A9301

$65.00

AGRICULTURAL COMMISSIONER
COUNTY OF RIVERSIDE
4080 Lemon Street, Room 19
Riverside, California 92501-4299

ARNOLD MALLIS

Mallis is the author of the Handbook of Pest Control, American Entomologists and numerous papers on household insects, household insecticides and the history of entomology. The Handbook of Pest Control has become the standard reference book of the pest control industry and, like Arnold himself, has played a vital role in helping people understand and respect pests and deal with pest problems in a rational manner.

—1982—

HANDBOOK
of
PEST CONTROL

The Behavior, Life History, and Control of Household Pests

Sixth Edition

By ARNOLD MALLIS

Associate Professor (Retired)
Department of Entomology
The Pennsylvania State University
University Park, Pennsylvania

Editorial Director
Keith Story

Production Editor
Dan Moreland

Art Director
Stuart Cantor

Contributing Editors
Bill Blasingame, Richard Carr, Norm Ehmann, Austin Frishman, Stanley
Green, Dariel Howell, Walter Howard, Harry Katz, Douglass Mampe, Rex
E. Marsh, George Okumura, Robert Snetsinger, David Shetlkar, Keith Story,
Vern Walter, Alfred Wheeler.

Franzak & Foster Company
Cleveland, Ohio 44113

PREFACE TO THE FIRST EDITION

This handbook concerns itself with household pests other than man, classical examples of which are the cockroach and bedbug. In numerous instances the author wanders from the immediate vicinity of the threshold to pay his respects to an ant, sowbug, or similar pest whose permanent abode is other than the home of its unhappy host.

Emphasis is laid on the control of household pests since this is the primary purpose of the work. During the last 10 years entomologists and chemists have devoted unusual attention to the control of household pests, and the author has taken particular pains to include these advances in the text. Various individuals have remarked to the author, "What we need is just ONE good method of control for each group." The author can merely reply that this is a Utopia as yet unattained, and some reasons for this are indicated below:

(a) No ONE good method of control is as yet known, e.g., against the black carpet beetle.

(b) No ONE good method of control is effective against all the species in a group, e.g., against both the German and the Oriental cockroaches.

(c) No ONE good method of control is effective against the same species in different localities, e.g., against the Argentine ant.

(d) No ONE good method of control is always the most applicable, e.g., fumigation against mice.

(e) Insecticides for the ONE good method of control are not always available, e.g., pyrethrum powder under present circumstances.

In some instances the author has separated the material on control into a concise introductory portion, for the benefit of those who are in a hurry, and a more detailed discussion, for those desiring additional information.

If the author has on occasion treated the introductory material lightly, it must be remembered that the entomologist and pest control operator in time spin about themselves a protective cocoon of humor to fend off the good-natured jests directed at their profession.

The author desires to express his gratitude to Professor E. O. Essig for constantly advising and encouraging him in the preparation of this work, and to Mr. A. E. Davie for his kind cooperation. Mr. Jack Schwartz, Dr. and Mrs. R. M. Bohart, and Dr. W. M. Hoskins, all rendered valuable assistance. Original photographs were furnished by Mr. R. J. Pence and Mr. J. C. Elmore. The following individuals and organizations have given the author permission to either use their illustrations or to quote from their works: W. S. Patton, Paul Griswold Howes, The British Museum of Natural History, The University of California Press, The Science Press, Smithsonian Institution Series, Inc., The Comstock Printing Co., The National Geographic Magazine, C. C. Thomas, Publisher, Little, Brown & Co., Charles Boni & Sons, Inc., American Cyanamid & Chemical Corporation, Zonite Products Corporation, Innis, Speiden & Co., E. I. du Pont de Nemours & Co. (Inc.).

ARNOLD MALLIS
1945

PREFACE TO THE SIXTH EDITION

In 1937, a salesman for a pest control firm in California mentioned the need for a comprehensive work on household pests and their control. This suggestion sparked the idea for the first edition of the *Handbook of Pest Control* which I started in 1938 and submitted to the original publisher, MacNair-Dorland Co., Inc., in 1942. Because of war-time conditions, publication of the first edition was delayed until 1945. Subsequently, four other editions appeared: the second edition in 1954, the third edition in 1960, the fourth edition in 1964, and the fifth edition in 1969.

For each edition I attempted to incorporate the latest information on pest biology and control. A comparison between the sixth edition and the fifth edition illustrates the remarkable progress made in the last decade in developing new products and techniques. But just as noteworthy have been the changes in pest control philosophies. The sixth edition reflects this evolution in our thinking and includes integrated pest management and pest eradication approaches, though for continuity I have kept the original title of the book.

For a variety of reasons I have not been able to revise the sixth edition entirely by my own efforts, and instead, I have been fortunate to have the cooperation of 16 other authors in the preparation of this edition. These authors, well-known in pest control, have updated the text and given it a modern, fresh look. Several chapters are new and have added a great deal of pertinent information to the book. For these reasons I wish to thank each of the authors, whose names head the chapters, for their contributions to the sixth edition.

I am especially indebted to Keith Story, who assembled the authors, edited and coordinated their efforts, and did all this with great competence in record time. I wish to acknowledge the initiative of Richard J. W. Foster and the work of Dan Moreland, Franzak & Foster, Co., and *Pest Control Technology* magazine for achieving publication of the sixth edition of the *Handbook of Pest Control*.

ARNOLD MALLIS
1982

TO

PROFESSOR E. O. ESSIG

and

MR. A. E. DAVIE

ISBN: 0-942588-00-2
Library of Congress Catalogue Card No. 81-71963
Manufactured in the United States of America.

TABLE OF CONTENTS

Credit for illustrations and photos appear on page 1059.

AUSTIN M. FRISHMAN

Dr. Austin M. Frishman gained his bachelor's and master's degrees from Cornell University and his Ph.D. from Purdue University. Frishman has had broad experience in the pest control industry. Prior to establishing his current consulting company in 1980, his work has ranged from pest control technician for a New York State PCO to a distinguished teaching career at the State University of New York (SUNY) at Farmingdale. For 14 years he was professor of biology at SUNY during which time he wrote several scientific and popular books on pest control and authored or co-authored more than 60 articles for scientific and trade journals. Currently he writes a column in the magazine Pest Control Technology magazine.

Frishman is well known for his entertaining lectures and during his career has addressed more than 50,000 people engaged in pest control.

Frishman has been active as a consultant in the food industry, developing pest control programs and manuals for several major food manufacturers and processors. In addition, he has worked closely with local, state and federal government agencies and played a major role in writing the state examination in three of the subcategories currently used in New York State.

Frishman is a member of the honor societies Phi Kappa Phi, Pi Chi Omega and Sigma Xi. He is a member of various professional associations including the National Pest Control Association and the Entomological Society of America. He is an honorary member of the Indiana, Long Island and Empire State Pest Control Associations.

CHAPTER ONE

Rats and Mice

Revised by Austin Frishman[1]

".......... Rats!
They fought the dogs, and killed the cats,
*   And bit the babies in the cradle,*
And ate the cheeses
*   out of the vats,*
*   And licked the soup from the cook's own ladles . . ."*

—Robert Browning

In Japan, the rat has the honor of having the first year of the oriental zodiac named after it. The rat is often associated with the God of Wealth — Daikoku, one of the seven gods of luck.

The Chichineca-Jonaz, in the Mexican state of Guanajuanto, were eating rats as part of their basic diet (Hirschhorn 1974) as late as 1950. Several veterans returning from Vietnam and others visiting the Philippines and other Asiatic countries have told the author of children bringing rats home for dinner. Rat meat brought a good price in 1798 in the French garrison at Malta (Zinsser 1935). Rats also serve as scavengers feeding on the garbage of man.

Both roof rats, *Rattus rattus*, and Norway rats, *Rattus norvegicus* serve as excellent research animals for a host of physiological, behavioral and other scientific studies. The acute oral toxicity of all pesticides is based on tests conducted with laboratory strains of Norway rats.

Also, rats can be conditioned so their brain waves show a response to TNT (Egelhof 1978). Thus rats have been trained to search out bombs. So much for the desirable contributions of these animals. The rest of this chapter is devoted to how to control rats.

The remarkable ability of the rat to adapt itself to nearly every environment, along with its natural cunning, have made it one of the most successful of all animals. Man continually must fight these animals to protect himself, his foodstuffs and his property from destruction. The war between man and rats is unrelenting since the victories achieved are of a temporary nature, only very rarely resulting in a long-lasting reduction of rat populations.

Rats as enemies of mankind. Rats are extremely important pests on the farm. Hamilton, Jr. (1947) states, "On the farm, rats eat incredible quantities

[1]*President, AMF Pest Management Services, Inc., Farmingdale, N.Y.; Extension Specialist, State University of New York, Farmingdale, N.Y.*

of foodstuffs, destroy poultry, lay waste the stored fruits and vegetables, and riddle buildings with their sharp teeth. Rats tear down growing corn, eat melons, pumpkins, and tomatoes on the vines, and even take an appreciable toll of cherries, climbing the tree in search of the fruit." Rats also are a major problem to the poultryman. They destroy feed, carry disease and kill chicks.

Hamilton, Jr. (1947), speaking of the dèpredations of rats on wildlife, notes, "In America, game keepers and conservationists might well look to the rat as one of the chief predators of game. Rats victimize the nests of robins and ground-nesting birds, insular colonies of terns have been completely destroyed by rats, and rat destruction of quail chicks in the South is well known. On game farms, the abundant supply of feed attracts hordes of rats, which in turn kill the young pheasants and other birds. Rats ate the legs off 40 young black ducks on a Long Island game farm. A single rat destroyed, in three nights, 120 pheasant chicks. In rural areas, rats are not restricted to the farm. The author has seen them in the dense woods of the Adirondacks, miles removed from habitation. In these areas it is obvious that some of the food consists of wildlife."

The young of pigs and lambs are not immune from their depredations, and it has been reported they will on occasion gnaw holes in the bellies of swine, tear the nipples from a farrowing sow, and otherwise attack animals. A circus owner in Germany was forced to kill three elephants because the rats gnawed the feet of these animals, and a pest control operator observed rats to gnaw on the hides of living alligators in a Los Angeles alligator farm.

Rats are responsible for damage in the home and warehouse. Here they gnaw upholstery, bolts of silk, papers, books, and like materials for nesting material. They gnaw hard substances such as bone, aluminum, lead, and similar materials (Fig. 1-1). It was believed this was to maintain their ever-growing incisors at the proper length. However, gnawing is not necessary for rats to keep teeth worn down; the incisors work like opposing chisels. In a hospital on the edge of the Mohave Desert, they gnawed for water through lead drains in the plumbing. Rats also can chew through plastic water pipes and garbage cans.

Accounts of the rat's ability to start fires by carrying matches to their hiding places, as well as by causing short circuits through the gnawing of electrical wires, are commonplace. Hamilton, Jr. (1947) notes rats "carry matches to their nests and gnaw the phosphorus or paraffin coating, causing the match to burn. Their nests of highly inflammable material are often composed of oily rags and other substances which encourage spontaneous combustion. By gnawing through electric wiring and lead gas pipes, a not uncommon practice, rats jeopardize human life through fire or asphyxiation."

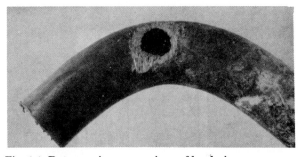

Fig. 1-1. Rat gnawings on a piece of lead pipe.

The mere presence of rats reduces the rental value of apartments and stores. The noise they make as they climb between the walls and floors of buildings often keep the inhabitants from sleeping. But historically their effects have been of little import compared to the diseases they carry. Richter (1946) studied the incidence of rat bites and rat bite fever in Baltimore. He concluded that in an area of less than two square miles at least 93 individuals were bitten by rats from 1939 to 1943. Seven of the 65 individuals treated in Johns Hopkins Hospital developed rat bite fever — none died. Sixty percent of those bitten were under one year of age. Most of the bites were mere punctures on the hands and face, although some were more serious. The rats regard the sleeping infant or adult as a source of food, and in most instances the first bite awakened the victim, and the rat ran away. Possibly, once a rat bites a human being it is apt to bite other human beings. Individuals dwelling in areas where housing and living conditions are poor are more apt to be bitten.

Nor are rats averse to attacking man as this report from Hogarth (1929) so well certifies: "In the Walker colliery, near Killingsworth, in which many horses were employed, the rats had accumulated in great multitudes. It was customary at holiday times to bring to the surface the horses and the fodder, and to close the pit for a time.

"On one occasion, when the holiday had extended to 10 days or a fortnight, during which the rats had been deprived of food, on reopening the pit, the first man who descended was attacked by the starving rats and speedily killed and devoured."

More recently, Newsday (1979) reported the following under the headline, RATS GNAWED ON MAN, 45.

"An indigent man who had lain in the garbage-strewn basement of a Chicago apartment building for two weeks while rats chewed on him was reported in fair condition yesterday after having both of his legs and two fingers amputated.

Cook County Hospital said David Hallman, 45, was transferred from Loretto Hospital early Saturday and underwent surgery later in the day. The hospital reported Hallman's legs were amputated below the knees, and that two fingers were removed from his left hand.

Police discovered Hallman in the basement of a West Side apartment building Friday night. He had rat bites on both legs and hands and one of his ankles was bare to the bone, police said.

A resident of the building said she had heard moans and screams for about two weeks, but didn't call authorities because 'she thought we would think she was crazy,' police said."

DISEASES CARRIED BY RATS

Although we have listed some of the depredations of the rat, these are minor when compared to the dread diseases conveyed and spread by this pest. Schwarz (1944), in evaluating the several species of rats as carriers of disease notes, "There are the true or exclusively domestic rats, the Black rat (*Rattus rattus rattus* Linnaeus) and the Greybellied rat (*R. r. alexandrinus* Geoffroy); these are the chief carriers of infection, because their contact with man is of the closest. The other two commensal rats, the White-bellied rat (*R. r. frugivorus* Rafinesque) and the Norway rat (*R. r. norvegicus* Berkenhout), although associated with man and dependent on him for their existence, to a certain extent may live with the wild rodents, or act as go-betweens between them and the domestic rats. Therefore, they are primarily responsible for animal epidemics (epizootics) and for the transport of infections over wide distances." Schwarz's taxonimic structure is no longer used. It is now standard practice to use only the terms *Rattus rattus* (with possible specific color phases) and *Rattus norvegicus*.

Plague. The great plague of London that killed more than half of the city's inhabitants, and the "black death" that devastated Europe for more than 50 years in the fourteenth century, killing some 25,000,000 individuals, were in part due to the abundance of rats. The plague-infected rats carry plague-infected fleas which in turn infect man. Fortunately, such epidemics no longer devastate Europe, yet it is estimated that from 1898 to 1923 11 million lives were lost from the plague in India, China, Mongolia, and other parts of Asia.

The thousands of miles of water separating us from the Asiatic and the European shores are not sufficient to spare us from this dread scourge, for plague outbreaks have occurred in San Francisco in 1900, in Oakland and San Francisco in 1907 and 1908, in New Orleans in 1914, in Galveston in 1920, in Los Angeles in 1924, and in other cities since then.

Plague is a bacterial disease of the circulatory and respiratory systems. The germ *Yersinia pestis,* which invades the body, was discovered independently in 1894 by the Japanese investigator Kitsato and by the French investigator Yersin. At this time it was established that rat plague and human plague were identical.

In man, plague may manifest itself in four ways:

- **Bubonic plague.** Here the blood is infected and the bacilli are arrested in the glands, particularly in those of the groin and under the armpits, resulting in inflamed glands or buboes which suppurate. This is the most common form of plague and results from the bite of a flea. The mortality may range from 40 to 70 percent. It should be noted bubonic plague also can be contracted by contact of the abraded skin with infected dust and dirt.
- **Septicemic plague.** In more serious cases the glands fail to arrest the bacilli which appear in large numbers in the blood. Numerous hemorrhages occur under the skin, which turns black, accounting for the name "Black Death." This form of plague is spread by the bite of an infected flea, and since the disease in this case is very virulent, death nearly always results.
- **Pneumonic plague.** Here, where the bacilli are in the lungs, we have the most dangerous form of plague from a public health standpoint, since it is spread so readily through contact and coughing, as well as by the consumption of contaminated food. This form of plague nearly always results in a mortality above 90 percent.
- **Sylvatic plague.** This is a form of plague wherein the virulence is greatly diminished. Ground squirrels contracted the plague in San Francisco in 1900, so now it is endemic or established in this country. This form of plague was first discovered in 1908 and is gradually spreading throughout the West. It is now found in ground squirrels, wood rats, deer, mice, and woodchucks. Silver (1927) states, "The sylvatic form of the plague is apparently not highly contagious to man, as an average of only about one human case each year has been reported. The menace, however, remains a most disturbing one because of the ever-present possibilities that house rats may become reinfested in the population centers and that human cases of bubonic plague contracted from native rodents may develop the secondary, or pneumonic form, which is highly contagious directly from person to per-

son." Sylvatic plague is conveyed by contact with dead or dying rodents and less often by flea bites.

Plague is primarily a rat's disease, and the black rat, *Rattus rattus rattus,* is particularly susceptible to this disease which periodically reduces its numbers. Other rodents such as mice, the brown rat, *Rattus norvegicus,* and the California ground squirrel, *Citellus beecheyi,* are readily infected by the plague. Animals like the horse, dog, pig, and man himself, often fall victim to this dread disease. Since the black rat lives in closest association with man, plague may be present wherever the black rat is a serious household pest. In fact, it has been observed that an epidemic of plague is often ushered in by a noticeable mortality among rats. The fact that plague is no longer important in such cities as London, Marseilles, and Rome may in part be attributed to the fact the black rat and its plague-carrying flea, *Xenopsylla cheopis* (Rothsch.), have been largely replaced by the brown rat and its flea, *Nosopsyllus fasciatus* (Bosc.).

The black rat has been driven away by the competition of the stronger, more ferocious, more adaptable, and more prolific brown rat. The black rat is now, for the most part, limited to seaports and ships, where because of its superior climbing ability it can successfully compete with the brown or Norway rat. Of late, the black rat has been making a "comeback" in England and other parts of the world, largely due to the ratproofing of buildings which successfully excludes the brown rat, but not the more agile black rat.

As long as rat and flea control is neglected, the menace of plague remains. The danger is greatly reduced because of the scientist's knowledge of the cause of plague, as well as by modern sanitation and medication. Moreover, introduction of plague-infected animals has been almost completely curtailed by the inspection and fumigation of ships.

The San Bruno Mountain region of California has been extensively studied in plague outbreaks (Hudson and Quan, 1975). Plague transmission can take place when wild rodents come into contact with urban rats. On one hog farm in which wild rodents and rats were present together, many animals were trapped including rats, voles, mice, a rabbit and two weasels. Close study of the animals showed flea exchange took place between the Norway rat and some wild rodents (deer mice and voles).

OTHER DISEASES WHICH MAY BE CONVEYED BY RATS

Murine typhus fever. USPHS (1948) states there are two kinds of typhus fever, "epidemic or European, and endemic or murine. The epidemic form is transmitted from person to person by body lice, while murine typhus is contracted from domestic rodents, probably both rats and mice; rats being the more active in spreading the infection. If louse-infested individuals contract murine typhus, the infection may then be transmitted by the patient's lice to other people."

Andrews and Link (1947) note the oriental rat flea is an important agent of transmission of the disease. "It must be emphasized, however, that rickettsiae have been found in the excrement of rat fleas and in the urine of rats. Thus, the possibilities of transfer to man by inhalation of dried flea feces in dust, or by the consumption of food or drink contaminated by flea feces or rat urine must be considered as well."

In the United States, in the 1940's, murine typhus was most prevalent in the South and Southeast including parts of Texas. Eskey (1943) notes 3,700 cases were reported in 1942, and he says he believes the disease is much more common than is indicated by official records.

Infectious jaundice or Weil's disease. Although this is a common disease in the Orient, what is not so well known is its prevalence in the United States. The disease is caused by the spirochaete, *Leptospira icterohaemorrhagiae,* which is found in the blood and urine of the rat. Human beings may become infected "by handling or eating things contaminated with rat urine. It is also contracted by swimming and wading in contaminated water." USPHS (1948) states the disease "does not usually cause death, but is very debilitating, confining the patient to his home for a week or longer. 'Yellow jaundice' may be caused by a number of conditions, but rats are probably responsible for many undiagnosed cases." Storer (1948) notes the disease has caused epidemics among city dogs in California.

Rat-bite fever. As was previously noted, Richter (1946) showed seven of 65 cases treated for rat bites in Johns Hopkins Hospital in Baltimore developed rat-bite fever. The symptoms of this disease may develop after the wound has healed. The infected individual may have a relapsing type of fever for weeks or months. Larson (1941) showed in Washington, D.C., that rat-bite fever is due to the two bacterial organisms, *Spirillum minus* and *Streptobacillus moniliformis.*

Jellison, et al. (1949), have the following to say in discussing a case in Montana where a girl was bitten by mice: "Rat-bite fever is most frequently communicated to man by the bite of rats, *Rattus* spp., occasionally by the bite of other rodents, and rarely by the bite of dogs, cats, or ferrets which presumably have become contaminated by eating infected rodents." The disease is particularly dangerous to babies and small children since they are the ones most frequently bitten by rats.

Trichinosis. This disease of man is caused by a minute worm, *Trichinella spiralis.* Rats and mice are the principal agents in the dissemination and the perpetuation of the disease. Large numbers of *Trichinella* in the adult or sexual state are most commonly present in the intestine of man, pigs, and rats. The worms may be found encysted in the muscles of mammals and birds. It has been estimated the flesh of an infected human being contained 100,000,000 encysted worms.

While encysted, the worms suspend animation and undergo no further development. Further development of the encysted and sexless worms will only take place if the infected flesh is eaten by another animal in which the worm is capable of living, e.g., man, pig, or rat. Once this is done, the cysts are dissolved by the digestive juices, the worms escape, become sexually mature, mate, and migrate, producing the disease again.

Rats become infected by feeding upon excrement or meat infected with these worms. Pigs eat the rats and mice, or food fouled by the excrement of the rodents. Man eats the trichinous pork and becomes infected when the meat is not properly cooked.

Food poisoning. USPHS (1948) notes both "rats and mice suffer from intestinal infections that are communicated to man, who is infected from eating foods contaminated by the excreta of infected rodents. Acute food poisoning of this type is probably much more common than generally realized, and many involve a large number of persons at one time." This same source states it is possible for man to contract amoebic dysentery by eating food contaminated with rat excreta.

Chorio-lepto meningitis. USPHS (1948) notes this acute infectious disease of the nervous system "is caused by the excreta of mice contaminating food and dishes. The infection is a rather mild meningitis that causes disability for a number of days. It is caused by a virus."

Poliomyelitis. According to USPHS (1948), two different types of poliomyelitis may be contracted through domestic rodents. "Recently, cases of this disease have been definitely associated with an infection of mice found on the premises of the sick person. The more common type of poliomyelitis, or infantile paralysis, is believed to result from rats contaminating food with organisms from human excreta with which they may come in contact in privies, sewers, or on the ground. In this case, the rat is simply a mechanical carrier of the infection and is not infected with the disease. It is possible that rats may act also as mechanical agents in the transmission of typhoid fever and dysentery."

Rabies. USPHS (1948) notes "the possibility that rats may play an important part in the dissemination of rabies or hydrophobia. Rats have not been found infected with rabies in nature, but they are susceptible to the infection. Recently an individual developed rabies a couple of weeks after being bitten by a rat, and two incidents have been reported of dogs developing hydrophobia after being bitten on the nose by rats. These dogs were shut up and had no known contact with other dogs.

To date, however, we have no documented cases rats carry the rabies virus.

HISTORY OF THE RAT

There is apparently great disagreement among pest control historians as to just when the rat arrived in Europe. Zinsser (1935) states, "De L'Isle believes that the black rat (Fig. 1-2), *Rattus rattus rattus,* originated from *Rattus rattus alexandrinus* and that it did not become parasitic on human beings until the seventh century, and that prior to this time, it lived a wild existence, possibly in the Arabian deserts. Moreover, De L'Isle was of the opinion that the brown rat became domesticated at the time of the Crusades, when it accompanied man everywhere by ship, and thus spread through the Mediterranean ports.

Fig. 1-2. The black rat, *Rattus rattus rattus.*

Fig. 1-3. The brown or Norway rat, *Rattus norvegicus*.

"The brown rat (Fig. 1-3), *Rattus norvegicus*, which most authorities agree originated in Central Asia, was supposed to have reached western Europe through the agency of the Russian fleet which visited Copenhagen. And in 1727, Pallas observed the rats to emigrate from Russia and to swim across the Volga in vast hordes, probably in search of food. They thereupon overran Europe."

Sambon (1924) heartily disagrees with the above and his criticisms are herewith reprinted verbatim: "From recent publications I notice that naturalists and archaeologists continue to perpetuate the erroneous notion that the black rat *(Rattus rattus)* was unknown to the ancient Greeks and Romans, and that it came to Europe during the twelfth century in the ships of returning Crusaders. On several occasions, during the last 20 years, I have endeavored to show that the black rat has inhabited Europe from time immemorial. To the ancients the rat was merely a mouse, and indeed, even the early medieval Bestiaries describe a *Mus major* (rat) and a *Mus minor* (mouse). I have drawn attention to the innumerable and excellent representations of rats in ancient Greek, Etruscan, and Roman works of art, and especially to works portraying them in actions more appropriate to rat than mouse, such as gnawing ship cordage, as on Etruscan bronze votive boats (Etruscan Museum, Florence); feeding on mussel-beds, as on the silver coins of Cumae (cir. 490-480 B.C.); spreading the bubonic plague, as on a Roman colonial coin struck at Pergamum (between A.D. 161 and 169). I laid particular stress on the raging of plague in Rome three centuries before our era, because, in the light of modern knowledge, it clearly reveals the presence of the rat. Indeed, outside the permanent Asiatic plague-area in which the Bobak marmot stands as reservoir, plague and rat necessarily are inseparable.

"As to the brown rat *(Rattus norvegicus)*, a similar error survives. Naturalists persist in asserting that it did not reach Europe before the eighteenth century, and give figures purporting to be exact dates of the first arrival of brown rats in different countries, Prussia, 1750; Norway, 1762; Faroe Island, 1768; Sweden, 1790; Switzerland, 1808, as if they had actually stamped their passports. We know from Pallas that in 1727 — a 'mouse year' in the Caspian region — vast hordes crossed the Volga and swarmed into Astrakan, thence spreading westward across Russia; but the fact that while travelling in Southern Russia, a distinguished naturalist had witnessed the migration of a great rat army does not prove that the brown rat first came to Europe at that particular moment. Aelian, in his work, *'De Natura Animalium,'* written in the second century A.D., undoubtedly refers to the brown rat when he states that 'Caspian rats' at times migrate in countless hosts and bridge the rivers, forming live rafts, each rat holding by teeth to the tail of the rat in front. Among ancient bronze representations of rats found in Italy, while some show the large ears, sharp muzzle, slender build and long tail of the black rat and, therefore, often are confounded with mice, others portray most faithfully and unmistakably the small ears, blunt muzzle, heavy build and shorter tail of the brown rat, proving that both species were available as models to the Italian sculptor of at least 20 centuries ago."

In any event, it is believed the black rat reached the United States with the first ships that sailed from European ports. Although some American authorities are of the opinion the brown rat reached the United States in 1775, Dr. Sambon's remarks about the brown rat leave this question open to debate. The progress of the black rat across the country was rather slow due to the high altitudes and the great stretches of arid regions in the West. With man pioneering the way, and providing food and shelter, the rat finally "thumbed" its way into California in 1851; it did not arrive in Montana until 1923. Silver stated in 1927 that rats at that time were present in every one of the United States.

In the United States the brown rat has replaced the black rat everywhere in the more temperate areas except in the southern states, especially in Florida and the Gulf States, and here in some sections the black rat may actually predominate. Silver (1927) notes black rats are to be found in small colonies in practically all seaport towns because of their presence on seagoing vessels. In England it was the black rat that was largely responsible for the plague epidemics and in the last few centuries it has been largely replaced by the brown rat. However, as was previously mentioned, at the present time the black rat is returning since ratproofing effectively keeps out the brown rat, but not the more agile and better climbing black rat.

Pratt and Brown (1976) describe the geographic range of domestic rats as follows: "Roof Rat: The roof rat is a native of Southeast Asia. It followed the caravan routes across India into the eastern Mediterranean region and entered Europe about the time of the Crusades. During medieval times it was the common house rat in Europe during the outbreaks of plague known as the Black Death. In the European area the roof rat has two distinct color phases: the black rat of Western Europe and the brown alexandrine rat common around the Mediterranean. When this species was carried to the Americas, however, this situation changed. These introductions into North America began well before 1750, and roof rats were well known throughout the French, English and Spanish colonies. Here the color phases from all parts of Europe were dumped together in the same ports, where they interbred freely. As a result, today in North America all the color phases can crop up in one population. Often a single litter of young roof rats will contain both black and brown animals.

"Norway Rat: There is evidence that the Norway rat is a later, more highly developed species originating in or near the center of origin of the *Rattus* group. This comparatively late comer is adapted to the dry, grassy plains of Central

Asia. It is characteristic among mammals that the most advanced species of a group are found closest to the center of origin, where they replace the more primitive forms. So it appears to be with the rats. As the more highly developed, more aggressive Norway rat spread outward from Asia, the more primitive roof rat disappeared over much of its original range.

The Norway rat first appeared in Europe in the 1700's. It spread so rapidly that the Europeans called it the "Wanderatte" or migratory rat. Soon after the Norway rat reached Western Europe, it was carried to the New World. Here it quickly began spreading outward from the seaports, especially along the east coast of North America.

"The present distribution of the Norway and roof rats appears related to two factors, competition between the two species and the reaction of both to different climates. When the aggressive Norway rat and the roof rat compete for the same areas, the Norway rat frequently becomes dominant, and the roof rat soon disappears. Only under special conditions do both species live in the same area. In one eastern seaport, roof rats live in the top of a grain elevator and Norway rats live in the bottom. This is probably because roof rats are better climbers than Norways. It is generally only in such situations as these that roof rats are found living in Norway rat territory.

"As the spread of the Norway rat approaches tropical regions, the picture is altered by its reaction to the warmer climate. It appears that the Norway rat is definitely an animal of the temperate climates. In its original range in Asia it is restricted to temperate regions. It is found in the tropics only in seaport areas. On the other hand, the roof rat is most common today throughout the tropics. This is true both in its native area and in areas where it has been introduced. In these areas roof rats commonly inhabit regions quite far removed from man's activities."

KINDS OF RATS

Although zoologists place mice and rats in different genera, viz. *Rattus* for the rats, and *Mus* for the mouse, Zinsser (1935) differs with them and states, "Rats and mice belong to the same genus, and the closeness of the relationship is attested by the experiment of Ivanoff, who artificially inseminated a white mouse with the sperm of a white rat, and obtained two hybrids after a pregnancy of 27 days. Mice may have developed out of rats under circumstances which made it less desirable to be large and ferocious than to be able to get into a smaller hole." Eaton and Cabell (1949) place both rats and mice in the genus *Mus*. However, most zoologists still retain the genus *Mus* for mice and *Rattus* for rats.

Until recently, it was generally believed we had only three house rats in the United States: *Rattus norvegicus* (Erxleben)[*], commonly referred to as the brown or Norway rat, but also known as the barn rat, wharf rat, sewer rat, water rat, and gray rat; *Rattus rattus rattus* (Linnaeus), the black rat or ship rat; and the subspecies *Rattus rattus alexandrinus* (Geoffroy), the roof rat, gray-bellied rat, or Alexandrine rat. Schwarz (1942) makes the important observation that what is very often called the roof rat is actually a fourth and distinct subspecies, *Rattus rattus frugivorus* Rafinesque, the white-bellied rat. Milmore (1943) notes the black rat may only be a color variant of the roof rat and the

[*]Authorities differ as to whether Erxleben or Berkenhout is the original describer.

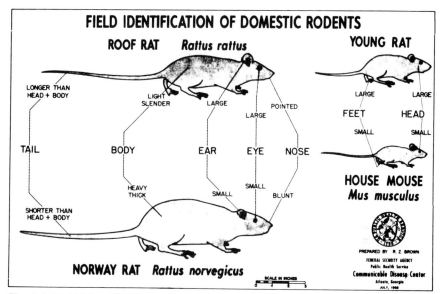

Fig. 1-4. Field identification of domestic rats.

habits of the two species do not differ materially. Green (1950) observes that *R. r. frugivorus* and *R. r. alexandrinus* have been "found in the same colony and even in the same litter." Today, however, subspecific designation is no longer used. Color variant or color phase is more valid from a technical standpoint.

Because of individual variations, the rats very often cannot be separated by color. The brown rat is usually a grayish brown, but it may vary from a pure gray to a blackish or reddish brown. The roof rat and the white-bellied rat may be distinguished from the black rat through their lighter coloration, which is believed to have been developed while living in the warm climate of northern Africa. The roof rat has a brown back and a grey belly, and the two colors gradually pass into one another without any line of demarcation. The white-bellied rat, like the roof rat, has a brown back, but the belly is white or lemon yellow and distinctly set off from the back. The black rat is a uniform dusky black with a white or cream-colored belly. Since the color of the black rat, roof rat, and white-bellied rat vary and often blend, the use of color alone may not be sufficient for the separation of these rats.

A few of the more obvious characteristics by which the adult black rat can be separated from the adult brown rat as adapted from Hinton (1931) and Pratt and Brown (1976) are shown in table 1-1.

Distribution. The Norway rat is the most common of rats and occurs practically everywhere. The black rat is found most commonly in seaports and in the Gulf States, and Dykstra (1950) notes this rat outnumbers the Norway rat nine to one in some parts of Texas. The black rat and the roof rat are primarily house rats, with the black rat having a more northern range than the roof rat. The roof rat is found most often in the southern states and along the Pacific Coast. The white-bellied rat is closely associated with the roof rat, and occurs in the southern states and possibly California. The white-bellied rat is primarily an outdoor rat that lives in trees, shrubs and in vines on the outside

TABLE 1-1

Some characteristics separating the adults of the black rat from the adults of the brown rat.

Characteristics	Black Rat	Brown Rat
Size	Body and head approximately 8 inches/200 mm	Body and head approximately 10 inches/250 mm
Muzzle	Sharp.	Blunt.
Ears	Large and almost naked.	Small and densely covered with short hairs.
Tail	Slender, and often longer than head and body. Unicolored.	Stout, usually not as long as head and body. Bicolored.
Mammary Glands	Usually 10 in number.	Usually 12 in number.
Weight	Eight to 12 ounces/227 to 340 g	16 ounces/454 g or more.
Fur	Somewhat stiff.	Soft.

walls of houses. As a rule, it does not live too close to man, although occasionally, especially during the winter, it will enter barns and the upper stories of houses. Where the white-bellied rat has been able to maintain itself, despite the presence of the Norway rat, it has been due to its ability to live and nest in trees. In some of the warmer parts of the United States, all the rats have been found in the same locality. However, some experts think the characteristics attributed to different color phases are not valid.

Brooks (1964) notes roof rats occur in sanitary sewers in California. He also states (1966) dense growth of trees, shrubs, and vines, woodpiles, sheds and accumulated yard rubbish all contribute to infestation by this rat.

The brown rat thrives best in temperate climates and Hinton (1931) notes that in hot countries it may be found living in cold storage buildings: "At Tammefors, a little town of wooden buildings in western Finland, and a very cold place, it has according to Zuschlag, invaded the houses and completely replaced the house mouse. Here it has become most impudent and bold; it is to be found even in the beds of the inhabitants."

HABITS OF RATS

Since rats can adapt themselves to a great many environments, their behavior also varies, and for this reason one must be careful when attributing certain set habits to one species or the other. USPHS (1948) illustrates this by noting that although the roof rat is a better climber than the Norway rat, the latter may at times be trapped on roofs, and roof rats may be trapped in sewers.

Night and day activities. Hamilton (1947) observes that if rats were not largely nocturnal animals, we would soon be apprised of their tremendous numbers. They usually become active "one-half hour after dusk, after the premises have become quiet." Williams (1948) adds, "Rats are not truly nocturnal, their apparent nocturnalism being an expression of their mode of self-defense, that is immediate retirement in the face of danger, into inaccessible and hidden retreats. When unmolested, rats roam about, seeking food, during any part of the day or night. That they are more frequently abroad at night is because their principal enemies are relatively inactive during the night hours." Chitty and Shorten (1946) observe well-fed rats are predominantly nocturnal, and hungry rats may feed during the day.

Harborage and nesting. Williams (1948) notes rats "require harborage for two major purposes, one as a protective refuge for the adults and the other as protection to the young." According to Williams, harborage "generally means retired spaces wherein rats are at least out of sight. Good harborage consists of retired spaces wherein the rat is not only out of sight but cannot be directly reached by its enemies.

"Rat's nests are made up of bits of any kind of material available. Soft materials are preferred such as bits of paper, rags, burlap, straw, string, chips, etc. These are carried into the selected harborage, packed loosely and built up around the side, until there appears a crudely constructed nest resembling a bird's nest. In the nests the young are deposited and cared for by the mothers." Hamilton (1947) notes "both bank notes and corn husks are acceptable for the nest, and that more than one innocent victim has been made out a thief, only to be pardoned when a piece of paper currency has been recovered from a rat's nest."

According to Williams (1948), when a nest containing young rats is invaded, "the young rats unless killed at once, will be removed by the mother at the first opportunity. If such a nest is exposed and then left for a half hour or longer, often it will be found empty on return." The bucks or male rats will feed on the young if they are given the opportunity. Frantz (1979) found in some cases, males can be left with the female even after the birth of young.

USPHS (1948) observes one of the favorite nesting places for rats in buildings is the closed space under the ground floor of buildings. "All other closed spaces of buildings may be utilized by rats if they can gain access to them. Nests will often be found in piles of rubbish and merchandise if the latter is stored over a month without being moved."

Pratt and Brown (1976) report a rat's nest is bowl-shaped and about eight inches/200 mm in diameter. Sometimes they are completely roofed over.

Emlen (1947) found an average of approximately 75 Norway rats per block in the downtown residential section of Baltimore. "About three-fourths of the rats in residential areas live outdoors in yards, garages, and sheds, and one-fourth indoors; very few rats are found in sewers." On the other hand, Worcester (1959) regularly baits for rats in sewers in South Dakota.

The senses of rats and mice. Storer (1948) discusses the senses of rats as follows: "Rats and mice have rather poor vision, but the sense of smell, taste, hearing, and touch are keenly developed. Their frequent sniffing movements tell them much about their surroundings through odors received. Their choice in foods is undoubtedly based upon taste preferences. They are frightened by unusual sounds, which may cause them either to stop abruptly or to hurry to safety. They become used to ordinary noise, however, and are often active where people, domestic animals, or machines are close by.

"The long 'whiskers,' or vibrissae, on the nose, and other long hairs above the eyes, serve the sense of touch. There are sensory nerves about the base of each

Fig. 1-5. Rat burrows beneath concrete slab porch.

hair. It is the habit of a house rat or mouse to run close beside a wall, against which these sensory hairs touch to give the animal information about its surroundings. In the laboratory, rats with the vibrissae removed have been found less skillful in running and finding their way."

Exterior nesting places and burrows. According to USPHS (1948), "Norway rats will nest in underground burrows even when nesting places are available within adjacent buildings. In fact, it sometimes appears as though this species preferred to nest underground rather than inside buildings. Blocking Norway rats out of buildings will have little effect in reducing their numbers when they have access to garbage. On the other hand, *Rattus rattus* prefer to nest inside and under buildings or in piles of rock, rubbish, boxes, lumber, and other materials in the open. However, roof rats may nest in underground burrows near buildings when Norway rats are not present, and they often have to live underground in fields. Storer (1948) notes roof rats in California may nest in palms and other trees, as well as in dense hedges and vines on fences. In this

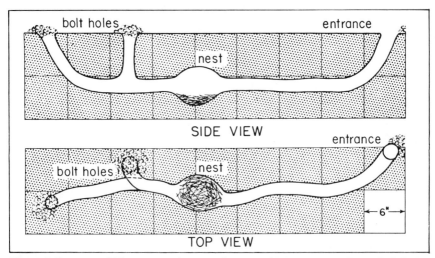

Fig. 1-6. Burrow of Norway rat in a poultry pen. Emergency exits or bolt holes have a light covering of soil and are not easily seen.

respect, Brown (1960) states that in Hawaii and Florida roof rats sometimes "build large closed nests in trees much as squirrels do."

According to Pisano and Storer (1948), the tunnels of Norway rats are two to three inches/50 to 76 mm in diameter, of varying lengths, with two or more entrance tunnels, a central den, and very often with a "bolt-hole" lightly plugged with earth at its outer end permitting emergency escapes by the rat. These bolt-holes may be hidden under grass or boards. The rat digs the earth in the tunnels with its front feet and shoves the loose soil back under its belly. The rat then kicks the soil back with powerful strokes of its hind feet. Then it turns around and pushes the soil along with the forepart of its body.

USPHS (1948) notes these burrows rarely extend 18 inches/45 cm below the ground. However, the rats may burrow (four feet/1.2 m) or more to get in or out of buildings. "Even *Rattus rattus* have tunneled under a foundation extending over (30 inches/75 cm) below the surface of the ground."

Running, climbing, jumping. Williams (1948) discusses these characteristics of rats as follows: "The ship rat *(Rattus rattus)* is not, compared with many other animals, a good runner, an athletic deficiency that probably plays a considerable part in its dependence on harborage. It is not nearly so fast a runner as man and in the open is readily overtaken and, by the expert, may be caught with bare hands. In the open, its efforts to dodge are clumsy so that even on fairly rough ground it is easy prey to an active dog. In familiar surroundings it gives the impression of speed by the rapid and expert utilization of cover and by the fact that it seldom stops at the point where it disappears, unless that is into a well protected retreat, suddenly reappearing at some distant spot only to immediately disappear once more.

"If the ship rat is not a good runner it makes up for the deficiency in climbing ability. It will positively climb anything that it can reach halfway around as well as a great many structures that it cannot. It will go straight up or down the edge of a steel door or an angle iron almost as fast as it can run. It easily climbs a one-and-a-half-inch pipe. If the walls are rough (a frequent result of repeated painting of metal surfaces on ships) it will go up either the outside or inside of a 90 degree angle. It has been seen to run along telephone wires. While a ship's mooring line is to it a broad highway. On land it climbs almost any kind

Fig. 1-7. The hind foot of a Norway rat.

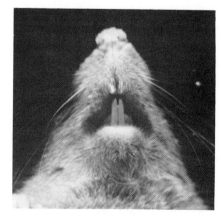

Fig. 1-8. Roof rat incisors.

Fig. 1-9. A Norway rat doing a "tightrope" walk on a wire.

of wooden structure. It will climb trees and readily climbs cocoanut or other palms, often making its nests in the tops.

"A full grown *Rattus rattus* can jump straight up very nearly two feet. Its horizontal flight has not been directly measured, but is generally estimated at four or five feet. It can jump down great distances without serious injury. This once received an unusual test when two *Rattus rattus* were intentionally dropped from the window of a grain elevator 183 feet above the ground. Both of them spread out their legs, very much as a flying squirrel does, caught the air and planed off to one side. They literally hit the ground running, and promptly disappeared under cover. This rat, however, does not ordinarily make much use of its jumping abilities, preferring a circuitous climb instead of a direct jump. Sometimes regularly traveled runways are found involving a jump of 10 to 15 inches, but it is when cornered or closely pursued that it displays its real prowess as a leaper."

Frantz (1979) reports roof rats, when able to hit a vertical surface, can easily clear four ft/1.2 m with a running start. They hit the surface part way up, then push off again vertically. Frantz also reports their ability to climb at the junction of smooth, painted metal walls.

USPHS (1948) notes Norway rats "are not inclined to climb wires, pipes, trees and other objects, as they prefer to travel over flat surfaces or up stairways and across roofs. They do climb to some extent, however, when necessary to gain access to buildings or to food." Johnson (1946) makes the following notes on the climbing ability of rats in a study of their ability to board ships: "Experiments were conducted to obtain data on factors limiting the ability of a rat to jump in horizontal and vertical directions. It is pointed out that rats can jump upwards by rapid ricochetting against smooth surfaces. They can learn to jump over and across properly placed ratguards.

"Observations are given that indicate rats are able to climb wherever they can get a toehold or can apply pressure. Their climbing ability was studied and results showed that they were able to move freely on vertical surfaces of rough lumber, along wire lines using the tail to assist in balancing, and to ascend sheet metal edges by gripping both sides of a free edge. Furthermore, they could climb galvanized pipe by reaching partly around for a better grip and could climb reasonably well on smooth glass tubing. Cloth wrapped around wire and thickly smeared with various adherent substances failed to prevent rats from climbing."

Bacon (1945) discusses the ability of the rat to climb vertical pipes as follows: "Rats can easily climb one-, two- and three-inch pipes. Above this size, his ability to climb the pipe will depend to a great measure upon whether the surface of the pipe is smooth or rough, and also upon the height he must climb to gain access to openings. If the pipe is smooth, he probably will not climb a four-inch pipe for any height; but if rough, it definitely should be considered an avenue of entrance to the building. It should also be borne in mind that rats will climb large pipes which are close to buildings by forcing their bodies between the pipe and the building." Norway rats can also slide down a rope with the aid of their tail wrapped around the rope.

Since rats are extremely wary creatures, they usually run close to the skirting or along pipes that are close to the wall, or behind objects next to the wall. These runways in the house can be readily recognized by the black spindle-shaped droppings, and by the dirty-gray markings their bodies leave around holes and along the skirting of their habitual passages. The rat markings along runways may only be distinct where these runways have been used for some time.

Water. The Norway rat is a moisture-loving creature, and of necessity must live near an available supply of water. It infests all situations where water is present, and is particularly abundant along the banks and watercourses, as well as in drains and sewers. Here it not only obtains water, but also food. In certain instances it has been known to climb to the roofs of buildings in order to obtain its moisture from the tanks situated there.

Norway rats can obtain moisture by licking the dew off grass or chewing on ice. Rats drink water by lapping directly with their tongue or cupping it in their front feet. Under conditions of severe water deprivation (i.e. grain warehouse in the summer, in tropical climates, etc.), rats have been observed drinking undiluted human urine (Frantz, 1979).

Fig. 1-10. Roof rat perched on a ledge. Note the large ears.

Williams (1948) states *Rattus rattus* obtains moisture in ships by drinking the water that condenses in ship holds. He also notes this rat has been shown capable of existing for weeks without water. If fruits, vegetables and other moist foods are present, it can forego any other water supply.

Not only do Norway rats drink water, but they swim and dive in it, and are generally quite at home in this medium. The black rat is also a good swimmer, but it is usually not as closely associated with water as the Norway rat. Cottam (1948) found Norway rats in a fish hatchery competing with fish for food (horse meat) thrown into the water. The rats swam well and rapidly during the day-time. They were even in the water when the outside temperature was below freezing. Not only did the Norway rats feed on the horse meat, but they also preyed on fingerling fish in the hatchery.

Food. Although rats, as previously mentioned, are omnivorous creatures, they do have specific food preferences. Garlough (1946) states the white-bellied, roof, and black rats are fond of succulent seeds, and fresh vegetables or fruits are preferred to cereals and meats. Of all the foods, fresh sweet potatoes are the most readily accepted. On the other hand, the brown rat prefers those foodstuffs high in fat content. Williams (1948) notes *Rattus rattus* shows a preference for grain such as wheat and corn, as well as for potatoes, fruit (except citrus fruit) and eggs. Williams further observes: "When preferred foods are absent, the rat eats what is there: when pushed by starvation, it will literally eat anything. The clothes of members of the ship's crew are often cut by rats, apparently in their efforts to eat out grease spots. This also appears as the probable explanation of the constant cutting off of coat buttons, attributed to rats."

At times rats appear to feed upon the strangest materials such as leather, bone, lead and plastic pipes, sash-weights, cement, as well as wood. Some feel rats tend to gnaw on a wide variety of materials in their environment as a test of local resources.

Hamilton (1947) discusses the feeding habits of Norway rats on the farm as follows: "The appetite of a rat is prodigious. Rats apparently feed twice during the night; once shortly after dark and again in the early morning. An average rat, weighing one-half pound, will eat or destroy two-thirds as much mash as will a hen. A pair of rats will eat the equivalent of a 100-pound sack of mash

Fig. 1-11. Wild Norway rat found dead. Note the tremendous growth of the top left incisor.

or grain in a year. The average farm may readily support 50 rats. If half their food be waste, they yet eat enough to support 16 fowls for a year."

Rohe and Stone (1962) note Norway rats eat meat meal, tankage, copra, and blood meal unless stored in rat proof containers. "Rats of the same species living in the same community may show considerable variation in their appetites for different foods."

Although some authorities claim cannibalism among rats is a common practice, USPHS (1948) notes that in captivity cannibalism ordinarily does not occur among well-fed, caged domestic rats. In many instances where trapped rats are found eaten, this may be due to cats, mice, and other animals that feed upon them. Rats of the same species are usually friendly to one another, yet they do not hesitate to devour their injured or weakened brethren.

Mills (1945) makes these additional notes regarding the feeding habits of rats: "Rats tend to carry food and other articles to their nest. They select the best or choice parts of the food when possible. They may have a regular hour for feeding and eat like squirrels, holding a morsel of food in their paws. After eating, they clean their mouth and foreparts with their paws. An average adult rat may consume one ounce/28 g food and drink two ounces/59 ml of water a day."

The behavior of rats in relationship to rat-baiting is considered in greater detail in the section on control.

In a study conducted in our laboratory with Albino Norway rats and reported here for the first time, it was found the color of rat fecal pellets is dependent upon the diets consumed over the previous three days. Rodents fed carrots produced orange-colored fecal material, those fed meat (Alpo® all beef) were dark brown, a macaroni diet produced a dark black color, and a fruit diet of apples and oranges resulted in reddish brown droppings.

Migration and range. Whelan (1944) discusses the migratory habits of rats as follows: "Twice a year there is a restlessness among the rats resulting in a

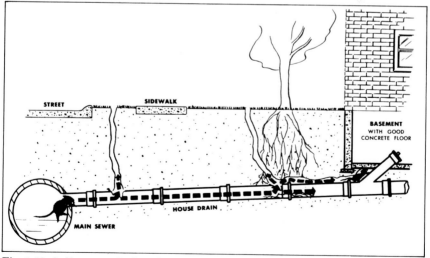

Fig. 1-12. A diagramatic sketch showing how rats may enter a house through the sewer.

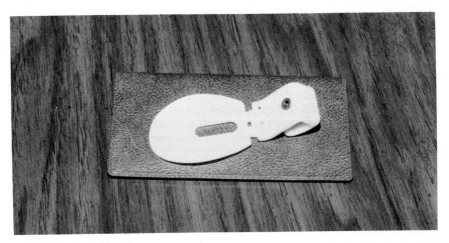

Fig. 1-13. Pictured above is a barrier for preventing rats from coming up through the toilet bowl.

mass movement, sometimes manifested by a general migration and, it takes place slowly over a longer period of time. In general, there are two such migrations each year. One takes place in the spring when there is an urge for more succulent food or for places where there are fewer rats. This spring-fever is usually away from the farm buildings to the open field or from old buildings in town to the country and to vacant lots where rubbish has accumulated. It is an urge to find better places to eat and more commodious places to bring up their families without too much competition. The fall migration is from the fields and vacant lots back to the farm and city buildings. This is a protective urge seeking shelter from the coming cold months to places where they can find stored up food. They must eat during the winter because they cannot hibernate. If no food is available they will not remain."

Davis, et al. (1948) studied the home range in the Norway rat. They define *home range* as "the term commonly used to designate the area regularly frequented by an individual. Its size may vary from season to season, according to sex, or to population density."

"The distance a rat regularly moves is probably largely dependent upon the relation between suitable harborage and the food supply. If the two are close together the rats travel little; if they are far apart the rats travel much." If the source of food as well as their shelter is shifted or destroyed, this likewise causes them to alter their usual movements. In cities rats may live for months in an area 60 feet in diameter." The above authors did not observe truly seasonal movements although some local readjustments did occur in spring or fall due to environmental changes.

Davis (1948) notes when rats are liberated in strange places they may wander far. Creel (1915) captured wild Norway rats, marked, and released them. Some of these were captured as far as four miles from the point of release. Stewart (1946) tagged rats in Los Angeles and captured them nine days later 11 miles from where they were released.

Cohabitation. According to Williams (1948), the black rat preys upon mice and may drive them from a building in which they are unable to protect themselves. "The Norway rat, being heavier and stronger, preys upon both the black rat and the mouse. Not being as good a climber as the black rat, however, it cannot drive it out of overhead locations." Thus, it is possible for both rats and mice to be present in the same building. Storer (1948) states a reduction of rats in a building will often result in an increase of mice. USPHS (1948) notes both roof rats and Norway rats may occur in the same building and even in the same rooms. This source also notes the Norway rat and *Rattus rattus* often infest certain districts and buildings in a city. Milmore (1943) states although the roof rat and the Norway rat may cohabit buildings, in such an event only the Norway rat is to be found in burrows. He also observes the roof rat may be found under buildings, as well as in basements and sewers, besides its usual places in the upper parts of structures.

Life history of the rat. Hinton (1931) and Silver (1927) studied the life history of rats and much of the following is from their papers.

The black rat becomes sexually mature in four months and continues to grow until it is 18 months old. The brown rat may reach sexual maturity in two months. Several authorities claim the life span of a rat is between three and five years. Davis (1948), whose studies were primarily with Norway rats, notes only about five out of 100 rats live as long as 12 months and the average life of rats in a population is about six months. The adult females usually live longer than the adult males.

Rats are capable of breeding every month of the year. Davis (1948) found reproduction to reach a peak about May and then again in September. Perry (1945) notes in England the Norway rat showed the greatest period of breeding from March to June. According to Davis, "Studies conducted in Baltimore blocks and markets show that if the rat population is reduced in the fall, it will return to the original level in about 12 months and if reduced in the spring it will reach the original level in about six months." Williams (1948) notes the status of a rat colony may be judged from the number of young rats, and a colony with 20 percent or less young is probably decreasing in size.

The number of young in a litter of Norway rats may vary from six to 22, and Williams (1948), who examined thousands of *Rattus rattus,* states the average number of embryos in pregnant females is six to eight. Perry (1945) notes the average number of young in a litter of Norway rats is seven to eight. There may be from three to 12 litters of Norway rats in a single year with an average of three to six per annum. There is a record of seven litters of Norway rats in seven months. The number of litters per annum, and the number of young per litter, are dependent on the food supply, harborage, age and condition of the female competition, temperature, climate, and other factors. Calhoun (1948b) observed rats living under poor environmental conditions grew slowly, and in some instances did not give birth to young. In warm temperate climates, the black rat is more fecund, whereas the Norway rat is more fertile in cold temperate regions.

The males are capable of pairing at all times, but the females can only pair during heat, periods of which may extend over nine months. The estrous cycle extends four to five days. There are four stages. In the first stage, which lasts about 12 hours, heat may occur. In the second stage, also 12 hours, there is heat and copulation. In the third stage, which lasts 15 to 18 hours, the animal is not in heat. In the fourth period, about six hours, eggs are in the oviduct. (Farris and Griffith, 1949). Gestation is from 21 to 25 days. The female may be impregnated a few hours after birth of a litter. Hinton (1931) states Lataste found "after effective coitus the vagina is plugged by a stopper, the purpose of which is apparently to prevent the escape of semen from the female before impregnation has taken place. This vaginal stopper is a joint production of the sexes, its larger, central and quickly coagulating portion being furnished by the male. The stopper remains in place for some hours and then is expelled from the vagina."

The pink young are helpless, blind and naked at birth with their external ears sealed down. According to Brown (1960), fine hair appears on the body in about a week and they open their eyes in 12 to 14 days. The young depend on the mother for food for about three weeks and then feed on solid food.

The largest wild rat personally viewed by the author was a 1.36 lb/617 g adult male. The rat was caught on a hog farm in Goldsboro, North Carolina. The rat was weighed in the research facilities at ICI's Biological Research Center in Goldsboro, North Carolina. The largest wild Norway rat known to the author was weighed by Dr. Stephen Frantz (Frantz, 1979). He recorded one adult male Norway rat weighing 22 oz/620 g from Washington D.C. Salmon et al. (1977) reported capturing three male Norway rats weighing 20½ oz/620 g, 19 oz/581 g and 18½ oz/536 g each. These three were the largest of 28 rats caught in California in 1976. Fifty percent of the 28 males trapped exceeded 1 lb/454 g in weight.

Natural enemies. Stewart (1946) notes cats will catch only small or half-grown rats as the majority of them are afraid to tackle a full-grown Norway rat. Terriers are good rat catchers. However, since rats can escape into holes, terriers cannot catch all rats in infested buildings. Lynch (1941), writing about New York in the period between 1879-1880, states rat-baiting was a form of so-called sport that achieved a minor degree of popularity. Rat terriers were especially bred and trained for this purpose.

From one to several dozen rats were dropped into zinc-lined or tin-lined pits where they were permitted to run around for a short period in order to "warm up." The dog was then placed into the ring and bets were made as to the time necessary for the dog to annihilate the rats. A good dog would kill as many as 100 rats in 30 or 40 minutes. One champion killed 100 of the squealing, nipping rats in 11½ minutes. Stotik (1967) reports the use of dogs to control rodents. This was portrayed in the "Great Train Robbery," a 1979 Columbia-Universal production.

Plummer (1978) devotes an entire book on the art of rat hunting with dogs and ferrets. Hamilton (1947) discusses the natural enemies of rats as follows: "Owls are among the most useful of all rat predators. These birds inflict very little damage upon the poulterer, but do confer vast benefits upon the farmer. Some misguided agriculturists omit no opportunity to deprive nature's means of holding in check the vermin (rats) that ruin his crops and despoil his harvest. Observers have counted the undigested parts of 172 rats under the nest of a pair of barn owls, and 113 rats were once counted under the nest of a great horned owl. The writer once entered the tower of a church in the city limits of New York and found no end of rats which had been brought by a pair of useful barn owls to their six growing young."

Ferrets are used to this day, particularly by a few pest control operators in the East. These ferrets drive the rats from their burrows. The mongoose was introduced into Jamaica, Hawaii, and other areas where it became an important enemy of rats. Unfortunately, it may itself become a pest by feeding on fruit and beneficial animals, including rare groundnesting birds.

Other enemies of rats are weasels, skunks, and reptiles, the latter being especially important. Beal (1943) observed a boa constrictor that kept a residence in South America free of rats.

The use of microbes for control of rodents in the field was actively pursued in the very early 1900s. A product called Danysz Virus was used to control "field mice" in France, rats in London subways and rats and mice in department stores in Hoboken, New Jersey. It was even used on Riker's Island, once a 600-acre garbage dump in New York City, but now a prison. From an old flyer presented to this author, Virus Limited Inc., New York, New York, called for the use of one dozen tubes of the product to take care of about 25 to 30 rat holes. A small house could be cleared of mice with one tube and six tubes for an ordinary dwelling-house containing rats. This product is no longer used or registered.

Rat fable. The stories regarding rats and eggs are many, and Gudger (1935) has traced them back to 1291 A.D. According to Gudger, many stories have risen concerning the rats' ability to remove eggs intact. There are even authentic eye-witnesses to verify the rats' ability in this respect. A modern version of the rat-egg-story is presented herewith: "I've heard the tale three times from different men: each swears it is true. All three claim to have seen rodents carry hen eggs from the shelf to the floor in this fashion; the rat climbs to the shelf grasps an egg with its front legs and jumps to the floor landing on its back. 'Brother Rat' then carries the unbroken egg to the nearest hole."

Fig. 1-14. An imaginative drawing of rats transporting an egg without breaking it. After LaFontaine, *Fables* — 1759.

Said the mouse
"Much protection you are to a house!
I go and come at will, I eat my fill;
Early and late.
What sort of strength is this you think you prove
And never move?
Said the trap, "I wait."

                                                    —*Charles Malam*

THE HOUSE MOUSE

Garlough and Spencer (1944) note the word "mouse" can be traced to the Sanskrit word *musha* which is derived from a word "to steal." Mice were well known to the ancient Greeks and Romans and were featured prominently in their art and literature. This cosmopolitan rodent is believed to have come originally from Central Asia.

Several centuries ago two subspecies of house mice came to the United States on ships from Europe. The English, French and Dutch ships brought *Mus musculus domesticus* which occupied the northern United States and southern Canada. The Spanish and Portugese ships brought *Mus musculus brevirostris* which occupied the southern states, California and Latin America (Schwarz and Schwarz, 1943). In the Midwest, the mice interbred. Today there are more than 300 separate varieties of house mice native to the United States. Mice are found throughout the world from the tropics to the Arctic regions.

House mice should not be confused with young rats. A young rat has a head and feet which look too large for the body. The mouse's head and feet are in proportion to its body. Adult house mice are about 3½ inches/89 mm long. The ears are moderately large and distinct. Mice are dusky-gray, but some may be light brown to dark gray with the belly slightly lighter or a very light cream color. Their tail is semi-naked and about as long as the body and head combined (2½ to 3¾ inches/65 to 90 mm). An adult mouse weighs ½ to one ounce/12 to 30 g.

Fig. 1-15. The house mouse, *Mus musculus.*

The albino form is bred and used for laboratory studies.

House mice can live outdoors as a field rodent. However, they should not be confused with pine mice, meadow mice, moles, and shrews, all of which are entirely different animals.

Life history of the house mouse. House mice are extremely prolific breeders. At 35 days of age they mature and carry for 18 to 21 days. If a mouse aborts, it can become pregnant again within 48 hours. The average litter size is about six. Variations in genetics, food supply, and temperatures can affect these numbers. At low temperatures, mice produce fewer offspring and they are smaller in size. A female with young suckles her young for about four weeks. Therefore if all goes well, a female can have another young litter approximately once every 50 days. The female may become pregnant while still lactating. In such cases, the gestation period will be lengthened. Under optimum conditions (often found indoors) mice can breed throughout the year. Outdoors, mice are more seasonal breeders, peaking in the spring and fall.

Newborn mice are extremely small, blind, pink and naked except for short vibrissae. They weigh between 0.02 and

Fig. 1-16. This mouse spent the night licking the syrupy side of this empty soda bottle and was trapped by its waistline.

0.03 ounces/0.5 and 0.8 g. After two weeks, the eyes and ears open and the young mouse is covered with fine hair. After about three weeks, the young mouse is fully covered with hair, makes short trips from the nest, and begins feeding on solid food (Marsh and Howard, 1976). At four months of age an adult mouse weighs about 25 grams, slightly less than an ounce. Eaton and Cabell (1949), on studying laboratory mice, state, "Young mice may be moved from their dam at three weeks of age and the dam rebred. A female is not usually productive after 15 months, but may live much longer. Male mice have been known to live as long as 2½ to three years."

Garlough and Spence (1944) note mice usually live for 15 to 18 months and some have lived up to six years. However, considering their natural enemies and the diseases to which they are susceptible, it is generally agreed their life expectancy is less than a year (Truman, Bennett and Butts, 1976).

Storer (1931), Orr (1944), and others have observed "community nests of mice wherein several females may share the same nest with their accumulated brood. Orr observed as many as three females to use the same nest. Storer found one nest to harbor 36 young. "Of these one brood was newly born, another of a size ready to leave the nest, and the remainder of intermediate development, with the eyes still unopened."

Feeding patterns. Garlough and Spencer (1944) note house mice "eat about the same kinds of food as do human beings, including meats, grains, cereals, seeds, fruits, and vegetables. They prefer sweet liquids to pure water for drinking." According to Mills (1947 a & b), house mice feed 15 to 20 times a day, consuming 100 to 200 milligrams of food at each feeding. Ives (1948), speaking of the feeding habits of mice, says, "They do not sit down and eat a large amount of food at one sitting like a rat does. Instead, they duck in and out and dart around picking up a morsel of food here and there."

The same author notes that four seeds suitably treated with a poison will kill a mouse, and a mouse will eat that many at one time.

While mice are nibblers and feed many times in many places, they have two main feeding periods. They eat at dusk and just before dawn, interspersed with many other feeding bursts approximately ¾ to 1¼ hours apart (Marsh and Howard, 1976). They have to consume about 10 to 15 percent of their body weight every 24 hours. When water is available, they may drink it.

Dysktra (1950), quoting from the laboratory reports of H. J. Spencer, states "mice are able to exist for periods exceeding four months without water, and that food preference varies from dry cereals (corn meal, rolled oats, etc.) in the presence of ample water to moist foods (apples, etc.) in the absence of water."

Southern and Laurie (1946), who studied the house mouse in grain stacks in the field, make the following notes about the water requirements of house mice: "Unlike rats, mice do not need to forage outside a rick for water; in captivity they live in good health on a diet of wheat, which itself contains about 15 percent of water, though they will take a small amount more if it is provided. This is generally less than one ml per day, and this quantity would be easily obtained on the outside of the rick in the form of dew or rain. This probably accounts for most mice in a rick visiting the outside, and for the readiness with which they take poison bait mixed with water."

The required water uptake in wild house mice, collected from a California salt marsh, was studied by Fertig and Edmonds (1969). They reported: "In their jars, where they lost little water by evaporation, the mice maintained themselves at full body weight on a 'natural' diet consisting only of dry seeds. Even in the lower humidity of the cages they lost little weight on this diet. When the diet was changed, however, to one with a high content of protein or roughage, the mice, if deprived of drinking water, reduced their caloric intake of food. The metabolic processing of large quantities of protein increases the excretion of urine, and the processing of roughage increases the loss of water by way of feces. Hence, in order to conserve water the mice on such a diet ate less, thereby subjecting themselves to slow starvation. Thus, for a house mouse unable to obtain drinking water a high-protein or high-roughage diet is essentially lethal.

"On the 'natural' diet of dry seeds, without drinking water, the mice were able to maintain their body weight at a stable level about 10 percent below normal. They were unable, however, to stabilize body weight on the Purina Chow diet usually fed laboratory mice unless this was supplemented with water. With this diet they needed about half a milliliter of water per day, about 12 percent of the amount they would consume when allowed to drink freely. When they were given the small water allowance, they ate substantially more Purina Chow than when they were deprived of water entirely.

"In the laboratory, with no intake except dry seeds, the mice were able to survive for months in apparent good health. No doubt the house mouse rarely has to endure a regime as severe as this in its usual habitats. It appears, therefore, that the mouse can live through almost any naturally occurring drought or dry season if it has access to a stored supply of seeds or grain and can establish a den where limited air circulation keeps the relative humidity high. We found that dehydrated mice could tolerate a temporary loss of body weight of about 40 percent. When they were returned to a full ration of water, they recovered full body weight in a short time-within 40 hours on the average. In this respect they resembled other small animals that have been studied and somewhat resembled the camel, which can make a dramatic recovery from its emaciated appearance with a single long drink."

Therefore, the success of liquid baits for mouse control may depend upon the diet of the mice. Mice feeding on high protein diets must supplement their diet with free liquid.

Once mice find a food to their liking, they may avoid all other items. It is quite common for mice to select one specific brand of candy in a supermarket and bypass anything else you attempt to use to achieve control.

Klimstra (1968), when testing feral mice, found two of their favorite liquid and solid form items were prunes and pineapples. Other attractive solid foods included salted peanuts, "Cheez-it" crackers, cornmeal, and wheat germ.

Social behavior. Women's liberation has a long way to go in the world of house mice. Each male mouse stakes out a territory and guards it. Within the territory can be several females and lower ranking males. While the dominant male is busy defending his territory, the female mice may be "getting accquainted" with the lower ranking males. Female mice will often mate with more than one male. A mouse's territory depends upon a number of factors including number of mice in the entire structure and arrangement of materials within the structure. The more mice present, the less territory each has. Some mice can remain in a desk or pallet for their entire lifetime, but keep in mind that the mouse is climbing up and down within the materials stored on the pallet or in the desk. Mice entering a territory already occupied are not welcome and are driven off. When mouse populations swell, the mice will seek out rodent bait stations, exposed window ledges, and any other area where they can hide from other more aggressive mice.

Mice are cannibilistic and will feed on each other when hungry. The author has seen hundreds of mice caught on glue boards partially eaten by other mice. Mice in multiple live traps will often be eaten by other mice caught in the same trap.

Because mice scurry from place to place and deposit fecal droppings wherever they please, the easiest way to determine if mice are present is to locate their droppings. Other signs include gnaw marks, small holes in walls and doors, and a pungent odor of their urine. The easiest way to discern active infestations is to sweep up the droppings and see if new ones appear the next day.

Emlen (1950) made a study of the distances mice travel by capturing 1,572 mice, marking them, and then releasing them at the point where they were caught. He found the average distance was 12 feet/3.6 m and that 90 percent moved less than 30 feet/9.1 m and 70 percent less than 10 feet/3.0 m. From this he concludes mice are "stay-at-homes" and do not move around unless disturbed. As a result of his studies, Emlen recommends the distribution of a large number of poison baits spaced at least 20 feet apart. Southern (1946) notes the average range of mice in cellars is not more than 48 square feet in 24 hours.

INTERESTING PHYSICAL AND BIOLOGICAL ABILITIES

- Mice are excellent climbers and can run up almost any roughened wall without breaking stride.
- Mice, although preferring not to swim, can do so. More than once a live

mouse flushed down a toilet has resurfaced a minute later.
- They can jump a vertical distance of 12 inches/30.5 cm from the floor onto an elevated flat surface.
- Are capable of jumping from a height of 8 feet/2.5 m to the floor without injury.
- Survive and thrive in cold storage facilities at 14° F/−10° C.
- Capable of squeezing through an opening slightly larger than ¼-inch/six mm in diameter.
- Mice can run horizontally along pipes, wires, and ropes.
- Develop thick long coats of hair when living in a cold environment (i.e. in cooler boxes).
- In six months, one pair of mice can eat about four pounds/1.8 kg of food and during that same period produce some 18,000 fecal droppings.
- Mice feeding on colored crayons will produce droppings based on the color of the crayon they feed on.
- Mice chew on electrical wires and thereby are capable of starting fires.
- Mature mice are not blind, but have poor vision and cannot clearly see beyond about six inches/15 cm.

Occasionally we hear of mouse plagues in which the mice become so numerous they overrun the country. These usually develop in the summer, following a mild winter and a plentiful food supply. Such plagues are usually terminated by a disease epidemic which quickly decimates the mice, reducing them to normal numbers.

Hall (1927) observed such an outbreak in Kern County, California. The mice were for the most part fully grown. He computed the number of mice per acre at 82,280. He notes: "Truly the number of mice was almost unbelievable, and one who has not seen this or a similar outbreak can scarcely comprehend the vast numbers that can occur in a given area of limited extent. Certainly the numbers were to be reckoned in the tens, and possibly in the hundreds of millions."

The mice damaged grain, reduced sacks of straw to chaff, destroyed foodstuffs, clothing, bedding, linen, and other articles in houses, and gnawed holes through floors and walls of frame buildings. Hall arrives at the following conclusion in regard to this unusual outbreak: "The causes of this overabundance of house mice may therefore be stated as: favorable meteorological conditions, abundant food and shelter, and removal of the principal natural enemies of small rodents that normally hold their numbers in check. The factor determining the time of the spectacular emigration of the mice was, probably, the destruction of their food and shelter."

Hinton (1931), in discussing the great mouse plagues of 1916 and 1917 in Australia, notes a farmer who baited for mice and picked up 28,000 dead on his veranda the following morning, and only stopped "because he was tired." Besides feeding on grain, the mice ate such unusual articles as lead pencils, leaden bullets out of cartridges, and seaweed on the beach.

Klimstra (1968) best describes the behavior of a mouse as it enters a structure in the following manner: ". . .he moves in this new territory for investigation and exploration, he tends to go along the edges. He will go a short distance and come back to where he made his entrance and then maybe he'll go a little farther and then, after a bit, a little farther. After a period of time he has investigated most of the edges of this facility. Now he has developed some confidence, but in addition, he has established patterns or blueprints which he is able to remember kinaesthetically, that is, by subconscious recordings of muscular reactions which allow him to respond in various ways on the basis of what the past experience has been. Once he explores the edges, then he starts moving into the interior. The amount of time he is going to spend in this interior is going to depend on what is in the interior; and, exactly where he is going to be will depend on what's there. If you have various kinds of structures in this interior, he is going to start investigating and when he does, it's much the same as he did for the perimeter of the room. He is going to go into the interior areas at a given point and will explore around the edges of any structure. Again he is developing memory patterns and as he does so he may move to another part of the interior. Pretty soon he will have several routes that he knows will take him back to his entrance place; he returns there because this is the place with which he has the greatest familiarity.

"We have to appreciate that when a mouse moves into a new environment he goes through a process of developing memory paths of use of this area, and this, of course, is an important principle which has to be considered in any program of control. If you are dealing with a new mouse coming into a building, you know that he is going to be around the edge, but after he has been there for awhile (which may take no more than about an hour) he may learn other patterns of movement. As I say, the diversity of the interior of this area is going to determine a great deal how simple its going to be to be able to trap him or bait him. But, if something happens to his home base, that point where he came in, he suddenly becomes all excited and starts scurrying around because that is the one thing with which he had greatest familiarity. In about 15 to 20 or 30 minutes, he will have established some other place as a home base. It may be one of those items that is in the interior of the room.

"Another thing that we see in conjunction with house mice is that if you make changes in this area, if you move any of the facilities, immediately it activates the mice. This is important to them because they operate in the fashion that they do, that is, on the basis of the subconscious recording of patterns of muscular movement. They have to find their way around and become totally familiar with it again or otherwise if they are running after being frightened or scared and trying to return to their home base, they won't know where these obstacles are and may run into them. You must remember that mice are relatively nearsighted. Also, even though they may be able to establish scent trails, when they are moving fast, they don't have time to pick up these trails. The end result is that they do not have a familiarity as to where these objects are, which will cause them to certainly run into them.

"In a stored food area the food is stacked on skids and there is some readjustment of these stacks, this is how you bring these mice out of their confined areas; they are now subject to trapping because they are out to investigate the rearrangement."

Disease carriers. Cameron (1949) discusses mice as carriers of disease and the following is largely from his paper. According to Cameron, mice may transmit disease to man in a number of ways including:

- By biting man.
- By infecting human food with its droppings.
- By infecting human food with its urine.
- By being eaten.
- Indirectly via the cat or dog.
- Indirectly via the blood-sucking insects.
- Indirectly via bloodsucking mites.
- Indirectly by dying in a water supply and contaminating it with organisms

contained in its body at death. There is no evidence this has happened, although it is a possibility which must be considered.

Although rat-bite fever is usually transmitted by the bite of a rat, it may be likewise transmitted by a mouse.

Aside from the spread of food poisoning, the importance of house mice as carriers of disease and parasites does not rank as a major threat to man. However, the potential is there.

Mice may infect foodstuffs with their droppings which may harbor such food-poisoning organisms as the *Salmonella* bacteria, or the microscopic eggs of the tapeworms of the genus *Hymenolepis.*

Leptospirosis (Weil's Disease) in man is caused by the contamination of water or food by rat or mouse urine which is infected with the causative organisms, *Leptospira,* a spirochete.

Mice carry *Favus,* a fungus disease of the scalp which results eventually in the formation of bald patches on the head. It can be transported to man by mice or contracted indirectly from mice through cats (Davis-Rowe, 1963). Plague and epidemic or murine typhus are carried by infected fleas on rats, mice, and other rodents.

Rickettsial pox is transmitted from mouse to man by the mite, *Liponyssoides sanguineus.* The disease is caused by the organism, *Rickettsia akari.* In the United States, cases have been reported from Boston, New York City, Cleveland, Philadelphia and West Hartford, Connecticut (NPCA, 1971). Huebner, et al. (1946), showed the mite was responsible for the outbreak of this disease among apartment dwellers in New York, New York. It should also be noted rats carry other mites which may infest and cause nervous and eczematic reactions in humans.

Mice may be the vectors of such diseases as lymphocytic choriomeningitis, histoplasmosis, and tularemia. "LCM (lymphocytic choriomeningitis) is a virus disease transmitted from rodents to man. Since the principal symptoms are fever, headache, and muscle pains, it may be impossible to distinguish it from influenza or many other diseases on clinical grounds. Up to 50 percent of people infected may have no symptoms at all. On the other hand, as many as one in five clinical cases may require hospitalization because of central nervous system involvement (encephalitis or meningitis). Onset is from one week to two months after exposure to an infected rodent, and the duration of the illness is usually two to three weeks, sometimes with a remission. Fatalities are rare. Convalescence may be prolonged, but no permanent damage has been recorded." (Woodall, 1975).

The white footed deer mouse *Peromyscus* will invade structures and can even be found in attics. This rodent has a distinct bi-colored tail and when viewed from the side, a distinct line appears created by the difference between the tawny brown top and white underside. Deer mice are the same size and larger than house mice. They are seed feeders. In the West meadow mice or voles (*Microtus* sp.) invade homes for short periods of time. It has a chunky larger body and weighs more than twice as much as a house mouse. Mice often leave fields in groups in the fall or winter and move directly into heated buildings.

RODENT CONTROL

The world still has much to learn on how to control rodents effectively. As recently as 1979, in some parts of the world, the bounty method was still a primary approach to rodent control. UPI (1979) reports:

"Jakarta, Indonesia (UPI) — Local authorities in central Java have decided the price for love and hate will be 10 and 23 rats, respectively.

In a recent decision, the authorities of Boyolai decided not to marry or divorce any couples unless the partners could hand over 10 rats if they wanted to wed and 23 rats if they were filing for divorce.

It's all part of a campaign to eradicate rats, which are far too numerous in the area, a Boyolai official explained Thursday.

The new regulation is an apparent success: 22,683 rats have been brought to authorities in the five days of the anti-rat campaign."

With such situations still existing, this portion of the text provides the necessary information to program rodent control from an integrated pest management approach.

The author recognizes the usefulness of the concept of pest management in rodent control. Therefore, this section on control strives to cover a balance between chemical and non-chemical controls. From the viewpoints of efficiency and environmental soundness, a balance between the use of rodenticides and non-chemical approaches is advisable. To avoid repetition, the control section for both rats and mice are combined. However, where appropriate, specific sections are outlined for each species.

RODENT SURVEY

A survey of the building when properly conducted will shed light on the type and extent of infestation, the breeding places or harborages, openings whereby the rodents enter the building, and the method and control most suitable for the premises. When making a rodent inspection a form is useful for recording information.

Williams (1948) made a detailed study of the signs of rat infestation on shipboard, and the results of this investigation, as well as those of the U. S. Public Health Service (1948) are incorporated in the following discussion:

According to USPHS (1948), the most important signs of rat infestation are droppings, runways, tracks, and gnawings; rat burrows are occasionally useful for estimating the degree of rat infestation. Other signs of rat infestation are, of course, the presence of live or dead rats, rat nests, rat odor, and undue excitement of domestic pets such as cats and dogs.

Droppings. According to Williams, this is a sign on which most inspectors rely: "The excreta of the rat is in small firm masses. These are rod-shaped, straight or slightly curved, with rounded ends. In size they vary from one-fourth inch long by one-sixteenth in diameter to three-fourths inch long by one-fourth in diameter. Nearly always they are quite dark or black in color. Where freshly passed, they are soft enough to be squeezed out of shape and often have a glistening wet appearance. Within two or three days they dry and become hard. Later, the surface becomes dull. Very old ones are dust or dirt covered, and may be discolored." Rodent hairs often are embedded in rodent feces. Mills (1945) notes fresh and old feces indicate the presence of a long standing infestation. The droppings from the Norway rat are longer than those from the black rat, and are to be found in greater numbers along runways, near the harborage, and in secluded corners It is possible in the survey work to learn three valuable things through the examination of excreta: (1) whether the infestation is an old one or of recent origin, (2) from the quantity and location, the probable extent of the infestation, and (3) from the various sizes of excreta, where families of rats are being reared. (Fig. 1-17).

SUGGESTED RAT SURVEY FORM

Note: The following might be considered as representing an ultimate "Form" card, 4 × 6 inches, printed on both sides of the card, easily carried by the inspector and convenient for permanent filing.

(A) Front Side of Form

PROPERTY INSPECTION FOR RATS—GENERAL INFORMATION

(Residential and Business Areas)

1—Location ——————————————————— Code —————————
2—Owner and Address ———————————————————————
3—Occupant —————————————————— Used For —————————
4—Type of Structure: Wood ——————— Brick ——— Stucco ——— Other———
5— Food Supply: Debris ——————— Exposed Foods ————— Garbage ———
   Partial Basement ———————————
6—Condition: Good ———————— Fair ———————— Poor ——————
7—Other Structures on premises: Garage ———— Shed ———— Other————
8— Previous History: New——————— Occasional———————— Frequent————
   Continuous ————————————
9— Previous Control: Baits ————— Traps ——— Gases ——— Proofing ——
   Self-Help ———————— PCO ——————
10—Past year's damage estimated $ ————— Illness, bodily harm —————
Date———————————— Inspector's Name————————————

(B) Reverse Side of Form

REASONS FOR PRESENT RAT INFESTATION

1—Rat Evidence, Feces ————— Gnawings ——— Holes ——— Odors ————
   Rats Seen ———— Runways———— Tracks ————
2—Entrance By: Adjoining Premises ————— Burrows ——— Coal Shutes ———
   Conduits ——— Doors——— Foundation Walls ——— Pipes ——— Pits ———
   Unexcavated Spaces ——————— Windows——————— Other—————
3—Interior Passages: Conduits———————————— Plumbing ——————————
   Structural Openings —————————————— Other————————
4—Inside Harborages: Rubbish ——————— Stored Goods ——————— Enclosures ———————
   Double Floors ——————— Double Walls ——————— Closed Showcases ———————
   Enclosed Toilets——————————— Fixtures ——————————— Other—————————
5—Food Supply: Debris —————— Exposed Foods —————— Garbage ————
   Improper Storage ——————————— Water ———————————
6—Damage to: Food ————— Fixtures ——— Structure ——— Other Materials ———
7—Outside Conditions: Debris ———————————— Garbage ————————— Rubbish —————
   Alleys——————— Other————————
8—Environmental Factors: — Explain ———————————————————
   ——

This form and instructions for its use have been prepared by Ernest M. Mills, U.S. Fish and Wildlife Service, in cooperation with the National Pest Control Association.

Fig. 1-17. Rat feces.

Runways. Runways are frequently traveled routes. The dirty and greasy body of the rat leaves characteristic marks on surfaces contacted by it. USPHS (1948) notes: "Generally, the runways along beams and pipes are indicated by darkly discolored, greasy markings that are most evident where the rats have had to swing under or around obstructions in their pathways, thus producing semi-circular markings at the points of obstruction. In bakeries and grain mills where rats get white flour on their bodies, the runs may be outlined in white. The discolorations of runways vary from a very faint shading that require careful scrutiny to delineate to definitely outlined ones that require little or no search to find." Runways frequently occur along the inner sides of pipes and wires. Where rat runs are present along floors, stairs, shelves, and similar places, these are made evident by the presence of rat droppings and rat tracks and dusty surfaces. According to Williams, "Runways are of the utmost importance to the rat-proofer since they show him where to place a barrier and where the harborage that must be closed or removed is located."

Tracks. Next to the absence of droppings, the absence of rat tracks is indicative of no rats. In dust, the trail of the tail and the four-toed front paws and five-toed rear paws can be seen. Usually, the rat drags its tail only when moving slowly. Rat tracks are most readily seen when the light is from the side. The separate toes can be seen only on a thin layer of dust over a hard surface; in thick dust only pits are visible.

According to USPHS (1948), "Various sized tracks point to the presence of rats of different sizes or different ages. Rat tracks not only indicate presence of these rodents, but furnish information regarding places they travel or frequent, and should always be sought for when making rat infestation inspections. Sometimes tracks will be found on the tops of large bottles in drug stores or on tin cans on grocery shelves. They may be found on the tops of desks, along shelves, on the tops of crates and boxes as well as along the floors."

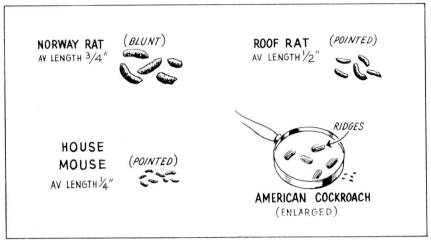

Fig. 1-18. Fecal pellets of rats, mice and cockroaches.

USPHS (1948) recommends testing for tracks by sprinkling a thin layer of non-toxic dust, not over $1/16$-inch thick, over a small area on floors, shelves, beams, and any other place where rats have been known to travel or are likely to use as runs. The dust can be applied with a large can having several openings punched in its top. If there are any rats on the premises, positive evidence should be found the next morning by inspecting the spots covered by dust for tracks. Powdered chalk or talc have proved suitable for testing for tracks and are preferable to flour which may attract insects. Unscented "dry" spray deodorants also may be used and these work especially well inside plastic bait boxes.

Gnawed materials. Gnawed materials in certain instances reveal the time when the rats were present. Gnawed fresh foods change their appearance within 24 hours. Wood recently gnawed presents a fresh appearance for about one week. Badly damaged merchandise usually indicates the presence of large numbers of rats. Tin and lead pipe may show the tooth marks of rats.

Live rodents. The presence of live rodents, particularly where there are many good rat-hiding places, indicates a heavy infestation.

Dead rodents. The presence of dead rodents may be due to poison or disease.

Nests. Nests prove the presence of rodents, breeding, and the establishment of colonies. Fresh droppings near a nest indicate a fresh nest. Old droppings indicate an abandoned nest.

Rodent odor. Individuals experienced in rodent control who have a keen sense of smell may detect the odor of rodents. The occurrence of this odor in a large room usually indicates the presence of many rodents.

Undue excitement of domestic pets. Cats and dogs may excitedly probe an area where rats or mice are present, and often indicate this by scratching or making unusual sounds.

USPHS (1948) states that in general few rats are found in nonfood establishments such as shoe, dry goods, furniture, or jewelry stores, whereas food establishments such as restaurants, grocery stores, or grain warehouses, particularly where sanitary conditions are poor, usually serve as feeding places for rats.

The U.S. Public Health Services uses such criteria for estimating the number

Fig. 1-19. Rat smears on beams caused by the rats swinging around each joist.

of rats in food establishments as evidence observed by occupants, losses caused by rats, degree of cleanliness, harborage, and evidence of rats such as those previously mentioned. Although all of the above criteria are used by the experienced inspector, one of the more useful methods for estimating rat infestations is through careful study of rat droppings. According to USPHS (1948), rat infestations may be estimated as 20 rats or less, 20 to 50 rats, and over 50 rats. The following is from their handbook:

1) **Twenty rats or less.** "Even in establishments that are frequently cleaned, some fresh droppings should be found either on the tops of contents, in out-of-the-way corners, or shelves, or tops of installations. Very few droppings should be found on recently delivered packages and sacks. Evidence of young rats, as determined by finding droppings smaller than those of adults, must be given careful attention. If droppings of more than two different sizes are found, the number of rats will probably exceed 20. Not over 15 to 30 fresh droppings should be found in any one place, and the number of places in which fresh droppings are found should not exceed a half dozen."

2) **Twenty to 50 rats.** "In estimating an infestation of 20 to 50 rats probably more reliance should be placed on droppings than any other rat sign.

"The amount of old droppings will naturally depend on how frequently the places where rats travel and congregate are cleaned. Fresh, dark, shiny, or soft droppings should be encountered in anywhere from 10 to 30 places if not found in very large quantities in any one location. On packages and sacks that have been stored a week or less, depending on their contents, some fresh droppings may be found in fairly large quantities.

"Active multiplication will take place among rats in a moderately infested building, but if over three different sizes of fresh droppings are found indicating the presence of adults and two families of young rats, the infestation will probably exceed 50 adult and young. This may be the case even when the total number of droppings is small in amount. If the proportion of adult and young droppings is about equal and are found widely scattered through the building the infestation will be greater than moderate."

3) More than 50 Rats. "Many fresh droppings will be found along floors where rats travel, in dark corners, on contents and other places where the rodents may congregate, if the infestation exceeds 50. If old or new droppings are found of several different sizes, indicating considerable multiplication among the rat population, the infestation will usually be very heavy."

Another method of estimating rodent numbers and an important preliminary in some baiting programs is the use of unpoisoned bait, or pre-baiting. Once the unpoisoned bait is being accepted, a rat population can be estimated by measuring total consumption over a 24-hour period and dividing by ½ oz/15 g. A mouse population can be estimated by dividing by $^1/_{20}$ oz/1.5 g.

Rodent proofing buildings. Rodent proofing, also known as "rat stoppage," "rat blocking," "vent-stopping," and "exclusion" involves the construction and maintenance of a building so rats and mice are unable to gain entrance to or maintain harborage in a building. Rat proofing by keeping rats from having access to food and by eliminating most of their living and breeding areas results in a lasting reduction in the number of rats. Silver, et al. (1942) note all the methods for killing rats "have utterly failed to reduce materially the total number of rats in the world. Rat proofing, however, is at last making definite headway against the age-old enemy of mankind, and it is upon this that the ultimate solution of the rat problem will depend." The same authors note modern buildings with their concrete and steel materials, fire stopping in double walls and floors, reduction of dead spaces in which rats can hide, and the sanitary storage of foods are, essentially, rat-proof buildings.

According to Silver, et al. (1942), two principles must be kept in mind in the rat proofing of buildings:

Fig. 1-20. Rat tracks.

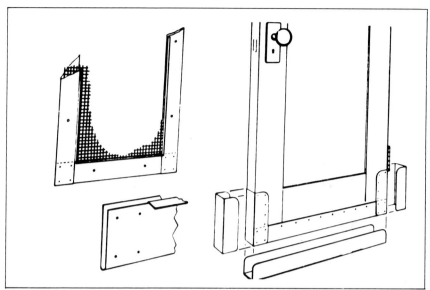

Fig. 1-21. Ratproofing of windows and doors with galvanized sheet metal.

"First, the exterior of those parts of the structure accessible to rats, including porches or other appurtenances, must be constructed of materials resistant to the gnawing of rats, and all openings must be either permanently closed or protected with doors, gratings, or screens; second, the interior of the building must provide no dead spaces, such as double walls, spaces between ceiling and floors, staircases, and boxed-in piping, or any other places where a rat might find safe harborage, unless they are permanently sealed with impervious materials."

Most contact between domestic rodents and people occurs inside buildings so it's important *not* to seal rodents indoors by proofing until baiting or trapping inside is also conducted.

Where a building is rat proofed, constant vigilance must be maintained to keep it rat proofed. USPHS (1948) notes rat proofing, "regardless of its cost or thoroughness, cannot be insured for a week or even for a day. Within this time a door may be broken by a truck backing into it, a basement window may be broken, a workman may remove the rat proofing or make a new opening, or in some other mechanical way a leak may be made through which rats may reinfest a building that has been stopped. Furthermore, rats may gnaw new openings or tunnel into buildings that have been rat stopped. This biological pressure is greatest during the first week or two after rodents have been sealed out of places in which they have been accustomed to feed."

A modern concrete and brick building, as was previously mentioned, can be rat proofed with a minimum of alterations whereas a poorly constructed wooden building with the wooden floor upon the ground may require expensive reconstruction. The following information, largely from a circular issued by the Georgia Department of Public Health (1944), lists the more important avenues of entry by rats into buildings, as well as the means by which such openings may be closed.

1) Rats often burrow beneath the foundations of buildings lacking basements. Such buildings may be protected from rat entry by placing a "curtain wall" or barrier of metal, concrete, or brick around and below the foundation of the building. Vertical curtain walls two feet below the surface of the ground with an eight-inch horizontal L or flange directed away from the building are usually effective. Although rats may burrow more than two feet below the ground, the horizontal flange will discourage them from doing so. Corrugated iron sheets, 29 gauge, as well as concrete, may be used for this purpose.

2) Rats may gnaw their way through buildings with wooden floors. These floors should be replaced with concrete where it is economically feasible, or the building should be enclosed in curtain walls as was previously discussed.

3) Openings around pipes passing through wood siding can be covered with 24-gauge, galvanized, sheet iron flashing. Bricks and cement can be used to close the vents around pipes passing through brick, stone, or cement walls.

4) Ventilator grills should be covered with 18-gauge, ½-inch mesh, galvanized, expanded metal or 16-gauge, ½-inch mesh, galvanized wire cloth. Low windows can be protected with 19-gauge, ½-inch mesh, galvanized, wire cloth.

5) Wooden doors are often gnawed through by rats. Galvanized, 24-gauge, sheet iron flashing around the bottom of the door, door sills, and jambs will prevent the entry of rats through tight-fitting doors. There should not be more than ⅜-inch clearance between door and sill. Doors should be kept closed when not in use.

6) Sidewalk gratings should be closed with 18-gauge, ½-inch mesh, galvanized, expanded metal.

7) Wooden cellar doors, particularly along the edges, should be flashed with 24-gauge, galvanized sheet iron.

8) Openings into hollow walls, particularly between the floor and floor-sills should be closed with 19-gauge, ½-inch mesh, galvanized, wire cloth, or 24-gauge, galvanized, sheet iron.

9) Rats often travel inside defective drain pipes or burrow along such pipes. A metal cover with small perforations should be cemented over the drain pipe

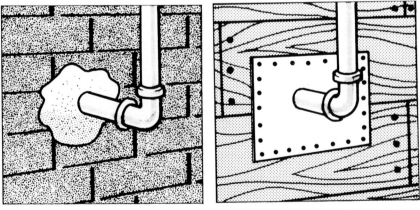

Fig. 1-22. Concrete (left) and sheet metal (right) can be used to "build-out" rodents when instituting a comprehensive rodent control program.

in the floor, and openings around the drain where it enters the building should be patched with cement mortar.

10) Rats may burrow through large sidewalk cracks into the building. These should be repaired with concrete.

11) Circular rat guards of galvanized metal should be placed around vertical wires and pipes to prevent rats from climbing upwards. Rats may wedge themselves between a drain and the wall and thus reach the upper stories of the house; a circular rat guard on the drain will prevent this.

A very common avenue of entry for rats into the open eaves of the attic is a tree limb or branch touching the roof of the house. Prune such limbs.

In structures, the most permanent form of rodent control is to build them out, thereby decreasing the chance of any reoccurrence.

Mouse proofing. Rodent proofing is usually confined to comments about rats. However, the concept is just as important for mice. The simple act of tightly ramming course steel wool in a hole leading to the outdoors (for temporary blockage until permanent construction changes can be made) may very well stop a mouse influx. Plugging holes and confining mice to smaller areas can enhance trapping techniques. An opening slightly more than a ¼-inch/6 mm in diameter is extremely small, thereby requiring time to find and patch all possible holes. Even if all such holes are plugged, mice will enter hidden in boxes or "scoot in" when somebody opens an exterior door. Building materials used for mouse proofing are discussed in the next section with rats.

Rat proofing. Rats are extremely intelligent animals when it comes to getting into a structure. So called "rat proofed buildings" frequently are infested by rats hiding in pallets carried on forklifts or by rats waiting in hidden grass

Kind	Quality
Sheet metal	— Twenty-six gauge thickness or heavier, galvanized.
Expanded metal	— Twenty-eight gauge or heavier, not larger than ¼ inch/ 6 mm mesh, rust-resistant coating, or preferably galvanized, unless made of nonrusting metal.
Perforated metal	— Twenty-four gauge thickness or greater, perforations not to exceed ¼ inch/6 mm in width.
Iron grills	— Sufficiently heavy to be equivalent to above materials; slots in grill not to exceed ¼ inch/6 mm
Hardware cloth	— Nineteen gauge or heavier, galvanized or other rust-protective coating, with no opening larger than ¼ inch (4 × 4 meshes per square inch or 6 mm).
Cement mortar	— Cement mortar should be 1:3 mixture or richer; concrete should be a 1:2:4 mixture or richer.

JEDCO of Omaha, Nebraska, produces a Homogard Rodent Barrier which is used in toilets to prevent rats from entering structures through this manner. (See figure 22).

Truman, Bennett and Butts 1976, Scott and Borum 1965 and Peterson 1978, have more details on rat proofing.

ready to rush forward as a door is opened. Nevertheless, the need for rodent proofing a building is essential for good control of rodents, particularly where structures are in neighborhoods where rats are present. Howard and Marsh (1974) outline the kind and quality of rat proofing materials needed.

Sanitation within structures. For both mice and rats sanitation within structures means cleanliness and looking to clean up "quiet zones." Mice and rats will move to areas least frequented by people. With rats, keeping a tight lid on garbage cans and picking up spills and debris goes a long way to reducing populations. Mice are another story. Someone's leftovers from lunch, the cookies used for coffee breaks, or even crayons can keep a mouse alive.

Moving all items off the floor at least six to 18 inches/15 to 45 cm and away from walls 12 to 18 inches/30 to 45 cm does two important things: A) It allows an inspector to quickly detect when new rodents are in the area. B) It increases the efficiency of baits and traps because the rodent is free to run along the walls and find the material placed there.

Stock rotation in large warehouses is most critical. First in, first out prevents rodents from reproducing without being disturbed for long time periods. This is particularly important with mice which can harbor undetected in cardboard boxes.

Blacklight. This is another tool used in the overall pest management approach. In large food processing areas, food warehouses and similar situations, the shipment of contaminated materials into a structure bypasses the best of rodent-proof jobs. Therefore, a systematic blacklight program whereby one out of every "x" number of boxes is examined with a blacklight can stop trouble before it enters a building. How many are blacklighted depends upon the items handled, available manpower, and the economics of the situation. Blacklight lamps are harmless to the skin and eyes because they utilize long wave ultraviolet energy (320-380 nm). Dried rodent urine will fluoresce when exposed to this type of light. It will glow blue/white if fresh, yellow/white if old. Rodent hair glows blue/white. The lights work best in the dark. Often, the presence of rodent droppings and the smell of rodent urine reveal the situation long before the blacklight is turned on. There are many other materials which fluoresce under blacklight including:

- Sulfide waste matter — blue/white
- Bleached sack fibers — blue/white
- Pitches and tars — yellow
- Optical bleaches — blue/white

Rodent urine will appear in a droplet pattern rather than as large blotches. This is a verification test for assuring what is being looked at is truly rodent urine. The object to be inspected is placed on Urease-Brom Thymol-Blue test paper, moistened with water, and covered with a cover glass. If a bluish spot appears after three to five minutes, it is rodent urine.

Measures around the exterior or structures. Short grass, neatly trimmed shrubs, paved roads, and proper drainage are all important in detracting rodents. The building foundation should be surrounded with an 18 to 24 inch/46 to 61 cm strip of ⅛ inch/3 mm pebbled rock in a trench four inches/10 cm deep, with the bottom of the trench lined with roofing paper to prevent weeds from growing. Any rodents moving around the exterior of a structure will seek shelter inside a rodent station, thereby achieving the desired result, mainly getting the rodent to eat outside before it ever enters a building. In residential areas it may not be possible to bait outside because the risk of children getting into the bait is too great. Loose garbage, rubbish and other debris should not be allowed to accumulate in any good sanitation program.

ADVANTAGES OF MECHANICAL CONTROL OVER RODENTICIDES INSIDE FOOD ESTABLISHMENTS

- Rodenticides in food establishments should be placed in bait stations. In order to kill the rodent, it must walk into the bait station and feed. Once inside, why let it out?
- It can take seven to 10 days to kill a mouse with anticoagulants and a rat three to five days. During that time the rodents continue to contaminate the area with 25 to 100 plus fecal pellets per day, urinating freely, dropping rodent hairs, and gnawing on numerous items. With mechanical means the rodent is caught once and is immediately prevented from any future contaminating.
- Rodenticides are toxic and can be accidently spilled, carried away by rodents or, in some cases, blown in the wind. In each case, the pesticide is no longer where you want it and could result in contaminating food.
- Rodent baits left for any length of time lose their palatability and enhance the chance of being contaminated with grain insects. These insects can then migrate to other foods.
- When used properly, rodents caught in glue or traps, do not die in walls, thereby alleviating odor and fly problems.

Mice, in most instances, can be eliminated without the use of rodenticides within a structure. Rats on the other hand pose a bigger problem. Some rats become "combat wise" and learn to avoid traps and glue boards, thereby making it sometimes necessary to use a rodenticide.

Around the exterior of a structure, rodenticides still play an important role in the control of rodents. In urban areas where children can readily get at any outside bait stations, the bait has to be placed indoors where the rodent feeds on the material after it enters the premises.

Glueboards. The use of glue to catch rodents is not new. Unfortunately, in the past many of the glues were very inconvenient to work with. During the summer months, prepared boards would melt in the trunk of a vehicle and create a mess. It was difficult to pull the cardboard apart to expose the glue. Further, anticoagulants worked so well that relatively few glue boards were used. During the last several years, several changes have occurred that have revitalized the need and use of glue boards for rodent control. These include greater restriction of pesticides in food plants, anticoagulant resistance, and the development of more convenient-to-use glues. Here are several tips to enhance the efficiency of glue:

- Place glue boards inside bait stations so they avoid getting stepped on and becoming dusty.
- Make hollow tubes out of cardboard by bending the glue board into a cylinder and taping. These can then be taped to pipes hanging near ceilings, on narrow ledges, and other areas where rodents are running.
- Some glue boards have proven extremely effective for house mouse control in areas as cold as 14° F/−10° C and as warm as 100 ° F/38° C or higher.

The success of glue boards is now well accepted. The author witnessed a control program consisting of glue boards remove more than 3,000 mice out of a cold storage facility in less than 30 days. One of the largest rodent control jobs ever undertaken in a public facility resulted in the removal of more than 120,000 rodents during a period of nine months. The majority of those rodents were removed with multiple live traps and rodent glue.

The last edition of this handbook dedicated five lines to glue boards, three of which outlined negative remarks. Times have changed and the use of glue now plays a major role in the control of rodents, particularly for mice in food establishments. Glue, along with proper sanitation, is often credited with totally eliminating mouse problems.

Rats, on the other hand, are much harder to catch, keep, and eliminate with glue. Some rats can be caught with glue. However, after awhile the remaining rodents will throw dirt on top of the glue, drag the boards away, or simply avoid them. When using a glue board for rats, the board must be secured to the ground or floor covering. Otherwise the rat may drag the board away and die in the wall of a structure.

Today the success of glue boards can be attested to by the surge in the number of companies producing and selling large quantities of glue for rodent control.

Multiple live traps. These traps have a winding mechanism which, with one setting, permits the capture of numerous mice (Fig. 1-24). To achieve best results it is important to use the trap properly. Key points to remember in maximizing the efficiency of the trap are:

- Items should be pulled away from walls. This will encourage mice to scurry along their pathway and investigate appropriately placed traps.
- Boxes of material or pallets of material containing mouse droppings and other rodent signs should be dismantled. This will force the mice to leave their protected areas and visit the trap.
- If the top inside frame of the trap is waxed the lid will slide more easily. This increases the chance of technicians cleaning the inside of the trap when it gets dirty and/or dead rodents are found inside.
- The trap should be placed either parallel to wall and two to three inches/50 to 76 mm from each wall or perpendicular and flush to the wall.
- It the traps are wound half-way the mice are "gently" flicked into the holding area. This affords a better chance the mouse will remain alive. Although no scientific evidence has proven this, it appears a live mouse in the trap leads to more mice being caught. Winding traps too much can break the winding mechanism rendering the trap ineffective.

Fig. 1-23. Typical trigger snap trap.

- Traps should be located where mouse droppings are found and near doorways.

In large food processing structures, a great deal of success in the control of mice has been achieved by placing multiple live traps outdoors near entrances. The mice are trapped before they enter the building.

TRAPPED MICE   MOUSE ENTERING TRAP

Fig. 1-24. Multiple live trap for mouse control.

Cleaning the inside of a multiple live trap can be a problem. A putty knife and wire brush have proved effective. In commercial food establishments, dead rodents, droppings, and/or hair left in traps constitute a problem and such traps should be checked daily.

Snap-traps for mice. The homeowner has long counted on the snap-trap as an effective tool for mouse control. Because it is readily available in hardware stores throughout the United States many professional pest control people are hesitant to use it. In actuality it is an excellent tool, and when in the hands of a professional does a good job. Here are several tips to enhance the art of using snap-traps.

- Tying the bait securely to the trigger prevents the rodent from licking or nibbling the bait without setting off the trigger. If a bait is used that cannot be tied (e.g. peanut butter), it is best to use small amounts.
- Sensitivity of the trap is adjusted by bending the long, narrow metal prong that touches the trigger. In areas where vibrations are a problem, the trap must be made less sensitive.
- The bait can be food for the rodent or small pieces of cotton. Cotton does not spoil and the mice pull at it for possible nesting material.
- Many traps should be used, flooding the area so the rodent has no choice but to walk into one of them.
- Traps can be used again and again. The rodent is not repelled by the odor of a previous rodent.
- Use of a small piece of cardboard to expand the trigger greatly increases the "catchable surface". Care should be taken not to expand the trigger too far. If it expands over the edge of the trap, the rodent can release the trigger without getting caught. Commercial models with expanded triggers are also available.

■ The actual bait used on the trap can be quite varied and the success of any material depends largely upon how much other food the mouse has to eat and what it is accustomed to eating. Peanut butter, salami, freshly fried bacon, and peanuts are but a few of the items which have proved successful.

Traps have two important drawbacks. First, they cannot be placed where children, pets or unknowing feet wander. Second, if not secured to a surface, the rodent will on occasion drag the trap into another area while caught by the tail or one foot.

Snap-traps for rats. Trapping is applicable to those situations where the use of poisons would be too dangerous. Moreover, it can be used where the odor of the decomposing bodies of the rats may become a very conspicuous annoyance. An evident disadvantage of trapping is it is a form of rat control where the experience and skill of the trapper is of paramount importance. Moreover, trapping may require more time and effort than some of the other methods of control.

In the United States, the snap trap ("guillotine" "spring," or "break back") is the most efficient trap. Cage traps may also be used, but as a whole, they are below the snap trap in effectiveness. When using the snap trap, it should be remembered a large treadle is preferred since it is often possible to trap the rat without using bait merely by placing the treadle in a position next to the wall where the rat may step on it.

It is usually recommended that a variety of baits be used in successive traps. Meat in one trap, vegetable in another, and cereal in another gives the rat a choice. It is advisable to use nuts in a meat packing establishment and meat in a nut or candy house, so the trap is baited with some food that is not otherwise available.

Traps are most effective when set next to the wall beneath a runway made by leaning a board against the wall. The trigger end of the trap should be situated next to the wall. Do not set the tray directly in front of the hole where the rat gains entrance to the room since it is likely to become suspicious. Where the rat becomes "wise," it may be necessary to bury the trap in a shallow pan of meal, sawdust, or grain, with the trigger protruding. Make sure the action of the trigger will not be clogged by the material beneath it. The baits should be tied to the treadle with a thread or rubberband, thus making certain of pressure on the treadle. Old traps should be boiled, scraped, and kept clean. Dipping traps in melted paraffin lengthens wear, deodorizes them, and may make them spring more readily.

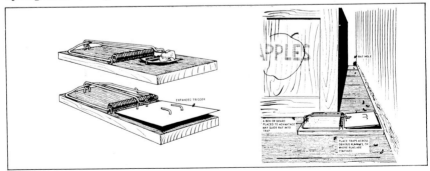

Fig. 1-25. Trap with expanded trigger and one way to use it.

In Berkeley, California, great success was experienced by placing the traps in discarded paper bags in which students had kept their lunches. An enterprising pest control operator in this same city baited the rats with frozen pudding flavored with sherry: "I got those rats so tight I could catch them with my bare hands."

The U. S. Public Health Service has had a tremendous amount of experience in trapping rats, and the following material is from their Rodent Control Manual (USPHS, 1948):

The trigger of a set trap should be parallel to the body of a trap when in its most sensitive position.

Do not use petroleum oils to prevent rusting of traps since rats do not like such oils. Lard or some other animal fat should be used. Thus, a trap whose metal parts are wiped with bacon rind will be protected from rust for some time and also will be attractive to rats.

A dirty trap may be more effective for catching rats than a clean one. Contrary to the usual belief, there is evidence that a trap in which a rat has been caught and is smeared with rat blood is more effective than a clean unused trap. It would seem that rats may be attracted to traps having the odor of other rats on them.

More rats are trapped on the first night than on any other night. After three to four nights of trapping, the catch may drop to zero. For this reason, it is essential to distribute a large number of traps on the first night of trapping, and in business establishments the number of traps may vary from 20 to 100 traps, depending upon the degree of infestation and the facilities for locating the traps.

Where there is a hole in the floor used by the rats, place several traps in a semi-circle two, or three feet from the opening.

When unbaited traps are employed to trap runways on floors, shelves, or other wide surfaces, the traps should be set in the run, placing them about one or two inches away from the wall or other surface the run follows.

Baited traps should be set eight to 12 inches from the surface followed by rat runs so trapped rats will not be directly in the line of travel. It is generally advisable to use baited and unbaited traps alternately along floor runs with the distance between them varying in accordance with the total traps required to catch the number of rats that are using the runway.

Where only a few rats remain after the control of a large population, flour smoothed out around the runways will often indicate the best places to set the traps.

At least three or four kinds of bait should be used when traps are set the first time in any building. Those that fail to attract rats should be replaced by others.

Trapping roof rats. Trapping roof rats is no easy task. Roof rats are more difficult to trap than Norway Rats, particularly at floor level. More success is achieved by nailing the snap trap to the beams where signs of roof rats (grease marks) are located. Pratt, Bjornson and Littig (1976) diagram how to do this.

Electromagnetic and ultrasound devices. Electromagnetic devices are supposed to work on the principle that a magnetic field produces a barrier which has a stunning effect on rodents. It is in no way associated with ultra-sound devices. *Pest Control* magazine (Anonymous, 1977) reported after interviewing some of the top experts in the country that no scientific evidence is available to show these devices work. *Pest Control* magazine (Anonymous 1978) also indicated three different electromagnetic devices tested by the Nevada Department of Agriculture failed to control pocket gopher populations. In 1979, *Pest*

Control magazine (Anonymous, 1979a) further reported that EPA stopped the sale of several electromagnetic devices because they did not work.

Ultrasonics work on the principle that certain high frequencies and amplitudes of sound are irritating. Rodents are capable of hearing ultrasonic sounds. Anderson (1954) showed laboratory rats are capable of emitting such sounds. *Pest Control* magazine (Anonymous, 1977) reported, "Tests performed by G. Keith LaVoie and J.F. Glahn at the Wildlife Research Center, Fish and Wildlife Service, in Denver, Colorado, indicated that ultrasonic devices did not expel or prevent feeding by Norway rats in an experimental warehouse setting. The tests were performed with an established colony of rats. Data showed that the rats became accustomed to the sound in less than three weeks."

Other experts describe "shadows" of ineffectiveness when conditions aren't perfect, as they usually aren't in typical warehouses.

The author's personal experience with ultrasonics for rodent control is somewhat limited. The sound emitted from several of the units resulted in a headache when he remained in the protected area for any length of time. A food warehouse in New Jersey, which used ultrasonic devices had many rats within the structure. It appears the rodents rode in on pallets and then could not escape because of the sound devices over each door. Ultrasonic sounds are directional which allows rodents to seek shelter behind solid objects. In such cases there should be complementary use of baits or traps in these "shadow" zones.

On the favorable side, such sound devices, if properly installed, cause rodents to move about and enhance the efficiency of a baiting or trapping program.

Jackson (1980) reported that the installation of ultrasonic units at an egg farm contributed significantly to the reduction of rat damage and rat activity. Jackson (1980) concluded: "Clearly ultrasonics is not the simple solution to rodent infestation problems but is a factor to be considered in planning pest management activities which of course may call for positive sanitation steps, access control, baiting, etc."

Bait stations. Before considering chemical approaches we must first consider bait stations. Regardless of how good a rodenticide is, once it is placed inside a bait station, it is not effective unless the rodent enters. Therefore, the location of the bait station is critical and the previous discussions on rodent behavior will aid in knowing where to make placements. Another factor which is critical is how "appealing" is the station for the rodent to enter and remain long enough to feed. Metal bait stations in cold areas are not particularly attractive. Keep in mind that the bottom of a rodent's feet are bare and the floor of the metal bait stations can be extremely cold in the winter. The placement of a piece of cardboard along the runway inside the bait station solves this problem. The insertion of small sections of PVC piping into each end of a bait station will minimize dust and debris that can contaminate the bait. It also prevents birds from entering. The PVC piping about six inches/15 cm long is fitted snuggly into the bait station by using glue or wrapping a small piece of carpeting around the piping and ramming it into position. In hot areas where the sun beats down all day, a light-colored top will decrease the temperature inside the bait station. Even with these precautions, when first placed it may take several days to a week for the rats to become acclimated to bait stations before entering to feed.

Pest control technicians who have had problems with cockroaches eating their rodent baits outdoors in bait stations have solved the problem by treating the inside runway with bendiocarb insecticide. It appears to have no detrimental effect on bait acceptance by the rodents.

Norway rats, when eating cornmeal baits, can eat more easily if they are able to hunch up and scoop the food with their front feet. If a bait station is too low, the rat can only eat by lapping and thus eats less. Most commercial bait stations are high enough to avoid this problem.

A chemical approach. Ingestion of rodenticides resulting in the rapid death of rodents (within 24 hours) are referred to as single-dose rodenticides, acute, or "quick-killers." Anticoagulants are often referred to as chronic or multiple dose rodenticides. With the development of non-anticoagulant, single-feeding rodenticides that provide delayed kill, and new anticoagulants that kill with a single feeding but over a period of days, the ease of classifying rodenticides in this manner is no longer so clear cut. This portion of the text attempts to cover rodenticides as non-anticoagulants, fumigants and anticoagulants. Attention is paid to those materials available, or until recently available, in the United States. No reference is made to products such as the narcotic chloralose used as a rodenticide in other countries.

THE NON ANTICOAGULANTS

Antu. The rat-killing properties of ANTU or alphanaphthyl thiourea were discovered by Dr. C. P. Richter of Johns Hopkins University. Chemists of E. I. du Pont de Nemours & Company synthesized the compound. Dr. Richter found a related compound to be highly toxic to rats used in taste reaction studies, and the more rat palatable ANTU was synthesized by Du Pont scientists.

ANTU is a fine greyish powder that is insoluble in water, but soluble in some organic solvents such as ethyl alcohol and acetic acid. It is a stable compound that is nonvolatile. This chemical is relatively nontoxic and nonirritating to the human skin. ANTU is highly toxic to Norway rats, dogs, and pigs. It is much less toxic to roof rats, mice, and cats. Anon. (1946) discusses ANTU as follows: "Disadvantages inherent in ANTU include the rapid build-up of tolerance that follows the eating of less than a killing dose, its nonsolubility in water and its lack of toxicity for rodents other than Norway rats. The development of tolerance to the poison by Norway rats is so rapid that ANTU cannot be used most effectively in a locality at intervals of less than one month; thus if additional control measures are necessary, some other effective rodent poison should be employed. A tolerance to many lethal doses can be built up within a few days. Likewise, a rat poisoned but not killed by ANTU may subsequently refuse to touch bait containing ANTU for a period of up to six months.

"Insolubility in water makes it more difficult to obtain even mixing of ANTU in bait and decreases its effectiveness as a water poison.

"ANTU is acceptable to rats in two to three percent concentrations. In these concentrations it may be considered more acceptable than red squill, less acceptable than thallium sulfate, and approximately equal in acceptability to 1080 and zinc phosphide. ANTU has a tendency to absorb odors, which will reduce its acceptability if the odor absorbed is unpleasant or gives warning to the rat.

"Because dogs may eat lethal amounts of ANTU-poisoned grain or poisoned meat, care must be taken not to allow dogs to have free access to poisoned baits. Some evidence indicates that the same precautions may also have to be observed with pigs. ANTU has a much lower toxicity for cats so that these animals rarely are fatally poisoned. ANTU probably is not toxic to man but in the absence of definite knowledge, all necessary precautions should be observed.

"ANTU kills rats by causing an acute lung dropsy and an accumulation of fluid in the chest cavity. Death usually occurs within 12 to 48 hours." ANTU should be thoroughly mixed into bait.

ANTU is used against Norway rats as follows:

1) "In 0.75 to three percent concentration (12 ounces to 48 ounces of ANTU per 100 lbs. of bait) in baits of the usual type. The higher concentration decreases acceptance, but may be necessary to kill young rats.

2) "Dusted lightly on and around freshly cut fruit or vegetables.

3) "Dusted lightly on the surface of water or shaken up in water in one to two percent concentration (1.3 to 2.6 ounces of ANTU per gallon of water).

4) "As a contact poison, using ANTU mixed 50-50 with an equal amount of flour or pyrophyllite and placed in patches on floors and in rat runways. This method is expensive, but has been particularly successful in eliminating the last survivors in ratproofed buildings.

5) "As a contact poison, by blowing ANTU into burrows and holes with a standard dust pump."

Since rats that survive an ANTU baiting may become "bait shy" for months, it is essential great care is taken in the preparation of the baits, and sufficient bait be used to cope with the estimated rat population. For best results with ANTU, it is recommended the pre-baiting system be used. Once the baits become stale be sure to remove them.

Munch (1947), in summarizing the studies of several papers, states that the 50 percent lethal dosage value for Norway rats falls between seven and 35 mg/kg going up to 58 mg/kg for the very young rats who are apparently not as susceptible to the poison. Latven, et al., (1948) showed there was a seasonal variation in the susceptibility of laboratory rats to ANTU; the rats being most susceptible in September and least susceptible in January.

Arsenic. White arsenic (AS_2O_3), also known as arsenic trioxide and arsenious acid, is a widely used and effective rat poison. This chemical is practically odorless and tasteless. White arsenic is sold on the market as commercial white aresenic, which is a white powder of rather large particle size. Micronized or microfine arsenic of 98 percent to 99 percent purity, which is a very fine powder and as soluble arsenic trioxide. The commercial white arsenic is less toxic to rats and less soluble in water than the very finely ground arsenicals.

Sennewald (1948) notes rats may live up to 48 hours or more when poisoned by baits containing five percent to 10 percent commercial white arsenic, whereas rats fed on 1½ percent finely ground arsenic died within 24 hours. Sennewald also observes rats on consuming a commercial soluble arsenic solution "show signs of distress within two to 10 minutes after drinking the diluted solution; or eating baits made from the solution. Rats generally die within 30 to 90 minutes after drinking the solution. Mice are generally killed within 25 minutes."

Due to the relatively odorless and tasteless properties of white arsenic, great care must be used in preparing and distributing baits so human beings and domestic animals are not injured. Since the water soluble arsenic is readily absorbed, skin contact with the solution should be avoided.

As rats become quickly suspicious of an arsenical bait, it is essential the concentration of arsenic in the bait be relatively large so the rat will not have to consume too much of it. USPHS (1948) states arsenical baits can be stored for some time without spoiling.

Barium carbonate. Barium carbonate is a tasteless and odorless white mineral salt and one of the so-called weaker rodenticides. Technical barium carbonate of at least 98 percent purity is used in rat baits. It is fairly effective and rather inexpensive. It may be fatal to chickens, dogs, cats, and larger animals if eaten in sufficient quantities. Since this chemical is comparatively weak, much bait must be eaten, and thus the food should be doubly attractive.

USPHS (1948) notes: "Unless the presence of barium carbonate is well disguised by the bait, rats soon discover that there is something wrong with it, and may stop eating it before they have consumed a fatal amount. The toxicity of this poison is not constant as some rats die after eating very small quantities while other rats may recover, or not even become sick, after having consumed an equal amount. In spite of these imperfections, barium carbonate is a fairly effective poison for killing rats."

Barium carbonate may be used with cereals such as bread, corn meal, rolled oats, with meats, such as hamburger, sausage, and fruits and vegetables, such as apples, melons, and tomatoes, as well as sardines and eggs. Rats prefer wet baits when barium is used. Garlough (1946) states the toxic action "of this poison on the animal is quite different from that of arsenic. Barium carbonate stimulates most of the muscles of the heart, arteries, and intestines, thereby producing powerful heart action and strong constriction of the blood vessels, increasing greatly the peristaltic movements of the intestines. There is finally an excessive stimulation of the central nervous system which brings on paralysis and death."

DDT. Because of legislation this product can no longer be used for rodent control in the United States. As a 10 percent tracking powder it was effective against house mice and killed young rats. It was never effective against adult rats, but was used to dust rat burrows for the control of fleas.

Norbormide. This product is specific for Norway rats, but achieving constant

TABLE 1-2

Toxicities to man of rodent baits prepared with common non-anticoagulant rodenticides*

Poison	Accepted Lethal Dose to Man (mg/kg)	Concentrations Used in Bait	Ounces of Bait Containing a Lethal Dose for a 150 lb. Man
Arsenious acid (AS$_2$O$_3$)	1.5-15.0	3.0% 1 part in 33 parts	0.12-1.22 oz.
Strychnine	1.0	0.3% 1 part in 320 parts	0.8 oz.
Sodium fluoroacetate (1080)	5.0	0.4% 1 part in 256 parts	3.15 oz.
Thallium sulfate** (T1$_2$SO$_4$)	20.0	1.5% 1 part in 65 parts	3.2 oz.
Zinc phosphide (Zn$_3$P$_2$)	40.0	2.0% 1 part in 50 parts	4.9 oz.
Barium carbonate (BaCO$_3$)	800.0	20.0% 1 part in 5 parts	9.9 oz.
ANTU (alphanaphthyl thiourea)	Unknown	5.0% 1 part in 20 parts	Probably very large

*Ward (1947) prepared this table.
**No longer used.

TABLE 1-3

Time to death and cause of death*

Note: The time to death depends upon various factors, such as amount of poison ingested, resistance of the animal, age, sex, etc. Most poisons cause disturbances to the various organs and systems of the animal. Only the final cause of death is given.

Name of poison	Minimum Time	Average Time	Maximum	Final cause of death
ANTU	12 hours	24 hours	36 hours	Plural effusion
Arsenic	5 hours	24 hours	48 hours	General paralysis
Barium carbonate	2 hours	12 hours	24 hours	Respiratory failure
Sodium fluoroacetate	1 hours	12 hours	72 hours	Heart paralysis
Squill	6 hours	24 hours	120 hours	Respiratory failure
Strychnine	¼ hour	½ hour	2 hours	Respiratory failure
Thallium sulfate**	12 hours	24 hours	120 hours	Respiratory failure
Zinc phosphide	12 hours	24 hours	120 hours	Heart paralysis

*From unpublished manuscript by E. M. Mills
**No longer used.

good bait acceptance is a problem. Rats die within 15 minutes to one hour by impairment of blood circulation. It is non-toxic to mice and not very effective against roof rats. Poor bait acceptance has resulted in norbormide never becoming a major rodenticide.

Phosphorus paste. Phosphorus pastes usually contain two percent white phosphorus, with corn starch and glucose as bait. This material is then spread on bread and small crackers. The phosphorus paste is placed between two slices of bread in the form of a sandwich which is then cut into squares about ½-inch in diameter.

The taste and odor of phosphorus are characteristic of the poison, and anyone who has spread the bait for roaches beneath a house or in the dark has noticed its luminous properties. Since phosphorus is dissolved in fats, it is usually used in greasy types of baits.

The preparation of phosphorus paste should not be attempted by the average individual since white phosphorus ignites spontaneously in air, and must be stored and worked with under water. Phosphorus burns are extremely serious. However, when used with care, phosphorus paste is usually a satisfactory product.

Red squill. This rodenticide is noteworthy since it is toxic to rats and mice, and yet is relatively safe to man and other animals. Red squill, also known as "scilla or sea onion," grows wild in the hills bordering the Mediterranean and is a perennial plant that belongs to the lily family. The bulbs average three to six inches/76 to 152 mm across and weigh up to six pounds/2.7 kg According to a Federal specification, "red squill powder shall consist of the dry powdered fleshy inner bulb scales of the red variety of *Urginea maritima* fortified when necessary with the alcohol soluble extract of the same, and free from any other modifying agents, diluents or adulterants."

Red squill should be distinguished from white squill which is used in medicine as an emetic, heart tonic, etc. Roques (1946) notes "the toxic principle is the scilliroside of Stoll and Renz. Fresh white squill is as toxic to rats as red squill, but loses its toxicity upon drying and aging. In red squill the red tannin pigment protects the toxic principle from destruction by oxidation upon drying."

Red squill is available in the form of oven-dried powder, liquid red squill, and red squill paste. Liquid red squill is found on the market in a concentrated form that must be diluted with water for use, or is sold in an already usable state. Liquid red squill may be either an extract of the bulb or of the powder. The extract has a somewhat disagreeable odor.

When handled, squill produces a needle-like stinging sensation due to the needle-like crystals of calcium oxalate which it contains and which irritate the skin. These needle-like crystals or raphides give it an acrid, prickly taste. Due to its objectionable taste and emetic effect, most animals other than rats avoid squill or regurgitate it upon consumption. Although red squill may make other animals very sick, they usually recover upon vomiting. Crabtree (1944) states work conducted by the former Bureau of Biological Survey, and now the Fish and Wildlife Service, showed slices of red squill when dried in an oven "at a temperature not in excess of 180° F made powder which was more toxic than powder prepared from sun dried slices."

The Fish and Wildlife workers in 1940 devised a method for strengthening or fortifying raw red squill so that squill of poor quality was improved in toxicity. According to Crabtree (1944), squill is fortified by adding varying amounts of red squill extract to one pound of the unextracted powder to form the final product. "Thus a two to one degree of fortification means that the extract from two lbs. of squill powder is impregnated upon one pound of unextracted powder taken from the same lot." The Fish and Wildlife Laboratory at Denver has set a standard for the minimum toxicity of red squill requiring that "red squill powders or extracts have an average lethal dose for male rats not exceeding 600 mg/kg in order to be considered as an effective lethal agent for rats."

Crabtree states in using a 500-600 mg-kg squill, "effective results can be obtained if such a squill is used at a concentration of 10 percent in baits composed of ingredients that are specially attractive to rats. Since a rat may normally be expected to eat at one feeding a quantity of food equal to one percent of his body weight, an adult rat weighing 250 grams would consume 2.5 grams of bait or 250 mg of squill. This amount would constitute a 1,000 mg/kg dose and produce lethal results."

Dykstra (1050), in discussing red squill, makes the following interesting note: "Other observers report that while a squill toxicity of 500 mg/kg has been accepted as the minimum toxicity suitable for commerical preparations, uniformly better results are obtained when products of 250 to 300 mg/kg are employed. This is particularly true in the case of black rats which are 20 to 25 percent more resistant to squill than the Norway form. There is also evidence of increased hazard to livestock when fortified products are employed, so even squill baits should be carefully exposed." In general, red squill is not recommended for roof rats or mice because of poor acceptance.

According to Garlough (1941), "Red squill depresses the muscular activity of the body, and so produces a slower pulse rate, labored breathing, and a rise in temperature. It stimulates the nervous system, causing paralysis of the hind limbs, convulsions, and peculiar gyrations. Finally, it depresses breathing so excessively that the animal dies of respiratory failure."

In the use of squill baits, the choice of food is an extremely important item, for if the rats are not materially reduced by the first baiting they may become "wise" to subsequent baitings.

Red squill has been found to be a fairly effective poison, usable in a great variety of baits, and relatively harmless to human beings and domestic animals. It is a slow-killing poison that will permit the rats to leave the premises, and thus reduces the possibility of odor.

One disadvantage of squill is its great variability in toxicity. This factor has been largely eliminated due to the fortification of squill to standard toxicities. Squill that is permitted to age will lose much of its toxicity in time. Rats may become extremely wary of squill baits if they survive the first baiting.

Sodium fluoroacetate (1080). Although highly restricted to how and by whom it can be used, this technical editor wishes to retain most of the information compiled by Mallis in previous editions. *"READERS SHOULD BE AWARE THAT 1080 CANNOT BE USED WITHOUT PROPER GOVERNMENT APPROVAL.*

Kalmbach (1945) states that R. Treichler, a chemist at the Patuxent Research Refuge in Maryland, first demonstrated the toxicity of this chemical to rats in the laboratory: and J. C. Ward and D. A. Spencer of the Fish and Wildlife Service, Denver, Colorado, definitely proved its effectiveness as a rodenticide both in the field and laboratory.

Finger and Reed (1949) note 1080 occurs in a South African plant, and that the natives "have been aware of its poisonous properties to cattle and rodents for many years.

"Compound 1080 is a stable compound chemically and is not corrosive to metals in general. It is very soluble in water and relatively insoluble in organic solvents such as kerosene, alcohol or acetone. Neither is it soluble in animal or vegetable fats and oils. When the dry pure powdered 1080 is exposed to air, it rapidly takes up water from the atmosphere and may become sticky.

"The favorable features of 1080 appear to be high toxicity to all species of rodents tested, excellent acceptance, absence of significantly objectionable taste and odor, non-volatility, non-toxicity on the skin and non-irritation to the unbroken skin of workers. In addition, domestic rodents apparently do not develop any significant tolerance to 1080 on ingestion of sub-lethal amounts nor, in general, are they able to detect the poison except after the ingestion of lethal amounts. Although they may tend to develop a slight aversion to 1080 bait during a poisoning operation, it is not sufficiently pronounced to affect the efficiency of the operations.

"Disadvantages inherent in 1080 include its high solubility in water which may result in the poison being washed out of baits by rain. The rapidity of absorption of 1080 by the gastro-intestinal tract, may in the case of some field rodents, cause symptoms which warn the animal before it has ingested a lethal dose and lead to hazardous amounts of bait left uneaten. This ease of absorption is a disadvantage in the treatment of accidental poisoning in man and other animals. A major drawback to the use of 1080 is the extreme susceptibility of dogs and cats. The hazard of accidental poisoning of humans is heightened by reason of 1080 being a white powder without odor and with only a slight salty taste. It must be emphasized that there is also a complete lack of any specific therapy or any antidote for 1080".

Simmons and Nicholson (1947) discuss 1080 as follows: "An important factor contributing to the success of 1080 is its acceptance by rats and mice. It has no objectionable odor or taste, and some investigators claim that rats accept it in water solution in preference to pure water. Certainly the repellent effect, if present, is exceedingly slight.

"An important quality of 1080 is its relatively quick action in comparison with other rodenticides. When used in the recommended dosage, rats die within one to several hours after taking a lethal amount. Since the first symptoms of illness usually immobilize them, dead rats are frequently found in the open near the poison cups where they may be picked up and destroyed. The likelihood of odors developing in inaccesible places is thus minimized. Since death occurs over such a short span of time, all odors that develop usually appear within a few days after poisoning. Tolerance of a significant degree is not developed under practical conditions on the ingestion of the sub-lethal amounts, so that 1080 poisoning may be repeated satisfactorily. Rats apparently develop only slight, if any aversion to 1080 solution, although they probably learn to avoid a particular type of solution container if used repeatedly over a short period of time in a single establishment. This reaction may be overcome by refraining from using 1080 solution for an adequate period of time following its first use or perhaps changing the type of solution container."

Instructions for the use of 1080 for the control of rats and mice are as follows:

1) "In poisoned water use ½ ounce of 1080 per gallon of water. In food baits use one ounce of 1080 in 28 pounds of bait.

"Compound 1080 does not appear to deteriorate when mixed with bait or with water in which it is very soluble.

2) "Do not use more 1080 in the bait than is necessary. Hazards to other animals with 1080 are lessened by using the recommended concentrations.

"If good control of rats does not result from the use of 1080 in the recommended concentrations, the trouble lies with the bait or the way it is applied, not with the 1080. Rats are cautious in the selection of baits and control methods must be adapted to the prevailing local conditions.

3) "Compound 1080 should not be used inside dwellings or anywhere in residential areas. It should be used only in and around commercial and business districts by carefully instructed, reliable, and competently supervised personnel.

4) "Remove all pets from the area to be poisoned, and keep them out for at least five days. Remove and destroy by burning if possible or by burying all surface kill of rats and mice before releasing pets which might feed on and be poisoned by them.

5) "Place baits carefully. Baits may be placed adjacent to burrows and along runways, preferably behind boards or boxes in specially prepared bait stations and other places frequented by rats and out of reach of irresponsible persons, pets, and other animals.

"The poisoned water may be placed in shallow ¾-ounce paper souffle cups and similarly distributed.

6) "Do not expose baits or water containing 1080 under conditions that might result in the contamination of food supplies.

7) "At the conclusion of operations, remove and burn if possible or bury any uneaten bait and all water containers."

Ward (1947) presents the following details on the use of 1080 water solutions: "Early in the field testing program it was found that a simple 1080 solution could be used for rat and mouse control. The best way for exposing the water was in small flat containers — and the squat-style, ¾-ounce souffle cup was most convenient. Experimental use of a concentration of ½-ounce of 1080 in one gallon of water gave the best control of rats and mice which had ever been obtained in the southern part of the country. In using the water, the best system was to place the containers at frequent intervals along the rat runways and then to pour about ½-ounce of liquid into each souffle cup. A supplemental system which is gaining favor is to expose the solution in standard or modified baby-chick water fountains. The use of such devices has been found to be most valuable in premises where the rats have been removed once, since the presence of the semi-permanent stations will catch any rats as they start to reinfest the area."

Since 1080 is so toxic and is used in such minute amounts, it is one of the most economical rodenticides. Nicholson et al. (1947) recommend 1080 be used at the rate of 12 grams per gallon of water. At this dosage, one pound/0.45 kg of 1080 is sufficient to prepare approximately 38 gallons/144 l of poisoned solution.

Nigrosine black is a dye for making water solutions of 1080 unattractive to humans. One fl. oz/29 ml is used per gallon/3.8 l of 1080. Twenty percent of 190 proof alcohol and 10 percent glycerine can be added to 1080 solutions to prevent freezing.

Lynch (1958) controlled rats in sewers with barley poisoned with 1080. Brooks (1961) prepared paraffin blocks containing 1080 baits for rat control in sewers. Brooks (1964) eliminated rats from sewers through the use of sodium fluoroacetate-barley baits in twice-a-year campaigns.

Although 1080 is an extremely effective rodenticide because of its toxicity to man and other animals, it is necessary to dwell as much on safety precautions in its use as on the toxicant itself. Some precautions are previously mentioned. Additional ones include:

- Some sources recommend 1080 be used in water solutions alone since rats may carry solid baits to other areas where they may possibly poison foodstuffs.
- The 1080 solution should be carried on the job in special rubber or plastic containers which will not break if they are dropped. Soft drink or milk bottles should never be used to store the poison solution.
- The souffle cups should bear the word "POISON" and be illustrated with skull and cross bones. It is preferable they be placed in special "bait boxes" made for this purpose. All used souffle cups should be burned. Children have been killed by chewing on souffle cups from which 1080 solution has evaporated.

Interestingly enough, rodents can develop chemical resistance even to a chemical as toxic as sodium monofluoroacetate. Australian scientists (Anonymous, 1979b) report some populations of the bush rat *Rattus fuscipes* from western Australia have developed a 30-fold threshold before death occurs. They found the rats can defluorinate the toxic compound which appears naturally in a few species of plants the rodent feeds upon.

Fluoroacetamide. (Fluoroacetamide, 1081, FCH_2CONH_2) is related to sodium fluoroacetate and like it is a very toxic rodenticide (oral toxicity — 15 mg/kg). It is a white crystalline material which is readily soluble in water and sold as a purplish black powder. It can be used in solid and water baits. Brooks (1963) used it for the control of sewer rats at a concentration of one and two percent in baits. Rohe (1966) obtained excellent control of roof rats in sewers in California with one percent fluoroacetamide in paraffin bait blocks. The use of this rodenticide is restricted to qualified individuals. A thorough discussion of 1081 in paraffin blocks is covered by Ebeling (1975).

Strychnine. Strychnine is a complex nitrogen compound and the sulfuric acid salt, strychnine sulfate, is used generally in place of the alkaloid. Strychnine sulfate is extremely poisonous and characterized by very rapid action. It also is soluble in water, with a very bitter taste. This rodenticide is not well-taken by rats in buildings. Its bitter taste must be disguised to be consumed by rats. Because of its rapid action, its use in residences is discouraged. Sennewald (1948) notes mice often die within 10 minutes after feeding on strychnine baits. Usually it is incorporated around seeds, and it may be used for control of rats in the field, as around city dumps. When used outdoors great caution should be employed to see the baits are placed in such situations that domestic animals and birds do not feed on them.

Strychnine alkaloid and strychnine sulfate are used in powder or crystalline form. The sulfate is much more soluble in water, one ounce/28 g in about one quart/0.9 l.

With the advent of anticoagulant baits, particularly in seed forms, strychnine no longer plays a major role in mouse control.

Thallium sulfate. Here is another rodenticide banned by the government and perhaps rightfully so. If ingested at non-lethal doses humans can lose their hair and nails; at slightly higher concentrations they lose control of their muscles. A few accidents have occurred as a result of improper use.

Zinc phosphide. Zinc phosphide is a quick killing rodenticide. As a concentrate, it is a dark, gray powder which is insoluble in water. Zinc phosphide possesses a strong garlic-like odor caused by the release of phosphine gas. It is more effective against roof rats than Norway rats, but is also effective against mice. Rats die within 17 minutes to several hours.

After consuming low dosages, rats may survive for several days. The LD_{50} for Norway rats is 35 to 48 mg/kg.

Secondary poisoning is minimal as the rodenticide breaks down rapidly within the rodent's stomach. Rodents die from heart failure and for those that linger, kidney failure also can occur. Rats killed with zinc phosphide often are found on their belly with their legs and tail extended. The antidote for zinc phosphide is copper sulfate before emetic, cathartic and water. Fats and oil are to be avoided.

Between 1917 and 1965, 26 persons died from zinc phosphide poisoning, but 18 of these were suicides (Schoof, 1970). Schoof, in citing Blisnakov and Iskrov, 1961, reports:

"One murder case was somewhat unique in that the victim was stabbed with knives coated with zinc phosphide, an approach that indicated that the murderer was somewhat unsure of his weapon. Despite immaculate surgical attention to the wounds in the muscle, gut, liver and lung, the victim died of renal and liver failure within 60 hours."

For rodent control this material is used in several formulations.

As a 10 percent tracking powder it is registered for the control of mice indoors. However, it must stay dry to remain effective. One drawback is that if left in an area where there is a draft, some of the powder will float away from the target area. Label directions call for it to be placed inside tamper proof bait stations with a spoon, *not* with a hand duster. Using a hand duster results in some of the product drifting. The dust should be sprinkled in locations not accessible to children, pets or wildlife.

Tracking powder containing zinc phosphide kills anticoagulant resistant mice. The normal phosphine odor associated with zinc phospide has been eliminated. Label application directions call for "evenly sprinkling two to four grams (one to two level teaspoonful amounts) in three by 24 inch/7.6 by 60 cm patches. Maintain in area for at least 20 days. Collect and dispose of all dead animals and used powder properly. Do not treat the same area with this product at less than 30 day intervals." Do not apply to wet or highly porous surfaces. Mice, after walking through the powder, groom and thereby ingest the toxic material. Mortality occurs in less than 24 hours.

A prepared commercial two percent zinc phosphide bait is registered for the control of rats, house mice, ground squirrels, pocket gophers, meadow voles, and deer mice. Indoors, it must be placed in a protected bait station.

The following six safety precautions should be kept in mind when mixing your own baits:

- Work in a well ventilated room.
- Wear gloves.
- Wear a respirator.
- Make very small pieces of bait so it isn't eaten by larger pets and humans.
- Avoid mixing with water.
- Add a tartar emetic.

Adding oils and fats in the bait increases the rate of absorption in rats. However, it also increases the hazard of human skin absorption if a cut is present.

Moist baits deteriorate rapidly, but baits kept dry remain potent for long periods. Elmore and Roth (1943) found grain baits coated with 0.6 percent zinc phosphide lost only 29 percent of their poison content after exposure to weather for 29 days.

Two recommended baits are:
1) Chop apples into ½-inch cubes.
2) Add one percent toxicant — enough to coat apple cubes.
3) Place in bucket.
4) Add flour at ratio of 1:2.
5) Rotate bucket (Note: if adding a tartar emetic, use at three parts to eight parts zinc phosphide.)

To prepare a second bait, grind meat, bacon, canned fish, grains, cereals or a combination of these mixed with zinc phosphide at the ratio of 99 parts bait to one part zinc phosphide. For example, in 25 lbs./11.3 kg of bait only four ounces/113 g of zinc phosphide concentrate would be needed. In all cases follow label directions.

Recently, a bait formulation of zinc phosphide was registered as a restricted-use pesticide for control of rats, mice, and other rodents outdoors, in crop and non-crop areas.

Fumigants. With the highly successful calcium cyanide dust no longer produced, gassing rodent burrows outdoors is greatly restricted.

Chloropicrin (tear gas). This gas is still registered for rodent control. It is often used in ridding large buildings, warehouses, and other structures of severe rat infestations. It is usually not necessary to seal the building when using this gas, but all doors and windows, as well as other openings should be closed.

The manufacturers of chloropicrin find a dosage of 2½ lbs./1.15 kg chloropicrin per 1,000 sq ft/91 sq m for rats, and 1½ lbs/0.7 kg per 1,000 sq ft/91 sq m for mice is a satisfactory dosage in small warehouses, storehouses, corn dryers, and similar buildings. If the building is of loose construction or largely filled with stored materials, the dosage should be increased by ½ lb/0.23 kg per 1,000 sq ft/91 sq m. The chloropicrin is poured on crumpled burlap sacks; use three sacks for each pound of fumigant.

The manufacturers of chloropicrin (Anon., no date) recommend the fumigant be used as follows: "Start from rear and work toward exit. Proceed rapidly, but without undue haste. Pour approximately ⅓ of bottle onto each sack, with head well away from bottle. Close exit door tightly and post fumigation notice. Keep building closed overnight. Open up an hour or so before starting work the next morning." For further details on the use of this gas, see chloropicrin in the chapter on fumigation. Chloropicrin can also be used in rodent burrows. Be sure to pack dirt on all exit holes to confine the gas in the burrows. Ebeling (1975) recommends, "It may be mixed with motor oil and poured into rat burrows, after which the burrows are plugged with soil. The gas adheres tenaciously to soil particles, and remains active in the burrow for a month or more, not only killing the rats, but repelling others that might otherwise take over the uninhabited burrow."

Methy bromide. According to Dow Chemical Co., "All rodents may be effectively controlled in warehouses by using four ounces (¼ pound) of Methyl Bromide for each 1,000 cubic feet of storage space for an overnight exposure (12-18 hours). Dosages must be increased in proportion to the leakage factor.

"In fumigating cold storage rooms for rodent control, use ¼ pound of Methyl Bromide for each 1,000 cubic feet of storage space for an exposure period of five hours at temperature of 32-45° F. Circulating fans previously placed should be on for at least 30 minutes to properly mix the air and gas. A quick and complete change of air after fumigating is necessary." Do not use methyl bromide in burrows near plants as this gas is phytotoxic.

Carbon monoxide. Carbon monoxide, the exhaust of gasoline engines, is a deadly poisonous gas which can be used to fumigate rat burrows. Haynes, et al. (1945) used carbon monoxide in the following manner: "The carburetor mixture should be enriched and the engine allowed to run at moderate speed, forcing fumes into the hole for about 10 minutes. All holes from which the smoky exhaust is seen emerging should at once be closed with sod to confine the gas. This helps force it to all parts of the burrow and fewer rats escape." Storer (1948) notes carbon monoxide remains in the burrow for some time since it is not readily absorbed in the soil or soil moisture.

Carbon bisulfide. Silver and Garlough (1941) recommend use of this fumigant in the following manner. "Carbon bisulfide is only fairly effective in destroying rats, but under favorable conditions it may be used in fields or in holes in dirt floors in chicken houses, cellars, and similar places. The gas is more effective in heavy damp soils and during wet weather. A wad of cotton or other absorbent material should be saturated with one ounce (about 2 tablespoonfuls) of carbon bisulfide and pushed as far as possible into each burrow entrance or the gas may be forced into the burrows by a special type of applicator now marketed by manufacturers of carbon bisulfide. Entrances should then be closed with moist earth to prevent escape of gas. Long forceps are convenient for handling the absorbent materials.

"Carbon bisulfide is highly inflammable and explosive and should be kept away from fire. As a gas, mixed with air, it may be discharged even by an accidental spark. Since it evaporates rapidly, it should be kept in an air-tight container.

Howard and Marsh (1974) state this gas is used for gassing ground squirrels, but it can also be used for rats.

It should be noted any fumigant used within a structure to control insects will destroy any rodents in the area at the time of fumigation.

ANTICOAGULANTS

Warfarin. Warfarin or Compound 42 whose chemical name is 3-(alpha-ace-tonylbenzyl)-4-hydroxycoumarin, is an anticoagulant rodenticide developed through the efforts of the University of Wisconsin research workers, Link, Ikawa, and Stahmann.

Anon. (1949) notes warfarin "is a stable, colorless crystalline solid m.p. 161° C. It is acidic and has a low solubility in water, but the sodium salt is readily soluble. Both are odorless and tasteless. The free acid is used in solid baits, the sodium salt for water solution (fountain) baiting." This compound "kills by destroying the coagulating powers of the blood and by causing capillary damage. The stricken rodents die a peaceful death from internal bleeding. Single large doses are not effective, but small daily exposures over a short period give a high percentage of kill (90%-100%). A daily intake of about 0.025 to 0.0125 milligram per gram of bait for six to 10 days will usually kill the average rat and mouse. Bait shyness and tolerance do not develop; both sexes are equally vulnerable to the product and seasonal effects are nil."

Dykstra (1950), in discussing warfarin, makes some notes that may be of interest to the reader: Although male and female rats show no difference in susceptibility, "pregnant females and young usually die first.

"The toxicity of warfarin varies for the different species of rodents. While daily ingestion of one mg/kg of the poison for five days is fatal to 85 percent of Norway rats, black rats are approximately three to four times as resistant to the chemical, and house mice are even more so. This species difference partially explains the variation in degree of control sometimes attained . . ." Hayes and Gaines (1959) show wild Norway rats are more susceptible than roof rats to warfarin, coumachlor, diphacinone, pindone and isovaleryl indandione.

One pound/0.45 kg of bait for the average home and two to three/0.9 to 1.4 kg pounds for the usual business establishment, are the amounts recommended.

Laboratory studies indicate birds as a group exhibit a high tolerance to warfarin. On the other hand, dogs have succumbed to single doses of about one pound/0.45 kg of bait, and cats, to about 1½ oz/43 g of bait. Dykstra (1951) states cats may be killed if they eat too many warfarin-poisoned rats and mice which are readily caught. Hogs are highly susceptible and sheep are highly resistant. For these reasons, the lower concentrations of warfarin are recommended. Be sure to remove stale baits to avoid unforeseen problems.

McAlindin (1956) showed warfarin baits to be effective against Norway rats when diluted one to 38. Link and Ross (1956), and Spencer (1946) note such dilution of the bait will make it ineffective against black rats and house mice.

Oderkirk (1951) found the following warfarin bait did not readily develop rancidity on storage:

by weight	
Ground oats — Quaker Oats Imperial A	45%
(steel cut groat)	
Non-degerminated corn meal	45%
Crude corn oil	5%
Warfarin	5%
	100%

Place baits in dry locations to prevent molding. Also, place baits in bait boxes to keep them away from pets. Ehmann (1960) states care should be taken to keep the baits from picking up insecticide odors.

One commercial formulation consists of 0.5 percent warfarin and 0.5 percent sulfaquinoxaline which destroys a strain of bacteria in the rodent that produces vitamin K. Vitamin K is the antidote for warfarin (Derse, 1963).

Wax base, weather-proof rodent baits are not as readily accepted by rodents as are other types of baits. EPA's minimum acceptance/kill standards call for 25 percent acceptance and 80 percent kill. Non-weatherproof baits call for 33 percent acceptance and 90 percent kill.

Other multiple dose anticoagulants. Besides warfarin, there are a number of other anticoagulants on the market that are effective in the control of rats and mice. In general, they have properties similar to warfarin.

Coumafuryl. 3-(alpha-acetonylfurfuryl)-4-hydroxy coumarin. This antico-agulant is sold as a 0.5 percent concentrate, and diluted 19:1 with bait. A water soluble sodium salt of coumafuryl is used at the rate of one tablespoon to one quart water plus two tablespoons of sugar.

Pindone. 2 pivalyl-1,3 indandione. Brooks and Scott (1961) showed baits made with 0.025 percent pindone are less subject to insect infestation and to molding. A sodium salt of pindone is marketed which is water soluble. The ad-dition of two tablespoons of sugar to each quart of water increases the accept-ability of the solution. Mackie (1964) used pindone baits to control roof rats in sewers in San Diego, California. He also recommends ratproofing sewer vents on roofs since rats enter these.

Bjornson and Brooks (1962) recommend a paraffin bait for rat control in sew-ers. It consists of 28½ oz/0.8 kg chicken scratch feed (cracked corn and wheat), 1½ oz/43 g 0.5 percent pindone, one lb/0.45 kg paraffin and 25 ml mineral oil. First add oil to the bait material and then mix thoroughly in a #10 tin can; then add pindone. Next, pour melted paraffin over bait. The can should be tilted to form a gradual slope at the bottom of the container. Then the cans are nailed to two by six inch/five by 15 cm boards. These are lowered by means of a wire to bottom of manhole.

Isovaleryl indandione. 2-isovaleryl-1,3-indandione. This anticoagulant is sold as a one percent concentrate. A mixture of one part of the concentrate to 19 parts bait results in a 0.05 percent bait. Isovaleryl indandione is also rec-ommended as an anticoagulant tracking powder and is dusted in runways and burrows. Water soluble isovaleryl indandione is used at the rate of four table-spoons per quart.

Diphacinone. 2-diphenylacetyl-1,3-indandione is recommended in the fin-ished bait at 0.005 percent.

A commercial pelletized 0.005 percent diaphacinone bait is produced pri-marily for outdoor rat control. However, this product is registered for rat control both indoors and outdoors. When wet, it becomes succulent and results in good bait acceptance. If left untouched for several weeks, it becomes moldy and in-effective. Different flavors do not appear to play a major role in bait acceptance. Indoors, it presents a problem in that mice and rats often carry the material away from the area where it is initially placed. Particularly with mice, they may drop the diphacinone pellets in areas where children could easily come in contact with them. In large food facilities grain pests can easily infest the prod-uct if left for a month or longer.

Outdoors in rodent burrows, diphacinone pellets work well. Uhlarik (1974) writes, "In one of the tests, run at the manufacturer's testing facility at Ed-wardsville, Kansas, diphacinone pellets were left to soak in water — completely immersed — for more than 48 hours. In another test the bait was weathered 15 days and then fed to rats and mice. While there was slight softening and distention noted when the pellets were removed, they returned to their original shape and hardness upon drying. They were then used in government approved acceptance and kill tests and performed above EPA minimum standards."

The diphacinone pellets kill more rapidly than warfarin at 250 ppm and during tests, the bait killed rats in as few as three days, with an average of 4.7 days. With mice, death resulted in an average of 5.5 days. However, the product is no longer registered for control of mice.

The commercial diphacinone bait is available in three flavors: apple (brown), meat (red) and fish (green).

Chlorophacinone. 2-[(p-chlorophenyl) phenylacetyl]-1,3-indandione is said to be the most active multiple dose anticoagulant available with an acute oral LD50 to the rat of 2.1 mg/kg and to the mouse 1.06 mg/kg. This level of activity makes it much more active than warfarin and even more active than the more modern multiple dose anticoagulants. It is extensively used against all three domestic rodents, especially because of the wide range of formulations available. These include 0.005 percent baits, 0.2 percent tracking powder, a dry concentrate, and a mineral oil concentrate. Baits based on chlorophacinone include a canary seed bait especially successful in mouse control, paraffin blocks, and pellets for use against rats and mice in damp situations indoors and outdoors.

When preparing chlorophacinone baits using the mineral oil concentrate the manufacturer finds the following formulations acceptable:

Oat groats (ground rather coarse)	64 lbs.
Yellow corn (ground rather coarse)	30 lbs.
Powdered sugar	4 lbs.
Chlorophacinone oil concentrate (about one quart)	1 lb. 12 oz.
Additional amount of oat groats	
	100 lbs.

OR

48% coarse ground yellow corn meal
48% coarse ground crimped oats
 4% confectionary sugar
      chlorophacinone oil is added at 20 cc/kg bait base mixture

The sugar is added after the grain has been mixed with the oil concentrate and then remixed. The last portion of oat groats is added to arrive at the total weight of 100 lbs. These formulae have been kept purposely simple for the PCO trade. However, the addition of other accepted ingredients may increase palatability.

If the bait is to be used soon:

Grain: If cereal is preferred use coarse cracked grain rather than fine ground meal. Pour the concentrate slowly over the bait and mix, turning continuously with a ladle or metal spoon to obtain a thorough impregnation. Allow to settle and dry for about 30 minutes during which time the oil should penetrate the bait. If the bait is not for immediate use it should be left in an open, ventilated place for a few hours before bagging.

Fruit: The concentrate may be used for impregnating fruit or vegetables such as apples, potatoes, carrots, roots, etc. Cut fruits or vegetables in pieces and dip them in the oil concentrate. Stir with a ladle until they become impregnated with the concentrate. Allow the fruit to drip excess liquid for about 30 minutes. The bait is now ready for use.

SINGLE-DOSE ANTICOAGULANTS

In the last two years two so-called "second-generation" rodenticides have been introduced in the United States — brodifacoum and bromadiolone. These coumarin-type anticoagulants represent a major breakthrough since they are better at controlling warfarin-resistant rodents than any previously marketed anticoagulants. Warfarin and other multiple-dose or first-generation anticoagulants are used as chronic rodenticides, causing death after several feedings by gradual inhibition of prothrombin and increased hemorrhaging. With the second generation rodenticides a single small dose has the same result, although the effect is delayed as with the conventional anticoagulants. Vitamin K_1, as with the conventional anticoagulants, is the antidote to these new single-dose anticoagulants.

Brodifacoum is labeled for indoor and outdoor control of rodents and is available as a 0.005 percent pelleted bait which has shown good acceptance under field conditions. Laboratory tests indicated that as little as one feeding of two grams of 0.005 percent pelleted bait will kill a Norway rat (Anon. 1978).

Bromadiolone is also labeled for indoor and outdoor control of rats and mice and is available as a 0.005 percent meal bait. Marsh et al. (1980) reported, ". . . a single overnight feeding often has resulted in 100 percent mortality of test groups of 20 wild Norway rats and 20 house mice." They reported successful field results and also stated, "Norway rats that have consumed a lethal acute dose of bromadiolone generally start dying on the third day, which is quite similar to reactions to other anticoagulants. Bait acceptance of this compound is exceptionally good and bait shyness is not evident."

Of special note is the field trial in Chicago's resistant-rat area, where control with bromadiolone was measured at the very high level of 88 percent (Ashton and Jackson, 1979).

ANTICOAGULANT RESISTANCE

MAJOR CHRONOLOGY OF ANTICOAGULANT RESISTANCE IN DOMESTIC RODENTS

1950—Warfarin first came into wide usage.

1958—Scotland, near Glasgow — Norway rat resistance to warfarin and diphacinone encompass at least 155 square miles of Western Scotland.

1959—England and Wales-Norway Rats — Three widely separated areas, Wales area most widespread.

1961—England-House mouse resistance to warfarin.

1962—Denmark-Norway rat resistance in southeastern Jutland.

1968—Netherlands-Resistant Norway rats were totally exterminated with other type rodenticides.

1970—Idaho near Maridian, small dairy farm, resistant to warfarin and pindone.

1971—Raleigh, North Carolina — First case in United States in Norway rats.

1972—New York, New York — Small pocket covering several blocks.

1972—Liverpool, England — Black rats.

1975—California — Two areas — Roof rat — First incidence in United States.

1975—California — San Francisco wharf area — First reported case of resistance in mice in United States.

1977—Lake Erie-Buffalo region, New York — Mice.
1979—Long Island, New York — Isolated areas — Mice.
    In April of 1977, a new anticoagulant rodenticide resistance surveillance program was initiated through The Center for Disease Control, Atlanta, Georgia (Frantz, 1977). This program involves a sampling procedure that provides relatively accurate information on the status of resistance in urban rat populations in the United States. A recent summary of study results by the New York State Rodent Evaluation Laboratory in Troy, New York, (Frantz, 1979a) indicates the following: "Here is the best data available on resistance in rats; that is, it is the only statistically valid data and it is the most up-to-date (the sampling began on April 1, 1977 and the table summarizes all data through March 31, 1979). Only 26 cities have been sampled by both labs thus far and a significant problem occurs in 13 of them or 50 percent. However, some resistance was found in 22 of the cities sampled, or in 84.6 percent."

TABLE 1-4

**Summary of anticoagulant rodenticide resistance test data
for rats — standard samples[1]
(1 April 1977 — 31 March 1979)[2]**

Project Location	Anticoagulant resistance in rats[3]	
	Identified	Significant problem[4]
ALABAMA		
Mobile	Yes	No
CALIFORNIA		
Los Angeles	Yes	No
Oakland	Yes	No
CONNECTICUT		
Hartford	Yes	Yes
DISTRICT OF COLUMBIA		
Washington	Yes	No
FLORIDA		
Ft. Lauderdale	No	No
Miami	Yes	No
Tampa (Roof rats, *R. rattus*)	Yes	No
GEORGIA		
DeKalb Co.	Yes	Yes
ILLINOIS		
Chicago	Yes	Yes
KENTUCKY		
Louisville	Yes	Yes
MARYLAND		
Baltimore	Yes	Yes

TABLE 1-4

Project Location	Anticoagulant resistance in rats[3]	
	Identified	Significant problem[4]
MASSACHUSETTS		
Boston	Yes	Yes
MICHIGAN		
Detroit	Yes	No
MISSOURI		
St. Louis	No	No
NEW JERSEY		
Jersey City	Yes	No
NEW YORK		
New York City	No	No
OHIO		
Cincinnati	No	No
Cleveland	Yes	Yes
Columbus	Yes	Yes
PENNSYLVANIA		
Chester	Yes	Yes
Clairton	Yes	Yes
PUERTO RICO		
San Juan	Yes	Yes
TENNESSEE		
Memphis	Yes	Yes
VIRGINIA		
Chesapeake	Yes	No
Portsmouth	Yes	Yes

[1]S. C. Frantz. 1977.
[2]Data provided by: The Rodent Control Evaluation Laboratory, New York State Department Health, Troy, NY and the Environmental Studies Center, Bowling Green State Univ., Bowling Green, OH.
[3]Norway rats *(Rattus norvegicus)* unless specified otherwise.
[4]S. C. Frantz. 1979b.

Tables five through seven summarize through March 31, 1979, the extent of warfarin resistance in groups of rodents tested from the United States and Canada. Note that in some cases very few rodents were tested. Thus, the data may not be statistically valid. The presence of resistance does not necessarily indicate the situation throughout a given city unless the rodents were trapped in a well-distributed fashion in the appropriate numbers.

Anticoagulant resistance occurs in areas where sanitation is poor and rodents have been exposed to anticoagulants for many generations. In Europe, almost all early cases occurred in rural areas. In the United States, it has surfaced in urban areas.

TABLE 1-5

Warfarin Resistance
10% or Greater Levels*
House Mouse

		Number Tested	Number Resistant
U.S.	Buffalo, N.Y.	88	69
	Decatur, Ala.	2	2
	Detroit, Mich.	6	5
	Ft. Worth, Texas	1	1
	Hoboken, N.J.	20	2
	Holland, Mich.	10	7
	San Francisco, Calif.	10	3
	Troy, N.Y.	31	23
CANADA	Vancouver, B.C.	75	60

(*10 of 15 cities tested, equals 67%)
Twenty-one day no-choice feeding on 0.025% warfarin.

TABLE 1-6

Warfarin Resistance
10% or Greater Levels*
Norway Rat

	Number Tested	Number Resistant
Akron, Ohio	52	5
Atlanta, Ga.	76	10
Baltimore, Md.	151	15
Boston, Mass.	175	20
Charlotte, N.C.	19	5
Chester, Pa.	108	11
Chicago, Ill.	369	242
Clairton, Pa.	86	15
E. Orange, N.J.	13	2
Hartford, Conn.	140	17
Houston, Texas	136	16
Memphis, Tenn.	59	8
Pittsburgh, Pa.	127	17
Portsmouth, Va.	70	20
Poughkeepsie, N.Y.	345	116
San Juan, Puerto Rico	85	16
Seattle, Wash.	5	1
Selma, N.C.	160	81
Sioux City, Iowa	18	14

(*19 of 93 cities tested, equals 20%)
Six-day no-choice feeding on 0.005% warfarin.

TABLE 1-7

Warfarin Resistance
5% or Greater Levels*
Roof Rat

	Number Tested	Number Resistant
Fresno, Calif.	27	3
Houston, Texas	14	4
Los Angeles, Calif.	23	8
Ontario, Calif.	54	3
Santa Ana, Calif.	19	2
San Diego, Calif.	22	1
San Juan, Puerto Rico	12	1
Seattle, Wash.	14	1

(*8 of 24 cities tested, equals 33%)
Twelve-day no-choice feeding on 0.025% warfarin.

With the exception of the Idaho and Holland population, none of the resistant rodent populations have been totally eliminated. Different non-anticoagulant baits have successfully controlled resistant rats in different situations including.
- Norbormide in Idaho (Brothers, 1972).
- Thallium baits in Denmark. (No longer available in the United States).
- Racumin in Scotland and other areas (however, in England and Denmark some rats are also resistant to Racumin); (not available in the United States).
- DLP 787 baits in New York City. (No longer marketed).

Jackson (1969) predicted eradication of resistant rodents is unlikely but frequency of occurrence should be reduced. Zinc phosphide and other acute poisons are effective and the second-generation single-dose anticoagulants have also been successful against resistant rats.

Rodents resistant to one multiple-dose anticoagulant exhibit cross resistance to most others, both hydroxycoumarins and indandiones.

Genetic resistance in a population is selected over a series of generations. It does not occur overnight. Where large numbers of rodents exist that are subjected to heavy rodenticide pressure, a few may survive who are more resistant than the average. During the next several generations, under constant exposure to rodenticide, a larger and larger percent of rodents survive that exhibit resistance. The anticoagulant eliminates the genetically susceptible rodents. Physically, the animal itself looks no different than non-resistant animals.

Resistance does not arise from exposure to sub-lethal doses of anti-coagulant. However, by improving sanitation, the chance of good bait acceptance both with poison and on traps is enhanced. If all rodents are eliminated there can be no resistance. In addition, it must be emphasized rodents cannot become resistant to the lack of food and shelter resulting from environmental sanitation and from the rodent proofing of structures.

Resistant Norway rats found in the United States can be divided into two camps, "hard resistant" and "soft resistant" warfarin rats. "Hard resistant" rats can feed on the 0.005 percent bait for six days and show no adverse symptoms. They are also referred to as "resistant-non-affected". "Soft resistance" or "resistant-affected" rodents, after feeding on the test bait for a few days, reduce their food intake, become sluggish and occasionally bleed slightly from the mouth and anus. Animals which show one or more of these symptoms, but subsequently recover and remain alive for at least 10 more days can be classified as soft resistant. Bowerman and Brooks (1974) showed when two resistant rats are crossed their offspring exhibit survival rates of 38 to 100 percent when fed warfarin test baits. The survival rates for offspring of a susceptible and a resistant parent range from zero to 77 percent. The exact genetic mechanism involved is still not understood. It appears that a single dominant gene is involved (Price-Evans and Sheppard 1966, Greaves and Ayres 1967, Bowerman and Brooks 1974). The development of house mice resistance is more complex and appears to involve several genes, according to Brooks (1973).

House mice have always been more tolerant to warfarin than rats. Jackson (1975b) reports, "In any house mouse population there likely will be individuals which can tolerate a dose of warfarin that is more than tenfold (on a per-unit body-weight basis) what is lethal for susceptible Norway rats. House mice, under normal conditions, can survive on commercially prepared warfarin baits for seven to 10 days before death occurs.

"For testing purposes, house mice which survive 21 days of feeding on 0.025 percent warfarin are designated resistant" (Jackson, 1975a & b). At a pest control symposium in Tampa, Florida in 1978, Jackson presented previously unpublished data showing house mice resistance now occurs in Decatur, Alabama, Cleveland, Ohio, Detroit and Battle Creek, Michigan, St. Paul, Minnesota, Hoboken, New Jersey, Buffalo, New York, and Fort Worth, Texas.

Although in several instances less than 20 animals were tested per location, it does show resistance is widespread. Undoubtedly, as mice are captured in other locations, the phenomenon of anticoagulant resistance will become more widespread. As this publication goes to press the technical editor has found resistance in house mice in Farmingdale, Long Island, New York. This resistance was confirmed by the New York State Health Department Rodent Evaluation Laboratory, Troy, New York.

Brooks (1973) summarizes anticoagulant resistance as "the metabolism of vitamin K in the liver is affected by warfarin. In normal rats, warfarin prevents the conversion of vitamin K and the subsequent formation of blood clotting proteins, leading to death of the animals through internal bleeding. In resistant rats, the chemical conversion of vitamin K takes place even in the presence of anticoagulants."

In the United States, only two facilities have been federally supported to evaluate the extent of anticoagulant resistance. Those were the laboratories at Bowling Green State University Bowling Green, Ohio, and the New York State Health Department Rodent Evaluation Laboratory, in Troy, New York. Through February of 1978, they showed 62 of 86 urban areas studied had Norway rats exhibiting at least low levels of anticoagulant resistance. Currently, no federal funds are provided for investigating the level and extent of anticoagulant resistance in mice. In 1979 federal funds for the Bowling Green Laboratory were terminated leaving only one laboratory in the United States to continue rodent resistant studies.

According to the World Health Organization, rodents which survive a six-day no choice feeding test with 0.005 percent warfarin bait are considered resistant.

TABLE 1-8

Situations where rodents are incorrectly accused of being resistant.

1) Rodents are not eating the bait.	— This is created by unpalatable bait or the presence of other food sources. This is not genetic chemical resistance.
2) Rodents are eating the bait. A once a week count shows all bait is gone, but the population is not declining.	— The rodent population is too large for the amount of bait used. For the anticoagulants to work, the rodent must be exposed to daily continual feedings. Once the anticoagulant is depleted, the rodent switches to another food.
3) Some rodents are dying but new rodents keep showing up. There are many areas where mice are in boxes of food against the wall.	— The mice are breeding within the boxes and are sending out new mice into other areas. The use of rodenticides without proper sanitation will not diminish the problem.

Pest control operators use commercial warfarin baits for rats at 0.025 percent and for mice from 0.025 percent to 0.05 percent. Although the World Health Organization technique uses a lower concentration, it gives the rodent no alternate food source. Jackson (1978) reports neither switching currently available anticoagulants nor increasing their concentration in baits results in good control where resistance occurs.

A compilation of 160 different references on resistance in rodents was completed by the Warfarin Resistance Laboratory in Bowling Green, Ohio, and the Rodent Control Evaluation Laboratory in Troy, New York. It is complete through October 1974. (Anonymous, 1974). This was revised by Kaukeinen in 1979.

Chemical repellents. Repellents do not kill. Therefore the best that can be hoped for them is that they will move rodents from one area to another. To date, repellents are seldom practical (Howard and Marsh, 1974).

RAT ODORS

Dead mice, because of their relatively small size, present little problems in reference to creating odors bothersome to man. Rats present another story. The use of poisons in a building may result in the rat dying in wall, floor, ceiling or other enclosed spaces. Needless to say, the stench from the decaying rats is extremely offensive. To offset this evident disadvantage from the use of poisons, some early merchandisers of rat poisons made exaggerated claims as to the ability of their poisons to embalm without odor, or to drive the poisoned rat from the building. There is no published information to substantiate these claims, and now in this more enlightened age, the legend persists only in our native folklore, viz.

> *Little Johnny and the other brats*
> *Ate all the Tuff-on-Ratz*
> *Father said, as mother cried*
> *Never mind, they'll die outside*

The rat odor problem is a serious one to the pest control operator. Maguire and Cook (1948) discuss this problem in some detail. They are of the opinion that the mercaptans and free fatty acids in the decaying rats are responsible for the odor. "These find their way into the air by volatilizing and mere traces of them give the impression that the air is saturated with their foulness." The odor is most prominent when the rat dies in a warm humid area as around a steam pipe.

Maguire and Cook (1948) note: "As the putrefaction progresses the odor changes from day to day since some chemicals are produced in the early part and others in the latter stages. It should, therefore, be obvious why a chemical of a given type might be expected to neutralize a three–day dead rat odor, but would be incapable of helping the situation after the rat was dead a week.

"At room temperature and with 50 percent relative humidity most of the bacterial action will be completed after 72 to 98 hours. The carcass will still continue to give off odor for upwards of a week while further bacterial, fungicidal, and possible chemical action together with desiccation and in some instances, insect life, complete nature's cycle."

Of course, the odor can be quickly abated if the dead animal is found and removed. Where this is not possible, Maguire and Cook (1948) recommend fans be used to ventilate the area, and cracks through which the odor enters the room be sealed off.

Rohm and Haas Hyamin 1622, a pure quaternary ammonium salt, has also been useful as a deodorant, as has Fritzche Bros. Neutroleum Alpha, and Du Pont Isobornyl Acetate. Nilodor, a proprietary deodorant, has been especially effective against a variety of odors. Turgasept is also an effective product (Frishman, 1974).

There are now deodorant aerosol bombs containing such materials as Metazene and Meelium which will offer temporary relief.

Sometimes the precise spot where the rat died can be located by releasing several blue or green bottle flies in the room where the dead rat is believed to be present. The bottle flies will settle directly over the locality where the dead rat is located. This author has found many a dead rat by following bottle flies. For a detailed discussion on odor control, see Buslik (1965).

Students interested in further information on the biology and control of rodents should obtain the comprehensive bibliography compiled jointly by the World Health Organization and the Food and Agriculture Organization of the United Nations (FAO and WHO, 1973a, 1973b, 1977).

LITERATURE

ANDERSON, J.W. — 1954. The production of ultrasonic sounds by laboratory rats and other mammals. Science 119:808-809.

ANDREWS, J.M. and V.B. LINK — 1947. The murine typhus fever problem in the United States. Pests 15(1):12-20.

ANONYMOUS — 1946. Instructions for using ANTU (Alphanaphthyl-Thiourea) Pests 14 (7):22,24.

ANONYMOUS — 1949. New rodenticides available. Soap S.C. 25 (12):143.

ANONYMOUS — 1974. Bibliography on anticoagulant resistance in rodents. (Preliminary-October 1974). Warfarin Resistance Laboratory Environmental Studies Center, Bowling Green, Ohio and Rodent Control Evaluation Laboratory, Bureau of Rodent Control, Troy, New York.

ANONYMOUS — 1977. Rodent repellers: More proof is needed. Pest Control Magazine 45 (8):26-27.

ANONYMOUS — 1978. Talon Technical Information, ICI Americas, Inc. 4 pp.

ANONYMOUS — 1978. 3 Electromagnetic devices fail to control in NV Test. Pest Control 46 (11):51.

ANONYMOUS — 1979a. EPA Stops Sale of Electro-Magnetic Repellers. Pest Control 47(5):54-55.

ANONYMOUS — 1979b. Rats thrive on poison diet. New Scientist, June 7, p. 812.

ASHTON, A.D. and W.B. JACKSON — 1979. Field testing of rodenticides in a resistant-rat area of Chicago. Pest Control 47(8): 14-16.

BACON, V.W. — 1945. Methods of estimating rat proofing and rat eradication changes. U.S.P.H.S. Lecture at Houston "P.C.O." school.

BEAL, C. — 1943. Rio Grande to Cape Horn (Houghton Mifflin).

BJORNSON, B.F. and A.J. BROOKS — 1962. Rat bait acceptability in sewers. Pest Control 30(10):24.

BOWERMAN, A.M. and J.E. BROOKS — 1974. Anticoagulant resistance studies in urban Norway rats. N.Y.S.D.H. Rodent Control Bureau.

BROOKS, J.E. — 1961. Baits for sewer rat control. Calif. Vector Views 8 (6):30-32.

1963. Fluoroacetamide, Calif. Vector Views 10(1):1-3.

1964. The presence of the roof rat *Rattus rattus* in sewers in California. Calif. Vector Views 11(11):71-72.

1966. Roof rats in residential areas. The ecology of invasion. Calif. Vector Views 13(9):69-73.

1973. A Review of Commensal Rodents and their control. Critical Reviews in Environmental Control 3(4):405-453.

BROOKS, A.J. and H.S. SCOTT — 1961. Preliminary studies of insect mold infestations of anticoagulant rodenticide baits. Fla. Entomol. 44(1):15-23.

BROTHERS, D.R. — 1972. A case of anticoagulant rodenticide resistance in an Idaho Norway rat *(Rattus norvegicus)* population. Calif. Vector Views 19(6):41.

BROWN, R.Z. — 1960. Biological factors in domestic rodent control. USPHS Publication No. 773.

BUSLIK, D. — 1965. Notes on odor control. Pest Control 33(7):14-15, 42 and 33 (8):20, 22-24, 26.

CALHOUN, J.B. — 1948a. Mortality and movement of brown rats *(Rattus norvegicus)* in artificially supersaturated populations. J. of Wildlife Management 12(2):167-172.

1948b. The development and role of social status among wild Norway rats. Anatomical Record 101(4):694.

CAMERON, T.M.W. — 1949. Diseases carried by house mice. Pest Control 17(9):9-11.

COTTAM, C. — 1948. Aquatic habits of the Norway rat. J. Mammal 29(3):299.

CRABTREE, D.G. — 1944. Fortified red squill; its development and application in rat control. Pests 12(12):24-26.

CREEL, R.H. — 1915. The migratory habits of rats with special reference to the spread of plague. U.S.D.H.S. Rats 30 (23):1679-1683.

DAVIS, D.E. — 1948. Principles of rat management. Pests 16 (11):9-10, 12.

DAVIS, D.E., J.T. EMLEN, JR., and A.W. STOKES — 1948. Studies on home range in the brown rat. J. Mammalogy 29(33):207-225.

DAVIS, R.A. and F.P. ROWE — 1963. The ubiquitous mouse. New Scientist 19:127-130.

DERSE, P. — 1963. Anti K factor in anticoagulant rodenticides. Soap & Chem Spec. 39(3):82-84.

DYKSTRA, W.W. — 1950a. A review of the history, ecology and economic importance of house mice. Pest Control 18(8):9-12, 14.

1950b. New techniques and developments in rodent control. Pest Control 18(12):9-10, 12, 24, 39.

1951. Toxicity of Warfarin, Pest Control 19(3):16.

EATON, O. and C.A. CABELL — 1949. Raising laboratory mice and rats. U.S.D.A. leaflet No. 253.

EBELING — 1975. Urban Entomology, Univ. of California 695 pp.

EGELHOF, J. — 1978. Bomb-sniffing rodents trained to smell a rat. Chicago Tribune, Section 1, Sept. 24, 1978. p. 4.

EHMANN, N.R. — 1960. Do's and don'ts with anticoagulants. P.C.O. News 20(5):6, 9, 10, 11.

ELMORE, J.W. and F.J. ROTH — 1943. Analysis and stability of zinc phosphide. J. of the Assn. of Official Agric. Chemists 26:559-564.

EMLEN, J.T., JR. — 1947. Baltimore's community rat control program. Pest 15 (10) 30, 32, 34, 38.

1950. How far will a mouse travel to a poisoned bait? Pest Control 18(8):16, 18, 20.

ESKEY, C.R. — 1943. Murine typhus fever control. U.S.P.H.S. Rpts. 58(16):631-639.

FAO and WHO, — 1973a. Bibliography on rodent pest biology and control 1950-1959. Part I and II.

1973b. Bibliography on rodent pest biology and control 1960-1969. Part I, II and III.

1977. Bibliography on rodent pest biology and control 1970-1974. United Nations, Via delle Terme di Caracalla, 00100 Rome, Italy.

FARRIS, E.J. and J.Q. GRIFFITH. — 1949. The rat in laboratory investigation. J.B. Lippincott Co., Philadelphia. 542 pp.

FERTIG, D.S. and V.W. EDMONDS — 1969. The Physiology of the house mouse. Sc. Am. 221 (4):103-108, 110, 148.

FINGER, G.C. and F.H. REED — 1949. Fluorine in industry. Part 2: Industrial fluorine chemicals. Chemical Industries 64(1):51-56, 149.

FRANTZ, S.C. — 1977. Procedures for collecting rats for anticoagulant resistance studies — Urban rat control projects. U.S. Dept. HEW, Public Health Serv., CDC, Atlanta, GA. 16 pp.

1979a. Personal communication

1979b. Procedures for sampling urban rat populations for anticoagulant resistance evaluation. Pages 20-28 in J.R. Beck, Ed. Vertebrate Pest Control and Management Materials. ASTM STP 680. American Society for Testing and Materials, Philadelphia. 323 pp.

FRISHMAN, A.M. — 1974. The Rodent Handbook, Frishman Publisher, Farmingdale, NY 164 pp.

GARLOUGH, F.E. — 1946. Need for permanent city rat control project. Pests 14 (9): 10, 12, 13.

GARLOUGH, F.E. and D.A. SPENCER — 1944. Control of destructive mice. U.S. Dept. Int. Fish & Wildlife Service. Conservation Bull. 36.

GREAVES, J.H. and P. AYRES — 1967. Heritable resistance to Warfarin in rats. Nature (Lond). 215, 877.

GREEN, D.D. — 1950. Answer to question on *R.r. frugivorus*. Pest Control 18 (12): 6.

HALL, E.R. — 1927. An outbreak of house mice in Kern Country, California. Univ. Calif. Publ. Zool. 30 (7): 189-203.

HAMILTON, W.J., JR. — 1947. Rats and their control. N.Y. Agric. Col. Ext. Cornell Ext. B. 353 rev. 34 p.

HAYES, W.J. and T.B. GAINES — 1959. Laboratory studies of five anticoagulant rodenticides. U.S.P.H.S. Rpts. 74(2): 105-113.

HAYNES, D.W., M.D. PIRNIE, and C.H. JEFFERSON — 1945. Controlling rats and house mice. Mich. Agr. Expt. Sta. Circ. Bull. 167.

HINTON, M.A.C. — 1931. Rats and mice as enemies of mankind. British Museum. Econ. Ser. 8.

HIRSCHHORN, H. — 1974. All about rats. T.F.H. Publications, Inc. 76 pp.

HOGARTH, A.M. — 1929. The rat, a world menace. London, J. Bale, Sons and Danielsson, Ltd. 112 p.

HOWARD, W.E. and R.E. MARSH. — 1974. Rodent control manual. Pest Control Magazine. August.

HUDSON, B.W. and T.J. QUAN SEREOLOGIC — 1975. Observations during an outbreak of rat borne plague in the San Francisco Bay Area in California. J. of Wildlife Diseases 11(3):431-436.

HUEBNER, R.J., W.L. TELLISON and C. POMERANTZ — 1946. Rickettsialpox — A newly recognized disease. IV. Isolation of a *Rickettsia* apparently identical with the causative agent of Rickettsialpox from *Allodermanyssus sanguineus,* a rodent mite. U.S.P.H.S. Rpts. 61(47):1677-1682.

IVES, H.B. — 1948. House mice and their control. Pests 16(10):11-12, 14, 16, 18, 54.

JACKSON, W.B. — 1969. Anticoagulant resistance in Europe. Pest Control 37(3):51-55.

JACKSON, W.B. — 1980. Ultrasonics protect egg farm. Pest Control 48(8):5 pp.

JACKSON, W.B., ET. AL. — 1975a. Anticoagulant resistance in Norway rats as found in U.S. Cities. Part I. Pest Control 43(4):12-16.

1975b. Anticoagulant resistance in Norway rats as found in U.S. Cities. Part II. Pest Control 43(5): 14-24.

KAUKEINEN, D.C. — 1977. Bibliography of rodent resistance to anticoagulant rodenticides. Environmental studies center, Bowling Green State Univ., Bowling Green, OH 23 pp.

JELLISON, W.L. et al. — 1949. Rat bite fever in Montana. U.S.P.H.S. Rpts. 64(52):1661-1665.

JOHNSON, M.S. — 1946. Notes on preventing rats from boarding ships. Abst. in Abst. Bull. N.S. No. 4 Insect Control Comm. Nat'l. Res. Council.

KALMBACH, E.R. — 1945. "Ten-eighty" a war-produced rodenticide. Science 102(2644):232-233.

KLIMSTRA, W.D. — 1968. Biology and behavior of the house mouse, NPCA Service letter #1220-Talk presented at the 35th Annual Convention. NPCA. Salt Lake City, Utah. 1968.

LATUEN, A.R., A.B. LOANE and J.C. MUNCH — 1948. Antu, the lethal dose. Soap and S.C. 24 (1):127, 129, 131.

LINK, K.P. and W. ROSS — 1956. Anticoagulants. Pest Control 24(8):22, 24.

LYNCH, D.T. — 1941. The wild seventies. Vol. II Appleton — Century Co., N.Y.,

Chapter 34, pp. 301-311. Reprinted by Kinnikat Press. Port Washington, N.Y.

LYNCH, F.G. — 1958. Berkeley's sewer poisoning program for rodent control. Calif. Vector Views 5 (11):71.

MACKIE, R.A. — 1964. Control of the roof rat *Rattus rattus,* in the sewers of San Diego, CA. Vector Views 11(2):7-10.

MAGUIRE, E.G. and K.C. COOK — 1948. Handling dead rat odors. Pest Control and Sanitation 3(4):26-27.

MARSH. R.E. and W.E. HOWARD — 1976. House mouse control manual. Part one. Pest Control 44(8):23, 24, 26, 29, 30, 32, 33, 62, 64.

MARSH, R.E., W.E. HOWARD and W.B. JACKSON — 1980. Bromadiolone: A new toxicant for rodent control. Pest Control 48(8):22, 24, 26.

MCALINDIN, D.P. — 1956. Anticoagulants. Pest Control 24(8):24, 27.

MILLS, E.M. — 1945. The detection of rats by their droppings. Pests 13(6):18.

1947a. Analysis of "Questionnaire on House Mouse Control". N.P.C.A. Service Letter No. 475, Appendix No. 1, Page No. 1.

1947b. House mouse control with poisoned baits. U.S. Fish and Wildlife mimeo. Leaflet.

MUNCH, J.C. — 1947. Antu. Soap & S.C. 23(4):147, 149, 169.

NEWSDAY — 1979. Rats gnawed on man, 45, March 5.

NICHOLSON, H.P. et. al. — 1947. Laboratory toxicity studies on dosages of 1080 in aqueous solution against *Rattus norvegicus,* U.S.P.H.S. mimeograph. Dec. 1947.

ODERKIRK, G.C. — 1951. Successful bait, Pest Control 19(3):16.

ORR, R.T. — 1944. Communal nests of the house mouse *(Mus musculus* Linnaeus).* Wasmann Collector 6(2):35-37. Biol. Abst. 19:1635, 1945.

PERRY, J.S. — 1945. The reproduction of the wild brown rat. Zool. Soc. London Proc. 114(1-2)19-46.

PISANO, R.G. and T.I. STORER — 1948. Burrows and feeding of the Norway rat. J. Mammal 29(4): 374-383.

PETERSON, G.N. — 1978. Rodent proof design/construction. Waverly Press, 151 pp.

PLUMMER, B. — 1978. Tales of a Rat-hunting man. Robin Clark, Ltd. 141 pp.

PRATT, H.D., B.F. BJORNSON and K.S. LITTIG, — 1976. Control of domestic rats, 2 mice. U.S.D. H.E.W. PHS HEW Publication No. (CDC) 76-8141. 47 pp.

PRATT, H.D. and R.Z. BROWN — 1976. Biological factors in domestic rodent control. U.S.D. H.E.W. PHS HEW Publication No. (CDC) 77-8144. 30 pp.

PRICE-EVANS, D.A. and P.M. SHEPPARD — 1966. Some preliminary data on the genetics of resistance to anticoagulants in the Norway rat. W.H.O. Seminar on rodents and rodent ectoparasites. Oct. 24-28, WHO/VC 66:217, Geneva 1966, 155.

RICHTER, C.P. — 1946. Incidence of rat bites and rat fever in Baltimore, Md. Abst. in Abst. Bull. N.S. No. 4 Insect control com Nat'l. Res. Council, Washington, D.C. July.

ROHE, D.L. — 1966. Field evaluation of the rodenticides, fluoroacetamide and norbormide against roof rats in sewers. Calif. Vector Views 13(11):79-82.

ROHE, D.L. and R.S. STONE — 1962. Feeding preferences of Norway rats for selected food industry by-products. Calif. Vector Views 9(3):14-15.

ROQUES, H. — 1946. Toxic principal of squill, Soap & S.C. 22 (1):78.

SALMON, T.P. and R.E. MARSH and K. WHITE — 1977. Record weight wild

Norway rats. 24(1-2):6-9.

SCHOOF, H.F. — 1970. Zinc phosphide as a rodenticide. Pest Control 38(5):38, 42-44.

SCHWARZ, E. — 1942. Origin of the Japanese waltzing mouse. Science 95:46.

SCHWARZ, E. and H.K. SCHWARZ — 1943. The wild and commensal stocks of the house mouse *Mus musculus* Linnaeus. J. Mammalogy 24(1):59-72.

SCHWARTZ, E. — 1944. Premises for rat control. Pests 12(9):7.

SCOTT, H.G. and M.R. BOROM — 1965. Rodent-borne disease control through rodent stoppage. U.S.D. H.E.W., PHS 82 pp.

SENNEWALD, E.F. — 1948. Alkaloids. Senco News 14(3):2-3.

SILVER, J. — 1927. The introduction and spread of house rats in the United States. J. Mammal 8(1):58-60.

SILVER, J. and F.E. GARLOUGH — 1941. Rat Control U.S. Fish Wildlife Services Conservation Bull. 8.

SIMMONS, S.W. and H.P. Nicholson — 1947. The use of rodenticides 1080 and Antu. Pest 15(5):8, 10.

SOUTHERN, H.N. — 1946. The normal daily range of wild house mice (*Mus musculus*) Abst. in Abst. Bull. N.S. No. 4 Insect Control Com Nat'l Res. Council, Washington 25, D.C.

SOUTHERN, H.N. and E.M.O. LAURIE — 1946. The house mouse (*Mus musculus*) in corn ricks. J. of Animal Ecology 15(2):139-149.

SPENCER, H.J. — 1946. Anticoagulants. Pest Control 24(8):27, 28, 50.

STEWART, C.K. — 1946. Aspect of plague control. Pest Control & Sanitation 1(4):14-15.

STORER, T.I. — 1931. Community nests of the house mouse. J. of Mammal. 12:317.

1948. Control of rats and mice. Calif. Agric. Ext Service Circ. 142.

STOTIK, A.M. — 1967. Using dogs for control of rodents (in Russian) Zhivot-novodstov 9:86-88.

TRUMAN, L.W. BUTTS and G. BENNETT — 1976. Scientific Guide to Pest Control Operations. Harvest Publications.

UHLARIK, J. — 1974. All weather rodenticide. Soap Cosmetics Chemical Specialties. March.

UPI — 1979. With this rat, I thee wed, St. Louis Post Dispatch, Jan. 26., p. 9Ag.

US-NY. — 1979. Anticoagulant Resistance Testing — Quarterly Report — January 1-March 31, 1979. Rodent Control Evaluation Lab, N.Y. Dept. Health, Troy, 17 pp, Mimeo.

US-OHIO — 1977-1979. Anticoagulant Resistance Study — Quarterly Report. Bowling Green State University, Environmental Studies Center, Mimeo, January 1977 — early 1979.

U.S.P.H.S. — 1948. Rodent Control Manual. Typhus Control Unit. U.S.P.H.S.

WARD, J.C. — 1947. Certain new techniques in rodent control. Pests 15(3):12, 14.

WHELAN, D.B. — 1944. Best time of year to kill rats. Nebraska Municipal Review, June.

WILLIAMS, C.L. — 1948. The ship rat. Pest Control and Sanitation 3(4):8-13.

WOODALL, J.P. — 1975. LCM — Lymphocytic Choriomeningitis. Proceedings of the Seminar on Environmental Pests and Disease Vector Control. pp. 1-15.

ZINSSER, H. — 1935. Rats, lice and history, Bantam Science and Mathematics. 228 pp.

ARNOLD MALLIS
RICHARD V. CARR

Arnold Mallis was born in New York City on October 15, 1910, the son of Russiam immigrants to the United States. He attended grade school and high school in Brooklyn and Long Island. In 1927, at the age of 16, he moved with his parents to Los Angeles, California, where he completed his last year of high school.

In 1929, he entered the University of Southern California and began pre-dental courses. Since the Great Depression began at about this time, he dropped out of college and worked in a Los Angeles garment factory. After two years on the job he decided to return to college and, because of his interest in trees and natural history, began a two year course in forestry at Pasadena Junior College (now Pasadena City College).

Because he was hard-of-hearing, his forestry teacher told him he would have great difficulty in getting a job as a forest ranger and that instead he should enroll in forest entomology and forest pathology in Berkeley. He entered the University of California, Berkeley in 1932

After two years graduate work he left the University in 1936 and sought employment as an entomologist in southern California.

Mallis worked two years for pest control firms in Los Angeles, Hollywood and Bakersfield and then in 1938 obtained a job as a field aide in the USDA Bureau of Entomology, doing research on vegetable insects in southern California. This job lasted for six months. He then returned to Berkeley for two purposes, to complete his Masters degree on the ants of California and to collect information for the first edition of the Handbook of Pest Control.

Dr. Richard Carr's career reflects a varied background of experience in urban pest control. As a service technician in Tucson, Arizona, Carr served his apprenticeship learning the trade at its most important level. He also gained valuable experience with his own firm, experiencing the business at the level where the "buck stops."

His master's degree, earned at the University of Arizona, Tucson, offered him the opportunity to study the behavior of the honey bee and resulted in his interest in another social insect, the termite. During the years Carr spent working on his Ph.D., he investigated the mating behavior and morphology of sex attractant glands in termites.

In the early 1970s, Carr joined Velsicol Chemical Corporation in Chicago, where he became their first research and development investigator to concentrate his efforts on the company's specialty product line.

In 1979, Carr worked for Armak Company, Chicago, Illinois, acquiring experience in polymers for use in slow releasing pesticides. Then in 1980, he assumed his current position as director of research at the National Pest Control Association.

CHAPTER TWO

Silverfish

Revised by Arnold Mallis[1] and Richard V. Carr[2]

Preferring books to other dishes
  Are our friends the silverfishes
Alas for them! All their lore
  Is filed on the library floor!

A.M.

SILVERFISH ARE AMONG the most common insects in the home and are of concern because they eat paper, fabrics, and get into cereals. They remove the sizing of paper in books and magazines, damage etchings and prints, nibble on book bindings, and feed on the glue and paste in the binding. At times, silverfish are important pests in libraries and other places where books, documents, and papers are stored and filed. Silverfish feed on fabrics made from plant fibers such as linen, rayon, lisle, and cotton, particularly starched linen or cotton. Sometimes they infest flour and other cereal products.

Wygodzinsky (1972), in his study of silverfish, found 13 species in the United States. The following are pests indoors: silverfish, *Lepisma saccharina* L.; firebrat, *Thermobia domestica* (Packard); *Thermobia campbelli* (Barnhart); gray silverfish, *Ctenolepisma longicaudata* Escherich; fourlined silverfish, *Ctenolepisma lineata pilifera* (Lucas); and *Acrotelsa collaris* (Fabricius). Two of these silverfish, *Thermobia campbelli* (Barnhart) and *Acrotelsa collaris* (Fabricius), are not as well known as the others.

Silverfish are among the most primitive of all insects and evolved before the cockroach. These insects have a distinct carrot-shaped form, long and slender, broad at the fore end, and gradually tapering to the rear. Superficially, they have a fish-like appearance and hence the name silverfish or "fishmoths." They have short legs, long slender antennae, and three jointed anal cerci, which are tail-like appendages at the end of the body and have given rise to the common name "bristletails." Silverfish are wingless and characteristically scale-covered. Their development is simple because the young are very similar in appearance to the adults except for size. Silverfish are long-lived insects, capable

[1]*Associate Professor (Retired) Extension Entomology, The Pennsylvania State University, University Park, Pa.*
[2]*Director of Research, National Pest Control Association, Vienna, Va.*

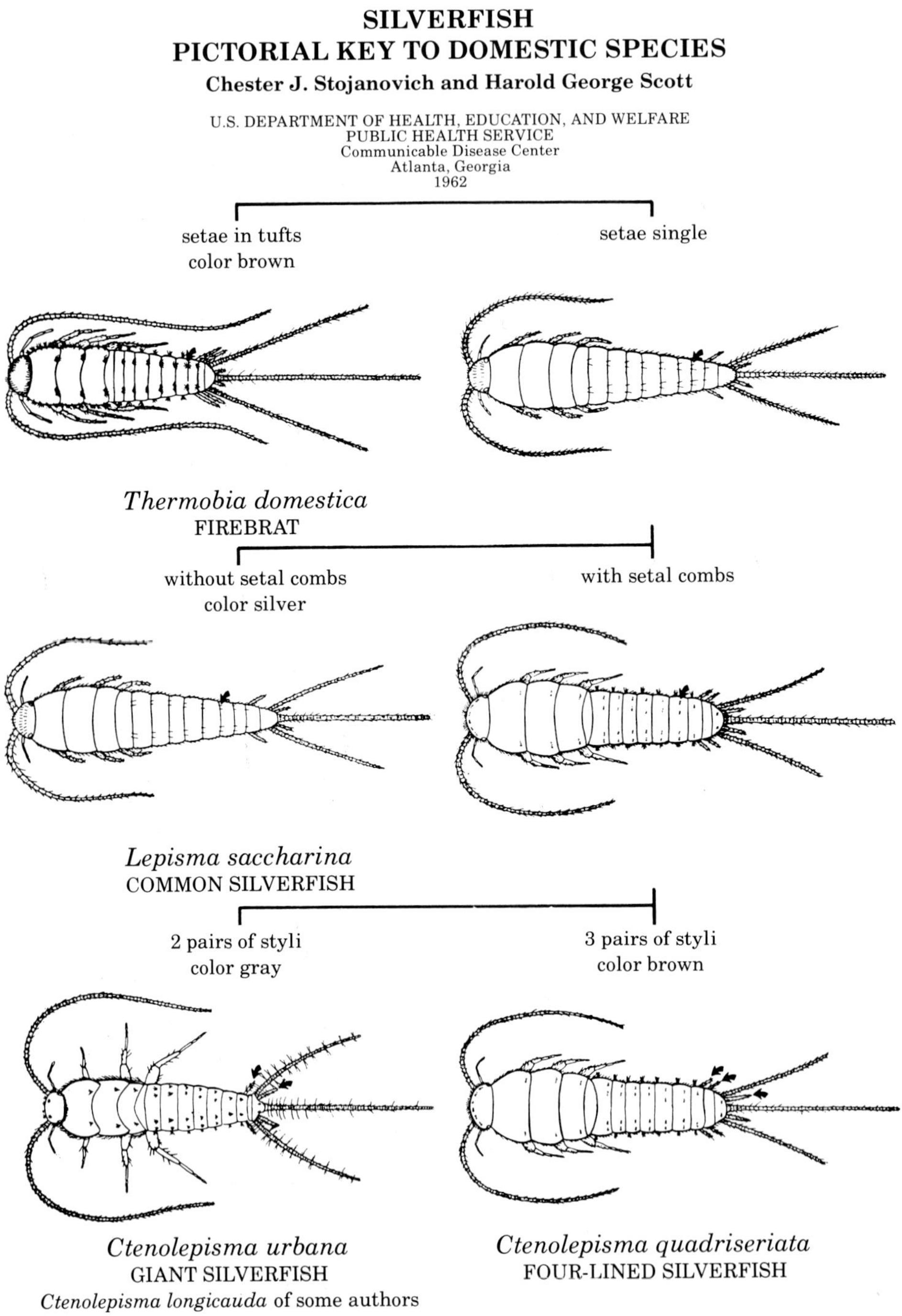

SILVERFISH
PICTORIAL KEY TO DOMESTIC SPECIES
Chester J. Stojanovich and Harold George Scott

U.S. DEPARTMENT OF HEALTH, EDUCATION, AND WELFARE
PUBLIC HEALTH SERVICE
Communicable Disease Center
Atlanta, Georgia
1962

setae in tufts
color brown
                setae single

Thermobia domestica
FIREBRAT

without setal combs
color silver
                with setal combs

Lepisma saccharina
COMMON SILVERFISH

2 pairs of styli
color gray
                3 pairs of styli
                color brown

Ctenolepisma urbana
GIANT SILVERFISH
Ctenolepisma longicauda of some authors
                Ctenolepisma quadriseriata
                FOUR-LINED SILVERFISH

Fig. 2-1. Pictorial key to silverfish.

of living for several years. They are dark-loving insects and avoid direct sunlight.

These insects are often entrapped in the smooth bowl of the wash basin or in the bathtub. Because of this some people have inferred that silverfish have a definite preference for the bathroom. In the author's opinion silverfish are to be found throughout the house, but are observed most frequently in the bathroom because there is little molding behind which they can hide. Furthermore, their dark bodies are most prominent against the background of light bathroom tile. Moreover, they are often trapped in the wash basin or the bathtub because they fall in and cannot climb out on the smooth surface of these fixtures.

Some species of silverfish are found beneath rocks, the bark of trees, and in leaf mold in forested areas. A number of species live in the nests of ants and termites, along with bird nests. Linsley (1944) notes that the silverfish and firebrat have been found outdoors in the nests of insects, birds and mammals, and under the bark of trees.

Feeding habits of silverfish. The feeding habits of silverfish in homes are very similar. These insects may roam for quite some distance in search of food, but once they have found a satisfactory source, they remain close to it.

Silverfish consume both carbohydrates and proteins. Adams (1933) showed *Thermobia domestica* (Pack.) was extremely fond of rolled oats and ground raw dried beef. Sweetman (1939) noted *Lepisma saccharina* L. preferred proteins to carbohydrates. He states, "Animal products such as fresh dried beef, are eaten eagerly and may be an essential portion of the diet. If fed only whole wheat flour or Mead's cereal for about two weeks and then offered fresh dried beef, they eat the latter ravenously, fighting among themselves for the food." Moreover, he found this insect to be cannibalistic, eating cast skins and dead and injured individuals. Rustin and Richardson studied the damage done by *Thermobia domestica* to fabrics and papers. They observed the greatest damage was done to medium typewriter bond paper, regenerated cellulose, either knitted or plain weave, and linen. Cotton and silk were attacked when the texture was suitable for feeding. Mallis (1941), working with *Ctenolepisma longicaudata* Esch., found this silverfish fed on materials of both animal and vegetable origin, including wheat flour, whether or not it contained sugar or salt. Silverfish were also extremely attracted to beef extract, when spread as a paste on paper.

Lindsay (1940) observed the silverfish, *Ctenolepisma longicaudata* Esch. had small sensory papillae on the labial palpi, which are sensitive to taste. The palpi determine the palatability of the silverfish food.

Silverfish are pests of paper, particularly paper that has a glaze upon it. They are especially fond of the sizing in paper, which may consist of starch, dextrin, casein, gum, and glue. These insects often attack wallpaper, in which they may eat holes, or may remove the paste from behind the wallpaper, which eventually causes it to become detached from the wall. According to Lindsay (1940), only the starch and dextrin sizes in the wallpaper are attacked extensively and the pigment has little effect on the nature of the attack. Sweetman et al. (1944) studied the firebrat as a pest on wallpaper and found this insect may be drawn by the paper itself, but ordinarily the paste, sizing, and dyes are the attracting ingredients.

Mallis (1941) found in experimenting with the paper preferences of *Ctenolepisma longicaudata* Esch. that such paper as cleansing tissue, onion skin, and cellophane are preferred by this silverfish, and papers such as newsprint, printed or unprinted, cardboard, and brown wrapping paper are not eaten. Lindsay (1940), who worked with the silverfish, *Ctenolepisma longicaudata* Esch., as a pest on wallpaper in Australia, found that papers of pure chemical pulp were more likely to be attacked than those consisting in part of mechanical pulp. Her conclusions are born out by the experiments of Mallis (1941) who observed the papers that are extremely well eaten are all highly refined chemical papers, whereas those that are not fed upon have a high mechanical pulp content. It should be noted, however, that silverfish may even nibble on papers they ordinarily avoid if other food is not available.

Silverfish are extremely fond of flour and starch; this in part accounts for their presence in breakfast cereals.

As a rule, books and papers that are in constant use are damaged little, although even these may show ragged edges and

Fig. 2-2. Silverfish damage to a map.

markings on the bindings. Back (1931) notes the silverfish "frequently eats off the gold lettering to get at the paste beneath and gnaws off the label slips glued on the backs of books."

Slabaugh (1939) found the firebrat *Thermobia domestica* (Pack.) fed on rayon, preferring the finishing agents such as starch, gums, etc., to the rayon itself. Lindsay (1940) noted the silverfish, *Ctenolepisma longicaudata* Esch., readily ate artificial silk and cotton, but did not touch wools or true silks. Mallis (1941) found under natural conditions the silverfish, *C. longicaudata* Esch. will feed on textiles of vegetable origin, such as linen, rayon, cotton, and lisle. Linen is by far the preferred vegetable fiber. Wall (1953) noted the firebrat and the four-lined silverfish to feed on carpets made with viscose rayon. Brett (1962) showed the firebrat preferred rayon to acetate. The feeding by *C. longicaudata* Esch. on fibers of animal origin, such as silk or wool, is practically negligible. Damage by silverfish to textiles may be recognized by the presence of feces, scales, the irregular feeding on the individual fibers of the textile, and in certain instances, especially in the case of linens, by yellowish stains. Where damage to textiles is believed to be due to silverfish, a card coated with flour paste may be placed in the vicinity. Subsequent examination for feeding marks will reveal whether or not silverfish are present.

The linen wrapping once used around insulated pipes beneath homes is known to be a very common source of food for the silverfish. They are especially injurious to materials that have been starched. In view of the fact that silverfish are so fond of rayon and cellophane, it is possible that with the introduction of new synthetic materials, the silverfish may become more important pests than they are at present.

Lindsay (1940) states *Ctenolepisma longicaudata* Esch. does not drink liquids, but instead obtains its moisture from food and oxidation of foodstuffs. This insect derives part of its moisture from the atmosphere and excretes dry feces.

Wall and Swift (1954) note the firebrat has starch, fat, and protein digesting enzymes in the digestive tract. Lasker (1957) showed the fourlined silverfish, which is found under the bark of *Eucalyptus* trees in California, produces the enzyme cellulase in its midgut by means of which it can digest cellulose.

Accumulations behind moldings apparently provide the insect with a varied diet. Where silverfish are found in and around rugs, carpets, etc., they are probably feeding on bread crumbs and similar foods that may be in the vicinity.

Silverfish are extremely resistant to starvation and may be kept alive in a glass jar without food or water for weeks. Lindsay (1940) found one adult out of 20 survived for 307 days without food. She also observed cellulose-digesting bacteria, as well as fungus hyphae in the crop of the silverfish, *Ctenolepisma longicaudata* Esch.

A TABULATION OF THE SPECIES OF SILVERFISH IN BUILDINGS

- The firebrat, *Thermobia domestica* (Packard), is grayish in color with numerous dark markings. This cosmopolitan insect attains a size slightly larger than ½inch/12 mm. It prefers localities where the temperature is 90° F/32.2° C and above.
- *Thermobia campbelli* (Barnhart) is known from Ohio and Pennsylvania (Wygodzinsky, 1972) and differs from the firebrat by being uniformly gray.

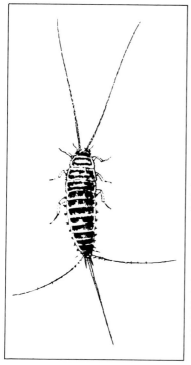

Fig. 2-3. Adult firebrat.   Fig. 2-4.Adult gray silverfish with eggs.

- The silverfish, *Lepisma saccharina* L., has a silvery sheen and attains a length of about ½ inch/12 mm. This cosmopolitan insect prefers lower temperatures than the firebrat.
- The fourlined silverfish, *Ctenolepisma lineata pilifera* (Lucas) (=*quadriseriata* Packard) has four dark lines extending longitudinally on its body and attains a size of almost ⅝ inch/16 mm.
- The gray silverfish, *Ctenolepisma longicaudata* Escherich (=*urbana* Slabaugh) is silvery to gunmetal in coloration. It attains a size of ¾ inch/18 mm.
- *Acrotelsa collaris* (Fabricius) has been reported from Florida by Wygodzinsky (1972). It is a tropical species of economic importance.

Sweetman, in correspondence with the author, states "the species of *Lepisma* can be separated from *Ctenolepisma* by the setal combs on the abdominal sternites of *Lepisma* and the absence of those combs on *Ctenolepisma*. With living specimens I have found it very convenient to place them in shallow vials about ¾ inch/19 mm deep and wide. By placing such a vial on a mirror, in a slanting position, one can examine the ventral side of the specimens by the reflection in the mirror, and thus can readily separate the species. This is also a convenient method for separating the sexes of the species where the ovipositor is not visible from the dorsal view."

THE SILVERFISH

Lepisma saccharina L.

The silvery sheen of this common pest which darts so unexpectedly from papers, its elongated shape, as well as the side-sway of the moving insect, are responsible for its common name, the silverfish. It is also referred to as the bristletail and the slicker, and less commonly by such interesting titles as sugarlouse, silverwitch, sugar fish, wood fish, and paper moth. The adult insect attains a length of approximately ½ inch/13 mm.

Distribution. The silverfish has been reported from Europe, Asia, Japan, Hawaii, Australia, and North America. It is believed to be originally a warm-country insect. It has adapted itself to living in temperate climates by dwelling in warm and humid areas, especially in basements, around water pipes, etc. Although knowledge of the distribution of the silverfish in the United States is incomplete, this insect is apparently most common in the eastern states, but does occur in the middle west and on the Pacific coast.

L. saccharina L. is often brought into new homes in cardboard cartons and books and papers that have come from infested sites. The silverfish themselves need not be transported, since their eggs can be the source of fresh silverfish hordes. The silverfish is primarily a house-infesting insect and is uncommon outdoors. Back (1931) states the silverfish may be very common in newly-built establishments before the masonry has completely dried. He notes, "In large apartment houses and hotels, silverfish may become so abundant in rooms on the lower floors as to make it difficult to keep them rented." The silverfish prefers cooler and moister areas than does the firebrat, and in the eastern United States may become less active during the winter.

Life history. Sweetman (1939) has studied the life history of the insect in detail and much of the following material is from his work.

The eggs are soft and white when first laid. They then turn yellow, and in a few hours, light brown. The eggs are laid singly or in two's or three's. The female may lay from one to three eggs per day, or at irregular intervals of days or even weeks. The eggs may be laid in crevices or under objects, or at times they may be merely dropped haphazardly. The eggs hatch at temperatures from 72° to 90° F/22° to 32° C. The majority hatch at 72° F/22° C. The average incubation period of the eggs is 43 days at 72° F/22° C and 19 days at 90° F/32° C.

Although some investigators have stated that the life cycle from egg to adult is two to three years, Sweetman found they may reproduce at three to four months. The nymphs become clothed with scales at the third molt, and continue to grow after the first reproduction. All the nymphs die at or above 98° F/37° C. The first instar is from seven to 10 days, and the later instars under favorable conditions are two to three weeks long.

The adults lay an average of 100 eggs at temperatures of 72° to 80° F/22° to 27° C. They are long-lived insects and some were still alive and reproducing after 3½ years at 72° F/22° C. The majority live about two years at a temperature of 84° F/29° C. The best development of this insect in every stage is at 72° to 80° F/22° to 27° C, and at a relative humidity of from 75 to 97 percent. These environmental conditions are most favorable for reproduction of the silverfish. Wigglesworth (1964) notes the female silverfish alternately molts and lays eggs and may molt up to 50 times after the adult stage is reached. Silverfish are unique because they keep their ventral glands so that molting can continue.

THE FIREBRAT

Thermobia domestica (Pack.).

Much of the information that is presented herein is from the works of Adams (1933) and Sweetman (1938), particularly the latter, who completed an excellent monograph on this insect.

This cosmopolitan silverfish probably originally came from some warm country, and in temperate climates it has adapted itself to heated buildings. According to Wygodzinsky (1972), secondary free-living populations have become established in the West and Southwest. The sobriquet, "firebrat," is applied to this insect since it commonly occurs in and around ovens, bakeries, and other extremely warm areas. Quick, active, and dark-loving, one often becomes aware of this insect only upon turning on the light, when it scurries with great alacrity into the security of some crevice. The firebrat attains a size of approximately ½ inch/12 mm, and is grayish in color with numerous dark markings. Spencer (1930) reports a race of a dark color pattern in Canada. Paper and paper products, such as books with glazed finish, are especially subject to injury by *Thermobia*. Sweetman (1941) found a temperature of 90° to 106° F/32° to 40° C is optimum.

Life history. The female deposits the eggs in crevices. She may oviposit when 1½ to 4½ months old at temperatures of 90° to 106° F/32° to 40° C. The female lays from one to 195 eggs, but averages 50. Only one batch of eggs is laid during an instar and fertilization must take place for each batch. Adams (1933) states there is a reduction in the number of eggs laid when the females are disturbed.

The eggs, which are soft, white, and opaque when laid, are one mm long and 0.7 mm wide. The young nymph emerges from the egg with the aid of an egg-burster which is situated on its so-called "forehead," the pressure of which breaks the enclosing chorion. The incubation period under optimum conditions is 12 to 13 days.

Sweetman (1938) states "the newly hatched nymphs are about two mm long, omitting the appendages. They are white, opaque, and free from scales and the ventral abdominal styli. The first instar is more plump and sluggish than later instars." He continues, "One day or less is spent in the first instar, about four in the second, about six in the third and fourth, approximately eight days in the fifth to the eleventh, with the period gradually increasing to about 12 to 13 days in later instars. The number of instars depends entirely on the length of the life of the individual. Many pass through from 45 to 60 instars before death." The young will feign death in the first few instars, but discontinue this when they become older. Injured appendages on the insect are regenerated throughout life.

Since these insects are ametabolous and molt throughout their life, it is difficult to determine when they are mature. Sweetman (1938) has designated an arbitrary means for the female only. He considers the female mature when the ovipositor becomes visible from a dorsal view. There is wide variation in the number of days in which nymphs mature. Under optimum conditions, the nymphs will mature in two to four months, and there may be several generations a year. Adams (1933) notes the cycle from egg to egg at 96.8° F/36° C is 11 to 12 weeks; Sweetman finds it to be somewhat longer under the same temperature. At temperatures ranging from 90° to 98°F/32° to 37° C, the insect will live from one to 2½ years. Under favorable conditions the firebrat completes its life history in from two to four months. Temperatures below 32° F/0° C, and above 112° F/44° C readily kill the nymphs. Spencer (1930) and Sweetman (1938) have observed the primitive and interesting manner in which this insect mates. There is no copulation between the sexes; fertilization occurs when the female picks up a sperm bundle from the floor which is deposited by the male during a "love dance."

According to Sweetman (1938), the sexes may be differentiated by the fact "the ventral abdominal styli usually appear in the fifth to tenth instars. The females typically have three pairs and this can be used as a partial means of determining the sexes. An occasional female may have only two pairs of styli. The males typically have two pairs, but a third pair frequently develops."

Thermobia campbelli (Barnhart)

Barnhart (1951) described this species as *Ctenolepisma campbelli* from specimens which were pests on book covers in Columbus, Ohio. Wygodzinsky (1972) placed this species in the genus *Thermobia*. J. E. Luke found large numbers of this insect on the campus of Pennsylvania State University, University Park, Pa.

GRAY SILVERFISH

Ctenolepisma longicaudata Escherich (=*urbana* Slabaugh)

This silverfish is reported from the South, Midwest, and southern California. Mallis (1941) studied the habits of these insects under the name *C. urbana* Sla. They are important pests in southern California and can be found throughout the house from basement to attic, but not outdoors. They are at times found in great numbers in relatively new homes. Ventilators and warm air ducts originating in the basement facilitate the spread of this insect throughout the building. The gray silverfish may live in very dry areas such as the crawl spaces beneath the house and the attic. It also occurs around water pipes in the bathroom.

Lindsay (1940), working with the gray silverfish, found this species reached sexual maturity in two to three years and it may continue to grow for another five years. This silverfish will molt three to five times per year. The female lays her eggs in lots of two to 20. They usually are pushed by her ovipositor into some crack so that the eggs are rarely seen. The first instar is without hairs or scales. The scales appear in the fourth instar and genitalia in the fourteenth instar. The insect may molt for several years after sexual maturity.

The gray silverfish has a silvery sheen which takes on the darker cast of "gunmetal" immediately after molting. The author, in a casual examination of this species in southern California, found the female to lay about 30 eggs in March, several at a time. The eggs were laid in a mass and hatched two months later at room temperature. Eggs of the same species were observed also in June.

FOURLINED SILVERFISH

Ctenolepisma lineata pilifera (Lucas) (=*quadriseriata* Packard)

This species can be recognized by the four dark lines extending lengthwise on the body. It reaches a length of ⅝ inch/16 mm, and is common in the East and much of the West. Ziegler (1955) notes it is an important pest in homes in Oklahoma. It also occurs in the mulch of flower beds and often infests the attic, especially in houses roofed with wooden shingles. Sweetman (1941) observes that in New England this insect may be seen during the summer in attics, on walls, and outdoors on the roofs and walls. They also may live outside the building. The author has collected this species outdoors in California. Sweetman and Kulash (1944) noted in the laboratory that the fourlined silverfish did more damage to fabrics and paper than either the silverfish or the firebrat.

Acrotelsa collaris (Fabricius)

This tropical species has been found in Florida and was first reported in the United States by Wygodzinsky (1972). He notes that this insect is of economic importance.

CONTROL OF SILVERFISH AND FIREBRATS

Insecticides such as DDT, chlordane, dieldrin, and lindane formerly available for the control of silverfish and firebrats are no longer registered for these uses in the United States. Other insecticides, however, are currently registered for control of both of these insect pests.

Although liquids, dusts or baits can be used for control of silverfish and firebrats, liquids are usually preferred in visible or exposed areas of the home where dusts or baits may present an undue hazard to small children or pets. Dusts can be used with good effect in attics, dry crawl spaces, basements, and in other places where their use is not potentially hazardous.

Regulations governing the use of pesticides are subject to constant change, therefore, it is important to use the insecticide *only as specified on the label.* In controlling the firebrat or silverfish the insecticide will be most effective when applied where the insects are most commonly seen. For the firebrat this may be alongside of heating ducts or outlets, furnace areas, and other warm places. The silverfish commonly occurs in damp areas such as behind or beneath kitchen or bathroom sinks, beneath or behind cabinet shelves, and inside cabinets such as cupboards or linen closets. Both the silverfish and the firebrat display marked thigmotactic behavior, that is, they prefer to hide or rest where there are tight cracks or crevices. Particular attention should be paid to injecting small amounts of liquids or dusts into cracks and crevices formed by shelving, loose moldings or floor tiles, loose fitting drawer components or drawer glides. Beneath heater housings is a favorite hiding place for firebrats and in this case only dusts should be injected into the accessible cracks and crevices due to the potential safety hazard created when liquids are used around electrical wiring or electrical motors.

Silverfish are often commonly found around shelves of books. Therefore, the back of the shelves should receive crack and crevice injections or small spot applications of liquid insecticides. In this situation, even though dusts would provide adequate control the risk of contaminating books and thus the book users is too high to use dusts. Caution must be exercised in the use of liquids since insecticides carelessly applied may stain or otherwise damage the books, especially when water-based sprays are used. When neither a spray application nor the use of a dust is the appropriate treatment for book shelving, then baits can often provide effective control of silverfish. A small amount of bait, bait dispenser or even a single bait pellet may be placed in the corners of book shelves. Baits should *not* be used where children or pets have access to them. A *public* library would *not* be an appropriate place to use baits. In public facilities an alternative insect control device such as "sticky traps" may provide an adequate measure of control (Sastry, in press). If the infestation in an attic, basement or crawl space is severe it may be necessary to select an insecticide where labeling will permit a more general broadcast application. Ziegler (1955) controlled the fourlined silverfish by spraying the underside of shingles, the eaves along the sheathing board outside, as well as the outer walls, the mulched flower beds adjacent to house walls, and unattached garages. Approximately 10 days may be required after spraying before satisfactory results are obtained.

Integrated concepts of pest control emphasize the integration of non-chemical with chemical control measures. If only non chemical techniques of pest control are used they should be *both* practical *and* effective.

For control of silverfish and firebrats three techniques of non chemical control should be considered.

1) Attempt to bring about a change in the physical environment in the immediate area of infestation. For example, controlling or eliminating moisture (e.g. leaking plumbing, around laundry areas, etc.) where a silverfish population is thriving can be quite effective in significantly reducing the level of infestation.

2) Reduce the potential sites of harborage. A simple caulking gun or can of patching plaster and a spatula may well prove to be the most effective tools in your arsenal of pest control devices. Seal up obvious and easily accessible cracks and crevices. Do not leave silverfish and firebrats a preferred place to hide and breed.

3) Of course, a good practice in almost all pest control strategies is to remove potential food supplies whenever and wherever possible. This is extremely important in the control of silverfish and firebrats. As stated earlier these insect pests seek out starch containing materials such as paper, book bindings, starched linens, and organic debris. If these materials cannot be secured in tightly sealed containers or cabinets, make sure your pest control application cuts off access of these pests to potential food sources.

INSECTICIDES FOR CONTROL OF SILVERFISH AND FIREBRATS

Liquids. Residual insecticides usually provide 15 to 45 days of control and should be applied to the areas where the silverfish and firebrats are most commonly seen. The effectiveness of residual sprays often can be enhanced by using them in conjunction with the short lived contact sprays that cause "flushing" activity to occur. These "flushing" agents, although usually insecticides themselves, can, in sublethal doses, irritate the insects enough to cause them to run to areas free of the irritant. In their search for a more acceptable harborage silverfish and firebrats will often run over an area on which a residual spray was just previously applied. This behavior increases the likelihood of more rapid and effective chemical control of these pests.

Currently registered liquid insecticides and their rates of use for silverfish and firebrat control include the following: propoxur 0.5%, chlorpyrifos 0.5%, bendiocarb 0.25%, diazinon 0.5%, microencapsulated diazinon 1.0%, malathion 3.0%, ronnel 1.0% and propetamphos 0.5%. Luke and Snetsinger (1972) evaluated six insecticides against firebrats in the laboratory and found chlorpyrifos effective.

The contact or short-lived liquid insecticides can be effective either used alone or in combination as "flushing" agents with the residual insecticides. The effective residual life of contact materials is usually less than 10 days and in the case of some pyrethrins less than two days. DDVP at 0.5% is a proven material that can be very effective in controlling silverfish or firebrats hiding in less accessible areas such as within wall voids and other void spaces beneath kitchen or bathroom cabinets. Several pyrethrin formulations are effective contact insecticides at 0.25 - 0.50% concentrations. Resmethrin at 0.5% has been effective against silverfish.

Dusts. Dusts provide exceptional control of silverfish and firebrats, although they can be more visible and have the potential of moving from where they were originally applied. A common "standard" dust used prior to World War II was a mixture of sodium fluoride, pyrethrum powder, and corn starch. Such a dust was used against a number of household insects including silverfish. These dusts were "shot" (predecessor of crack and crevice injection) behind baseboards, under pipe flanges, beneath and in back of bookcases, and in crevices that harbored silverfish and firebrats. Later, dusts containing the chlorinated insecticides took the place of the earlier dusts and in turn were replaced by the newer organophosphate and carbamate dusts.

The commonly used dusts that are currently registered for control of silverfish and firebrats are insecticides such as boric acid, diazinon, bendiocarb, malathion, and sodium fluoride dusts, all of which provide some residual control, especially when applied in dry areas. Pyrethrin dusts can effect good control but have a much shorter residual effectiveness. A proprietary dust of proven value to the professional pest control operator is a combination of pyrethrins, piperonyl butoxide (synergist) and silica aerogels. Tarshis (1958 and 1961) and Ebeling and Wagner (1959 a,b) investigated and uncovered the unique insecticidal properties of the sorptive silica aerogels. The aerogel absorbs some of the protective wax layers covering the cuticle or external "skin" of insects. These wax layers protect the insect from water loss. The sorption of the wax by the silica aerogel causes the cuticle to release the internal water of the insect eventually bringing about its death from dehydration. When applied in dry areas, the silica aerogel goes on working long after the more acutely toxic pyrethrin has lost its effective potency.

Recently, a new product was introduced for the control of silverfish which contains diatomaceous earth and pyrethrins. Unlike the sorptive silica aerogels, the diatomaceous earth abrades the waxy covering or cuticle of the insect. This insecticidal action brings about both dehydration and increases the speed of penetration of the acutely toxic pyrethrin. As with the former products, dusts such as this should be applied in areas that are dry.

Baits. Baits will slowly control localized infestations of silverfish or firebrats. However, due to the time factor and the effort required to make proper bait placement, this type of formulation has been largely replaced by sprays and dusts. In the past, some of the generally used baits consisted of mixtures of sodium fluoride and wheat flour. Mallis (1944) demonstrated that a bait consisting by weight of five parts sodium fluoride and 95 parts flour was effective against the gray silverfish in California. The sodium fluoride and the flour were very thoroughly mixed. This bait was distributed in "pinches" at one to three-foot/0.3 to 0.9 m intervals behind baseboards, in the corners of shelves in bookcases, and in crevices in general.

It appears the attractancy of baits for firebrats is important only at very close ranges. Ebeling and Reierson (1974) investigated the odor of rolled oats and its attractancy for *Thermobia*. They observed that direct antennal contact with the bait was needed to trap this pest. "Empty traps trapped as many firebrats as jars with wheat, flour, and sugar, or wheat, flour, sugar, and chipped beef." Bait placement and not odor is apparently the most important element for successful trapping of silverfish and firebrats. Mallis (1941) showed the gray silverfish was repelled by flour pastes containing white arsenic, tartar emetic, sodium fluoride, and sodium fluosilicate, but were not repelled by pastes containing barium fluosilicate or barium carbonate.

Most labels of commercially formulated baits most commonly used for insect control do not at this writing cite silverfish and firebrats.

Pest control operators may not formulate toxic baits for insect control unless the EPA registered label of the toxicant allows for such formulation. Pre-baiting with non-toxic food substances such as wheat paste painted onto paper is allowable and may be an effective strategy if used just prior to a dust or spray application.

DETECTING DAMAGE CAUSED BY SILVERFISH OR FIREBRATS

- Book bindings will show minute scrapings. The sizing of paper will be removed in an irregular fashion and the edge of the paper will present a notched appearance. Where the injury is severe, irregular holes will be eaten directly through the paper.

- The small, dark feces of silverfish are visible to the eye, and silverfish scales can be readily seen with a hand lens.

- Other signs of silverfish damage are the irregular chewing marks on individual fibers of textiles, as well as minute yellowish stains associated with such feeding.

- The presence of silverfish in any locality may be determined by coating paper with flour paste. After exposure for one week, this coated paper will show the feeding marks made by silverfish if they are in the vicinity.

SUMMARY OF IMPORTANT ELEMENTS OF CONTROL

1) Conduct a thorough inspection.
2) Note places where pest populations are most abundant.
3) Note any environmental conditions that can be corrected, e.g. excessive moisture from leaking pipes, cracks and crevices providing abundant harborage. Correct these conditions.
4) Note excessive food sources. Correct these conditions with appropriate sanitation measures whenever possible.
5) Choose the insecticide formulation most appropriate considering:
    a. Surface upon which application will be made.
    b. Exposure of the surface to children and pets.
    c. Length of control desired.
    d. Type of application required (crack and crevice injection vs. spot or general application).
    e. The potential for staining or otherwise damaging and/or contaminating household effects.
6) Apply the insecticide only according to the label instructions.
7) Make the application to places where the insects are most abundant.
8) Note the relative success of your treatment for future reference.

LITERATURE

ADAMS, J. A. — 1933. Biological notes upon the firebrat, *Thermobia domestica* Pack. J. N.Y. Entomol. Soc. 41:557-562.

BACK, E.A. — 1931. The silverfish as a pest of the household. USDA Farmers' Bull. 1665.

BARNHART, C. S. — 1951. A new silverfish of economic importance found in the United States. Ohio J. Sci. 51:184-186.

BRETT, C. H. — 1962. Damage and control of silverfish and firebrats. Pest Control 30(10):75-76.

EBELING, W., and D. A. REIERSON — 1974. Bait trapping silverfish, cockroaches, and earwigs. Pest Control 42(4):24,36-39.

EBELING, W. and R. E. WAGNER — 1959a. Rapid desiccation of drywood termites with inert sorptive dusts and other substances. J. of Econ. Entomol. 52:190-207.

EBELING, W. and R. E. WAGNER — 1959b. Control of drywood termites. California Agriculture 13(1):7-9.

LASKER, R. — 1957. Silverfish, a paper-eating insect. Scientific Monthly 84(3):123-127.

LINDSAY, E. — 1940. The biology of the silverfish, *Ctenolepisma longicaudata* Esch. with particular reference to its feeding habits. Royal Soc. of Victoria 52(1):35-83.

LINSLEY, E. G. — 1944. Natural sources, habitats, and reservoirs of insects associated with stored food products. Hilgardia 16(4):187-224.

LUKE, J. E. and R. SNETSINGER — 1972. Laboratory evaluation of six insecticides for control of the firebrat. J. Econ. Entomol. 65:917-918.

MALLIS, A. — 1941. Preliminary experiments on the silverfish, *Ctenolepisma urbani* Slabaugh. J. Econ. Entomol. 34:787-791.

1944. Concentrations of sodium fluoride-flour mixtures for silverfish control. J. Econ. Entomol. 37:842.

SASTRY, K. S. S. — 1981. Role of 'Mr. Sticky' as an effective tool in structural pest management. Pest Control Magazine (in Press).

SLABAUGH, R. E. — 1939. The silverfish in the new role. Trans. Ill. Acad. Sci. 32:227-228.

SPENCER, G. J. — 1930. The firebrat, *Thermobia domestica* Packard (Lepismatidae) in Canada. Canad. Entomol. 62:1-2.

SWEETMAN, H. L. — 1938. Physical ecology of the firebrat, *Thermobia domestica* (Packard). Ecological Monogr. 8:285-311.

1939. Responses of the silverfish, *Lepisma saccharina* L. to its physical environment. J. Econ. Entomol. 32:698-700.

1941. The pest thysanurans of New England. Pests 9(6):8-9.

SWEETMAN, H. L., and W. M. KULASH — 1944. The distribution of *Ctenolepisma urbana* Slabaugh and certain other Lepismatidae. J. Econ. Entomol. 37:444.

SWEETMAN, H. L., F. E. MORSE, and W. J. WALL, Jr. — 1944. The influence of color and finish on the attractiveness of papers to Thysanurans. Pests 12(10):16,18.

TARSHIS, I. B. — 1958. The use of sorptive dusts for the control of cockroaches. Syllabus 6th Annual Cal-Poly Pest Control Conference. Dec. 12 & 13, 1958, pp. 1-25.

1961. Laboratory and field tests with sorptive dusts for the control of arthropods affecting man and animals. Experimental Parasitology 11:10-33.

WALL, W. J., Jr. — 1953. Damage to carpet materials. J. Econ. Entomol. 46:1121-1122.

WALL, W. J., Jr., and A. H. P. SWIFT — 1954. The digestive enzymes of the firebrat. J. Econ. Entomol. 47:187-188.

WIGGLESWORTH, V. B. — 1964. The Life of Insects. World Publ. Co.

WYGODZINSKY, P. — 1972. A review of the silverfish (*Lepismatidae, Thysanura*) of the United States and the Caribbean Area. Amer. Museum Novitates No. 2481.

ZIEGLER, T. W. — 1955. Silverfish control in dwellings. Pest Control 23(6):9-12.

ARNOLD MALLIS

Arnold Mallis was born in New York City on October 15, 1910, the son of Russian immigrants to the United States. He attended grade school and high school in Brooklyn and Long Island. In 1927, at the age of 16, he moved with his parents to Los Angeles, California, where he completed his last year of high school.

In 1929, he entered the University of Southern California and began pre-dental courses. Since the Great Depression began at about this time, he dropped out of college and worked in a Los Angeles garment factory. After two years on the job he decided to return to college and, because of his interest in trees and natural history, began a two year course in forestry at Pasadena Junior College (now Pasadena City College).

He entered the University of California at Berkeley in 1932, where he studied under such outstanding entomologists as Professor E. O. Essig in economic entomology, Dr. E. C. Van Dyke in forest entomology, Professor W. B. Herms in medical entomology and insect ecology, and Dr. W. M. Hoskins in insect toxicology. These teachers and others instilled in him a life-long interest in insects and their control. He received a bachelor's degree in entomology in 1934.

After two years graduate work he left the University in 1936 and sought employment as an entomologist in southern California. Since there were no jobs open in entomology he took the state examinations for a license in structural pest control and received a Class A license (all categories, including fumigation).

Mallis worked two years for pest control firms in Los Angeles, Hollywood and Bakersfield and then in 1938 obtained a job as a field aide in the USDA Bureau of Entomology, doing research on vegetable insects in southern California. This job lasted for six months. He then returned to Berkeley for two purposes, to complete his master's degree on the ants of California and to collect information for the first edition of the Handbook of Pest Control. In 1939, he returned to southern California and became entomologist-in-charge of pest control for the Buildings and Grounds Department, UCLA. In 1942, shortly after the United States entered WWII, he became associated with the USPHS, implementing malaria control programs around military camps in Louisiana. In 1944, he began to screen compounds as pyrethrins substitutes for Hercules Powder Company at the University of Delaware. The screening resulted in the discovery of Compound 3956, later known as toxaphene.

In 1945, Mallis commenced his employment in the entomology laboratory of Gulf Oil Corporation in Harmarville, Pennsylvania. Here he worked as an entomologist on household insects and household insecticides for 23 years until the company closed its entomology laboratory in 1968. He then became an extension entomologist for The Pennsylvania State University and retired as an associate professor in 1975, at the age of 65.

CHAPTER THREE

Springtails

By Arnold Mallis[1]

THESE VERY SMALL INSECTS, generally measuring one to two mm. in length, are occasional invaders of homes and particularly prominent in the basement, bathroom and kitchen of structures. According to Scott (1966), springtail infestations in buildings are usually associated with dampness, organic debris and mold.

Springtails are widely distributed and among the most common soil insects. They usually occur in the soil of, as well as beneath, potted plants and in vegetable matter such as decaying plant bulbs. Buildings with constant high humidity may be overrun with springtails. In one particular case in western Pennsylvania, an unknown species of springtails infested the floor of a living room in the thousands after heavy rains in September. Two days later, when the weather was dry, only three springtails were seen in the house. On another occasion in central Pennsylvania, springtails (snowfleas) were present in enormous numbers on snow covering a lawn and invaded the porch of the house. In Europe they have at times been so numerous as to impede the progress of railway trains by preventing the wheels from gripping the rails.

Curran (1947) discusses their presence in the home as follows: "Most people who complain about the presence of springtails find them in the cellar. This does not indicate that the whole cellar is unduly damp. The insects may be breeding in a drain, along a wall through which water seeps, or on the surface of floor or paneling that is too damp for the good of the building. Or, they may occur in water and heating system risers, escaping from these into bathrooms and other portions of the house. At times the insulation in steam and water pipes may be "alive" with them. Their presence in rooms above the cellar is evidence that there is probably a leak in the piping system and that there should be an immediate examination to determine the source of the insects and the necessary repairs.

[1]Associate Professor (Retired) Extension Entomology, The Pennsylvania State University, University Park, Pa.

COLLEMBOLA: PICTORIAL KEY
TO COMMON DOMESTIC SPECIES
Harold George Scott, Ph.D. — 1961

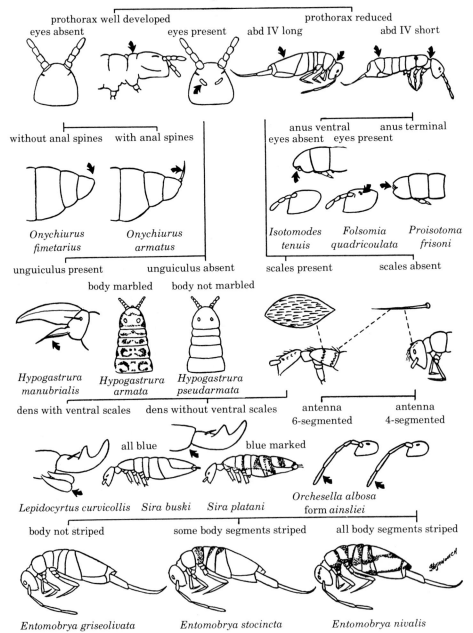

prothorax well developed
eyes absent · eyes present · prothorax reduced · abd IV long · abd IV short

without anal spines · with anal spines · anus ventral eyes absent · anus terminal eyes present

Onychiurus fimetarius · *Onychiurus armatus* · *Isotomodes tenuis* · *Folsomia quadricoulata* · *Proisotoma frisoni*

unguiculus present · unguiculus absent · scales present · scales absent

body marbled · body not marbled

Hypogastrura manubrialis · *Hypogastrura armata* · *Hypogastrura pseudarmata*

dens with ventral scales · dens without ventral scales · antenna 6-segmented · antenna 4-segmented

all blue · blue marked

Lepidocyrtus curvicollis · *Sira buski* · *Sira platani* · *Orchesella albosa* form *ainsliei*

body not striped · some body segments striped · all body segments striped

Entomobrya griseolivata · *Entomobrya stocincta* · *Entomobrya nivalis*

Fig. 3-1. Pictorial key for springtails.

"The springtails that get into the house usually are whitish or grayish in color. They do not bite, lack wings and undergo a simple metamorphosis. Springtails lay spherical eggs and the young greatly resemble the adults except for size, color and sexual maturity. They are equipped with a forked muscular appendage at the end of the abdomen, which when suddenly released, enables them to jump in the air, and hence the name springtails."

Mills (1950) notes that they move ordinarily "by short runs, separated by periods of rest." Springtails, for the most part, feed on algae, fungi, and fungus spores, pollen and decaying vegetable matter. Some species attack living plants, and a number act as scavengers on dead animal matter. Springtails, as mentioned previously, can withstand low temperatures and at times, in Arctic zones, are so numerous as to almost cover the snow.

Mills (1950) notes that *Sira buski* Lubbock and *S. platani* (Nic.) occur in the home and due to their protective scales can exist in a drier environment than most springtails. Scott (1966) states there are about 500 species of springtails in North America, of which 19 have been found infesting buildings. He also notes that springtails do not transmit human disease. *Entomobrya nivalis* (worldwide) and *Entomobrya tenuicauda* (Australasian) reportedly cause an itching type of dermatitis in man. *Orchesella albosa* (North American and European) has been recorded as infesting, without dermatitis, the head and pubic areas of man. *Entomobrya atrocincta* (worldwide) is a pest of dried milk powder, according to Scott.

The author is also acquainted with an unidentified species of springtail that was a pest in a home in Los Angeles in 1941. This insect invaded the house during hot weather, apparently in search of moisture, since it was most commonly found around the kitchen and bathroom sinks. These insects hid in minute cracks and crevices and large numbers were observed climbing the stucco on the outside of the house. Many dead springtails were also found in a globe around an electric bulb in the ceiling of the bathroom. Limited control was obtained by spraying crevices around sinks with a pyrethrum fly spray, and by dusting the sills above sinks with pyrethrum powder.

Scott (1966) notes that springtails are attracted to lights "and may pass under lighted doorways at night. Presence of plants and mulch under windows causes buildup of large outdoor populations."

Fig. 3-2. Springtails. Left, dorsal view; middle, side view showing spring situated beneath body; right, side view showing spring released.

Fig. 3-3. Building-infesting springtails.

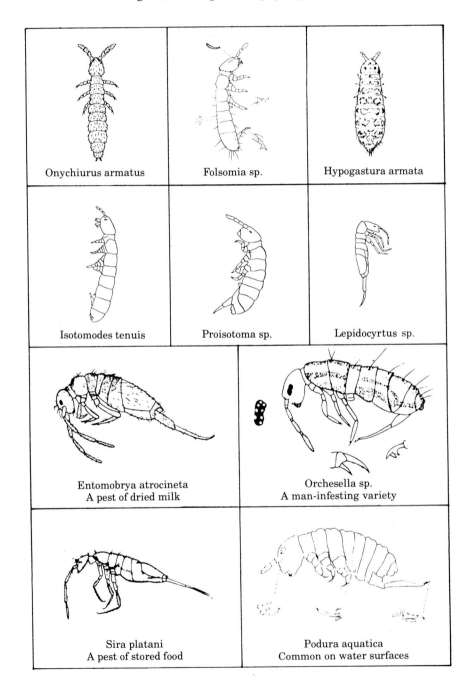

Onychiurus armatus   Folsomia sp.   Hypogastura armata

Isotomodes tenuis   Proisotoma sp.   Lepidocyrtus sp.

Entomobrya atrocineta
A pest of dried milk

Orchesella sp.
A man-infesting variety

Sira platani
A pest of stored food

Podura aquatica
Common on water surfaces

Control. Since springtails occur in greatest numbers under moist, humid conditions, infestations can be reduced by drying out the places where the insects occur. Areas in the house wetted by leaking plumbing, damp basements, and moist flower pots are common sources of springtail infestations. Sprays containing 0.5 percent diazinon or two percent malathion are effective for several weeks. Contact sprays containing 0.25 percent synergized pyrethrins, or 0.5 percent DDVP, or 0.25 percent resmethrin quickly control springtails.

At times, springtails invade structures from outside sources. Therefore, mulch, moist leaves and other damp materials near the foundation of structures should be monitored. If possible, these materials should be removed or sprayed with 0.5 percent diazinon or two percent malathion.

LITERATURE

CURRAN, C. H. — 1947. Insects in the house; springtails (Collembola) and snowfleas. Nat. Hist. 56:476, Dec.

FOLSOM, J. W. — 1933. The economic importance of Collembola. J. Econ. Entomol. 26:934-939.

MACNAMARA, C. — 1924. The food of Collembola. Canad. Entomol. 56:99-105.

MILLS, H. B. — 1950. Springtails. Pest Control Technology. National Pest Control Association.

SANDERS, G. E. — 1940. The snowflea as a household pest. Pests 9(12):20.

SCOTT, H. G. — 1966. Insect pests: Part I: Springtails. Modern Maintenance Management 18(9):19-21.

SCOTT, H. G., J. S. WISEMAN, and C. J. STOJANOVICH — 1962. Collembola infesting man. Ann. Entomol. Soc. Amer. 55(4):428-430.

AUSTIN M. FRISHMAN

Dr. Austin M. Frishman gained his bachelor's and master's degrees from Cornell University and his Ph.D. from Purdue University. Frishman has had broad experience in the pest control industry. Prior to establishing his current consulting company in 1980, his work has ranged from pest control technician for a New York State PCO to a distinguished teaching career at the State University of New York (SUNY) at Farmingdale. For 14 years he was professor of biology at SUNY during which time he wrote several scientific and popular books on pest control and authored or co-authored more than 60 articles for scientific and trade journals. Currently he writes a column in the magazine Pest Control Technology magazine.

Frishman is well known for his entertaining lectures and during his career has addressed more than 50,000 people engaged in pest control.

Frishman has been active as a consultant in the food industry, developing pest control programs and manuals for several major food manufacturers and processors. In addition, he has worked closely with local, state and federal government agencies and played a major role in writing the state examination in three of the subcategories currently used in New York State.

Frishman is a member of the honor societies Phi Kappa Phi, Pi Chi Omega and Sigma Xi. He is a member of various professional associations including the National Pest Control Association and the Entomological Society of America. He is an honorary member of the Indiana, Long Island and Empire State Pest Control Associations.

CHAPTER FOUR

Cockroaches

Revised by Austin M. Frishman[1]

Timid roach why be so shy?
We are brothers, thou and I.
In the midnight, like thyself,
I explore the pantry shelf!
        Christopher Morley.

IT WAS IN THE geologically distant Carboniferous Period, in the moist and humid Age of Cockroaches, that these animals reached their zenith, and so abundant were they, that the mere thought brings a wave of warmth to the pest control operator's heart. In a sense, domestic cockroaches continue to carry on the legacy of their ancestors. No matter how many times these creatures are "eliminated" from a structure, they continue to reappear. The backbone of the structural pest control industry pivots around the need for continual cockroach control. Cockroaches, despite man's advancements in technology and pesticide development, continue to maintain an admirable position on the face of this earth. Their chance of ever appearing on the endangered species list seems dim indeed.

The Romans called the cockroach *lucifuga* and *blatta* for its habit of fleeing from the light, and Miall and Denny (1886) state the term "cockroach" can be traced to the Spanish *cucaracha*. European nationals name their cockroaches after their neighbors across the border, an honor which is always reciprocated.

There is little agreement as to the country of origin of house-infesting cock-roaches other than they were almost all tropical, and to a great extent followed commerce, as stowaways on ships, to all those portions of the earth where conditions were suitable.

Rehn (1945) has shown the American, Australian and brown-banded roaches, as well as the Madeira roach, *Leucophaea maderae* (Fabr.), "reached America by the slave ship route from West African sources," whereas the oriental roach and German roach "reached America directly from Europe, which, however, represented a way station on the long trek of these originally north or northeast African types."

---

[1]*President, AMF Pest Management Services, Inc., Farmingdale, N.Y.; Extension Specialist, State University of New York, Farmingdale, N.Y.*

101

What is certain is wherever man can survive, so can cockroaches. For the polyglot wishing to broaden his/her horizons, the following should be of interest:

THE NAME FOR COCKROACH IN:

Norwegian	kakerlakk
Swedish	mort
Danish	kakerlak
Spanish	cucaracha
German	kuchenschabe
French	blatte, cafard
Dutch	kakkerlak
Portuguese	bicho de conta
Polish	karaluch
Italian	blatta
Russian	tarakan amerikanski (American) turusak (German)
Chinese	chang-lang
Japanese	abula mushi

Let us give the roaches the courtesy of airing their rare virtues. These insects are born scavengers and are known assassins of that odious roommate, the bedbug. Foster (1855), in his interesting "Voyage," has this to say about them: "Cockroaches, those nuisances to ships, are plentiful at St. Helena, and yet, bad as they are they are more endurable than the bugs. Previous to our arrival here in the *Chanticleer,* we had suffered great inconveniences from the latter, but the cockroaches no sooner made their appearances than the bugs entirely disappeared. The fact is that the cockroach preys upon them and leaves no sign or vestige of where they have been. So that it is a most valuable insect." Laboratory tests by Mallis (1969) show hungry American cockroaches will eat bedbugs when confined with them.

Rehn (1945) has a kind word for roaches in general as the following note indicates: "It often takes some effort to convince the 'doubting Thomases' that the number of species of cockroaches which are domiciliary pests is greatly limited — in fact, less than one percent of the known forms — and that cockroaches of many kinds are diurnal, with hundreds of species tropical forest foliage forms, others semi-aquatic, some in one sex living in the ground, a few wood-boring, while a dozen or so genera will be found, in a state of either known of suspected commensalism, in the nests of ants, wasps, or termites." To date, about 4,000 species of cockroaches have been described, of which only a dozen or two are serious pests of buildings. Rehn has indicated that roaches, in view of their omnivorous eating habits, probably fed on materials used by early man, and thus some species were transferred into his abode, and finding an abundance of food and a congenial atmosphere therein, became cohabitants with him.

Nor are all cockroaches maltreated on the first provocation. In fact, many of them are tenderly nursed from infancy on specially prepared diets since laboratories find them most useful animals in research. Rau (1946) notes that fishermen use the oriental roach as bait for bream, a bluegill fish.

The profitable business of rearing cockroaches for chicken feed is suggested by Revel (1978). For medicinal purposes, Ebeling (1975) points out that the 1907 edition of *Merck's Index* suggested how oriental cockroaches could play a useful role for certain internal and external problems. He reports: "Constituents: Blattaric acid; antihydropin; fetid, fatty oil. Uses: internal, in dropsy, Bright's disease, whooping cough, etc. External: as an oily decoction for warts, ulcers, boils, etc. Doses: 10-15 grains in dropsy, as powder or pills, or four fluid drams decoction."

As this eulogy of the roach has now exhausted itself, let us look on the red side of the ledger. Cockroaches are prime pests because they devour food, which it is believed they discover primarily through the sense of smell. They are of extreme economic importance in hotels, restaurants, butcher shops, bakeries, private homes, on ships, and many other places. In these localities they abound near hot water pipes, moist kitchen sinks, behind stoves and refrigerators, under meat chopping blocks, in cracks or wooden store fixtures, and inside drains.

Cockroaches are omnivorous. They are especially fond of starchy materials such as cereals, sweetened or sugary substances, and meat products. A few of the substances upon which they feed are cheese, beer, leather (such as that found in upholstered furniture), hair, wallpaper, and dead animals. They eat books, especially those soiled with perspiration, and may feed on the binding of books in order to get to the paste beneath the binding. In some countries, parchment cannot be used for legal documents because of the roach's fondness for this material. Linnaeus tells of a species of cockroach, *Blatta lapponica* Linn., which in company with another insect, *Silpha lapponica* Herbst, has been known to devour the whole stock of dried and unsalted fish of a Lapland village in one day. Frankly, we advise you to take this statement with one or more grains of salt. Cockroaches may also be found in greenhouses where they often damage plants.

Wheeler (1910) discusses the interesting behavior of tiny cockroaches of the genus *Attaphila* that live in the nests of the fungus-growing ants, *Atta texana* Buckley. These roaches climb on the backs of the large *Atta* soldiers and lick the surface for exudates. The ants tolerate the roaches and show no signs of hostility, although occasionally they may clip off a portion of their antennae by accident. Cleveland, et al., (1934) note that the roaches, *Cryptocercus punctulatus* Scudder, are wood-eating roaches which honeycomb wood with their galleries, and may be ancestors of the termites. For further information on the relationship between the two see the chapter on termites.

It is not sufficient for roaches to partake of our food, but they must also apprise us of their presence with odor and fecal pellets. The "attar of roaches," that unpleasant stench that is so well associated with these animals, is the combined product of their excrement, the fluid which they exude from their abdominal scent glands, and a dark-colored fluid they regurgitate from their mouths while feeding, and with which they stain their runways.

Herbert H. Smith, in his interesting letter to Dr. L. O. Howard (Howard and Marlatt, 1902), presents the roach in an entirely different character: "At Corumba, on the upper Paraguay, I came across the cockroach in a new role. In the house where we were staying there were nearly a dozen children, and everyone of them had their eyelashes more or less eaten off by cockroaches — a large brown species, one of the commonest kind throughout Brazil. The eyelashes were bitten off irregularly, in some places quite close to the lid. Like most Brazilians, these children had very long, black eyelashes, and their appearance thus defaced was odd enough. The trouble was confined to children, I suppose because they are heavy sleepers, and do not disturb the insects at work. My wife and I sometimes brushed cockroaches from our faces at night, but thought nothing more of the matter. The roaches also bite off bits of the toe nails. Brazilians very properly encourage the large house spiders, because they tend to rid the house of other insect pests." There is a possibility that the roach described by Smith was *Periplaneta americana* (Linn.) since this species has similar habits and is common in South America.

The technical editor, while a service technician, observed young children with eyebrows apparently eaten by German cockroaches in Monticello, NY. One particular apartment contained more than 10,000 cockroaches. Incredible as it may seem, the adult tenant was not anxious to have her premises treated.

DISEASES ASSOCIATED WITH COCKROACHES

Cockroaches often have been held suspect as disease carriers, and they are believed to be as suitable for the conveyance of common germs as the house fly. Jettmar (1935) observed *Blattella germanica* (Linn.) to overrun the bodies of men who died of plague in Manchuria, where it was found feeding on the infested secretions. These plague organisms were later recovered in their excreta in a viable and highly virulent condition. Barber (1914) states that they may also act as vectors of cholera vibrios. In San Antonio, Texas, the health department cited the case of a dairy which found its bottled milk contained coliform bacilli, despite pasteurization and all known sanitary precautions. Eventually the infection was traced to the bottle caps. When the caps were stored in a place where cockroaches could not enter, the contamination disappeared. Mackerras and Pope (1948) and Olson and Rueger (1950) conducted laboratory tests that showed cockroaches harbor food-poisoning *Salmonella* organisms. These studies indicated that if the roaches deposit their excreta on food or dishes, the *Salmonella* organisms will remain alive for weeks. Moiser (1946) believes cockroaches are a source of infection for leprosy because they excrete leprosy bacilli in their feces. Roth and Willis (1957 and 1960) have written monographs on the medical importance and biotic associations of cockroaches. They list 18 species of cockroaches which have been incriminated in the transmission of pathogenic agents. Tarshis (1962) believes they may spread hepatitis.

Tarshis (1962) reported a six-year study (1956-1962) where a strong relationship between cockroaches and infectious hepatitis existed. As control of cockroaches improved, the incidence of hepatitis declined despite the fact that in other areas, the number of cases increased. The study area involved 124 buildings containing 582 apartments which were inhabited by about 2,800 persons. Ninety-five percent of the apartments had cockroaches. There was a severe infestation of German cockroaches and a lesser infestation of brown-banded and oriental cockroaches.

Cardone and Gauthier (1979) reported *Salmonella* could not survive in the intestinal tract of German cockroaches for more than 11 days.

At a cockroach seminar in 1977 sponsored by the New York State Department of Health, Ebeling (1977) summarized the relationship of *Salmonella* and cockroaches as follows:

"Food Poisoning. Cockroaches have been most commonly implicated as carriers of *Salmonella* bacteria, the cause of food poisoning (Roth and Willis, 1957, 1960; Rueger and Olson, 1969). When feces of American cockroaches infected with *Salmonella oranienburg* were spread on human foods and on glass, the bacteria survived for the following periods: on corn flakes, 3.5 years; on crackers, over 4.25 years; and on glass slides, 3.67 years. Mice in jars with a minute quantity of infected cockroach feces were infected with *Salmonella* in one day.

Salmonella remained viable in the gut of oriental cockroaches for as long as six weeks (Olson and Rueger, 1950), in the digestive tract of the German cockroach for nine days (Janssen and Wedberg, 1952), and on its exoskeleton for at least 10 days (Graffar and Mertens, 1950), and in another experiment, for 78 days (Olson and Rueger, 1950). A food poisoning epidemic in the children's ward of a Belgian hospital was terminated when the German cockroaches were controlled (Graffar and Mertens, 1950). *Salmonella oranienburg* and *S. panama,* which also causes food poisoning, were taken from the intestinal tracts of the American cockroach taken from manholes."

On cockroach allergies, he reported, "Some people are allergic to cockroaches. In one investigation among 253 normal persons, 7.5 percent showed positive skin tests with extracts of cockroaches (*Periplaneta americana* and *Blatta orientalis*), compared with 28 percent of an unselected group of 114 allergic patients. Skin-sensitizing antibodies were present in the blood sera of positive reactors. The allergen is thermostable. There is evidence in some instances symptoms attributable to food allergy may be caused by food contaminated by cockroach allergen (Bernton and Brown, 1964, 1970*a, b*).

"In an investigation to test the age of onset of skin reactivity, 38 out of 102 allergic children, ranging from infants to 12 years old, gave positive cutaneous reactions to body extract of the German cockroach, *Blattella germanica,* compared with only five out of 100 non-allergic children. A four-year-old asthmatic child was the youngest to give a positive reaction (Bernton and Brown, 1970*a*). In a later investigation, it was found that an extract of *B. germanica* caused an attack of asthma in 10 asthmatic persons with skin hypersensitivity to the extract and other allergens, but not in asthmatics without such skin hypersensitivity (Bernton *et al.,* 1972).

"An allergen in the feces of *B. germanica* acts as an ingestant when it contaminates food and as an inhalant when dried fecal particles become incorporated with house dust (Bernton and Brown, 1970*b*).

"In a study of four ethnic groups in New York City, the percentage of their members with positive reactions to German cockroach allergen injected subcutaneously ranged from 59 percent in the group with the highest incidence of cockroach infestation to five percent in the group with the lowest (Bernton and Brown, 1967). In another study made in New York City, the German cockroach was again implicated as a potent sensitizing agent in exposed populations. Sensitivity to the allergen injected intradermally was determined in children from three groups coming from areas that were (1) overcrowded and heavily infested, (2) uncrowded and lightly infested and (3) suburban, uncrowded and uninfested. Among males, positive reactions were obtained in 75 percent, 31 percent and 10 percent, and among females, 85 percent, 28 percent and zero percent, respectively. Children obtained relief from symptoms when moved from heavily-infested homes. Sensitization was believed most likely obtained by inhalation of cockroach emanations or disintegrating parts, or possibly by ingestion of the excretions or parts of the cockroach itself (Schulaner, 1970).

"Injections with house-dust extract obtained from homes of allergic patients and with cockroach extract (a blend from several species) resulted in 87.5 percent and 77.5 percent positive skin reactions (weals larger than four mm) among patients with bronchial asthma, allergic rhinitis, or both, in Thailand (Choovivathanavanich *et al.,* 1970). These and most other investigators have concluded that hyposensitization to cockroach allergen may be beneficial to allergic patients."

Frishman and Alcamo (1977) spent several years collecting live cockroaches from structures in an attempt to determine to what degree they transport disease organisms. Additional studies by these authors are summarized in the accompanying two tables.

TABLE 4-1

Summary of incidence of bacteria associated with American cockroaches, *Periplaneta americana* L. and brown-banded cockroaches, *Supella longipalpa*

Cockroach Species	# tested		Type of premises	Bacterial Concentrations on Cockroaches					
	Nymphs	Adults		Staphylo-coccus	Strepto-coccus	Coli-form	E. Coli	Fungus	Bacillus
Brown banded *Supella longipalpa*	1	—	Kitchen-private home	None	None	None	None	None	None
	4	—	House-bedroom	Med.	Hevy	Heavy	None	Heavy	None
American cockroaches *Periplaneta Americana*	—	1	Kitchen-private home	None	None	None	None	None	None
	3	1	Birdhouse-zoo	None	Low	Low	None	None	None
	4	—	Barn area-outdoors	Med.	None	Low	None	Hevy	None
	—	3	Boiler room-Hospital	Med.	None	Med.	None	Low	None
	2	—	Hospital	Med.	Low	Heavy	None	None	Heavy
	1	3	Reared colony	None	Heavy	Heavy	None	None	Low

Light = 1 to 20 bacterial colonies per media plate
Medium = 21-100 bacterial colonies per media plate
Heavy = Over 100 bacterial colonies per media plate

TABLE 4-2

Summary of incidence of bacteria associated with German cockroaches *Blattella germinica* (L.)

Type of Location	Total No. of Specimens Tested		Number of Locations & Bacterial Concentrations on Cockroaches					
	Nymphs	Adults	Staphylococcus	Streptococcus	Coliform	E. Coli	Bacillus	Clostridium
Laboratory reared (3 sites)	1	22	1 Light 2 medium	3 Light	1 Light 1 Medium 1 Heavy	—	—	1 Medium 2 None
Restaurants, Bakeries, Cafeterias (12 Sites)	34	42	2 Heavy 2 Medium 4 Low 4 None	5 Heavy 3 Medium 2 None 2 None	6 Heavy 4 Medium 2 None	10 None 2 Light	4 Heavy 2 Medium 2 Light 4 None	1 Medium 11 None
Hospitals and Mental Institutions (19 Sites)	41	70	5 Heavy 3 Medium 2 Low 9 None	8 Heavy 3 Medium 2 Low 6 None	13 Heavy 1 Medium 4 Low 1 None	1 Heavy 1 Medium 2 Low 15 None	1 Heavy 3 Medium 6 Low 9 None	1 Medium 18 None
Private homes and Apartments (9 Sites)	22	6	5 Heavy 1 Low 3 None	4 Heavy 4 Medium 1 Low	8 Heavy 1 None	9 None	7 Heavy 2 None	9 None
Animal Facilities Zoological Parks (5 Sites)	5	13	1 Heavy 1 Medium 2 Low 1 None	3 Heavy 2 Low	1 Heavy 3 Medium 1 Low	5 None	2 Low 3 None	5 None
Prison (1 Site)	3	0	None	High	Medium	None	Medium	None
TOTAL NO.	106	153						

Light = 1 to 20 bacterial colonies per media plate
Medium = 21-100 bacterial colonies per media plate
Heavy = Over 100 bacterial colonies per media plate

FIG. 4-1
PICTORIAL KEY TO SOME COMMON ADULT COCKROACHES

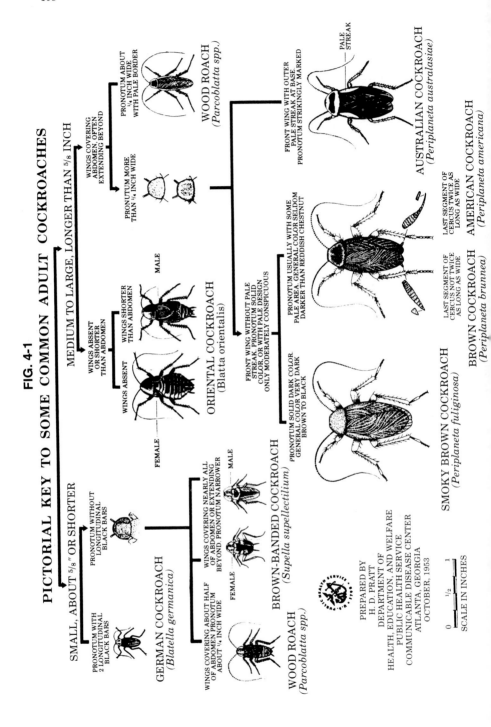

FIG. 4-2

COCKROACHES: KEY TO EGG CASES
OF COMMON DOMESTIC SPECIES
Harold George Scott, Ph.D. and Margery R. Borom

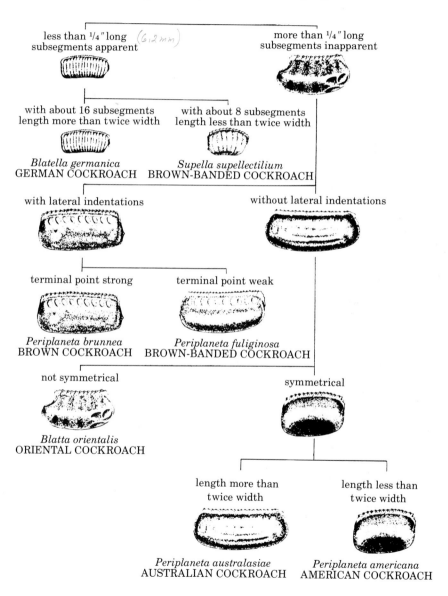

less than ¼″ long *(6.2 mm)*
subsegments apparent

more than ¼″ long
subsegments inapparent

with about 16 subsegments
length more than twice width

with about 8 subsegments
length less than twice width

Blatella germanica
GERMAN COCKROACH

Supella supellectilium
BROWN-BANDED COCKROACH

with lateral indentations

without lateral indentations

terminal point strong

terminal point weak

Periplaneta brunnea
BROWN COCKROACH

Periplaneta fuliginosa
BROWN-BANDED COCKROACH

not symmetrical

symmetrical

Blatta orientalis
ORIENTAL COCKROACH

length more than
twice width

length less than
twice width

Periplaneta australasiae
AUSTRALIAN COCKROACH

Periplaneta americana
AMERICAN COCKROACH

Stek et al (1978) showed cockroaches can harbor internally *Proteus, Klebsiella, Shigella* and *Salmonella.*

Comprehensive reviews on cockroach biology are available from Cornwell (1968) and Guthrie and Tindall (1968).

THE GERMAN COCKROACH

Blattella germanica (Linn.)

The German cockroach is known as the "Steam Fly" in England. This worldwide species is about ½-inch in length, brown in color, with two dark streaks on the thorax. The male is light brown and somewhat boat-shaped. The female is darker in color with broader and rounded posterior.

Not only is this pest prominent in and around homes and restaurants, but ships, too, are often badly infested. This roach breeds throughout the year, but favors a humid atmosphere and an average temperature of approximately 70° F/21° C.

The German roach is a more wary and active species than the others; produces more eggs per capsule; and its young complete their growth in a shorter period of time. According to Noland, et al. (1949) the German roach has matured in as little as 36 days at 86° F/30° C. However, in laboratories where the average temperature is approximately 80° F/27° C, with a 40 percent relative humidity, 50 to 60 days are required for it to reach the adult stage.

The female carries the egg capsule until hatching time, and the capsule, in time, becomes as large as the abdomen. Gould and Deay (1940) state, "At the time of hatching the seam edge of the capsule, while still held in the ovipostor of the female, may be opened by the hatching nymphs, or the capsule may be dropped several hours or perhaps a day previous to the escape of the young. Capsules removed from the female do not hatch unless this act is performed

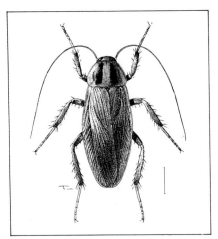 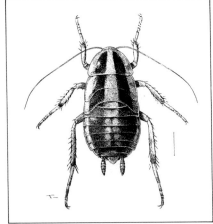

Fig. 4-3. Adult German cockroach. Note two dark vertical streaks on the thorax which are characteristic of the adult.

Fig. 4-4. Nymph of the German cockroach. The homeowner may see more nymphs than adults.

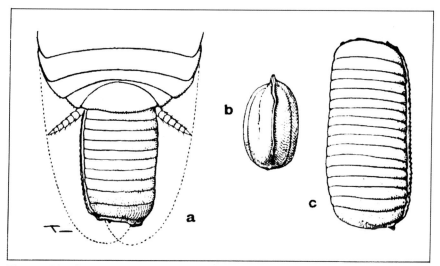

Fig. 4-5. The German cockroach.(a) Egg capsule projecting from the abdomen of the female as seen from above. (b and c) Egg capsule from the end and from the side.

only a day or so before hatching." According to Gould and Deay (1938a), the egg capsule, in hatching, split open along the seam, and the white heads of the numerous young roaches then appeared. Then, after "much struggling and squirming, about 12 worked themselves out and were soon pulling their legs and antennae free from their individual egg shells." According to Laing (1938), "in this species the mother roach is interested in her young and may aid them to escape from their swaddling clothes. The larvae run actively as soon as they are born. It appears as though the females far outnumber the males." Ross (1928) and Rau (1944a) observed the young to emerge from the egg capsules while they were still being carried by the mother, and they observed some females carrying the egg capsule even after the young emerged.

Ross (1928) studied the life history of this roach at a temperature of 95° F/ 35° C and a 90 to 95 percent relative humidity. A life cycle was completed in about three months. The egg stage required 14 days, and the seven nymphal instars 60 days for males and 65 days for the females. The number of eggs in the egg cases varied from 18 to 50 and an ample moisture supply was required for their complete development. The number of days for development, from hatching to the adult stage, varied from 55 to 68. He also observed, unlike Gould and Deay (1938b), that the males and females ordinarily were present in approximately equal numbers. Ross makes this interesting observation, "It is curious to note that the younger females, even when carrying an oötheca, can walk upside down on dry glass with the utmost ease; but shortly after the first oötheca has been dropped, the females lose this power and cannot walk up even a vertical surface of glass." W. C. Easterlin has observed that first and second instar American and oriental roaches are incapable of climbing vertical glass surfaces, whereas the older instars have no difficulty in doing so.

Gould and Deay (1940) found the dropped capsules hatch in a period of three weeks, and apparently were unaffected by the usual roach powders. The number of eggs in a capsule usually is between 30 and 40, with a maximum of 48. The average number of nymphs hatching from 59 capsules was 29.9. The average incubation period at 76° F/24° C was 28.4 days. At a temperature of 76° F/24° C, the developmental period varied from 54 to 215 days with an average of 103. Individual females at room temperature may produce an average of four to five capsules. The female may live for more than 200 days.

Roth and Willis (1952 and 1954) made careful studies of the sexual behavior of German cockroaches. Twomey (1966) reviewed the biology and control of this cockroach.

Howard and Marlatt (1902) noted that this was the species that caused much annoyance in the Department of Agriculture in Washington, D.C., and in several educational institutions, by eating and scraping the covers of cloth-bound reports.

This roach is disseminated through the various avenues of commerce and transportation, as well as by its migrations. Howard (1895) provides the following interesting detailed account of this phenomenon:

"On a dark drizzly day in September, 1893, Mr. P. H. Dorsett came to me and stated that he had just seen a remarkable sight on D Street, near the Department grounds. A vast army of cockroaches, according to his story, were crossing the street. A few hours later I visited the spot with Mr. Marlatt and found that the bulk of the army had disappeared, but that many stragglers still remained. Mr. Dorsett is an assistant in the Division of Vegetable Pathology, and according to his statement the army issued from the rear of an old restaurant fronting Pennsylvania Avenue and marched across the muddy street, undeterred by pools of water, ash heaps, and other barriers, directly south to the building opposite.

"This building was a machine shop and at the direction of the foreman several of the men took brooms and swept back the advancing horde. They swept until their arms were tired, but were unable to stem the advancing tide. The foreman then directed that a line of hot ashes from the furnace be laid along the brick sidewalk. This proved an effective barricade. The foremost cockroaches burned their antennae and their front legs and the army divided to either side and scurried down into the area ways of adjoining buildings in which they disappeared. The march is said to have continued for two to three hours and many thousands of the insects crossed in this way.

"A moment's glance, after arriving at the spot, showed me that the insect was the Croton bug [German Cockroach] and that nearly all of the individuals were females carrying eggcases.

"I called at the restaurant and found to my surprise that no house cleaning had been going on and that no special effort had been made by the application of insecticides to rid the establishment of the roaches.

"It seems then to have been a true migration, a development of the true migratory instinct in the Croton bug. The restaurant had become overpopulated, perhaps not for its actual denizens, but certainly for the thousands of about-to-be-born young. The maternal instinct originated the migratory instinct and the army by one common impulse started on its journey for more commodious quarters.

"The darkness of the day is significant, and there is no reason to suppose that similar migrations do not frequently occur, but undoubtedly under ordinary circumstances at night. This is the way that new houses become infested."

Many homes and business establishments become infested with roaches when these insects or their capsules are introduced in infested cartons, foodstuffs, and other materials. This subject is discussed at greater length in the section on control. It may be of interest at this point to describe a home in Austin, Texas, that had an unusually high population of German roaches. This four-room apartment was sprayed on August 1, 1947, with approximately 3.5 quarts of an experimental contact spray. The roach population was estimated to be 50,000 to 100,000 roaches, the greatest number by far being German roaches. For the most part they were present in the kitchen. It may be an understatement when we note that the housekeeping was poor. On August 25 and September 17, 1947, the apartment was examined for roaches, and a pyrethrum aerosol spray was used to flush the roaches out of their hiding places. No effort was made to kill the roaches. An estimated 400 to 1,000 German roaches were found on these respective dates.

On January 21, 1948, approximately six months after the initial spraying, the house once again was inspected for roaches. Although this was a sunny day, the interior of the house was dark due to the drawn window shades. On approaching the wood shelving over the kitchen sink in order to flush the roaches out with the aerosol spray, it was observed that the shelves were covered with plates in an upright position, and German roaches of all stages were clustered on the walls in great numbers, particularly in the corners where the shelves joined the sides of the open cupboard. Apparently there were so many roaches in the cracks and crevices between and behind the shelving that the roach population overflowed on the walls into the open, which is unusual for German roaches. On directing the aerosol into the cracks and crevices, the German roaches began to emerge immediately in enormous numbers and scurried frantically over the walls and ceilings. Upon lifting the oil cloth of the kitchen table, the edge of the table was found to be encrusted by a great mass of German roaches. These began to fall like rain drops and frantically scatter upon contacting the mist from the aerosol container. It is estimated that there were ap-

Fig. 4-6. Tree infested with German cockroaches being treated.

Fig. 4-7. The Board of Health in Schenectady, N.Y., ordered this two-family home destroyed as a result of a massive German cockroach infestation.

proximately 15,000 to 25,000 German roaches in this infested kitchen six months after treatment with a contact spray.

During the summer of 1979, Linindoll Pest Control took on what was perhaps the largest German cockroach population ever within a two family dwelling. Called in by the Board of Health in Schenectady, New York, the company was faced with so many cockroaches that the wall of every room was saturated. It was estimated that over a million cockroaches were present. The population had overflowed outdoors and a large tree, (Fig. 4-4) the lawn, and walls on surrounding buildings were heavily infested with additional specimens. Sewers within a three block area were also heavily infested with German cockroaches. The house in question harbored a population of 24 dogs, 20 cats, two mice and a parrot. Many of the animals were badly bitten by cockroaches and fleas. The house was ordered destroyed. Because there were so many cockroaches present, standby power sprayers were necessary to kill cockroaches as the walls of the building were destroyed (Fig. 4-5).

Although uncommon, German cockroaches can survive outside. In 1966, cockroaches were seen surviving on rotten pears outdoors in Indianapolis, Indiana. In the early 1970's, a heavily-infested home in Nashville, produced an excess population in the surrounding grass and on the exterior of structures.

On one occasion, thousands of German cockroaches were observed harboring inside a large radio located in a patients' lounge of a hospital. There was no moisture source within 60 feet. Upon questioning some of the staff, it was learned that the cockroaches often rushed out to drink human urine produced by patients who could not control themselves.

COCKROACH PHEROMONES

Immature German cockroaches are especially responsive to an aggregation pheromone. This pheromone is found in the fecal material, Reierson (1977). Wood surfaces lend themselves to the adherence of the fecal material, thereby resulting in the tendency of German cockroaches to congregate on wooden surfaces in areas where there is little air movement, darkness, and large quantities of fecal material.

THE AMERICAN COCKROACH

Periplaneta americana (Linn.)

The American cockroach, waterbug, or Bombay canary as it is known to the English dockhands, is the largest of the house-infesting roaches, being 1½-inches long with fully developed reddish/brown wings, and light markings on the thorax. The cosmopolitan American roach likes the seafaring life, and is one of the most common roaches on ships. Kellogg (1908) says a friend of his in Mazatlan, Mexico, sent him quarts of large native American roaches which he scooped up from his bedroom floor. He goes on to state that ships came into San Francisco with the sailors wearing gloves on their hands while asleep to keep the hordes of roaches from gnawing off their fingernails. This roach is commonly found in basements, particularly around pipes. Schoof and Siverly (1954), as well as other investigators, show this roach to be a common inhabitant of sewage systems. Observations after heavy rain show that basements in numerous cities throughout the United States have increased populations of American cockroaches.

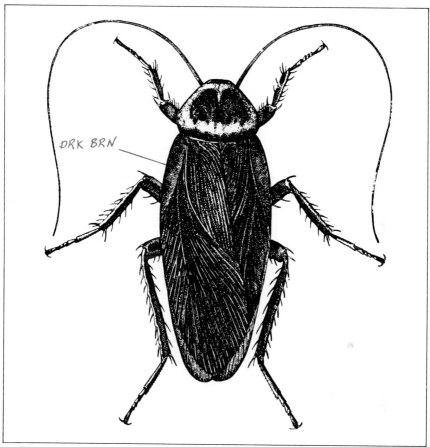

DRK BRN

Fig. 4-8. The American cockroach, *Periplaneta americana* **(Linn.)**

Much of the following material on the life history of the American roach is from Gould and Deay (1940) who made a detailed study of this species. Rau (1940a) notes the capsule (Fig. 4-7) is not dropped as soon as completed, but may protrude from the body from a few hours up to four days. Capsules are often glued to surfaces. In Florida, capsules are often found covered with paint or other building material, blending well with the surroundings. According to Rau (1943), roaches "do not drop the egg cases indiscriminately, but usually hide them with great care in crevices, or bury them in soft wood or workable material." This clearly has survival value since pest control technicians routinely squash egg capsules easily spotted around exterior door frames. He showed that the incubation period was from 38 to 49 days, and the number of egg cases produced by the female was from six to 14, with an average of 9.5.

The eggs in the capsules are in two parallel rows, and the number of eggs in a capsule is variable. The first few capsules produced by a female may be perfect, and contain 16 eggs. However, the average number of young to emerge from 511 fertile capsules was 13.6. According to Gier (1947), "Eclosion, or hatching, occurs when all embryos are developed enough to exert sufficient pressure to open the ootheca. As soon as the cement on the lips of the ootheca is broken, all nymphs wriggle out, moulting their embryonic skins as they do so. If one or two

fail to emerge when the others do, they perish, as they are not able to force the ootheca open again." The whitish, newly-emerged nymphs begin to run around actively in about 10 minutes.

The first few molts of the nymph occur at approximately monthly intervals, but thereafter may vary from one to six months. Gould and Deay (1940) state, "The American cockroach apparently molts 13 times before reaching maturity. The duration of the nymphal period from records now available varies from 285 to 616 days, with an average of 409, although the final average will be over 450 days. One individual hatching on January 11, 1936, is still in the eleventh instar, being 786 days old. Growth is slower during the winter months, although maturity is reached any month in the year.

Fig. 4-9. Egg capsules of five common cockroaches. (Top to bottom) *Periplaneta brunnea* **Burm.;** *Blatta orientalis* **L.;** *Periplaneta americana* **(L.),** *Blattella germanica* **(L.); and** *Supella longipalpa* **(Serv.).**

The average development period of six male nymphs was 426 days and that of 11 females was 396 days." Gier (1947) showed that at 86° F/30° C this roach normally has ten instars which gradually increase from 18 days for the first, to 50 days for the tenth. He also found that body weight approximately doubles between molts and the roaches may reach sexual maturity in seven months under optimum conditions. In the author's laboratory approximately six months were required to reach the adult stage from the egg stage at 82° F/27° C and R.H. of 30 percent or higher.

The wing pads become evident in the third or fourth instar. The sexes in the early instars are distinguished by the fact the female has a sharp medial notch in the caudal margin of the ninth sternum which is either lacking or very slightly indented in the male.

Gier (1947) states the American roach may reach sexual maturity in seven months under optimum conditions. The sexes are almost identical in size, the females averaging 1⅓ inches/34.7 mm and the males 33.6 mm. The female has a broader abdomen than the male. However, only the male has both cerci and stylets. The wings of the male extend from four to eight mm over the end of the abdomen, while in the female they are equal to or only slightly larger than the abdomen.

None of the females reproduced parthenogenetically. Roth and Willis (1956) showed parthenogenesis does occur in the American, oriental, German, and brown-banded cockroaches, as well as in *Nauphoeta cinerea*. A female roach need mate but once to produce more than one capsule. In Indiana, during the months of June, July, and August, a female may produce capsules at intervals of approximately four days. These intervals may lengthen to 12 days during the winter. One female produced 58 capsules, of which 33 were fertile. The maximum number of capsules produced was 90.

The sexes are approximately equal in number. Adult females lived from 102 to 588 days, with an average of 440 days. The life spans for three females were 783, 793, and 913 days. Gould and Deay observed the mating procedure of the roaches. "In pairs found *in copula* the roaches were usually headed in opposite directions. Amatory actions of the males are a common sight at night. They walk around with their abdomen distended, legs stiffened and their wings erect, and attempt to back under the female. Copulation was of short duration when there were many individuals in a jar, but where single pairs were in a jar, it may last for 30 minutes or longer." Willis, et al. (1958) discuss in some detail the reproduction and nymphal development of the roaches of the genus *Periplaneta*.

Both German and American roaches occur together in some structures. American roaches of varying sizes may live together in perfect harmony. They are active throughout the year where the temperature is 70° F/21° C or higher. Temperatures of 15° to 20° F/−8° to −6° C will kill them.

The American roach is found most commonly in restaurants, grocery stores, bakeries, and where food is prepared or stored. During the summer months alley ways and yards may be badly infested.

According to Gould and Deay (1940), American roaches have been found flying around street lights in Texas. Here they are reported capable of long flights, but their flight in the North is more of a gliding type.

However, Frishman has observed American cockroaches flying in heated steam tunnels on the campus of the State University of New York at Farmingdale.

From many reports in literature, it is evident that cockroaches, particularly the American roaches, are born inebriates, and their desire for fermenting liquid is often so strong that it leads to their undoing. Thus, bread saturated with beer is often employed to trap cockroaches. While scanning some stenographic reports of a pest control operator's meeting, Mallis (1969) came upon this discussion about roaches and alcoholic spirits that should be of interest to readers. "There was a beer parlor located in Los Angeles, we have a few, that had been bothered with roaches to the extent that they didn't know what they were going to do about it. I talked to this particular manager, and he said, 'You know, it is a funny thing; up until a few weeks ago this place was overrun with cockroaches of all types and we were about ready to call you fellows to see if you could do something about it. Now the roaches have all practically disappeared.' I asked, 'What did you do about it?' He said, 'We didn't do anything. We had been selling a large amount of bottled beer and one of the bartenders was getting out of some work by just taking the bottles and dumping them down in the basement. There was some beer left in the bottles which apparently attracted these roaches. Upon inspecting the basement we found hundreds of these bottles packed full of dead roaches: they got into the bottles apparently for the beer: they went in there and couldn't get out. I don't know whether they got drunk, but they worked into those bottles until they practically exterminated themselves.' "

The ability of American cockroaches to survive the winter outside under certain conditions was confirmed by observations at an open dump in Waltham, Massachusetts. Numerous American cockroaches were found thriving under several inches of snow in smoldering refuse when the outside air temperature was well below freezing. As material under the snow was turned over, American cockroaches scurried in all directions.

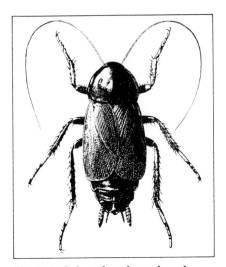

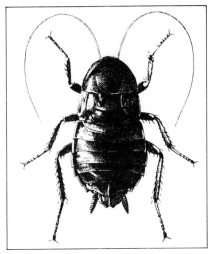

Fig. 4-10. Oriental cockroach male.   **Fig. 4-11. Oriental cockroach female. Note the rudimentary wings.**

THE ORIENTAL COCKROACH

Blatta orientalis (Linn.)

Hebard (1917), in discussing this insect, notes: "In infested houses about Philadelphia it appears in swarms during the month of May, coincident with the arrival of the shad in the Delaware river, and in consequence is locally known as the Shad Roach." Very often this roach, as well as other species, is called a "water bug," possibly because roaches inhabit damp places, but it should be remembered that a water bug is merely a roach. In fact, the author recalls that when he lived in Los Angeles the oriental roach was a "cockroach" in the poorer parts of the city, but a "water bug" in Beverly Hills. In some parts of the world the oriental roach is referred to as the "Black Beetle."

This species, which is less wary and more sluggish than the others, is the most disliked of all roaches since it often travels through sewer pipes and lives on filth (Anon., 1957). It may enter the home in food packages and laundry, or merely come in under the door or through air-ducts or ventilators. The oriental roach is most common in dark, damp basements, but is known to ascend water pipes to the upper floors of apartment houses.

Since high-rise apartments no longer allow the burning of garbage, oriental cockroaches have a free highway to climb to upper floors via garbage chutes.

At times large numbers occur in one great mass around leaks in the basement or under-area of the home. By simply trimming shrubbery and permitting the sunlight to enter through the ventilators, it is possible to decrease the dampness beneath homes without basements, and thereby greatly reduce infestations. This is a notably gregarious species which, according to Howard and Marlatt (1902), lives "together in colonies in the most amicable way, the small ones being allowed by the larger ones to sit on them, run over them, and nestle beneath them without any resentment being shown."

The oriental roach is so dark brown in color that, as was previously noted, it is often referred to as the "Black Beetle" in both the United States and England.

Fig. 4-13. Three views of egg capsule of the oriental cockroach. Note the 16 eggs in the split capsule.

Fig. 4-12. Female oriental cockroach with egg case.

The adult is one inch/25 mm in length. The male has fully-developed wings, but apparently does not fly. The female has rudimentary wings which are reduced to mere lobes. The mature female may be distinguished from the large nymph by the fact the wing stubs have a definite venation. Laing (1938) states the abdomen of the female is broader than the male and appears to be dragged along the floor when the insect is in motion, while the male keeps its body clear of the ground when running. The male can be recognized by the styli between its pairs

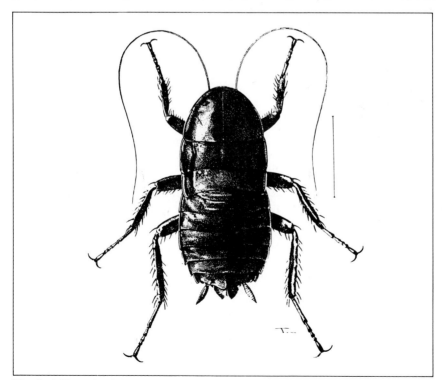

Fig. 4-14. Nymph of the oriental cockroach.

of jointed cerci; the females are far more numerous, and pairing takes place at any season.

The female may carry the brown egg capsule for 30 hours. There are 16 eggs in the perfect capsule, which consists of two rows of eight eggs each. Gould and Deay (1940), who studied the life history of this species, found the average number of nymphs to hatch from a capsule to be 14.4. The female may deposit from one to 18 capsules with an average of eight per female. The egg capsule is carried from 12 hours to five days and then deposited in some warm sheltered spot where food is readily available. In this species the female roach gives no assistance to her newly born young. The larvae of *Dermestes* often infest the egg cases, and fowls are known to feed upon them. At room temperature, the incubation period was 42 to 81 days, with an average of 60 days. Rau (1945) showed this roach prefers to feed upon starchy foods.

In England, Zabinski (1929) observed the developmental period to be 12 months, during which interval the insect undergoes seven molts. He found there is a seasonal history for this species since the adults appear in May, June, and July. Gould and Deay (1940), who studied this roach in Indiana, noted the smaller roaches became more and more evident during the summer months. In southern California the oriental roach was observed on sidewalks during the summer months. The time it took the oriental cockroach to complete development in the laboratory varied from 311 to 800 days. Capsules were produced from December to August. No adults matured during the months of October, November, and December. The adult females may live from 34 to 181 days.

To help track down pockets of these insects in basements, a careful search of spider webs will often lead to the source of infestation.

THE BROWN-BANDED COCKROACH

Supella longipalpa (Serville) (= *S. supellectilium* [Serville])

The name "brown-banded" roach is much more appropriate for the nymphs than for the adults. This roach is of African origin, and according to Rehn (1945), was introduced from Cuba to Miami where he first collected it in 1903. It has been reported in many states throughout the United States and in Canada, but is most prominent as a pest in southern cities. In regard to the distribution of this insect, Gould and Deay (1940) have this to say: "This species is gregarious and hides in cupboards, pantries or even in other rooms of the house. Preferences are shown for high locations such as shelves in closets, behind pictures and picture moldings, and the like. Egg capsules may be deposited about the kitchen sink, desks, tables and other furniture, and even in bedding. This habit of hiding capsules in furniture probably accounts for its spread northward, for it has been found that infested premises follow a trip to the South, or transportation of goods from a southern state."

In a survey made in the cities of Houston, San Antonio, San Marcos, and Austin, Texas, by Mallis (1969), it was observed that the brown-banded roach was far less prevalent than the German roach. Notes on the brown-banded roach from this survey follow:

"The brown-banded roach, although not too common, was widely distributed throughout the apartment and probably was the most common roach to be seen in the bedroom. Its favorite dwelling places were beneath and behind braces in the corners of kitchen chairs, and underneath tables, as well as behind pictures and other objects on walls, and in shower stalls. The egg capsules were as widely distributed as the adults and were commonly fastened on walls and ceilings. German and brown-banded roaches often were found in the same crevice. It is of interest to note that a male brown-banded roach was observed flying indoors."

The brown-banded cockroach, to some extent, replaced the German cockroach as a household pest in Southern California. In recent years it has, on occasion, become an important pest in some parts of the North.

This roach, which is in some respects similar to the German roach, may be distinguished from it by the following characteristics, according to Back (1937):

1. There is a greater difference in form between the male and the female than between the two sexes of the German roach.

2) The brown-banded roach lacks the two dark stripes on the thorax found on the German roach.

3) The wings are twice-banded with brownish/yellow stripes.

4) The egg capsule is smaller than that of the German roach, being but $3/16$ inch/5mm long, and having about half as many eggs.

Gould and Deay (1940) studied the biology of this roach. They found the insects are active and fly readily when disturbed. The brown-banded cockroach may be found with other roaches in buildings. The egg capsule is yellowish or reddish/brown in color. $3/16$ inch/5 mm in length. The female carries the capsule for 24 to 36 hours and then attaches it to some object. The maximum number of eggs to be found in a capsule was 18. The average number of young to emerge from a capsule was 13.2 at room temperature. The flattened virgin female develops an enormously enlarged abdomen when she becomes gravid. At 77° F-25° C, the average incubation period is 69.7 days. The females produce a greater number of capsules during the summer.

The time for complete development of the roach varied from 95 to 276 days, with an average of 161 days. The males have a shorter development period. The brown-banded roach prefers temperatures over 80° F/27° C; temperatures below 75° F/24° C retard its development. The brown-banded cockroach (Cornwall 1968) is now officially reported throughout the United States except in Ver-

Fig. 4-15. The brown-banded cockroach. Upper left, female; upper right, large nymph; below, male.

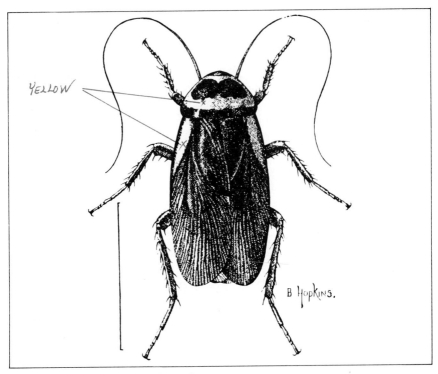

YELLOW

B Hopkins.

Fig. 4-16. The Australian cockroach.

mont, where a careful survey probably would indicate that it also exists.

When a few German cockroaches are placed in a culture of brown-banded roaches, the German species will overrun the colony within nine months.

THE AUSTRALIAN COCKROACH

Periplaneta australasiae (Fabr.)

NOT COMMON IN
SO. CALIFORNIA

The Australian cockroach closely resembles the American roach, but can be separated from it by its slightly smaller size (over one inch/25 mm), the yellow margin on the thorax, and the light yellow streaks on the sides at the base of the wing covers or elytra. Late instar nymphs possess distinct bright yellow spots along the margins of their abdomen.

This world-wide species has become established in many greenhouses and is apparently more vegetarian than the others. Nevertheless, it is also a pest in homes where it may eat holes in clothing and feed upon book covers. Hebard (1917) noted that although this roach has been recorded as far north as Canada, it is firmly established only "near the borders of towns in peninsular Florida." Hebard further states, "It is constantly being introduced north of the regions in which it has become established, but is evidently much more decidedly affected by cold than *P. americana,* and in consequence has never become permanently established in the United States north of the areas in which the winter climate is comparatively mild. In the colder regions of the United States, when it appeared in greenhouses and such artificially heated places, it has been found to breed and increase in numbers with great rapidity, temporarily becoming a dangerous pest, so that vigorous efforts have been found necessary to exterminate such a colony."

This insect's life cycle was studied by Cornwell (1968). It takes 40 days for the ootheca to develop and hatch. There are 24 eggs per egg capsule, 16 of which hatch. Each female produces 20 to 30 oothecae. Nymphs take about one year to develop. Some eggs produced parthenogenetically hatch, but the nymphs do not mature. In the United States, it is most abundant in Florida and the southern states, and in California it ranges as far north as San Francisco. Control may be realized by the same methods used for the American roach.

One of the greenhouses at the State University of New York at Farmingdale had a severe population of Australian cockroaches. All of the young seedbeds were decimated shortly after germination. As a class project, the premises were treated using crack and crevice applications of chlorpyrifos. Less than 50 cockroaches were visible at the time of treatment, the next morning the floor was covered with hundreds of dead and dying Australian cockroaches.

There is an increased chance Australian cockroaches will become a more important pest in dwellings as interest in imported plants continues to grow.

Fig. 4-17. Numerous Australian cockroaches.

THE SMOKYBROWN COCKROACH

Periplaneta fuliginosa (Serville)

Since the previous revision of this book, the smokybrown cockroach has taken on increased importance in several southern states. In Houston, Texas, it has become a major pest and is the dominant pest of residences in New Orleans, Louisiana.

Although reported in Florida about 140 years ago (Serville, 1839), little on its biology was recorded. Gould and Deay (1940), Rau (1945) and Willis et. al (1958) reported on its life history, but it has only been recently that a two-and-a-half-year laboratory study was completed at North Carolina State University (Wright 1979). It was found that at temperatures of approximately 80° F/26.7° C and 55 percent relative humidity, egg incubation averaged 45 days with a mean of 20 nymphs per ootheca. It took females 320 days to mature and males 388 days. Egg capsules are produced 17 days after mating and are carried for one day. Various researchers report that after egg capsules are deposited they hatch within 24 to 70 days.

Zuben (1955), in a comprehensive study of cockroaches in municipal sewers, showed that smoky brown cockroaches can survive in this habitat along with *Parcoblatta pennsylvanica* and *Blatta orientalis*.

THE BROWN COCKROACH

Periplaneta brunnea Burmeister

The brown cockroach is often mistaken for an American cockroach. To an untrained eye, both species look alike. Often, the adults of *P. brunnea* are darker than *P. americana*, but this is not always true. The different stages differ as follows:

TABLE 4-3

Structure	Stage	P. brunnea	P. americana
Cerci-last segment	Adult	Stubby, triangular, and less than twice as long as wide.	More elongate, and more than twice as long as wide.
Wings	Adult male	Barely extend beyond tip of abdomen.	Extend well beyond tip of abdomen.
Ootheca	Eggs	Securely glued when deposited.	Not so securely glued when deposited.
		12-16 mm long.	8 mm long.
		Average 24 eggs.	Average 16 eggs.
Antennae	First stage nymph.	The first eight and last four antennal segments conspicuously white. The intermediate ones are brown.	Antennal segments uniformly brown.
Mesothorax		A median translucent area allows light to pass through.	This area absent.
Abdomen		Faint cream colored spots on dorso-lateral margins of first and second segments.	These segments entirely brown.

P. brunnea was first reported in the United States in 1907 in Illinois, but is actually more confined in its distribution than *P. americana*. One of the so called Southern species, it is well established in numerous states throughout the Southeast.

Edmunds (1957) notes this roach is found in the Southeast to Texas and as far north as Philadelphia, Pennsylvania and Columbus, Ohio. It occurs indoors, as well as under the bark of trees, and in sewers. He found an average of 24 eggs in an ootheca. Dow (1955) trapped *P. brunnea* mostly in houses and *P. americana* mostly in privies, in south Texas.

Mallis (1968) reports that it is particularly prevalent in San Antonio, Texas. Accidental introductions have occurred twice in California (Reierson and Ebeling 1970, Waldron and Hall 1972).

The cockroach is capable of producing offspring without mating (Willis et al 1958), although the female normally copulates within a few hours after the final molt (Edmunds 1957).

This author observed brown cockroaches congregating at night in a ground floor garbage room in a condominium in Hollywood, Florida. During the day, the cockroaches remained hidden outdoors under nearby vegetation. Residual treatments of bendiocarb provided excellent control.

It is suspected the brown cockroach is more prevalent than reported because it is, in many cases, misidentified as an American cockroach. However, while its distribution has grown during this century, it is still less widely spread than *P. americana*.

THE FIELD COCKROACH

Blattella vaga Hebard

This is believed to be an introduced cockroach. It is very similar in appearance to *Blattella germanica* (Linn.). Flock (1941) states it can be distinguished from the German cockroach "by the blackish/brown area on the face from mouth parts to between the eyes. This species is slightly smaller and more olivaceous in coloration than is the German roach."

This cockroach, unlike the German cockroach, is not repelled by light and can often be seen during the day. It is most common in irrigated regions of southern Arizona and adjacent areas in southern California. According to Gurney, *Blattella vaga* is known in Arizona, California, and Texas. The field cockroach is associated with and feeds largely on decomposing vegetation; it also occurs under stones, clumps of earth, and similar objects.

During the drier part of the year, it temporarily may come into the house in search of moisture. Palermo (1960) notes control can be obtained by removing "decomposing plant material from around the home. Spray foundation area, edges of lawn, outdoor flower boxes, and similar areas where moisture collects with a water base spray. . ." Buxton and Freeman (1968) reported a heavy infestation in Orville, California.

THE PENNSYLVANIA WOOD-ROACH

Parcoblatta pennsylvanica (DeGeer)

Rau (1940b) notes this roach "is usually found in hollow trees, under loose bark, and often in wood piles and in crevices in rural buildings. The wood-roach is a trim, pretty creature, with its chestnut brown color, and thorax and wing pads edged in white. It does not possess the repulsive odor so characteristic of roaches of other species that infest the dwellings of man. In the adult form the males are fully winged, while the females have only the conspicuous wing-pads (actually short wings like that of the female oriental roach) which are functionless. The males fly swiftly, but have not the ability to sustain themselves in the air for long periods." The male is one inch/25 mm and the female is ³/₄ inch/19 mm in length. The two sexes differ so greatly in appearance that they once were described as two species. This roach is widely distributed in the eastern, southern, and midwestern states, up into Canada.

According to Gould and Deay (1940), in rural areas "where a house is surrounded by or near a woods, these roaches are common invaders. The males are capable of long flights, while the females may migrate some distance by crawling. During the past summer a trip was made to investigate an infested farm house. Many mature males and a single female were found on an enclosed porch, and a few males were in the house, but there were no young anywhere. About dusk the roaches began flying in short flights around an old building used for a garage and chicken house. By watching carefully, an occasional roach was observed to fly to the house which was about 100 feet in the other direction."

Fig. 4-18. The Pennsylvania wood roach, *Parcoblatta pennsylvanica* (DeGeer).

This species is the one frequently encountered by motorists driving in the woods at dusk. In nature, the wood-roach passes the winter as a partially grown nymph, and even in sub zero weather is quite active when exposed by pulling away bark. Adults are present from May until early October.

Gould and Deay (1940) investigated the life history of this species. The capsule is yellowish/brown and approximately two times longer than wide. The maximum number of eggs in a capsule is 32. An average of 26.1 nymphs hatch from each capsule. The capsules are deposited only during the summer. At 80.6° F/27° C the average incubation period was 34.2 days.

In Indiana the nymphs hatch in summer and mature the following May or June. In some instances the life cycle may take two years. The adults may remain alive for several months. Although most roaches are repelled by light, the males of this species are attracted to light, and it is this sex commonly encountered in the house at night. *P. virginica* (Brunner) is another wood-roach occasionally found in houses.

Although not numerous, both female and male Pennsylvania wood-roaches have been found indoors surviving under shingles on the inside of a garage in Long Island, New York. However, with the increased use of firewood, the popularity of cedar shake shingles, and the continued building of homes in wooded areas, an increase in problems of *P. pennsylvanica* seems inevitable.

LESS COMMONLY ENCOUNTERED COCKROACHES

Of the 55 species of cockroaches in the United States, most live in outdoor habitats. However, circumstances sometimes result in these outdoor species becoming a household pest. The increased use of firewood and live plant material indoors, along with the movement of large numbers of people to suburbia enhances the chance of new species invading homes.

Aglaopteryx gemma Hebard

Aglaopteryx gemma Hebard, the little gem cockroach, inhabits tree trunks in the Gulf states including Texas, Louisiana, Mississippi, Alabama, Georgia and Florida, and favors resting behind signs nailed to trees (Hebard 1917). These cockroaches were first reported indoors in Thomasville, Georgia. (Gorham, et. al. 1971). The cockroaches were found both indoors and outdoors, including the attic area. To the inexperienced, they resemble brown-banded cockroaches, but they have vestigial wings.

Control was achieved with indoor spot applications of one percent diazinon and treatment of the attic and outside perimeter with a dust of 0.2 percent pyrethrum and two percent sulfoxide.

Ectobius livens (Turton)

Gurney (1953) notes that *Ectobius livens* (Turton), from the Mediterranean region, is a household pest in eastern Massachusetts. It was found under outdoor conditions and entering homes at night. It also occurs on fresh vegetables.

Eurycotis floridana (Walker)

Eurycotis floridana (Walker), according to Creighton (1943), "is also frequently called the 'woods' roach. Often it may be discovered in stacked lumber and firewood. It is large, dark brown to almost black, averaging 1.5 to 1.75

inches in length. It is an infrequent visitor in homes." Willis et al. (1958) studied the life history of this Florida species in some detail.

The species can multiply parthenogenetically (Cornwell 1968). When touched, these cockroaches emit an oily liquid which has a strong odor. In Florida, they may be found hiding under palmetto leaves and other debris alongside buildings.

Leucophaea maderae (Fabricius)

The Madeira cockroach is four cm to five cm long. They are easy to distinguish by their large size and the fishnet-like pattern or mottled appearance on their front wings. The author has watched numbers of them in alleys on the island of Nassau (Bahamas). One night he was able to pick several off a drunk sleeping in an alley. The wall and ground nearby were alive with hundreds more, in addition to several other species not collected. This cockroach is tropical and frequents the western Mediterranean areas, various parts of South America, the West Indies, Southern Africa, Bahama Islands and Hawaii.

In the United States, it was first reported in a section of New York City occupied by people originating from Puerto Rico (Gurney, 1953 and Cornwell, 1968). The cockroach gives off a strong odor when approached or handled. In 1972, it was found indoors in Maywood, California (Ebeling, 1975).

This cockroach is capable of living under extremely crowded conditions in laboratory containers. Its "body odor" is more repulsive than that exhibited by most cockroaches.

Nauphoeta cinerea (Oliv.)

Nauphoeta cinerea (Oliv.), according to Ratcliffe (1952), is established in some food handling establishments in Tampa, Florida. Since it has a lobster-like design on the prothorax (the section immediately behind the head), it is referred to as the lobster roach. It is also called the Cinereous cockroach, which best describes its color.

In 1970, it was found indoors in San Francisco, California. The origin of the infestation is unknown, but may have been associated with vermicelli flour from the Orient. The species is known to occur in East Africa, Germany, Australia, Hawaii and Florida. It infests grain, fruit and vegetables. In Hawaii, it also feeds on other cockroaches.

The adults are 1¼ inches/32 mm in length, and in the female the wings are shorter than the body, exposing the abdomen. Roth and Willis (1954) note the oothecum has 26 to 40 eggs.

Panchlora nivea (L.)

Panchlora nivea (L.) is a pale green species that is brought into various ports in bananas from Central America. Essig (1926) calls it the "green Cuban roach." According to Gurney (1955), *P. cubensis* Saussure is a synonym for *P. nivea* (L.) These cockroaches move about more rapidly than German cockroaches, are capable of flight, and difficult to catch. When placed in alcohol, they quickly loose their pale green color. Nymphs are dark brown. Detailed biology is reported by Roth and Willis (1958).

Currently they are found in many areas of Florida and attracted to lighted doorways or windows. The standard pesticides used to control domestic cockroaches have proved effective against these species (Burden and Madden 1975).

Pseudomops septentrionalis

Pseudomops septentrionalis, the September roach, was first collected by Paul Adams of Adams Pest Control Company in Alexandria, Louisiana. He notes, "In June of 1967 I did collect a September roach in a men's clothing store here in Alexandria.

"In the fall of 1967, I collected my second one. It was crawling on a wooden yard gate. The next year our service technicians and I collected several, and since that time they have gradually become more numerous.

"In the late spring of this year I covered two areas of my yard with thick blocks of St. Augustine sod. Each time I watered this sod I could depend on chasing eight or 10 Septembers out of each of the two areas. If I watered it first thing in the morning it would be double or triple that number.

"Over the years we have found very few of these inside structures. I have collected three in my own residence. All were at night and they were spotted crawling across the floor or on the wall. We have had two or three occasions where homeowners would kill them and save them for us to identify, all of which were found at night and in situations similar to the ones I found in my own home. Other than the original case in the men's clothing store, our technicians have never flushed one out in their normal service routine. The situation could be correctly summarized by saying that even though the insect now occurs quite commonly in this area, it is not known as a pest and when encountered indoors is usually just incidental." (Adams 1978).

It is primarily an outdoor species and is ordinarily found in the western portion of Texas and the northern and southern edges of Oklahoma to northern Mexico. (Anonymous 1968).

Pycnoscelus surinamensis (L.)

Pycnoscelus surinamensis (L.), the bicolored or Surinam cockroach, is a burrowing insect which is capable of destroying various plants.

Although this cockroach is not in the strict sense a household pest, it is nevertheless a source of much annoyance in related structures such as greenhouses, and has been reported in the reptile house of the New York Zoological Society and the bird house of the Philadelphia Zoo.

They are reported from the Brownsville area of Texas, New Orleans, Lousiana and San Antonio, Texas (Hebard, 1917). A visit to the Toronto Zoo in Toronto, Canada, showed thousands of these cockroaches living inside one of the walk-in bird buildings. The insects are extremely sensitive to cold temperatures. Laboratory-reared colonies were wiped out when the heat went off for one night in the laboratory. The pipes did not freeze and all other species survived except the Surinam cockroaches. The egg capsule is retained within the abdomen. This is also true of Madeira and Lobster cockroaches.

Schwabe (1949) notes the bicolored cockroach is the intermediate host of the chicken eyeworm.

Zappe (1917), who made an extensive study of these roaches, found that in greenhouses they hid during the day under the soil in the benches, on the sides of the benches, under boards, barrels, in holes and crevices in the walls of buildings and wherever it was dark and possible for them to conceal themselves. At night, they came up in great numbers and gnawed stems of plants such as roses and Easter lilies.

The number of eggs in each capsule varied from 14 to 42, the average being 24. Although Zappe examined 1,000 specimens, he was unable to find a single male, and thus thought there was a possibility this form may reproduce parthenogenetically. Roth and Willis (1956) state this species is parthenogenetic in North America and in Europe and bisexual in Indo-Malaysia.

CONTROL OF COCKROACHES

The need to "think like a cockroach" in order to successfully control these creatures has been punctuated by three major events:

- The control once achieved by chlordane is now history. Chemical resistance and governmental regulations have made chlordane obsolete for cockroach control.

- Substitute pesticides in food processing areas are greatly restricted as to their application.

- In the increasingly busy world of transporting materials from structure to structure, cockroaches continue to travel undetected.

Therefore, to achieve proper levels of control, the "roach approach" must be taken to beat them at their own game. But what levels of cockroach control are realistic? In structural pest control one normally thinks of achieving a zero population level, first, through exterminating the existing cockroaches, and thereafter through implementing a pest control program which prevents re-infestation. This concept starts with the premise that the technicians doing the control can reach all necessary areas with pesticide. However, today's apartment residents frequently place double and triple locks on their exterior doors, thereby preventing access. Under such conditions, zero levels of cockroaches become improbable, if not impossible. Even in commercial food establishments, lack of access to liquor cabinets, tool boxes and other areas make control difficult to achieve. Even the good environmental practice disallowing the incineration of garbage in large cities works to the cockroach's advantage. These insects now can breed and move unobstructed throughout the incinerator chutes of apartment complexes. Even when access is possible, Gupta et al. (1973) showed control is rather uncertain and/or limited if sanitation is poor.

The successful control of cockroaches forces adoption of a pest management approach wherein pesticides play an important, but no longer predominant role. Additionally, the recent marketing of new pesticide formulations has necessitated fewer applications because of their longer residual activity. To eliminate existing conditions before deciding what control methods to use, it is most important an inspection be conducted. A good working flashlight, a mirror to reflect light back into cracks, and a willingness to climb high and low are still the hallmarks of a good cockroach control technician. Some use of pyrethrins or other contact sprays are helpful in flushing cockroaches from machinery. Cornwell (1976) presents an exhaustive review on cockroaches and cockroach control in his definitive work, "The Cockroach Volume II." It contains 921 references on this important subject.

Construction changes to eliminate hiding areas and proper sanitation to decrease food and water sources also play an important role in the success of a cockroach control program.

Before World War II, German cockroaches were controlled by sodium fluoride powders, or combinations of sodium fluoride and pyrethrum powders. It was essential when using these powders to dust every crevice and hiding place in order to obtain control.

According to *Soap and Sanitary Chemicals* (Anonymous, 1940), the mixture preferred by most insecticide manufacturers was 75 percent sodium fluoride and 25 percent pyrethrum. Their next choice was a 50-50 mixture. For the most part, the pyrethrum powder ranged from 0.6 to 0.9 pyrethrins and the sodium fluoride was usually a finely ground product of at least 95 percent purity. The pyrethrins in the mixture acted as both stimulants and toxicants and caused the roaches to run around excitedly and pick up more of the mixture. The pyrethrins in the powder decomposed on exposure to air and sunlight after a few days. The sodium fluoride did not deteriorate and acted as the long-killing toxicant in the mixture. Since sodium fluoride is highly toxic to human beings, it is essential that formulations be applied carefully. Gibson (1960) uses a duster with a modified nozzle for his mixtures of 50 percent sodium fluoride/50 percent pyrophyllite.

Bedingfield (1952), a pest control operator in Corpus Christi, Texas, wrote a letter to the editor of *Pest Control* magazine stating that two percent chlordane sprays did not control German cockroaches in that city in 1951. He concluded his letter by noting, "We have a breed of German cockroaches immune to the residual of chlordane." Heal et al. (1953) confirmed Bedingfield's observation, and found the German cockroaches from Corpus Christi to be more than 100 times resistant to chlordane, 10 to 12 times resistant to lindane, and five to six times resistant to DDT as non-resistant roaches. Fisk and Isert (1953), and many others, have conducted numerous tests with resistant German cockroaches and shown them resistant to DDT, chordane, dieldrin and other chlorinated insecticides, as well as malathion, diazinon, fenthion and other organic phosphates.

However, if resistant German cockroaches are no longer exposed to diazinon for six to seven generations, resistance is lost (NPCA, 1961). Cornwell (1976) reports one documented case of resistance to the carbamate propoxur in 1969.

The relatively short life cycle and greater number of offspring per generation make German cockroaches more likely to develop resistance to pesticides than any other domestic cockroach.

The fear that German cockroaches would exhibit, throughout the United States, widespread resistance to all registered organophosphates and carbamates has not materialized. Parker (1977) reports pest control operators having trouble controlling German cockroaches in Georgia, and a few pockets of resistance of German cockroaches to commonly used organophosphates and carbamates have been reported from cities as far apart as Baltimore (Wood 1980) and Los Angeles (Rust and Reierson 1978). However, although resistance was first reported to diazinon more than 20 years ago, this product continues to be effectively used over a wide area. This indicates cockroach resistance to organophosphates and perhaps also carbamates will not develop as steeply or extensively as in the case of chlordane. Nonetheless, pest control technicians should be watchful for signs of resistance and wherever possible avoid repeated partial elimination of cockroaches since this may hasten the selection of resistant strains.

The history and development of insecticides for cockroach control through 1972 is best summarized by Cornwell (1976).

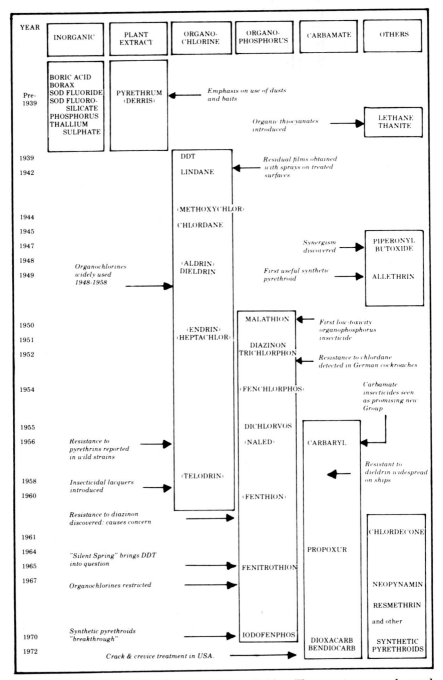

Fig. 4-19. History of the development of insecticides. Those not commonly used for cockroach control are in parentheses.

CHLORINATED HYDROCARBONS IN RETROSPECT

Prior to the advent of chlordane, DDT was used as both a residual spray and as a dust against the several species of common household cockroaches. When chlordane appeared on the market, it soon began to prove itself more effective than DDT against cockroaches. Chlordane was first designated "1068" in reference to its empirical formula $C_{10} H_6 Cl_8$. Strangely enough, in the laboratory, DDT showed longer residual effectiveness against cockroaches than chlordane, e.g. Knipling (1947), a fact that has been demonstrated in a number of laboratories concerned with the problem. In later work, however, chlordane also displayed its effectiveness in the laboratory.

Chlordane is one of the major reasons the practice of treating accounts once a month was implemented. Its residual activity was good for at least 30 days.

Resistance and federal regulations have now eliminated use of chlorinated hydocarbons for cockroach control in the United States, though chlordane, dieldrin and other chlorinated hydrocarbons continue to be used successfully in some other countries.

Nowadays, a wide range of pesticides are available to the professional for cockroach control, but the prime emphasis is on organic phosphate (also known as organophosphates) and carbamate insecticides. These groups of insecticides act by inhibition of the enzyme cholinesterase which occurs both in insects and non-target organisms. While the newer organophosphates and carbamates can provide superb control, many pest control specialists report they do not perform as spectacularly as chlordane first did.

Tables 3 and 4 indicate which pesticides are currently registered for cockroach control and how they may be applied. It is readily apparent from these tables that general or overall applications of long-lasting residual insecticides are no longer permitted. Prior to 1973, the initial or "clean-out" treatment commonly involved general application of residual insecticides, but a federal regulation issued in that year abruptly put a stop to that practice.

RESIDUAL AND CONTACT PESTICIDES

Sprayable Products. Acephate is one of the newest organophosphate insecticides. It is formulated as a liquid concentrate for dilution with water. At recommended rates acephate provides many weeks of residual control, particularly when applied on non-porous surfaces. It has proven very effective against resistant strains of German cockroaches, including the Baltimore strain. It has unusually low toxicity to non-target organisms.

Bendiocarb is a carbamate insecticide introduced in the mid 1970s specifically for control of cockroaches and other household pests. It is formulated as a wettable powder. This type of formulation has an advantage over most liquid insecticides on porous surfaces because the insecticide remains on the surface and does not become absorbed and unavailable to the insects walking on such surfaces. Story (1972) noted under practical field conditions the length of residual control at the labeled dose rates ranged from three to 23 weeks, with a mean of 10 weeks at the 0.25 percent rate of use. Since it is highly concentrated and effective at low doses, the spray deposits are less visible than with most wettable powder products. Bendiocarb has no odor and low volatility and this, together with good residual control and good labeling, has made it one of the most widely used products in sensitive situations such as restaurants, offices, modes of transport and hospitals. Bendiocarb has an excellent record of safety though some non-target vertebrates are highly susceptible. Grayson (1976) found bendiocarb less repellent than diazinon against German and American cockroaches.

Carbaryl is another odorless carbamate insecticide available both as a wettable powder formulation and as a newly developed flowable suspension formulation. It is seldom used for indoor cockroach control because of poor residual action and visible spray deposits. However, low cost, broad labeling and low environmental impact have resulted in its extensive use around the exterior of buildings. Grayson and Messersmith (1959) found 2.5 percent and five percent suspensions of carbaryl to be effective against German cockroaches.

Chlorpyrifos is an organophosphate insecticide which has become the dominant insecticide among PCOs for cockroach control. It is primarily formulated as an emulsifiable concentrate dilutable in oil or water and available in two strengths, both of which have an obnoxious mercaptan odor. Chlorpyrifos has longer residual activity than diazinon, but kills more slowly (Cornwell 1976). Excellent labeling, including clearance for use as a spot treatment, as well as crack and crevice treatment in food handling establishments, has contributed largely to its success. Chlorpyrifos lacks good flushing action, but vapor kill can contribute to achieving good control in confined spaces such as kitchen cabinets. Various formulations are available containing dichlorvos to achieve faster knockdown of pests. Mixtures containing pyrethrins and chlorpyrifos may cause unacceptable scattering of infestations. Reierson and Rust (1979) discuss field results using 0.5 percent chlorpyrifos in foam which still achieved 87 percent control of German cockroaches in apartments after 12 weeks.

Diazinon is second only to chlorpyrifos for general pest control use by PCOs. It is formulated as an emulsifiable concentrate with separate formulations for oil or water dilution. Compared with the newer organophosphate and carbamate insecticides it has only a moderate residual action and is particularly short-lived at high temperatures, on exposure to ultra violet light and on stainless steel. Familiarity and cost-effectiveness contribute to its continuing popularity in cockroach control. However, restriction of diazinon to crack and crevice treatments in food handling establishments has resulted in products with superior labeling ousting diazinon from this sector. Rapid degradation of diazinon when mixed with water in stainless steel compressed air sprayers makes it essential to use the spray on the day of mixing. Like the other leading cockroach insecticides, diazinon is moderately toxic to mammals and should be handled strictly in accordance with label directions.

Malathion is used alone and in combination with other insecticides in the control of German cockroaches. It is an organophosphate insecticide characterized by a highly objectionable odor and poor residual action. Lofgren et al. (1957) and Grayson and Jarvis (1958) showed malathion to be effective against resistant German cockroaches. Though available in a wide variety of formulations, including oil base, emulsifiable concentrate and wettable powder formulations, malathion is now seldom used by professionals for cockroach control. For less skilled users of pesticides, a big advantage of malathion is its very low mammalian toxicity.

Propetamphos is a new-generation organophosphate insecticide characterized by its excellent stability in water and long residual action against cockroaches. It is formulated as a low-odor liquid concentrate for dilution in oil or water. It is particularly effective against German cockroaches and has performed well against this species in urban housing projects.

Propoxur is a carbamate insecticide available as an emulsifiable concentrate

TABLE 4-4
A Guide to Some Pesticides and Type of Use
(Check and Follow Labels of Product being Used)

Insecticides	Crack & Crevice	Inside — Food Handling Establishments		Non-Food Areas Limited Area — may include spot
		Food Areas		
		Spot	General	
acephate	no	no	no	yes
bendiocarb	yes	yes	no	yes
boric acid	yes	no	no	yes
carbaryl	yes	no	no	yes
chlorpyrifos	yes	yes	no	yes
diazinon	yes	no	no	yes
dichlorvos (DDVP)	yes	yes	yes	yes
malathion	yes	yes	yes	yes
MGK 264	yes	yes	yes	yes
piperonyl butoxide	yes	yes	yes	yes
propetamphos	no	no	no	yes
propoxur	yes	no	no	yes
pyrethrins	yes	yes	yes	yes
resmethrin	yes	yes	yes	yes
silica gel	yes	yes	no	yes
sodium fluoride	no	no	no	no
trichlorfon	yes	no	no	yes

Adapted from New York State Pesticide Applicator Training Manual Category 7: Industrial, Institutional, Structural and Health Related Pest Control. Subcategory: Structural and Rodent p. 136.

and a wettable powder formulation for spray application. Of all the residual insecticides propoxur has the fastest knockdown action on cockroaches, typically five to 10 minutes in the case of German cockroaches. Propoxur also has a distinct flushing action, though this is not as pronounced as pyrethrum and may cause undesirable scattering of cockroaches. Flynn and Schoof (1966) note propoxur has two to six weeks residual activity. The solvents in the emulsifiable concentrate formulation may cause damage to treated surfaces and sprayer components. Therefore, label precautions must be followed carefully. In addition, the spray emulsion breaks at high and low temperatures so care must be taken to maintain a moderate tank temperature. These problems do not occur with the wettable powder formulation, though highly visible spray deposits from this formulation make it more suitable for use in basements than living rooms. Propoxur is moderately toxic to most mammals, though like other carbamates, some non-target animals are particularly susceptible. Particular care should be taken to avoid dermal exposure when handling the emulsifiable concentrate since the solvents increase its skin toxicity.

Trichlorfon is an odorless organophosphate insecticide available for cockroach control as a soluble powder formulation. Its mode of action is attributed to its metabolic conversion within the insect to dichlorvos. Trichlorfon has poor residual action against cockroaches, but its low toxicity and convenient formulation contribute to its use.

TABLE 4-5
A Guide to Some Pesticides and Type of Use
(Check and Follow Labels of Products being Used)

Insecticides	Inside—Non-food Handling Establishments Residual General	Spot	Space and/or Contact	Outside Residual	Space
acephate	no	yes	no		
bendiocarb	no	yes	no	yes	no
boric acid	no	yes	no	no	no
carbaryl	yes	yes	no	yes	no
chlorpyrifos	no	yes	no	yes	no
diazinon	no	yes	no	yes	no
dichlorvos	no	no	yes	no	yes
malathion	yes	yes	yes	yes	yes
MGK 264	no	no	yes	no	yes
piperonyl butoxide	yes	yes	yes	yes	yes
propetamphos	no	yes	no	yes	no
propoxur	no	yes	no	yes	yes
pyrethrins	yes	yes	yes	yes	yes
resmethrin	no	no	yes	no	yes
silica gel	no	yes	no	no	no
sodium fluoride	yes	no	no	no	
trichlorfon	no	yes	no	yes	no

Adapted from New York State Pesticide Applicator Training Manual Category 7: Industrial, Institutional, Structural and Health Related Pest Control.
Subcategory: Structural and Rodent p. 137.

Dusts. Now that the use of chlordane for cockroach control is illegal in the United States, and more stress is placed on crack and crevice application of pesticides, pest control operators are returning with renewed interest to dusts. Dusts offer the advantage of floating back further and coating voids in cracks and crevices. If left dry, some remain effective for years.

Borax alone is an old roach remedy, but has long been considered too slow in action, since it takes four to seven days or longer (Gould 1945) to kill cockroaches. Walter Ebeling, with good cause, is one of the individuals that has revitalized and promoted the current use of boric acid for cockroach control (Ebeling et. al. 1966 a & b, Ebeling and Reierson 1969, Ebeling 1975).

Klostermeyer (1943) showed that in the laboratory, borax, sodium fluoride and sodium fluosilicate were equally toxic to German cockroaches. Boric acid found some usage when sodium fluoride and pyrethrum were in short supply during World War II.

Since borax and boric acid work slowly on cockroaches (Moore 1972), there is little use for it on initial clean-outs by professional pest control operators. Application during construction has proven extremely effective and currently is one of the primary uses for this material. Ebeling (1975) advocates using a water-type fire extinguisher for application.

Ebeling et al. (1966 a & b) showed boric acid powder was less repellent and remained effective for longer periods than most powders. It was used by a housing authority at the rate of one pound per one- or two-bedroom apartment. Small amounts of the dust were applied with a bulb duster with a six-inch spout ($^3/_{16}$ I.D.). Cockroaches succumbed by cuticular penetration, as well as by ingestion through cleaning themselves. A special boric acid powder was used because it did not lump in the duster. The addition of flour or sugar was of no value.

Boric acid, although long used as a medication, can cause death in humans if considerable quantities are accidently ingested. Ebeling (1975) reports death occurred within 46 hours after adults ate 75 grams. Other reports indicate that less than ½-teaspoon of boric acid may kill a child. Therefore, it is important to exercise care when applying boric acid and confine its use to inaccessible sites.

Pest Control magazine (Feb. 1979) clarifies the difference between Boric acid and Borax. "Boric acid is derived from borax (found naturally, and called sodium tetraborate decahydrate) by treatment with hydrochloric acid or sulfuric acid. Borax is not the same thing as boric acid. Borax, too, in the past, has been included in the pesticidal arsenal as a slow acting toxicant. A recipe for borax (36 percent by weight) in a mixture of flour, powdered sugar and corn meal as a roach bait was found to be very slow acting.

Boric acid, when used as a medicine, such as eyewash, is prepared as a 2.2 percent isotonic solution. Boric acid when used as a pesticide is used at a higher, sometimes undiluted concentration."

Cornwell (1976) reviews the practice of washing floors with boric acid as follows. "A solution of boric acid (1.5 percent) in 'vials' in infested premises is said to 'clean-up' the roach population. This seems doubtful, although boric acid dissolved in warm washing water (five percent) for cleaning floors produces a thin film of boric acid crystals when dry, which may kill cockroaches by contact over a period of weeks. The crystals are more difficult for the insects to pick up than when the insecticide is applied as a dust."

Tarshis (1958) demonstrated the effectiveness of silica aerogel. This dust is a chemically inert, odorless, non-crystalline white powder. The particles are extremely small, averaging three microns in size, and the dust is very light, having a density of 4.5 pounds per cubic foot.

Silica aerogels kill cockroaches and other insects by removing the waterproof fatty layer through absorbing the fatty or lipid coating, or by abrading the fatty coating. This results in the loss of water from the body of the insect so it dries up and dies. According to Tarshis (1958), silica aerogel acts primarily by dehydration and secondarily by a physical/chemical reaction.

In homes and apartments ½- to ¾-pound/227g to 340g of silica aerogel is sufficient to control cockroaches. It is extremely important a careful and thorough application be made to all harborages. For this reason, application of silica aerogels requires more time than the application of sprays. In severe infestations, an additional pound of dust per 1000 square feet/454 grams per 90 square meters should be blown into the attic and sub-floor spaces. When applying the dusts in such areas, be sure openings into the house are closed to keep the dust from travelling throughout the house. Ebeling and Wagner (1964 a and b) obtained long-lasting control of several species of cockroaches by applying three grams of silica aerogel per interstud void.

It is claimed that silica aerogels are relatively non-toxic, nevertheless, pest control operators are advised to wear dust masks because the dusts are irritating to the nose and mouth. The dusts are so fine that the many dust masks are not effective. It is believed the Mine Safety model CM876668 is adequate.

The control of cockroaches by silica aerogels is slow. In fact, during the first week after application, there may appear to be an increase in cockroaches because the dust flushes out the cockroaches from their hiding places; then there is a decrease of cockroaches over the next few weeks. When the dust becomes wet, it apparently does not act as quickly upon the cockroaches. The dust does not affect the oothecae or egg cases. However, the newly emerged cockroaches are very susceptible. For this reason, it may be necessary to re-dust several weeks after initial application to kill emerging nymphs, especially if dust was removed shortly after the original application.

Ordinary dust guns and dust blowers can be used to apply silica aerogels. The dust is so fine it may even be applied with a clean paint brush.

Because silica aerogels may stain dark surfaces, care should be taken in applying them to such areas. Silica aerogel dust may drift from the place of application because it is so light. A combination of 0.9 pyrethrum dust and silica aerogel on a 50:50 weight basis results in a much more manageable bulk, and also provides the fast knockdown of pyrethrum. There are now several commercial mixtures of pyrethrins and silica aerogel dusts on the market. One of the most popular contains one percent pyrethrins and 10 percent piperonyl butoxide, plus silica aerogel, ammonium fluosilicate and petroleum hydrocarbons. Wagner et al. (1966) controlled American cockroaches in sewers by blowing 136 g of this mixture into a manhole. The same procedure was successful in Yakima, Washington and other Pacific Northwest cities (Ebeling, 1975).

It is believed cockroaches cannot become immune to silica aerogels because of their combined physical and chemical action. Because cockroaches are developing an immunity to some insecticides, dusts such as sodium fluoride, pyrethrum, and the silica aerogels are becoming more important.

Next to boric acid and silica aerogel, sodium fluoride is the most important of the inorganic dusts. At first, it was thought that sodium fluoride acted as a toxicant on roaches because of the "cleaning-up" habits of insects. Roaches and other insects frequently clean their antennae and legs by drawing them through their mouthparts. In this way the poison enters the stomach, even if it is distasteful to the cockroach. The conclusion was drawn it was unnecessary to place an attractive bait in the powder, since the roach consumed the powder in the process of cleaning itself.

A number of later investigators were not wholly in accord with the "cleaning-up" theory. Hockenyos (1933 and 1936) studied the problem in the laboratory and found sodium fluoride was absorbed through the body integument of the oriental roach. This absorption was greatest where the outer covering of integument was thinnest and most flexible. His studies also showed the finer the sodium fluoride dust, the more readily it was absorbed. Moreover, the cockroach on cleaning its antennae, tarsi, and palpi did not necessarily consume a fatal amount of sodium fluoride. Sweetman and Laudani (1942) and Griffiths and Tauber (1943) observed that under practical field conditions, sodium fluoride acts primarily as a contact poison. Marcovitch (1945) obtained a patent on the use of dextrin as a synergist in pyrethrum/sodium fluoride mixtures, where at least 60 percent dextrin was present in the mixture. In discussing his patent application he stated, "As soon as a roach becomes dusty with this combination, the pyrethrum paralyzes the roach, so that it cannot use its mouthparts effectively to remove the powder. The pyrethrum also interferes with the normal permeability of the tissues and causes water droplets to appear over the body and induces salivation from the mouth. The water thus produced serves to make a gum of the dextrin, which sticks the sodium fluoride onto the roach so tightly that the roach cannot remove it by the movements of its body. This gives the sodium fluoride time to dissolve in the water produced and then to enter the tissues through the pore canals of the chitin, resulting in the final knockout. All three materials thus act in cooperation; the pyrethrum as a paralyzer and water producer, the dextrin as a sticker, and the sodium fluoride as the final killing agent."

Hutzel (1942 a and b), who studied the effect of pyrethrum on cockroaches, found the pyrethrins diffuse through the oily film on the surface of the roach, thereby affecting the nervous system.

Recently another dust has been introduced based on diatomaceous earth and pyrethrum. This combines the long-lasting desiccant action of the diatomaceous earth with the knockdown and flushing action of pyrethrum.

Dusts based on carbaryl are also available for cockroach control and are broadly labeled for consumer or professional use. Five percent carbaryl dust blown in manholes for control of American cockroaches has proved very effective (Ebeling, 1975).

The most commonly used dust for professional use based on a synthetic organic insecticide is diazinon dust. As a one percent formulation this dust provides several months of residual control in dry situations. Eastin and Burden (1961) showed one percent diazinon dusts gave better control over 30 days than silica aerogels. When formulated with pyrethrum it provides a valuable combination of residual and contact/flushing action against cockroaches.

In the past three years a one percent bendiocarb dust has become available for consumer and professional use against cockroaches. It has broader indoor labeling than other residual dusts and provides longer residual control than diazinon or carbaryl dusts. These factors, combined with excellent handling properties, make it one of the most popular dusts for professional control of household pests.

Baits. Insecticide baits are often used to supplement other residual treatments in the overall pest management approach. Care should be exercised not to spray the bait with a repellent insecticide. To achieve effective control with baits, three principles must be followed:

• Place small amounts in many locations.
• Reduce amount of other food readily available for the cockroaches.
• Use a registered bait.

There are several baits recommended for cockroach control. These include two percent propoxur, chlordecone paste and pellets, and 0.5 percent chlorpyrifos bait. Cockroach control tests conducted in apartments in Indianapolis, Indiana indicated that chlorpyrifos bait gave good control (Lund and Bennett 1978).

With the production of chlordecone bait eliminated, two percent propoxur bait has become the major bait material used for cockroach control. It is formulated small enough to be forced into cracks and can be applied with a hand bellows duster. The propoxur bait stays effective so long because it remains dry. Baygon bait aged six weeks was as effective as fresh material (Cornwell 1976). Ninety-five percent kill of German cockroaches was achieved after exposure to bait for nine days (Anonymous 1966). Care should be taken with propoxur bait in pet shops. The vapor can kill fish.

Borax — boric acid. One popular bait utilizing borax consists of the following ingredients by weight:

	Percent
Borax	36
Flour	16
Powdered sugar	10
Cornmeal	38
Total	100

Bare (1945) studied the effect of boric acid baits on the German cockroach. He recommended use of a bait consisting of 10 percent powdered boric acid and 90 percent powdered confectioner's sugar. He found this preparation was slowly effective, non-repellent and of relatively low toxicity to human beings. He notes that baits with sodium fluoride were repellent.

Chlordecone bait. Chlordecone is a chlorinated hydrocarbon which until recently was available in pelleted and paste form. The cereal pellet contained 0.125 percent chlordecone in peanut oil and peanut meal. Its stability in water allowed the manufacture of a gel-like paste. Control with chlordecone was slow, but proved most successful in pet shops and in areas where food for cockroaches was the limiting factor. Cornwell (1976), in his definitive work on cockroach control, predicted chlordecone would play a significant role in the 1980s. The following footnote accompanied this statement which summarizes the abrupt end of chlordecone use by PCOs: "This statement was written before the closure of manufacture in the U.S.A., in late 1975, due to inappropriate precautions in manufacturing and having nothing whatever to do with the safe use, or otherwise, of the insecticide."

With the loss of chlordecone we saw the end to one of the most effective tools for cockroach control in pet shops.

Phosphorus paste. Phosphorus paste is the poison bait previously used for the control of the Oriental roach, the American roach, and the brown-banded roach. Cowan (1865) mentions phosphorus paste was used for the control of cockroaches and rats in London as early as 1858. These commercial phosphorus pastes consist usually of two percent white phosphorus in the form of minute particles well dispersed in some attractive food material such as flour, glucose and honey. The commercial phosphorus pastes do not harden rapidly, and are not a fire hazard when made properly and not diluted with water. Cheng and Campbell (1940) found the particles of phosphorus were surrounded by a 98 percent non-flammable material, and the phosphorus particles did not produce sufficient heat to catch fire. They confirmed this "by heating the paste on kerosene-saturated filter paper to 250° F., and also by rubbing the paste on glass with a gasoline-saturated piece of waste." As a result of these experiments, they concluded phosphorus paste was not a fire hazard. Their studies with a two percent phosphorus paste showed the minimum lethal dosage for the American roach was 0.02 milligrams per gram. The minimum lethal dosage for the German cockroach was 0.13 milligrams per gram. Moreover, the German cockroach apparently does not consume baits with phosphorus paste as readily as the larger roaches. This may be another reason for the relative ineffectiveness of phosphorus pastes in German cockroach control. It is believed phosphorus paste acts as both a stomach and contact poison, and primarily as a stomach poison under field conditions.

American cockroaches receiving minimal lethal amounts may linger a month or more and become sluggish before they die (Cornwell, 1976). With the advent of chlordecone and propoxur baits, use of phosphorus paste declined.

It should be noted that "phosphorous is a deadly poison, and it should not be used where children or pets can find it, nor should the paste be placed where there is a possibility of food contamination. Phosphorous paste has been known to drip on food in warm weather."

NEW TECHNOLOGIES: LACQUERS, TAPES, ETC.

Insecticides in lacquer is not a new concept. The United Kingdom did research in this area in the mid-1950s (Cornwell 1976).

In the United States, a plastic lacquer-based material is available in two formulations, one containing one percent chlorpyrifos and the other two percent chlorpyrifos. The manufacturer claims residual activities of six months and 12 months respectively. The materials are applied undiluted with a four-inch paint brush or in a pin-stream apparatus. Early tests conducted in Arizona with one percent chlorpyrifos gave good control for at least three months (Olson, 1975). Reports from pest control operators (personal communication) show favorable results using chlorpyrifos lacquer in previously hard to control areas. Researchers at Rentokil (Cornwell 1976) performed their own tests using lacquers combined separately with diazinon, propoxur and chlorpyrifos. Two percent chlorpyrifos gave 80 percent control of German cockroaches exposed to surfaces treated and aged four weeks. Eight percent levels of the compound in lacquer gave 100 percent control after the same period of time.

Cornwell (1976) reported on other work in Denmark where different concentrations of diazinon and dieldrin were effective for about nine months and chlorpyrifos for five months. This same author advises that lacquer-base pesticides should not be applied on exposed surfaces which require the cockroaches to walk over the treated area. Instead, it should be placed where the cockroaches hide.

Several pest control operators report difficulty in applying chlorpyrifos lacquer in private homes. Any surface containing dust results in visible smear marks. Treating marble, glass and stainless steel surfaces with chlorpyrifos lacquer should be avoided.

Use of present pesticide insect tapes relies on placing the tape in areas where cockroaches come in contact with it. In laboratory tests, Kydonieus, et. al. (1976) reported 100 percent kill (after three days) of German cockroaches after a two-second contact with the tape containing propoxur. Moore (1976) substantiated this data and reported 100 percent knockdown of German cockroaches after 50 minutes and 100 percent mortality after three days with only one two-second exposure time. He also showed there is minimal fumigation action, thereby indicating the cockroaches must come in contact with the material for it to work. His tests showed the propoxur tape to be less repellent than two percent propuxur bait. Field trials by Moore (1976) show tapes to be less effective than residual sprays in apartments which initially had populations of more than 50 German cockroaches.

A personal survey showed several pest control operators were able to obtain good control of German cockroaches in nightstands in hospitals and nursing homes. The PCOs placed one strip in the top corner of cabinets to achieve control.

However, there are three major drawbacks with insect tape including:

1) It cannot be used in edible product areas of food processing plants, restaurants or other areas where food is commercially prepared or processed.

2) A significant amount of tape is needed to achieve control in an entire room.

3) The tape does not adhere well to surfaces that readily get wet, even when package directions are followed.

Once the tapes are removed from the original package, there is no label on the actual pesticide strip. Therefore, extra caution is needed when these strips are placed.

A recently marketed microencapsulated synergized pyrethrum allows pyrethrum to exhibit residual activity for 30 to 60 days. Pest control operators using the material have reported satisfactory results in food processing areas, though in routine service work it has not performed as well as standard residual insecticide sprays. The use of microencapsulation provides a way for contact pesticides to become residual materials. Bennett and Lund (1977) reported on the effectiveness of microencapsulated pyrethrum and found residual activity against German cockroaches for at least one month. It did leave a visible residue on some surfaces. This product diluted with water at various ratios, 4:1 down to 20:1, and aged for 21 months in closed bottles is still capable of killing German cockroaches. The higher concentration (4:1) was effective for a full five weeks in laboratory tests (Bennett and Runstrom, 1979).

A much longer lasting microencapsulated product is based on diazinon. The active ingredient is placed in 30 to 50 micron polymer particles. The formulation contains two pounds of diazinon per gallon (23 percent). Toxicity of the product to humans is greatly reduced, but the residual activity is maintained. Reierson and Rust (1979) report that according to the manufacturer, the oral toxicity is more than 21,000 mg per kg. Theoretically this would mean a small child could accidently ingest an entire gallon of a one percent mixture and not die. These same authors report applications of microencapsulated diazinon at one percent in apartments gave 85 percent control of German cockroaches after 12 weeks, but at 0.5 percent only 43 percent control was achieved.

The crack and crevice aerosol application technique was refined in the mid 70s after numerous years of development. This development provided an answer in the period when residual insecticides were restricted to crack and crevice applications in food handling establishments. In the late 70s, spot treatments with residual insecticides were again labeled and there was less need to change to crack and crevice technology. Jackson and Wright (1975), applying diazinon and chlorpyrifos using crack and crevice aerosols, showed the pesticide remained where it is applied. Blow (1976) substantiated this when propoxur was used in the same manner.

The most popular crack and crevice aerosol system incorporates the use of gas (carbon dioxide) to carry the pesticide to the target area. No solvents, water or oil are left on treated surfaces. Only the technical pesticide remains. On first use there is a period of adjustment necessary in that the technician tends to apply more than needed. Training is definitely necessary to use the material properly. Training manuals and cassette tapes are available from manufacturers to aid new users. Manuals are currently available for control in institutional kitchens (Whitmire, 1976), hospitals and nursing homes (Whitmire, 1978a), hotels, motels, and resorts (Whitmire, 1978b). These references also serve as excellent training guides.

Flushing agents. These are materials used to irritate cockroaches and bring them out of their hiding places. Ideally, a flushing agent should also knock down and kill those that are flushed. Pyrethrin, resmethrin and dichlorvos are the three major materials used for this purpose. Care should be taken to use this technique only when necessary. Technicians have a tendency to lean too heavily on this technique and end up chasing and scattering cockroaches throughout a room. Ebeling (1975), in citing Grothaus et al. 1972, states, "Resmethrin does not seem to flush cockroaches out of their hiding places as quickly as pyrethrins and the insects appear to be less agitated. However, within 15 to 20 minutes, a greater proportion of the cockroach population is said to be flushed out."

According to Campbell (1942), the pyrethrum acts directly on the nervous system and the pyrethrins go "into solution in the outermost oily/waxy layer and penetrate the cuticle through the pore and gland canals, thus reaching the nerve endings in the epidermis." Woodbury (1938) found pyrethrum sprays may cause the female German cockroach to drop its ootheca prematurely. Parker and Campbell (1940) found the pyrethrins caused a majority of the females to drop their oothecae, whereas thiocyanate had no such effect. In their tests, the adult females and large nymphs of the German roach were less susceptible than other stages to pyrethrum sprays, and the egg capsules were even more resistant than the adult females. McGovran (1943), using sprays with high pyrethrins content, showed the German cockroach was paralyzed more rapidly, but less easily killed than the American roach. Tests in the author's laboratory indicate where the three species of roaches are sprayed with an equal dosage at the same pyrethrum concentration, the German cockroach is the most susceptible, followed by the American cockroach and oriental cockroach.

In a similar series of tests, Miller et al. (1954), it was observed that German cockroaches were the most susceptible to the petroleum distillate base alone, which accounts for the greater susceptibility of German cockroaches to oil-base pyrethrum sprays. Regarding the dropped capsules, Campbell (1942) makes the following observation: "Although some of these capsules may hatch later, the percent hatch is less than that which would have occurred if the capsules had remained attached to the female. Dropped capsules lose water and shrink. The younger the eggs in these capsules, the more likely it is that the embryos in them will fail to mature and hatch. The dropping of the capsules is brought about through the nervous system by contraction of the abdomen."

Pyrethrum-synergist sprays are applied commonly by pest control operators for control in specific situations, the final formulation resulting in a more effective insecticide for the control of a number of insects, particularly flies. Against cockroaches, this activation of the pyrethrins is not so noticeable. Some of the more common pyrethrum synergists that have been on the market are sesame oil extractives, 264, sulfoxide, n-propyl isome, piperonyl cyclonene and piperonyl butoxide, the latter being one of the most commonly used. The manufacturers recommend the pyrethrins be combined with the piperonyl butoxide in ratios of one to 10 and one to 20, although one to five combinations may be applied for longer-lasting control. These pyrethrin-piperonyl butoxide sprays are formulated in oil-base concentrates, which can be diluted with petroleum base oils of the household type, or in emulsion concentrates which can be diluted with water.

Keller et al. (1956) showed a number of strains from different parts of the South to be resistant to contact sprays consisting of pyrethrins alone, and combinations of pyrethrins plus piperonyl butoxide. The level of resistance was much less than usually observed in chlordane resistance.

It was long felt that flushing cockroaches and watching them run from their hiding places was both spectacular and efficient. However, the addition of pyrethrins or synthetic flushing agents added to residual insecticides does not enhance the efficiency of the residual material (Ebeling and Reierson, 1973). Yet some pesticide formulations are sold that include both types of pesticides. With one material the insect is being irritated and with the other it is desirable to have it walk over the treated surface. Better control will be achieved if a good residual pesticide is first put in cracks and crevices with spot treatments where appropriate; this procedure then should be complemented with a flushing-killing action pesticide in an ultra low dosage apparatus.

Use of ULV (Ultra Low Volume) or ULD (Ultra Low Dosage) machines to dispense flushing and knockdown insecticides for cockroach control is well recognized. ULV machines employ the principle that the pesticide is much more concentrated and the particle size is more ideal for reaching and hitting the insect. Recent advances in application equipment technology, particularly the availability of small portable units, has made ULV application of flushing and knockdown products more convenient for pest control personnel. More on ULV is discussed in the mosquito and equipment section of this book.

Flushing and knockdown insecticides are also available in the form of total release aerosols for cockroach control. These products are based on pyrethrins or dichlorvos, sometimes in combination with propoxur. Such total release aerosols have advantages and disadvantages. On the positive side, they provide a fast and easy way to penetrate hard-to-reach areas such as cluttered basements. For such uses they are an excellent complement to residual sprays. On the negative side, a pesticide is being left unattended, cockroaches may be flushed into adjacent areas, and people cannot remain in the area while a total release aerosol is in use.

With this type of treatment the user should be sure to:
- Get everyone out of the room (pets and people).
- Turn off all vents, fans and air conditioners.
- Allow the pesticide to reach room temperature before using. If too cold, they will sputter and be ineffective.

Cockroach repellents. R-11 (2, 3, 4, 5 bis [2 butylene] tetrahydrofural) R-326 (di-n-propyl isocinchomeronate), R-874 (2-hydroxyethyl-n-octyl sulfide) as well as Tabatrex (dibutyl succinate) have been found effective in keeping cockroaches out of beverage cases, dispensing machines and similar places. These repellents are combined with pyrethrins for quick kill. Mallis et al. (1961) showed a Fumol mixture of one percent R-11 and three percent MGK-264 at about 12 ml per case repels cockroaches from beer cases for five weeks or more. DeLong (1962) also studied these repellents in cases and cartons.

Sterilization and radiation. Mortality after exposure to UV radiation was exhibited in five species of cockroaches (Cohen, Sousa and Roach, 1973). Burden and Smittle (1963) first reported on the effect of chemosterilants against German cockroaches.

Frishman (1968) was able to sterilize both sexes of German cockroaches with oral, dermal and residual deposits of TEPA, tris (1-aziridinyl) phosphine oxide, a chemosterilant. The relative danger of this compound and related compounds to humans and other animals has curtailed further research in this area.

Kenaga (1965a, 1965b) showed several triphenyl tin compounds partially suppressed reproduction in German cockroaches when fed to fourth instar nymphs in a dog food-yeast mixture.

Experimentation with ionization radiation on German cockroaches also was attempted. German cockroaches can be killed by a dosage of 105,000 r (Cole, et al., 1959). Ross and Cochran (1963), in an attempt to induce mutations in German cockroaches, exposed late instar nymphs to ionizing radiation. Dosages ranging from 100 r to 9,600 r produced a decrease in egg production and fertility, abnormal ovaries, cessations of spermatogenesis, undeveloped ovaries, sterility and mortality. The authors concluded there was a progressive increase in dormant lethals with an increase in exposure to radiation.

Smittle (1964) reported for the first time that C.S. Lofgren, G.S. Burden and P.H. Clark in 1956 had sterilized female and male German cockroaches with a dose of 1,500 r and 5,500 r respectively.

The use of massive releases of sterile male cockroaches to achieve control has been the subject of research on naval vessels. Such techniques offer long-term hope of success, but only in situations where tolerance of cockroaches is high or can be enforced.

Fumigation. Use of the common fumigants will result in cockroach control providing the building can be tightly sealed. Fumigation has been replaced almost entirely by residual sprays 'and dusts because these ordinarily provide long-lasting control which fumigants do not. Cockroaches and other vermin are controlled aboard ships with fumigants.

Dichlorvos or DDVP is now commonly formulated as 20 percent concentrations in resin strips. Russell and Frishman (1965) and Smittle and Burden (1965) showed that DDVP resin strips are not very effective against German cockroaches in well-ventilated and air-conditioned rooms. DDVP vapors do not effectively penetrate the hiding places of cockroaches when used in strip formulations.

NON-CHEMICAL APPROACHES

Traps. Use of traps to totally eliminate a heavy cockroach population within a large structure is not a realistic solution, nor have the authors ever witnessed this.

Haber (1919) recognized the importance of trapping more than 60 years ago, but even then he stated such methods are often over-emphasized. He described one of several trapping methods as follows: "At night put old cloths dampened with dish water in the sink or near their runways and places of seclusion. Darken the room and leave it. At half- or three-quarter hour intervals return with a liberal supply of scalding hot water and dash it upon the cloths, thus destroying many cockroaches which have secluded themselves in the folds of the cloth or beneath it. The dead cockroaches should be collected and burned before the cloths are rearranged to trap more."

More than a quarter of a century ago, Gould (1943) reported use of traps as a recent development for cockroach control. However, today both the public and professionals are finding them a useful tool in monitoring cockroach infestations. They allow an individual to be alerted to the fact that cockroaches are intruding into an area and additional control is needed. In "sensitive" areas where all pesticides are prohibited (eg. insect rearing rooms and certain research facilities), these traps may be the only item allowed.

Personal observations in apartments heavily infested with German cockroaches revealed traps almost completely covered with cockroaches, with thousands more in boxes adjacent to the traps. On some occasions, mice get caught in the cockroach traps.

Electrical traps as reported by Burgess, McDermott and Blanch (1974) corralled 12,000 German cockroaches in a hospital canteen during one year.

Barak, Shinkle and Burkholder (1977), using traps as a method of control, found greater success with oriental cockroaches than German cockroaches.

With trapping, the greater the reproductive capacity of the cockroach, the less significant the impact in reducing the population. Where the cockroaches have numerous cracks and crevices to hide, traps also proved less effective.

The authors believe traps could be used for the following purposes:

• To detect low-level populations. The existence of a potential problem can be confirmed before a population explosion takes place.

• To locate problem areas or harborages. This can greatly enhance control efforts, allowing the operator to intensify treatments in certain areas and perhaps solve continuing problems.

• To monitor population increases. Thus, the need for frequent and expensive applications or treatments can be minimized.

• To reduce or control infestations. As a primary method in certain instances, but more commonly, integrated with current methods of chemical treatment to improve efficacy.

Live trapping. The ecological and physiological mechanism influencing food-finding in cockroaches is complex and well documented by Miesch (1964). He found a semi-solid bait of dehydrated potatoes, sucrose and water was the most attractive material for German cockroaches. Dr. Charles Wright (personal correspondence) found baited raisins attracted German cockroaches. Reierson (1977) found that in the laboratory Coca-Cola® syrup, stale beer, and fresh bananas are good attractants for German cockroaches, but white bread is more convenient for field research.

Mallis (1969) successfully trapped American and German roaches through the use of empty gallon tin cans that are clean on the inside. The upper inner surface of the can is lightly greased with Vaseline® one to two inches/25-50 mm beneath the rim. Then, a piece of white bread is wetted with beer. The can is then placed in a dark corner so that its outer surface is in contact with two walls. The number of traps used depends upon the infestation. These traps are usually set up in the evening and the roaches are collected in the morning. Carbon dioxide from a small CO_2 tire-filling cylinder can be used to anesthetize cockroaches so they can be transferred easily. Wright (1966) developed a technique of collecting live German and brown banded cockroaches via a modified vacuum cleaner.

Biological control. This facet of pest management holds little promise for domestic cockroaches, particularly indoors. In most cases a zero population level is sought. Biological agents reduce populations to lower levels, but if the biological agent completely eliminates its host, it would also eliminate itself.

More than 60 years ago, Shipley (1916) reported the intentional introduction of hedgehogs to reduce cockroach populations. Marlatt (1908) found it worthy to report that three frogs left in a room overnight would effectively rid the premises. Monkeys and turtles have been reported eating cockroaches at zoological parks.

The hymenopterous insect parasites of the genus *Evania* lay their eggs in the egg capsules of both American and oriental roaches. Mites of the genus *Pimeliaphilus* also have been known to destroy colonies of cockroaches in the laboratory (Field et. al. 1966). One predator, the Surinam toad *(Pipa pipa)*, which has been introduced into a number of roach-infested islands, has shown both a palate and appetite to become an effective natural enemy of the cockroach. However, the great reproductive potential of the cockroach, its ability to adapt and protect itself, its omnivorous appetite, and the scarcity of effective natural enemies, guarantees its survival despite the efforts of PCOs.

LITERATURE

ADAMS, P.K. — 1978. Personal Communication. Adams Pest Control, Inc.

ANONYMOUS — 1940. What combination for roach powders? Part II. Soap & S.C. 16(11):96, 99.

1957. Oriental roaches from sewers. Pest Control 25(2):26, 46.

1966. Public Health Pesticides. Pest Control 34(3):10-34.

1968. Louisiana Association Newsletter.

BACK, E.A. — 1937. The increasing importance of the cockroach *(Supella supellectilium* Serv.) as a pest in the United States. Proc. Entomol. Soc. Wash. 39:205-213.

BARAK, A.V., M. SHINKLE and W. E. BURKHOLDER — 1977. Using attractant traps to help detect and control cockroaches. Pest Control 45 (10):14-16, 18-20.

BARBER, M.A. — 1914. Cockroaches and ants as carriers of the vibrios of Asiatic cholera. Philippine J. Sci. Sec. B., 9:1-4.

BARE, O.S. — 1945. Boric acid as a stomach poison for the German cockroach. J. Econ. Entomol. 38(3):407.

BEDINGFIELD, W.D. — 1952. Insecticide resistant roaches. Pest Control 20 (4):6.

BENNETT, G.W. and R.D. LUND — 1977. Evaluation of encapsulated pyrethrins (Sectrol® insecticide) for German cockroach and cat flea control. Pest Control 45(9):44, 46, 48-50.

BENNETT, G.W. and E.S. RUNSTROM — 1979. New developments in pest control insecticides. Pest Control 47(6):14-16, 18, 20.

BERNTON, H.S. and H. BROWN — 1964. Insect allergy: Preliminary studies of the cockroach. J. Allergy 35:506-513.

1967. Cockroach Allergy. II. The relation of infestation to sensitization. So. Med. J. 60:852.

1970a. Age of onset of skin reactivity. Ann. Allergy 28:420-422.

1970b. Insect allergy: the allergenicity of the excrement of the cockroach *Blattella germanica* Ann. Allergy 28:543-547.

BERNTON, H.S., T.F. MCMAHON and H. BROWN — 1972. Cockroach asthma. British J. of Diseases of the Chest 66:61-66.

BLOW, D.P. — 1976. England's Protim Ltd. tests. Whitmire system on cockroaches. Pest Control 44(6):46-47.

BURDEN, G.S. and B.J. SMITTLE — 1963. Chemosterilant studies with the German cockroach. Florida Entomol. 46:229-234.

BURDEN, G.S. and E.E. MADDEN — 1975. *Periplaneta americana:* Comparative susceptibility to residuals. Pest Control 43(1):20.

BURGESS, N.R.H., S.N. MCDERMOTT, A.P. BLANCH — 1974. An electrical trap for the control of cockroaches and other domestic pests. J.R. Army Med. Cps. 120, 173-175.

BUXTON, G.M. and T.J. FREEMAN — 1968. Positive separation of *Blattella vaga* and *Blattella germanica* (Orthoptera: Blattidae) Pan-Pacific Entomologist 44:168-169.

CAMPBELL, F.L. — 1940. Toxicity of phosphorus to cockroaches. Pests 8(4):12.

1942. Pyrethrum vs. roaches. Soap & S.C. Parts I & II 18(5):90-93.

CARDONE, R.V. and J.J. GAUTHIER — 1979. How long will *Salmonella* bacteria survive in German cockroach intestines? Pest Control 47(6):28-30.

CHENG, T.H. and F.L. CAMPBELL — 1940. Toxicity of phosphorus to cockroaches. J. Econ. Entomol. 33:193-199.

CHOOVIVATHANAVANICH, P.P. SUWANPRATECEP and N. KANTHAVICHITRA — 1970. Cockroach sensitivity in allergic Thais. Lancet 2:1362.

CLEVELAND, L.D., S.R. HALL, E.P. SANDERS and J. COLLIER — 1934. The wood-feeding roach *Cryptocereus*, its protozoa, and the symbiosis between protozoa and roach. Mem. Amer. Acad, Arts & Sci. 17:185-342.

COHEN, S.H., J.A. SOUSA and F. ROACH — 1973. Effects of UV irradiation on nymphs of five species of cockroaches. J. Econ. Entomol. 66(4):859-862.

COLE, M.M., G.C. LABRECQUE and G.S. BURDEN — 1959. Effects of gamma radiation on some insects affecting man. J. Econ. Entomol. 52:448-450.

CORNWELL, P.B. — 1968. The Cockroach. Vol. I. Hutchinson, London. 391 pp.

1976. The Cockroach. Vol. II. Hutchinson, London. 557 pp.

CREIGHTON, J.T. — 1943. Household Pests. Fla. Agr. Ext. Serv. Bull. 122

DELONG, D. — 1962. Beer cases and soft drink cartons as insect distributors. Pest Control 30(7):14, 16, 18.

DOW, R.P. — 1955. A note on domestic cockroaches in South Texas. J. Econ. Entomol. 48(1):106.

EASTIN, J.L. and G.S. BURDEN — 1961. Tests with five silica dusts against German cockroaches. Fla. Entomol. 43(3):99-102.

EBELING, W. — 1975. Urban Entomology. Univ. of California. 695 pp.

1977. The cockroach as implicated in public health. Proceedings of the 1977 Seminar on Cockroach Control sponsored by the N.Y.S. Dept. of Health.

EBELING, W. and D.A. REIERSON — 1969. The cockroach learns to avoid insecticides. California Agriculture 23(2):12-15.

1973. Should flushing agents be added to blatticides? Pest Control 41(6):24, 46, 48, 50-51.

EBELING, W. and R.E. WAGNER — 1964a. "Built in" pest control for wall and cabinet voids in houses and other buildings under construction. Calif. Agr. 18(11):8.

1964b. The treatment of voids under cabinets. P.C.O. News 24(4):8-11.

EBELING, W., R.E. WAGNER, and D.A. REIERSON — 1966a. Influence of repellancy on the efficacy of blatticides. I. Learned modification of behavior of the German cockroach. J. Econ. Entomol. 59:1374-1388.

1966b. Influence of repellancy on the efficacy of blatticides. II. Laboratory experiments with German cockroaches. J. Econ. Entomol. 60:1375-1390.

EDMUNDS, L.R. — 1957. Observations on the biology and life history of the brown cockroach *Periplaneta brunnea* Burmeister. Proc. Entomol. Soc. Wash. 59 (6):283-286.

ESSIG, E.O. — 1926. Insects of Western North America. Macmillan Publ.

FIELD, G., L.B. SAVAGE and R.J. DUPLESSIS — 1966. Note on the cockroach mite *Pimeliaphilus cunliffei* (Acarina: Pterygosomidae) infesting oriental, German and American cockroaches. J. Econ. Entomol. 59(6):1532.

FISK, F.W. and J.A. ISERT — 1954. Comparative toxicants of certain organic insecticides to resistant and non-resistant strains of the German cockroach, *Blattella germanica* (L.) J. Econ. Entomol. 46 (6):1059-1062.

FLOCK, A.A. — 1941. The field roach, *Blattella vaga*. J. Econ. Entomol. 34:121.

FLYNN, A.D. and H.F. SCHOOF — 1966. Evaluation of toxicants as residues against *Blattella germanica (L.)* J. Econ. Entomol. 59(5):1270-1274.

FOSTER, E. — 1855. Foster's voyage. Proc. Entomol. Soc. London. p. 77.

FRISHMAN, A.M. — 1968. A comparison and evaluation of residual, oral and topical treatments of tepa, Tris (1-aziridinyl) phosphine oxide, used to sterilize german cockroaches *Blattella germanica* (Linnaeus) Ph.D. Thesis. Purdue University, Lafayette, Indiana.

FRISHMAN, A.M. and I.E. ALCAMO — 1977. Domestic cockroaches and human bacterial disease. Pest Control 45(6):16, 18, 20, 46.

GIBSON, A.W. — 1960. How we apply powders for insect control. Pest Control 28(1):18, 20, 22, 24.

GIER, H.T. — 1947. Growth rate in the cockroach *Periplaneta americana* (Linn) Ann. Entomol. Soc. Amer. 40 (2):303-317.

GORHAM, J.R., KARL P. CONRADI and K. PAGE CONRADI — 1971. Household infestation by the cockroach *Aglaopteryx gemma* in Georgia J. of the Georgia Entomol. Soc. 6(2):133-135.

GOULD, G.E. — 1943. Recent developments in roach control. Pests 11(12)12-13, 22-24.

1945. Roach control tests. Soap & S.C., N.Y. 21(2):113-115, 121.

GOULD, G.E. and H.O. DEAY — 1938a. The biology of the American cockroach. Ann. Entomol. Soc. Amer. 31(4):489-498.

1938b. Notes on the bionomics of roaches inhabiting houses. Proc. Ind. Acad. Sci. 47:281-284.

1940. The biology of six species of cockroaches which inhabit buildings. Purdue Univ. Agric. Exp. Sta. Bull. 451.

GRAFFAR, M. and S. MERTENS — 1950. Le rôle des blattes dans la transmission de salmonellosis. Ann. Inst. Pasteur. 79:654-660.

GRAYSON, J.M. — 1965. Resistance to three organophosphorus insecticides in strains of the German cockroach. J. Econ. Entomol. 58(5):956-958.

1976. Cockroach control research in 1975: comparative effectiveness of various insecticides against cockroaches. Pest Control 44(2):30-32, 39.

GRAYSON, J.M. and D.H. MESSERSMITH — 1959. Resistant roach control research at VPI, Pest Control 27(2):26-27.

GRAYSON, J.M. and F.E. JARVIS JR. — 1958. Residual effectiveness of certain insecticides in German cockroach control. Soap & S. C. 34(3):91, 92, 133.

GRIFFITHS, J.T. and O.E. TAUBER — 1943. Evaluation of sodium fluoride as a stomach poison and as a contact insecticide against the roach. *Periplaneta americana* L. J. Econ. Entomol. 36(4):536-540.

GROTHAUS, R.H., W.N. SULLIVAN, M.S. SCHECTER and E.L. COX. — 1972. Resmethrin odor and performance improved. Soap & S. C. (September) pp. 54, 56, 58.

GUPTA, A.P., et al. — 1973. Effectiveness of spray-dust-bait combination and the importance of sanitation in the control of German cockroaches in an inner-city area. Pest Control 41(9):20, 22, 24, 26, 58, 60-62.

GURNEY, A.B. — 1953. Distribution, general bionomics and recognition characters of two cockroaches recently established in the United States. Proc. U.S. National Museum 103(3315):39-55.

1955. Notes on the Cuban cockroach *Panchlora nivea* (L.) Orthoptera, Blattidae) Proc. Entomol. Soc. Wash. 57:285-286.

GUTHRIE, D.M. and A.R. TINDALL — 1968. St. Martin's Press, N.Y. 408 pp.

HABER, V.R. — 1919. Cockroach pests in Minnesota with special reference to the German cockroach. The Univ. of Minn. Agricultural Experiment Station Bulletin 186. 16 pp.

HEAL, R.E., K.B. NASH and M. WILLIAMS — 1953. An insecticide-resistant strain of the German cockroach from Corpus Christi, TX J. Econ. Entomol.

46(2):385-386.

HEBARD, M. — 1917. The Blattidae of North America, North of the Mexican Boundary, Mem. Amer. Entomol. Soc. 2:1-284.

HOCKENYOS, G.L. — 1933. The mechanism of absorption of sodium fluoride by roaches. J. Econ. Entomol. 26:1162-1169.

1936. Mechanism of absorption of pyrethrum powder by roaches. J. Econ. Entomol. 29:433-437.

HOWARD, L.O. — 1895. Migration of cockroaches. Insect Life 7:349.

HOWARD, L.O. and C.L. MARLATT — 1902. The principal household insects of the United States. U.S.D.A. Bull. #4.

HUTZEL, J.M. — 1942a. The activating effect of pyrethrum upon the German cockroach. J. Econ. Entomol. 35(6):929-933.

1942b. Action of pyrethrum upon the German cockroach. J. Econ. Entomol. 35 (6):933-937.

JACKSON, M.D. and C.G. WRIGHT — 1975. Diazinon and Chlorpyrifos residues in food after insecticidal treatment in rooms. Bull. Environ. Contam. Toxicol. 13(5):593-595.

JANSSEN, W.A. and S.E. WEDBERG — 1952. The common house roach, *Blattella germanica* Linn., as a potential vector of *Salmonella typhimurium* and *Salmonella typhosa*. Amer. J. Trop. Med. Hyg. 1:337-343.

JETTMAR, H.M. — 1935. Kuchenschaben als Krangheitsubertrager. (Cockroaches as vectors of disease). Wein. Klin. Wechr. 48(20):700-704.

KELLER, J.C., P.H. CLARK, C.S. LOFGREN — 1956. Susceptibility of insecticide-resistant cockroaches to pyrethrins. Pest Control 24(11):14-15, 30.

KELLOGG, V.L. — 1908. American insects. Henry Holt and Co. 694 pp.

KENAGA, E.E. — 1965a. Triphenyl tin compounds as insect reproduction inhibitors. J. Econ. Entomol. 5:64-68.

1965b. Triphenyl tin compounds as insect reproduction inhibitors. Proceedings of the XIIth Int. Congr. Entomol. p. 517.

KLOSTERMEYER, E.C. — 1943. Roach powders: Study of comparative effectiveness of insecticidal powder mixtures against the German cockroach. Soap & S.C. 19 (2):98-99.

KNIPLING, E.F. — 1947. Newer synthetic insecticides. Soap & S.C. 23(7):127, 129, 131.

KYDONIEUS, A.F., A.R. QUISUMBING, I.K. SMITH, S. BALDWIN, R.A. CONROY — 1976. Hercon Technical Bull. No. 26. 4 pp.

LAING, F. — 1938. The cockroach, its life-history and how to deal with it. British Mus. (Nat. Hist.) 23 pp. 3rd. Ed.

LOFGREN, C., G.S. BURDEN and P.H. CLARK — 1957. Experiments with insecticides for the control of German roaches. Pest Control 25(7):9-10, 12 47.

LUND, D. and G.W. BENNETT — 1978. Evaluation of Bolt® roach bait. Insecticide and Acaricide Test Bulletin, Section 3, 176-177. Entomological Society of America.

MACKERRAS, I.M. and P. POPE — 1948. Experimental *Salmonella* infections in Australian cockroaches. Aust. J. Expt. Biol. Med. Sci. 26(6):465-470. Rev. Appl. Entomol. B 38(1):15, 1950.

MALLIS, A. — 1969. Handbook of Pest Control, 5th edition, MacNair-Dorland Co., 1158 pp.

MALLIS, A., W.C. EASTERLIN, and A.C. MILLER — 1961. Keeping cockroaches out of beer cases. Pest Control 29(6):32-35.

MARCOVITCH, S. — 1945. U.S. Patent 2,377,798. June 1945. Univ. of Tenn. Research Corp.

MARLATT, C.L. — 1908. Cockroaches. U.S.D. A. Div. Entomol. Circ. No. 51.

MCGOVRAN, E.R. — 1943. The relative resistance of roaches to pyrethrum spray. Proc. Entomol. Soc. Wash. 45(2):55-56.

MIALL, L.C. and A. DENNY — 1886. Structure and Life History of the Cockroach.

MIESCH, M.D. — 1964. Ecological and physiological mechanisms influencing food finding in Blattaria. Ph.D. Thesis. Oklahoma State Univ., Stillwater, OK.

MILLER, A.C., A. MALLIS and W.C. EASTERLIN — 1954. A testing procedure for evaluating liquid sprays against cockroaches. J. Econ. Entomol. 47(1):23-26.

MOORE, R.C. — 1972. Boric acid. Silica dusts for control of German cockroaches. J. Econ. Entomol. 65. (2):458-461.

1976. Efficacy of Hercon Roach tape, Pest Control 44(6):37, 38, 40, 42.

N.P.C.A. (National Pest Control Association) — 1961. German roach resistance and control. Tech. Release No. 8-61.

NOLAND, J.E., J.H. LILLY and C.A. BAUMAN — 1949. A laboratory method for rearing cockroaches and its application of dietary studies on the German roach. Ann. Entomol. Soc. Amer. 42 (1):63-70.

OLSON, G.S. — 1975. Slow-release formulation shows promise. Pest Control 43(10):20, 53.

OLSON, T.A. and M.E. RUEGER — 1950. Experimental transmission of *Salmonella oranienburg* through cockroaches. U.S. P.H.S. Rpts. 65(16):531-540.

PALERMO, M.T. — 1960. The cockroach twins. Pest Control 28(6):12.

PARKER, T. — 1977. Pi Chi Omega Newsletter.

PEST CONTROL MAGAZINE — 1979. Basically Borax? Vol. 47(2):9-10.

RATCLIFFE, J. — 1952. Lobster roach. Pest Control 20(5):44, 54.

RAU, P. — 1940a. The life history of the American roach. Entomol. News 51:121-124, 151-155, 186-189, 222-227, 273-278.

1940b. The life history of the wood roach *Parcoblatta pennsylvanica* DeGeer, (Orthoptera: Blattidae) Entomol. News 51:4-9, 33, 35.

1943. How the cockroach deposits its egg-case. A study in insect behavior. Ann. Entomol. Soc. Amer. 36(2):221-226.

1944a. A note on the period of incubation of eggs of the cockroach *Blattella germanica* L. Can. Entomol. 76(10):212.

1944b. Another use for the cockroach *Blatta orientalis* Entomol. News. 55(2):49-50.

1945. Food preferences of the cockroach *Blatta orientalis* Linn. Entomol. News 56(10):276-278.

REHN, J.A. G. — 1945. Man's uninvited fellow traveler- the cockroach Sci. Mo. 61(4):265-276.

REIERSON, D.A. — 1977. Methods for evaluating cockroach infestations. Proceedings of: The 1977 Seminar on Cockroach Control. N.Y.S. Dept. of Health. pp. 21-24.

REIERSON, D.A. and W. EBELING — 1970. The brown cockroach *Periplaneta brunnea* Burmeister, found in California PCO News 30(7):6-9.

REIERSON, D.A., M.K. RUST — 1979. New sprays for cockroach control. Pest Control Technology 7(3):14, 16, 17.

REVEL, C. — 1978. Hottest New Business Ideas. Baron book Publishing Co.,

Santa Monica, Calif.

ROSS, H.H. — 1928. The life history of the German cockroach, *Blattella germanica* Trans. Ill. State Acad. Sci. 21:84-93.

ROSS, M.H. and D.G. COCHRAN — 1963. Some early effects of ionizing radiation on the German cockroach, *Blattella germanica*. Ann. Entomol. Soc. Amer. 56:256-261.

ROTH, L.M. and E.R. WILLIS — 1952. A study of cockroach behavior. Amer. Midland Naturalist 47(1):66-129.

1954. The reproduction of cockroaches. Smithsonian. Misc. Coll. 122(12):1-47.

1956. Parthenogensis in cockroaches. Annals. Entomol. Soc. Amer. 49(3):195-204.

1957. The medical and veterinary importance of cockroaches. Smithsonian Misc. Coll. 134(10):1-147.

1958. The biology of *Panchlora nivea* with observations on the eggs of other Blattaria. Transactions of the American Entomol. Soc. 83:195-207.

1960. The biotic association of cockroaches. Smithsonian Institute. Washington. 439 pp.

RUEGER, M.E. and T.A. OLSON — 1969. Cockroaches (Blattaria) as vectors of food poisoning and food infection organisms. J. Med. Entomol. 6:185-189.

RUSSELL, M.P. and A.M. FRISHMAN — 1965. Effectiveness of dichlorvos in resin strips for the control of the German cockroach *Blattella germanica*. J. Econ. Entomol. 58(3):570-572.

RUST, M.K. and D.A. REIERSON — 1978. Comparison and field efficacy of insecticides used for German cockroach control. J. Econ. Entomol. 71:704-708.

SCHOOF, H.F. and R.E. SIVERLY — 1954. The occurrence and movement of *Periplaneta americana* within an urban sewerage system. Amer. Trop. Med. and Hyg. 3(2):367-371.

SCHULANER, F.A. — 1970. Sensitivity to cockroaches in three groups of allergic children. Pediatrics 45:465.

SCHWABE, C.W. — 1949. Observations on the life history of *Pycnoselus surinamensis* (Linn.), the intermediate host of the chicken eyeworm, *Oxyspirura mansoni* in Hawaii. Hawaii Entomol. Soc. Proc. 13:433-436.

SERVILLE, J.G.A. — 1839. Histoire naturelle des insects, Orthopteres, xviii 777 pp.

SHIPLEY, A.E. — 1916. More Minor Horrors. Smith Elder, London. 163 pp.

SMITTLE, B.J. — 1964. The effects of tepa on the embryogeny and reproductive organs of the German cockroach *Blattella germanica* (L.) Ph.D. thesis. Rutgers Univ. 83 p.

SMITTLE, B.J. and G.S. BURDEN — 1965. Dichlorvos as a vapor toxicant for control of roaches, bedbugs, fleas. Pest Control 33(10):26-32.

STEK M., R.V. PETERSON and R.L. ALEXANDER — 1979. Retention of bacteria in the alimentary tract of the cockroach, *Blattella germanica,* J. Environ. Health 41(4):212-213.

STORY, K.O. — 1972. Control of cockroaches and other domestic pests with a new carbamate insecticide. International Pest Control. 14(6):6-10.

SWEETMAN, H.L. and H. LAUDANI — 1942. Sodium fluoride. Soap & S.C. 18 (4):90, 93.

TARSHIS, I.B. — 1958. The use of sorptive dusts for the control of cockroaches. Syllabus 6th Annual Cal-Poly Pest Control Conference Dec. 12 & 13, 1958.

1962. The cockroach — a new suspect in the spread of infectious hepatitis. Amer.

J. Trop. Med. & Hyg. II (5):705-711.

TWOMEY, N.R. — 1966. A review of the biology and control of the German cockroach, *Blattella germanica* (L.) in California. Calif. Vector Views 13(4):27-37.

WAGNER, R.E., W. EBELING and D.A. REIERSON — 1966. Control of cockroaches in sewers. Public Works 97(1):82-84.

WALDRON, W.G. and F. HALL — 1972. Mode of entry into Los Angeles County, Calif., of the brown cockroach, *Periplaneta brunnea* Burmeister. California Vector Views 19:1-2.

WHEELER, W.M. — 1910. Ants. Columbia Univ. Press. 663 pp.

WHITMIRE RESEARCH LABORATORIES — 1976. Prescription Treatment. Pest Management System, Insect Treatment Manual for Institutional Kitchens. 19 pp.

1978a. Prescription Treatment Systems. Pest Management Manual for Hospitals and Nursing Homes. 80 pp.

1978b. Prescription Treatment System Pest Management Manual for Hotels, Motels and Resorts. 45 pp.

WILLIS, E.R., G.R. RISER, and L.M. ROTH — 1958. Observations on reproduction and development in cockroaches. Annals Entomol. Soc. Amer. 5(1):53-69.

WOOD, F.E. — 1980. Cockroach control in public housing: Is it an overwhelming problem? Pest Control. 48(6):14.

WOODBURY, E.N. — 1938. Test methods on roaches. Soap 14(8):86.

WRIGHT, C.G. — 1966. Modification of a vacuum cleaner for capturing German and brownbanded cockroaches. J. Econ. Entomol. 59(3):759-760.

1979. Life history of the smoky brown cockroach. J. of Georgia Entomol. Soc. 14(1):70-75.

ZABINSKI, J. — 1929. The growth of black beetles and cockroaches on artificial and incomplete diets. Part 1. Brit. J. Expt. Biol. 6:36-85.

ZAPPE, M.P. — 1917. A cockroach pest in greenhouse *Pycnoscelus (Leucophaea) surinamensis* Linn. Conn. Agr. Expt. Sta. Bull. 203:302-303.

ZUBEN, F.J. VON — 1955. Cockroaches in municipal sewers. Pest Control 23 (5):14-16.

RICHARD V. CARR

Dr. Richard Carr's career reflects a varied background of experience in urban pest control. As a service technician in Tucson, Arizona, Carr served his apprenticeship learning the trade at its most important level. He also gained valuable experience with his own firm, experiencing the business at the level where the "buck stops."

His master's degree, earned at the University of Arizona, Tucson, offered him the opportunity to study the behavior of the honey bee and resulted in his interest in another social insect, the termite. During the years Carr spent working on his Ph.D., he investigated the mating behavior and morphology of sex attractant glands in termites.

It was at this time that the International Biological Program (IBP), Desert Biome, came into existence. Dr. William Nutting, Carr's major professor was senior investigator for the termite research to be conducted in the Sonoran Desert for the IBP. This program exposed Carr to the largest single research effort ever conducted on termite behavior, ecology and biology.

Following his tenure as research assistant to Nutting, he participated in a brief post-doctoral tour at the University of Arizona investigating a systems approach to predicting infestations of the cotton pink bollworm, an IPM study. In the early 1970s, Carr joined Velsicol Chemical Corporation in Chicago. He became their first research and development investigator to concentrate his efforts on the company's specialty product line. His work included the development of products for structural and household pest control in addition to those for professional turf care. By the mid 1970s he was promoted to regional development manager and then to director of commercial development for all of Velsicol's pesticide products.

In 1979, Carr worked for Armak Company, Chicago, Illinois, acquiring experience in polymers for use in slow releasing pesticides. Then in 1980, he assumed his current position as director of research at the National Pest Control Association.

Carr is the author of many scientific publications and popular articles on urban pest control and has lectured widely at numerous training and certification workshops. Carr is now actively involved in expanding the research base, both in technical and commercial sectors, to help the professional sector meet today's pest control needs.

Carr is a member of Pi Chi Omega, Entomological Society of America, an honorary member of the Wisconsin Pest Control Association and an entomologist on the American Registry of Professional Entomologists.

CHAPTER FIVE

Crickets

Revised Richard V. Carr[1]

The cricket, there beside the fire,
Who twangs upon a tuneless lyre,
Who simply sits and stridulates
The while he loudly cogitates.
Go sing his praises you who will;
In me he fails to rouse one thrill.
In fact, he makes my hackles twitch —
He's just a cockroach with an itch.

Hal Borland

THE HOUSE CRICKET, *Acheta domesticus* (L.), and the field cricket, *Gryllus* spp., are the cricket pests which sometimes invade homes. Aside from their occasional annoying entry into the home, they may injure clothes and other materials. Enormous populations may explode and congregate around lights at night making roads slick and public places unattractive.

Crickets are members of the order Orthoptera, which includes grasshoppers and roaches. However, they belong to separate families. According to Marlatt (1896), the word "cricket" is derived from the imitative French common name "cricri," and is indicative of the cricket's chirping sound. The crickets can be separated into several distinct groups known as:

1) Tree crickets
2) Ground crickets
3) House and field crickets
4) Camel crickets
5) Jerusalem crickets
6) Mole crickets

The tree crickets, ground, house and field crickets are usually placed in the family Gryllidae. The camel crickets and Jerusalem crickets are placed in the family Gryllacrididae and the mole crickets are in the family Gryllotalpidae.

[1]*Director of Research, National Pest Control Association, Vienna, Va.*

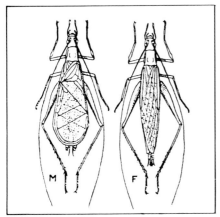 

Fig. 5-1. The snowy tree cricket, *O. ni-*   Fig. 5-2. *Ecanthus fasciatus*, female.
vecus. Left, male; right, female.

The tree crickets, such as the snowy tree cricket (Fig. 5-1 and 5-2), *Oecanthus fultoni*, live in trees and shrubs where they chirp at a regular rate which varies with the temperature. In fact, a good approximation of the temperature in degrees Fahrenheit is to add 40 to the number of chirps in 15 seconds. These are also the crickets commonly heard in the background noise of movies and television. Most tree crickets are not pests, but some damage fruit bearing plants and ornamentals by making holes in stems and branches during egg laying.

The ground crickets (Fig. 5-3) are common insects in pastures, meadows and in wooded areas. They look like the house cricket but are much smaller, usually less than ½ inch/13mm in length. Their songs are often soft, high-pitched, pulsating trills or buzzes. they are not often pests, though they may be abundant around homes in the fall after harvest or hay cutting.

The house and field crickets (Fig. 5-4 and 5-5) are common everywhere. They usually live on plants, but are not strictly vegetarians. Sometimes they feed on each other or other insects. The eggs are laid in the fall, usually in sandy soil, and hatch the following spring. Most field crickets chirp and may sing both day and night.

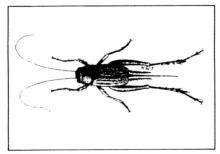  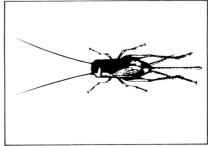

Fig. 5-3. *Nemobius fasciatus*, form *vitta-*   Fig. 5-4. *Gryllus abbreviatus* Serville.
tus, female.

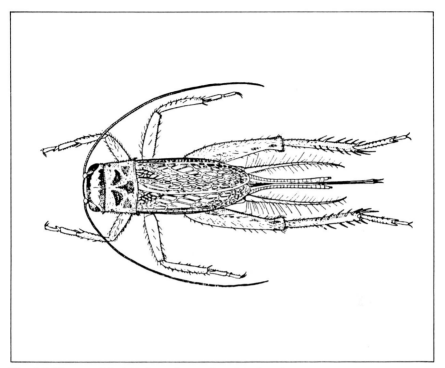

Fig. 5-5. House cricket (Acheta domesticus L.), female.

The camel crickets are wingless hunch-backed creatures with long antennae. They are sometimes called cave or cellar crickets as they prefer to live in dark moist places. Camel crickets are often found under porches or in basements and have been known to damage textiles.

The Jerusalem cricket or stone cricket is another wingless cricket with a very large head. The group occurs in the western states where it is an occasional pest of root crops. These crickets sometimes wander into homes where their large size and conspicuous coloration may cause concern. Though their bite is powerful, they do not warrant chemical controls.

The mole crickets are so named because of their striking resemblance to moles. These crickets burrow in the ground with their front legs which are broad and shaped like the front feet of a mole. The mole crickets feed on the tender roots of plants. They are not often pests throughout most of their range, but are serious pests of turf in the Southeast, especially Florida. They also fly to lights during their spring mating period.

Laufer (1927) adds the following information on this group: "Like their near relatives, (mole) crickets have biting mouth parts, and like the grasshoppers and katydids, long hind legs which render them fit for jumping. Although many of them have wings when full grown, they move about mainly by jumping or hopping. When the young cricket emerges from the egg, it strongly resembles the adult, but it lacks wings and wing-covers, which gradually appear as the insect grows older and larger. The final development of wings and wing-covers

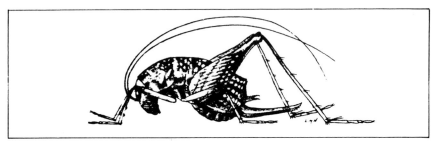

Fig. 5-6. *Ceuthophilus lapidicolus* **Burmeister, female.**

furnishes the means whereby the male cricket can produce his familiar chirping sound. It is only the adult male that sings; the young and the females cannot chirp."

The cricket's chirp song, by which it supposedly, unceasingly declares its love, is produced by the friction of the upper wings on each other. The stridulation or vibrant sound produced by the male cricket results from the scraping of the file-like under surface of one wing over the roughened veins of the other wing. The crickets supposedly hear the vibrations caused by this stridulation through the so-called "insect ears" on the fore tibiae. Thus, the Chinese and Japanese place crickets in beautiful ornate cages so they may liven the room with their cheerful chirping. The pugnacious males are placed together in cages to fight. Essig (1942) noted good fighters were worth up to $100,000 each and $90,000 was actually wagered on one champion appropriately named "Gengis Khan."

THE HOUSE CRICKET

Acheta domesticus (L.)

This cricket was introduced into Canada and the United States in the eighteenth century, and attracted the attention of the early chroniclers by its seranades and whimisical habit of chewing on clothes. The house cricket is 18 to 20 mm long, light yellowish-brown, with three dark bands on the head and long thin antennae. The female has a long slender ovipositor.

Since these crickets are fond of warmth, they are often present in the vicinity of the fireplace, kitchen and basement, where they conceal themselves in cracks and crevices, behind baseboards and may even burrow into the mortar of walls. Bakeries, because of their warmth, are frequently overrun by them. In a cold

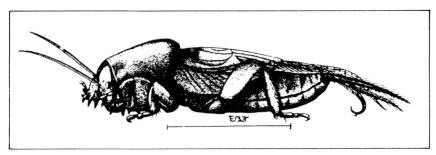

Fig. 5-7. *Gryllotalpa columbia.*

room, the crickets are torpid, but with the hearth merrily ablaze, the crickets amaze all with their saltatorial talents and musical renderings. Modern buildings provide fewer hiding places than those of more than a century ago.

Back (1936) presents this interesting description of a plague of crickets that migrated from a waste paper dump and descended on nearby homes in the city of Baltimore. "No crickets were in evidence throughout the day except as they were exposed during a careful search in the rubbish, but upon the approach of dusk hordes in all stages of growth began emerging from the debris in the dumps and swarmed upon tree trunks and over the ground between the dump and the nearest dwellings, a distance of several hundred feet. The winged forms flew readily, particularly after the city lights were turned on, and telephone and electric light poles along the streets became covered with crickets. Crickets crawled in countless numbers up the sides of the houses seemingly looking for openings through which to enter. They entered the second and third story windows, and even the skylights on the roofs. Scarcely any article of furniture or other object could be moved anywhere in the house without uncovering one to a dozen crickets. Crickets got into the food on the tables, dropped from the ceilings, were found in the beds and flew onto the residents as they sat on porches or in the rooms, and their chirping was incessant, loud and distracting. Owners sat with fly swatters or brooms in hand. At one house the housewife displayed several dustpanfuls of dead crickets swept from her floors as a result of her warfare during the previous evening. Garments of all sorts, but especially thin fabrics, were badly eaten. It is difficult to express the exasperation of occupants of houses subject to cricket invasion."

The cricket is especially destructive to silk and woolens. Caesar and Dustan (1938) record a case in Ontario, Canada where a woman living close by a badly infested dump swept up some 2,000 crickets each morning, for several days, and fought a losing battle to keep them from eating her garments. Anon. (1956) records damage to nylon, rayon and wood by the house cricket in houses in North Carolina. Janjua (1939) tells of swarms of *Acheta domesticus* (L.) in India that were 220 yards/189 m long and 100 yards/91 m wide. When these swarms reached a village, the nymphs entered the houses and attacked food, clothing and other household articles.

Habits. In warm weather this cricket lives outdoors, especially in garbage dumps. It may also be seen in these dumps during the winter, as well as in houses, sheds and other shelters. With the coming of cold weather it enters homes and, as previously mentioned, is torpid when the house is cold and active when the house is heated.

The house cricket is apparently nocturnal and usually first makes itself evident at dusk when it begins to seek food in the home. These insects are omnivorous and frequently drown in liquid nourishment due to their overanxiety. The house crickets are pugnacious little animals and they will bite when captured. They are predacious on other insects as well as on one another.

Life history. In Europe, all stages are found in the home, and it is here they deposit their eggs and undergo their nymphal growth. The eggs are supposedly deposited singly in crevices in dark places and behind baseboards.

Kemper (1937) studied them in some detail. He found they prefer raw or cooked juicy fruits and vegetables, and soft dough products of flour, cereal products, cooked meat and dead or living insects. Textiles and papers were gnawed but not infested, or only to a slight extent. The females laid 40 to 170 eggs with an average of 103.6 at room temperature. Ghouri and McFarlane (1958), who studied the life history of this cricket, found them to lay about 728 eggs at 82° F. The egg stage lasts eight to 12 weeks, the nymphal stage 30 to 33 weeks, with nine to 11 molts. Their ability to stand cold is shown by the fact they survived an exposure of 16 hours to temperatures as low as 16.7 to 23.9° F/ −8.5° to −4.4° C.

THE FIELD CRICKETS

Gryllus spp.

These crickets appear under many synonyms, the most common being *Gryllus assimilis*. F. Alexander (1963) separated *assimilis* from the North American fauna and defined six distinct species indigenous to North America. The genus is distributed widely throughout North America, Central America and the northern part of South America. The field crickets are usually black in color, larger and more robust than the house cricket and the wings project back beyond the front wings like pointed tails. Many of the species can only be separated on the basis of their song. Mickle and Walker (1974) provide a useful key to field crickets of the Southeast.

The field crickets are, at times, very injurious to field crops, especially alfalfa, wheat, oats, rye, etc. They also consume dead or weakened crickets, grasshoppers, cutworms and other insects. When these crickets invade the home they may attack textiles of cotton, linen, wool, silk, as well as furs. Clothing and paper if stained with perspiration, greasy foods, milk, syrup, etc., are liable to injury. According to Lintner (1893), they have been known to ruin a suit of clothes in one night.

Howell and Hensley (1953) recorded a tremendous outbreak of the field cricket throughout Oklahoma, Texas, Louisiana, Arkansas and Kansas. They noted that immense swarms invaded the well-lighted areas of these cities. "During warm nights the streets beneath bright lights were black with crickets, sides of buildings were completely covered with tremendous numbers of the pests and some streets were hazardous for driving due to the slipperiness caused by the crushed cricket." Articles made from nylon, wood, plastic fabrics, thin rubber goods and leather were most often damaged in houses and other buildings. Hutchins and Langston (1953) suggests these outbreaks of field crickets occur when rainfall follows a period of drought.

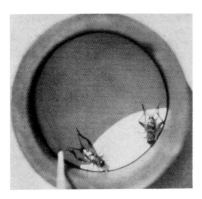

Life history. Severin (1926) notes there is one generation a year. Ninety-five percent of these insects hibernate in the egg stage, five percent as nymphs in the fifth and sixth instars. Crickets that hibernate in the egg stage become adults in July and August. They then mate and usually die in September, although some may live until the first freeze.

Two weeks after becoming adult, the female may intermittently lay eggs for two months or longer, until death ensues. The eggs are deposited in the soil at

Fig. 5-8. Two fighting crickets in the ring. The Chinese often wager large sums on the contenders.

depths of from ¼ inch/six mm to one inch/25mm. Fifty or more eggs may be laid in an area covering not more than two square inches/13 square cm. The eggs are laid singly with no protective secretion. The number of eggs laid range from 150 to 400. These usually hatch in May though temperature and rainfall affect the time of hatching.

The newly emerged cricket can walk, run and jump immediately after shedding its viteline membrane. It passes through eight to 10 instars before becoming an adult. The females usually have nine instars, the males eight. Thus, the females become adults later than the males. The crickets mature in 78 to 90 days, although the range may be from 65 to 102 days.

Mackie (1938) notes in California the conenose bug, *Rasahus thoracius* Stal., is predacious on *Gryllus,* and in the vicinity of homes, a number of cases were reported where the conenose bit man. *Calasoma* beetles and American cockroaches also have been observed feeding on dead and injured crickets.

CAMEL CRICKETS AND GREENHOUSE STONE CRICKET

Tachycines asynamorus Adelung

This greenhouse stone cricket is one of the most wide-spread of the camel crickets, being found in greenhouses in Europe and the United States. Most of the other camel crickets which are found around homes belong to the genus *Ceuthophilus*. These camel crickets may be found outside in soil, under logs or stones. Crampton (1923) found specimens crawling about open privies and on pantry shelves of a nearby cabin. He suggested they might serve as vectors of disease. These crickets often move inside dwellings during dry, hot weather and have been observed feeding on clothes and lace curtains (Hubbell, 1936 and Townsend, 1893).

CONTROL OF CRICKETS

The Jerusalem, mole and ground crickets are not normally pests of human habitations. Whereas, the house and camel crickets can both live and breed in homes, offices, food establishments or in industrial buildings such as warehouses. The field cricket is a frequent interloper in human habitations, especially in late summer and early fall and becomes more of an indoor pest problem in the mid-south and southwestern United States.

Tree crickets are not found indoors and most often confine their activities to trees and woody ornamental plants outdoors.

It is important to note that with the exception of the mole cricket these insects are not considered to be among the most serious of urban pests. The field, ground and Jerusalem crickets can be listed as serious pests of some agricultural crops, but may present more or less of a problem from season to season.

The behavior of the house cricket, *Acheta domesticus* (L.) the field crickets *Gryllus* spp. and the camel crickets *Ceuthophilus* spp. is an important key to acquiring and maintaining control of these insects.

The crickets as a group are generally nocturnal, display a preference for space-limited shelter such as cracks and crevices and most often invade homes seeking moisture. Porch and street lights or strongly lighted windows readily attract the field, house and occasional ground crickets, but lights are not attractive to the camel cricket which prefers a damp, darkened basement to a roost near the living room fireplace.

The life cycle of these insects are usually limited to one generation per year. Mole and Jerusalem crickets often overwinter in the soil as adults. Field and ground crickets overwinter in the egg stage. House and camel crickets can overwinter indoors as adults, but often die in late fall or early winter having by then lived out their annual life cycles.

The first perimeter of defense in a cricket control strategy should consist of limiting harborage. Well mowed lawns and regular weeding of ornamental shrubbery surrounding a house are good approaches to limiting the potential for house, camel, field or ground cricket invasions.

The next perimeter of defense is to utilize outdoor lighting such as yard lights and porch or patio lights as little as possible. However, when lighting is required use the "insect repellant" yellow lights which are less attractive to insects in the place of white, neon or mercury vapor lamps.

Ground level access to structures should be inspected to determine whether there are repairable openings that might admit insects pests. Tight-fitting doors, clothes dryer vents, basement window frames or the caulking of holes around plumbing where it enters the structure can help prevent cricket entry. Garbage storage areas should be inspected and cleaned up if required. Garbage cans should never sit directly on the ground. Raise them above ground level using a pallet or stand them on bricks. Do not allow debris to collect beneath these containers.

If firewood is stacked against the house, remove it for a distance of at least one foot/0.3 m from the foundation or outside wall. If the wood is not being rapidly used, especially during summer months, turn the pile over every three to four weeks or apply a six-inch/15 cm band of ashes around the base of the wood pile adding a fresh covering of ashes when the previous layer becomes damp and compacted.

Indoors, crickets such as the house and occasional field and ground crickets seek harborages much like their cousins the cockroaches. These crickets are most often in search of water, but when their thirst is quenched they will then seek out food and harborage. With this behavior in mind the beginning of control for these pests indoors starts with repairing leaking pipes or other damp indoor conditions, then reducing harborage sites and practicing sound sanitation principles.

The caulking gun, patching plaster, spatula, a hammer to tighten loose moldings, cabinet frames or the vacuum cleaner may be the most useful tools in a pest control arsenal.

When chemical control is required, several formulations of pesticides should be considered, making sure to fit the formulations to the problem. Maximum consideration should be given to selecting the chemical that will do the job and provide the safest possible application. Several pesticides for the control of crickets are available and are currently registered with the United States Environmental Protection Agency as emulsifiable concentrates, wettable powders, dusts, granules, aerosol sprays, "lacquer" or "paint-on" and bait formulations.

It is very important to *carefully* read the labels of pesticides registered for control of crickets. Most "indoor use" pesticides will simply describe the control of crickets by using a phrase such as ". . . and for the control of crickets." The labels usually will not specify whether the rate of application and the site to be treated is for control of camel, house, field, ground crickets, etc. Labels that do specify field, Jerusalem, ground or mole crickets are usually restricting the pesticide's use to *outdoor application only*, e.g., turf, around trash piles, dump sites or in field crop uses. *Always take special precaution* so that you *do not* use an outdoor rate of pesticide application to treat an indoor infestation of crickets.

Table 5-1 lists the formulations and generic chemical names of many of these pesticides commonly known to be effective agents for the control of crickets.

Selecting the appropriate formulation to fit the conditions where applications are to be made is the keystone to successful chemical control of crickets. Dusts should not be used in areas where moisture is likely to occur and cannot be eliminated. This precaution includes dusts based on silica aerogel or diatomaceous earth. Granules on the other hand are very effective under moist conditions and where they are used in turfgrass to control mole crickets may require "watering in" in order to activate the release of the toxicant from the clay base.

Since crickets can often be found in food storage, linen cabinets or even on book shelves, a crack and crevice injection of an emulsifiable concentrate or wettable powder or the use of an injectable aerosol or a few bait granules may be preferred instead of dusts.

TABLE 5-1

Commonly used pesticides currently registered for indoor and/or outdoor use in the control of crickets.

	PESTICIDE FORMULATIONS					
Pesticides	Emulsifiable Concentrate	Wettable Powder	Dust	Granular	Bait	"Paint On"
Bendiocarb		X	X			
Boric Acid			[1]			
Carbaryl		X	X	X	X	
Chlorpyrifos	X			X	X	X
Diatect®			X			
Diazinon	X	X	X	X		
Diazinon (encapsulated)	X					
Dichlorvos	[2]					
Dri Die®			X			
Drione®			X			
Fenthion	X					
Malathion	X	X	X			
Propoxur	X	X	X		X	
Pyrethrin	[3]					
Resmethrin	X					
Ronnel	X					
Sodium Fluoride			X[4]			

[1] Not currently registered for this use but known to be effective.
[2] Used alone but more often in combination with other emulsifiable concentrates or in aerosol formulations.
[3] Pyrethrins are common ingredients in aerosol sprays.
[4] Caution, very toxic to plants.

It cannot be emphasized too strongly that all pesticide labels should be read carefully, especially for control of crickets. Although ground crickets do not often invade homes, the outdoor application rate of diazinon, for example, is far too high to safely use indoors when this pest suddenly appears in a broom closet.

Important Elements of Cricket Control.

- Do a thorough inspection.
- Be sure to identify the cricket. This is an important key to developing your pest control strategy.
- Note the likely points of entry, especially where the crickets involved are basically outdoor pests (e.g. field or ground crickets). Suggest or carry out necessary repairs to restrict or eliminate entry of crickets to the structure.
- Crickets such as field, house and ground species are attracted to lights. Be sure lights are not used unnecessarily at night. If night lighting must be used, utilize yellow rather than white, neon or mercury vapor lamps.
- In basements where camel crickets are a problem, correction of excessive moisture conditions will often effect complete control without the use of a pesticide.
- If a pesticide is used, choose the insecticide formulation most appropriate considering:
  a) Surface upon which application will be made.
  b) Exposure of the surface to children and pets.
  c) Length of control desired.
  d) Type of application required (crack and crevice injection vs. spot application).
  e) The potential for staining or otherwise damaging and/or contaminating household effects.
- Remedial applications of pesticides may be required indoors. However, sound cultural pest management outdoors can often result in 100 percent control of house-invading crickets. For example, frequent mowing of lawns, cutting of weeds, cleaning up of garbage collection areas and firewood piles greatly limits the potential attraction and routes for cricket invasions.
- Finally, be sure to make notes on the relative success of your treatment for future reference.

LITERATURE

ALEXANDER, R. D. — 1957. The taxonomy of the field crickets of the eastern United States (Orthoptera: Gryllidae, *Acheta*). Ann. Entomol. Soc. Amer. 50(6):584-602.

ANON. — 1956. House cricket. Coop. Econ. Insect Rpt. 6(16):335.

1963. Crickets. Coop. Econ. Insect Rpt. - Sept. 27.

BACK, E. A. — 1936. *Gryllus domesticus* L. and city dumps. J. Econ. Entomol. 29:198-202.

CRAMPTON, G. C. — 1923. The cave cricket, *Ceuthophilus*, as a possible vector of pathogenic organisms. J. Econ. Entomol. 16:460.

CAESAR, L. & G. G. DUSTON — 1938. Control of the house cricket. Ann. rept. Entomol. Soc. Ontario. 69:101-105.

ESSIG, E. O. — 1942. College Entomology. Macmillan Co.

GHOURI, A. S. K. & J. E. MCFARLANE — 1958. Observations on the development of crickets. Canada. Entomol. 90:158-165.

HOWELL, D. E. & S. D. HENSLEY — 1955. Field cricket control in buildings. Proc. Okla. Acad. Sci. 34:105-107.

HUBBELL, T. H. — 1936. A monographic revision of the genus *Ceuthophilus*. Univ. Florida Publ. Biol. Sci. Ser., 2(1):551 pp.

HUTCHINS, R. T. & J. M. LANGSTON — 1953. An unusual occurence of the field cricket. J. Econ. Entomol. 46(1):169.

JANJUA, N. A. — 1939. A preliminary note on the bionomics and control of the black headed cricket (*Gryllus domesticus* L). in Usta Colony (Sidi District) of Baluchistan. Rev. Appl. Entomol. A. 28:436, 1940.

KEMPER, H. — 1937. Observations on the biology of the house cricket. Rev. Appl. Entomol. (A) 26:37, 1938.

LAUFER, B. — 1927. Insect musicians and cricket champions of China. Field Mus. Nat. Hist. Anthrop. Leaflet 22:1-26.

LINTNER, J. A. — 1893. The common black cricket. Eighth Rept. N. Y. Insects.

MACKIE, D. B. — 1938. Entomological Service Bull. Dept. Agr. Calif. 27:645-668.

MARLATT, C. L. — 1896. The house cricket. U.S.D.A. Div. Entomol. Bull: 4.

MICKLE, D. A. & T. J. WALKER — 1974. A morphological key to field crickets of Southeastern United States (Orthoptera:Gryllidae:*Gryllus*). Florida Entomologist 57(1):8-12.

SEVERIN, H. C. — 1926. The common black field cricket, *Gryllus assimilis* (Fab.) and its control. J. Econ. Entomol. 19:218-227.

SHORT, D. E. & P. G. KOEHLER — 1977. Control of mole crickets in turf. Fla. Entomol. 62(2):147-148.

TOWNSEND, C. H. T. — 1893. Note on *Ceuthophilus* eating curtains and other fabrics. Insect Life. 6:58.

DAVID J. SHETLAR

Dr. David Shetlar gained his bachelor's and master's degrees in zoology from the University of Oklahoma and his Ph.D. from The Pennsylvania State University. After various teaching assignments at the University of Oklahoma, Shetlar began his research and teaching career at The Pennsylvania State University in 1971.

Shetlar's research activities have included pest management studies for ornamental plants, turf and greenhouses. His teaching responsibilities cover the whole field of entomology, including medical entomology, economic entomology and use of pesticides.

In addition to his research and teaching experience, Shetlar is an accomplished illustrator of insects and has taught scientific illustration.

Shetlar has authored or co-authored numerous articles in scientific publications and has lectured widely at seminars of professional associations and civic groups.

Shetlar has held office in the Phi Sigma Biological Research Society (University of Oklahoma Chapter) and is an associate member of Sigma Xi, the Scientific Research Society of North America. Shetlar is active in several professional societies and is a member of the Pennsylvania Academy of Sciences, the Entomological Society of Pennsylvania and the Entomological Society of America.

CHAPTER SIX

Earwigs

Revised by David J. Shetlar[1]

ACCORDING TO FULTON (1924), the word earwig is derived from the Anglo-Saxon *earwicga*, which literally means "ear creature." The term undoubtedly originated from the widespread superstition that earwigs purposely crawl into the ears of sleeping persons. More gullible individuals actually believed that once the insect gained access to the human ear, it could bore into the brain.

Earwigs are like crickets and roaches in that they are characterized by biting mouthparts and a simple metamorphosis. The characteristic which distinguishes them most readily is the forceps (cerci) on the end of the abdomen. In the female, the forceps are fairly straight-sided, whereas in the male, they are often caliper-like. The forcep-like cerci are both defensive and offensive weapons, and are used by the earwig to capture prey.

Earwigs are active at night and hide during the day in cracks or crevices. They are mainly scavengers, but occasionally feed on plants. The eggs are laid in burrows in the ground and are usually guarded by the female. Most species overwinter in the adult form. Some species have repugnatory glands from which they can squirt a foul-smelling, yellowish-brown liquid.

There are about 15 species of earwigs in the United States, but only four or five of these are common pests. Plate 6-1 illustrates some of the common earwigs that invade homes.

THE EUROPEAN EARWIG

Forficula auricularia L.

The adults (Fig. 6-1) are 5/8 inch/16 mm long and dark reddish brown in color. They are also characterized by a reddish head and pale, yellowish-brown legs. Some adult males may grow considerably larger. The males also have two forms of forceps. Some have forceps approximately 3/16 inch/5 mm long, while others have forceps 3/8 inch/9.5 mm long.

[1]Assistant Professor, The Pennsylvania State University, University Park, Pa.

EARWIGS: PICTORIAL KEY TO COMMON DOMESTIC SPECIES

Chester J. Stojanovich and Harold George Scott

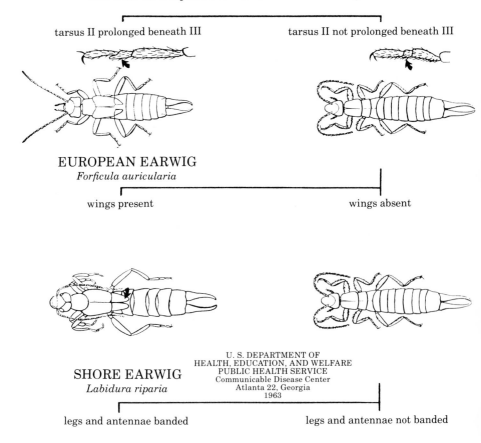

tarsus II prolonged beneath III   tarsus II not prolonged beneath III

EUROPEAN EARWIG
Forficula auricularia

wings present   wings absent

SHORE EARWIG
Labidura riparia

U. S. DEPARTMENT OF
HEALTH, EDUCATION, AND WELFARE
PUBLIC HEALTH SERVICE
Communicable Disease Center
Atlanta 22, Georgia
1963

legs and antennae banded   legs and antennae not banded

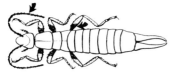

RING-LEGGED EARWIG
Euborellia annulipes

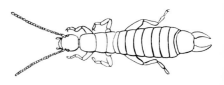

SEASIDE EARWIG
Anisolabis maritima

Fig. 6-1. Pictorial key to common earwigs found in the United States.

This cosmopolitan insect is believed to have been first observed in the United States in 1907 in Seattle, Washington. It is now found throughout much of the United States and parts of Canada.

The European earwig is an onmivorous feeder, very fond of plant food, and for a time it was feared it would become a prominent plant pest. However, because earwigs feed on a variety of plants, the injury they cause is widely distributed. As a result, they are usually considered minor pests of plants. Because it often invades the home in large numbers, the European earwig has also become a notorious household pest in some areas.

The European earwig is disseminated largely through man's activities. In this respect, Crumb et al. (1941) note: "The earwig rarely flies and is not inclined to travel very extensively by crawling, but is admirably adapted for transportation by man. Wandering at night, it crawls into any available hiding place at the approach of day and thus may be carried long distances in bundles of newspaper, the luggage of travelers, cut flowers, packages and crates of merchandise, lumber and shingles, automobiles, and even rarely in letters. Ships often are infested, and their cargoes are likely to carry earwigs. The female can deposit fertile eggs several months after mating, and the insect is able to survive under a variety of environmental conditions."

This insect readily avails itself of any dark, moist crevice, such as those found in balled plants, heaps of manure, boards, and similar locations. Morgan (1926) notes that the adults will float in water for 24 hours, and then may resume immediate activity on reaching a dry surface. It is in this way they are distributed on debris floating in streams and rivers.

Fulton (1924) gives the following graphic description of these insects as noxious pests in the home: "At night they swarm over porches in such numbers that many people prefer to remain inside on a summer evening, rather than spend it with such unwelcome guests.

"In the morning earwigs are found by the handful under rugs and cushions which have been left outside. They crawl into basements and hide in the laundry which is waiting to be ironed. In the bedrooms they sometimes find their way into the clothing hanging in closets. A man reported that he pulled on a sock one morning which contained three earwigs. A correspondent in a letter appealing for information describes the situation as follows: 'Literally thousands and thousands of these bugs inhabited my premises last summer. They made it almost impossible to live in my home; they inhabited the sleeping porch till we had to leave it.' They work mostly at night, but in the daytime might be found in kitchen drawers and often burrow an inch into a loaf of bread. They crawl over the ceiling and drop on the bed, or inhabit themselves in a person's clothing during the night, and while their bite has never proved serious it is entirely uncomfortable. No part of the house seems to be entirely free from earwigs, not even the roof, and it is almost impossible to keep them out by the use of screens. The pests have become so annoying in some districts that property values have depreciated considerably."

These insects also have a foul odor. Apparently, the only favorable thing we can say about them is that they attack and devour other insect pests.

As mentioned previously, the European earwig is omnivorous, and eats practically anything that is not too hard for its mandibles. Earwigs normally feed on green plants. Dimick and Mote (1934), upon examining the insects digestive tracts, found the contents to be largely lichens and pollen. When these insects feed on plants, they make small irregular holes in the leaves, and may even

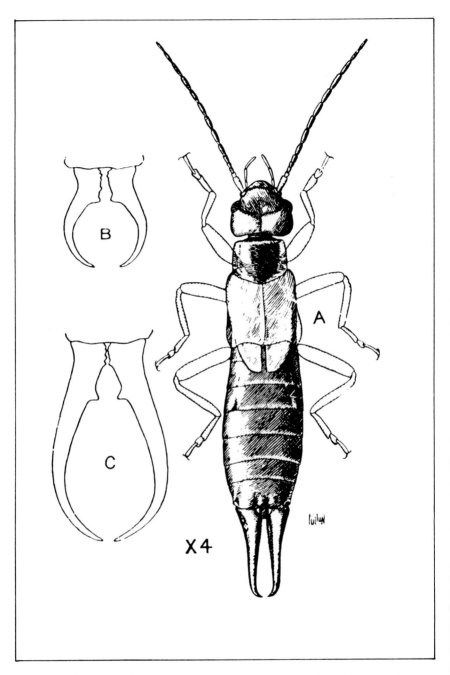

Fig. 6-2. A. Adult female earwig. B. and C. Forceps of small and large males.

Fig. 6-3. Female earwig in winter quarters with eggs.

skeletonize them. If it were not for an absence of slime, their work could be mistaken for that of slugs.

Life history and habits. Fulton (1924b) and Crumb et al. (1941) studied the life history of *Forficula auricularia* L., in Oregon in detail. They noted that the adults are more or less active during the colder months of the year, and overwinter beneath boards, stones, and similar materials, singly, and in little cells. The overwintering eggs are deposited in the fall or on warm winter days. In the early spring, the males may leave their hibernating quarters and congregate in clusters in dark crevices above the ground.

Crumb et al., (1941) found that each female laid an average of about 30 eggs for the first lot and considerably less for the second batch. "The duration of the egg and nymphal stages is influenced strongly by the temperature. The winter eggs require about 72.8 days from deposition to hatching, whereas the spring eggs require only about 20 days." The nymphal stages require about 68 days or more for their completion in nature.

The European earwig has four nymphal instars. The first instar only ventures a short distance from the nest. The second instar is much more venturesome. It forages for food. Should it fail to return to its own nest, and seek the shelter of another nest, it appears to be entirely welcome. The third instar leaves home, and from then on is on its own. The majority of earwigs which hatch from overwintering eggs mature in the latter part of June. Some of the females may lay a second and smaller batch of eggs from April to June. Although there may be two broods, there is but one generation a year in the United States.

The female European earwig has earned some measure of fame because of her maternal care for the young. This trait in the earwig can be observed readily by placing a pregnant female earwig in a Petri dish partially filled with moist sand. The female earwig gathers her eggs together in a pile, and stores them in a shallow hole, over which she stands guard. The mother frequently moves the eggs from one part of the cell to the other, rolling and cleaning them in her mouth. Hatching usually begins in April, but varies according to external conditions.

Fulton offers the following description of the mother earwig and her young-sters: "For a few days after the young earwigs appear the female keeps the nest tightly closed to prevent their escape. A nest of young earwigs is a most inter-esting sight; they seem to fill the cavity, a squirming, writhing mass. The old earwig standing in their midst with the nymphs crawling under, over and about her, reminds one of an old hen with a brood of chicks.

"After a few days the nest is opened to the outside, but the female guards it against all intruders by using her forceps as a weapon. Occasionally, she brings in some food for the young earwigs, who remain in or close to the nest until they have moulted."

THE RING-LEGGED EARWIG

Euborellia annulipes (Lucas)

The ring-legged earwig occurs rather generally throughout the United States and is most common in the South and Southwest. It is reported also in British Columbia. This insect has habits similar to that of the European earwig, feeding on both plants and insect life. Klostermeyer (1942) observed that the ring-legged earwig feeds on grain and grain insects in packing plants in various parts of the country. Gould (1948) states that it is one of the most common in-sects in corn processing plants in Indiana, being particularly fond of starch. For this reason, it is known locally as a "starch bug."

The adults may reach a length of approximately one inch/25 mm. They vary from brown to black in color, with a yellowish brown undersurface. The femur and tibia of the yellowish legs are ringed with brown stripes. "The antennae are 16 jointed and black in color except for the third and usually the fourth segment from the apex, which are white," according to Gould (1948).

This insect is nocturnal, avoids light, and is an omnivorous feeder. Outdoors it lives under litter, rocks, and similar places where it excavates shallow nests.

Life history and habits. Klostermeyer (1942) studied the life history of the ring-legged earwig at temperatures ranging from 65° to 85° F/18° to 29° C, and at high humidities. He found that the average life cycle took 10.8 days for pre-oviposition; 14.3 days for the egg to hatch; 10.6 days for the first instar; 13.5 days for the second instar; 16.2 days for the third instar; 20 days for the fourth instar; and 29.5 days for the fifth instar. Thus, the average number of days re-quired for completion of the nymphal stage was 90 days, and from egg to adult 104 days. Some of the adults remained alive in the laboratory for more than seven months. Bharadwaj (1966) also made a detailed study of the life cycle of this insect. He pointed out that the nymphs start with eight segments in the antennae and added segments at each molt. Though the females may produce up to six clutches of eggs, most produce one to four.

This earwig is considered a nuisance pest because it invades houses, espe-cially in Gulf Coast states, and attacks stored foods.

The biology of a closely related earwig, *Euborellia cincticollis* (Gerstaecker) has been described by Knabke and Grigarick (1971) based on its occurrence in California.

SHORE EARWIG

Labidura riparia (Pallas)

The shore or riparian earwig is also commonly called the striped earwig. The adults are usually larger than the European earwig, measuring about one inch/ 25 mm long. Their color varies from pale brown to chestnut or reddish brown, with black markings. The abdomen is usually banded.

The shore earwig can be found in tropical and subtropical areas worldwide. The species was first reported in Texas (Hebard, 1942) and is now found from California through Arizona and the Gulf Coast states.

Life history and habits. The shore earwig generally lives in burrows beneath debris and thatches. The females make a brood chamber like the European earwig, where they tend and wash the eggs which hatch in about a week. The nymphs remain in the chamber and are groomed until after their first molt, which occurs about a week later. The female then opens the burrow and soon thereafter considers her own offspring as fair game for food. Clements and Kerr (1969) found that the nymphs take 49 to 60 days to mature.

This earwig is predatory and a scavenger. It has not been known to damage plants, but commonly seeks shelter in buildings. This species is also attracted to lights at night and has an odor if disturbed. Bishop (1962) notes it has caused minor skin abrasions in humans.

CONTROL OF EARWIGS

In the past, earwigs have been effectively controlled through the use of poisoned fish-bran baits such as the one described by Crumb et al. (1941):

Wheat bran .... 12 pounds
Sodium fluosilicate .... 1 pound
Fish oil .... 1 quart

They suggested the bran and toxicant be thoroughly mixed, and then the fish oil added until evenly distributed. No water was used in the bait. This quantity of bait was sufficient to cover 5,000 to 8,000 square feet.

Prior to the severe restrictions on their use various chlorinated hydrocarbon insecticides were highly effective against earwigs. Thus, Legner and David (1963) recommended the use of granule formulations of aldrin, chlordane and dieldrin. Lyons (1952) recommended dust and sprayable formulations of DDT and chlordane for earwig control.

Today, a variety of insecticides are labeled for control of earwigs. However, the first step in control is to remove unessential plant debris, mulch, and boards from around buildings. The object is to establish a zone of bare concrete or soil, exposed to sunlight, which will dry out and be disagreeable to earwigs. Such sanitation measures can then be supplemented by applications of insecticides in a three- to six-foot band around the building adjacent to the foundation.

Perimeter sprays of bendiocarb, chlorpyrifos, diazinon and propetamphos applied according to label directions have provided good residual control of earwigs. When spraying turf or shrubs in the band area, care should be taken to avoid phytotoxicity.

Good residual control has also been obtained using exterior applications of two percent propoxur bait or granular formulations of carbaryl, chlorpyrifos and diazinon. For crawl spaces, dust formulations of bendiocarb, carbaryl, diazinon and silica gel have also been useful.

Whether using sprays, baits or granules, special attention should be paid to the areas most frequented by earwigs. In addition to around building foundations, this includes areas along fences and walks, around trees and utility poles and around wood piles and rocks.

For severe infestations resulting in heavy invasions it may be necessary to treat a much wider outdoor area. This can be achieved using products labeled for whole yard treatment such as carbaryl, chlorpyrifos, diazinon and malathion. Such measures may be particularly appropriate in the fall when more earwigs may seek shelter indoors.

In general, indoor treatments are only supplemental to outdoor control measures. Many of the residual insecticides labeled for outdoor use also have inside recommendations. In addition, for quick kill of earwigs indoors, various products based on pyrethrins, resmethrin and DDVP have appropriate labeling.

It is also worth noting that a parasitic fly of the tachinid family, *Digonochaeta setipennis* Fall., from Europe, has been of some value in reducing earwig populations.

LITERATURE

ANONYMOUS — 1949. The European earwig *(Forficula auricularia)*. Agric. Gaz. N.S.W. 60 (4): 200-202. Rev. Appl. Ent. A. 39(2): 36, 1951.

BHARADWAJ, R. K. — 1966. Observations on the bionomics of *Eurborellia annulipes*. Ann. Entomol. Soc. Amer. 59(3): 441-450.

BISHOP, F. C. — 1962. Injury to men by earwigs (Dermaptera). Proc. Entomol. Soc. Wash. 63(2): 114.

CLEMENTS, R. H. and S. H. KERR — 1969. Earwigs. Pest Control. 37(10): 26, 28, 30.

CRUMB, S. E., P. M. EIDE and A. E. BONN — 1941. The European earwig. U.S.D.A. Tech. Bull. 766: 76 pp.

DIMICK, R. E. and D. C. MOTE — 1934. The present status of the European earwig. Calif. Dept. Agr. Monthly Bull. 23: 298-300.

EBELING, W. and D. A. REIERSON — 1974. Earwigs of California. National Pest Control Operator News. 34(3): 24-27.

FULTON, B. B. — 1924a. Some habits of earwigs. Annals Entomol. Soc. Amer. 17: 357-367.

1924b. The European earwig. Oreg. Agr. Exp. Sta. Bull. 207.

GOULD, G. E. 1948. Insect problems in corn processing plants. Jour. Econ. Entomol. 41(5): 774-778.

HEBARD, M. — 1942. The Dermaptera and orthopterous families Blattidae, Mantidae, and Phasmidae of Texas. Trans. Amer. Entomol. Soc. 68: 239-319.

KLOSTERMEYER, E. C. — 1942. The life history and habits of the ring-legged earwig, *Euborellia annulipes* (Lucas) (Order Dermaptera). Jour. Kans. Entomol. Soc. 15(1): 13-18.

LANGSTON, R. L. and J. A. POWELL — 1975. The earwigs of California. Bull. of California Insect Survey.

LARSON, A. W. — 1948. The European earwig. Pest Control & Sanitation 3(5): 16-17.

1951. The European earwig and its control. U.S.D.A. EC-25.

LEGNER, E. F. and D. W. DAVIS — 1963. Some effects of aldrin, chlordane, dieldrin, and heptachlor on the European earwig, *Forficula auricularia*. Jour. Econ. Entomol. 56(1): 29-31.

LYONS, C. — 1952. The European earwig and its control. U.S.D.A.B.E. & P.Q. EC-25, 5 pp.

MORGAN, W. P. — 1926. A note on the mode of distribution of earwigs. Proc. Ind. Acad. Sci. 26: 311-333.

C. DOUGLASS MAMPE

Dr. C. Douglass Mampe received degrees in entomology from Iowa State University, North Dakota State University, and a Ph.D. from North Carolina State University. Upon completing his university education, he joined the staff of the National Pest Control Association (NPCA). During his 10 years there he was responsible for keeping abreast of all technical and regulatory actions related to the structural pest control industry. This included reviewing research proposals and overseeing research projects related to the biology and control of termites, cockroaches, commensal rodents, and other related pests.

Mampe left NPCA as its technical director and joined Western Termite and Pest Control, Inc., and its sister company Residex, a pest control industry supply house. He served as Western's technical director for six years and developed practical procedures for field operations. He eventually became general manager of Residex, which included developing new pesticide registrations and industry training programs, including certification training programs for a number of the northeastern and middle Atlantic states.

In 1980, Mampe started his own consulting firm for the urban and structural pest control industry. His firm provides technical consultation, training and training programs, evaluates new pesticides and equipment, custom develops personnel management programs and provides expert testimony for legal cases.

Mampe served as editor of _NPCA's Approved Reference Procedures for Control of Subterranean Termites_, and editor and co-author of the _Manual for Structural Wood Decay_. He writes a monthly column called "Answers" for _Pest Control_ magazine. This column, which deals with a variety of questions, has consistently been one of the best read sections of the magazine since the column began at the beginning of 1975.

Mampe is a member of the NPCA, Pi Chi Omega (a fraternity of industry professionals involved in training), the Entomological Society of America, and a number of state pest control associations. He is currently licensed as a certified pesticide applicator in a number of northeastern states. In 1979, he was chosen "Pest Control Operator of the Year" by the New Jersey Pest Control Association for his contributions to the industry in that state. He is currently serving as a resource person for the United States House of Representatives Committee on Agriculture.

CHAPTER SEVEN

Termites

Revised by C. Douglass Mampe[1]

Some primal termite knocked on wood
And tasted it and found it good
And that is why your Cousin May
Fell through the parlor floor today
                    —Ogden Nash

THE ROMANS referred to these insects as "Termes" which aptly means "woodworm." Their fossilized remains have been found in formations some 55 million years old, and day in and day out their diet has consisted, monotonously enough, of wood and other cellulose materials. Thus, the impression entertained by some individuals that termites, like some ravenous horde from Mars, have suddenly descended upon the scene with but one avowed purpose, namely to consume and digest their homes, is evidently doing these insects an injustice. The 45 to 50 species of termites in the United States are mostly native American stock. Man very obligingly concentrates finished timber products in wooden structures and makes them readily available to termites.

Isherwood (1956) reproduced an essay by Dr. T. W. Harris which showed termites were pests of homes in New England prior to 1849 and a scourge of valuable documents in New Orleans at that time. Urquhart (1953) believes the eastern subterranean termite, *Reticulitermes flavipes* (Kollar) was introduced into Toronto, Canada, from the United States between 1935 and 1938. The increase in the termite problem of late partially may be attributed to the modern system of central heating which provides the warmth of summer throughout the year and enables the termites to feed without pause. Kofoid, et al. (1934), Snyder (1935, 1948 and 1954b), Harris (1961) and Hegh (1922) have written interesting and informative books on termites, the latter in the French language. Weesner (1965) prepared a very useful publication on the termites of the United States. The most recent and comprehensive work on termite biology was prepared by Krishna & Weesner (1969).

Why are termites so destructive? Light et al. (1930) answer this question by stating the individuals are long lived, the colony is self-perpetuating, the insects have a constantly available supply of food, they are social insects and are

---

[1]*President, D.M. Associates, Westfield, N.J.*

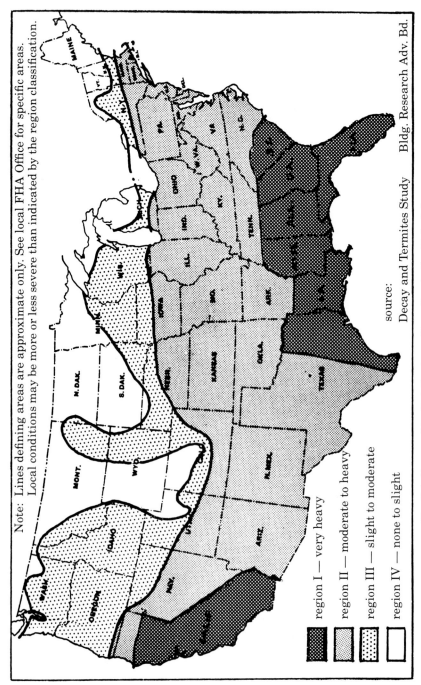

Note: Lines defining areas are approximate only. See local FHA Office for specific areas. Local conditions may be more or less severe than indicated by the region classification.

source:
Decay and Termites Study   Bldg. Research Adv. Bd.

region I — very heavy

region II — moderate to heavy

region III — slight to moderate

region IV — none to slight

Fig. 7-1. Geographic distribution of termites in the United States.

protected from enemies and extremes of heat and cold, storms, etc., all of which is provided for by their "cryptobiotic or hidden mode of life." This results in a constant increase in numbers with a consequent increase in destructive powers.

Termites were not created for the sole purpose of giving a livelihood to pest control operators. The termites in our forests are undoubtedly beneficial, for they, with the aid of other organisms, reduce fallen and decaying wood into the organic material whence they originally arose. Moreover, through aeons of time, the earth and wood through which they burrow, and which pass through their bodies, help enrich the soil. Nor should it be forgotten there are some inhabitants of the globe who have a material fondness for these "woodworms." Thus, the palates of many aborigines are susceptible to living termites which supposedly have a flavor very similar to Brazil nuts. Incidentally, Reynolds (1963) reports a very unusual case where the intestinal tract of a human being was infested with termites.

What is a termite? Those unhappy people who, through no desire on their part, make the acquaintance of the termite, often plaintively ask, "Just what is a termite?" A termite is an insect with six legs and a body divided into three main constrictions, namely head, thorax and abdomen. On the head is situated a pair of antennae made up of many bead-like segments. Also, on the head are mandibles that bite the wood, as well as black compound eyes which are usually present. The thorax or chest consists of three segments, to each of which are attached a pair of legs. And in those termites that are winged, a pair of wings is attached to each of the last two segments of the thorax.

When an infested piece of wood is broken open, one observes what appears to be a white maggot with legs. This is the blind or almost blind and ever faithful "worker" termite. Since the outer covering of this worker is thinner than in most insects, it is less effective in conserving body moisture, and for this reason the worker termite must dwell in the dark, away from the desiccating rays of the sun.

Termites are sensitive to stimulation by touch, odor, taste and pressure. And this sensitivity in turn is due to the tactile sensory hairs and specialized sense organs which are on various parts of their bodies. Turner regards their behavior as bordering on intelligence and has noted they have in a number of well-built homes found the only possible point of entry. This is the result of their faculty of constantly constructing exploratory tubes. Kofoid believes the individual termite can benefit by experience and it may behave in a manner closely akin to intelligence. There are those who retain a similar opinion in regard to ants.

Termites and roaches. Termites are among our most ancient insects and are, in fact, related to such primitive insects as roaches. Thus, Snyder (1935) states the most primitive living termite, *Masotermes darwiniensis* Frogg., "has wing structure similar to roaches as well as an egg mass similar to the egg capsule of roaches indicating at least a common ancestry." Moreover, there is a large, primitive wingless brown roach, *Cryptocercus punctulatus* Scudder, that burrows into the fairly sound wood of decayed logs which serves the roach for both shelter and food. Interestingly enough, this roach has Protozoa in its digestive tract and termites inhabiting the same log may have obtained their Protozoa from the roaches.

Termites as social insects. Termites live in colonies, are social insects and divide their work among specialized members. The members of these colonies move in passageways that are hidden from the sun's rays and are protected from undue moisture loss. According to Kofoid (1934), some marked instincts accom-

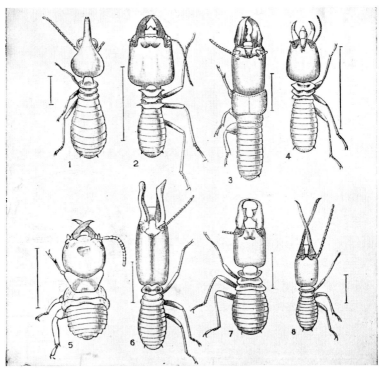

Fig. 7-2. Head forms of different species of termites.

pany their social way of life. Thus, they must recognize members of the colony and act accordingly. Their habit of grooming each other not only cleanses them, but also results in the consumption of secretions from glandular cells situated in their body covering. They cooperate in the construction and in the cultivation of fungus gardens. Finally, Kofoid makes the following interesting observation: "Another instinct which is basic in the social habit of these insects is their ceaseless industry. They have never been observed to rest for any considerable length of time except when chilled or when molting and under normal conditions of moisture and temperature the workers or nymphs toil ceaselessly in the various industries of the colony."

Each colony is a separate cosmos, distinct from any other colony of termites. In fact, when two colonies meet, the members of each slaughter one another. The colonies, as was previously mentioned, are hidden or cryptobiotic and this enables them to conserve the necessary water supply, and also protects them from animal enemies.

The cryptobiotic habit, or the habit of living in darkness, is of fundamental importance, and has resulted in such noteworthy characteristics in termites as poor or absent eyes, lack of wings in some adults and a thin body covering. According to Light et al. (1930), "Termites live in darkness, in narrow passageways, where the temperature, the moisture and probably the oxygen pressure are to some extent under their control." Their elaborate systems of tubes and runways are constructed either to obtain food or as outlets for the winged sexual forms. In order to protect themselves from desiccation they consume the wood until only a thin outer shell is left, and this may not be evident from the exterior.

REACTION OF TERMITES TO THEIR ENVIRONMENT

Temperature. In those parts of the world where the winter is mild, termites are active throughout the year. In heated buildings, in the colder areas of the world, the house-infesting termites also are active throughout the year. Colonies situated in the field become inactive during cold weather. To a limited extent termites are capable of regulating temperature conditions to suit themselves. Thus, their burrowings often are situated so some run above and some run below the ground. Therefore, during extremes of cold or hot weather, the termites will be found in the burrows situated some distance below the surface of the ground where temperature conditions are more equable.

Moisture. The amount of moisture required by different species of termites is extremely variable. Subterranean termites need a constant supply of moisture. Part of this moisture is procured from the products of their own metabolism and part from the soil moisture which diffuses through their tunnels or tubes. The rate of diffusion of moisture through ground connections of subterranean termites partially determines the height of such tubes. Fungi, when present in the wood, serve as another source of moisture. Moreover, being consumed directly, these fungi aid in the regulation of humidity in the passages. The plugs of partially masticated wood, feces, etc., placed by the termites in the passageways also assist in the regulation of their moisture content. Brown (1936), speaking of termites, states, "It is possible, however, for them to colonize

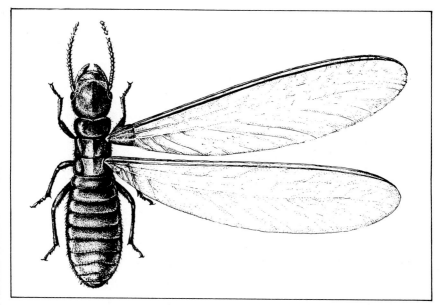

Fig. 7-3. **Winged form of *Reticulitermes tibialis* Banks; wings shown on one side only.**

without ground connections if they have other means of obtaining moisture. For example, an ocean liner at the dock in San Francisco was found to be infested by an oriental subterranean termite. In this instance they had colonized in the woodwork of a cabin and were obtaining their moisture supply from a leaky coffee urn." He also notes dry-wood termites are able to colonize in dry-seasoned wood. This wood absorbs some moisture from the air. The damp-wood termites are extremely tolerant to moisture and have been found in water-logged wood, staves of water tanks and in harbor structures near and over the sea, as well as in water tanks on high buildings.

Ratner (1963) notes house timbers on the second floor wetted by rain are often infested by subterranean termites. The termites build tubes up to these timbers.

Soil conditions. The subterranean termite, since it originally obtains its moisture from the soil, must of necessity be greatly dependent on soil types. The moisture in clay soils is tightly bound to the particles and so not readily available to the termite, whereas sandy soils, although having a much lower percentage of total moisture, nevertheless have more free capillary moisture. Thus, the moisture in sand is more available to the termite than is the moisture in clay. For a more detailed discussion of moisture and soil, see the section on termite control and Lee & Wood (1971).

Termites and sound. Termites can detect vibrations through their legs. They are unable to hear a noise near the nest, but are immediately aroused when the nest is tapped. When alarmed, the soldier termites rattle their hard heads against the walls, thereby initiating the vibrations which instantaneously warn the entire colony. Human beings can detect these vibrating sounds by rapping the infested pole or timber and applying their ears to the termite-ridden wood. It is believed termites, like ants, can communicate with one another through the tactile movements of their antennae.

THE FOOD OF TERMITES

Termites probably have arrived at such highly developed social instincts because in cellulose (the hard part of plants) they have a constant supply of food. Subterranean termites obtain their cellulose primarily from wood that has been previously attacked by fungi. Goetsch and Gruger (1941) note *Kalotermes* and *Reticulitermes* "feed primarily on wood and wood products, use various fungi as additional food and can live on these a long time." Although termites are soft-bodied insects, their hard, saw-toothed jaws, which work like blades of horizontally-held shears, are able to bite off extremely small fragments of wood a piece at a time.

The common dry-wood termite feeds on both spring and summer wood, whereas the western subterranean termite avoids summer wood. It also prefers sapwood to heartwood, but will attack both.

Wolcott (1946) notes the dry-wood termite, *Cryptotermes brevis*, in Puerto Rico, prefers some woods to others and some woods are practically immune to termite attack. "Of the major constituents of wood, cellulose is readily digested by the enzymes of protozoa living in the digestive tract of termites, whereas lignin is entirely indigestible, so that woods with a high lignin content are invariably avoided. In the case of all timbers tested, the sapwood, which contains starch and sugar, was more attractive than the heartwood which contains more lignin in many tropical hardwoods." Whatever resistance to termites coniferous woods of the temperate zones possess, it may be attributed for the most part to the resinous gums that are present. Behr et al. (1972) found subterranean termites prefer soft woods.

No wood is wholly resistant to the termite. In the United States, heartwood of redwood, *Sequoia sempervirens* of California, and southern cypress, *Taxodium distichum* of the southern United States, as well as some species of junipers are resistant to termite attack. All these so-called resistant woods are subject to attack once the repellent ingredients leach out. These chemical constituents which render wood termite-resistant are the sesquiterpene alcohols, oils, alkaloids, gums, resins, as well as silica. Termites do not ordinarily eat mortar or plaster, but are able to use chemical means to penetrate these materials. Snyder (1935) has the following to say regarding this: "By means of acidulous secretions from the frontal gland, certain tropical termites (species of *Coptotermes*) are able to dissolve lime mortar." Pence (1957) showed subterranean termites under laboratory conditions will eat stucco and cement. Snyder (1955b) notes some termites penetrate many plastics other than cellulose acetate, as well as neoprene and pure rubber insulation.

Termites and protozoa. It will be found that in all but the highest family of termites, Termitidae, the lower intestine of the termite contains a swarming mass of one-celled flagellated Protozoa. In addition, there are numerous amoebae, bacteria, spirochaetes and fungi of diverse sorts. These unique organisms contain enzymes which digest the wood fragments. If these organisms are removed, the termites die of starvation.

The highly specialized king, queen, soldiers and first instars depend upon nymphs and workers for their feeding, and if these dependent individuals are isolated, they starve since they are unable to feed themselves. Kennedy (1947) notes these stages cannot feed themselves because of toothless jaws or very poorly developed jaw muscles. As the young have Protozoa some 24 hours after hatching, they soon are able to eat wood. The young feed on fecal pellets and on the fluid droplets from the rectums of other termites and are also fed from the mouths of their fellow termites. It is in this manner the young termites are readily inoculated with the intestinal Protozoa as well as with the other intestinal fauna and flora. McMahon (1963) showed transfer of food from the hindgut was the usual method of food exchange in *Cryptotermes brevis*. Kennedy (1947) states the "flight queen is limited to a few protozoa since she cannot carry many and be light enough for easy flight; but she has to carry a few to infect her first young."

Cleveland et al. (1931) believe the primitive termites obtained their Protozoa from the older wood-boring roaches which have similar Protozoa and it is possible the more specialized termites obtained their Protozoa from the lower termites since many types of termites may live in close association. This is all the more possible when we remember termites readily feed on materials of anal or oral origin and at times they may be cannibalistic.

Termites and fungi. As was mentioned before, many species of termites enter wood previously invaded by fungi. They carry the spores on their bodies and drop them in their feces. In the opinion of some researchers, it is thought wood alone does not supply all the nutritional needs of the termite and nitrogen is obtained from fungus mycelium. The fungus provides moisture for the termite and may regulate the moisture content in the passageways. Pence (1957) and Lund (1959, 1960) showed the subterranean termite can feed on sterile wood without any fungus whatsoever on the wood. Using moisture gradient tubes made of plaster, Pence found this termite preferred a relative humidity of 97.5 percent.

Kofoid et al. (1934) note the galleries "of termites are kept closed so the air contained therein is humid up to the point of saturation from the moisture of respiration of the colony and from the wood itself. The fungi may be of value to termites in the preliminary softening of the wood, particularly during the incipient stages of development of the colony, but their most important function is probably the fulfillment of protein and vitamin requirements of the termites." Even the galleries of the dry-wood termite contain fungi. Although termites may be dependent upon fungi, fungi are not dependent upon termites for they may persist long after the termite colony has vanished. However, termites do aid in the dissemination of fungi and provide shelter and the humid atmosphere so necessary for the fungi.

The higher family of termites, Termitidae, lack Protozoa and gather vegetation upon which mushrooms are cultivated. The termites then cut these mushrooms and thereby obtain cellulose which has been prepared for them by a fungus. This is the case with the mound-building termites in Africa and Asia. According to Snyder (1935), other foraging or harvesting termites in Africa and South America proceed in files in the open during the day to collect grass stems or other vegetation. These termites are dark-colored and have well-developed eyes. There are a few such species of nasutiform daylight-foragers in the United States, which when out foraging are eagerly devoured by many animals.

These harvester termites are of great economic importance in some parts of the world. An example of this is cited by Coaton (1946) on harvester termite damage in South Africa. "Although the Harvesters do not attack timber, the type of damage most commonly associated with termites, they are nevertheless

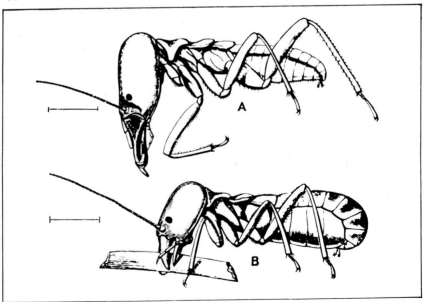

Fig. 7-4. A. Soldier termite. B. Worker of eyed grass-cutting termite from South Africa. The piece of grass is carried in a vertical position.

of great economic importance in the Union. The townsman knows them well and as a result of their denudation of lawns and stripping of hedges, shrubs and plants in gardens. Their habit of burrowing through and undermining walls constructed of raw-brick, of destroying thatched roofs, wallpaper, cotton materials, books and other materials of a similar type containing cellulose, also frequently brings them into the public eye. It is, however, in agricultural and pastoral areas where they play their major role. In the first place they bring about widespread denudation of grass and bush needed as grazing for live-stock; apart from the considerable lowering of the carrying capacity of such veld, the removal of the grass cover decreases normal reseeding and results in soil desiccation and erosion. Secondly, the Harvesters destroy or seriously decrease the yield from large areas under wheat, beans, lucerne, teff, and oats in many parts of the Union."

Esenther, et al. (1961) showed one fungus to have a chemical that was highly attractive to some termites. Smythe, et al. (1971) found subterranean termites are attracted to wood infested by decay and survive better in such wood. Termites do, however, attack and survive in sound wood in the absence of decay organisms. Lund (1962) showed a metabolite of the fungus *Lentinus lepideus* was highly toxic to subterranean termites.

Grooming. Termites clean themselves by licking or grooming one another. This results in each termite consuming the secretions of the other. While the grooming is going on, the termites exchange nourishment by mouth. This exchange of nourishment which is akin to that in ants, is known as *trophallaxis*. Castle (1934) describes the grooming of *Zootermopsis angusticollis* (Hagen) as follows: "The groomed individual stands quietly while the groomer carefully cleans the legs, thorax and abdomen. The mandibles and palps are used in the process. Most attention is given to the white areas between the abdominal segments. The termite being groomed assists in the process by turning the body and flexing its abdomen so the groomer has better access to the covered portions. At times the termite turns on its sides so the lower portion of the body can be groomed." If the termite accidentally bites, causing a droplet of body fluid to exude, it may excite the groomer to cannibalism and the groomed individual may be eaten, as are weak, diseased and dead individuals. Grooming is important in the control of dry-wood termites since it is in this fashion they disseminate the poison dusts. Castle further notes, "If individuals weakened or killed by toxic powder are eaten by healthy ones the poison may thus be spread further through the colony."

TERMITE CASTES

The caste system. Termites have a rigid caste system and are adapted in form and instinct to carry on their special duties. The castes of termites are primarily three, namely, the reproductive, worker and soldier. Some termite species lack the workers, their nymphs or immature forms performing the duties ordinarily fulfilled by workers. In time, these nymphs become reproductives or soldiers.

The reproductive caste. The male termite commands a great deal more respect in the termite world than does the male ant in his particular cosmos. The male termite is an extremely important individual in the life of the colony for the termite male, unlike the ant, mates throughout his life with the female or queen.

At certain times of the year, usually immediately prior to and immediately after the rains, the winged sexual forms of subterranean termites are extremely common. These winged forms are both male and female, and entirely unlike the popular concept of the termite. They now have two long narrow wings, dark-colored, flattened bodies and big compound eyes. Thus, they are radically different from the whitish soft workers or nymphs which are often mistakenly called "white ants."

Fossil remains show the worker evolved later than did the winged-sexual form. Thus, Snyder (1935) believes the winged sexual forms existed as individual males and females, which only associated in order to mate and rear young. These young, in turn, developed into winged sexual forms and their social phase in the termites came later.

The winged or sexual forms emerge from the colonies at specific times, usually after the first rains, or when temperature conditions are suitable. The reproductives associate in pairs, break off their wings and attempt to start a new colony. These are the primary and permanent queens and kings.

With time, the body of the queen becomes extended or enlarged because of the great development of the ovaries and glands. Because of this extension of the abdomen, tropical queens often reach a length of three inches/76 mm. Although our queens have enlarged abdomens, they do not approach the tropical queens in size. It will be noted the abdomen of the queen presents a striped appearance. This is due to the fact that as the abdomen stretches, the white membranes, between the dark sclerites or plates which cover her body, become

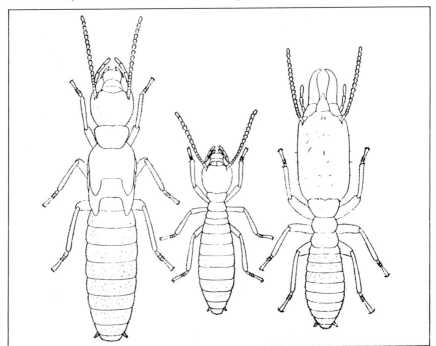

Fig. 7-5. (Left) Supplementary queen; (center) worker; and (right) soldier of the eastern subterranean termite.

extended. Thus, the plates between the white inter-segmental areas present a striped appearance.

The original or primary queen attains maturity in about two years. The average worker or soldier lives only two or three years, but queens are believed to live up to 25 years. Emerson and Fish (1937), in discussing a queen in British Guiana, have these interesting things to say concerning her. Speaking of her egg-laying capacity they note, "She laid them constantly, at the rate of six eggs a minute. That adds up to 360 eggs an hour, 8,460 eggs a day and 3,153,000 eggs a year. Each egg, as it came from the queen's body, was covered with a clear fluid which the termites liked to drink, so each worker was eager to care for the eggs. The workers were not only feeding the queen, but a score of them were busy rubbing and massaging her. A delicious oily liquid flowed from her pores. The workers were in constant attendance not from any idea of duty or loyalty, but because they found this liquid pleasant to taste and stimulating. The queen was large because her egg-making organs, or ovaries, and the glands which produced the secretions had grown and grown. She weighed at least 100 times as much as a worker or soldier."

Coming closer to home, the queens of our own *Reticulitermes* lay their eggs singly, six to 12 in the first batch. The eggs in large colonies of this genus may occur by the thousands. The queens of native termites, unlike the tropical queens, are active, may move about and are rarely discovered.

Supplementary queens. Besides the primary queen, there may be supplementary queens. These supplementary queens are, according to Light et al. (1930), "nymphs in various stages whose reproductive development is suddenly speeded up while their normal development into soldiers or alates is inhibited save for a certain increase of pigmentation. They are, therefore, as a rule, darker than other nymphs but much less heavily pigmented and chitinized than are the first-born kings and queens and have at the most but rudiments of eyes and wings."

The original or primary queens are called Macropterous queens or long-winged females. In the supplementary queens, we have the Brachypterous or short-winged queen and the Apterous or wingless queen. The Brachypterous queen is much more active than the Macropterous queen. These supplementary queens readily replace the primary queens when the latter die or are injured.

The supplementaries are important because they rapidly increase the termite population. Moreover, the supplementaries with a consort of a few soldiers and workers may migrate and initiate new colonies. For all practical purposes, a colony never dies since the supplementaries replace the injured or dead queens. The rapidity with which a new structure may be infested is explained by Light et al. (1930) as "the invasion of foraging groups which form subcolonies headed by supplementary reproductives. Workers and nymphs, when separated from the mother colony, exist as a new colony, with supplementary queens which may be produced in from six to eight weeks. What is more, as many as three queens may develop in the same cavity and live peaceably together, with the eggs from all three deposited on a single pile by the workers. A supplementary queen attains approximately the physogastric size in a few weeks and at the height of egg-laying (the complete group of supplementary queens) will deposit more eggs in a day than the primary queen lays in the first two years of colonial development. Some of the largest supplementary queens can lay 60 to 80 eggs a day."

According to Light et al. (1930), the primary reproductives after breaking off their well-formed wings, are characterized by wing scales, fully formed compound eyes and the completely chitinized and pigmented chitinous exoskeleton. The supplementary reproductives "lack the wing-scales and show all degrees of size, pigmentation and development of eye and wing intermediate between conditions found in nymphs of the various older instars and the fully developed primary reproductives."

BIOLOGICAL INFORMATION ON TERMITES

The swarming of termites. As a rule, termites avoid light and seek narrow cracks and crevices. The winged reproductives before swarming may wait three months or more for the arrival of correct atmospheric conditions, especially as far as temperature and humidity are concerned. The alates or winged reproductives now completely reverse their instincts and seek the light and open spaces. They fairly burst from the tubes constructed by the workers and nymphs. These workers and nymphs also may show a reversal of instinct by being present outside the opening. However, it is possible they are merely carried out by the rush of the winged reproductives. The apertures which are opened in the ground later may be closed so there is no indication where the opening occurred.

The flight is short and weak. The termite wings break off at the "fracture point" in the air, the termites then spiralling to the earth, or the termites may land upon the earth and then remove their wings. According to Light (1942), the male and female are attracted to one another only after the loss of the wings, and their flight cannot be considered a mating flight.

Fig. 7-6. Subterranean termites. Note swarm of winged forms on right.

In the tropics, the late evenings and nights are sufficiently warm for the swarming flights, whereas in the temperate zone, swarming occurs during the day because of the cool nights. In both areas, however, swarming commonly occurs after a rain. This is understandable when we know the ground must be sufficiently soft to enable the colonizing termites to excavate it.

On the Pacific Coast, the greatest swarms occur after the rains in autumn and lesser swarms occur in the spring. Although in areas far under the house swarms may emerge much later. This sudden emergence of winged forms simultaneously from many different colonies is favorable to cross-breeding. In the East, the greatest swarms are to be seen with the rise of temperature in the spring.

Light et al. (1930) make the following interesting comments on the swarming of termites in the tropics: "Their numbers are almost incredibly great and, in the case of certain ground-dwelling termites, they arise in great clouds which darken the sky of late afternoon or early evening. Attracted in enormous numbers by the lights, the alates of certain species make life a burden to mankind on the evening of their swarming. Eating is impossible, since they drop in countless numbers into food and water. Indeed, the only chance for comfort is to sit in the dark, since they enter the clothing in great numbers if one sits near a light, while quite unable to do any real harm, their excited crawling over the skin is not pleasant."

Courtship and pairing. There is a reason for the enormous numbers of winged forms produced, for once their wings are gone, termites are among the most helpless of all animals. They are now a prey to predacious insects, especially ants, who celebrate a Roman holiday with the swarming of the termites. The writer has observed the Argentine ant, *Iridomyrmex humilis* Mayr, to individually pack off a hapless termite, with practically every other ant in the trail bearing a termite in its jaws. Even the spider beneath the house has an opportunity to replete its larder since many of the emerging termites are entangled in its web. Birds and lizards gorge themselves with this winged delicacy. Rare, indeed, is the termite couple that survives.

Snyder (1935) describes the courtship and pairing of the surviving dealated termites as follows: "The surviving deälates pair off in couples, the male closely and tirelessly following the female, with head close to her abdomen and his feelers or antennae in constant touch with her. This rather rapid pursuit of running about after the female occurs for some time after the swarm and is called "amatory procedure." The male and female together found a new colony "by excavating a shallow cell in the earth, in the earth under wood or other vegetation, or in crevices or under bark in moist wood." Nor are they safe even now, for the adverse conditions in the soil, as well as temperature, humidity and food conditions may foil them. Snyder further states, "It is about one week after the swarm that copulation takes place and the female is fertilized." They now prepare the "royal cell." It is believed the male continues to mate with the female for life.

In regard to the rearing of the young, we once again borrow from Snyder's excellent work. It will be found the first-born in a colony are smaller than normal because of a lack of sufficient nourishment. "The male and female termites are equally active in caring for the young and both young parent adults eat wood. This condition is quite strikingly in contrast to the lonely life of the young queen of the "carpenter ant" (species of *Camponotus*). Here the consort male is dead and the queen, without taking any external nourishment, has to feed and

Fig. 7-7. Workers and soldiers of the western subterranean termite. Note the elongated head of the soldiers.

rear the first brood by her own efforts alone. Her food for periods of six months or more is obtained entirely from internal metabolic processes (or internal body changes), a breaking down of muscle now useless and of the reserve fat bodies into food." When the termite colony is in the incipient stage the king and queen take care of the young and keep the place clean. Later, all care of the young in the nests will be assigned to the workers or nymphs.

The young are fed on a prepared and predigested food, and only later are they capable of eating wood. The cellulose in the wood is digested and the lignin in the wood is excreted. Once the workers and nymphs are produced, the king and queen are fed from the mouths of the workers or nymphs on a specially prepared diet. They then cease feeding on wood. Some species of termites, particularly those that are tropical in distribution, raise fungi and feed their young on this plant material.

How are the supplementary kings and queens produced? There are two main theories in regard to this question. The first that the supplementary kings and queens are selected on a basis of heredity. The second as advanced by Pickens (1932) suggests the main or primary king and queen in the colony secrete a substance that inhibits the development of the reproductive organs in other members of the colony. This substance is distributed throughout the colony by the grooming habit. Should the original reproductives die or be injured, those forms that are most sexually mature become the supplementary kings and queens, and these in turn secrete the substance that inhibits the sexual development of the other members of the colony. Keene and Light (1944) note although there is no evidence to prove it, the inhibition theory still remains a plausible proposal. Light (1942) prepared an excellent review of the subject of caste determination in social insects.

The worker caste. The worker is the form most commonly encountered when a piece of infested wood is picked up from the ground or when a tree stump is broken apart. These creamy-white, thin-skinned, maggot-like insects are the most numerous form in the colony, and the one that does the actual damage to wood and cellulose products. It is this caste that undertakes all the labor in the colony, such as constructing the tunnels and excavating chambers, obtaining food, cultivating fungus gardens, and feeding the young, soldiers and king and queen.

It is believed the workers evolved from the soldier since workers are absent in the more primitive termites. Thus, dry-wood termites have soldiers and reproductives but no workers. Instead, the nymphs, which later become the soldiers and reproductives, do the work in the colony. Termite workers develop to maturity within a year. In ants, the worker is a sterile female. In termites the worker is either a male or a female.

The soldier caste. The soldier is the grotesque individual with the enormously elongated head. The head is brownish, hard and equipped with two large jaws or mandibles. The soldier is not nearly as common as the workers or nymphs in the colony. It also is wingless, blind and practically eyeless. Except for the head, he is as softbodied as the worker or nymph. The soldiers must be fed by the workers or nymphs since they are incapable of feeding themselves.

Apparently their sole purpose in life is that of defense, particularly against their inveterate enemy, the ant. When a hole is made in a termite tube or carton, if the hole is small, the soldiers attempt to block the opening by "using their heads." If the opening is large, they form a guard around the entrance, their large mandibles wide open, ready to shear in two any insect enemy having the

Fig. 7-8. Newly molted drywood soldiers side-by-side with soldiers with pigmented head and body. Various instars of nymphs or workers are also present.

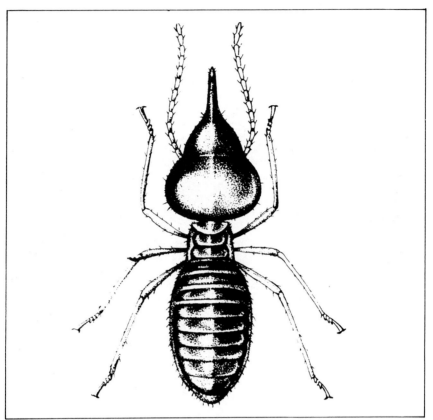

Fig. 7-9. A soldier of *Nasutitermes*. Head has small jaws and a long snoutlike horn through which a gummy liquid is ejected for defense purposes, particularly against ants.

temerity to pass close by. However, should an insect, such as the ant, attack them from the rear, their soft bodies are quickly rended by the aggressive and swifter ant. The workers with their smaller jaws are often as effective in protecting the colony as the soldier. There have been lawsuits in the United States concerning imaginary injuries suffered from the bites of termites. Such suits have little success since the bites of our native termites are weak and may not even be felt.

Some termite soldiers are living spray-machines. These are most common in tropical areas. They either have a frontal gland, or a tube or *nasus* which extends from the head and from which is expelled a milky-white and glue-like material. When an insect, such as an ant attacks, the termites are capable of shooting this liquefied material a distance up to one inch/25 mm. Not only does it entangle the ants, but it is supposed to have a powerful chemical effect on them. This secretion from the head is acidic in nature and may be responsible for the etching of metals by termites. It even may enable the insects to penetrate through lime mortar.

Distribution of termites. Termites are world-wide in distribution, but are absent for the arctic and antarctic areas. They occur in greatest preponderance

Fig. 7-10. Western subterranean termites construct a tube from the ground to the upright at the extreme right of this photo. The other tubes were then started from the top and built down.

in the tropics where they are supposedly one of the dominant animal types. In fact, it is believed by some individuals the backwardness of the tropical areas is due to termites, since it is extremely difficult to maintain libraries because of their depredations. Some 1,800 species of termites have been described and T. E. Snyder believes that with the species yet to be found, the number will reach 5,000. Termites are known from every state in the Union, with approximately 45 species occurring in the United States.

In 1965, E. Dale McCullough, a pest control operator in Houston, Texas, found the Asiatic termite, *Coptotermes formosanus* Shiraki infesting a warehouse in Houston. Since then this termite has been found established in Galveston, Texas, as well as in New Orleans and Lake Charles, Louisiana, and in parts of South Carolina and Florida. Padget (1960) describes the control of *Coptotermes crassus* in a drydock in Houston, Texas, by methyl bromide fumigation under tarps. Emerson (1936) states, "The environmental requirements seem so rigid that it is unlikely a given species can become established in an environment very different from that in which it is a native. A possible exception to the general rule that introduced termites have not spread is the case of *Heterotermes tenuis* (Hagen), a native of Brazil, the Guianas and Panama. It is reported to have been introduced into St. Helena in 1840 and to have damaged houses and furniture, but there is no evidence that it spread to wild habitats in the Islands."

Termites in the United States are most common in the South and Southwest. Nasutiform termites occur in Texas and Arizona, but not in California or Florida. Members of the genus *Reticulitermes* are extremely widespread throughout the United States, but they do not occur at elevations higher than 7,000 to 8,000 feet/2,138 to 2,438 meters in the western states. The termites of this genus are the ones so prominent in wood injury throughout the United States.

Fig. 7-11. Drywood termites in an infested piece of wood. Note the characteristic
well-formed fecal pellets, and the two soldiers with their long dark mandibles.

Nesting sites of termites. Termites may be divided into two groups as far
as their nesting sites are concerned. The earth-dwelling termites that extend
their tunnels in the earth, or prolong their tubes above the surface of the earth
are one of these groups. Of this type, the subterranean termite of the genus
Reticulitermes is an excellent example.

The second major nesting type of termite is the wood-dwelling termite in
which the colony is situated in the wood and has no contact with the earth.
There are several groupings of the wood-dwelling type of termites. Thus, those
termites that attack only dry-sound wood are known as dry-wood termites. The
Kalotermes and *Incisitermes* are typical of the dry-wood termites. The termites
that attack damp and decaying wood are the damp-wood termites. Termites of
the genus *Zootermopsis*, occuring on the Pacific Coast, are representatives of
this group. Another group are the so-called "powder-post" termites of tropical
regions which occur in very dry, seasoned wood. The genus *Cryptotermes* is representative of the powder-post termites. These may actually form small colonies
in woodwork, furniture, picture frames, etc. They are often carried from the
tropics into temperate regions, and usually die out in the cooler climates. The
author has personally seen *Cryptotermes* infestations in Washington, D.C. and
northern New Jersey that had spread from an imported wood member and were
well established in numerous structural members of the buildings involved.

The nests of termites may be found in the soil under buildings or in the wood
itself. The subterranean termites move up or down in the soil to meet changing
moisture and humidity conditions. During the winter they may go below the
frost line and in areas where drought prevails the termites may go down to
regions of greater moisture content. McCauley and Flint (1938), discussing
Reticulitermes flavipes Kollar, state, "they may be found underground to depths

Fig. 7-12. Drywood termite laying eggs.

of five feet, feeding on plant products which are present. On the other hand, they may sometimes be found quite a distance above the ground, the workers foraging 100 feet or more from the nest." It is in the tropics that termites make their huge mound-nests (Fig. 7-14), as well as carton nests which are suspended from trees.

Termite guests. An interesting sidelight on the social life of termites are the "termitophiles" or termite loving guests that dwell with termites. They include insects of many kinds especially beetles and flies. These insects live with the termites for numerous reasons, but particularly because the termites and their environs offer them food, shelter, and security. These guests of termites are very similar to the guests of that other common social insect, the ant, and some of these guests are beneficial, some neutral dwellers, others predacious and even parasitic on the termites. If one may be permitted to view the termites from an anthropomorphic viewpoint, the feelings of the termites toward their guests may run the gamut of emotions from love, through indifference, to hate. The relationship between the guests and the termites rests on the exchange of nourishment and the grooming habit. In some respects, these guests have taken on an appearance similar to that of the termites.

How to distinguish ants from termites. Probably the most prevalent statement to be gleaned from general works on termites is "that termites are called white ants because they are neither white nor ants." People confuse them with ants because of the similar social life. Moreover, termites are found in the earth, where they move in ant-like files. Termites have broad waists and straight, bead-like antennae. Ants have narrow waists and elbowed antennae.

In discussing ants as natural enemies of termites, Kofoid et al. (1934) make the following interesting statement: "The habit of termites living in closed bur-

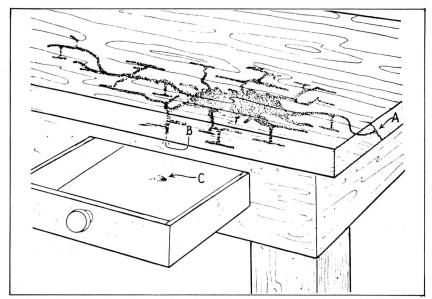

Fig. 7-13. Schematic diagram showing the work of a powder-post termite, *Cryptotermes brevis* **(Walker). A. Point of entry. B. Opening from which a tiny pile of fecal pellets C. is deposited.**

rows is the mechanism by which they successfully ward off destruction by these, their hereditary enemies. The ants constitute an ever-present menace to the existence of the termites, but the evolution of the burrowing habit is the latter's successful answer to this menace." Snyder (1935) furnishes this inevitable exception to the rule. "An ant, *Camponotus novegranadense* Mayr, inhabits the carton tree nest of termites in Panama. This ant is pacific, has its own galleries and is not molested by these termites." Hatfield (1944) notes ants and termites can exist in close proximity. "It is true that ants are deadly enemies of termites, but termites can and do establish themselves in areas occupied by ants and can actually carry on destruction in wood, a portion of which is occupied by ants."

ECONOMICALLY IMPORTANT TERMITES IN THE U.S.

1) Make termite tubes, pellets not present . . . genus *Reticulitermes*.
   a) *Reticulitermes flavipes* (Kollar). The common Eastern subterranean termite in the East.
   b) *Reticulitermes viginicus* (Banks). Smaller and darker than above. Occurs most commonly in mid-Atlantic states.
   c) *Recticulitermes tibialis* (Banks). Occurs in the West, not nearly as common a pest in houses as *R. hesperus* (Banks); head of soldier short, broad and dark.
   d) *Reticulitermes hesperus* (Banks). Occurs on the Pacific Coast; head of soldier pale, long, and narrow.
   e) *Recticulitermes humilis* (Banks). Occurs in Arizona and New Mexico.
2) Do not make termite tubes, pellets present.
Occur on the Pacific Coast; winged forms more than one inch/25 mm in length;

pellets oval-shaped . . . *Zootermopsis angusticollis* (Hagen). Winged forms smaller than above, about ½ inch/12 mm in length; pellets well formed, hectagonal . . . genus *Incisitermes or Marginitermes.*

> a) *Incisitermes minor* (Hagen). Occurs on Pacific Coast, especially in California.
> b) *Marginitermes hubbardi* (Banks). Occurs in Arizona.

Snyder (1948) lists the termites and has prepared keys to the genera and subgenera of termites in the United States. Snyder (1949) catalogued termites of the world and (1954) prepared a bibliography.

THE DAMPWOOD TERMITES

Zootermopsis

These are our largest termites and since they require much moisture, their colonies are commonly found in logs and damp or decaying wood. The damp-wood termites are distributed from British Columbia to lower California.

Fig. 7-14. **Huge termite nest in Port Darwin, N. Australia.**

According to Light et al. (1930), the winged forms vary from reddish/brown to black/brown and are about one inch/25 mm or more in length with wings twice as long as the body. The soldiers have large reddish/brown heads with black mandibles and are ¾ inch/19 mm in length. The light yellowish gray/ brown nymphs have a mottled gray/brown abdomen which is wider than the head and the head is wider than the thorax. The feces of these termites have the appearance of large oval pellets.

The termites of this genus are all situated in the West and are most widespread on the Pacific Coast. Castle (1934) states three species of the genus *Zootermopsis,* formerly the genus *Termopsis,* are known. They include *Zootermopsis angusticollis* (Hagen), the common damp-wood termite; *Zootermopsis nevadensis* (Hagen), the small damp-wood termite, and *Zootermopsis laticeps* (Banks), the Arizona damp-wood termite. *Zootermopsis angusticollis* (Hagen) and *Zootermopsis nevadensis* (Hagen) are very similar in many respects and are difficult to tell apart.

THE PACIFIC DAMPWOOD TERMITE

Zootermopsis angusticollis (Hagen)

Since *Zootermopsis angusticollis* (Hagen) is the most important dampwood termite from an economic viewpoint, and since more is known about this species than about the others, it will be discussed in greater detail. This is the largest Pacific Coast termite, and according to Castle (1934) is among the most primitive of the termites occuring in America as evidenced by their "less highly

Fig. 7-15. Soldier of the common dampwood termite.

evolved social system and many primitive morphological characters." There are but two well defined castes, the soldier and the reproductive caste. The worker caste is absent. The work of the colony of these termites is carried on by the immature soldiers or reproductives which are known as nymphs.

Distribution. This species is distributed from British Columbia to Mexico and Baja California. In the southern part of their range they have been found up to 6,000 feet/1,829 meters. The dampwood termite occurs most commonly in the cool and humid areas along the coast in the region cited.

Snyder (1954) and David (1955) record the shipment of this termite in lumber made from Douglas Fir. According to Snyder, infested lumber has been received in the Midwest, South and East. When it is discovered in these regions it usually is destroyed by burning in the lumber yard.

The common dampwood termite attacks wood of all types throughout the Pacific Coast. Wood buried in the ground and subject to much moisture and decay is most likely to be attacked. The tolerance of this species to moisture is very great as is evidenced by reports that it has been found in wood piling at low tide in buildings situated on the San Francisco waterfront. This termite remains in decaying wood as long as it can control the fungus growth and then moves to sounder wood when it cannot.

Description of castes. There are three kinds of reproductives in the colony of which the primary reproductives are the most common and most important. The winged forms are light brown with dark brown leathery wings which are approximately one inch/25 mm in length. After mating, the female drops her wings and merely retains the wing stubs.

The nymphs are whitish to cream-colored with a darker abdomen due to the intestinal contents. They are from ⅛ to ½ inch/three to 13 mm long. These immature forms or nymphs are by far the most common individuals in the colony. The nymphs that are to become primary reproductives have wing pads in their more advanced stages.

The pugnacious soldiers vary from ⅜ to ¾ inch/nine to 19 mm or more in length. Castle (1934) describes them as follows: "The head is large and is armed with a pair of long, black, toothed mandibles. The anterior portion is black, generally shading to a dark reddish brown in the posterior position. The thorax and abdomen are a light caramel color, the abdomen being slightly darker, due to the intestinal content."

Swarming. Although swarming may occur throughout the year, the common dampwood termite most often swarms from July through October. The reproductives usually emerge in the late afternoon or early evening on warm sultry days, particularly in September. They may then be present in some localities in great numbers and are commonly referred to as "rainbugs" since they so often appear after the early rains. The reproductives are greatly attracted to electric lights. They apparently are strong fliers for they have been known to attack the wet wood of a water tank situated on the seventh story of a building.

Life history. Castle (1939) studied the life history of this termite, the results of which are presented herewith: The female excavates an opening and later the male enters. The opening is then closed with pellets and chips of wood cemented together with liquid feces. They copulate within two weeks of the founding of the colony. Fourteen to 18 days after the founding, the long slender and beanshaped eggs are laid. These are ¹/₁₆ inch/1.6 mm long and less than one-half as wide, and are white and translucent. The queens lay from six to 22 eggs with an average of 12. The second lot is laid the next spring.

Although the queen is not a great egg-layer, nevertheless colonies of several thousand individuals may result. This is probably due to the numerous supplementary reproductives that are present. Well established colonies in decaying wood give rise to reproductives which attack nearby susceptible woods. The colonies vary greatly in numbers and are known to have up to 4,000 individuals.

Tunnels and pellets. The tunnels are of many sizes and shapes, and according to Castle, "the surface of the runway has a velvety appearance and sometimes is covered with dried liquid feces." Castle further notes pellets are found throughout the tunnels and are small, hard, oval objects about ¹/₂₅ inch/one mm long. The color of the pellets varies according to the wood being eaten and these may have a slightly hexagonal shape due to being compressed in the rectum prior to ejection.

When attacking sound wood, the softer spring wood is preferred. Like other termites, the damp-wood termites aid in spreading wood-decaying fungi, the spores of which they carry in their intestines.

THE SMALL OR DARK DAMPWOOD TERMITE

Zootermopsis nevadensis (Hagen)

This termite is found from British Columbia to central California and ranges into Nevada and Montana. It occurs in the drier and higher altitudes, ranging up to 9,000 feet/2,743 meters and is primarily a forest-dwelling termite. This is the common termite found in the stumps in the mountains of California. It is very similar in habits and morphology to the common damp-wood termite. The soldier of *Zootermopsis nevadensis* (Hagen) has long mandibles and a longer head with straighter sides than does *Zootermopsis angusticollis* (Hagen). The winged reproductive of the small damp-wood termite is a darker brown than that of the common damp-wood termite. The reproductive of *Zootermopsis nevadensis* (Hagen) is smaller in size and the nymphs have larger and more numerous hairs on the head and body parts.

This species is of potential economic importance. Records of its attacks on man-made structures are few. The reason for this is it occurs in less densely populated areas. It has been reported from California in railroad ties, poles, docks, some houses and one bridge.

THE DAMPWOOD TERMITES

Prorhinotermes

Miller (1949) studied the habits and distribution of *Prorhinotermes simplex* (Hagen) in Florida. The following notes are from his paper. This species has not been collected north of Fort Lauderdale nor west of Homestead, Fla., and its distribution is limited to a few miles along the east coast where it may do considerable damage to structural timber. No contact with the ground is necessary if the timbers are sufficiently moist. The eaten portion of the wood is replaced with a clay-like material.

The winged forms may be confused with those of *Cryptotermes brevis*. They swarm at dusk from October through January and are attracted to lights.

THE DRYWOOD TERMITES

Kalotermes, Incisitermes and other genera.

The drywood termites are so-called because they establish themselves in wood that is not decayed, nor in contact with ground moisture and which is to all appearances perfectly dry. Gunn (1952) notes this species can exist in wood with as little as 2.8 to three percent moisture. The termites of the genus, *Incisitermes*, formerly *Kalotermes*, the common drywood termites in southern California, are pests in the attics of buildings where they maintain themselves without any soil connections whatsoever. Krishna (1961) recently split the genus *Kalotermes* into a number of genera.

Drywood termites are larger than subterranean termites and smaller than dampwood termites. The winged form and soldiers are up to ½ inch/13 mm long. The pellets are regularly formed and when in abundance, are to be found in dry sawdust-like piles. The cavities in the wood are clean and smooth as though the surface has been gone over with sandpaper.

THE WESTERN DRYWOOD TERMITE

Incisitermes (formerly *Kalotermes*) *minor* (Hagen)

The western drywood termite is an extremely important termite pest in the Southwest. Hunt (1949) notes 75 percent of a total of 186 buildings in California treated for termites in 1948 by his company had *Kalotermes minor* infestations. This species also occurs in Arizona, with a single report of its presence in Washington. In California, its range extends north of Sacramento. Small piles of well-formed fecal pellets are ready evidence of infestation by the drywood termite.

On rare occasions this termite may be found in areas outside its usual range. Gross (1949) notes an infestation in a church in Ohio and he quotes the following observation from a letter by T. E. Snyder. "Such termites have been carried from the tropics or more southern parts of this country in wooden articles and have infested furniture and the woodwork of buildings, as far north as Ontario, Canada, Madison, Wisconsin, Connecticut, and California. However, they can only live in buildings and have not been established as a native pest. They cannot live outdoors in northern climates."

Purdy (1951) notes an infestation by drywood termites in a lumber yard in Ohio in fresh lumber from the West Coast. Heal (1952) found a *Kalotermes* infestation in a New York home. The termites had been transported in several cases of California grapes and had then established themselves in the home. Katz (1955) records an infestation in a home in Pittsburgh, Pennsylvania which is of unknown origin. This infestation persisted for several years. A similar infestation was reported in wood paneling in a home in Philadelphia.

Swarming and flight. According to Harvey (1934), who made a detailed study of this insect, "Swarming is stimulated by brilliant sunlight when the temperature is 80° Fahrenheit or higher. Scattered swarming may, however,

Fig. 7-16. Workers, soldiers and sexual forms of the drywood termite. A queen is in the lower left hand corner and a king can be seen near the upper margin of the photo.

occur at lower temperatures and when the sky is overcast. Precipitation stimulates swarming of the western subterranean termite, but not the common drywood termite or common dampwood termite. Swarming of the common drywood termite reaches a peak and then tapers off. It occurs with almost clocklike regularity for each locality. It is during June and July in northern California and from September through October (until December according to Hunt, 1949) in southern California, but the exact date varies from one locality to another." These termites are to be seen during the day, and according to Brown (1936), in Golden Gate Park in San Francisco they swarmed from 11 a.m. to 3 p.m., the peak of their flight occurring at noon.

Harvey notes they may be able to fly more than a mile when aided by air currents. However, they often do not fly more than a few feet before settling down. After mating they seek cracks or knotholes in nearby wood whether it is a roof, lumber pile or telephone pole.

After they have mated, the royal pair, known as the king and queen, find a crack or crevice or crawl between boards and gnaw a small tunnel which they then plug up with a brownish material. Here they "excavate a pear-shaped royal cell where the queen lays the first egg." In this respect Hunt (1949) states: "It was Harvey's observation that, although they often land on wood-shingle roofs they seldom enter the cracks and crevices in the shingles but seem to prefer to crawl under the eaves and enter the wood through joints between rafters and sheathing, or junctures between rafters and ridge or studding and plates. Other typical places of entry are mortises in window sashes, cracks between sashes and frames, under casings, or even through checks in knots, which may occur in siding or the eave extension of rafters."

Quoting Harvey, "The nymphs all the while aid the royal pair in making passages in the wood and in extending the limits of the colony which gradually penetrate throughout the infested wood and may riddle it." The nymph passes through seven instars.

There is no worker caste in the colony. The nymphs eventually become sexual forms or soldiers. Harvey states, "an alate or a soldier may emerge from a seventh instar nymph. A soldier may emerge also from a fourth, fifth, or sixth instar nymph, so these individuals will vary in size according to the instars from which they were formed, while all the alates are of one size because they come only from seventh instar nymphs."

The nymphs make emergence holes during the swarming season. These holes, which are 1/16 to 1/8 inch/1.6 to 3.2 mm in diameter, are guarded by the soldiers with their long mandibles. Harvey further notes the soldiers "may even plug up the holes with their enlarged amber-colored heads until it is safe for the alates to emerge and may perform a similar service when a crack appears in the wood. The nymphs repair these places immediately with plugs of cement, feces and partially masticated wood."

Whereas the alates require seven instars, the supplementary reproductives may arise from instars four to seven inclusive. These supplementaries, as is the case with other termites, may form a colony within a colony. They also make the termite colony practically immortal since they replace the primary reproductives if they die or are injured.

Colony growth. Harvey has this to say regarding colony growth: "A pair of primary reproductives may remain dormant for nearly a year or produce up to five eggs, 20 nymphs and one soldier in this length of time. The colony consists of from six to 40 nymphs and one soldier, toward the close of the second year. Three-year-old colonies have a total of from 40 to 165 individuals; four-year-old colonies have a total of from 70 to 700 individuals. The ratio of soldiers to nymphs in three-year-old and four-year-old colonies varies from 1:15 to 1:60. The age and population of colonies beyond four years is not definitely known, but some colonies have been estimated to be 15 years old. The total population of a colony of this age may reach 2,750, which is estimated as the maximum for a colony of the common drywood termite. This is considerably smaller than the estimated population of the western subterranean termite.

"The first alates are released when the colony is four years old. These are replaced by newly hatched nymphs, and as the colony continues to grow the number of nymphs exceeds the total of alates that emerge during the swarming season. Passages are slowly and insidiously tunneled in the wood in the meantime. It is riddled eventually and may separate from the outside. The colony declines then and dies, or its ant enemies break through unrepaired cracks in the shell and destroy it. However, this should not be taken as an important factor in control."

These termites attack poles, posts, lumber yards, etc. In homes they are found in rafters, studs, window frames, door and window jambs, door sills, sheathing and places of a similar nature. They have also been found in bridge piles and marine piling, as well as oil derricks. According to Hunt (1949), "Dry-wood termites are general feeders, but they seem to prefer the sappy or spring-growth wood. Unlike the subterranean termites, they attack redwood freely. Piles of new lumber and new houses, in the framing stage, seem to be especially attractive to this species. In addition to structural timbers in general, they often attack English walnut, eucalyptus and citrus trees, gaining entrance through injuries or sunburned areas. Infestations have been found in apricot, avocado, alder, almond, cherry, California laurel, Monterey cypress, oak, peach, pear, plum, sycamore, willow and a number of other ornamental and fruit trees and shrubs." For further information on this species, see the section on drywood termite control.

THE DESERT DRYWOOD TERMITE
Marginitermes hubbardi (Banks)

The desert drywood termite, *Marginitermes hubbardi* (Banks), is known from Arizona, where it infests poles, posts and houses. Light et al. (1930) note in western Mexico "its abandoned and exposed workings are conspicuous sights in the wood of shutters, doors, door and window frames and jambs, floor boards, furniture, etc. It is also known to occur in the southeast part of California. This species is in very many respects similar to *Incisitermes minor* (Hagen), but it is found in localities having higher temperatures, and has lower moisture requirements than *Incisitermes minor* (Hagen). It is almost a true "house termite" and one that is very rare in natural wood. In the northern part of its range the most important natural reservoirs of the desert drywood termite are the giant cacti of the desert and the cottonwood of the canyons and stream beds.

Marginitermes hubbardi (Banks) may be separated from *Incisitermes minor* (Hagen) by the fact the third segment in the antenna of the soldier is nearly as long as all the following segments. The third segment in *Incisitermes minor* (Hagen) is enlarged, but only as large as the four following segments. Moreover, the alates of *Marginitermes hubbardi* are much lighter than those of *Incisitermes minor,* being light yellow with light wings. The alates, including their wings, are ½ inch/13 mm long. This termite flies at night.

OTHER SPECIES OF DRYWOOD TERMITES

Kalotermes approximatus (Snyder) occurs in Florida, New Orleans and Virginia, in dead trees, logs, stumps and buildings. Hetrick (1961) notes it infests living rosaceous trees in Florida.

Neotermes jouteli (Banks) is found in the southern Florida Keys, Cuba, Panama, West Indies and Mexico. Miller (1949) notes this species is largely limited to coastal zones, usually within 10 miles/16 km of the shore line. This termite requires more moisture than some of the other drywood termites. Snyder states, "This termite has seldom been found to be injurious except in a few cases in the moist foundation timbers of buildings. In nature it lives in dead trees and in logs and branches lying on the ground."

Incisitermes schwarzi (Banks). According to Snyder (1935), "This is one of the common termites of the genus *Kalotermes* in southern Florida Pensacola and its northern limit. This species also occurs in Cuba, Jamaica and the Bahamas in the West Indies, and Yucatan, Mexico. It attacks both the woodwork of buildings and the bases and tops of telegraph and telephone poles, including cross arms and insulating pegs. In nature this termite is found in dead trees, logs and stumps."

Incisitermes snyderi (Light). This is the most widespread and injurious species in the East where it damages the woodwork of buildings and poles. The damage is somewhat similar to that of *Cryptotermes brevis* (Walker).

Kalotermes marginipennis (Latrelle). Light (1943) notes the alates of this species are unusually large, being about ¾ inch/19 mm long, including the wings. They are found on the tablelands of Mexico where they are injurious to buildings and plants.

THE POWDERPOST OR FURNITURE TERMITES

Cryptotermes

These termites develop colonies in small pieces of dry wood and in the home they attack furniture, woodwork and floors. They are called powderpost termites because of their tiny fecal pellets, which are in themselves an annoyance through their constant dropping. Light et al. (1930) state termites of this group are "easily distinguished from other species of the genus *Kalotermes* by 1) the small size of their pellets; 2) the small size of all castes of the colony; 3) the fact the soldier has a short, high, strongly truncated head, often excavated in front and usually black or dark brown in color."

The common powderpost termite, *Cryptotermes brevis* (Walker), is widely distributed throughout Central and South America, Mexico and West Indies, and in some localities in the Gulf states such as Tampa and Miami, Florida and New Orleans, Louisiana. It has been found as far north as New Jersey. It is only found in buildings, never outdoors. It infests and destroys woodwork, floors and furniture. Anon. (1958) records it in a television set in California which was shipped from Hawaii.

According to Snyder (1935), the powderpost termite may have been introduced in the United States. It was first found in this country in furniture in Key West, Florida, in 1919, and Miller (1949) notes this termite may invade new areas through the exchange of used furniture. There apparently has been no natural spread. In the United States, colonizing flights occur in May or June.

Fig. 7-17. Damage to furniture by drywood termites in Key West, Fla.

Light (1934) has the following to say about this termite: "The colonies, while usually small, are numerous and in heavily infested areas the evidences of their attacks are to be seen on every hand in furniture, in isolated boards, and in timbers of houses. They are the true house termites and are found in great numbers in wood structures, even when absent or very rare in natural wood.

"Where abundant, *Cryptotermes brevis* is responsible for a great deal of damage. Its colonies are commonly found in shelves of stores, where they often eat into books, stationery, dry goods, etc. A single small drygoods establishment in Colima, Mexico, reports an annual damage of $500 from this species, which was found riddling drygoods and books, and was reported to have destroyed books left on the counter overnight." These insects badly infested a hotel in Florida. The damage was most severe to the woodwork on the third, fourth, fifth and sixth floors. The building was sealed and fumigated with hydrocyanic acid gas at a dose of 12 ounces/0.34 kg of sodium cyanide per 1,000 cu. ft./28 cubic meters for 2½ days. The results were satisfactory.

Anon. (1962) notes that *C. cavifrons* was found in a building in Sarasota, Florida.

THE DESERT DAMPWOOD TERMITE
Paraneotermes simplicicornis (Banks)

Light (1934) has the following to say about this species. The desert damp-wood termite is "common on the border of the desert, along the washes and river valleys, and in the semi-arid portions of the southern border states and Mexico." Light further notes this species "is occasionally, if not typically, of semi-sub-terranean or temporary subterranean habit. It differs from drywood termites in that colonizing pairs cannot enter wood above the ground and in that its colonies are always located in wood which is partly or entirely buried; and from the subterranean termites in that it does not build covered runways to reach wood above the ground.

"The abdomen of the nymph of *Paraneotermes* ordinarily has a spotted appearance due to the fact that the dark contents of the intestine show through in places. The soldiers have low, flat heads, brown or yellowish brown in color. The mandibles are short and are thick near the base, but with narrow tips. The third segment of the antenna shows little if any enlargement or special chitin-ization. The alate is dark brown throughout, including the wings.

"This species breeds in such semi-desert plants as mesquite, and in the Colorado and Mohave deserts and in Arizona it is already of some economic importance and it may be expected to become increasingly so with the Boulder Dam developments." It damages untreated poles and posts, attacking them just below the ground level. Butt-treatment has been found satisfactory for the control of this insect. Bynum (1951) reports this termite damages the root system of young citrus trees in the Lower Rio Grande Valley of Texas. Anon. (1953) found this species attacking hardwood floors in Needles, California, and Anon. (1964) reports it heavily infesting and damaging a home in Idaho.

SUBTERRANEAN TERMITES
Reticulitermes

The Subterranean termites of the genus *Reticulitermes* are the termites most commonly injuring structural timbers in the United States. They are found in practically every state in the Union. These termites, as compared to the dry-wood and dampwood termites, are small. The winged forms are only ½ inch/13 mm long and the workers and soldiers are approximately ¼ inch/six mm long. The soldier has an elongated, quadrangular, light yellow head equipped with mandibles which are straight except for the curved tips. The thorax of the worker is narrower than the head and abdomen.

The subterranean termites are ground-dwelling termites which nest in the soil or in wood or vegetable material in contact with the ground. They are able to reach wood or cellulose material above the ground level by means of the earthen tubes which they build. Holway (1941) notes, however, termite tubes in the great majority of cases are dropped down from infested wood. He states, "With few exceptions tubes found on foundation walls have been traced to col-

Fig. 7-18. Winged sexuals of the eastern subterranean termite. Winged termites in the house indicate an infestation in or in the immediate vicinity of the house.

onies which had become established in the building at some points, usually hidden, where wood was in ground contact or where wood could be reached through hollows, voids or mortar joints." Moreover, they rarely construct tubes over foundations in the open. Since these termites are dependent on soil moisture, they almost always have a ground connection. There are rare exceptions where they have been able to maintain themselves in damp wood in humid areas even without a ground connection.

Unlike the drywood and dampwood termites, their feces consist of liquid drops, not pellets, which characteristically spot their excavations in wood. In their galleries is found a "frass" which is also characteristic. The frass consists of triturated wood and earth cemented together with saliva and liquid feces to form a "mud-like" cement. Mampe (1976) notes from a study made by W.F. Exner that subterranean termites rebuild tubes fast, 2½ inches/63 mm in 65 minutes.

THE WESTERN SUBTERRANEAN TERMITE

Reticulitermes hesperus Banks

This species will be discussed in somewhat greater detail than the others, since it is a termite of prime economic importance on which much research has been done. It is this species that is so commonly encountered in wood in contact with the ground along the entire Pacific Coast from the lower slopes of the mountains to the seashore. It ranges all along the coast from British Columbia into lower California and eastward into Idaho and Nevada.

Swarming. On the first sunny day following the first rains in autumn, these termites will be found erupting from small exit holes in the ground and from

Fig. 7-19. Worker of the western subterranean termite in an exploratory tube.

cracks and crevices in buildings. They often appear in great clusters around these exit holes. If they infest a home, they fly around in the rooms, and it is at this time many individuals mistakenly refer to them as "flying ants." These winged forms have black bodies with light gray wings. Light (1934) observes, "If it is warm enough they fly some distance, toward the sun if not carried elsewhere by the wind, drop to the ground, lose their wings at a preformed basal joint and assort in pairs." He further notes, "If the temperature is 64° F or higher, the insects readily remain outside and may fly aloft even with a slightly clouded sky above them." These dropped wings, as well as the feeble fluttering of the termites in the air, are all familiar sights to those in any way acquainted with termites. The importance of the previous rains may well be understood when we realize the moisture has softened the earth and dampened the wood, thereby permitting the termites to chew out an excavation for a nest. Sporadic flights occur throughout the fall. In the spring of the year, on the first clear days after a long period of rainfall, there are some minor mating flights. Brown (1936) states since these termites live in the ground at varying depths, "the time required for moisture to reach the colony and stimulate the mating flight will vary." This partially explains why winged forms of this species may be encountered in basements and cellars in May and June. Light (1934) notes, "If the fall is very dry the majority may stay thus penned up until late in the season." This may account for the presence of the winged forms in their galleries up until December and even January.

Pickens (1934) gives us the following details about the flight and mating of these termites: "The flight is weak and the direction is largely determined by the prevailing breeze; most of the insects soon sink to earth in the immediate vicinity, though a few may flutter upward for 60 feet and be borne along for as much as 200 yards. Many do not fly aloft at all, but drop their wings while running about on the ground." After settling to the earth and breaking off their wings, the females "now take up a passive position with the abdomen raised high in the air, increasing the area of visibility and probably emitting an odor which is attractive to the males. The latter rush about the surface hurriedly, one by one coming in contact with a female. When a pair thus come together the female depresses the abdomen and the male follows as she leads, the two proceeding in tandem, sometimes joined by a third and even a fourth insect, the whole moving along like a train of cars. At times, in turning the leader curves so far to the right or left that she finds herself behind the last one of the train and the group revolves for some time in a circle. Again, the male may become separated from the female leading him, in which case she stops, re-elevates the abdomen, and waits until he or another male joins her before proceeding again."

They now seek the dark. Once a suitable cavity has been found by the female, the male aids her in the founding of the new colony. They feed and groom each other and copulate within a day of excavating their cavity. According to Pickens (1932), "Copulation has been observed to be repeated at intervals for some months and probably actually occurs at intervals throughout the life of the royal pairs."

Whereas the drywood termites are capable of living in wood without any contact with the soil, the subterranean termites are not as able to utilize the moisture in the wood and therefore they almost always have tubular connections with the moisture in the soil.

These winged sexual forms are produced in tremendous numbers and well they may be, for now they are the prey of every animal that is in any way predacious on insects. Spiders under buildings construct their webs directly over the exit holes of the emerging reproductives, and thus immediately account for some of them. Ants now consume them with great gusto. One may now observe every other ant in a trail hustling off one of these reproductives to a banquet. Birds, lizards, toads, etc., now all help themselves to this feast provided through the bounty of nature. The mortality is terrific, for besides the animal predators, the termites may be killed by dust, dry soil, or they may drown in pools of water.

According to Light (1934), a colony of subterranean termites may vary from three to four to many thousands, the number being dependent upon their age and location. He further notes, "Each colony has at least one queen, but there are often several, depending on the size of the colony. The queen is usually attended by a male, but seems to lay fertile eggs for a considerable period after her consort's death. She lives in a large chamber with the male, attended by a squad of soldiers, and from time to time shifts her quarters to some neighboring cavity of similar size."

Light further notes, "An average of less than 10 eggs is produced as a first clutch. The number may be as low as four to six, or as high as 15, or in rare cases 20. The eggs are laid at fairly regular intervals. A period of from one to three days elapses between the deposition of two successive eggs, except for the last two or three, which are laid at increasingly long intervals, of as much as a week or 10 days." The termites must clean the eggs to prevent mold from growing upon them. The eggs hatch in from 30 to 90 days, with an average of 50 to 56 days.

Pickens (1932) observes the growth of a colony from a primary pair is slow. Few eggs are laid the first year, and these hatch in little more than a month after laying, the periods for the others increasing gradually until the last egg may require almost four months." Pickens also notes the eggs of soldiers appear different and are laid at considerable intervals after the main group of eggs are laid. They also require longer periods for hatching. It appears as though there is a separate soldier group and worker reproductive group. Weesner (1956) prepared a monograph on colony foundation in this species.

Growth. Pickens (1932), in his studies on the biology of *Reticulitermes hesperus* found the first instar requires 14 to 18 days. In the second instar, the termite already contains intestinal Protozoa. The second instar extends from 14 to 18 days. The third instar lasts for approximately a month and the fourth instar requires two months for completion. "In young colonies started by a swarming pair the fourth instar, or a small fifth may be the dominant instar in the colony for as much as two years. Sixth instars result in the development of large, well-matured workers and of reproductive nymphs, but only in colonies sufficiently large to supply abundant food and a large amount of fraternal feeding, which seems necessary to such full development." There may also be a seventh instar of still larger workers. In this instar the perfect reproductive stage is attained in the reproductive caste.

"The development of reproductives requires several years of growth and an abundance of inhabitants, depending apparently on the abundance of material supplied for interfeeding. Swarms, even under highly favorable conditions, cannot be expected before the third or fourth year, at the earliest." The workers are long-lived and Pickens notes, "Workers that do not fall victim to cannibalism may reasonably be expected to live from three to five years, and queens probably live much longer." In pieces of infested material the insects may maintain themselves for weeks after the wood has been lifted from the earth by sealing all openings.

THE ARID LAND SUBTERRANEAN TERMITE

Reticulitermes tibialis (Banks)

This is probably the most widely distributed species of *Recticulitermes* in the United States. It appears in sand dunes to locations as high as 7,000 feet/ 2,133 m in Colorado in a great variety of soils and plant locations. Snyder (1935) states, "It is found in moist river bottom land and along streams in canyons, but essentially it is a desert or prairie species." It is distributed from the Pacific to Mississippi and from Montana to lower California.

The soldier of *Reticulitermes tibialis* has a short, broad, yellow head, while the head of the soldier of *Reticulitermes hesperus* is long, narrow and pale. The arid land subterranean termite is found in the creosote bush and greasewood, which is not a common locality for *Reticulitermes hesperus*. *Reticulitermes tibialis* dwells in drier areas. Thus, it does less damage than the western subterranean termite merely because there are fewer habitations to attack. They have been known to be quite destructive in the Mohave Desert areas.

Pickens (1932) has this to say regarding this species and the western subterranean termite: "Periods of swarming in the eastern humid states appear to be controlled by rise of temperature in the spring, and in the dryer western states by rise in the amount of moisture during autumn. *Reticulitermes hesperus* and its congener *R. tibialis* are strikingly similar in markings and general habits of the colony. The chief difference lies in geographical distribution, reaction to moisture content of soil and in the degree of pugnacity in the various castes in the two species.

"Fertile crossings of *Reticulitermes hesperus* males with *Reticulitermes tibialis* females have been accomplished in the laboratory, as has also the fertile crossings of the unpigmented primary with the pigmented in *Reticulitermes hesperus.*"

THE EASTERN SUBTERRANEAN TERMITE

Reticulitermes flavipes (Kollar)

This species occurs in the Eastern part of the United States up to the Mississippi. Snyder (1934) states it extends its range into Mexico and is present in southern France. Paulson and Boulanger also (1958) record it from Maine. This is the most destructive termite in the East, besides injuring wooden structures and lumber products in the manner of the western subterranean termite, it is also of some importance as a pest of living trees, shrubbery, flowers and crops.

This insect is somewhat similar in habits to the western subterranean termite. According to Snyder, the outdoor colonies are dormant during the winter and the first signs of activity in Washington, D.C., are observed in February or March. The reproductives usually emerge in the vicinity of Washington, D.C., in April or early May. McCauley and Flint (1938) note in Illinois the first swarms "emerge within heated buildings, most often during February. The later swarms emerge out of doors during late May or early June in the northern part of the state and earlier in the southern part." According to Snyder, "Often several swarms emerge from the same colony in the same day, sometimes as many as four separate swarms extending over a period of one month. In size, however, the first swarm, from the writer's observations, is usually the largest." According to information collected by Weesner (1965), this termite has been found in flight in the East from January through August.

Snyder notes there is no definite nor permanent royal cell in this species. He also observes, "The queens of the species of *Reticulitermes* do not reach extraor-

Fig. 7-20. Queen, workers and soldiers of the subterranean termite.

dinary size and never lose the power of locomotion. They probably change their location in the colony to conform to the most favorable conditions of temperature and moisture and go below the frost line in the ground in winter." The rate of egglaying is slow and the period of maximum production is from the middle of May to early September. Large colonies may contain hundreds of thousands of individuals. Turner and Townsend (1936) are of the opinion that individual workers may live for five years.

Reticulitermes hageni (Banks)
Reticulitermes virginicus (Banks)

According to Snyder (1935), these two species occur in the southeastern United States and are very similar in appearance. They swarm from May to October. Snyder (1955) records *R. virginicus* from Long Island, New York, and from Philadelphia.

Reticulitermes humilis (Banks)

This species occurs in Arizona, New Mexico and Mexico. It swarms in June or July.

THE DESERT SUBTERRANEAN TERMITE

Heterotermes aureus (Snyder)

The desert subterranean termite is a pest of some economic importance in the Imperial Valley of California and portions of Arizona where it has attacked hardwoods, particularly oak veneering. It is also very destructive along the west coast of Mexico. According to Light et al. (1930) in Arizona this species has "as its natural reservoir the wood of dead giant cacti and cholla cacti."

Pickens and Light (1934) find the workers of *Heterotermes aureus* (Snyder) are very similar to those of *Reticulitermes* in appearance and habits. The soldier of *Heterotermes* has relatively stronger and more slender mandibles. "The reproductives, while similar in form and size to those of *Reticulitermes* are very light in color, with transparent white wings and yellow or yellow/brown bodies, in contrast to the black of the *Reticulitermes* alate." Moreover, the reproductives of the genus *Heterotermes,* have night-flying habits, unlike those of *Reticulitermes*. The fecal spots and frass of *Heterotermes* are light yellow in color and those of *Recticulitermes* are a dirty light brown.

Pickens and Light are of the opinion "*Heterotermes* is favored by the advance of civilization into the desert. Irrigation brings more moisture and artificially heated basements are quite as attractive to it as they are to *Reticulitermes hesperus*. The transportation of dead wood for fuel possibly leads to the distribution of *Heterotermes* into areas not previously inhabited."

According to Pickens and Light, this species does serious damage to sound drywood. It prefers hardwoods, building runways over impervious materials in order to reach the wood. In homes it is especially common around furnaces and fireplaces as was observed in Phoenix, Arizona. It is also of some consequence in Arizona because of its damage to the butts of poles. Beneath buildings these termites may build tubes that are 24 inches/0.6 meters or more high.

FORMOSAN SUBTERRANEAN TERMITE

Coptotermes formosanus Shiraki

This termite invader of the United States is of great concern as a potentially important pest. E. Dale McCullough, a pest control operator in Houston, Texas, found it in 1965 in a shipyard warehouse in Houston. In 1966 it was also found established in Galveston, Texas, as well as in New Orleans and Lake Charles, Louisiana. It also was found in 1967 in Charleston, South Carolina. In 1980 it was reported established in a condominium in Hallandale, Florida (Koehler, 1980). It is capable of doing serious damage in as little as three months. This termite is common in China and Taiwan as well as Guam and Hawaii.

Beal (1967) notes the carton nests made of masticated wood and termite saliva "may be several cubic feet in size and a single colony may contain several hundred thousand individuals. Tunnels have been found extending as far as 10 feet beneath the soil. They are normally much larger in diameter than those constructed by our native species."

The queen has a reproductive capacity as high as 1,000 eggs a day. The aggressive soldiers have a frontal gland which secretes an acidic substance that can go through such materials as asphaltum, lead, plastics, mortar, plaster, etc., to enable these termites to attack wood. In Hawaii, this termite has caused short circuits by damaging electric cables. Gentry (1966) observes the oriental subterranean termite is largely limited to warmer areas with adequate rainfall.

This termite is a fairly important pest in Hawaii. Ehrhorn (1934) states this is a subterranean and moisture-loving termite. It nests in the ground, at the base of poles in old tree stumps, etc. The adults fly in the evening and are attracted to lights. It can become established in the upper stories of buildings when the moisture conditions are satisfactory. It is believed to have come to Honolulu in the soil of potted plants or in wood from China or Taiwan. It attacks wharfs, buildings and packing cases, as well as the woodwork of steamers.

Light and Pickens (1934) note, "Like other subterranean termites they ordinarily have a ground connection and live partly in ground and partly in wood. *Coptotermes formosanus* has the ability, however, not known to be possessed by other subterranean termites, of living without ground connection where there is a suitably located constant water supply. A ship arriving in San Francisco from the tropics, for example, was found to contain a very large colony with probably hundreds of thousands of nymphs, workers and soldiers, and several supplementary reproductives, living under the heavy flooring and wainscoting of three cabins, and centering under a leaking percolater. Without such constant water supply, however, the colonies of this species cannot persist without ground connection."

THE DESERT TERMITES

Amitermes sp.

According to Light et al. (1930), the termites of this genus extend their range from Mexico into the United States, chiefly into the desert and semi-arid regions of California, Arizona and Texas. *Amitermes arizonensis* (Banks), which is common in southeastern California, has eaten the bark of young citrus trees. *Amitermes wheeleri* (Desneux) has been reported doing damage in Texas. At present they are of little economic importance and for the most part feed on desert vegetation.

TERMITE CONTROL

Introduction. Our interest in termites as economic pests is almost entirely due to their use of finished lumber for food. Termites are responsible for damage of wood structures in the United States amounting to millions of dollars annually, for which the subterranean termites are for the most part responsible.

Termites destroy products other than wood such as paper, books, clothing, shoes and leather materials, and are often injurious to living trees and shrubs.

In a survey, Anon. (1951c) made by the author's laboratory of termite infestations in the United States, the Pest Control Operators of California organization estimated $8 million was spent on 60,000 jobs in the state for the year 1950. State authorities in Illinois reported the same year, 10,000 termite jobs at an estimated cost of $2 million. According to this survey, the cost of the average termite job in the United States during 1950 was $147.

Heal (1964) brings some of the above information up-to-date. He notes Georgia had 34,000 termite control jobs in 1963, Oklahoma had 16,145 jobs in 1963 and California had 112,177 jobs in 1962. In 1981 it was estimated that in the United States termites accounted for over $500 million in property damage each year.

The question has been raised as to how long it takes for subterranean termites in the continental United States to cause severe damage. This is of importance because some individuals have attempted to "write off" such damage as a tax loss. In general, authorities agree it would take three to eight years of termite infestation to cause extensive damage and even under very severe conditions it would take one to six years.

Termite control can be accomplished most effectively by individuals experienced in this type of work such as professional pest control operators. These specialists must understand building construction, soil and related insecticides, the species and habits of termites and other wood-boring insects. The species of termites, the intensity of infestation, building construction, the nature of the soil and other factors differ greatly from one part of the country to the other, and for these reasons control methods also vary. Finally, the value of the structure as well as the economic status of the owner must be considered. The owner

Fig. 7-21. Damage by drywood termites. Note the fecal pellets.

who can afford it may be able to undertake extensive alterations to correct the conditions which make the wooden portions of his buildings susceptible to termite attack. On the other hand, where cost is a primary factor, the termite operator may resort to piecemeal repairs and localized or "spot" chemical treatment.

The homeowner who is not acquainted with termites and termite control can obtain literature on this subject from the United States Department of Agriculture, Washington, D.C., as well as from the Agricultural Experiment Station in the state college. It is possible for the intelligent homeowner to cope with his termite infestation where it is of a limited nature. However, where the infestation is severe and widespread throughout the house, an experienced and reliable termite operator is essential for control of these pests.

Fig. 7-22. Damage to door frame by eastern subterranean termites.

In the following pages, the author outlines some of the factors to be considered, and some of the methods to be used in combating termites. The National Pest Control Association has published a comprehensive work on subterranean termite control which is available to members of their organization. The Building Research Institute (1956) released a very useful publication on decay and termite control in houses. Du Chanois (1961) discusses some shortcomings in termite control by pest control operators. Hunt (1966) prepared a very useful manual on the art of inspecting for termites and other wood-destroying organisms.

Termites and fungi. Much of the wood failure in buildings is caused either by termites or by wood-destroying fungi or by a combination of both of these factors. Termites and fungi have a very close interrelationship since the fungi provide the nitrogenous portion of the termite diet which is essential for their growth and development. And the termites for their part aid in spreading the fungi. As a matter of fact, termite burrows provide a moist atmosphere rich in carbon dioxide that is very favorable for the growth of fungi. Conditions conducive to the growth of fungi are usually favorable for termite infestation and *vice versa*. Thus, efforts undertaken to control one will usually control the other. It is believed when the termite operator utilizes such fundamental precepts as insulation of the wood from the ground, good cross-ventilation, the use of chemically-treated wood, etc., he controls the termites in part by making conditions unfavorable for fungus growth. Incidentally, wood in advanced stages of decay may often be valueless as food for termites since the fungus has assimilated so much of the nutritive materials in the wood.

Hatfield (1944) has the following to say in regard to the relationship between termites and decay: "Our results, based upon the examination of over 1,600 specimens, show that decay is not prerequisite for attack by subterranean termites. Conversely, we found no evidence that termite attack necessarily induces decay. It can be stated, however, that if conditions are suitable for decay, termite attack is likely to occur more readily and progress more rapidly. Also termites often carry soil and humus into the wood. This increases the moisture-holding capacity of the wood, thereby improving conditions necessary for decay.

"Another misconception is that termites require shade, poor ventilation and poor drainage as adjuncts to tube building."

Termites as native pests. Many individuals harbor the false belief the termite is a comparatively new insect that has but recently invaded their immediate vicinity. Termites have always been with us, but in the past they applied

Courtesy University of California Agricultural Experiment Station

Fig. 7-23. Subterranean termites reach the wood by means of the earthen runways they construct.

their appetites to stumps, dead logs and other plant life. When man commenced to clean the forests and other areas and build homes of "choice cuts" of timber, and then warmed these homes throughout the year, a termite's dream of paradise was realized. There are some individuals who are of the opinion that at times there may be a tremendous increase in the number of termites and these increases are cyclic in nature. However, further work is necessary to confirm this belief.

The home that collapses because of termite injury is indeed a rarity (Turner, 1961). Nevertheless it has been shown some homes badly damaged in an earthquake in California were initially weakened by termite injury to the foundation timbers. There can be no doubt that at times termites do severe damage to structural timbers in the home. Hagen (1876), speaking of *Reticulitermes lucifugus*, a European subterranean termite, states in France a whole dinner party suddenly fell through the floor from the third story, down to the cellar. One cannot be certain of the truth of this tale, since it is believed Hagen received this information from some secondary source. However, every experienced termite operator is aware of instances where some heavy piece of furniture, such as a piano, has cracked floor boards previously weakened by termites.

Distinguishing termite work from other wood-infesting insects. Wood infested by subterranean termites will usually show their earthen tubes attached to the wood or the galleries of infested wood lined and spotted with hard adobe-like material. Drywood and dampwood termites have galleries packed with pellets.

Since powderpost beetles bore in wood, too, their work is commonly confused with that of termites. Powder post work can be distinguished by the small circular holes ($1/16$ of an inch/1.6 mm or less in diameter) formed by the emerging beetles, as well as by the extremely fine borings which issue from the holes and is often mistaken for spilled face powder.

The work of the death-watch beetles of the family Anobiidae is often confused with that of termites. They honeycomb wood with their tunnelings. Old infested pieces of lumber show numerous emergence holes. The pellets of anobiids may be distinguished from the pellets of the drywood termite by the long narrow shape and pointed ends.

The tunnelings caused by the larvae of the true wood-boring beetles of the families of Cerambycidae and Buprestidae may also be confused with termite galleries. The burrows are wide, flat and plugged here and there with a fibrous coarse sawdust-like material.

Carpenter ants, unlike termites, excavate wood not as a source of food, but merely to provide a home. The ant chambers, unlike those of the termites, are always free of chaff and refuse. These chambers resemble the clean chambers of the drywood termite, but lack the pellets present in the chambers of the latter.

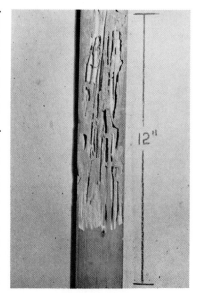

Fig. 7-24. Damage to door stop by eastern subterranean termite.

CONTROL OF SUBTERRANEAN TERMITES

Evidence of subterranean termite infestation. Evidence of infestation by these termites may be recognized by a number of signs:

- Earthen tubes extend from the ground to the wood and at times may pass for long distances over concrete foundations, etc. These tubes are earthlike in appearance, often flattened and composed of tiny particles of earth and of partly digested wood, glued together to form the tube.
- Wood attacked by subterranean termites comes apart readily when badly infested, revealing termite tunnels, as well as earthlike frass or fecal pellets. If the infestation is a fresh one, the white termites themselves may be seen.
- The presence of the winged termite males or females, particularly when they fly in large numbers inside the home, usually indicates an infestation in the structure. According to Light et al. (1930) in the western United States these flights very commonly occur "on the first clear days after the first real rain of the autumn; lesser swarms emerge under similar conditions in the spring. These male and female subterranean termites are jet-black with gray wings and are commonly mistaken for flying ants."

Snyder (1948) notes in the East the time of swarming varies with the species, the locality and the seasons. In the vicinity of Washington, D.C., *Reticulitermes flavipes* (Kollar) swarms from the last of April; *R. virginicus* the first of June; and *R. hageni* in July or August.

In the eastern states information as to the extent of the termite population in the ground may be obtained by setting blocks or stocks of unpainted white pine in the ground and then examining them after 60 days. This procedure is unnecessary in many of the southern and western states since here subterranean termites are widespread.

Termite inspections. The presence or absence of termites in a wooden structure can best be determined through a thorough inspection by a competent pest control operator or experienced entomologist, Denny (1961), Hunt (1966).

In order to make a satisfactory inspection, the inspector should be provided with a pair of coveralls, either a flashlight or an electric light bulb in a cage, attached to at least 100 feet/30 m of extension cord. The operator equips himself with a hammer or a tool such as an icepick, screwdriver, or geologist's pick so he can probe the wood. Graph paper and drawing board are used to outline the plan of the house to indicate the location of termites and the necessary repairs. In some parts of the United States trained beagles are used to assist inspectors. They are reported to be of particular value where infestations are not readily accesible to humans. (Caruba, 1981).

The job of inspecting a building is usually a difficult and arduous one. It often consists of crawling on one's stomach in the dirt, mud and litter beneath the home, brushing aside spider webs, avoiding broken bottles, boards with nails and like objects. Where the attic is inspected, one may encounter very high temperatures and great humidity in narrowly confined and extremely dusty places. Here one must often lay flat and wriggle over innumerable rafters that are not especially down-like in texture. Thus, not only is skill required to make a thorough inspection, but stamina as well. Howell (1952) and Zimmern (1950 and 1952) have prepared a list of safety precautions to be observed by the pest control operator in doing termite work. Isherwood (1950) described and illustrated building terms for use by the pest control operator.

Termite operators in the South, in addition to the above hazards in termite inspections, are subject to infestation by the hookworm, *Ancylostoma brazil-*

iense, known as "creeping eruption." These cause intense itching and irritation in the skin for prolonged periods. Anon. (1951b) states the larvae "from this hookworm of pets, swarm in the droppings. They crawl from the dropped manure and hide in damp soil until an animal or human victim shows up. Then they quickly pierce the skin and enter. Most cases are contracted in the southern states in areas where soil is sandy and moist. The season lasts from May to November."

The above article recommends the following preventive measures:
- "Postpone annual termite inspections until after the danger season.
- Avoid inspections (initial), during the danger season, of houses where crawl spaces are open to pets.
- Postpone treatment of suspicious houses until worst part of season is past.
- Take special precautions at all times under houses where there has been a known history of infection."

According to Guion (1951), the infested soil can be disinfected by raking salt into it to a depth of five inches/13 cm or more or fumigating it with ethylene dibromide. The wearing of gloves and suitable clothing to prevent the bare skin from contacting the soil also is helpful. Walters (1957) claims an ointment, Kerodex No. 71, prevents the itch when applied to the skin before creeping under a building.

Fuller (1966) recommends the use of ethyl chloride spray for the dermatitis. He also suggests the use of sodium borate as a larvicide at the rate of 10 lbs. per 100 sq. ft./4.5 kg per nine square meters of soil. The use of a plastic sheet over the soil, long-sleeved shirts and trouser legs secured at the ankles also are helpful. Showering after work also is recommended.

In seeking evidence of subterranean termites in a home, one should thoroughly examine the exterior and interior surfaces of the foundation, particularly where wood construction is on or near the soil. A careful inspection should be made of the wood construction in the basement or underarea of the house for evidence of termite tubes, tunnels or termite damage. These tubes, if present, are readily seen when they pass over the outer surface of the foundation or on the outside of the wooden understructure, but very often they may come up through a hollow block or a crack in the cement or brick construction, an expansion joint, etc. The tube may be hidden in a not readily perceived crevice such as very often occurs around the chimney. Sills, girders, joists, wood columns and basement window frames of wood should be probed carefully. Scrap wood on the ground should be examined for the presence of termites. Where fills are in contact with portions of the wooden understructure, evidence of termite attack may not be visible. Hollow masonry, the voids in cement blocks, siding and stucco close to the ground will also hide evidence of termite infestation. Needless to say, special

Fig. 7-25. Typical tubes of subterranean termites.

note must be taken of any wood in contact with the ground.

At the same time the pest control operator inspects a building for termites and other wood-boring insects, he also examines the structure for evidence of or conditions likely to promote fungus infestation. This is because alterations that correct one condition normally aid in rectifying the other.

How inspections are performed and what is reported varies somewhat from area to area, due primarily to local, state and FHA & VA requirements. The inspector puts his integrity and his company's liability on the line with each inspection. All of the issues are not currently resolved, but generally a thorough inspection should be made of all accessible areas where termites are likely to occur, including attics in drywood "country." Damage found should be disclosed in the report. Many state associations of pest control operators have developed useful forms that meet local conditions and regulations.

STRUCTURAL ALTERATIONS FOR TERMITE CONTROL

How termites infest buildings. Subterranean termites may infest the wood structure of a home through direct contact between the wood and the soil or by means of their earthen tubes, which often may pass along hidden crevices in the masonry. In very rare instances termites have entered homes from infested firewood stored in the basement. Formerly it was believed winged sexual forms of subterranean termites cannot initiate a colony directly in the wood of a house, but instead begin their new colony in the soil. Now we are not so sure for some students of subterranean termites believe the sexual forms may be able to infest consistently damp wood even without ground contacts. More research is needed here. In any event, severe damage ordinarily is caused after one or more years of infestation. See FHA (1958) for their recommendations against termites and decay and note the following:

1) No wood contact with the ground. It has been estimated that 90 percent of termite infestations in the home can be traced to contact of the wood with the ground. Where wood is used that is within six inches/15 cm of the ground it should be either coal-tar creosote or chromated zinc chloride pressure-treated wood, or wood protected by some equally effective chemical. Where the ground contact or tubes and tunnels of the subterranean termites have been broken, the termites in the house (except where the wood is damp) eventually die since they cannot maintain the necessary moisture in their galleries.

According to Howell (1951), when the wood is moist as in humid areas in the South, subterranean termites may continue to live in the timbers even when ground contact is broken. The supplementary queens keep the colony alive and in one or two years damage to timbers may be observed once again. Schendel (1952) and Buswell (1955) found subterranean termites established in wet wood 50 to 70 feet/15 and 21 m above the ground, respectively. There was no evidence of tubes connecting the infestation with the ground. Delaplane (1953) discovered a colony of subterranean termites with a primary reproductive functioning as a queen in the damp wood of a leaky attic in Illinois. Such infestation of subterranean termites, without any connections to the ground, are the exception to the rule.

Wood siding, shakes, or shingles should not be closer than six inches/15 cm to the ground. Piers should be of solid concrete and should extend at least six inches/15 cm above the ground level. Wooden steps should be supported on a concrete base at least six inches/15 cm above the ground and the steps should be separated from the remainder of the building by metal shielding.

2) **Foundation.** The foundation should be of solid concrete and preferably reinforced to prevent cracking since cracks $^1/_{32}$ inch/0.8 mm will permit the entry of termites. Where the foundation walls meet and intersect, they should be joined by steel rods to prevent opening of the joints or cracks due to shrinkage. Where stone, brick, or hollow unit masonry is used, it should have a solid capping of four inches/10 cm reinforced concrete or a metal shield. The exterior foundation walls should extend at least six inches/15 cm above the outside grade line. Where exterior foundation walls are level with or below the exterior grade line, the foundation must be raised to a point at least four inches/10 cm above the exterior grade or a three inch/eight cm concrete wall installed against the building at all points where the foundation is found to be in this condition, with the top of the curb extending at least six inches/15 cm above grade and being sloped from the building and properly flashed with the bottom extending to a point at least six inches/15 cm below the top of the foundation. The concrete should consist of the following mixture: one part Portland cement, three parts sand and where hydrated lime is used it should not exceed three percent of weight of cement that is to be used.

The U.S. Bureau of Entomology and Plant Quarantine, (Anon., 1949) recommends where wooden piers or posts are used for foundations that these be impregnated with an approved chemical preservative by a standard pressure process or heartwood of naturally resistant species should be required. As an additional safeguard, metal termite shields may be installed on top of such piers or posts to prevent termites from tubing up through checks or cracks and thus gaining hidden access to the building.

3) **Debris beneath the home.** All wood from beneath the building whether

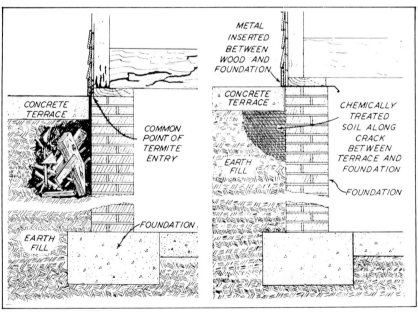

Fig. 7-26. (Left) An earth-filled terrace is a common source of termite attack. (Right) The same terrace pictured at left after being treated chemically and provided with a metal shield.

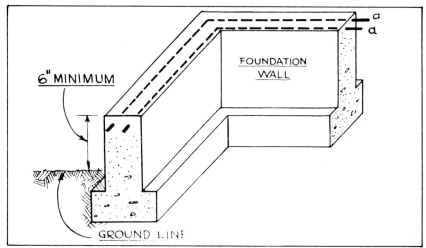

Fig. 7-27. A well-built foundation wall with $^1/_4$-inch square or $^3/_8$-inch round rods to prevent cracking.

stumps, tree roots, scrap wood, form boards, wood chips or sawdust should be removed. No wood should be buried in any of the fills. See Russell (1958).

4) **Ground area.** The ground beneath a building should be leveled. It should provide at least 18 inches/46 cm clear space between all horizontal timbers and the ground. According to Anon. (1949), 24 and 30 inches/60 and 76 cm clearance is to be preferred in the South and other humid sections of the country. Moreover, 30 inches/76 cm clearance facilitates regular inspection and the application of control measures when necessary. Where the ground level is lowered to as little as 12 inches/30 cm clearance, each 100 square feet/nine square meters

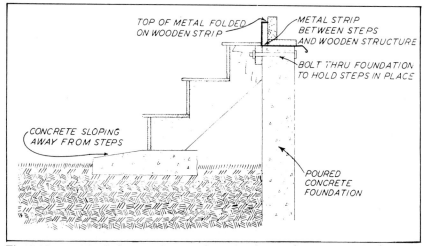

Fig. 7-28. A method of protecting wooden steps from subterranean termites.

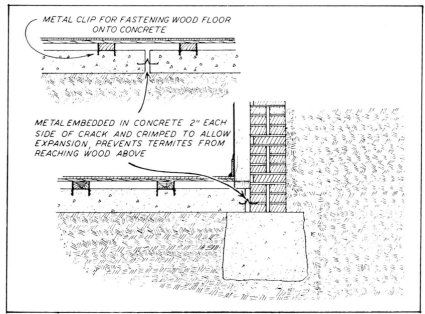

Fig. 7-29. Metal expansion joints installed between concrete floors and walls, and between two sections of concrete floor, will prevent entrance of termites.

of ground area should then be treated with at least five gallons/19 l of toxic solution. The ground level of the outer wall should be at least six inches/15 cm from the top of the foundation and the ground so graded that moisture drains away from the building.

5) **Basement.** This is an important area for termites and fungus injury. Wood should be used as sparingly as possible here, and where used, should preferably be of the pressure-treated type. No wood should extend into the concrete or masonry, nor should timbers be permitted to remain where concrete is poured around them. Timber resting on concrete should be treated with coal-tar creosote. Since cracks or openings in the concrete floors and between the concrete floor and side walls are a common source of termite infestation, all these should be sealed with a noncorrosive metal expansion joint or with such materials as coal-tar pitch, or a black roofing cement. Wooden partitions are to be placed on concrete supports raised at least four inches/10 cm above the floor level. Studding against unfinished cellar walls should preferably be of pressure-treated wood. Stored material like books, letters, files, etc., should be raised on shielded supports.

6) **Cellar hatchways.** The cellar hatchways should be part of the main foundation, constructed of solid concrete and of sufficient height so no wood is closer than six inches/15 cm to the ground. The doors and casings should be made of treated wood where the hatchway is not part of the main foundation and should be separated from the building proper by metal shielding.

Anon. (1949) makes these additional suggestions for the protection of wood in basements from termite attack. "A waterproofing coating, such as roofer's felt mopped on with asphalt, should be applied before the wood floor is laid.

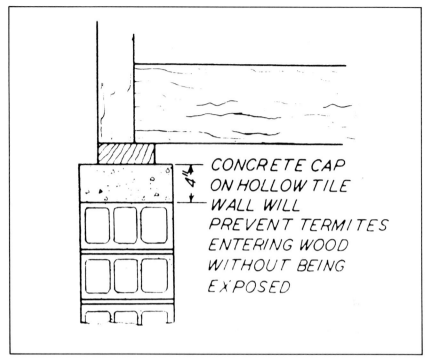

Fig. 7-30. A method for preventing termites from passing through or between hollow masonry.

Every effort should be made to prevent the formation of concealed cracks in the concrete.

"Wooden girders, sills or joints in or on foundation walls should not be placed below the outside grade levels because termites may find hidden access to this wood. Decay also will be a serious problem under such conditions. Floor joists and girders set in masonry or concrete walls should have an air space of at least one inch around the sides and ends.

7) **Cellar windows.** The frames of cellar windows should be made of either metal or pressure-treated wood. The openings of the cellar windows and doors should be finished with cement mortar so there is no possibility of the termites entering through cracks into the frames of windows and doors. The bottom of the well should be at least six inches/15 cm below any wood.

8) **Vents.** A sufficient number of vents should be constructed to provide cross-ventilation under structures having no finished basement. The vents should furnish two square feet/0.19 square meters of air space for each 25 lineal feet/ 7.6 meters of foundation wall. One vent should be within five feet/1.5 meters of each exterior corner of the building. Provisions should be made so there are no dead air pockets anywhere beneath the building. Some of the vents should be of sufficient size to permit a man to crawl beneath a house. All openings should be screened with ¼-inch/6 mm galvanized iron mesh screen to keep rodents and other animals outside. The frame of the vent should not be in contact with the ground. The top of concrete under all vents should be at least three

inches/7.6 cm above the finish grade. Where the vents are beneath the ground level they should be protected by solid concrete window casements. A three-inch/7.6 cm concrete (or four-inch/10 cm brick laid in cement mortar) curb extending four inches/10 cm above the ground level and six inches/15 cm below the bottom of the vent with an earth or gravel bottom should be installed around each vent or opening which is less than three inches/7.6 cm above grade. This precaution is recommended to prevent entrance of surface water to the subfloor area. Roof drainage should be diverted from vents or openings by galvanized iron gutters at the eave line. Plants should not be permitted to grow over the vents and thereby prevent cross ventilation of air. See Anon. (1959).

9) Drainage. Anon. (1949) recommends precautions be taken to prevent the accumulation of moisture in the soil beneath a building. "The soil surface should be sloped so that surface water will drain away from the building. Eaves and downspouts connected to storm sewer systems are very helpful. Buildings with basements should have drainage tile around the outside of the foundation footings if the site is low or wet." This same source advocates the use of roofing paper to cover the soil beneath buildings without basements. This practice reduces evaporation from the soil as well as condensation of moisture on floor timbers. The roofing paper used should weigh 55 pounds/24.9 kg per roll of 108 square feet/10 square meters and should be lapped for two or three inches/five to seven cm without being fastened together.

10) Porches. The foundation of the house should be extended under the porch and the wood used for the porch framework should be pressure-treated. A metal shield or apron should separate the porch from the house proper. Ample cross ventilation should be provided beneath the porch and wood lattices should be at least two inches clear of the soil. The foundation adjoining an earth-filled masonry porch should be of solid and preferably reinforced concrete at least four inches/10 cm thick. In no case should timber be permitted to be in contact with the earth fill.

Anon. (1949) notes concrete or masonry porches, terraces and steps are responsible for a large part of the termite infestations in homes since such construction almost always joins the exterior wall of the building above the top of the foundation. Thus, termites obtain hidden access to the building. The above source discusses the protection of porches as follows. "Protection against such infestation can be provided by the use of a properly designed and installed metal barrier or apron. This apron must effectively isolate the soil and slab from the woodwork of the building and make an impervious barrier to termites. An important feature that should be embodied in all such aprons is a vertical extension to serve as a flashing to prevent moisture from reaching the sill and causing decay. Painting the apron with asphalt after the apron is in place will help to prolong the life of the metal."

It is recommended porches, terraces and steps should not be filled since such fills definitely promote termite infestations where wood is exposed to termite attack. "Where such structures are not filled, the slab or floor should be adequately reinforced and an access door should be left in the foundation so that the form boards can be removed and periodic inspections made. Ventilation openings must be provided.

"The lower or outer step and the platform supports should rest upon poured or solid concrete bases or aprons at least six inches above grade."

Wooden skirting between piers should not touch the ground and preferably should rest on a low concrete wall. Where skirting is used, it is preferable to use pressure-treated wood and to periodically inspect the exterior and interior surfaces of the skirting for signs of termites.

The Pest Control Operators of California (Anon., 1950) recommended the use of concrete barriers where the earth-filled porch presents a termite hazard to the house proper: "Where earth fills of porches and terraces are not properly isolated from the wood of the building, such fills shall be isolated by a barrier wall not less than four inches thick of either poured concrete or masonry set in cement mortar."

Scott (1951) discusses in some detail the skilled work necessary to make a concrete "seal-off" or barrier for dirt-filled porches in Southern California.

11) Buttresses, pilaster, porch and patio walls, etc. The framework of all buttresses, porch columns, pilasters, archways and porch or patio walls should rest on concrete slabs or bases which should be at least three inches/7.6 cm thick and in no instance less than four inches/10 cm above the exterior or interior grade line.

12) Concrete slabs on frame construction. Exterior porches and patios or interior hallways, sunrooms, solariums, etc., having concrete floor slabs on frame construction should have sufficient clearance between the frame supporting the slab and the earth below to permit complete inspection at all times. The area beneath construction of this type should have cross-ventilation and under no circumstances should there be dead air-space.

13) Pipes. Pipes should be set tightly in concrete floors and surrounded with coal tar pitch where they pass through basement floors or walks. The pipes should be sufficiently distant from the walls to allow for inspection. Care must be exercised to see that condensed moisture from sweating pipes does not wet nearby walls. Pipes should not be elevated from the ground by wooden supports since these permit the termites to build their tubes to nearby structural timbers.

14) Miscellaneous items.

- Joists should not be inserted in walls, but hung on metal stirrups or supported on metal hangers.
- Stucco on an exterior wall should be anchored to the foundation wall so termites cannot build runways between the stucco and the wall. Kick the side of the stucco to see if it is hollow below ground.
- Where creosote-impregnated wood is cut, it should be coated with at least two coats of hot coal-tar creosote.
- Structurally unsound timber should preferably be replaced with pressure-treated timber.
- Where there is evidence of subterranean termites or wood-destroying fungi in the ground, the earth should be raked and loosened up. Trenches not less than four inches/10 cm deep should be dug around all interior foundation walls, intermediate footings and piers. Then apply chemicals as indicated in the section on soil poisons.
- Retaining-wall flower boxes are often termite problems; see Anon. (1963).

15) Reinspections. Reinspections of a building should be made once or twice a year, preferably the latter.

The following table prepared by Kofoid and Chase from "Termites and Termite Control" presents a diagnosis of 1,000 cases of attack by *Reticulitermes* in the western states:

1. Direct ground contacts 768
2. Concrete foundation trouble 92

GROUND TREATMENT WITH SOIL INSECTICIDES

The purpose of ground treatment is to make the soil repellent, distasteful or poisonous so termites are prevented from penetrating through the treated layer. Most authorities recommend methods of correct construction be resorted to in the first place and chemical control be resorted to as a secondary measure. The cost of labor and the effectiveness of modern termiticides, however, have made chemical soil treatment the favored method of controlling subterranean termites. Although soil insecticides have been used for many years in the control of subterranean termites, our knowledge of their effect on termites, on plants, their reaction with soils, the duration of their effectiveness and other factors is far from complete. Soil insecticides are used for the protection of slabs, foundation piers, cellar, porch, terrace, steps, drives, walks and other areas where the soil treatment can prevent termite infestations. The preconstruction soil treatment, applied beneath and immediately adjacent to the foundation before the rest of the house is constructed, is now recommended (Heal, 1957). Recognized soil insecticides properly applied will give 20 or more years protection against subterranean termites (Johnston, et al. 1971). St. George (1952) describes methods used for testing the effectiveness of soil insecticides.

Berger (1947), in his bulletin on termite control in Illinois, has the following to say concerning the use of soil insecticides for subterranean termite control: "The length of time a soil poison is effective varies with the locality and the climate, the kind of soil, the amount of poison used, how well it is mixed with the soil, and the location of the water table. Sometimes the application of a soil poison exterminates the only termite colony in the vicinity, and the fact that termites do not reappear in the treated structure is credited to the lasting effectiveness of the poison used; consideration is not given to the possibility that, if there has been a survival of termites in or near the structure after the original application, the effective period would have been very limited. Chemical treatments are usually guaranteed for five years by professional termite control operators. The term of five years is considered to be a safe average period of effectiveness for good soil poisons."

Professional termite control companies employ many labor-saving devices for applying the necessary chemicals. These include electric, drills, pumps and other special pieces of equipment. Smith (1956 and 1957), Spitz (1958), Anon. (1960) and Jones (1966) review some of the equipment useful in their work.

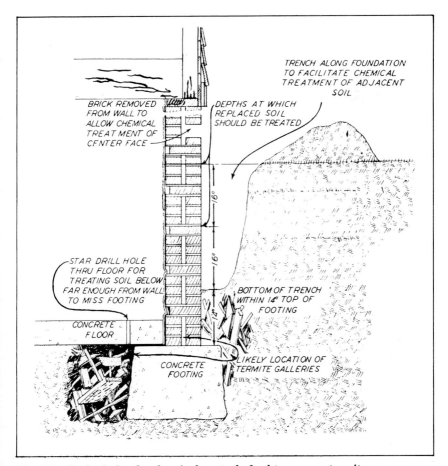

Fig. 7-31. Methods for the chemical control of subterranean termites.

FHA recommendations. The Federal Housing Administration recommends the following measures for soil treatment:
a) Chemicals and concentrations.
Apply to soil areas to be treated, one of the following chemicals at not less than the designated concentration:

CHEMICALS	CONCENTRATIONS
Aldrin	0.5% applied in water emulsion.
Benzene hexachloride (BHC)	0.8% of gamma isomer applied in water emulsion.
Chlordane	1.0% applied in water emulsion.
Dieldrin	0.3% applied in water emulsion.
Heptachlor	0.5% in water emulsion.

Notes:
1) Other materials may be used provided they contain at least one of the above-mentioned chemicals in the concentration recommended.
2) The listed chemicals are toxic to plant and animal life. They should be applied only with caution by an experienced person.
3) Oil solutions shall not be used under concrete slabs or in other locations where the solution may come in contact with vapor barriers.
4) Where individual water supply systems are used, soil treatment is acceptable only when compliance with 1102-3.4 is assured. See 1102-3.5 for additional restrictions (FHA specifications).
b) Application — General.
Treatment shall not be made when the soil or fill is excessively wet or immediately after heavy rains, to avoid surface flow of the toxicant from application site. Surface flow of toxicants toward sources of individual water supply shall be avoided. Unless the treated areas are to be immediately covered, precautions shall be taken to prevent disturbance of the treatment by human hands or animal contact with the treated soil.
c) Application — Under Slabs.
1) Apply an overall treatment under entire surface of floor slab including porch floors and entrance platforms. Apply at rate of one gallon per 10 square feet/3.8 l per 0.9 square meters, except that if fill under slab is gravel or other coarse absorbent material, apply at rate of 1½ gallons per 10 square feet/5.7 l per 0.9 square meters.
2) Where finished rooms are provided in basements or below-grade areas, apply an overall treatment under entire surface of basement floor at rate of one gallon per 10 square feet/3.8 l per 0.9 square meters except that if fill under slab is gravel or other coarse absorbent material, apply at rate of 1½ gallons per 10 square feet/5.7 l per 0.9 square meters.
d) Application — Foundations.
1) Apply at a rate of four gallons per 10 lineal feet/15 l per three lineal meters to critical areas along both sides of foundation walls, interior foundation walls, around plumbing, piers, etc.
   a) Where concrete foundations are used or where construction is slab-on-grade, apply to a depth of one foot/30 cm by any convenient method, acceptable to the FHA field office, that will provide a uniformly treated soil barrier.
   b) Where masonry foundations are used, increase rate of application by multiplying basic rate by depth of foundation in feet.

2) Voids of unit masonry foundation walls and piers, apply to void at or near bottom of foundation at rate of two gallons per 10 lineal feet/7.6 l per three lined meters.
3) Application may be by either of the following methods when acceptable to the FHA field office:
   a) Chemical shall be mixed with the soil as it is being replaced in the trench. Replace in approximately one foot/30 cm lifts, or
   b) Chemical shall be applied by the rodding method in which a soil injector rod is inserted into the soil around the foundation, at approximately 12-inch/30 cm intervals, at a distance six inches/15 cm from the wall. Penetration of the rod shall be to within six inches/15 cm of the top of the footing. Chemicals shall be dispersed, under pressure, through the rod as it is being inserted into the soil. As a guide, the following pressures should be provided to assure dispersal of the toxicants through various types of soil:

SOIL TYPES	PRESSURE (psi)
SW, SP, GW, GP	30 to 50
GM, GC, ML, SM, SC, OL, CL	50 to 150
MH, CH, OH	150 to 300

e) Guarantee.
   1) A written guarantee shall be furnished to the homeowner providing:
      a) That the chemical having at least the required concentration and the rate and method of application complies in every respect with the standards contained herein, and
      b) That the soil treatment firm guarantees the effectiveness of the treatment against termite infestation for a period of not less than five years from date of treatment. Any evidence of infestation within the guarantee period will require treatment, without cost to the owner, in accordance with FHA standards, and
      c) The name and state license number of the soil treatment firm, when required by state regulation.
   2) The guarantee shall be drawn in favor of the owner, successor or assignee.
   Table 7-1 shall be used in establishing the minimum acceptable distances between wells and sources of pollution located on either the same or adjoining lots. These distances may be increased by either the health authority having jurisdiction or the FHA field office when site conditions warrant.

TABLE 7-1
MINIMUM DISTANCE IN FEET

Source of Pollution	Minimum Distance (feet)
Septic tank	50
Absorption field	100 (1)
Seepage pit	100 (1)
Absorption bed	100 (1)
Sewer lines with permanent watertight joints	10
Other sewer lines	50
Chemical poisoned soil	100 (1)
Dry well	50
Other	– (2)

1) The horizontal distance between the sewage absorption system and the well, or chemically poisoned soil for termite treatment and the well may be reduced to 50 feet only where the ground surface is effectively separated from the water bearing formation by an extensive, continuous impervious strata of clay, hardpan, rock, etc. The well shall be constructed so as to prevent the entrance of surface water and sewage as effectively as the undistributed impervious soil prior to the well construction.

2) Recommendation of health authority.

Individual water supply systems are not acceptable in areas where chemical soil poisoning is practiced if the overburden of soil between the ground surface and the water bearing formation is coarse-grained sand, gravel or creviced or channeled rock which will permit the recharge water to carry the toxicants into the zone of saturation.

Soil treatment along foundations. The soil on either side of a foundation should be treated to prevent termites from tubing up the foundation wall or entering cracks or other openings in the foundation and moving upward to attack wood members above. Current labels call for approximately two gallons/ 7.6 l of termiticide per five linear feet/1.5 linear meters per foot/30 cm of depth. Thus, if the footing was three feet/0.9 m deep, six gallons/23 l per five linear feet/1.5 meters are called for.

Earlier methods for accomplishing this treatment called for digging a trench and treating the back fill as it was replaced. The most common method of treating is to create a very shallow trench and insert rods along the foundation to inject termiticide into the soil. High pressures are not needed for this work, 50 to 60 pounds per square inch are sufficient. The rods are usually inserted at 12 to 18 inch/30 to 46 cm intervals.

Marshall (1966) and O'Brien et al. (1965) discuss rodding techniques and insecticide injection into soils in some detail.

Under slabs. Smith (1952), in an informative article considers termite control in the latest type of construction which uses concrete slab floors and eliminates basements and underareas. Smith discusses this problem as follows: "Inspection of slab construction for termite infestation often reveals termite activity only in super-structure such as ceiling joists, window casing, door jambs and roof rafters. In conventional construction it is usually foundation timbers on the first floor which show the first sign.

"There is generally a pattern with each type of slab whereby termites gain access to the woodwork of the building. Usual hazards which accompany this construction are expansion joints, precast door or window lintels, air space behind brick veneer, shrinkage of concrete around plumbing, grade stakes left within the slab and shrinkage cracks which develop between concrete pours. If cracks develop in the floor from settling there is no definite pattern."

Smith notes there are a number of classes of slab construction and an intimate knowledge of these is essential if the treatment is to be successful and damage to the building is to be avoided. In this respect he states: "Several years ago when termites first developed in hardwood flooring laid over a concrete base, standard procedure was to drill numerous holes through the wood floor and lightly flood the area between wood floor and concrete. It develops, however, that moisture proofing compounds such as tar, pitch and asphalt are often applied over the top of the concrete floor. In this event the oil carrier for termite chemicals used may cause discoloration for hardwood flooring from tar compounds in solution. On the other hand, when water soluble chemicals are used to flood over this slab there is risk of buckling the wood floor from moisture."

In addition to the above dangers inherent in the treatment of slab-type structures, radiant heating pipes and service lines embedded in the concrete offer other hazards to the termite operator.

The chemicals may be applied through the expansion joint or crack between slab and foundation floor cracks, the void space between masonry foundation and brick veneer, and holes drilled horizontally through the foundation walls. In the latter case, the treatment should be as near the top of the fill as possible.

Termiticides are injected beneath slabs, usually until the chemical emerges through the drilled holes adjacent to the one being treated. The holes are usually 18 inches/four to six cm apart. For pre-construction application, labels currently call for one gallon of termiticide per 10 square feet/3.8 l per 0.9 square meters. When existing construction is treated, less chemical is used and is usually applied beneath critical areas of the slab such as beneath partition walls, where it adjoins a foundation, and around utility lines. The entire area beneath the slab is seldom treated.

Smith has treated wood floors over the slab floor by drilling a one inch/25 cm hole "through the wood floor and a smaller hole drilled through the concrete floor. A short piece of rubber hose inserted into the concrete floor or a wood dowel driven into the concrete floor with a small hole drilled through it serves as an excellent pressure fitting for the standard tapered nozzle.

"Hardwood floors may be repaired by using a standard size wood dowel of the same wood as the floor and finished as the floor. While this type of floor repair is not unsightly, some customers may object to its use. In this case, since we are forcing the chemical under the concrete floor, we can usually place our holes inconspicuously.

Smith (1953) discussed the problem of staining in slab treatment. He notes solvents in the chemicals used may dissolve tar, pitch, or the mastic coating between wood floor and concrete slab, as well as creosote from treated wall studs. Solvents mixed with diatomaceous earth or powdered talc are used to remove such stains.

Denny (1953), Kowal (1954), Smith, M. W. (1957), Spitz (1958) and Anon. (1959) have prepared informative articles on the inspection and treatment of slab houses.

Osmun and Pfendler (1955) describe an injection unit, a sub-slab nozzle, which is useful in the injection under pressure of insecticides for sub-slab treatment. Johnston (1954) discusses tests in the application of soil insecticides beneath slabs and a nozzle that delivers a fan-shaped spray of 180 degrees beneath the slab.

Anon. (1949) says the following regarding use of soil poisons in porch and ground slab construction: "To treat and maintain an enclosed unfilled porch, it is necessary to make an opening in the wall at each end so that the interior can be inspected. Any form boards or other wood present should be removed, existing termite tubes destroyed, and the space ventilated. One of these openings should be large enough to serve as an access door. The soil poison should be applied in a trench along the foundation and the walls supporting the porch.

"In a dirt-filled porch or other similar area adjacent to the outside foundation wall, a metal apron should be inserted between the enclosed area and the wall . . . If such procedure is impractical, a poison can be applied to the soil in a trench adjacent to the foundation after openings have been made through the side walls and after debris and the filled earth have been removed down to the outside grade. In some types of construction, where the slab is long and not well attached to the main wall, it may be necessary to install a supporting wall of piers to prevent cracking of the concrete or masonry slab.

"Where an entrance ground slab is next to a masonry foundation wall surrounding a basement, it often is more convenient to work from the basement than to excavate beneath the slab from outside. From two to three feet of the foundation wall, parallel with and slightly below the lower inner edge of the slab, should be removed so soil can be removed along the wall to provide for future inspection.

"If the foundation wall is of poured concrete, it is usually easier to apply the soil poison from the outside. This would mean excavating a shallow trench along the foundation wall and beneath the slab, working from one or both sides of the slab, removing any debris present, and applying the chemical in this trench.

"In some buildings it is almost impossible to trench under a ground slab abutting a wall. In such case, a strip of the slab along the foundation wall may have to be removed. After the wood debris has been taken out and the soil chemically treated, the slab can be repaired. In other buildings, it may be desirable to bore holes through the slab, and into the ground about 18 inches apart, near the foundation wall. After the treatment with the chemical the holes in the concrete can be plugged."

Basement treatment. Berger (1947) discusses this phase of termite control as follows; see Anon. (1960) for illustrations.

"It is not uncommon during early spring for termites to be seen swarming from cracks in the basement floor around the furnace. For a thorough job of termite control, the basement floor, as well as the foundation walls and outside soils, must be treated with a suitable poison. If wood such as that in door casings, supporting wooden columns and step runners extends through the floor to the soil, this wood should be sawed just above the floor surface; the wood extending through the floor to the soil should be removed, and the hole treated with chemical and then filled with concrete.

"In unexcavated areas beneath the building, the soil inside the foundation wall must be treated the same as that on the outside, except that a relatively odorless poison should be used. If the soil is less than 18 inches from the floor joists, enough soil should be removed to make this clearance. Termites are able to construct tubes from infested floor joists to the soil if the clearance is less than 18 inches; by so doing they can make contact with soil moisture without coming in contact with the treatment applied for their control. If, after a soil poison has been applied to the inside foundation wall, inspection shows that termites have been successful in dropping a tube from the floor to the soil, this tube should be broken and the surface of the soil in this area treated with a poison that has little or no odor.

"Concrete basement floors require considerable effort for termite control treatment. Holes one-half inch to an inch in diameter should be made with a star drill along the foundation walls, along both sides of supporting walls, around the foundation walls to miss the footing (usually six inches is sufficient), and they should be about two to three feet apart. Concrete floors may be treated with the trichlorobenzene oil solution, usually at the rate of about one or two quarts to each hole. Holes that will readily take more should have more added. The holes should be filled with concrete after the solution has soaked into the soil."

CHEMICALS USED IN SOIL TREATMENT

The soil insecticides most commonly used are chlordane, dieldrin, heptachlor and aldrin. Other soil insecticides that have been used are lindane, benzene hexachloride, DDT, sodium arsenite, trichlorobenzene and orthodichlorobenzene. The sodium arsenite is used in water solutions and the organic soil insecticides are either diluted with fuel oil, kerosene and other petroleum bases, or they may be emulsified with water.

The commonly used termiticides for soil treating (chlordane, dieldrin, heptachlor and aldrin) are usually applied as water emulsions. Johnston et al. (1971) found these materials to protect wood in contact with treated soil for more than 20 years. The tests have continued and still provide protection after 25 years or more.

Smith (1968) found DDT applied to soil for subterranean termite control moved less than 12 inches/30 cm after nearly 20 years in soil subjected to heavy rainfall. He believes the same is true for the above chlorinated hydrocarbons.

These four termiticides do not damage vegetation when soil treatments are made under normal circumstances. And if label directions are followed, there should be no harm to the applicator.

Anon. (1949) notes: "Where valuable shrubs and flowers are near the area to be treated (one to two feet) and it is not desirable to remove them temporarily, the plants may be protected by lining the side of the trench next to the shrubbery with tar paper, paraffined canvas, or copper-coated kraft paper. The last material is preferable where orthodichlorobenzene, trichlorobenzene, creosote or light petroleum oils are used, since these chemicals have a solvent action on tar products. Special care must be exercised when poisoning the soil with sodium arsenite because it is water-soluble and is readily absorbed by plants and kills them. Although oils are not absorbed by the roots, they burn any parts they touch.

"Soil poisons should be applied when the ground is dry and warm. When the earth is soaked with water, the chemicals are less able to penetrate through the spaces in the soil."

A newsletter of the National Pest Control Association makes the following remarks on this subject: "Where vegetation is a problem it should be remembered that plants are more subject to injury when the soil is dry than when it is wet; also that at times of high temperature and active growth they are more readily damaged than when temperatures are low and plants are dormant. There is evidence that organic chemical toxicants tend to injure and destroy those roots which they contact, while water soluble inorganic toxicants may be absorbed by part of the root system and by translocation in the plant destroy the entire plant. It is possible that in hot weather and in close quarters the vapor of volatile chemicals such as orthodichlorobenzene may cause injury to evergreen planting through the action of the vapors on the foliage of the plant."

St. George protected yews from termite damage with a one percent chlordane emulsion applied at the rate of ½ to one pint per sq ft/0.24 to 0.47 l per 0.09 sq m.

Handling soil insecticides. Where water soluble salts like sodium arsenite are used, a clean towel should be available to wipe hands and arms. A clean handkerchief should be used to wipe away any of the toxic material that may be splashed into the eye, and the eye should be washed thoroughly. Use rubberized work gloves. The hands may be covered with vaseline if the spray solution for some reason or other comes in contact with the skin. Special note should be made of the fact the oil-soluble soil insecticides are for the most part volatile and will result in irritation of the eyes and nose on close confinement with the fumes. The skin and eyes are sensitive to these soil insecticides, and contact with them may result in severe irritation. Where such contact does occur, the skin should be thoroughly washed with soap and water and the eyes thoroughly rinsed with water. Rubber gloves, specifically made for handling these chemicals, as well as goggles, should be used while working with soil insecticides. Clothes that become wet with soil insecticides should be removed and exchanged for clean clothes.

Chlordane. Shelford (1952), Hetrick (1950, 1952, 1957), Delaplane (1951), Snyder (1951), St. George (1952), and other workers were impressed with the effectiveness of water emulsions of chlordane as soil treatments for subterranean termites, as well as their lack of toxicity to plants, as previously discussed.

Delaplane (1951) notes he has used one to two percent chlordane water emulsions or soil solutions since 1946 in subterranean termite control. "Using 200 pounds psi on the sprayer at all times, we applied two gallons per linear foot for trenching, void treatment and flooding. One gallon per 10 square feet was applied where the top soil was saturated." Water emulsions were used primarily to protect vegetation; however, oil solutions were preferred in damp situations. Where emulsions were applied, it was essential to use a sprayer with a constant agitator unless stable emulsions were used. Delaplane did not observe any difference in effectiveness between water emulsions or soil solutions at the one and two percent chlordane concentrations.

The Velsicol Corporation (Anon., 1952) makes the following recommendations for the use of chlordane soil insecticides. "Chlordane water emulsion type sprays should be used for this method. They provide excellent soil penetration, and also minimize the possibility of injury being done to vegetation, shrubbery and ornamental plants."

Dieldrin, aldrin and heptachlor. Johnston (1958) showed these three compounds are effective against subterranean termites. Dieldrin and aldrin are now widely used in soil treatment and in preconstruction treatment. The Federal Housing Administration (1958) recommends dieldrin or aldrin at 0.5 percent in oil solution or water emulsion. Lichtenstein and Schulz (1959) discuss the breakdown of aldrin to dieldrin in the soil. Bess et al. (1966) showed aldrin and dieldrin to be very persistent soil poisons. Allen et al. (1964) note the use of dieldrin in concrete for termite control.

Other chemicals. Chlorpyrifos was registered in 1981 for use as a soil treatment against subterranean termites after 13 years of testing. These tests showed that when used at the labeled rate of one percent chlorpyrifos provided a minimum of five years protection. Chlorpyrifos is less persistent in the environment than chlorinated hydrocarbon soil termiticides but as an organophosphate insecticide with moderate acute toxicity, monitoring of cholinesterase levels of applicators is recommended with extensive use.

Bendiocarb was also registered in 1981 for application as a 0.25 percent suspension in water to channels in damaged wooden members of a structure or to cracks, bearing joints, etc. in locations vulnerable to termite attack. The label states the purpose of such applications "is to kill workers or winged reproductive forms which may be present in the treated channels and spaces at the time of treatment and to impart a temporary resistance at these places to termite attack which may aid, in some part, in the prevention of immediate reinfestation by termites. Such applications are not a substitute for mechanical alteration, soil treatment or foundation treatment but merely a supplement, particularly where use of other insecticides may pose an odor problem.

CHEMICALS RARELY USED OR NOT CURRENTLY LABELIED

Lindane. Hetrick (1950 and 1952), Snyder (1951), St. George (1952) note the effectiveness of gamma benzene hexachloride as a soil poison for subterranean termite control. The California Spray Chemical Corporation (Anon. 1951) has prepared directions for the use of its 20 percent water or oil emulsifiable lindane concentrate. The concentration of lindane used varies with the source and is from 0.2 to 0.8 percent. The California Spray Chemical Corporation recommends use of its 20 percent emulsifiable lindane concentrate to be diluted at the rate of one gallon to 100 gallons water. This lindane emulsion is then applied in six inches trenches, or bar holes two to three feet deep and about two feet apart, around the foundation. One hundred gallons of this emulsion is sufficient to treat a two-room bungalow and increased amounts are necessary for larger buildings.

Hetrick (1952) notes "the gamma isomer of benzene hexachloride is more quickly toxic than chlordane at comparable dilutions (one-10,000 and one-20,000). Both of these chemicals are extremely toxic to the test insect. Substantiating proof of their effectiveness has been obtained by the use of these insecticides on structures naturally infested with the test insect, *Reticulitermes flavipes* (Kollar)."

Feytaud (1950) showed DDT requires two to three hours to affect *Reticulitermes lucifugus* Rossi, whereas benzene hexachloride is much more rapid.

The FHA recommended benzene hexachloride at 0.8 percent gamma isomer in water emulsion.

Lead arsenate. Headlee and Jobbins (1939) found when acid lead arsenate was worked into the soil it was effective against termites, and T. E. Snyder suggests it be used at the rate of five pounds per 1,000 square feet/2.3 kg per 93 square meters of surface. The lead arsenate is mixed with the top layer of soil. Kowal and St. George (1948) recommend lead arsenate be mixed with soil at the rate of ¼ to one pound per cubic foot/0.11 to 0.45 kg per 0.028 cubic meters.

Sodium arsenite. Ten percent water solutions of sodium arsenite have been widely used for the chemical treatment of soil for the control of subterranean termites.

Sodium arsenite solutions may leach through the soil and it is for this reason they cannot be used where there is any possibility of contamination of wells or streams. Care must be used in applying this material about foundations for the solution is very toxic to plants. Sodium arsenite, when properly applied, is an effective soil insecticide. It was often used for interior applications since it did not present odor or fire hazard problems as in the case of the oil-base soil poisons. It was recommended that one gallon/3.8 l of the 10 percent solution be used per linear foot/30 cm for deep trenches and ½ gallon per linear foot/1.9 l per 30 cm for shallow trenches. Where the surface soil was treated underneath buildings without basements, the solution is applied at the rate of five gallons per 100 square feet/19 l per nine square meters. Beal recommended that the 10 percent sodium arsenite solution be used at the rate of ¹/₅ gallon per linear foot/0.8 l per 30 cm for a foundation depth of less than two feet/0.6 m and ²/₅ gallon/1.5 l for a depth of two to five feet/0.6 to 1.5 meters. At one time it was feared the volatile gas generated by the action of molds and fungi on arsenical solutions was the dangerous arsine. Snyder (1952) notes this is not so since the gas produced is "Gosio gas" which is not harmful to human beings.

Kowal and St. George (1948) found sodium arsenite used dry and in 10 percent solution was satisfactory as a soil insecticide after an exposure of three years.

Ball et al. (1936) give the following directions for the preparation of a 50 percent sodium arsenite solution:

> Mix in dry condition:
> White arsenic .... 20 pounds.
> Sodium hydroxide .... 5 pounds.
> Then weigh out
> Water .... 15 pounds.

Half of the water is added to the mixed chemicals. This is then stirred to loosen the arsenic and lye. The mixture heats rapidly and as it approaches the boiling point, the arsenic goes into solution, forming sodium arsenite. If the boiling becomes too violent, a little cold water is added. The mixture is constantly stirred to loosen the chemicals from the bottom until they are entirely free in the solution. When boiling ceases, any arsenic that remains undissolved is mashed against the container with a paddle. A piece of wire screen can be used to hold the lumps of arsenic against the side during the mashing. After the chemicals are completely dissolved, the remaining water is added and the mixture is stirred thoroughly and strained through doubled cheese cloth. This stock solution can be kept for a few weeks, but will crystallize if the moisture is permitted to evaporate. It can then be redissolved by boiling.

Pentachlorophenol. This oil soluble crystalline compound is dissolved in kerosene or fuel oil and is used at a five percent concentration. Although its odor may not be quite as penetrating as some of the other oil-soluble soil poisons, it is quite noticeable when used in the home and may linger for several weeks. At room temperatures this chemical is only soluble to the extent of three percent in kerosene; and may be dissolved in fuel oil to the extent of five or six percent, but only with great difficulty. Usually pine oil, linseed oil or trichlorobenzene is used to dissolve the crystals, and then the kerosene or fuel oil is used as the diluent. Commercial five percent solutions are being offered for sale, as well as 40 percent concentrates which are diluted by volume one part concentrate to 10 parts fuel oil or kerosene to give five percent pentachlorophenol by weight.

Pentachlorophenol solutions have found wide usage not only for the control of termites and wood-boring insects, but also for protection against dry rots and other wood-destroying fungi.

Behr (1949) recommends a five percent pentachlorophenol solution "should be added to the soil around the foundation, pipes or steps at the rate of two gallons per five cubic feet of soil or two gallons per five lineal feet of trench. These recommendations are for a trench 10 inches deep. For deeper trenches, which are necessary if cracks exist in foundation walls, more preservative is necessary. Thus, for back-filling a 30-inch deep trench every six inches, 10 gallons would be required for five lineal feet of trench. Even 2.5 gallons per 10 cubic feet will give protection, but it is best to have a safety factor present. In tropical locations it may be necessary to use seven gallons per five cubic feet." Beal (1951) recommends five percent pentachlorophenol be used per linear foot/ 30 cm at ½ gallon/1.9 l for depths less than two feet/0.6 m and one gallon/3.8 l from two to five feet/0.6 to 1.5 meters.

Sodium pentachlorophenate, a water soluble salt of pentachlorophenol, is used in place of pentachlorophenol where odor is a factor and where leaching is not likely to occur.

Where it is desirable not to stain the wood, the pentachlorophenol is first dissolved in pine oil or linseed oil and then diluted with a petroleum base of the type used in fly sprays.

Johnston (1958) shows in field tests pentachlorophenol does not give as good protection against termites as some of the other soil poisons.

Wayne K. Davis, a California termite control operator, patented a mayonnaise-like oil-in-water pentachlorophenol emulsion which has some penetrating ability when applied as a coating on wood. Hatfield and Van Allen (1956) describe this product as a semi-stable, highly viscous, self-sustaining emulsion which forms a heavy coating on the treated surfaces. The product consists of 87 percent by weight solvents and carriers, including 10 percent pentachlorophenol. The remaining 13 percent is water and emulsion stabilizer. The oils and pentachlorophenol penetrate the wood, the water evaporates, and the emulsion stabilizers remain on the outside.

Orthodichlorobenzene. This chemical is used as a soil poison. Orthodichlorobenzene is a colored liquid which has a strong odor. It is soluble in oil and miscible with most organic solvents, as well as being a good solvent itself. This material is somewhat volatile and thus may give control for only a comparatively short time. St. George (1939) recommends the use of orthodichlorobenzene at the rate of four gallons per 10 linear feet/15 l per three meters. He states that instead of trenching one may make two-inch/five cm bar holes 30 inches/76 cm deep and not more than 18 inches/46 cm apart. However, other sources recommend orthodichlorobenzene be used in the trenches as noted under pentachlorophenol. This material may be applied with a sprinkling can from which the rose has been removed. One should avoid prolonged breathing of the fumes. Care should be observed that the chemical does not come in contact with plants.

Trichlorobenzene. This relatively colorless liquid is used in the same manner as orthodichlorobenzene. Due to its persistent naphthalene-like odor it should not be used for interior treatment. Some pest control operators use this chemical in water emulsion. It is usually used with fuel or furnace oil and is diluted one part trichlorobenzene to three parts oil. Beal (1951) recommends the trichlorobenzene mixture be applied per lineal foot/30 cm at $^1/_5$ gallon/0.8 l at less than two-foot/0.6 m depth and -2$^2/_5$ gallon/1.5 l at two to five foot/0.6 to 1.5 m depth.

Creosote. Coaltar creosote is an old and well established material used in the preservation of wood. It also is used in the soil as a repellent and as a contact soil insecticide against termites. The composition of coaltar creosote is dependent on the coal from which it is derived. Creosote has a strong and persistent odor which is disliked by most people. Moreover, it stains surfaces. Creosote is slightly heavier than water. It is soluble in oils and is miscible with orthodichlorobenzene and most of the other organic materials used in termite work. Coal-tar creosote is usually mixed with kerosene at the rate of one part creosote to two parts kerosene or fuel oil. A trench is dug 30 inches/76 cm deep and 12 inches/30 cm wide around the foundation walls and piers. Four gallons/15 l of the mixture is then used for each 10 linear feet/three meters. Some of the liquid mixture should be saved to treat the top layer of the trench. Beal (1951) recommends coal tar creosote be used per linear foot/30 cm at a rate of $^3/_5$ gallon/

2.3 l at a depth of less than two feet/0.6 m and 1-$\frac{1}{5}$ gallons/4.6 l at a depth of two to five feet/0.6 to 1.5 meters.

Berger (1947) states coaltar creosote when applied to soil is usually ineffective after one year. Coal tar acid or tar acid oil is used in similar fashion to creosote.

With the advent of chlorinated insecticides, coaltar creosote is no longer widely used as a soil poison. However, it is employed in pressure-treating wood against termites and decay fungi.

Nature of soil in relationship to soil treatment. Soils range from sand to clay with various gradations. Hockenyos and Hyndman (1937) made a study of this factor. "The amount of liquid poison a soil can take up depends largely on the total air space minus the water present. Air space is important also in that liquid poisons are probably lost largely by evaporation. Thus, evaporation is a factor not only on the surface of the soil but also from the particles below the surface. Where soil particles are very fine as in the case in soils containing much organic matter, force of capillarity is greater and liquids disappear much more rapidly from a soil of this type than they do from pure sand. Absorption is another factor that must be kept in mind; the smaller the particles of a soil are, the more firmly liquids and gases will be held on the surface of the particles. Normally soil particles are coated with a film of adsorbed moisture that persists long after the gravitational water filling the air spaces is drained or evaporated away.

"One other property of soil is of interest in considering poisons and that is base exchange. Divalent bases such as calcium, barium, copper or magnesium will displace univalent bases such as sodium or potassium. Thus, copper salts might be expected to displace sodium or potassium salts and themselves be fixed by the acid colloids of the soil so long as soluble salts of sodium or potassium were not added, to in turn displace the copper. Acid ions, however, such as arsenate, sulfate, etc., are not fixed by the soil and remain leachable."

Harris (1964) notes inactiviation of insecticides in moist soils "is proportional to the organic content of the soil, while in dry sôils inactivation is related to the absorptive capacity of the mineral fraction."

According to Hockenyos and Hyndman (1937), a clay soil "may not be penetrated by road oil or an asphalt solution that will barely pour at ordinary temperatures. Too thin a material will go down too far and be washed. The liquid should be applied in several applications at different depths with soil filled in and thoroughly tamped with a post hole or similar tamper. This tamping fills in loose soil and packs the soil against the foundation. The soil should be built up above the ground level and finally capped with as heavy a layer of asphalt or pitch as possible, so as to run water away from the foundation and to retard evaporation. In tight soils it is preferable to apply a very thin and fluid solution first and a heavier one later, rather than to apply the heavier fluid alone. Better penetration of the heavier material is obtained this way."

Zimmern (1957) and Lichtenstein (1959) discuss the penetration and persistance of chemicals in soil. Osmun (1958) reviews the importance of soil moisture in competing with the movement of soil fumigants. Ebeling and Pence (1957) show sand, cinders and slag when of certain sizes are definite barriers to penetration by subterranean termites.

Foundation treatment. Berger (1947) and Hockenyos (1949) studied this phase of termite control and the following discussion is from their papers. Void treatment involves flushing or injecting soil insecticides "into all spaces in the foundation walls (beams), piers and posts where termites might travel from soil to wood." Such treatment usually involves extensive drilling of the foundation and is necessarily laborious, time-consuming and expensive. Wall void treatment is usually employed in those areas where termite infestations in homes are of common occurrence. The most common foundation types are poured concrete, brick, cement block, hollow tile, brick veneer, stone and rubble. Since the voids or spaces in the foundation necessarily vary with the construction, one must know the arrangements of the hollow spaces in the foundation for the proper treatment of such areas. Berger (1947) illustrates several types of foundation walls.

When the foundations with hollow spaces inside are treated, it is necessary to drill holes through the foundation, the distances varying with the type of foundation, so every inch of soil at the bottom of the void is treated with a soil insecticide. The drilling is accomplished by electric drills. The drills have carbide cutting edges especially made for working with masonry. An effort is made to drill the holes cleanly so that nozzles can be inserted where it is necessary to use pressure to flood the voids. In structures where the foundation plate does not cover the voids completely, the chemical may be poured through these spaces into the foundation without drilling any holes. See Anon. (1960) for drilling sites.

Butts (1958) demonstrated that chemicals penetrate deeper when injected near the footing than when injected higher up in the void of the block. The purpose of such treating is to get the chemical into openings that might occur in the footing and provide termite entry points. This would be favored by injecting near the footing. Many choose to drill and inject high, however, because it may be done outside with less mess indoors and there can be no accusations by the homeowner concerning moisture seepage during rainfall if the holes are above grade.

Any of the chemicals mentioned under soil treatment may be effective in void treatment. However, only those currently registered for such use should be applied. Pentachlorophenol, trichlorobenzene, orthodichlorobenzene, and coaltar creosote are not favored for this type of treatment due to their persistent odor. Hockenyos (1949) has the following to say about NC crystals (nitrochlorotoluene): "It would seem that the most effective form of wall void treatment would be to have deposited in the wall voids a crystalline mass of a slowly volatile chemical, toxic to the termites and not too offensive or toxic to human beings if some breakage occurred in the operation of pumping the voids. At present, this appears to be rather well accomplished by the use of NC crystals, especially in the form of a strong concentration in organic solvent and emulsified in water."

The holes are bored at intervals of from one to several feet depending upon the type of voids in the foundation. An idea of the spread of the chemical can often be gained by observing leakage at cracks and other places and then drilling the holes accordingly. Berger states generally less soil insecticide is needed for the treatment of voids than for outside treatments, and usually one-fourth of the amount used for outside treatment will suffice for void treatment.

Hockenyos makes the following remarks on the removal of excess soil poisons from the basement. "Materials spilled should be flushed away as soon as possible to avoid staining. Where it is known a wall is leaky and may result in spillage, it is well to have an absorbent material ready to apply to the floor at the needed points. For water base solutions, a bag of Kieselguhr (diatomaceous earth) will cause prompt absorption and will dry readily if spread in the sun. For oil solutions, pyrophyllite or sawdust may be used. If an odor persists after the solution has been removed, this may usually be taken care of by covering the spot with burlap and then sprinkling a layer of activated charcoal over the burlap." All drill holes should be closed with cement, mortar, or black roofing cement.

The FHA recommends voids of unit masonry walls and piers be treated at the rate of one gallon per five lineal feet/3.8 l per 1.5 meters.

Allen et al. (1964) studied the effectiveness of concrete mixed with dieldrin to prevent termites from tubing up foundations. While effective, the residual period was too short to encourage this as a practical method.

Wood treating. Termites will tube over treated wood (Lund, 1959 and 1961), but it is common practice to flush existing galleries with a termiticide to prevent post-treatment swarms and to give immediate control of the above ground members of the colony. This should be done as product labels permit.

Termite baits. Beard (1974) tested the idea of placing blocks of wood in the soil around a structure to control termites. The wood was treated with a decay fungus to make it as attractive as possible. The wood was also impregnated with chlordecone, an organic insecticide. It was hoped termite workers would find the wood, feed, return to the nest and spread the chlordecone throughout.

While the results were not encouraging, they were inconclusive. Manufacture of chlordecone has since been terminated. This idea has merit if the proper attractant and toxicant can be found.

Termite barriers. According to Turner and Townsend (1936), "Termites can penetrate 1) lime mortar; 2) ordinary masonry work of brick, stone or hollow blocks; 3) tar and asphalt compounds; 4) poisoned paint films; 5) roofing felt and 6) cracks in solid concrete. Termites cannot penetrate: 1) solid concrete without cracks or solid unit masonry laid in cement mortar, or 2) wood treated by a standard pressure process, using standard or equivalent wood preservatives. They can, however, construct shelter tubes over either of these. Termites cannot penetrate non-corroding sheet metal, and they cannot construct shelter tubes up over the edges of properly installed termite shields."

H. O. Lund (1959 and 1961) showed subterranean termites will build tubes over pine wood chemically treated for rot control. A. Lund (1958) found arsenical water-borne preservatives in wood to be more effective than those with copper, zinc or chromium in protecting wood from subterranean termites.

Termite shields. Termite shields are metal barriers which are inserted in or on foundations, piers, pipes and other structures to prevent termites from gaining access to the wood construction in a building. Moreover, such shields usually force the subterranean termites to build their tubes where they may readily be seen. Termite shields may be recommended for new construction, but they are difficult and expensive to install in old construction.

Unless such shields are properly installed they are of little value in preventing termite infestation. Some of the usual defects encountered in the installation of termite shields are listed by the U.S.D.A. (Anon., 1949), Johnston and Osmun (1960):

- "Loose joints between sections of metal.
- Improperly cut and soldered corners or angles where walls intersect.
- Strip shield placed on top of foundation instead of being embedded in or attached to the side of the wall.
- Anchor-bolt holes cut in bread-pan shields and not sealed with coal-tar pitch.

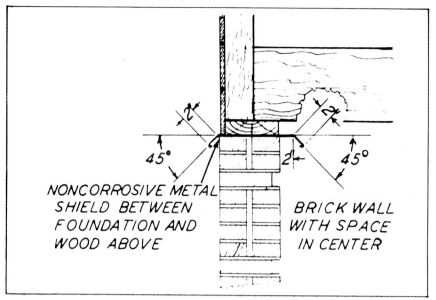

NONCORROSIVE METAL
SHIELD BETWEEN
FOUNDATION AND
WOOD ABOVE

BRICK WALL
WITH SPACE
IN CENTER

45° 2" 45°

Fig. 7-32. In the eastern states, pest control operators at times resort to the use of metal termite shields to protect structural timbers from subterranean termites. Note that the shield is bent down at a 45 degree angle.

- Insufficient clearance between the outer edge of the shield and adjacent woodwork or piping.
- Shields less than 12 inches above grade line, sometimes even buried by grading operations.
- Projecting edge of shields battered and bent out of shape, often flattened against piers or foundation wall.
- Shields installed on section of a foundation where there was little danger of termites attempting to gain entrance to the building, whereas the points of greatest danger, such as filled porches, were left unprotected.
- Shields constructed of material subject to rapid corrosion or to being easily torn or bent out of shape."

There are a number of different kinds of shields in use, several of which are patented. Two shields that are widely used are the "breadpan" and "strip" shields. The bread-pan shield consists of a strip of metal which extends across the top of the wall or pier for two inches/five cm or more with the edges bent down at a 45° angle. Anchor bolts in the bread-pan shield should be covered with coaltar pitch or black roofing cement. The strip shield consists of a straight piece of metal which is usually embedded in the foundation or pier.

R. A. St. George (1939) states, "For walls made up of masonry units, such as tile, hollow blocks of various kinds, brick, or stone, it is best to cap them with high-grade mortar and to use the 'bread-pan' type of shield which extends entirely across the foundation. On the inside, it should project horizontally for a distance of two inches before being bent downward at a 45° angle, for a similar distance. On the outside, if there is no shrubbery adjacent to the building and if no woodwork comes close to the ground and the foundation can be readily inspected periodically, the shield can either be extended flush with the exterior surface of the building or it can be projected as indicated for the interior wall. Where such a projection is objectionable because of appearance, a compromise can be made as to the amount of the extension without materially lessening the effectiveness of the shield.

"For solid walls composed of a good grade of poured concrete, a six-inch strip type of shield may be sufficient. Such a strip should be securely embedded in the wall, especially on the inside, for a distance of two inches, project horizontally for a like distance and be bent downward as described above. In all cases, the ends of the metal strips must be firmly joined, either by soldering or by an interlocking mechanical joint. Simply lapping and riveting the pieces together is entirely unsatisfactory, as the joint is most likely to open if the building settles or if much contraction or expansion of the metal takes place with changes in the weather. Particular attention should be given to such shields in only partial excavated portions of buildings and around the inside of excavated porches and sun parlors."

Johnston (1943) notes shields do not guarantee complete control of termites and periodic inspections for the presence of tubes are necessary. His studies showed a flat or horizontal shield of 26-gauge galvanized iron projecting two inches/five cm from the foundations affords as much protection against the eastern subterranean termite as the standard 45° shield.

Clements (1952) discusses termite shields and shows failures result from improper installation of the shields. Isherwood (1957) believes pressure-treated wood sills have certain advantages over metal shields.

Naturally termite-resistant woods. The sapwood of most trees is more susceptible to attack than is the heartwood. The heartwood of California redwood, tidewater red cypress and western red cedar is fairly termite resistant. It is believed the natural resistance of these woods is due to certain extractives present in the wood which are either distasteful or injurious to the termites or their protozoa. When with time the toxic ingredients of these extractives are diminished or lost, the termite resistance of the woods is likewise lessened.

CONTROL OF DRYWOOD TERMITES

Drywood termites are the cause of extensive damage to timber in homes and to poles outdoors in the South and South-

Fig. 7-33. The work of two colonies of drywood termites in the wood in an attic. A and B. One avenue of entry. C and D. Another avenue of entry.

west. Snyder (1950) discusses these insects in some detail in his U.S.D.A. bulletin. The colonizing sexual winged forms can enter the wood at all levels and are the termites most commonly found in roofs, attics, window frames and less often in the substructure. These termites may riddle the wood until there is merely an outer protective shell. Dry-wood termites prefer to enter the wood through checks, through cracks, and through broken knots. They have definite wood preferences and it is very likely such factors as the chemical content of the wood, its hardness, age in the tree, etc., determine these preferences.

Hunt (1949) notes houses less than one year old in Southern California are often infested with drywood termites and he lists some of the factors which are responsible for the increase in drywood termite infestations: "Close proximity of homes caused by the general acceleration in residence construction in recent years; lack of knowledge of means of prevention due to the fact that national building codes and termite specifications have stressed the prevention of infestation by subterranean termites, but have not given information on prevention of the drywood species; the extensive practice of subdividing orchards into home sites leaving a few trees which may serve as foci of infestation; inadequate protection of piled lumber in lumber yards during the season of termite flights; the use of the poorer grades of lumber necessitated by the general shortage of building materials in recent years; lack of vigilance on the part of both the layman and the termite operators which has delayed the discovery of the infestations and the employment of inadequate control methods after the termites were known to be present; and, the presence of numbers of heavily infested utility poles throughout the older residence districts."

Evidence of the presence of drywood termites. The presence of these insects is most readily recognized by the characteristic fecal pellets which are small, seed-like and usually straw colored. The pellets may be pushed up from the wood in small piles or scattered by falling from chinks or cracks in the infested wood situated overhead. Timber attacked by drywood termites is characterized by tunnels or workings in the wood which may be crowded with fecal pellets. Drywood termites consume the wood up to the paint itself, forming what appear to be paint blisters, which on the slightest pressure break and spill forth fecal pellets. One does not find the earth-like frass or earthen tunnels in or on wood attacked by drywood termites. Where there is infested wood in the home, in California the winged forms may be found swarming in the home from June until December, the time varying with the locality as well as with other conditions. The winged form or alate is dark brown and about ½ inch/13 mm long. The biology of the drywood termite is discussed in an earlier section of this chapter.

Inspecting for drywood termites. Prior to control one must estimate the extent of the infestation. The procedure followed is similar to that described for the inspection fo subterranean termites. All exposed timber beneath the house proper and in the attic, as well as window frames, floors, etc., should be carefully observed for signs of termites. The inspector searches for the pellets which are so characteristic of infestation and the wood is scratched or probed with some instrument such as a screwdriver, hammer, geologist's pick, etc.; see Hunt (1966). Packard (1951) notes, "As a general rule, it is not advisable to break open the infested wood with a prospecting pick or screwdriver as this disturbs the colony, causing the termites to spread and also making treatment more difficult as it is hard to dust or apply liquid effectively in open galleries." Infested wood has a hollow sound and usually when probed breaks, spilling forth the seed-like pellets. Some pest control operators use such methods as the removal of plaster from the walls in order to examine the framework, although others do not deem this essential. In some instances the pest control operator may drill into the framework directly through stucco or plaster in order to determine if these hidden timbers have been hollowed out by termites.

Packard (1951) describes the method for determining the presence and distribution of drywood termites in homes in southern California: "Attics are attacked as readily as other portions of the buildings and so must be inspected. Infestations are usually found around vents, in rafters, ridge poles and in general around the perimeter except where the infestations are heavy and of long standing so that the center sections of the attic may become involved. During the swarming season, many reproductives fly out of the parent colonies and infest other portions of the attic until even the wooden lath and shingles become infested. On flat-roofed structures, this method of infestation often results in the spread of drywood termites into the very low sections which are inaccessible to the inspector.

"Any inaccessible areas should be noted on the diagram prepared with the inspection report so that liability for such areas will be understood between the owner and the pest control operator.

"Substructure infestations are usually found around the outside walls with the pellets appearing on the mudsills or in piles on the ground. Sometimes the foundations are high and the floor joists rest directly on the mudsills, leaving no room for the inspector to look on top of the sills next to the outer wall. Some inspectors have found a mirror to be of great help in getting into close and out of the way places. If the pellets appear to be falling from behind the framing next to the wall, rapping on the wood may cause more pellets to drop, indicating that the infestations are probably in the wall above the floor. Again, the probable extent of the infestations can only be approximately determined, as the outer wall covering would have to be removed to establish the true extent of the termite activity. However, if the pellets appear in the attic, in the windows and substructure directly below, it may be safely assumed that the wall studs and braces are infested.

"Exterior exposed woodwork is subject to drywood termite attack and a good inspection will include an examination around the entire building. Pellets will be found on flat surfaces, in spider webs or on the ground. If a search is made directly above the place where the pellets have accumulated, the tell-tale 'kick-out' holes will be found stuffed with pellets to keep out insect intruders. These pellet plugs are plastered together with a material exuded by the termites and when the holes are opened to throw out pellets which have accumulated in the galleries, soldiers stand guard around the opening to repel any invaders.

"Interiors of structures are subject to infestation and examination of all windows, casings and outside doors is necessary. Exposed floors and woodwork are checked and the people occupying the building are questioned to see if they have found any pellets which they might have cleaned up or swept away and would not otherwise be discovered.

"Garages and other buildings on the premises subject to normal occupancy are examined and very often are found to be infested with drywood termites. Control service on these termites has become a substantial business in southern California and many operators have made a practice of examining trees, poles and fences on the property to determine the potential hazards which may exist to the buildings under service. Many times the electric light and telephone companies have been requested to remove badly infested poles which have served as a source of infestation for the entire neighborhood."

The grooming habit of termites in relation to control. Termites constantly lick or groom their own bodies, as well as one another. This habit is of extreme importance from the viewpoint of control since it is in this manner that the insecticide dusts used in the control of drywood termites are disseminated. Thus, Light et al. (1930) note when a single individual was dusted with white arsenic and placed in a dish four inches/10 cm in diameter with 149 other termites, all were dead within 30 hours due to the grooming habit. The sensory hairs on the antennae and legs of the termite are stimulated by contact with the dust and this induces them to groom themselves and one another. Moreover, it is likely termites lick one another for the exudations that are present on the surface of their bodies as in the case of ants.

Preventing attack by drywood termites. Until the advent of the silica aerogel dusts, it was practically impossible to devise any method whereby drywood termites could be prevented from initiating a colony in the wooden portion of a structure. Pressure-treated timbers if used throughout the framework would undoubtedly serve this purpose, but the expense would be so great as to make this method impractical. Where exposed wood is coated with several layers of paint, this to a limited extent discourages infestation. Vents screened with 20 mesh, non-corroding metal screening will aid in keeping the termites from the framework as long as there is no other means of ingress. Pest control operators have noticed that drywood termites spread in the direction of the prevailing winds from infested homes, poles, tree stumps, etc.

Silica aerogel dusts are very light and penetrate attics well, coating exposed wood members. The dust is not toxic to mammals and is effective indefinitely if not wetted. By treating attics and similar areas with silica aerogels, the likelihood of drywood termite infestation is reduced. The dust also covers existing pellets, swarmers and wings, making it easier for the inspector to determine new activity during reinspections.

Control with toxic dusts. This method was once commonly used to control drywood termite infestations. It is slow, laborious, has limited effectiveness and its legality is presently unclear. It is presented here as a possible consideration for limited infestations and for the future when this technique might be resurrected.

Once the infested area is determined, the infested timber, as well as nearby wood is drilled with an electric drill. A ½-inch/13 mm drill is used for larger timber, a ¼-inch/six mm or ⅛-inch/three mm drill is used for smaller timbers and an even smaller drill is used for finished woodwork and interior trim. The size and the number of holes drilled are dependent on the extent of the infestation. Pest control operators bore at least half way into the timber at approximately one-foot intervals. Practice will enable the operator to determine when he strikes hollow galleries with the electric drill. Because drilling is time consuming and a painstaking job many pest control operators prefer to fumigate whenever possible.

The dusts most commonly used in drywood termite control were Paris green, sodium fluosilicate, sodium fluoride and calcium arsenate. Certain pest control operators diluted these insecticides with such carriers as corn starch to make the dusts lighter and fluffier. Sodium fluosilicate is less toxic to human beings than Paris green, but it has a tendency to cake. It should be remembered the smaller or more finely ground the dust particle, the more efficient the treatment, since the finer dust is more likely to adhere to and irritate the minute sensory hairs of the termite. One ounce/28 g of dust is sufficient to treat from 15 to 30 holes. The dust may not be effective in wet wood, since it becomes caked in the moist runways.

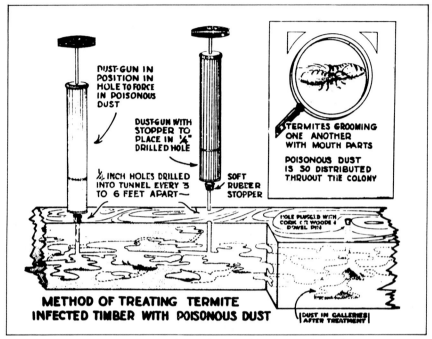

DUST-GUN IN
POSITION IN
HOLE TO FORCE
IN POISONOUS
DUST

DUST-GUN WITH
STOPPER TO
PLACE IN ½"
DRILLED HOLE

¼ INCH HOLES DRILLED
INTO TUNNEL EVERY 3
TO 6 FEET APART—

SOFT
RUBBER
STOPPER

TERMITES GROOMING
ONE ANOTHER
WITH MOUTH PARTS

POISONOUS DUST
IS SO DISTRIBUTED
THRUOUT THE COLONY

HOLE PLUGGED WITH
CORK OR WOODEN
DOWEL PIN

METHOD OF TREATING TERMITE
INFECTED TIMBER WITH POISONOUS DUST

DUST IN GALLERIES
AFTER TREATMENT

Fig. 7-34. Method of treating timber infested with drywood termites.

Once the dust is blown into holes, the opening is closed with wooden dowels, corks, putty or plastic wood. It is believed if the hole is sealed, the termites may continue to use the treated tunnels, otherwise they may block off the dusted galleries.

Pest control operators usually use a special pressure duster, the "kaligun." A small hand duster, as illustrated, may also be used for the dusting. Randall and Doody (1934) give the following directions for the preparation of such a hand duster: "The nozzle of this hand gun should be changed for the application of dust poisons for termite control. A small rubber suction cup of the kind used for holding ash trays, show cards, etc., on glass should have a hole bored through it and be placed on the ½-inch pipe nozzle of the gun. Place the rubber shoulder on the back of the suction cup firmly against the top of the dust chamber. The pipe nozzle of the dust gun should then be cut off, leaving only about ¼-inch (0.6 cm) of it projecting beyond the inside surface of the suction cup." This makes an air-tight seal and prevents dust from blowing into the face of the operator. R. H. Smith (1930) used a No. 79 De Vilbiss atomizer as a duster, to which was attached a piece of flexible copper tubing $5/64$ inch/two mm in diameter. This tube was used to apply the dust into very small holes as in the wood trim. Where dust has been applied in wood, control has been obtained as far as 15 feet/4.6 m from the point of application. It may be necessary to reinspect treated wood several months later since isolated colonies may have escaped the dust. Poison dusts do not kill immediately, but act slowly and several weeks in many instances transpire before they take full effect.

Some operators force liquid under as much as 75 pounds pressure into the wood. These liquids may have an oil or kerosene base with one of several toxic ingredients other than the base. Thus, a five percent solution of pentachlorophenol in pine oil and/or a base oil may be used, as well as orthodichlorobenzene, paradichlorobenzene, chlordane or lindane solutions. Great care must be observed to prevent seepage of the liquid through the treated wood.

Ebeling and Wagner (1959) obtained control of drywood termites with a treatment of 10 percent pentachlorophenol in patented oil emulsion, previously described in the section on the control of subterranean termites. "Painting the emulsion over the surface of the infested area enables the toxicant to penetrate the galleries inhabited by the termites and cause the extermination of the entire colony." This penetration is assisted by the fact many of the galleries are very close to the surface of the wood.

Ebeling and Wagner (1959) have also demonstrated the effectiveness of silica aerogel dusts in killing drywood termites and preventing reinfestation. According to the authors, a silica aerogel is impregnated with a small percentage of silicofluoride. These silica aerogels remove the wax-coating and dry out the insect and also affect it chemically by means of water-soluble fluorides. The dust is blown into the attic and other areas by an electric blower at the rate of one to two lbs. per 1000 sq. ft./0.45 to 0.9 kg per 93 square meters (see Ebeling and Wagner (1963)). The dust is also blown under the house to prevent infestation in the substructure. Because silica aerogels may be irritating, a dust respirator especially suitable for so fine a dust, should be worn by the operator. Reierson (1966) showed in the laboratory drywood termites are susceptible to wood dusted with boric acid at 1¾ lbs. per 1000 square feet/0.8 kg per 93 square meters.

Hunt (1949) and Packard (1951) both note the use of the injection of volatile liquid fumigants as a successful means of combating drywood termites. Ethylene dibromide in a petroleum distillate was the primary material used for this purpose. About one-third of a teaspoonful of the mixture per hole was the usual dosage. Its current legality is questionable.

Control by fumigation. Pencille (1947) and Hodel (1949) have shown fumigation is an effective method of drywood termite control. However, if it is believed the infestation is localized, consideration should be given to some of the previously mentioned control methods, since fumigation procedures may not be necessary for small infestations. Fumigation does not prevent reinfestation of the treated building. It should be undertaken only by *very experienced fumigators*. See Chapter 26 for additional information on fumigation.

Young (1955) recommends the use of a heat-exchanger which warms the methyl bromide as it comes from the cylinder. This results in faster dispersion of the gas as well as counteracting the cooling effect on the piping system caused by evaporation of the gas.

The Florida Pest Control Association (1963) states fumigant dosages "will vary with such factors as size of building, temperature, wind, length of fumigation period, etc.; however, the following dosages would be considered minimal for an average building of approximately 25,000 cu. ft. under ideal conditions and temperatures above 60° F.

Methyl bromide — 2-lb. per 1000 cu. ft. — 18 hrs. exp.
Sulfuryl fluoride — 1-lb. per 1000 cu. ft. — 18 hrs. exp.
Acrylonitrile-carbon tet. — 5-lb. per 1000 cu. ft. — 36 hrs. exp.
Hassler (1956) discusses the characteristics of fumigation tarps used in fu-

Fig. 7-35. A drywood termite infested building sealed with a plastic covering in preparation for a methyl bromide fumigation.

migation. Wolcott (1956) notes the effectiveness of pentachlorophenol in preventing drywood termite infestations, and Stewart (1957) describes the fumigant, sulfuryl fluoride for use against drywood termites, and Gray (1960) recommends its use at two lbs. per 1000 cu. ft./0.9 kg per 28 cubic meters.

Stewart (1966) recommends six, eight and 16 ounces/0.17, 0.23 and 0.45 kg of Vikane at 80°, 70° and 55° F/27°, 21° and 13° C, respectively, for a 20-hour exposure. See Stewart (1962 and 1966) for further information on fumigation with Vikane.

Protection of stored lumber. Where the protection of lumber piles is desired, the wood should be raised on concrete piers, using creosote impregnated wood as a platform. All debris should be removed from the vicinity to discourage breeding.

It has been found if the tops and sides of lumber piles are sprayed with a sodium fluosilicate solution using one pound/0.45 kg sodium fluosilicate to 10 gallons/38 l of water, it discourages infestation by drywood termites. Termites already in the wood may be killed with kerosene or orthodichlorobenzene or by placing lumber which is two inches or less in thickness in a dry kiln at temperatures of 125° to 150° F/42° to 66° C. Lindane water-base sprays are effective against wood-boring insects and may offer protection against drywood termites. Lumber dusted with silica aerogels may also give some protection.

CONTROL OF THE POWDER POST TERMITE

The powder post termite *Kalotermes (Cryptotermes) brevis* was controlled in Hawaii by fumigation with sulfuryl fluoride for 1.5 hours at two lbs. per 1000 cu. ft./0.9 kg per 28 cubic meters. (Bess and Ota, 1960). It was more effective than methyl bromide for 15 to 19 hours at 2.5 lbs. per 1,000 cu. ft./1.1 kg per 28 cubic meters. Wolcott recommends the introduction of pentachlorophenol solutions into the galleries through the holes they make to eject excreta for the control of active colonies. He also notes lindane at 0.05 percent made wood repellent to this species. E. M. Miller (1949) notes fumigation of furniture or buildings with HCN, chloropicrin or methyl bromide will often give satisfactory control of these termites. Sulfuryl fluoride is also effective here.

CONTROL OF DAMPWOOD TERMITES

(Zootermopsis)

Dampwood termites are of economic importance as timber pests in northern California, Oregon and Washington. These termites produce pellets, unlike the subterranean termites. As a whole, the dampwood termites are of minor economic importance, but in certain localities they may become serious pests. Thus, Brown (1936) notes dampwood termites are very destructive to property in San Francisco and are present in approximately 50 percent of the structures that were inspected and found infested. On the average these termites do much greater damage than subterranean termites since they often extend their activities from the foundation up to the roof rafters and on one occasion were found on the fifth floor of a fireproof structure.

They usually attack wood that is constantly dampened, such as water tanks, pilings, wood in contact with the damp ground or wood moistened by leaking plumbing, steam pipes or sweating pipes, as well as marine structures above water. In the Pacific Northwest they are economically noteworthy because of their attacks on poles. These dampwood termites almost always work in conjunction with fungi. Although dampwood termites usually occur in decaying wood, they may extend their workings into the sound wood.

The colonizing pair enter directly into the dampened wood. Brown (1936) states, "We found many instances where dampwood termites had been present but appeared to have died out due to the excessive amount of decay that had taken place in the wood. On more careful examination, however, the investigator found that these termites had not died, but had abandoned the decayed wood and moved to sound wood where conditions were more favorable to their mode of living and where the growth of fungi could be controlled. These observations probably explain both the failure of workmen to locate such termite colonies and the finding of termites in structures immediately following the completion of attempts at eradication." Dampwood termites are controlled by methods recommended for the subterranean termites.

Chamberlain and Hoskins (1949) studied the toxicity and repellency of chemicals to dampwood termites. They showed hexachlorocyclohexane (BHC), and 3,5-dinitro-*o*-cresol to be especially effective. These chemicals when used in wax "prevented the insects from boring through the treated papers for more than six months."

Furniss (1953) found a ¼ to one percent chlordane solution or pentachlorophenol effective against dampwood termites. He controlled them in the same way carpenter ants are handled, or by structural modification to eliminate moisture. Chlordane is now commonly used to control these termites.

LITERATURE

ADAMSON, A.M. — 1941. Laboratory technique for the study of living termites. Ecology 22(4):411-414.
ALLEN, T. C., G. R. ESENTHER, & P. LICHTENSTEIN, — 1964. Toxicity of

dieldrin-concrete mixtures to termites. J. Econ. Entomol. 57(1):26-29.
ANONYMOUS. — 1947 Tentative rules for fumigating structures for dry wood
termites. Pests 15(9):28, 1948. Decay and termite damage in houses. U.S.D.A.
Farmers' Bull. 1993, 1949. Preventing damage to buildings by subterranean
termites and their control. U.S.D.A. Farmers' Bull. No. 1911. 1950. Ap-
proved reference procedures for subterranean termite control. National Pest
Control Assoc., Inc. 1950. Termite operators accepted standards for inspec-
tions and recommendations. Pest Control Operators of California. Rev. Aug.
1950. DDT, another soil poison for subterranean termites control U.S.D.A.,
B.E. & P.Q. Mimeog. Sept. 19, 1950. 1950. "NC" Soil compound. Mimeog E.
I. Du Pont de Nemours. 1950. Chlordane for subterranean termite control.
Velsicol Bull. No. 40, Nov. 1951a. Control of termite operator's itch (Creeping
eruption). Pest Control. 19(6) 24, 32. 1951b. Termite operator's itch. Pest
Control. 19(5):26, 46. 1951c. Summary of subterranean termite control survey undertaken in 1950. Gulf Research & Development Co. Mimeog. 1951d.
Soil fumigation can eliminate termite infestation in cold joint failure. Pest
Control. 19(8):29. 1951e. How would you treat infested timber beneath heavy
electric converters. Pest Control. 19(7):29. 1951f. Isotox Spray No. 200 for
termite control. Ortho. Field News. Nov. 1, 1951. 1952. Control with chlordane. Form 317. Velsicol Corp. 1953. A new termite attacking hardwood
floors. P.C.O. News, May, 1953, p. 4. 1958. *Coptotermes formosanus* in television set. Coop Econ. Insect Rept. 8(12):215. March 21, 1959. Ventilators
— How to buy them; where and how to install them. Pest Control. 27(5):54.
1959a. How Savannah TO Oliver pretreats 1,100 unit navy project to protect
against termite infestation. Pest Control. 27(2):19. 1960. Getz pretreats
basement type house during construction. Pest Control 28(2):23. 1960a. Concrete block foundation voids: Should they be treated at grade level or near
footing. Pest Control. 28(11):34. 1960b. Equipment and methods for drilling
masonry. Pest Control. 28(5):62. 1962. Fumigator's page. P.C.O. News 22(6):10.
1962. A powder-post termite *(Cryptotermes cavifrons)*. Coop. Econ. Insect
Rpt. 12(24):641. 1963. Retaining-wall "flower boxes" are termite heavens.
Pest Control. 31(5):92. 1963. Accepted and recommended procedures for the
treatment of buildings. Florida Pest Control Assoc. 1964. Dampwood termite. Coop. Econ. Insect Rpt. 14(4):300, 1964. Formosan subterranean termite. Coop. Econ. Insect Rpt. 14(10):150 1967. South Carolina included in
Formosan termite quarantine hearing. U.S.D.A. 1924-67.
BALL, W. S., A. S. CRAFTS, B. A. MADISON, W. W. ROBBINS, and R. N. RAY-
NOR. — 1940. Weed control. Calif. Agr. Ext. Serv. Circ. 97.
BEAL, J. A. — 1951. National Pest Control Assoc. Service Letter. Feb. 28, 1951.
1967. Formosan invader. Pest Control. 35(2):13-17.
BEAL, R. H. & V. K. SMITH. — 1964. Progress report on granular formulations
of insecticides. J. Econ. Entomol. 57(5):771. 1965. Are granular formulations
of insecticides effective in subterranean termite control? Pest Control.
33(5):78.
BEARD, R. L. — 1974. Termite biology and bait-block method of control. Conn.
Agri. Expt. Sta. Bull. 748.
BEHR, E. A. — 1949. Application of pentachlorophenol as soil poison and wood
preservative. Pests. 17(3):19-20, 22.
BEHR, E. A. et al. — 1972. Influence of wood hardness on feeding by the eastern
subterranean termite, *Reticulitermes flavipes* (Isoptera: Rhinotermitidae).
Annals ESA 65(2):457-460.

BERGER, B. G. — 1947. How to recognize and control termites in Illinois, Ill. Nat. Hist. Survey, C41.

BESS, H. A. & A. K. OTA. — 1960. Fumigation of buildings to control dry wood termite, *Cryptotermes brevis*. J. Econ. Entomol. 53(4):503-510.

BESS, H. A., A. K. OTA & C. KAWANISHI, — 1966. Persistence of soil insecticides for control of subterranean termites. J. Econ. Entomol. 59(4):911-915.

BROWN, A. A. — 1936. Report of the San Francisco termite survey. City & County of S. F. Dept. Publ. Works Proc.

BUILDING RESEARCH INSTITUTE. — 1956. Protection against decay and termites in residential construction. Building Research Advisory Board, FHA. National Research Council Publ. 448. Addendum, 1958.

BUSWELL, W. S., Jr. — 1955. Termites on a wet tin roof. Pest Control 23(9):44.

BUTTS, W. L. — 1958. Termite chemical dispersion in concrete blocks. Pest Control 26(3):32.

BYNUM, W. M. — 1951. The desert dampwood termite in the lower Rio Grande Valley of Texas. J. Econ. Entomol. 44(6):996-997.

CARUBA, A. — 1981. This beagle sniffs out termites. Pest Control. 49(2):15, 16.

CASTLE, G. B. — 1934. Termites and Termite Control. Univ. of Calif. Press.

CHAMBERLAIN, W. F. & W. M. HOSKINS. — 1949. The toxicity and repellence of organic chemicals toward termites, and their use in termite-proofing food packages. Hilgardia 19(9):285-307.

CLEMENTS, W. B. — 1952. Termite shields . . . Are they effective? Pest Control 20(11):29-30.

CLEVELAND, L. R., E. P. SANDERS & S. R. HALL. — 1931. The relation of Protozoa of *Cryptocercus* to the Protozoa of termites and bearing of this relationship on the evolution of termites from roaches. Anat. Rec. 15, supplement :92.

COATON, W. G. H. — 1946. The harvester-termite problem in South Africa. Dept. Agr. Bull. 292. Union of South Africa.

CROSS, J. C. — 1942. A simple method of controlling termites. Science 95 (2469):433.

DAVIS, J. J. — 1950. The prevention and control of termites. Purdue Univ. Agr. Ext. B. 225. 1955. Dampwood termites "hitchhike" ride in lumber shipped east from West Coast. Pest Control 23(4):38, 57.

DELAPLANE, W. K. JR. — 1951. New chemicals we've tried for termite control. Pest Control. 19(9):31, 32. 1953. Subs don't always come from Soil. Pest Control 21(7):39, 40.

DELONG, D. M. & R. J. KEAGY. — 1949. Termite cultures in the laboratory. Turtox News 27:114-116.

DENNY, C. — 1953. How to inspect slab-type homes for termites. Pest Control 21(2):14, 16. 1961. First draft of NPCA's approved inspection procedures. Pest Control 29(2):26, 28, 30, 32, 34.

DEWS, S. C. & A. W. MORRILL, JR. — 1946. DDT for insect control at army installations in the Fourth Service Command. J. Econ. Entomol. 39(3):347-355.

DuCHANOIS, F. R. — 1961. Shortcomings in termite control work in Florida. Pest Control 29(11):42, 44, 46, 55.

EBELING, W. & R. J. PENCE. — 1956. UCLA entomologists evaluate research data on drywood and subterranean termite control. Pest Control 24(10):46, 50, 52, 54-58, 62, 64. 1957. Relation of particle size to the penetration of

subterranean termites through barriers of sand and cinders. J. Econ. Entomol. 50(5):690-692. 1958. Laboratory evaluation of insecticide-treated soils against the Western Subterranean termite. J. Econ. Entomol. 51(2):207-211.
EBELING, W. & R. E. WAGNER. — 1959. Two newer weapons for drywood termite control and prevention . . . Woodtreat-TC and Silica aerogels. Pest Control 27(2):40, 42, 44-45. 1959. Rapid desiccation of drywood termites with inert sorptive dusts and other substances. J. Econ. Entomol. 52(2):190-207. 1963. Methods and equipment for treating houses with Dri-Die 67 during construction. P.C.O. News. 23(6):14-17.
EHRHORN, E. M. — 1934. In Termites and Termite Control. Univ. of Calif. Press. pp. 321-333.
EMERSON, A. E. — 1936. Distribution of termites. Science. 83(2157): 410-411.
EMERSON, E. E. & E. FISH. — 1937. Termite City. Rand McNally & Co.
ESENTHER, G. R., T. C. ALLEN, J. E. CASIDA & R. D. SHENEFELT. — 1961. Termite attractant from fungus-infected wood. Science 134(3471):50.
FEDERAL HOUSING ADMINISTRATION. — 1958. Minimum property standards for one and two living units. FHA No. 300 Nov. 1.
FEYTAUD, J. — 1950. Comparison of the effect of HCH and DDT on termites *(Reticulitermes lucifugus)*. Chem. Abst. 44:10247.
FULLER, C. E. — 1966. A common source outbreak of cutaneous larva migrans. USPHS Rpts. 81(2):186-190.
FURNISS, R. L. — 1953. Damp-wood termite control. Pest Control. 21(6):20.
GENTRY, J. W. — 1966. Formosan subterranean termite. Agr. Chem. 21(11):63-64.
GOETSCH, W. & R. GRUGER — 1942. Pilzzucht and Pilznahrung staatenbildenden Insekten. (Fungiculture and fungal nutrition of social insects.) Biol. en. (Vienna) 16¹/₃):41-112. Biol. Abst. 21(3):544-545, 1947.
GOSSWALD, K. — 1943. Directions for the rearing of termites. Z. angew. Entomol. 30(2):297-316. Rev. Appl. Entomol. A., 35(10):333, 1947.
GRAY, H. E. — 1960. Vikane: A new fumigant for control of drywood termites. Pest Control. 28(10):43-46.
GROSS, J. R. — 1949. On *Kalotermes minor* (Hagen). Pests. 17(3):14.
GUION, M. J. — 1951. Creeping eruption. Pest Control. 19(9):6, 50.
GUNN, J. W. — 1952. Pest Control 20(4):28. 1953. Termite Control in slabs with ethylene dibromide. Pest Control 21(6):20.
GUNN, J. W., H. SMITH, BOB LOIBL, JR., & C. W. PENCILLE. — 1947. Report on the semi-annual convention of pest control operators in California. Pests. 15(9):8, 10, 11.
HAGEN, H. — 1876. The probable danger from white ants. Amer. Naturalist. 10(7):401.
HARRIS, C. R. — 1964. Influence of soil type and soil moisture on the toxicity of insecticides in soils to insects. Nature. 202(4933):724.
HARRIS, W. V. — 1961. Termites: Their Recognition and Control. Longman, Green and Co., Ltd. 187 pp.
HARVEY, P. A. — 1934. Life history of *Kalotermes minor*. In Termites and Termite Control. Univ. Calif. Press.
HASSLER, K. — 1953. E.D.B. — A new approach to Sub Control. Pest Control 31(3):37, 38. 1955. Termite fumigation in California. Pest Control 23(2):14-16. 1956. Some facts about fumigation tarps. Pest Control 24(2):25-26. 1960. Procedure for precautions to observe with sub-slab fumigation for subterranean termites. Pest Control 28(7):36.

HATFIELD, I. — 1944. Research tests on soil-poisoning chemicals for the control of subterranean termites. Pests. 12(3):10-14.

HATFIELD, I. & R. VAN ALLEN. — 1956. Termite-repelling wood preservative. Pest Control 24(10):32, 34, 78.

HEADLEE, T. J. & D. M. JOBBINS. — 1930. Some effects of acid arsenate of lead used against the termite, *Reticulitermes flavipes* (Kollar). J. Econ. Entomol. 32:638-640.

HEAL, R. E. — 1952. Drywood termites as occasional invaders in the east. Pest Control. 20(4):28, 1957. Chemicals in termite control. Soap & San. Chemicals. 33(8):73-76, 109. 1964. Termite jobs(?). Pest Control. 32(5):82.

HEGH, E. — 1922. Les Termites (Bruxelles, Impr. Industrielle & Financiere).

HENDEE, E. C. — 1933. The association of the termites, *Kalotermes minor, Reticulitermes hesperus and Zootermopsis angusticollis,* with fungi. Univ. Calif. Publ. Zool. 39:111-134.

HETRICK, L. A. — 1950. The toxicity of some organic insecticides to the eastern subterranean termite. J. Econ. Entomol. 43(1):57-59. 1952. The comparative toxicity of some organic insecticides as termite soil poisons. J. Econ. Entomol. 45(2):235-237. 1957. Ten years of testing organic insecticides as soil poisons against the Eastern subterranean termite. J. Econ. Entomol. 50(3):316-317. 1961. *Kalotermes approximatus* Snyder infest rosacious trees. Fla. Entomol. 44(1):53-54. 1962. Effectiveness of insecticides in soil against termites after 15 years. J. Econ. Entomol. 55(2):270-271.

HOCKENYOS, G. L. — 1940. Properties of soils related to termite control. Pests. 8(3):10-12. 1949. Wall void treatment. Pests. 17(3):9-13, 38.

HOCKENYOS, G. L. & H. HYNDMAN. — 1937. Soil poisons for termite control. National Pest Control Assoc. Communication #115. Appendix #1, p. 1.

HODEL, C. — 1949. Drywood termite control in southern California. Pests. 17(3):30,32.

HOLWAY, R. T. — 1941. Tube-building habits of the eastern subterranean termite. J. Econ. Entomol. 34:389-94.

HOWELL, J. L. — 1951. Importance of soils in termite control. II. Pest Control 19(12):26. 1952. Safety is part of every termite job, too! Pest Control. 20(1):27.

HUNT, R. W. — 1949. The common drywood termite as a pest. J. Econ. Entomol. 42(6):959-962. 1959. Wood preservatives as deterrents to drywood termites in the Southwest. J. Econ. Entomol. 52(6):1211-1212, 1966. Inspection for wood-destroying organisms. Pest Control Operators of Calif., Inc.

ISHERWOOD, H. R. — 1950. Definitions of building terms. Pest Control. 18(2):18-19. 1956. Over 100 years . . . A forgotten essay on termites by Dr. Thaddeus W. Harris. Antimite Co., St. Louis, Mo. 1957. Pressure-treated wood tops shields in home mechanical termite control. Pest Control. 25(11):32, 34.

JOHNSTON, H. R. — 1943. Laboratory tests of termite shields. J. Econ. Entomol. 36(3):386-392. 1954. Soil poisons under concrete slabs. Pest Control. 22(2):24, 28, 46. 1958. Tests with Soil Poisons for controlling subterranean termites. Pest Control. 26(2):9, 11-16. 1960. Soil treatments for subterranean termites. U.S. Forest Serv. So. Forest Expt. Sta. Occas. Paper 152.

JOHNSTON, H. R. & J. V. OSMUN. — 1960. Good-bye termite control. Pest Control. 28(5):62-63.

JOHNSTON, H. R., V. K. SMITH & R. BEAL. — 1971. Chemicals for subterranean termite control: results of long-term tests. J. Econ. Entomol. 64(3):745-748.

JONES, E. — 1966. Thorough treatment prevent TO call-backs. Pest Control. 34(8):46.

KATZ, H. — 1955. TO's memory pegs second story job. Pest Control. 23(10):86.

KEENE, E. A. & S. F. LIGHT. — 1944. Results of feeding ether extracts of male supplementary reproductives to groups of nymphal termites. Univ. California Publ. Zool. 49(9):283-290. Biol. Abst. 19:1956, 1945.

KENNEDY, C. H. — 1957. Child labor of the termite society versus adult labor of the ant society. Sci. Monthly. 65:309-324. Oct.

KOEHLER, P. — 1980. Formosan termite now in Florida. Pest Control. 48(11):20.

KOFOID, C. A., et al. — 1934. Termites and Termite Control. Univ. of Calif. Press., Berkeley.

KOWAL, R. D. — 1954. Termite control in slab construction. Pest Control, 22(2):12, 14, 16, 18.

KOWAL, R. J. & R. A. ST. GEORGE. — 1948. Preliminary results of termite soil-poisoning tests. J. Econ. Entomol. 41(1):112-113.

KRISHNA, KOMAR. — 1961. A generic revision and phylogenetic study of the family Kalotermitidae (Isoptera). Amer. Mus. Nat. Hist. Bull. 122:400-408.

KRISHNA, K. & F.M. WEESNER. — 1969. Biology of termites. Vol. 1 & 2. Academic Press, N.Y.

LEE, K. E. & T. G. WOOD. — 1971. Termites and soils. Academic Press, N.Y. 252 pp.

LESSER, M. A. — 1947. Wood preservatives. Ag. Chem. 2(2):22-25, 65.

LICHTENSTEIN, E. P. & K. R. SCHULZ. — 1959. Factors affecting insecticide persistence in various soils. Pest Control. 27(8):40, 42, 56.

LICHTENSTEN, E. P. & K. R. SCHULZ. — 1959. Breakdown of lindane and aldrin in soils. J. Econ. Entomol. 52(1):119-124.

LIGHT, S. F. — 1934. Drywood termites, their classification and distribution. In Termites and Termite Control. Univ. of Calif. Press. 1942. The determination of the castes of social insects. Quart. Rev. Biol. 17(4):312-326.

LIGHT, S. F. & A. L. PICKENS. — 1934. In Termites and Termite Control. Univ. of Calif. Press. pp. 150-156.

LIGHT, S. F., M. RANDALL & F. G. WHITE. — 1930. Termites and termite damage. Calif. Agr. Expt. Sta. Circ. 318.

LIGHT, S. F. & F. M. WEESNER. — 1947. Methods of culturing termites. Science. 106(2745):131-132.

LINSLEY, E. G. — 1948. Some habits of the western subterranean termite with special reference to factors which must be considered in relation to control. Pest Control & Sanitation. 3(2):8-9, 11.

LUND, A. E. — 1958. The relation of subterranean termite attack to varying retentions of water-borne preservatives. Proc. Amer. Wood Preserver's Assoc. 54:4-52. 1959. Subterranean termites and fungi; mutualism or environmental association. Forest Prod. J. 9(9):320:321. 1960. Termites and wood-destroying fungi. Pest Control. 28(2):26-28. 1962. Subterraneans and their environment. Pest Control. 30(2):30-34, 36, 60.

LUND, H. O. — 1959. Tests of the ability of *Reticulitermes flavipes* (Kollar) to build tubes over pine wood chemically treated for rot control. J. Econ. Entomol. 52(3):533-534. 1961. Will wood preservatives block termite tubing. Pest Control. 29(4):58-60.

MAMPE, C. D. 1976. Answers. Pest Control 44(2):16.

MARSHALL, C. W. — 1966. A digest of termite rodding techniques. Pest Control. 34(5):72, 74, 76.

McCAULEY, W. E. & W. P. FLINT, — 1938. Outwitting termites in Illinois. Ill. Nat. Hist. Surv. Circ. 30.

McMAHON, E. A. — 1963. New radioactive tests show how termites feed. Pest Control. 31(2):32-34, 36.

MILLER, E. M. — 1949. Florida termites. Editor, Univ. Miami Press, Coral Gables 34, Fla.

MILLER, J. M. — 1941. Public relations. Pests. 9(4):6-7.

NATIONAL PEST CONTROL ASSOCIATION. — 1951. Approved reference procedures for subterranean termite control, by the Wood Destroying Organisms Committees, 1948-1951. Edited by Ralph E. Heal, technical director. 1957. NPCA now recognizes EDB as a soil fumigant; gives use instructions; warns against MB. Pest Control. 25(8):30, 32, 34.

O'BRIEN, R. E., J. K. REED & R. C. FOX. — 1965. Insecticide distribution in soils following application by soil injector rods. Pest Control. 33(2):14-15, 42.

OSMUN, J. V. — 1958. Factors affecting sub-slab dispersion in soil. Pest Control. 26(2):23-24, 56.

OSMUN, J. V. & D. C. PFENDLER. — 1955. A device for sub-slab pressure injection of insecticides. J. Econ. Entomol. 48(4): 479-480.

PACKARD, H. R., JR. — 1951. The control of drywood termites in Southern California. Pest Control. 19(2):9-10.

PADGET, L.J. — 1960. Program for eradication of *Coptotermes crassus,* a subterranean termite new to the United States, at Todd Shipyards, Houston, Tex. Down to Earth, 16(2):11-14. Fall.

PAULSON, R. W. & L. W. BOULANGER. — 1958. Summary of insect conditions — 1957. Coop. Econ. Insect Report, 8(1):10.

PENCE, R. J. — 1957. The prolonged maintenance of the western subterranean termite in the laboratory with moisture gradient tubes. J. Econ. Entomol. 50(3): 238-240.

PENCILLE, C. — 1947. Structural fumigation to eliminate drywood termites *(Kalotermes)* with liquid hydrocyanic acid gas. Pest Control & Sanitation 2(8):10-12

PICKENS, A. L. — 1932. Observations on the genus *Reticulitermes* (Holmgren). Pan-Pacific Ent. 8:178-180. 1934. The biology and economic significance of the western subterranean termite, *Reticulitermes hesperus.* In Termites and Termite Control. Univ. of Calif. Press.

PICKENS, A. L. & S. F. LIGHT. — 1934. The desert subterranean termite, *Heterotermes aureus.* In Termites and Termite Control. Univ. of Calif. Press.

PURDY, J. L. — 1951. *Kalotermes* in shipped lumber. Pest Control. 19(10):40.

RANDALL, M. & T. C. DOODY. — 1934. Poison dusts. In Termites and Termite Control. Univ. of Calif. Press.

RATNER, H. — 1963. Spot and stop moisture conditions; avoid second-story termite attack. Pest Control. 31(7):38, 40, 42.

REIERSON, D. A. — 1966. Feeding preference of drywood termites and termiticidal effect of boric acid. PCO News. 26(11):14-15.

REYNOLDS, W. B. — 1963. Human infection with termites. J. Amer. Med. Assoc. 186(4):426.

RUSSELL, R. M. — 1958. Relationship of construction to sub-termite infestations. Pest Control. 26(10):74.

SCHENDEL, R. R. — 1952. Sub tubes 50 feet above ground level. Pest Control. 20(10):38, 40.

SCOTT, K. G. — 1951. How to make a "seal-off" for dirt-filled porches in Southern California. Pest Control. 19(10):35, 37, 40.

SHELFORD, V. E. — 1952. Termite treatment with aqueous solution of chlordane. J. Econ. Entomol. 45(1):127.

SMITH, H. C. — 1952. ABC's of safety on fumigating a house with methyl bromide. Pest Control. 20(8):20, 28.

SMITH, M. K. — 1952. Termite control in slab-type construction. Pest Control. 20(2):9, 10, 12.

SMITH, M. W. — 1953. Staining of hardwood floors. Pest Control. 21(8):41. 1956. Where are we going in our control methods for subterranean termites? Pest Control. 24(11):36, 38, 40. 1957. New approaches to "sub" treatment of slab houses. Pest Control. 25(7):36, 38, 40.

SMITH, R. H. — 1930. Experiments with Paris green in the control of termites. Mon. Bull., Calif. Dept. Agr. 19:557-60.

SMITH, V. K. — 1968. Long term movement of DDT applied to soil for termite control. Pesticides Monitoring. J. 21:55-57.

SMYTHE, R. B., et al. — 1971. Influence of wood decay on feeding and survival of the eastern subterranean termite, *Reticulitermes flavipes* (Isoptera: Rhinotermitdae). Annals ESA. 64(1):59.

SNYDER, T. E. — 1915. Biology of the termites of the eastern United States with preventive and remedial measures. U.S.D.A. Bur. Ent. Bull. 94:13-95. 1935. Our Enemy the Termite. Comstock Publ. Co. Revised 1948. 1949. Catalog of the termites of the world. Smithsonian Inst. Misc. Collect. 112. 1950. Control of nonsubterranean termites. U.S.D.A. Farmers' Bull No. 2018. 1951. Roadblocks for sub control. Pest Control. 19(2):28. 1952. The arsenic hazard in termite control. Pest Control. 20(3):34, 48. 1954a. Large Pacific Coast dampwood termite introduced into East. Pest Control. 22(3):33, 34. 1954b. Order Isoptera — The termites of the United States and Canada. National Pest Control Assoc. 64 pp. 1955a. Northward termite migration. Pest Control. 23(9):28, 30. 1955b. Termite attack on plastics and fabrics. Pest Control. 23(3):48, 56. 1956. Annotated, subject-heading bibliography of termites, 1350 B.C. to A.D. 1954. Smithsonian Misc. Collect. Vol. 130. 350 p. Publication 4258.

SNYDER, T. E. & J. ZETEK. — 1943. Effectiveness of wood preservatives in preventing attack by termites. Circ. U.S.D.A. No. 683.

SPITZ, W. J. — 1958. How we get complete coverage of sub-slab soil with termite chemicals. Pest Control. 26(2):38, 40, 43, 51.

ST. GEORGE, R. A. — 1939. Termite control in buildings. Pests. 7(2):13. 1941. Protection of log cabins, rustic work, and unseasoned wood from injurious insects. U.S.D.A. Farmers' Bull. 1582. 1944. Test of DDT against ants and termites. J. Econ. Entomol. 37(1):140. 1952. Test with new insecticides for termite control. Pest Control. 20(2):20. 1952. Testing soil poisons for termite control. Pest Control. 20(4):36. 1957. Protecting yews from termite damage. Pest Control. 25(2):38, 40.

STEWART, D. — 1957. Sulfuryl fluoride — a new fumigant for the control of the drywood termite. *Kalotermes minor* (Hagen). J. Econ. Entomol. 50(1):7-11. 1962. Precision fumigation for drywood termites with Vikane. Pest Control. 30(2):24, 26, 28. 1966. Balanced fumigation for better termite control. Down to Earth. (Dow). 22(2):8-10.

TURNER, N. — 1939. Construction of metal termite shields. Pests. 7(2):16. 1961. Termites in buildings. Conn. Agr. Sta. Circ. 218.

TURNER, N. & J. F. TOWNSEND. — 1936. Termite control in buildings in Connecticut. Conn. Agr. Exp. Sta. Bull. 382.

URQUHART, F. A. — 1953. The introduction of the termite into Ontario, Canad. Entomol. 85(8):292-293.

WALTERS, D. T. — 1957. "Kerodex" No. 71 stops termite operator's itch. Pest Control. 25(5):42.

WEESNER, F. M. — 1956. The biology of colony foundation in *Recticulitermes hesperus* (Banks), Univ. Calif. Publ. in Zool. 61(5):253-314. 1965. The termites of the United States. NPCA.

WOLCOTT, G. N. — 1945. DDT as a termite repellent. J. Econ. Entomol. 38(4):493. 1946. Factors in the natural resistance of woods to termite attack. Caribb. Forester. 7(2):121-134. Rev. Appl. Entomol. A. 35(11):375. 1947. 1947. Termite repellent: a summary of laboratory tests. Puerto Rico Agr. Expt. Sta. Bull. 73. 1950. Benzene hexachloride as a termite repellent. J. Agr. Puerto Rico. 1956. One-percent pentachlorophenol protects woods against dry-wood termite attack for more than 11 years. J. Agric. Univ. Puerto Rico. 40(1):85-86. Biol. Abst. 30(12):3584, 1956.

YOUNG, T. R. JR. — 1955. Inexpensive heat-exchanger for methyl bromide dry-wood termite fumigations. Pest Control. 23(3):45.

ZIMMERN, A. — 1950. Safety engineering in termite control. Pest Control. 18(2):30, 32, 34. 1952. Precautions in the use of oil products with termite chemicals. Pest Control. 20(5):29. 1957. Soil penetration. Pest Control. 25(2):32, 34, 36, 50.

C. DOUGLASS MAMPE

Dr. C. Douglass Mampe received degrees in entomology from Iowa State University, North Dakota State University, and a Ph.D. from North Carolina State University. Upon completing his university education, he joined the staff of the National Pest Control Association (NPCA). During his 10 years there he was responsible for keeping abreast of all technical and regulatory actions related to the structural pest control industry. This included reviewing research proposals and overseeing research projects related to the biology and control of termites, cockroaches, commensal rodents, and other related pests.

Mampe left NPCA as its technical director and joined Western Termite and Pest Control, Inc., and its sister company Residex, a pest control industry supply house. He served as Western's technical director for six years and developed practical procedures for field operations. He eventually became general manager of Residex, which included developing new pesticide registrations and industry training programs, including certification training programs for a number of the northeastern and middle Atlantic states.

In 1980, Mampe started his own consulting firm for the urban and structural pest control industry. His firm provides technical consultation, training and training programs, evaluates new pesticides and equipment, custom develops personnel management programs and provides expert testimony for legal cases.

Mampe served as editor of NPCA's Approved Reference Procedures for Control of Subterranean Termites, and editor and co-author of the Manual for Structural Wood Decay. He writes a monthly column called "Answers" for Pest Control magazine. This column, which deals with a variety of questions, has consistently been one of the best read sections of the magazine since the column began at the beginning of 1975.

Mampe is a member of the NPCA, Pi Chi Omega (a fraternity of industry professionals involved in training), the Entomological Society of America, and a number of state pest control associations. He is currently licensed as a certified pesticide applicator in a number of northeastern states. In 1979, he was chosen "Pest Control Operator of the Year" by the New Jersey Pest Control Association for his contributions to the industry in that state. He is currently serving as a resource person for the United States House of Representatives Committee on Agriculture.

CHAPTER EIGHT

Decay Fungi

Revised by C. Douglass Mampe[1]

MUCH OF THE DAMAGE to structural timbers attributed to termites actually is caused by fungus attack, and some individuals believe fungi may be the more important of the two as structural pests. As a rule, conditions that favor the growth and development of fungi are also favorable for infestation by subterranean and dampwood termites, and vice versa.

The decay "dry-rot" fungi, which are the ones involved in structural damage, are plants that lack chlorophyll, and since they cannot produce food carbohydrates, they live on dead trees, dead limbs or lumber products. The fungus consists of thread-like strands known as hyphae which aggregate into masses called mycelium.

The mycelium, under suitable conditions, forms fan-shaped sheets, especially when developing in a very moist locality. These may give rise to the fruiting body of the fungus, which in the case of the dry-rot fungi is relatively flat. The fruiting bodies bear enormous numbers of microscopic spores that are similar in function to seeds in the higher plants. Spores are readily distributed by water, air currents, or man and animals. If the spore contacts a suitable material in an agreeable environment, it germinates and gives rise to hyphae which penetrate into wood or other cellulose products such as paper, cardboard, books, etc. The fungus may also be distributed by means of hyphae which spread from decayed material to sound material.

Lumber attacked by a decay fungus has a brownish discoloration and presents a crumbly appearance, with square and rectangular pieces of wood. Although the dry, crumbly appearance of the decayed wood has given rise to the term "dry-rot," this term is a misnomer, since the decay took place originally in the presence of a noticeable amount of moisture.

[1]President, D.M. Associates, Westfield, N.J.

Fig. 8-1. Decay fungus on joists and beams of poorly ventilated underarea of house.

In the United States, the building poria, *Poria incrassata,* is the fungus that is most destructive to structural timbers in homes and other buildings. Richards (1933) notes this dry-rot is widespread "along the Pacific Coast, in Atlantic Coast States, as far north as Pennsylvania, and thoughout the Gulf States . . ." Boyce (1938) observes that *Poria incrassata* is serious only on coniferous wood. According to Richards, the tear fungus, *Merulius lacrymans,* which is so destructive to buildings in Europe, occasionally causes much damage to structural timbers in the northern part of the United States and Canada. *Merulius lacrymans* has been designated the "tear fungus" because of the drops of moisture that often accumulate on the fruiting body. Boyce states the higher summer temperatures and the greater heat maintained in buildings during the winter in the United States do not favor the development of the tear fungus. There are a number of other fungi that attack lumber in buildings such as *Poria microspora* (Hirt and Lowe, 1945), but as a whole they are not as important as the previously mentioned species.

Both *Poria incrassata* and *Merulius lacrymans* produce a similar decay on coniferous wood. The sporophore or fruiting body of the *Merulius* fungus is orange or light brown when dry. The *Poria* fungus varies from orange through olivaceous to deep purplish. In *Merulius,* the fruiting body is waxy, and the spore-bearing layer is shallow and net-like. Whereas, in *Poria* the fruiting body is membranous and has spore-bearing tubes which appear wart-like.

Fig. 8-2. Decay fungus showing whitish mycelia and checked appearance of rotted wood.

Humphrey (1923) notes the *Poria* fungus "works very rapidly and for this reason has occasioned serious concern in more than one household. In some instances, tenants have complained of the floors breaking through in walking about, or the post of a bed has crashed through, or a door has fallen from its hinges."

Fig. 8-3. Flooring destroyed by decay fungus.

Food of the fungus. The hyphae of the fungus bore through the cell walls of the wood by means of chemical agents they secrete known as enzymes. These enzymes reduce the complex chemicals that make up the cellulose and lignin of the cell walls into simpler substances that are more readily consumed. The fungus also feeds on the cellular contents of the cell, such as sugars and starches. With the destruction of the cell walls and their contents the wood disintegrates and decay becomes evident.

Effects of moisture. The decay fungi thrive in a damp and humid environment. Wood is most subject to decay where the moisture accumulates and does not evaporate readily. Decay does not occur where the wood is very dry or where it is completely saturated with moisture. In the latter case, the growth of the fungus is inhibited by lack of air. Wood with a moisture content below 20 percent inhibits the growth of the fungus. According to Verall (1954), *Poria* dies in 24 hours in wood with eight percent moisture, and in 12 days or less in wood with 13 percent to 21 percent moisture. A piece of infected lumber, when dried, may show no external evidence of the fungus, but should the piece of lumber be returned to a suitably moist location, the fungus may become active and grow. Richards (1933) notes some of the wood-destroying fungi can remain dormant for years.

Decay fungi are somewhat unique since they very often flourish many feet from the source of moisture. In such instances, the water is transported by the fungus to normally dry wood by thick strands of mycelium which are root-like in appearance, white or brown, and vary from ⅛-inch to one-inch in thickness. These water-transporting strands may extend for 30 feet/nine m or more feet across brick, concrete, and material of a similar nature.

Decay occurs most commonly when wood is buried or in contact with the moist ground. In this respect, Humphrey (1923) notes, "It is not necessary, however, that the sub-floor timbers and floors be close to the ground, for in the moist climate of the Gulf States houses built over apparently dry sandy soil and well ventilated beneath have suffered heavy losses. In frame buildings the fungus spreads from the subfloor timbers up the walls and usually first becomes evident by the rotting and shrinkage of baseboards, panels, wainscoting, door casings, etc. The shrinkage, and consequent opening of the joints, expose the whitish mycelium behind."

Serious injury from decay also occurs in wooden basement floors laid over concrete resulting in a dead-air space without ventilation. Timbers embedded in concrete walls, and wood moistened by defective plumbing or a "sweating" of water pipes, also provide an environment suitable for the development of decay fungi. Factories with high humidity and wooden roofs often find these roofs attacked by fungi since moisture is very apt to condense on the roofs during the cold weather of winter. The above are but a few of the structural conditions that provide suitable localities for the growth of timber-destroying fungi.

If the air is sufficiently humid and other conditions satisfactory, the decay fungi may attack objects in contact with the infected wood, such as books, paper documents, and permanent fixtures. Edgerton (1924) mentions several valuable electrical measuring instruments stored in the basement of a physics laboratory that were ruined by the building poria.

Effects of temperature. Cartwright and Findlay (1946) note the optimum temperature for growth of *Merulius lacrymans* is about 74° F/23° C and the maximum temperature is about 79° F/26° C. Low temperatures retard the growth of the fungus, but the dry-rots resume growth with the coming of warmer weather. Temperatures above 115° F/46° C, if sustained for some time, usually kill these fungi. The heat in the kiln drying of lumber kills the fungus and reduces the water content. Thus, the decay fungi cannot reestablish themselves until the wood reabsorbs the necessary amount of moisture.

Natural resistance of wood to fungi. The heartwood of lumber is, as a rule, much more durable than the sapwood. Hunt and Garratt (1952) note that in the living tree the above condition is reversed since the moisture content of the sapwood is so high as to reduce the amount of air below the minimum required for fungus growth. The decay resistance of heartwood lumber is due largely to such chemicals as water-soluble extractives, resins, pine oils, turpentines, essential oils and tannins. It must be remembered that a fungus is a living plant that is nourished by liquid food materials, and toxic substances that are water soluble will injure or kill the fungus. Thus, when such a toxic material as zinc chloride is added to lumber, it inhibits or prevents fungus growth.

Conifers that are naturally very durable are Pacific yew, juniper, redwood, bald cypress, and western red cedar. Hardwoods that are very durable are osage orange, black locust, red mulberry, catalpa, and black walnut.

Verall (1954) notes *Poria* can decay the heartwood of durable species such as the heartwood of bald cypress, western red cedar, and redwood. Nevertheless, this fungus shows no unusual resistance to such fungicides as creosote, zinc chloride, or pentachlorophenol. It does show some resistance to copper fungicides.

CONTROL OF DECAY FUNGI

The decay fungi may be prevented from establishing themselves, or in any event, controlled, by making unfavorable all the conditions essential for their growth. The air and temperature requirements of the fungus cannot be as readily regulated as their moisture and food requirements, and thus man directs his control efforts largely towards the latter two. Hicken and Levy discuss the problem of decay fungi in some detail. Amburgey (1978) and NPCA (Anon., 1974), discuss practical methods of controlling wood decay in structures. However, where wood is constantly subjected to conditions favoring decay, naturally-resistant woods such as bald cypress, western red cedar, redwood or pressure-treated timbers should be used. In certain instances, the dry rots, by means of their water-transporting mycelia, may cross resistant materials in order to attack susceptible lumber. Verall (1954) is of the opinion that *Poria* is often introduced into a building in infected lumber rather than by airborne spores of the fungus.

The prevention of decay in structural timbers. Before listing the various conditions to be observed in the prevention of infection by dry-rot fungus, it is worth noting that these same factors are also applicable in the control of subterranean and damp-wood termites, and are discussed in greater detail in the section on termites.

Repair of decayed buildings. Structural timbers can be protected by periodic wetting (at least once a year) with five percent pentachlorophenol solutions. The crevices formed where wood joins wood, a favorite point of fungus infection, can be wetted with a solid stream from a force oiler. For wetting large surfaces, a 3½-gallon garden spray outfit with an oil-resistant hose is satisfactory for small jobs. Anon. (1949) recommends the following when pentachlorophenol is applied with a brush: "Flood the solution on the surface. Actually it is best to slop it on in order to get a thorough coverage and as much penetration as possible. Use one gallon to every 100 to 250 square feet of surface. At least two heavy coats should be applied. Allow sufficient time between coats to let the wood soak up the pentachlorophenol. Two to three times longer life for wood can be obtained if enough solution is properly brushed on." It is likely a cream emulsion of pentachlorophenol would also be effective against decay fungi. It should be noted that pentachlorophenol is odorous and the odor is persistent.

Creosote and other coal tar oils may be used in place of pentachlorophenol for the treatment of foundation timbers. There are highly refined creosote oils suitable for this purpose currently on the market. Some of the more refined creosote oils impart a rich brown color to the wood. Several coats should be applied to the dry wood. Anon. (1948) advises the use of 1½ gallons/5.71 per 100 sq. ft/nine sq. m for each coat.

Cooley (1934) recommends creosote, as well as a five percent solution of zinc chloride that is prepared by mixing five pounds/2.3 kg of the chemical with 11½ gallons/43.5 l of water.

All of the above treatments by spraying or brushing result in very slight penetration of the wood surface and must be repeated. In some instances, the ground itself may be treated for the control of dry-rots with some of the soil toxicants mentioned in the chapter on termites. Anon. (1941) says the following in regard to the repair of decayed buildings: "In repairing a building damaged by decay, the primary job is to determine the source of the moisture and remove it. Ordinarily, if adequate ventilation and soil drainage are provided and if all contacts of untreated wood with the soil or moist concrete or masonry are broken, the decayed wood will dry out and further decay will be prevented. In making replacement, it is a good plan to cut out at least a foot beyond the rotten area, because wood is usually infected beyond the point where the rot is apparent. New, green, untreated lumber should never be nailed against old infected material, since this exposes the new wood to immediate attack, with the result that decay may be much more rapid than it was in the original construction. Because of the need for effectively remedying causes of decay that may be inherent in the form of construction used, the replacement of decayed parts of a structure frequently should assume the character of a remodeling job to provide more ventilation, or otherwise improve on the design, rather than superficial replacement of decayed lumber."

The U.S. Forest Products Laboratory, Madison, Wisconsin (Anon., 1941) lists the following principles for the prevention of decay in building timbers: "The cardinal principles of good building practice to avoid decay can be summarized in the following rules:

1) "Build on a well-drained site. This requires the avoidance not only of marshy locations where the water table is at or near the surface, but also of the more common error of poor grading which, especially in the case of houses without basements, causes drainage from the home site or from surrounding areas to seep under the house. Rain water and melting snow should be drained from the building and the drainage of the general area should be sufficient to keep the ground beneath and around the structure dry. Moisture sources, such as fish ponds close to the house, should be recognized as dangerous.

2) "From lumberyards where the stock is kept off the ground and protected from rain, select only decay-free lumber that is dry, and keep it dry between delivery and installation. Lumberyards should deliver lumber at the building site at such a moisture content as to prevent decay and give minimum trouble from shrinking and swelling after installation. Once delivered, lumber is often handled with an apparent disregard not only of its value, but also of the swelling, shrinking, and decay that may result if it is allowed to become wet. During temporary storage on the building site, all lumber should be protected from rain or other moisture sources, and should never be piled directly on the ground. For the parts of the building in which the decay hazard is high, select the heartwood of decay-resistant species, such as bald cypress, cedar or redwood, or use wood that has been properly treated with a good preservative.

3) "Maintain sanitary conditions with respect to foundation, basement, and masonry. All wood scrap and debris that might furnish food for fungi should be removed. A tree stump left beneath a building without a basement has been known to furnish an entrance point for fungus infection resulting in hundreds of dollars worth of damage to floors and woodwork. For the same reason, all concrete forms and form stakes should be removed.

"If dirt-filled porches or terraces are used, the wooden sill back of the fill should be completely isolated from the soil by a non-corrosive metal flashing extending between the sill and the foundation upon which it rests, bent upwards over the outside face of the sill, and extending upwards beyond the porch floor and under the siding or other surfacing of the building.

4) "As a rule, place no untreated wood within 18 inches of the ground. The wide variation in temperature, rainfall, prevalence of extremely destructive fungi, and availability of the more decay-resistant woods in different sections of the country permit considerable latitude in the application of this rule with respect to contact with the ground. In warm, humid regions, 18 inches may not be enough. In the colder and drier parts of the country, however, wood may be placed considerably closer to the ground. The 18-inch clearance should always be observed, however, unless ample local experience over a long period has definitely demonstrated that there is no risk in violating it. In some localities, so

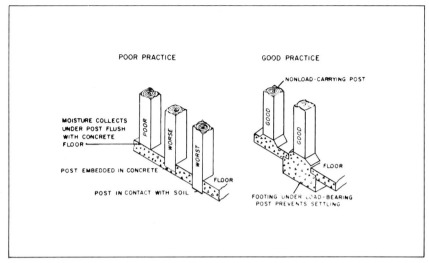

Fig. 8-4. Wooden posts on concrete floors.

far as decay is concerned, buildings may be safely supported on piers or posts of decay-resistant wood, such as the heartwood of bald cypress, cedar or redwood, if they are provided with concrete footings extending above the soil. Whenever there is uncertainty as to the safety of using unprotected material in any particular location in a building, the lumber should be thoroughly impregnated with a suitable preservative.

"Although decay does not always follow, it is bad practice to lay a wood floor or any untreated wood directly on a brick, cinder or concrete base at or below the soil grade line, because the wood may absorb sufficient moisture to bring about rapid decay. Good practice necessitates treating the floor sleepers with a preservative; any concrete sub-floor should be thoroughly dried before the wood is laid. As a further precaution, the concrete or other base should be waterproofed.

"Partition plates, stair carriages, and wood pillars should be on concrete bases and preferably separated from the concrete by some water-resisting materials, such as roofer's felt mopped on with asphalt.

"Similarly, embedding the ends of girders in masonry or concrete walls is not good practice unless the point of contact is well above the outside grade line so that the wall does not become damp and transmit dampness to the wood. Where necessary to seal wood in a wall close to the grade line, preservative-treated wood or all-heartwood stock of a naturally decay-resistant wood should be used.

5) "Beneath all buildings that are not provided with basements or in which the basements are so damp that exposed woodwork will absorb considerable quantities of moisture from the air, provide adequate cross ventilation so that no dead air pockets exist. Buildings without basements should be supported on foundations of adequate height with at least two square feet of opening per 25 linear feet, to insure ample air circulation. The ventilators may be grilled holes left in otherwise solid foundation, latticed brick in brick walls, unenclosed or wood-latticed spaces between supporting masonry piers or between the ends of floor joists above the foundation plate. Dense bushes or other plants should not

be placed directly in front of ventilators, as they will greatly reduce the effectiveness of these openings. In cold climates it is desirable to install special vents which may be closed during the winter months to avoid unnecessary cooling of the ground floor. However, the vents should always be opened in the spring. Porches elevated above the ground should be so built as to insure ample circulation of air beneath them.

6) "Make all exterior joints tight enough to keep moisture from accumulating in the adjacent wood. The most critical places are at the corners of the building and around windows, doors, and porches. Unless shutters and garage doors are made from naturally decay-resistant heartwood or properly preserved wood, avoid outside battens and cross rails, which frequently create decay hazards. Provide drainage through bases of porch columns and at the bottom rails of porch screens to avoid trapping water behind these members.

"In general, architectural frills or novel forms of construction should be studied carefully to determine whether they provide entrance points or pockets in which moisture may remain long enough to make wood susceptible to decay. Avoid all forms of construction that will trap moisture in the wood.

7) "Avoid the accumulation of moisture condensed from the atmosphere. In many parts of the United States, the water vapor in the air within the walls of a house may condense on the back of the sheathing or the under side of the roof during cold weather and even freeze there. When this happens, more water vapor moves into the wall space from within the house, and this, in turn, is condensed until, if the process is continued long enough, the amount of moisture taken up by the wood may be sufficient to permit decay as well as to cause swelling and paint difficulties. This condition will be aggravated if the air in the house is artificially humidified and if the walls are insulated.

"The most positive and least expensive method of preventing condensation within the wall structure of new houses is the use of vapor-resistant barriers at or near the inner face of the wall. Among the materials that are highly resistant to the passage of water vapor are: 1) light-weight asphalt roofing materials; 2) asphalt-impregnated and surface-coated sheating paper, glossy surfaced, weighing 35 to 50 pounds per roll of 500 square feet; 3) laminated paper made of two or more sheets of kraft paper cemented together with asphalt, 36-60-30 grade; and 4) double-faced reflective insulation mounted on paper.

"In houses already erected, painting the inner plastered surfaces of exterior walls with aluminum paint will provide a moisture barrier, although not quite so effectively as the use of the papers recommended for new construction.

"During warm, humid weather, there may be so much condensation of moisture on cold-water pipes that a considerable quantity of water drips on the woodwork. This may raise the moisture content of the wood to such a point that the wood is liable to serious localized decay. Such difficulty may be avoided either by insulating the cold surfaces or by making some provision for preventing the condensed moisture from reaching the wood. During cold weather, water may condense on window panes in sufficient quantity to run down over the sash or sills and soak into the wood. The installation of storm windows will greatly reduce the danger of decay in such cases. Reduction of the humidity within the building will also help materially. Chemically treated sash is available commercially and should be used unless the sash can be kept dry."

Verall (1954) lists the causes of rot by decay in the South as follows: wet soil, leaky plumbing, rain leaks, green lumber, leaky drainspouts. He notes that wood-soil contacts contributing to decay include joists or sills, dirt-filled porches,

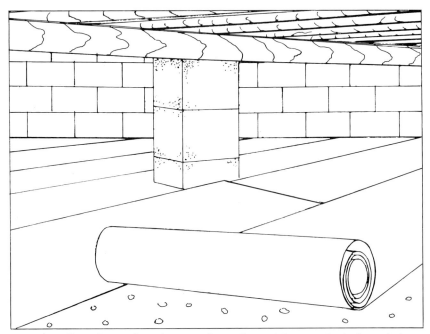

Fig. 8-5. Roll roofing under a house reduces the moisture content of wood above the roofing.

foundation forms, siding, trim, sheathing, lattice work, wood foundation, wood on ground line concrete slab, as well as inadequate ventilation. Conditions such as these favor decay in all parts of the country.

Hartley (1947) was concerned with the encouragement of decay in structural timbers of homes lacking suitable ventilation and situated on moist sites. The moisture condenses on the inner faces of sills, on joists near the outside foundation or throughout the substructure of unheated buildings. Hartley discusses this situation as follows: "Evidence was obtained indicating that vapor from soil was an important source of this condensation moisture, and a simple method was devised for shutting it off. Asphalt roll roofing was laid on the soil under buildings that were too moist. There was a surprising decline in the moisture content of the wood; no more condensation has been observed during the 3½ years since the cover was put on, although nearby houses that had the same moisture content before the beginning of the test continued to show high moisture, and in some condensation. Many houses, even on moist sites and with mineral-board-skirting completely closed around the base during the winter, do not reach the moisture danger point. The houses that do become moist enough for decay can apparently be safeguarded sufficiently by the use of roofing of the grade weighing 55 pounds per 108 square feet. . . ." Diller (1950) recommends the roofing be laid with a two-inch overlap, but without fastening. Amburgey (1978) and NPCA (Anon., 1974) recommended four-mil polyethylene sheeting as an effective vapor barrier in crawl spaces to prevent decay. In addition, if the substructural wood is saturated with water, the drying-out process should be slow to reduce warpage, shrinkage, and splitting. No more than 80 percent of the exposed soil initially should be covered with a vapor barrier. As the wood dries, additional soil may be covered, but there is no need to cover 100 percent of the soil surface nor to overlap the sheeting.

The use of moisture meters to determine the moisture content of wood, and the likelihood of decay occurring is useful for both inspecting buildings initially and on reinspections to determine if the moisture control techniques are adequate. Cassens (1979) reviews the use of such meters and lists four sources for them. The readings produced by moisture meters are usually reliable, but wood treated with a salt type preservative or fire retardants will produce erroneous readings (Anon., 1976). Temperature also affects readings. A correction table relating temperature to readings is presented by Cassens (1979).

WOOD PRESERVATION PROCESSES

St. George (1939) notes there are three common methods of wood preservation including 1) standard pressure process treatments in a closed steel cylinder, 2) open tank or hot- and cold-dip treatments, and 3) plain dip or cold soak, brush, or spray treatment. "The first method gives approximately 25 years of protection, the second around 15 years, and the third prolongs the life of the wood from two to five years."

Standard pressure process. It should be noted that in any wood preservation process the sapwood absorbs the preservative much more readily than does the heartwood. According to Randall and Doody (1934), pressure treatments are of two types, *full-cell* and the *empty-cell* process. "The preservative is applied by pressure-treating processes in large steel cylinders or retorts. The wood is placed in the treating cylinder, which is then filled with the preservative, and a pressure of 100 to 175 pounds per inch/seven to 12 kg/sq. cm is applied to force the preservative into the wood. In the full-cell treatment a vacuum is applied to the retort to remove as much air as possible from the cells of the wood structure before the preservative is run into the retort. This allows the preservative to flow readily into the wood cells, where it remains at the conclusion of the treatment.

"With the Lowry empty-cell process, the cells of the wood are allowed to remain full of air at atmospheric pressure when the preservative is run into the retort. In the Reuping empty-cell treatment the cells of the wood are filled with air under pressure (from 30 to 110 pounds per square inch/two to eight kg per sq. cm, is used depending on the type of treatment desired) before the preservative is applied. The preservative is then forced into the wood under pressure until the desired absorption is obtained. In both empty-cell processes the final step is the application of a drying vacuum. The air in the cells forces the excess preservative out, leaving the cells empty and the wood relatively dry at the conclusion of the treatment."

Any pressure-treating process results in an outer shell of varying thickness of treated wood and an inner core of untreated wood. It is the treated shell that protects the wood.

Open-tank method. This method permits full penetration of the sapwood of poles when properly done. Randall and Doody (1934) note, "In the application of coal-tar creosote by this treatment the poles are set in a vat of the hot preservative held at 220° to 270° F/104° to 132° C for four to eight hours and then allowed to cool. The poles are usually in the preservative bath for a period of 16 to 24 hours."

Anon. (1949) refers to the hot- and cold-dip method as the "hot-cold soak," particularly as it applies to the use of pentachlorophenol. In this process, the wood is seasoned to reduce the moisture content. Then it is soaked in a hot pentachlorophenol oil solution which opens the pores and cells. It may then be transferred immediately to a cold solution of pentachlorophenol, or, as was mentioned previously, be left in the hot solution until it cools.

Plain dip or cold soak method. In this method, the lumber is seasoned before being treated. Anon. (1949), referring to pentachlorophenol, notes: "Cold soaking can be done in a discarded watering tank, or in oil drum halves welded together. If the full length of the wood is to be treated, the tank should be long enough to accommodate lumber of that length. An upright tank made by welding three drums together, ends removed except for the bottom drum, is useful for treating fence posts."

The wood is immersed completely in the solution, and the length of time depends upon "the species, density of the wood, its cut thickness or how it is to be used."

Brush-treatment method. The use of pentachlorophenol by this method was described previously. Three coats of coal-tar creosote may be applied, allowing each to soak in thoroughly before the next application. Applying preservatives to the surface of the wood by means of brush or spray are not nearly as satisfactory as impregnation by other methods, since this treatment merely covers the surface, and any check, break or crack in the surface permits dry-rot and termites to find means of ingress. In such cases, use of resilient caulking to fill cracks, checks, and other gaps including wood borer damage, will be a valuable supplement to surface application of preservatives.

Be careful when spraying preservatives in a petroleum base since the mist is inflammable. The presence of open flames and pilot lights may result in an explosion. The operator, when spraying with preservatives, should wear a mask, and use rubber gloves, goggles, and protective (non oil-absorbing) clothing.

WOOD PRESERVATIVES

In order to protect a house against fungi and subterranean termites, there should be no contact between wood structures and the ground. In certain instances it may be necessary for the wood to be close to the ground, if not in actual contact with it. Where this is the case, the use of pressure-treated wood is essential. Pressure-treated wood, as was mentioned previously, consists of an outer shell of wood preservative forced into the wood cells under pressure. The preservative must be of a type that is resistant to leaching by ground or atmospheric moisture.

Hunt and Garratt (1952) separate wood preservatives into three classes, namely the toxic oils, such as creosote; the salts that are injected into wood in the form of water solutions, e.g., zinc chloride; and preservatives that consist of a highly toxic chemical in a solvent, or mixture of solvents other than water, e.g., beta naphthol or pentachlorophenol.

Coal tar creosote. This is our oldest wood preservative and was experimented with as early as 1756. It came into common usage in 1865 due to the establishment of a pressure-treating plant in Massachusetts. This wood preservative has proved satisfactory and is fairly economical. Coal tar creosote has a marked toxicity to wood destroying fungi, and since it is not readily soluble in water, it is retained in the wood for a long time.

However, wood preserved with coal tar creosote has several objectionable features which prevent the use of the treated wood on interior finish and wood work. Freshly creosoted wood presents somewhat of a fire hazard, it has a black color that is very difficult to paint over and is characterized by an odor that is objectionable to many people, and which may taint nearby food. Moreover, it is difficult to work with since it soils clothing and may cause skin irritation. When creosote is applied as a fine spray it is combustible and explosive. For these reasons the use of creosote-impregnated wood in the home is usually limited to the foundation and framework of the house.

Liquid coal tar creosote may be applied to wood by brushing, spraying or dipping. When warmed it penetrates more readily. When the preservative is applied in this manner it does not penetrate as deeply as by pressure treatment, and its service life is not as long as in pressure-treated timber. It has been found one can paint creosote-treated wood with aluminum paint that contains a fairly hard drying spar varnish-type of vehicle. Where the treated wood has been cut through, it is necessary to treat the fresh surface by soaking or brushing it with hot coal tar creosote. It is claimed that wood treated with certain refined coal tar products, is paintable.

Anthracene oils or carbolineums which are also coaltar distillates are used in the same manner as liquid coal tar creosote.

Zinc chloride and its derivatives. Zinc chloride was prepared as a wood preservative by Thomas Wade in 1815. Although creosote is a more permanent preservative, zinc chloride is used where the odor, color, and staining of creosote is objectionable. Wood treated with zinc chloride is subject to leaching, so it is used for above ground construction. Various other derivatives of zinc chloride are on the market, such as chromated zinc chloride. Chromated zinc chloride consists of 80 or more percent of zinc chloride, the remainder being sodium dichromate. Chromated zinc chloride and copperized chromated zinc chloride are claimed to be more resistant to leaching than zinc chloride. Wood treated with zinc chloride shows considerable resistance to fire. Such preservatives usually are applied to the lumber by the full cell pressure-impregnation process. Other water soluble preservatives which are not used widely are sodium fluoride, mercuric chloride, copper sulfate, and arsenicals.

Pentachlorophenol. Blew (1946, 1948, and 1949) reviews the use of pentachlorophenol. The following is largely from his papers.

Solutions of pentachlorophenol in light petroleum oil solvents have been used widely in recent years for the treatment of window sashes and frames. It has been used for treating other products where a clean, odorless, and non-swelling treatment is required, particularly when the wood must be painted after treatment.

Pentachlorophenol is available in dry flakes or crystals which can be dissolved in such solvents as pine oil. However, due to the difficulty in dissolving the crystals, and also to the irritation to the eyes, ears, and nose caused by the fine dusts from the crystals, this material is usually purchased in solution in a petroleum base. Pentachlorophenol solutions may be purchased in concentrate form which is usually diluted to five percent pentachlorophenol with fly spray base oil, kerosene, or Stoddard's solvent, when it is desirable to have clean, paintable lumber. Otherwise, a No. 1, 2, or 3 light fuel oil may be used. Brushes, force oilers, and even garden-type sprayers may be used for surface application of pentachlorophenol. Such applications are usually of a superficial nature since the spray does not penetrate too deeply. Thus, they must be repeated. When using pentachlorophenol in the house, it should be remembered oil base solutions have a persistent odor that is objectionable to some people.

Blew (1948) notes: "The penetrating properties of the pentachlorophenol solution and the paintability of the treated wood are influenced chiefly by the solvents used in the solution. Light oils usually penetrate wood better than viscous oils. Oils with a viscosity as high as creosote or higher, however, cannot be expected to excel creosote in penetrating properties. Such oils are not likely to leave the wood in a paintable condition.

"There is evidence that treating solutions made with lighter fuel oils will not perform so well as those containing heavier petroleum oils."

The preservative properties of pentachlorophenol were studied by Hatfield (1935) and the U.S.D.A. Forest Products Laboratories in Madison, Wisconsin. Further, Hunt and Garratt (1952) state that the Western Pine Association developed several preservative formulas. Although this association has copyrighted the name *Permatol,* the formulas are apparently not patented.

Permatol A

Pentachlorophenol 5   pounds
Pine oil or other solvents 1   gallon
Refined kerosene .... 1¼ gallons
Penetrant (Stoddard solvent) .... 10¾ gallons

Blew (1949) states pentachlorophenol "in the dry form and in solution irritates the skin of workers, but with careful handling and the use of suitable protective clothing it is possible to avoid harmful effects. The use of "bloom" preservatives, such as ester gum, is required in pentachlorophenol solutions with volatile solvents to prevent the formation of irritating crystals of the preservative on the surface of the wood after treatment."

According to the Monsanto Chemical Co., resins, synthetic resins, oils of vegetable origin (linseed or tung oil), as well as Monsanto Aroclors and Hercules Powder Co. resins such as Abalyn and Hercolyn, are suitable anti-blooming agents. Such anti-blooming agents are often used at five to 10 percent by weight in the finished pentachlorophenol formulations.

Woodtreat-TC, a creamed penta emulsion, is used for decay and termite control because of its penetrating ability when brushed on wood. The product consists of 87 percent by weight solvents and carriers, including 10 percent pentachlorophenol. The remaining 13 percent is water and emulsion stabilizers.

The U.S. Forest Products Laboratory has developed several wood preservatives other than pentachlorophenol which are used with highly refined petroleum distillates as a carrier, and with pine oil, as an additional solvent for the toxicant. Beta naphthol, chlorbetanaphthol, tetrachlorophenol, and chlororthophenylphenol are usually used as five percent solutions, and are clean, rapid drying preservatives that can be used in finished furniture. Copper naphthenate and chlorinated naphthalene are used in a fashion similar to the above.

According to Verrall (1965), pentachlorophenol and copper naphthenate in water-repellent solutions are highly effective. These water repellent solutions contain about one percent paraffin, ceresin, or other inexpensive waxes.

Proprietary preservatives. There are many patented preservatives on the market. Among the most widely used products is "Bruce preservative," which consists of five percent beta naphthol in an organic solvent. The Wolman Salts were developed in Germany. Two types of these salts used in pressure-treated wood are *Triolith,* which contains sodium fluoride, potassium bichromate, and dinitrophenol, and *Tanolith,* which is based on sodium fluoride, sodium chromate, anhydrous disodium arsenate, and dinitrophenol. Zinc meta-arsenite (ZMA), according to Snyder and Zetek (1943), is prepared "by mixing a solution of arsenious acid with a water solution of zinc acetate which contains an excess of acetic acid. Still another commercial product is "Celcure," a copper sulfate, potassium bichromate, and acetic acid preservative. Copper naphthenates in the form of ready-to-use solutions and concentrates for application in petroleum oils are being used for the control of decay fungi. Boron compounds are used in a proprietary treatment called "Timborize." Bis (Tri-n-butyl-tin) oxide is the active ingredient of the increasingly popular "Cuprinol" products marketed for consumer use as water repellent and rot resisting stains.

CHANGES IN LEGISLATION GOVERNING WOOD PRESERVATIVES

Preservatives. The Federal Register (2/19/81) reports the EPA has released its preliminary RPAR determination for wood preservative pesticides containing coal tar, creosote and coal tar neutral oil, inorganic arsenicals, and pentachlorophenol (PCP) (Anon., 1981).

The agency also decided to cancel the spray method of application of pentachlorophenol products which are now available for retail sale in concentrations of five percent or less. It will retain all other wood preservative uses, but classify them as restricted use pesticides and add further label restrictions and safety precautions.

Here are the restrictions for creosote and pentachlorophenol for home and farm use:

- Restricted use for all creosote and PCO products where concentration is greater than five percent.
- Strict use of protective clothing. Even "noncertified applicators" using less than five percent PCP must use long-sleeved cotton coveralls.
- Strict pesticide container disposal.
- Eating, smoking, and drinking prohibited during application.
- Careful application to prevent direct exposure to livestock or domestic animals, or food and feed contamination, or drinking or irrigation water pollution.
- No creosote or PCP can be used indoors except for support structures (foundation timbers, pole supports, or bottom six inches of stall skirtboards in barns or stables) and millwork which has outdoor surfaces (door frames, windows, and patio frames).

LITERATURE

AMBURGEY, T.L. — 1978. Prevent Wood Decay. Pest Control, 46(11): 32-26.
ANONYMOUS. — 1941. Prevention and control of decay in dwellings. U.S. Forest Products Lab. Tech. Note No. 251. 1948. Decay and termite damage in houses. U.S.D.A. Farmer's' Bull. No. 1993. 1949. Penta to preserve wood on the farm. Monsanto Chemical Co. 26 pp. 1949. — Wood preservative offered as a concentrate. Down to Earth (Dow Chem. Co.) 4(4); 10-11. 1974. Wood decay and its control in structures. NPCA. 1976. Effects of Wood pre-

servatives on electric moisture-meter readings. U.S. Forest Serv. Res. Note FLP-0106. 1981. Wood Preservatives Restricted. Pest Control. 49(6): 53.

BLEW, J. O. — 1946. Preservatives for wood poles. Purchasing. April. — 1948. Treating wood in pentachlorophenol solutions by the cold-soaking method. U. S. Forest Products Lab. No. R1445. Rev. 1949. Development in wood preservatives. Chem. Indus. 64:218-223.

BOYCE, J. S. — 1938. Forest Pathology. McGraw Hill. 600 pp.

CARTRIGHT, K. ST. G. & W. P. K. FINDLAY. — 1946. Decay of Timber and Its Preservation. Dept. of Sci & Indust. Research. Great Britain, 294 pp.

CASSENS, D. L. — 1979. Using electric moisture meters. Pest Control, 47(2): 15-16, 46.

COOLEY, R.H. — 1934. Termites and Termite Control. Univ. Calif. Press.

DILLER, J. D. — 1950. Reduction of decay hazard in basementless houses on wet sites. USDA Forest Pathology Special Release No. 30.

EDGERTON, C. W. — 1924. Dry rot in buildings and building materials. La Agr. Exp. Sta. Bull. 190: 1-42.

FRITZ, E. — 1938. The decay of wood. — Its cause and control. Ext. Log. 6:14-16, 1948. Cause and control of decay. Pest Control & Sanitation. 3(5): 8-10.

HARTLEY, C. — 1947. Fungi in forest products U.S.D.A. Yearbook 1943/1947 883-889.

HATFIELD, I. — 1935. Toxicity in relation to the position and number of chlorine atoms in certain chlorinated benzene derivatives. Proceedings American Wood-Preservers' Association. 31:57-66. 1944. Information on pentachlorophenol as a wood preserving chemical. Reprint of American Wood-Preservers' Association. pp. 1-19.

HENRY, W. T. & R. J. KEPFER, — 1949. Copperized chromated zinc chloride Proc. Am. Wood-Preservers' Assoc. 45:66-74.

HIRT, R. R. & J. L. LOWE, — 1945. *Poria microspora* in house timbers. Phytopathology 35(3): 317-318. Abst. Expt. Sta. Rec. 93(1): 50-1.

HICKIN, N. E. — 1967. The Conservation of Building Timbers. Rentokil Library. Great Britian.

HUBERT, E. E. — 1924. The effect of kiln drying, steaming and air seasoning on certain wood fungi. U.S.D.A. Bul. 1262.

HUMPHREY, C. J. — 1923. Decay of lumber and building timbers due to *Poria incrassata* (B & C) Burt Mycologia 15: 258-277. 1923. The destruction of the fungus *Poria incrassata* of coniferous timber in storage and when used in the construction of buildings. A.W.P.A. Proc. 19th Ann. Meeting: 188-201.

HUNT, G. M. & G. A. GARRATT. — 1952. Wood Preservation. McGraw-Hill.

LEVY, M.P. — A guide to the inspection of new houses and houses under construction for conditions which favor attack by wood-inhabiting fungi and insects. U.S. Dept. Housing Urban Development.

RANDALL, M. & T. C. DOODY. — 1934. Termites and Termite control. Univ. of California Press.

RICHARDS, C. A. — 1933. Decay in buildings. Report presented at 29th Annual meeting of American Wood-Preservers Association, Jan.

SNYDER, T. E. & J. ZETEK — 1943. Effectiveness of wood preservatives in preventing attack by termites. U.S.D.A. Circular No. 683.

ST. GEORGE, R. A. — 1939. Termite control in buildings. Pests 7(2): 13.

VERRALL, A. F. — 1953. Factors leading to possible decay in wood siding in the South, USDA. Forest Pathology Special Release No. 39, 1953. Decay prevention in wooden steps and porches through proper design and protective treatments. J. Forest Prods. Research Soc. Nov. pp. 54-60. 1954. Preventing and controlling water-conducting rot in buildings. USDA Forest Service Occasional Paper 1933. 1965. Preserving wood by brush, dip and short-soak methods. USDA Tech. Bull. No. 1334.

C. DOUGLASS MAMPE

Dr. C. Douglass Mampe received degrees in entomology from Iowa State University, North Dakota State University, and a Ph.D. from North Carolina State University. Upon completing his university education, he joined the staff of the National Pest Control Association (NPCA). During his 10 years there he was responsible for keeping abreast of all technical and regulatory actions related to the structural pest control industry. This included reviewing research proposals and overseeing research projects related to the biology and control of termites, cockroaches, commensal rodents, and other related pests.

Mampe left NPCA as its technical director and joined Western Termite and Pest Control, Inc., and its sister company Residex, a pest control industry supply house. He served as Western's technical director for six years and developed practical procedures for field operations. He eventually became general manager of Residex, which included developing new pesticide registrations and industry training programs, including certification training programs for a number of the northeastern and middle Atlantic states.

In 1980, Mampe started his own consulting firm for the urban and structural pest control industry. His firm provides technical consultation, training and training programs, evaluates new pesticides and equipment, custom develops personnel management programs and provides expert testimony for legal cases.

Mampe served as editor of NPCA's _Approved Reference Procedures for Control of Subterranean Termites_, and editor and co-author of the _Manual for Structural Wood Decay_. He writes a monthly column called "Answers" for _Pest Control_ magazine. This column, which deals with a variety of questions, has consistently been one of the best read sections of the magazine since the column began at the beginning of 1975.

Mampe is a member of the NPCA, Pi Chi Omega (a fraternity of industry professionals involved in training), the Entomological Society of America, and a number of state pest control associations. He is currently licensed as a certified pesticide applicator in a number of northeastern states. In 1979, he was chosen "Pest Control Operator of the Year" by the New Jersey Pest Control Association for his contributions to the industry in that state. He is currently serving as a resource person for the United States House of Representatives Committee on Agriculture.

CHAPTER NINE

Wood-Boring, Book-Boring, and Related Beetles

Revised by C. Douglass Mampe[1]

Through and through the inspired leaves
  Ye maggots, make your windings;
But, oh! respect his lordship's taste,
  And spare his golden bindings.
            — Robert Burns

POWDER POST BEETLES and other wood-boring beetles are second only to termites in destruction of wood articles and structural timbers. Doane et al. (1936), Chamberlin (1948), Craighead (1950), and Hickin (1963) have prepared comprehensive works concerned with insects injurious to wood. Osmun (1955) and Becker (1956) illustrate wood damaged by various species of insects.

POWDER POST BEETLES

Family Lyctidae

For the most part, the name "powder post beetle" is restricted to the members of the family Lyctidae. These insects are so named because of the powder-like dust produced as a result of their working in wood. This fine dust is often mistaken by women for face powder that has been inadvertently dropped. Gerberg (1957) completed a well-illustrated taxonomic study of powder post beetles in the Family Lyctidae. Hickin (1960) has prepared a similar work on British Lyctidae.

[1]*President, DM Associates, Westfield, NJ*

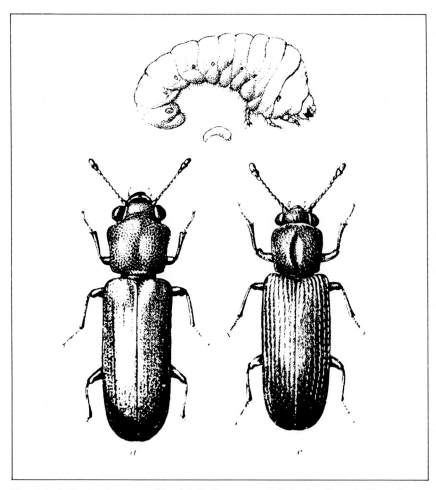

Fig. 9-1. A. *Lyctus brunneus* (Steph.); its larvae; and **C.** *Lyctus linearis* (Goeze).

Powder post beetles are from ¹/₁₂ to ¹/₃ inch/two to 7.5 mm in length and brownish to reddish in color. Their short, 11-segmented antennae, with two-segmented terminal clubs, are inserted before the eyes. The tibiae have distinct spurs.

The work of these insects may be recognized by the small, round holes which appear in hardwood floors, furniture, etc., as well as by the presence of fine dust. This dust results from the boring of the larvae and the holes are the openings through which the adult beetles emerge. Since generation after generation may work in the same piece of wood, in time, the interior may be reduced to a fine powder.

According to Christian (1940), the *Lyctus* larvae attack only the seasoned or partly seasoned sapwood of hardwoods, since the sapwood is richest in starch, and starch is believed to be the principal food of the *Lyctus* larvae. Oak, ash, hickory, maple, and walnut are preferred woods, and bamboo is a favorite in the Orient. Other hardwoods such as magnolia, sweet gum, black gum, birch, persimmon, locust, elm, poplar, sycamore, and cherry at times are damaged. It is believed the softwoods or the conifers are not attacked by the powder post beetles.

Under natural conditions, these beetles breed in old and dried wood (e.g., the dead branches and limbs of trees) which they aid in reducing to fine particles. It is only when they are present in stored lumber, finished wood, and furniture products that we become cognizant of their importance as pests. As a rule, they enter lumber while it is being stored and cured and later emerge from the finished product. Very often, old pieces of furniture, such as fine antiques, are reduced to worthless hulks because of the activities of powder post beetles.

One should note when these insects are carried into the home in a piece of infested furniture or timber, the beetles may multiply and then attack other hardwoods in the vicinity. Since these beetles breed in such common commercial commodities as tool handles, furniture, etc., they are subject to world-wide trade, and in this manner have become cosmopolitan in distribution.

A TABULATION OF SOME OF THE COMMON *LYCTUS* BEETLES

(In part after Gahan and Laing (1932))

- *Lyctus brunneus* (Steph.)
  The color is reddish brown and the prothorax is wider in front than behind and has a shallow depression along the middle. The hairs on the wing cases have no definite arrangement.

- *Lyctus linearis* (Goeze)
  This insect is brownish in color. It is more parallel-sided with a deep elongate, elliptical pit in the middle of the elytra. The elytra are punctured with a single row of large, shallow, circular punctures, and the hairs on the wing cases are in longitudinal rows.

- *Lyctus parallelopipedus* (Melsh.)
  The color of this species varies from reddish brown to very dark brown and the lateral margins of the prothorax converge behind. The body is covered with yellow pubescence not arranged in longitudinal rows. According to Gerberg (1957), the correct generic name of this species is *Trogoxylon*.

- *Lyctus planicollis* (LeC.)
  This powder post beetle is black when mature and the prothorax usually has a broad shallow median depression. The front part of the thorax is as wide as the elytra at the base.

- *Lyctus cavicollis* (LeC.)
  This insect is reddish brown in coloration and somewhat similar to *Lyctus linearis* (Goeze), but the punctures on the striae are in double rows, and the anterior portion of the prothorax is narrower than the elytra at the base.

Life histories, The life histories of the various species are somewhat similar. When breeding outdoors, the winter is passed in the larval stage. With the coming of spring, the powder post beetle bores closer to the surface and pupates. Then, it emerges from the wood in two or three weeks and either flies or crawls

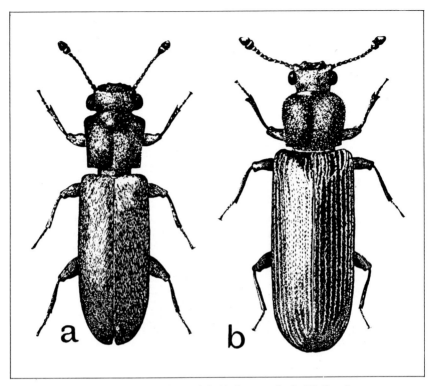

Fig. 9-2. A. *Lyctus parallelopipedus* **Melsh. B.** *Lyctus planicollis* **(Lec.)**

away. The adult lays her eggs one or two days after mating. She may oviposit one or several eggs deep into the pores of the wood, as well as in old emergence holes. She oviposits in the spring when outdoors, but in a heated building she may lay her eggs at an earlier date.

Gahan and Laing (1932) state the larvae have two short, four-jointed antennae on the head, and this readily distinguishes them from the larvae of *Anobium*, whose antennae are only two-jointed and barely visible under the microscope. The lyctid larvae are also without spinules.

The burrows follow the grain of the wood and are usually somewhat parallel and do not branch. Outdoors the larvae are probably torpid in cold weather, but indoors in heated buildings they may be active all through the winter. Christian (1940 and 1941) made a detailed study in Louisiana of the biology of two species of powder post beetles, *Lyctus parallelopipedus* (Melsh) and *Lyctus planicollis* (LeC.) He notes these powder post beetles are the most common *Lyctus* beetles in the United States, and the following is largely from his studies: "The life cycle of *Lyctus* beetles is completed in four distinct stages, namely, the egg, the larva, the pupa, and the adult. The beetles lay their eggs within the surface wood pores to a depth of several millimeters, and the larvae which hatch from them burrow inside the wood and reduce it to fine powder-like dust. Pupation occurs in the larval tunnels, usually near the surface. A few days after trans-forming to the adult stage, the beetles cut small, round holes directly to the surface through which they emerge. The adults feed very scantily on the surface of the wood, mating occurs, and the cycle is repeated. The complete cycle ranges from three months to a year, or longer, depending on the species of *Lyctus* and the condition of the host species of wood, such as starch content, moisture, and prevailing temperatures.

"Females of *L. parallelopidedus* insert their eggs into the wood pores to a depth of one to three millimeters, while those of *L. planicollis* lay their eggs somewhat deeper, ranging from four to 7.5 millimeters. Because of this fact, certain larval stomach poisons, such as borax, when applied as 10-second hot dips to lumber, give effective protection against *L. parallelopipedus,* but such treated wood may become slightly infested by *L. planicollis.* The latter beetles may lay some of their eggs in the wood below the point penetrated by the toxic chemical."

The egg laying is accomplished "by the female inserting her long ovipositor within the open surface timber pores. When fully extended, the ovipositor some-times approaches in length that of the entire beetle's body, but normally it is almost completely concealed within the abdomen." The white cylindrical-shaped eggs hatch in six to 15 days.

Smith (1956b) studied the oviposition habits of *L. planicollis* in some detail. He notes approximately 45 seconds are required for the oviposition of a single egg. After 10 to 12 days the larva emerges from the egg and begins to bore into solid wood.

The whitish larvae of *Lyctus* beetles are characterized by the presence of an enlarged spiracle on either side of the abdomen near the end of the body. The dust produced by the boring larvae is like a fine powder.

"With the exception of the exit hole made by the adult, all the boring is caused by the larvae, and in most cases, infested wood cannot be detected by external examination until some of the beetles reach maturity and bore their way out.

"The grown larvae usually bore near the wood surface (approximately ⅛ inch) where they prepare small cylindrical pupal chambers, in which they change to the pupal stage.

"The larval stage covers a period varying from two to nine months or longer, depending on the species of beetles and the condition of the wood host. Under ideal conditions, *L. parallelopipedus* larvae may mature in 60 days; under sim-ilar conditions, *L. planicollis* larvae require a much longer period, usually four to nine months.

"When the larvae complete the pupal chambers, they remain quiet and undergo a gradual change into the pupal stage. The pupae remain inactive in their cells for a period of 12 days to three weeks, when they transform to the adult beetles.

"The beetles remain in the pupal cavities until their bodies become suffi-ciently hardened, then they commence to bore their way out to the surface." In the process of emerging, dust is usually pushed in front of the body, which is often the first external sign of infestation.

"The exit holes are circular in shape, and their average diameters vary some-what with the two species. Those by *L. parallelopipedus* average about one millimeter in diameter, while those of the larger species, *L. planicollis,* average slightly less than two millimeters.

"The adult beetles are small (two to six millimeters in length) and somewhat flattened in shape, the latter feature being more pronounced in the *L. planicollis* adult. The average size of *L. parallelopipedus* is less than that of *L. planicollis,* and specimens of the smaller species are usually reddish brown in color, while those of *L. planicollis* are usually black or approaching black."

The beetles mate and begin to lay eggs shortly after emergence. Wright (1960) showed they laid an average of 51 eggs which incubated in about eight days. The beetles are less active during the day and most of their activities take place beginning at dusk and continue during the night. Adult beetles consume very little wood. Thus, their feeding signs after emergence are negligible.

"Adults of *L. parallelopipedus* are much more rapid crawlers than members of the larger species, which are almost sluggish in their movements. Both are strong fliers and many travel some distance from their original host. The beetles live about three or four weeks. The longest record of adult life was that of *L. parallelopipedus,* which lived for 50 days. Generally, there appeared to be little difference in the length of adult life of the two species."

Lyctus beetles may be separated from Anobiid beetles by the nature of their work. Thus, *Lyctus* beetles make a flour-like dust, whereas Anobiids make small pellets which are granular in appearance. *Lyctus* beetles confine their work to sapwood, but the adult may penetrate the heartwood when emerging.

The brown powder post beetles, *Lyctus brunneus* (Steph.). This species, which is believed originally to have come from Europe, is now cosmopolitan in distribution. It is $1/6$ to $1/5$ inches/four to five mm long, dark brown, with glossy, reddish brown wing cases. Essig (1926) notes this cosmopolitan species is found frequently in bamboo art goods and Mah Jongg sets from China and Japan which the beetles may reduce to powder-filled shells. Control is accomplished by dipping the articles into a bath of kerosene.

The European Lyctus, *Lyctus linearis* (Goeze). This is the common European species that is now cosmopolitan in distribution. It is also dark brown with glossy, reddish brown wing cases, and measures $1/8$ to $1/4$ inch/three to seven mm long. Essig (1926) states it is a common pest of commercial products of ash, wild cherry, hickory, locust, oak, poplar, and walnut. Essig also states it breeds in dead limbs of oaks, orange, and sycamore. It is very similar to *L. cavicollis* (LeC.), but can be distingiushed by the single row of large round shallow punctures on the wing covers.

The southern lyctus, *Lyctus planicollis* (LeC.) This *Lyctus* is black and approximately $1/5$ inch/five mm in length. The fine punctures on the wing covers are to be seen in double rows. It is a native species which is most commonly encountered in the South. It is also found in the West as well as in England, where it was imported from the United States in oak, ash, and hickory. In the South Atlantic and Gulf States, the adults are active from the middle of February until late September, according to Hopkins and Snyder (1917). They make the interesting note that as a result "of continued breeding in the same wood for several generations in confinement, the beetles were found to decrease in size." Christian's studies (1940 and 1941) on the life history of this beetle were noted previously. Smith (1956a) describes a technique for rearing the Southern lyctus. The same author (1955) notes production of adults was increased when using wood with a moisture content of 15 to 18 percent as compared with an eight percent moisture content.

The western lyctus, *Lyctus cavicollis* (LeC.). This rusty, red-brown species is $1/8$ inch/2.5 to three mm long with fine punctures in two double rows on the dorsum. It is a native California insect commonly encountered in oak firewood.

It is possible it spreads from the firewood to the hardwood furniture in the home. Essig (1926) notes this insect attacks seasoned eucalyptus, hickory, oak, orange, as well as furniture and building materials in California and Oregon.

The velvety lyctus, *Lyctus parallelopipedus* (Melsh.) The velvety *Lyctus* is rusty, red-brown to black in color and from two to four millimeters long. The punctures on the wing covers are not in rows and the fine yellowish hairs give it a velvety appearance. Originally a pest in the United States, it is now present in England. It occurs in ash, hickory, and oak timbers. Christian (1940 and 1941), as previously noted, studied the life history of this insect. Gerberg (1957) notes the correct genus for this species is *Trogoxylon. Trogoxylon prostomoides* is recorded in oak flooring in Illinois.

THE ANOBIIDS

This family includes some of the most important furniture and timber beetles. Important members of this family (e.g., *Lasioderma serricorne* (F.) and *Stegobium paniceum* (L.)) are treated in the chapter on stored product insects. White (1962) has written a well-illustrated work on the Anobiidae of Ohio and a key to the North American genera (1971).

The wood infesting anobiids breed in old, dry limbs, but also are found in girders, beams, foundation timbers, as well as in furniture. They have become cosmopolitan in distribution, since they are readily introduced in antique furniture and other old woods. Smith and Forbes (1944) note *Anobium* chiefly attacks sapwood in softwoods, while *Lyctus* confines its attacks to the sapwood of hardwoods.

Christian (1945) states that unlike *Lyctus,* anobiids attack coniferous, as well as hardwood timbers, and the damage may extend into the heartwood. Kelsey et al. (1945) found the larvae infesting woodfiber wallboard and the cardboard filler of a leather suitcase. Hatfield (1949) observes anobiids usually are found in old wood rather than new lumber and thus may be referred to as "old furniture beetles." According to Bletchly (1951), more larvae become established in partly decayed wood than in sound wood.

The furniture beetle, *Anobium punctatum* (DeG.) (*A striatum* Oliv.). Gahan and Laing (1932) and Kelsey et al. (1945) studied this insect and much of the information presented here is from their works. Hickin (1961 and 1963) wrote very useful reviews of this insect.

This cylindrical beetle is 1/6 to 1/4 inch/four to six mm long, reddish brown to dark brown in color with punctures on the dorsum in longitudinal rows. The last three joints of the 11-segmented antennae are distinctive in length and shape. The furniture beetle is a common pest in Europe and is present in the United States.

In England, these beetles emerge in June from cells immediately below the surface of the wood and are found occasionally crawling or flying inside the house. They occur in floors and rafters and furniture of pine, and also have been reported from oak, beech, alder, willow, etc. Hicken (1953) states the common furniture beetle usually infests sapwood indoors and heartwood outdoors.

According to Gahan and Laing, they mate shortly after emerging from the wood, and one day later the female lays her eggs in a crevice in the wood or in the mouth of an old exit hole. She may oviposit one or more eggs in the crevice if it is sufficiently large. Eggs are usually not laid on a smooth surface and wood with pores filled with paint, varnish, and other materials is less likely to be

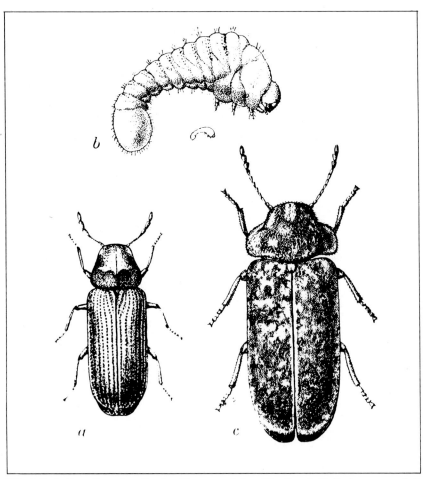

Fig. 9-3. A. Furniture beetle, *Anobium punctatum* (De Geer); B. its larvae; and C.
death watch beetle, *Xestobium rufovillosum* (De Geer).

infested than unfinished rough wood. Kelsey et al. (1945), at a temperature of
71° F/22.5° C and 75 percent relative humidity observed the female to deposit
an average of 18 eggs, but the number is quite variable. Spiller (1964) showed
the average number of eggs in the laboratory was 54.8. The white oval, lemon-
shaped egg is less than one mm long and shows fine honeycombed sculpturing
at one end. The remainder of the egg is smooth. Spiller (1949) showed that no
eggs hatched, or hatching was impaired at humidities below 60 percent.

The eggs hatch in six to 10 days. The full grown larva is about ¼ inch/six
mm long, white with nearly black jaws, and characterized by three pairs of five-
jointed legs. On the dorsum are small brown spinules which are set in double
rows. The presence of these spinules readily separates the anobiid larva from

the *Lyctus* larva. The frass, produced while boring, consists for the most part of small oval pellets. The larva pupates for two or three weeks in a cell immediately below the surface of the wood.

Kelsey et al. (1945) found the ratio of males to females was approximately equal, although at times the percentage of females captured was low. According to their observations, the explanation "lies in the behavior of the beetles, which mate normally soon after their emergence from wood, and while still *in copula,* retreat into exit holes and crevices in the timber. After an hour or so, males disengage from females and return to the wood surface, while the latter remain in the tunnels, etc., to lay some of their eggs. The females later reappear to go through the same procedure, and the time occupied in ovipositing is greater than that spent on the surface. As males spend most of their time on the surface of timber, it is apparent that at any one time the percentage of males will be high."

Outdoors, in dead branches, the life cycle is completed in ony year. However, in dry antique furniture two or more years may be necessary before the beetle emerges. Thus, it is possible for several generations to be inside the same piece of wood at one time.

The emergence holes of the beetle are approximately two millimeters in diameter, varying with the size of the beetle. The adults, when collected, will play "possum." Gahan and Laing have never heard them give forth a tapping sound and so question their right to be called a "death watch" beetle.

Desch (1950) discusses the furniture beetle and the conditions that favor it in England as follows: "Abundance of food supplies, which are more likely to occur in the sapwood than in the heartwood, are obviously a factor favoring rapid development of the pest: there appears to be considerable variation in these supplies, not only in different timbers, but in different parts of the same piece. At the same time, the more plentiful the sapwood in the vicinity of detected infestation, the more likely it is that the attack will be widespread, calling for really thorough curative steps. Rising moisture contents of wood in service, and high prevailing atmospheric humidities and mild temperatures, are also conditions favorable to the pest, and consequently to a short life cycle. It follows that timber in cellars, cupboards under stairs, pantries, and unused attics are places where conditions most favorable to the common furniture beetle are likely to develop, if they do not already exist. Unfortunately, except for pantries, it is in just such places that infected timber is most likely to be introduced: logs, old packing cases and timber from garden sheds are very liable to be infected, and they are very commonly dumped in cellars with a view to their conversion to firewood at a later date — there is little doubt that many cases of infestation can be traced to this decidedly risky practice. Again, outdoor games apparatus has a way of being brought in at the end of summer, and stored in cupboards under stairs and other equally inaccessible storage places, and is unlikely to be taken out again until well in May. If infected when put away, the adults will emerge from these wooden articles inside such cupboards, since they appear from the middle of April onwards, and the unpainted and rough-finished timber of which the cupboards are constructed make ideal breeding grounds for launching a new generation of furniture beetles. Finally, the odds are further weighted in favor of the furniture beetle, in that attics and other little used rooms are just those we tend to furnish from 'bargains' purchased from junk shops and sales, and such pieces may well be infested when acquired. Since the rooms are little used, the first two or three generations of furniture beetles may be produced before the presence of any infestation is detected, when it is probable that attack has spread over the whole house.

"By comparison with fortuitous infestation arising in any of the ways suggested above, the genuine piece of antique furniture, purchased through a reputable dealer or firm, is a much less likely source of infection. It is almost certain to have been attacked by the furniture beetle at some time or another, but if the piece is really old, and the attack is still active and of long standing, its presence can hardly fail to escape notice. Moreover, if the piece has passed through the hands of a specialist in the trade, he is likely to have taken all reasonable precaution to minimize the risk of infestation of his premises, and deterioration, while in his possession, of his assets. The gift of antiques, by way of legacies and the like, from other private homes is a very different matter. Such pieces are very likely to be attacked, and if the infestation is still active, it may well give rise to widespread attack in the new home.

"Finally, 'bargains' however cheaply acquired, and gifts from friends and relations, should never be given the benefit of the doubt: it is always worth while to give them a very thorough dressing with one of the solvent wood preservatives, paying particular attention to all hidden surfaces, joints, crevices and the like, before admitting the new acquisitions among our existing possessions."

Spiller (1949a) notes a five percent pentachlorophenol treatment will control the beetles emerging from the wood for three to four years. Pentachlorophenol treatment is discussed in the section on control.

The deathwatch beetle, *Xestobium rufovillosum* (DeG.). Imagine the hush and stillness of the home suddenly disturbed by a series of eerie ticks, and then a period of ominous quiet, one alternating with the other. Is it any wonder superstitious individuals attributed the sound to the shrouded figure — a foreboding of dire things to come!

Science has found a less exciting reason for this sound, which resembles the ticking of an old-fashioned clock. This tapping is merely a love-call between adult deathwatch beetles during the mating season. Blake (1926) states: "The noise is produced by the beetle when calling to its prospective mate, by erecting itself firmly on its legs and jerking its body forward rapidly from eight to 10 times, striking its head sharply against the surface on which it stands at the same time, each succession of taps being given out in the space of about one second.

"If another beetle is within hearing, and the call is accepted, it is immediately answered in a similar manner, and thus a convenient means of communication is established between the two. The tapping or ticking is generally heard in the late spring when the beetles emerge from wood and their instincts warn them that pairing time is at hand."

Gahan and Laing (1932) note: "A female of this species, captured when it had just come out of the wood at the end of March, 1917, was placed in a small box where it continued to live for 10 weeks; and at almost any moment throughout the whole of that time was ready to respond by tapping its head against the bottom or side of the box, to a sound made by tapping at the same rate with a pencil on anything within a few yards of its prison."

The deathwatch beetle is approximately ¼ inch/seven mm long, reddish to dark brown, and spotted with patches of short, yellow-gray hair. The holes made by this beetle are approximately twice the diameter made by the furniture beetle, or ⅙ inch/four mm in diameter. The pellets or grains of wood are "bun-shaped." According to Gahan and Laing, in England they begin to appear in

April and May, when their tapping most commonly is heard. The adults, unlike the furniture beetle, may not bore out immediately, but may remain in their pupal cells until the following spring, thus having a two-year life cycle. They mate shortly after emergence from the pupal cells and then die in a few weeks. The female may lay up to 70 eggs. The white oval eggs are smooth throughout and twice as large as those of the furniture beetle.

The larvae measure up to $7/16$ inch/11 mm. They have a yellowish head with mandibles that are almost black. Gahan and Laing state the larva differs from the larva of the furniture beetle by having two small black spots on each side of the head, whereas the furniture beetle has but one. The larva has been known to live three years. The pupal stage may last two to three weeks, and as stated previously, pupation takes place in late summer or autumn. The beetle may remain in the pupal cell until the following spring. In old buildings, the life cycle may extend for at least three years.

This beetle ranges throughout Europe, as well as in Corsica, Algeria, New Caledonia, and the United States. Craighead (1950) states it is "occasionally found in the woodwork of moist cellars in the New England states." According to Fisher (1937), the deathwatch beetle is an important pest in England, since so many of the buildings are constructed of oak, the principal timber infested by it. The building timbers most likely to be attacked are those constantly soaked by rain or those situated in poorly ventilated localities. Heavy furniture and fixtures that are more or less permanently situated, particularly if made of oak or chestnut, very often are infested. The deathwatch beetle also is reported as a pest in books.

Normally, the deathwatch beetle dwells outdoors in dead and dying trunks and large branches of such hardwoods as oak, chestnut, and willow, as well as in beech, hawthorn, etc. Softwoods of conifers are attacked when they are in the immediate vicinity of hardwoods.

▪ *Ptilinus pectinicornis* (L.). As a whole, this beetle is not of very great economic importance, since it occurs most commonly in dead trees, gates, posts, etc. It has been reported damaging furniture in England and the woodwork of a house in Seattle.

The beetle is dark brown, ¼ inch/seven mm long, with very typical antennae. Infested furniture is usually made of beech or maple. However, willow and sycamore may also be subject to infestation. The holes made in wood are slightly larger than those made by *Anobium*. In one instance the larvae have been known to reduce a new bedstead to powder in three years. The result of their borings consists of finely divided particles of wood, which are not aggregated into fecal pellets.

▪ *Trypopitys punctatus* (LeC.). This anobiid has a superficial resemblance to *Anobium punctatum* (DeG.). In southern California, it has damaged floors and doorcasings of oak, as well as maple wainscottings. It breeds outdoors in Monterey cypress and in pine.

▪ *Hadrobregmus spp.* Several species of this anobiid have reportedly damaged wood products and timbers. Hatch (1946) notes its presence in the "weathered spruce boards of the flooring of a porch" which was tunneled by the larvae and beetles. Linsley (1943a) makes the following comments on this species: "The California deathwatch beetle, *Hadrobregmus gibbicollis* (LeConte), is an elongate, subparallel brown to dark brown beetle about four to six millimeters long. The antennae are 11-segmented, with the last three segments distinctly longer than the preceding three segments. The pronotum is convex and distinctly narrower than the base of the wing covers. The striae of the wing covers are not deep, and the punctures are regular. It breeds in old, well-seasoned Douglas fir ("Oregon Pine") studs, joists, and supporting timbers along the Pacific Coast. It is especially common in basement timbers of buildings 20 or more years old, and is the most serious pest among the native powder post beetles in the area which it occupies. It breeds year after year in supporting timbers, reducing the contents to powder while leaving the surface intact, thus providing a dangerous hazard. The adults emerge through round holes nearly ⅛-inch in diameter and frequently fly to windows congregating on the sills. However, because the beetles work commonly in basements in unfinished wood, emergence holes and signs of infestation are not conspicuous, often obscure and inaccessible. Much serious damage attributed to termites should be credited to this species.

Linsley discusses the eastern deathwatch beetle as follows: "The eastern death watch beetle, *Hadrobregmus carinatus* (Say), is an elongate, subparallel, reddish brown to dark brown beetle 3.5 to 6.5 millimeters long. It differs from the western deathwatch beetle by having the antennae 10-segmented and the pronotum about as broad as the base of the wing covers. It is widely distributed in eastern North America and breeds in both new and old wood. It has been recorded from ash and basswood flooring in Ohio and Michigan, maple and beech in Indiana, sills and floors in Connecticut, elm floor joists in Minnesota, etc." Payne (1936), in discussing the infested elm joints in a house in Minnesota, notes that the joists, which supported the basement floor were completely eaten out except for a thin outer shell. The injury was similar to that of *Lyctus sp.*, with larger emergence holes and coarser frass.

Fisher (1938) reports *Hadrobregmus destructor* Fisher as damaging wooden articles in a museum in Sitka, Alaska, as well as the unpainted surface of the supporting columns of a building.

- *Ernobius mollis* (L.). According to Kelsey (1946) and Craighead (1950), this small brown beetle (⅙ to ⅕ inch/four to five mm) is a native of Europe, but has become established in the United States. It often damages coniferous flooring, as well as lumber covered with bark. The life cycle is usually one year, but may be longer under adverse conditions.

Craighead (1950) notes the following anobiids are injurious to structural timbers of furniture: *Xyletinus peltatus* (Harr.), *Trypopitys sericeus* (Say), *Trichodesma gibbosa* (Say), *Ptilinus ruficornis* (Say), *Ptilinus pruinosus* (Casey), and *Nicobium hirtum* (Ill.). Hetrick (1955) and Wright (1959, 1960a) note *Xyletinus peltatus* (Harris) is one of the most important anobiid beetles in the Southeast. "It infests sapwood and heartwood of coniferous and hardwood species and is most commonly associated with sills and floor joists." Spink et al. (1966) obtained control of this insect with treatments of 0.5 percent and one percent dieldrin or one percent or two percent chlordane in either water or kerosene.

Anobids injurious to books. Back (1939a) reports the anobiid, *Neogastrallus librinocens* Fisher as injuring books in Florida that originally came with the beetle from Havana. Watson (1943) records it from Florida and Louisiana. This dark reddish-brown insect with short recumbent grayish pubescence averages ¹⁄₁₀ inch/2.4 mm in length and ¹⁄₂₀ inch/1.2 mm in width. The whitish larva, which is ¹⁄₁₂ to ⅛ inch/two to three mm long when fully grown, makes burrows a little more than ¹⁄₂₅ inch/one mm in diameter.

Both old and new books are subject to injury, and Back states, "The feeding burrows of the larvae cut the binding materials, causing the pages to fall out. The pages themselves may be riddled and cut by burrowing larvae in constructing their pupal chambers." This pest was controlled by fumigating the library with HCN at the rate of one pound/10.45 kg. sodium cyanide to each 100 cubic feet/2.8 cubic meters.

Herrick (1936) notes *Nicobium hirtum* (Ill.), a native of southern Europe, and known to be established in the southern United States, has injured books in the State Library in Baton Rouge, Louisiana. This insect is also injurious to timber.

According to Taylor (1928), a tropical anobiid of the genus, *Catorama,* was found infesting a stack of books that came to Boston from Honolulu. The infested books were fumigated with carbon bisulfide. It is possible that was the *Catorama bibliothecarum* (Poey), mentioned by Back (1939) in his paper on bookworms. White (1963) notes the Mexican book beetle, *Catorama herbarium,* which is established in the United States, is also a pest of seeds, furniture, leather, chocolate, and Cayenne pepper. In this same paper, Back lists the following insects, excluding those already mentioned, as pests of books: The cigarette beetle, *Lasioderma serricorne* (F.), the drugstore beetle, *Stegobium paniceum* (L.), the Dermestidae, *Trogoderma* sp., *Attagenus* sp., and *Anthrenus* sp., as well as the American cockroach, silverfish, and termites.

Back (1939) notes the National Archives in Washington, D.C., has fumigation vaults which are used to fumigate every lot of books received prior to their being unpacked. The vaults are evacuated until a vacuum of 29.9 inches of mercury is reached. A mixture of ethylene oxide and carbon dioxide is released into the vault until the vacuum falls to 21 inches of mercury. "The gas is then agitated for 15 minutes by pumping it out at the top and in the sides of the chamber. After the records have been exposed for a total of three hours, the chamber is re-evacuated to 29.8 inches of mercury, the vacuum is broken with air, and the fumigated materials are removed."

Weiss and Carruthers (1945) and Back (1939) discuss the insect pests of books in some detail.

FAMILY BOSTRICHIDAE

Several members of this family may at times be pests in furniture and other wooden commodities in the household. For the most part, the bostrichids occur in the warmer areas of the world. The adults are from ⅛- to one inch/three to 25 mm in length with an enlarged thorax which gives the beetle a decidedly hump-backed appearance. Hatfield (1949) discusses this group of wood-boring insects as follows:

"The eggs of Bostrichid beetles are not laid in the surface pores of wood as in the case of the Lyctids, nor are the eggs laid in surface cracks or crevices as in the case of Anobiid beetles. In the case of Bostrichid beetles, the adults bore into the wood, and the female lays in egg tunnels prepared by adult beetles or in pores leading from the egg tunnels. The larvae, which hatch from the eggs, are also capable of producing tunnels in the wood. Since both the adult beetles and the larvae bore into the wood, the galleries formed vary in size and shape."

■ *Polycaon stouti* (LeC.). This completely black insect is ⅝ to ¾ inch/15 to 20 mm long with numerous fine punctures on the wing covers and coarse punctures on the head and prothorax. It is distributed in the western United States and

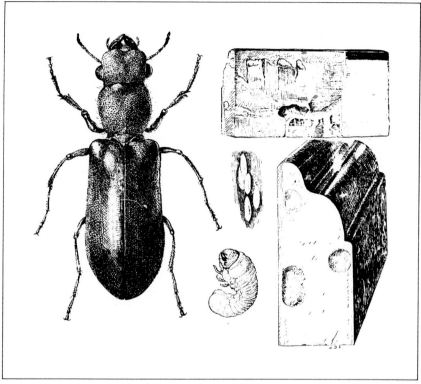

Fig. 9-4. *Polycaon stouti* (Lec.), a bostrichid. Adult, larva, eggs, along with damage in molding.

is of some economic importance, since it attacks cured hardwoods in lumber yards, buildings in mountainous areas, as well as furniture and other wood products.

Essig (1926) notes under natural conditions it breeds in such trees as dead eucalyptus, maple, oak, fruit trees, California laurel, madrone, manzanita, sycamore, etc. Doane et al. (1936) state this insect infests three-ply panel used in making desks, such as basswood with an outer layer of either oak or mahogany. The adults attack these panels at night in the warehouses, establishing numerous colonies. The larvae work in the inner panel and may require some time to reach maturity. They then bore out of the outer veneer.

The bamboo borer, *Dinoderus minutus* (F.). This bostrichid is a cylindrical brown beetle, ¹/₈ to ¹/₆ inch/three to four mm long and a native of Asia. It breeds in bamboo, reducing the inner area into a fine powder. Thus, much of the bamboo furniture, ornaments, poles, etc., is rendered worthless. It occurs also in drugs, spices, cacao, corn, rice, stored grain, dried bananas, and flour.

Plank (1948) made a detailed study of the biology of this beetle in Puerto Rico. Eggs hatch in from three to seven days. The larvae feed for an average period of 41 days and molt four times. The pupal period extends for four days. The emerged adult waits about three days before gnawing its way out of the pupal cell and the bamboo. The entire life cycle from egg to emerged adult averages 51 days. Plank and Hageman (1951) showed the severity of attack was correlated with the amount of starch present. Duval obtained control with 0.75 to three percent oil solutions of gamma BHC.

The leadcable borer or shortcircuit beetle, *Scobicia declivis* (LeC.). This is a cylindrical, reddish brown beetle, ¹/₅ to ¹/₄ inch/five to six mm in length, whose major claim to fame is its ability to bore through the lead sheathing of aerial telephone cables. The holes it bores are ¹/₁₀ inch/2.5 mm in diameter and may extend into the paper insulation. The attack is usually made where the

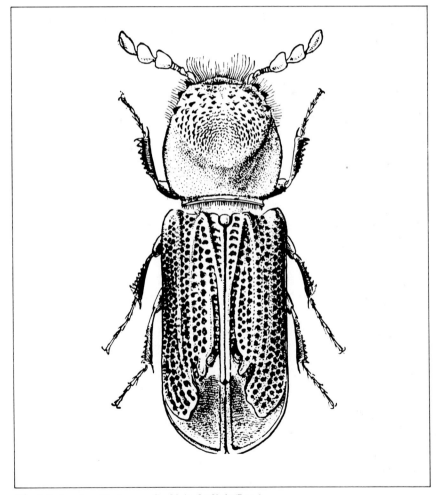

Fig. 9-5. Lead-cable borer, *Scobicia declivis* (Lec.)

suspending ring supports the cable. The beetle attacks at this point, since here it can obtain a foothold to bore into the lead.

Where the sheathing is penetrated, moisture enters and a short circuit results. According to Burke, et. al. (1922), "The number of holes may vary from one to 125 a span (section of cable between two poles, the distance averaging 100 feet). As one hole may put from 50 to 600 or more telephones out of use for from one to 10 days, the damage is rather extensive." Most of the damage occurs from June to August, when the beetles are emerging in the greatest numbers. However, the injury does not become apparent until the first rain, when all the short circuits occur at one time.

The leadcable borer may swarm about dwellings that have been newly painted or where a fire has started. R. M. Bohart states after a fire where oak trees were abundant, many leadcable borers departed from the burned-over area and attacked lead cables, causing great damage. Rivers (1886) reports the beetles severely damaged wine casks in California. Ebeling and Reierson (1973) in California and Anon. (1966) note that in Oregon this insect damaged plasterboard wall, plaster casts, hardwood paneling, and floors. They are known also to be injurious to living trees and plastic conduits.

Life cycle. According to Burke et al. (1922), the female bores into solid wood and lays her eggs. The eggs hatch in 21 days. The larvae mine in the wood for about nine months and molt about six times. The pre-pupal stage lasts for about six days and the pupal stage for 14 days. The beetle does not emerge for 30 days, hardening and maturing in the meantime. The females, upon emerging, often attack the same piece of wood. When they mate they are ready to start their galleries. There is one generation a year.

Experiments in the control of this insect conducted by Burke et al. show the beetle can penetrate any lead alloy or cable sheathing used, as well as any poison placed on these materials. It probably can penetrate the poisoned surface since it does not feed while boring through. These same authors note: "Theoretically if any grease or grease compound, which will soften in the sun when the beetle is most active, is placed on the rings, it will stick to the beetle and suffocate it when it tries to bore into the cable. Beef tallow appears to be the best for this purpose."

Other insects injurious to lead. Insects of several Orders are known to attack metals. Burke et al. (1922) state beetles are the most numerous pests of lead, and include the families Anobiidae, Anthribidae, Bostrichidae, Buprestidae, Bruchidae, Cerambycidae, Curculionidae, Dermestidae, Lyctidae, Ptinidae and Tenebrionidae. Insects of other Orders, which are known to injure metals, are the Cossidae in the Lepidoptera, the termites or Isoptera, and the horntail *Sirex*, as well as a wasp in the Hymenoptera.

Other metals attacked by insects include tin, zinc, silverplate service, and the quicksilver linings of mirrors. In most instances, the metal is attacked since it is in the way of the emerging adult or burrowing larva. However, some insects make direct attacks on the metals, e.g., *Scobicia declivis* (LeC.) and the apple twig borer. *Amphicerus hamatus* (F.) (Craighead, 1950). Snyder (1956b) also discusses some of the insects that bore into metals.

Back (1945) reports the bostrichid, *Prostephanus punctatus* (Say) as a pest of barrel staves in Georgia.

G. E. Wallace and the author have several records of the bostrichid, *Stephanopachys rugosus* (Oliv.), occuring in building timbers in homes in Pittsburgh, Pennsylvania. One infestation occurred in bark-covered joists in the basement of a two-year-old house. *S. substriatus* is a pest of wood in California and is known to have emerged from a kitchen cabinet. The oriental wood borer, *Heterobostrychus aequalis* (Waterhouse), is now established in Florida (Anon., 1967) and found in paneling.

CONTROL OF POWDER POST AND ANOBIID BEETLES

The control of powder post, anobiid beetles, and other wood-boring insects present problems which in many instances may be handled in a similar fashion. As a whole, anobiids are much more important pests in Europe than they are in the United States.

The methods utilized by the pest control operator are either liquid treatment or fumigation. The liquids currently being used are pentachlorophenol (as an oil base formulation or a proprietary gel) and lindane (in oil or water).

Discussion of control methods. Lyctus, anobiid, and related beetles are usually introduced into the home in building materials, since these may have become infested in lumber stock piles. These insects may also enter the home as larvae or adults in finished wood products such as oak flooring, paneling, furniture, etc. Hatfield (1949) notes rustic furniture purchased along the roadside is particularly likely to be infested by *Lyctus* beetles. These wood-boring insects breed in the wood and riddle and reduce it to a powder-like dust or sawdust-like frass which becomes readily evident as the infestation in the wood increases. Some wood-boring insects spread from the infested source to other susceptible wood in the home. There is always the possibility that the dead limbs of trees, which harbor many of these wood-boring insects, may serve as a source from which the flying adults may enter the house. However, there is a dearth of published material on this aspect of the problem.

The author has observed in California that where the beetle *Lyctus cavicollis* (LeC.) is present in a home, one often finds oak logs, a host of the beetle, have been used as firewood. It is possible these beetles have spread from the firewood into flooring, furniture, and paneling in the home. The larvae can burrow from one piece of wood into another, but are unable to disperse widely by their own movements.

Lehker (1947) notes the "parts of buildings damaged most frequently are joists, sub-flooring, hardwood flooring, sills, plates, studding, rafters, and interior trim." Many household articles such as implement handles, ladders, and others may be infested.

The sapwood of hardwoods is the wood that is almost always attacked by *Lyctus* larvae. The wood of conifers is tunneled only when adjacent to such hardwoods as oak, walnut, etc. According to Hunt and Garratt (1953), injury is confined to the sapwood because sapwood is rich in starch and starch is the principal food of the *Lyctus* larvae. Where the amount of starch in the wood is insufficient for the normal rate of development, the life cycle of the insects may be extended for two or more years. Unlike *Lyctus,* the anobiids attack coniferous, as well as hardwood timber. Further, they may even bore into the heartwood.

The size of the pores in the wood is another factor that limits *Lyctus* larvae in attacks on wood. Christian (1940) notes large pores in the wood aid in the process of egg-laying, permitting ease of oviposition. For this reason, woods such as ash, oak, and pecan, having exceptionally large pores, are very often infested. Smith (1956b) notes the average size of the egg of *L. planicollis* is 0.87 mm long and 0.16 mm wide.

Liquid treatment. A five percent solution of pentachlorophenol in oil or one percent lindane in oil or water is the formulation applied most commonly in the home for the control of wood-boring insects, such as *Lyctus,* anobiid beetles, and other beetles. Creamed penta emulsions are also effective, especially where rough timber is infested.

Hatfield (1944) notes chemically pure pentachlorophenol is a white needle-shaped crystalline solid. However, the non-pure or technical grade of pentachlorophenol is used for wood preservation purposes. The product, as sold commercially, has approximately six to eight percent tetrachlorophenol. Pentachlorophenol, as mentioned previously, is sold as five percent solution in a base oil. Lehker (1947) observes that an "ordinary light fuel oil (stove or furnace oil) is the best diluent, but kerosene can be used if no finished surfaces are to be treated." There are on the market 40 percent concentrates of pentachlorophenol, which when mixed one part by volume with 10 parts by volume of base oil will result in a five percent by weight pentachlorophenol solution. These concentrates are more economical to use than ready-made

Fig. 9-6. Powder post beetle emergence holes in a piece of badly infested wood.

Fig. 9-7. Burrows of powder post beetles in infested wood.

five percent pentachlorophenol solutions and are preferred by the professional pest control operator.

Since the vapors of the chemical and the solutions are irritants to the skin, precautions must be taken not to inhale the vapors or wet the skin. Where the solution contacts the skin, it should be washed off with soap and water. The professional operator who is subject to prolonged exposure utilizes goggles, rubber gloves, and a respirator.

Lehker (1947) offers the following suggestions for using pentachlorophenol solutions: "The chemical is applied by spraying it directly upon infested wood. A three-gallon compressed air-type sprayer is quite satisfactory, and the best penetration can be obtained if the nozzle is held about three inches from the surface being treated. A fog-type sprayer is not satisfactory since the object is to get the chemical on the wood and not in the air. The use of a brush can be used if spray equipment is not available or if the material is applied to small areas next to white paint or wall paper."

Hatfield (1949) amplifies the above remarks with his notes on the subject of spraying: "If the choice of a treating method lies between spraying, brushing or mopping, it will generally be advantageous to use a spray, as a more thorough job can be done without adversely affecting the finish. Whereas varnished and painted surfaces might be ruined by brushing-pentachlorophenol solutions onto wood, spray treatments, when applied carefully, penetrate the wood without 'roughing' the finish.

"Where sub-floors, floors, joists, or girders are sprayed, it has been found that a nozzle which delivers a conical-type spray is most useful. Likewise, it has been found that the spray nozzle should be held about three inches from the surface being sprayed. If held closer than three inches, the solution 'bounces off' the wood; if held farther away than three inches, the spray fogs and much is lost. Where hardwood floors are to be treated, the operator should start in a corner or a side of a room away from the door and work toward the door so that he will not have to walk the treated areas. In general, 24 to 48 hours should elapse before anyone walks on the treated floors. Applications should be made so that the surfaces are adequately 'wet' with the solution, but there should not be so much solution applied that the material 'puddles' on the floor surface."

Lehker recommends the use of one gallon of the five percent pentachlorophenol spray to "250 square feet of rough surface, such as joists and sub-flooring, and up to 450 square feet of finished surface." He also notes: "When floors are being treated, it is important to spray right up to the baseboard, but care should be exercised not to get the material upon white paint, plaster, or wallpaper. A brush can be used for treating small areas such as interior trim. In a few instances, hardwood floors are backed up with a tar paper material between the hardwood and sub-flooring. Ordinary spraying will not harm such floors, but the use of excessive amounts of chemical should be avoided, since it may soak up the paper, dissolve the tar and allow it to work out of the cracks or stain the ceiling below."

Hatfield adds to these precautions with the following remarks: "It is always desirable, if a finished floor is to be treated, to try a small floor section before the entire area is treated. Such an area may be located in and under the rug' area or some other inconspicuous place. It must be recognized that there are almost as many floor finishes (varnishes, lacquers, paints, waxes, stains, etc.) as there are floors, and it is impossible to make a definite statement relative to the possible damage that might be done by a spray solution. Likewise, care must be exercised that a spray material is not applied so heavily that it runs down through floor cracks and stains plaster or other areas.

"It should be mentioned that the spraying of oil solutions in basements or other areas where ventilators are present in the foundation, or the treatment of floors in the house with the windows open, may make it possible for the tiny oil droplets to 'drift' out of the openings and onto the shrubbery outside the house. If five percent pentachlorophenol in oil hits such shrubbery, it can easily be the cause of leaf drop or even the killing of certain plants. Extreme care in preventing this 'drifting' should be exercised by the person making the application.

"Any petroleum solvent, regardless whether it has a low or high flash point, presents a certain degree of fire hazard. Smoking should not be allowed on the premises and care should be taken that open flames are not present. The mist from the spray should be kept off lighted electric bulbs, as the droplets that fall on the hot glass make pin-point holes in the bulbs and thus burn them out quickly." The odor of pentachlorophenol will be apparent at first, but usually disappears within a few days.

Johnston et al. (1958) recommend the use of five percent DDT in a highly refined oil base for the control of infestations of powder post beetles in hardwood floors, interior woodwork, and furniture. They note that more than one application may be necessary and best penetration is obtained where the finish is removed before treatment. Do not handle furniture or walk on floors until they are thoroughly dried. These workers also recommend 0.5 percent lindane in a refined kerosene as a dip for infested dry hardwood products. A three-minute dip of tool handles and other rough materials will result in five or more years of protection. Now that DDT is no longer available for controlling wood-boring beetles, lindane has become the standard material for dipping or applying by spray or brush onto surfaces of infested wood. Coleman (1979) partially attributes control by lindane to repellency preventing adult emergence from infested wood which has been treated.

When treating finished wood surfaces, the precautions suggested for pentachlorophenol should be followed for lindane. Penetration is greater if the surface finish is first removed, especially if the newer polyurethanes are present, since they virtually prohibit any penetration.

Pence (1956) controlled powder post beetle in flooring by means of an infrared radiation unit.

Prior to publication of this handbook Dow Chemical Company reports that clearance will soon be obtained for use of chloropyrifos for powder post beetle control.

Weiss and Carruthers (1945) and Back (1939) consider insect pests of books in some detail.

Dipping. Christian (1940) states a three-minute dip in a solution consisting of three percent pentachlorophenol and 97 percent fuel oil will kill all stages of *Lyctus* in one inch/25 mm and thinner material, and renders the wood immune from further attack for an unknown period. However, if the treated wood is reworked, it may be necessary to retreat the exposed areas. Hatfield (1949) recommends a five percent pentachlorophenol solution be used as a five-minute dip on stock four inches/101 mm in diameter. Such dips can be used effectively on infested implement handles and similar materials.

Heat treatment. This method of control is utilized in the lumber yard rather than in the home. Nevertheless, it is an effective means of coping with wood-boring insects.

Snyder and St. George (1924) found that kilns operated by live steam, whereby wood up to one inch/25 mm thick was heated to a temperature of 130° F/54° C for 1.5 hours, killed the insects. The humidity was at the saturation point. Thicker wood requires a longer period of treatment. This method of control does not render the wood immune from future attack. The steaming under high pressure may weaken and discolor the wood, and for this reason, this process should not be used for wood requiring great structural strength or having a fine finish.

In general, the use of heat for the control of wood-boring insects in household furniture is not recommended, since the heat may warp the wood, loosen the joints, and injure the fabric or finish.

Fumigation. Infested furniture often can be treated effectively by vacuum and vault fumigation using such fumigants as methyl bromide and sulfuryl fluoride. In the lumber yard, Christian (1940) found methyl bromide at a dosage "of two cubic centimeters of the liquid gas per cubic foot of chamber space for 18 hours at 80° F will kill the stages inside the wood in loosely piled one-inch and thinner boards." Ehmann (1961) recommends two to three lbs./0.9 to 1.4 kg of methyl bromide per 1,000 cu ft./28 cubic meters for 24 to 48 hours where the infestation is widespread in a house. Tarpaulins are used to confine the gas. Details on such fumigations are presented in the chapter on fumigation.

The use of fillers. If the pores in the wood are filled with some material, the *Lyctus* female is unable to oviposit. A coating of shellac, varnish, furniture oil, or wax will serve this purpose.

Gahan and Laing (1932) impart the following information on a method of strengthening furniture weakened by wood-infesting insects. Parchment size "is prepared by dissolving clippings of parchment in boiling water. A proportionately small amount of water (about a pint to a pound of clippings) should be used, and it should be allowed to simmer for several hours until the parchment is all dissolved. The solution, while hot, may be worked into the 'wormholes' by means of a camel's hair brush or may be injected with a fine syringe, which should be warmed first. When the size has dried and hardened, what remains of it on the surface may be removed with warm water.

Prevention and control of lyctus beetles in the mill. Although the author primarily is concerned with the control of these insects in the home, it is appropriate to discuss briefly some of the control methods, other than those previously mentioned, that can be utilized where rough lumber is sawed and stored. If wood in the lumber yard can be protected from *Lyctus* and other wood-infesting insects, injury to finished products will be less likely, and introduction of the beetles into homes will be a less common occurrence.

Snyder (1938) recommends the lumber dealer institute the following program to safeguard his stock from attacks by *Lyctus*. This program undoubtedly is applicable to other wood-infesting insects:

- Inspect material annually, preferably in November and February, especially material two or more years old; burn material showing evidence of powder post beetles.
- Burn all useless sapwood and prevent accumulation of wastewood.
- If sapwood material is arranged by species (hickory, ash, oak, etc.) and by age, this will make it easier to examine the stock more readily.
- Sell the oldest stock first, since this is the most susceptible to attack.
- Inspect all newly arrived stock to prevent the introduction of powder post beetles.

Christian (1939), working with unfinished lumber, found that an immersion of green oak for 10 seconds in a five percent water solution of borax protected it from *L. parallelopipedus* (Melsh.), but not from *L. planicollis* (LeC.). This material also prevents sap stain. In 1940, the same author found that a one or two percent water solution of sulfocide (sodium pentasulfide in liquid form), as well as other finely divided sulfurs, effectively protected wood that was immersed for 10 seconds at 190° F/88° C. The beetles are repelled by the treated wood and are apparently killed by irritation as a result of contact with the sulfur particles.

Kowal (1949) recommends 14 pounds of 12 percent gamma benzene hexachloride in 50 gallons No. 2 fuel oil, as well as water emulsions of benzene hexachloride for the protection of green lumber against ambrosia beetles and other wood-boring insects. The U.S.D.A. (Anon., 1952) recommends the application on logs of 1¾ lbs/0.8 kg gamma benzene hexachloride in 50 gallons No. 2 fuel oil at the rate of one gallon/3.8 l for each 100 sq. ft./9.3 sq. meters bark surface for the control of ambrosia and other beetles. Hetrick and Moses (1953) showed an 0.5 percent emulsion of benzene hexachloride to be more effective than other chlorinated insecticides in protecting pine pulpwood from forest insects. Becker (1955), Becker et al. (1956 and 1962) showed 0.1 to 0.4 percent lindane emulsion to effectively protect white pine saw logs.

OTHER WOOD-BORING INSECTS

The beetles to be considered in this section often damage rustic furniture and buildings and are particularly important in wooded and mountainous areas. They are at times introduced into the home in firewood. The larvae and adult stages are responsible for the injury. These insects attract attention because of the frass (chewed and/or excreted wood) they produce, the galleries and holes they make, or by their size, coloration, and distinctive appearance. Snyder (1952) describes how to distinguish wood boring insects by the frass they produce.

The bark beetles or scolytidae. The adults are cylindrical beetles from $^1/_{16}$ to $^1/_4$ inch/1.5 to six mm long and are red, brown, or black in coloration. The white, legless larvae or grubs usually bore between the inner bark and the phloem. This causes the bark to fall off and then much frass is evident.

As an example of an injurious scolytid, mention may be made of the California hardwood bark beetle, *Micracis hirtellus* (LeC.). This dark reddish-brown beetle is $^1/_8$ inch/three mm long. It attacks the twigs of many shrubs and trees in the San Francisco Bay region, northward along the coast to Oregon. It is a source of much injury to lead cables in California (Struble and Hall, 1954).

According to Thompson (1932), another scolytid, *Phleosinus dentatus* (Say), damages furniture and log cabins of white cedar in New Jersey.

 The flatheaded borers or buprestidae. These are various-sized, metallic-colored, flattened, and stream-lined beetles. They attack dead and living shrubs and plants, as well as fences, log cabins, rustic furniture, etc. The whitish or cream-colored larvae are called "flat-headed" borers because of the enlargement of the thorax immediately behind the head. The larval mines are packed tightly with their sawdust-like borings.

Linsley (1943b) notes the buprestid beetles of the genus *Melanophila* are attracted "over long distances by smoke from a variety of burning materials, including wood, oil, mill refuse, smelter products, and possibly tobacco. In nature, this attraction leads them to forest fires, where they normally oviposit in scorched coniferous wood. The beetles are also stimulated by heat, and in the vicinity of the source they fly rapidly and run about over hot surfaces. Light probably plays little if any role in their attraction to fires." These beetles become annoying around sugar refineries, sawmills, etc., where they may even bite people. They are attracted to football games, possibly due to cigarette smoke.

According to Linsley (1943c), the golden buprestid, *Buprestis aurulenta* (L.), which may oviposit on green lumber, is capable of prolonged existence in cured wood, such as varnished stairways and handrails. There is a record of the emergence of an adult golden buprestid from wood 26 years after it had been installed.

The roundheaded borers or cerambycidae. The adult beetles are usually large, $^1/_4$ to three inches/six to 76 mm attractive-appearing insects with antennae which may be in some species much longer than the body. This accounts for the adults being referred to as "long horned" beetles. The larvae or grubs are long and narrow (from $^1/_2$ to four inches/12 to 101 mm), cream-colored insects with dark-brown jaws. They are, for the most part, legless, noticeably constricted by rings or segments, and often covered with "warts" and other protuberances. The larvae commonly infest dead and living trees and their galleries are packed with frass, which varies in appearance according to the species from a fine dust to a coarse sawdust-like material. Firewood, lawn furniture, and even building materials may harbor the living larvae.

On one occasion, the author was summoned to a house where a woman insisted beetles were attacking her curtains. Truly enough, the cerambycid, *Xylotrechus nauticus* (Mann.) was found on the curtains. Investigation revealed the firewood from where they had emerged. Their presence on the curtains was merely accidental, for upon emerging, they flew toward the windows.

The spined pine borer, *Ergates spiculatus* (LeC.). This huge, reddish-brown beetle may reach 2.25 inches/57 mm in length. It occurs throughout the Pacific Slope where it breeds in conifers. The full grown, enormous larvae, which are two to three inches/50 to 75 mm long, are often found in lumber long after it is milled. On one occasion, the larva was found boring into a lead cable from an outer casing of wood.

The old house borer, *Hylotrupes bajulus* (L.). This cerambycid is a black to dark brown beetle with grey pubescence and with a white patch on its back. Houghton (1939) states, "Linnaeus gave it the specific name *bajulus,* which means "porter" because of this patch. It is a very important pest in Norway, Denmark, and Germany. McIntyre and St. George (1968) have reported this insect as causing serious damage in the Atlantic Coastal states, as well as Mississippi, Louisiana, and Texas. The old house borer is a pest of both old and new houses. Korting (1962) showed that lumber in newer houses was more likely to be infested than in older houses. Becker (1954) notes the old house borer is widespread in Massachusetts, and Simeone and McAndrews (1955) find it well distributed throughout New York state. St. George et al. (1957) state it occurs along much of the East Coast, as well as along the Gulf of Mexico to Texas. In Pennsylvania it is recorded as far west as Pittsburgh. Moore (1978) presents a current review of this pest as it occurs in the United States. He also discusses current control recommendations.

According to Craighead (1950), the larva can be recognized "by the thin texture of the skin and the fact that the head is wider than long, the apex of the mandible is rounded, and there are three ocelli on each side of the head. The prothorax is smooth and shining and the ampullae reticulated, approaching tuberculate. Legs are present. It feeds in dry, seasoned, coniferous woods, filling the extensive galleries with loose granular frass."

According to St. George et al. (1957), presence of the old house borer in a building can be determined in five ways. "(1) By the rasping or ticking sound made by the larvae (borer) while boring; (2 and 3), by the presence of powdery borings in the sapwood or by the larva in its tunnel, both of which may be seen when the wood surface is probed with a sharp instrument; (4) by the presence of ¼-inch, broadly oval holes made by the adult when it emerges; or (5) by the presence of the beetle. The last two evidences of attack will be found only in buildings that are five or more years old."

Patton (1931) observes this insect attacks timbers in houses , especially the roof timbers. He notes that in Copenhagen, no less than 10 churches, several factory buildings, and more than 1,000 dwellings have had their roof timbers damaged seriously. In some cases, the infestation was so bad there was a fear of the roof falling in. Roofs with timbers covered with metals and more than 20 years old were the ones likely to be attacked. Houghton also found these beetles prefer attic and roof timbers in the United States, but they have been reported in flooring too. In one instance, wood in a bridge in Pennsylvania, had to be replaced because of the attacks of the old house-borer. This beetle prefers pine, spruce, hemlock, and fir. Becker (1944) notes it is very injurious to wood of con-

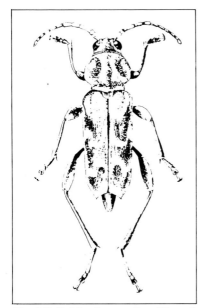 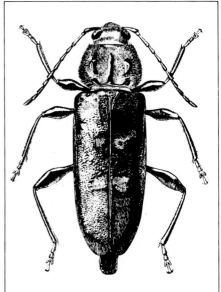

Fig. 9-8. *Xylotrechus nauticus* (Mann.),   **Fig. 9-9. Adult old house borer.**
a cerambycid beetle.

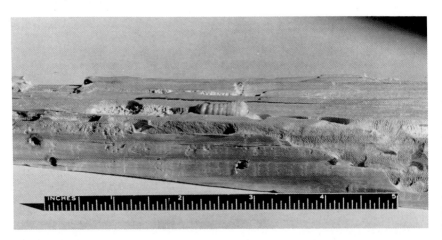

Fig. 9-10. Old house borer damage. Note the larva on top of the piece of wood.

ifers in Germany, but is unable to develop in deciduous trees. Duffy (1949) states the larvae feed mainly in the sapwood and the infestation decreases as soon as the heartwood is reached.

According to Patton (1931), in England, the female lays her eggs in the cracks and crevices in the wood; the larvae hatch in a few days and bore into the wood. the larval stage may last from three to 11 years, in which time the interior of the wood may be reduced to a powdery consistency, but there may be no external evidence of attack. The pupal stage is about one week, but the adults may remain in the mines from five to seven months before emerging. The emergence holes of the adults may be the only evidence of attack.

Craighead (1950) discusses the life history of the old house borer in the eastern United States as follows: "Adults appear late in summer, and deposit the eggs in season checks or irregularities of the wood. The resulting larvae feed from a few to many years in the dry sapwood, until it is destroyed completely. The mines are loosely filled with granular frass, some of which may fall out and reveal their presence. The frass forms a distinguishing characteristic, being composed of tiny pellets and fine, powdery material. The usual larval period is three years, but in many instances it may extend to five years or more. Pine and spruce woodwork and possibly the woods of other conifers, in rather dry situations are subject to attack and destruction by these larvae." Cannon and Robinson (1981) found wood consumption is reduced where temperature and relative humidity fluctuate more than 10°.

Houghton (1939) notes, "There are two theories as to why this insect has not increased to a greater extent in the United States. First, the construction of homes and buildings in the northern section is entirely different than found in Europe at approximately the same latitude. Along with this, the climate in the New England states has a wider variance in temperature the year around. Second, the soft woods which this insect infests are exported from this country rather than imported. The insect is most likely to be carried from one area to another in lumber." Snyder (1955a) notes the old house borer has not been found infesting logs and stumps and that it is strictly a structural pest. But unpublished research ongoing at Virginia Polytechnic Institute and State University shows the old house borer is capable of surviving all the seasons in stumps. Therefore, its capability of becoming a major pest is real.

Craighead (1950) stated the old house borer was becoming increasingly important in the United States. Moore (1978) also indicates this and the author's personal experience supports this. Conditions in the Middle Atlantic states appear especially attractive to the old house borer with some structures sustaining heavy and wide-spread damage due to this beetle.

In the Carolinas, the infestations are most common in crawl spaces, but attics are the common infestation site in Pennsylvania, New Jersey, and states to the north. The practice of decorating with used barn siding and beams has also brought this pest into structures, as these wood items are often infested.

When the adults emerge, they will penetrate virtually any covering, including hard asbestos shingles, aluminum siding, and ceramic tile in baths. It is indeed a destructive insect.

Control ranges from waiting for the infestation to run its course to fumigation. Moore (1978) discusses the alternates and puts them in good perspective. Pentachlorophenol and lindane are the only two currently registered liquids for controlling this beetle. Both are limited in that they will probably not prevent emergence, nor kill larvae in the wood furniture and are useful only when infested wood is accessible. Fumigation, while not always necessary to eliminate an infestation, is often resorted to for the homeowner's peace of mind or when a PCO must issue a clearance on a building for a real estate transaction. Methyl bromide or sulfuryl fluoride are the fumigants most commonly employed. Sulfuryl fluoride fumigations are costly as the dosage is four times that required for drywood termites. This is to ensure kill of old house borer eggs.

Durr (1956) studied the morphology and biology of the old house borer in South Africa and prepared a monograph on the insect.

The new-house borer, *Arhopalus productus.* (LeC.). This cerambycid beetle is an important pest of green lumber in the western United States. Although it causes only minor structural damage to new wood, the beetles and their clear-cut ¼-inch/six mm oval emergence holes are quite upsetting to the new owner. As many as 50 holes have been reported in a single residence. Although the new-house borer attacks only coniferous woods, its exit holes may appear in adjoining hardwood floors, linoleum, rugs, plasterboard, and roof coverings. Subflooring of Douglas fir is a common source of infestation by this insect.

The adult is a narrow black beetle measuring ¾ to 1¼ inches/19 to 32 mm long. The cream-colored larva may reach a length of 1½ inches/38 mm. The larvae at first bore through the sapwood and later the heartwood leaving behind tunnels of tightly-packed frass.

In the forest, this beetle attacks trees killed by insects or fire as well as green logs. It does not reinfest wood in the house. The life cycle requires at least two years and the beetle usually emerges in the first year of the new house's existence.

There is no practical way to control this pest other than by fumigation and ordinarily the insect is not a serious enough pest to require such treatment. Eaton and Lyon (1955), Ebeling (1963) and Wood (1964) have made a study of this insect.

Thompson (1932) found that insects most prominent in the injury of rustic furniture and log cabins were mainly the Cerambycidae, *Hylotrupes ligneus* (F.), *Oeme rigida* (Say), and *Callidium antennatum* (Newm.). R. M. Bohart reports *Phymatodes nitidus* (LeC.) as a pest of rustic furniture in a mountain resort in southern California, and Leech (1944) found *P. dimidiatus* (Kby.) emerging from structural timbers in British Columbia. Hatfield (1949) lists some additional roundheaded borers such as the flat oak borer, *Smodicum cucujiforme* (Say), the grey banded ash borer, *Neoclytus caprea* (Say), and the roundheaded borer, *Megacyllene antennatus* (White). Gerberg (1951) notes adults of the southern pine sawyer, *Monochamus titillator* (F.), were found in a show window

Fig. 9-11. Red-headed ash borer,
Neoclytus acuminatus **(F.).**

in Maryland feeding on rayon dresses. Davis (1952) notes fireplace logs are often an annoying source of insects. Two cerambycids emerging from such logs are the hickory borer, *Megacyllene caryae* (Gah.) and *Phymatodes testaceus* var., *variabilis.*

Orthosoma brunneum (Forst.) is a pest of structural timber and Craighead states this insect may damage cross ties, telephone poles, and all structural timbers in contact with the ground or in moist, exposed places.

Oedemeridae, the wharf borer, *Nacerdes melanura* (L.). G. E. Wallace of the Carnegie Museum notes the wharf borer, *Nacerdes melanura* (L), is a pest in dwellings in the Pittsburgh, Pennsylvania area. Since this insect is found in very moist wood, it is his belief the occurrence of the adult insect indicates the presence of rotting timbers and even unsatisfactory plumbing.

According to Drooz (1953), this insect promotes the extension of rot into wood and is responsible for very costly damage to foundation piling underneath buildings in Milwaukee, Wisconsin. Pomerantz (1954) reports several cases where it infested foundation timbers and was a great nuisance to tenants in apartment houses in New York. According to Anon. (1950), a severe infestation of this insect was found in greenhouse benches in Ohio.

Snyder notes the insect was introduced from Europe and that it has a life cycle of one or more years.

Balch (1937) studied the habits of this insect in St. John, New Brunswick, and

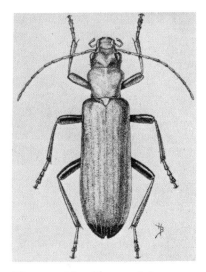

Fig. 9-12. Wharf borer adult.

notes this insect is recorded throughout much of the United States and Canada, particularly along coastal areas. Larvae of the wharf borer are found in timbers and posts adjacent to salt water, in cellars, floor timbers, old boxes, and lumber yards. Balch found the larvae in wharf timbers between flooding and the high water level which were badly decayed and well riddled. Wood wetted by dog urine also attracts the beetles.

"The adult is somewhat narrow and depressed, about 10 millimeters long, brownish to reddish-yellow above with the tips of the elytra, the eyes, side of the thorax, legs, and ventral parts generally blackish, the whole body covered with a rather dense yellow pubescence. The elytra are pointed, and each has four more or less evident raised lines. The antennae are rather more than half the length of the body." The adults were seen in large numbers in St. John during July and August, which is apparently the period of mating and oviposition. H. Katz and J. Becker note this beetle is found in Pittsburgh, Pennsylvania in May and June, and even earlier.

The mature larvae are about 1¹/₅ inch/30 mm long and covered with brown hairs or setae. The body is cream-colored with the mouthparts brown and the mandibles almost black at the tips. The larvae overwinter in many different stages. The wharf borer larvae breed in rotten wood that is kept moist. Creosote-impregnated timber must be substituted for susceptible wood to control them where they are serious pests. Spencer (1947) recorded an infestation in buried piling in Vancouver, British Columbia.

Pomerantz (1954) controlled an infestation of the wharf borer in an apartment house by injecting five percent pentachlorophenol and 10 percent orthodichlorobenzene oil solution into the infested subfloor area. Egleton (1955) saturated the infested area with five percent chlordane solutions.

With many redevelopment projects underway, timbers from old structures are often bulldozed beneath slabs and into backfills of the new construction. This buried wood — often in direct contact with the subsoil water table, provides an ideal environment for wharf borers. Annual swarming of adults into the new structure is severe enough to disrupt business and even close it down for several days. The injection of one percent chlordane into the backfill and beneath slabs as a preventive termite treatment has usually proven effective in preventing such swarms in future years.

Snyder (1954) discusses another insect, *Micromalthus debilis* (LeC.), which also bores into rotten wood. This insect belongs to the Family Micromalthidae.

Curculionidae. Turner and Townsend (1936) report the weevil, *Hexarthrum ulkei* (Horn), has been found in several buildings in Connecticut. The damage is similar to that of powder post beetles and the control measures are the same. G. E. Wallace reports this species as damaging wood in Forbes Field, the home of the Pittsburgh Pirates.

CONTROL OF ROUNDHEADED AND FLATHEADED BORERS

St. George (1956) recommends that the surface of the bark or wood be thoroughly sprayed with an 0.5 percent oil solution of gamma benzene hexachloride. He notes that one gallon/3.8 l is sufficient to treat approximately 50 square feet/ 4.65 sq. meters of bark surface. "When dressed timbers in a building are being treated, one gallon of the material will cover approximately 150 square feet of surface."

Becker (1959) found 0.4 percent benzene hexachloride emulsions to be effective. Allen and Rudinsky (1959) showed that endosulfan, carbaryl and lindane, each, when used at one to three lbs./0.45 to 1.4 kg wettable powders to 100 gallons/378 l of water, protected freshly cut logs for at least eight weeks.

LITERATURE

ALLEN, D. G. & J. A. RUDINSKY, — 1959. Effectiveness of Thiodane, Sevin, and Lindane on insects attacking freshly cut Douglas fir logs. J. Econ. Entomol. 52(3): 482-484.

ANONYMOUS — 1951. A diagnostic key to the more common structural wood-destroying pests of the Pacific Coast. Pest Control 19(2) :26. 1952. Research-developed treatment protects felled logs from insect attack. U.S.D.A. Mimeog. 805-52. 1954. Powder-post beetles in buildings; what to do about them. USDA Leaflet 358. 1960. Wharf borer. Coop. Econ. Insect Rpt. 10(8) :10. 1964. *Stephanopachys substriatus.* Coop. Econ. Insect Rpt. 14(24) :627. 1966. Powder post beetle, *Trogoxylon prostomoides.* Coop. Econ. Insect Rpt. 16(14) :280. 1966. Lead-cable borer, *Scobicio declivis.* Coop. Econ. Insect Rpt. 16(29) :709. 1967. Oriental wood borers now established in Florida. Pest Control 35(4) :34.

BACK, E. A. — 1939. Bookworms. Smiths. Inst. Ann. Rept. 365-74. 1939a. A new pest of books, *Neogastrallus librinocens* Fisher. J. Econ. Entomol., 32 :642-645. 1945. Minutes of the 553rd regular meeting of the Entomological Society of Washington. Proc. Ent. Soc. Wash. 47(6) :182-184.

BALCH, R. E. — 1937. Notes on the wharf borer *Nacerda melanura* Linn. Canadian Ent. 69(1) :1-5.

BECKER, G. — 1944. The natural resistance of wood of deciduous trees to larvae of *Hylotrupes bajulus* L., and its cause. Z. angew. Ent. 30(2) 319-417. Rev. Appl. Ent. A. 35(11) :350-351, 1947.

BECKER, W. B. — 1942. *Prionus laticollis* (Drury) in a subterranean wooden duct for telephone cables. J. Econ. Entomol. 35 :608. 1954. The old house borer in Massachusetts. J. Econ. Entomol. 47(2) :362-363. 1955. Tests with BHC emulsion sprays to keep boring insects out of pine logs in Massachusetts. J. Econ. Entomol. 48(2) : 163-167. 1956. Common insect damage to wood seen in buildings in the Northeast. Pest Control 24(2) :9-10, 12, 14, 16, 19. 1959. Further tests with BHC emulsion sprays to keep boring insects out of pine logs in Massachusetts. J. Econ. Entomol. 52(1) :173-174. 1962. Autumn versus spring spraying of unseasoned pine logs with BHC. J. Econ. Entomol. 55(6) :1020-1021.

BECKER, W. B., H. G. ABBOTT, & J. H. RICH. — 1956. Effect of lindane emulsion sprays on the insect invasion of white pine sawlogs and the grade yield of the resulting lumber. J. Econ. Entomol. 49(5) :664-666.

BLAKE, E. G. - 1926. Enemies of Timber. D. Van Nostrand Co., N. Y. 206 pp.

BLETCHLY, J. D. — 1951. A summary of some recent work on the factors affecting egg-laying and hatching in *Anobium punctatum* DeG. (Coleoptera — Anobiidae). Rev. Appl. Entomol. A 41(8) 231.

BURKE, H. E., R. D. HARTMAN & T. E. SNYDER — 1922. The lead cable borer or "short-circuit beetle" in California. USDA Bull. 1107.

CANNON, K. F. & W. H. ROBINSON, — 1981. Old house borer larvae: factors affecting wood consumption and growth. Pest Control 49(2) :25-28.

CHAMBERLIN, W. J. — 1948. Insects Affecting Forest Products and Other Materials. Oregon State College Coop. Assoc. Corvallis, Ore. 159 p.

CHRISTIAN, M. B. — 1939. Southern Lumbermen. Dec. 15, 1939. 1940. Biology of the powder-post beetles, *Lyctus planicollis* LeConte and *Lyctus parallelopipedus* (Melsh.). Part I. Louisiana Conservation Review 9(4) :56-59. 1941. Biology of the powder-post beetles, *Lyctus planicollis* LeConte and *Lyctus parallelopipedus* (Melsh.) Part II. Louisiana Conservation Review 10(1) :40-42. 1945. Powder-post beetles and their control and notes on chlorinated phenols as soil poisons for termite control. Pests 13(6) :20.

COLEMAN, G. R. — 1979. British Wood Preserving Assn. News Sheet No. 155, May.

CRAIGHEAD, F. C. — 1950. Insect Enemies of Eastern Forests. U.S.D.A. Miscellaneous Publications 657.

DAVIS, J. J. — 1952. Fireplace logs an annoying source of insects. Pest Control 20(10) :19-20.

DESCH, H. E. — 1950. The common furniture beetle as a household pest today. Roy. Inst. Chartered Surveyors, J. 30 140-144. Aug.

DOANE, R. W., E. C. VAN DYKE, W. J. CHAMBERLIN & H. E. BURKE — 1936. Forest Insects. McGraw-Hill

DROOZ, A. T. — 1953. The wharf borer problem in Milwaukee, Wisc. Proc. 8th Annual Meeting No. Central States Br. Ent. Soc. Amer. 22-23.

DUFFY, E. A. J. — 1949. Recent research on long-horned timber beetles. British Sci. News 3(25) :19-21.

DURR, H. J. R. — 1956. The morphology and bionomics of the European houseborer, *Hylotrupes bajulus* (Coleoptera: Cerambycidae). Union of So. Africa Dept. Agric. Ent. Mem. 4(1) :1-136. Biol. Abst. 31(2) :607, 1957.

EATON, C. B. & LYONS, R. L. — 1955. *Arhopalus productus* (LeC.), a borer in the new buildings. USDA Forest Service, Calif. Forest and Range Exp. Sta. Tech. Paper No. 11.

EBELING, W. — 1963. Damage from new house borer increasing. P.C.O. News 23(10) : 18-19.

EBELING, W. & D. A. REIERSON — 1973. Leadcable borers infesting houses.

National PCO News 33(8) :16-17.

EGLETON, S. P. — 1955. English PCO reports wharf borer British problem, too. Pest Control 23(3) :6.

EHMANN, N. R. — 1961. A new method for control of powder post beetles. Pest Control 29(6) :42, 46, 46b.

ESSIG, E. O. — 1926. Insects of Western North America. Macmillan Co.

FISHER, R. C. — 1937. Studies of the biology of the death-watch beetle, *Xestobium rufovillosum* DeG. I. A summary of past work and a brief account of the development stages. Ann. Appl. Biol. 24:600-13.

FISHER, W. S. — 1938. A new anobiid beetle from Alaska. J. Wash. Acad. Sci. 28:26-7.

GAHAN, C. J. & F. LAING, — 1932. Furniture beetles. British Econ. Ser. 11.

GERBERG. E. J. — 1951. An unusual food habit of *Monochamus titillator* Fab. J. Econ. Entomol. 44(3) :317/1957. A revision of the new world species of powder-post beetles belonging to the family Lyctidae. USDA Tech. Bull. No. 1157.

HATCH, M. H. — 1946. *Hadrobregmus gibbicollis* infesting woodwork J. Econ. Entomol. 39(2) :274.

HATFIELD, I. — 1944. The present status of the powder post beetles and effective control. Proc. 23rd Ann. Meeting No. Central States Entomologists. Univ. Ill. pp. 138-142. 1948. Powder post beetles and their control. Pests 16(5) :23-32. 1949. How to recognize and control powder post beetles in wood. Forest Products Research Society 33rd Ann. National Meeting. pp. 1-13.

HERRICK, G. W. — 1939. Insects Injurious to the Household and Annoying to Man. Macmillan.

HETRICK, L.A. — 1955. Personal communication.

HETRICK, L. A. & P. J. MOSES — 1953. Value of insecticides for protection of pine pulpwood. J. Econ. Entomol. 46(1) :160-161.

HICKIN, N. E. — 1953. The common furniture beetle, *Anobium punctatum* (DeGeer) (Col., Anobiidae): Some notes on its outdoor occurrence. Entomologist 86(9) :216-217. 1960. An introduction to the study of the British Lyctidae. Brit. Wood Preserv. Assoc. Rec. Annu. Conv. 1960 :57-96. 1961. An introduction to the study of *Anobium punctatum*. Pest Control Conf. Rentokil Group. 1963. The Insect Factor in Wood Decay. Hutchinson. 336 pages.

HOPKINS, A. D. & T. E. SNYDER — 1917. Powder-post damage by *Lyctus* beetles to seasoned hardwood. USDA Farmers' Bull. 778.

HOUGHTON, C. W. — 1939. The porter beetle, *Hylotrupes bajulus* L. Pests 7(10) :19.

HUNT, G. M. & G. A. GARRATT — 1953. Wood Preservation. McGraw-Hill.

JOHNSTON, H. R., R. H. SMITH & R. A. ST. GEORGE — 1958. Control of lyctus powder-post beetles in lumber yards and processing plants. Pest Control 26(1) :39-42.

JUDD, W. W. — 1948. Powder-post beetles in imported bamboo. J. Econ. Entomol. 4(1) :113.

KELSEY, J. M. — 1946. A preliminary report on timber preservation with Wolman Tanalith in New Zealand. N. Z. J. Sci. & Tech. 28(3) :136-44. 1946. Tests with timber preservatives in New Zealand. N. Z. J. Sci & Tech. 27(6) Sec. B: 446-457. 1946. A note on rearing *Anobium punctatum* DeGeer N. Z. J. Sci & Tech. 27(4) :329-335. 1946. Insects attacking milled timber, poles, and posts in New Zealand. New Zealand J. Sci. & Tech. 28(2) :(Sec. B): 65-100.

KELSEY, J. M., D. SPILLER & R. W. DEENE — 1945. Biology of *Anobium punctatum*. Progress Report. N. Z. J. Sci. Tech. 27(b)1 :59-68.

KORTING, A. — 1962. Development and destructivity of *Hylotrupes bajulus* in rafters of varying ages. Biol. Abst. 43(3) :1003, 1963.

KOWAL, R. J. — 1949. Benzene hexachloride for the control of insects attacking green logs and lumber. Down to Earth. Dow Chem. Co. 5(2) :12-13.

LEECH, H. B. — 1944. The cerambycid beetle, *Phymatodes dimidiatus,* in cedar structural timbers. Canad. Ent. 76:211.

LEHKER, G. E. — 1947. Powder post beetles and their control. Purdue Univ. Ext. Bull. 314.

LINSLEY, G. E. — 1943a. The recognition and control of deathwatch, powder-post and false powder post beetles. Pests. 11(3) :11-14. 1943b. Attraction of *Melanophila* beetles by fire and smoke. J. Econ. Entomol. 36(2) :341-342. 1943c. Delayed emergence of *Buprestis aurulenta* from structural timbers. J. Econ. Entomol. 36(2) :348.

McINTYRE, T. & R. A. ST. GEORGE — 1968. The old house borer. U.S.D.A. Leaflet No. 501.

MOORE, H. B. — 1978. The old house borer — an update. Pest Control 46(3) :14-17, 46(4) :28-30, 46(5) :26-28.

MUIRHEAD, D. M. — 1911. A beetle control problem in timbers of the old South Meeting House. J. Econ. Entomol. 34 :381-383.

OSMUN, J. V. — 1955. Recognition of insect damage. Section Three: Wood and wood products. Pest Control 23(7) :20-22.

PARKIN, E. A. — 1943. The moisture content of timber in relation to attack by *Lyctus* powder-post beetles. Ann. Appl. Biol. 30(2) :136-142. Rev. Appl. Ent. 32:35-36, 1944.

PATTON, W. S. — 1931. Insects, Ticks, Mites and Venomous Animals II. H. R. Grubb, Ltd.

PAYNE, N. M. — 1936. Injury to lumber by *Hadrobregmus carinatus* Say. J. Econ. Entomol. 29 :1027.

PENCE, R. J. — 1956. UCLA's infrared radiation unit to control lyctus powder-post beetle infestations in hardwood floors can be built by PCO's. Pest Control 24(7) :30, 32, 34, 36.

PLANK, H. K. — 1948. Biology of the bamboo powder-post beetle in Puerto Rico. Bull. No. 44. Federal Experiment Station in Puerto Rico; Mayaguez, Puerto Rico.

PLANK H. K. & R. H. HAGEMAN — 1951. Starch and other carbohydrates in relation to powder-post beetle *(Dinoderus minutus)* infestations in freshly harvested bamboo. J. Econ. Entomol. 44(1) :73-75.

POMERANTZ, C. — 1954. The way we handled a wharf borer infestation in a N.Y. apartment building. Pest Control 22(10) :35-36,40.

RIVERS, J. J. — 1886. Contributions to the larval history of Pacific Coast Coleoptera. Bull. Calif. Acad. Sci. 2(5) :63-72.

ST. GEORGE, R. A. — 1956. Protecting log cabins, rustic work, and unseasoned wood from injurious insects in Eastern United States. USDA Farmers' Bull. 2104.

ST. GEORGE, R. A., H. R. JOHNSTON & T. MCINTYRE — 1957. Old house borer. Pest Control 25(2) :29-31.

SIMEONE, J. B. & A. H. MacANDREWS — 1955. The old house borer in New York State. J. Econ. Entomol. 48(6) :753-754.

SMITH, J. H. & A. C. FORBES — 1944. How to defeat wood-borers. N. Z. J.

Agr. 68(2) :83, 85-88. Rev. Appl. Ent., A. 32:385-387, 1944.

SMITH, R. H. — 1955. The effect of wood moisture content on the emergence of the southern lyctus beetle. J. Econ. Entomol. 48(6) :770-771. 1956a. The rearing of *Lyctus planicollis* and the preparation of wood for control tests. J. Econ. Entomol. 49(1) :127-129. 1956b. A technique for studying the oviposition habits of the Southern lyctus beetle and its egg and early larval stages. J. Econ. Entomol. 49(2) :263-264. 1956c. Lyctus powder-post beetle control by surface applications of oil preparations and solvents. Pest Control 24(4) :42, 45.

SNYDER, T. E. — 1923. High temperatures as a remedy for *Lyctus* powder-post beetles. J. Forestry 21 :810-814. 1938. Preventing damage by *Lyctus* powderpost beetles, USDA Farmers' Bull. 1477. 1944. Powder-post beetles and their control. Pests 12(4) :8. 1952. How to distinguish wood-boring insects by their frass. Pest Control 20(1) 28. 1954. Two borers of unusual habits. Pest Control 22(9) :30, 32, 34. 1955a. Introduced wood borers. Pest Control 23(1) :28. 1956b. Borers through metal. Pest Control 23(5) :28,30.

SNYDER, T. E. & R. A. ST. GEORGE, — 1924. Determination of temperatures fatal to the powder-post beetle, *Lyctus planicollis* LeConte, by steaming infested ash and oak lumber in a kiln. J. Agr. Res. 28:1033-8.

SPENCER, G. J. — 1947. The status of *Anobium punctatum,* the death watch beetle in the lower Fraser Valley in 1946. (Coleoptera: Anobiidae). Entomol. Soc. Brit. Columbia, Proc. 43:9-10. Feb. 1947. An unusual record of the wharf borer. *Nacerda melanura,* in buried piling at Vancouver, British Columbia (Coleoptera: Oedemeridae). Ent. Soc. Br. Columbua Proc. 43:7-8.

SPILLER, D. — 1949a. Toxicity of pentachlorophenol to the common house borer *Anobium punctatum* DeGeer. 1. Residual contact and ovicidal action. 1949b. Toxicity of boric acid to the common house borer. New Zealand J. Science Technol. 30B :22-23. Chem. Abst. 43:8598. 1949. Effect of humidity on hatching of eggs of the common house borer, *Anobium punctatum.* Rev. Appl. Entomol. A. 38(6) :263. 1964. Number of eggs laid by *Anobium punctatum* (DeGeer), Bull. Entomol. Res. 55(2) :305.

SPINK, W. T., H. R. GROSS, & L. D. KIRST — 1966. *Xyletinus peltatus* in structural timbers. Pest Control 34(2) :12, 13, 15, 44.

STRUBLE, G. R. & R. C. HALL — 1954. Telephone cables invaded by shrub bark beetle in Pacific Coastal Region. J. Econ. Entomol. 47(5) :933-934.

TAYLOR, R. L. — 1928. A foreign book pest enters Boston. J. Econ. Entomol. 21:626-627.

THOMPSON, F. M. — 1932. Pine oils as agents for protecting rustic furniture and log cabins from various wood borers. J. Econ. Entomol. 25:347-351.

TURNER, N. & J. F. TOWNSEND — 1936. Termite control in buildings in Connecticut. Conn. Agr. Exp. St. Bull. 382.

WATSON, E. B. — 1948. Powder-post beetles in Canada. Pests 16(30) 24, 26.

WATSON, J. R. — 1943. A tropical book worm in Florida. *Neogastrallus librinocens* Fisher in Florida. Fla. Entomol. 26(4) :61-63. Dec.

WEISS, H. B. & R. H. CARRUTHERS — 1945. Insect enemies of books. N. Y. Public Library.

WHITE, R. E. — 1962. The Anobiidae of Ohio. Ohio Biol. Survey 1(4) — pp. 58. 1963. The Mexican book beetle, *Catorama herbarium,* established in the United States. Ann. Ent. Soc. Amer. 56(3) :280-285. 1971. Key to North American genera of Anobiidae, with phylogenetic and synonymic notes. Annals ESA. 64(1) :177-190.

WOOD, D. L. — 1964. *Arhopalus productus* (LeConte) a pest of increasing importance in new home construction. P.C.O. News 24(3) :22-23.
WRIGHT, C. G. — 1959. Beetles found in yellow pine floor joists of buildings in North Carolina. J. Econ. Entomol. 52(3) :452. 1960. Biology of the Southern lyctus beetle, *Lyctus planicollis*. Ann. Entomol. Amer. 53(3) :285-292. 1960A. A technique useful in comparing powderpost beetle population in the structures of buildings. J. Econ. Entomol. 53(2) :329-330.

DAVID J. SHETLAR

Dr. David Shetlar gained his bachelor's and master's degrees in zoology from the University of Oklahoma and his Ph.D. from The Pennsylvania State University. After various teaching assignments at the University of Oklahoma, Shetlar began his research and teaching career at The Pennsylvania State University in 1971.

Shetlar's research activities have included pest management studies for ornamental plants, turf and greenhouses. His teaching responsibilities cover the whole field of entomology, including medical entomology, economic entomology and use of pesticides.

In addition to his research and teaching experience, Shetlar is an accomplished illustrator of insects and has taught scientific illustration.

Shetlar has authored or co-authored numerous articles in scientific publications and has lectured widely at seminars of professional associations and civic groups.

Shetlar has held office in the Phi Sigma Biological Research Society (University of Oklahoma Chapter) and is an associate member of Sigma Xi, the Scientific Research Society of North America. Shetlar is active in several professional societies and is a member of the Pennsylvania Academy of Sciences, the Entomological Society of Pennsylvania and the Entomological Society of America.

CHAPTER TEN

Psocids or Book Lice

Revised by David J. Shetlar[1]

THOUGH THERE ARE approximately 150 species of psocids in the United States, only a few of these are pests in homes and stored products. The more common house infesting psocids in the United States are the booklice, *Liposcelis* species, and the larger pale trogiid, *Trogium pulsatorium* (L.). Also, *T. pulsatorium* is commonly called the "deathwatch" because of its habit of tapping its abdomen. Gurney (1950) lists other psocids that are pests in the home, such as the cosmopolitan grain psocid *Lachesilla pedicularia* (L.), *Psyllipsocus ramburii* (Selys), *Psocathropos lachlani* Ribaga and *Lepinotus patruelis* Pearman.

The booklice and larger pale trogiid are worldwide in distribution, and usually range from 1/25- to 1/13-inch/ one to two mm in length. These insects have a simple development. The adult booklice are wingless and the trogiid has small scale-like wings, although some of the other species of psocids have two pairs of wings which are held roof-like over the body. Booklice vary from an almost colorless appearance to grayish or brown. The trogiid has brownish spots on its pale-colored body. These psocids are almost colorless when young, but obtain some coloring as they grow older, possibly due to the food they consume.

Some psocids are called booklice because they superficially resemble the bird lice and because they are commonly found scurrying about books and papers, especially in damp locations.

Many psocids occur outdoors on or under bark, grass, leaves, damp wood and similar places. This is where they get the common name of "barklice." Barklice often occur in large numbers on the bark of trees where they feed on mosses and lichens. Some species, such as the southern *Archipsocus*, spin webs along the trunks of trees or along the walls of buildings (Mockford, 1957). Though they

[1]*Assistant Professor, The Pennsylvania State University, University Park, Pa.*

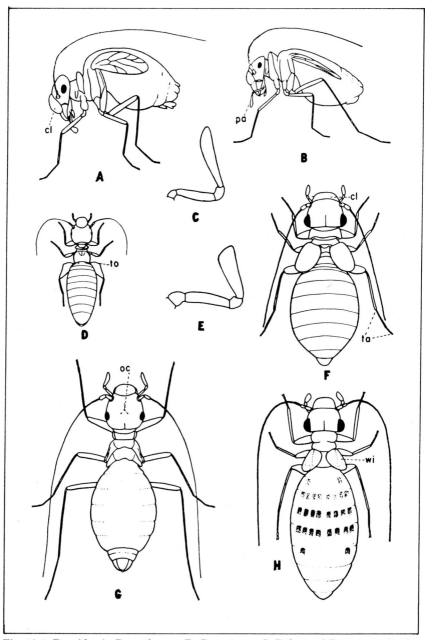

Fig. 10-1. Psocids. A. *Psocathropos.* B. *Dorypteryx.* C. Palpus of *Dorypteryx.* D. *Liposcelis.* E. Palpus of *Psocathropos.* F. *Lepinotus.* G. *Psyllipsocus* (short winged adult). H. *Trogium.* cl. clypeus; oc, ocelli; pa, palpus; ta, tarsus; to femoral tooth; wi, wing.

do not damage plants, a large mass of barklice may cause concern for the home-owner who does not like the accumulation of animated legs and antennae.

Broadhead and Hobby (1944) note the booklouse has been recorded in houses, warehouses, herbaria, insect collections, libraries and in stored foods. Linsley (1942) found this species feeding on pollen in bees' nests in southern California. Counts taken from sticky slides near Pittsburgh, Pennsylvania indicated numerous specimens of the booklouse *L. divinatorius* (Mull.) which were floating in the air in August 1967.

Although psocids may become extremely annoying by crawling over everything in the home and by contaminating foodstuffs by their presence, they usually cause negligible damage to commodities. Back (1939) states damage attributed to these insects should actually be charged against silverfish and cockroaches.

Psocids prefer damp, warm, undisturbed situations. They become most numerous in homes during the spring and summer. Hartnack (1939) reports in Chicago they are most common during the month of September and practically disappear during the winter months. This may be attributed to the cold weather and to the artificial heat in structures that reduces the dampness and fungi upon which they feed. However, Pearman (1928b) states that in England all stages of the booklouse may be found during the winter.

Psocids, for the most part, feed on microscopic molds. Thus, any manufactured material of plant origin, such as furniture, paper, or books, when stored in a damp locality is likely to support a profuse growth of these mildews, which in turn, encourage infestation by psocids. These insects are also known to overrun cereals and materials of a starchy nature; here they feed on the material itself as well as on any mold present. *Liposcelis entomophilus* (End.) preferred to feed on flour, bark and dead insects rather than straw and paper (Strivastava and Sinha, 1975). Thus, this species probably feeds on the starchy paste and glue of books and wallpaper when found in these habitats (Finlayson, 1932).

Hawkins (1939) reports them infesting ground feed, and Finlayson (1932) observed them in large numbers in soft wheat, where they fed on the eggs of *Sitotroga cerealella* (Oliv.).

Normally, psocids live outdoors, and when they enter the home, their small size makes them inconspicuous. Occasionally they may infest just one material and then appear in alarming numbers throughout the house.

There are a number of records where these insects have become extremely abundant in old-fashioned straw mattresses and made life unbearable by creeping over the walls, food, cupboards and throughout the house. Furniture upholstered with tow and Spanish moss is a frequent source of infestation. Back (1939) explains the unusual interest in psocids as "due to heavy losses sustained by property owners resulting, not from the destruction of buildings and furnishings, but from broken leases, lawsuits to recover damages for fancied loss to health and furnishings, and the payment of fees for the services of pest control operators."

Psocids have been pests of prime importance in newly constructed apartment houses, although they also occur in older buildings. Back (1939) states that complaints concerning the psocids commence four to 12 months after the opening of a new house. The conditions initiating such infestations are better understood when one considers the fact that new buildings, because of insulation, hollow walls and wrappings around pipes and electrical fixtures, incorporate moisture which is not readily evaporated. The psocids are readily introduced on building materials, as well as on furniture, boxes, books and paper, all of which may have microscopic molds upon which they feed. Once in the building, they will be found on the walls, in cupboards where they occur in cereals and sugar, behind molding and baseboards, between floors and in the wallspace, behind electrical and plumbing fixtures, as well as in vegetable wrappings on the latter.

In some instances, tenants have attempted to move *en masse,* and were refused transportation by the van companies for fear of infesting the vans. Hartnack (1939) notes infestations are commonly the worst on the top floors of the new buildings since these are the dampest. As a rule, these pests are much less abundant in rooms that are bright and sunny.

BOOKLICE

Liposcelis sp.

The common booklouse is wingless as an adult and is most commonly found under flat objects such as books, flowerpots, boards, loose wallpaper and cardboard boxes.

In the literature, *Liposcelis divinatorius* (Muller) is most commonly referred to as the "booklouse". However, there is some controversy as to the validity of this name and most older references probably were referring to any of several species of *Liposcelis.* However, most of the species in this genus are similar in form and habits.

Life history. Rosewall (1930) studied the booklouse in corn meal. During the several years he had this psocid under observation, he never found a male booklouse since reproduction is parthenogenetic. The eggs are white and oval-shaped and, according to Pearman (1928b), covered with a crusty coating. Booklouse nymphs upon hatching are white, motionless and difficult to find. After each molt, the nymph becomes slightly grayer, the degree of darkness being variable. Ghani and Sweetman (1951) report there are usually four molts and occasionally three. From October to January with temperatures ranging from 50° to 87° F/10° to 30° C, the preoviposition period averages 45 days; the number of eggs deposited averages 20; the incubation period averages 21 days; the life cycle averages 110 days; and the post oviposition period averages nine days. However, the life history from June to August, at temperatures ranging from 60° to 90° F/15° to 32° C, has an average incubation period of 6.91 days, an average life cycle of 24.39 days and an average deposition of 57 eggs. One female deposited a total of 98 eggs. During cold weather the adults die and the eggs hatch in the spring.

Finlayson (1932) and Candura (1932) also made a detailed study of the life history of this species, and the latter notes there may be six to eight generations a year. Pearman (1928b) notes in England the booklouse breeds irregularly and continuously, and it may be found in all stages during the winter.

THE LARGER PALE TROGIID OR DEATHWATCH

Trogium pulsatorium (L.)

This larger psocid is referred to as the "deathwatch" since it, like an anobiid beetle, produces a ticking sound by banging its abdomen against paper and similar objects. Others have expressed doubt that such small, soft bodied insects are actually capable of making sounds and attribute these noises to the true deathwatch or anobiid beetles, which infest wood overrun by the psocids. Pearman (1928a) made a study of the deathwatch psocid and definitely observed this species to tap with its abdomen. There may be as many as five or six taps a second. These may continue for as long as a minute with the number of taps per second gradually decreasing. The taps are most audible when the insect is on paper, since this is a thin and resilient material. Only the female produces these sounds and they are believed to be mating calls. They are most noticeable from July to October. *Lepinotus inquilinus* (Heyd) also makes a tapping noise.

The deathwatch does not seem to breed continuously, but has one generation per year (Pearman, 1928a). In England, this pest overwinters in the nymphal stage. Finlayson (1949) mentioned the deathwatch preferred dark areas in storage buildings such as under sacking. In captivity, the eggs are laid in crevices in food and debris. There are five nymphal instars which can be distinguished by head and body size, the size of the wing rudiments, and the number of antennal segments which increase with each molt.

OTHER PSOCIDS

There are many other species of psocids which can be found in the home or around stored foods, especially under humid conditions. However, several groups of semitropical psocids spin webbing over the surface of tree trunks or outside house walls. Psocids in the genus *Archipsocus* commonly spin heavy webs and have been considered pests in Gulf Coast states from Florida to Texas (Mockford, 1957). Species in this genus may form webs up to one foot/30cm in length. They have been found on tree trunks, the leaves of some plants and the walls of brick and wooden buildings. These psocids seem to make webs where lichens grow.

Nymphs of *Cerastipsocus* have an interesting habit of living in dense herds on tree trunks, stones, or occasionally building walls, where their food may occur (Mockford, 1957). Though these herds of animated legs and antennae may alarm a homeowner, they tend to disperse after reaching adulthood.

PSOCID CONTROL

Booklice and the deathwatch are among the most difficult household pests to cope with for there is no simple way they can be prevented from entering a house.

A number of researchers contend the house-invading psocids feed primarily on microscopic molds and the reason for the presence of psocids in damp places is due to the prevalence of the molds in these areas. The utilization of methods to reduce dampness, or the use of fungicides, both serve to kill molds that the psocids consume. Anon. (1904) recommends the use of two percent formalin in a household oil base spray to kill these molds. An infestation of psocids in a newly built home was controlled with a spray of one percent formalin and one percent lindane in an oil base (Mallis, 1969).

Psocids also have been found to be very sensitive to changes in relative humidities. If the relative humidity is reduced below the "critical equilibrium" of the psocid, the insects desiccate and eventually die. Finlayson (1932) demonstrated that some *Liposcelis* would lose up to 50 percent of their water in 11 days when subjected to 77° F/25° C and 33 percent relative humidity. However, these insects recovered their water loss in six to seven hours when transferred to a 58 percent relative humidity environment. Knülle and Spadafora (1969)

noted *L. knullei* died within one week at all humidities below their critical level
and *L. bostrychophilis* survived only ten days. Thus, reduction of household
humidities below 50 percent relative humidity should help to control psocids.
Drying agents or dessicants, such as silica aerogel, should also be effective control materials.

When psocid infestations are localized, direct methods of control can be taken.
Infested cereals or stored foods can be frozen or thrown away. Furniture up-
holstering having tow or Spanish moss can be exposed to drying in sunlight or
sent away for vacuum or vault fumigation.

In the past, Mallis (1969) mentions a pest control operator who successfully
used a two percent chlordane oil base spray to control psocids in a six month-
old house, which was built with green lumber. Jensen and Holdaway (1946)
obtained control of booklice on rabbit hides by spraying the hides with three to
five percent DDT in kerosene. Back (1946) pointed out that fumigation with
hydrocyanic acid gas has often failed. He recommends the use of one pound/0.45
kg of flake naphthalene or paradichlorobenzene for each 100 cubic feet/2.8 cubic
meters to kill psocids in enclosed areas such as closets. Archer (1952) controlled
herbarium pests with a pyrethrin and DDT aerosol.

At present, only pyrethrins are registered for psocid control. However, many
of the longer lasting household materials, such as chlorpyrifos, diazinon, permethrin, propetamphos and propoxur, should be effective if they are labeled for
application to the sites infested by psocids.

LITERATURE

ANONYMOUS — 1904. Psocids, book lice, dust lice, etc. British Museum (Natural History) Econ. Leaflet No. 4
ARCHER, W. A. — 1952. Aerosol for controlling herbarium pests. Science
116(3009): 233-234.
BACK, E. A. — 1939. Psocids in dwellings. J. Econ. Entomol. 32: 419-423.
1946. Psocids in dwellings. Pests 14(11): 26-27.
BROADHEAD, E. and B. M. HOBBY — 1944. Studies on a species of *Liposcelis*
occuring in stored products in Britain. Entomol. Monthly Mag. Part I, 80:
45-59; Part II, 80: 163-173.
CANDURA, G. S. — 1932. Contributo alla conoscenza biologica del *Troctes di-
vinatorius* (Müller). Bollettino di Zoologia. 3: 177-184.
FINLAYSON, L. H. — 1949. The life history and anatomy of *Lepinotus patruelis*
(Psocoptera: Atropidae). Proc. Zool. Soc. London 119: 301-323.
FINLAYSON, L. R. — 1932. Some notes on the biology and life history of pso-
cids. Entomol. Soc. Ontario Annual Rept. 63: 56-58.
GHANI, M. A. and H. L. SWEETMAN — 1951. Ecological studies of the book
louse, *Liposcelis divinatorius* (Müll.). Ecology 32: 230-244.
GURNEY, A. B. — 1950. Corrodentia. In: Pest Control Technology. N.P.C.A.,
Inc. Elizabeth, N.J. pp. 129-163.
HARTNACK, H. — 1939. 202 Common Household Pests in North America.
Hartnack Publishing Co., Chicago. 319 pp.
HAWKINS, J. — 1939. Corrodentia as pest of ground feed. J. Econ. Entomol.
32: 467.

JENSEN, D. D. and F. G. HOLDAWAY — 1946. DDT for control of a book louse. J. Econ. Entomol. 39(2): 274.

KNULLE, W. and R. R. SPADAFORA — 1969. Water vapor sorption and humidity relationships in *Liposcelis* (Insecta: Psocoptera). J. Stored Prod. Res. 5: 49-55.

LINSLEY, E. G. — 1942. Insect food caches as reservoirs and original sources of some stored product pests. J. Econ. Entomol. 35(3): 434-439.

MALLIS, A. — 1969. Handbook of Pest Control. (5th Edition) MacNair Dorland, N.Y. 1158 pp.

MOCKFORD, E. L. — 1957. Life history studies on some Florida insects of the genus *Archipsocus* (Psocoptera). Bull. Florida State Mus., Bio. Sci. 1(5): 253-274.

PEARMAN, J. V. — 1928a. On sound production in the Psocoptera and on a presumed stridulatory organ. Entomol. Monthly Mag. 64: 179-186.

1928b. Biological observation on British Psocoptera. Entomol. Monthly Mag. 64: 209-218, 239-243, 263-268.

ROSEWALL, O. W. — 1930. The biology of the book-louse, *Troctes divinatoris* Müll. Ann. Entomol. Soc. Amer. 23: 192-194.

SRIVASTAVA, D. C. and T. B. SINHA — 1975. Food preference of a common psocid *Liposcelis entomophilus* (End.) Psocoptera: Liposcelidae. Indian J. Ecol. 2(1): 102-104.

ALFRED G. WHEELER, JR.

Alfred G. Wheeler, Jr., a native of Nebraska, received his bachelor's degree in biology from Grinnell College in 1966 and his Ph.D. in entomology from Cornell University in 1971.

Since June 1971, he has been an entomologist with the Bureau of Plant Industry, in Harrisburg, Pa., with primary responsibility for insect survey and detection. He also has held an adjunct appointment with the Department of Entomology, The Pennsylvania State University, since 1973, and currently is an adjunct associate professor.

Wheeler's special interests are biology of the Hemiptera-Heteroptera, especially Berytidae and Miridae; life histories of insects associated with ornamentals; and history of entomology. His more than 80 publications deal mainly with insect life history and distribution. With a thorough understanding of biological aspects and the principles of control, he is particularly well qualified to revise the chapter on "bugs" in the 6th Edition of the Handbook of Pest Control.

Wheeler has been elected to membership in Phi Kappa Phi and Sigma Xi and is a member of the Coleopterists Society, Entomological Society of America, Entomological Society of Pennsylvania, Entomological Society of Washington, and Pennsylvania Academy of Science. He serves as editor of the history issue of the Melsheimer Entomological Series and of Regulatory Horticulture, a grower's magazine published biannually by the Pennsylvania Department of Agriculture.

CHAPTER ELEVEN

Bed Bugs and Other Bugs

Revised by A. G. Wheeler, Jr.[1]

Most democratic is the bedbug chappie
Who prefers red blood to blue
And nothing makes him quite so happy
As sharing — your blood with you!
—A.M.

To THE LAYMAN, a bug is any insect, regardless of whether it is a fly, beetle or wasp. But an entomologist would usually reserve the word "bug" for "true bugs" of the Hemiptera-Heteroptera. Thus, he may refer to plant bugs, squash bugs or stink bugs. True bugs make up a diverse group of insects and, as Slater and Baranowski (1978) pointed out, it is difficult to find one set of characteristics that will fit all species. They note that in most text books the Hemiptera are generally said to have a segmented beak (proboscis) that arises from the front of the head and extends backward underneath the head. They are also characterized by front wings that lie over the abdomen. The wings at their tips (apical portion) overlap one another and are more membranous than the basal area. Literally, the name Hemiptera means "half wings," which refers to the different *textures* (thin and membranous versus more thickened or leathery) rather than to an insect having only half a wing.

The word bug apparently was first applied to the familiar bed bug in 17th Century England. Although the origin of the word is in doubt, it may have been derived from the old English "bogy" or "hobgoblin," signifying a "terror in the dark." Usinger (1966) suggests another possibility: derivation from the Arabic "buk," an "old and widespread name for *Cimex lectularius* throughout the Arab world" that "could have been picked up by travelers and brought back to England."

[1]*Entomologist, Bureau of Plant Industry, Pennsylvania Department of Agriculture, Harrisburg, Pa.*

Regardless of the actual derivation, it seems appropriate to begin the chapter on true bugs with the first "bugs" — the bed bugs. After all, they have been associated with man since the beginning of civilization. Although bed bugs are less important than they were before the advent of DDT, these pests provided targets for some of the first professional exterminators. The successful control of these bloodsucking insects can be regarded as one of the hallmarks of the pest control industry.

FAMILY CIMICIDAE

The human bed bugs and their relatives form a rather small group (70+ species) of bloodsucking ectoparasites. Hosts, in addition to man and his domesticated animals, include bats and birds. These are strongly flattened bugs, often reddish brown, with forewings reduced to mere pads. The hind wings are absent. Although there are three cimicid species that qualify as human bed bugs, the Entomological Society of America recognizes "bed bug" as the official common name for only a single species.

THE BED BUG
Cimex lectularius (L.)

History. This pest has plagued man since the dawn of civilization. It is believed the bed bug originally was associated with bats living in tree holes and in caves of the Middle East. It was in caves that these parasites became associated with man. According to Usinger (1966), "man's use of fire definitely made him a more suitable host for the bed bug, especially in the temperate regions in winter." As man gradually moved from cave to village, the bugs followed and became permanent associates. As an alternative hypothesis, Usinger suggests the bugs may have moved onto man when bats roosted in his houses. Busvine (1976) notes as man was leaving caves for villages, the climate of the Middle East was becoming drier, so "forest-living bats must have begun to vanish. But the bed bug was pre-adapted to take advantage of the new habitats provided by human dwellings. A few colonies must have remained associated with bats in caves and it is claimed that some may even yet persist."

The bed bug has a rich tradition in man's recorded history, having "contributed not only to his misery, but also to his folklore, pharmacopoeia and literature" (Usinger, 1966). The Greeks were well acquainted with these pests. Usinger quotes from *The Clouds* by Aristophanes in 423 B.C.: "Socrates: Where is Strepsiades? Come forth with your couch. Strepsiades (from within): The bugs do not permit me to bring it forth."

Likewise, the Romans were aware of these parasites and used them in their primitive medicines, possibly because the bed bug's "peculiar smell made it especially revolting" (Busvine, 1976). Remedies using bed bugs, according to Busvine, can be traced to an army surgeon employed during the reign of Nero (54-68 A.D.). Writing at about this same period, Busvine cites Pliny: "The bed bug, a most foul creature and nauseating even to speak of, said to be effective against the bites of serpents, especially asps, as also against all poisons." Pliny believed these bugs also could be used to loosen the hold of leeches. It should be noted the bed bug remained a part of folklore medicine well into the 20th Century in parts of Europe and North America. Riley and Johanssen (1938) state the 5th (1896) Edition of the *American Homeopathic Pharmacopoeia* included directions for preparing a tincture of *Cimex* to be used as a malaria remedy.

The German worker Kemper, as cited by Busvine, traced the spread of the bed bug to northern Europe. This pest was first recorded from Germany in the 11th Century, from France in the 13th Century and from England in the 16th Century. Busvine says the first English record, cited as 1503, more likely was 1583.

Marlatt (1896) believes the bed bug came to North America with the early colonists, noting that Kalm, in his writings of 1748-49, stated the bed bug was plentiful in the English colonies and in Canada, though unknown among the Indians. In the New World this now cosmopolitan pest followed man on his excursions by ship and rail. Infestations were especially obnoxious on steamboats that plied rivers of the Mississippi Valley (Walsh and Riley, 1869). As noted by Marlatt, the "old-fashioned heavy wooden bedsteads" favored the breeding and concealment of this insect. Snodgrass (1944) remarks that although the bed bugs had been around long before beds were invented, "it might as well be said that beds were made to accommodate the bugs." In different parts of the United States bed bugs, according to Marlatt, were given different names: "chintzes" or "chinches" (Boston), "mahogany flats" (Baltimore), "red coats" (New York) and "crimson ramblers" (western states).

Occurrence and detection. Although this democratic creature draws no line between the impoverished or the wealthy, its presence is more evident in poorer quarters owing to conditions more favorable for its survival. The early writers (16th Century and earlier) realized these bugs were not as abundant in beds of the rich in which the linen and straw were changed regularly. Busvine (1976) notes, however, these observers "were right for the wrong reason," because they believed the bugs arose spontaneously from "dirt and sweat."

As Marlatt points out, although many considered bed bugs to inhabit only "houses of the meaner sort" and the "careful housekeeper would feel it a signal disgrace to have her chambers invaded by this insect," these pests can enter even the most immaculately kept homes. He described the capacity of bed bugs

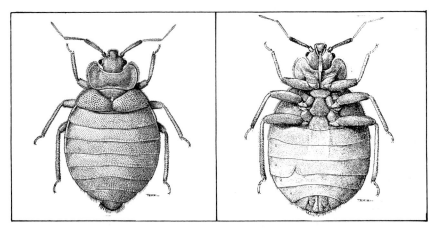

Fig. 11-1. (Left) Dorsal view of male bedbug. The abdomen of the adult male is not as broad as that of the adult female. (Right) Ventral view of female bedbug. Note the sharp pointed beak used for piercing the skin and sucking the blood.

for infesting new quarters: "Escaping through windows, they pass along walls, water pipes, or gutters, and thus gain entrance to adjoining houses." The bed bug is distributed readily in laundry and on clothes and baggage of individuals who have visited infested premises. This pest is disseminated primarily from one house or apartment to another by stowing away in furniture and bedding that is being moved, or by attaching itself to articles placed in an infested moving van. Second hand furniture, old books and lumber salvaged from demolished houses offer other means of ingress.

Once in the home, bed bugs become established in any convenient crack or crevice, particularly along the seams or in the buttons of mattresses, in the coils of the bedspring, wooden bedsteads, upholstered furniture, the backing of pictures, behind wallpaper and calendars, behind skirting boards and between floor boards. Theatre seats often are found infested. In some countries, including Rumania, homeland of Count Dracula, bed bugs infest airplane seats and are thus a 20th Century form of flying bloodsucker.

Bed bugs are exceedingly wary and cautious, and their hiding places tax the patience and ingenuity of man. Thus, there are many records of persons who have prepared themselves for a night of undisturbed sleep by placing the legs of the bed in crocks of kerosene, only to find bed bugs dropping down on them from the ceiling. This behavior, according to Usinger, has been known since the time of the European entomologist Latreille in 1829, but it should not be considered purposeful. He notes, however, bugs often will drop when alarmed. Usinger also states the means by which bed bugs locate their host is a controversial topic. Certain researchers have concluded a host is found strictly through random wanderings, whereas others have suggested the bugs are attracted to odor or to heat.

Bed bug-infested homes often can be recognized by their distinctive "buggy" smell, an odor that has been described as "obnoxious sweetness" (Usinger, 1966). R. C. Hill, who raised bed bugs in the laboratory, describes the odor of bed begs as that of fresh red raspberries. Busvine (1976) feels this characteristic smell may result as much from generally unhygienic conditions as from the bugs themselves. The odor, attributed by the ancients to the notion these bugs lacked vents, is produced by scent glands, which in the adult are situated between the coxae of the hind legs. These glands secrete a clear, oily volatile liquid.

Infestations also can be detected by the blood stains on walls or linen and by the characteristic spots of excrement. In regard to the repulsiveness of bed bug infestations, I cannot resist quoting Busvine (1976): "In addition to their generally disgusting appearance, bed bugs have two other unpleasant characteristics: they make a mess and they stink. The mess is due to their frequent excretion, which causes brownish, yellowish or black spots on the wall near crevices where they hide. The dark marks are due to the presence of partly digested blood in their feces excreted to make way for a fresh meal."

External appearance. The oval body of the bed bug is flattened from top to bottom, allowing ready access to narrow crevices. The adult is $^1/_6$ to $^1/_5$ inch/four to five mm long, reddish brown and becomes blood-colored when fully fed or replete. When engorged, its abdominal segments are stretched, which give the abdomen a somewhat elongated appearance. Engorgement results in such changes in size, shape and color that it is possible to think bugs in various stages of distention might represent different species. The membranous areas of the adult abdomen, which allow for enormous expansion during feeding, have been called "hunger folds." Nymphs, after feeding, look like animated drops of blood.

The head bears four-segmented antennae and beak-like mouthparts consisting of two pairs of stylets which pierce the skin of the host. "The outer pair of stylets corresponds to the mandibles in other insects and is provided with barbs which exert a sawing action; the inner pair, which correspond to the maxillae, forms two tubes, one for sucking up the blood and a second which functions as a channel for the injection of the saliva into the wound" (Wright, 1944). The bed bug, because of its parasitic mode of life, has lost the power of flight and its wings have degenerated into simple, functionless, oval structures. The adult female can be separated from the adult male by having the end of the abdomen broadly rounded. The female also has a cleft on the right side of the posterior margin of the fourth apparent abdominal segment, indicating the opening of the pouch which receives sperm during copulation. It should be noted insemination in the Cimicidae differs from that in other insects. During copulation, the male punctures the body wall of the female to inject sperm into her abdomen. This unusual manner of insemination, often referred to as "traumatic" or "extragenital," has received intensive study by European and American workers (Usinger, 1966).

The bite and host reaction. The German worker A. Hase, as quoted by Usinger (1966), gives a good description of biting behavior. "The bug secures itself with its claws on the skin, with the forelegs reaching quite far forward, in order to have leverage when introducing the stylets . . . First the beak is touched vertically to the surface and the skin is tested repeatedly with its tip . . . The antennae are no longer pointed forward, but rather backward on a line level with the eyes. At this point — while the insect makes rather energetic pushing movements with the head and the entire body may be brought into sway with the abdomen moving up and down — the introduction of the stylets begins." Girault (1905) found first-instar nymphs became engorged after only three minutes, whereas adults required 10 to 15 minutes for engorgement. Busvine (1976) says an egg-laying female has the greatest demand for blood, about eight mg at a full meal. According to Busvine, the bed bug's food requirements do not normally have an adverse effect on host tissues.

As a well-adapted parasite, the bed bug inflicts a painless bite. Some persons are quite sensitive to the bites; others are not. Individuals who are not affected by the bites have been known to become carriers of bed bugs and have infested new surroundings simply because they were unaware that bugs were present in their belongings. Many people, however, become immediately aware of the bug's presence because of severe itching and the large inflamed spot that is produced. Usinger notes the eyes of a colleague were once swollen shut because of a bed bug encounter. The typical swelling, sometimes called a wheal, is the result of an allergic reaction. It is doubtful if the distinction between the bite of a bed bug and that of a flea lessens the annoyance, but it is at least of value to the pest control operator. Generally, the lump or swelling contains no red area in the center; fleas produce bites that leave a red central area surrounded by a reddish halo (Patton, 1931).

Life history. Numerous workers have dealt with various aspects of the biology of *Cimex lectularius*. For a summary of these studies, the reader should consult the excellent monograph of Usinger (1966). The brief sketch of life history that follows should serve the needs of most pest control specialists.

Eggs of the bed bug are about 1/25 inch/one mm long, white, elongate and slightly curved. There is a rim at the anterior end, which the insect pushes off upon emerging. A quick drying cement, which is exuded when the eggs are laid, fastens them in cracks and crevices, on rough surfaces, behind woodwork or nearly anywhere the adults conceal themselves.

During warm weather or in heated buildings, eggs may hatch in six to 10 days, but the incubation period is lengthened considerably under cooler conditions. Johnson (1940) found eggs did not hatch below 55.4° F/13° C or above 98.6° F/37° C. Humidity did not significantly affect the duration of the egg stage.

The nymph, immediately after forcing off the rim of the egg, is white to straw colored. Upon feeding, the nymph turns red or purple from the engorged blood; older nymphs take on a darker color. The bed bug has five nymphal stages with a nymphal period lasting 35 to 48 days under ideal conditions. Girault (1912) reported poorly fed nymphs might require as long as 156 days to complete their development. If the bed bug cannot obtain food in any of its nymphal stages, it may remain unchanged for an indefinite period. Ordinarily, only one meal is taken between molts. Girault (1912) stated three or four generations are produced annually.

According to Cragg (1923), both sexes of *C. lectularius* are fully mature soon after the final molt, and mating may occur before either sex has fed. Girault (1914) noted in mating the male climbs on the female's back or they mate end to end. Few eggs are laid until a female has had a blood meal. A male is capable of fertilizing several females in 24 hours. Females deposit an average of one to five eggs a day during a period of two months until some 200 eggs are laid. Maximum egg production occurs between 70° and 82° F/21° and 28° C.

Adults are able to survive long periods in unoccupied houses. In the absence of man, feeding can take place on a great variety of hosts: poultry, canaries, English sparrows, mice, rats and guinea pigs. Thus, the survival of bed bugs in long vacant houses may be explained in part by their resistance to starvation and because other animals can substitute for man, their preferred host. According to Girault, adults that are well fed may live six to seven months, or even up to 1½ years.

It should be emphasized that bed bugs do not obtain nourishment from moistened wood or from dust, as stated in some early 20th century text books of parasitology. Patton (1931) refutes the notion that bed bugs can subsist on food other than blood: "A reference to the structure of the mouth parts of the bed bugs will clearly show that it is *impossible* for it to ingest *solid* particles, or to obtain moisture from wood. No food, other than blood, has ever been found in the alimentary tract of many hundreds of bugs the writer has dissected."

The bed bug and disease. Investigators and medical men have tried for a long time to attach the stigma of "disease-conveyor" upon the bed bug. According to Girault (1906), this pest was implicated in the transmission of human diseases as early as 1887. In the laboratory, as Wright (1944) pointed out, the bed bug can be infected experimentally with the causative agents of numerous diseases. He was forced to conclude, however, that in spite of all evidence offered, "it seems probable that the bed bug plays only an insignificant part in the carriage of disease to man." After reviewing all experiments on disease transmission, Burton (1963) also stated natural transmission had not been proven.

Disregarding their possible role as vectors of disease organisms, bed bugs are capable of inducing nervous and digestive disorders in sensitive people. McKenny-Hughes (1937) observed, "In infested areas it is often possible to pick out children from buggy homes by their pasty faces, listless appearance and general lack of vigor. It can be argued that the house in which bugs are tolerated also will be the home of malnutrition, dirt and other causes of physical inferiority. Such causes cannot be solely responsible, and sleepless nights with constant irritation due to injection of the minute doses of bed bug saliva into the blood are likely to contribute to the ill-health of children and even adults." It should be mentioned the mere presence (or even suspected presence) can have severe psychological effects, bringing about loss of sleep and much apprehension. Busvine (1976) believes man's natural revulsion toward bed bugs and other crawling creatures is a "cultural tradition rather than an instinct." This feeling of revulsion is an outgrowth of a society preoccupied with an "abhorrence of refuse or excrement" and a desire for cleanliness. As will be discussed later, a pest control specialist must always bear in mind the psychological aspects of a bed bug infestation.

Enemies of the bed bug. That certain insects would prey on bed bugs was known to Linnaeus in 1758. The habits of the insect Linnaeus observed, *Reduvius personatus,* are discussed later in this chapter under the family Reduviidae.

The entomologist T. Pergande, who was a Union soldier in the Civil War, related to Marlatt (1896)that the "little red ant," or Pharaoh ant, *Monomorium pharaonis* (L.), invaded bed bug-infested quarters at Meridian, Mississippi. Pergande witnessed the dismembering of the bugs as part of an attack that in a single day eliminated this pest so only the odor remained to remind him of his late inhabitants. The Argentine ant,*Iridomyrmex humilis* (Mayr), has likewise given short-shrift to the bed bug wherever it has encountered this insect.

Usinger (1966) reports in other countries spiders, pseudoscorpions, mites and reduviids will feed on bed bugs. The American cockroach, *Periplaneta americana* (L.), has received attention as another possible natural enemy of *Cimex lectularius.* Johnson and Mellanby (1939) tested this cockroach and reported it exercised no real control over bed bug populations. More recently, Mallis (1971) reported on tests conducted at the Gulf Research Laboratory at Harmarville, Pennsylvania, near Pittsburgh. He concluded the "American cockroach will eat bed bugs when it has no other food available. This is not surprising in view of the omnivorous appetite of the American cockroach." I should add that Mallis (1981) has described the "inner works" of the insecticide laboratory at Gulf in an entertaining and informative article.

RELATED SPECIES ATTACKING MAN

In North America there is one additional human bed bug and a number of species associated with bats and birds that are occasional household pests. In recent years, a species of bat bug seems to be encountered as often (or even more so) than the bed bug in houses of the eastern states.

Cimex hemipterus (F.). The so-called tropical human bed bug is mainly a parasite of man, but also has been found in association with bats. This species is widespread throughout the warmer areas of Africa, Asia and the American tropics. The only definite record from the United States is that of Hixson (1943) who reported *C hemipterus* from five localities in Florida. This cimicid closely resembles *C. lectularius* but lacks the wing-like pronotal expansions of the latter. Perhaps the most diagnostic character is that *C. hemipterus* has the pronotum less than $2\frac{1}{2}$ times as wide as long at the middle. In *C. lectularius*, the pronotal width is greater than $2\frac{1}{2}$ times the median length.

Cimex pilosellus (Horvath). This western bat bug is known from British Columbia south to California and from Idaho, Montana and Nevada (Usinger, 1966). *C. pilosellus* may be distinguished superficially from the bed bug, *C. lectularius,* by the presence of long hairs on its body. A more technical character is that the hind femur of *pilosellus* is less than 2.6 times as long as its greatest width; in the bed bug the hind femora are usually more than 2.6 times as long as wide.

Spencer (1935) notes in British Columbia he found this bug mainly in roosting places of bats rather than on the animals themselves. Although *C. pilosellus* is not common in human habitations, Spencer reported an outbreak in a hotel: "Another record shows a very well known summer hotel in Dry Belt whose log construction afforded splendid hiding for bats. Up to the time of our visit in July, no less than 72 bats had been destroyed because they harbored the bugs which swarmed into the neighboring rooms through cracks in the plaster, especially in one of the bathrooms. Although human beings had not actually been bitten by the bugs, the guests seemed to resent their presence and the management was much concerned over the situation." Many of the reports of *C. pilosellus* from the East, e.g., Stearns (1937), probably should refer to the next species to be discussed, *Cimex adjunctus.*

Cimex adjunctus Barber. This eastern bat bug, closely related to the western *C. pilosellus,* occurs throughout the East and Midwest. It is known to range as far west as Colorado. The same characters discussed as helping separate *C. pilosellus* from *C. lectularius* apply to the eastern bat bug. Smith and Chao (1956) gave a table listing characters a pest control operator can use to distinguish the human bed bug from *C. adjunctus.* They point out the hairs fringing the body of *lectularius* are mostly shorter than the width of an eye, whereas in *adjunctus* the hairs are longer than an eye width.

C. adjunctus uses bats as its primary host and may build up large populations in bat-infested attics. In Delaware, Rice (1936) described this bug as overrunning an attic, biting family members and causing a "severe annoyance." Back (1940) reported "hordes" from an attic in Ohio where the bugs were migrating to the living rooms.

Oeciacus vicarius Horvath. Some of the early North American references to *C. pipistrelli* (Jenryns), a European bed bug, may refer to a cimicid whose primary hosts are cliff swallows. The swallow bug, *O. vicarius,* also has been recorded from nests of the barn swallow. This bug is known to occur throughout much of the country but, according to Slater and Baranowski (1978), is relatively uncommon in the Northeast. The pest control specialist may distinguish this species from the human bed bugs and bat bugs by the long, pale hair-like bristles that clothe the body and the smaller size. Technical characteristics for separating swallow bugs from other cimicids may be found in Usinger (1966) or Slater and Baranowski (1978).

Myers (1928) studied the life history and Spencer (1930) prepared a note on *O. vicarius* as a pest in homes. "The insects breed freely all summer in the birds' nests and in early autumn when migration occurs, the bugs scatter from the birds' nesting areas and invade dwellings where they may attack human beings for a few days: the infestation then dies down. In certain instances, apparently of very heavy infestation, invasion of dwellings from nesting areas may occur some weeks before the birds leave the nests." A correspondent from Alberta, Canada, addressed a letter to Spencer, of which the following reveals the importance of this pest in infested homes: "About the end of July we discovered these insects in the house and on knocking down the swallows nests (about 200 of them) we found they were simply swarming with them. After this they came into the house in thousands and bit us so much we had to move into tents to sleep. They bite just as badly in the day only of course there isn't the same chance of them getting on one. They seem to be practically all gone now although there are still odd ones."

Mail (1940) notes an infestation of this insect in a high school in Alberta, British Columbia, and Mills and Pletsch (1941) reported this pest from a Montana school building. Mallis, in earlier editions of this handbook, noted the presence of *O. vicarius* in a Los Angeles home where pyrethrum fly spray brought the infestation under control. Eads et al. (1980) noted an increased interest in this pest following the recent laboratory isolation of two viruses from bugs collected in cliff swallow nests. One of the viruses was related to western equine encephalitis and the other to the Venezuelan equine complex. These authors note even though swallow bugs have not been established as transmitting disease organisms to humans, they will bite man and have caused a recent nuisance in both rural and urban localities in Colorado and Wyoming.

Cimexopsis nyctalis List. Doner (1945) discussed another external parasite of birds, a chimney swift associate, that is a rare pest in homes. This small, elliptical species is about ⅛ inch/three mm long and occurs from Maine south to Florida and west to Nebraska and Arkansas (Slater and Baranowski 1978). Smith and Chao (1956) describe the characters useful in making a "sight" identification of this bug: hind coxae nearly touching; body smooth, lacking hairs or bristles; and last two antennal segments thin, segments two and three about equal in length, four a little shorter. Nettles (1961) reported at South Carolina's Clemson University he had encountered infestations of a chimney swift bug in several houses on campus; an infant was bitten by this cimicid.

Haematosiphon inodorus (Dugés). The poultry bug, recognized by its long rostrum, or proboscis, and its small size, is the only cimicid known to have only four rather than five nymphal stages. Also referred to as the "Mexican chicken bug," or in New Mexico as the "coruco," this southwestern and California bug is a parasite of various birds, among them eagles, owls, the California condor and chickens. Lee (1955) states *H. inodorus*

Fig. 11-2. The American barn-swallow bug, *Oeciacus vicarius* Horv., will often attack human beings in the home.

"shows no reluctance to feed on human subjects" and cited Townsend in 1893 as saying these bugs once spread from roosts to dwelling-houses where they proved more difficult to control than the bed bug. Hall and Wehr (1949) note the poultry bug lives "in the nests or about roosting places, hiding in cracks during the day and coming out at night to suck the blood of the fowls. They especially annoy sitting hens, sometimes causing them to desert their nests." The poultry bug may invade homes and become an even more serious pest on man than on birds. According to Essig (1926), the Indians and Mexicans have been so plagued

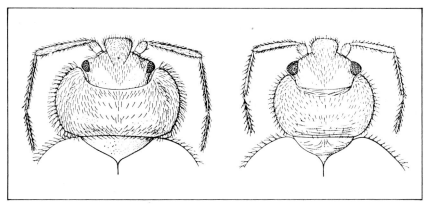

Fig. 11-3. Head and prothorax of two bedbugs. (Left) *Cimex lectularius* L. (Right) *Cimex hemipterus* (F.).

by this pest that they have been forced to burn and abandon their homes. Town-send (1894) described one such serious outbreak: "They have been known . . . to swarm in military posts in southern New Mexico to such an extent that the soldiers were ordered out and formed in two lines, one line with brooms to sweep the corucos en masse up against an adobe wall, where the other line stood ready with trowels and mud and plastered them into the wall alive — a novel but effective means of riddance!"

KEY TO THE COMMON BED BUGS AND RELATED SPECIES

1) Bases (coxae) of middle and hind legs nearly touching; body lacking hairs . . chimney swift bug, *Cimexopsis nyctalis*/Bases of middle and hind legs widely separated; body distinctly hairy 2

2) Beak or proboscis reaching coxae of hind legs . . . poultry bug, *Haemato-siphon inodorus*/Beak or proboscis not reaching beyond front coxae 3

3) Front margin of pronotum shallowly concave, not enveloping head; 3rd and 4th antennal segments about equal in length; body covered with long pale hairs . . . swallow bug, *Oeciacus vicarius*/Front margin of pronotum deeply con-cave to receive head; 3rd antennal segment distinctly longer than 4th segment; body covered with short or long, distinctly golden, hairs 4

4) Upper surface of body thickly covered with long hairs ($1\frac{1}{2}$ or more times the diameter of 2nd antennal segments) . . . bat bugs, *Cimex adjunctus* and *Cimex pilosellus*/Upper surface of body sparsely covered with short hairs (usu-ally less than the diameter of 2nd antennal segment) 5

5) Pronotum broad with winglike expansions extending laterally well be-yond eyes bed bug, *Cimex lectularius*/Pronotum much more narrow, sides not greatly expanded beyond eyes. . . . tropical bed bug, *Cimex hemipterus.*

CONTROL OF BED BUGS

Early attempts at controlling bed bugs can be traced at least as far back as the Greek philosopher Democritus in the late 5th to early 4th Century B.C. According to Cowan (1865), Democritus recommended hanging "the feet of a hare or of a stag at the foot of the bed." Cowan also notes the ancients resorted

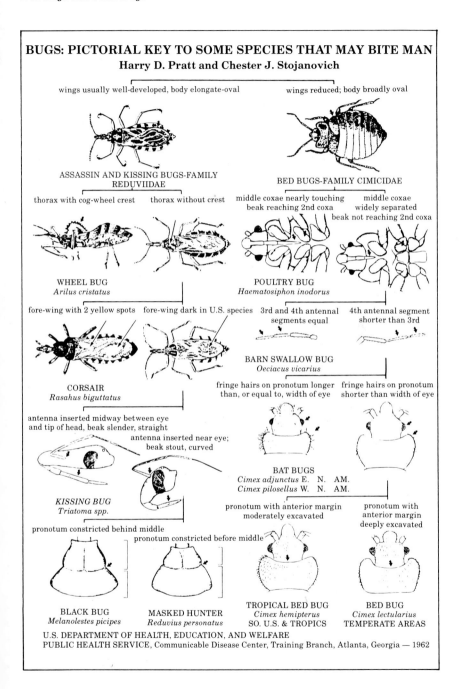

BUGS: PICTORIAL KEY TO SOME SPECIES THAT MAY BITE MAN
Harry D. Pratt and Chester J. Stojanovich

wings usually well-developed, body elongate-oval

wings reduced; body broadly oval

ASSASSIN AND KISSING BUGS-FAMILY REDUVIIDAE

BED BUGS-FAMILY CIMICIDAE

thorax with cog-wheel crest

thorax without crest

middle coxae nearly touching beak reaching 2nd coxa

middle coxae widely separated beak not reaching 2nd coxa

WHEEL BUG
Arilus cristatus

POULTRY BUG
Haematosiphon inodorus

fore-wing with 2 yellow spots

fore-wing dark in U.S. species

3rd and 4th antennal segments equal

4th antennal segment shorter than 3rd

BARN SWALLOW BUG
Oeciacus vicarius

CORSAIR
Rasahus biguttatus

fringe hairs on pronotum longer than, or equal to, width of eye

fringe hairs on pronotum shorter than width of eye

antenna inserted midway between eye and tip of head, beak slender, straight

antenna inserted near eye; beak stout, curved

BAT BUGS
Cimex adjunctus E. N. AM.
Cimex pilosellus W. N. AM.

KISSING BUG
Triatoma spp.

pronotum with anterior margin moderately excavated

pronotum with anterior margin deeply excavated

pronotum constricted behind middle

pronotum constricted before middle

BLACK BUG
Melanolestes picipes

MASKED HUNTER
Reduvius personatus

TROPICAL BED BUG
Cimex hemipterus
SO. U.S. & TROPICS

BED BUG
Cimex lectularius
TEMPERATE AREAS

U.S. DEPARTMENT OF HEALTH, EDUCATION, AND WELFARE
PUBLIC HEALTH SERVICE, Communicable Disease Center, Training Branch, Atlanta, Georgia — 1962

Fig. 11-4. Pictorial key to common species that may attack man.

to "train-oil, linseed and hempseed, crushed up all together, and the bugs were to eat it till they burst." The modern pest control operator does not have to wait for mass suicide among bed bug populations, but can hasten the process through the use of several chemicals.

It should be remembered effective insecticides did not become available for hundreds of years. Professional exterminators in 18th Century England had to rely on their own secret concoctions. In the United States, even as recently as the mid- to late 19th Century, entomologists were using methods that by today's standards seem crude. Marlatt (1896) described a "fire and brimstone" technique advocated by J. A. Lintner, state entomologist of New York: "Place in the center of the room a dish containing about four ounces of brimstone, within a larger vessel, so that the possible overflowing of the burning mass may not injure the carpet or set fire to the floor. After removing from the room all such metallic sufaces as might be affected by the fumes, close every aperture, even the keyholes, and set fire to the brimstone. When four or five hours have elapsed, the room may be entered and the windows opened for a thorough airing."

As noted by Marlatt (1896), bed bug control was made somewhat easier by the late 19th Century when iron and brass rather than heavy, wooden bedsteads came into frequent use. With benzene or kerosene applied liberally to infested beds, eradication of the bugs was "comparatively easy." The widespread use of central heating helped return the bed bug to prominence, and in the early 20th Century exterminators relied on fumigation with highly poisonous hydrocyanic-acid gas (Knowlton, 1935a). Such fumigation was about the only method available for eradicating infestations, but it had to be carried out by trained operators. In the typical home the commercial poison being used was a "saturated solution of corrosive sublimate in alcohol" (Riley and Johanssen, 1938).

The post-World War II era brought what Matheson (1950) termed "an almost perfect method for control" — DDT. This chlorinated hydrocarbon, widely heralded as a miracle insecticide, largely replaced all other methods of controlling bed bugs. A five percent DDT-oil base spray gave excellent control. Where white spray deposits were unobjectionable, for example in barns and rough buildings, DDT wettable powders and water emulsion were used. According to Mallis (1981), the arrival of DDT revolutionized bed bug control, almost eliminating the insects and the markets for bed bug sprays. "When a customer sprayed a bed, mattress, springs, and the immediate surfaces surrounding an infested bed with DDT-petroleum base spray, he effectively controlled the bed bugs for a year or more." Sailer (1952) presents a fascinating account of the bed bug's "rise and fall."

But it was not long before Johnson and Hill (1948) reported bed bugs had become resistant to DDT, both in the field and in the laboratory. These authors observed several instances in the summer of 1947 where DDT deposits failed to control the bed bug in barracks of the Naval Receiving Station in Pearl Harbor. In the Pittsburgh area, Katz (1957) remarked he had begun to encounter problems with DDT resistance in 1950. Within a few years, a bed bug population in Israel was found to have developed a resistance to lindane (Cwilich et al. 1957). In 1956, the National Pest Control Association began recommending malathion sprays be used against resistant populations.

Despite the development of insecticide resistance in certain populations, bed bugs are no longer regarded as common household pests. Most pest control specialists probably make fewer than a dozen bed bug calls each year, and some have never seen an infestation. Traynor (1978) considers bed bugs as one of the "occasional pests" confronting the pest control operator, and Mallis (1979) acknowledges "bed bug jobs now-a-days do not approach pre-war numbers."

Bed bug problems, however, still occur, even in the most luxurious of homes, and the pest control operator must be prepared to treat an insect with which he has had little or no experience. Pratt (1958) cautions that because most homeowners still think of this pest as a sign of slovenly housekeeping, the control specialist should use considerable tact when investigating problems, especially in more affluent surroundings.

No matter what chemical is chosen for the job, effective control, as Smith (1958) stressed, depends on "searching out all their hiding places." The various refuges mentioned earlier in this chapter — upholstery of chairs and couches, loose wallpaper, picture frames, openings of water pipes, etc. — should be examined. In locating bed bug infestations, Truman et al. (1976) offer this sound advice: "The pest control specialist must look for them in any place which offers darkness, isolation, and protection." They note even when the bugs themselves cannot be found, spots of fecal material reveal their hiding places.

Truman et al. suggest using sprays of one percent malathion, one percent fenchlorphos, 0.5 percent DDVP, and 0.5 percent synergized pyrethrins. Mattresses should be treated only at the seams, folds and buttons and should never be soaked with spray. Mattresses should be allowed to dry thoroughly and should be covered before being used. These authors list several residual sprays, used as water emulsions or oil-base solutions, that may be applied away from beds: 0.5 percent diazinon, two percent malathion and one percent fenchlorphos. They point out a hand sprayer usually is adequate and spraying should be done early in the day so the insecticide can dry before the room is used for sleeping.

Burden (1980) suggests lindane, malathion, pyrethrins or fenchlorphos can be used to spray mattresses. Pyrethrum treatments, because of their limited residual action, should be made several times at weekly intervals.

It should be re-emphasized mattresses and upholstery should not be soaked with spray; a light application should be sufficient. One should never sleep on a freshly sprayed mattress; it should be allowed to dry thoroughly. Treatment of infant bedding and cribs should be avoided. Care should be taken to spray *all* possible hiding places of the bugs. On beds, these areas include slats, springs and frames. Baseboards, wall crevices and other cracks should receive spray. In addition to sprays, insecticide dusts may be useful for control of bed bugs. Recently, a dust based on diatomaceous earth and synergized pyrethrins became available for treatment of "baseboards, moldings, cracks, bedsteads and both sides of mattress" for bed bug control.

The preceeding suggestions and precautions, although written primarily for *Cimex lectularius*, the "human" bed bug, apply generally to bat bugs. In some parts of the country, bat bugs may be encountered more often than the bed bug. The two common bat associates, *C. adjunctus* and *C. pilosellus*, have life cycles and habits similar to *C. lectularius,* hiding by day in cracks and crevices. Bat bugs, however, tend to be more abundant in areas of the house adjacent to attics where bats are roosting. If a bat population is successfully excluded from the house, bat bugs will seek other hosts, with man usually being the most convenient alternate host. Mampe (1978) suggests primary efforts at control should be directed toward the attic. He says large areas generally do not require spraying, but controls should be concentrated on living areas below the attic by treating "cracks and crevices around window frames, light fixtures, ceiling light fixtures and any other areas where the bat bugs that have migrated from the attic may have been hiding." Mampe states the bed and bedding probably will not need to be sprayed. He also suggests using a total release aerosol in the attic to supplement the use of residual sprays.

It should be pointed out that whether bats are to be controlled in the home is an emotional issue in the field of pest control. For a review of the controversy and both environmental and human sides of the question, the reader should consult the article by Pitchon (1980). He concludes by saying the best solution, although not always successful or practical, is to exclude bats mechanically from infested dwellings.

Control of the swallow bug, *Oeciacus vicarius,* presents a somewhat different problem. Eads et al. (1980) note federal and state laws protect cliff swallows and their nests. These regulations must be kept in mind when any management of swallow colonies is attempted.

FAMILY REDUVIIDAE

Assassin bugs represent a large family of true bugs or Hemiptera-Heteroptera. Most of the species are active predators of other insects, but members of the subfamily Triatominae feed on vertebrate blood and are important disease vectors. Many reduviids are capable of inflicting painful bites, but it is the conenoses, or triatomines, that pose the greatest threat to man's health. These reduviids have a cone-shaped head, with a three-jointed beak, or proboscis, that has a membranous connection between segments two and three. This allows flexibility of the proboscis during bloodsucking.

Conenoses. In 1909 the Brazilian scientist Carlos Chagas discovered a sometimes fatal, trypanosomyiasis disease of humans. Although the most obvious symptoms are a swelling of the eyelids and face, there also may be a loss of nervous control, high fever, anemia and destruction of cardiac and skeletal muscles. The causal agent of this disease, now widely referred to as Chagas' disease, was shown to be a flagellate protozoan, *Trypanosoma cruzi.* Conenose bugs have been established as the only important vectors of *T. cruzi.* According to Lent and Wygodzinsky (1979), the fever is mainly a parasitic disease of wild animals and is transmitted by sylvatic or "wild" species of triatomines, which serve as reservoirs of the parasite. The fact that certain of the conenose vectors have become associated with human habitation is strictly a secondary adaptation. The disease is especially prevalent in Central and South America where several common conenose bugs colonize human habitations. In 1960 the World Health Organization estimated that some seven million South Americans suffered from Chagas' disease. Zeledon and Rabinovich (1981) would place the number of infected persons at 13 to 14 million.

Although nymphs of conenose bugs may feed on other insects, all triatomines require bloodmeals to complete their development. Conenoses, or "kissing bugs" as they sometimes are called, feed on infected mammals or rodents and then are capable of transmitting the parasitic protozoan to man and his domestic animals. In their thorough review of triatomine taxonomy and vector relationships, Lent and Wygodzinsky (1979) note immature bugs defecate shortly after taking a bloodmeal. These authors emphasize conenoses cannot transmit the causal agent, *T. cruzi,* in their saliva. Rather, the protozoan parasite is transmitted to man only through the bug's feces. According to Lent and Wygodzinsky, "Trypanosomes penetrate into the host's tissues through skin and abrasions, through the mucosa of the mouth, and, in humans, especially through the con-

Fig. 11-5. These and similar insects are often called "kissing bugs."

juctiva of the eyes. It is obvious that transmission of the parasite to the vertebrate cannot occur in this way when defecation takes place after the vector has
left the host; hence the importance of the time span between feeding and the
first subsequent defecation." Compared to South American species, several
North American triatomines do not defecate or urinate as promptly after a
bloodmeal (Wood, 1951).

Although no truly domestic Triatominae occur in the United States, at least
four actual or potential vector species have been reported as household pests.
Wood (1975) has discussed the invasion of homes by conenose bugs. Ryckman
and Casdin (1976) present a checklist and bibliography of triatomines occurring
in western North America.

Grundemann (1947) quotes Morrill as stating "that in many parts of Arizona,
Triatoma species have taken the place of the common bed bug as a household
pest." Mead (1965) notes Chagas' disease was not known to occur in the United
States until two cases were found at Corpus Christi, Texas in 1955. Since then,
Farrar at al. (1963) tested blood of a few hundred Georgians and found that
some gave a positive serological reaction to the Texas strain of the parasite. The
principal wild reservoirs of the disease carrier *T. cruzi* in the United States are
raccoons and oppossums.

The triatomine reduviid most often reported indoors in this country is *Triatoma sanguisuga* (LeConte), the bloodsucking conenose. This is a medium-sized
bug, dark brown to black, with reddish-orange or yellowish markings. *T. sanguisuga* ranges from Maryland south to Florida and west to Kansas and Arizona
(Lent and Wygodzinsky, 1979). Walsh and Riley (1869) were the first to report
this bug from human dwellings, although its bloodsucking habits were known
to LeConte in 1855. This conenose also is known to be infected with equine
encephalitis virus.

Marlatt (1896) discussed the habits of *T. sanguisuga,* a bug sometimes called the "Mexican" or "Texas" bed bug or the "big bed bug." He points out that until recently this conenose had rarely been encountered inside, but it was becoming more common in country homes of the Mississippi Valley. He states this insect's "buggy" odor is more intense than that of the bed bug. Marlatt also says these bugs will prey on bed bugs and occasionally on cockroaches. The bugs become active at night and sometimes come to light. Once inside, they will pierce human skin to feed on blood, frequently while the victim is asleep. Blood meals of conenoses, according to Lent and Wygodzinsky, last from 20 to 30 minutes. Their saliva contains an anesthetizing agent so persons rarely are aware they are being attacked. Grundemann (1947) describes the bite of *T. sanguisuga*: "To check the details of engorgement, the author allowed a number of laboratory-reared *Triatoma* to feed upon the back of the hand. The bite was found to be somewhat anesthetic . . . The only indication of the presence of the insect was a slight tickling sensation when the proboscis was first introduced, and no reaction followed the bites. However, severe reactions have been observed and reported from several individuals sensitive to the bites of this species." Griffith (1947) notes a case where *T. sanguisuga* was a pest in Oklahoma City, Oklahoma causing severe irritation to the inhabitants. In Georgia, Scott (1958) reported an infestation in an "upper income home." More recently, in the "Infestation Reports" of *Pest Control* magazine, *T. sanguisuga* has been a nuisance in homes in Alabama, Florida, Indiana, North Carolina and Texas.

Lent and Wygodzinsky stress the bloodsucking conenose does not colonize houses and its natural habitats are "hollow trees inhabited by various invertebrates and vertebrates, e.g., raccoons and oppossums . . . The bugs frequently are found in or near wood rat nests . . ." Truman et al. (1976) suggest when this bug does come indoors it is in substandard structures having "the area immediately surrounding and the crawl space beneath . . . open to the activities of domestic fowl or other animals." In domestic situations, *T. sanguisuga* may occur "in bedding, cracks in floors and walls, under furniture, in hen houses, outhouses, barns, and doghouses" (Mead, 1965).

Triatoma protracta (Uhler) appears to be the second most common conenose that invades houses in the United States. Its range, as defined by Lent and Wygodzinsky, is Colorado and Utah, west to California, and south to Arizona, New Mexico and Texas. The western bloodsucking conenose occurs in nests of the wood rat (*Neotoma* spp.) and has been found naturally infected with the parasite *Trypanosoma cruzi*. It occasionally enters (but does not colonize) houses. Wood (1953) describes an allergic reaction to the bite of *T. protracta*, which included nausea, heart palpitation, breathlessness and violent itching. Anon (1970b) reported in Utah several sleeping children were bitten and became unconscious; others who were bitten developed rapid heartbeat. Wood (1950) noted California had at least 10 localities where populations of *T. protracta* were naturally infected with the causal agent of Chagas' disease. As Truman et al. (1976) pointed out, this species is a poor flyer, but in the West Coast foothills it will fly downhill and sometimes create a problem around swimming pools. Apparently it is hunger that causes these insects to disperse. Zeledon and Rabinovich (1981) cite studies showing that *T. protracta* requires more food during the hot summer months. It is then that adults leave wood rat nests to seek new hosts in human habitations.

Conenose bugs reported less frequently as invading houses are *Triatoma gerstaeckeri* (Stal) and *T. lecticularia* (Stal). Both are known Chagas' disease vectors that are associated with nests of wood rats (Lent and Wygodzinsky, 1979).

In the United States, *T. gerstaeckeri* is found only in New Mexico and Texas. According to Lent and Wygodzinsky, it occurs in "chickenhouses, pigsties, horse stables and horse and cattle corrals." They note large numbers may be found indoors. As recorded in *Pest Control's* "Infestation Reports," this conenose was "unusually heavy" in Texas during 1967; several people were bitten, with one person requiring hospitalization. *T. lecticularia* ranges from Pennsylvania south to Florida and west to California and Arizona. Packchanian (1940) recorded nymphs of this species living in beds, behind wallpaper and in cracks in wood. In their review of arthropods of medical importance, Travis et al. (1969) also list *T. rubida* (Uhler) (*=uhleri* Neiva) as a household pest. This species, known in the United States from Arizona, California, New Mexico and Texas, was studied by Wehrle (1939).

Control. Chagas' disease, as noted by Lent and Wygodzinsky, has no known cure. Therefore, control of the disease depends largely on controlling the triatomine vectors. Marlatt (1896) realized that controlling conenose bugs in their natural habitats is "manifestly out of the question." Instead, all potential breeding areas near houses should be eliminated, including trash piles and bird nests (Rachesky, 1970). In California, Michelbacher et al. (1961) suggested wood rat nests be removed from around dwellings. They note these rats often have nests in piles of cactus pads in desert regions and in piles of sticks in forested areas. Homeowners also should try to keep domestic animals out of dwellings and crawl spaces. Because these insects fly at night and are attracted to artificial light, adequate screening should be used. Wehrle (1939) suggests if these insects cannot be excluded from the house, sleeping persons should be protected by mosquito netting. If a conenose should invade and alight on one's face or hand, the bug should be brushed away gently because it is likely to bite if pinched or crushed.

At one time, water emulsions of DDT (Randolph, 1946) and DDT-pyrethrum aerosols (Griffith, 1947) provided adequate control of *Triatoma* species. With this use of DDT no longer permitted, Truman et al. (1976) recommend a residual application of two percent malathion.

Other reduviids. Triatomine reduviids, which require a blood meal for their development, may actively seek man to obtain additional food. In contrast, other assassin bugs prey on insects and bite humans only incidentally, most often in self defense when roughly handled or crushed. For more detailed information on the habits of these other nuisance species of Reduviidae, the reader should consult the work on biology by Readio (1927).

In the summer of 1899 some of these non-bloodsucking reduviids aroused widespread alarm. Following several attacks on residents of Washington, D.C., the newspapers helped inspire a "kissing-bug craze" that soon spread throughout most of the country. Howard (1900) felt compelled to prepare a matter-of-fact account of reduviid habits to help allay the nation's fears.

The two assassin bug species implicated in the "scare" of 1899 were the masked hunter, *Reduvius personatus* (L.), and a dark, superficially similar bug, *Melanolestes picipes* (H.-S.). The former species, native to Europe, was accidentally introduced to North America. Howard suggests that *R. personatus* may have followed closely in the "wake of the bed bug," upon which the masked hunter feeds in Europe. This assassin bug now occurs throughout much of the United States, from New England to Florida and west to Kansas (Slater and Baranowski, 1978). Scudder (1961) reported a population living in "dockside warehouses" in Vancouver, British Columbia.

The specific name, *personatus,* meaning "masked," was given because of this bug's curious habits. Butler (1923), writing in England, described a nymph as "covering itself with fragments of foreign matter, which might collectively be called 'dust'; its appearance is thereby greatly altered . . . These particles are attached not so much to the body itself as to the long and fine hairs with which it is covered, and which appear to have some adhesive power." Another most appropriate description of these well-camouflaged nymphs is "small animated dust bundles" (Slater and Baranowski, 1978). As Readio has pointed out, nymphs usually occur in houses and other buildings, where they live in dusty corners and other less used areas.

Adults are known to inflict painful bites. Howard cites Lintner's description of the bite as resembling a bee sting and being followed by numbness and considerable swelling. Judging from the "Infestation Reports" in *Pest Control* magazine, the masked hunter each year inflicts a few bites that require medical attention.

Melanolestes picipes is a mainly ground-dwelling reduviid which, according to Readio, flies to light at certain times of the year. In Kansas, he noted most flights took place during May and June. It is the fully-winged males that usually come to light, a habit that gave rise to the popular name "kissing bug" and resulted in notoriety for this insect during the late 19th Century.

Howard notes that Glover, in 1875, may have been the first to record the bite of *M. picipes.* Howard further states the pain, initially like that resulting from a bee sting, has been described in some cases as becoming "ten times more painful" and inducing "a feeling of weakness followed with vomiting." Like the masked hunter, this wide-ranging assassin bug is a household nuisance that is recorded annually as biting a few persons in or around the home.

Other reduviids at times attract attention, though perhaps less than the two "kissing bugs" just described. A large assassin bug (nearly 1½ inches/38 mm long) that nearly always impresses a first-time observer is the bizarre wheel bug, *Arilus cristatus* (L.), so named because of the semicircular crest on its thorax. Mead (1974) describes the thoracic teeth or tubercles as numbering from eight to 12 and resembling a "cog-wheel" or "chicken's comb." This insect's bite has been described by Barber (1919), Hall (1924), and Smith et al. (1958). The latter workers concluded the prolonged effects from this bite generally are the result of individual hypersensitivity or a secondary infection. Wheel bugs, voracious predators on other insects, are capable of capturing large caterpillars, beetles, bees and wasps. In Pennsylvania, Mallis (1979) identified the wheel bug as the species causing alarm when large numbers appeared on a woman's porch. The bugs had been attracted to nearby grape vines where they were feeding on Japanese beetles.

In California, Anon. (1963a) reported the leafhopper assassin bug, *Zelus renardii* Kolenati, as biting man. Species of the genus *Rasahus* also have been recorded as biting several persons in California (Anon., 1964a). In Oregon, a young child was painfully bitten by *Rhynocoris ventralis* (Say) (Anon., 1968). Thread-legged bugs of the reduviid subfamily Emesinae sometimes occur in domestic situations, but these slender, elongate insects are not noted for inflicting bites. Wygodzinsky (1966) quotes Uhler as saying the large elongate bug, *Emesaya brevipennis* (Say) "takes shelter in sheds, outhouses and barns." This thread-legged bug often occurs in spider webs where it feeds on entrapped insects, or perhaps on the spider inhabitants. In Georgia, Snoddy et al. (1976) discussed the habits of another emesine reduviid, the spider predator *Stenole-*

mus lanipes Wygodzinsky, which is found in webs of the house spider, *Achaeranea tepidariorum* (C.L. Koch). "The adult bugs were commonly observed in and resting near the spider webs. The nymphs and eggs were abundant in protected eaves, downspouts and gutter attachments . . ."

KEY TO THE ASSASSIN BUGS OCCURRING IN HOUSEHOLDS

1) Small bugs, less than ½ inch/13 mm long; legs slender, thread like . . . thread-legged bugs, Emesinae/Larger bugs, length ¾ inch/19 mm or more; legs rather stout 2

2) Large species, length more than 1¼ inches/32 mm; pronotum produced into a distinct serrate or toothed crest . . . wheel bug, *Arilus cristatus* — Smaller species, lengths not more than one inch/25 mm; pronotum not produced into a distinct crest 3

3) First antennal segment longer than 2nd antennal segment 4
First antennal segment shorter than 2nd antennal segment 5

4) Front femora as long as and about equally thickened as hind femora; body (except wings), including basal portion of pronotum, light brown . . . *Zelus renardii*/Front femora shorter than and distinctly more thickened than hind femora; body usually reddish, basal portion of pronotum black . . . *Rhynocoris ventralis*

5) Membrane of wings with a large yellowish spot at middle . . . *Rasahus biguttatus* and *R. thoracicus*/Membrane of wings uniformly colored, distinct yellowish spot absent 6

6) Head produced into an elongate cone, tip reaching beyond tips of 1st antennal segments; body dark, accented with red and/or yellowish markings . . . *Triatoma* species/Head not produced into an elongate cone, tip not reaching tips of first antennal segments; body uniformly dark brown to black 7

7) Body and legs uniformly black; front legs stout, apical ⅓ of front tibiae broadly dilated and flattened beneath . . . *Melanolestes picipes*/Body and legs dark brown with knees paler; front legs slender, apical area of front tibiae little thicker than base . . . *Reduvius personatus*

FAMILY RHOPALIDAE (= CORIZIDAE)

Rhopalids, or scentless plant bugs, have been referred to as corizids or coreids in some of the older literature. Closely related to squash bugs (family Coreidae), members of this small family show a preference for fruits and seeds of their host plants. Although some species are quite common on weeds in old fields and along roadsides, many entomologists are unfamiliar with these insects. For identifying scentless plant bugs, Hoebeke and Wheeler (1982) have given a key to the nine genera known to occur in the United States and the 16 species that are found in the eastern states. Scentless plant bugs are of little economic importance except for the boxelder bugs, which frequently become household pests during the fall, winter and spring months.

THE BOXELDER BUG

Leptocoris trivittatus (Say)

Boxelder bug is the accepted common name for this household nuisance, but in the Midwest the names "pop," "populist" and "democrat" bug have been used, and in the South these bugs mistakenly have been called "cotton stainers." The scientific name long used for the boxelder bug is *Leptocoris trivittatus,* but a European entomologist who specializes in the taxonomy of rhopalids has placed this species in the genus *Boisea* (Göllner-Scheiding, 1980). It is unfortunate a name so familiar to pest control specialists and entomologists must be changed. Until it is clear that American workers accept this name change, it seems best to retain the older name.

Slater and Schaefer (1963) have shown the boxelder bug is native to the western states where it breeds mainly on the boxelder tree, *Acer negundo* L. Schaefer (1975) would restrict the native range to the southwestern states. In Midwestern farmyards and towns this large black bug with red markings had become a familiar household pest by the 1930's. Boxelder bugs leave their host trees during fall to seek overwintering sites, frequently around foundations and windows. As Henry (1977) describes, they often congregate on the south and west sides of buildings where the sun warms them during the day and early evening. Adults are capable of flying two miles or more in search of suitable hibernation quarters (Knowlton, 1944), although they usually overwinter near their host trees. These insects do not sting and seldom bite, but become annoying when they invade houses during warm days in fall, winter and spring. As noted by Swenk (1929), they are capable of staining draperies.

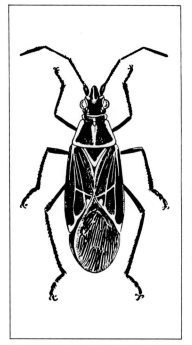

Periodic outbreaks are known to occur, such as the one in North Dakota in 1894. Lintner (1894) noted the nymphs, occurring in patches four or five to 60 feet in diameter, formed "a deep, writhing mass."

Fig. 11-6. Boxelder bug, *Leptocoris trivittatus* (Say).

Smith and Shepherd (1937) recorded outbreaks in Kansas during the 1930's, and in Wisconsin populations had been building for several summers where, according to Chambers (1934), infestations near seed-bearing trees were nearly "unbearable." He remarked some villages were removing boxelder trees and some were even trying to outlaw the tree. In Michigan, McDaniel (1933) suggested the "elimination of boxelder in the vicinity of houses would settle the local question of control measures for all time."

In recent years the boxelder bug has expanded its range to include most of the eastern United States and southern Canada. Slater and Schaefer (1963), after reviewing the economic literature and checking for boxelder bug specimens in major collections, were able to trace the spread of this insect. It was not until the late 1940's and early 1950's that it appeared as a pest in New England. According to Slater and Schaefer, boxelder is not native east of the Alleghenies. It is likely that the boxelder bug spread eastward as its host was planted in towns of the Northeast. Because adults have been found in a vehicle several hundred miles from where it had been parked under boxelder trees (Wollerman, 1965), it is possible that man's commerce has aided in the dispersal of this pest. The boxelder bug has now become a common household pest in urban and suburban environments. Wollerman (1965) predicted the "current emphasis on outdoor recreation and the growing number of homes and parks in sites where boxelder is among the few trees adaptable" will increase the demand for control information. According to Ascerno (1981), the boxelder bug ranked first among insects submitted to the University of Minnesota's diagnostic clinic in 1977(2,501 inquiries) and second in 1978(1,742). The large populations in those years were correlated with the long, hot, dry summers of 1976 and 1977.

Life history. Smith and Shepherd (1937), noting information on habits of the boxelder bug was limited mainly to accounts in farm journals and newspapers, provided the first details of life history. In Kansas, they observed adults began to hibernate around October 1 and leave their overwintering sites about the last week of March. Mating and oviposition took place after adults had fed for about two weeks. Eggs, which are "light straw-yellow" when first deposited and eventually a "dark reddish-brown," were laid mainly from late April to May 10. Eggs were placed on stones, leaves, grasses and on shrubs and trees, especially in bark crevices. Smith and Shepherd found eggs hatched after 11 to 19 days, and the emerging first-stage nymphs (immature bugs) were bright red, sparsely covered with bristles, and about $1/16$ inch/1.6 mm long. There are five nymphal instars with later-stage nymphs darker red; the last-stage nymph has "slate-black wing pads" that cover about $1/4$ of the abdomen. In the laboratory the nymphal period averaged 60 days. These authors believed two generations were produced each season in Kansas.

According to Smith and Shepherd, boxelder bugs fed on a "wide variety of plants." Although these bugs are known primarily as household pests, earlier authors have noted injury to strawberries (Hutson, 1932) and various tree fruits (Knowlton, 1951). In Washington State, Webster (1926) reported these bugs became so numerous on apple that they resembled bee swarms. In Utah, Knowlton (1935b) recorded boxelder bugs causing prunes to shrivel and the outer layer of apples to become corky. Knowlton (1944) found several thousand boxelder bugs around hives where they were feeding on dead honeybees. These insects also will cannibalize, particularly during molting (Abbott, 1948) and occasionally will bite man. Knowlton (1947a) was bitten by a nymph that attempted to feed and then withdrew its stylets. He also notes several individuals have been bitten in bed. The bite may cause an irritation and produce a red spot similar to a small ulcer (Brannon, 1961).

Tinker (1952) further clarified the habits of the boxelder bug. He observed nymphs and adults of the season's first generation were found on the ground or on low vegetation. The bugs fed primarily on fallen seeds of boxelder. The occurrence of boxelder bugs on low-growing plants in spring and early summer may help explain the wide host range reported by Smith and Shepherd (1937). Once seeds began to form on boxelder trees during mid-summer, Tinker found the bugs moved into the pistillate or female trees. Long (1928) already had observed that staminate (or male) trees usually were avoided, even when their branches were intertwined with those of pistillate trees. It should be noted that in Pennsylvania the seed-bearing trees of silver maple, *Acer saccharinum* L., sometimes are important boxelder bug hosts.

Tinker made additional observations that are helpful in understanding and controlling this pest. Because only adults survive a normal winter, the size of the following spring's population depends on the proportion of adults entering hibernation the previous fall. Two complete generations are more likely to be produced in seasons having a greater number of day-degrees available. Thus, a longer growing season would give rise to a larger proportion of hibernating adults.

Other rhopalids and related species. Barber (1956) described a new western boxelder bug, a species he named *Leptocoris rubrolineatus*. If Göllner-Scheiding's concepts of classification are accepted, this rhopalid would be called *Boisea rubrolineata*. The western and the "common" boxelder bug are closely related. Schaefer (1975) found the only consistent character distinguishing the two was the greater amount of red on the outer forewings of the western species.

The western boxelder bug has a limited distribution. Schaefer records it from the California coast and in Idaho, with scattered populations known from Arizona, Nevada and Texas. Scudder (1979) notes it also occurs in British Columbia. The life history has not been studied, but large numbers have been reported from bigleaf maple in California (Horn, 1973). According to Horn, the western boxelder bug occasionally feeds on larvae of the California oakworm inhabiting coast live oak growing near infested maples. Schaefer suggests the California boxelder eventually may prove to be a host plant. Like the more familiar boxelder bug, the western species sometimes attacks fruit (Schaefer, 1975) and invades houses during the winter. It has entered houses in California (Anon. 1964b), Nevada (Anon., 1974a) and British Columbia (Scudder, 1979).

A rhopalid recently has become troublesome in houses is *Arhyssus crassus* Harris. This western species invaded residences in California during the mid-1960's. It has been "attaining increasing importance" as a household pest by "defecating on drapes, bedspreads and rugs" (Anon., 1967). It invaded a California air base, amassing by the thousands on sides of buildings and creating "a definite morale problem" (Anon., 1966a). In the pest control literature this insect has been referred to as a "grass bug." North American rhopalids, however, rarely feed on grasses, and other plants may be the preferred hosts. The life history is unknown, but Chopra (1968) recorded *A. crassus* from alfalfa, wild gooseberry and antelope bitterbrush *(Purshia tridentata)*. In the "Infestation Reports" of *Pest Control* magazine, *Arhyssus scutatus* (Stål) also has been reported to occur in California houses. It should be noted the species of *Arhyssus* are difficult for the non-specialist to identify and some of the literature records may be based on misidentifications.

Adults of the handsome rhopalid *Jadera haematoloma* (H.-S.) occasionally enter houses. This strikingly colored insect is mainly black with contrasting red on the head and pronotum. The life history has not been studied, but this species has been recorded in Oklahoma as invading homes from nearby Chinaberry plants (Anon., 1966c). Anon. (1964e) notes *J. haematoloma* has been a nuisance insect in Texas. Other rhopalids that may enter houses are species of the genus *Stictopleurus* (Anon., 1963b, 1964f).

Members of the squash bug family Coreidae, related to the rhopalids, also become occasional pests. In Pennsylvania, the familiar squash bug, *Anasa tristis* (De Geer), has been found overwintering in basements. Squash bugs normally overwinter under loose bark and in various kinds of debris, but Beard (1940) notes it is not unusual to find them in buildings. According to Beard, they "may occur in the basements of dwellings, or elsewhere in less furnished buildings. The insects are commonly wedged in crevices or between boards."

Another coreid, *Leptoglossus occidentalis* Heidemann, has invaded homes in British Columbia. Spencer (1945) described this insect as creating "great alarm" when large numbers of adults suddenly began "crowding into houses for hibernation." This large leaf-footed bug is found from British Columbia, east to Nebraska, and south to California and New Mexico. Koerber (1963) found this coreid feeds on seeds of Douglas fir and various pines, thus lowering the quality of seed crops.

Alydus sp., belonging to the broad headed bugs of the family Alydidae, has been mentioned as a nuisance insect in Georgia. Anon. (1964e) reported large numbers of these slender-bodied bugs entering houses from nearby groundsel fields.

Control. Outbreaks of the boxelder bug in the 1890's prompted the use of kerosene emulsion, whale oil soap, tobacco decoctions and other compounds to control this pest. Gillette (1898) noted these materials merely made the bugs "uncomfortable" for awhile. Smith and Shepherd (1937) found applying hot water or kerosene to adults and drenching nymphs with cold water from a hose would provide adequate control. Later, Knowlton (1948) recommended the use of ordinary fly sprays, three to five percent chlordane, or five percent DDT oil-base sprays. The insects required a thorough wetting. Bruce (1950) and Rogoff (1950) found dieldrin sprays gave very satisfactory control, and Monro and Post (1949) obtained good results on sunny sides of buildings with DDT, chlordane, lindane and toxaphene emulsions.

More recently, Truman et al. (1976) have suggested residual sprays of one percent diazinon and propoxur, 0.5 percent chlorpyrifos, two percent malathion, or one to two percent carbaryl for indoor use. They also recommended space sprays of 0.5 percent DDVP, 0.25 percent synergized pyrethrins, or 0.15 percent resmethrin. The entire room should be filled with space spray and closed off for several hours. A vacuum sweeper may be used to pick up and destroy the bugs. Truman et al. suggest dusts of five to 10 percent carbaryl for treating wall voids and other sites that may be used for overwintering.

As noted by Mampe (1979), the chlorinated hydrocarbons such as chlordane worked well against boxelder bugs "but sadly, may no longer be used for this purpose." In his opinion, with organophosphates sometimes falling short of good control, only the carbamates remain to provide adequate control for entering bugs. He mentions carbaryl, which is labeled for this pest, and propoxur as possibilities. Once the bugs are inside, Mampe suggests carbaryl dust (a formulation labeled for indoor use) gives good control. Diazinon-based dusts or silica gels also work well, according to Mampe. If sprays are necessary, he would suggest propoxur.

Outside, Truman et al. (1976) note the same materials and formulations as used for entering bugs will provide control. The bugs should be wetted thoroughly, as well as the surfaces over which they crawl. Host trees, of course, can be sprayed. Because boxelders are not generally considered prized ornamentals or shade trees, extension entomologists often note a more permanent solution to the boxelder bug problem can be obtained only through removal of the female or pistillate trees. If nearby boxelder trees are allowed to remain, it is of course possible for the bugs to re-invade.

OCCASIONAL INVADERS

As Mallis (1979) pointed out, "Anyone who has worked in the pest control industry for a period of time comes up with some unusual household pests that make life interesting." These pests are refered to variously as "miscellaneous pests," "occasional invaders" or "minor pests." Traynor (1978) stresses many pest control operations concentrate on the more profitable, recurring problems — ants, cockroaches and rodents. In his words, this "leaves little, if any, leeway for handling the unusual pest or pest occurrence." Several species of Hemiptera not discussed in previous sections qualify as unusual problems. Although they are met with infrequently and could be dismissed as insignificant, Mallis noted that what a pest control operator or entomologist might consider minor "is a 'major' pest to the homeowner or tenant involved."

Hemiptera-Heteroptera. True bugs other than bed bugs, conenoses and boxelder bugs and their relatives have been reported to invade houses. Many of the entering species are attracted to artificial light. The majority of these occasional pests are stink bugs (family Pentatomidae) or seed bugs (Lygaeidae).

The stink bug, *Brochymena affinis* Van Duzee, becomes a nuisance when it overwinters in houses in British Columbia (Scudder, 1979). In New Mexico, Ruckes (1946) notes hibernating adults of *B. sulcata* Van Duzee have been found "under breeding cages, clapboards of hen houses and dilapidated dwellings, where the bugs appear in sufficient abundance to become nuisances." Anon. (1965a) mentioned *Brochymena* sp. as a household pest in Oklahoma. In Idaho, the Say stink bug, *Chlorochroa (=Pitedia) sayi* Stal, created a nuisance when large numbers came to light (Anon., 1958), and in Nebraska, *Thyanta custator* (F.) swarmed under street lights and entered houses (Swenk, 1925). The green stink bug, *Acrosternum hilare* (Say), a well-known pest of fruit crops, has been taken in a South Carolina residence (Anon., 1974b). Anon. (1965a) reported the two-spotted stink bug, *Perillus bioculatus* (F.), from a house in Delaware. It should be noted this two-spotted species is a predacious rather than a plant-feeding stink bug and an enemy of the Colorado potato beetle. In South Carolina, a negro bug of the related family Corimelaenidae (or Thyreocoridae) has been found indoors. *Corimelaena pulicaria* (Germar), a small black insect with a pale stripe laterally on the forewings, invaded from nearby grain fields (Anon., 1970a). Westcott (1973) records *Pangaeus bilineatus* (Say), a burrower bug of the family Cydnidae, as a household pest in Pennsylvania.

The lygaeid bugs, forming the second largest family of true bugs, were once considered to be predators. Sweet (1960), however, has demonstrated that many species are seed feeders that show a preference for fallen seeds; hence the common name seed bugs. False chinch bugs of the genus *Nysius* may enter the household, sometimes in alarming numbers. Parshley (1919) recorded an outbreak of *N. ericae* (Schilling) in British Columbia. "For a few days the walls of the houses inside and out were covered with the insects and complaints were made that they bit children — altogether a most unusual occurrence." More recently, *Nysius* spp. have been found on furnishings in a New Mexico house (Anon., 1964e), found creating a "considerable nuisance" in residences in California (Anon., 1966b), invading houses adjacent to alfalfa fields in New Jersey (Anon., 1962a), swarming at light in Idaho (Anon., 1962b), moving into houses from cut and drying grain fields in Oregon (Anon., 1959), entering yards and homes in Nevada (Anon., 1966b), and invading a hospital in Arizona (Anon., 1964c). The species most often mentioned are *N. raphanus* Howard and *N. ericae*. Identification of these bugs is difficult and specimens frequently are misidentified. Ashlock (1977) has shown most records pertaining to *N. ericae* in North America should refer to *N. niger* Baker.

Other seed bugs recorded indoors are *Ischnodemus falicus* (Say), which swarmed about houses during early November in North Carolina (Farrier, 1958) and the notorious chinch bugs, which are pests of corn, small grains and turf. According to Leonard (1966), chinch bugs at one time threatened the economy of American agriculture. "Because of this threat probably more has been written about the chinch bug [*Blissus leucopterus* (Say)] than any other American insect." Truman et al. (1976) noted *Blissus* spp. will crawl over sidewalks and sides of houses, and sometimes will enter houses. In discussing migrations of these bugs during days of "Indian Summer" in Missouri, Stedman (1902) remarked: "Very frequently they occur in villages at this time of year in such immense numbers as to attract a great deal of attention from the village people, who do not understand the invasion or the name of the insect."

In Pennsylvania, an unusual infestation of the so-called birch catkin bug, *Kleidocerys resedae* (Panzer), plagued a homeowner who had white birches in his yard. Wheeler (1975) related how large numbers of this lygaeid migrated from host trees, aggregated on shrubs and on sidewalks and were carried into the house. The odor of crushed bugs was particularly objectionable. As noted by Wheeler (1976), birch catkin bug often breeds in the seed capsules of azaleas, rhododendrons and Japanese andromeda *(Pieris japonica)*. Anon. (1963c) recorded another lygaeid outbreak. In New Mexico, a large red and black species, *Neacoryphus (=Lygaeus) lateralis* (Say), became a nuisance on and in buildings. Knowlton (1960) reported on a diverse group of true bugs that motel owners in St. George, Utah regarded as a "plague." The bugs "entered cabins, beds and clothing, often 'biting' the inhabitants."

Hemiptera attracted to light also may create problems indoors by biting man. In fact, members of several normally plant-feeding families are known to feed on human blood (Usinger, 1934). Several plant bugs, members of the largest family of true bugs (Miridae), will bite humans. The azalea plant bug, *Rhinocapsus vanduzeei* Uhler, is especially troublesome to gardeners as they work near ornamental azaleas (Wheeler and Herring, 1979). This attractive reddish-orange and black species is attracted to perspiration and may be brought inside on clothing or on skin. The bite of this plant bug feels like a pin prick on the skin and at the site of penetration, it produces a small welt, somewhat similar to that left by a chigger.

Hemiptera-Homoptera. Aphids, leafhoppers and their relatives also become occasional household pests. Compared to the true bugs, or Heteroptera, fewer homopteran species have been recorded as nuisances. Hackberry psyllids, jumping plantlice of the family Psyllidae, are the most consistently reported pests in this group of insects. They seem to be more persistent pests in the Midwest where hackberry is common. Chambers (1940) noted the hackberry nipplegall maker, *Pachypsylla celtidismamma* (Riley), is a considerable nuisance in Wisconsin, where large numbers of overwintered adults swarmed into houses in spring. According to Tuthill (1943), adults usually emerge in the fall from galls on hackberry leaves and overwinter in bark crevices. Also, he stated that "large numbers collect on screens, which barrier they readily penetrate and move into the household to hide for the winter." Truman et al. (1976) acknowledge this insect is best controlled while the nymphs are still developing on their hosts, but they admit that it is the *adult* stage that the pest control specialist is asked to control. They recommend residual sprays of one percent diazinon or propoxur, 0.5 percent chlorpyrifos, or two percent malathion applied to screens and window frames to help keep these insects from coming inside. Once psyllids have entered the house, they suggest using aerosols of 0.5 percent DDVP, 0.25

percent synergized pyrethrins, or 0.15 percent resmethrin. Eliminating nearby hackberry trees may be a practical solution to the problem. Mallis (1979) offers some of the best advice when he notes infestations of hackberry psyllids are so short lived that a homeowner may simply choose to wait for the problem to go away. The same advice might be given for many of these "occasional invaders," particularly those that do not threaten man's health.

Among aphid species, Britton (1920) reported *Calaphis betulaecolens* (Fitch) swarming from birch trees in Connecticut so "tops of automobiles and clothes of persons were literally covered." In Utah, Knowlton (1962) described an unusually heavy flight of a white, wax-covered aphid, *Drepanaphis utahensis* Smith & Knowlton, which alighted on cars and otherwise drew attention. More recently, several aphids have been mentioned in the "Infestation Reports" of *Pest Control* magazine as causing inconveniences to homeowners. These species include the cotton or melon aphid, *Aphis gossypii* Glover; the mealy plum aphid, *Hyalopterus pruni* (Geoffroy); and the so-called giant willow aphid, *Tuberolachnus (= Lachnus) salignus* (Gmelin).

Leafhoppers attracted to artificial lighting also may enter houses where they create a nuisance and, according to Tucker (1911) and Knowlton (1947b), occasionally bite humans. In *Pest Control's* "Infestation Reports," *Keonolla confluens* (Uhler) was described as troublesome indoors. Certain species sometimes swarm in enormous numbers. A seldom-collected leafhopper, *Xerophloea viridis* (F.), invaded a Nebraska town in 1920, as graphically described in a newspaper article by M. H. Swenk and quoted by Lawson (1931). "The eleventh day of September was unique in that the city of Neligh was visited by a cloud of the little green bugs, and windows other than mine were clouded with them and houses and stores were closed to shut them out . . . All the stores were closed and it was quite a chore next morning to sweep up the thousands that were caught when the stores were closed." In another instance, Osborn (1912) reported that in Missouri, *Draeculacephala mollipes* (Say) became so abundant at light that this species could be "gathered up by the bushel." Because so many leafhopper species are attracted to light, Osborn concluded this habit might be generally distributed throughout the family.

INSECTS IN SWIMMING POOLS

Many insects living in water are naturally attracted to swimming pools. Many of these aquatic insects also may be attracted to lights in the vicinity of pools. Among the true bugs, the backswimmers (family Notonectidae), giant water bugs (Belostomatidae), and water boatmen (Corixidae) are the most common swimming pool pests.

The backswimmers are medium-sized Hemiptera that are so named because they swim on their backs. Their long hind legs have fringes of hairs that serve as oars that propel them through the water. Backswimmers are predators that feed on other insects, tadpoles and even small fish. Their painful bite is said to resemble a bee sting, and Slater and Baranowski (1978) note that early German writers referred to these aquatic bugs as "water bees" (Wasserbienen). In California, *Notonecta kirbyi* Hungerford has been an especially troublesome pest in swimming pools (Anon., 1964g).

The giant water bugs are among the largest of our North American insects; some species are more than two inches/five cm long. These dull brown or yellowish bugs have enlarged front legs that enable them to capture aquatic insects and small fish. In some parts of the couantry, belostomatids are called "electric light bugs" or "toe biters." In Michigan, *Lethocerus americanus* (Leidy) is reported to cause painful bites accompanied by local swelling (Anon., 1965b). It also has been a nuisance around homes in North Dakota (Anon., 1964d).

The water boatmen, or corixids, differ from the notonectids and belostomatids in feeding on algae and other microorganisms rather than preying on other animals. Corixids also do not bite. Members of this large family often come to light. It should be noted that corixid mouthparts are so highly modified, allowing solid particles to be ingested, some specialists have considered placing the water boatmen in an order or suborder separate from other Hemiptera. Hungerford (1948) summarized the uniqueness of these insects: "The ingestion of multicellular organisms and the packing of the stomach with skeins of filamentous algae is indeed unique amongst the sucking insects of the order Hemiptera."

Control. Because insecticides cannot safely be applied, other means must be used to remove aquatic bugs from pools or keep them from invading. Rachesky (1973) suggests pool dip nets be used to skim insects from the water surface. Lighting should be used sparingly and lights should be turned off when not needed. Yellow lights may be less attractive to some insect species. If bright lights are necessary, they should be set 20 to 30 feet/six to nine m from the pool. Rachesky also suggests that 0.1 percent pyrethrin, in a pressurized spray, can be used to mist around lights for a temporary knock-down of flying insects. The use of insect electrocuters might have a similar effect. He stresses that good sanitation practices prevent slime from building up and may reduce the attractiveness to certain aquatic insects. "Small children's pools should be emptied when not in use. If the owner is planning to be away for two weeks or longer, either the pool should be drained or someone should look after its maintenance."

It should be noted adult flies and their larvae (maggots), and certain beetles, moths and leafhoppers occasionally pose a problem in and around pools. Mallis (1973) discussed an unusual infestation of a beetle in Pennsylvania. A pool owner submitted larvae of a water scavenger beetle (family Hydrophilidae) that was identified as a species of *Tropisterna*. The long, sharp jaws of the larvae enabled them to pierce the plastic liner of the pool.

LITERATURE

ABBOTT, C. E. — 1948. Cannibalism in *Leptocoris trivittatus* Say. Bull. Brooklyn Entomol. Soc. 43:112-113.

ANONYMOUS — 1958. Pest Control 26(10):86. 1959. *Ibid.* 27(9):46. 1962a. *Ibid.* 30(10):110. 1962b. *Ibid.* 30(11):40. 1963a. *Ibid.* 31(3):86. 1963b. *Ibid.* 31(5):104. 1963c. *Ibid.* 31(8):68. 1964a *Ibid.* 32(2):60. 1964b. *Ibid.* 32(3):94. 1964c.*Ibid.* 32(4):72. 1964d. *Ibid.* 32(7):66. 1964e.*Ibid.* 32(9):62. 1964f.*Ibid.* 32(11):52. 1964g.*Ibid.* 32(12):56. 1965a.*Ibid.* 33(5):90. 1965b.*Ibid.* 33(10):55. 1966a. *Ibid.* 34(7):70. 1966b. *Ibid.* 34(8):54-55. 1966c. *Ibid.* 34(9):54. 1967. *Ibid.* 35(2):56. 1968.*Ibid.* 36(9):35. 1970a.*Ibid.* 38(9):40. 1970b.*Ibid.* 38(12):31. 1974a. *Ibid.* 42(5):24. 1974b. *Ibid.* 42(7):28.

ASCERNO, M. E. — 1981. Diagnostic clinics: more than a public service? Bull. Entomol. Soc. Am. 27:97-101.

ASHLOCK, P. D. — 1977. New records and name changes of North American Lygaeidae (Hemiptera:Heteroptera:Lygaeidae). Proc. Entomol. Soc. Wash. 79:575-582.

BACK, E. A. — 1940. Bat bug (*Cimex pilosellus* Horv.). Insect Pest Surv. Bull. 20(6):359.

BARBER, G. W. — 1919. On the bite of *Arilus cristatus*. J. Econ. Entomol. 12:466.

BARBER, H. G. — 1956. A new species of *Leptocoris* (Coreidae:Leptocorini) Pan-Pac. Entomol. 32:9-11.

BEARD, R. L. — 1940. The biology of *Anasa tristis* DeGeer with particular reference to the tachinid parasite, *Trichopoda pennipes* Fabr. Conn. Agric. Exp. Stn. Bull. 440. pp. 595-679.

BRANNON, D. H. — 1961. Boxelder bug (*Leptocoris trivittatus*). Coop. Econ. Insect Rep. 11(49):1100.

BRITTON, W. E. — 1920. Swarms of aphids. P. 203 *in* 19th Rep. State Entomol. Conn., 1919 (Bull. Conn. Agric. Exp. Stn. 218).

BRUCE, W. N. — 1950. Boxelder bug control. P. 9 *in* Proc. 5th Ann. Mtg. No. Centr. States Branch, Am. Assoc. Econ. Entomol.

BURDEN, G. S. — 1980. Household pests. pp. 741-747 *in* Guidelines for the control of insect and mite pests of foods, fibers, feeds, ornamentals, livestock, households, forests, and forest products. U.S. Dept. Agric. Handb. 571.

BURTON, G. J. — 1963. Bedbugs in relation to transmission of human diseases. U.S. Pub. Health Service Rep. 78:513-524.

BUSVINE, J. R. — 1976. Insects, Hygiene and History. Athlone Press of London Univ., London. 262 pp.

BUTLER, E. A. — 1923. A Biology of the British Hemiptera-Heteroptera. H. F. & G. Witherby, London, 682 pp.

CHAMBERS, E. L. — 1934. Boxelder bug (*Leptocoris trivittatus* Say). Insect Pest Surv. Bull. 14(8):275.

CHAMBERS, E. L. — 1940. Hackberry nipple gall (*Pachypsylla celtidis-mamma* Riley). Insect Pest Surv. Bull. 20(8):467.

CHOPRA, N. P. — 1968. A revision of the genus *Arhyssus* Stål. Ann. Entomol. Soc. Am. 61:629-655.

COWAN, F. — 1865. Curious Facts in the History of Insects. J. B. Lippincott, Philadelphia. 396 pp.

CRAGG, F. W. — 1923. Observations on the bionomics of the bedbug (*Cimex lectularius* L.), with special reference to the relations of the sexes. Ind. J. Med. Res. 11:449-473.

CWILICH, R., G. G. NIER, and A. V. MERON — 1957. Bedbugs resistant to gamma BHC (lindane) in Israel. Nature 179(4560):636-637.

DONER, M. H. — 1945. Biology and identification of bedbugs. Pests 13(7):5-7.

EADS, R. B., D. B. FRANCY, and G. C. SMITH — 1980. The swallow bug, *Oeciacus vicarius* Horvath (Hemiptera:Cimicidae), a human household pest. Proc. Entomol. Soc. Wash. 82:81-85.

ESSIG, E. O. — 1926. Insects of Western North America. Macmillan Co. 1035 pp.

FARRAR, W. E., JR., I. G. KAGAN, F. D. EVERTON, and T. F. SELLERS, JR. — 1963. Serologic evidence of human infection with *Trypanosoma cruzi* in Georgia. Am. J. Hygiene. 78:166-172.

FARRIER, M. H. — 1958. Summary of insect conditions — 1957: North Carolina. Coop. Econ. Insect Rep. 8(13):240-245.

GILLETTE, C. P. — 1898. Colorado's worst insect pests and their remedies. Col. Agric. Exp. Stn. Bull. 47; 64 pp.

GIRAULT, A. A. — 1905. The bedbug, *Clinocoris* (= *Cimex* = *Acanthia* = *Kli-*

nophilos) lectularia Linnaeus. Part I. Life-history at Paris, Texas, with biological notes, and some considerations on the present state of our knowledge concerning it. Psyche 12:61-74.

GIRAULT. A. A. — 1906. The bedbug, *Cimex lectularius* Linnaeus. Part II. Critical remarks on its literature, with a history and bibliography of pathogenic relations. Psyche 13:42-58.

GIRAULT, A. A. — 1912. Preliminary studies on the biology of the bedbug, *Cimex lectularius.* II. Facts obtained concerning the duration of its different stages. J. Econ. Biol. 7:163-188.

GIRAULT, A. A. — 1914. Preliminary studies on the biology of the bed bug, *Cimex lectularius,* Linn. J. Econ. Biol. 9:25-45.

GOLLNER-SCHEIDING, U. — 1980. Revision der Afrikanischen Arten sowie Bermerkungen zu weiteren Arten der Gattungen *Leptocoris* Hahn, 1833, und *Boisea* Kirkaldy, 1910 (Het., Rhopalidae). Dtsch. Entomol. Z.27:103-148.

GRIFFITH, M. E. — 1947. The blood-sucking conenose, or "big bedbug," *Triatoma sanguisuga* (LeConte), in an Oklahoma City household. Proc. Okla. Acad. Sci. 28:24-27.

GRUNDEMANN, A. W. — 1947. Studies on the biology of *Triatoma sanguisuga* (LeConte) in Kansas, (Reduviidae, Hemiptera). J. Kans. Entomol. Soc. 20:77-85.

HALL, M. C. — 1924. Lesions due to the bite of the wheel bug, *Arilus cristatus* (Hemiptera; Reduviidae). Arch. Int. Med. 33:513-515.

HALL, W. J. and E. E. WEHR — 1949. Diseases and parasites of poultry. U.S. Dep. Agric. Farmers' Bull. 1652. pp. 94-95.

HENRY, T. J. — 1977. Boxelder bug, *Leptocoris trivittatus* (Say) (Hemiptera:Rhopalidae). Reg. Hort. 3(2):19-20.

HIXSON, H. — 1943. The tropical bed bug established in Florida. Fla. Entomol. 26:47.

HOEBEKE, E. R. and A. G. WHEELER, JR. — 1982. *Rhopalus (Brachycarenus) tigrinus,* recently established in North America, with a key to genera and species of Rhopalidae in eastern North America (Hemiptera-Heteroptera). Proc. Entomol. Soc. Wash. in press.

HORN, D. J. — 1973. *Leptocoris rubrolineatus,* an occasional predator of the California oakworm, *Phryganidia californica* (Hemiptera:Rhopalidae: Lepidoptera:Dioptidae). Pan-Pac. Entomol. 49:196.

HOWARD, L. O. — 1900. The insects to which the name "kissing-bug" became applied during the summer of 1899. U. S. Dep. Agric. Div. Entomol. Bull. 22:24-30.

HUNGERFORD, H. B. — 1948. The Corixidae of the Western Hemisphere (Hemiptera). Univ. Kans. Sci. Bull. 32:5-827.

HUTSON, R. — 1932. Boxelder bug on strawberries. J. Econ. Entomol. 25:1107.

JOHNSON, C. G. — 1940. Development, hatching and mortality of the eggs of *Cimex lectularius* L. (Hemiptera) in relation to climate, with observations on the effects of preconditioning to temperature. Parasitology 32:127-173.

JOHNSON, C. G. and K. MELLANBY — 1939. Bed bugs and cockroaches. Proc. R. Entomol. Soc. London (A) 14(2-3):50.

JOHNSON, M. S. and A. J. HILL — 1948. Partial resistance of a strain of bed bugs to DDT residual. Med. News Letter 12(1):26-28.

KATZ, H. — 1957. Bedbug haven. Pest Control 25(10):102.

KNOWLTON, G. F. — 1935a. Bedbugs and cockroaches. Utah Agric. Exp. Stn. Leaflet 68. 4 pp.

KNOWLTON, G. F. — 1935b. Boxelder bug (*Leptocoris trivittatus* Say). Insect Pest Surv. Bull. 15(9):402.
KNOWLTON, G. F. — 1944. Boxelder bug observations. J. Econ. Entomol. 37:443.
KNOWLTON, G. F. — 1947a. Boxelder bug "bites" man. Bull. Brooklyn Entomol. Soc. 42:33.
KNOWLTON, G. F. — 1947b. Leafhopper "bites" man. Bull. Brooklyn Entomol. Soc. 42:169.
KNOWLTON, G. F. — 1948. Boxelder bug control. Utah Agric. Exp. Stn. M. S. 772 (mimeo).
KNOWLTON, G. F. — 1951. Boxelder bug damage to crops. J. Econ. Entomol. 44:994.
KNOWLTON, G. F. — 1960. An unusual flight of Hemiptera in southern Utah. Proc. Utah Acad. Sci. 37:53-54.
KNOWLTON, G. F. — 1962. An unusual fall flight of a maple aphid. Utah State Univ. Extension Serv. Mimeo. Ser. 201: 2 pp.
KOERBER, T. W. — 1963. *Leptoglossus occidentalis* (Hemiptera, Coreidae), a newly discovered pest of coniferous seed. Ann. Entomol. Soc. Am. 56:229-234.
LAWSON, P. B. — 1931. The genus *Xerophloea* in North America (Homoptera, Cicadellidae). Pan-Pac. Entomol. 7:159-169.
LEE, R. D. — 1955. The biology of the Mexican chicken bug, *Haematosiphon inodorus* (Duges) (Hemiptera:Cimicidae). Pan-Pac. Entomol. 31:47-61.
LENT, H. and P. WYGODZINSKY — 1979. Revision of the Triatominae (Hemiptera, Reduviidae), and their significance as vectors of Chagas' disease. Bull. Am. Mus. Nat. Hist. 163(3):123-520.
LEONARD, D. E. — 1966. Biosystematics of the "*leucopterus* complex" of the genus *Blissus* (Heteroptera:Lygaeidae). Conn. Agric. Exp. Stn. Bull. 677: 47 pp.
LINTNER, J. A. — 1894. North Dakota's new bug. Country Gentleman 59:841.
LONG, W. H. — 1928. Why only staminate box-elders should be used for shade trees. J. Econ. Entomol. 21:433-434.
McDANIEL, E. I. — 1933. The box elder bug as a household pest. Q. Bull. Mich. Agric. Exp. Stn. 15(4):226-227.
McKENNY-HUGHES, A. W. — 1937. The bedbug: its habits and life-history and how to deal with it. Brit. Mus. (Nat. Hist.) Econ. Ser. 5: 19 pp.
MAIL, G. A. — 1940. Infestation of a high school by *Oeciacus vicarius* Horv. J. Econ. Entomol. 33:949.
MALLIS, A. — 1971. Do cockroaches eat bed bugs? Pa. Pest Control Q. 14(1):5.
MALLIS, A. — 1973. Unusual insect problems in Pennsylvania. Pa. Pest Control Q. 16(2):3,6.
MALLIS, A. — 1979. Some unusual household pests in Pennsylvania. Pest Control 47(10):36-37, 44-46.
MALLIS, A. — 1981. The inner works of an industrial insecticide laboratory. Melsheimer Entomol. Ser. 30:30-38
MAMPE, C. D. — 1978. Answers. Pest Control 46(8):45-46.
MAMPE, C. D. — 1979. Control boxelder bugs? Pest Control 47(1):38-39.
MARLATT, C. L. — 1896. The bedbug and cone-nose. Pp. 32-42 *in* Howard, L. O. and C. L. Marlatt. The principal household insects of the United States. U.S. Dep. Agric. Div. Entomol. Bull. 4.
MATHESON, R. — 1950. Medical Entomology. 2nd ed. Comstock Publishing

Co., Ithaca, N.Y. 612 pp.

MEAD, F. W. — 1965. The blood-sucking conenose (Hemiptera:Reduviidae). Fla. Dept. Agric. Entomol. Circ. 33: 2 pp.

MEAD, F. W. — 1974. The wheel bug, *Arilus cristatus* (Linnaeus) (Hemiptera:Reduviidae). Fla. Dept. Agric. Consum. Serv. Entomol. Circ. 143: 2 pp.

MICHELBACHER, A. E., D. P. FURMAN, C. S. DAVIS, J. E. SWIFT, and I. B. TARSHIS — 1961. Control of household insects and related pests. Calif. Agric. Exp. Stn. Circ. 498: 40 pp.

MILLS, H. B. and D. J. PLETSCH — 1941. Another infestation of a school building by *Oeciacus vicarius* Horvath. J. Econ. Entomol. 34:575.

MUNRO, J. A. and R. L. POST — 1949. Control of boxelder bugs. J. Econ. Entomol. 42:994.

MYERS, L. E. — 1928. The American swallow bug, *Oeciacus vicarius* Horvath (Hemiptera, Cimicidae). Parasitology 20:159-172.

NETTLES, W. C. — 1961. Household insects. Clemson Coll. Exten. Serv. Bull. 101: 54 pp.

OSBORN, H. — 1912. Leafhoppers affecting cereals, grasses, and forage crops. U.S. Dep. Agric. Bur. Entomol. Bull. 108: 123 pp.

PACKCHANIAN, A. — 1940. Natural infection of *Triatoma heidemanni* with *Trypanosoma cruzi* in Texas. Pub. Health Rep. 55:1300-1306.

PARSHLEY, H. M. — 1919. On some Hemiptera from western Canada. Occ. Papers Mus. Zool., Univ. Mich. 71:1-35.

PATTON, W. S. — 1931. Insects, Ticks, Mites and Venomous Animals of Medical and Veterinary Importance. Part II. Public Health. H. R. Grubb Ltd., Croydon, England. 737 pp.

PITCHON, S. — 1980. Control bats? Pest Control 48(9):34-35.

PRATT, H. — 1958. Ectoparasites of birds, bats and rodents and their control. Pest Control 26(10):55-56, 58, 60, 94, 96.

RACHESKY, S. — 1970. A list of recommendations for common insect problems. Pest Control 38(1):32, 34, 36-37.

RACHESKY, S — 1973. Swimming pool pests. Pest Control 41(8):39-40.

RANDOLPH, N. M. — 1946. DDT for the control of *Triatoma*. J. Econ. Entomol. 39:419.

READIO, P. A. — 1927. Studies on the biology of the Reduviidae of America north of Mexico. Univ. Kans. Sci. Bull. 17:5-291.

RICE, P. L. — 1936. Bat bedbug (*Cimex pilosellus* Horv.). Insect Pest Surv. Bull. 16(8):388.

RILEY, W. A. and O. A. JOHANNSEN — 1938. Medical Entomology. 2nd ed. McGraw-Hill, New York. 483 pp.

ROGOFF, W. M. — 1950. Control of boxelder bugs. P. 19 *in* Proc. 5th Ann. Mtg. No. Centr. States Branch, Am. Assoc. Econ. Entomol.

RUCKES, H. — 1946. Notes and keys on the genus *Brochymena* (Pentatomidae, Heteroptera). Entomol. Am. 26:143-238.

RYCKMAN, R. E. and M. A. CASDIN — 1976. The Triatominae of western North America, a checklist and bibliography. Calif. Vector Views 23(9-10):35-52.

SAILER, R. I. — 1952. The bedbug: an odd bedfellow that's still with us. Pest Control 20(10):22, 24, 70, 72.

SCHAEFER, C. W. — 1975. A re-assessment of North American *Leptocoris* (Hemiptera-Heteroptera:Rhopalidae). Ann. Entomol. Soc. Am. 68:537-541.

SCOTT, H. G. — 1958. *Triatoma sanguisuga* infesting a bedroom in Decatur, Georgia. J. Econ. Entomol. 51:549.

SCUDDER, G. G. E. — 1961. Some Heteroptera new to British Columbia. Proc. Entomol. Soc. Br. Columb. 58:26-29.

SCUDDER, G. G. E. — 1979. Hemiptera. Pp. 329-348 *in* Danks, H. V. ed. Canada and its insect fauna. Mem. Entomol. Soc. Can. No. 108.

SLATER, J. A. and R. M. BARANOWSKI — 1978. How to Know the True Bugs (Hemiptera-Heteroptera). Wm. C. Brown, Dubuque, Iowa. 256 pp.

SLATER, J. A. and C. W. SCHAEFER — 1963. *Leptocoris trivittatus* (Say) and *Coriomeris humilis* Uhl. in New England (Hemiptera:Coreidae). Bull. Brooklyn Entomol. Soc. 58:114-117.

SMITH, C. N. — 1958. Control of bedbugs and human lice. Pest Control 26(11):9-10,12.

SMITH, F. D., N. G. MILLER, S. J. CARNAZZO, and W. B. EATON — 1958. Insect bite by *Arilus cristatus,* a North American reduviid. Am. Med. Assoc. Arch. Dermatol. 77:324-330.

SMITH, M. R. and H. F. CHAO — 1956. Bedbug, batbug and chimney swift bug. Pest Control 24(2):58.

SMITH, R. C. and B. L. SHEPHERD — 1937. The life history and control of the boxelder bug in Kansas. Trans. Kans. Acad. Sci. 40:143-159.

SNODDY, E. L., W. J. HUMPHREYS, and M. S. BLUM — 1976. Observations on the behavior and morphology of the spider predator, *Stenolemus lanipes* (Hemiptera:Reduviidae). J. Ga. Entomol. Soc. 11:55-58.

SNODGRASS, R. E. — 1944. The feeding apparatus of biting and sucking insects affecting man and animals. Smithsonian Misc. Coll. 104(7):1-112.

SPENCER, G. J. — 1930. The status of the barn swallow bug, *Oeciacus vicarius* Horvath. Can. Entomol. 62:20-21.

SPENCER, G. J. — 1935. The bedbugs of British Columbia. Proc. Entomol. Soc. Br. Columb. 31:43-45.

SPENCER, G. J. — 1945. On the incidence, density and decline of certain insects in British Columbia. Proc. Entomol. Soc. Br. Columb. 42:19-23.

STEARNS, L. A. — 1937. Important insects of the year. Bull. Del. Agric. Exp. Stn.207

STEDMAN, J. M. — 1902. The chinch bug. Univ. Mo. Agric. Exp. Stn. Bull. 51. 28 pp.

SWEET, M. H. — 1960. The seed bugs: a contribution to the feeding habits of the Lygaeidae (Hemiptera:Heteroptera). Ann. Entomol. Soc. Am. 53:317-321.

SWENK, M. H. — 1925. A stink bug (*Thyanta custator* Fab.). Insect Pest Surv. Bull. 5(7):372.

SWENK, M. H. — 1929. Boxelder bug (*Leptocoris trivittatus* Say). Insect Pest Surv. Bull. 9(3):87.

TINKER, M. E. — 1952. The seasonal behavior and ecology of the boxelder bug *Leptocoris trivittatus* in Minnesota. Ecology 33:407-414.

TOWNSEND, C. H. T. — 1894. Note on the coruco, a hemipterous insect which infests poultry in southern New Mexico. Proc. Entomol. Soc. Wash. 3:40-41.

TRAVIS, B. V., H. H. LEE, and R. M. LABADAN — 1969. Arthropods of Medical Importance in America North of Mexico. U.S. Army Natick Lab., Natick, Mass. Tech. Rep. 69-2-ES. 335 pp.

TRAYNOR, P. C. — 1978. Pest Control Technol. 6(12):24-31.

TRUMAN, L. C., G. W. BENNETT, and W. L. BUTTS — 1976. Scientific Guide to Pest Control Operations. 3rd ed. Harvest Publ. Co., Cleveland, Ohio. 276 pp.

TUCKER, E. S. — 1911. Random notes on entomological field work. Can. Entomol. 43:22-32.

TUTHILL, L. D. — 1943. The psyllids of America north of Mexico (Psyllidae:Homoptera). Iowa State Coll. J. Sci. 17:443-660.

USINGER, R. L. — 1934. Blood sucking among phytophagous Hemiptera. Can. Entomol. 66:97-100.

USINGER, R. L. — 1966. Monograph of Cimicidae (Hemiptera-Heteroptera). Thomas Say Foundation 7:1-585.

WALSH, B. D. and C. V. RILEY — 1869. The parasites of the human animal. Am. Entomol. 1:84-88.

WEBSTER, R. L. — 1926. Boxelder bug (*Leptocoris trivittatus* Say). Insect Pest Surv. Bull. 6(6):200.

WEHRLE, L. P. — 1939. Observations on three species of *Triatoma* (Hemiptera:Reduviidae). Bull. Brooklyn Entomol. Soc. 34:145-154.

WESTCOTT, C. — 1973. The Gardener's Bug Book. 4th ed. Doubleday & Co. 689 pp.

WHEELER, A. G., JR. — 1975. Birch catkin bug — a nuisance insect. Pa. Pest Control Q. spring issue, p.4.

WHEELER, A. G., JR. — 1976. Life history of *Kleidocerys resedae* on European white birch and ericaceous shrubs. Ann. Entomol. Soc. Am. 69:459-463.

WHEELER, A. G., JR. and J. L. HERRING. 1979. A potential insect pest of azaleas. Q. Bull. Am. Rhododendron Soc. 33(1):12-14, 34.

WOLLERMAN, E. H. — 1965. The boxelder bug. U.S. Dep. Agric. For. Pest Leaflet 95: 6 pp.

WOOD, S. F. — 1950. The distribution of California insect vectors harboring *Trypanosoma cruzi* Chagas. Bull. So. Calif. Acad. Sci. 49:98-100.

WOOD, S. F. — 1951. Importance of feeding and defecation times of insect vectors in transmission of Chagas' disease. J. Econ. Entomol. 44:52-54.

WOOD, S. F. — 1953. Conenose bug annoyance and *Trypanosoma cruzi* Chagas in Griffith Park, Los Angeles, Calif. Bull. So. Calif. Acad. Sci. 52:105-109.

WOOD, S. F. — 1975. Home invasions of conenose bugs (Hemiptera:Reduviidae) and their control. Nat. Pest Control Operator News 35(3):16-18.

WRIGHT, W. H. — 1944. The bedbug — its habits and life history and methods of control. U.S. Pub. Health Service Rep. Suppl. No. 175: 9 pp.

WYGODZINSKY, P. W. — A monograph of the Emesinae (Reduviidae, Hemiptera). Bull. Am. Mus. Nat. Hist. 133:1-614.

ZELEDON, R. and J. E. RABINOVICH — 1981. Chagas' disease: an ecological appraisal with special emphasis on its insect vectors. Ann. Rev. Entomol. 26:101-133.

ARNOLD MALLIS

Arnold Mallis was born in New York City on October 15, 1910, the son of Russian immigrants to the United States. He attended grade school and high school in Brooklyn and Long Island. In 1927, at the age of 16, he moved with his parents to Los Angeles, California, where he completed his last year of high school.

In 1929, he entered the University of Southern California and began pre-dental courses. Since the Great Depression began at about this time, he dropped out of college and worked in a Los Angeles garment factory. After two years on the job he decided to return to college and, because of his interest in trees and natural history, began a two year course in forestry at Pasadena Junior College (now Pasadena City College).

He entered the University of California at Berkeley in 1932, where he studied under such outstanding entomologists as Professor E. O. Essig in economic entomology, Dr. E. C. Van Dyke in forest entomology, Professor W. B. Herms in medical entomology and insect ecology, and Dr. W. M. Hoskins in insect toxicology. These teachers and others instilled in him a life-long interest in insects and their control. He received a bachelor's degree in entomology in 1934.

After two years graduate work he left the University in 1936 and sought employment as an entomologist in southern California. Since there were no jobs open in entomology he took the state examinations for a license in structural pest control and received a Class A license (all categories, including fumigation).

Mallis worked two years for pest control firms in Los Angeles, Hollywood and Bakersfield and then in 1938 obtained a job as a field aide in the USDA Bureau of Entomology, doing research on vegetable insects in southern California. This job lasted for six months. He then returned to Berkeley for two purposes, to complete his master's degree on the ants of California and to collect information for the first edition of the <u>Handbook of Pest Control</u>. In 1939, he returned to southern California and became entomologist-in-charge of pest control for the Buildings and Grounds Department, UCLA. In 1942, shortly after the United States entered WWII, he became associated with the USPHS, implementing malaria control programs around military camps in Louisiana. In 1944, he began to screen compounds as pyrethrins substitutes for Hercules Powder Company at the University of Delaware. The screening resulted in the discovery of Compound 3956, later known as toxaphene.

In 1945, Mallis commenced his employment in the entomology laboratory of Gulf Oil Corporation in Harmarville, Pennsylvania. Here he worked as an entomologist on household insects and household insecticides for 23 years until the company closed its entomology laboratory in 1968. He then became an extension entomologist for The Pennsylvania State University and retired as an associate professor in 1975, at the age of 65.

CHAPTER TWELVE

Clothes Moths

Revised by Arnold Mallis[1]

The clothes moth is a creature rarę
That lives and even thrives on air
And has a light and buoyant soul
Since when it feeds it eats a hole.

—A.M.

THE QUESTION HAS BEEN RAISED as to just what the clothes moth ate before man started to wear clothes. The clothes moth, undoubtedly, fed on animal carcasses, as well as the furs and feathers found in the caves of early man. Linsley (1944) notes the natural food of the clothes moth is pollen, hair, feathers, wool, fur, dead insects, and dried animal remains.

The ancestors of the clothes moth may have come over on the Mayflower, and Peter Kalm, keen observer of insects and a traveler in the Colonies, has this to say of moths in what was then the village of Philadelphia: "Moths, or *Tineae,* which eat the clothes, are likewise abundant here. I have seen cloth, worsted gloves, and other woolen stuffs, which have hung all the summer locked up in a shrine, and had not been taken care of, quite cut through by these worms, so that whole pieces fell out." However, in view of Linsley's studies showing that the webbing clothes moth lives outdoors in bird nests, bee cells, and similar places, it is possible the webbing clothes moth was here before the arrival of the white man.

The word moth is derived from the early English *mothe* or from the German *motte.* According to Colton (1927), the word "moth" comes from the same root as mouth, and has come to mean something that devours, such as a maggot. Thus, certain carpet beetles, specifically their larvae, are called buffalo-moths, and silverfish are called "fish moths." Materials damaged by clothes moths are called *moth-eaten,* but the clothes moth adult, because of its imperfect mouthparts, is incapable of eating clothes. Naturally, neither the egg nor the pupa can feed, leaving the only remaining stage, the larva, with its chewing or biting mandibles.

The larvae of clothes moths feed on woolens and other animal products and

[1]*Associate Professor (Retired) Extension Entomology, The Pennsylvania State University, University Park, Pa.*

for this reason are of concern. The adult clothes moths are characterized by narrow wings which are fringed with long hairs. They are yellowish or golden-colored and are less than ½ inch/12 mm in length. At times they are confused with the Angoumois grain moth, a pest of cereals, that may be found flying during the day. The adults of the Angoumois grain moth have pointed hind wings and can be separated from clothes moths by this characteristic. Clothes moths are often confused with many other moths that feed on vegetation and enter the home.

It should be noted clothes moths rarely hover around a light as do the larger outdoor moths, and in this respect, Michelbacher and Furman (1951) state: "They prefer darkness, but will flit about the margins of lighted areas. When infested fabrics are disturbed, the moths run rapidly or fly to conceal themselves in folds or other secluded places. This ability to disappear is characteristic." The life cycle of the clothes moth is complex, having an egg, larva, pupa, and adult or moth stage. In the United States, concern for the most part is with the webbing clothes moth, *Tineola bisselliella* (Hum.). Occasionally, the case-making clothes moth, *Tinea pellionella* (L.), and rarely the tapestry moth, *Trichophaga tapetzella* (L.), are responsible for moth damage, There are several other species of moths that feed on clothing, but these are uncommon, and some of them are discussed in the chapter on stored product pests.

Ferguson (1950), with the aid of extension entomologists, made a survey of clothes moth and carpet beetle infestations in the United States. He makes the statement that "in the area east of the Rocky Mountains, the bulk of the damage in the southern states is due to clothes moth; whereas in the northern states, the carpet beetle is much more important than the clothes moth." He attributes this difference in distribution, in part, to the fact the higher humidities of the South are more favorable to the development of the clothes moth, and the carpet beetles can better withstand the colder weather in the North.

Food of the clothes moth. As noted previously, the clothes moth probably plagued the prehistoric housewife by residing in her cave and feeding on the animal parts to be found there. That they were originally scavengers is borne out by the fact they attack not only furs, but imperfectly cleaned animal skeletons, mammal and bird skins, insect specimens, etc. Moncrieff (1950) notes that "the clothes moth and the carpet beetle attack cadavers when they are in a very advanced stage of decomposition known as mummified. . . ." Linsley (1944) observed the webbing clothes moth in California to breed in the nests of the English sparrow, in the cells of a wild bee, as well as under loose bark. Here they were probably living on animal products, insect remains, and possibly stored pollen in the case of the beehive.

According to Waterhouse (1958), insects are the only animals capable of digesting keratin, a protein present in wool, hair, fur, horns, hoofs, and feathers.

There are those who have attempted to analyze in a scientific way the unique ability of the clothes moth and the carpet beetle to feed on such dry, dead things like fur, hair, wool, etc. McTavish (1938), using the work of Linderstrom and Duspiva (1935) as a basis of his interpretations of this problem, summarizes the subject. The clothes moth digests keratin through a unique enzyme or a series of enzymes called keratinases, which hydrolyze the keratin in the digestive system of the moth. McTavish states, "It would appear that among the insects the clothes moth larva is a nuisance largely because of the intestinal secretion of a thiol compound which is a great reducing agent. Thiol as a reducing agent breaks up the disulfide linkages in the cystine molecule (which makes up seven to eight percent of the wool molecule)." McTavish then notes: "It would appear that among the insects, the clothes moth larva is a nuisance largely because of the intestinal secretion of a thiol compound which renders wool susceptible to attack by proteolytic enzymes." Waterhouse (1958) prepared a comprehensive summary of the biochemistry of wool digestion by fabric pests.

Vollmer (1931) observed the larva feeding on the cast skin of snakes as well as on the eggs and living larvae of ticks. It is his opinion that in hot countries the larvae of *Tineola bisselliella* (Hum.) were originally predacious. The larvae have been found infesting beef meal, fish meal, pemmican, casein, milk products, and have even consumed fingernail clippings. Back (1935) notes, "In experimental work clothes moths have thrived best upon a diet of bristles, hairs, feathers, fur, or raw wool, and not so well upon ordinary woolen cloth used in the making of wearing apparel." Griswold (1944) proved fish meal to be the most satisfactory food of a great many foods tested for the webbing clothes moth, and the larvae were unable to complete their growth on clean woolen fabrics.

In the home, clothes, carpets, rugs, upholstered furniture, the felts in pianos, as well as brush animal bristles are all prime sources of moth infestation. Lint from rugs, as well as the hair of pets accumulate behind baseboards and other inaccessible places, where they may serve as a source of moth infestation. Spencer (1931) found lint from carpets and other sources had accumulated in parts of heating units and served as important breeding places for the clothes moth. Ferguson (1950) showed that in the United States, the types of goods damaged in the order of decreasing importance are carpets and rugs, woolen clothes, blankets, upholstery, and stored wool.

Although it is often stated clothes moths do not feed on plant products, several individuals such as Laing (1932) and Swenk (1922) claim that *Tineola bisselliella* (Hum.) will consume injured stored grain such as oats and barley. Nagel (1920) found the clothes moth larva will feed on cotton, linen, and silk when forced to do so, and that the larva attains a size only one-half as large as the larva fed on woolens. However, Waterhouse (1958) states neither clothes moth nor carpet beetle larvae can digest cellulose fibers or synthetic fibers that are nonprotein in origin. Griswold (1944) noted the larvae could not live on raw silk alone. Back (1931) states, "In rare instances, the larva may eat holes in paper or in cotton and linen goods, but this, when it occurs, usually results from excessive infestation in close quarters, or the cutting of the goods by the larva that it may obtain bits of material to build into its cocoon. For all practical purposes, moth larvae do not feed upon vegetable products." Moncrieff (1950) notes that reports of clothes moth "attacking silk or other fibers which are not animal hairs have been due to the real attack being on a size or finish on the yarn or fabric. It seems to be fairly well established that silk is immune from its attack." Curran (1949) found the webbing clothes moth infesting stored cheesecloth insect screening which harbored dead insects. Dingler (1928) reports the clothes moth larva in sugar, probably the accidental occurrence of a migrating larva.

Mallis et al. (1958) showed the webbing clothes moth does not feed on such synthetic fabrics as nylon, Dynel, Dacron, Orlon, Vicara, acetate rayon or viscose rayon, but readily feeds on mixtures of wool and some of these synthetics. Reumuth (1946) notes that "in a mixed fabric, the larva destroys all other fibers. These fibers, although they cannot be digested, are carried into the alimentary canal somewhat like ballast and are excreted, chemically unchanged, with the pellets.

Kemper (1936) came to the conclusion the clothes moth larva cannot complete its development on furs such as rabbit or hair such as calf. He found rabbit and wildcat furs and calf and goat hair were subject to slight damage, whereas skunk, marmot, and opossum furs were damaged severely. Moncrieff (1950), on the other hand, finds rabbit fur attractive to moths and readily subject to their attacks. The larva consumes the hide near the base of the fur, thus causing the fur to fall out.

According to Back (1931), in upholstered furniture "the larva of clothes moths fed upon woolen fibers in covers, the feathers in cushions and pads, and to a limited extent upon hair fillings. In advanced cases of infestation the dead moths and larvae resulting from the high rate of mortality suffered by moth colonies are important sources of food." He further states that such vegetable products in upholstered furniture as cotton batting, burlap, cotton and linen fabrics, Spanish moss, flax straw or tow, palm fiber, and sea moss are not eaten, even though these materials may be found full of dead worms, larvae, and excrement. The larvae invade such substances in order to rest in them. Curwen (1932) found woolen coverings of wire in a telephone exchange in London to be extensively damaged by clothes moths, and that "on reaching maturity the larvae attack material of other than animal origin from which to construct the cocoon, one formed of cotton yarn being obtained." Ott (1955) and Pence (1966) describe photomicrograph methods for determining the insects responsible for damaging woolen fibers.

Colman (1932) showed that Vitamin B as supplied in dry brewer's yeast was essential to full development of clothes moth larvae, and Billings (1936) also observed that the clothes moth larva cannot be bred on thoroughly clean woolen goods. In this respect he notes, "Eggs laid on such woolens hatch normally, but the little larvae feed slightly or not at all, and death from apparent starvation follows in about two weeks." He also found food stains stimulate larval growth, as did ordinary house dust but with varying effect. He had to sprinkle dry brewer's yeast or fish meal on the clean woolen goods in order to get the young larvae to live.

Hoskins et al. (1940), commenting on Vitamin B states, "It is very probable that the well-known tendency of moth larvae to seek out soiled portions of woolen fabrics is due to their need for Vitamin B and not to any need for fats. An interesting sidelight of this work is the suggestion that human sweat contains considerable Vitamin B." Fraenkel and Blewett (1945) found certain constituents of Vitamin B complex such as pantothenic acid and nicotinic acid are essential for the growth of the larva of the webbing clothes moth. These authors note thiamin, riboflavin, nicotinic acid and pantothenic acid occur in human sweat.

Mallis et al. (1959 and 1962) showed the larvae of the webbing clothes moth are attracted to such common stains as human sweat, human urine, tomato juice, milk, coffee, and beef gravy. Salts are essential nutrients to certain fabric insects and this is one reason they seek out spots stained with fruit juices, urine, perspiration, etc. Some of the salts especially attractive to the larvae are KCl, K_2HPO_4, NaCl and Na_2HPO_4.

The question may well be asked, just how does the larva of the clothes moth, which is very susceptible to the least bit of drying, obtain its moisture from a source such as hair, fur, etc? Colton (1927) has the following to say in this regard: "Although the caterpillar never drinks a drop of water in its life, yet, as everybody knows it is a very juicy animal. In most animals, the elimination of the nitrogen in the body waste is in the form of urea, which is dissolved in water and passed out of the animal. The clothes moth, in common with birds and certain desert animals, excretes these wastes as uric acid in little crystals which pass out almost dry." Thus does the larva conserve its water supply.

THE MORE COMMON CLOTHES MOTH ADULTS (AFTER AUSTEN)

1. Head and forewings shiny, golden-buff or putty colored, unspotted. The Webbi.ng Clothes Moth, *Tineola bisselliella* (Hum).
2. Head and forewings dusty brownish, the latter with three dark spots, sometimes indistinct. Larvae within a case. The Casemaking Clothes Moth, *Tinea pellionella* (L.).
3. Head white; forewings black at the base, then snowy-white, with the extreme tips mottled with black. The Tapestry Moth, Carpet Moth, *Trichophaga tapetzella*. (L.).

THE WEBBING CLOTHES MOTH
Tineola bisselliella (Hum.)

The webbing clothes moth, *Tineola bisselliella* (Hum.), is worldwide in distribution. Some individuals are of the opinion that Europe is its original home, but Austen et al. (1935) state that in all probability it is native to Africa. This species is undoubtedly the most common moth in the United States, since it is the one most often referred to in papers on moth damage. Benedict (1917), in order to produce clothes moths in some quantity, bought old woolen rags from a dealer who obtained his material from every part of the country, yet in these rags he found only the species *Tineola bisselliella* (Hum.)

Adults. The adults of *Tineola bisselliella* (Hum.) have a body that is covered with shiny golden scales with the top of the head bearing a fluffy pompadour of reddish golden hairs. In 1971 the author examined adult clothes moths in fish food from West Germany. The adults did not have the reddish golden hairs on top of the head. The compound eyes are black and the antennae are darker than the rest of the body. The wings are without spots and have an expanse of approximately ½ inch/12 mm, with that of the female being slightly larger. When resting, the expanse of the folded wings is from ¼ to ⅓ inch/six to eight mm, the larger size being that of the female.

This moth flies, stays in the darker areas of the room, and ordinarily does not flit around the lamps like the larger outdoor species of moths. The clothes moths that do fly are ordinarily males, since the females, usually heavy with eggs, prefer to walk or run. However, the female may be found on the wing after she has laid her eggs, and at times may fly even when she is gravid. Austen et al. (1935) explain the preponderance of flying males as follows: "It will be seen then that if the eggs are to be fertilized, the males must find the females and not *vice versa*, which is one of the reasons why so many males and so few females are found on the wing." Herrick (1933) found the adults to be capable of flying fairly long distances. Thus, houses 150 to 300 feet from a warehouse containing infested raw wool were soon infested by the flying clothes moths. According to Griswold (1931 and 1933), the sexes are about evenly divided in number. Studies by Titschack (1936) showed that when food is scarce males predominate, when abundant, females are most common.

The moth, upon emerging, leaves its pupal skin partly protruding from the cocoon. It may mate and lay eggs before it is one day old, and in a laboratory

Fig. 12-1. Webbing clothes moth adults.

culture most of the eggs are deposited the first few days after emergence. During copulation, the moths rest with their bodies in opposite directions. Austen et al. (1935) observed the average life of the adult was 28.3 days for the male and 16 days for the female at a temperature of 68° to 77° F/20° to 25° C. During the winter at a lower temperature, the time was extended to 34 days. Whitfield and Cole (1958) completed detailed studies on the biology of this moth.

Egg. The egg is large, oval, ivory white, about $1/24$ inch/one mm long, and barely visible against a black background. Griswold (1944) notes that under the binocular microscope the egg has many narrow ridges. According to Austen et al. (1935), the eggs are laid singly or in groups of two or more. Although the eggs are fragile, they actually are harder than carpet beetle eggs and can be

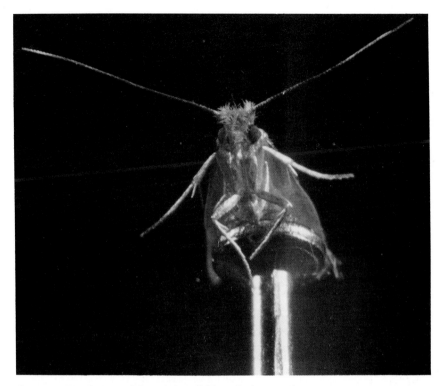

Fig. 12-2. Head view of the adult webbing clothes moth.

shaken in a sieve without harming them. Benedict (1917), who did some careful work on the egglaying habits of this moth, observed the female to lay eggs 24 hours after maturing.

The average number of eggs laid by a female was between 40 and 50. The number, however, ranged from 30 to 160, and Austen et al. (1935) had one female that deposited 221 eggs. Griswold (1944) notes the female may deposit eggs before 24 hours have elapsed. The eggs may be laid in one night or over a period of two to three weeks. According to Nagel (1920), the maximum deposition of eggs is between the seventh and eleventh days of adult life. The female dies after laying all her eggs. The males often live several weeks longer, during which time they continue to mate.

Benedict (1917) noted that "the eggs are carefully placed among the threads of the clothes and fastened by some gelatinous material so that they do not readily shake off. If the cloth has a ravelled edge the female will generally lay most of the eggs deep among the loose threads." The material which aids the sticking qualities of the eggs is a secretion of the accessory glands. Back (1935) observed the females to place their eggs deep in the meshes, where they are held by the threads of such loosely-woven materials as yarns, carpets, rugs, woolens, etc. On closely-woven materials such as serges, the eggs may be laid on the threads themselves. Benedict found the moths to lay their eggs as readily on cotton and silk as on wool.

The time necessary for the hatching of the eggs varies according to the environmental conditions. Various authors find the eggs to hatch in from four days to three weeks, with an average of four to 10 days during the summer, the longer periods occurring during the winter. In artificially heated buildings, the moths will emerge, and the larvae will feed all through the cold season. The insect reportedly does not overwinter in the egg, since the embryo is unable to survive long periods of incubation. Individuals often mistake the fine, hard, and gritty particles of excrement for eggs.

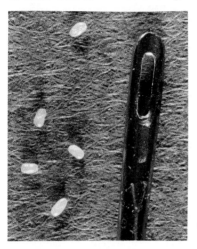

Larva. The larva upon emerging from the egg is active, white, translucent, and about one mm long. The average weight of the newly-emerged larva is 0.0273 mg, and the average of the full-grown larva is 3.08 mg. The healthy young larva is shiny and greasy-appearing and has three pairs of jointed legs, and four pairs of pseudopods or larval legs. The last segment has a pair of appendages, the claspers.

Fig. 12-3. Eggs of the webbing clothes moth when compared to the eye of a needle.

Colman (1940) found the small size of the newly-born larva permits it to enter anything with an opening greater than .01 mm. Since the female may lay her eggs in the crevices of what may otherwise be a very tight container, it can readily be perceived how the clothes moth later appears in such a repository. The emerging larva commences to feed almost immediately where suitable food is available.

The size of the full-grown larva is dependent on a number of conditions (e.g. food, temperature, moisture, etc.) In any event, it ordinarily does not attain a size greater than ½ inch/12 mm. With regard to the effect of food on larval growth, Austen et al. (1935) note that starving the larvae results in a reduction in size, and when they become adults, they are not capable of depositing fertile eggs. The measure of control based upon starvation of the larvae may not control the clothes moth. Larvae, especially the older ones, may live without food for some time.

Benedict (1917) ran some interesting experiments utilizing the fact that ingested foods are seen through the transparent body. He fed the larvae on felts of diverse colors and obtained larvae with median streaks of red, blue, green, etc. Since the dyes of the various materials are unaffected by the digestive processes, the excreta took on the color of the larval food.

The larva may spin a feeding tunnel of silk with which is incorporated some of the fibers on which it is feeding, as well as its excrement. A portable case is not made by the webbing clothes moth larva. Benedict (1917) found the larva may feed for some distance in this tube, and in certain instances it will abandon the tube, build another, and commence to feed once again. Some of the larvae merely spin a small silken patch here and there and graze as they go along.

Reumuth (1946) notes the silk-forming organs "are long tubular glands alongside the center of the intestine. The spinneret is a pointed appendage of the middle lower jaw (submentum)."

The larva is more or less stationary in the tunnel, and it is its concentration on its feeding area that does the most damage. Feeding tubes occur commonly in dark, secreted areas of clothes, such as under collars or cuffs. The larvae often are very active and may be found crawling on clothes or on the floor beneath badly infested upholstered furniture. Curran (1949) states if the larva finds the food satisfactory, it may finish development within one inch/ 25 mm of where the egg was deposited. Otherwise, it moves on. When the larva feeds between the carpet and the floor, it may extend its tubes along and in the floor cracks. The presence of the larva is indicated by the webbing beneath the carpet.

Fig. 12-4. Newly emerged larva of a webbing clothes moth on a pin head.

As was previously mentioned, food stains as well as other stains are especially subject to damage.

According to Back (1935), many of the larvae that hatch during the summer do not become moths until the following spring. At times, the larva, without apparent cause, may enter a resting stage or cocoon. It becomes inactive and ceases feeding and growing, only later, especially if disturbed, continuing its normal activities. These dormant periods may extend from eight to 24 months, and it is believed that in nature such cessation of activity may tide it over unfavorable periods. Apparently, the number of molts varies greatly, depending especially upon the duration of the larval period. Thus, Titschack (1936) found the larva may molt from five to 41 times. He also noted the time for the larva to complete the larval stage is of short duration on raw wool, cow hair, and

Fig. 12-5. Larva of webbing clothes moths are attracted to stains. The word "stain" was treated with tomato juice and exposed to a culture of the larvae.

Fig. 12-6. Larvae of webbing clothes moths on a buttonhole.

rabbit fur, but is somewhat extended on woolen clothing. Nagel (1920) also noted a varied number of molts with a maximum of 11. He determined that on an average the first molt occurred in nine to 10 days, the second after 12 to 13 days, the third 15 to 20 days later, and the remaining molts at intervals of 25 days. Back (1935) observed the shortest record for the completion of the larval stage was 40 days at 85° F/29.5° C. Griswold (1944) observed the males require approximately 30 days and the females 35 days for the duration of the larval stage.

The great variation in the time to complete the larval development does not appear to be entirely dependent on food, temperature, and humidity and may be very short or up to 29 months. Griswold notes: "The adult of *Tineola bisselliella* has rudimentary mouth parts and cannot eat; hence a sufficient supply of nutrients must be consumed by each larva to carry the individual not only through the pupal stage, but through adult life as well."

Pupa. The larva, having reached full development, spins a pupal case of silk, which is smooth on the inside, but to the outside, bits of the textile on which it is feeding as well as its excrement may be incorporated. The size of the case varies from ⅙ to ¼ inch/four to six mm in length. Investigators differ greatly on the time necessary for pupation, probably because their observations were made under varied environmental conditions. Back states that in the warm summer months the pupal stage may be completed in eight to 10 days, and during the winter in a steam-heated building, in from 21 to 28 days. Nagel

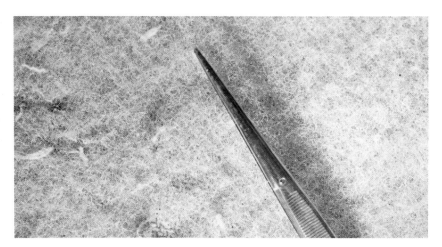

Fig. 12-7. Typical webbing of webbing clothes moth.

(1920) reported the pupal stage was completed in 18 to 19 days, and Titschack (1936) found there was a variation of between 14 and 44 days. Griswold (1944) observed the pupal stage required approximately 12 days at 75° F/24° C and 23 days at 65° F/18° C.

Seasonal history. The adults may be found in artificially-warmed buildings at any time of the year. During the summer, the adults are on the wing from May through July. Nagel (1920) states that in Germany there are two chief flight periods, in the spring and autumn, but that there are overlapping broods due to the prolonged life of some of the larvae. He found the time from egg to egg was about five to nine months. Back (1935) recorded a development in an incubator, from the day the egg was laid to emergence from the pupal stage, of 55 days at a temperature of 85° F/29° C. He concluded the variation in the length of the life history may be from 55 days to four years, but under optimum conditions averages 65 to 90 days. Griswold (1944) recorded a development period from egg to adult of 35 days for five webbing clothes moths. Notini (1939), in Sweden, found this insect may produce four generations a year in the warmed parts of buildings, but only one generation in unheated areas. Griswold et al. (1936) noted the most favorable humidity was 75 percent and lower relative humidities increased the time required for development. They were of the opinion the dry atmosphere of the house during the winter may retard development and the damp atmosphere of spring and summer may hasten development. Griswold (1944) observed females that were kept at approximately 50° F/10° C laid eggs and crawled about but did not attempt to fly. She states, "Adults have been found in the breeding jars at all times of the year, but they have been somewhat less abundant during the winter. A series of experiments showed that the females will mate and lay eggs during every month of the year and that some of these eggs will develop into adult moths . . ." She found mated females may live as long as 30 days at 65° F/18° C and unmated females who are capable of laying infertile eggs live up to 32 days at the same temperature. According to Roth and Willis (1952), the male is attracted to the female by a scent from the abdomen of the female. This scent also induces courtship behavior.

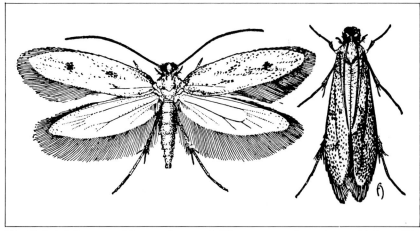

Fig. 12-8. Adult casemaking clothes moth. Note spots on forewings which readily separate it from webbing clothes moth.

THE CASEMAKING CLOTHES MOTH

Tinea pellionella (L.)

The casemaking or casebearing clothes moth is not nearly as common as the webbing clothes moth and therefore is not as important from an economic viewpoint. According to Curran (1949), in the American Museum of Natural History in New York City, not a single casemaking clothes moth has been brought to him in moth-eaten clothes in the last 20 years. However, where it occurs, it may become a serious pest. Herrick (1916) notes that Riley believed this moth to be more common in the North than the webbing clothes moth. The National Pest Control Association (1954) disagrees with this and states the casemaking clothes moth is more common in the South, especially in Georgia and Florida. MacNay (1947) notes the casemaking clothes moth destroyed a large shipment of Persian rugs arriving at Victoria, British Columbia.

This moth has more of a brownish hue than the webbing clothes moth and three dark spots, often indistinct, on the wings. These spots may be rubbed off on the older moths. It is smaller than *Tineola bisselliella* (Hum.), being 10 to 14 mm in wing expanse. The name "casemaking" clothes moth is derived from the habit of the brown-headed larva spinning a case of silk and interweaving in this some of the fibers on which it is feeding. When it moves, this case is carried around with it. The case is from ¼ to just over ⅓ inch/six to nine mm in length. The larva dies if separated from its case. When the larva moves, it thrusts its head and thoracic legs out of the case and drags the case along. Unlike the webbing clothes moth, the larva of the casemaking clothes moth rarely spins a web on the material on which it is feeding; it crawls around and grazes hither and yon, damaging the textile according to the time it feeds in one place. The case may be attached to the garment by silken threads and in this manner it may be carried for some distance.

Marlatt (1896) describes the method whereby the larva enlarges the case as follows: "Without leaving its case, the larva makes a slit half way down one side

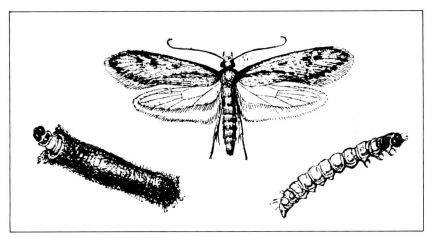

Fig. 12-9. Adult and larvae of casemaking clothes moth.

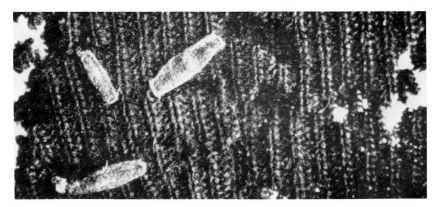

Fig. 12-10. Cases of the casemaking clothes moth.

and inserts a triangular gore of new material. A similar insertion is made on the opposite side, and the larva reverses itself without leaving the case and makes corresponding slits and additions in the other half. The case is lengthened by successive additions to either end. Exteriorly, the case appears to be a matted mass of small particles of wool; interiorly, it is lined with soft whitish silk. By transferring the larva from time to time to fabrics of different colors, the case may be made to assume as varied a pattern as the experimenter desires and will illustrate in its coloring the peculiar method of making the enlargements and additions described."

When the larva is ready to pupate, it seeks some protected place, such as a crevice on the wall or ceiling, withdraws completely into the case, and fastens both ends with silk. Marlatt (1896) found that in the northern United States there is one generation per year, with the moths in evidence from June to August. During the winter the larvae are not to be found even in heated homes. In the South there are two generations per year, and in centrally heated homes in this section of the country *Tinea pellionella* (L.) lays eggs throughout the year. The life history of the casemaking clothes moth, with the exception of some of the aforementioned points, is somewhat similar to that of the webbing clothes moth. Cheema (1956) made a detailed study of the life history of *Tinea pellionella* (L.). This insect, like the webbing clothes moth, is believed to be worldwide in distribution.

The casemaking clothes moth is a pest of feather-filled upholstery and cushions. Back (1931) states; "Feathers and down cushions are sometimes reduced completely to a mass of frass, dead insects and their cases, and feather quills." It, of course, attacks woolens, rugs, and furs, and is recorded in felts in pianos, as well as in stored tobacco. Austen et al. (1935) mentioned a number of drugs, some of which are actually poisonous to human beings, that were consumed by the larvae of this moth. The materials eaten were *"Aconitum* root, Cayenne pepper, horse-radish, *Strophanthus* (used as an arrow poison in East Africa), common hemp *Cannabis sativa,* cherry-laurel leaf, black mustard seed, ginger, orris root, laurel leaves, poppy capsules, linseed, almonds, saffron, and a monkey's skin used as a bag for the import of bitter aloes."

Meeuse (1952) noted an infestation of the casemaking clothes moth and the tapestry moth originated in owl pellets deposited in a church tower.

THE CARPET MOTH

Trichophaga tapetzella (L.)

This species is rare in the United States and only of minor importance in the British Isles. It is recorded from tapestries, old carpets, upholstered seats in autos, as well as in the furs and feathers of stuffed animals and birds. Patton (1931) reports it has been found in Germany in mousehair casts ejected by owls. Burbutis (1966) found this moth infesting stored goose feathers.

The male and female, with the wings folded, are approximately $^1/_3$ to $^5/_{12}$ inch/ eight to 11 mm in length, the female usually approaching the maximum size. This moth is therefore larger than the two previously mentioned species of clothes moths. The tapestry moth may be recognized by its white head, as well as by the fact that the first third of the forewing is black, and the remainder of the wing is white, mottled with black and gray.

Richardson (1897) observes that the larvae makes no case, instead it fashions a silken tube or burrows through the material. This tube, combined with the feeding, causes much damage to infested material. The larva may feed in its silken tunnel throughout the winter.

The life history of this species in the British Isles, according to Richardson (1897) and Austen et al. (1935), is as follows: The adults are found flying from April to June and mate almost immediately after emergence. The female lays her eggs in batches and may deposit from 60 to 100 eggs. The larvae molt four to six times and the larval stage persists through the summer months. The larva then constructs a rough cocoon in which silk is interwoven with bits of its food and pupates in 10 to 14 days. Unfavorable environmental conditions may prolong the larval period. Adults may emerge a second time from August throughout October, mate, and then lay eggs. The resultant larvae emerge and overwinter, pupating in the spring of the following year. There are one or two generations each year.

Other moth pests of fabrics. The brown house moth. *Hofmannophila pseudospretella* (Staint.), the white shouldered house moth, *Endrosis sarcitrella* (L.),

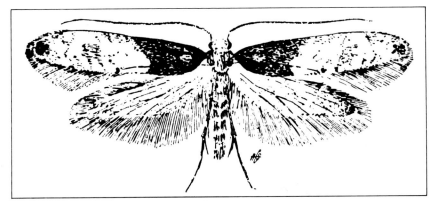

Fig. 12-11. The tapestry or carpet moth.

the plaster bagworm, *Tineola walsinghami* (Busck), and the fan palm caterpillar, *Litoprosopus coachellae* Hill, are moths that occasionally damage woolens and other fabrics.

The brown house moth, *Hofmannophila pseudospretella* (Staint.), formerly *(Borkhausenia),* is common in homes and warehouses in Britain. Woodroffe (1951a) and Cole (1962) made detailed studies of this moth. Although this moth is a pest on clothes, it also feeds on cereals and cereal products. Chrystal (1932) found it damaging book bindings. Anderson (1956) observed it to be a serious pest of the insulation of racetrack cables. Cereal rat baits near the cables were infested by the moths. Cole (1962) reports it in bird nests and dry organic debris. According to P. E. S. Whalley, it can eat its way through nylon and plastic materials.

The life history is extremely variable, primarily due to temperature changes. The incubation period varies from 8.5 days to 110 days, the larval stage from 71 to 145 days. The larva may enter a desiccation-resistant stage prior to pupation, which may last for months. Under laboratory conditions, the life history varies from 192 to 440 days, but in the field it is about 12 months.

The whiteshouldered moth, *Endrosis sarcitrella* (L.), will attack clothes on rare occasions. However, it is primarily a warehouse pest of stored seeds such as peas and beans. According to Woodroffe (1915b), the incubation period for the eggs may require 10.4 to 58 days, the larval stage 38 to 133 days, and the egg to adult stage 62 to 235 days. As in the case of the brown house moth, temperature is largely responsible for the great variation in completion of the various stages.

Watson (1939 and 1946) found the plaster bagworm, *Tineola walsinghami* (Busck), feeding on woolen goods, rugs, carpets, etc. in houses in southern Florida. It is originally supposed to have been a feeder on dead insects in spider webs (a precarious way to make a living). Watson (1946) observed that pyrethrum and DDT sprays and dusts controlled the larvae under rugs and carpets. The latter no longer can be used in the United States.

According to Flock (1951), the fan palm caterpillar, *Litoprosopus coachellae* Hill, a feeder on the flowers and fruit of the fan palm in California, may occasionally remove the pile from rugs, drapes, clothes, and other fabrics in the process of making its cocoon. The caterpillars are often blown to the ground by

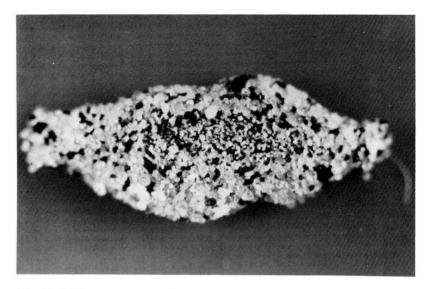

Fig. 12-12. The case of the plaster bagworm.

the wind and may then enter buildings prior to pupation. Flock notes that if lighted doors near infested palm trees are closed tightly at night, and especially if weather stripping is used, large caterpillars will be kept out of the house. Anon. (1965) reports *L. futilis* migrates in homes in Florida where it pupates in rugs, draperies, and stuffed furniture and infests sofas and damages rugs of synthetic fibers.

According to Pence (1955), the fan palm caterpillar is pink in color and reaches a length of 1½ inches/38 mm. The adult is a buff-colored moth with dark lines on the front wings. These pests are especially noticeable when a fan palm is "skinned."

Another unusual clothes moth is the brown-dotted moth, *Acedes fuscipunctella*. It has been reported from California and from North Carolina.

Some of the above moths, as well as the cigarette beetle and other fabric pests, are discussed in the chapter on stored products pests.

CONTROL OF THE CLOTHES MOTH

Entomologists and pest control operators interested in fabric pests note that since World War II the webbing clothes moth, *Tineola bisselliella* (Hummel), is not as important a pest in many parts of the Unites States as it was in the past. One proof of this is the disappearance from the marketplace of many insecticides previously formulated specifically for the control of clothes moths. The reason for this is not known, several factors may account for this including:

■ DDT, dieldrin, chlordane, lindane, Perthane® and similar insecticides were highly effective and in some cases very persistent.

- Synthetics have replaced much of the wool formerly used in clothes and home furnishings.
- Air conditioning and central heating and insulation have changed temperature and humidity conditions in the home. The lower humidity is not favorable to the clothes moth.
- House construction differs greatly from pre-war construction. For example, instead of hardwood flooring with its many crevices, plywood flooring is now used or concrete covered with tightly-fitted squares of asphalt tile, linoleum, or similar materials that do not have crevices so favorable to moth infestations.
- A good deal of raw wool came to the Unites States from Australia where sheep were treated for blow fly infestations. Dieldrin and other insecticides were applied to the sheep to protect them. Dieldrin, in particular, had an affinity for the wool and was a good mothproofer.
- It is believed the newer vacuum cleaners are more efficient than some of the earlier models and do a better job of pulling up the larvae of the clothes moth.

Katz (1975) also discusses some of the reasons for the decline of clothes moths as fabric pests.

Although the webbing clothes moth and some of the related moths may be less common than they were, they still are a problem and must be controlled. Since the last edition of this book (1969), control methods for the clothes moth have changed drastically. The most important insecticides used for the control of fabric insects such as DDT, dieldrin, chlordane, and lindane are no longer registered for the control of clothes moths. Using DDT as an example, a claim could be made on the label that a formulation of five percent DDT would protect woolens from clothes moths for a year if the spray was used according to directions on the label. There are, at this time, few registered insecticides that can make long-lasting claims. In fact, many of them are only effective against clothes moths for a few weeks, if that long.

Even though the older mothproofers can no longer be used, a brief review of them is pertinent from a historical viewpoint.

DDT. The insecticidal properties of DDT were discovered largely through the interest of the Geigy Corporation in mothproofing chemicals. DDT was marketed as a moth spray at a five or six percent concentration in a petroleum distillate. It was necessary to use an aromatic solvent to keep the DDT in solution. The odor of the aromatic was unpleasant to most people and for this reason many pest control operators used a three percent DDT concentration which did not require the addition of the aromatic.

One quart of a DDT-petroleum distillate spray consisting of five grams DDT per 100 ml (6% DDT by weight) would be sufficient to spray from five to eight men's winter suits. On the same basis, one quart/0.95 l of the above spray could treat 3.5 square yards/2.9 square meters of carpeting that weighed approximately six pounds per square yard/3.2 kg per square meter. As mentioned previously, clothes or carpeting treated with DDT would be protected for about one year. Adults or larvae of fabric pests hit by the spray would die in a few hours to a few days. After the petroleum distillate and the aromatic dried (the time for these to volatilize depends on temperature, humidity, air currents, and the fabric itself), only DDT crystals remained on the sprayed material. Larvae of the clothes moth placed on treated, but dried fabric would be affected in a few hours to about one day. Larvae of the black carpet beetle placed on a similarly

treated surface would live for weeks and months. The reason for the ineffectiveness of DDT to the larva of the black carpet beetle is not known. A guess is
the hairs on the larva prevented contact with the DDT-treated surface. Nevertheless, the DDT crystals protected the surface because the larvae of black carpet beetles did not feed on the treated material, even if confined to it for months.

Dieldrin. This insecticide was used as a 0.5 percent solution for the protection
of woolens or in the "spot" treatment of premises. Because of its toxicity to
warm-blooded animals, it was recommended that all sprayed articles be dry
cleaned before being used for clothing or bedding. Dieldrin was also used as a
dye-bath mothproofer by textile manufacturers.

Chlordane. Chlordane, like dieldrin, was widely used by pest control operators for the control of fabric pests. It was usually applied as a two percent spray.
Because of its toxicity to warm-blooded animals it was recommended that
sprayed clothes be dry-cleaned before being worn.

Methoxychlor. This is another DDT-related insecticide that was effective
against larvae of the clothes moth, but was not widely used because it was less
soluble in petroleum distillates than DDT.

Lindane. This chlorinated insecticide was effective against fabric insects,
but was not as widely used as DDT because it did not protect woolens as long
as DDT.

Silicofluorides or fluosilicates. Prior to World War II, water-soluble products containing sodium or magnesium silicofluoride were widely marketed for
the control of the larvae of clothes moths. These gradually disappeared with the
advent of the more effective organic compounds previously mentioned.

USE OF SPRAYS AND FUMIGANTS

When clothes moth infestations are severe, a great amount of skill may be
required in their control. Pest control operators usually use one of two methods,
spraying or fumigation.

1) **Spraying.** At this writing, pyrethrins, resmethrin, allethrin, DDVP, malathion, bendiocarb, and Perthane® may be used as sprays against clothes moths
and their larvae. *In any event, it is important to follow directions on current
labels before applying insecticides against clothes moths and their larvae.*

2) **Fumigation.** Sulfuryl fluoride is registered for fumigation in houses
against clothes moths. Sulfuryl fluoride, ethylene dichloride, and ethylene dibromide are used in vault fumigation. In addition, napthalene and paradichlorobenzene remain available as fumigants.

*It should be noted that many of the control measures recommended for clothes
moths are applicable to carpet beetles.*

DISCUSSION OF CONTROL MEASURES

Source of infestation. When clothes moths become noticeable in a home,
there may be one or several sources of infestation. A thorough search may reveal
this source. It can be an old woolen sock left in a drawer, the inside cuff of flannel
pants, or a fur piece in a cardboard box. Clothes moths and carpet beetles often
breed in felt-like accumulations that collect behind baseboards, under door
jambs, treads, etc., and even in bird nests (Linsley, 1944). Spencer (1931) found
these accumulations in the rectangular cold-air shafts, running horizontally
along the floor joist to the furnace, in homes warmed by hot air furnaces. Spencer

notes: "Especially is this likely to be the case in a much carpeted house with gratings flat on the floor, where a careless housemaid may save herself the trouble of using a dust pan or where the job of sweeping is sometimes entrusted to the man of the house." The writer found a source of infestation in felt knee and shoulder pads and hair-filled ice hockey shin guards in school gyms. Fish meal used for fish food may be another source of infestation. Upholstered furniture sometimes serves as a place for the breeding of clothes moths. Occasionally, piano felts may be infested by clothes moths or carpet beetles. NPCA (1958) recommends piano tuners or technicians be consulted if the felts have to be treated.

It should be recalled that the newly-hatched larvae are so tiny they can crawl into any crack or crevice. Thus, once the female has laid her eggs on the cover of woolen upholstered furniture or in the crevices or seams, the larvae can readily enter the furniture proper. Moreover, the female may creep inside the furniture through any opening she can find. Slipcovers encourage infestation because they provide darkness for the adults and larvae.

Back (1931), who studied the feeding habits of the clothes moth, found that the larvae, when feeding on upholstered furniture, may stay and feed on the outside cover as surface feeders, or they may feed from under the covers. Some of the material on which they are feeding may be incorporated in the tubes. The larvae, when older, then build the tubes over the pile. Surface feeding occurs most commonly in the shaded parts of chairs which receive little usage, such as behind pillows and the backside of furniture turned toward the wall. The larvae may also work beneath the covers. Their presence may be unsuspected until the pile begins to fall out in patches when brushed.

SPRAYING OR TEMPORARY MOTHPROOFING

When a material is treated with a spray or temporary mothproofer, it is wetted by some chemical in water or in an organic solvent such as deodorized kerosene (petroleum distillate). The insecticide is deposited on or combined with the fibers in the fabric and protects the treated material from attack by the larvae for days to months. Thus, the larvae either do not feed upon the fibers or if they do, they are quickly poisoned by the insecticide. In any event, they cannot complete their growth on the fiber. Since carpet beetles or dermestids are at times referred to as "buffalo moths" by the layman, the word "mothproof" actually designates a material that is resistant to feeding by carpet beetles, as well as clothes moths.

Tests for the effectiveness of insecticides against fabric pests are conducted in a variety of ways (see Cole and Whitefield [1962]). In general, a treated piece and an untreated piece of woolen cloth are exposed to larvae of the clothes moth or carpet beetle. The effectiveness of the moth insecticide is determined in a number of ways such as visual estimation of damage, weight of the excrement, loss of weight of the cloth, the number of larvae dead or alive, the number of head capsules of the larvae of the clothes moth, the number of cast skins of the larvae of the carpet beetle, and by the microscopic examination of the individual fibers. The CSMA Textile Resistance Test (1969) goes into great detail regarding test methods and the rearing of test insects to determine the effectiveness of insecticides against fabric pests.

A review of a few of the pertinent habits of the larva in regard to moth protection against fabric insects should be helpful at this point. The larva is the only stage in clothes moths and carpet beetles that feeds on animal products. The larva digests animal fibers and excretes synthetics as undigested pellets. The larva of the clothes moth, even if it does not digest the fibers, may cut them to make a pupal case. In this case, a repellent type of spray may be satisfactory, whereas a stomach poison may not keep the larva from cutting some of the fibers.

SPRAYS OR TEMPORARY MOTHPROOFERS

These chemicals consist of inorganic salts or emulsifiable concentrates that are soluble in water or organic compounds soluble in petroleum distillates. Thousands of patents have been issued on mothproofing chemicals, and of these, only a few have made it to the marketplace.

By far the best mothproofing for an unfinished textile is obtained by submerging the material in a bath containing the mothproofing ingredients and thereby insuring the most thorough penetration.

There are currently on the market a number of temporary mothproofing insecticides that are applied as a spray, and this is the method used by the homeowner and the pest control operator. In employing these sprays it is necessary to wet the fibers thoroughly. When using water-soluble insecticides, wetting agents that aid the water in penetrating the fabrics are combined with the mothproofing chemicals. The organic solvents (petroleum distillates) penetrate the textile fibers much more readily than do the water solvents.

The staining problem. Among those individuals interested in sprays or temporary mothproofers there is a continuous verbal battle as to the respective merits of water solvents versus organic solvents (petroleum distillates). One of the most important topics of discussion, but not the only one, is the problem of staining. In fact, it is often the knowledge of this factor that makes the difference between the experienced mothproofer and the beginner. Jackson and Wassell (1927), who worked on cinchona alkaloids in organic solvents, present the following criticisms of water as a solvent: "Water causes wool to shrink; loosens glued joints in wooden furniture; water-spots silk linings of wool or fur garments; and destroys the curl and shape of such feather articles as plumes; it does not have the penetrating properties of most organic solvents; and lastly, it is devoid of the specific insecticidal (larvicidal) properties of most organic solvents; as, for example, carbon tetrachloride or petroleum naphtha."

In addition to the above, water-soluble mothproofing insecticides are removed from fabrics by washing or exposure to large amounts of water. It should be noted the lack of penetration of water solutions have been overcome to some extent through the use of wetting agents that reduce the surface tension of water and permit penetration of the insecticide into the fabric.

On the other hand, some organic solvents may be a fire hazard and may stain some fabrics, especially silks. Oil-soluble insecticides are usually removed during dry cleaning. Insecticide base oils (petroleum distillates) may become a hazard by making floors slippery. They may also mar asphalt tile and some plastic surfaces and may dissolve the mastic beneath tile and other floor covering. Rubber on the bottom of some carpeting and the rubber and synthetic pads beneath rugs and carpeting may be damaged by oil-base sprays. Moreover, insecticide base oils are relatively expensive, whereas one merely turns the tap for a water solvent.

In view of the fact synthetic fibers are widely used in combination with wool fibers, it may be worthwhile to note that tests conducted in the author's laboratory showed oil base sprays badly spotted celanese acetate and slightly spotted nylon, silk crepe, silk shantung, Dacron®, Orlon®, and rayon. Wool and Dynel® were not affected by the oil base. Water badly spotted celanese acetate and slightly spotted nylon, silk crepe, silk shantung, viscose rayon, and Orlon®. Dacron®, Dynel®, and Vicara® were not stained by water. Most of the stains from oil-base sprays could be removed by immersing the stained material in Stoddard's solvent.

CHEMICALS CURRENTLY LABELED FOR CONTROL

At this time, the following insecticides are labelled for the control of the clothes moth: Pyrethrum (fog or aerosol), resmethrin (SBP-1382) (aerosol), allethrin (aerosol), DDVP (dichlovos) (emulsifiable concentrate, aerosol, and plastic strips), malathion, bendiocarb, Perthane®, sulfuryl fluoride, ethylene dibromide, ethylene dichloride.

Pyrethrum. Pyrethrum, in combination with the synergist piperonyl butoxide, may be used as 0.5 percent space spray or 0.1 percent contact spray. The piperonyl butoxide is usually used at two to five times the concentration of pyrethrins. In pressurized sprays, Synergist 264 may be included in addition to piperonyl butoxide.

One label for a pyrethrins-synergist combination makes the following claim: "Spray all articles to be protected giving particular attention to folds and seams. Spray interior of storage containers such as closets, chests, and trunks regularly. Treatment should be repeated every 30 days unless the articles are stored in a moth-tight container."

Tittanen (1971) showed woolen test fabrics treated with an aerosol containing 1.2 percent pyrethrins and five percent piperonyl butoxide and placed in dark wardrobes protected the fabrics from larvae of the webbing cloth moth for six months. For further information of the effectiveness of synergized pyrethrins against fabric insects, see Bry & Simonaitis (1975).

Although the pyrethrum extracts currently being used are largely free of the impurities responsible for allergies, it may still be a good idea to recommend that treated garments be dry cleaned before being worn again. This will aid individuals allergic to pyrethrins.

Resmethrin (SBP-1382). Resmethrin is effective against clothes moths when used at 0.25 percent in an aqueous pressurized spray. Penick Corporation notes that SBP-1382 "is effective both as a direct contact spray against these insects and as a protectant for as long as six months when applied to fabrics according to label directions."

For indoor application for the protection of woolens, the SBP-1382 label reads as follows: "Apply only to garments, blankets, carpets, and other woolen fabrics before placing in storage for protection from larvae of webbing clothing moths and black carpet beetles.

"Dry clean and wash garments and blankets before treating. Clean carpets, draperies and other fabrics by thoroughly brushing before treating.

"Hold the container 14 to 18 inches from the fabric and apply the spray while moving evenly in a forward and back motion across the fabric while covering one foot per second. Spray the entire fabric surface until it is *slightly moist or damp to the touch.* When possible, treat both sides of the fabric. For maximum protection, place treated fabrics after drying in plastic or paper bags. Also treat cracks and crevices of closets, chests, and trunks where these fabrics are being stored. Treat the cracks and crevices at the rate of two seconds per linear foot.

"When applied as directed on fabric and in storage areas this treatment will offer protection for six months. To insure protection inspect the fabric in the storage area once every three months. Repeat application only when necessary. Avoid heavy applications. When possible make applications outdoors. If application is made indoors do not remain in treated areas. Ventilate the area after treatment is complete."

Permethrin. This synthetic pyrethroid-like insecticide is *not* cleared at this time for use against fabric insects. However, because of its low warm-blooded animal toxicity and its long-lasting effectiveness against both adults and larvae of clothes moths and carpet beetles, it is possible it will be labeled for such use in the future.

Bry *et al.* (1979) and Bry *et al.* (1980) discuss in some detail the effectiveness of this pyrethroid against both adults and larvae of the webbing clothes moth and the black carpet beetle. Permethrin, as mentioned previously, has residual properties and protects sprayed woolens for weeks to months. The latter paper showed sprayed cloth to be lethal to the eggs of the webbing clothes moth, but not to the eggs of the black carpet beetle.

Allethrin. This insecticide, d-trans allethrin (allyl homolog of Cinerin I), is formulated for flying moths in a pressurized house and garden spray at 0.25 percent in combination with piperonyl butoxide and Synergist 264. The label reads as follows: "To kill flies, mosquitoes, gnats, and flying moths: close all doors and windows. Cover or remove all exposed food and utensils. Direct spray upward into center of the room with a slow, sweeping motion. Spray five to 10 seconds for average room. Keep nozzle three feet from surfaces. Vacate premises. Keep room closed for 15 minutes after spraying. Ventilate room before re-entry. Repeat as needed."

DDVP (dichlorvos). Shell Chemical Company has a label on its two pounds per gallon DDVP product recommending an 0.5 percent DDVP spray to control "flying moths" in household pest control. The label for use in households, theaters, industrial plants, and non-food warehouses reads as follows: "Apply 0.5 percent spray with low pressure sprayer to localized areas insects may infest. Treat infested areas around baseboards, cracks, walls, door and window frames, and localized areas of floors. Do not apply to areas which will be contacted frequently by children and do not allow them in treated areas until surfaces are dry. Do not treat animals." Note the above does *not* say anything about applying the insecticide to garments or carpeting.

 Shell's vinyl strip formulation of DDVP which contains 20 percent technical DDVP may be used against "flies, gnats, mosquitoes, and other small flying insects." Apparently the category "other small flying insects" includes "flying moths." Thus, the strip may be placed in clothes closets. Some of the cautions on the label read as follows: "Do not place in hospital or clinic rooms, such as patients' rooms, wards, nurseries, operating or emergency areas. Do not use in any room where infants, the sick or aged are or will be present for any extended period of confinement. Do not use in kitchens, restaurants, or areas where food is prepared or served."

Batth and Singh (1974) showed the vapors of DDVP to be more effective against the larvae of the webbing clothes moth than against the larvae of the black carpet beetle.

Malathion. This insecticide is used against clothes moths and their larvae at a three percent concentration in water or deodorized kerosene. The label reads as follows: "Apply to baseboards, floors (including areas under carpets, along margins of carpets, and in closets), behind radiators and other lint accumulation areas, closet shelves and walls, and infested surface areas of carpeting. Application to these will also control clothes moth larvae."

The insecticide is *not* to be used on clothing Since malathion has a strong odor, open the windows and air the sprayed areas if the weather permits.

Bendiocarb. This carbamate insecticide is formulated as a 76 percent wettable powder and is labeled for use by professional applicators against webbing clothes moths as a 0.5 percent spray. The label reads as follows: ". . . apply spray as spot applications in closets and other storage areas where webbing clothes moths are found. This treatment is an adjunct treatment and will not control moth larvae already on the clothes. Do not apply to clothing." An advantage of bendiocarb over malathion is its lack of odor.

Perthane. At this writing, Perthane is no longer available because the manufacturer, Rohm & Haas, has ceased production. However, in view of past experience, some other manufacturer may make it available.

This insecticide was used as five percent technical Perthane in a pressurized moth spray. It is effective against clothes moths and carpet beetles. When applied to woolens before storage it prevented damage for 12 months when used as directed.

The directions on the label read as follows: "Clean and brush woolen articles, removing persistent spots and stains by dry cleaning. Spray each article evenly from approximately 18 inches distance until the surface is thoroughly moistened. Avoid overspraying and excessive application. Pay particular attention to folds, seams, cuffs, etc., and spray on both sides whenever possible. For maximum protection, spray insides of closets or containers before storing treated articles. Air mothproofed articles before using.

"Repeat application if fabrics are washed or dry cleaned. Flying moths and carpet beetles are killed by spraying directly. Spray all folds and crevices where insects tend to hide and spray directly as they emerge."

Application of sprays. When applying insecticides for the control of fabric pests, adjust the nozzle of the sprayer to a fine stream for crevices in baseboards and similar places and to a coarser spray when wetting surfaces such as the edge of carpets and like surfaces.

Sprayed surfaces should not be walked on when wet. Some pest control operators cover their shoes with heavy cotton or wool socks so the soles do not react with the spray and cause stains. If possible, spray early in the day in good weather so the windows can be opened and the rooms aired out. Do not spray radiators, baseboard heaters, and other heat sources because this will result in rapid volatilization of the insecticide and the petroleum distillate in oil-base sprays.

Carpets vary greatly in sizing, dyes, and backing which may consist of synthetic materials, cotton or jute. These in turn may be affected by the water or oil-base carriers of the insecticide. It may be a good idea to test the spray on an inconspicuous area of the carpeting to make sure it is not affected by the spray. Furniture should not be replaced on sprayed carpet until the carpet is thoroughly dry. Usually those areas that have daily traffic need not be sprayed.

FUMIGATION FOR CLOTHES MOTHS

Where infestations are heavy, the pest control operator may resort to fumigation. Sulfuryl fluoride is the fumigant of choice. At times, even fumigants can fail because the fumigant may not penetrate an infested area in sufficient concentration to kill the larvae.

Since clothes moths and carpet beetles may occur in the same building, the minimum dosage used is that needed to kill the more resistant carpet beetles. As a rule, the larvae of carpet beetles are much more resistant to fumigants than many other household insects. The dosage for carpet beetles is usually twice that for other insect pests in the home.

Some pest control operators have airtight fumigation vaults. In these they place infested upholstered furniture, carpets, etc., and since the concentration of fumigant may be retained for some time, the results are usually satisfactory. Besides sulfuryl fluoride, ethylene dichloride and ethylene dibromide are some of the other fumigants that may be used in vault fumigation.

Fumigants should be used only by licensed and experienced individuals.

Naphthalene and paradichlorobenzene. Clothes moths are immediately associated with these materials. Practically every housewife has one or the other in the home. Unfortunately, these insecticides tend to lull the housekeeper into a false sense of security, since it is believed the mere presence of them, often in insufficient quantities, will keep the moths away. Billings (1934) and Abbott and Billings (1935) showed paradichlorobenzene, naphthalene, and the cedar oils do not repel the moths or keep them from laying eggs, nor the larvae from feeding on articles in the presence of these materials. Billings concluded the materials have no repellent action, but they do have some fumigation value. Since the worth of these materials rests on the fumigation effect when in heavy concentrations, it is essential these insecticides be enclosed in containers that are practically airtight. Trunks, satchels, chests, etc., when sealed from the outside, are effective when used with these fumigants. Such containers which retain high concentrations of the gases emanating from these fumigants may keep the larvae from feeding and death may finally result through starvation or the toxic effect of the fumigants.

Herrick and Griswold (1933) found napthalene flakes at the rate of two to three ounces per five cubic feet/57 to 85 g per 0.14 cubic meters were toxic to eggs and larvae of *Tineola bisselliella* (Hum.) when confined for some time in a tight box. Mothballs of naphthalene at the rate of eight ounces to five cubic feet/227g per 0.14 cubic meters were also toxic to the larvae, when confined in the box from three to four weeks. The naphthalene flakes were more effective when scattered in the clothing, preferably between newspapers. The dermestid or carpet beetle larvae, *Attagenus megatoma* (F.) and *Anthrenus verbasci* (L.), were much more resistant.

Frey (1939) found that with naphthalene, 40 percent of the larvae died in 15 days; 100 percent of the adults died in five days, and 99.6 percent of the eggs were killed in four days, all at a temperature of 68° to 73° F/20° to 23° C at a dosage of 15 ounces per 100 cubic feet/0.42 kg per 2.8 cubic meters.

The U.S.D.A. Production and Marketing Administration discussed naphthalene preparations for the control of clothes moths as follows: "Naphthalene is slowly volatile at ordinary temperatures and under proper conditions acts as a fumigant (gas) to kill moths. *To be effective it must be used in tight containers, and dosage must be adequate, and the period of exposure must be sufficient to kill.* Naphthalene does not repel moths and is of no practical value against such household insects as flies, roaches, ants, and bedbugs.

"Tightness of containers: The articles to be treated must be in a container which will permit the naphthalene to build up a sufficient concentration to be effective. Well-built closets *with all cracks sealed with tape,* tight wooden storage cupboards or chests, and well constructed trunks have been found satisfactory for this purpose. Unsealed closets with cracks under or around the doors, loosely built cupboards or chests, and many types of corrugated cardboard containers will not retain sufficient gas to be effective.

"Dosage required: The dosage of naphthalene required depends on the articles to be protected, the temperature, and the physical form of the product. At temperatures below 70° F naphthalene volatilizes very slowly, and if the temperature is too low, the confined air, even when saturated with the gas, may not hold enough to be effective. The form of the naphthalene, whether fine crystals, balls or cakes, and the form of the holder, when one is used, will affect the rate of volatilization. Other conditions being equal, the amount of gas given off in a definite period varies directly with the surface of the naphthalene exposed. Under proper conditions, in gas-tight containers where the gas is given off freely and absorption does not take place, a dosage of one pound of naphthalene crystals to 100 cubic feet will kill moths in all stages with sufficient time for exposure.

"Length of exposure and necessity for cleaning articles before storage: Adult moths, their eggs, and very young larvae are killed within two or three days by the above-mentioned dosage of the fumigant. On the older larvae, however, it acts very slowly and, even when used at high dosages, will not kill them in less than three or four days; meanwhile, they will continue feeding on the fabric and may seriously damage it. In order to avoid damage, all articles must be freed of the older larvae by thoroughly brushing them, unless they have just been washed or dry cleaned before they are packed away. Any eggs or very young larvae that may be missed will be killed by the naphthalene vapor before material damage is done."

Paradichlorobenzene. This insecticide, which comes in flakes and more commonly in crystals, is used in a manner similar to naphthalene. Freedman (1948) notes paradichlorobenzene or "PDB" is in great demand, as evidenced by department store sales. Frey (1939) found the mortality of *Tineola bisselliella* (Hum.) at 68° to 73° F/20° to 23° C, at a dosage of 15 ounces per 100 cubic feet/ 0.42 kg per 2.8 cubic meters was 100 percent of the larvae in four days; 100 percent of the adults in one day; and 100 percent of the eggs in four days. If these figures are compared with those offered previously by Frey for naphthalene, it will be seen that paradichlorobenzene is a more rapid-killing fumigant than naphthalene. However, Doner and Thomssen (1943) note that although naphthalene releases its fumes more slowly than paradichlorobenzene, it is supposedly more toxic to insects than paradichlorobenzene. It is recommended that 12 to 16 ounces/0.34 to 0.45 kg be used to 100 cubic feet/2.8 cubic meters of a very tight container. According to Back (1935), two ounces/ 56 g of paradichlorobenzene per 100 cubic feet/2.8 cubic meters in a tight container quickly stops the feeding by the clothes moth larva.

The U.S.D.A. Production and Marketing Administration discusses paradichlorobenzene preparations for the control of clothes moths as follows: "Paradichlorobenzene is slowly volatile at ordinary temperature and acts as a gas (fumigant) in killing moths, but does not repel them. Its effectiveness against moths depends mainly on three factors: the tightness and contents of the space treated, the length of exposure, and the amount of gas present. The first two conditions can be controlled readily, but the amount of gas given off is affected by several factors, the most important one being temperature. At temperatures of 70° F and above paradichlorobenzene crystals, when exposed to air, vaporize rather rapidly, and confined air at this temperature will hold enough of the gas to kill moths. At temperatures considerably below 70° F, the confined air may not hold enough of the gas to be effective, but the limit has not been determined accurately.

"The form in which the paradichlorobenzene is used (fine crystals, lumps, or cakes), the surface exposed to the air and circulation of the air also greatly influence the rate of gas production. Other conditions being equal, the amount of gas given off in a definite period varies directly with the surface of the paradichlorobenzene exposed.

"Under ordinary conditions, in tight closets, trunks, and other tight containers, where the gas is freely given off, a dose of one pound of paradichlorobenzene crystals to 100 cubic feet of confined space should be effective. In an absolutely air-tight chamber, at a temperature of 77° F, 0.5 pounds of paradichlorobenzene will saturate 1,000 cubic feet of air, but this dosage would not be satisfactory under the usual household conditions.

"The length of exposure necessary to kill moth larvae varies with the temperature, the concentration of the gas, and the age of the larvae, but at least three days should be allowed. If the form of the paradichlorobenzene or its container is such that the evolution of the gas is retarded, a greater dosage may be necessary, but this can be determined by actual tests only." Many users feel the odor of PDB is not retained as long on clothes as naphthalene.

Arnold (1953) recommends the use of fans or heat-vaporizing units to produce a high vapor concentration of PDB in closets. He notes that vacuum cleaners forcing warm air through the crystals are especially effective. Arnold (1956) found PDB in plastic garment bags to vary in effectiveness because of the differences in permeability of bags to the PDB vapors. Approximately 12 ozs/340 g of crystals should be used per 7.6 cu. ft/0.2 cubic meter bag. "Crystals should not be placed on the floor of the bag; aside from the low position, retarding vapor movement the PDB may react with the plastic film, softening it." The clothing in the bag should be loosely separated, and the bag should remain closed from one to two weeks.

Cedar products. The housewife has long attributed special "mystical powers" to cedar oil, cedar wood, and cedar chips, with respect to their "supposed" ability to drive away clothes moths. The heartwood alone of red cedar *(Juniperus virginiana)* has a volatile oil, the principal constituents of which, according to Back (1935), are "the alcohol cedrol or cedrol camphor, the sesquiterpene alcohol cedrenol, and the sesquiterpene cedrene. The characteristic odor of cedar chests is probably due to the first two compounds mentioned." The heartwood contains from two to four percent of these volatile oils.

As a rule, red cedar closets are not sufficiently tight or kept closed long enough to retain the volatile oil, and thus are of little use in controlling the clothes moth. Since chests are tighter, they are usually slightly more effective in this respect. Back (1935) states that "chests with the sides, ends, and bottoms made of red cedar heartwood at least three-fourths inch thick and the cover of solid red cedar or of a neutral wood lined with red cedar veneer will kill all the newly hatched or young larvae of the clothes moth." It should be recalled the volatile oil of cedar is not a repellent.

Huddle and Mills (1952) demonstrated that "cedar oil vapor at a sufficient concentration and exposure time will kill the half grown larvae" of the webbing clothes moth. Back and Rabek (1923) found the larvae of *Tineola bisselliella* (Hum.), upon hatching from eggs in a chest, died for the most part in two or three days, and all were dead in two weeks. Larvae that were three to four months old, or more than half grown, as well as the eggs, pupae, and adult stages, were not affected. Back and Rabek state that "a chest of ordinary wood, as tightly constructed, would be just as effective, provided the clothing were thoroughly cleaned, brushed and aired in the sun and from one to two pounds of good grade naphthalene enclosed with it."

Laudani and Clark (1954) and Laudani (1957) show the vapor from cedar wood is greatly limited in its toxicity to fabric pests, and after cedar chests are 36 months old, they are practically useless in killing fabric insects.

Very tight garment bags or cabinets will aid in keeping adult moths from laying eggs on clothes. However, if the larvae are already on the clothes, the bags and cabinets are useless. When these containers are fairly tight, the use of naphthalene or paradichlorobenzene in them will be helpful. Arnold (1956) showed that many of the plastic bags do not retain PDB vapors long enough to kill fabric pests. Bags scented with various oils, although pleasing to the nostrils, annoy the moth very little, if at all. Freedman (1948) observes from department store sales that tar paper bags alone are being replaced by plastic film bags such as polyethylene or vinyl. Some vinyl bags are affected by paradichlorobenzene, whereas the polyethylene bags are not. He also notes cedar oil preparations are losing their popularity. Here, again, it should be mentioned that oil base sprays may spot some types of plastic bags. Pimentel and Weiden (1959) found tightly-sealed plastic bags without any insecticides whatsoever protected clothes from fabric pests. Batth (1972) tested vinyl garment bags for paradichlorobenzene (PDB) fumigation of fabric insects. Mortality of the larvae of the webbing clothes moth in the bags was 100 percent in 10 to 28 days at temperatures of 80° F/27° C and 50 percent relative humidity, depending on the type of bag. The results with larvae of black carpet beetles showed 100 percent kill in 14 to 28 days in some bags and 11 percent to 67 percent in other bags at the end of 28 days. PDB fumigation in tight metal boxes killed 100 percent of the larvae of black carpet beetles in three to six days. Lower temperatures greatly reduced the effectiveness of PDB, especially against the larvae of the black carpet beetle. Apparently, fumigation with PDB at lower temperatures inhibits feeding, but may not kill the larvae.

MISCELLANEOUS CONTROL METHODS

Dye-bath mothproofers. Some of the permanent mothproofers consist of colorless dyes which are applied by the manufacturer in the dye-bath by impregnation or padding and by exhaustion. Anon. (1948) states, "Padding is a textile mill operation in which piece goods are run over rollers in and out of a tank solution. The goods are usually dry when they go into the tank and often travel at a speed of 30 yards a minute. Exhaustion is a method in which the material absorbs the solution while being worked around in a washer, vat, or kettle. Yarns are most conveniently processed by this method." It is claimed such mothproofers are effective for years, are fast to washing, dry cleaning, and to light, and do not interfere with subsequent working of the goods. Crossley (1946), Moncrieff (1950) and Waterhouse (1958) consider in some detail the chemistry of dye-bath mothproofers. Some of the dye-bath mothproofers of the past were Eulan, Mitin, Bocon, and Lanoc. It is interesting to note studies by Paul Müller, a research chemist of the Ciba-Geigy Corporation, on mothproofing chemicals led to the discovery of DDT.

Furs. Individuals are not capable of mothproofing their furs since the application of insecticides to furs requires the services of an expert. Furs are treated in dry cleaning establishments equipped with sawdust tumblers to aid in removing the matting caused by immersion in a mothproofing bath. Anon. (1955) notes that furs are drummed in tumblers containing about ½ pound/0.23 kg per garment of ground nutshells, corncobs, or sawdust, depending on the color or class of fur, and then dampened with dry-cleaning solvent. Another tumbler shakes the dust out so the fur is clean. The use of cold storage for the protection of furs is discussed in the following section.

Cold storage. Large department stores, wholesale dealers, warehouses, etc., depend upon cold storage to protect articles subject to attack by clothes moths and carpet beetles. Back (1935) has this to say regarding cold storage: "Articles will be protected from injury in storage at temperatures ranging from 40° to 42° F. A number of years ago, a manager of a large storage-warehouse company in Washington, D.C., conducted certain experiments at the instance of the Chief of the Bureau of Entomology, with the result it was found that larvae of the webbing clothes moth and of the black carpet beetle can withstand for a considerable time a temperature of 18° F. It has been discovered it is not so much the cold that kills, it is the sudden change from a cold to a warmer temperature and back to cold temperature that most quickly results fatally. Thus, it was learned if articles infested with clothes moths were refrigerated at 18° F for several days, then suddenly exposed for a short time to 50°, and then returned to 18°, and finally held permanently at about 40°, all moth life in them would be killed.

"If storage concerns intend to destroy the clothes moths in articles entrusted to them, as well as to protect from injury during the period of storage, it is recommended the article be exposed to changes of temperature, as noted above, before they are placed at 40° to 42° F. A lower temperature is needless and wasteful."

Since clothes moths are fairly resistant to cold, it is possible to have had clothes in cold storage and still find living, well-grown larvae on the garments a short time after removing them from cold storage. Cold storage mey prevent the larvae from feeding, but does not necessarily kill them. Regarding this, Back notes, "well-grown larvae of the webbing clothes moth in fur and wool held in commercial cold storage at a temperature said to fluctuate between 24° and 48° F, but held mostly at about 40° were found by the writer to be alive after storage for six, eight, 10, 11 and 12 months." The resultant adults laid many eggs which hatched normally. During the winter if furniture is placed outdoors at 0° F/18° C for several hours, it often results in good control.

The use of heat. Dry heat as a method of control of the clothes moth is not ordinarily used. It is known that when clothes are sunned and the larvae exposed to high temperatures, some control may result. According to Wallace (1925), the larvae of *Tinea pellionella* (L.) and *Tineola bisselliella* (Hum.) may be controlled by subjecting infested material to a temperature of 130° F/54° C for 24 hours or to a higher temperature for a shorter period. Back (1935) observed that all larvae "exposed in an incubator to 128° F, 120° F, and 110° F died in six, 11 and 31 minutes, respectively. Exposure to the sun at 110° and 105° for 31 and 11 minutes, respectively, did not kill the eggs. Well-grown larvae of clothes moths in garments, exposed for several hours to the hottest rays of the sun usually become restless and spin down from the garment. The old-time custom of sunning clothing to kill moths is based on experience."

Rawle (1951) conducted temperature studies on the webbing clothes moth at 70 percent relative humidity. All stages were susceptible when exposed for four hours to 105.8° F/41° C. The egg was the most resistant stage and some were able to survive an exposure of 104° F/40° C for four hours. However, all stages are capable of developing at 91.4° F/33° C. Rawle draws the following conclusion: "As far as can be ascertained from the literature *T. bisselliella* does not occur in tropical climates, though it is otherwise distributed widely throughout the world. This fact may be related to the inability of the moth to withstand temperatures in excess of 91.4° F/33° C or 93.2° F/34° C for prolonged periods."

It is possible to treat clothes moths by placing articles in heat vaults where the temperature may rise to 160° to 170° F/71° to 76° C. In regard to furniture, this method has certain disadvantages. Thus the finish of the furniture may blister, the glue in the joints may liquefy and fail to hold, and some of the wood may even warp.

Brushing. The "good old" method of removing woolens and like materials and brushing and beating them at intervals of once or twice a month is very effective. Wearing apparel and blankets, etc., in constant use are rarely damaged by clothes moths. By beating and brushing clothes, the eggs and possibly even the younger larvae are crushed or dislodged. The use of a vacuum cleaner on the clothes will facilitate this. Regarding fur, Back (1935) states: "Furs cannot always be rid of the older worms by merely brushing or beating. Clothes moth worms often lie hidden next to the skin and are so firmly established by the webbing they spin that they are not dislodged from the fur by brushing or shaking. For this reason furs suspected of being infested should be combed out with a very fine comb or should be fumigated." Further, Back (1931) notes that on upholstered furniture the larvae feed on the outer surface between the rows of pile and the underside. This surface feeding can be curbed through constant brushing and the use of a vacuum cleaner. It should be remembered that upholstered furniture in constant use, unlike clothes, may be badly infested. The larvae that are beneath the foundation warp cut the fibers, and bare spots result in the mohair after the cut threads fall out.

Trapping. Wilson (1940) found clothes moths and carpet beetles are attracted into box traps containing cloth pads treated with fish meal or an alcoholic extract of fish meal. He also found sticky fly paper treated with various animal substances was useful in trapping clothes moths. The adults are attracted to such traps and lay their eggs there. Moreover, such traps also attract the feeding larvae.

LITERATURE

ABBOTT, W.S., and S.C. BILLINGS — 1935. Further work showing that paradichlorobenzene, naphthalene, and cedar oils are ineffective as repellents against clothes moths. J. Econ. Entomol. 28:493-495.
ANDERSON, J.G. — 1956. Odds out, moths in. Pest Control 24(6):50.
ANONYMOUS — 1948. Of moths and men. Chem. Indust. 63(1):29-30.

1955. Fur-bearing market. Chem. Week 76(17):68.

1965. *Litoprosopus futilis.* Coop. Econ. Ins. Rpt. 15(15):328.

ARNOLD, J.W. — 1953. Para and naphtha as closet fumigants. Soap & Chem. Spec 29(8):134-135, 137, 139, 141, 155.

1956. Effectiveness of paradichlorobenzene in plastic garment bags. Soap & Chem. Spec. 32(2):121-124, 167, 169.

AUSTEN, E.E., A.W. McKENNY HUGHES, and H. STRINGER — 1935. Clothes moths and house moths. British Mus. Econ. Series 14.

BATTH, S.S. — 1972. Evaluation of vinyl garment bags as chambers for paradichlorobenzene fumigation of fabric-insect pests. J. Econ. Entomol. 65:1074-1080.

BATTH, S.S. and J. SINGH, 1974. Evaluation of dichlorvos vaporizing solids for controlling insects. Can. Entomol. 106:31-37.

BACK, E.A. — 1931. The control of clothes moths in upholstered furniture. USDA Farmers' Bull. 1655.

1935. Clothes moths and their control. USDA Farmers Bull. 1353.

BACK, E.A. and F. RABEK — 1923. Red cedar chests as protection against moth damage. USDA Bull. 1051.

BENEDICT, R.C. — 1917. An outline of the life history of the clothes moth, *Tineola bisselliella* Hummel. Science 46:464-66.

BILLINGS, S.C. — 1934. Paradichlorobenzene, naphthalene and the cedar oils inefficient as repellents against clothes moth adult. J. Econ. Entomol. 27:401-5.

1936. Notes on clothes moth breeding. J. Econ. Entomol. 29:1014-1016.

BRY, R.E., and R.A. SIMONAITIS — 1975. Synergized pyrethrins effective against fabric pests. Soap/Cosmet./Chem. Spec. 51(5):34-36,76.

BRY, R.E., R.E. BOATRIGHT, J.H. LANG, and R.A. SIMONAITIS. — 1979. Spray application of permethrin against fabric pests. Pest Control 47(4):14-17.

BRY, R.E., R.E. BOATRIGHT, and J.H. LANG — 1980. Ovicidal effect of permethrin against the black carpet beetle and the webbing clothes moth. J. Econ. Entomol. 73:449-450.

CHEEMA, P.S. — 1956. Studies on the bionomics of the case-bearing clothes moth *Tinea pellionella* (L.). Bull Entomol. Res. 47(1):167-182.

CHRYSTAL, R. — 1932. An oecophorid moth, *Borkhausenia pseudospretella* Stainton, attacking bookbindings. Entomol. Mon. Mag. 68:9-10.

COLE, J.H. — 1962, *Hofmannophila pseudosprettella* (Stnt.) (Lep. Oecophoridae), its status as a pest of woolen textiles, its laboratory culture and susceptibility to moth proofers. Bull. Entomol. Res. 53(1):83-89.

COLE, J.H., and F.G.S. WHITFIELD — 1962. A comparison of the principal test methods for evaluating mothproofing agents. J. Textile Inst. 53(3):326-51.

COLMAN, W. — 1932. Effect of yeast on clothes moth larvae. J. Econ. Entomol. 25:1242.

1940. Minimum size of openings through which clothes moth larvae can pass. J. Econ. Entomol. 33:582.

COLTON, H.S. — 1927. The unnatural history of the clothes moth. Sci. Monthly 24:47-58.

CROSSLEY, M.L. — 1946. Protection of furs, wool, silk, and related materials from destruction by moths and other insects. Mothproofing of woolen materials in Europe. Textile Res. Inst., Inc. pp. 4-6.

CSMA. — 1969. Chemical Specialty Manufacturers' Association Textile Resistance Test. Soap Blue Book 44(4a):197-200.

CURRAN, C.H. — 1949. Clothes moths. Natural History 58:324-31.

CURWEN, B.S. — 1932. Infestation of telephone exchange wiring by *Tineola bisselliella*. Proc. S. London Entomol. Nat. Hist. Soc. 1931-32:47-48.

DINGLER, M. — 1928. Remarkable occurrences of home household pests. Rev. Appl. Entomol. (A)17:11.

DONER, M.H., and E.G. THOMSSEN — 1943. Clothes moths and their practical control. Soap & Sanitary Chem. 19(10):102-05.

FERGUSON, G.R. — 1950. Survey on incidence of fabric pest damage. Soap & Sanitary Chem. Official Proc. 36th Mid-Year Meet. pp. 77-79.

FLOCK, R.A. — 1951. Damage to household goods by the fan palm caterpillar. J. Econ. Entomol. 44:260-61.

FRAENKEL, G., and M. BLEWETT — 1945. The dietetics of the clothes moth, *Tineola bisselliella* Hum. J. Experimental Biology 22:156-61.

FREEDMAN, E. — 1948. Trends in consumer demand for household sanitary products. Soap & Sanitary Chem. 24(7):123-25.

FREY, W. — 1939. Ueber die wirksamkeit von naphthalin, paradichlorobenzol and hexachlorathum als kliedermottenbekamp fungsmittel. Arb. Physiol. Angew. Ent. Ber. 6:189-198.

GRISWOLD, G.H. — 1931. On the length of the adult life in the webbing clothes moth. *Tineola bisselliella* Hum. Ann. Antomol, Soc. Amer. 24:761-74.

1933. Fish meal as a food for clothes moths J. Econ. Entomol. 26:720-22.

1944. Studies on the biology of the webbing clothes moth (*Tineola bisselliella* Hum.). N.Y. (Cornell) Agr. Experiment Memoir 262.

GRISWOLD, G.H., and M.F. CROWELL — 1936. The effect of humidity on the development of the webbing clothes moth (*Tineola bisselliella* Hum.). Ecology 17:241-50.

HERRICK, G.W. — 1916. Insects injurious to the household and annoying to man. Macmillan.

1933. An unusual invasion of the clothes moth, *Tineola bussellliella*. Entomol. News 44:99-101.

HERRICK, G.W., and G.H. GRISWOLD — 1933. Naphthalene as a fumigant for the immature stages of clothes moths and carpet beetles. J. Econ. Entomol. 26:446-51.

HOSKINS, W.M., and M.J. VAN ESS — 1940. Protection of fabrics and furs from clothes moths and carpet beetles. Pests 8(4):8-12.

HUDDLE, H.B., and A.P. MILLS — 1952. The toxicity of cedar oil vapor to clothes moths. J. Econ. Entomol. 45:40-43.

JACKSON, L.E., and H.E. WASSELL — 1927. Cinchona alkaloids. J. Indus. Engin. Chem. 19:1175.

KATZ, H.L. — 1972. Why carpet beetle has declined in U.S. International Pest Control Jan./Feb. p. 25.

KEMPER, H. — 1936. Ueber die anfalligkeit verschiedener pelzsorten gegenuber mottenfrass. Anz. Schadlingsk 12:1-6.

LAING, F. — 1932. *Borkhausenia pseudospretella* and other moths. Entom. Mon. Mag. 68:77-80.

LAUDANI, H. — 1957. PCOs should treat cedar chests and closets for moths. Pest Control 25(10):39-40,98.

LAUDANI, H., and P.H. CLARK — 1954. The effects of red, white, and South American cedar chests on the various stages of the webbing clothes moth and the black carpet beetle. J. Econ. Entomol. 47:1107-11.

LINDERSTROM-LANG, K., and F. DUSPIVA — 1935. Nature 134:1039-40.

LINSLEY, E.G. — 1944. Natural sources, habitats, and reservoirs of insects associated with stored food products. Hilgardia 16(4):187-214.

MACNAY, C.G. — 1947. A summary of the more important insect infestations and occurrences in Canada in 1947. Entomol. Soc. Ontario 78th Ann. Rpt.

MALLIS, A., A.C. MILLER, and R.C. HILL — 1958. Feeding of four species of fabric pests on natural and synthetic textiles. J. Econ. Entomol. 51:248-49. 1959. The attraction of stains to three species of fabric pests. J. Econ. Entomol. 52:382-84.

MALLIS, A., B.T. BURTON, and A.C. MILLER — 1962. The attraction of salts and other nutrients to the larvae of fabric insects. J. Econ. Entomol. 55:351-55.

MARLATT, C.L. — 1896. USDA Bull. 4, N.S.

McTAVISH, W.C. — 1938. Mothproofing problems. Soap & Sanitary Chem. 14(1):103.

MEEUSE, A.D.J. — 1952. On the origin of clothes moths, carpet beetles and similar household pests. Beaufortia 15:1-8. March 18.

MICHELBACHER, A.E., and D.P. FURMAN — 1951. Control of household insects and related pests. Calif. Agr. Expt. Sta. Circ. 172.

MONCRIEFF, R.W. — 1950. Mothproofing. Leonard Hill Ltd. London.

NAGEL, I. — 1920. Contributions to the biology of *Tineola bisselliella*. Rev. Appl. Entomol. (A) 9:1, 1921.

NPCA (National Pest Control Association) — 1954. Biology and control of clothes moths and carpet beetles. NPCA Technical Release 5-54. 1958. Don't treat pianos. NPCA Technical Release 14-58.

NOTINI, G. — 1939. Kladesmalen. Medd. St Vaxtskyddanst No. 28.

OTT, D.J. — 1955. Identification of insect damage on wool and related animal fibers. Amer. Dyestuff Reporter 44(16):515-20.

PENCE, R.J. — 1955. The fan palm caterpillar as a household pest P.C.O. News June, p. 7. 1966. Analyzing fur damage with a microscope. Calif. Agr. Ext. Serv. Circ. 541.

PIMENTEL, D., and M.H.J. WEIDEN — 1959. Protection of stored woolens from insect damage. J. Econ. Entomol. 52:457-60.

RAWLE, S.G. — 1951. The effects of high temperatures on the common clothes moth, *Tineola bisselliella* (Hum.) Bul. Entomol. Res. 42(1):29-40.

REUMUTH, H. — 1946. The major textile pest — "The Moth." Moth proofing of woolen materials in Europe. Textile Research Institute, Inc. pp. 7-24.

RICHARDSON, N.M. — 1897. Dorset clothes moths and their habits. Proc. Dorset Nat. Hist. Soc. 18:138-49.

ROTH, J.M., and E.R. WILLIS — 1952. Observations on the behavior of the webbing clothes moth. J. Econ. Entomol. 45:20-25.

SPENCER, G.J. — 1931. An important breeding place of clothes moths in homes. Can. Entomol. 63:199-200.

TITTANEN, K. — 1971. The efficiency of a pyrethrins aerosol against the larvae of the clothes moth *(Tineola bisselliella)*. Pyrethrum Post 11(1):15-17.

TITSCHACK, E. — 1936. Experimentelle Untersuchungen Uber den Einfluss der Massenzucht auf das Einzeltier. A. Angew. Entomol. 23:1-64.

VOLLMER, O. — 1931. Kleidermottem als Fresser Lebender Zecken. Z. Angw. Entomol. 18:161-74.

WALLACE, F.N. — 1925. Report of the Division of Entomology. 7th Ann. Rpt.

Indiana Dept. Conserv.

WATERHOUSE, D.F. — 1958. Wool digestion and moth-proofing. Advances in Pest Control Research. Vol II. Interscience, N.Y. pp. 207-262.

WATSON, J.R. — 1939. Control of four household pests. Fla. Agr. Expt. Sta. Press Bull. 536.

1946. Control of three household insects. Fla. Agr. Expt. Sta. Press Bull. 619.

WHITFIELD, F.G.S., and J.H. COLE — 1958. The bionomics of *Tineola bisselliella* (Hum.) under laboratory culture and its behavior in biological assay. Rev. Appl Entomol. (A) 47(9):301-302, 1959.

WILSON, H.F. — 1940. Lures and traps to control clothes moths and carpet beetles. J. Econ. Entomol. 33:651-53.

WOODROFFE, G.E. — 1951. A life-history study of the brown house moth *Hoffmanophila pseudospretella* (Staint.) (Lep. Oecophoridae). Bull. Entomol. Res. 41:529-53.

1951a. A life-history study of *Endrosis lactella* (Schiff) (Lep. Oecophoridae) Bull. Entomol. Res. 41:749-760.

ARNOLD MALLIS

Arnold Mallis was born in New York City on October 15, 1910, the son of Russian immigrants to the United States. He attended grade school and high school in Brooklyn and Long Island. In 1927, at the age of 16, he moved with his parents to Los Angeles, California, where he completed his last year of high school.

In 1929, he entered the University of Southern California and began pre-dental courses. Since the Great Depression began at about this time, he dropped out of college and worked in a Los Angeles garment factory. After two years on the job he decided to return to college and, because of his interest in trees and natural history, began a two year course in forestry at Pasadena Junior College (now Pasadena City College).

He entered the University of California at Berkeley in 1932, where he studied under such outstanding entomologists as Professor E. O. Essig in economic entomology, Dr. E. C. Van Dyke in forest entomology, Professor W. B. Herms in medical entomology and insect ecology, and Dr. W. M. Hoskins in insect toxicology. These teachers and others instilled in him a life-long interest in insects and their control. He received a bachelor's degree in entomology in 1934.

After two years graduate work he left the University in 1936 and sought employment as an entomologist in southern California. Since there were no jobs open in entomology he took the state examinations for a license in structural pest control and received a Class A license (all categories, including fumigation).

Mallis worked two years for pest control firms in Los Angeles, Hollywood and Bakersfield and then in 1938 obtained a job as a field aide in the USDA Bureau of Entomology, doing research on vegetable insects in southern California. This job lasted for six months. He then returned to Berkeley for two purposes, to complete his master's degree on the ants of California and to collect information for the first edition of the <u>Handbook of Pest Control</u>. In 1939, he returned to southern California and became entomologist-in-charge of pest control for the Buildings and Grounds Department, UCLA. In 1942, shortly after the United States entered WWII, he became associated with the USPHS, implementing malaria control programs around military camps in Louisiana. In 1944, he began to screen compounds as pyrethrins substitutes for Hercules Powder Company at the University of Delaware. The screening resulted in the discovery of Compound 3956, later known as toxaphene.

In 1945, Mallis commenced his employment in the entomology laboratory of Gulf Oil Corporation in Harmarville, Pennsylvania. Here he worked as an entomologist on household insects and household insecticides for 23 years until the company closed its entomology laboratory in 1968. He then became an extension entomologist for The Pennsylvania State University and retired as an associate professor in 1975, at the age of 65.

CHAPTER THIRTEEN

Hide and Carpet Beetles

The lowest kind of vermin
And the one I most abhor
Is the bug that ate my wife's mink
When it was only half paid for.

— A.M.

IT IS SAID THAT the Egyptians held these beetles in high esteem and even
embalmed them in their mummies. However, Mallis (1978) believes the beetles
were attracted to the bodies and embalmed by accident. In any event, man's
affection for hide and carpet beetles has worn thin with time, as these are the
insects which the taxidermist, the museum caretaker, and the housewife, to
name but a few, constantly have to contend with. The hide beetles, as their
common name indicates, are special pests of hides, meat, and similar animal
products, and are often referred to as skin beetles, larder beetles, tallow beetles,
and dermestids. Carpet beetles are smaller in size than hide beetles, and are
the common pests of woolens, rugs, carpets, upholstered furniture, museum
specimens, and materials of a like nature. In the case of carpet beetles, the
adults feed largely on pollen and nectar. The larvae alone are responsible for
damage to commodities.

Where the carcass of a dead animal is encountered in the field, hide beetles
will commonly be found feeding on the remains. Here they are of value as scavengers; unfortunately for man, these beetles consider taxidermic specimens in
the museum, or a cured hide in the warehouse, as fair prey. Their meat-eating
tendencies have been put to work by museum preparators for removing the flesh
from skeletons, a delicate job for which they have no peer.

Hinton (1945), in discussing these insects, notes: "From the point of view of

[1]*Associate Professor (Retired) Extension Entomology, The Pennsylvania State University,
University Park, Pa.*

the type of food eaten, dermestids may be divided into three groups: (1) species which are able to maintain themselves only on animal matter or materials containing animal proteins; (2) species which normally live on animal matter, but are also able to breed successfully on vegetable matter exclusively; and (3) one species (*Trogoderma granarium* Everts) which appears normally to be restricted to grain and cereal products. Most of the species of the family belong to the first group, and it seems certain that feeding on dry and decomposing animal matter is the primitive habit in the Dermestidae, and that the ability to survive on vegetable matter alone is a secondary specialization acquired more recently in the history of the family."

The household-invading carpet beetles very often frequent flowers, and it is a moot question whether or not they fly from the flowers to the homeowner's cherished possessions. The late Ronald Hunt saw the varied carpet beetle in southern California fly from crepe myrtle and spiraea into houses, and he said when the crepe myrtle and spiraea are in flower, varied carpet beetles can be shaken from the flowers in great numbers. Kemper (1939), who studied this insect in Germany, believes that *Anthrenus scrophulariae* (L.) has two varieties; one that breeds indoors and does not visit flowers, and another that breeds outdoors. "This view was supported by constant differences in coloration of the adults of *A. scrophulariae* taken on flowers and in dwellings in the same locality in the summer of 1935, and by experiments in which house strains of *A. scrophulariae* and *A. verbasci* could be easily reared in the laboratory at any time of the year, while field strains could not. Eggs of both species laid by beetles taken on flowers gave rise to adults in autumn when reared in the laboratory, but these failed to emerge from the larval skin and died, whereas when the pupae of *A. verbasci* were exposed to cold, they overwintered and did not give rise to adults until the spring. Some eggs were obtained from the latter. It thus appears that a diapause is obligatory in field strains of *Anthrenus spp.* and not in house strains."

Linsley (1944) studied the natural habitat of several species of carpet beetles in California and came to the following conclusions: "Nests of birds, rodents, insects, and spiders harbor large numbers of carpet beetles and clothes moths. Nests about houses and other buildings are more apt to be heavily infested than those in the open. The protected environment provides drier conditions and favors survival of the insects. Such nests provide foci for household and warehouse infestation and may invalidate control efforts applied within the structure.

"Control programs should include destruction of nests at the close of the breeding season before the onset of cool weather, which may drive the insects indoors.

"A number of the carpet beetles require pollen for successful production of eggs, and care should be taken not to bring these insects into the house on cut flowers. Such individuals, having mated and fed, are more likely to provide the source of a serious infestation than individuals which have reached maturity within the confines of the building."

In some areas, carpet beetles are responsible for more injury to carpets, rugs, and other woolens than the clothes moth. Ferguson (1950) concludes from his survey that "in the area east of the Rocky Mountains, the bulk of the damage in the southern states is due to clothes moths; whereas in the northern states, the carpet beetle is much more important than the clothes moth. Florida reports that there are no important economic species of carpet beetles in the state, and that they are not a problem." A limited survey conducted by the author among experienced pest control operators and entomologists from coast to coast showed carpet beetles were definitely responsible for more fabric damage in the North than clothes moths. In the author's laboratory in Pittsburgh, Pennsylvania, approximately 10 calls are made for information about carpet beetles for every one on clothes moths. This confirms Ferguson's conclusions.

According to Ferguson (1950), the explanation for the comparative distribution of clothes moths and carpet beetles in the United States is as follows: "It is the writer's opinion that the differences in geographical incidence of injury are due primarily to humidity conditions. In northern states, the humidity in the average home is extremely low during the winter months. Griswold and Crowell (1936) reported that under high humidity conditions, a high percentage of eggs of the black carpet beetle became moldy. The greater resistance of carpet beetle larvae to cold temperatures could account for their greater importance under wool storage conditions in the northeast, but not for their greater importance under heated household conditions."

Ferguson's survey also shows that the black carpet beetle is the most important pest species of carpet beetle in the eastern states. He further notes: "The common carpet beetle was mentioned particularly by Rocky Mountain and far mid-western states. The varied carpet beetle was stated to be most important by two of the three Pacific Coast states, and *Trogoderma sp.* the most important in Idaho."

Mallis et al. (1958) showed that carpet beetles and clothes moths readily fed on combinations of wool and synthetic fibers. Bry (1975) notes that only the wool is digested and the other fibers are passed unchanged. According to Mallis et al. (1959 and 1962), young and mature larvae of the clothes moth, and young larvae of the black and furniture carpet beetles are attracted to such stains as tomato juice, human sweat, human urine, etc., because of the salts they contain. Older larvae of the furniture carpet beetles are less attracted to these stains, and the older larvae of the black carpet beetle are not attracted at all. Baker (1974) notes that the addition of minerals to test cloth greatly increases the rate of consumption of wool.

COMMON HIDE AND CARPET BEETLES

A. Adults five or more mm in length.
1. The larder beetle, *Dermestes lardarius* (L.) (seven to nine mm)
   The adult is dark brown with pale grayish-yellow pubescence, forming a yellow band at the base of the wing covers, containing approximately six black spots. The larva is 11 to 13 mm long, brown and hairy, with a pair of stiff curved spines on the ninth abdominal segment. —CURVED CAUDAD
2. The hide beetle, *Dermestes maculatus* De Geer (five to 10 mm)
   The shape of the adult is similar to that of the larder beetle. It is black in color with white hairs on the sides and undersurface. The apex of each wing cover comes to a fine point. CURVED SPINES ON 9TH CURVED CEPHALAD
3. The black larder beetle (incinerator beetle), *Dermestes ater* De Geer (seven to nine mm)
   The adult is black in color with a yellowish-gray pubescence. There are black rounded spots and hook-shaped spots on the ventral side of the abdomen. The mature larva is similar in appearance to that of other *Dermestes* larvae and is brownish to black in color and up to 14 mm in length.
B. Adults less than five millimeters in length.

FOR LARVAL DETERMINATIONS SEE A. PETERSON-1957

C-45

1. The black carpet beetle, *Attagenus megatoma* (F.) (2.8 to five mm)
   The adult is dark brown or black and twice as long as wide. The larva has an elongated body with a long tail of hairs. It varies from light brown to dark brown in color, and is seven to eight mm long when mature.
2. The common carpet beetle, *Anthrenus scrophulariae* (L.) (2.2 to 3.75 mm)
   The adult is blackish in color, speckled with white, and characterized by a longitudinal band of orange red scales, with three lateral projections of the same color down the center of the back. The undersurface of the body is covered with white and orange scales. The mature larva is five mm long, brown, with long hairs at the tail end.
3. The furniture carpet beetle, *Anthrenus flavipes* LeConte (two to 2.5 mm)
   The adult is blackish in color, mottled with white, black, and brown. These areas are outlined by yellow scales. The ventral side is covered with white scales. On each side of the ventral segments there is a spot of yellow and black scales. The femora are clothed with yellow scales. The larva is similar in size and appearance to that of the common carpet beetle.
4. The varied carpet beetle, *Anthrenus verbasci* (L.) (two to three mm)
   The thorax of the adult is black with yellow scales along the base and white scales along the sides. The black elytra have two transverse zigzag bands of white scales which are bordered by yellow scales. The ventral side of the body is covered with grayish yellow scales. The larva is similar to that of the common carpet beetle.
5. The larger cabinet beetle, *Trogoderma versicolor* (Creutz) (two to five mm)
   This is an oval, blackish beetle with a mottled reddish-brown appearance and covered with brown and gray scale-like hairs. There are a number of species of *Trogoderma* requiring a specialist for classification. The larva is reddish brown on the dorsal surface and lighter underneath. The body is covered with short yellowish brown hairs and has a tuft of short hairs at the posterior. *Trogoderma* larvae usually are much lighter (light brown) than the larvae of most carpet beetles.

THE LARDER BEETLE

Dermestes lardarius (L.)

The name *Dermestes* is derived from Greek and means "to devour a skin," a habit which is typical of this genus. This cosmopolitan species is a pest of cured meats in Europe, the United States, and Canada. The adult is a dark-brown beetle, some seven to nine mm in length, with a pale yellow six-spotted band on the elytra. The undersurface of the body, as well as the legs, are covered with fine yellow hairs. The larva is brownish in color, 11 to 13 mm in length, and characterized by two curved spines on the last visible body segment.

It has been recorded from such stored provisions as ham, bacon, meats, cheese, dried museum specimens of all kinds, stored tobacco, dried fish, etc. In the United States, larval infestations have been associated with the presence of dead cluster flies and dead face flies. Patton (1931) also reports that the larva have attacked newly hatched chickens and ducklings in poultry houses in Germany. The infestations originated from pigeon lofts above the poultry houses.

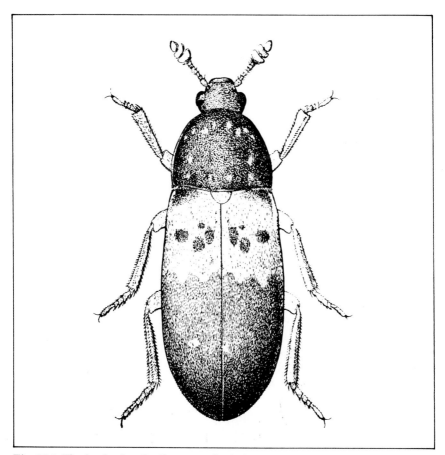

Fig. 13-1. The larder beetle, *Dermestes lardarius* (L.)

All stages of the beetle have been found in boxes of dog biscuits. Schwarz (1936) found the beetle in enormous numbers in a building in Hamburg, West Germany where several tons of crabs, which were to be ground into manure, were stored. The beetles infested the roof beams and other timbers to such an extent as to threaten the safety of the building. They had bored into the spruce to a depth of ½ inch/13 mm. Bauer and Vollenbruck (1930) record from Berlin that the larvae of *Dermestes lardarius* (L.) and the related species, *Dermestes peruvianus* (Lap.), made holes in felt-covered lead water pipes in a cellar used for storing smoked hams and sausages. Later, experiments proved they could penetrate lead with ease, tin with some difficulty, and were unable to perforate zinc or aluminum. Canzanelli (1937) reports that the beetle is a pest in Italy in the rearing of silkworms, where it feeds on eggs, pupae, and dead moths, and oviposits on the cocoons.

Spencer (1948) notes that in British Columbia the larder beetle "is best collected under old bacon rind or an old ham bone to which they generally arrive in pairs. In the same way, a male and female often arrive during the night to the bread box into which a freshly-baked loaf has been put and has fermented or soured a little."

Life history. According to Herrick (1936), the beetles are often found overwintering in crevices of bark. In May and June the beetles enter the house and seek food on which to oviposit, and if the beetle, like Old Mother Hubbard, should find the cupboard bare, she deposits her eggs in cracks and crevices in the vicinity of the pantry. Thus, the hatching larvae are not far from food.

Kreyenberg (1928) observed the females to lay from 102 to 174 eggs. The eggs were laid from June through August, and the incubation period was 12 days or less. According to some authors, the larvae seem to prefer the fatty portions more than the lean muscular parts, whereas Kreyenberg notes that fresh lard or fat clogs the mouthparts of the larvae and is avoided. The larvae apparently prefer hams that are beginning to spoil. According to Kreyenberg, the larva molts up to six times, the male five times, and the female six times. The larva eats constantly until the next to the last molt, when it begins to wander and seek a suitable place for pupation.

When ready to pupate, the larva burrows into ham or bacon, or as previously stated, it may leave the meat, wander, and then bore into anything in the immediate vicinity. It is this tendency that accounts for the recorded injury to wood, lead, etc. The burrowing into these strange materials is entirely for protection and not for food. If nothing is available, the larva will pupate in its last larval skin. The pupal stage extends from three days to a week or longer, depending on environmental conditions, and a complete generation may be completed in 40 to 50 days under suitable conditions. During the period of pairing and oviposition, both sexes avoid light, although they will venture into it when hungry. Hinton (1945) states that there is usually one generation a year, but in some localities as many as five per year have been observed. Kreyenberg found that the optimum temperature for development is from 64.4° to 68° F/ 17.7° to 20°C.

Control. Gray (1947) says the following about the occurrence of the larder beetle in the home: "The general use of refrigeration, the purchase of meat in small quantities, and the fact the home curing of meat is practically a lost art have resulted in a great decrease in the damage caused by this insect. It is now more likely to be introduced with dog or cat food than otherwise. A careful inspection of these commodities should always be made to avoid such a happening."

However, of late, larder beetles are once again becoming a pest of some importance, especially in suburban areas. The reason for this is that fertilized lawns in the suburbs produce large numbers cluster flies, which are parasites of earthworms. The cluster flies, along with face flies, bluebottle flies and other flies accumulate in the fall in attics and other void spaces in the home. Many of these flies die and their bodies attract larder beetles who lay eggs on the dead flies. The larvae feed and grow on the dead flies and eventually produce more larder beetles.

DDVP, pyrethrins, resmethrin, malathion sprays or aerosols should be effective against the larder beetle in attic areas. DDVP strips in such areas may also slowly affect the adults and larvae of this insect.

THE HIDE OR LEATHER BEETLE

Dermestes maculatus (DeG.)

This beetle, which is five to 10 mm in length, is similar in habits and appearance to *Dermestes lardarius* (L.) and, like it, is cosmopolitan in distribution.

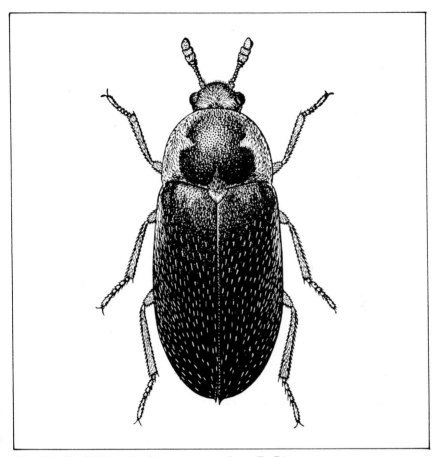

Fig. 13-2. The hide beetle, *Dermestes maculatus* (DeG.)

In entomological literature it is often referred to as *D. vulpinus* (F.). This insect has been distributed widely through the shipping trade. It is believed to have come to California in the 1840's on the old schooners that carried hides around the Horn.

It may be distinguished from the larder beetle by the fact that its wing covers or elytra are dark all over, whereas the anterior portion of the elytra of *Dermestes lardarius* (L.) is a pale yellow/brown. The undersurface of *Dermestes maculatus* (DeG.) is for the most part white, while the undersurface of *Dermestes lardarius* (L.) is clothed with a fine yellow pubescence. Moreover, in *Dermestes maculatus* (DeG.) the apex of each wing cover tapers into a fine point.

Normally, this insect is beneficial since it is an important scavenger; it has been used by Hall and Russell (1933) and Case (1959) to remove the flesh from delicate museum specimens. However, its habit of feeding on carcasses of stored skins, hides, etc., makes it an important pest.

Kreyenberg (1928) states that the larvae of the hide and leather beetle prefer hides and skins, while the larvae of *Dermestes lardarius* (L.) prefer ham, bacon, cheese, meats, etc. Von Dobkiewicz (1928) noted that the larvae and adults of *Dermestes frischii* (Kug.) and *Dermestes maculatus* (DeG.) thrived on smoked meat, cakes, and dried cheese, but could not survive on fat alone. Because the mature larva has the habit of boring into various hard substances in order to pupate, there have been many records of injury to such materials as woolen goods, hair brushes, cork, tea chests, and woodwork. Brimblecombe (1938), in Australia, observed this beetle do severe damage to the rafters of Oregon pine in a mill in Queensland, where bones and offal were received. The larvae climbed some 24 to 36 feet and bored into the rafters. They preferred to mine the spring wood since this is softer. Some pieces of wood were so badly injured that they presented a laminated appearance consisting of alternate layers of hardwood and mined areas of soft wood.

Life history. A number of individuals have studied the life history of this species. Because of differences of locality and conditions under which the studies were undertaken, the results vary.

The eggs, which are two mm long and creamy in color, are laid singly or in batches of two to 20 in cracks, skins, hides, etc. If such shelters are not available, mating and oviposition cease. Kreyenberg (1928) states that two females laid 648 and 845 eggs, respectively. The eggs hatch in two to 12 days.

The extremely hairy and active larvae undergo their first molt in two days after hatching. At 82.4° to 86° F/28° to 30° C five molts occur at intervals of five days. Grady (1928) observed seven molts to be average. There may be up to 11 molts if the larva is reared under adverse conditions. Smit (1931) found that the larval period ranged in length from a minimum of 35 days in summer to a maximum of 238 days the rest of the year. Most of the damage caused by this species is due to the larva, which is a much more voracious feeder than the adult.

The larvae cease to feed four days prior to pupation and wander in search of shelter in which to pupate. Hinton (1945) says the following about this species: "The larvae are very active and strongly negatively phototropic. When full grown they leave their food to find a suitable place in which to pupate. They will bore a pupal chamber in almost any compact substance that happens to be near at hand, and this habit of indiscriminate boring into various materials which they do not use for food has frequently been noted. Probably the earliest record of this kind of injury is that referred to in 'The Last Voyage of Thomas Cavendish', where there is an account of a ship in 1593 carrying a cargo of dead penguins being nearly sunk because of the honeycombing of its sides and bottom by Dermestes larvae. . . ."

Walker (1944) found the life cycle to require 60 to 70 days at average temperatures and humidity. There may be six generations per year under favorable conditions. Russell (1947) studied the life history of this species at 85° F/29° C. He found the average length of the various stages to be: egg — three days; larval — 30 days; pupal — seven days; and adult before oviposition — five days. Bellemare and Brunelle (1950) found the number of molts decreased with increases in relative humidity from 70 to 100 percent. The pupal stage was affected only by a rise in temperature.

Walker (1944), in discussing an infestation in Massachusetts, notes that the "adult stage lasts for about 60 to 90 days, during which time they feed on skins. They are not negatively phototropic and may be found at windows trying to get outdoors to the flowers and shrubs. They are often found burrowing in the wool on the skins, but no damage to the wool could be found. They mate and deposit their eggs in this stage. Actual oviposition was not seen. The beetles are strong, active fliers and feed on a great variety of flowers and shrubs. They frequently gain entrance to buildings by flying through windows and other openings, or they may be brought indoors on flowers. This beetle may be found in some heated buildings at any time during the year, but it is usually more abundant during spring and summer."

Locally, this insect spreads by the wanderings of the newly emerged and mature larvae and by the flights of the adults. Von Dobkiewicz (1928) notes that this species is very cannibalistic, since the parents eat the younger larvae, and the older larvae eat the fresh pupae. Roche and Smith (1974) used plastic boxes to prevent cannibalism in raising the larvae. Von Dobkiewicz notes that hybrids from *Dermestes frischii* (Kug.) and *Dermestes vulpinus* (F.) also have been obtained.

Control. Since the highly effective sprays and dusts of DDT, lindane, and chlordane no longer can be used against these insects, we must resort to such insecticides as pyrethrins plus synergist, resmethrin, or organic phosphate and carbamate insecticides like DDVP, malathion, chlorpyrifos, bendiocarb, propoxur and diazinon. In the author's laboratory it was shown that adults of *Dermestes* are susceptible to sprays containing the petroleum fly-spray base oil alone. Dried pet foods packaged in paper bags are at times infested with *Dermestes* larvae and may serve as a source of infestation in the home. Such infested material should be removed immediately.

THE BLACK LARDER BEETLE (INCINERATOR BEETLE)

Dermestes ater (DeG.)

This insect is recorded in the literature as *Dermestes cadaverinus* (F.), which Hinton (1945) notes is a synonym of *D. ater*. The cosmopolitan black larder beetle has been found infesting a wide variety of materials somewhat similar to that recorded for *D. maculatus* (DeG.). Pet foods in bags are often infested.

Roth and Willis (1950) studied this insect in the laboratory and fed the adults moist canned beef to stimulate oviposition. The larvae were reared on fish meal and a supply of drinking water hastened their development. The egg to adult stage required six weeks at approximately (80° to 82° F/27° to 28° C). The female may oviposit over two months and lay up to 400 eggs during this period.

Johnson (1945) notes that this black beetle with black rounded and hook-shaped spots on the bottom of the abdomen, and with yellowish/grey pubescence, is a pest in Connecticut and commonly found in incinerators. The adults have been found flying in a Connecticut apartment building during the months of November, December, and January. It is believed mouse cadavers in the building may have served as a source of infestation. Johnson describes the annoyance as follows: "One apartment on the first floor, directly above one of the incinerator rooms, experienced intermittent flights of the insects from late November into January. In the evening, when the temperature was high, the beetles were very active. They would fly around the lights and annoy the family members when reading. They would fall into food, annoying dinner guests, or be found crawling around in general. A nightly catch often numbered 25 or more beetles. The adults were often observed coming out of crevices, from behind baseboards and moldings. Some larvae were also found emerging from behind the baseboards or on the floor in the kitchen. This indicated that the insects were present in the partitions. As a number of crevices and a few holes were observed in the basement ceiling directly beneath the apartment, it was possible for the insects to infest the apartment directly from the incinerator room."

Johnson describes the causes of infestations in incinerators as follows: "Complete combustion, reducing all waste material to ash, is not obtained in the average small incinerator. This is due to insufficient dry, combustible material in the substance being burnt. As a result, partially burnt garbage remains in the ashes.

"The odors from the burning waste materials or from the unburnt garbage probably attract the adult insects. Sometimes several weeks elapse before the incinerator is cleaned out, providing ample time for breeding. Careless cleaning, such as leaving debris in the corners of the ashpit, leaves material available for continuous breeding. Small ledges in the chimney or disposal chutes formed by protruding bricks, stones, or disposal door frames might protect the insects sufficiently to form breeding areas.

"General infestations in incinerators may be avoided by thorough removal of all ashes and unburnt material at least every two weeks. Weekly removal would reduce the possibility of an infestation to a minimum."

OTHER SPECIES OF *DERMESTES*

There are a number of other species of *Dermestes* which are, for the most part, similar in habits and life history to those discussed previously. A few of the more important species are:

1) *Dermestes frischii* (Kug.). This insect occurs in the United States, Canada, Europe, Asia, and Africa. The adults and larvae have been reported in Ohio where they were feeding on bird feathers. Von Dobkiewicz (1928) found this dermestid with *Dermestes maculatus* (DeG.) in a case of skins imported into Bavaria from Canada.

The life history and feeding habits are similar to those of *Dermestes maculatus* (DeG.). Howe (1953) showed that the quickest larval development, 23.4 days, occurred at 91.4° F/33° C and 70 percent relative humidity, and the fastest pupal development, 4.4 days took place at 99.5° F/37° C and 70 percent relative humidity. This species may be distinguished from the hide or leather beetle by the fact that the apical margin of the elytra is not serrate.

2) *Dermestes peruvianus* (Castelnau). Tragardh (1934) noted that the larvae excavated pupal cells in boards of a box of preserved meat that originally came from South America and ultimately arrived in Sweden. A piece of wood measuring 18 mm by 24 cm harbored more than 230 holes.

3) *Dermestes carnivorus* (F.) is reported from North and South America, Europe, and India. This insect is not very important from an economic viewpoint.

4) *Dermestes marmoratus* (Say.). According to Essig (1926), this cosmopolitan species, which feeds on carrion and dry animal products, was distributed throughout California by hide and tallow traders in pioneer times, and now occurs throughout the West.

Dermestes in their relation to disease. Since *Dermestes* are carcass and hide feeders, there is always the possibility they may spread the bacilli or spores of anthrax, and in fact, the anthrax bacilli have been recovered from the feces of a dermestid.

Patton (1931) notes, ". . . the hairs on the larvae of *Dermestes* are known to produce allergic conditions. A case of this nature is recorded by Loir and Langangneux. A number of dock laborers were unloading a cargo of bones at Havre, and they soon began to suffer from urticaria, conjunctivitis, nausea, and irritation of the respiratory tracts; some of the men were ill for 15 to 20 days. Examination of the holds showed among the debris, hairs of the larvae of *Der-*

mestes, which filled the air with a fine powder at the slightest movement. Large numbers of larvae of at least two species of *Dermestes* were found, as well as those of *Necrobia rufipes, Necrobia ruficollis,* and *Tenebrio molitor."*

THE BLACK CARPET BEETLE

Attagenus megatoma (F.) [*=piceus* (Oliv.)]

According to Back and Cotton (1938), the black carpet beetle is the most destructive and widespread carpet beetle in the country. This was the carpet beetle most commonly referred to the U.S. Bureau of Entomology and Plant Quarantine by housewives. It is not as important a pest in Europe as it is in the United States. Beal (1970) has reviewed the taxonomy of these beetles.

The adult is 2.8 to five mm long, and dark brown or black in color. Griswold (1941) notes the terminal antennal segment in the male is twice as long as the same segment in the female. The black carpet beetle is found outdoors in the sunlight, very often feeding on the pollen of flowers, particularly that of *Spiraea.* These dermestids occur on windows and screens, so in all probability they enter the house while they are on the wing. Klein (1940) found the larvae in the sparrow nests in ventilators above an infested apartment. Here they apparently were feeding on feathers, dead birds, and bird manure. In the vicinity of Washington, D. C., the adults are very rarely found after July except in very warm buildings, where they may continue to feed throughout the winter. Spen-, cer (1948) notes that in British Columbia these beetles may infest nearly dry horse and cow carcasses.

Rees (1943) states that in "rare instances the larvae of *Attagenus piceus* (Oliv.) have caused physical discomfort to human beings by their presence in the nasal passages and sinuses."

At one time it was believed that this pest came from Europe in the baggage of the early colonists. It was first recorded in the United States in 1806, by Melsheimer, a chaplain to the Hessian soldiers. However, Linsley (1942a,

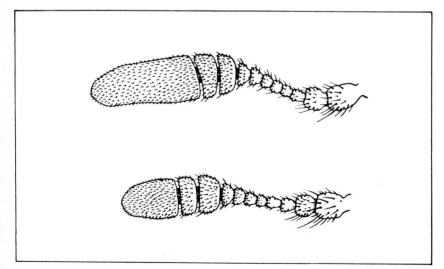

Fig. 13-3. Antennae of the black carpet beetle. (Top) Male and (bottom) female.

Fig. 13-4. (Left to right) Adult and larva of black carpet beetle; adult and larva of furniture carpet beetle.

1942b, 1944 and 1946) studied the distribution of this insect in California and showed it breeds in bird and rodent nests in California, where it passes the winter in the larval state. Studies such as these cast some doubt on the theory that this insect came over from Europe with the early settlers. It is now distributed widely throughout the United States and Canada, as well as Mexico.

The black carpet beetle is an important pest of many plant materials, particularly seeds, grain, and cereals, as well as a notable pest of animal products. Back and Cotton (1938) list woolen rugs, clothing, silk, carpeting, felts, furs, skins, yarn, velvet, feathers, hair-filled mattresses, upholstered furniture, wool-filled blankets, house insulations containing sheep wool and cattle hair, meat and insect meal, kid leather, milk powders, casein, books, birds' nests, cayenne peppers, and many seeds and grains. Takahashi and Uchiumi (1934) state that this insect causes considerable damage to raw silk in Japan, where it feeds on the dried pupae of silkworms, as well as on artificial silk. In the author's laboratory, the black carpet beetle was fed on the Chemical Specialties Manufacturers Association media consisting of 95 parts (by weight) ground dog meal and five parts brewer's yeast. Since the larvae bore into food containers and make them vulnerable to infestation by insect pests that ordinarily cannot penetrate the uninjured container, the larvae of this beetle may actually be responsible for much more damage than they are accredited. The larvae of the black and varied carpet beetles are often found in spilled flour in cupboards.

Life history. The small, pearly-white egg which is rarely seen by PCOs is deposited in the lint around baseboards, in ducts of hot-air furnace systems, etc.

Fig. 13-5. Cross section of wool carpet showing black carpet beetle feeding.

It hatches in six to 11 days in warm weather, but may require from five to 16 days under other temperature conditions. In laboratories where the black carpet beetle is raised, the eggs must be handled with great care since they are very fragile. In view of this, vigorous cleaning and brushing of clothes by housewives should kill many of the eggs. The eggs of clothes moths can take much more abuse.

The larva is long and narrow, up to seven to eight mm in length, excluding the long characteristic tuft of hairs. It has short, stiff hairs covering the body, and is dark brown to almost golden in color. Moore and Moore (1942) showed there may be two species of the so-called black carpet beetle, since they observed larvae that were chestnut brown and larvae that were silky yellow in color. Reproduction did not occur between the adults of the black vs. yellow larvae, whereas matings of pairs from similar larvae resulted in offspring. Apparently the yellow larva is typical. The authors note that in "living adults of the yellow species, the antennae are yellow throughout. The antennal clubs of the other species are black." Beal (1970) is of the opinion that the light-colored larvae are a related species, *A. elongatus Casey.*

When disturbed, the larva curls up and "plays possum." Griswold (1941) studied the behavior of the larvae and found that they are repelled by light, burrowing into the nap of a thick rug. They move so slowly that they appear to be gliding. At room temperature, the length of the larval life ranged from 258 to 639 days. This variation was to a great extent due to changes in temperature, food, and humidity. Back and Cotton (1938) found the larva to molt five to 11 times, and up to 20 times when conditions were unfavorable. The larval skins are often mistaken for the grubs themselves. The larva pupates in the last larval skin, and the pupal period may extend from six to 24 days. In Washington, D.C., the insects pupate from February through July, the large majority pupating from April to June. The adults may remain in the partially shed pupal skin from two to 20 days before emerging.

Takahashi and Uchiumi (1934) note the female adult lives for 36 days, and the male adult for 38 days at 77° F/25° C. Back and Cotton (1938) observed some of the females to commence egg-laying in less than one week after emergence, and to deposit an average of 53 eggs within another week's time. Griswold (1941) found that the females laid from 42 to 114 eggs, generally dying a few days after oviposition.

The Chemical Specialties Manufacturers Association (Anon., 1969) has described in great detail methods for rearing and testing the black and furniture carpet beetles, as well as the webbing clothes moth.

THE COMMON CARPET BEETLE

Anthrenus scrophulariae (L.)

Greenwald (1941) made a detailed study of the life history and behavior of the common carpet beetle, also known as the "carpet beetle," and much of the

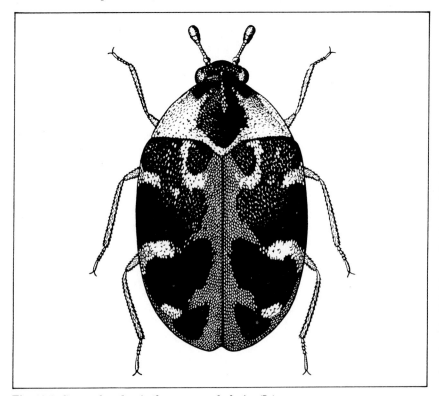

Fig. 13-6. Carpet beetle, *Anthrenus scrophulariae* (L.)

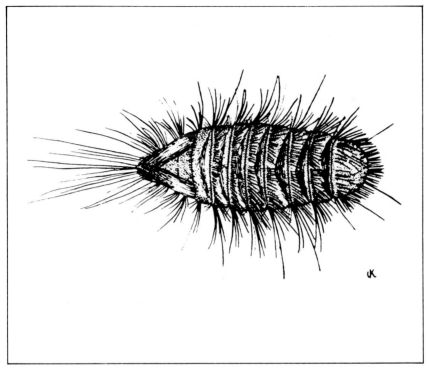

Fig. 13-7. Larva of the common carpet beetle, *Anthrenus scrophulariae* (Linn.)

following is from her observations. The common carpet beetle is a special pest of carpets. Back (1938) notes the larvae "eat irregular holes in fabrics, but in carpets tacked to floors they are more likely to eat slits following cracks. They never cause a webbing on the fabric." Besides carpets, this pest attacks woolen and other animal products that are left undisturbed, such as feathers, leather, furs, hairbrushes, silks, mounted museum specimens, as well as pressed plants. Linsley (1942b) found *A. scrophulariae* infesting chipmunk nests in the mountains of California.

The common carpet beetle is widespread throughout much of the world. Although it is most common in the northern states, it occurs as far south as Texas and Florida.

Since this was the injurious carpet beetle in Buffalo, New York, in the 1870's, it received the common names of "buffalo moth" and "buffalo bug." Lintner (1884) was of the opinion it was so named because of its habit of infesting the then widespread buffalo robes. It is now referred to as the "carpet beetle," "buffalo bug," and "old-fashioned carpet beetle."

The adult is a small oval beetle about three mm long, blackish in color, with minute whitish scales and a longitudinal band of orange/red scales down the middle of the back. The head is black with a few orange/red scales around the eyes and on the clypeus. In regard to the adults, Greenwald (1941) notes they were collected "in late May and early June, on blossoms, where they were found feeding on the nectar and the pollen. Toward dusk and on cold days they were found on the underside of flowers, although in a few cases beetles that were still feeding were collected late in the evening. When disturbed, the beetles usually 'played possum,' drew in their appendages, and dropped to the ground; some however, merely slid around to the underside of the flower cluster where they remained motionless with their heads and antennae drawn in." These day-flying beetles were most common on bushes during sunny days, particularly on such plants as *spiraea, ceanothus,* wild buckwheat, etc., which have white or cream-colored flowers. According to McQueen (1965), in Alabama the adults are "numerous on annuals, especially daisies and wild asters throughout the state; entering homes with cut flowers." L. O. Howard observed the adults in the East in heated houses throughout winter and the ensuing spring.

Life history. The small white eggs have projections at one end which apparently aid in anchoring them to textiles. Greenwald (1941) states that some of the eggs "are dropped upon the surface of the flannel, some are laid so that one or two fibers girdle them, and some are thrust so deeply into the fabric that only one end remains visible." One female was observed by L. O. Howard to deposit 36 white eggs in the fold of a carpet during April. The eggs hatched in 10 to 18 days.

The mature larva is reddish/brown, clothed with numerous black or brown hairs, and about 2.5 to 3.5 mm in length. The active larva appears to run rather than crawl. Greenwald (1941) found the larva underwent six instars for an average of 66 days at room temperature. The larva then pupated in the last larval skin. If the carpet is infested, pupation often occurs in cracks in the floor beneath the carpet. Length of the pupal stage averages 13.5 days.

The adult beetle, upon emerging from the pupal stage, lays quiescent for approximately 18 days in the old larval skin, and then becomes an active adult for approximately one month. The entire period of development at room temperature averages 94.5 days, with a range of 89 to 108 days. Griswold found the developmental period for one of the common carpet beetles to extend for 439 days.

Kunicke (1941), who studied the common carpet beetle in Germany, notes that beetles of the genus *Anthrenus* are changing gradually from a diet of natural foods in nature to finished animal products, such as wool and carpets. He notes that the common carpet beetle copulates and feeds on flowers, and then the female may fly into the home from May through July and oviposit on suitable larval food. This species apparently requires nectar and pollen to stimulate oviposition. The larvae feed until the end of August, when approximately 75 percent pupate, with the remainder overwintering as larvae. The beetles hibernate in unheated rooms or in the open.

Sweetman (1956) succeeded in rearing this beetle on brewer's yeast, soybean flour, honey and pollen. Cormia (1967) found this insect to be responsible for dermatitis.

THE FURNITURE CARPET BEETLE

Anthrenus flavipes LeC. [= *vorax* (Waterh.)]

Back and Cotton (1936) and Griswold (1941) concerned themselves with the life history and behavior of the furniture carpet beetle. Ayappa et al. (1957) and Patel (1958) studied the life history of this species in India. This beetle is one of the most destructive household pests in Washington, D. C. It has been re-

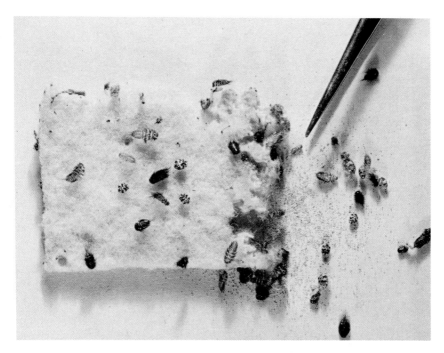

Fig. 13-8. Test cloth damaged by furniture carpet beetle. Note adults, larvae, cast skins and feces.

ported from many cities along the eastern seaboard and is a pest in the warmer parts of the South and West, as well as in heated buildings throughout the country. Barber (1951) states that the correct name for this species is *A. flavipes* LeConte.

The presence of this cosmopolitan species was first established in the United States in Augusta, Georgia in 1911. It appeared in upholstered furniture stuffed with horsehair imported from Russia. At first, it attracted attention as a pest of hair-filled furniture and was accordingly designated "the furniture carpet beetle." It is destructive to animal products of all kinds, particularly upholstered furniture, brushes, and carpets. Back and Cotton (1936) state the following materials and commodities are damaged by the larvae of this beetle: wool, hair, fur, feathers, bristles, horn, tortoise shell, and silk. When cellulose material such as linen, cotton, rayon, jute, softwood, as well as leather, are stained with animal excreta, or when paper and the above-mentioned materials enclose animal products, the larvae will gnaw through them. Herfs (1932) found spores also serve as an item of diet. Takio (1937) reports the larva of this beetle is a pest on dried silkworm pupae and cocoons in Japan. Back and Cotton (1936) note the larvae skeletonize dead mice, eat insects, dried cheese, old grain, casein, dried blood, and the glue of book bindings.

The adult is 2.0 to 3.5 mm long, blackish, and because of the presence of yellow and white scales, presents a mottled yellow/white and black appearance on the dorsal surface, and a white appearance on the ventral surface. The femora of the legs are thickly clothed with yellow scales. Hinton (1945) states this species

Fig. 13-9. Color variations in the adult furniture carpet beetle.

is closely related to *A. pimpinellae,* and may be distinguished from it by having an antennal club with the "first segment distinctly shorter than the second instead of about as long."

Life history. The white eggs, although small, can be seen with the naked eye. The female beetle commences laying eggs one to three days after leaving the larval skin. The female may lay 37 to 96 eggs in one to three batches, ranging from one to 57 eggs per batch. At room temperature the eggs hatch in approximately three weeks.

The adults are the overwintering stage, and no eggs are laid in cold weather. Herfs (1932) states the eggs of this insect failed to hatch at 104° F/40° C and development was slow at 68° F/20° C. Eggs laid before the coming of cold weather will ultimately die and shrivel. The female usually lays her eggs in the pile of mohair chairs, in the nap, or loosely on the surface of clothes, as well as in numerous crevices. Small pellets of larval excrement are often mistaken for eggs.

The larva of *A. flavipes* LeC. is so similar to that of *A. scrophulariae* (L.) that it is difficult to separate the two species in the larval stage. Rees (1943) and Hinton (1945) illustrate the antennae and peculiar spearheaded hairs from the posterior tufts of the larvae. These and other anatomical details are used to classify the larvae of *Anthrenus* and *Trogoderma.* Griswold (1941) notes: "As the larva of *A. vorax* moves about, the long pencil of hairs projecting backward from the end of its body is in constant vibration. If the larva is touched with a camel's-hair brush or disturbed otherwise, it opens up the tufts of hairs near the end of its body." Griswold found the duration of the larval life at room temperatures to be from 112 to 378 days for the females. "In general, larvae that developed rapidly had fewer instars than did those that developed more slowly. None of the larvae that eventually pupated had less than six instars." At room temperature, the number of instars varied from six to 12. Herfs (1932) found one larva that molted 29 times at a temperature of 68° F/20° C.

The full grown larva is generally chestnut brown in color and about five mm long. Since the larvae feed in a limited radius, their cast larval skins accumulate, often giving the appearance of more larvae than is actually the case. The larvae vary with age from white, light yellow, to dark red in color their color changes with the type of food consumed. It should be noted that this carpet beetle, as is the case with all pest dermestids, passes the greatest part of its life cycle as a destructive larva, which is usually not seen until the damage itself becomes evident.

The white pupa develops in the last larval skin, and Griswold (1941) found at room temperature the pupal stage extends from 14 to 19 days. The adult life is divided into a quiescent phase, during which it dwells in the old larval skin, and an active stage. At room temperature, the beetle has remained in the quiescent phase from six to 71 days. The time from egg to adult varies from 149 to 422 days.

The adults emerge for the most part in June, July, and August, but may be seen as early as March. Griswold observed that individuals that developed rapidly had a long adult life, and individuals that developed slowly had a short adult life.

Where the furniture carpet beetle is raised under methods recommended by the CSMA Scientific Committee, the larval period is completed in 10 to 13 weeks. The entire life cycle requires 12 to 15 weeks. The temperature of the rearing room is 80° F/26.7° C plus or minus 2° F/1.1° C, and a relative humidity of 55 percent plus or minus five percent. The food consists of 30 grams wool cloth, over which has been spread one teaspoonful of brewers' yeast.

THE VARIED CARPET BEETLE

Anthrenus verbasci (L.)

It has been noted, in some of the literature on this cosmopolitan species, that' it is of European origin. However, in view of Linsley's (1946) studies which show these beetles occur naturally in the nests of bees and wasps in California, it is quite possible it is a native insect. The adult is two to three mm long, and the dorsal side of its body is for the most part blackish in the center, with a variable, irregular arrangement of white, brownish, and yellowish scales. The ventral surface is clothed with fine, long, grayish/yellow scales. There are a number of varieties of this species whose scales differ in shape, size, color, and pattern. According to Hinton (1945), the scales on *A. verbasci* are 2½ to four times as long as broad, whereas the scales on *A. flavipes* are broadly oval, and two times or less long as broad.

This carpet beetle has been recorded feeding on a great variety of animal and plant products, such as carpets, woolen goods, skins, furs, stuffed animals, leather book bindings, feathers, horns, whalebone, hair, silk, fish manure, and dried silkworm pupae. Such plant products as rye meal, cacao, corn, and red pepper are also attacked. The varied carpet beetle is a museum pest, particularly in insect collections. Larvae often live as scavengers in nests of birds and bees.

In this respect, Linsley (1946), discussing this insect in California, has the following to say: "After mating, the females search out the nests of bees, wasps, spiders and other favorable habitats for oviposition. The favored larval food is dead insects and spiders, a fact which accounts for their ravages in insect collections. The nests of *Sceliphron* (Author's note: a mud dauber, a species of wasp), which stocks its cells with spiders and spider webs, are common natural habitats, as are various other places where dead insects are available. Fur and feathers undoubtedly serve as elements in their natural diet, as does the pollen stored by bees. (This latter product provides a satisfactory laboratory diet, bringing the beetles to maturity in a little more than 300 days. Milled cereals such as whole wheat flour may require over 400 days development.)"

Linsley (1946) notes that in California the varied carpet beetle is the most

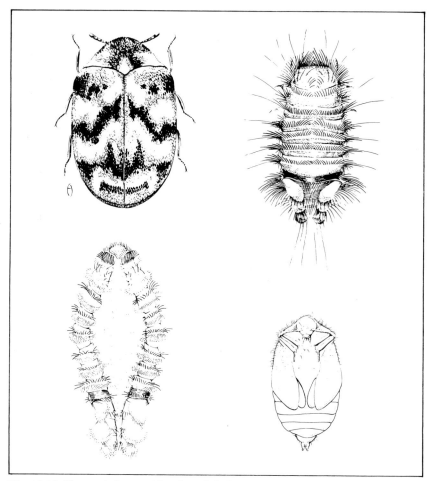

Fig. 13-10. The varied carpet beetle, adult, larva, pupal skin and pupa.

injurious dermestid to fabrics. This can probably be attributed to the wasp nests about homes which these beetles infest. H. Katz has found this and other carpet beetle infestations in Pittsburgh associated with the presence of bird nests on homes. Woodroffe (1953) and Woodroffe and Southgate (1954) have shown *A. verbasci* to be present in sparrow nests in England. Bird nests in attics resulted in heavy infestations. The larvae migrated from the bird nests in autumn to the rest of the house.

Metcalf (1944) records an epidemic of varied carpet beetle infestations in northern Illinois shortly after the United States entered World War II. Servicemen returned with infested coats, trunks, etc., and the carpet beetles spread from the infested articles to the rest of the house. The author is aware of an infestation of these insects in the carpeting in the rear seat of a car. Patton (1945) notes *A. verbasci* damaged nylon hosiery in the laboratory.

Life history. Griswold (1941) made a study of the varied carpet beetle, and much of the following is from her paper. The average size of the eggs is 0.27 mm wide and 0.55 mm long. The surface of the egg appears to be rough, with short spine-like projections at one end. The egg is white when first laid, but later becomes cream-colored. At room temperature, the egg hatches in 17 to 18 days.

The larva begins to feed almost as soon as it leaves the egg. The number of instars varies from five to 16, averages seven or eight, and is correlated with the length of the larval life. In this respect, Griswold notes: "The length of the larval life varies considerably and is probably influenced by a number of factors such as temperature, humidity, and the quality of the food available. In data obtained from 47 individuals, the larval periods for all but two ranged from 222 to 323 days. Of the two exceptional individuals, one required 604 days to attain its larval growth, and the other required 630 days."

The mature larva is four to five mm long, and its body seems to consist of a series of light and dark brown transverse stripes. Back (1938) states: "When unmutilated, they possess on each side at the end of the body three dense tufts of bristles and hair. If suddenly alarmed, the larvae erect these tufts and spread the bristles and hairs out so as to form beautiful round balls."

Pupation takes place in the last larval skin. At room temperatures the pupal period extends from 10 to 13 days. Griswold found the development period from egg to adult at room temperatures ordinarily requires 251 to 351 days.

The adult carpet beetle remains quiescent in the last larval skin for some days, finally emerging as an active adult. The males lived 13 to 28 days, and the females 14 to 44 days. Back (1938) notes that in Washington, D. C., the adults reach their greatest numbers in late spring and summer and are found at times feeding on the pollen of such plants as white roses, *viburnum spiraea,* and about 30 species of flowers. Griswold observes: "Some writers have stated that the beetles probably lay most of their eggs indoors before they go outside to feed on pollen. However, beetles collected outdoors at Ithaca have laid large numbers of eggs when brought into the laboratory, and colonies have been started in this way. In several instances, pairs of mating adults have been found on flower clusters of spiraea." Kunicke (1941) notes that the female does not require special food to induce oviposition, and Spencer (1948) also showed that this beetle can reproduce inside the house without having to fly outside to feed on flowers.

Kiritani (1958a) notes that adults bred indoors show a negative attraction to light on emergence from the last larval skin, but after laying most of their eggs they become positively attracted to light. Most of the outdoor adults show a positive attraction to light.

Linsley (1946), who studied this insect in California, states: "In the field, as in the laboratory, this species has but a single generation a year. The winter is passed in the larval stages, and when temperatures are favorable, feeding continues during this period. In early spring (about the middle of March in central California) the larvae pupate and transform to adults. In the field, the pupal period usually requires between two and three weeks, although as little as nine days may be involved under laboratory conditions. After transformation, the adults remain in the larval skin for a day or two and then emerge, seeking flowers for mating and feeding."

Kiritani (1958b) believes the two or three year life cycle as compared with a one year life cycle, is caused by unfavorable diet. Blake (1958, 1959 and 1961) is of the opinion that the number of diapauses (resting stages) depends on temperature. *A. verbasci* has one diapause at 77° F/25° C and two diapauses at 59° F/15° C. Where there is one diapause, the life cycle is completed in one year, and where there are two diapauses, it takes two years to complete the life cycle. A one-year cycle is more typical.

Since the beetles fly fairly high, they readily can enter houses through open windows. Should they oviposit in the home, they do not necessarily lay their eggs on the food material of the larvae. Imamura (1935) noted the adults are attracted to blue and white colors.

THE LESSER MUSEUM BEETLE

Anthrenus museorum (L.)

This beetle is very similar in appearance to *Anthrenus verbasci* (L.), having yellow and white spots. Patton (1931) states that the two species are best distinguished by the antennae, which in *Anthrenus museorum* (L.) consist of eight segments, with the club of two closely-joined segments, whereas the antenna of *Anthrenus verbasci* (L.) consists of 11 segments with a club of three closely-joined segments.

Anthrenus museorum (L.) is common in Europe, and although distributed widely in North America, it is not seen as often as the other carpet beetles. The life history and habits of this species have not been investigated. The larvae have been recorded from grain, wool, woolen articles, silk, and museum specimens. It is important as a pest of museum specimens. Spencer (1928) found this beetle infesting a house in Toronto, Canada, in 1928. The larvae were breeding in the attic on dead cluster flies, *Pollenia rudis* (F.). Kunicke (1939) records this species in birds' nests and dovecots in Germany.

THE BIRDNEST CARPET BEETLE

Anthrenus pimpinellae lepidus (LeC.)

This carpet beetle is a fabric pest, particularly in California. Linsley (1946) studied the life history of this insect. The adults emerge in March and April and feed and mate on flowers. They then oviposit in bird nests. The larvae feed for the most part on feathers and animal hairs, but require pollen and nectar to deposit viable eggs. Pupation occurs in the fall and unlike *A. verbasci,* the adult overwinters in the last larval skin.

Recently, two other carpet beetles have been recognized as potential pests, namely *Anthrenus coloratus* Reitter and *Novelsis aequalis* (Sharp). Kingsolver (1969) notes that *Anthrenus coloratus* Reitter has been recorded in Virginia, Washington D.C., Maryland, Illinois and California. It also has been found infesting insect collections, seeds and other plant products. The adult is 1.5 to 2.25 mm long and resembles *A. verbasci* (L.). It can be distinguished from the latter by the nine segments in the antenna, whereas *A. verbasci* (L.) has 11 segments.

Kingsolver and Fales (1974) record *Novelsis aequalis* (Sharp) throughout the eastern United States, as well as Mexico, Texas, and Oklahoma. The adult and larva appear somewhat similar to the black carpet beetle. The adult can be distinguished from the black carpet beetle by the three grayish bands it has on the elytra.

DERMESTIDS OF THE GENUS TROGODERMA

Trogoderma

This genus has assumed great importance in the United States because of the discovery of the khapra beetle. The khapra beetle, *Trogoderma granarium* Everts, is a major granary pest in many parts of the world. The khapra beetle is discussed in the chapter on stored product pests. Dermestids of the genus *Trogoderma* are serious pests of cereals and dried plant products such as cocoa, dried soups, etc. as well as milk powder, grains and seeds. Okumura (1972) states that one of the most important pests of dried foods is *Trogoderma variabile* Ballion, the warehouse beetle.

Beal (1954) notes that *Trogoderma* are ordinarily scavengers in "bird and mammal nests, tent caterpillar nests, spider webs, and old wasp and bee nests provisioned with spiders, insects, or pollen."

Okumura (1967) reports enteric irritation in two infants that swallowed the larvae of *Trogoderma* in infested cereal. Very likely this was caused by the larval hairs (hastisetae) in the cereal.

Incidentally, larvae of *Trogoderma* often can be spotted by the fact they are often light tan in appearance, much lighter than the usual wool-feeding carpet beetles.

Beal (1954, 1956 and 1960) and Strong and Okumura (1966) prepared taxonomic reviews that are of interest to students of *Trogoderma*.

THE WAREHOUSE BEETLE

Trogoderma variabile Ballion (=*parabile* Beal)

Okumura (1972) believes this beetle, which is a voracious feeder, is one of the most important dermestid pests in warehouses. A few of the many materials it feeds on are seeds of all kinds, dead animals, cereals, candy, cocoa, cookies, corn, corn meal, dog food (dried and 'burgers'), fish meal, flour, dead insects, milk powder, nut meats, dried peas, potato chips, noodles, spaghetti, and dried spices. Partida and Strong (1975) found it preferred such foods as barley, wheat, mixed animal feeds, processed grains, some grocery products, and pollen. The warehouse beetle occurs throughout most of the United States and is one of the most common stored product pests in seaports throughout the world.

According to Okumura (1972), the larva is approximately ¼ inch/6.3 mm long and varies in color from a yellow/white to dark brown, depending on its age. Okumura notes the larva "possesses two major types of setae: hastisetae, the spear-headed shafts with numerous barbs; spicisetae, the slender, elongate structures resembling a rat tail in shape and bearing many sharp pointed hairs. The structure of these setae and their abundance may be a source of irritation to sensitive individuals who contact or ingest the larva. The warehouse beetle has about 1,706 hastisetae and about 2,196 spicisetae."

The adult beetle is brownish black and about ⅛ inch/3.2 mm long. The warehouse beetle, unlike the khapra beetle, can fly. Loschiavo (1960), Partida and Strong (1975) studied the life history of this insect. The male has five molts and the female six molts before pupation. At 90° F/32.2° C and 50 percent relative humidity the eggs hatched in six days. Jt took 32.1 and 36.6 days for the eggs to hatch and become adults. Pupation to the adult stage required about five days.

Vincent and Lindgren (1975) recommend both aluminum phosphide and methyl bromide as fumigants for the control of this insect. Methyl bromide is the fumigant of choice in most cases. Strong (1970) found of the commonly used insecticides, DDVP was most toxic to *Trogoderma*.

Trogoderma ornatum (Say)

Almost all the information in the following pages is from the work of Wodsedalek (1912), who studied the life history of this species under the name of

Trogoderma tarsalis (Melsh.) This is a common and injurious museum pest, and is especially injurious in mounted collections of insects.

The beetle, which is distributed throughout the United States and is most common in the North, occurs outdoors in the cracks of hollow trees and similar situations, where it feeds on dead insects. It is even more common in the deserted cells of various Hymenoptera that store their cells with spiders and insects, as well as in bird nests.

The adults vary greatly in size and are from 1.6 to four mm in length. They are oval, somewhat oblong, black beetles. The elytra have four sinuous and confluent red bands, which are clothed with a whitish pubescence. The larvae are up to 10 mm in length, including the two caudal appendages. They are reddish/brown above and whitish below, and are covered all over with short, soft, yellowish/brown hairs.

The larvae thrive on dried insects and fish. They have been reported infesting such animal matter as wool, feathers, furs, skins, bee glue, cocoons, and on such vegetable matter as grain, hickory nuts, peanuts and peanut meal, wheat, corn, malt, cayenne pepper, old bulbs, pumpkin seeds, flax seed, castor beans, and tobacco. Hunt (1958) reports the larvae feed on dead drywood termites.

Life history. This insect occurs in all stages throughout the year in warm buildings, with two and a partial third generation each year. The females mate one day after crawling from the pupal skin. The male may mate with several females and *vice versa*. The female may lay from five to 62 small, oblong, white eggs, from three to seven days after emerging. The eggs hatch in 10 to 16 days with an average of 12 days. The larvae, upon emerging, feed on the commodity upon which they hatched, if edible, and wander away only if food is not available. The young will not leave the source of food upon which their parents are feeding until it is almost entirely consumed.

Wodsedalek (1912) did some interesting work on the larva, and his results are well worth reporting in some detail. He found there is great variation in the amount of feeding on the same food material, despite the fact the larvae are all feeding on the identical food under the same environment. He states, "Very often some specimens attain full size, metamorphose, and produce young long before others are half-grown, but not infrequently do these young overtake the other members of their parent group and even reach maturity much sooner under the same conditions." Some larvae which are active, feeding, and normal in every respect will not enter the pupal stage for an extremely long period. The majority of larvae molt eight to 12 times during a larval period of five months. Thirty-two molts were recorded for a larva with a prolonged life history, and Wodsedalek notes they probably undergo even more molts. Under normal conditions the larvae molt about every two weeks. The great number of larval skins give the impression of a whole colony of larvae.

Wodsedalek had one larva that lived for five years, one month, and 29 days, or 1,884 days without any food whatsoever. He notes: "Many of the largest larvae, which were about eight mm in length, dwindled down to practically the hatching length of one mm before dying." Speaking in terms of reduction in size, it is astonishing to note that some of the largest larvae have been reduced to about 1/600th of their maximum larval mass. Another, and even more interesting phenomenon, is the fact that when the starved specimens almost reach the smallest size possible and are then given plenty of food, they will again begin growing in size. A number of the larvae which were half-grown when placed under starvation for the first time, have through alternating periods of 'feasting and fasting' attained that size three times and are now on the way to their fourth 'childhood,' and even some of the large specimens have started dwindling down to their third 'childhood' after having twice attained the practically maximum larval size." Beck (1971) discusses retrogression in larvae of *T. glabrum* (Herbst).

The larvae pupate in their last larval skin, and the pupal period lasts from 11 to 17 days. "Should a specimen be forced out of the larval case when not fully matured though capable of locomotion, it invariably returns to its former position within the protective larval skin upon coming in contact with it."

Of particular interest is Wodsedalek's observation on the light and dark habits of this insect. They are dark-loving in the larval stage. The female adults become light-loving from one hour to several days after egg-laying. Wodsedalek dissected the females after they began to appear in the light and found no eggs. Thus, control when adults are found in the light may be of little use.

THE LARGER CABINET BEETLE

Trogoderma inclusum LeC. [= *versicolor* (Creutz.)]

This cosmopolitan beetle, which is two to five mm in length, is often a serious pest of seed collections in the United States. It has been recorded in stored wheat, rice, and other grain, as well as woolen clothing, grain, dried insects, dried casein, and corn meal. Beyer (1922), who worked out its life history as *Trogoderma inclusa* (LeC.), was concerned greatly with this dermestid because the larvae would live on grain in straw used in filling horse-collars and then bore through the leather. In one factory, these beetles damaged 50 percent of the collars.

Anon. (1954) and Marzke (1955) show the larger cabinet beetle to be important pests of dry milk solids. Fletcher (1946) records this species as an important pest of insect collections.

Beyer's results on the life history of this species duplicate almost exactly those of Wodsedalek on *Trogoderma ornatum* (Say.). The adults mate a day or two after emerging from the pupal skin. They lay 10 to 45 whitish and translucent eggs four to six days after copulation. The eggs are about 0.5 mm long, and one end of the egg has hair or threadlike projections which adhere to any object it contacts. The eggs hatch at room temperature in eight to 12 days. The larval life is five months, and under normal conditions the larvae molt twice in two weeks. They pupate 10 to 14 days, and the adult lives eight to 25 days.

Hadaway (1956) has prepared a detailed study comparing the life history of the khapra beetle and the larger cabinet beetle.

Beyer found that a temperature of 119° to 120° F/49° C was fatal to all stages. "A small room, 12-feet long, eight wide and 12 high, was constructed, and lined on the interior with asbestos. Two large steam-heat radiators were installed, and the collars hung on brackets about the room. One large thermometer was placed in the room and several smaller ones inserted in collars." The results were satisfactory when the temperature was maintained for several hours. At temperatures of 2° below zero the larvae appeared dead, but revived as the temperature was raised.

Other Trogoderma. There are 16 species of *Trogoderma* and Okumura (1972) notes they can be distinguished only by an expert. Besides those mentioned previously, several other important species include *T. glabrum* (Herbst), which is a pest of dried foods, and cigarettes (Fletcher et al. 1969), *T. simplex* Jayne and *T. grassmani* Beal.

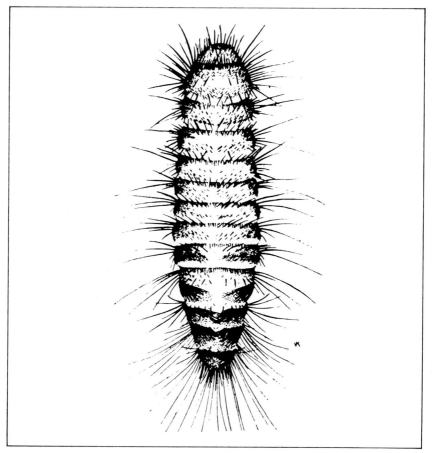

Fig. 13-11. Larva of the larger cabinet beetle, *Trogoderma inclusum* (Lec.)

THE ODD BEETLE

Thylodrias contractus (Mots.)

The male and female are utterly unlike in appearance, and the identification of this species was somewhat of an enigma to entomologists at the beginning of this century. Annie T. Slosson (1903 and 1908) wrote some interesting papers on this species. Some time previous to 1903 she mailed specimens of the peculiar female to a number of outstanding authorities. All were puzzled and declared they had never seen anything like it. Mrs. Slosson then authored a humorous article on this species and in fun called it *Ignotus aenigmaticus*. Since at that time it was thought to be a new species, the name stuck, and it was a number of years before its true identity became known.

This insect was first recorded in the United States from New York City in 1902. The odd beetle has been found in clean garments, clean muslin bedding, china closets, book cases, and white tissue paper. At one time, this insect was referred to as the "tissue paper bug" since an infestation occurred in such paper. However, Barber (1947) has shown tissue paper is of no importance in its life history, and its normal food is dry animal matter. However, tissue paper is penetrated as a barrier or may be eaten if the larvae are starved. MacNay (1950) notes it as destroying valuable prints in the National Gallery, Ottawa, Canada. Petrakis (1939) states the odd beetle has been known to injure silk hose and make small holes in garments. Numerous individuals have reported them as pests of dried insects mounted in insect collections. In Pittsburgh, Pennsylvania, it was found feeding on the feathers of a pheasant in a bureau drawer.

The adults are two to three mm long, yellowish/brown, and thinly covered with pale hairs. It is noteworthy that a single ocellus is found between the eyes of both sexes.

Twinn (1932) describes these insects as follows: "The male is elongate, has long slender antennae and legs, and possesses elytra, but no wings. The female is larviform and has neither wings nor elytra. It is stouter and its antennae and legs are much shorter and weaker than in the male. The larva is a stout brownish grub, somewhat similar in appearance to the larva of the buffalo carpet beetle, but smaller, and lacks the caudal hair tufts of the latter species and bears on its abdominal dorsum a dense covering of short clavate bristles." Barber (1947) notes the male has well developed underwings which it uses for flight. Metcalf (1933) states the larva is smaller than that of the buffalo carpet beetle and can roll itself into a ball when disturbed. This insect can be distinguished from other dermestid larvae by the single transverse row of erect club-like and spiny hairs that cross the dorsum of each principal segment. As mentioned previously, although the larvae resemble those of the dermestids, the adults are quite unlike typical dermestids in appearance. The larvae may live three or four years without food, and the life cycle may take one year.

CONTROL OF CARPET BEETLES

Since the 5th edition of this book appeared in 1969, the control of carpet beetles has undergone drastic changes. The reason for this is that the most effective insecticides against fabric pests such as DDT, dieldrin, chlordane, lindane and others can no longer be used in the United States. None of the later residual

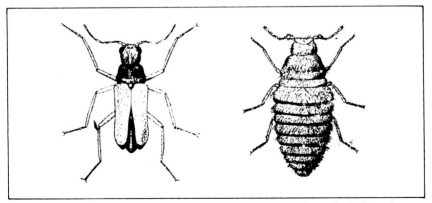

Fig. 13-12. The odd beetle, *Thylodrias contractus* (Mots.). The male (left) and female (right) are pictured above.

insecticides such as propoxur, diazinon, chlorpyrifos, malathion or bendiocarb have the long residual effectiveness of DDT and some of the other chlorinated insecticides. This means more frequent application of insecticides is required for control than in the past. Moreover, since most currently available residual insecticides including diazinon, chlorpyrifos and propoxur are limited to crack and crevice or spot treatment, the problem of controlling carpet beetles becomes more difficult than ever.

The first step in the control of carpet beetles is to inspect the premises for the source of infestation. Animal products of all kinds are infested and occasionally flour and other cereals. Rugs and the mats beneath rugs, upholstered furniture stuffed with mohair, trunks packed with woolens, mounted animal specimens, piano felts, accumulations of lint in floor cracks, ventilators and areas behind baseboards all may harbor dermestids. Badly infested materials should be burned or otherwise destroyed.

One should very carefully ascertain if insulating or soundproofing materials of plant or animal origin have been used in the construction of the building. If this is the case, and if larvae and adults are seen crawling from such sources, the infestation may then be considered general. Jennings (1940) notes where hair felt insulation in a building is infested by carpet beetles, control can be realized only by removing the insulation from the building.

Turner (1940) reported finding *Dermestes sp.* in insulating board, with most of the larvae coming from the fireplace. Investigation revealed the presence of a dead squirrel on a shelf back of the damper.

An infestation of black carpet beetles in a home in Pittsburgh, may furnish a clue to similar problems elsewhere. Numerous black carpet beetle adults were found around the windows of the home. It was some time before the source was discovered in a blocked-off fireplace. Several dead birds heavily infested with larvae were found in the flue when the fireplace was opened for inspection. Pieces of toast, suet, and cherry seeds were also present, and these apparently had been dropped in the flue by the birds. It appears as though the birds fell in the flue, possibly when numbed by cold weather, and after they died, they became a haven for the carpet beetles.

Zappe (1945) records an occasion where insects were falling on girls operating sewing machines in a shirt factory. "We investigated this complaint and found that the previous tenant had made woolen garments, and large amounts of lint had been left in the building, particularly between the floors. This material was breeding numbers of larvae of the black carpet beetles, which were dropping from the ceiling, annoying the girls. A few adult beetles were also found to be present." Davis (1947) notes an infestation of the black carpet beetle originating in accumulations of lint and cast-skins "some 14 inches back from under the edge of kitchen linoleum that had been laid 15 to 20 years previously." When this was removed, the infestation was terminated.

Hinton (1943 and 1945) and Linsley, in several previously mentioned papers, studied the natural resting places of carpet beetles as sources of infestation in the home. The latter also concerned himself with the natural habitat of clothes moths, and much of the following is from his studies.

According to Linsley (1946), the household is not as favorable an environment as such natural habitats as the nests of birds, rodents, insects, and spiders. Such nests, when on a house, show heavier infestations than those in the open, and serve as sources of infestation in the home. Linsley recommends the nests be destroyed "before the onset of cool weather, which may drive the insects indoors." He also notes that some "carpet beetles require pollen for successful production of eggs, and care should be taken not to bring these insects into the house on cut flowers. Such individuals, having mated and fed are more likely to provide the source of a serious infestation than individuals which have reached maturity within the confines of the building.

"In the last few years, a number of cases of extremely persistent infestations of carpet beetles or clothes moths have been called to our attention by homeowners or pest control operators. In most of these cases repeated fumigation with unusually heavy dosages of hydrocyanic acid gas had provided little more than temporary relief. Further investigation revealed that these infestations were arising from bird nests under eaves or in tile roofs, honeybee nests in walls or chimneys, deserted hornet nests (*Dolichovespula arenaria* Fab.), and in one case, nests of the yellow and black mud dauber (*Sceliphron servillei* Sauss.) in an attic."

In California, *Anthrenus verbasci* (L.) and *Trogoderma variabile* Ballion are the most injurious carpet beetles to fabrics and foodstuffs, respectively, and are common in wasp nests and spider webs in and around buildings. Farinaceous materials, as well as dried seeds of various kinds, also are attacked by some species of carpet beetles and their larvae. Woodroffe (1953) showed the varied carpet beetle to be a common inhabitant of sparrow nests.

Of all the household pests, carpet beetles are among the most difficult to kill. Where the infestation is of long duration, the larvae and adults become very widely distributed, wandering from one food source to another, or they may find some hidden and obscure crevice that protects them from sprays and toxic gases. The larvae may then reappear a day or so after the control measures have been applied, much to the anger of the housewife and the chagrin of the pest control operator. It is for this reason that extra large quantities of sprays or fumigants must be applied in an especially thorough manner. However, very often the cast skins of the larvae accumulate in one locality, giving the impression of a more extensive infestation than is the case.

Sprays for carpet beetles. Oil base and water base sprays are used for the control of carpet beetles and clothes moths and it should be noted that *many of the materials and methods recommended for the control of clothes moths are applicable to the control of carpet beetles.*

As mentioned previously, the insecticides currently available for the most part do not have the long residual effectiveness of DDT, dieldrin, chlordane and lindane. This means that insecticides such as bendiocarb, chlorpyrifos, diazinon, propoxur and others with a shorter insecticidal life will have to be applied more frequently. The following insecticides are now commonly used for carpet beetle control:

- BENDIOCARB: Apply a 0.25 percent bendiocarb spray as spot applications on and under edges of floor coverings, under rugs and furniture, and in closets or other localities where these insects are found. This material is one of the few residual insecticides labeled for treatment of entire rugs.
- CHLORPYRIFOS: Thoroughly apply a 0.5 percent spray as a spot treatment along baseboards and edges of carpeting, under carpeting, rugs and furniture, in closets and on shelving, and wherever else these insects are seen or suspected.
- DDVP (dichlorvos): DDVP at a 0.5 percent concentration by weight has been found effective as a contact spray against carpet beetles. Apply it as directed for diazinon. Boles et al. (1974) found DDVP resin strips effective against

furniture carpet beetles after 24 hours exposure. Zettler and LeCato (1974) showed that dosages of DDVP or malathion that did not kill black carpet beetles reduced their fecundity. According to Strong (1970), DDVP was highly effective against larvae of *Trogoderma.*

- DIAZINON: Apply a 0.5 percent diazinon spray in spot applications along baseboards and edges of carpeting, under carpeting and rugs and furniture, in closets and on shelving, and wherever these insects are seen or suspected. Miller et al. (1958) showed 0.5 percent diazinon in a petroleum distillate was highly effective as a contact spray against larvae of the black carpet beetle.

- MALATHION: Use malathion as a three percent water or oil based spray. Apply to baseboards, floors (including areas under carpets, along margins of carpets and in closets), behind radiators and other lint-accumulation areas, on closet shelves and walls, and on infested surface areas of carpeting. Spray surfaces until wet and take care to treat all cracks and crevices. Malathion, as well as some of the other insecticides, are odorous and this should be taken into consideration when applying. Malathion has a shorter residual life than diazinon. Miller et al. (1958) showed that malathion at three percent by weight in a petroleum distillate was highly effective as a contact spray against the larvae of the black carpet beetle. According to Butler (1974), wetting agents such as Tween, Triton and others used with water-based malathion sprays increased the effectiveness of malathion against black carpet beetles.

- PERMETHRIN: See Clothes Moths chapter.

- PERTHANE: Until recently, this insecticide at a five percent concentration was available in a pressurized aerosol. The manufacturer of perthane has discontinued production, but it is possible someone else may produce it. It was applied against carpet beetles as directed for diazinon. This pressurized aerosol at one time was widely used by the public for treatment of clothes, blankets, drapes and upholstery. However, at this writing it is not certain perthane can be used for the treatment of woolens as mentioned above. Bry et al. (1973b) found perthane effective against the larvae of the black carpet beetle, but they also noted that it caused some soiling.

- PROPETAMPHOS: Use a rate of 0.5 percent for light infestations and a rate of one percent for heavy infestations. Apply as a spot application along baseboards and edges of carpeting, under carpeting, rugs and furniture, and in closets or shelving where these insects are seen or suspected. Where fleas are present entire rugs may be treated at the 0.5 percent rate.

- PROPOXUR: Apply as a 0.5 percent spray as specified for diazinon.

- PYRETHRUM (pyrethrins): Pyrethrins plus synergists such as piperonyl butoxide or MGK 264 are used as contact sprays against carpet beetles. The pyrethrins are usually applied at 0.1 percent, but a higher concentration is preferable against the hard-to-kill larvae of carpet beetles. The effectiveness of pyrethrins is usually of short duration, although Bry (1975) showed in experimental studies that pyrethrins and piperonyl butoxide in acetone solutions protected test cloth against larvae of the black carpet beetle for six months.

  Where clothing or blankets are sprayed with pyrethrins, it may be a good idea to dry clean them before use to prevent allergic reactions in susceptible individuals.

- RESMETHRIN (SBP-1382): This pyrethroid insecticide has been highly recommended by Bry et al. (1973a and 1973c) as an effective insecticide against clothes moths and carpet beetles. They claim it protected woolen cloth from feeding of the webbing clothes moth and the black carpet beetle for six months.

Resmethrin is used at 0.25 percent by weight in a pressurized aerosol. The manufacturer recommends its application as follows: "Apply only to garments, blankets, carpets and other woolen fabrics before placing in storage for protection from larvae of webbing clothes moths and black carpet beetles.

"Dry clean or wash garments and blankets before treating. Clean carpets, draperies and other fabrics by thoroughly brushing before treating.

"Hold the container 14 to 18 inches from the fabric and apply the spray while moving evenly in a forward and back motion across the fabric while covering one foot per second. Spray the entire fabric surface until it is *slightly moist or damp to the touch*. When possible, treat both sides of the fabric. Do not soak or wet fabric. For maximum protection place treated fabrics after drying in plastic or paper bags. Also treat cracks and crevices of closets, chests and trunks where these fabrics are being stored. Treat the cracks and crevices at the rate of two seconds per linear foot.

"When applied as directed on fabric and storage areas this treatment will offer protection for ˚six months. To insure protection inspect the fabric in the storage area once every three months. Repeat application only when necessary. Avoid heavy applications. When possible make treatment outdoors. If application is made indoors do not remain in treated areas. Ventilate the area after the treatment is complete."

Anon. (1949), in discussing the control of carpet beetles and clothes moths with oil solutions or water emulsions, recommends:

1) "Apply the spray annually through the premises with special attention being given to treatment of attics, basements, wall cavities, floor margins, window sills, behind wainscoting ledges, behind quarter round, behind picture trim, in and around incinerators (only when fire is extinguished should oil sprays be used), behind drawers and cabinets where susceptible products are stored, in plumbers' closets, behind and around built-in tubs, basins, sinks, lights, heat pipes, air ducts, cleaning and ventilating systems, or in a word, throughout the building where insect fragments, lint, crumbs, feathers or life sustaining wastes accumulate."

2) Susceptible floor coverings are treated in a thorough manner so that "all materials in the protected areas, such as margins beneath or near quarter rounds, stairkeepers, treads, low-set furniture, and in similar situations are exposed to the control." Infestations in carpeting or rugs rarely occur in areas constantly used.

3) The spray should be applied "in closets and storage areas, interior surfaces, including floors of clothes closets, cabinets, drawers and storage compartments."

4) The effects of oil or water on sprayed material should be determined. "Soiled rugs or tapestries are likely to show spotting following application of any liquid which dissolves the dirt. Solubility of rug dyes in a spray solution can be determined by rubbing the rug briskly with a clean white cloth wetted with the solution. If the dye is readily removed by rubbing, the solution *should not* be used on the rug." *(Author's note: Do not apply sprays on linens, rayons or other synthetic textiles, or on plastic materials since they may be spotted or otherwise affected. Moreover, such materials are rarely damaged by clothes moths or carpet beetles.)*

Dusts for carpet beetles. Dusts are at times effective for carpet beetles, especially in areas such as attics, wall spaces and similar sites. Diazinon and silica gel plus pyrethrins are two of the dusts that have been used against carpet beetles. Unfortunately, many individuals object to unsightly dusts so their use is limited. Moreover, people with respiratory problems may be adversely affected by dusts and this fact should be considered before they are used.

Fumigation. Where a carpet beetle infestation is severe, pest control operators may resort to fumigation. Sulfuryl fluoride is used for household fumigation. For further details see the chapter on Fumigation. Methyl bromide is highly effective against dermestids, but ordinarily is not used in household fumigation because of the odorous effect it has on leather and certain other household goods. Pence and Morganroth (1962) showed methyl bromide to be effective against all stages of the furniture carpet beetle including the eggs. Unfortunately, fumigation does not always work because at times the gas does not penetrate some infested pockets in sufficient concentrations to kill larvae and adults. Moreover, fumigation has no residual effect and reinfestation may occur shortly after fumigation.

Some pest control operators have fumigation vaults for fumigation of infested household furnishings. They may use such fumigants as sulfuryl fluoride or ethylene dichloride in such vaults.

Some pest control operators find a combined fumigation and spray treatment is most satisfactory, since the fumigation quickly kills most of the insects, and the spray kills the remaining pests.

Paradichlorobenzene. Paradichlorobenzene (PDB) is commonly used by homeowners against clothes moths and carpet beetles. Colman (1940) found low concentrations of PDB vapors inhibited the feeding of the black carpet beetle. Arnold (1957) showed under practical conditions, PDB is not an effective repellent against the black carpet beetle. For further information on the use of PDB see the section featuring PDB under clothes moths.

LITERATURE

ANONYMOUS — 1949. Direct control of clothes moths, carpet beetles and other pests attacking stored products. Julius Hyman & Co. Tech. Suppl. 206, April.

ANONYMOUS — 1954. Procedure for insect prevention and control in plants processing nonfat dry milk solids. USDA multilith.

ANONYMOUS — 1960. U.S.D.A., PCO, work together to intercept, fumigate khapra beetle in French ship in Cleveland. Pest Control 28 (9):24-26, 34.

ANONYMOUS — 1969. CSMA Textile Resistance Test. Soap Blue Book 44(4A):197-200.

ARCHER, T. L., and R. G. STRONG — 1975. Comparative studies on the biologies of six species of *Trogoderma: T. glabrum.* Annals Entomol. Soc. Amer. 68:105-114.

ARNOLD, J. W. — 1957. Toxicity and repellency of paradichlorobenzene to larvae of the black carpet beetle. J. Econ. Entomol 50:469-471.

AYAPPA, P. K., P. S. CHEEMA, and S. L. PERTI — 1957. A life history of *Anthrenus flavipes* LeC. (Col., Dermestidae). Bull. Entomol. Res. 48(1):185-198.

BACK, E. A. — 1938. Carpet beetles. USDA Leaflet 150.

BACK, E. A., and R. T. COTTON — 1936. The furniture carpet beetle (*Anthrenus vorax* Waterh.) a pest of increasing importance in the United States. Proc. Entomol. Soc. Wash. 38:191-198.

1938. The black carpet beetle, *Attagenus piceus* (Oliv.). J. Econ. Entomol. 31:280-286.

BAKER, J. E. — 1974. Influence of nutrients on the utilization of woolen fabrics as a food for larvae of *Attagenus megatoma* (F.) (Coleoptera:Dermestidae). J. Stored Prod. Res. 10:155-160.

BARBER, H. S. — 1947. On the odd or tissue paper beetle supposed to be *Thylodrias contractus*. Ann. Entomol. Soc. Amer. 40:344-349.

1948. Postscript on the "odd beetle." Ann. Entomol. Soc. Amer. 41:478.

1951. Another name for the furniture carpet beetle. Coleopterists' Bull. 5:44-45.

BAUER, O., and O. VOLLENBRUCK — 1930. Uber den Angriff von Metallen durch Insecten. Z. Metalk. 22(7):230-233.

BECK, S. D. — 1971. Growth and retrogression in larvae of *Trogoderma glabrum* (Coleoptera:Dermestidae). Characteristics under feeding and starvation conditions. Annals Entomol. Soc. Amer. 64:149-155.

BEAL, R. S. — 1954. Biology and taxonomy of the Nearctic species of *Trogoderma* (Coleoptera:Dermestidae). Calif. Univ. Publ. Entomol. 10:35-101.

1956. Synopsis of the economic species of *Trogoderma* occurring in the United States with descriptions of a new species (Coleoptera:Dermestidae). Ann. Entomol. Soc. Amer. 49:559-566.

1960. Descriptions, biology, and notes on the identification of some *Trogoderma* larvae (Coleoptera:Dermestidae). USDA Tech. B. 1228.

1970. A taxonomic and biological study of the species of *Attagenini* (Coleoptera:Dermestidae) in the United States and Canada. Entomol. Americana 45:141-235.

BELLEMARE, E. R., and L. BRUNELLE — 1950. Larval and pupal development of *Dermestes maculatus* (DeGeer) under controlled conditions of temperature and relative humidity. Canad. Entomol. 82(1):22-24.

BEYER, A. H. — 1922. A brief resume of investigations made in 1912 on *Trogoderma inclusa* (LeC.), (a dermestid). Kans. Univ. Bull. 14(15):373-387.

BLAKE, G. M. — 1958. Diapause and the regulation of development in *Anthrenus verbasci* (L.) (Col., Dermestidae) Bull. Entomol. Res. 40(4):751-775.

1959. Control of diapause by an "internal clock" in *Anthrenus verbasci* (L.) (Col., Dermestidae). Nature 183 (4654):126-127.

1961. Length of life, fecundity, and oviposition cycle in *Anthrenus verbasci* (L.) (Col., Dermestidae) as affected by adult diet. Bull. Entomol. Res. 52(3):459-472.

BOLES, H. P., R. E. BRY, and L. L. McDONALD — 1974. Dichlorvos vapors: Toxicity to larvae of the furniture carpet beetle. J. Econ. Entomol. 67:308-309.

BRIMBLECOMBE, A. R. — 1938. Hide beetle damage to Oregon pine. J. Aust. Inst. Agr. Sci. 4(1):49-50.

BRY, R. E. — 1975a. Synergized pyrethrins as a short-term protectant of woolens against larvae of the black carpet beetle (Coleoptera,Dermestidae). J. Ga. Entomol. Soc. 10(1):14-17.

1975b. Feeding studies of larvae of the black carpet beetle (Coleoptera, Dermestidae) on wool/synthetic blend fabrics. J. Ga. Entomol. Soc. 10(4):284-286.

BRY, R. E., R. E. BOATRIGHT, J. H. LANG, and R. S. CAIL — 1973a. Protecting woolen fabric against insect damage with Resmethrin. Soap, Cosmetics, Chemical Specialties 49(3):40,42,44.

BRY, R. E., J. H. LANG, and S. A. BROWN — 1973b. Protection of woolen carpeting against black carpet beetle larval damage: investigation of six compounds. J. Econ. Entomol. 66:546-548.

BRY, R. E., J. H. LANG, and R. E. BOATRIGHT — 1973c. Toxicity of resmethrin to carpet beetles and clothes moths. Pest Control 41(11):32,47.

BUTLER, L. — 1974. Comparative study of adjuvants for increasing mortality of malathion-treated black carpet beetles. J. Econ. Entomol. 67:571-573.

CANZANELLI, A. — 1937. *Dermestes lardarius* (L.). Boll. Sez. ent. Oss. fitopat. 6:19-65. Rev. Appl. Entomol. A. 26:153,1938.

CASE, L. D. — 1950. Preparing mummified specimens for cleaning by dermestid beetles. J. Mammal. 40(4):620.

COLMAN, W. — 1940. Effect of paradichlorobenzene on the feeding of the black carpet beetle. J. Econ. Entomol. 33:816-817.

CORMIA, F. E. — 1967. Carpet beetle dermatitis. J. Amer. Med. Assoc. 200(9):799.

DAVIS, J. J. — 1947. Carpet beetles. Pests. 15(7):38.

ESSIG, E. O. — 1926. Insects of Western North America. Macmillan.

FERGUSON, G. R. — 1950. Survey on the incidence of fabric pest damage. Soap and Sanitary Chemicals Off. Proc. 36th Mid-Year Meeting, pp. 77-79.

FLETCHER, F. C. — 1946. DDT and the insect collection. Ward's Nat. Sci. Bull. 20:32.

FLETCHER, L. W., D. P. CHILDS, and J. S. LONG — 1969. An infestation of *Trogoderma glabrum* (Herbst) in cigarettes. USDA Coop. Econ. Ins. Rpt. 19(15):270.

GRADY, A. G. — 1928. Studies in breeding insects throughout the year for insecticide tests. II. Leather beetle (*Dermestes vulpinus* Fab.). J. Econ. Entomol. 21:604-608.

GRAY, H. E. — 1947. The common dermestid beetles. Pests. 15(8):8,10,12,14.

GREENWALD, M. — 1941. Studies on the biology of four common carpet beetles. Cornell Agr. Exp. Sta. Memoir 240.

GRISWOLD, G. H. — 1941. Studies on the biology of four common carpet beetles. Cornell Agr. Exp. Sta. Memoir 240.

GRISWOLD, G. H., and M. F. CROWELL — 1936. The effect of humidity on the development of the webbing clothes moth (*Tineola bisselliella* Hum.). Ecology:17:241-250.

HADAWAY, A. B. — 1956. The biology of the dermestid beetles, *Trogoderma granarium* Everts and *Trogoderma versicolor* (Creutz.). Bull. Entomol. Res. 46(4):781-796.

HALL, E., and C. RUSSELL — 1933. Dermestid beetles as an aid in cleaning bones. J. Mammal. 14:372-374.

HERFS, A. — 1932. Untersuchungen Zur Oekologie und Physiologie von *Anthrenus fasceatus* Herbst. Cong. Int. Entomol. Paris, 1932. pp:295-302.

HERRICK, G. W. — 1936. Insects Injurious to the Household and Annoying to Man. Macmillan.

HINTON, H. E. — 1943. Natural reservoirs of some beetles of the family Dermestidae known to infest stored products, with notes on those found in spider webs. Proc. R. Entomol. Soc. London (A) 18:33-42.

1945. A Monograph of the Beetles Associated with Stored Products. British

Museum (Natural History) 443pp.

HOWE, R. W. — 1953. The effects of temperature and humidity on the length of the life cycle of *Dermestes frischii* Kug. (Col., Dermestidae). Biol. Abst. 28(5):1191,1954.

HOWE, R. W., and H. W. BURGES, — 1956. *Trogoderma arfum* Priesner, a synonym of *T. granarium* Everts and a comparison with *T. versicolor* (Cruetz.). Bull. Entomol. Res. 46(4):773-780.

HUNT, R. — 1958. Dermestids on termite diet. Pest Control 26(2):58.

IMAMURA, S. — 1935. On the olfactory and visual senses of *Anthrenus verbasci* (L.). Bull. Seric. Exp. Sta. Japan 9(1):1-21.

JENNINGS, H. E. — 1940. Carpet beetle control. Pests 8:10-12.

JOHNSON, J. P. — 1945. The incinerator beetle, *Dermestes cadaverinus* Fabr. Conn. Agr. Exp. Sta. Bull. 488, June, pp. 411-415.

KEMPER, H. — 1939. Oekologisch-biologische Beobachtungen an schadlichen Dermestiden. Verh. 7 Int. Kongr, Entomol. Berlin, 1938. 4:2825-2832. Rev. appl. Entomol. A. 33:229-230,1945.

KINGSOLVER, J. M. — 1969. *Anthrenus coloratus* Reitter: a dermestid new to North America (Coleoptera). USDA Coop. Econ. Ins. Rpt. 19(5):61-62.

KINGSOLVER, J. M., and J. H. FALES, — 1974. *Novelsis aequalis* (Sharp) (Coleoptera: Dermestidae), a potential household insect in the eastern United States. USDA Coop. Econ. Ins. Rpt. 24(42):818-820.

KIRITANI, K. — 1958a. The ecological study of adult *Anthrenus verbasci* L. Botyu-Kagaku 23:92-98.

1958b. Factors influencing the development of *Anthrenus verbasci* L. Botyu-Kagaku 23:137-146.

KLEIN, H. W. — 1940. An unusual source for black carpet beetles. Pests 8(11):19.

KREYENBERG, J. — 1928. Experimental investigation on the biology of *D. lardarius* and *D. vulpinus*. A contribution to the inconsistency in the number of moults in Coleoptera. Z. Angew. Entomol. 14(1):140-178.

KUNICKE, G. — 1939. Beitrage sur kenntnis der Gattung *Anthrenus* (Coleoptera, Dermestidae). Verh. 7. Int. Kongr. Entomol. Berlin, 1938. 4:2833-2839. Rev. appl. Entomol. A. 33:230, 1945.

1941. Investigations of the biology of the carpet beetles. Biol. Abst. 24(10):2929, 1950.

LINSLEY, E. G. — 1942a. A natural habitat for the black carpet beetle. J. Econ. Entomol. 35:452.

1942b. Insect food caches as reservoirs and original sources of some stored product pests. J. Econ. Entomol. 35:434-439.

1944. Natural sources, habitats, and reservoirs of insects associated with stored food products. Hilgardia 16(4):187-214.

1946. Some ecological factors influencing the control of carpet beetles and clothes moths. Pests 14(7):10,12,14,16,18.

LINTNER, J. A. — 1884. The carpet beetle *Anthrenus scrophulariae*. Cult. and Country Gent. 49:676.

LOSCHIAVO, S. R. — 1960. Life history and behaviour of *Trogoderma parabile* Beal (Coleoptera:Dermestidae). Canad. Entomol. 92(8):611-618. Rev. Appl. Entomol. A. 49(10):531-532,1961.

MacNAY, C. G. — 1950. Odd beetle. 80th Ann. Rpt. Entomol. Soc. Ontario, 1949, p. 77.

MALLIS, A. — 1978. Sacred scarabs-sacred dermestids? Melsheimer Entomol. Ser. 24:9-11.

MALLIS, A., A. C., MILLER, and R. C. HILL — 1958. Feeding of four species of fabric pests on natural and synthetic textiles. J. Econ. Entomol. 51:248-249.

1959. The attraction of stains to three species of fabric pests. J. Econ. Entomol. 52:382-384.

MALLIS, A., B. T. BURTON, and A. C. MILLER — 1962. The attraction of salts and other nutrients to the larvae of fabric insects. J. Econ. Entomol. 55:351-355.

MARZKE, F. O. — 1955. Mites and insects in dairy products. Proc. Tenth Ann. Meet. No. Central State Branch Entomol. Soc. Amer. p. 70.

McQUEEN, H. F. — 1965. *Anthrenus scrophulariae.* USDA Coop. Econ. Ins. Rpt. 15(9):456.

METCALF, C. L. — 1933. *Thylodrias contractus* Mots. J. Econ. Entomol. 26:509-510.

1944. Records of new insect pests found in various stages not known to have occurred there previous to the present world war. Proc. 23rd Ann. Meet. No. Central States Entomologists. Univ. Illinois p. 133.

MILLER, A. C., A. MALLIS, and W. C. EASTERLIN — 1958. Oil base sprays against the larvae of the black carpet beetle. 51:249-250.

MOORE, W., and M. B. MOORE — 1942. Two species of black carpet beetle. J. Econ. Entomol. 35:288.

OKUMURA, G. T. — 1967. A report of canthariasis and allergy caused by *Trogoderma.* Calif. Vector Views 14(3): 14-22.

1972. Warehouse beetle-A major pest of stored food. National PCO News 32(1):4,5,24.

PARTIDA, G. J., and R. G. STRONG — 1975. Comparative studies on the biologies of six species of *Trogoderma:T. variabile.* Ann. Entomol. Soc. Amer. 68:115-125.

PATEL, H. K. — 1958. The furniture carpet beetle (*Anthrenus vorax* Waterhouse). Memoir Entomol. Soc. India No. 6. Rev. Appl. Entomol. 47(9):303,1959.

PATTON, W. S. — 1931. Insects, Ticks, Mites and Venomous Animals. II. H. R. Grubb, Ltd.

PENCE, R. J. — 1966. Analyzing fur damage with a microscope. Calif. Agr. Expt. Sta. Circ. 541.

PENCE, R. J., and J. MORGANROTH — 1962. Field effects of methyl bromide on carpet beetle eggs. Pest Control 30(7):20-24.

PETRAKIS, M. M. — 1939. The tissue paper bug. Pests 7(3):7.

REES, B. E. — 1943. Classification of the dermestidae (larder, hide, and carpet beetles) based on larval characteristics, with a key to the North American genera. USDA Misc. Publ. 511.

ROTH, L. M., and E. R. WILLIS — 1950. The oviposition of *Dermestes ater* DeGeer, with notes on bionomics under laboratory conditions. Amer. Mid-Nat. 44(2):427-447. Biol. Abst. 25(9):2611,1951.

RUSSELL, W. C. — 1947. Biology of the dermestid beetle with reference to skull cleaning. J. Mammal. 28(3):284-287.

SCHWARZ, L. — 1936. Erhebliche holzzerstorung durch speckkafer. Anz. Schadlingsk 12(4):46.

SLOSSON, A. T. — 1903. A coleopterous conundrum. Canad. Entomol. 35:183-187.

1908. A bit of contemporary history. Canad. Entomol. 40:213-219.

SMIT, B. — 1931. Insect damage to hides and skins. Farming S. Afr. 1931, reprint No. 27, May, Pretoria.

SPANGLER, P. J. — 1961. Notes and pictorial key for separating khapra beetle (*Trogoderma granarium*) larvae from all other Nearctic species of the genus. USDA Coop. Econ. Ins. Rpt. 11(6):61-62.

SPENCER, G. J. — 1928. Dead *Pollenia rudis* (Fabr.) as hosts of dermestids. Canad. Entomol. 60:283.

1948. Notes on some dermestids of British Columbia (Coleoptera). Entomol. Soc. British Columbia Proc. 44:6-9.

STRONG, R. G. — 1970. Relative susceptibility of larvae of six species of *Trogoderma* to ten organophosphorus insecticides. J. Econ. Entomol. 63:1836-1838.

1975. Comparative studies on the biologies of six species of *Trogoderma: T. inclusum*. Ann. Entomol. Soc. Amer. 68:91-104.

STRONG, R. G., and G. T. OKUMURA — 1966. *Trogoderma* species found in California. Bull. Calif. Dept. Agr. 55(1):23-30.

SWEETMAN, H. L. — 1956. Rearing successive generations of the carpet beetle under controlled conditions. J. Econ. Entomol. 49:277-278.

TAKAHASHI, S., and M. UCHIUMI — 1934. Studies on *Attagenus piceus* Oliv., a pest of raw silk. First report. Rev. Appl. Entomol. A. 23:209,1935.

TAKIO, M. — 1937. Morphology and ecology of *Dermestes vorax* Motschulsky (Dermestidae). Bull. Seric. Exp. Sta. Japan 9(3):167-184.

TRAGARDH, L. — 1934. En flaskanger som skadegorare i labrader. Skogen 23:514-515.

TURNER, N. — 1940. *Dermestes* sp. Unusual problems. Pests 8(1):17.

TWINN, C. R. — 1932. The occurrence of the odd beetle and a brief note on other dermestid species in Canada, Canad. Entomol. 64:163-165.

VINCENT, L. E., and D. L. LINDGREN — 1975. Toxicity of phosphine and methyl bromide at various temperatures and exposure periods to the four metamorphic stages of *Trogoderma variabile*. J. Econ. Entomol. 68:53-56.

VON DOBKIEWICZ, L. — 1928. A contribution to the knowledge of the bacon beetles. Rev. Appl. Entomol. (A) 17:133,1929.

WALKER, F. A., JR. — 1944. Life histories and control tests on three insect pests of skins stored in the tannery. J. Kansas Entomol. Soc. pp. 7-14.

WODSEDALEK, J. E. — 1912. Life history and habits of *Trogoderma tarsale* (Melsh.), a museum pest. Ann. Entomol. Soc. Amer. 5:367-381.

WOODROFFE, G. E. — 1953. An ecological study of the insects and mites in the nests of certain birds of Britain. Bull. Entomol. Res. 44(4):739-772.

WOODROFFE, G. E., and B. J. SOUTHGATE — 1954. An investigation of the distribution and field habits of the varied carpet beetle, *Anthrenus verbasci* (L.) (Col. Dermestidae) in Britain with comparative notes on *A. fuscus* Ol. and *A. museorum* (L.). Bull. Entomol. Res. 45(3):575-583.

ZAPPE, M. P. — 1945. Black carpet beetle. Conn Agr. Exp. Sta. Bull. 488, p. 421.

ZETTLER, J. L., and G. L. LECATO — 1974. Sublethal doses of malathion and dichlorvos: effects on the fecundity of the black carpet beetle. J. Econ. Entomol. 67:19-21.

DAVID J. SHETLAR
VERNON E. WALTER

Dr. David Shetlar gained his bachelor's and master's degrees in zoology from the University of Oklahoma and his Ph.D. from The Pennsylvania State University. After various teaching assignments at the University of Oklahoma, Shetlar began his research and teaching career at The Pennsylvania State University in 1971.

Shetlar's research activities have included pest management studies for ornamental plants, turf and greenhouses. His teaching responsibilities cover the whole field of entomology, including medical entomology, economic entomology and use of pesticides.

In addition to his research and teaching experience, Shetlar is an accomplished illustrator of insects and has taught scientific illustration.

Shetlar has authored or co-authored numerous articles in scientific publications and has lectured widely at seminars of professional associations and civic groups.

Shetlar has held office in the Phi Sigma Biological Research Society (University of Oklahoma Chapter) and is an associate member of Sigma Xi, the Scientific Research Society of North America. Shetlar is active in several professional societies and is a member of the Pennsylvania Academy of Sciences, the Entomological Society of Pennsylvania and the Entomological Society of America.

Vern Walter, with more than 30 years of experience in pest control, is one of the most respected professionals in the industy. He graduated from Purdue University in 1950 with a degree in entomology.

Walter worked for 12 years with Dr. Lee Truman, one of the leading pest control operators and pest control educators in the United States. He then joined Terminix International Inc., one of the largest and most prestigious pest control companies in the world, progressing in a 15-year career to the key position of technical director.

Walter knows every aspect of pest control from personal experience. He has been a service technician, salesman, service manager, owner/operator, consulting sanitarian and technical director. For the past four years he has worked with The Industrial Fumigant Company as a professional fumigator and consultant.

It is appropriate that Walter should have contributed to the chapter on Ants, especially the section on control methods. His father was a noted authority on ants and Walter has had a longtime interest in this subject.

CHAPTER FOURTEEN

Ants

Revised by D.J. Shetlar[1] and V.E. Walter[2]

The ant has made himself illustrious
Through constant industry industrious
So what?
Would you be calm and placid
If you were full of formic acid?

　　　　　　　—Ogden Nash

Scientists, unlike homeowners, have a rather favorable opinion of ants. This may be due to the very successful social habits of these insects which are comparable to that of man, even unto his faults. The late Dr. William Morton Wheeler, an authority on ants and other social insects, believed ants outnumber any other class of terrestrial animals. Although students of aphids (plant lice) have challenged him on this point, one must admit, using numbers as a criterion, that ants are extremely successful animals. According to Wheeler (1910), ants have become such dominant animals because of their habit of dwelling in the earth and through their ability to range from the arctic regions into the tropics, from sea level to timberline and from very moist to dry desert areas. Their colonies are long lived and "may outlast a generation of man." Worker ants may live from four to seven years and the queen as long as 15 years. Ants, as a group, feed upon practically everything consumed by human beings, as well as on a variety of other things. Ant nests are usually terrestrial and are adapted to a great variety of climatic and soil conditions. One of the most prominent assets of the group is its ability to adapt itself to a varying environment.

Anatomy of ants. Ants are distinguished from other insects primarily by the narrow pedicel, consisting of one or two joints, situated between the thorax and the abdomen. Moreover, ants are characterized by elbowed antennae. Termites may be distinguished readily from ants by the broad connection between the thorax and the abdomen, the straight and bead-like antennae, and the four wings that are almost equal in size.

The integument of ants differs greatly in color, sculpture and pilosity. Ants are usually either blackish, brownish, yellowish or reddish in coloration, or a

[1]*Assistant Professor, The Pennsylvania State University, University Park, Pa.*
[2]*Regional Coordinator, The Industrial Fumigant Co., Olathe, Kan.*

425

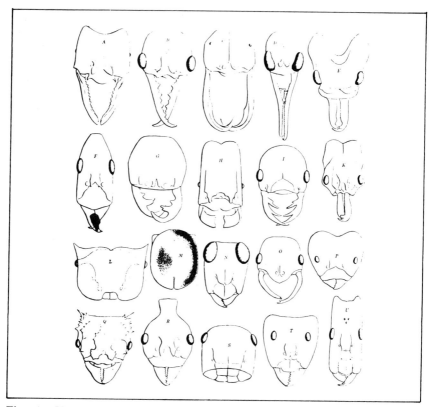

Fig. 14-1. Heads of various ants.

combination of these. There are some species of tropical ants with beautiful metallic coloring. The fact ants are extremely sensitive to "touch" may be correlated with the presence of the many hairs covering their bodies.

Ant heads vary greatly in form. The mandibles on the head are extremely important organs. Forel (1930) calls them the "hands" of the ants and states, "Ants use their mandibles for almost everything: biting, pricking, piercing, cutting off heads, building, sawing, gnawing, cutting, carrying, leaping and even bounding, but never for eating."

The antennae of ants are extremely important organs. It is the antenna that harbors the many sensory cells, particularly those of touch and smell. The antenna of the ant is made up of a scape and funiculus (whip), the latter being much more mobile than the former. Morley (1949) notes the funiculus vibrates very rapidly and this rapid vibration is associated with the high degree of development of the olfactory sense in ants. Other parts of the body may have similar sensory cells, but these are mostly concentrated in the antennae. The removal of the antennae is a calamity comparable to that of a man going deaf, dumb and blind, for it is principally through the antennae that ants are aware of their environment and orientate themsleves accordingly.

Forel (1930) is of the opinion the sense of smell in the ant is radically different from that of humans. Since both the sense of smell and touch are in the antennae, ants can "recall smells as round, square, elongated, hard, soft, etc., and as having a certain height, and being in a certain direction." Forel refers to this type of smell as "contact or topochemical smell." Ants have lateral compound eyes; the queen, male and workers of some species have three simple eyes or ocelli. However, even the best developed sight in ants is relatively poor. The eyes detect movement, but are not believed capable of seeing objects distinctly. Ocelli are adapted for seeing light or dark only.

Spurs are usually present on the legs of ants and those on the forelegs are especially large and comb-like. The ant removes dust from the antennae and legs by drawing these through the comb of the tarsus and the spur of the tibia. Moreover, the tarsal hairs are lubricated by the tarsal glands. In this regard, Forel states, "The comb and brush are never absent from the fore-legs. The secretion of the glands of the tarsus causes the grains of dust and other impurities to adhere to each other, and this makes it easier for the ant to dispose of them with its comb and brush."

The smallest ant is 0.8 mm long and the largest ant, a female of *Dorylus (Anomma) wilverthi* Emery, attains a length up to four cm Ant eggs are small and whitish. The helpless larvae are whitish to cream-colored, soft-bodied and

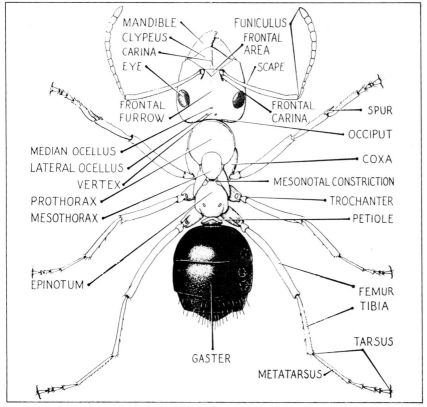

Fig. 14-2. Diagrammatic drawing of worker ant.

legless. Some species of ants have pupae with cocoons; in other species, these cocoons are lacking.

Ant bites and stings. In some ants the sting has practically disappeared, but poison may be injected into a wound made from a bite. Some ants actually produce a toxin secreted by glands in the head which is deposited in the bite. Most stingless ants which inject poisons curve their abdomens foreward as they bite. This is thought to be stinging, but actually is not. Some ants are capable of ejecting their poison up to 20 inches/50 cm in a fine spray.

Ants produce a large number of materials for their poisons. In general, ants are unusual in that their toxins contain large amounts of fatty acids instead of short chain protein molecules. In general, primitive ants have wasp-like proteinaceous venom, while the more advanced groups utilizes more formic acid. Martin (1971) summarized the makeup of many ant poisons. He pointed out many ants had formic acid, histamines and some hemolytic factors in their toxins. On the other hand, fire ants (*Solenopsis* ssp.) have venom with unique alkaloids. The most prominent alkaloid, *trans*-2-methyl-6*n*-undecylpiperidine (Solenopsin A), is a potent necrotoxin (MacConnell *et al.*, 1971).

The materials produced by ant poison glands are not always used as toxins. Wheeler (1910) notes in the army ants (*Eciton* sp.), the poison material contains the chemical leucine which has a fecal odor whereby these blind ants are able to follow one another. These scent trails are common among the ants and seems to direct the travel of the ants to and from the nest. Other odors also are released from the mouth, anal glands and sting glands which are used by ants to communicate messages. These chemicals are called pheromones and are discussed by Wilson (1963 and 1965).

Some of our native ants have severe stings, e.g., *Solenopsis xyloni* (McCook), as do the agricultural ants of the genus *Pogonomyrmex*. In California, the agricultural ants have been known to kill young pigs and injure children. Brett (1950) notes the Texas harvester ant was responsible for the death of a child in Oklahoma and Coarsey (1952) states an eight-month-old baby died from the stings of *Solenopsis xyloni* (McCook) in Mississippi. However, the stings of these ants do not even compare with the formidable stings of some tropical species, especially the Ponerine ant, *Paraponera clavata* (Fabr.), which is so dreaded in the high tropical rain forests of South America. The sting of this ant results in large blisters, which later become suppurating sores (Weber, 1939), and may even force its victims off their feet and into bed. Weber (1937) states some of the Indians use Ponerine ant poisons in the treatment of rheumatism. In other Indian tribes, the ants are applied to various parts of the bodies of young men to test their hardihood, whereupon they celebrate their "coming of age" party in a hammock for a week.

Methods of establishing the colony. With the great majority of ants, a new colony is initiated when the newly mated queen rids herself of her wings, digs a nest or seeks a cavity under a stone or piece of bark. Wheeler (1923) gives the following description of colony formation after the queen has established herself: "She then closes the opening of the cell and remains a voluntary prisoner for weeks or even months while the eggs are growing in her ovaries. The loss of the wings has a peculiar effect on the voluminous wing muscles in her thorax, causing them to break down and dissolve in the blood plasma. Their substance is carried by the circulation to the ovaries and utilized in building up the yolk of the eggs. As soon as the eggs mature, they are laid and the queen nurses the hatching larvae and feeds them with her saliva till they pupate.

"Since the queen never leaves the cell during all this time and has access to no food, except the fat stored in her abdomen during her larval life and her dissolved wing muscles, the workers that emerge from the pupae are all abnormally small. They are, in fact, always minimae in species which have a polymorphic workers' caste. They dig their way out through the soil, thus establishing a communication between the cell and the outside world, collect food for themselves and their mother, and thus enable her to lay more eggs. They take charge of the second brood of eggs and larvae, which, being more abundantly fed, develop into larger workers. The population of the colony now increases rapidly, new chambers and galleries are added to the nest and the queen devotes herself to digesting the food received from the workers and to laying more eggs. In the course of a few years, numerous males and queens are reared and on some meteorologically favorable day, the fertile forms from all the nests of the same species over a wide expanse of country escape simultaneously into the air and celebrate their marriage flight. This flight provides not only for the mating of the sexes but also for the dissemination of the species, since the daughter

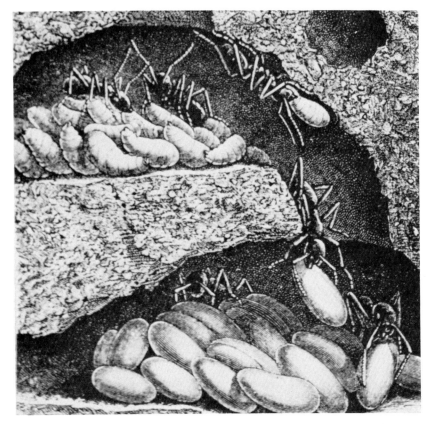

Fig. 14-3. Interior of a nest showing ants placing eggs, larvae and pupae in specific piles according to stage of development.

queens, on descending to the ground usually establish their nests some distance from where the parental colony is located."

Flanders (1945) notes that in such insects as ants, "the higher the rate of egg deposition the higher the proportion of eggs escaping fertilization and therefore the higher the proportion of males.

"It follows that if differences in the rate of egg deposition by the queen ant determine the occurrences of the various castes, the males and queens will be produced when the rate is high, that is, when the ripe eggs are retained in the ovary for a relatively short time, while the sterile female castes and associated anomalies will be produced when the rate is low, that is, when the ripe eggs are retained in the ovary for a relatively long time."

The worker. As a rule, the castes of ants are made up of three forms — worker, male and female. The workers are sterile females and may vary in size from forms almost as large as the queen to very small ants. The larger ants may defend the nest or use their large jaws to crush seeds, as in some of our seed-eating forms, e.g., *Pheidole*. When there are not intermediate forms, but only two classes of workers, small and large, the large workers are called "soldiers." In many species of ants, there may be only one size for all the workers. The workers perform all of the labor in the ant colony such as nest building, nursing of the young, procuring the food and duties of a similar nature. At times the worker may take over the egg-laying duties of the queen as Leutert (1963) reports.

The queen. The queen is almost always the largest individual in the colony. In species where the queen is winged, virgin queens found in the nest still retain their wings, unlike the mated queen, who removes hers. Once the queen has reared her first brood, she usually becomes an "egg-laying machine" and is cleaned and fed by workers. Many colonies have more than one queen in the nest. Should all the queens in a nest die or be killed, specially fed workers may undertake the egg-laying function. The queen mates only once, but may produce offspring until she dies.

Some ant colonies are 30 to 40 years old, the original queen having been replaced a number of times. The workers may live up to seven years, but usually exist under natural conditions for a much shorter period of time. Sir John Lubbock reared a queen for 15 years and when she died, a Paris newspaper noted his loss "of a relative."

The male. There are some people who deem the male an unfortunate creature because he dies a day or two after the mating flight. There are the others who think him thrice-blessed to forego a society completely dominated by females. His are the largest eyes so as better to see the ladies, particularly the queen. His huge thorax harbors great wing muscles.

Feeding habits of ants. Ants as a group eat an extremely wide range of foods, feeding on sweet materials, greasy materials, starchy materials, and plant and animal materials of all kinds. Ants somewhat higher in the scale of ant-evolution are liquid-sugar imbibers and obtain much of their nourishment either from the sweet exudations of plants or insects.

These sugar-eating ants attend the nectaries on leaves and in flowers, as well as collect the "honey dew" deposits of aphids, whiteflies, scale insects, mealybugs and other insects. This honey dew is the excess juice left over after the insects have assimilated the nutrients which they can use. These honey dew producers have changed the cane sugar to invert sugars which is similar to honey.

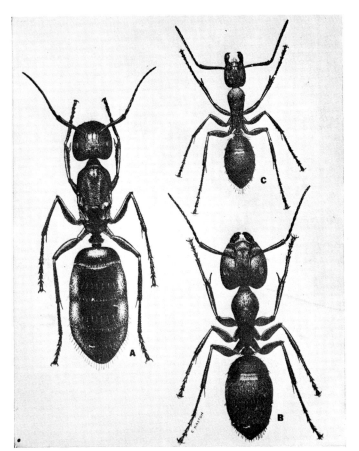

Fig. 14-4. Black carpenter ant, *Camponotus herculeanus pennsyl-*
vanicus **(DeGeer). A. De-alated queen. B. Larger worker or worker**
major. C. Small worker or worker minor.

Ants lap up liquids and large solid particles do not enter the digestive tract.
When found feeding on solid material, they are merely squeezing the liquid
juices from this food. Although the larvae of most ants feed on liquids, the larvae
of a few species may swallow solid food.

Ants and honeydew-secreting insects. The relationship between ants and
honeydew-secreting insects such as aphids, scale insects, mealybugs, etc., is one
of great importance, due to the fact the insects under question become very
injurious to plants because of the care shown them by the ants.

Ants protect aphids or "ant-cows" from their natural enemies. It is known
that certain species of ants store the eggs of aphids in their nests during the
winter. The ants treat their charges with the utmost care. For example, Forbes
(1908) states, "An ant coming up with a young aphid in its mandibles, carried
this about two feet and placed it on a smartweed near the ground. In about an
hour and a half one of the ants returned for its aphid and took it to the nest,
and 35 minutes later all had been carried back. One of these ants, which was

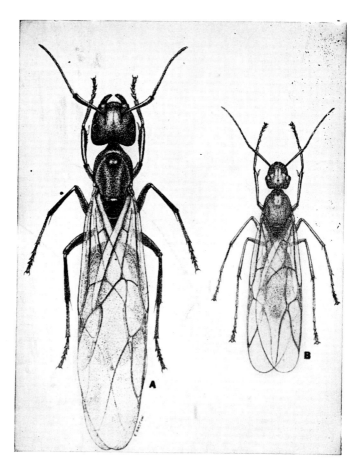

Fig. 14-5. The black carpenter ant. A. Adult winged female. B. Adult winged male.

so marked that it could be recognized on its return, carried to the nest the same aphid which it had previously brought out." Flanders (1951) and DeBach (1951 and 1952) and other researchers have shown that scale insects, mealybugs, as well as other important orchard pests often can be effectively controlled by their natural enemies in the absence of ants. When ants gather honeydew, they automatically protect the honeydew-producing insects from their natural enemies. Bartlett (1961) showed in the orchards of California, the Argentine ants interfered in the activities of most insect predators.

When disturbed, many ants will pick up their aphids or ant-cows and depart. Other ants will wage war ferociously when possession of their ant cows are disputed by some "ant-rustler." In some instances, aphids have become so dependent on ants they cannot establish themselves on plants without the aid of ants. Beyer (1924) noted the citrus aphid becomes entangled in its own honeydew and soon dies if it does not have an ant-attendant to consume its honey-dew. In any

Fig. 14-6. Ants attending their "ant cows" or aphids on a plant stem.

event, it should be remembered not all ants attend aphids, nor are all aphids attended by ants.

Some species of ants build earthen sheds over aphids. In the opinion of a number of individuals, these sheds are built to protect the aphids from rain and heat so the honeydew production of the aphids will not be curtailed; others believe the sheds protect the ants themselves from the heat of the sun. In any event, these sheds protect the aphids from their natural enemies, although the ants do not necessarily need the sheds for this purpose.

Where the honeydew falls on leaves, it forms a coating on which a sooty fungus grows and which cover the photosynthetic parts of the plant. Ants also are important disseminators of such plant diseases as fire blight, curcurbit blight, etc.

Honey ants. Some species of ants, especially the genus *Myrmecocystus* in the arid southwestern United States, have the remarkable habit of collecting the secretions from the nectaries of plants and the honey-dew of aphids and scale insects and storing them in specially adapted castes. These castes, the so-called "honey-casks," have crops which have become so enlarged they encompass almost all of the abdomen. The honey-casks are filled in times of plenty against periods of dearth. These honey-casks are attached by their tarsi to the roof of the nest chamber, and whenever a hungry ant desires a little liquid refreshment it palpates the honey-cask with its antennae and receives a regurgitated droplet.

Meat-eating ants. Several students of ants have observed the most primitive ants, such as the army ants, are mainly carnivorous. The army ants *(Eciton)* of the Americas and the legionary ants *(Dorylus)* of Africa are well known for their lengthy columns, military phalanxes and for their predatory habits. No living thing, including man, is able to resist these blind hordes of marching ants. Animals many times their size, such as tethered cattle, are quickly stripped of their flesh and even man must leave his home when these marching ants invade it. However, inhabitants of homes invaded by these hordes often welcome them despite the momentary inconvenience they cause, since these army and legionary ants depart with the unlamented remains of innumerable vermin.

Seed-eating ants. There are quite a few species of ants whose diet consists exclusively of plant products. Thus, the agricultural or harvesting ants of our arid and semi-arid regions harvest and store seeds, which they use for food. Ants of the genus, *Atta,* the "leaf-cutting" or "parasol" ants, actually raise fungi upon the leaves stored in the ant nests. Weber (1966) wrote a very interesting article on fungus-growing ants.

Ant guests. Insects of many orders live in the nests of ants, primarily because of the food and shelter to be found in the subterranean chambers. The relationships between the ants and their guests are varied and remarkable. Some guests are persecuted, but escape annihilation either through their speed or hardness of integument; others are tolerated without arousing any special feeling; and still others that favor the ants with secretions from their bodies are fondled and treated with evident liking. Many of these guests have a remarkable resemblance to the ants they associate with in color, form and other respects.

Beneficial aspects of ants. Ants are undoubtedly most useful to man as the tireless scavengers that help to reduce dead and decaying organic materials. Forel observed a large colony of ants and found they would bring into the nest 28 insects per minute and 100,000 on a day when they were especially active. In Germany, according to Gosswald (1944), the ant *Formica rufa* Linn, is considered beneficial in forests and anyone found disturbing the nests or collecting the cocoons is fined. In Hawaii, the ant *Pheidole megacephala* (Fabr.), is believed responsible for reductions in house flies and blowflies, since it preys on the immature stages of the fly. Pimentel (1955) showed 91 percent of a potential fly population in Puerto Rice was destroyed by ants, especially by the fire ant *Solenopsis geminata* (F.). According to Baerg (1937), in Mexico, lice-infested garments are placed on ant hills by the natives to rid the clothes of lice. This procedure also had been followed by soldiers. Naturalists at times clean delicate skeletons and birds eggs by placing them on the nests of ants. The cocoons of certain large ants of the genus *Formica* in Europe are gathered and sold as food for goldfish and birds. These are the "ant eggs" of commerce. Ants are considered beneficial because they aerate the soil like earthworms, by bringing up earth from their nests.

Wheeler (1907) quotes the use of the head of Atta soldiers for surgical purposes by Indians: "They use a certain species of said ants, because they bite severely, for closing wounds instead of stitching them with a needle. This is done in the following manner: They bring the skin of the two sides or lips together and then they cut off the insects' heads which remains attached to the wound with their mouths or mandibles as firmly closed as they were in life."

According to Wheeler (1908), the peasant women and children in Mexico assiduously seek honey ants. If these are to be eaten immediately, the sugar portion is sucked out. However, if the ants are to be carried back, the head and thorax are bitten off and the abdomens are placed in a dish. Otherwise the ants in desperation may bite each other, thus permitting the honey to run out. In New Mexico, the natives concoct a drink from these honey ants and in some instances use a mixture of water and honey ants as a cure for fever and eye diseases.

Quayle (1938) has some interesting notes on the cultured citrus ant in Kwangtung province in south China. This ant makes nests from leaves which are fastened together by silk. The adult ant takes a larva into its jaw and then uses it as a shuttle, since this is the stage that produces the silk. However, the unfortunate larva that has its silk employed in this manner must pupate without a cocoon.

Citrus growers buy the nests of this ant and places them in the crotches of trees during the late winter before the ants become active again. With the coming of spring the ants either enlarge the old nest or build a new one. The growers distribute bamboo rods from one tree to another in order to provide the ants with bridges. It is believed this may be the first practical application of biological control. Cultured citrus ants kill and attack caterpillars and larger insects. However, the scale insects are not molested and even occur in the nest.

In certain tropical countries, the nests of ants are placed in various parts of a building since the ants act as six-legged pest control operators to reduce the vermin in the building.

Ants are important predators on insects and the literature cites their importance in this respect on such pests as bedbugs, termites, screwworm larvae (Lindquest, 1942) and others.

Harmful aspects of ants. Ants have to be curbed when their behavior conflicts with that of man, as is frequently the case. Once inside the home they mingle with the provisions therein, thereby ruining them as food. Moreover, they transmit disease organisms to food they contact. Such ants as *Solenopsis* sp. may injure textiles by feeding on the soiled portions. Carpenter ants hollow out wooden structures. Ants frequently are annoying and sometimes dangerous because of their bites and stings. They annoy the farmer, since they may feed upon his plants. The large mounds of the imported fire ant break farm machinery and their stings keep the workers out of the fields. Agricultural ants spoil much range land for grazing by clearing sites for their nests. Many species of ants attend and care for honeydew-secreting insects of plants, protecting them from their natural parasites and thereby increasing the propensity of these pests for injuring the plants. A more detailed discussion of the injurious aspects of ants is found in the following pages. M.R. Smith (1965) has prepared a very useful and well-illustrated work on ant pests.

Ant identification. Most of the common species of house-invading ants are not difficult to identify, but the use of a hand lens or microscope is most useful to identify species. The following pictorial key includes most of the common genera of house-infesting ants found in the United States.

KEY TO COMMON HOUSE ANTS

1a) Ants with one (1) segment in petiole 2
  b) Ants with two (2) segments in petiole 8
2a) Ants with hair ring around anus (Subfamily Formicinae) 3
  b) No hair ring around anus (Subfamily Dolichoderinae) 6
3a) Workers small, antenna base close to clypeus 4
  b) Workers large, 7 mm or larger, antenna base far from
clypeus .... *Camponotus*
4a) Petiolar scale inclined foreward 5
  b) Petiolar scale erect, ocelli absent .... *Lasius*
5a) Mesothorax strongly constricted .... *Prenolepis*
  b) Mesothorax not constricted .... *Paratrechina*
6a) Thorax without projection or spine 7
  b) Thorax with pyramid-like projection .... *Dorymyrmex*
7a) Petiole very small, not distinct .... *Tapinoma*
  b) Petiole distinct, erect .... *Iridomyrmex*
8a) Eyes lacking or extremely small (Subfamily Dorylinae) not normally
found in household
  b) Eyes present, more normal (Subfamily Myrmicinae) 9
9a) Antennae without distinct club 10
  b) Antennae with 2-segmented club (see 9c) .... *Solenopsis*
  c) Antennae with 3-segmented club 12
10a) Petiole attached to front end of abdomen 11
  b) Petiole attached to dorsum (top) of heart-shaped
adbomen .... *Crematogaster*
11a) With three (3) or more pairs of spines on dorsum of abdomen ..... *Atta*
  b) Without paired spines on thorax, small ants .... *Tetramorium*
12a) Thorax with teeth or spines .... *Monomorium*
  b) Thorax without teeth or spines .... *Pheidole*

THE ARGENTINE ANT

Iridomyrmex humilis (Mayr)

Range. The Argentine ant probably made its entry into the United States at New Orleans, via the coffee ships from Brazil. It was first noticed in 1891 by Edward Foster (1908), although it was undoubtedly present some years before Foster's observation.

This insect is now established throughout the southern states and in California. Isolated infestations have been reported from Arizona, Missouri, Illinois, Maryland, Oregon and Washington. It also is known from Portugal, the Union of South Africa, France, Australia and from many islands including Hawaii, where it first became established in 1940, according to Zimmerman (1941).

Description. The worker or sterile female is 2.2 to 2.8 mm long, but it may appear larger when the abdomen is distended with food. In color it varies from a light to a dark brown, with the thorax, scapes and legs somewhat lighter. The

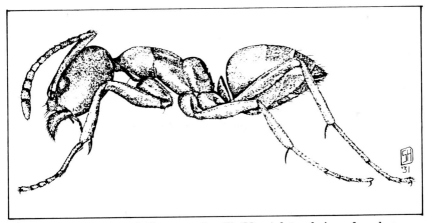

Fig. 14-7. Argentine ant, *Iridomyrmex humilis* (Mayr), lateral view of worker.

mandibles are yellowish and dentate. Often, when the worker is descending from a tree, the abdomen will appear honey-colored, due to distension of the abdomen by liquid nourishment. The worker does all the work, such as excavating the nest, obtaining food, feeding the young, protecting the colony and annoying the homeowner.

The brownish queen is from four to six mm in length and is by far the largest ant in the colony. There are usually a number of queens in one colony. One may readily demonstrate this by pouring a milk-bottle full of water onto a nesting site; within a short time, the workers with young in the mandibles, and very often a number of queens, will emerge. Since a colony may consist of a maze of trails ramifying throughout an extensive area, queens of the Argentine ant may number in the hundreds.

Newell found the winged queen on rare occasion in an outdoor nest in Louisiana and encountered the alate female much more frequently in artificial nests. It was his opinion the queen mates in the nest, making wings unnecessary.

Hertzer (1930) notes the queen is not a mere egg-laying machine, but takes an active part in the feeding and grooming of the young. Moreover, in this highly successful species, the queen feeds and cleans herself. The winged male is a small-headed, big-chested nonentity who shortly after mating departs from this vale of tears. Like other gentlemen we know, he is very fond of bright lights on warm summer nights.

Life history. Newell and Barber (1913) made a detailed study in Louisiana of the life history of the Argentine ant and it is from this source that the following was gleaned; Markin (1970 and 1970a) studied the seasonal life cycle and foraging habits of this ant in Southern California.

The eggs are white and 0.3 by 0.2 mm in size. Although the eggs are deposited throughout the year, the vast majority of them are laid during the summer. "In outdoor colonies, oviposition ceases when the daily mean temperature drops below 65° F, but is usually begun again when the mean temperatures rise above this point, regardless of the time of the year." The shortest incubation period was 12 days, the longest was 55 days and the average incubation period was 28 days. The longer periods for the incubation of the eggs was in part due to cooler weather.

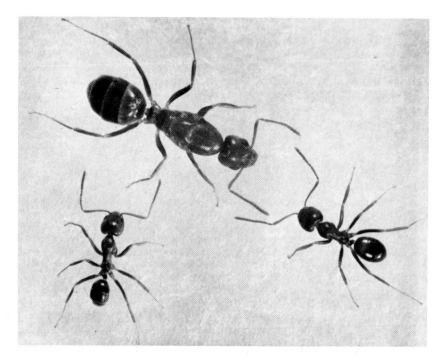

Fig. 14-8. Argentine ant queen and two workers.

The larvae are white, helpless creatures that are dependent entirely on the workers. The latter "care for their charges with the greatest solicitude. They feed and groom the young larvae continually and transport them from place to place whenever necessary. In case of danger, their first instinct appears to be to remove the young to a place of safety and they readily sacrifice their own lives in order to accomplish this."

The larvae are fed by the workers on predigested and regurgitated food. The workers constantly groom and lick the larvae and pupae. During the larval period, nothing is voided from the alimentary canal since the stomach has no connection with the intestine. The undigested mass of material or meconium is voided through the intestine when the larvae enters the prepupal stage. The larval stage varies from 11 to 61 days, with an average of 31.4 days. The pupal stage ranges from 12½ to 25 days with an average of 15 days.

Immediately after pupation, the ant is colorless and transparent and it's commonly designated as a "callow." It is a clumsy creature and walks in a stumbling and uncertain manner. The integument of the callow darkens in color in 48 to 72 hours after emergence from the pupal stage. The time required for completion of the egg to adult stage varies from 33 to 141 days with an average of 74 days.

In the spring, the nest will often be found in open ground with small piles of excavated earth a short distance from the nest holes. Form boards along walks and wooden objects of any kind are preferred as nesting sites and permanent runways, as are cracks and crevices in concrete walks. The area beneath a plant infested with "ant cows" often will be honey-combed with their tunnels. The ants may be encountered in enormous numbers in and under dead and decaying stumps. During warm weather they are partial to the underareas of houses and may use the mudsills as their runways. The nests even may be established in the house proper. The queen and her workers have been observed to invade the second floor of a house and to establish themselves beneath a damp rag on a kitchen sink.

The nests during the summer are usually very shallow, only one to two inches/ 2.5 to five cm beneath the surface. On one occasion, a nest was found that was more then 12 inches/30 cm in depth about the roots of a tree, situated on a slope.

In the autumn, the insects aggregate into a virtual ant metropolis in which there are hundreds of queens. The huge nests may be found beneath boards, sheets of tin, buildings, etc., as well as in accumulations of dead plant material with its attendant heat of decay. The author has found the Argentine ant overwintering in enormous numbers in tunnels containing hot conduit pipes. Smith (1936) noted in northern localities this ant may become established in buildings and spread from one structure to another, thereby avoiding the cold of winter. With the coming of spring, these colonies break up into smaller units consisting of one to several queens, with a large number of workers, all of whom migrate and establish themselves elsewhere. In fact, during the spring and summer months, an individual queen may take a stroll in the open and upon acquiring a retinue of workers, establish her own colony.

Damage. The Argentine ant is one of the most common household pests in the wide area where it is distributed. For example, in those sections of southern California where the insect is prevalent, practically every household utilizes some measure to combat it. In homes where its control is neglected, hordes of this persistent creature with its ever present trails may be found in practically any sweet food. It is an extremely persistent pest, especially during the warm summer months. In the winter, immediately after rain, it may invade the home seeking protection from inclement weather.

Documentation that the Argentine ant was a most serious household pest in New Orleans may well be realized when we peruse Newell's (1908) remarks on this ant: "The species does not sting, but can bite severely when so inclined and sometimes becomes an annoyance to human beings. I have known of several cases where people have had to place their beds, during the summer months, upon panes of glass covered with vaseline in order to pass the night in peace. There have been rumored cases of infants being killed by these ants, but so extreme a case has not come within my observations. That such might easily occur is not at all improbable."

Barber (1916) was given this rather graphic description of the attack of the ants on a four-week-old baby. "We were awakened in the night by a weak cry from the baby and when the light was turned on the baby's face was black with ants. They were in the baby's nose, ears and mouth. We hurriedly carried the baby to the bathtub and started to wash off the ants. It took us nearly an hour and a half to get the last ant off the baby. I feel sure that if we had not heard the cry, in a few hours the child would have perished."

The farmer is another victim of the ant's depredations. In the citrus orchards of Louisiana, some 35 years ago, the ant was incriminated for eating the buds on the trees and thereby ruining the yield. Fig crops were likewise damaged. Moreover, this hexapod fostered aphids, mealybugs and scale insects on plants, so they increased enormously in numbers. The Argentine ant was an especially prominent pest in the citrus orchards and sugar cane fields in Louisiana. Flanders (1943 and 1945), Bartlett (1961) and other workers have shown this ant interferes with the activities of scale parasites in citrus orchards in California.

Food of the argentine ant. The Argentine ant has a decided "sweet tooth" in the presence of sugars, syrups, fruit juices and materials of a like nature. It feeds with great avidity on the floral and extrafloral secretions of plants. The honeydew excreted by aphids is at times also very attractive to it. One often may see large numbers of ants on a fresh bone. In Berkeley, California, the insect was found to invade the university hospital and feed on blood smears. It also is recorded as feeding on cereal products, especially on corn meal.

When once on the trail of food, the ant is extremely persistent. It raids refrigerators with abandon and like Eliza will even cross ice to attain its objective. The ant has been observed to drop from the ceiling on a table which had its legs wrapped in cloths soaked with coal oil and has entered tightly screwed Mason jars (lacking a rubber gasket) by following the spiral thread between the cover and the glass.

There are certain plants which almost are certain to attract myriads of the insect. At times, the ants will be found ascending these plants in trails that are several inches wide. Cherry laurel is one of the "ant-plants" and the Argentine ant is particularly fond of the secretions from the nectaries on the leaves. During the summer, when the insect ascends this plant, it will not be turned aside by the most seductive of baits. Fig trees, with their fresh figs, are an ant heaven. Bamboo, many species of pine, as well as numerous other plants infested with aphids and other Homopterous insects, constantly shower the hungry ant with their "manna."

The harassed homeowner is apt to attribute certain occult powers to this insect, so quickly does it locate food in the home. The ant accomplishes this feat despite relatively poor eyesight and a limited olfactory sense. It practically "bumps" into the food before discovering it. Newell and Barber (1913) found the ant can travel 29 inches/74 cm a minute or 145 feet/44 m in an hour. Thus, the scouts who forage day and night and patrol every square inch of an infested area are bound to contact some stray food. They thereupon report the bonanza to the home base and the characteristic safari of ants ensues. In hot weather, the rate of locomotion of the ants, as with other insects, is greatly accelerated.

Means of spreading. The Argentine ant has been distributed to all parts of the country, mainly, by "riding the rails," in vehicles of all kinds. Its distribution by self movement is rather slow and Smith found this natural spread to be a few hundred feet per day. The Argentine ant often is introduced into new areas in balled nursery stock.

The fact water may also serve as a transporting medium is noted by Barber, who is quoted as follows: "Lumber, rotting trees, uprooted shrubs, cane growth, fruit, vegetables and all manner of refuse contribute to the mass of matter borne on the crest of flood water, and in this the ants seek refuge and are involuntarily transported. Nature has endowed this species with a remarkable habit of self-preservation from drowning in times of floods, for when rising water floods their nests and no other means of escape are presented, they cluster together and form a compact ball. The immature stages form the center of this ball, with the queens and workers as the outer portion. As the ball enlarges from the addition of other workers which have been struggling alone in the water it gradually revolves. It is kept revolving slowly by the outside workers continually striving to reach the top of the ball, thus permitting air to reach the interior."

Reasons for the success of the Argentine ant. What are some of the factors contributing to the prominence of this species?

- Unlike ants of most other species, the Argentine ants, even from widely separated colonies, are friendly to one another. The queens are likewise so.
- In an infested area, the widely separated runways of the ants eventually come together. Moreover, there are a great number of queens in an infested area, having a high reproductive potential, thus building large colonies.
- The ant is readily adaptable and nests in a great variety of situations.
- There is usually no dearth of food for the omnivorous worker. The worker, moreover, is persistent, courageous and tenacious. These qualities and their great numbers result in the defeat of practically every insect that opposes them.
- The female is not a mere egg-laying machine, but partakes in the grooming and feeding of the young. She mates in the nest and is therefore not subject to the perils that attend the mating flight of other species of ants. The mated female can initiate a new colony merely by walking off with a following of workers.
- There is no important natural enemy of the ant in the United States.
- Finally, they are readily distributed through natural and man-made facilities.

Feeding habits. Ants do not eat solid food in the adult state, but instead suck, lick or crush the solid food and then lap up the resultant liquids with a ridged, protrusible, pad-like tongue. This liquid material may be stored in the crop or "social stomach" as Wheeler named it. When an ant is quite turgid with honeydew, and is stopped and palpated by the antennae of a fellow worker, she may regurgitate a tiny droplet of liquid to this soliciting worker. Several other workers, as well as the larvae and the queen, will be fed in a like manner. Thus, a liquid containing a slowly acting poison may be conveyed to other members of the colony.

Certain plants encourage ants in a locality. Thus, fig trees with ripe figs become badly infested. Bamboo always causes trouble because of the scale insects thereon, as does oleander. Ants feed very eagerly on the secretions from the nectaries of cherry laurel and several other plants. When such ant-plants are present, one can usually be sure of having an ant problem.

Reasons for the Argentine ant entering the home. During the spring and summer, when the number of young and the size of the colony are increasing tremendously, the ants seek all possible sources of food and thus forage in the home. Moreover, summer showers may wash the honeydew from the plants, forcing the homeowner to share the contents of the cupboard with the hungry ants. In cold and rainy weather, these ants seek the warmth and the shelter of the home. On warm days they often enter the house to avoid the desiccating effect of sunlight and to obtain moisture.

Control. This species probably responds better to baits than any other control. If these baits are legally available, the sweet syrup baits will appeal to the "sweet tooth" of this ant. It is known to feed on many different foods and has a limited ability to see or smell so baits must be put in many areas so that the ants will find it.

The control of the Argentine ant with poisoned baits is based on the theory that if the queens are killed or made ill, the reproductive capacity of the colony is diminished Thus, there will eventually be a curtailment in the number of workers. The poisoning of the young and workers is a secondary consideration.

The ants that enter homes are almost always the workers or neuters who, incidentally, are sterile females. Merely killing these workers will give temporary relief, but is not a permanent solution to the problem. Newell and Barber (1913) showed one percent of the workers can keep the remainder of the colony including the queens and immature stages supplied with food. This reveals the futility of killing the foraging workers. Moreover, since there are so many queens in the nests of the compound colonies of the Argentine ant, these may produce eggs as rapidly as one kills the workers.

Barrier treatments outside will help since the ant is often nesting outside. Treatments of trails inside have also been helpful but not always completely successful for the reasons mentioned above.

Removal of ant-attracting plants such as Cherry laurel, bamboo and fig trees may reduce the incidence of this pest.

Treatment for this ant may require multiple visits that can be efficiently scheduled with contract pest control. Relatively few problems occur when a good program is followed consistently.

Time to control ants with poisoned baits. With the coming of autumn, colder weather and, for many localities, increasingly rainy weather arrive. Nectar in flowers and nectaries decreases or disappears and the honeydew is washed from the plants by the rain. Since these natural sources of food are greatly curtailed, the ants tend to feed much more readily on the ant poisons. The fall of the year, therefore, is one of the most appropriate periods in which to start an ant control campaign. Since the ants have an absolute minimum of natural foods during the winter, the size of the colony is greatly reduced, and since the colonies are concentrated in limited areas, one should continue to keep the poison distributed during this season. With the coming of spring, prior to the blooming of plants and just before the ants become more widely distributed, the ant control campaign should be reinitiated. One should look for the ants on warm sunny days and not when the weather is cold, moist or windy, for they may not forage on such days.

SOUTHERN FIRE ANT

Solenopsis xyloni McCook

The ants in the genus *Solenopsis* are called "fire" ants because of their fiery stings. Smith (1936 and 1965) studied this species in detail. The ant is found in the Gulf Coast region from South Carolina to southern California. In Arizona and California, the ant is lighter in color and is usually displaced from Florida by its cousin the fire ant, *S. geminata,* and imported species of *Solenopsis.*

The workers are of various sizes, ranging from 1.6 to 5.8 mm They are brownish red, with a brown to black abdomen and head. Part of the thorax is a yellow color.

The nests occur usually in exposed places under boards, stones, at the base of plants and tufts of grass, etc. The nests also are found in rotten wood, cracks in concrete work, beneath houses and especially around fire hearths, where the artificial heat stimulates them to year round activity (Mallis, 1938). The nests consist of loose soil, with many craters often scattered over extensive areas. The average size of the nests is an area of two to four square feet/0.19 to 0.37 sq. meters.

People react differently to the sting, but generally adults have little aftereffect. However, babies and children may retain sores. The ants have been known to attack people in bed and some women have reported the ants attack new-born babies that have been oiled.

Around the house and farm, the ant is an omnivorous feeder (Smith, 1965) and is known to eat meats, grease, butter, seeds, grains, nut meats and similar products. Also, it readily chews clothing, especially if soiled with foods. Farmers claim young chicks and other domestic animals suffer from attacks. The ants also attack the bases of plants, especially if the nest is so placed. Further, they bore into vegetables and fruits. The workers tend aphids, scales and mealybugs on ornamental and fruit plants. This ant also has been noted to remove the rubber insulation from telephone and electrical wires.

CHEWS CLOTHING [handwritten annotation]

Travis (1941) studied the life history of this species. The captive mother queens lay as many as 1,123 eggs in 24 hours. These eggs hatch in 14 to 30 days and the adult stage is reached in as little as 44 days during the summer.

RED IMPORTED FIRE ANT

Solenopsis invicta Buren

The red imported fire ant apparently was brought to the Mobile, Alabama or Pensacola, Florida area from central Brazil between 1933 and 1945. Originally it was believed to be a species, *S. saevissima richteri* Forel, with two color forms, red and black. However, Buren (1972) described the black form to be a distinct species, *S. richteri* Forel, which was different than the red form, *S. invicta*. Literature prior to 1972 used the older name. The black imported fire ant was apparently introduced before the red species and is localized around the Mobile area. The red fire ant has now spread from southern North Carolina through Texas.

The red fire ants build mounds in all types of soil except swampland and dense forest. The mounds may reach sizes of one to two feet/0.3 to 0.6 m in diameter and 1½ feet/0.46 m in height. There may be 30 to 50 mounds per acre/0.4 ha and in some soils the mounds may be very hard. These mounds are unsightly and can damage farm equipment.

Control. Ideally, fire ants should be attacked on a community or nationwide basis rather than on individual sites. This probably will not occur until fire ants become the dominant pest in certain key buildings in Washington D.C.

Although the imported fire ant has received much of the publicity, native fire ants have existed in this country longer than recorded history. There probably always will be some fire ants in the United States. They can be considered beneficial if you are not the one being stung.

Fire ant control may need to be done inside the home, but this fortunately is rare. On these occasions, the ants will normally be found in damaged woodwork or in masonry voids, particularly those near chimneys. Control here is similar to any other void-infesting ant.

Most fire ant control will occur outside. The large mounds are usually easily seen and control should be done very carefully to avoid painful stings.

When and if specific baits are available, they should be used because good control of the entire colony can be achieved. We PCOs have lost, probably permanently, mirex and chlordecone baits, and related materials have had trouble getting registration. Some new materials show promise and the reader should consult the state entomologist to see what is currently available and directions for its use.

Residual pesticides have been used very effectively on the nests when only one yard or other limited area is concerned. An entire large farm would be a different problem. Dusts, granules and sprays have all been used. Granules around the entrance holes and for an area around the hole are the easiest to apply. Dusts should be used with caution since they may be blown where not desired.

Emulsion sprays can be used to try to soak deep down into the nest. Some have suggested probing the nest first with a rod to loosen the soil, but we doubt these authorities tried this very often. This ant is very sensitive to vibration and quite willing to defend its nest. Kicking the mound open is rarely tried more than once. We suggest power spraying from a distance and the use of large amounts of water to try to gain penetration. If you are trying to coat the soil and vegetation, wettable powders will stay on the surface longer and are less likely to harm desirable vegetation. Most labels will require that children be kept off the treated area until it has dried or been washed in by rain or other water source. In our experience, children have already learned to avoid these areas.

Fumigation of the areas where the nests are located is possible if permitted by the label, but would require care for the surrounding vegetation, as well as the workers.

THE THIEF ANT

Solenopsis molesta (Say)

This is one of the smallest of all ants. In fact, it is so small (one to 1.5 mm) that its presence in the home may escape the eye.

It is a yellowish ant with two segments in the pedicel and 10-segmented antennae with a two-segmented club. It has vestigial eyes and a sting that is very rarely used. Like its larger cousins, the fire ants, there may be many queens in one nest. The queen, which is about five mm long, is considerably larger than the worker. The winged males and females have been observed in flight during June in Los Angeles.

This species is generally distributed throughout the United States. Thief ants

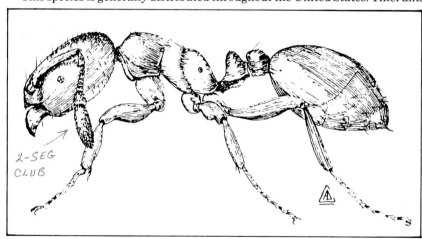

Fig. 14-9. Thief ant, *Solenopsis molesta* (Say), lateral view of worker.

often live in association with the nests of larger ants. They enter the other ants nests by means of tiny passages that are too narrow for the larger ants. The "thiefs" move about the chambers and kill and eat the immature forms of the host ant.

This insect is one of the most important of our house-infesting ants and is apt to be prevalent around the kitchen sink and in the cupboards. It nests in cracks and crevices in cupboards or wall. It is a very persistent creature and is controlled with much difficulty.

The thief ant is attracted to greasy materials such as cheese and animal matter. Some individuals are of the opinion these ants are not attracted to sweets, but McCulloch and Hayes (1916) state they feed on sweetened food in the house. Eckert and Mallis (1937) noted this ant often feeds "upon dead rats and mice and thus might convey disease-inducing organisms to human food." The ants also may serve as intermediate hosts for poultry tapeworms because the workers have been seen carrying egg segments into the nest (Smith, 1965).

Thief ants are of some importance to the poultry-man since they attack young chicks. At times the thief ants are responsible for injury to sprouting seeds and vegetables. These ants are believed to hollow out the seeds for the oil content. They are also predacious on many insects.

McCulloch and Hayes (1916), who studied the life history of *Solenopsis molesta* (Say), found the eggs have an incubation period of 16 to 28 days. The queens lay 27 to 387 eggs, generally averaging about 105. During the summer, the larval stage may be as short as 21 days. However, the larval stage may be greatly extended when the ants overwinter as larvae. The prepupal stage extends from two to 11 days during the summer and the pupal stage is 13 to 27 days. The larvae are fed upon regurgitated food, insect parts and small pieces of seed. The nests occur in a great variety of locations outside the home, particularly under rocks. MacNamara (1945) notes the queens in their mating flights may carry some of the tiny workers and it is possible the latter may aid in founding new nests.

Control. Thief ants are more likely to have an outside nest than Pharaoh ants so an outside barrier strip should be part of a control program under most conditions.

Sanitation is important in all insect control, but sanitation by itself would probably leave enough food for this very small omnivorous feeder. Therefore, pesticides are required for control.

Although this ant may nest inside all year long, it is thought by some to nest inside more frequently in hot weather. Following the trails to the nest is reportedly difficult, but should be attempted.

Control should be centered around the pantry and other food sources, as well as around sinks and other moisture sources. Natural voids such as the area under kitchen cabinets or nearby wall voids should be dusted with a residual dust.

Baits may be difficult to use since this ant will eat almost anything. It seems to prefer grease or high protein food over sweets. In the past, bacon grease has been used as an attractant.

PHARAOH ANT

Monomorium pharaonis (Linn.)

If good things come in small packages, then the Pharaoh ants are an exception to the rule. The worker of the Pharaoh ant is 1.5 to two mm long. Although

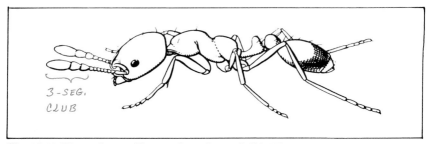

Fig. 14-10. Pharaoh ant, *Monomorium pharaonis* (Linn.).

small in size, they loom large as important ant pests in the home, particularly because of their persistence and the difficulty with which they are controlled.

The Pharoah ant is distributed widely throughout the United States and Canada. It is believed to have come to the United States from Europe and may have originated in the Old World tropics. Riley is of the opinion it received the "scientific name of 'Pharaoh's ant' on account of a defective knowledge of Scripture on the part of its describer, who doubtless imagined that ants formed one of the plagues of Egypt in the time of Pharaoh, whereas the only entomological plagues mentioned were lice, fleas and locusts." Linnaeus described this ant in 1767 and named its habitat as Egypt.

This ant varies in color from yellow to red and may be distinguished from the thief ant with which it is commonly confused by the fact it has three segments in the antennal club, rather than two segments as in the case of the thief ant.

Riley (1889) found this ant is not a nuisance "from the actual loss which it causes by consuming food products, but from its inordinate faculty of *getting into things.*" These ants have been observed feeding on such sweets as sugar syrups, fruit juices, jellies, cakes and fruit pies, as well as greases, shoe polish, bath sponges, insect collections and on the juices of recently deceased cockroaches. Lintner observed them running all over the dining table and getting in the food, and he made the interesting note that "a large number of these insects in food can impart an agreeable acid flavor." Although this ant feeds on sweets, it is believed to prefer fatty foods. It is known to be predacious on a wide variety of insects among which are bedbugs, white grubs, cotton boll weevils, etc., and Martini (1934) saw ants of the same genus attack children in a hospital. Ploschke (1943) found these ants to be serious pests on a hospital ship, where the ants frequently entered the wounds, and were suspected of transmitting germs by contact.

Beatson (1972) reported more than a dozen pathogenic bacteria found on Pharaoh ants collected from hospitals. These included *Streptococcus pyogenes, Pseudomonas aeruginosa* and *Staphylococcus epidermidis.* Keith Story (personal communication) reported particular problems of Pharaoh ant infestations in operating theaters and intensive care units where pest control measures often are discouraged by medical staff who are unaware of the disease transmitting potential of these ants. Story further reported Pharaoh ants have penetrated the security of recombinant DNA laboratories. Rachesky (1974) incriminated this pest with the death of baby reptiles in a zoo.

The nests were observed by Riley to occur "in almost any secluded spot, between the walls or under the floors or behind the baseboards, or among trash in some old box or trunk, or in lawn or garden walk just outside the door." R.C.

Smith (1934) studied the ants in Kansas, and found them nesting "in some cracks or crevices, behind baseboards, wainscoting, in cement or stone walls, on foundation walls, between flooring or in furniture, or outside of buildings, as under stones, heaps of trash and in boxes." In Illinois, Flint and McCauley (1936) observed the ants to nest in such unusual places as clean linens, electric irons and a can of modeling clay. Armand (1942) noted the ants have a preference for nesting places having a temperature between 80° and 86° F/27° and 30° C. Their trails usually were restricted to areas adjacent to hot water pipes, and the nests were most readily discovered by examining areas adjoining heating systems.

Zappe (1917) observed their apparent affinity for the bathroom or kitchen, where they were seen drinking water. Since these ants may nest in relatively dry situations in the house, the popularity of the kitchen and bathroom faucets may be accounted for by their need for moisture, especially for rearing their young. Armand's studies support this view since he found ant trails to earth beneath the house did not necessarily indicate the presence of a nest, rather the desire for soil moisture.

The winged or sexual forms mate in nuptial flights during the summer months. The male soon dies and the female bites off her wings and starts the serious business of founding the colony. She lays her eggs and cares for the first-born from the egg, larval and pupal through the callow or immature adult stage. These first workers aid in the growth of the colony by caring for the new progeny, enlarging the nest, foraging for food and protecting the colony from enemies. The number of eggs produced by a queen is relatively large and in a few months a colony may contain several hundred workers. Riley noted in time each of the nests contained several females, "each laying her hundreds of eggs and attended by a retinue of workers caring for the larvae and starting out from dawn till dawn on foraging expeditions in long single files like Indians on the war path." Bellevoye (1889), in Reims, France, was of the opinion the ants mated in the nest. He set out pieces of liver and trapped, in one room in six months' time, some 572 females and 239 males, and about 349,500 workers, the latter being computed on a weight basis. These ants were nesting in the walls of the room. Not all of the ants were collected at the termination of the experiment.

Peacock et al. (1959) have published the results of an extensive study on the life history and control of this ant. Under laboratory conditions of 80° F/27° C and a relative humidity of 80 percent, the life history of the Pharaoh ant is approximately as follows:

"Period from egg-laying to emergence of worker .... 38 days
Incubation period of eggs .... 7½ days
Larval period .... 18½ days
Prepupal period .... 3 days
Pupal period .... 9 days

"Regarding females and males, there is evidence that the life history period is about four days more, presumably because their larger size as larvae necessitates a longer period of feeding and growth.

"Females are capable of producing 400 or more eggs, though it does not follow that all females give this yield. Normally the rate of laying appears to be 10 to 12 eggs in a batch during the first few days of egg production, but the later batches comprise only about four to seven eggs."

Flint and McCauley (1936) found the colonies "may be spread from one place to another by being carried in groceries, laundry bundles or by similar means." They also note in Illinois it is the only species known to be active throughout the year in houses.

Control. Most pest control operators will agree this is the hardest ant to control. If there is any doubt about the identification of an ant infestation, no control program or price should be given until the presence or absence of Pharaoh ants has been established.

Pharaoh ant control should be considered at least a one-year project. Even in the heyday of the strongest of baits and the liberal use of chlorinated hydrocarbons, even a graduate entomologist with years of practical experience might spend a year or two eliminating this pest from a large hospital complex.

Pharaoh ant control is extremely difficult in large buildings with many wall voids. Hospitals and food plants are the most difficult, but even a small residence may be a challenge. Outside treatment is rarely needed, but an inspection is always wise.

If good baits are available the program should start with baiting before other treatments are commenced. There is still a need for more research on the best attractants. Time honored materials include liver, sponge cake, vanilla wafers, honey, bacon grease and many others. Combinations are often used. Although some researchers think fresh pork liver is best, this may vary with the location. One clue would be to study what food is being infested and then set out a series of non-toxic baits of this material. Select the one that draws the most ants. At the time of this writing there have been successful tests with everything from old-fashioned boric acid to the latest insect growth regulators. However, none are registered and commercially available at this moment. Sugar-based baits laced with thallium sulfate or sodium fluoride (Davidson, 1950) were used successfully in the past.

However, there probably will never be a miracle bait for this ant. We will always have to study the feeding and nesting habits and attack it with a multiple type of treatment. Incidentally, commercial baits, especially pellets, should be crushed because the ants are too small to carry larger pieces.

After a baiting program has had sufficient time to allow ants to carry a quantity of toxicant back to the nest, residual insecticide treatments can be used to further suppress the colonies. At this point it is very important to be thorough and to start at a point far beyond any known infestation (which has been determined by use of non-toxic baits). All residual insecticides can scatter the infestation (though not as readily as pyrethrins) and this ant has no problem starting new colonies with new queens.

Ideally, residual insecticides will consist of dusts for all void areas and sprays for travel paths, plumbing areas and pantry feeding areas. Retreatments should be based on observation and the expected life of the residual insecticide.

Overseas, especially in Europe, where the public health importance of this pest is better appreciated than in the United States, paint-on insecticide lacquers containing dieldrin are used with great success.

THE LITTLE BLACK ANT

Monomorium minimum (Buckley)

This jet black ant differs from the Pharaoh ant by its color and its smaller size, being 1.5 mm long. It is distributed widely throughout the United States. The nests are found beneath rocks, in lawns or in areas free of vegetation. Nests

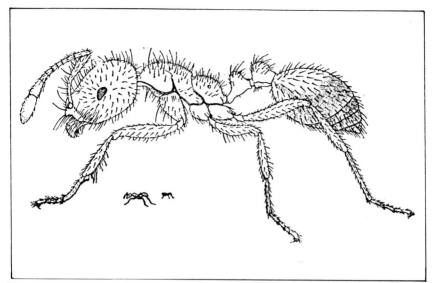

Fig. 14-11. Worker of the California fire ant. The 2-segmented antennal club is characteristic of this species.

in the ground may be detected by the very small craters of fine soil.

According to Blake (1940), the natural food of this ant is honeydew and the sweet secretions of plants. At times it invades the house and feeds on the food therein. It may also establish its nest in the woodwork or masonry of buildings.

Blake states, "mating occurs in July and August in New England. The colonies are large and each contains a number of queens."

Monomorium destructor (Jerdon)

Wheeler (1906) believes this ant was introduced from India into the United States. It is extremely common in India where it lives in nests in the ground, as well as in the walls of buildings. This dark red ant, which is about two mm in length prefers animal material to vegetable matter, but rarely feeds on sugar. Clark (1922) reports on ships sailing between the United States and Central American ports, the ants attacked the passengers and the crew, getting into the beds and inflicting painful bites which resulted in local swelling and inflammation. The ants also made a nuisance of themselves by mingling with the food and did not become active until the ship entered the tropics, whereupon they appeared in large numbers. They were found in the pantries, galleys, storerooms and were present in vast numbers in the ventilator shafts. Live steam was run into these shafts destroying large numbers of this pest.

Kalshoven (1939) reports this ant damaging electric cables. He also observed it doing serious damage to clothes and fabrics.

Control. These two "cousins" of the pharaoh ant also have more than one queen and are also omnivorous. They can nest inside or outside, but their small size precludes serious structural damage when nesting inside. They may sometimes eat the insulation off some wires.

Outside, nest and barrier treatments should be done first to prevent reinfes-

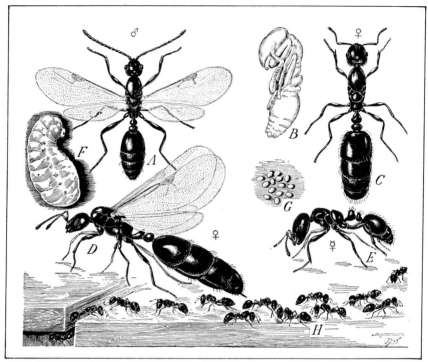

Fig. 14-12. The little black ant. A. Male; B. Pupa; C. Female; D. Female with wings; E. Worker. F. Larva; G. Eggs; H. Workers in a trail.

tation. Nests should be located with non-toxic sweet or grease baits or by following the workers. Both the nest and the normal pathways should be treated. Control is usually much easier than with Pharaoh ant.

THE BIG-HEADED ANTS

Pheidole spp.

On occasion, ants of this genus nest in and around the house and become pests very much like the fire ants. These ants are found in the warmer and more arid regions of the United States. The big-headed ant may be distinguished from the fire ant by its 12-segmented antenna and three-segmented antennal club. This genus is remarkable for the large heads of the soldiers, which when viewed under a magnifying glass or binocular microscope, present a very comical appearance. It has been stated that in some instances, when the soldiers inadvertently tumble over on their huge heads, they are unable to right themselves and starve to death. In some species of *Pheidole,* the huge heads of the soldiers are removed by the workers prior to the winter season since it is simpler apparently to breed new soldiers than feed old soldiers. Mallis (1969) found *Pheidole hyatti* Emery dwelling in the steps of a home some distance below a colony of fire ants. Both species invaded the home, but the fire ants were much more active in this respect. The *Pheidole* invaded the house to a much greater extent after the fire ants were controlled, thus necessitating further control measures.

Smith (1965) records *P. bicarinata vinelandica* Forel, *P. floridana* Emery and
P. dentata Mayr. as house-invading ants in the eastern U.S.A.

Control. Control of this ant is not normally needed, but it may invade the
home and can be an intermediate host of poultry tapeworms.

It prefers sweets or high protein foods. A mixture of peanut butter and honey
has served as a bait for a number of similar ants. Toxicants only can be added
if permitted by the label.

Outside barrier treatments with a residual insecticide and spot spraying of
probable travel paths inside is recommended.

THE PAVEMENT ANT

Tetramorium caespitum (Linn.)

Brown (1957) believes this ant is of European or Asiatic origin. According to
Eckert and Mallis (1937), the pavement ant "is a small blackish-brown ant,
some two to three mm in length with pale legs and antennae and with a black
abdomen. The head and thorax are furrowed by parallel lines. A pair of small
spines adorn the posterior portion of the thorax. Hairs are thickly distributed
over the entire body." This ant has two segments in the pedicel and 12-seg-
mented antennae.

The pavement ant is of interest to the pest control operator since it invades
the home for food. It is distributed in New England, sporadically in the Middle
West and is found in California in the San Joaquin and Sacramento Valleys.

Gross (1948) notes this species occurs 25 times as commonly as any other
house-invading ant in Cleveland, Ohio, and Mallis (1969) found most of the
complaints about ants in the Pittsburgh, Pennsylvania, area are due to the presence of the pavement ant. Incidentally, this ant will be found foraging in the
home throughout the year, although it is observed in greatest numbers during
summer. The nests are to be found outdoors under stones, along the edges of
curbing and in cracks in the pavement, especially when the latter is next to the

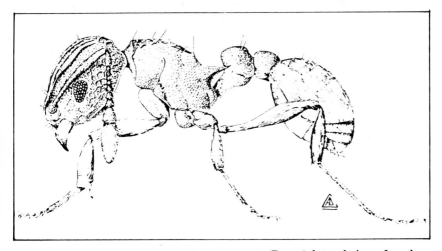

Fig. 14-13. Little fire ant, *Wasmannia auropunctata* (Roger), lateral view of worker.

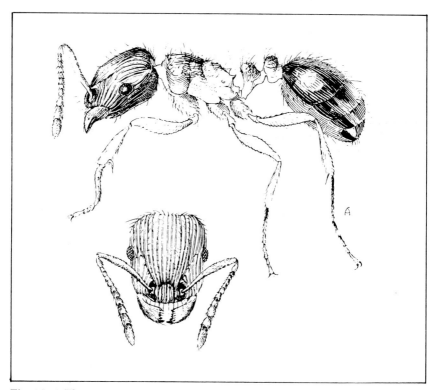

Fig. 14-14. The pavement ant, *Tetramorium caespitum* **(Linn.). Side view of worker and front view of head.**

lawn. In some vicinities, the nest openings are surrounded by small craters. In other vicinities, these craters are lacking. During the winter, the ants often nest in the home in a crevice in the immediate vicinity of some heat source such as a radiator. These slow-moving ants have been found tending aphids on willow by B. E. White. Blake (1940) states at times they store seeds in the nest chambers, as well as feed on insect remains and greasy materials. They have been recorded as pests on eggplants, peanuts and strawberries.

The workers and all the winged stages were observed emerging from the cork floor of a machine shop during the month of July. Burbutis and Conrad (1958) found it swarming in buildings in Delaware in January. Although the workers can bite and sting, they are not as aggressive or painful as the fire ants in this respect. Bruder and Gupta (1972) studied the life history of this ant in the laboratory at 70° to 75° F/21° to 24° C. The newly dealated queens begin oviposition within 48 to 72 hours after swarming. The queen produced five to 20 eggs per day. The brood developed in 36 to 45 days. The worker development required 43 to 63 days. There were three larval instars in the growth of the larva.

Control. The pavement ant is occuring increasingly in basements and slab-on-grade homes. It often is found entering the home through heating ducts which are hazardous to treat with pesticides because of the potential movement of pesticide vapors. Outdoors, it also lives under sidewalks and other areas.

Whenever practical, residual sprays should be forced into the entry point with drilling if necessary. Where heating ducts are the entry point, it may be necessary to use baits which will be more effective in the ducts than they will be in other parts of the structure. Sticky traps such as glue boards may be helpful here, but they will soon get a coating of dirt and become ineffective.

Regular treatments of the mounds as seen along the sidewalk and pavement outside can reduce the likelihood of an interior infestation.

THE LITTLE FIRE ANT

Wasmannia auropunctata (Roger)

This tiny pale yellow ant is very common throughout central and northern South America, the West Indies and the warmer portions of Mexico. Wheeler received specimens of this ant from Miami, Florida in 1924, and Fernald (1947) stated that it was fairly well distributed throughout the peninsular portion of the state. This ant has been reported in Los Angeles County, California (Osburn, 1948), and Mallis (1969) examined specimens collected in a San Francisco warehouse in 1936. The little fire ant, which is $^1/_{16}$ inch/1.6 mm in length, has an 11-segmented antenna and the antennal club is two or three segmented. The posterior dorsal part of the thorax bears two spines. These ants characteristically move very slowly. When little fire ants are present in a locality, their stings soon apprise one of the fact, and according to Spencer (1941), this is "another sure means of identification, since no other small grove ant stings as severely as the *Wasmannia*." The queens of little fire ants are rare and about ¼ inch/six mm in length.

Spencer (1941) observes this ant is a serious household pest, since "it contaminates food, is attracted to dirty and sweaty work clothing and even infests beds and stings severely when rolled upon or touched." Speaking of its effect on fruit pickers in the field, he notes: "A single ant may sting three or four times before it can be dislodged. Each stung place gets red and swells and finally a spot larger than a silver dollar is affected. It may become whiter than the normal skin or redder. For some people a few stings give only slight passing discomfort, but for others the sting lasts for three days, aching painfully at first and later itching intensely by spells. If stings are numerous, say a dozen or two within a short time, the victim gets pale and becomes "shaky" and unnerved, and when a picker reaches this stage he quits work."

Fernald (1947) notes although in nature this species attends honeydew-secreting insects, in the home it is attracted to fats, peanut butter, cooked fat meats and oily materials. He also stated, "These ants are sensitive to cold and do not appear in spring until the weather has become quite warm, even at night, and disappear with the first touch of cool weather in the fall, though if this is followed by a brief period of warm weather they may reappear for a short time." According to Fernald, "The general color of a group is dark gray, though under a lens the red of the body becomes evident. They do not seem to form definite nests underground but clusters of them may sometimes be found under bricks and stones partly covered by earth or grass. More often these clusters occur in cracks or crevices; under pieces of wood or even under dead leaves or rubbish on the ground. Such clusters often appear to be connected with others nearby and these conditions seem to imply a colony divided into two or more foci."

Control. Since this ant prefers to nest outdoors, controls seem most effective if applied to the surrounding soil and trees where aphids may be present (Osburn, 1948).

HARVESTING ANTS OR AGRICULTURAL ANTS

Pogonomyrmex spp.

These are the ants of the warmer and drier regions of the South and West, famous for their habit of collecting seeds and their vicious stings.

They are comparatively large-sized ants, varying from red to dark brown, and usually ⅕ to ¼ inch/five to six mm or more in length. Many of the species in this genus have long hairs (the psammophore) on their chins with which they clean their legs and antennae, carry water and remove sand during the excavation of the nest. Though never observed to invade the home, they nest in the lawn, around doorsteps, paths, etc. Entomologists are primarily interested in harvesting ants because of their habit of clearing large areas of cultivated fields, orchards and range land. Cole (1968) gives the most complete information on the taxonomy and biology of this genus.

There are several common species of harvester ants found in the United States:

- **Florida harvester ant.** *(Pogonomyrmex badius* (Laf.) This is the only eastern species and is found from Louisiana to North Carolina. The species tends to nest in open woodlands and grassy fields. The adults are a dark rust red color.

- **Red harvester ant.** *(Pogonomyrmex barbatus* (F. Smith) Wildermuth and Davis (1931) paid special attention to the red harvester ant, since it is an important pest in agricultural regions. These ants make circular, bare areas, 25 to 35 feet/7.6 to 10.7 m across and are especially injurious in alfalfa and grain fields. They are known also to destroy trees in citrus orchards. According to these workers, the red harvester ant "is also a great annoyance around dooryards and is especially troublesome when in city lawns. In such localities, the ant is a nuisance not only because of its har-

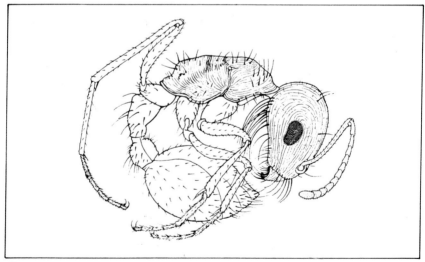

Fig. 14-15. Worker of the California harvester ant, *Pogonomyrmex californicus.* These ants often coil themselves up when picked up. Note the long hairs on the underside of the head.

vesting activities, but also because it has the unpleasant habit of inflicting painful stings, as almost all residents of a district where this ant occurs can testify." Its presence in yards and grounds make these places unsuitable for recreation. Children, adults, and animals may all be severely stung. Garden (1943) observed the red harvester ant to damage airplane runways in Texas.

The red harvester ant and a closely related species, *P. rugosus* Emery, are about 4.5 to 7.2 mm long. *P. barbatus* is often more red in color and occurs from Kansas through Texas and into Arizona. *P. rugosus* is brownish and is found from west Texas into California.

- **Western harvester ant.** *(Pogonomyrmex occidentalis* (Cresson) This red ant may reach up to 10 mm in length and is found at higher altitudes in most of the western states as far north as Idaho and Wyoming. Fritz and Vickers (1942) found the western harvester ant to be responsible for considerable damage to highways in Kansas. The ants remove the soil-binding vegetation at the edge of the road and tunneled beneath the surface of the road, thereby abetting erosion and damage to the roads during periods of rainfall.

  Lavigne (1969) found this ant may build mounds 4 to 12 inches/10 to 30 cm high and up to 36 inches/0.9 m in diameter. Excavations of nests found that galleries may go nine ft/2.7 m deep and contain from 400 to over 8,000 workers.

- **California harvester ant.** *(Pogonomyrmex californicus* (Buckley) The California harvester ant is light red with somewhat lighter legs, and is from 5.5 to 6.0 mm long; the thorax lacks spines. This insect is in the same category as *Pogonomyrmex barbatus* (F. Smith), for when it occurs immediately in the vicinity of the home, children and small animals playing thereabouts may be stung. Herms (1939) states, "These ants will readily attack humans and smaller animals. Hog raisers in the Imperial Valley, California, report many young pigs are killed by ants, particularly by the stings of *P. californicus* (Buckley). It is a matter of common observation to see a small pig walk leisurely upon an ant mound and suddenly begin to kick and squeal, due to the terrific attack of the myriads of ants rushing forth from the nest. The animals commonly topple over with the legs outstretched and death may result."

  Michener (1942) studied the habits of these ants and made the following comments on this species: "Observations on a colony of harvester ants *(Pogonomyrmex californicus)* in Pasadena, California, have shown that for three or four months during the winter the nest is continually closed and that during the remainder of the year it is closed every night. Outside activity goes on only during the warmer parts of the days. The actions of the ants are well correlated with temperature; they are sluggish at 70° F, exhibit maximum foraging activities with temperatures at the surface of the ground between 90° and 115° F, and are driven into the nest except for very brief excursions by temperatures over 120° F. Swarming occurs more than once each season, during the late mornings of certain clear, hot days in June and July."

- **Pugnacious ants.** When molested, the workers become very pugnacious. Wildermuth and Davis (1931) noted. "It is only when something interferes with their activities that they become hostile. The colony stubbornly resists interference and the pugnacious habits of its members have caused them

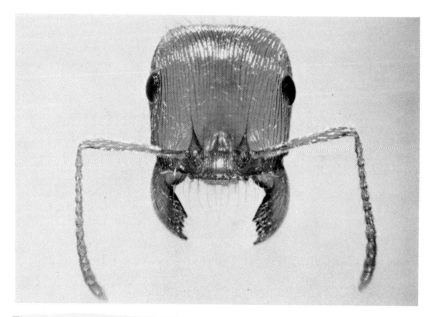

Fig.. 14-16. Head of the California harvester ant.

to be known as among the most ferocious of all American ants. When molested, they not only turn to give battle, but actually run about in search
of the intruder. When once they have set their powerful jaws in an object
there is no way to remove them without tearing the head from the rest of
the body, and even then the jaws remain locked." The sting is even a more
efficient defensive than offensive weapon. The poison they inject is very
irritating to man and animals and may cause pain for a long period of time,
as well as some swelling. Brett (1950) states one child is known to have
been killed in Oklahoma by the stings of, *P. barbatus.*

Habits of the ants. These ants swarm from June to October, the swarms
being most common in August and September. The swarms occur most commonly in the afternoon after a rain. Although the winged forms emerge in
enormous numbers, very few survive, since birds, toads, lizards and other ants
search them out for food. The male dies soon afer mating and the female removes
her wings and starts a new colony. The colonies may survive for a number of
years and V. L. Wildermuth has observed one for 19 years. The principal food
of the ants is seed, which is collected and stored in the fall of the year. Should
the colony be greatly disturbed, it may hole up and live on stored seeds for
months.

The ant nest. The nest consists of a flat, vegetation-free circular area about
the entrance hole or holes. This area averages about 12 feet/3.6 mm in diameter,
but may range from three to 35 feet/0.9 to 10.7 m. One or more paths may lead
from the entrance hole to the surrounding area, and these pathways may range
up to 200 feet/61 m. The nest consists of many subterranean tunnels and chambers. The ants store food in these chambers and also hibernate in them. E. G.
Davis excavated one nest which extended to a depth of 15 feet/4.6 mm, contained

436 chambers and a total of 12,358 ants. The ants remove the vegetation around the nest in order to permit the sun to shine on it. This, in turn, prevents excessive moisture in and around the nest and permits the nurseries to be warmed.

Control. Harvester ants are not a problem in homes, but may cause problems in or around the lawn.

They do not seem to be a problem in well maintained St. Augustine lawns which may be due to the lack of seed on this grass or possibly to its vigorous growth when well watered. Bare areas occur most often when lawns are neglected and are a real problem in areas with limited care such as right-of-ways.

The older control chemicals used against the harvester ants were chlordane, dieldrin, aldrin and heptachlor sprays and dusts. Race (1964) indicated baits containing mirex were effective, while Lavigne (1966) recommended chlordecone bait.

Currently available insecticides should be applied around the entrance hole and out to the edge of the mound area so foraging ants will carry the pesticide into the nest. Fast-acting pesticides or irritants such as pyrethrin may merely cause the ants to make new entrances (Brett, 1950).

Harvester ants seem to readily accept coated seeds or cracked grains within 50 feet/15 m of the nest and where the label permits baiting with rye grass or cracked grains should be effective. Burrow fumigants such as the liquid fumigants if labeled should be effective.

Since most foraging is within 50 feet/15 m of the nest, a treatment of a residual chemical within this area and reapplied as often as necessary should be effective. This ant is highly temperature responsive. It would be interesting to see if the nest area could be artificially heated with black plastic blankets or other solar means to raise soil temperature enough to reduce foraging and thus reduce and weaken the colony.

Rodding the nest with a residual chemical probably would be very effective, but we do not know of a label at the present time with this site.

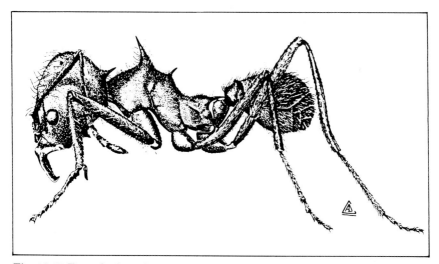

Fig. 14-17. Texas leaf-cutting ant, *Atta texana* (Buckley), lateral view of worker.

THE LEAF-CUTTING OR FUNGUS-GROWING ANT

Atta spp.

According to Walter, et al. (1938), the injurious fungus-growing ants are distributed from northern Texas to central Argentina. These ants are injurious since they cut the green vegetation from trees and shrubs and carry it into the nest, where they cultivate fungi on it. They have been known to denude a tree in one night. It has been estimated they do $1 billion damage per year in North and South America. Although primarily an agricultural pest, this insect on occasion may invade the home for cereals.

In the United States, the Texas leaf-cutting ant, *Atta texana* Buckley, occurs in Texas, Louisiana and some other southern states. It is present also in Mexico. This ant is believed to cause a total yearly loss of $5 million in the United States.

These ants are highly polymorphic, that is, they vary from very small to very large workers. Their queens are among the largest of ants. The workers usually have three or more pairs of spines upon the thorax. The larger workers bite ferociously and the natives of South America use them to sew wounds together. They force the ant to bite both edges of a wound, then sever the body of the ant from the head. In death, the jaws of the ants firmly hold the edges together. These ants are sometimes referred to as "parasol" ants, as a result of their habit of cutting vegetation and carrying it to the nest so the cut material extends over the head like a parasol.

Control. This ant can be very destructive around a home, and control is both needed and difficult. This ant will usually continue feeding on the leaves of one bush or tree until it is totally stripped. Spraying this particular plant and others around it with a residual may help, but nest treatment is needed. According to Byars (1949), weak solutions of chlordane, when applied in the nest, appear to show good results against leaf-cutting ants. Autori (1950) obtained excellent control of leaf-cutting ants in Brazil in 30 days using methyl bromide and 10 percent trichlorobenzol. Echols (1966a and 1966b) controlled the Texas leaf-cutting ant with 0.45 percent mirex baits.

The nest may cover 3,000 to 4,500 square feet/279 to 418 sq. m and may be eight feet/2.4 m or more deep. Fumigants with heavy molecular weights, such as carbon disulphide, have produced good results (Walters et al. 1938, Smith, 1939 and Johnston, 1944). Light gases such as cyanide have not done as well probably due to poor downward penetration. The liquid fumigants would be the ones of choice but few, if any, labels still permit this treatment. Fumigants registered for soil treatment with mixtures of ethylene dibromide, methyl bromide and chloropicrin should give good results with deep injection and covering. Johnson (1944) and Little (1950) show an application of one lb/0.45 kg methyl bromide will destroy an ant colony of ordinary size. Little applied the methyl bromide with a special "rodent" gun.

THE ODOROUS HOUSE ANT

Tapinoma sessile (Say)

This native ant is a common house-invading pest which produces a foul odor when crushed. Its habits are greatly similar to those of the Argentine ant, but is of lesser economic importance. The odorous house ant has a much wider range than the Argentine ant, being distributed widely throughout the United States from Canada to Mexico. The workers forage tirelessly night and day in trails reminiscent of their arch-rival, the Argentine ant. The ants are especially likely to invade the home during rainy weather, since their natural food supply, honeydew, is washed from the trees.

Prior to the invasion of the Argentine ant, the odorous house ant was one of the country's most successful native ants. Like the Argentine ant, its colonies contain numerous queens, who are friendly to one another. Although the colonies are large, as a rule, they are separate entities and do not anastomose as do those of the Argentine ant. The colonies range in size from 100 to 10,000 ants. Since the colonies of the odorous house ant are numerically inferior to the compound colonies of the Argentine ant, the latter eventually succeed in driving the odorous house ants from any area that they invade.

Description of the worker. The worker of the odorous house ant is from two to three mm in length with a brownish-to-black body. It may be distinguished from the Argentine ant by its darker color, unpleasant odor on being crushed and the overhanging abdomen which hides a vestigial scale. Kerpen (1961) separates the two species by the 17 teeth on the jaw of the Argentine ant versus the 14 teeth on the jaw of the odorous house ant.

Nesting sites. The nests of this species are found in a great variety of situations from sea level to 10,000 feet/3,500 mm. The nests are usually shallow and in the field often located beneath a board, stone walk, etc. In the home, the odorous house ant often is found nesting in the walls, sills or beneath the floor. On one occasion, a nest was found on the subfloor of a home, close to a copper hot water pipe. Another nest was discovered in the abandoned excavation made by subterranean termites in a porch floor. The odorous house ant is very common in apiaries, where its large colonies including numerous dealated queens, nest beneath the top and inner covers of the hive. Strangely enough, although this ant is largely a sweet-eater, it is of little consequence to the bees and apparently nests in the hive because of the warmth given off by the colony.

Life history. Smith (1928) observed the ant to be active in homes throughout the year. Breeding takes place continuously in suitably heated homes. Outdoors

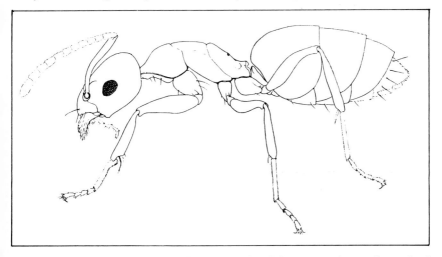

Fig. 14-18. Worker of the odorous house ant. The abdomen overhangs the scale of the petiole in such a manner that the scale is often difficult to see.

the ant overwinters as workers, dealate females and partly grown larvae. Smith observed this hardy ant to forage in Illinois at temperatures below 50° F/10° C. The workers commence to seek food early in March. Egg-laying and uniform development of the brood is continuous from April to November.

According to Smith, the white eggs have an incubation period of from 11 to 26 days, varying with the temperature. Although there are numerous females in a nest, each individual female may deposit only one egg a day. The workers, at times, lay eggs which develop into castes other than the male. The larval stage extends from 13 to 29 days, the prepupal from two to three days and the pupal stage from eight to 25 days, with an average of 14 days in midsummer. The odorous house ant has a complete life cycle of six to seven months during the winter, and five to nine weeks during late spring and summer. The alate females and males first appear in June and July, the males emerging first. They probably mate both inside and outside the nest and there may be four to five generations a year. The workers and females, under natural conditions, live for several years, whereas the males die within a few days after emergence. Wang and Brook (1970) studied the life cycle of this ant in Mississippi.

Control. These ants have food habits that are similar to those of the Argentine ant. The colonies of the odorous house ant are much more localized than those of the Argentine ant, and fortunately so, for the control of the odorous house ant by poison baits and syrups is very often unsatisfactory. As a result of this situation, one must follow carefully the trail of the ants and attempt to discover the nesting site. This may be in the woodwork of the house, floors or sills or under the subfloor, near hot water pipes.

W. S. Creighton once suggested the best control was to bring in a colony of Argentine ants which would drive out the odorous house ant. The Argentine ant is often easier to control. Some PCOs reported failures even when they were able to use chlordane for power spraying around mouldings and cupboards indoors.

Locating the nest is well worth the effort particularly if previous attempts have not been successful. First search for a nest outside. Look under boards and stones. If found, as the source of the problem, control is a simple matter of nest treatment with a residual pesticide. If there is no sign of this ant outside, the nest is probably inside.

THE VELVETY TREE-ANT

Liometopum spp.

The California velvety tree-ant, *Liometopum occidentale* Emery, is distributed widely throughout the state of California. It occurs in the foothills and mountains of southern California, at lower levels in northern California and east to Colorado and south to central Mexico (Ebeling, 1975).

The worker is 2.5 to six mm in length, with a glistening velvety-black abdomen, red thorax and brownish-black head. It dwells under the bark and in the cavities of trees and is found constantly ascending and descending trees such as oaks and poplars. This ant occurs commonly along the banks of streams.

The tree-ants attend honeydew-secreting insects and also are predacious. At times they become pests in homes and cabins. Their trails may extend for several hundred feet from the nesting site. Their huge colonies nest in trees, tree stumps and beneath stones in the ground. The ants are very pugnacious and when disturbed, they bite and inject a poison in the wound. They occur often on

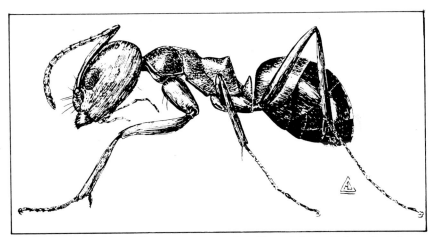

Fig. 14-19. Pyramid ant, *Dorymyrmex pyramicus* (Roger), lateral view of worker.

clothes and will sometimes allow some time to pass before attacking the wearer.

Wheeler (1905) quotes Emery as follows: "The rapidity with which the *Liometopum* communicate with one another and assemble in force for the purpose of overwhelming an enemy is truly wonderful. On such occasions a faint crackling sound is heard. Hardly any other animal dares venture up into trees inhabited by *Liometopum*.

Only long-legged ants that can get over the ground very rapidly endeavor to run the gauntlet in order to reach the plant-lice, which are heartily despised by the *Liometopum*." However, in the United States the ants eagerly attend aphids and coccids. These ants, like *Tapinoma sp.*, have the rank "*Tapinoma*" odor when crushed.

The species, *Liometpum apiculatum luctuosum* Wheeler, which is blackish in appearance, is found usually in the upper parts of pine trees and has habits similar to those of the previously mentioned species.

Control. The tree-ants invade homes not only for food, but also for the insects that are lodged there. Their long columns are most commonly encountered entering the house towards evening.

If the trees they are using for nesting are nearby, then tree spraying can be effective. Since this ant may wander great distances, the infested tree may not be on property that can be sprayed.

The columns themselves can be sprayed with residual or contact sprays by the homeowner. Barrier treatments may be effective, but will usually lose their effectiveness quickly outdoors. Inside, preventative spraying would be minimally effective.

THE PYRAMID ANT

Dorymyrmex pyramicus Roger and

ESA/ACN
Conomyrma insana

THE BICOLORED PYRAMID ANT

Dorymyrmex bicolor Wheeler

Pyramid ants are found throughout the southern states and California. At times they invade homes and become pests in the garden where they are wont to attend aphids on ornamentals. These ants are also predacious, attacking ants and other insects. In California they were formerly common pests in homes, but now have been replaced largely by the Argentine ant.

The worker is from 1.5 to two mm long with a definite pyramid on the thorax. The ants vary in color, either being uniformly dark brown as is *Dorymyrmex pyramicus* Roger, or characterized by a reddish tint on the head and abdomen as in the case of *Dorymyrmex bicolor* Wheeler. Pyramid ants construct small cone-shaped nests. Snelling (1973) placed these ants in the genus *Conomyrma*.

Control. The Pyramid ant builds obvious mounds in the yard or soil around a structure. These nests are relatively shallow with the chamber near the surface and should respond well to heavy spraying with emulsions. Diluting the

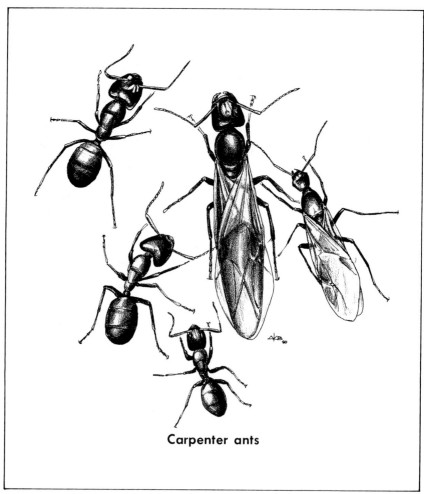

Carpenter ants

Fig. 14-20. Carpenter ants, *Camponotus* spp., are distributed widely throughout the United States.

pesticide more and using higher spray volumes so it will be carried deeper in the soil may be helpful. Control should always include outside treatments to avoid reinfestation.

Inside structures, the pyramid ant will seek out our food or feed on available dead insects. Spot spraying in or around the pantry may be valuable to protect food.

Baits, if available, also could be helpful. Sweet type baits are usually the best choice.

CARPENTER ANTS

Camponotus spp.

Carpenter ants, because of their large size and biting ability, attract more than a passing glance from even the neophyte nature lover. It is a common experience of hikers, who seek the comfort of a woodland log, to find themselves suddenly beset by large numbers of these biting ants. Carpenter ants are distributed widely throughout the United States and range from sea level to well above 9,000 feet/2,740 m in the western mountain ranges. They very often are pests in lawns as well as homes. Since theirs is the habit of dwelling in and excavating wood, they were given the common name of "carpenter ant." These long-legged, swiftly-moving emmets are among our largest ants. The workers are highly polymorphic, that is, they vary greatly in size, ranging from six to more than 10 mm, the queens may be from 13 to 15 mm long. Ants of the genus *Camponotus* apparently have better eyesight than ants of most other genera.

There are many species of carpenter ants which may occasionally invade homes foraging for food or for nest building. The black carpenter ant, *Camponotus pennsylvanicus* (DeGeer), is the most common carpenter ant in the eastern states. It is completely black and foraging workers may be from ¼ to ½ inch/six to 13 mm or more in length. Other common eastern species include the red carpenter ant, *C. ferrugineus* (Fab.), and the Florida carpenter and, *C. abdominalis*. A species found throughout most of the United States is *C. nearcticus* Emery which is smaller with the workers being 4.5 to 7.5 mm.

In the western states, species of carpenter ants may be: black, *C. laevigatus* (F. Smith); brownish with black, *C. hyatti* Henry, *C. clarithorax* Emery, and *C. modoc* Wheeler; or red and black, *C. vicinus* Mayr.

The carpenter ants attend aphids and other honeydew-secreting insects. At times they are predacious, lapping the juices from the insects they have captured and killed.

The colony is ordinarily initiated by one queen, who begins the nest beneath a rock or in the soil, or in an insect-bored tunnel in a tree, etc. The queen lays only a few eggs and these hatch into very small workers. These *minimae* or small workers then go forth to forage. The small workers then feed the young and queen, whose sole interest in life is the production of eggs. If the environment is propitious, the colony thrives. The pupae are enclosed in cocoons, which are referred to by most individuals as "ant eggs." When the nesting site is in wood, it often resembles an ornate carving due to the multiplicity of the galleries. These galleries are so smooth, they appear to have been sandpapered. The carpenter ants ordinarily excavate that portion of the wood softened by decay or by the attacks of other insects. However, Furniss (1944) and Brown (1950) note they can and will tunnel sound wood.

Pricer (1908) made a detailed study of the life history of the black carpenter ant in Illinois at a temperature of 70° to 90° F/21° to 32° C. The female laid 22 eggs in 15 days. The egg stage took 24 days, larval stage 21 days, and pupal stage 21 days (66 days from egg to adult). Under natural conditions, the larval stage may be of much longer duration during the winter. The winged sexual forms were observed to emerge about the first of July. Large colonies may be characterized by winged males and females during the winter. The colony does not produce the winged forms until it is more than two years old. These winged sexuals, which may be produced during one summer, overwinter in the parental nest and emerge for their marriage flight from May to July. One colony had as many as 3,212 worker ants.

Blake (1940) notes there may be more than one queen in a colony and the largest workers may lay eggs which will produce only male ants. "After mating, the queen makes or finds a smaller chamber in which she remains, sealed up, until her first brood of workers is adult. This requires at least two months and sometimes as long as 10 months." Most of the work of the carpenter ant is done by smaller workers.

Anon. (1959) reports these ants are able to withstand cold because they generate glycerol in their bodies whenever the temperature falls below a certain point. This is nature's way of manufacturing "antifreeze."

Carpenter ants in the home. Furniss (1944), who studied the control of carpenter ants in Oregon, discusses their infestations in homes as follows: "Where an infestation is of long standing and the colony is a large one consisting of several thousand ants, structural damage is frequently extensive enough to require major repairs. If the infestation is noted at an early stage, however, all that may be necessary is to get rid of the ants.

"Carpenter ant colonies become established in new situations either through invasion by a fertile queen and development of her progeny, or through immigration of all or part of an existing colony. The latter seems to be the more common way in which houses become infested. Evidence gathered over a period of several years shows that houses near wooded areas, 'stump' land, or brush-

Fig. 14-21. Wood damaged by carpenter ants. Tunnels lack the earthy appearance characteristic of timber damaged by subterranean termites.

covered vacant lots are most likely to become infested, although it is by no means uncommon for carpenter ants to invade dwellings in thickly populated districts of a city. Ants from any neighboring colony may move into a house, especially when seriously disturbed, as often happens in the clearing of building sites: in fact, cases are on record where disturbed carpenter ants became established in new houses before the home owners moved in. Usually they take the course of least resistance and enter any available openings about the foundations, but occasionally they exhibit considerable ingenuity in gaining access to a house, and have even been known to enter along telephone and electric wires. *(Author's note: They may enter the home by crawling on branches that contact the roof and other parts of the house.)* Often the point, or points, of entry are a considerable distance from the place where the brood galleries are excavated.

"All kinds of houses, from the oldest to the newest and from the most poorly constructed to the best, may become infested. In general, the houses most subject to attack are frame buildings without basements or with only part basements, those with very low foundations, those with open rambling porches, and those of loose construction, such as rustic cabins. The ants show some preference for moist rotting timbers about the foundations, but readily mine sound dry wood any place in a house. Among the commonly mined portions are porch pillars and supporting timbers, sills, girders, joists, studs and casings of houses, garages and other buildings.

"When a house becomes thoroughly infested there is little likelihood that the fact will be long unnoted by the human occupants. The continued presence of numerous workers, whether they are attracted to food or are merely running around the rooms, is strong though not conclusive evidence that a colony is established in the house. On warm days early in the spring, many people first become aware of carpenter ants in their homes when swarms of large winged ants emerge from the walls and try to escape through the windows. The appearance of these winged forms is an almost certain sign of continued trouble, for the main part of the colony remains behind and continues to develop, unless controlled. A faint, rustling sound in walls, floors and woodwork is another common clue to the presence of carpenter ants. Often the workers make slitlike openings through the surface of infested wood and through these openings expel their borings, which accumulate beneath in characteristic piles of fibrous 'sawdust'. Such refuse piles can be found most frequently in basements, in dark closets, under porches, and in similar out-of-the-way places. They are a sure sign that a colony is established. . . ."

Control. The extent of the treatment for carpenter ants varies in various parts of the country due to several factors such as the species encountered and, more importantly, the relative abundance of ant colonies inside or outside of the structure. Where there are many colonies indoors, or the chance for reinfestation is high, a very thorough job must be done. This seems to occur most often in the extreme northern parts of the United States. In other areas with infestations that normally contain only a single colony and relatively low rates of reinfestation, the treatment is often centered on locating and destroying the single nest.

The owner of the structure can and should take a series of steps to reduce the chance of reinfestation and to cooperate in the control program. These include:

- The homeowner should provide the PCO with detailed information about where they have seen the ants, witnessed signs of damage or moisture problems.

- Trim all trees and bushes so no branches touch or come close to any part of the house.
- Correct any moisture problem such as leaking roofs, leaking chimney flashing or plumbing, poorly ventilated attics or crawl spaces and blocked gutters.
- Consider replacing all rotted or water damaged wooden parts of the structure and eliminate wood/soil contacts.
- Remove dead stumps that are on their property and within 50 feet of the structure if practical.
- Use standard methods to repair trees with damage at the crotches, broken limbs and any type of hole.
- Store firewood up off the ground and well away from the structure.
- Bring only firewood that will be used quickly into the structure. Examine it and if necessary remove bark or discard infested logs.
- Consider non-organic mulches near the house in heavily infested areas.
- Homeowners should expect control to take a period of time and to occasionally see wandering carpenter ants reenter the structure. If large numbers are seen or damage is apparent, the PCO should be called back to evaluate the need for more treatement.

Treatment for carpenter ants should include both inside and outside inspections. Inspection should include looking for the ants themselves, nest, potential food sources and any conditions that are favorable to carpenter ant development or entry into the structure.

The nest itself is generally recognized first by the sawdust-like bits of wood that are sometimes removed in enlarging the nest area. There will often be partial ant bodies that also are discarded. The nest will be above the area where the wood fibers are seen.

The damaged wood differs from wood damaged by other pests in several important ways.

- There will be no "mud" in the galleries and the damage is irrespective of wood grain. This is in contrast to subterranean termites.
- There will be no minute bun-shaped fecal pellets as there would be with drywood termites.
- There will not be matched chisel-shaped marks as would occur with rodent gnawing.
- Cavities made by carpenter bees would have a nearly perfectly round entrance hole opening into a narrow gallery at right angles. Carpenter ants might infest such a hole if it was abandoned, but would enlarge the galleries.
- Buprestid beetles would have an oval opening with very narrow galleries. Carpenter ants have small openings in the wood, but larger cavities beyond.
- Round-headed borers will have a more circular opening that will only be approximately the size of the insect and again narrow wandering galleries instead of the larger carpenter ant galleries.
- Rotted wood may be associated with carpenter ants or may exist by itself. Even advanced rots will show a general disintegration rather than an excavated chamber.

Nests outside are usually associated with parts of a tree that are rotting or at least retaining extra moisture. They also can be found under boards, in leaves or pine needles or even under stones. They have been known to forage up to 300 feet/91 m from their nest.

Inside structures, ant nests can be found ranging from the mud sill to the shingle roof and anywhere in-between. Contrary to popular opinion, this ant is actually lazy. It will occupy an already created void such as a hollow core door if available and create its nest from softer damp and rotting wood in preference to dry hard wood. This seems rather a practical approach.

Spider webs may give clues, but the primary source of information should be the resident who has had a chance to observe this pest on a daily basis.

If the nest is found, treatment is usually easy with either a dust or a spray. Dusts are preferred because of its better float and coating ability in voids. However, good control has been achieved with various sprays or even aerosols. The needle tip on the B&G sprayer or the small tube available as an injection tool with some pressurized aerosol containers is helpful.

Where extensive treatment is needed, a perimeter spray around the structure is advisable and this should include spraying at least the trunks of nearby trees or stumps.

It may be necessary to drill and treat the voids between each stud with a residual chemical. Treatment around all window frames and door frames and along the sill plate also may be advisable.

Attics and crawl spaces are often important areas to treat with either dusts, sprays or space treatments. In living areas vulnerable locations such as under dishwashers, bathtubs and sinks, should be spot treated with residual sprays or dusts.

Extremely difficult infestations have been fumigated by covering the house and introducing methyl bromide or other gas. This should eliminate any existing infestation, but would need to be accompanied by other treatments to reduce the chance of reinfestation. In particular, unless all moisture problems in the structure are corrected, reinfestation can be expected. Indeed some PCOs have found that eliminating moisture problems has eliminated infestations of carpenter ants without any use of insecticides.

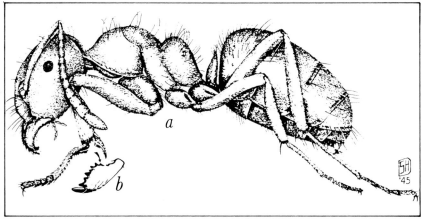

Fig. 14-22. Larger yellow ant, *Acanthomyops interjectus* (Mayr). A. Lateral view of worker. B. Left mandible showing small tooth on the superior border near the junction of the masticatory and superior border.

THE LARGER YELLOW ANT

Acanthomyops interjectus (Mayr.)

This is a common ant throughout New England and the Midwest. The worker is five mm in length, and the female is six to seven mm long. According to Flint and McCauley (1936), the large winged forms swarm, from cracks and crevices in the walls and floors of houses or from the basement and outside foundations of the house during the late winter or spring. For this reason, they are often mistaken for termites.

The large yellow ant is an aphid or honey-dew-attending insect which is found nesting outdoors in old logs and under stones. When crushed, the worker has an agreeable citronella odor. Smith (1928) states they occasionally carry soil into the house or annoy merely by being present, but otherwise they do no damage.

Blake (1940), speaking of the ant in New England, states, "The winged forms are discharged from the colony in mid or late summer and, after mating the queens (as in the cornfield ant) seek places to spend the first winter. New colonies are started the following spring. These newly mated queens sometimes take shelter in houses. Since the colonies of the large yellow ant and its near relatives are not uncommon in gardens the winged forms are occasionally found in buildings or on their walls."

The Larger Yellow Ant is often encountered by the PCO in response to a "Termite" call. This ant lives in the same areas where termites travel. It also is found when the termite operator is digging his trench along the foundation to do a termite treatment. Since they forage at night, they are seldom seen by the homeowner other than at swarming time.

Control. Since they live in soil next to a building foundation, under basement floors or even in concrete block voids, they are often killed by a thorough termite treatment. Control will usually depend on getting a toxicant to the nest which may require treatments that mimic a full or partial termite treatment. (A preventative termite treatment would be legal if the homeowner realized all of the facts). A barrier treatment might kill off enough of the workers to eliminate the colony. Treatment should be made to the area next to the foundation and to plantings likely to support honey-dew producing pests. Control often is not economically justifiable based on potential damage.

FIELD ANTS

Formica spp.

There are a great many species, subspecies and varieties of this common genus in the United States. They are moderately sized ants, usually varying from three to seven mm in length. They may be brown, black, red or various combinations of these colors. Ants of this genus are encountered most everywhere. They are fond of sweets and attend honeydew-secreting insects on plants. They also are predacious on other insects. Gosswald (1944) and other authors have noted *Formica rufa* L. is a valuable predator on forest insects in Germany.

In California, the brown field ants, *Formica cinerea* subspecies and varieties, are often found nesting in cracks in sidewalks, along the sides of buildings, at the base of trees, etc. They attend plant lice and occasionally invade the home. These common ants may often be observed running in long trails along fences, sidewalks or buildings. Undoubtedly, other ants of this genus enter homes occasionally in different parts of the country.

The moundbuilding ant of the northeastern United States, *Formica exsectoides* Forel, kills small trees and shrubs. These ants slowly girdle the trees by biting into the bark, and squirting formic acid into the wounds. Haviland (1945) states the poison of this species is "stored in a sac in the posterior dorsal part of the gaster and is ejected by the workers, as a means of defense, for anaesthesia or, according to Peirson, to kill seedlings." Acree et al. (1946), showed the venom is formic acid.

Field ant control. Since this group of ants apparently do not nest inside homes, minimal treatment would be all that is required inside. The primary treatment should be outside with a barrier strip of insecticide spray or granules around the structure as wide as permitted by the label.

Treating the outside trails along the foundation may help, but they will sometimes avoid a treated area if another pathway is available.

Nests, when found, can be treated with insecticide sprays, but this ant may wander a fair distance and thus nest far from the structure. Since this is an outside dwelling ant, baits are rarely effective by themselves.

THE CRAZY ANT

Paratrechina longicornis (Latr.)

The crazy ant is believed to be a native of India that has now become cosmopolitan. Since it presents the appearance of running aimlessly about a room, it was given the name "crazy ant." Its very long legs and antennae are its most obvious physical characteristic. The worker is 2.3 mm long, dark brown, with one segment in the pedicel.

Marlatt (1930) observes it is a pest in Florida and the Gulf states and notes it has been found even on top floors of large apartment buildings in New York City, as well as in hotels and flats in Boston. The author determined this as the species that was present in a hotel kitchen in San Francisco, California. It also has been reported from the District of Columbia and in Europe it is a common pest in greenhouses.

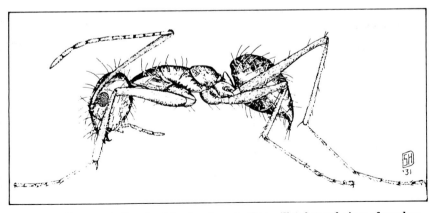

Fig. 14-23. Crazy ant, *Paratrechina longicornis,* (Latreille), lateral view of worker.

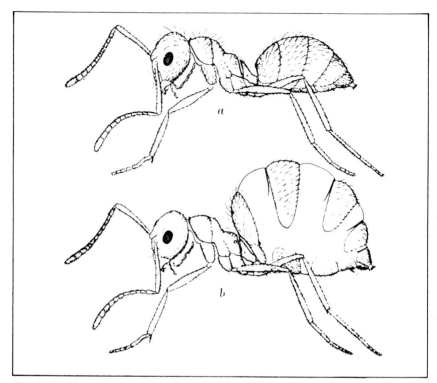

Fig. 14-24. The small honey ant, *Prenolepis imparis* (Say). A. Worker. B. Worker with gaster enlarged from imbibing a large quantity of liquid nourishment.

According to Blake (1940), the crazy ant nests in crevices, and although it prefers to feed on animal matter and insects, it will feed on sweets and kitchen scraps. He further states, "The colonies are rather small, containing up to 2,000 workers. There are commonly from eight to 40 queens in each colony. An entire colony may desert its nest and migrate to another site." This ant has been called *Prenolepis longicornis*, and *Nylanderia longicornis*. Fox and Garcia-Moll (1961) report it as a predator of fleas in Puerto Rico. Pimentel (1955) observed the workers preying on fly larvae and adults.

The crazy ant was noted by V. Walter to be the most common ant in food processing plants in south Texas. It also was a frequent invader of homes and was often seen after a termite treatment apparently feeding on dead termites.

Control. Baits have been minimally successful against this ant since it has a varied diet and the baits have to compete with so many other potential food sources.

It more or less follows a pathway and has relatively few members in a colony so it usually responds well to repeated spraying of trails. Nests do occur outside so a barrier treatment around the building should be considered part of a treatment.

In a few cases it may be necessary to drill and treat a colony that is living in a wall void or other sheltered area. Doorway mouldings and window trim have sometimes sheltered nests.

THE SMALL HONEY ANT

Prenloepis imparis (Say)

This ant and its subspecies and varieties are distributed throughout the United States. The workers vary from light to dark brown, are very shiny, have a somewhat triangular abdomen and are about three to four mm long. A few of the subspecies are slightly smaller. As a rule, it is only of slight importance in the home.

Prenolepis imparis (Say) may be observed ascending plants and feeding on ant cans. When the ants are turgid with nourishment, the gasters appear greatly enlarged, almost like the repletes or "honey casks" of the true honey ants. In fact, Wheeler (1930) states they actually have repletes that store food. This ant nests in damp soil in shaded places.

Talbot (1943) notes this ant is a "Cold weather ant, beginning activity above ground at temperatures just above freezing and reaching its peak of foraging at temperatures between 45° and 60° F." There is actually a decrease in the numbers of the small honey ants above ground at temperatures from 60° to 75° F/10° to 24° C. This species can move slowly at 32° F/0° C, and has no definite hibernation period. "In contrast, there is maintained a midsummer aestivation period of one to two months, in which no ants appear above ground even though temperatures are favorable. Flights of the males and females occur on the first warm days of spring when other species of ants are just finishing hibernation. *Prenolepis* responds positively to high humidity and, during favorable temperatures, reaches peaks of activity at humidities of 80 to 100 percent."

Control. Control is only occasionally needed. Marion R. Smith reports this species may occasionally nest inside a home during the winter. When this occurs, baits containing honey, sugar or very ripe fruits should attract the ants. If toxicants are counterindicated, it may be possible to place baits on sticky rodent glue boards and achieve control of the limited numbers of ants that occur in the winter with this species.

Nests located outdoors have a single entrance hole and if this can be found, control could be achieved by rodding the soil with any insecticide emulsion that would permit this treatment with its labeling. A barrier strip treated with a residual insecticide would probably be effective and baits could also be used outside if permitted by label.

THE ACROBAT ANTS

Crematogaster spp.

The acrobat ants may be recognized by the heart-shaped abdomen, which is flattened on the upper surface and curved below. They are black, brown or yellow in color and have two segments in the pedicel. At times, these ants invade the home for food. Snyder (1957) notes these ants may tunnel and nest in wood.

The slow-moving acrobat ants are found throughout the country, where they dwell beneath stones and in old stumps, etc. Their peculiar habit of raising the abdomen over the head and thorax has given rise to the name "acrobat ant." These ants are known to construct "cowsheds" of plant or earthen material over aphids and coccids which they attend.

Wheeler (1906) notes *Crematogaster lineolata* (Say), which has many subspecies and varieties, "ranges over the whole country from the Atlantic and Pacific seaboards to an altitude of about 7,000 feet in the Rocky Mountains."

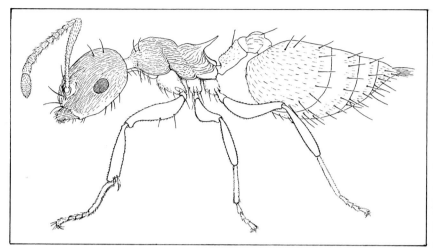

Fig. 14-25. Worker of the acrobat ant. Note the shape of the abdomen, which is often held up over the thorax.

Where the colonies are large, the ants are courageous and sting and bite, but where the colonies are small, the members are timid.

The various species of *Crematogaster* often live in decaying tree stumps outside and can live in woodwork, particularly doorframes or windowframes, inside the home. They will also take over areas hollowed out by other insects.

They may use tree limbs or power lines for entrance into the building and these should be considered in control.

Control Treating likely nesting areas outside plus a barrier treatment around buildings is the accepted form of treatment. Tree trimming or spraying is indicated where the branches touch the house.

Inside treatment, if needed, should include all doorframes, windowframes and any damp wood or previously infested wood. Wooden roof shingles may need to be treated. Baits of sweet syrups or ground dried meats may be helpful in some situations.

THE CORNFIELD ANT

Lasius alienus (Foerster)

Ants of the genus *Lasius,* including the above ant, are important house pests in the northern states. This ant also has been named *Lasius niger americanus* Emery. The cornfield ant is distributed widely throughout the United States and according to Wheeler it is the most abundant of ants, and hence of all our insects. Forbes (1908) made a detailed study of this ant, particularly in its relationship to the corn root-aphid. He found the cornfield ant collects the eggs of the corn root-aphid and stores them in its burrows during the winter. Then, in the spring, the ant places the aphids on various plants suitable for their growth.

The cornfield ant enters the home for sweets and for this reason is a common pest. Moreover, it attends aphids and other homopterous insects on ornamental

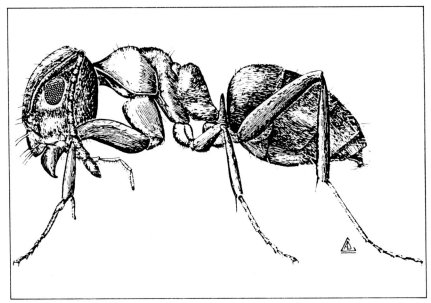

Fig. 14-27. Cornfield ant, *Lasius alienus* (Foerster), laterel view of worker.

plants. The numerous mounds of its nests in the lawn also are a common sight in those areas where it is abundant. This ant is predacious on other insects and after killing them it laps up their juices. The nests occur very commonly in the fields and in and around the home they may be found between bricks in the walk, beneath rocks, in cracks in the pavement, in the lawn, etc. The nests may become extended and form a group of mounds.

Forbes (1908) studied the life history of the cornfield ant and he found in Illinois the sexual forms are in flight from August to September. The females enter the ground and lay their eggs during the summer and fall, although some of the females may not lay their eggs until May of the following year. The queen feeds and cares for her first young. The first workers emerge in July. The egg stage extends from 22 to 28 days. The larval stage is of 16 to 23 days duration. During the summer, a period of four months may elapse for completion of the egg to the adult stage. Davis (1949) recommended plowing to a depth of 6½ to seven inches/16.5 to 17.8 cm in the spring for the control of this insect in corn fields.

Control. This ant looks so much like its close relative *Lasius neoniger* that even the best of authorities have occasionally been confused. Apparently *L. neoniger* is more common in open fields, building many small mounds in yards and on golf courses. *L. alienus* is more common in or near wooded areas. We will treat them together since control will only depend on the location of the nest in both cases.

Total lawn spraying, where permitted by the residual insecticide label, is the preferred treatment. Little if any treatment is needed inside unless the ants are well established. However, this rarely occurs. They will feed on sweets or meats and a "combination bait" of both grease and sweets might help inside.

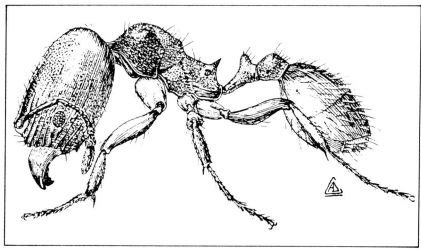

Fig. 14-26. *Pheidole flavens floridanus* Emery.

MISCELLANEOUS ANTS

Marlatt (1930) states, "Four other species of ants from tropical America have gained, through the agency of commerce, some foothold for considerable periods of time in northern heated houses." These ants are *Prenolepis fulva pubens* Forel, *Neoponera villosa* F. Smith, *Camponotus abdominalis floridanus* Buckley, and *Pheidole flavens floridanus* Emery. In addition to the above, the ant *Iridomyrmex iniquus* var. *nigellus* Emery, which has been introduced into the United States with soil from tropical plants, occurs in greenhouses in Illinois and Massachusetts.

GUIDELINES AND PRODUCTS FOR CONTROL OF ANTS

Ant control has changed throughout the years to reflect the efficacy of various materials that were available at that time. Earlier editions of the *Handbook of Pest Control* listed various sprays dusts and baits that were legal at that time and which were effective.

Baits with slow acting toxicants that can be carried by the foraging ants back to the nest to kill the queen and some of the workers and young will always be a good concept. As of this writing, arsenic, chlordecone (Kepone®) mirex or thallium sulphate are no longer available. These products were all used effectively for years in ant baits, but have all been removed from the market by the Environmental Protection Agency. There are still some premixed ant baits available, particularly to the homeowner. These do not have the efficacy of the former baits, but still should be considered. It is hoped that other toxicants will be available for mixing specialized baits for special ants.

During the heyday of chlorinated hydrocarbons, ant control was relatively easy on most species because materials such as chlordane lasted for a long time and were regularly used to control other insects. Ant control was often an inexpensive and sometimes unintentional by-product of other pest control measures. Baseboard spraying was of questionable value in roach control, but probably did do a good job on some species of ants. Broadcast surface spraying of chlorinated hydrocarbons around the house is not permitted now in the United States. Replacement materials do not have the very long residual life of the chlorinated hydrocarbons.

Ants can be beneficial as was discussed earlier, but there are valid reasons for controlling some ants under certain conditions including:

- They may destroy seeds that have been planted by homeowners.
- They may directly destroy plants as occurs with the leaf-cutting ants.
- They may indirectly injure some crops through their care of honeydew-producing insects.
- They may injure or kill livestock as can occur with fire ants.
- Their mounds can damage farm machinery as occurs with fire ants.
- They may inflict painful and even serious stings or bites on people.
- They can contaminate food to the point that it is legally adulterated and cannot be sold.
- Food contamination by ants can result in an allergic reaction by certain sensitive people.
- They can carry disease-producing organisms to patients in hospitals or other areas.
- They can be an annoyance at picnics and at home.
- Some ants can damage wooden portions of structures.
- They can interfere with enjoyment of yards or golf courses.
- Some ants can undermine a slab causing clean-up problems now and possibly structural problems later.

The pest control operator may have lost some effective ant control tools, but ant control is still possible. However, a complete understanding of ant identification is necessary before a successful control program can be initiated.

USDA (1980) recommends the following insecticides for ant control: bendiocarb, carbaryl, chlorpyrifos, dichlorvos, diazinon, fenthion, lindane, malathion (premium grade), propoxur and pyrethrins.

Ant control can be divided into a series of treatment situations. These include:

- Treatments of nests in open land.
- Treatment of nests in yards or adjacent to a structure when the ants are not entering the structure.
- Treatment of nests in yards or adjacent to a structure when the ants are entering the structure.
- Treatment of ants under concrete slabs that are a part of a structure.
- Treatment of nests in walls and other voids.
- Treatment of barrier strips around the structure.
- Treatment of travel paths of ants in order to kill by attrition of workers.
- Sanitation to remove food favored by ants.
- Tree trimming to discourage entrance to roof or window areas by carpenter ants and others.
- Changing landscaping or crops to discourage ants or ant honeydew-producing insect relationships.
- Structural changes and the reduction of hollow core doors and other favored nesting areas of carpenter ants.

DETAILS OF TREATMENT SITUATIONS

Treatments of nests outside. Outside nests can be treated in a variety of ways if permitted by the label.

The interior of the nest can be treated with fumigants or with sprays with large volumes of water. Rodding or probing can be used to aid penetration.

The area around the nest can be ringed with toxicant so all foraging ants must travel through this lethal barrier if this is permitted by the label. This could concentrate considerable toxicant in a small space and ants will sometimes tunnel under the toxicant or just wait until it is no longer effective. However, this type of treatment is usually effective.

Overall spraying of the yard or other area may be necessary when there are many nests. Chlorpyrifos and diazinon have been effective in the past. Check labels for current permission and precautions. This treatment will usually require power spraying because of the large amount of material necessary.

Treatment of nests under slabs. Treatment is often effective with dusts and sprays through the cracks ants are using. The ants will often excavate the area under these cracks and treatment is easy. If not, drilling and treating as in termite control will be necessary.

Treatment of nests in walls and other voids. Access to void areas can sometimes be obtained by removing switch plates and other receptacle covers. Hand dusting with plastic-tipped dusters is a favorite control method but other techniques also are successful. Drilling, treating and then plugging is needed in some instances and requires the owner's permission.

Treatment of barrier strips around the structure. When ants may be entering from the outside, this can be a very effective treatment. A strip as wide as permitted on the label, preferably 10 feet/three m, should be made with a residual chemical. The structure should be sprayed to window height if permitted by labels. Trees and bushes in the area should be sprayed to the same height if there is no problem of phytotoxicity and it is permitted.

Treatment of travel paths. This is largely a treatment of the floor wall junction or other area frequented by ants. It must extend beyond the observed travel paths of ants to assure complete control.

Ants can be controlled by using one or more of the treatment situations depending on what would be most cost effective.

The toxicants used in ant control can eliminate ants in many different ways including:
- Fumigants have been used on ant nests in the ground.
- Contact insecticides have been used in the nest.
- Residual insecticides have been placed in wide strips all around all nest exits to kill foraging workers.
- Baits may be designed to kill foraging workers and gradually eliminate the colony by attrition.
- Baits are usually more effective when they are designed to be carried back to the nests and eliminate the queens and the nest workers.
- Baits have been used at least experimentally that keep the ants at a "juvenile" stage where they cannot reproduce.
- Sprays can be designed to kill aphids and other honeydew-producing insects that are the main source of food for some ants through their honeydew excretions.
- Baits can be used to attract the ants away from a point where protection is desired (e.g. certain types of hospital cases or allergy cases).

Locating ant nests. Most successful ant control recognizes that ants are a social insect that must live as part of a colony. We can kill many individuals and the colony will rebuild to its original numbers. Some species of ants are said to survive the death of 99 percent of the foraging workers and still rebuild.

Control of the colony is easier if we know where it is located. In many cases, this may be difficult or impossible but certain techniques can help.

Outside, the colony location often is obvious due to the large mounds of dirt or the complete removal of vegetation around an entrance hole. It is important to inspect further because there may be many entrance holes and the subterranean colony may cover a vast area. Treating only one entrance may be ineffective.

Some ants will actually nest inside our structures instead of merely entering in search of food. This could include: carpenter ants, crazy ants, odorous house ants, pavement ants, Pharaoh ants, thief ants and California fire ants.

Carpenter ant nests can sometimes be found directly over piles of "wood shavings" that were tossed out as the ants enlarged a nest. There usually will be some parts of dead ants mixed with the wood fragments.

Some have reported they could hear carpenter ants with the aid of amplifying devices such as a stethoscope. This probably works best when the colony is chewing on the wood to enlarge the nest.

Some ants will follow definite trails from the nest to the food source. The trails sometimes can be followed from the food source back to the nest. It may be necessary to place a type of food that is preferred by that ant in order to create a trail of ants. The ants, unfortunately, may enter a wall void well before they get to the true location of the nest.

Ants that usually travel in trails include: Argentine ants, black harvester ants (outside species), crazy ants, California fire ants (outside species), field ants (outside species), odorous house ants, velvety tree ants, and Pharaoh ants.

Some ants often nest under slabs and this of course gives us a clue to where their nest may be located. They also may kick up mounds of earth which can indicate their presence. The piles of dirt do not always occur. This could include pavement ants, Argentine ants and yellow ants.

FOOD PREFERENCE OF SOME ANTS

Ants that normally prefer sweets:
- Honey ant.
- Carpenter ant.
- Field ant.
- Odorous house ant.
- Pavement ant.
- Velvety tree ant.

Ants that normally prefer fats:
- Western thief ant.

Ants attracted to both sweets and fats:
- Argentine ant.
- California fire ant.
- Pharaoh ant.
- Tiny black ant.

Ants that eat mainly seeds:
- Black harvester ant.
- Red harvester ant.
- California fire ant.

Ants that cut and remove leaves:
- Texas leaf-cutting ant.

Sprays for ant control. Sprays are the most common form of pesticide used in ant control. They usually are readily available since they are used in other pest control work. The sprays are usually not a problem in staining since the applicator can choose between oil-based and water-based materials and even wettable powders such as bendiocarb, carbaryl and propoxur.

There have been some indications that sprays such as dieldrin and possibly others achieved control more from a repellancy reaction than from their residual action. The ants avoided the treated areas and if no others were available, they might eventually starve to death.

Ant baits. Ant baits also can be a valuable tool in a control program. Non-toxic baits can be used to attract ants so their trails can be followed to the nest. Non-toxic baits also can be used to help identify the type of ants since some ants have very definite feeding habits. We may be able to merely observe what they are already eating to determine their food preference. Non-toxic baits also have been used to draw ants away from another area that needed protection.

Toxic fast-acting baits have been used that kill the foraging workers quickly, but these are not as effective as those that are slow acting and will be taken back to the nest for consumption. Baits based on mirex and chlordecone are excellent in this respect. Consumption in the nest can be from original bait or from regurgitated food.

Some new baits that are now being used experimentally and should soon be on the market are growth regulators which prevent the ants from reaching full maturity and sexual development. They show promise in some situations because of low mamalian toxicity.

Dusts for ant control. Many people fail to use dusts for ant control since heavy amounts of dust are repellant as well as unsightly if placed in the open. Dusts also can be blown from the original placement by winds or air currents and could become a contaminant in some situations.

Dusts do have advantages and should be considered an important tool in ant control. Ants "comb" their bodies and may be killed by dusts that adhere to their bodies.

Dusts can mechanically float back into crevices such as wall voids where the nest may be located and can easily cover vast areas of attics or crawl spaces. Dusts also can easily coat large areas of vegetation and some of the dust will float down to the soil where many ants wander.

Dusts already are sorbed or impregnated into particles that are chemically compatible with its chemistry and will therefore often have a longer residue than a spray of the same chemical.

Dust should be applied in light amounts in places where they can be expected to contact the ants without causing drifting, staining or residue problems.

Fumigants for ant nests in the ground. Throughout the years several fumigants have been labeled as pesticides for control of ants in nests in the soil. These included calcium cyanide, carbon disulphide and methyl bromide. They cannot be used under current laws unless the label you possess states the ant nest as a site.

E. V. Walter reported success in killing established colonies of the Texas leaf-cutting ant with carbon disulphide which he mixed with water to extend the volume. Calcium cyanide was often used to treat nests. It was sometimes packaged in a can with an "oil can" spout which made it easier to get the material into the nest. Where the nest is extensive, the fumigant would have to be injected at many points.

All fumigants are dangerous to man and should only be used by persons with both certification and thorough training. Some fumigants such as chloropicrin would probably be effective, but would kill vegetation in the area exposed.

Pre-plant soil fumigation undoubtedly kills ant colonies as well as other insects. However, structural fumigation for ants is justified in only a few cases where the costs or problem of other types of treatments are excessive.

Granules for ant control. Granules like dusts are probably not used often enough in ant control programs. They will roll down to the base of vegetation better than dusts and can be effective against some ants. They can be placed around or in ant mounds and may kill some of the ants that tend the queen and the young, or perhaps even the queen herself. Granules can be consumed by birds and this should be considered in placement.

LITERATURE

ACREE, F., JR., E. E. HAVILAND & H. L. HALLER, — 1946. The nature of the venom of *Formica exsectoides*. J. Econ. Entomol. 39(5):661-662.

ANON. — 1949. How to control ants in houses and lawns. Purdue U. Agr. Ext. Mimeog. E-22, 2 p. Mar. Purdue Univ. Agr. Ext. Serv. 1951. The red harvester ant. U.S.D.A. EC-18. 1953. Ants in the home and garden, how to control them. U.S.D.A. Home & Garden Bull. 28. 1954. The imported fire ant. How to control it. U.S.D.A. Leaflet 350. 1957. Keys to the species of fire ants. Coop. Econ. Insect Rpts. 7(48):901-902. 1958. Observations on the biology of the imported fire ant. U.S.D.A. ARS-33-49. 1959. Insects' "antifreeze." Agr. & Food Chem. 7(12):806.

ARMAND, J. E. — 1942. Thermal preference by Pharaoh's ant as a guide in control work. Pests 10(2):18-19. Also 72nd Annual Rpt. Entomol. Soc. Ontario pp. 30-32.

AUTORI, M. — 1950. M. M. 33, a new formicide with methyl-bromide base in the combat against the sauba ant. Biol. Abst. 25(9):2581, 1951.

BACK, E. A. — 1946. House ants. U.S.D.A. Leaflet No. 147. Rev.

BAERG, W. J. — 1937. Ants in Mexico. Introd. to Econ. Entom. p. 71.

BARBER, E. R. — 1916. The Argentine ant. Distribution and control in the United States. U.S.D.A. Bull. 377.

BARNES, O. L. & N. J. NERNEY — 1953. The red harvester ant and how to subdue it. U.S.D.A. Farmers' Bull No. 1668.

BARTLETT, B. R. — 1961. The influence of ants upon parasites, predators and scale insects. Ann. Entomol. Soc. Amer. 54(4):543-551.

BARTLETT, F. J. & C. S. LOFGREN — 1964. Control of a native fire ant, *Solenopsis geminata,* with Mirex bait J. Econ. Entomol. 57(4):602.

BEATSON, S. H. — 1972. Pharaoh's ants as pathogen vectors in hospitals. The Lancet. February 19. 425-427.

BELLEVOYE, M. A. — 1889. Observations on *M. pharaonis* Latr. Insect Life, 2:230-233.

BELLINGER, F., R. E. DYER, R. KING & R. B. PLATT — 1965. A review of the problems of the imported fire ant. Georgia Acad. Sci. 23(1):—reprint.

BEYER, A. H. — 1924. Life history of the new citrus aphid. Fla. Entomol. 8:8-13.

BLAKE, C. H. — 1940. Notes on Economic Ants. Pests 8(11):16-18. 1940. Notes on Economic Ants. Part II. Pests 8(12):8-10.

BLAKE, G. H., JR., W. G. EDEN, & K. L.HAYS — 1959. Residual effectiveness of chlorinated hydrocarbons for control of the imported fire ant. J. Econ. Entomol. 52(1):1-3.

BLUM, M. S., J. R. WALKER, P. S. CALLAHAN & A. F. NOVAK — 1958. Chemical insecticidal, and antibiotic properties of fire ant venom. Science 128:306-307; Aug. 8.

BRETT, C. H. — 1950. The Texas harvester ant. Okla. Agr. Expt. Sta Bull. No. B-353. July.

BROWN W. L., JR. — 1950. The status of two common North American carpenter ants. Ent. News 61(6):157-161. 1957. Is the ant genus *Tetramorium* native in North America? Breviora-Mus. Comp. Zool. (Harvard) 72:1-8.

BRUDER, K. W. & A. P. GUPTA — 1972. Biology of the pavement ant, *Tetramorium caespitum* (Hymenoptera: Formicidae). Annals Entomol. Soc. Amer. 65(2):358-367.

BURBUTIS, P. R. & M. S. CONRAD — 1958. Pavement ant. Coop. Econ. Insect Rpt. 8(51):1015, Dec. 19.

BUREN, W. F., 1972. — Revisionary studies on the taxonomy of the imported fire ant. J. Georgia Entomol. Soc. 7:1-26.

BURNETT, D., JR. — 1952. Pharaoh's ant a special problem. Pest Control. 20(2):22.

BYARS, L. F. — 1949. The Mexican leaf-cutting ant *(Atta mexicana)* in the United States. J. Econ. Entomol. 42(3):545.

CARTHY, J. D. — 1950. Odour trails of *Acanthomyops fuliginosus*. Nature 166 (4212):154.

CAVILL, G. W. K. & P. L. ROBERTSON, — 1965. Ant venoms, attractants and repellents. Science 149(3690):1337-1345.

CLARK, W. T. — 1922. Ant control on ship board. J. Econ. Entomol. 15:329-333.

COARSEY, J. M. — 1952. Southern fire ant, *Solenopsis xyloni* (Death of a child in Mississippi). Coop. Econ. Insect Report. Sept. 19, 1952.

COLE, A. C. — 1968. *Pogonomyrmex* Harvester Ants. A study of the Genus in North America. Knoxville: Univ. Tennessee Press. x + 222 pp.

COTTON, R. T. & G. W. ELLINGTON — 1930. A simple and effective ant trap for household use. J. Econ. Entomol. 23:463.

DAHMS, R. G. & F. A. FENTON — 1939. Methods of destroying red harvester ant nests. Pests. 7(4):12-14.

DAVIDSON, R. H. — 1950. An effective ant bait formula. J. Econ. Entomol. 43(4):565.

DAVIS, J. J. — 1949. The corn root aphid and methods of controlling it. U.S.D.A. Farmers' Bull. 891.

DeBACH, P. — 1951. The necessity for an ecological approach to pest control on citrus in California. J. Econ. Entomol. 44(4):443-447. 1952. California red scale: Study of prospects for biological control of pests in orange and lemon groves of San Diego County. Calif. Agr. 6(3):8,12.

DENNIS, C. A. — 1941. Some notes on the nest of the ant *Prenolepis imparis.* Annals Entomol. Soc. Amer. 34:82-84.

DENNY, C. — 1943. The control of pharaoh's ant. NPCA Service Letter 288.

DOWNES, D. — 1939. Derris for ants and wasps. J. Econ. Entomol. 32:883-884.

DURR, H. J. R., C. J. JOUBERT, & S. WALTERS — 1955. A biological evaluation of the effects, two years after application to the soil of aldrin, chlordane and dieldrin, on workers of the Argentine ant, *Iridomyrmex humilis* (Mayr.) J. Entomol. Soc. So. Afr. 18(2):235-237. Biol. Abst. 3(1):273, 1957. 1958. The effect of application to the soil of aldrin, dieldrin and chlordane on infestations of the Argentine ant *(I. humilis)* during a period of four years. S. Afr. J. Agr. Sci. 1(1):75-82. R.A.E. A. 47(5):175, 1959.

EAGLESON, C. — 1940. Fire ants causing damage to telephone equipment. J. Econ. Entomol. 33:700.

EBELING, W., — 1975. Urban Entomology. Univ. of Calif. Div. of Agri. Sci. 695 pp.

ECHOLS, H. W. — 1966a. Texas leaf-cutting ant controlled with pelleted Mirex bait. J. Econ. Entomol. 59(3):628-631. 1966b. Assimilation and transfer of Mirex in colonies of Texas leaf-cutting ants. J. Econ. Entomol. 59(6):1336-1338.

ECKERT, J. E. & A. MALLIS — 1937. Ants and their control in California. Calif. Agr. Exp. Sta. Circ. 342.

ENZMANN, J. — 1946. A new house-invading ant from Massachusetts. J. N. Y. Entomol. Soc. 54:47-49. *Crematogaster lineolata cerasi,* the cherry ant of Asa Fitch; with a survey of the American forms of *Crematogaster,* subgenus *Acrocoelia.* J. N.Y. Entomol. Soc. 54(2):89-97.

ESSIG, E. O. — 1921. The Argentine ant builds earthen protectors for mealybugs. J. Econ. Entomol. 14:506-508. 1926. Insects of Western North America. 1035 p. The Macmillan Company. 1939. The fire ant. Texas Citriculture. p. 15.

FAVORITE, F. G. — 1958. The imported fire ant. USPHS Rpts. 73(5):445-448.

FERNALD, H. J. — 1947. The little fire ant as a house pest J. Econ. Entomol. 40:428.

FLANDERS, S. E. — 1943. The Argentine ant versus the parasites of the black scale. Calif. Citrograph 28(5):117, 128, 137. 1945. Coincident infestations of *Aonidiella citrina* and *Coccus hesperidium,* a result of ant activity. J. Econ. Entomol. 38(6):711-712. 1945. Is caste differentiation in ants a function of the rate of egg deposition? Science 101 (1619):245-246. 1951. The role of the ant in the biological control of homopterous insects. Can. Entomol. 88(4):93-98. 1952. Ovisorption as the mechanism causing worker development in ants. J. Econ. Entomol. 45(1):37-39.

FLINT, W. P. & W. E. McCAULEY — 1936. Ants—How to combat them Ill. Agr. Exp. Sta. Circ. 456.

FORBES, S.A. — 1908. Habits and behavior of the cornfield ant, *Lasius niger americanus.* Ill. Agr. Exp. Sta. Bull. 131.

FOREL, A. — 1930. The Social World of the Ants Compared with That of Man. Albert & Charles Boni.

FOSTER, E. 1908. The introduction of *Iridomyrmex humilis* (Mayr.) into New Orleans. J. Econ. Entomol. 1:289-293.

FOX, I. & I. GARCIA-MOLL — 1961. Ants attacking fleas in Puerto Rico. J.

Econ. Entomol. 54(5):1065-1066.

FRIEND, R. B. — 1942. The black carpenter ant. Pests 10(2):12.

FRIEND, R. B. and A. B. CARLSON — 1937. The control of carpenter ants in telephone poles. Conn. Agr. Exp. Sta. Bull. 403:913-929.

FRITZ, R.F. and W.A. VICKERS — 1942. Damage to highways by the mound-building prairie ant. J. Econ. Entomol. 35:725-727.

FURNISS, R. L. — 1944. Carpenter ant control in Oregon. Ore. Agr. Exp. Sta. Cir. No. 158, 1957. Ore. Ext. Circ. 627.

GERHARDT, P. D. — 1952. Ant control in citrus groves. California Agriculture. 6(5):13. 1953. Chlordane, dieldrin, aldrin, and heptachlor for control of the Argentine ant in California citrus orchards. J. Econ. Entomol. 46(6):1063-1066. 1954. Argentine ant control with sprays and bait in two California lemon groves. J. Econ. Entomol. 47(4):591-593.

GOODFELLOW, A. — 1948. If you are troubled with red ants *(Monomorium pharaonis)*, read how to attack this bakery pest. Confectioner, Baker & Restauranteur 77 (1027):28-29, 56, (1948):30-31. May & June.

GORDON, W. M. — 1943. Airplane runways damaged by ants. J. Econ. Entomol. 36(2):354.

GOSSWALD, K. — 1944. The effect on the various kinds of forest ants collecting the pupae. Z. angew. Ent. 30(3):317-335. Rev. Appl. Entomol. A. 35(11):348. 1947.

GREEN, H. B. — 1952. Biology and control of the imported fire ant in Mississippi. J. Econ. Entomol. 45(4):593-597. 1962. On the biology of the imported fire ant. J. Econ. Entomol. 55(6):1003-1004.

GROSS, J. R. — 1948. City ants and their country cousins. Pests 16(8):26, 45.

HACKLEY, R. E. — 1939. Highlights of ant clinic. Pests 7(8):12, 1940. Ants and electricity. Pests 8(10):16-17.

HAINES, G. C. — 1935. The small house ant. Rev. Appl. Entomol. 23:70.

HAVILAND, E. E. — 1945. The effect on the fingers of the poison of *Formica exsectoides*. J. Econ. Entomol. 38(5):607. 1947. Biology and control of Allegheny Mound Ant. J. Econ. Entomol. 40(3):413-419.

HAYS, S.B. & F.S. ARANT — 1960. Insecticidal baits for control of the imported fire ant. J. Econ. Entomol. 53(2):188-191.

HERBERT, F.B. — 1932. Effect of cold storage temperatures on the Argentine ant. J. Econ. Entomol. 25:832.

HERMS, W.B. — 1939. Medical Entomology. 582 p. The Macmillan Co.

HERTZER, LUCILE — 1930. Response of the Argentine ant *(Iridomyrmex humilis* Mayr.) to external conditions. Annals Entomol. Soc. of Amer. 23:599. 1930. Studies on the Argentine ant queen *(Iridomyrmex humilis* Mayr.) Annals Entomol. Soc. of America. 23(3):601-609.

HOCKENYOS, G.L. — 1940. Pharaoh's Ant. Pests 8(10):11.

HORTON, J.R. — 1918. The Argentine ant in relation to citrus groves. USDA Bull. 647.

JOHNSTON, H.R. — 1944. Control of the Texas leaf-cutting ant with methyl bromide. J. For. 42(2):130-132. Rev. Appl. Entomol. A. 32:295-296, 1944.

JONES, C.R. — 1929. Ants and their relation to aphids. Colo. Exp. Sta. Bull. 34.

KALSHOVEN, L.G.E. — 1937. Further notes on the house ant, *M. destructor*. Rev. Appl. Entomol. A. 26:162.

KERPEN, F. — 1961. Count ants' teeth. Pest Control. 29(10):6.

KERR, T.W. JR. — 1948. Control of the cornfield ant in golf greens. J. Econ.

Entomol. 41(1):48-52.

KESTERTON, B.C. — 1961. A programme of control of Pharaoh's ants. Pest Technology. 4(3):66-68.

LAUDANI, H. & H.T. VANDERFORD, — 1952. Control of little black, Pharaoh and Argentine ants. Pest Control 20(5):18, 20, 22.

LAVIGNE, R.J. — 1966. Individual mound treatments for control of the western harvester ant, *Pogonomyrmex occidentalis* in Wyoming. J. Econ. Entomol. 59(3):525-532.

1969. Bionomics and nest structure of *Pogonomyrmex occidentalis*. "Hymenoptera: Formicidae". Annals Entomol. Soc. Amer. 62:1166-1175.

LEUTERT, W. — 1963. Systematics of ants. Nature. 200(4905):496-497.

LINDQUIST, A. W. — 1942. Ants as predators of *Cochliomyia americana* C. & P. J. Econ. Entomol. 35(6):850-852.

LITTLE, V. A. — 1950. Methyl bromide controls the Texas leaf-cutting ant *(Atta texana)*. Down to Earth 6:15 Summer.

LOFGREN, C.S., V.E. ADLER, W.A. BANKS, & N. PIERCE — 1964. Control of imported fire ants with chlordane. J. Econ. Entomol. 57(3):331-333.

LOFGREN, C. S., F. J. BARTLETT, & C. E. STRINGER — 1963. Imported fire ant toxic bait studies evaluation of carriers for oil baits. J. Econ. Entomol. 56(1):62-66.

LYLE, CLAY — 1936. Challenge of the Argentine ant. J. Econ. Entomol. 29:965-67.

LYLE, C. & I. FORTUNE — 1948. Notes on an imported fire ant. J. Econ. Entomol. 41(5):833-834.

MacCONNELL, J. G., J. M. BRAND, M. S. BLUM, and H. M. FALES — 1971. Fire Ant Venoms: Comparative analysis of alkaloidal components. Toxicon. 10(3)259-271.

MacNAMARA, C. — 1945. A note on the swarming of *Solenopsis molesta*. Say (Hymenoptera) Can. Entom. 77(2):40.

MALLIS, A. — 1938a. The Argentine ant. Pest Control: 1:11-16. 1938b. Army ants in California. Sci. Mo. 47:220-226. 1938c. The California fire ant and its control. Pan Pacific Ent. 14:87-91. 1941. A list of the ants of California with notes on their habits and distribution. Bull. So. Calif. Acad. Sci. 40:61-100 1942. Half a century with the successful Argentine ant. Sci. Mo. 55:536-545, Dec. 1948. The Pharaoh and the thief ant. Pest Control & Sanitation. 3(3):8-9. 1969. *Handbook of Pest Control.* 5th Ed. MacNair-Dorland, New York 1158 pp.

MARICONI, F.A.M. — 1964. Heptachlor dust for control of the leaf-cutting ant "Suavo Limao" in Brazil. J. Econ. Entomol. 57(6):797-798.

MARKIN, G. P. — 1970. Foraging behavior of the Argentine ant in a California citrus grove. J. Econ. Entomol. 63(3):741-744. 1970a. The seasonal life cycle of the Argentine ant in Southern California. Annals Entomol. Soc. Amer. 63(5):1238-1242.

MARLATT, C. L. — 1930. House ants: Kinds and methods of control. U.S.D.A. Farmer's Bull. 740. Revised.

MARTIN, P. R. S. — 1971. The venomous ants of the genus *Solenopsis*. Chapter 46 in Venomous Animals and Their Venoms. Vol. 3 Venomous Invertebrates. Ed. Bucherl & Buckley. Academic Press, New York. 95-101.

MARTINI, E. — 1934. *Monomorium* in Krankenhausern. Rev. Appl. Entomol. B. 22:163.

McCAULEY, W. E. — 1947. Chlordane emulsion against ants.

MCCULLOCH, T. W. & W. P. HAYES — 1916. A preliminary report on the life economy of *Solenopsis molesta* Say. J. Econ. Entomol. 9:23-28.

METCALF, C. L. & W. P. FLINT — 1939. Destructive and Useful Insects. 2nd ed. 981 p., McGraw-Hill.

MICHELBACHER, A. E. — 1950. Principles of ant control. Pest Control 18:(7): 14, 16, 18. 1950. Ant control program; modern insecticides correctly applied achieve indoor and outdoor control. Calif. Agr. (Calif. Sta.) 4(48):11-12. Aug.

MICHENER, C. D. — 1942. The history and behavior of a colony of harvesting ants. Sci. Mo. 55:248-258. 1948. Observations on the mating behavior of harvesting ants. J.N.Y. Ent. Soc. 56:239-242.

MOREHOUSE, C. H. — 1949. Unusual reaction to ants bites. J. Amer. Med. Assoc. 141:193. Sept. 17.

MORLEY, D. W. — 1949. Vibration of the flagellum of the ant antennae. Nature (London) 164 (4174):749.

MURPHY, R. T., W. F. BARTHEL, & C.S. LOFGREN — 1962. Residual studies in connection with successive applications of heptachlor for imported fire ant eradication. Agr. & Food Chem. 10(1):5-7.

MUSGROVE, C. H. & G. E. CARMAN — 1965. Argentine ant control on citrus in California with granular formulations of certain chlorinated hydrocarbons. J. Econ. Entomol. 58(3):428-434.

NEGIS, P.S. — 1934. The small red ant, *Solenopsis geminata,* subsp. *rufa* Jerdon, and its usefulness to man. Rev. Appl. Entomol. B. 22:102, June.

NEWELL, W. — 1908. Notes on the habits of the Argentine or "New Orleans" ant, *Iridomyrmex humilis.* Mayr. J. Econ. Entomol. 1:21-34. 1909. Measures suggested against the Argentine ant as a household pest. J. Econ. Entomol. 2:324-332. 1909. The life history of the Argentine ant, *Iridomyrmex humilis* Mayr. J. Econ. Entomol. 2:174-192.

NEWELL, W. & T.C. BARBER — 1913. The Argentine ant. U.S.D.A. Bur. Entomol. Bul 122:1-98.

NIXON, G. E. J. — 1951. The association of ants with aphids and coccids. Rev. Appl. Entomol. A. 39(6):189, 1951.

NOVAK, V. — 1942. A fresh occurrence of the ant. *M. pharaonis,* in Prague, Acta Soc. Ent. Bohem. 39(2):135-136. Rev. Appl. Entomol. A. 34(11):328, 1946.

OLIVE, A. T. — 1960. Infestation of the imported fire ant, *Solenopsis saevissima* v. *richteri,* at Fort Benning, Georgia. J. Econ. Entomol. 54(4):646-648.

OSBURN, M. R. — 1945. DDT to control the little fire ant. J. Econ. Entomol. 38(2):167-168. 1946. Effect of DDT on the little fire ant *(Wasmannia auropunctata)* Fla. State Hort. Soc. Proc. (1945) 58:156-158. 1948. Comparison of DDT, chlordane, and chlorinated camphene for control of the little fire ant. Fla. Entom. 31(1):11-15. 1949. Tests of parathion for control of the little fire ant *(Wasmannia auropunctata).* J. Econ. Entomol. 42(3):542.

OSBURN, M. R. & N. STAHLER — 1946. Use of DDT to control the little fire ant. B.E. & P.Q.E.-683.

PAPWORTH, S. D. — 1958. Practical experience with the control of ants in Britain. Ann. Appl. Biol. 46(1):106-111.

PARKER, G.H. — 1947. The number of ants in ant colonies. Ann. Entomol. Soc. Amer. 35(3):363-365.

PEACOCK, A.D., D.W., HALL, I.C. SMITH & A. GOODFELLOW — 1959. The biology and control of the ant pest *Monomorium pharaonis* (L.) Dept. of Agr. of Scotland. Misc. Publ. No. 17.

PIMENTEL, D. — 1955. Relationship of ants to fly control in Puerto Rico. J. Econ. Entomol. 48(1):28-30.

PLOSCHKE, J. — 1943. Die Pharaoameise an Bord eines Schiffes. Deutsch. tropenmed Zeitschr. 47(12):302-309. Biol. Abst. 20:655, 1946.

POMERANTZ, C. — 1955. The fabulous and destructive carpenter ant, Pest Control. 23(10):9-10, 14, 64, 70-71.

POPENOE, E.A. — 1926. Thallium sulphate as a poison for ants. Science 65:525.

POTGEITER, J. T. — 1944. The Argentine ant. Farming in S. Africa. 19(223): 631-632, 664. Biol. Abst. 19:1954, 1945.

PRICE, W. A. — 1945. The Allegheny mound ant and its control J. Econ. Entomol. 38(6):706.

PRICER, J. L. — 1908. The life history of the carpenter ant. Biol. Bul. 14:177-217.

QUAYLE, H. J. — 1938. Insects of Citrus and Other Subtropical Fruits. Comstock Publ. Co. 583 p.

RACE, S. R. — 1964. Industrial colony control of the western harvester ant, *Pogonomyrmex occidentalis.* J. Econ. Entomol. 57(6)860-864.

RACHAESKY, S. — 1974. The Pharaoh Ant. Pest Control. Feb. 1974. p 43.

RILEY, C. V. — 1889. The little red ant. Insect Life, 2(4):106-108.

ROBINSON, F. A. & E. OERTEL — 1950. Chlordane for control of Argentine ants. Amer. Bee Journal 90(9):406-407.

SCHMITT, J. B. — 1947. A recommended cure of the ant problem. Pests 15(10):50, 52.

SCHREAD, J. C. — 1947. Progress report on chlordane for ant control. Pests 15(11):33. 1948. Control of soil insects. J. Econ. Entomol. 41(2):318-324. 1949. A new chlorinated insecticide for control of turf-inhabiting insects. J. Econ. Entomol. 42(3):499-502. 1949. Control of ants. Conn. Agr. Expt. Sta. C. 173. 8 p.

SCHREAD, J. C. & G. C. CHAPMAN — 1948. Control of ants in turf and soil. Conn. Agr. Expt. Sta. Bull. 515.

SIMEONE, J.B. — 1954. Carpenter ant control. College of Forestry, Syracuse University, N.Y. Bull. No. 34.

SMITH, F. — 1944. Nutritional requirements of *Camponotus* ants. Annals Entomol. Soc. Amer. 37:401-408.

SMITH, M. R. — 1928. The biology of *Tapinoma sessile* Say an important house-infesting ant. Annals of Entomol. Soc. Amer. 21:307-329. 1936. Distribution of the Argentine ant in the United States and suggestions for its control. U.S.D.A. Circ. 387. 1936a. Consideration of the fire ant. *Solenopsis xyloni,* as an important southern pest. J. Econ. Entomol. 29:120-122. 1939. The Texas leafcutting ant *(Atta texana* Buckley) and its control in the Kisatchie National Forest of Louisiana. Southern Forest Expt. Sta. Occasional Paper 84. 1943. Ants of the genus *Tetramorium* in the United States with the description of a new species. Proc. Entomol. Soc. Wash. 45:1-5. 1947. A generic and subgeneric synopsis of the United States ants, based on the workers (Hymenoptera: Formicidae). Amer. Midland Nat. 37:521-647. 1965. House-infesting ants of the Eastern States. U.S.D.A. Tech. Bull. No. 1326.

SMITH, R. C. — 1928. *Lasius interjectus* Mayr. (Formicidae) a household pest in Kansas. Kans. Entomol. Soc. J. 1(2):14-18. 1934. A summary of published information about Pharaoh's ant, with observations on the species in Kansas. Trans. Kans. Acad. of Sci. 37:139-149.

*PYRAMID ANTS
INCLUDED*

SNELLING, R. R. — 1973. The ant genus *Conomyrma* in the United States (Hymenoptera: Formicidae). Los Angeles City Museum Contrib. Sci. 238.

SNYDER, T. E. — 1957. Kinds of carpenter ants and their importance. Terminix tech. Paper #3.

SPENCER, H. — 1941. The small fire ant *Wasmannia* in citrus groves. A prelimliminary report, Florida Entomologist, 24(1):6-14.

SRIVASTAVA, B. G. & H. R. BRYSON — 1956. Insecticidal seed treatment for the control of the thief ant. J. Econ. Entomol. 49(3):329-333.

ST. GEORGE, R.A. — 1944. Tests of DDT against ants and termites. J. Econ. Entomol. 37:140.

SWEETMAN, H.L. — 1945. The residual toxicity of DDT, influence of moisture and temperature on the residual kill of DDT. Soap & S. C. 21(12):141, 143, 145, 147, 149.

TALBOT, M. — 1943. Population studies of the ant *Prenolepis imparis* Say. Ecology, 24:31-44. 1943. Responses of the ant *Prenolepis imparis* to temperature and humidity changes. Ecology 24(3):345-352.

TITUS, E. G. — 1905. Report on the "New Orleans" ant, *Iridomyrmex humilis* Mayr. U.S.D.A. Bur. of Entomol. Bull. 51:78-84.

TOWNSEND, L. H. — 1945. Literature of the black carpenter ant. Kentucky Agr. Exp. Sta. Circular 59. March.

TRAVIS, B. V. — 1939. Tests of soil treatments for the control of the fire ant, *Solenopsis geminata* (F.) J. Econ. Entomol. 32:645-50. 1939. Poisoned-bait tests against the fire ant, with special reference to thallium sulfate and thallium acetate. J. Econ. Entomol. 38:706-713. 1941. Notes on the biology of the fire ant *Solenopsis geminata* (F.) in Florida and Georgia. Fla. Entomol. 24(1):15-23. 1943. Further tests with thallium baits for control of the fire ant. J. Econ. Entomol. 36:56-58.

USDA (1980). Guidelines for the control of insect and mite pests of foods, fibers, feeds, ornamentals, livestock, households, forests, and forest products. USDA Agr. Handbook No. 571.

VAN BODEGON, A. H. — 1941. Ants that feed on rubber, a domestic pest Trop. Nature 30(10-11):161-163. Rev. Appl. Entomol. A 30:273, 1942.

WANG, J. S. & J. S. BROOK — 1970. Toxicological and biological studies of the odorous house ant, *Tapinoma sessile,* J. Econ. Entomol. 63(6):1971-1973.

WALKER, H. G. & L. D. ANDERSON — 1937. Control of the pavement ant attacking eggplants. J. Econ. Entomol. 30:312-314.

WALTER, E. V., L. SEATON & A. A. MATHEWSON — 1938. The Texas leaf-cutting ant and its control. U.S.D.A. Circ. 494.

WALTER, E.V. — 1959. Carpenter ants in a pantry door. Pest Control 27(10):96.

WEBER, N. A. — 1937. The sting of an ant. Amer. J. Trop. Med. 17:765-68. 1939. The sting of the ant, *Paraponera clavata.* Science 89(2302):127-128. 1966. Fungus-growing ants. Science 154(3736):587-604.

WHEELER, W. M. — 1905. The North American ants of the genus *Liometopum* Bull. Amer. Mus. Nat. His. 21:321-333. 1906. On certain tropical ants introduced into the United States. Ent. News 17:23-26. 1906. *Monomorium destructor.* Entomol. News 17:265. 1906. The habits of the tent-building ant *(Crematogaster lineolata* Say) Bull. Amer. Mus. Nat. Hist. 22:1-18. 1907. The fungus-growing ants of North America. Bull. Amer. Mus. Nat. Hist. 23:669-807. 1908. Honey ants, with a revision of the American Myrmecocysti. Bull. Amer. Mus. Nat. Hist. 24:345-397. 1910. Ants. Columbia Univ. Press, New York. pp. 663, 286 figs. 1923. Social Life Among the Insects. Harcourt, Brace

Amer. 23:1-26.

WILDERMUTH, V. L. & E. G. DAVIS — 1931. The red harvester ant and how to subdue it. U.S.D.A. Farmer's Bull. 1668.

WILSON, E. O. — 1953. On Flanders' hypothesis of caste determination in ants. Psyche 60(1):15-20. 1963. Pheromones. Sci. Amer. 208(5):100-106, 108, 110, 112, 114. 1965. Chemical communication in the social insects. Science 49:1964-1971.

STANLEY G. GREEN

Dr. Stanley Green received his bachelor's and master's degrees from the University of Colorado. For his disserttation on the taxonomy of the family _Oribatulidae_ he received his _Ph.D. from The Colorado State University, Fort Collins, Col. in 1969._

Green is responsible for vector control and urban pest control programs in Pennsylvania, providing training to public health professionals and pest control operators. Green teaches a number of broad-ranging courses, including basic entomology and structural pest control. In addition, he teaches in-depth courses for specialists, including mosquito and termite control courses. Green's courses and workshops have played an important role in developing professionalism in the pest control industry by encouraging a more scientific approach to pest control. A _Vector Control_ and _Pest Control_ newsletter are published by Green to help accomplish this goal.

As an extension entomologist, Green is responsible for the preparation of the popular _Extension Entomological Fact Sheets_ on household arthropod pests. These are aimed at the general public to help them better understand arthropod pests and how to deal with them.

Green is well known for his selfless attitude toward those who need advice on pest problems. Many professionals have benefitted form his wise counselling and no problems are too big or too small to capture his interest. Green's availability and ability to communicate have made him a key resource for the news media when pest stories need professional input.

Green's achievements have been recognized by numerous honor societies, including Sigma Xi, Epsilon Sigma Phi and Pi Chi Omega. In addition, he is active in many professional associations including the Entomological Society of America, the American Mosquito Control Association, the American Registry of Professional Entomologists and the American Associaiton for the Advancement of Science. Green is also active in many state and regional associations and is currently president of the Mosquito and Vector Control Association of Pennsylvania.

CHAPTER FIFTEEN

Bees and Wasps

Revised by Stan Green[1]

the honey bee is sad and cross
and wicked as a weasel
and when she perches on you boss
she leaves a little measle
　　　　—Don Marquis.

AS A RULE, bees and wasps are beneficial insects unmindful of the activities of man, as long as man makes it a point to disregard them. At times, however, the nests of these insects may be built in such close proximity to the home, or even in the home, as to make the area too confining for both insect and man. Although the stings of bees and wasps are usually a painful experience, for a few individuals the consequences may be much more serious, resulting in a severe reaction or even death.

The order Hymenoptera consists of sawflies, horntails, wasps, ants and bees. There are some 113,000 known species in this order and more than 17,000 are found in North America. The fossil record for these insects, according to Busvine (1980), goes back to the middle of the Triassic period (about 180 million years ago).

While many hymenopterans do not sting man, members of the Sphecidae (of which the cicada killer is a member), Vespidae (paper wasps, yellow jackets, hornets) and Apidae (honeybees and bumble bees) do.

THE HONEYBEE

Apis mellifera L. (Family Apidae)

It is to the honeybee, one of man's oldest insect friends, that we are indebted for honey, beeswax and most important of all, the fertilization of many of our cropbearing plants. The honeybee is a social insect living in large colonies of from 20,000 to 80,000 individuals (Ribbands, 1953). There are a number of races of bees, which vary in pugnacity. Italian bees (*Apis mellifera ligusta*) are generally gentle creatures, whereas German bees (*Apis mellifera mellifera*) are the

[1]*Associate professor of Entomology, The Pennsylvania State University; Cooperative Extension Service, Philadelphia, Pa.*

reverse, and the hybrid German-Italian bee is quite often "all stings." Nevertheless, it should be noted that even the normally gentle Italian bee, when provoked, will unsheathe its dagger. The weather often affects the temper of bees, and on windy, cloudy days when they are unable to forage for nectar, pollen, etc., they are somewhat cross, or "frustrated" as the psychologists put it, and they may "take it out" on some innocent passerby.

Another honey bee with a nasty disposition is the Africanized honey bee, *Apis mellifera adansonii*, also known as the Brazilian honey bee. This hybrid resulted when African bees brought to Brazil in 1956 escaped and bred with the native bees. The African bees were imported to improve production in the bee keeping industry. African bees are very industrious, foraging earlier in the day and working longer in the evening. They also can work at higher or lower temperatures and thus produce more honey per year than the European strains. However, they are very aggressive, sting with little provocation and pursue their victims up to 328 feet/100 m (Italian bees will normally only pursue about 33 feet/10 m). Presently the Africanized bees are widespread in South America. It is estimated they could reach northern Mexico in nine to 13 years. Opinions vary as to what will happen to the American bee keeping industry if and when this bee reaches us. It is hoped that in its migration northward the Africanized bee will breed with the gentler European bees and its aggressiveness will be reduced.

The castes. Three types of individuals or castes can at one time or another be found in a honeybee colony including the queen (a fertile female), worker (infertile female) and drone (male). There is only one egg-laying queen in a hive. The bulk of the colony consists of workers who build and repair the hive, forage for nectar and pollen, produce wax and honey, feed the young and protect the hive against enemies. The males have but one purpose in life and that is to mate with virgin queens. Once they have done this they are no longer permitted to return to the colony. Drones buzz ferociously, but lack a sting and are entirely harmless.

The bee sting. Most individuals who fear bees, do so because of their potent sting. When the bee stings, the sting, poison sac and several others parts of the bee's anatomy are torn from the bee's body. It soon dies, a fact that offers little relief to the individual who is stung. The action of the sting takes place almost instantaneously. The sting has barbs on it, and if it is not immediately removed, the reflex action of the muscles attached to the sting drive it deeper and deeper into the skin, thus permitting more time for the discharge of poison from the poison sac. The pain from the sting is augmented by the discharge of the toxin.

Bee venom is complex. Frazier (1969) says bee venom contains:

1) Histamine. Moreover, the venom also causes more histamine to be produced by the tissues of a person who is stung.

2) At least eight other components (fractions) which have been detected by chemical fractionating techniques. Two of these are very active. Fraction F_1, called melitin, contains 13 amino acids and is responsible for local pain and inflammation, lowering of blood pressure and a paralyzing effect on nerves. Fraction F_2 contains 18 amino acids, plus two enzymes, hyaluronidase and phospholipase. This fraction supplements the action of melitin and in addition causes the destruction of red blood cells.

Different individuals are affected in different ways by bee stings. Some of the factors that make for this variation are the part of the anatomy that is stung, the amount of poison that has entered into the system and the natural immunity

Fig. 15-1. (Left to right) Drone, queen and worker of the honey bee.

of the individual. The actual pain from the bee sting is of short duration and it is the after effects, the swelling and itching, that are the most disturbing. Some individuals are naturally immune and do not swell, while others are so badly affected by a bee sting they may be confined to bed for a number of days. In some instances, the sting of a bee may result in red blotches on the skin, nausea, fainting and even death.

The sting of the bee, as was previously mentioned, has barbs on it, and thus remains in the skin. At times, complications may result from the sting being embedded in the skin and for this reason an effort should be made to remove the entire sting.

Reactions to stings. When stung, most people suffer nothing more than initial pain at the sting site followed by the development of a wheal. Itching and heat may be experienced for a few hours.

Multiple stings may present a more serious problem. Frazier (1969) notes about 500 stings (within a short time) may be fatal. However, he notes people have survived many more stings. He reports on a 30 year old male who was viciously and persistently attacked by bees (probably African honey bees) while he was submerged in water. He put his shorts over his head for protection, but the bees stung through the material. He then plastered his shorts with mud, leaving a small hole through which to breathe. The bees found this hole. The only thing he could do was to bite the bees as they entered through the hole. Amazingly, this man (who was in the water for 4½ hours) survived having sustained 2,243 stings. He did not develop an allergy to bee venom, having been stung later with no ill effects.

Frazier (1969) classifies people's reactions to stings as follows:

1) **Normal Reaction** — There is pain for a few minutes at the sting site and swelling which subsides in a few hours followed by itching and heat.

2) **Local Reaction** — Here there is an unusual amount of swelling which may persist for several days. An entire limb may become swollen. Local reaction can be serious if the sting and subsequent swelling is on the throat. Since the sting is not sterile, secondary infection can also occur and complicate things.

3) **Toxic Reaction** This occurs with multiple stings. Symptoms may include any of the following: headache, fever, fatigue, diarrhea, vomiting, unconsciousness, muscle spasms and sometimes convulsions.

4) **Generalized Allergic Reaction** — Allergic reactions may range from slight to serious shock. Symptoms range from a dry hacking cough, sensation of constriction in the throat or chest, wheezing, rapid pulse, drop in blood pressure, paleness or bluish cast to the skin, skin rash, a sense of impending disaster, loss of bladder or bowel control, dizziness, confusion and loss of consciousness. Anyone who suffers allergic reaction should seek medical help and desensitization treatments, since death can occur within 15 to 30 minutes of being stung.

Parrish (1959) analyzed 215 deaths reported from venomous animals in the United States for the years 1950-1954. Eighty-six deaths were from the stings of Hymenoptera (40%); 39 from spiders (18%); and 71 from poisonous snakes (33%). The rest (19) were from other animals such as scorpions (5); coelenterate (probably a Portugese man-of-war (1); sting ray (1); and 12 deaths were from unknown animals.

Marshall (1957) notes "repeated stinging over a period of time produces an immunity in about 75 percent of individuals; the actual prick remains painful, but the local reaction is transient. About 10 percent of persons are immune from birth; 15 percent never develop immunity, but continue to react locally. Acquired immunity is not permanent; it disappears in a few months if fresh stings are not received."

Marshall continues, "In about two percent of persons, repeated stinging creates a state of hypersensitivity in which each fresh sting produces reactions of increasing severity. In some persons there is only an intensification of the customary local reaction, but in others the reaction is generalized in character, producing alarming symptoms akin to those of anaphylactic shock; in a few of these persons the final sting results in death. Some develop hypersensitivity after their first sting, but in others it arises after a series of normal reactions, or even after some degree of immunity has been acquired. In a few, hypersensitivity appears to be inborn; their first sting results in death.

"Reactions to the sting of a wasp appear to differ little from those of bees. The only significant difference is infection. Unlike bees, wasps and hornets are scavengers and likely to transmit infection along with their venom."

Marshall recommends the following treatment for those individuals who are very sensitive to the stings of bees and wasps.

"Adrenaline remains the best drug in the treatment of acute systemic symptoms. An intravenous dose of five to 15 minims (0.3 to 1 ml) of 1:1000 adrenaline should be injected slowly and repeated as is necessary. The use of antihistamine drugs is based upon the allergic interpretation of the generalized symptoms. Diphenydramine ('benadryl') by mouth has been used with good effect, but at other times it has failed. Absorption by this route is too slow to combat the rapid effects of the poison. A delay, of minutes, in the action of the drug can lead to a fatal outcome; effective drugs must be given parenterally. In any event, it is better that antihistamines be used only in addition to adrenaline."

In some cases it is possible to desensitize individuals who are allergic to the venom of bees and wasps. "This treatment consists of giving the patients shots of diluted venom, gradually increasing the strength over a period which may be as long as six months or more. Thus, the patient builds up a tolerance to the stings."

Treatment for a bee sting. Once the sting penetrates the skin, the pain may become intense, and the natural tendency is to grab the sting and yank it out. In this operation, the tyro seizes the bulb of the poison sac and squeezes it as though it were a syringe, thereby forcing more of the toxin into the wound. This may be avoided by using the finger nail or a knife blade to scrape the sting from the skin without mashing the poison sac. Under no circumstances should the affected part be rubbed, since this merely promotes the flow of venom into the blood, thereby increasing the area of irritation.

The wound should be cleaned with soap and water and an antiseptic then applied. A cold compress should relieve pain and reduce swelling.

CONTROL OF HONEYBEES

Since honeybees are beneficial insects they should not be indiscriminately destroyed. In the spring, swarms of bees are frequently seen on objects, trees, bushes, etc. The swarm consists of the original queen who founded the colony and many thousands of workers. Swarmers are mild in nature and desire only to find a place in which to relocate their nest. The swarm is merely resting temporarily when it is seen and will disperse in a short time (anywhere from 30 minutes, to a few days if the weather is adverse). Either leave swarms alone or contact a local bee keeper. He should be interested in removing the swarm and thus acquiring another colony. While bees are almost always mild during the swarming stage, in order to avoid unforeseen incidents, it is advisable to wear a bee veil and to tie the cuffs of the pants tightly to the ankles. Where the swarm has settled on a limb, it may be deposited in a cardboard box by giving the limb a sharp jerk, repeating this several times to remove all the bees from the limb. One should be sure the queen does not remain on the limb. The cardboard box is then closed and the opening sealed to prevent the bees from escaping. After removing the box to a point some distance from where the swarm originally occurred, it may be then safely opened, permitting the bees to emerge. If the bees are to be returned to a hive, the box is completely sealed and moved at night to where the hive is located, and the bees transferred to the hive on the following day.

If bees should nest in the wall voids of a home they should be controlled or they will enter the living areas. The removal of a bee colony once it is established inside a wall void, chimney or attic is difficult and time consuming. Essentially what is done is that worker bees are removed (via trapping) from the colony. They are induced to nest nearby the old nest site and given a new queen. The queen and developing brood still inside the old nest die and the workers now established in the new nest are permitted to return to the old nest where they will remove any remaining honey. This process, published by Stranger (1967), takes many weeks for completion and is described by Ebeling (1975).

Since the above method is so time consuming and it would be most difficult to find a bee keeper willing to attempt it, most bee colonies in the walls of buildings are treated with insecticides. A variety of insecticides are effective including bendiocarb, carbaryl, diazinon, malathion and propoxur. The dust formulation of these products is preferable to spray formulations when bee and wasp nests are in enclosed spaces. Dust has the advantage of being easily and widely distributed by the movement of the insects as they enter and move about in the nest. Sprays, on the other hand, kill only those individuals which contact the sprayed surfaces. Further, nest contamination does not occur. Usually one application is sufficient with activity ceasing in one to two days, but retreating should be carried out if necessary. Nests should be treated at night if possible to minimize the risk of being stung.

Once the bees are killed, the wall in which they have been nesting should be opened and the comb removed. If not, the untended honey will run down and through the walls and attract such insects as wax moths, carpet beetles, cockroaches, meal moths and various cereal beetles.

THE CARPENTER BEES

Xylocopa spp. (Family Anthophoridae)

These large, attractive-looking bees with a blue-black, green or purple metallic sheen, often burrow into the exposed dry wood of buildings, telephone poles, fence posts and bridges. Anon. (1958) notes carpenter bees bore holes in wooden water tanks resulting in costly leaks. Since the burrow of one bee may be more than 12 inches/30 cm long, and since the bees often colonize in the same piece of wood, the damage to timber can be quite extensive. Carpenter bees resemble bumble bees, but they are not social insects, and most of the top part of their abdomen is without hairs.

Xylocopa orpifex Smith, the mountain carpenter bee, is ½ to ⅔ inch/12 to 17 mm long, and is black in both sexes. It is recorded from the western United States and lower California. Essig (1926) notes these bees may colonize close

Fig. 15-2. Adult female of the valley carpenter bee.

Fig. 15-3. Damage done to wood infested by carpenter bees.

to one another. A single egg is deposited in a cell containing honey-pollen, or bee bread. The burrow is partitioned into a number of such cells. The larval period extends from 37 to 47 days and the pupal period requires 15 days. This bee was observed damaging a redwood beam in an outdoor fireplace in Los Angeles. Hurd (1955) records this bee will nest in structural timbers.

The valley carpenter bee, *Xylocopa varipuncta* Patton, is ³/₄ to ⁴/₅ inch/18 to 20 mm long, and the female is black with metallic reflections, while the males are tannish in coloration. This insect is reported from various trees, telephone poles and house timbers. The valley carpenter bee occurs at lower altitudes in Arizona, California and lower California.

Chandler (1958) made a study of the carpenter bees and the following notes are from his paper. The seven species of carpenter bees in the United States may be transported outside their normal range in lumber and large-stemmed plants. The males are at times annoying because they will fly around the heads of humans. Since they lack a sting, they are entirely harmless. The females possess a potent sting which they use very rarely. In Chandler's opinion, *Xylocopa virginica* (L.), an eastern species, is probably the most destructive carpenter bee in the United States. The males and females overwinter in old nest tunnels and survive the winter if it is not too severe. The adults emerge in spring, mate and then the female deposits her eggs. The larvae and pupae develop during the summer and the adult bees emerge in late summer.

"The nesting gallery of *X. virginica* may be excavated in either a vertical or horizontal plane. The female may use a previously existing gallery without further boring; she may lengthen an existing gallery; or, she may bore a new gallery, either in its entirety or from a common entrance tunnel."

Galleries usually average four to six inches/10 to 15 cm in length, but galleries used by a number of bees may go up to 10 feet/three m in length. The female excavates the gallery by means of her mandibles. She can excavate one inch/25 mm in six days.

The female furnishes her nest with "bee bread" (a mixture of pollen and regurgitated nectar); she lays an egg on top of it and closes the cell with chewed wood pulp. There may be six such sealed cells in a linear row in one gallery. Rau (1933) showed that *X. virginica* requires two days for the egg, 15 days as a larva, four days as a prepupa and 15 days as a pupa.

While the damage to wood from the drilling activities of a pair of carpenter bees is slight, the activities of numerous bees during a period of years can cause considerable damage. They often attack such objects as window sills, wooden siding, eaves, railings, outdoor furniture and fences. The entrance hole (about ½ inch/13 mm in diameter) is usually against the grain of the wood. When the tunnel is about one inch/25 mm deep, the bee turns at right angles to the initial hole and then tunnels with the grain.

A carpenter bee infestation is often first detected by finding large amounts of sawdust on the ground below the area being drilled.

CONTROL OF CARPENTER BEES

Puff one percent bendiocarb, five percent carbaryl or two percent diazinon into the entrance holes. Do not plug up these holes immediately. The bees should be allowed to pass freely through the entrance where they will contact the dust and distribute it into the tunnels. Any newly matured bees will emerge through the openings and contact the dust placed there. In the fall, the holes can be filled and the entire wood surface painted or varnished.

WASPS

Chipps et al. (1980) record that at least 40 deaths (due to wasp and bee stings) are reported each year in the United States and this is probably an underestimate.

Fluno (1961) discusses wasps as enemies of man. He states each year more than 10,000 requests come into the U.S. Department of Agriculture regarding wasps and their control. "About half of these inquiries are accompanied by statements asserting that either the inquirer or one or more members of his immediate family have suffered severe reactions from the sting of a wasp."

As in the case of bees, individuals stung by wasps may show a variety of reactions ranging from no evident effect to anaphylactic shock resulting in death. O'Connor et al. (1964) investigated 88 deaths from wasp stings. "In 36 reports with sufficient data, 17 victims knew of previous severe reactions of stings, 18 were positive of no previous severe reactions, and one was positive of no previous sting. Thirteen of the 88 victims were aware of some serious heart condition, and in 10 others, severe heart damage was noted in postmortem examinations." Thus, it is believed many human deaths may wrongly be attributed to "heart attacks" or "heat strokes" when actually they result from the stings of venomous insects.

Fluno (1961) is of the opinion many auto accidents result from wasp stings because either they distract the driver or they so affect him that he loses control of the car. Such reports appear frequently in newspaper reports.

Of the thousands of described wasp species only a relatively few are of concern to the pest control operator. He receives calls from the public about these wasps either because they are large and "vicious" looking, or they are nesting in the ground, under the eaves or in the walls of their homes. The species of concern to the pest control operator include three types of solitary wasps (mud daubers — various species; the digger wasps — *Campsomeris tolteca* and *Scolia dubia;* and the cicada killer — *Sphecius speciosus*) and a number of social wasps belonging to the family Vespidae (hornets, umbrella wasps and yellow jackets).

MUD DAUBERS (Family Sphecidae)

These wasps are so called because they construct "nests" or brood chambers from mud. These clusters of mud are attached to the walls of buildings. The female mud dauber collects spiders which she stings and paralyzes and then places inside the mud chambers. She then deposits an egg on one of the spiders and leaves and closes the chamber. The young larval wasp hatches and feeds on the spiders provided. It later pupates and changes to an adult wasp which emerges from its mud chamber.

People become concerned when they find the clusters of mud on their homes. If the mud nests have holes in them it means the wasps have completed their life cycle and have left. Control is not necessary since mud daubers rarely sting and are beneficial in getting rid of unwanted spiders.

DIGGER OR SCOLIID WASPS (Family Scoliidae)

Ebeling (1975) describes a scoliid *Campsomeris tolteca* (Saussure) common in southern California and also found in Arizona, Texas, Mexico and Haiti. This wasp flies above lawns which are infested with beetle grubs. These grubs are food for the wasp larvae. The wasps are about one inch/25 mm long with a black head and thorax. The abdomen is pale red with black markings. The wings are colorless.

Scolia dubia is common along the Eastern coast, but also found as far west as California. The wasp flies low over the lawn and can easily be identified by the following characteristics:;

- Wings are a blackish/purple color.
- Head and front part of the body are black.
- First half of the abdomen is black, the second half is reddish with a pair of yellow spots on the top of the abdomen separating the black half from the red half.

These wasps appear in the morning and fly over the lawn all day, leaving in the early evening.

Digger wasps are actually beneficial. They are looking for beetle grubs (*Scolia dubia* is looking for green June beetle larvae). When one is found, they sting and paralyze it, construct a chamber in the soil for it and deposit an egg on it. Their young feed on these paralyzed grubs.

These wasps generally do not attack people. In fact, you can walk safely through them as they hover over the lawn. They disappear at the end of the season and help to control the beetle grubs. If control is desired, the lawn can be sprayed with carbaryl or the grubs controlled with chlorpyrifos or diazinon.

CICADA KILLER (Family Sphecidae)

This is a very large wasp; up to 1 ^{9}/16 inches/40 mm long. It has a black abdomen with yellow markings on the first three abdominal segments. Adults are seen in late July and August. They dig holes in lawns and may tunnel as much as six inches/15 cm deep and another six inches/15 cm horizontally. Burrows have piles of dirt piled up at their entrances. The wasp then locates a cicada, stings it to paralyze it and brings it back to the burrow. One or two cicadas may be placed in the burrow and an egg deposited on one. The wasp larvae feed on the paralyzed but still living cicada.

Control may be accomplished by putting dust (1% bendiocarb, 5% carbaryl, or 2% diazinon) into each hole if the infestation is not too widespread. If the entire lawn is involved, a spray with the same insecticides would be more practical.

UMBRELLA WASPS *(Polistes* spp.) (Family Vespidae)

Polistes wasps commonly have been called paper or paper-making wasps. This name is not appropriate since yellow jackets and hornets also construct nests of paperlike material. Ebeling (1975) proposes they be called umbrella wasps because their nest has the shape of an inverted umbrella.

Nests are usually small (when compared to yellow jacket or hornet nests)

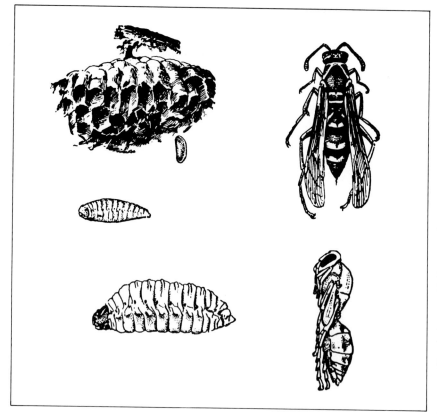

Fig. 15-4. Nest, larvae, pupa and adult of the umbrella wasp.

containing only up to about 250 wasps. The nest consists of a single comb with no paper envelope around it.

A good account of the life cycle of *Polistes* is given by Evans and West-Eberhard (1970). Mated females overwinter in sheltered areas. All females are potential queens — there is no worker caste. Which one actually becomes the queen, or nest initiator, seems to depend on which one begins laying eggs and building cells first. The other females take the subservient role of foraging for food and feeding the young.

Evans and West-Eberhard (1970) consider why one female becomes the dominant egg layer when more than one female begins laying eggs at the same time. It seems a contest is begun with females laying eggs and eating the eggs of the others, replacing eaten eggs with those of her own. The one which out-eats the others becomes the dominant female.

Nests are often built under the eaves of homes and while these wasps are not particularly aggressive, their proximity to the living area threatens many people. Nests are easily destroyed by spraying at night with any of the following: bendiocarb, carbaryl, chlorpyrifos, diazinon, dichlorvos, pyrethrins or resmethrin. Knocking down the nest without spraying is worthless since the wasps will only rebuild it.

YELLOW JACKETS AND HORNETS (Family Vespidae)

The European or giant hornet *(Vespa crabro)*, bald or white faced hornet *(Dolichovespula maculata)* and the yellow jacket *(Vespula* sp.) are the real problems of the wasp world, as far as humans are concerned. These social wasps live in colonies which number thousands of individuals. These beneficial insects would not anger or threaten man except that they have adapted themselves to living very closely with us. They take advantage of us by sometimes nesting in the wall voids and attics of our homes where they can go unnoticed all season. They leave the nest area flying outdoors in search of food. However, in the fall, when food becomes scarce and the temperature outdoors cools, they frequently find their way into the living areas of the home. Then they have become a health threat to the home's occupants and must be controlled.

Yellow jackets "like" to join us in our recreational activities (cookouts, camping, at swim clubs, picnics) where they feed in trash containers and attempt to share our food. They love our soft drinks, chicken and hamburgers apparently as much as we do, and they resent our efforts in keeping them from sharing our food.

Life cycle of social wasps. In the north temperate zone, wasps undergo an annual cycle. Only queens mated in the fall or early winter survive. They spend the winter in protected places such as under bark, stones, shingles and in abandoned rodent nests. In the spring, these queens establish a colony laying from 10 to 20 eggs. Since there are no workers present to help her raise her brood she alone has the burden of foraging for food, feeding the young and collecting wood from which she manufactures the paper used in nest construction. With the production of this first brood of workers, the queen gives up all her duties except that of egg laying. She remains in the nest and the workers forage for food and wood, feed the young, and enlarge, repair and defend the nest. At the end of the summer nests have multiple combs, thousands of cells and thousands of workers. Then, the colony begins to produce males and new queens. These fly away and mate. The males soon die as do the workers and the original founding queen.

Whereas the adults feed mainly on fruit juices and other sweet materials, the

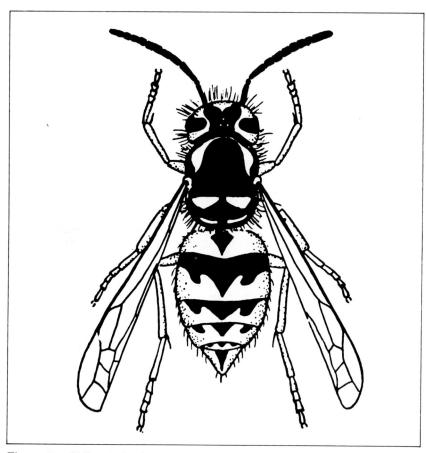

Figure 15-5. Yellow jacket, *Vespula pensylvanica* (Saussure).

larvae feed on animal materials such as insects. When the adults are able to obtain pieces of meat for the larvae, they do so. This flesh is chewed and conditioned by the workers prior to being fed to the larvae. The workers apparently do not feed the larvae out of "love." There is an exchange of food material — the larvae being given protein food by the adult and in return secreting a material containing sugar which is relished by the adult. This exchange of material was named trophallaxis by Wheeler (1918).

The yellow jackets and hornets belong to the family Vespidae (along with the umbrella wasps previously discussed). There are 19 species of these wasps found in North America. The identification and distribution of these species is given by Miller (1961) and Jacobson et al. (1978). Some of the most common and important species are:

- *Vespa crabro*. The European, brown or giant hornet is a large wasp nesting in attics and often attracted to light during summer evenings. It is found in 10 eastern states and Alaska.

- *Vespula consobrina.* A black and white "yellow jacket" recorded in 22 states.
- *Vespula squamosa.* Recorded in 24 states.
- *Vespula vidua.* Collected in 21 states.
- *Vespula vulgaris.* Collected in 26 states.
- *Vespula maculifrons.* Collected from 35 states.
- *Vespula pensylvanica.* The western yellow jacket recorded in 12 western states.
- *Vespula flavopilosa.* A recently described species (Jacobson et al. 1978) collected in 24 states.
- *Vespula germanica.* MacDonald et al. (1980) cites Morse et al. (1977) as saying this species was periodically introduced into northeastern North America and became established about 1968. It has thus far only been found in 10 states, but is spreading rapidly. This yellow jacket is a particular problem because it frequently feeds on human foods and prefers nesting inside structures. DeJong (1979) reports this wasp is a potential threat to the bee keeping industry. Thomas (1960) claims entire colonies of bees can be destroyed by overwintering workers. (In New Zealand they have perennial colonies).
- *Dolichovespula arctica.* This black and white wasp doesn't build its own nest but is a social parasite of *Dolichovespula arenaria.* No worker cast is produced. Queens and males are cared for by the *D. arenaria* workers. *D. arctica* has been found in 21 states.
- *Dolichovespula arenaria.* Sometimes called the yellow hornet, *Dolichovespula arenaria* has been found in 30 states. Their nests are usually constructed in shrubs or trees but they often place them under the eaves of buildings.
- *Dolichovespula maculata.* The bald or white faced hornet is widespread and

Fig. 15-6. Nest of the bald-faced hornet.

Fig. 15-7. Cross-section through nest of the bald-faced hornet.

has been found in 42 states. Their nests are found in trees or shrubs and
they become very large by the end of the summer.

MacDonald et al. (1976) recognizes two groups of typically ground nesting
yellow jackets: the *Vespula rufa* group (this includes *Vespula acadica, V. atro-
pilosa, V. consobrina, V. intermedia,* and *V. vidua*) and the *Vespula vulgaris*
group (this consists of *V. flavopilosa, V. germanica, V. maculifrons, V. pensylvanica* and *V. vulgaris*). The *Vespula rufa* group is not a problem to man since
their colonies are small and they only have between 75 and 350 workers. They
complete their life cycle by early September and feed strictly on insects. They
do not scavenge for human food. The *Vespula vulgaris* group does scavenge becoming a real nuisance at people's outdoor activities. They produce large populations, often having between 1,000 and 3,000 workers in a nest, and the colony
persists into late autumn.

CONTROL OF YELLOW JACKETS

If the nest can be found, control of its residents is simple. For the aerial nesters
such as the European hornet, white faced hornet and yellow hornet, simply
spray bendiocarb, carbaryl, chlorpyrifos, diazinon or resmethrin into the nest
opening, then wet the surface of the nest envelope. The nest can be removed,
if desired, within a day or two following treatment.

For ground nests or those nests within cavities such as wall voids, stone walls,
etc., a dust formulation of any of the above insecticides is preferable. Workers
entering the nest will track the dust into the nest and contaminate it.

The problem is that all too often the nest from which pest wasps are emerging
cannot be located. Certain yellow jackets have been shown to fly almost 3,000
feet/910 m from their nest in their search for food (Arnold, 1966, cited in Spradbery, 1973). It would be most difficult, if not impossible, to try and track a worker
wasp for this distance even if she returns directly to the nest. Spradbery (1973)
published a letter to a newspaper which gives a method used to locate nests:

"Sir, many years ago I recall being shown a certain method of finding
wasps' nests by the late General Prescott-Decie. The General caught a
wasp on the window pane, covered it with flour and released it out of doors,
saying that it would immediately return to its nest. This it did, zooming
in a straight line to the local river bank, with myself and a friend running
underneath the flour-coated insect.

"Today, 40 years later, I tried the same method. The wasp flew some 30
yards and disappeared into its nest. Possibly some of your country readers,
plagued by wasps, may care to try this. It's a certainty!"

Spradbery (1974) then gives the following replies from two persons who have
tried the above foolproof method:

"Sir, I caught one just now, devouring an expensive peach. I covered it,
and myself, with large quantities of flour. I descended from a great height
to street level and released the insect. It did not return directly to its nest.
It rose vertically and vanished towards the sun. I returned to my flat. That
same white wasp was once again lunching on that same peach. How can
one destroy a creature with a sense of humor?"

"Sir, I caught my wasp and coated it with flour. Then I ran down the garden beneath the whitened creature. Unfortunately it soared over a 10-foot-wall. I was not so agile. When my bruises are healed I shall try again, but this time will use self-raising flour, on myself as well as the wasp."

Recently, studies have been conducted to try and control yellow jackets over relatively wide areas using insecticide baits — Keh (1968), Perrott (1975) and Wagner and Reierson (1969 and 1971).

Wagner and Reierson (1969) rated the repellency index of various insecticides considered as candidates for bait formulations. They gave four categories of repellency. Those which were highly repellent included carbaryl, diazinon, dichlorvos, dieldrin, lindane, malathion and propoxur. Those which were moderately repellent included chlordane, chlordecone and ronnel. Showing slight repellency was trichlorfon, and one, mirex, was non-repellent. In fact mirex seemed to have some attractive value. Mirex worked well in controlling the western yellow jacket *(Vespula pensylvanica)*. However, mirex has since been banned and is no longer produced.

A microencapsulated diazinon product, mixed with a meat product such as tuna or mackerel, has been successfully used against some western yellow jacket species, but no success has yet been reported against eastern species. This product is presently registered in 11 western states for use in controlling yellow jackets.

It was once believed the best time to control wasp populations was to trap or kill the queens in the spring as they are flying about foraging for food. However, Richards (1961) mentions a campaign in Berkshire in 1901 where school children were paid to collect queen wasps. It was unsuccessful. He also cites in 1950 the Cyprus Department of Agriculture killed 243,394 queens, and that many queens also were killed by the public. Nineteen-fifty was one of the worst wasp years recorded. Spradbery (1973) states, "The mass destruction of these potential colony founders seems to have virtually no effect on subsequent wasp populations in the succeeding summer months." He further notes even if 99.9 percent of potential queens were eliminated the same number of annual colonies would be maintained.

It appears many queens coming out of hibernation are unsuccessful in establishing colonies. Richards (1961) cites Beirne as showing that the weather in the spring is the important factor in colony establishment. David (1978) cites Akre (1976) as reaching the same conclusion regarding yellow jacket populations in the Pullman area of southeastern Washington.

LITERATURE

BEIRNE, B. P. — 1944. The causes of the occasional abundance or scarcity of wasps *(Vespula* spp.) (Hym. Vespidae). Entomologists' Monthly Magazine. 80:121-4.

BUSVINE, J. R. — 1980. Insects and Hygiene. 3rd Ed. Chapman and Hall, London. 568 pp.

CHANDLER, L. — 1958. 7 species of carpenter bees are found in the United States. Pest Control 26(9):36, 38, 47. 1965. Coop. Econ. Ins. Rpt. 15:1080.

CHIPPS, B. E., M. D. VALENTINE, A. KAGEY-SOBOTKA, K. C. SCHUB-

ERTH and L. M. LICHTENSTEIN — 1980. Diagnosis and treatment of anaphylactic reactions to Hymenoptera stings in children. J. Pediatrics. 97(2):177-184.

DAVIS, H. G. — 1978. In Perspectives in Urban Entomology. Ed. G.W. Frankie and C.S. Koehler. Academic Press, NY. 417 pp.

DeJONG, D. — 1979. Social wasps, enemies of honey bees. Am. Bee J. 119:505-7, 529.

EBELING, W. — 1975. Urban Entomology. Univ. Calif. Press, Berkeley. 695 pp.

ESSIG, E. O. — 1926. Insects of Western North America. Macmillan. 1050 pp.

EVANS, H.E., and M. J. WEST-EBERHARD — 1970. The wasps. Univ. Michigan Press. Ann Arbor, 265 pp.

FLUNO, J. A., 1961. Wasps as enemies of man. ESA Bull. 7(3):117-119.

FRAZIER, C. A. — 1969. Insect Allergy. Warren H. Green, Inc. St. Louis. 493 pp.

HURD, P. D., JR. — 1955. The carpenter bees of California (Hymenoptera: Apoidea). Bull. Calif. Insect Surv. 4:35-72.

JACOBSON, R.S., R.W. MATTHEWS, and J. F. MacDONALD — 1978. A systematic study of the *Vespula vulgaris* group with a description of a new yellow jacket species in eastern North America (Hymenoptera: Vespidae). Ann. Entomol. Soc. Am. 71:299-312.

KEH, B., N. T. BROWNFIELD, and M. E. PERSON — 1968. Experimental use of bait with mirex lethal to both adult and immature *Vespula pensylvanica* (Hymenoptera: Vespidae). Calif. Vector Views. 15:115-118.

MacDONALD, J.R., R. D. AKRE and R. E. KEYEL — 1980. The german yellow jacket (*Vespula germanica*) problem in the United States (Hymenoptera: Vespidae). Bull. Entomol. Soc. Am. 26(4):436-442.

MacDONALD, J. R., R. D. AKRE, and R. W. MATTHEWS, 1976. Evaluation of yellow jacket abatement in the United States. Bull. Entomol. Soc. Am. 22:397-401.

MARSHALL, T. K. — 1957. Wasp and bee stings. The Practitioner. 178:712-722. June.

MILLER, C. D. F. — 1961. Taxonomy and distribution of Nearctic *Vespula*. Canad. Entomologist. Vol. XCIII (Supplement 22):52 pp.

MORSE, R. A., G. C. EICKWORT, and R. S. JACOBSON — 1977. The economic status of an immigrant yellow jacket, *Vespula germanica* (Hymenoptera: Vespidae), in the northeastern United States. Environ. Entomol. 6:109-10.

O'CONNOR et al. — 1964. Death from "wasp" sting. Ann. Allergy. 22:385.

PARRISH, H.M. — 1959. Deaths from bites and stings of venomous animals and insects in the United States. AMA Archives of Internal Medicine. 104:198-207.

PERROTT, D. C. F. — 1975. Factors affecting use of mirex-poisoned protein baits for control of European wasp (*Paravespula germanica*) in New Zealand. New Zealand J.

RAU, P. — 1933. Jungle Bees and Wasps of Barro Colorado Island. Published by Author. 324 pp.

RIBBANDS, C. R. — 1953. The behavior and social life of honeybees. Bee Research Association, Ltd. London. 352 pp.

RICHARDS, O. W., 1961. The social insects. Harper & Bros. NY. 219 pp.

SPRADBERY, J. P. — 1973. Wasps. Sidgewick & Jackson, London. 408 pp.

STANGER, W. — 1967. How to remove bees from buildings. Univ. Calif. Agr. Ext. Svce., One Sheet Answer. No. 161.

THOMAS, C. R. — 1960. The European wasp (*Vespula germanica* Fab.) in New Zealand. DSIR Information Series. 27:74 pp.

WAGNER, R. E., and D. A. REIERSON — 1969. Yellow jacket control by baiting. 1. Influence of toxicants and attractants on bait acceptance. J. Econ. Entomol. 62(5):1192-7. 1971. Yellow jacket control with a specific mirex-protein bait. Calif. Agr. 25(4):8-10.

WHEELER, W. M., 1918. A study of some ant larvae, with a consideration of the origin and meaning of the social habit among insects. Proc. Am. Phil. Soc. 57:293-343.

GEORGE T. OKUMURA

George Okumura is a graduate of California State University. Prior to founding his training institute, Okumura served for 28 years with the California Department of Food and Agriculture. In this Department, Okumura achieved the key position of chief of Laboratory Services with responsibilities in the areas of entomology, nematology, plant pathology, botany and seed.

Noted for his technical expertise on stored product pests, Okumura has been a consultant in this field for 15 years. He is particularly recognized for his knowledge of Khapra beetle and has played a role in helping insure this pest does not become established in the United States.

In addition, Okumura is a gifted educator. Thousands have attended training seminars at the Okumura Biological Institute or have benefitted from his lectures at conferences.

Okumura has authored more than 20 scientific articles and has co-authored several books including <u>Certification Training Manual</u> and <u>Community Pest and Related Vector Control</u>. Most recently he co-authored <u>Stored Product Insects</u>, a USDA publication soon to be released.

CHAPTER SIXTEEN

Stored Product Pests

Revised by George Okumura[1]

$\mathbf{F}$EW THINGS IRRITATE the homeowner as much as finding insects in their cereals and other foodstuffs. This starts a chain reaction which causes trouble for the retailer and, subsequently, the manufacturer of the infested foods.

These insects usually come into the home with some infested package of food. With favorable conditions, they greatly increase in number and may then crawl into hitherto uninfested food. They can enter through extremely small cracks, and in some instances bore directly through the wrapper and container.

An infestation of stored product pests may be recognized not only through the presence of the insects themselves, but by the holes in the package, by webbing in the food material, by insect feces, etc. Almost all of the insects considered in this chapter infest stored grain and cereal products, and some of them are responsible for millions of dollars' damage in the flour mill and warehouse.

Since it is the duty of the Food and Drug Administration to examine samples of foods and drugs for the detection of violations of the Food, Drug and Cosmetic Act, and since it is in their power to condemn contaminated foods (e.g., containing insects and insect excrement), it is essential for the manufacturer of food stuffs engaged in interstate commerce to prevent such foodstuffs from becoming infested. For those interested in the microanalysis of food and drug products for insect and other contaminants, the use of Food and Drug Circular No. 1 (Anon., 1944) is recommended, as well as publications by Harris (1943, 1950 and 1955), Harris et al. (1952), Kurtz et al. (1952 and 1955) and Heuerman (1955). Kurtz and Harris (1962) summarized their studies on insect fragments in food in a well-illustrated volume. Other useful books on stored product pests are the publications by Munro (1966) and Ebeling (1975).

Other stored product pests are discussed in the chapters on hide and carpet beetles, mites, etc.

[1]*Director of Okumura Biological Institute, Inc., Sacramento, Calif.*

CONTROL

The author is primarily concerned with the control of these stored product pests in the home and retail store. When insects are present in the vicinity of foodstuffs, or if the food is infested with insects or shows evidence of their presence, one should seek the major source of infestation. This may be a long-neglected box of meal, crackers, paprika, etc. Old lots of foodstuffs should be inspected frequently and commodities no longer useable should be destroyed, preferably by fire.

Where the infestation is serious, it may be necessary to resort to fumigation with one of the fumigants mentioned in the chapter on fumigation. This control measure should be undertaken only by professional pest control operators, entomologists and other authorized individuals. Appended to the discussion of important stored product insects is a consideration of control methods for mill or warehouse.

Dr. R. T. Cotton, an authority on stored product pests and their control, has discussed the problems of insects and dried food storage (1946), and the following summary is largely from his paper.

Food, moisture and temperature needs of stored-food insects. "Stored-food insects are predominantly of tropical or subtropical origin and thrive under warm, humid conditions. With few exceptions they are unable to resist exposure to low temperature and do not hibernate. They are all adapted to living on food that is low in moisture content, and some of them, for example, flour beetles of the genus *Tribolium,* are capable of breeding in milled cereal products that are practically devoid of moisture. Others have certain minimum moisture requirements below which they are unable to live and reproduce. The rice and granary weevils cannot breed in grain with a moisture content of eight percent or below, and they soon die if restricted to such grain for food. Mites, which are not true insects but are usually classed with this group of stored-food pests, require food of a comparatively high moisture content. They are not troublesome in foods unless the moisture content is above 12 percent." In this respect, Fraenkel & Blewett (1943) note the flour beetle, the saw-toothed grain beetle and the Mediterranean flour moth will grow fairly well at about six percent water content, and the latter will manage at even one percent. However, the drugstore and tobacco beetles, and *Ptinus* require at least 10 percent moisture content.

"Temperature and moisture are the two most important factors in the ecology of these insects. Up to certain limits their rate of reproduction increases in direct proportion to increase(s) in the temperature and moisture content of the food in which they are breeding. Temperatures above 95° F are not favorable for the development of most of these insects, however, and the growth of molds induced by moisture contents above 15 percent appears to interfere with the development of some species. Conversely, temperatures below 70° F greatly retard their development. Food temperatures of 65° F or below, entirely prevent the reproduction of insects such as flour beetles, since they do not lay eggs at these temperatures. The rice and granary weevils lay a few eggs at 60° F, but they develop so slowly at this temperature that they are of little consequence. On the other hand, tyroglyphid mites are reported to breed in foodstuffs at temperatures between 40° and 50° F when moisture conditions are favorable.

"Many of the insect pests of stored foodstuffs are general feeders and feed on nearly all types of dried vegetable or animal matter. Others are more restricted in their diet and have definite food preferences. The principal classes of dried foods, together with their more important insect pests and sources of infestation, are listed below:

Beans, peas, cowpeas, and similar foods. "Bruchid weevils, *Bruchidae*. Infestation starts in field, and with many species is continued in storage.

Cheese. "Cheese skipper, *Piophila casei* (L.); mites, *Tyroglyphus* spp.; ham beetle, *Necrobia rufipes* (DeG.); larder beetle, *Dermestes lardarius* (L.). Infestation may occur in curing rooms or in subsequent storage.

Dried fruit. "Indian meal moth, *Plodia interpunctella* (Hbn.); raisin moth, *Cadra figulilella* (Greg.); sawtoothed grain beetle, *Oryzaephilus surinamensis* (L.); dried fruit beetle, *Carpophilus hemipterus* (L.); mites. Infestation may occur in field while drying, or in storage.

Dried vegetables. "Indian meal moth and miscellaneous general feeders. Infestation occurs between dehydration and packing operations.

Flour and milled cereal products. "Flour beetles, *Tribolium* spp.; Mediterranean flour moth, *Anagasta kuehniella* (Zell.) Indian meal moth, sawtoothed grain beetle, cadelle, *Tenebroides mauritanicus* (L.); flat grain beetle, *Cryptolestes pusillus* (Schönherr); black carpet beetle, *Attagenus megatoma* (F.); mites, *Tyroglyphus* spp.; *Acarus siro* L. Infestation may occur in progress of manufacture, in storage, and during transportation.

Grain (wheat, rice, corn and other cereals.) Rice weevils, *Sitophilus oryza* (L.); granary weevil, *S. granarius* (L.); lesser grain borer, *Rhyzopertha dominica* (F.); cadelle, *Tenebroides mauritanicus;* flour beetles, *Tribolium* spp.; Angoumois grain moth, Sitotroga cerealella (Oliv.); Indian meal moth, *Plodia interpunctella;* mites, *Tyroglyphus* spp.; *Acarus siro;* and miscellaneous general feeders. Infestation occurs in the field in some cases, but more often in farms, warehouses, or elevator storage.

Macaroni and spaghetti. "Rice and granary weevils and milled cereal pests. Infestation occurs after manufacture, usually in drying and packing rooms.

Nut meats and candy. "Indian meal moth, sawtoothed grain beetles, mites. Infestation occurs during manufacture and processing and, in the case of nut meats, in storage before and after processing."

Inspection. It is essential large stocks of stored products, particularly if they are not used or moved at frequent intervals, be inspected regularly for signs of infestation, so control measures can be applied before the infestation becomes serious.

Sanitation. Periodical cleaning to prevent the accumulation of food materials will reduce the possibilities of infestation. Construction that eliminates or reduces crevices will aid in sanitation. The storage of materials on raised pallets or on movable trucks will facilitate cleaning. Screening of windows and doors, and rat and mouseproofing will reduce the entry of pests. Anon. (1956) prepared an inspection check list for grain elevators. Cotton et. al. (1953) and Michelbacher (1953) review the causes of outbreaks of stored grain insects. Wagner (1957) and Anon. (1958) discuss the role of sanitation for the control of insects in flour mills. Henderson (1955) is concerned with insect prevention in food plants. Evans and Porter (1965) consider the role of stored product pests on ships. Anon. (1972) provides a manual for sanitation inspection.

Prevention of insect infestation. "To preserve dried processed food from insect damage it is essential, first, to manufacture insect-free products and, second, to use insect-proof packages. During the processing of many foods, insects are removed mechanically or killed by heat or other agencies. Infestation of these products occurs between manufacture and packing, and results from infestation in the packing rooms or in machinery handling the finished product. Complete freedom from insect infestation in the processing plant should be sought by means of sanitation; by periodic fumigation of the entire plant with methyl bromide, by heat sterilization, and by the monthly local fumigation of machinery in which residues of foodstuffs occur." "When machinery containing residual stock is being fumigated, it is well to run off and set aside enough of the product, after the machinery is restarted, to avoid the immediate use of material containing residual fumigant. The material can be fed back into the machinery later, after it has become thoroughly aerated. Prompt packaging of processed food is essential." Shoenherr (1972) provides a guide to good manufacturing practices for the food industry.

Entoleter. In mills, a device known as the entoleter has been introduced for the mechanical destruction of insects. The manufacturer recommends it be installed at several points in the milling process. Freeman and Turtle (1947) explain the operation of the machine as follows: "Flour spouted to the machine is thrown by centrifugal force between two flat steel discs or plates that revolve on a central shaft at 2,900 r.p.m. (For whole wheat the speed should be 1,450 r.p.m.). Small round hardened steel posts are closely spaced in two concentric rings between the two discs. The impact of the flour against the revolving discs and posts, and against the housing of the machine is so great that all stages of insects and mites, including the egg, that may be in flour, are killed. The treated flour passes out through a spout at the base of the machine." Stateler (1943) notes tests by Cotton and Frankenfeld showed the centrifugal force disrupted the cells of insects, larvae and eggs of such test insects as the rice weevil, lesser grain borer, and flour beetle. Freeman and Turtle (1947) killed a high proportion of all stages of the mill moth, *Anagasta keuhniella;* the confused flour beetle, *Tribolium confusum;* the brown spider beetle, *Ptinus clavipes;* and the grain mite, *Acarus siro(Tyroglyphus farinae DeGeer)* infested flour and flour mixtures. For example, the mite infestation of flour was reduced by passage through the machine from about 60,000 mites per 200 c.c. of flour to only one mite per 200 c.c.

Whitney and Pederson (1962) review physical and mechanical methods of stored product insect control.

Heat treatment. Heat is often used in the "sterilization" of materials to free them of insects and mites. Grossman (1931) showed the red flour beetle, rice weevil, squarenecked grain beetle, Angoumois grain moth, and slenderhorned flour beetle are killed when exposed to 122° F/50°C for one hour. He notes, "The pupae resist a given temperature for a longer period of time than the larvae, which in turn, are able to resist longer exposures than the egg and adult stages. Young adults can resist more heat than older ones."

For the control of insects in cereal products in the home, the U.S. Bureau of Entomology and Plant Quarantine recommends the use of dry heat for the control of insects in packaged foods or in such loose materials as beans, whole grains, dried peppers, nut meats, etc. The packaged foods may be treated in their original containers and the loose materials should be spread in a shallow pan. These foodstuffs should be placed in an oven at 150° F/66° C for 20 minutes. "To keep the oven from getting too hot, prop the door open a few inches. It is a good idea to stir the materials in the trays every now and then."

There are heat sterilizing machines on the market that are used in large establishments that handle foodstuffs. Shepard (1940) says, "Heat is oldest, most common and most satisfactory method of 'sterilizing' products. It may be derived from steam, hot air, or electricity, and should be applied just before the material is placed in the package. Any type of machine that will keep the material in constant motion and heat all portions equally is satisfactory." The cereal should pass from the sterilizer directly to the packer. Sterilization will not prevent insects from re-entering.

Godkin and Cathcart (1949) investigated the effectiveness of heat in controlling insects infesting the surface of bakery products. Some of their conclusions are presented below: "Infra-red heat offers a rapid and effective means of eliminating surface infestations of all stages of stored product insects in finished packaged products, such as fruit cakes and similar items.

"Using an infra-red exposure temperature of approximately 500° F (260° C) it is possible to heat the surface of a finished package product like fruit cake to 140-150° F (60-66° C) within 20 seconds. This short interval of treatment is sufficient to kill any incipient insect infestation present without altering the food product or damaging the cellophane packaging material."

Kenaga and Fletcher (1942) studied the effects of high temperature on several household and stored grain pests, and the table from this paper on the following page summarizes the results of the test.

Cold treatment. Merchants often keep candy, chocolate and similar perishable materials in refrigerated containers to preserve them and prevent insect infestation. Housewives can control insects in cereals and other food products by placing them in the deep-freeze 0° F/−18° C for four days (French, 1964).

Fumigation. Fumigants such as aluminum phosphide gas, and methyl bromide are commonly used for the control of stored product pests. This subject is discussed in the chapter on Fumigation. Winburn (1952) Lindgren et al. (1954), Kenaga (1957) and Krohne and Lindgren (1958) evaluate grain fumigants against stored product insects. Monro and Upitis (1956) in the laboratory developed a strain of granary weevil resistant to methyl bromide. Barnes et al. (1956) discuss enclosures for fumigating dried fruit. Monro (1964 and 1969) wrote a manual on fumigation. Childs and Overby (1976) discuss fumigation of tobacco with ethylene oxide, and Bond and Buckland (1976) used a mixture of methyl bromide and acrylonitrile for fumigation at low temperatures.

Contact sprays including aerosols. Sprays containing pyrethrins and the pyrethrins-synergist combinations in a water or oil base can be successfully used against surface infestations of some insects. Pyrethrum aerosols can also be used for the same purpose. However, these methods give temporary control and must be repeated at frequent intervals. Cotton and Gray (1948) found a 0.8 percent pyrethrins spray in a highly refined oil base to effectively control surface infestations of Indian meal moth. The spray was applied by means of an electric sprayer above the grain in a closed top bin at the rate of approximately six ounces per 1,000 cubic feet/170g per 28 cubic meters. Freeman and Turtle (1947) obtained control of surface infestations in grain of the Mediterranean flour moth and the tobacco or chocolate moth with pyrethrin sprays. Sprays containing pyrethrins and piperonyl butoxide, of which there are a number on the market, are used as contact sprays for stored pests since they are effective and relatively non-toxic to human beings. These sprays are sold as water emulsion or oil base concentrates which must be diluted, or as ready-to-use oil base sprays.

Lloyd and Hewlett (1958) found oil sprays containing 1.3 percent pyrethrins or 0.3 percent pyrethrins plus three percent piperonyl butoxide to be equally effective. Holborn (1957) showed lab strains of the granary weevil to be more

TABLE 16-1

Effects of a temperature change from 80° to 105° F/26.7° to 40.6 °C

Insect	Egg	% Mortality (Approximate) Immature Stages	Adult	Remarks
Blattella germanica (L.)	0	0	0	All stages unaffected.
Periplaneta americana (L.)	—	40	99	1st and 2nd instars active, larger instars moribund. The larger the insect the greater the mortality.
Lioscelis divinatorius Mull.	—	0	0	All stages apparently unaffected. No counts made.
Cimex lectularius (L.)	0-?	0-?	0-?	Same as *L. divinatorius*.
Oncopeltus fasciatus Dall.	20	40	80	All stages desiccated. Eggs hatched 3 days early.
Attagenus megatoma (F.)	—	50	90	The larger the larva the greater the mortality. Cultures where crowding occurred showed a high percent kill.
Trogoderma versicolor Creutz.	—	70	90	Same as *A. megatoma*.
Dermestes sp.	—	50	50	Larvae slightly desiccated.
Oryzaephilus surinamensis (L.)	—	100	100	All stages killed.
Cryptolestes pusillus (Schönherr)	—	0	0	All stages unaffected.
Bruchus sp.	—	0	0	All stages unaffected.
Sitophilus oryza (L.)	100	100	100	All stages killed.
Cynaeus angustus LeC.	—	90	50	In the moist culture all stages were killed and in the dry culture only a few of the larvae were killed.
Tribolium confusum Duv.	0	0	0	All stages unaffected.
Rhyzopertha dominica (F.)	0	0	0	All stages unaffected.
Tineola bisselliella Hummel.	100	90	100	Only a few large larvae survived.
Anagasta kuehniella (Zell.)	100	100	100	All stages killed.

susceptible to pyrethrins than the wild strain. Dove and Schroeder (1955) discuss the protection of grain with emulsions of pyrethrins and piperonyl butoxide. Phillips (1959) used emulsion sprays of 0.3 percent pyrethrins and three percent piperonyl butoxide for wheat stored in ship's holds. Joubert (1962) recommends the use of concentrates containing five percent pyrethrins and 50 percent piperonyl butoxide for the protection of grain. Laudani et al. (1959) used pyrethrum-piperonyl butoxide wettable powders to protect citrus pulp.

Contact sprays are of value when applied in a vacant area before foodstuffs are moved into it since such spraying greatly reduces the potential of future infestations.

O'Farrell et al. (1949) showed highly concentrated pyrethrins sprays have persistent killing power on such materials as jute, cotton and wool.

Vincent and Lindgren (1957) showed malathion to be an effective contact spray against grain insects. Moore (1959) found a 1.5 percent malathion emulsion an effective bin spray. Other common contact sprays are dichlorvos and resmethrin. Spitler et al. (1976) showed malathion to protect walnuts in shells for eight to 12 months against the Indian meal moth and the red flour beetle. Anon. (1964) notes some stored product pests showed resistance to pyrethrins and malathion. Armstrong and Soderstrom (1975) demonstrated malathion resistance in stored product insects, and Bansode and Campbell (1979) found field strains of the red flour beetle resistant to malathion and other organophosphorus compounds.

Residual sprays. When insects are hiding in cracks and crevices or active on surfaces more long-lasting insecticides may be required. Available insecticides include bendiocarb, chlorpyrifos, diazinon and propoxur.

In 1972, amendments to the Federal Insecticide, Fungicide and Rodenticide Act resulted in severe restrictions on use of residual insecticides in food handling establishments. Applications of residual insecticides in food areas of food handling establishments (including food manufacturing, food processing and food service establishments) became, for a few years, limited to a "crack and crevice" treatment only. Food areas were defined to include places where food is exposed during receiving, storage, preparation and serving. Crack and crevice treatments were defined by the Environmental Protection Agency, the lead agency governing use of pesticides, as the application of small amounts of insecticides into cracks and crevices in which insects hide or through which they may enter buildings. Such openings were described as commonly occuring at expansion joints, between different elements of construction and between equipment and floors. Also included were openings leading to voids and hollow spaces in walls, equipment legs and bases. For crack and crevice treatments, applications are made at low pressure to avoid atomization, splashing or run-off which might lead to contamination of food or food-contacting surfaces. Any contamination of an exposed surface with an insecticide limited to crack and crevice use is a violation of the law.

The limitation of residual insecticides to crack and crevice applications proved impractical. Not only was it impossible with existing application equipment to avoid some contamination of exposed surfaces adjacent to treated cracks and crevices, but such applications often failed to control pests which constituted a greater health hazard than the low levels of insecticide residue. However, changes in the law allowed qualifying residual insecticides to be applied to exposed surfaces in food areas of food handling establishments as "spot" treatments. With characteristic attention to nonessential detail a spot was defined by the EPA as not exceeding two square feet, but spots could be any shape and could be adjacent to one another.

Residual insecticides which qualified to be used as spot applications were those which could be shown not to unduly contaminate non-target areas as a result of such spot treatments. To date products based on two residual insecticides, bendiocarb and chlorpyrifos, have qualified and are labeled for use as spot treatments in food areas of food handling establishments. Most other residual insecticides have the potential to qualify for such applications, though some manufacturers may consider the cost of obtaining a clearance for spot treatments in food areas too expensive. In any case, even when the EPA gives clearance for use of a residual insecticide as a spot or crack and crevice treatment in food areas, other agencies with jurisdiction over food handling establishments may disallow such use on purely arbitrary grounds. For instance, in food plants coming under the Meat and Poultry Inspection Program of the U.S. Department of Agriculture the use of any residual insecticides in food areas is not routinely allowed. Only under special circumstances will applications be permitted and then only as crack and crevice treatments.

Because of such inconsistencies in pesticide regulation between agencies and even within the same agency, the user of pesticides in food handling establishments is advised not only to abide by the instructions on the pesticide label, but also to consult with local representatives of the various regulatory agencies for additional guidelines which may further restrict use of pesticides in a particular situation.

Dusts. Chemically inert dusts such as silica or diatomite, etc., are used to protect seeds and stored grains from insects. These dusts cause breaks in the waterproof fatty covering of insects. Thus, evaporation increases and the insect eventually dies from desiccation. Wigglesworth (1947) studied the site of action of inert dusts on a number of stored product pests and he came to the following conclusions: "The abrasion of the protective wax layer in beetles kept in contact with fine alumina takes place chiefly in the articulations of the limbs, but to some extent also at other points where the soft cuticle of the moving insect is rubbed against the dust or where the dust gets into other moving joints. Normal insects living in flour or bran may show very slight amounts of abrasion with increasing age."

Strong and Sbur (1963) protected wheat seed with diatomaceous earth at the rate of six lbs./2.7 kg per ton of grain.

Parkin (1944) studied the effect of finely ground mineral dusts on the grain weevil and several other stored product pests. He found the grain weevil was the most susceptible to these dusts in the week following emergence. He used flint, felspar, limonite (a hydrated ferric oxide) and anhydrite (anhydrous calcium sulphate) in his tests. Flint was the most effective material. The most effective particle size was less than 10 microns in diameter and there was no correlation between the chemical composition of the mineral and its insecticidal action. Parkin notes "uninfested grain can be deposited on a floor thickly sprinkled with dust and the surface covered with a thick layer of it, but where infestation is suspected, the dust must be mixed intimately with the grain by means of a shovel or rotating drum mixer in the case of small quantities of grain, or in the case of large bulks, by introducing it at a measured rate into the grain stream just before its deposition on the storage floor. A trial showed that the dusts are readily removed by the ordinary cleaning process."

Alexander et al. (1945) used silica, carborundum, alumina and charcoal;

Wilson (1946), magnesite; Freeman and Turtle (1947), diatomite; and Cotton (1947), magnesium oxide and aluminum oxide. Cotton (1947) notes magnesium oxide is an inert, non-poisonous dust which is used medicinally in cases of stomach acidity. He recommended the mineral at a particle size of one micron or less to be used at the rate of one ounce per bushel/28 g per 35 l of seed with a moisture content of more than 12 percent. Magnesium oxide also has a repellent effect. Cotton and Frankenfeld (1949) found silica aerogel effective in protecting seed and milled cereal products from insects.

Lindgren et al. (1954) controlled grain insects with malathion dusts. Beckley (1948) recommends the use of finely ground pyrethrum powder at the rate of one pound/0.45 kg per bag of grain for protection against grain weevils. One great disadvantage of grains treated with chemically active dusts is they may be consumed by livestock and care must be taken to prevent this.

Watts and Berlin (1950) showed in the laboratory that dust containing pyrethrins/piperonyl butoxide combinations controlled rice weevils effectively in stored wheat for at least 30 days. There are now on the market pyrethrins/piperonyl butoxide grain and wheat protectants which, when mixed with the clean cereals, will effectively protect them from insect infestation for months. The wheat protectant is mixed with the wheat at the rate of approximately one pound per 10 bushels/0.45 kg per 352 l of wheat. In the case of the grain protectant, the pyrethrins and piperonyl butoxide are impregnated on a fibrous talc, and this mixture is dusted on southern stored corn. The piperonyl butoxide in these dust protectants is usually in the ratio of 10 to 20 parts to one part pyrethrins. Goodwin-Bailey and Holborn (1952) state pyrethrins-piperonyl butoxide dust mixtures containing 0.04 percent pyrethrins and 0.8 percent piperonyl butoxide when applied at the rate of one pound per 300 pounds/0.45 per 136 kg of wheat protect grain from attack.

Eden (1953), Osmun (1954) and Walkden and Nelson (1959) showed protectant dusts of pyrethrins and piperonyl butoxide to protect shelled corn in bins from nine months to two years.

Packaging. Stored product pests and other insects are a serious problem to those concerned with the packaging of food material. Annand (1944) reports an investigation of the insect proof qualities of packages for dried fruit, and found that newly hatched larvae of the Indian meal moth, *Plodia interpunctella,* and the confused flour beetle, *Tribolium confusum,* can penetrate crevices 0.12 mm in width. Larvae of the sawtoothed grain beetle, *Oryzaephilus surinamensis,* can pass through crevices 0.16 mm wide.

Back and Cotton (1926) showed waxed paper and cardboard can be penetrated readily by larvae of the cadelle. However, there can be no doubt good packaging, as described by Linsley (1944) as materials that are insect tight, with smooth surfaces that offer a minimum of folds, creases, seams, rough edges, etc., and are sufficiently thick walled to offer some mechanical resistance, can offer a large measure of protection against insect infestation.

Essig et al. (1943) conducted tests with various materials to determine the penetrating ability of a number of pests of stored products and they designate the larva of the cadelle, and the adults of the lesser grain borer and rice weevil as the best penetrators.

Kunike (1941) made a study on the protection of packaging materials against a wide variety of stored product pests and found the insects oviposited on the outside of the package and the larvae bored through the packing material. He found the females of the rice and grain weevils can gnaw through many paper wrappings and deposit their eggs on the contents. Munro (1942) investigated a packaged product containing malted cereals and wheat flour which was infested by adults and larvae of the red flour beetle and the sawtoothed grain beetle. He concluded the insects reached the food through an opening left in the sealing and not by gnawing through the cover. Collins (1963) noted that 75 percent of the infestation in packaging took place through folds of the overwrap, corners of a carton or a pinhole. Aluminum foil offered the best protection against insects.

Sweetman and Bourne (1944) showed a laminated Kraft paper containing an inner layer of asphalt was both pest and moisture proof and was resistant to the attacks of several species of cockroaches, silverfish and a species of termite. Essig et al. (1943) demonstrated that "Kraft-asphalt-lead foil-cellophane bags" did not prevent penetration by stored product pests, *Bemis* multiwall bags were more promising, but were readily penetrated by the cadelle, and the material offering the greatest protection were heavy cardboard cartons double-dipped in a thermoplastic material."

Anon. (1965) found polycarbonate film plastics to protect stored products from insects.

A number of workers have used chemically impregnated paper to prevent penetration by insects. Frings (1948) showed paper impregnated with ammonium chloride and ammonium nitrate repelled the American cockroach, mealworm and cadelle. Cotton and Frankenfeld (1949) recommended the use of pyrethrins-piperonyl butoxide insecticides to flour bags made from cotton for extended protection.

Gray (1952) made a thorough study of the chemical treatment of packages to prevent penetration by insects. He showed three species were by far the most common insects to occur in "ready-mixed" packages of biscuit flour. Namely, the sawtoothed grain beetle, confused flour beetle, Indian meal moth. Pyrethrins-piperonyl butoxide mixtures were found to be promising in protecting paper and fabric bags from insect penetration.

Gerhardt and Lindgren (1954) showed the lesser grain borer and cadelle to be the best penetrators of packaging films.

Incho et al. (1953) found a spray consisting of five mg pyrethrins and 50 mg piperonyl butoxide per square foot/0.09 square meters, when applied as a wettable powder, protected packages against the larvae of the cadelle for one year. Sivik and Kulash (1956) protected cloth bags with shelled corn with dusts of pyrethrins and piperonyl butoxide or ryania.

Laudani and Davis (1955) showed the primary mode of action of pyrethrins synergized with piperonyl butoxide was repellency. Laudani et al. (1958) review methods to reduce insect infestation by improving packaging, especially closures.

Warehousing susceptible foodstuffs. Cotton (1946) has the following to say on this subject: "It is unlikely that more than a small portion of susceptible food products will be packaged in insect proof containers for many years to come. Therefore, every precaution must be taken to guard against infestation in storage unless refrigerated storage is used. Warehouses should be of modern tight concrete or brick construction suitable for fumigation, particularly in warm climates. Floors should be concrete to facilitate cleaning and wooden partitions should be avoided. Foodstuffs should not be stored in the same warehouse with animal feeds or other products likely to harbor insects.

"Only such products as are guaranteed free from insect infestation should be purchased. Every incoming shipment should be carefully inspected on arrival and rejected or treated if found to be infested. Bagged material, such as flour, rice, beans, peas and similar products should be inspected to determine whether or not insects are crawling over the bags. If insects are found, the shipment should be rejected or held for further inspection or treatment. If no insects are seen on the outside of the bags, several bags should be selected at random from each shipment and their contents sifted with suitable screens. If no infestation is found, it is safe to accept the shipment.

"Regular monthly inspection of all stored foodstuffs for insect infestation should be conducted during periods when the temperature of the product is 60° F or above. Such inspection should reveal the presence of insects before serious damage can be done. Stocks to be used or shipped out should always be selected from lots that have been in storage the longest.

"Whenever food commodities are found to be infested, the warehouse should be fumigated. If the warehouse is not tight enough for fumigants, infested products can be removed to a fumigation vault or treated under a tarpaulin.

"Sprays are frequently suggested for use in warehouses. They may be utilized in clean-up operation when odd lots of food products are moved out of warehouses. It should be realized, however, that with sprays now available it is necessary to hit and soak stored-product insects with the spray in order to kill them.

"The need for strict sanitation in the warehouse cannot be emphasized too greatly. No accumulations of dust, flour, meal or other products in which insects can breed should be allowed in the warehouse. Broken bags should be repaired or disposed of. The use of lift platforms for storage and handling of bagged commodities is desirable when such equipment is available, since it will keep the bags off the floor and facilitate the removal of accumulations therefrom."

Transportation and receiving. Insects of stored products are introduced very often into the store or warehouse from infestations in trucks, railway cars, etc. Trucks transporting such susceptible products should be cleaned daily, preferably by vacuum cleaning. Cotton and Gray (1948) note railway cars can be treated by cleaning with compressed air by using space or residual sprays and by fumigation.

According to Cotton et al., (1950) it has been shown "cleaning cars with compressed air or fumigating them fail to destroy all the insects established in them. When infested cars are loaded with fresh flour, insects from the woodwork or from the accumulations of foodstuffs behind the car linings are quickly attracted to the flour."

Many milling companies apply residual sprays "to railway boxcars before they are lined with paper and loaded with flour."

A KEY TO THE MORE COMMON STORED PRODUCT PESTS FROM CHAPMAN AND SHEPARD (1932)

BEETLES (Coleoptera)

I) Adult beetles with long snouts. Larvae legless grubs inside of grains or other hard material.
 A) Adult — Red or blackish beetle without spots. Found in wheat, barley, corn, macaroni, or other hard products. Granary weevil, *Sitophilus granarius* (L.)
 B) Adult — Red or blackish beetle with four more or less obscure light spots on the back. Habits like those of the granary weevil. Rice weevil, *Sitophilus oryza* (L.)

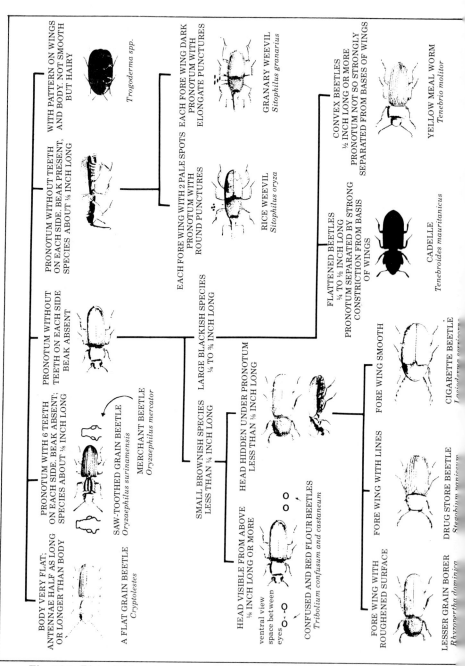

Fig. 16-1. Pictorial key to common beetles and weevils associated with stored foods.

II) Adult beetles without long snout. Larvae with legs. (For all beetles with long snouts see I.)

    A. Adults — Reddish brown.

        1) Adult — Half as broad as long. (For all other reddish brown beetles see 2.)

            a) Adult — Minute lines down back.
            Larvae — Grubs with few short hairs.
            Drugstore beetle — *Stegobium paniceum* (L.)

            b) Adult — No lines on back.
            Larvae — Grubs covered with long hair.
            Cigarette Beetle — *Lasioderma serricorne* (F.)

            c) Adult — Spider-like with indistinct patches of white on back.
            Larvae — White grubs in cases. Found in a wide variety of substances. White-marked spider beetle, *Ptinus fur* L.

        2) Adult — Less than half as broad as long. (For all other reddish brown beetles see 1.) Larvae all much alike.

            a) Adult — Mandibles projecting out in front like large horns. Broadhorned flour beetle, *Gnathocerus cornutus* (F.) Found in flour, meal, and other cereal products.

            b) Adult — Flat, no "horns", gives off a pungent odor when crushed between the fingers. Confused flour beetle, *Tribolium confusum* Duval.

            c) Adult — Very small flour beetles. Smalleyed flour beetle, *Palorus ratzeburgi* (Wissm.)

            d) Adult — Long, slender, "teeth" on sides of the thorax. Sawtoothed grain beetle, *Oryzaephilus surinamensis* (L.)

            e) Adult — Thorax straight sided, about as broad as the abdomen, no "teeth" on sides of thorax. *Cathartus quadricollis* (Guer.)

            f) Adult — Robust, antennae enlarged at end, thorax with nob on front angles. Foreign grain beetle. *Ahasverus advena* (Waltl.)

            g) Adult — Small and slender. Antennae long and slender. Flat grain beetle. *Cryptolestes pusillus* (Schönherr)

    B) Adults — Black, gray, bluish, or varied colors, never reddish brown. (For all reddish brown beetles without snouts see A.)

        1) Large black beetles more than ¼ inch/six mm long. Larvae white or yellowish, attaining a length of ½ inch/13 mm or longer.

            a) Adult — Beetle ½ inch/13 mm long. Larvae yellow, round, may become 1½ inches/38 mm long. Yellow mealworm, *Tenebrio molitor* L.

            b) Adult — Beetle a little more than ¼ inch/six mm long. Cadelle, *Tenebroides mauritanicus* (L.)

        2) Small beetles, less than ¼ inch long, color black, bluish, gray, or varied. (For all black beetles over ¼ inch/six mm long see 1.)

            a) Adults — Plump gray beetles found in peas and beans.

                1) Without small but distinct white spots — in peas. Pea weevil, *Bruchus pisorum* (L.)

                2) Without distinct white spots — in beans. Bean weevil, *Acanthoscelides obtectus* (Say)

            b) Adult — Black or dark brown or bluish or vari-colored beetles.

                1) Adult — Dark brown or black with a light band across the

LARVAL STAGES

Chester J. Stojanovich & Harold George Scott
U.S. DEPARTMENT OF HEALTH, EDUCATION, AND WELFARE
PUBLIC HEALTH SERVICE, Communicable Disease Center
Atlanta, Georgia
1962

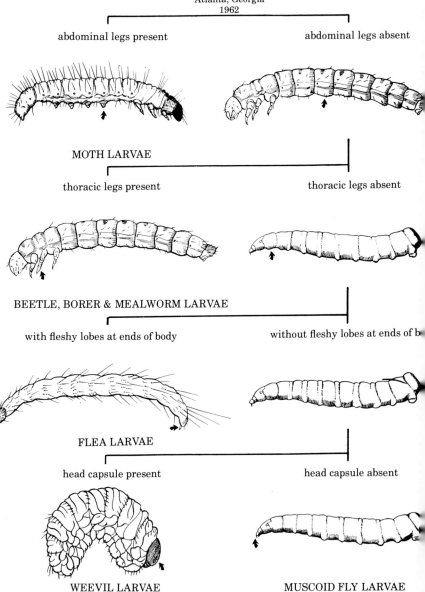

abdominal legs present   abdominal legs absent

MOTH LARVAE

thoracic legs present   thoracic legs absent

BEETLE, BORER & MEALWORM LARVAE

with fleshy lobes at ends of body   without fleshy lobes at ends of b

FLEA LARVAE

head capsule present   head capsule absent

WEEVIL LARVAE   MUSCOID FLY LARVAE

Fig. 16-2. Pictorial key to larval stages of common groups of household and stored food pests.

middle of its body. Larva covered with long hairs. Larder beetle, *Dermestes lardarius* L.

2) Adult — Steel-blue, legs reddish, ¹/₅ inch/five mm of an inch
long. Larva white with brown head. In cured meat. Red-legged
ham beetle, *Necrobia rufipes* (DeG.)

3) Adult — Small black beetle spotted with gray and light
brown. Larvae with long hairs at the posterior end. *Trogoderma* spp.

MOTHS (Lepidoptera)

I) Adult — Narrow pointed wings fringed with hair. Larvae found inside of
kernels of corn, wheat, etc. Emerge as adults through small round holes.
Angoumois grain moth, *Sitotroga cerealella* (Oliv.)

II) Adult — Inner half of front wings, light brown, outer half dark. Larvae
— yellowish, green, pink in color — living free-not in silk tube. Indian meal
moth, *Plodia interpunctella* (Hbn)

III) Adult — Base and outer portion of front wings brownish red middle portion
whitish. Larvae solid grayish, darker at ends, found webbing the food substances together, living within a silken tube in moist chaff or vegetable
debris. Meal moth. *Pyralis farinalis* (L).

IV) Adult — Dark grayish with blackish streaks across the wings resembling
the letter "V" or "W". Larvae yellowish white or pinkish, often with a dark
spot near the middle of the body. Found in flour and feed. Mediterranean
flour moth, *Anagasta kuehniella* (Zell.).

For additional keys and illustrations of stored product pests see Linsley and
Michelbacher (1943), Hinton and Corbet (1949), Okumura and Strong (1965)
and Munro (1966). LeCato and Flaherty (1974) describe the eggs of some stored
product insects.

THE GRANARY WEEVIL

Sitophilus granarius (L.) (Family Curculionidae)

This snout weevil has been known as a prominent pest for centuries and is
believed to be the *Curculio* of the Romans. The generic name *Sitophilus* means
"grain-loving" which is very descriptive of the habits of the insect. This weevil
prefers a cooler climate than does the rice weevil, and most commonly occurs
in the northern part of the United States, whereas the rice weevil is usually
encountered in the southern states. Since much of our detailed knowledge concerning this insect is due to the work of Back and Cotton (1926), the following
is largely from their paper.

The granary weevil breeds in grains of all kinds, and being an insect with
preferences for temperate areas, it is not often found breeding farther south
than North Carolina. Its vestigial wings are useless for flying, so it is chiefly
found where grain is stored and is primarily disseminated by man. Both the
larvae and adults feed upon grains and grain products of diverse types. The
adults do not lay their eggs in such finely ground material as semolina used in
the manufacture of macaroni, since the divided grain particles are too small to

serve as food for a single larva. However, if the material becomes caked, it may then be used for the rearing of the young. When the larvae are found in sacks of flour, it is probable they are merely migrating and not feeding. Birch (1946) observed the granary weevil to live and reproduce at a depth of five feet/1.5 m in wheat until the temperature became too high for breeding.

Adults. The adult weevil is resistant to cold and hibernates during the winter only to resume egg laying in the spring. The beetle is rather sluggish and draws its legs up to its body and remains motionless when disturbed. The adult is shiny reddish brown and about ⅛ to ³/₁₆ inch/3.1 to 4.8 mm long. Birch (1944) found one strain smaller than the other in stored wheat in separate parts of Australia. Cotton (1947) notes the size of the weevil "will depend to a considerable extent on the size of the kernel of grain. In small grains such as millet or milo maize, it will be small in size, but in corn it will attain its maximum growth. The great variation in size of these weevils resulting from the amount of food available has caused many people to consider the small and large forms to be different species. The contents of small grains are almost completely devoured by one weevil in its development, whereas a kernel of corn will provide food for several weevils." Surtees (1965) notes the smaller the weevil, the easier it is to kill.

According to Reddy (1951), the sexes in the adult grain and rice weevils can be distinguished by the fact the male rostrum or beak is shorter and wider, and has sharper and more defined punctures on the dorsal surface than is the case in the female. The granary weevil can be distinguished from the rice weevil by the elongated pits on the thorax, whereas the rice weevil has round or irregularly shaped pits on the thorax. Moreover, the wing covers of the granary weevil lack the four reddish spots to be found on the rice weevil.

Another weevil similar to the rice weevil, but larger, is the maize weevil, *Sitophilus zeamais* Mot.

Life history. The female excavates a hole in the grain, about the length of her beak, whereupon she deposits a single egg in this hole and then plugs it up with some gelatinous material. The period of egg laying is dependent upon the temperature. The female is capable of laying 36 to 254 eggs. Ewer (1945) showed this insect preferred to oviposit on large wheat grains in preference to small ones.

The larva is a small white legless grub that can breed in all the common grains as well as in acorns, chestnuts and sunflower seeds. The larval stage varies from 19 to 34 days and there are four instars. The larva makes its pupal cell in the grain and pupates in five to 16 days, depending on the temperature. The entire life cycle will extend from 30 to 40 days during the summer and 123 to 148 days during the winter. In Washington, D.C., there are four generations per year. With abundant food the adults can live from seven to eight months.

Defiel (1922) found the granary weevil has no poisonous qualities and horses and other stock can feed upon weevily grain without injury.

In Australia during World War I, the weevils became so numerous "that from every 2,500 bags treated, 200 to 300 pounds of weevils were removed, an enormous number, when one considers that there are about 442,000 weevils to the pound."

Control. Back and Cotton (1924) found superheating at temperatures of 120° F/49° C for one hour or 130° F/54°C for 30 minutes kills all stages. They also obtained control through the use of carbon disulphide at the rate of five to 15 pounds per 1000 bushels/2.3 to 6.8 kg per 35,000 liters

De Francolini (1935) obtained 100 percent mortality of the rice and grain weevils with methyl bromide at the rate of two ounces per 100 cubic feet/57 g per 2.8 cubic meters at a temperature of 67.1° F/19° C All the larvae in the grain were killed by an exposure to the gas for 24 hours.

Chiu (1939) noted crystalline silica was effective in the control of rice and granary weevil.

Samsoniya (1936) found the granary weevil and *Tenebrio molitor* L. were controlled in a period of 10 days with a dust consisting of two percent calcium carbonate by weight. He used both magnesium and calcium carbonate in his experiments at the rate of one percent the weight of the grain. The treated grain was freed almost completely from dust by air suction. He was of the opinion these salts release carbon dioxide under the influence of the acidity of the food during deglutination, and the gas dilates the midgut, destroys the epithelium and in this way kills the beetles. These dusts, moreover, act through desiccation and irritation on the insect. The dust interferes almost immediately with movement, especially of the mouthparts. Thus, there is an almost immediate cessation in the injury to grain.

THE RICE WEEVIL

Sitophilus oryza (L.) (Family Curculionidae)

This insect rogue is rated by many entomologists as our most important grain pest. The rice weevil is world-wide in distribution.

Cotton (1920) believes the rice weevil may have come originally from India and spread from there on commercial byways and highways. It was the rice weevil that played havoc with the cargoes of many of the old slow-going sailing ships, either destroying the cargo entirely or so polluting it that it was rendered worthless. The rice weevil is primarily a pest in warm countries and is important in the Gulf states and in the South. It apparently cannot establish itself to any great degree north of North Carolina. Hinds and Turner (1911), in referring to the then current conditions in the South, states, "The injury done by this species would seem to be one of the factors which has restricted the raising of corn and the production of livestock and tended to promote the 'one crop' system of cotton culture which has prevailed for a generation past."

A few of the seed and food products these insects have been reported from are rye, buckwheat, table beans, stored cotton, grapes, cashew nuts, cereals and wheat products of all kinds. The weevil will injure apples and pears by sucking the juice and gradually forming cavities, within which they then conceal themselves. Although the food of the larva is similar to the adult, it is more restricted since the larva must pass its entire larval period in a single seed, which must be large enough to permit it to feed until mature.

Life history. Hinds and Turner (1911) and Cotton (1920) made an intensive study of the life history of this insect. The egg is deposited in the kernel of some grain and requires three days for hatching at a temperature of 60° to 65° F/15.6° to 18.4° C. As a general rule, roughly 50 percent of the eggs produce no young. Hinds and Turner found "the female eats out a cavity large enough for the egg, occupying usually about 45 minutes in the operation when the corn is fairly hardened, then turns, locates the cavity with the tip of the abdomen and inserts the fleshy ovipositor. The deposition of the egg requires only about three minutes and the cavity is then sealed over as the ovipositor is withdrawn. The top of the egg is just below the surface of the grain. The female will then rest for a short period before starting another cavity." The rice weevil lays from zero to

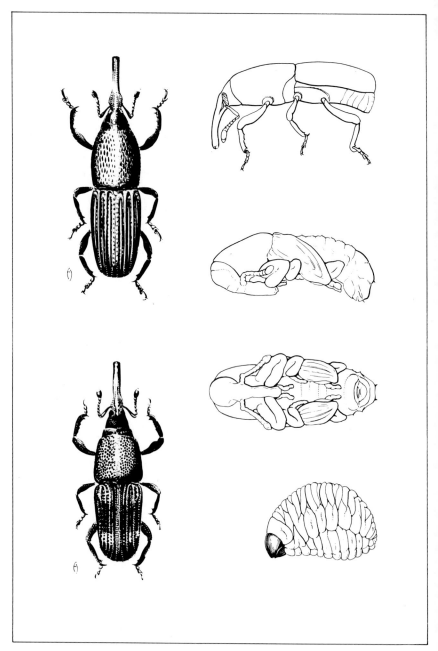

Fig. 16-3. (Top left) Granary weevil. (Bottom left) Rice weevil. (Top right) Lateral view of the rice weevil. (Middle and bottom right) Larvae and pupae of the granary weevil.

25 eggs per day, the average number being four. The female continues to oviposit until within a few days of death, and one weevil has been known to lay eggs over a period of 110 days. The female may deposit eggs at any period of the year, but egg-laying during the winter is sporadic and dependent on temperature.

The larva is creamy white with a brownish-black head. There are three molts, the average time for the entire larval period being 18 days. The pupal stage is from three to nine days with an average of six. Upon transforming to an adult, the insect will remain in the kernel from three to four days, where it hardens and matures.

Reddy (1950) concludes the optimum conditions for weevil activity are 80° to 86° F/27° to 30° C, 75 to 90 percent relative humidity and wheat having a moisture content of 13.5 to 17.6 percent.

The adults of the rice weevil measure from two to 2.8 mm, and are a dull reddish brown having the thorax pitted with deep punctures, and with four light spots on the wing covers. These punctures and light spots are lacking in *Sitophilus granarius*. Many of the weevils perish when trying to escape from the exit hole since they become wedged, with only the head, legs and prothorax free. During the warm weather of August, the life cycle may be only 32 days. The weevils will at times become so thick in corn they can be screened out in tons.

The female can usually be distinguished from the male by her thinner and less pitted proboscis; the two sexes are produced normally in approximately equal numbers. The rice weevils are both polygamous and polyandrous. The male and female will mate 24 hours after emerging from the kernel. There may be several matings. In the South, many of the weevils overwinter in the field. Floyd and Powell (1958) note that infestation in the field is largely due to bird damage. The weevils may live three to six months.

Hinds and Turner observed parthenogenesis (production of young without mating) and they note, "Unfertilized females deposit eggs occasionally but much more rarely than is normal. Many observations have shown that these eggs may hatch. We have bred a male and a female weevil from eggs deposited by a positively unfertilized female. Their development required about five months, from October to March, in a heated room."

Both the rice and the granary weevil feign death. This they do by drawing their legs up close to the body, falling and remaining quiet. Cotton has observed this habit of feigning death is much shorter in the rice weevil and he believes the rice weevil has functional wings with which it can escape, whereas the granary weevil must depend upon its histrionic ability of playing dead.

Control. The usual method of control consists of fumigation. The fumigants, as well as the methods of applying them, are discussed in greater detail in the chapter on Fumigation.

THE SAWTOOTHED GRAIN BEETLE

Oryzaephilus surinamensis (L.) (Family Cucujidae)

This insect is cosmopolitan in distribution and has long been known to naturalists. Linnaeus received his specimens from Surinam in Dutch Guiana and for this reason named the beetle *surinamensis*. The whole Latin name actually means "rice loving from Surinam." This beetle attacks such food as cereals, dried fruits, breakfast foods, macaroni, sugar, drugs, dried meats, chocolate, tobacco, snuff, etc. According to Back and Cotton (1926), the presence of this insect in cereal products such as flour, meal and breakfast foods makes them unsaleable and unpalatable. It readily penetrates packaged foodstuffs, where its presence rather than the food consumed makes it *persona non grata*. It will infest figs and other dried fruits after prolonged storage. The beetles and larvae have been known to enter every single package in a very badly infested grocery store. The insects soon became so abundant, it was necessary to suspend business. Herms (1915) records a case where the beetles entered a home and annoyed the individuals therein by nibbling and crawling upon them. The infestation lasted a number of days and evidently commenced in the stalls of an old barn where the insect had bred on grain in a manger. It is believed the dry heat of the California summer forced these insects to seek water in the bathroom and their nibbling of the occupants of the beds was incidental. Taschenberg recorded a similar infestation some 50 years ago in Germany, the beetles breeding in a brewery and then invading homes to the annoyance of individuals trying to sleep.

The adult. The adult is a small active brown beetle about $1/10$ inch/2.5 mm in length. The flattened body is well adapted for crawling into crevices. Its wings are well developed, but there is no record of the beetle being observed to fly. A closely related species, the merchant grain beetle, *Oryzaephilus mercator* (Fauvel), resembles the sawtoothed grain beetle and is known to fly. Loschiavo (1976) notes it frequently infests bran, rolled oats, brown rice and walnuts. The margins of the thorax are saw-like and bear six projections on each side. This is the reason for its popular name. The male can be distinguished from the female by the fact that it bears a tooth on the femur of the hind leg. Simmons et al. (1931) state adults are very long-lived, one beetle living three years and three months.

Life history. Back and Cotton (1926) made a detailed study of the life history of this insect and found the eggs which are laid singly or in small batches are deposited in some crevice in the food supply. They also may be deposited in such finely ground foods as flour or meal. In Washington, D.C., the females emerge in April and lay from 45 to 285 eggs.

The larva is yellowish white and has a brown head. When full grown it is a little less than $1/8$ inch/three mm in length. It has three pairs of legs and a pair of abdominal prolegs. According to Back and Cotton (1926), the larva thrives on foodstuffs of vegetable origin such as rice and wheat, breakfast foods, nutmeats, etc. It moves about nibbling hither and yon, and probably cannot feed on the whole grain. Thus, it is associated with other insects that feed on intact grain.

The larva molts from two to four times, depending upon temperature. The life cycle from egg to egg can range from 27 to 35 days in Washington, D.C. Simmons et al. (1931) reared a number of these beetles on raisins in Fresno, California, and found on an average the incubation period was 7.9 days; the larval period was 37.4 days, the pupal period 6.3 days, and the period from egg to egg 51.6 days. In the District of Columbia, there may be six to seven generations annually. Thomas and Shepard (1940) note the optimum temperature for development for this insect is between 86° and 95° F/30° and 35° C. In general, development is more rapid at the higher humidities. The egg and pupal stages appear to be affected little by atmospheric moisture conditions.

Back and Cotton state the insect will cease breeding during the winter unless the building is heated and moisture conditions are satisfactory. They may then breed at a slow rate throughout the winter. Since most heated buildings are fairly dry, little or no breeding takes place, but the adults remain active.

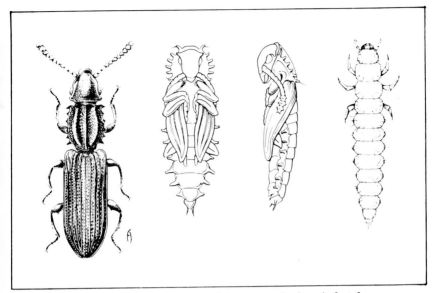

Fig. 16-4. The adult, pupae and larvae of the sawtoothed grain beetle.

Howe (1956) made a study of the biology of *O. surinamensis* and the related species *O. mercator*. The optimum temperature and humidity was approximately 90° F/32° C and 70 percent relative humidity or higher. Lergenmuller (1958) made ecological investigations with this species.

Control. Back and Cotton (1926) note that although this insect is resistant to fumigation it can be killed by fumigants. All stages succumb when exposed to temperatures of zero to 5° F/−18° to −15° C for one day and to a temperature of 125° F/52° C for one hour. Other methods of control of this insect are discussed in the introductory section.

THE CADELLE

Tenebroides mauritanicus (L.) (Family Trogositidae)

When Linnaeus described this insect in 1758, it was in all probability already cosmopolitan. The French called it the "cadelle" and this is the name by which it is commonly known at the present time. It is known also as the "bread beetle" because of its common occurrence in bakeries and as the "bolting cloth beetle" because of its habit of cutting the silk bolting in flour-mill machinery. Hatch (1942) notes in Europe the cadelle lives outdoors under bark and in rotten wood. Here it is predacious on wood-feeding insects.

This beetle is a pest of great importance in rice and flour mills, as well as in grain stored on the farm. It is found frequently in packages of ground cereals, breakfast foods, maize, oats, potatoes and in shelled and unshelled nuts, spices, fruits, cinnamon, nutmegs and ship biscuits. According to Bond and Monro (1954), the males can be distinguished from the females by the numerous and fine punctures on the ventral side of the abdomen, whereas the punctures on the female are less numerous and coarser.

The restless larvae and adults cause much damage by gnawing through sacks and paper packages and even through wooden boards. Since they are among the largest insects that feed on grain products, the holes they make on entering or escaping from these containers are sufficiently ample to enable the entrance of other insects.

Back and Cotton (1926) investigated the life history of the cadelle and the following is to a large extent from their studies. The adult beetles are shiny black in color and approximately ⅓ inch/eight mm in length. They avoid light and hide in the dark corners of the mills as well as in between the sacks. The food habits of the adults are similar to those of the larvae. The adults, however, readily attack any larva that they encounter, particularly *Plodia interpunctella* and *Oryzaephilus surinamensis*. There is some divergence of opinion among entomologists as to whether or not the cadelle makes any particular effort to hunt out these larvae.

The long-lived females deposit the eggs loosely in flour and the crevices of other food materials. They lay from 10 to 60 eggs, usually a batch at a time. The oviposition period may vary from two to 14 months. The maximum number of eggs deposited by the cadelle was 1,319, the minimum was 436. Bond and Monro (1954) reared the cadelle on oatmeal cultures containing molds or yeast. They found the female lays her eggs in clusters in the food or packs them in crevices. One female produced a maximum of 3,581 eggs.

The larvae feed on a great variety of grains, as well as on flour, meal, biscuits and bread, vegetables, dried fruits, etc. Upon migrating from its food source in order to pupate, it occurs accidentally in such unusual places as books, balls of twine, carpet rolls, rugs and in bottles of milk. The length of the larval period varies with the environment and may extend from 38 to 414 days. The cadelle may undergo from three to seven molts, the average for this insect being four.

One larva was found capable of destroying the germinating powers of 10,000 grains. The larvae, when attacking such grains as wheat and oats, usually confine themselves to the embryo, this being the softest part. These insects are also a serious pest in tobacco factories. Here they occur in bales of dry tobacco where it is claimed they bore into them in search of insect larvae. According to Candura (1943), the young larvae will feed on the tobacco, but will die unless they are able to feed on insects too. The larvae require two to three years to develop in the tobacco which they finally destroy through their borings. The larvae pupate or hibernate by hollowing out a cell in the adjacent wood. Thus, they may remain in the wood or empty bins for months and then emerge to reinfest fresh grain placed in these bins. This wood-boring habit of the larvae has resulted in the pupal chamber being excavated in such curious places as the corks of bottles and in books. The cadelle pupates in from eight to 25 days. In Washington, D.C., there may be two generations with a partial third. It is believed there are three generations in tropical countries.

All stages of the adults and larvae are active during the winter and only hibernate at low temperatures. The eggs and pupae, however, succumb to cold and are not observed during the winter. Both the adults and the larvae live in the open, especially in the mines of wood-boring insects upon which they feed. Adults emerging in spring and summer lived for six to seven months and those emerging at the end of the autumn lived for a year.

Control. Since a year may be required for the development of this insect from egg to adult, if the infested locality is cleaned several times during the year, there is little chance of the larva surviving this procedure.

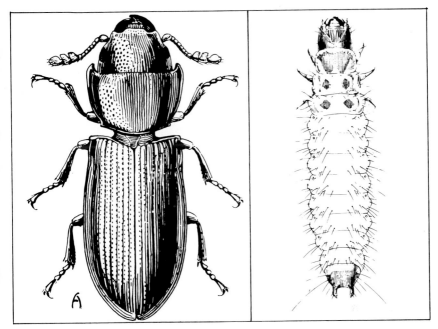

Fig. 16-5. (Left) The cadelle. (Right) Cadelle larva.

The eggs and pupae of the cadelle are easily killed by low temperatures, whereas the larvae and adults are rather resistant and can survive a temperature of 15° to 20° F/ −9° to −7° C for several weeks.

Badly tunneled timbers should be replaced with concrete, since the timbers act as a source of reinfestation. Control may be realized by superheating or fumigation.

Richardson (1945) studied the effectiveness of various fumigants on the cadelle in shelled corn. He found 1,1-dichloro-1-nitroethane and ethylene dibromide were most toxic. Carbon disulfide, methyl bromide-carbon tetrachloride mixture (19-90 parts by volume), B-methyl allyl chloride, were intermediate in toxicity. In his opinion, the cadelle beetle "is more resistant than *Tribolium castaneum, Oryzaephilus surinamensis* and *Sitophilus oryza* to carbon disulfide, B-methyl allyl chloride, carbon tetrachloride and ethylene dichloride." This same author recommends the usual fumigant dosage be doubled for cadelle control. Bond and Monro (1961) found acrylonitrile, HCN, chloropicrin and hydrogen phosphide to be the most toxic fumigants to the cadelle.

THE DRUGSTORE BEETLE

Stegobium paniceum (L.) (Family Anobiidae)

This common pest of the home and storehouse is world-wide in distribution. It is cylindrical in form and about 1/10 inch/2.5 mm long and is uniform brown in color. The legs and the antennae are appressed to the body when the insect is at rest. This enables it to escape detection. The adult beetle can be most read-

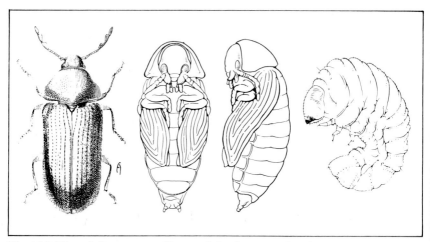

Fig. 16-6. The adult, pupae and larva of the drugstore beetle.

ily distinguished from the cigarette beetle which it resembles by the three-segmented club of the antenna. Griffith (1946), who studied the insect in dog biscuits, notes the infestation consisted of 46.2 percent males and 53.8 percent females, and that the variation in size was from 1.96 to 3.33 mm. The average length of the males was 2.39 mm and the females was 2.88 mm, but there is some overlapping in sizes between the sexes.

It was given the Latin name *paniceum* because it feeds on bread, but its appetite includes practically everything edible and many things nonedible to man. In the household, it feeds readily on flour, meal, breakfast foods and condiments, particularly red pepper, etc. During World War I it became a very serious pest of bread, as the crawling female deposited her eggs as she bored her way through the bread. Eichler (1943) records it as a pest in breadboards in Germany and DeOng (1948) notes it infesting green coffee beans from Colombia. This insect has infested even poisonous materials like strychnine, belladonna, aconite and wheat poisoned with strychnine.

 It is also a serious pest of books and manuscripts, and has been known to bore in a straight line through a whole shelf of books. Linsley (1942a and 1942b) states it has infested mah-jongg playing pieces and wooden forms in a home. It has been recorded feeding on a mummy and has been known to perforate tin foil and sheet lead. Apparently, some vexed entomologist must have made the statement that it will "eat anything except cast iron."

Life history. The eggs are laid singly in the foodstuffs. The larval period ranges from four to five months. The pupal stage lasts from 12 to 18 days. The complete life cycle requires seven months. Chittenden (1879) states in powdery substances the larva forms a little round ball or cell which becomes its cocoon, and in which it pupates. Chittenden succeeded in rearing the beetle from egg to larva in two months. He found there may be four broods per year in a fairly warm atmosphere. In a place as cool as the ordinary larder there will be but one generation per year.

Control. All heavily infested material should be burned. Heating infested

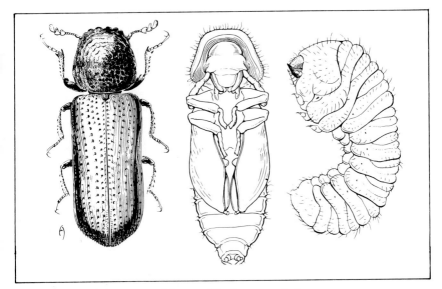

Fig. 16-7. The adult, pupa and larva of the lesser grain borer.

material at temperatures of 140 to 176° F/60° to 80° C for several hours will kill all stages. Brooks (1934) controlled these insects in the Huntington Library of San Marino, California, by means of vacuum fumigation with Carboxide, which is a mixture of ethylene oxide and carbon dioxide. Averin and Novinenko fumigated these insects in books with chloropicrin at the rate of one ounce per 50 cubic ft/28 g per 1.4 cubic m for 24 hours with satisfactory results.

THE LESSER GRAIN BORER

Rhyzopertha dominica (F.) (Family Bostrichidae)

This cosmpolitan beetle, which is approximately ⅛ inch/three mm long, dark brown or black, with small pits on the wing covers, is a very important United States pest of stored grain and is particularly prominent in the United States, southern Canada, Argentina, India, New South Wales and southeast Australia. It also attacks rice, wood and books. Its Latin name means "root destroyer from Dominica." Potter (1936), who has made an extensive study of the lesser grain borer, believes its original food was wood, possibly living trees. At times, this pest will reduce grain to a mere shell and several of the beetles may be found in one corn kernel. Potter states the general belief is the larva can attack only damaged grains, but he observed a first instar larva to enter an intact corn kernal through the hard testa. Hoffman (1933) found the beetles to be injuring the inner portions of the backs of books in a library in Puerto Rico, which resulted in the bindings becoming loose and less durable.

According to Potter, the eggs are laid singly or in clusters of from two to 30. The larva molts two to four times. The total time from egg to adult averaged

58 days. Dean (1947) found in the summertime this period can be reduced to 30 days.

Control. The pest may be controlled through the use of heat, dusts, or fumigants. Stracener (1943) showed that six ounces/170 g of derris dust mixed with 100 pounds/45 kg of rice before sacking killed all the insects present in the rice, as well as having a repellent effect on the beetles.

THE CIGARETTE BEETLE

Lasioderma serricorne (F.) (Family Anobiidae)

The fact the *Lasioderma serricorne* will hob-nob with royalty as well as with laity is revealed by its presence in the tomb of Tutankhamen. Alfieri (1931) believes it is probably native to Egypt and he notes it has shown practically no change in structure in the 3,500 years that have elapsed. Sivik et al. (1957) note this insect is the most destructive pest found in stored tobacco.

The cigarette beetle is a small squat beetle, and two to three mm in length. The antennae are saw like and the head

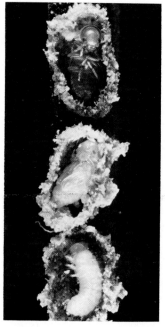

Fig. 16-8. The drugstore beetle in pupal case. Top, adult not yet emerged; middle, pupa; lower, pre-pupal larva.

retracted. This cosmopolitan species infests tobacco wherever it is stored. It is also a very serious pest of books where it causes much injury to the binding and the leaves. Furniture upholstered with flax tow or straw is also injured severely. Back (1939) states the larvae cause serious damage to sacks containing cottonseed meal. The adults fly on late afternoons and on dull cloudy days. Back (1939) notes, "At such times so many adults are on the wing in warehouses and about loading platforms that life may be made miserable for those who work or live within the neighborhood. The beetles do not bite, but get into the hair and beneath the clothing.

"These flying beetles invade residences in the vicinity of the meal establishments, usually for a limited number of days during the height of a flight, and cause no end of annoyance by disturbing the sleep of young children and ruffling the disposition of adults in addition to contaminating food supplies." The foods upon which it feeds are legion, and a few of these are rice, ginger, raisins, pepper, dried fish, dates, belladonna, drugs, seeds and pyrethrum powder strong enough to kill cockroaches. Dried straw flowers may be a source of infestation in the house. Merrill (1948) states this beetle is the chief offender in botanical circles where it is known as the "herbarium beetle." Flint and McCauley (1937) note the larvae feed on upholstered furniture, particularly stuffing. It may also damage silk. Mampe (1976) found place packets of rodent bait a source of this beetle.

Life history. Howe (1957) prepared an extensive review of the biology of the cigarette beetle. Bovingdon (1931) studied the life history of this insect on to-

HANDBOOK OF PEST CONTROL
COLOR IDENTIFICATION GUIDE

Alphabetical Listing of Color Illustrations

HANDBOOK OF PEST CONTROL
COLOR IDENTIFICATION GUIDE

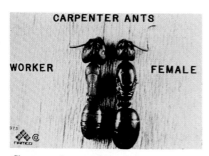

Carpenter ants.

Carpenter ant, queen.

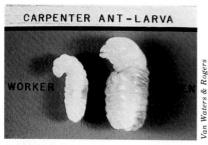

Carpenter ant larva.

Carpenter ant pupa.

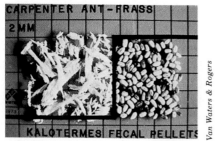

Carpenter ant frass (left) and
Kalotermes fecal pellets (right).

Argentine ant

Plate 1

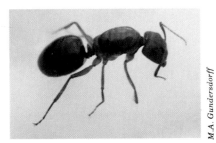

Carpenter ant

Carpenter ant damage

Carpenter ant

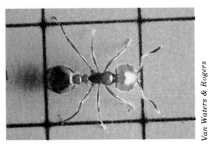

California fire ant

Fire ant on Jerusalem cricket.

Fire ant

Plate 2

Odorous house ant

Harvester ant

Pharaoh ant

Rough harvester ant

Velvet ant, female.

Thief ant

Plate 3

Clemson University

Black carpet beetle, adult (top), larva (bottom).

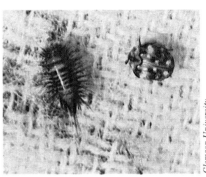

Clemson University

Furniture carpet beetle, adult (r), larva (l).

Degesch America Corp.

Cigarette beetle

Clemson University

Cigarette beetle on damaged cigar, pupa (l), larva (r).

Degesch America Corp.

Redlegged ham beetle on copra.

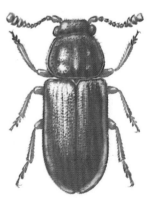

Degesch America Corp.

Redlegged ham beetle

Plate 4

Degesch America Corp.

Confused flour beetle on manioc.

Degesch America Corp.

Confused flour beetle

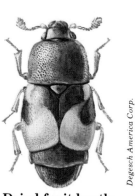

Dried fruit beetle

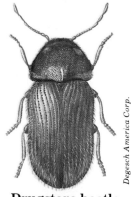

Degesch America Corp.

Drugstore beetle

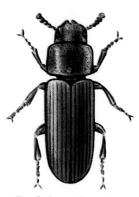

Degesch America Corp.

Red flour beetle

North Central States Extension Entomologists (USDA)

North Central States Extension Entomologists (USDA)

Flat grain beetle

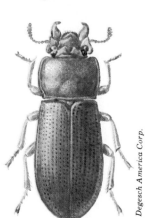

Degesch America Corp.

**Broad-horned
flour beetle**

Degesch America Corp.

Rusty grain beetle

Plate 5

Degesch America Corp.

Khapra beetle on cotton-seed cake.

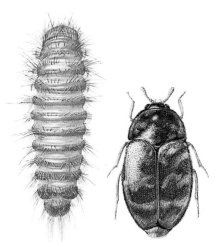

Degesch America Corp.

Khapra beetle

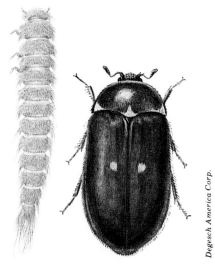

Degesch America Corp.

Fur beetle

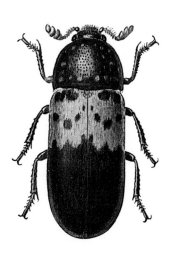

Degesch America Corp.

Larder beetle

Plate 6

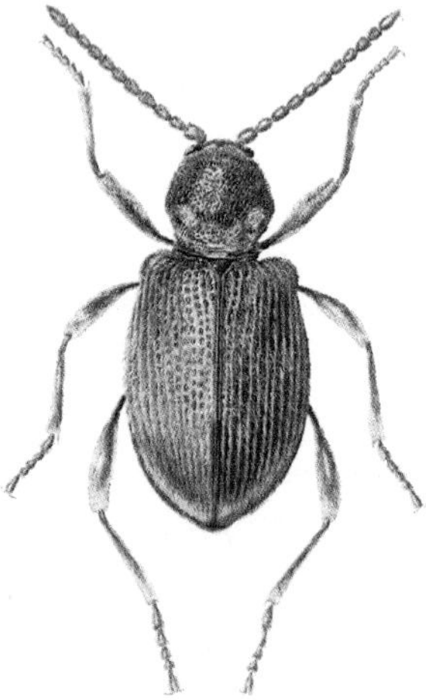

Degesch America Corp.

Australian spider beetle

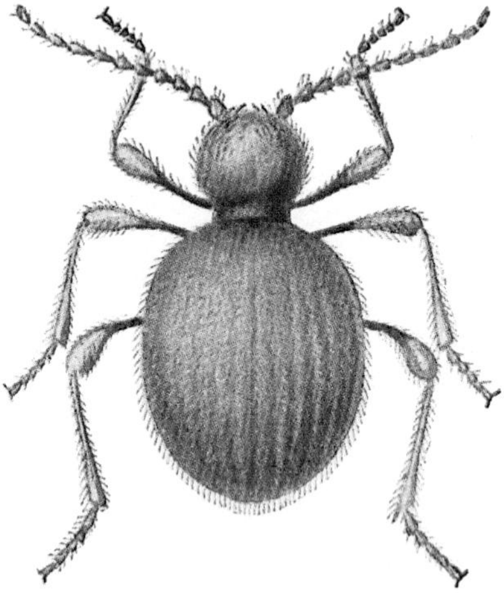

Golden spider beetle

Degesch America Corp.

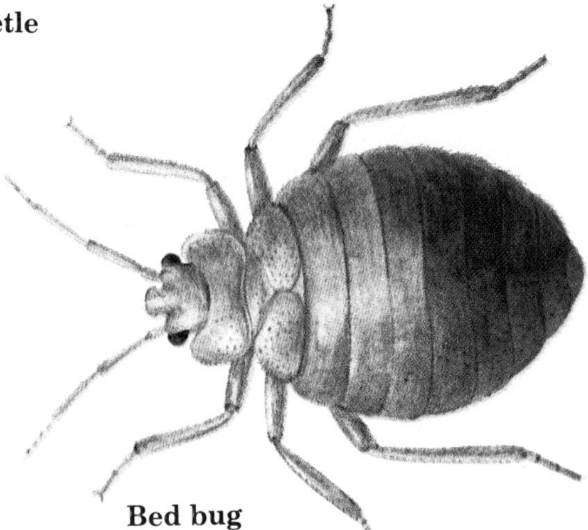

Bed bug

BFC Chemicals, Inc.

Plate 7

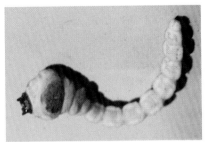

Flatheaded borer larva.

Metallic wood borer

Lesser grain borer

Old house borer larva

Old house borer

Plate 8

M.A. Gundersdorff

Saw-toothed grain beetle (l) and rice weevil (r).

Degesch America Corp.

Sawtoothed grain beetle

M.A. Gundersdorff

Casemaking clothes moth

Cadelle

Degesch America Corp.

Casemaking clothes moth larva

Plate 9

American cockroach

Australian cockroach egg

**Australian cockroach (r),
German cockroach (top, l)
and Australian cockroach
nymph (bottom, l).**

**Brown banded
cockroach**

Plate 10

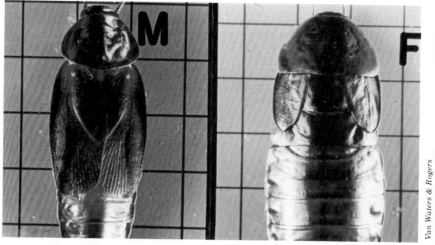

Van Waters & Rogers

Oriental Cockroach

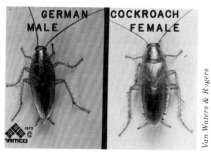

Van Waters & Rogers

German cockroach

Van Waters & Rogers

Smoky brown cockroach

Plate 11

Van Waters & Rogers

Earwigs

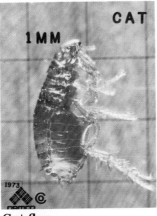

CAT FLEA

1MM

1973

Cat flea

Van Waters & Rogers

SQUIRREL FLEAS - MATING

FEMALE

Van Waters & Rogers

Squirrel fleas mating.

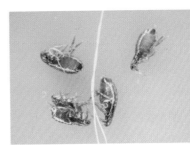

M.A. Gundersdorff

Dog fleas

Plate 12

Crane fly

Sandfly

Robber fly

Green bottle fly

Striped horse fly

Black blow fly

Plate 13

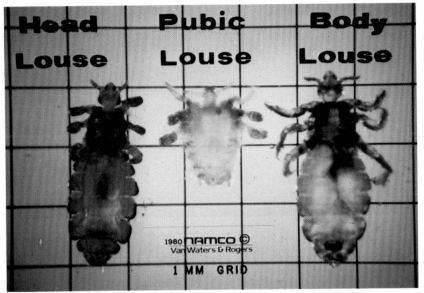

Van Waters & Rogers

Head louse (l), crab louse (c), and body louse (r).

Van Waters & Rogers

Bird louse

M.A. Gundersdorff

Book louse

Plate 14

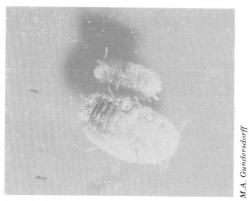

Mealybug

Midge

Adult millipede

Millipede

Plate 15

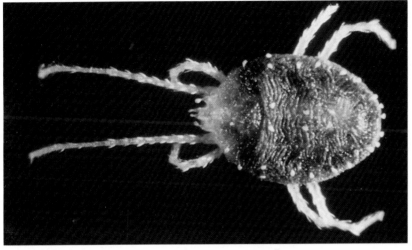

Van Waters & Rogers

Clover mite

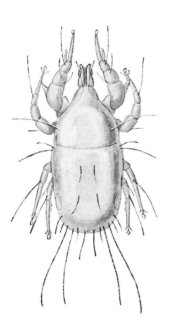

Degesch America Corp.

Common grain mite

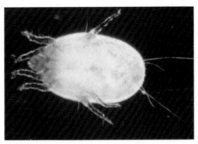

Van Waters & Rogers

Dust mite

Van Waters & Rogers

Tropical rat mite

Plate 16

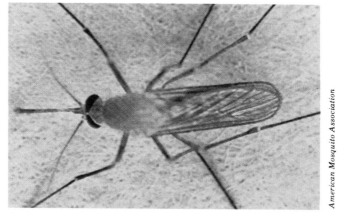

American Mosquito Association

Adult house mosquito

American Mosquito Association

House mosquito egg raft.

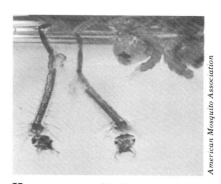

American Mosquito Association

House mosquito larvae and pupae.

Plate 17

Indian meal moth on dried apricots.

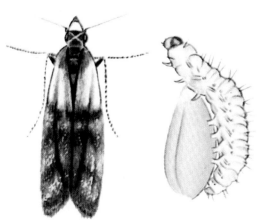

**Indian meal moth, adult (1),
larva (r).**

Plate 18

Mediterranean flour moth on sweepings.

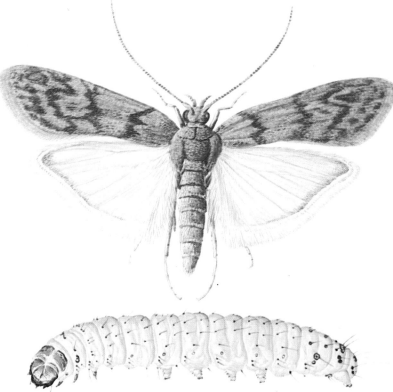

Mediterranean flour moth, adult (top), larva (bottom).

Plate 19

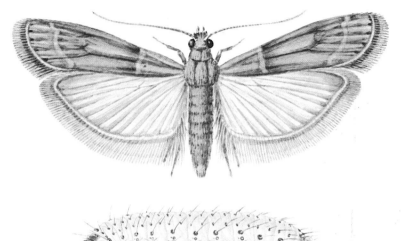

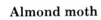

Almond moth

Angoumois grain moth

Plate 20

Warehouse (stored tobacco) moth

Webbing clothes moth

Plate 21

M.A. Gundersdorff

Sowbug

M.A. Gundersdorff

Pillbug

M.A. Gundersdorff

Pine sawyer

M.A. Gundersdorff

Saddleback

Plate 22

M.A. Gundersdorff

Brown recluse (fiddleback) spider

Van Waters & Rogers

Brown recluse (violin) spider

M.A. Gundersdorff

Crab spider with insect.

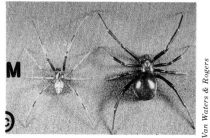

Van Waters & Rogers

Black widow spider

M.A. Gundersdorff

Black widow spider with egg.

Plate 23

M.A. Gundersdorff

Brown widow spider

M.A. Gundersdorff

Black and yellow argiope

M.A. Gundersdorff

Flower spider

M.A. Gundersdorff

Golden silk spider

Plate 24

Van Waters & Rogers

Tarantula

M.A. Gundersdorff

Steatoda grossa

M.A. Gundersdorff

Jumping spider

Van Waters & Rogers

Trap door spider

M.A. Gundersdorff

Wolf spider

Plate 25

M.A. Gundersdorff

Drywood termite

M.A. Gundersdorff

Drywood termite

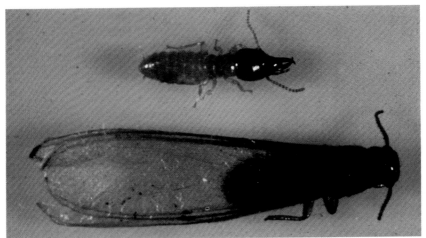

M.A. Gundersdorff

Coptotermes formosans

M.A. Gundersdorff

Subterranean termites

Plate 26

Brown wood tick

Brown wood tick

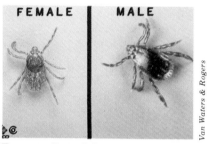

FEMALE   MALE

Brown dog tick

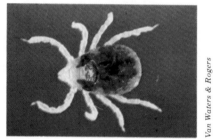

Brown dog tick nymph

EGGS

Brown dog tick eggs

American dog tick

Plate 27

Van Waters & Rogers

Bumble bee

M.A. Gundersdorff

Carpenter bee

M.A. Gundersdorff

Honey bee

Van Waters & Rogers

Yellow jacket

M.A. Gundersdorff

Cicada killer

M.A. Gundersdorff

Cicada killer

Plate 28

Wasp: mud dauber larva

Wasp: mud dauber larva

Paper wasp larva

Paper wasp

Cuckoo wasp

Plate 29

Degesch America Corp.

Granary weevil on wheat.

Degesch America Corp.

Granary weevil

Degesch America Corp.

Common bean weevil on dried beans.

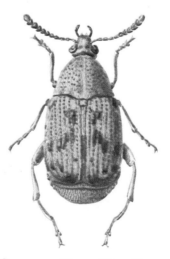

Degesch America Corp.

Common Bean Weevil

Plate 30

Rice weevil on corn.

Rice weevil

Coffee bean weevil on coffee.

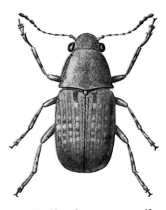

Coffee bean weevil

Plate 31

bacco. The eggs are laid in the folds of bundled tobacco in storage, never in the tobacco leaves in the field. The eggs may also be laid at the open end of a cigar, etc., where they are difficult to detect. The female lays about 30 eggs throughout a period of three weeks on newly harvested and baled tobacco. The beetles become distributed widely throughout the factory in the bales. They do not, as a rule, enter boxed cigars in order to oviposit.

The eggs hatch in about six to 10 days. The larvae make long cylindrical galleries through the leaves and feed on the edges and center of the leaves. The larval stage lasts from five to 10 weeks. The larvae shun the light. Although the larvae feed on a great variety of vegetable matter, they generally prefer cured tobacco leaves and manufactured tobaccos. At about 60° F/16° C the larvae become dormant and may then hibernate.

The pupal and prepupal periods last about two to three weeks and are passed in a cell. The life history lasts from 70 to 90 days and there may be about five to six overlapping generations per year in warm localities, but only one generation in more temperate regions. The adults are strong flyers, and are active in subdued light at temperatures above 65° F/18° C. The adult beetles may live from 23 to 28 days. In temperate climates, the beetles usually first swarm in May and then again in August. In the United States, there may be three generations per year, but in certain parts of the world there are as many as five or six. Runner (1922) notes in Virginia the winter is passed usually in the larval stage, but the adults, which are not too resistant to cold, may secrete themselves in some crevice.

Sivik et al. (1957) studied the life history of the cigarette beetle in warehouses. With flue-cured tobacco as a medium, the entire life cycle was completed in the summer in 52 days. The incubation period was seven days, larval 40 days and pupal five days. The adults lived in the summer from one to six weeks. The average number of eggs deposited was 42.

Control. Since these insects infest so many different objects, the methods of control must necessarily be varied. A few of the control methods are briefly reviewed, but it must be emphasized that some of these methods are not currently permitted in the United States.

1) Back (1939) controlled the insects in cotton seed meal by fumigating with sodium cyanide at the rate of one pound per 1,000 cubic feet/0.45 kg per 28 cubic meters. Childs (1958) showed HCN dosages of 32 ounces per 1,000 cu. ft./0.9 kg per 28 cu. m killed larvae in tobacco in a depth of five inches/13 cm.

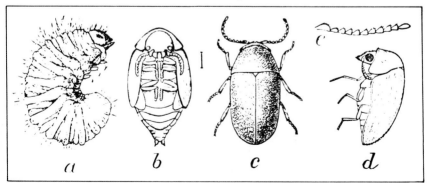

Fig. 16-9. The cigarette beetle. A. Larva. B. Pupa. C and D. Adult. E. Antenna.

2) Childs (1967) controlled the cigarette beetle with one HCN fumigation per year at the rate of 48 oz per 1,000 cu. ft/1.4 kg per 28 cu. m combined with the use of DDVP aerosols.

3) Stamatinis (1935) in Greece found a commercial fumigant consisting of a mixture of 90 percent ethylene oxide and 10 percent carbon dioxide at the rate of four ounces to 100 cubic feet for a period of 48 hours gives good control of the cigarette beetle. Bare (1948) found a 50-50 mixture of acrylonitrile-carbon tetrachloride to be effective in laboratory tests against this insect. Brubaker and Reed (1943) obtained satisfactory control of the cigarette beetle in tobacco through the use of methyl bromide and hydrocyanic acid gas in a vacuum chamber. Bare et al. (1946) found vacuum fumigation combined with a live steam treatment to be very effective. Bare and Tenhet (1950) controlled this insect, as well as the tobacco moth, with a 50-50 mixture of acrylonitrile-carbon tetrachloride as the fumigant. Tenhet (1961) recommended DDVP aerosols dispersed at the rate of 2 g/1,000 cu. ft. twice a week.

4) Cressman (1935) controlled these insects in a library in New Orleans through the use of heat. He used gas burners, which maintained a temperature of 140 to 145° F/60° to 63°C for six hours. He obtained even distribution of the temperature throughout the room through the use of electric fans. The books were loosened in order to allow the air to circulate around them.

5) Crumb and Chamberlin (1934) found the storage of cigars for 35 days at a temperature of 55° F/13°C gave good control since the eggs do not hatch at 50 to 60° F/10° to 16° C, and are nonviable after 35 days when removed to normal temperature.

6) Swingle (1938) found all stages of the insect are killed if infested material is kept at 36° F/2° C for 16 days and at 25° F/−4° C for seven days.

7) Tenhet (1947) showed the adult beetles are somewhat resistant to pyrethrins. However, Tenhet (1955) recommends periodical applications of pyrethrin sprays between 6 p.m. and midnight to coincide with the greatest activity of the cigarette beetle.

8) Reed et al. (1934) obtained satisfactory control of these insects in warehouses by means of an electric light suction trap. This trap was operated at the period of activity and migration, which is usually about sunset. Vinzant and Reed (1941) demonstrated an 18-mesh or 20-mesh screen was necessary to keep cigarette beetles from warehouses.

9) Merrill (1948) obtained effective control of the cigarette beetle in herbariums by trapping the gravid females on preferred plants which were placed in strategic locations as "bug traps." These were then picked up, fumigated and replaced at periodic intervals.

THE BEAN WEEVILS

The Bean Weevil, *Acanthoscelides obtectus* (Say)
The Cowpea Weevil, *Callosobruchus maculatus* (F.)
(Family Bruchidae)

We are concerned here for the most part with two weevils that breed in stored beans, cowpeas, peas, etc., *Acanthoscelides obtectus,* the common bean weevil, and *Callosobruchus maculatus,* the cowpea weevil. The bean weevil occurs throughout the United States and the cowpea weevil is limited for the most part

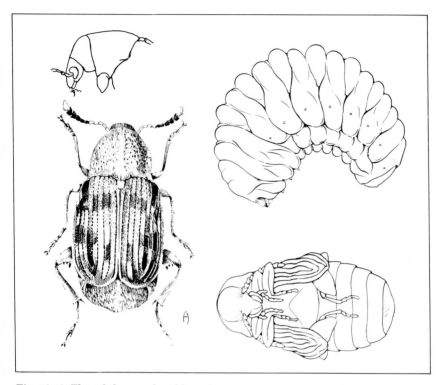

Fig. 16-10. The adult, pupal and larval stages of the bean weevil.

to the southern states. The bean and the cowpea weevil breed continuously in dried seeds if stored in a warm place. Thus, all stages are to be found during the winter. Metcalf and Flint (1939) separate the two species as follows: The bean weevil is "light olive-brown color, mottled with darker brown and gray. They are about ⅛ inch in length, the appendages are reddish and the body narrows evenly toward the small head. In the cowpea weevil the elytra have a large rounded spot at midlength and the tips are black. The bean weevil has smaller black mottlings and can also be distinguished from the former species by having one large and two small teeth at the apex of the hind femur, while the cowpea weevil has only one such tooth."

The adult bean weevil lays its white eggs on pod beans in the field, or in storage, and the larvae emerge in five to 20 days. The tiny legless grubs enter the beans and eat out a cavity, becoming mature in 11 to 42 days. They pupate in the beans for five to 18 days in cells near the surface. The life cycle is completed in 21 to 80 days. They continue to breed in storage, generation after generation. Larson and Fisher (1938) state when they placed the bean weevil in a bag containing 87 pounds/39 kg of red kidney beans, 250,000 adults were produced in 14 months. The adults hibernate in fields and warehouses. If the temperature is high enough, they will emerge during the winter and continue laying eggs. The adult weevils do not feed on the beans. Weiss (1944), who studied the death feint of the bean weevil, found only 72 out of 283 beetles could be induced to feign death and the feint varied from one to 300 seconds.

According to Nelson and Fisher (1952), the cowpea weevil is a pest of dried cowpeas and does not develop successfully in the common varieties of beans. The female attaches her eggs to the surface of cowpeas. The life cycle is similar to that of the bean weevil. The cowpea weevil is a strong flier. Jay et al. (1973) found this weevil to damage soybeans in storage.

Control. Linsley and Michelbacher (1943) note the bean weevil can be controlled in the home "by the destruction or protection of stored beans or peas since the species does not infest grains, cereals, or other stored food products."

Mayer and Nelson (1955) controlled insects by fumigating dry beans and cowpeas in the packaging line.

Nelson and Fisher (1952) recommended fumigation with one of the following fumigants: methyl bromide, chloropicrin, ethylene dibromide or combinations of these. The USDA (Anon., 1966) recommends the use of methyl bromide at two to four lbs. per 1,000 cu. ft./0.9 to 1.8 kg per 28 cu. m.

Marcovitch (1934) controlled the weevils in stored beans by mixing them with hydrated lime at the rate of one pound of lime to a bushel of seed. Chiu (1939) found an "inert" dust of bentonite could be used in a manner similar to hydrated lime. The inert materials have a desiccating effect on the weevils. He states, "it is believed that the fine particles of the 'inert' material, by means of their surface activities, draw water out from the insect body, which results in the death of the insect." Deay and Amos (1936) found dusting talc to be equally good, and Lathrop and Keirstead (1946) used tablegrade, ground black pepper.

Morgan and Pasfield (1943) killed the bean weevil adults in beans by using either copper oxychloride, copper carbonate or kaolin at the rate of one pound per bushel/0.45 kg per 35 l. This prevented further infestation and did not affect germination materially.

Thompson and Perry (1956) controlled insects in bagged beans with aerosol formulations containing one percent pyrethrins and 10 percent synergist.

Larson and Simmons (1924) found all stages could be killed if kept in cold storage at 32° F/0° C for 58 days. Essig (1926) recommends heating at 145° F/ 63° C for two hours.

Bean weevils may be common in the kitchen during the summer and may be found flying to the windows. The infested beans are undoubtedly in the cupboard and should either be discarded or heated in a stove, according to Essig's recommendations.

THE REDLEGGED HAM BEETLE OR COPRA BEETLE

Necrobia rufipes (DeG.) (Family Cleridae)

The adult of this cosmopolitan shiny blue to green beetle is 3.5 to seven mm long. Simmons and Ellington (1925) made a detailed study of this insect and the following material is largely from their paper. This insect infests meats which have become dried through evaporation or as long storage or as a result of prolonged smoking. Their injury is more or less sporadic, since they become only occasionally abundant enough to injure large quantities of meats.

This beetle and its larvae infest bacon, Egyptian mummies, fish and whale, guano, bone meal, garlic, coconuts, etc. The beetles live primarily on dead and decaying animal matter. The necessity of protecting cured meats with wrappings, sacks and washes is in great part due to the depredations of this insect. The larvae often infest bone and dog biscuit factories.

The adults, like the larvae, feed upon ham and cheese, are cannibalistic and readily destroy an infestation of maggots of the cheese skipper, *Piophila casei*. The redlegged ham beetle also infests baled cotton, woolen tops, rattan and salt. However, it does not feed upon these materials. The larvae cause most of the damage to stored meat. They bore holes into the fatty tissue near the rind, growing rapidly, and seem to assemble in the hollow of the bone at the butt end of ham. The adults are surface feeders.

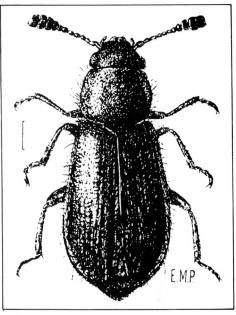

Fig. 16-11. The redlegged ham beetle.

Judd (1949) reports this insect in a shipment of copra from the Phillipines. Redlegged ham beetles invaded houses adjacent to the warehouse where they became a nuisance. These beetles "nipped the skin sharply" when disturbed. The beetles were not considered injurious to copra and no control measures were taken against them. The infestation soon died out. Gorman (1952) notes large numbers of these insects came into Tacoma, Washington with shipments of copra.

The adult beetle lays from 400 to 2,000 eggs on the exposed meat. The egg is smooth, shining, translucent and about one mm long. The newly hatched larva bores into the meat. The full-grown larva is 10 mm long with dark brown head and dorsal thoracic segments. The larva molts two or three times. Both the adults and the larvae are active and predatory. The larva, like the adult, is repelled by light, and after completion of feeding, the larva leaves the greasy material and seeks a dry spot to build its pupal case.

The period from egg to adult may be as short as 30 days, 17 days as a growing larva and 13 days in the cocoon. The adult may live more than 14 months and the female may deposit 2,100 eggs. There are several generations per year and in colder climates winter is passed in the larval stage. The first adults usually appear from the beginning to the middle of May. In hot weather, the adults may be seen in slow flights about the infested room, but their usual mode of movement is by running rapidly. The adults, when handled, emit a strong but ephemeral odor.

Control. The injuries caused by this insect may be attributed, in great measure, to careless packing of hams or to cracking or fraying of the protective covering on hams. The larvae also readily perforate grease-soaked paper wrappings.

Ordinarily, there is no infestation of newly smoked pork products by ham beetles. Where the meats are wrapped promptly there is no trouble, but long-stored hams that are unwrapped or stored in crates, may become infested during warm weather.

Anon. (1943), in a U.S. Extension Service Circular, recommends the following measures be observed to protect home-cured meat from insects and mites:

"An ounce of prevention will save all the pounds of meat cured.

"Eliminate breeding places of meat-house pests. Such insects feed and breed on any form of animal product. Even grease and crumbs of meat or cheese lodged in cracks in shelves, walls and floors harbor them.

"Brush and scrub thoroughly all places where meat has been stored. Smoke-houses free of meat for any length of time should be cleaned as soon as the last meat is removed.

"Keep in tight containers all scraps of meat until rendered. Keep insects out of meat storerooms, as they can fly and carry mites.

"Use 30-mesh screen or finer, fitting doors and windows tightly. Slaughter, cure and wrap meat before insects start to work in the spring.

"Wrap each piece separately and securely in waxed or other grease-proof paper. See that no insect life is on the meat when wrapped. Store meat in a clean, tight, well-ventilated dark smokehouse.

"Examine meat occasionally for grease-soaked bags that insects might penetrate and for holes chewed by rats.

"If meat becomes infested in spite of your precautions — remove from storeroom and trim out any infested parts, cutting deep enough to remove all larvae that may have tunneled into the meat. Uninfested parts of meat are safe, but should be eaten promptly.

"Protect the exposed lean of trimmed meat by greasing it with salad oil or melted fat to delay molding or drying.

"Store at temperatures below 45° F to prevent further growth of insects. Freezing will not spoil the meat.

"Render the trimmings to kill the insects present. The grease may be salvaged. Both dogs and chickens will like the scraps.

"The same pests that damage meat damage cheese. Most of the recommendations for control on meat apply also to cheese."

Anon. (1955) recommends application of sprays to the storage areas before the meat is brought in.

THE DRIEDFRUIT BEETLE

Carpophilus hemipterus (L.) (Family Nitidulidae)

These beetles attack fresh ripe fruit, but are of particular interest to us because of their infestation of dried fruits. They are cosmopolitan in distribution and especially common in fruit packing areas. We are not so much concerned with the actual amount of food these beetles consume as with the fact that a box of dried fruit containing but a few of these insects presents an extremely unsightly mess of larval excreta, pupal skins and adults. Such infestation is naturally a serious problem to both the retailer and the packer. The adult insect is for the most part attracted to moist rather than dried fruit. It has been recorded from dried figs, plums, peaches, apricots, bananas, drugs, nuts, bread,

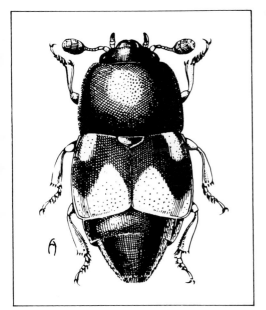

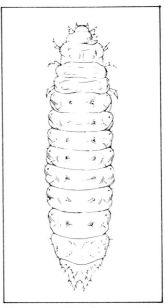

Fig. 16-12. The driedfruit beetle.

Fig. 16-13. The larva of the driedfruit beetle.

biscuits, grain, etc., especially when moist and decaying. It is a major pest of the dried fruit industry in California. Fresh figs which are being dried are very suitable for the development of driedfruit beetles. These insects carry into the ripening figs yeast cells and bacteria which cause a souring of the fruit and also are responsible for several fungus diseases of the fig and stone fruits.

Dobson (1954) has prepared a key to the species of *Carpophilus* associated with stored products. Lindgren and Lloyd (1953) studied the life histories and control of four species of nitidulid beetles infesting California dates: the corn sap beetle, *Carpophilus dimidiatus* (F.); the driedfruit beetle, *C. hemipterus*; the pineapple beetle, *Urophorus humeralis* (F.); and the yellow-brown sap beetle, *Haptoncus luteolus* Er. Okumura and Savage (1974) published an article concerning the larvae and adults of nitidulid beetles attacking dried fruits.

Description of adult. The driedfruit beetle is ⅛ inch/three mm in length. It is somewhat oval and black in color with two large conspicuous amber-brown spots at the posterior tips. These spots may run together to form one large area. The antennae or legs are reddish or amber in color. A characteristic that distinguishes these beetles from most other beetles is the very short wing covers which expose the tip of the abdomen and the antennal tip being knob-shaped.

Life history. According to Essig (1915), the eggs are laid on the outside of the fruit while it is still on the trees or while it is drying on the trays in the open before it reaches the packing sheds. Thus, the fruit is infested before actually arriving at the storehouses. The larval period extends from four weeks to four months; the pupal stage runs for approximately two weeks. The mature larvae are about ¼ inch/six mm long with the head and tip of the tail rich amber brown. The larvae are covered with long hairs and have two large tubercles at the extreme posterior end of the abdomen and two smaller tubercles just in front of the larger ones. The larvae are extremely active. There may be many broods a year since they breed continuously in warm storehouses.

Lindgren and Lloyd (1953) note that at 80° F/27° C the time for the various stages is egg 1.6 days, larva 4.9 days, prepupa 3.2 days, pupa 6.1 days and egg to adult 15.8 days. These authors recommend fumigation with methyl bromide or ethylene dibromide and the use of five percent malathion dusts to reduce and control dried fruit insects.

CORN SAP BEETLE

Carpophilus dimidiatus (F.) (Family Nitidulidae)

The corn sap beetle is two to 3.5 mm long, reddish brown in color and can be distinguished readily from the driedfruit beetle by the absence of the four yellowish areas that are found on the latter. Blazer (1942a) studied this insect as a pest in rice in storage in the South, and the following notes are from his investigations in Beaumont, Texas:

The corn sap beetle when "reared on cracked rough rice, may overwinter in both the pupal and adult stages. Oviposition commences early in March, females depositing from 175 to 225 eggs each. The beetles live about 63 days in summer and as long as 200 days in winter. First generation pupae may be found by the middle of March. Under optimum summer conditions, the life cycle may be completed in 18 days, whereas, in winter it may extend over 150 to 200 days. The beetle prefers food of 15 to 33 percent moisture content and can survive on food of higher or lower moisture content, but apparently it cannot mature to the adult stage in rice containing less than 10 percent of moisture. The insect is able to breed in putrefying food, but develops more rapidly in sterilized food. The beetle pupates in soil, but it can complete its life cycle in the food material it infests if no soil is available." Lindgren and Lloyd (1953) observed the length of development at 80° F/27° C to be: egg 2.2 days, larva 7.1 days, prepupa 4.1 days, pupa 7.3 days and egg to adult 20.7 days.

Gould (1948) found *C. nitens* Fall. common in a corn-processing plant in Indiana.

Osmun and Luckmann (1964) discuss another corn sap beetle, *Glischrochilus quadrisignatus* (Say) which flies from cornfields to picnic areas. It annoys people by crawling on them or by getting into their hair. Osmun and Luckmann have called this shiny black beetle with four yellowish-orange spots on the wing covers the "picnic" beetle. It overwinters as an adult and breeds in the spring. Because it is attracted to fermenting and decaying materials, it can be trapped in muskmelon slices. These are set out several hours before the picnic and then sprayed with DDVP.

Control. Simmons (1943), an authority on dried fruit insects, discusses the prevention of insect damage in home-dried fruits and the following is largely from his work: The Indian meal moth, *Plodia interpunctella;* the raisin moth, *Cadra figulilella,* in California and Arizona; the driedfruit beetle, *Carpophilus hemipterus;* the sawtoothed grain beetle, *Oryzaephilus surinamensis;* the pomace or vinegar flies, *Drosophila* spp., and other species of flies; and several species of mites are among the more common pests on home-dried fruits. In addition to the above, Donohoe (1946a) notes the dried fruit moth, *Vitula ed-*

mandsae serratilineella Rag., is an occasional pest of dried fruits in California, as is the stored nut moth, *Aphomia gularis* Zeller. Many of the insects listed in the chapter on stored product pests can infest dried fruit. Simmons (1960) recommends spray application to floors of packing plants of three to five pounds/ 1.4 to 2.3 kg malathion per 100 gallons/378 l of water.

Where the usual methods of sanitation have failed in preventing infestation, other methods such as fumigation must be undertaken.

Simmons (1943) recommends fumigation with sulphur, or ethylene dichloride and carbon tetrachloride. The latter mixture "should be applied at the rate of about two teaspoonfuls per cubic foot of space to be fumigated. It should be poured into a shallow pan placed above the fruit and the container should remain closed for at least 12 hours." The fumigation should be carried on in hot weather and preferably outdoors away from the house.

Besides proper sanitation, Zeck (1943) used ethyl formate in driedfruit packing sheds where it was "poured on top of the fruit in each box after packing, immediately before the lid is nailed on, at the rate of 10 to 14 cubic centimeters per 56 pound box, or applied to fruit in tins at seven to eight cubic centimeters per 56 pound tin, and subsequent applications are made every two months where the fruit is stored." This treatment did not harm the fruit or affect its flavor. Simmons and Fisher (1945) discuss the use of ethyl formate and isopropyl formate where it is pumped into dried fruit packages and sealed. "The usual dosage of ethyl formate for a 25-pound box of raisins (0.41 cubic foot) ranges from about four milliliters in hot weather to seven milliliters in cold weather . . ." Isopropyl formate is used in approximately the same way. Fisher (1945) and Armitage and Steinweden (1945) discuss the use of methyl bromide as a dried fruit fumigant. The latter authors used one pound/0.45 kg of methyl bromide per 1,000 cubic feet/28 cubic meters for 24 hours at 60° F/16° C for date fumigation. Brown (1937) used a 90 percent ethylene oxide and 10 percent carbon dioxide mixture for the fumigation of dried fruit in barges. He applied one pound/0.45 kg of the fumigant for 1¼ tons of produce. Barnes and Fisher found chloropicrin or a mixture of carbon tetrachloride and ethylene dichloride caused a high percentage of the insects to leave the fruit before death occurred. Balzer (1942b) has authored a detailed paper on the control of the corn sap beetle and other pests of rice. Lindgren and Lloyd (1953) controlled nitidulids with methyl bromide and ethylene dibromide fumigants and five percent malathion dusts.

Simmons (1943) has the following to say in regard to dried-fruit storage containers in the household: "The use of tight containers, such as glass fruit jars provided with rubber rings, friction-top metal cans of various sizes and moisture-vapor-tight containers made for use in freezing storage, will keep insects and mites from attacking dried fruits. Fruits should be dried and free of infestation before being sealed up. Cylindrical paper cartons, such as those used for ice cream, have tight covers and also make suitable storage units for dried fruits. As an added precaution, the lids may be sealed with cellulose tape. Heavy paper bags can be made insect-tight by gluing down the folds at the bottom and folding the top over several times before fastening it with paper clips. Cloth bags cannot be depended on to exclude insects or mites during prolonged storage. In storing dried fruits, it should be borne in mind that newly hatched storage insects are very small and that some are able to enter crevices only one-eighth a millimeter wide, or a space of about five one-thousandths of an inch.

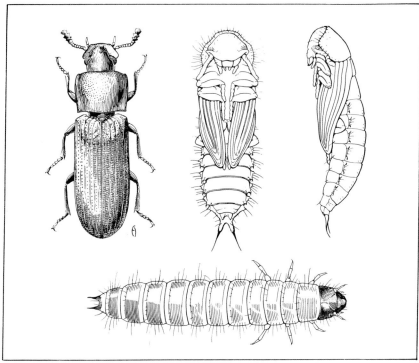

Fig. 16-14. The adult, pupae and larva of the confused flour beetle.

THE FLOUR BEETLES

The Confused Flour Beetle, *Tribolium confusum* Duval
The Red Flour Beetle, *Tribolium castaneum* (Hbst.)
(Family Tenebrionidae)

Tribolium confusum, the "confused flour beetle," apparently was so designated because of the confusion pertaining to its identity. *T. castaneum* is rust-red in color and hence its name. *T. ferrugineum* (Mot.) is a synonym to *T. castaneum* (Hbst.). Much of the information herein is from Good (1939), who made an intensive study of these insects.

This beetle was evidently a pest of the ancients, since it has been found in a jar which probably contained grains of flour in a tomb of the Pharaohs about 2,500 B.C. Prior to its presence in the homes of man, this beetle is believed to have lived under bark and in old logs, and probably scavenged for food, later becoming a flour-feeder. The flour beetles are cosmopolitan. The red flour beetle is essentially an insect of warm climates, whereas the confused flour beetle occurs commonly in the northern part of the United States. Hinton (1948) notes that *T. confusum* is of African origin and that *T. castaneum* is an Indo-Australian insect. The adults are small, reddish-brown beetles, about 3.5 mm in length. These active beetles quickly run to cover when disturbed. Because of their small size, they are able to force themselves into almost any container.

The adults of the two species are very similar in life history, habits and appearance, and this caused great confusion among earlier students of these insects. However, there are important differences in appearance between the two. The antenna of the red flour beetle is abruptly club-like, with a three-segmented club, whereas the antenna of the confused flour beetle is gradually club-like, the club consisting of four segments. Moreover, the sides of the thorax of the red flour beetle are curved, whereas those of the confused flour beetle are somewhat straighter. The beetles are omnivorous and are extremely important pests of flour. Cotton and Gray (1948) observed the "bran bugs," e.g., the above two species of *Tribolium* and the sawtoothed grain beetle, may constitute 80 percent or more of flour mill insects. Cotton (1947) notes the bran bugs "feed on grain dust and milled cereals, but are unable to attack sound and undamaged grain." Willis and Roth (1950) show *T. castaneum* is attracted to flour of high moisture content.

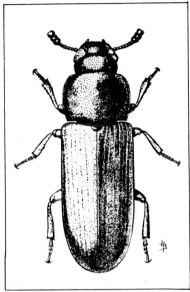

Fig. 16-15. The red flour beetle.

When the insects are present in great numbers, they may cause the flour to turn grayish, and to mold more quickly. Moreover, they impart a disagreeable taste and odor to the flour from a secretion of the scent glands. These insects also infest cereal products and thus become very annoying in grocery stores. The flour beetles are bad pests on ships, where they may infest all the foodstuffs. They are also brought into the house in infested cereals and flour, and may thereby contaminate the food materials in the cupboard.

The flour beetles are unable to feed on the undamaged grain and occur in common with *Sitophilus* and *Rhyzopertha,* where at first they act as scavengers. Later, as the grain develops cracks, they may do considerable injury through their own efforts. Other materials attacked are peas, beans, shelled nuts, dried fruits, spices, milk chocolate and herbarium, insect and other museum specimens. Mills and Pepper (1939) found human beings were not injured by the ingestion of confused flour beetles in cooked cereals. Cotton (1946) found these insects feeding on powdered hand soap containing corn meal. Bissell (1943) observed them thriving in a corn meal and cryolite base, and notes that Cotton found them to infest baits poisoned with arsenicals. Trehan and Rajarao (1945) observed the adult beetles to be predatory on the eggs of the rice moth, *Corcyra cephalonica* Staint.

Because of the ease with which these insects may be reared, Cotton states they are employed in laboratory experiments to determine the relative toxicity of poison gases, as well as in many studies of a biological nature.

Life history. The adults of the red flour beetle can fly short distances, but the adults of the confused flour beetle, although similarly provided with wings, have not been observed in flight. They are capable of breeding throughout the year where the building is warmed during the winter, but only the adults are found in unheated buildings. The adult insects are long-lived and may live for more than three years.

The average oviposition period was about eight months for that of *T. confusum,* and 5½ months for that of *T. castaneum.* The average number of eggs laid per day was two or three, with 300 to 400 eggs as an average total. The eggs are whitish or colorless and are covered by a sticky material to which particles of flour adhere, and thus are very difficult to detect. They are laid in the material in which the female happens to be at the time of oviposition. The incubation period for the egg is approximately nine days at room temperature. The larva, when mature, measures about ¼ inch/six mm in length, and is white to yellow in color. The larva of the flour beetle is distinguished by the two-pointed or forked termination of the last body segment.

The life periods vary with environmental conditions. The number of larval instars ranges from five to 18, Mickel and Standish (1946) noting the latter number for the beetles reared on edible soya products; the more common number is seven or eight. The length of the larval period varies from 22 to more than 100 days. The pupal period is approximately eight days for both species at room temperature. The life cycle may be completed in seven weeks, or it may require three months or longer.

Control. These insects are controlled by means of the fumigants commonly used against stored product pests, as well as through heat sterilization. Pepper and Strand (1935) controlled the flour beetles by maintaining a temperature of 120° F/49° C for several hours, after this temperature was recorded by themometers lying on the floor. DeCoursey (1931) trapped beetles in corrugated paper containing wheat flour as bait.

According to Green and Kane (1959), malathion was more effective in protecting bagged peanuts than either lindane or DDT. It was applied as a spray at 40 to 80 mg/sq. ft. at monthly intervals.

Bissell and DuPree (1946) studied the infestation of shelled peanuts in storage by *Tribolium* spp. and other insects, and showed jute bags became heavily infested in storage, whereas cotton bags were relatively lightly infested.

Saunders and Bay (1958) showed 0.2 percent pindone baits to prevent the development of the larvae of the confused flour beetle.

Anon. (1949a) recommends the use of a pyrethrum-piperonyl butoxide-water emulsion or oil base spray to be applied to clean surfaces for control of confused flour beetles. These sprays are sold as concentrates, usually in a ratio by weight of one part pyrethrins to 10 or more parts piperonyl butoxide. The emulsion concentrate is diluted with water or fly-spray base oil, respectively. The spray is applied every few weeks at the rate of approximately one gallon to 1,000 square feet.

THE MEALWORMS

The Yellow Mealworm, *Tenebrio molitor* L.
The Dark Mealworm, *Tenebrio obscurus* F.
(Family Tenebrionidae)

The common names are derived from the color of the wirewormlike larvae. Cotton notes *Tenebrio* means "darkness" and is appropriately applied to the larvae that are "nocturnal in habit and frequenters of dark places." The larvae, pupae and adults of these two beetles are very similar in form, size and color.

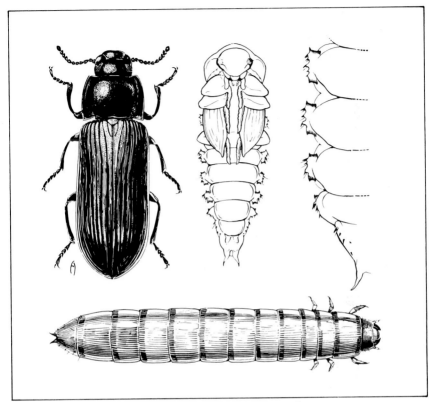

Fig. 16-16. the adult, pupae and larva of the yellow mealworm.

The adults are more than ½ inch/13 mm in length, but those of the dark meal worm are dull pitchy black in contrast to the polished and shiny dark brown or black of the yellow meal worm adult. The two species have well developed wings and are attracted to light.

These insects are cosmopolitan, and both are believed to be European in origin. *T. molitor* occurs in the cooler climates of the North, whereas *T. obscurus* is in practically all parts of the United States.

The mealworms are among the largest insects that infest stored products, but they are only moderately important in this respect. As mentioned previously, they are nocturnal in habit and hide in dark moist corners, refuse, grain, sacks, bins and other places that are not disturbed frequently. The active larvae wander about and occur in some of the most unsuspected places. The writer knows of an infestation of these insects in the ground of an unfinished cellar beneath a macaroni factory.

MacNay (1950) reports these insects in straw-stuffed furniture. Riley and Howard (1889) tell this interesting tale of a presumably haunted room: "A guest in a hotel in Rhode Island was awakened one night by a scratching sound that apparently emanated from a pincushion on the dressing table. After spending a sleepless night the guest reported to the landlord that the room was haunted.

Investigation resulted in the discovery that the pincushion was literally alive with beetles. The cushion had been made about four years earlier and had been filled with coarse shorts. The beetles, which proved to be *Tenebrio molitor,* had been breeding in the cushion until they became so abundant as to attract attention by their struggles."

Palmer (1946) notes this insect is probably the commonest beetle parasite of man in the United States. The eggs or the larvae of the insects are eaten in cereals and breakfast foods. A five-month-old child passed the larvae by the rectum. Beetles were found in the infant's precooked cereal. The author knows a man who while sleeping was bitten on the cheekbone by the beetle.

The mealworms are not considered pests by all individuals. In fact, some people make their living by raising them, and there are pamphlets that discuss the subject in technical detail. Anon. (1954) gives directions for rearing mealworms. These worms serve as food for animals in aquariums and zoological parks, and are very popular as fish bait. Stampfel (1944) notes *T. molitor* and *T. obscurus* are reared in large numbers in Czechoslovakia as food for fowl during the winter, when insect life is scarce.

Life history. Cotton and St. George (1929) made a detailed study of the life history of the mealworm. In Washington, D.C., mealworms overwinter as larvae. The adults emerge in the spring and early summer, when they live for two or three months and then die. There is usually but one generation a year, although some individuals may require two years to complete their development. The female lays bean-shaped white eggs, coated with a sticky secretion that causes flour and other particles to adhere to them. The eggs hatch in about two weeks. The female oviposits over a period of 22 to 137 days. The yellow mealworm lays an average of 276 eggs and the dark mealworm deposits an average of 463 eggs. The larvae, when full grown, are about 1¼ inches/32 mm in length. The larva of the dark mealworm is dark brown, and that of the yellow meal worm is bright yellow. There are usually 14 to 15 molts. The ninth abdominal segment of the larva has a pair of small, upturned, dark chitinous spines on the dorsal surface. The larval period for both species is often more than 600 days. The mealworms then pupate near the surface of the material on which they are feeding. The complete life cycle for both species of mealworm is 10 months to two years.

Control. This insect can be controlled through the use of fumigants, as well as by superheating and freezing. Anon. (1954) recommends sprays for the treatment of wooden floors or walls of empty granaries or warehouses. These sprays included emulsions containing pyrethrin alone or with a synergist. The dosage used was one to two gallons per 1,000 square feet/3.8 to 7.6 l per 93 square meters.

MEDITERRANEAN FLOUR MOTH

Anagasta kuehniella (Zell.) (Family Pyralidae)

Some students of insects are of the opinion this now cosmopolitan insect first came from the Mediterranean region. Others believe it originated in Central America. It infests a great variety of foodstuffs and is an important pest of flour. The Mediterranean flour moth occurs in nuts, chocolate, seeds, beans, biscuits, dried fruits and stored foods of many kinds, and is a common insect in the home. It is a pest of mills and warehouses, where it clogs the machinery with its webs, which at times necessitates a shutdown of the mill so the webs can be removed.

Patton (1931) notes an interesting case where the worms became so thick in the biscuits of soldiers that the hungry men could only eat the biscuits in the dark. Horn (1934) found one larva of this moth actually gnawing the edge of a lead sheet.

Description of the adults. The adult moth has a wing expanse of less than one inch/25 mm, with the hind wings a dirty white and the fore wings a pale gray with transverse black wavy bars. When at rest, the adult raises its forelegs, thus elevating the forepart of the body and thereby giving the wings a very distinct slope; this char-

Fig. 16-17. Yellow mealworm larvae.

acteristic is much more stable than the bars on the wings which may disappear when the wings become worn. The night-flying moth moves in a very characteristic rapid zig-zag fashion. According to Okumura (1966), this insect is often confused with the dried-fruit moth.

Life history. The eggs are laid wherever the food for the larva occurs and are usually fastened to the object. Ullyett (1945) proved chalk to be as suitable as flour for oviposition. He concluded "that the ultimate stimulus for oviposition is the texture of the medium, and not its olfactory attractiveness." One female is capable of infesting an entire mill and may lay from 116 to 678 eggs. At a temperature of 80 to 90° F/27° to 32° C the eggs may hatch in three days.

The larva attains full growth in 40 days. It is whitish or pinkish with a strongly chitinized head and is from 0.5 to 0.63 inch/12.7 to 16.0 mm in length. The young larva confines itself to silken tubes which it spins constantly. The silk forms balls of flour which clog the bolters and other parts of the machinery in flour mills, making it necessary to close the mills and clean the machinery. Since the larva migrates when mature, it may be found by homeowners in practically any part of the house.

The larva pupates in a cocoon in the flour or on the surface of the flour, or it may pupate in some crack or crevice either with or without a cocoon. In warm mills, the life history may be completed in from one to 1½ months and there may be six or more broods per year.

Essig (1940) found these insects breeding in the brood comb of honeybees situated in a chimney. These insects are controlled in mills and warehouses with the fumigants discussed in the chapter on Fumigation.

INDIAN MEAL MOTH

Plodia interpunctella (Hbn.) (Family Pyralidae)

The adults of this moth have a wing spread of about ⅝ inch/16 mm and have a broad grayish band across the bronzy appearing wings. They are very general feeders and feed upon grain and grain products, a variety of dried fruits, seeds, graham crackers, nuts, powdered milk, chocolate, candies, etc. Dried red peppers, bird seed and dehydrated dog food containing meat and cereal are some of the other sources of this insect in the house. This pest is very important and very destructive wherever dried fruits are stored. In this respect, it is interest-

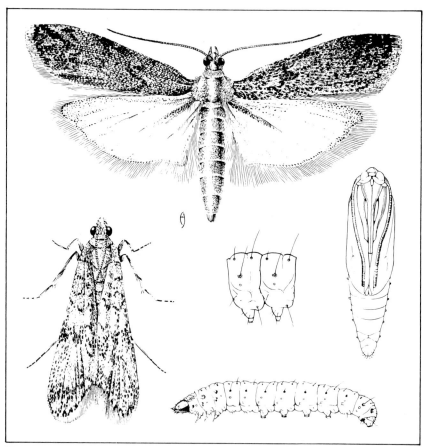

Fig. 16-18. Adult, pupa and larva of the Mediterranean flour moth.

ing to note that Candura (1943) found the larva attacking fresh apples and pears in a warehouse. It is a native of the Old World and presently cosmopolitan in distribution. This insect is an important pest of dried fruit since it lowers the quality and adds the extra expense of having the fruit processed before it is marketed. A few larvae in a package with their webbing and frass are very repulsive to the housewife, and costly to the firm marketing the material. The fig, of all the dried fruits, is the most readily infested.

Donohoe (1946a) writes of this insect as follows: "As for many other stored foods, the Indian meal moth is the number one lepidopterous pest of processed dried fruits in storage. It develops on such a varied assortment of foods and is of such world-wide distribution that without proper control methods, its presence in every food warehousing facility in the United States may be accepted as a foregone conclusion. Besides the industrial warehouses these include even the small storage facilities of local stores, institutions and the like, and the writer suspects, even most, if not all, of the home kitchens and pantries of the nation. As a sidelight, my own residence at this writing literally swarms with

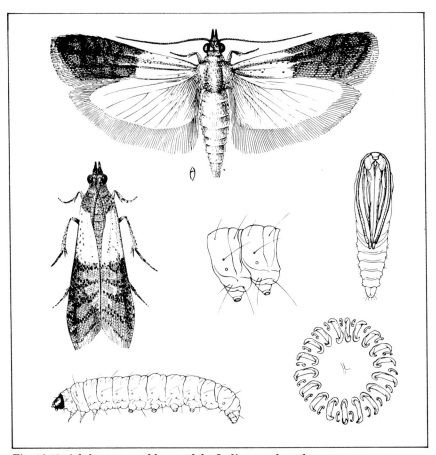

Fig. 16-19. Adult, pupa and larva of the Indian meal moth.

adults, despite the fact I have been unable to find any infested food supply. Curious enough, the adults are most abundant upstairs and not in the downstairs kitchen and food storage facilities. The 'white worms' so frequently reported by housewives in purchases of dried fruits probably always are of this species.

Although it is seldom a pest of milling machinery, like the Mediterranean flour moth, its webbing will in time become sufficiently abundant in previously injured grain to become noticeable. Since it prefers the coarser grades of flour, it is the most common insect in packages of whole wheat, graham flour and corn meal. The larva webs together the flour with strands of silk.

Hamlin et al. (1931) states the insects commence the year with a brood of overwintering larvae, which pupate mostly in March and emerge in April. The generations overlap as the seasons progress. The female usually commences to oviposit about three days after emergence and lays some 200 to 400 eggs in from one to 18 days. The eggs are deposited chiefly at night and range in number from 39 to 275, the average being 128. Upon hatching, the larvae begin to disperse. Within a few hours, they establish themselves on the crevices of the fruit and feed in or near a tunnel-like case of frass and silk they web together. Then they emerge from these hiding places and spin their cocoons in the crevices in the vicinity of their food. The range in the duration of the larval period is extremely great and may extend from 13 to 288 days. This is due principally to such environmental factors as food and temperatures. The shortest life cycle of this species was 27 days and the longest was 305 days. Five generations were raised on raisins in a single year and four on prunes.

The egg is grayish-white and from 0.3 to 0.5 mm in length. The larva is from nine to 19 mm in length, with an average of about 13 mm. Its color is usually dirty white, but may range from pink to brown to a greenish tinge.

Control. Strong (1936) reports that methyl bromide, three parts of ethylene dichloride and one part ethylene oxide, at the rate of four and seven ml respectively, per 25 pound/11.3 kg box, gave complete control of *Cadra figulilella.*, and *Plodia interpunctella.*

McLaine (1943) controlled the Indian meal moth and a species of *Ephestia* damaging the surface grain in a terminal elevator. Fumigation with hydrocyanic acid gas was first used. This was followed with a treatment of a proprietary fumigant to kill the larvae deeper in the bin and pyrethrum spray was used on the bin floor and around the ventilator boots. Such fumigants as methyl bromide and sulfuryl fluoride also are effective against this insect.

Quinlan and Miller (1958) controlled the Indian meal moth in shelled corn with synergized pyrethrins.

ANGOUMOIS GRAIN MOTH

Sitotroga cerealella (Oliv.) (Family Gelechiidae)

This cosmopolitan species is of European origin and was given the name Angoumois grain moth because it first was reported destroying grain in this French province. The Angoumois grain moth first established itself in this country in North Carolina in 1728. The adult is about the size of the webbing clothes moth and yellowish white, with pale yellow forewings and gray characteristically pointed hind wings.

This moth prefers to oviposit in barley, rye, corn, oats, rice, various seeds. It also chooses damp grain in preference to old dry grain. The presence of small holes in corn or wheat, as well as insects the size of clothes moths, is a good indication of an infestation by the Angoumois grain moth. It attacks dried grains in storage and also grains maturing in the field. It is considered to be an important pest of stored grain.

Since the Angoumois grain moth is active at low temperatures, it does much damage during the winter. Grain that is infested has a sickening smell and taste that makes it unpalatable. The moths, in their working on the grain, leave much debris on the top. Long storage will often lead to infestation.

Life history. The female may lay from 40 to 300 eggs. The egg is difficult to see without a magnifying glass. The eggs, which are at first white, turn red with age. During the warm weather of summer the eggs hatch after an incubation period of from four to eight days. The larva then bores into the grain and spins a silken web over the hole which it entered. The larval stage will last for about three weeks. The larva undergoes three molts.

It may hibernate in scattered grains or in bags of mill screenings and become full grown in the spring. Pupation may occur in May, taking place in a silken

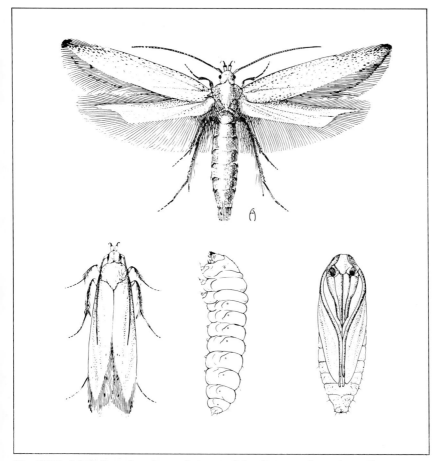

Fig. 16-20. Adult, larval and pupal stages of the Angoumois grain moth.

cocoon, in a cavity that results from feeding. The adults emerge through a small round hole. The pupal period is about 10 to 14 days, depending on the temperature. The full grown larva is about ¹/₅ inch/five mm long, white in color, with a yellowish-brown head. The pupa is reddish brown. It has small legs and four small pseudopods.

The life cycle in warm countries may require from five to seven weeks, whereas during the winter in colder climates, the larva is dormant for four to five months, and the life cycle may be fully six months long. As a rule, there are four or five generations per year, but in heated warehouses there may be as many as 10 to 12 generations.

Candura (1943) found *S. cerealella* attacking hulled chestnuts, red beans and buckwheat in storage. It was controlled by dusting the grain with a "two or three percent fossil farina (a mealy-looking infusorial or microphytal earth)." This dust protected the grain from attack and also destroyed the larvae.

SPIDER BEETLES OR PTINID BEETLES

Family Ptinidae

These are small oval or cylindrical beetles that often look like giant mites or small spiders. The larvae are similar to those of the scarabs. Spider beetles are cosmopolitan pests, feeding on dried animal and vegetable matter. At times, they become important pests in homes, warehouses, grain mills and museums. Howe (1959) notes these beetles are scavengers. They are attracted to moisture and excrement and are found in the nests of animals. The larvae have been found feeding on wool, hair, feathers, textile fabrics and old wood. Gray (1952) found the beetles to be active at freezing temperatures. Due to their resistance to cold, spider beetles are often of some consequence in cereal products in Canada and the northern United States. Brown (1940), Hinton (1941), Hall and Howe (1953) and Papp and Okumura (1959) made keys for the identification of these insects. Howe and Burges (1951, 1952 and 1953) made a comprehensive study of the biologies of *Niptus hololeucus* (Fald.), *Gibbium psylloides* (Czemp.), *Trigonogenius globulus* Solier, *Stethomezium squamosum* Hinton, *Eurostus hilleri* Reitt., *Ptinus ocellus* Brown (*=tectus* Boield), *P. fur* (L.) and *P. sexpunctatus* Panzer, on a variety of foods.

THE HAIRY SPIDER BEETLE

Ptinus villiger (Reit). These beetles are reddish brown in color, often with two irregular white patches on each wing cover. Moreover, the elytra or wing covers have several longitudinal rows of shallow pits. The adults are from 2.2 to four mm in length. They are distributed throughout Europe, Asia and North America. Gray (1942) states this beetle is an important pest of cereal products in Canada.

Gray (1934) made a detailed study of the life history of this insect in flour sheds. The adult appears in the spring and the female may deposit up to 40 eggs. The egg is spindle-shaped, pearly and 0.6 mm long. The eggs may be deposited on the outside of the bags, through the mesh of the sacks and in flour debris situated in cracks and corners. The larva is cream-colored with a light brown head capsule. When mature, the larva may reach a length of 3.8 mm. It completes its development in three months at a temperature of 82° to 86° F/28° to 39° C and pupates in the cereal debris of the flour, often burrowing into the wooden shed holding the flour. The pupal cell consists of a silky material which is secreted by the larva. This, in turn, is coated with the material in which the larva pupates. The larva often overwinters in the pupal cell, without transforming into a pupa until the following spring.

This insect infests stored products such as flour, farina, corn meal and wheat. Watters (1961) controlled these insects with a five percent methoxychlor water suspension.

THE WHITEMARKED SPIDER BEETLE

Ptinus fur L. The whitemarked spider beetle is cosmopolitan in distribution and was described by Linnaeus in 1766. The adults are spider-like beetles with reddish brown bodies covered with yellow hairs. The females, unlike the males,

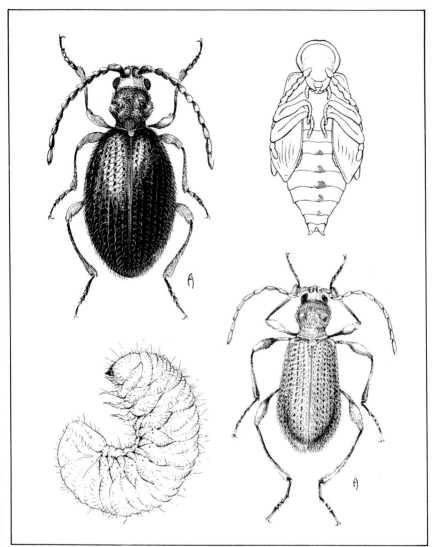

Fig. 16-21. The brown spider beetle, *Ptinus hirtellus* **Sturm. (Above) Adult female and pupa. (Below) Larva and adult male.**

have two white patches on each wing cover. These patches may join and form white bands. The adults are two to 4.3 mm long; the female being larger and rounder. The larva is very similar in appearance to that of the drugstore beetle and feeds in a globular cell formed from the material it is consuming.

Howe and Burges (1951) observed *P. fur* to complete its development at an optimum temperature of 73.4° F/23° C and 70 percent relative humidity in 32.1 days on fish meal.

In Washington, D. C., the life history is completed during the summer in 3½ months. In artificially heated buildings, there may be two generations a year. Chittenden (1896) found the adults in the dead of winter crawling up on walls of cellars and walls of unheated buildings.

The whitemarked spider beetle is a pest in warehouses, granaries, museums and libraries. It is reported from feathers, animal skins, stuffed birds, herbarium specimens, stored seeds, ginger, cacao, dates, paprika, rye bread, flour, stored cereals and insect specimens. Gross (1949) found the spider beetle breeding in fly manure. According to Chittenden (1896), in Concord, New Hampshire, the whitemarked spider beetle was found in a barn containing cotton seed in bags, and was feeding on both the seed and the bags. From here the beetles spread widely and even entered homes, where they ate the clothes. As a whole, the insect is of slight economic importance in the United States.

THE BROWN SPIDER BEETLE

Ptinus clavipes Panzer (*Ptinus hirtellus* Sturm.) This cosmopolitan insect is very similar in habits and appearance to *Ptinus fur* L. and may live together with it. It is uniformly brown in color and 2.3 to 3.2 mm in length. The brown spider beetle is omnivorous and is a scavenger in cellars, attics, storehouses, henhouses, etc. It feeds on books, feathers, skin, dried mushrooms, and excrement of rats and other animals, the powdered leaves of such drugs as senna and jaborandi, dried fruit, grains and sugar.

THE AUSTRALIAN SPIDER BEETLE

Ptinus ocellus Brown. The Australian spider beetle is a dark reddish brown spider beetle whose wing covers are covered with golden brown or yellowish hairs. It is 2.5 to four mm in length and is cosmopolitan in distribution.

Hatch (1933) lists it from "cayenne pepper, chocolate powder, desiccated soup, cacao, nutmegs, almonds, ginger, figs, sultanas, dried pears, dried apricots, beans, rye, fish food, maize, casein, stored hops, poultry food, paprika pepper and fish meal." In addition, Hinton (1941) reports this insect in dried insects and crushed crabs. Mackie (1932) observed it in a grain elevator where it not only infested the grain, but riddled the timbers of the elevators as well. The Australian spider beetle has been known to eat holes in carpets. Hatch (1943), speaking of damage by this insect to furs, notes "A fox skin looked as though perforated by a shot gun; the tail has been eaten off the skin of a white ermine, and much fur had been removed from an otter skin." Shapiro (1941) found this insect to infest ant pupae used for medicinal purposes. Howe (1950) notes this species is a pest in warehouses, since "it consumes the foodstuffs, adds excreta, silk and fragments to it and spoils the fabric of the rooms and sacks it invades, spilling the food in doing so." Duffy (1953) records this spider beetle damaging lead cable.

According to Patton (1931), the life history of the insect is as follows: The eggs are laid in batches in the larval food. The female may deposit from 70 to 120 or more eggs. The eggs are laid in early summer and there are four to five larval instars. The larvae pupate in September and October. They then bore through the food and pupate in some crack, and as was previously mentioned, may in some instances bore into wood in order to pupate.

Howe (1949) showed *P. ocellus* does not survive at temperatures above 82.4° F/28° C, can develop at 50° F/10° C and oviposit at 41° F/5° C. Optimum conditions for development of this species are temperatures of 71.6 to 77° F/22° to 25° C and 80 to 90 percent relative humidity. Howe and Burges (1953) completed a comprehensive study of the biology of this species.

Howe (1950) studied the behavior of this insect in a warehouse and found it to be most active during darkness. "Hence, it is fairly continuous in dark premises and periodic in light places. In the latter, beetles emerge from their hiding places at dark and return at dawn. If dark crevices are provided artificially they act as traps, for some of the beetles will use them as daytime hiding places. Activity is reduced by low temperatures but does not stop altogether until it is as low as 2° C."

Hickin (1942) showed hungry adult beetles and well-fed larvae can bore their way through transparent cellulose used for packaging. However, the larvae ordinarily bore out of packaged foodstuffs in order to find a more suitable place for pupation. Gunn and Knight (1945) had a female under observation that lived for approximately one year, during which time she laid almost 1,000 eggs.

THE AMERICAN SPIDER BEETLE

Mezium americanum (Lap.). This dark, reddish brown to nearly black insect is 1.5 to 3.5 mm in length and apparently cosmopolitan in distribution.

Montgomery (1936) found this beetle in bed blankets and around the edge of a carpet. They were treated there, but in a few days the infestation was renewed. The source of the infestation was finally discovered beneath the bathroom floor, where rats had nested. The nest contained old bones, rat excreta, dead rats, etc., and was alive with the beetles. Due to the inaccessibility of the nests, the area had to be treated with a powder containing a high percentage of sodium fluoride. This same infestation occurred in six different localities in San Francisco. Gerber (1946) observed these insects living in cracks and the accumulated dirt of a large baking establishment. The insects were most common in the receiving department and the basement storage room.

Swezey reports them from Hawaii in stored seed, dried sunflower heads and a sparrow nest. Cotton and Good (1937) note their presence in tobacco seed, cayenne pepper, opium and grain.

THE GOLDEN SPIDER BEETLE

Niptus hololeucus (Fald.). The golden spider beetle is three to 4.5 mm in length, with numerous long yellow silky hairs. Except for the tropics, this insect is cosmopolitan in distribution.

As a rule, it is a pest merely because of its presence, although at times it may be definitely injurious. This insect is found in homes, bakeries, flour mills, warehouses and granaries. Both the adults and larvae feed on woolens, linens and on natural silks, as well as dead animal matter. The textiles are much more likely to be damaged if they are soiled with grease. They have been reported from sponges, bones, feathers, casein, brushes, leather goods, cacao, spices, dead insects, rat and mouse excreta, books, paper, bran, grain flour, stored seeds and bread.

Patton (1931) notes the following life history for this insect. The female lays 25 to 30 eggs; the larva undergoes two molts. The complete life history from egg to adult is six to seven months, with two generations a year. Howe and Burges (1952) studied the life history of the species and noted it took 11.5 days from

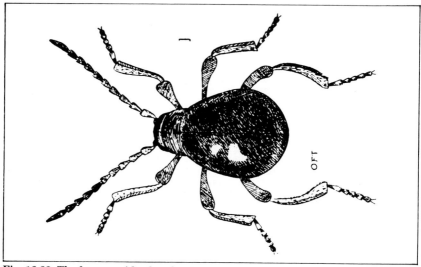

Fig. 16-22. The hump spider beetle, *Gibbium psylloides* (Czemp.)

egg to adult at 77° F/25° C. The adults are plentiful in June and July. They are wingless and can spread only by crawling.

THE HUMP SPIDER BEETLE

Gibbium psylloides (Czemp.). This ptinid is 1.7 to 3.2 mm long, brownish red to nearly black in color and cosmopolitan in distribution. In appearance it has been compared to a giant mite.

Kemper (1938) reports it from sponge rubber bath mats, woolens, towels, leather, paste and tallow. Hinton (1941) records it from houses, hotels, warehouses, mills, granaries, bakeries, latrines, cayenne pepper, stored seeds, wheat, bran, stale bread, decaying animal and vegetable refuse, stored wheat, cereals and opium cake. Gross (1948) found this insect breeding in rat excreta. Curran (1946) notes some individuals call this insect the "bathtub" beetle since it commonly falls in the bathtub and is trapped there. In addition to some of the above foods, Curran records it in baby food and dog biscuits. The life history is somewhat similar to that of *Niptus hololeucus*. Howe and Burges (1952) studied its biology. It grows from egg to adult at 91.4° F/33° C in 45 days.

Tarry (1967) controlled this insect with DDVP resin strips at the rate of one strip per 1,000 cu. ft./28 cubic meters.

MISCELLANEOUS SPIDER BEETLES

Some other spider beetles that are occasionally pests in the home are: *Ptinus raptor* Sturm from Europe, Russia, United States, Canada and Newfoundland; *Ptinus latro* F. with a world-wide distribution; *Sphaericus gibboides* Boieldieu from southern Europe, North Africa and California; *Trigonogenius globulus* Solier from Europe and North America; *Tipnus unicolor* P. & M. from Europe, Transcaucasia and Canada; and *Eurostus hilleri* (Reit.) from Japan, Great Britain and Canada. The latter insect breeds to a large extent on rat and mouse droppings.

Control. Since these insects feed on a wide variety of materials, it is essential the premises be cleaned thoroughly prior to the application of insecticides. Several species of spider beetles have been recorded breeding in and feeding on rat and mouse feces, and it may be necessary to control these rodents at the same time that measures are undertaken against the beetles. Residual DDT sprays containing five percent or more DDT have been used successfully against spider beetles, but no longer can be used. Other insecticides with residual properties also are of value. Tarry (1967) found DDVP resin strips to be effective against *Gibbium psylloides*.

Since spider beetles are important pests of cereal products in Canada, Canadian workers have devoted a great deal of attention to their control. Gray (1942) and Smallman (1948) studied these insects and the methods of control. Much of the following information is from their papers. Smallman notes spider beetles are of economic importance in Canada "due to their habit of laying eggs *through* the mesh of cotton flour sacks. In Manitoba, the egg-laying period extends from early May to the middle of July and may be assumed to begin during April in the milder climate of Alberta. There is an emergence of adults in the late summer, but no eggs are laid until the following spring. Accordingly, it is only during the four month period, April to July, that warehouse stocks require protection against infestation by spider beetles."

Gray states although spider beetles are especially important in Canada, they also are prominent cereal pests in the northern United States. He notes spider beetles "are primarily warehouse pests of cereal products and only invade mills when the premises are used for storage as well as production. The heaviest infestations are likely to be found in slow-selling lines which remain in storage for a long time. The materials attacked by spider beetles include all milling company products, as well as various seeds and certain whole grains (i.e., wheat, barley, rye and flax). The most serious infestations have been in flour and in feeds such as bran and shorts and various meal preparations.

"Spider beetle larvae cause a typical 'scarring' of the wood in buildings in the formation of pupal cells prior to pupation. This often serves as a valuable aid in warehouse surveys indicating spider beetle infestation in past seasons.

"Infestations vary considerably. In some cases, only a few larvae are present per sack. All the stages in one badly infested 49-pound sack of flour were counted and a total of 3,126 was secured or about 64 insects per pound of flour. As the sack contained many holes made by the larvae in leaving this particular bag, the population earlier in the season was even larger. All of this infestation occurred within a period of eight months in a country warehouse. Surely no one would envy a warehouseman the task of explaining to a feminine customer why she received such a 'prize' package."

Smallman (1948) controlled spider beetles in cereal warehouses by using either a five percent (weight per volume) DDT oil-base spray, or a DDT wettable powder at a five percent concentration or a 0.5 percent gamma benzene hexachloride oil-base spray. These sprays were applied to the floor and five feet/1.5 m up on the walls by means of a knapsack sprayer. Watters and Sellen (1956) controlled *Ptinus villiger* (Reit.) with five percent DDT emulsion at one gallon per 100 sq. ft./3.8 l per 9.3 sq. meters.

MacNay (1950) found five percent DDT suspensions or emulsions effective against these insects. Magnesium oxide and aluminum oxide dusts that were applied to the warehouse floors resulted in good control; although these dusts were not as effective as the chlorinated compounds. Cotton sacks impregnated with DDT or TDE protected the enclosed flour. Paper sacks also prevented infestations. Neither DDT or TDE may be used now.

Where residual sprays or dusts cannot be used for fear of unacceptable levels of food contamination, contact sprays employing pyrethrins, and pyrethrins and synergist combinations, give fairly good control at frequently repeated intervals.

Patton (1931) mentions a method of control whereby a cloth moistened with a solution of one drop amyl acetate in 3½ liquid ounces/103 ml of alcohol is placed in infested localities. The beetles are attracted to the treated cloth and then destroyed. Montgomery (1936) dusted with sodium fluoride to control *Mezium americanum* in a hotel.

STORED PRODUCT PESTS OF LESSER IMPORTANCE
THE ALMOND MOTH

Ephestia cautella (Wlkr.) (Family Pyralidae)

This cosmopolitan insect feeds and breeds on dried fruit, especially figs and dates. Often it infests shelled nuts, grain and seeds. The adult is a gray moth measuring 14 to 20 mm in wing expanse. It occurs in storerooms, warehouses and dwellings. The moths fly with a quick dart and with rapid vibration of the wings. Simmons et al. (1931) found that this insect laid an average of 114.1 eggs. These hatched in four to eight days during the summer. The larva is dirty white in color and when full-grown is tinged with brown or purple dots on its back, which give it a striped appearance. Ordinarily, the larva migrates out of the figs in order to pupate. When these insects are reared in raisins, they have an incubation period averaging eight days, a larval period averaging 63.6 days and pupal period averaging 10.4 days. In all, an average of 81.7 days is required for completion from the egg to the adult stage.

Davis (1947) records an infestation of adults in a theatre in Chicago. These moths flew in the spotlights and the actors referred to them as "B-29s." The larvae were found feeding on nuts, dried fruits and other foodstuffs lodged in the air-conditioning vents in the floor of the theatre. These infested areas were cleaned and then sprayed with a five percent DDT spray, which resulted in control of the nuisance.

Cotton (1950) notes the almond moth is a potentially serious pest of stored products, particularly in warehouses. He recommends fumigation or pyrethrum sprays where the building is not suitable for fumigation.

Donohoe (1946a) notes the singular fact that in California this moth does not infest drying or dried fruit, although it is a serious pest of almonds in the state.

THE TOBACCO MOTH

Ephestia elutella (Hbn.) (Family Pyralidae)

This insect is an important pest of chocolates, tobacco, cereals, seeds, etc. It is of lesser importance in grains and flour. Reed and Livingstone (1937) made a study of this insect on tobacco.

The cosmopolitan tobacco moth is often a severe pest on stored tobacco and may consume the entire leaf, with the exception of the larger veins. Tobacco that is not consumed is soiled or rendered worthless by the webbing and excrement adhering to the leaves. The moths attack the highest-priced tobacco grades, since these contain the greatest amount of sugar, principally levulose.

The tobacco moth is light, grayish-brown with two light-colored bands extending across each forewing. The hind wings are uniformly gray in color. The wings from tip to tip are ⅝ inch/16 mm in length. The larvae migrate when full grown and pupate in sheltered locations. The migration of the full grown larvae is most noticeable in the fall and the larvae may then overwinter and emerge in May and June.

The eggs are laid singly or in small clusters on or near tobacco. The larvae ordinarily molt five times. Prior to pupation, they construct loosely woven cocoons. The peak of emergence is during the spring. There are two complete generations in tobacco warehouses, with three periods of moth emergence.

Richards and Waloff (1946) studied the life history of the tobacco moth in bulk grain. The overwintering larvae spun their cocoons between the ceiling boards and in other cracks. Pupation began in May, although most of the insects pupated in June. The pupal stage lasted three to four weeks. The peak of adult emergence occurred in July. Mature larvae were very numerous in September. There may be one generation and also an overlapping generation per year. Overwintering adults were found to infest wheat.

Control. Reference should be made to the section on the control of the cigarette beetle, since these control methods are often applicable to the tobacco moth. Reed et al. (1933) controlled these insects with hydrocyanic acid gas fumigation in warehouses. They also used vacuum fumigation, applying a mixture of ethylene oxide and carbon dioxide (9:1), at 29 inches, when the temperature of the tobacco was 70° F/21° C with an exposure to the gas of three and one-half to four hours. Tenhet (1959) controlled this insect with weekly sprays of 0.2 percent pyrethrins in an oil base. Phillips et al. (1959) fumigated bulk stored cocoa beans with two lbs./0.9 kg methyl bromide per 1,000 cu. ft/28 cubic meters. Press and Childs (1966) controlled this pest in warehouses with DDVP applications.

THE RAISIN MOTH

Cadra figulilella Greg. (Family Pyralidae)

Cotton and Good (1937) note this cosmopolitan insect is found "in warehouses and granaries, feeding on grain, rice, meal, corn, oatmeal, dried fruits, etc." Donohoe et al. (1943 and 1949) studied the life history of this moth as a pest of commercial dried raisins, peaches, apricots, pears and figs in California.

"The drab-gray moths are about ⅜ inch long when at rest with wings folded. They are active chiefly in the early evening, spending the daylight hours in shaded, protected places. Their length of life is about two weeks, during which time the females produce about 350 eggs each. The minute white eggs are scattered about on the fruit. When full-grown, the larvae are one-half inch long, white, with four rows of purple spots along the back. In warm weather, the eggs hatch in about four days; the larvae feed and grow for about one month, then leave their food and enter the soil or some other dark retreat to transform and emerge as adults.

"The well grown larvae overwinter in the topsoil or under the loose bark of grapevines. Waste fruits, of which mulberries are the earliest to ripen, are abundantly available as larval food and breeding continues from April or May until November."

THE MEAL MOTH

Pyralis farinalis (L.) (Family Pyralidae)

This cosmopolitan moth is brownish in color and has a wing spread of about one inch/25 mm. It is a general feeder on cereals and webs, and binds together seed of all kinds. The larvae feed on flour, meal, damaged grain and seeds. They are fairly common in flour mills, but prefer damp and spoiled flour and grain.

The larva attains the length of one inch/25 mm when full grown. The head of the larva is black and the end of the body is tinged with orange. It very often spins tubes of silk, which contain particles of food material. The larva feeds from openings at the end of the tube. It spins a silken cocoon in which to pupate. The egg, larval and pupal stages may be completed in eight weeks.

The meal moth is primarily a pest of seeds in cool and damp localities. Berns (1958) found them infesting damp straw.

THE PINK SCAVENGER CATERPILLAR

Pyroderces rileyi (Wals.) (Family Cosmopterigidae)

The larva is $5/16$ inch/eight mm long when full grown, pink in color, with head and thoracic shield pale brown. The moth has a wing expanse of less than $\frac{1}{2}$ inch/13 mm. The forewings are banded and mottled with yellow, reddish brown and black. The hind wings are pale grayish, very slender and edged with long fringes.

In the South, the pink scavenger caterpillar causes injury to crops, both in the field and in storage. Infestation commences in the field and continues in storage. This insect is seldom an important grain pest. In Florida, the larva has the unusual habit of feeding on scale insects.

THE RICE MOTH

Corcyra cephalonica Stainton (Family Pyralidae)

This insect is a pale brownish moth which is 14 to 24 mm in width when the wings are spread. There are several generations in tropical countries, but usually one generation in temperate climates. The larva is dull white, with long fine hairs and a dark brown head. It spins a dense cocoon when full grown.

The rice moth, which is best adapted for living in a warm humid climate, has done considerable damage at times to stored rice and to chocolate products in the South. Besides feeding on rice and chocolate, the larva has been known to consume cocoa, nut meats and farinaceous materials.

THE BROWN HOUSE MOTH

Hofmannophila pseudospretella (Stainton) (Family Oecophoridae)

Although this insect is practically cosmopolitan in distribution, it is more common in England than elsewhere, if we are to judge by the available literature.

According to Austen et al. (1935), the male and female with their wings folded measure respectively $\frac{1}{3}$ and $7/12$ inch/eight mm and 15 mm in length. This species has dark brown forewings, across which there are several rows of dark brown spots.

In respect to the damage caused by this insect, Austen et al. (1935) quote N. M. Richardson that the larva "eats furs and skins, dried specimens of animals, birds, etc., including moths, amongst which it makes great havoc, if it gains admittance to the cabinet drawer, seeds of many kinds, corn, peas, etc., dried plants, live and dead chrysalides of butterflies and moths, figs, dates, groceries of many kinds and if I had the power of exterminating any one species of clothes moths, I think it is the one I would choose as the greatest general pest." It also damages clothing, mohair furniture and chairs stuffed with horse hair, and feeds upon herbs, some of which contain belladonna. It is partial to paper, as well as the corks in wine bottles. Book bindings, especially made of leather, are susceptible to attack.

The fully developed white larva is about ¾ inch/19 mm long with a tan head. This larva spins a cocoon with which it unites the material upon which it is feeding. The moths are found flying in England from May to October, where they occur not only in the house, but in the gardens as well.

Bender (1941) discusses the following moths as pests of corks in wine cellars: *Dryaduala pactolia* Meyr., *Oinophila v-flavum* Haw., *Nemapogon granella* (L.), *Endrosis sarcitrella* (L.), = *lactella* (Dennis & Schiff.), and *Hofmannophila pseudospretella* (Staint.).

The larvae of the above moths feed on corks and cause leakage. "The larvae of these moths were never found in the corks of bottles containing spirits. The damage to wine corks caused loss not only by leakage, but also by deterioration of the wine that remained in the bottles."

THE WHITESHOULDERED HOUSE MOTH

Endrosis sarcitrella (L.) (Family Oecophoridae)

Austen et al. (1935) state this insect is world-wide in distribution. The larva feeds on plant products, such as seed peas, seed potatoes and other types of dry seeds, as well as on rubbish in bird nests, the thatch on roofs, fungi on trees and also on all kinds of dry vegetable refuse. In the house, it may attack dry seeds, meal, carpets and may bore into wine corks, thus doing serious damage in wine cellars. Austin also notes, "It is often found floating on the top of milk jugs, etc. and also on water left in ewers and basins."

The male and female when at rest with wings folded, are ¼ inch/six mm and 5/12 inch/10 mm long, respectively. The head and thorax of this insect is white, with the forewing brownish in color. The white larva has a brownish head, and when reaching full development is ½ inch/13 mm long. It is found on the wing from May through October.

THE EUROPEAN GRAIN MOTH

Nemapogon granella (L.) (Family Tineidae)

This moth is of minor importance in the United States, Canada, and Europe on stored products. The adult with wings folded is six to seven mm in length and has a wing spread of nine to 14 mm. It has a white head, forewings splotched with brown and white, and gray hind wings with long hairs along the edges.

According to Patton (1931), it is a pest of stored cereals, cigars, etc. In wine cellars, the female lays her eggs in the cracks in the corks. The larva upon emerging from the egg, burrows into the cork. In Great Britain, it also feeds on rubbish out-of-doors.

The female lays her eggs singly on corn, corn cobs, barley, etc. The larvae complete their growth in 3½ to four months, then migrate from the food and pupate in a cocoon. There are two generations per year, adult moths being seen in April and May as well as August and September.

THE CORK MOTH

Nemapogon cloacella Haw. (Family Tineidae)

This cosmopolitan and minor pest is 10 to 17 mm with the wings spread and is so similar in appearance and habits to *N. granella* (Linn.) that Stellwaag (1924) made a distinct study of these two to show *T. cloacella* Haw. is not a variety of *N. granella* L.

Stellwaag states there are some biological differences between the two species, since *N. granella* L. prefers dry food such as grain, mushrooms, etc., whereas the larvae of *N. cloacella* Haw. die on too dry a diet. Since the cork moth may bore into corks in wine cellars, it is of interest to note Stellwaag did not find it in dry cellars, but rather damp bottle-stores and casks. Thus, one wonders if references of *N. granella* L. in corks may not actually be *N. cloacella,* and of *N. cloacella* Haw. in grain may not have been due to confusion of the two species. Patton (1931) states its biology is similar to that of *N. granella* L. and it breeds for the most part in rotten wood, fences and tree stumps. Stellwaag controlled this insect in wine cellars by fumigation.

Tinea fuscipunctella Haw. (Family Tineidae)

This species is 10 to 16 mm in wing expanse, with yellow gray forewings and bronzy-gray hind wings. It is cosmopolitan in distribution and feeds on animal matter. Reh found it destroying pigskin book bindings in Hamburg, Germany, and it has been reported injuring bedding in Norway. Patton (1931) states the larvae are to be found feeding on feces and feathers in bird nests. Mathlein (1941) made a study of a related species. *Nemapogon personella (=N. secalella* Zacher)as a pest of stored grain in Sweden.

THE GREATER WAX MOTH

Galleria mellonella (L.) (Family Pyralidae)

The wax moth occurs in the combs of abandoned or weakened bee colonies. If these combs are present in openings in the home, particularly where the colony has been killed, infestation of the comb by the wax moth is a common occurrence. It is at such times the adults may become unwelcome visitors in the home. Vansell (1943) notes in one house, the larvae of this moth bored through board walls and wallpaper and were found behind picture frames and other objects.

According to Eckert (1947), the larvae of the wax moth feed on the pollen and waste materials in the cell, spinning silken tunnels and burrowing through them.

This moth is approximately ⅝ inch/16 mm with a wing expanse of about 1¼ inches/23 mm. The folded wings appear ash gray in color with the basal third of each front wing being bronze colored. There are one or two generations a year, depending on the locality.

Where the moths are entering the house proper, the bee colony and comb should be located and removed as discussed in the chapter on bees.

Fig. 16-23. Foreign grain beetle.

THE FOREIGN GRAIN BEETLES

Ahasverus advena (Waltl)
(Family Cucujidae)

The foreign grain beetle is a small reddish-brown insect related to the square-necked beetle, *Cathartus quadricollis*. It has characteristic projections at the anterior corners of the prothorax.

It is a cosmopolitan species of minor economic importance and is attracted to damp and moldy grains, where it feeds on the molds, fungi and dead insects that occur in the grain. For this reason, it occurs in mills and warehouses, where farinaceous materials that are out of condition are stored. The author knows of one home in Pittsburgh, Pa., that was infested with it. This insect has been reported from beans, dates, figs, biscuits, coffee, yams, tobacco, etc. Strong and Okumura (1958) record it in dried animals and stale cake. Woodroffe (1962) and David and Mills (1975) made a study of this insect.

THE BROADNOSED GRAIN WEEVIL

Caulophilus oryzae (Gyllenhal) (Family Curculionidae)

The broad-nosed grain weevil is a small dark-brown weevil less than ⅛ inch/ three mm long, that is distributed generally throughout North America and Europe and is believed to be native to the United States.

This insect is widespread throughout the southeastern states and often is found infesting stored corn, dried peas and a great variety of seeds and cereals which are quickly reduced to a powdery mass through the combined efforts of the grubs and the adult weevils. It can attack only soft and injured grain, and is often associated with the rice weevil.

THE FLAT GRAIN BEETLES

Cryptolestes pusillus (Schon.) (formerly *Laemophloeus minutus* Oliv)
(Family Cucujidae)

The flat grain beetle is a very small, reddish brown beetle about ¹/₁₆ inch/1.6 mm long, with rather long antennae. This beetle can jump and fly. It is one of the smallest beetles found in stored grain in the United States. This cosmopolitan beetle is found in flour mills, warehouses, granaries, etc., where it infests

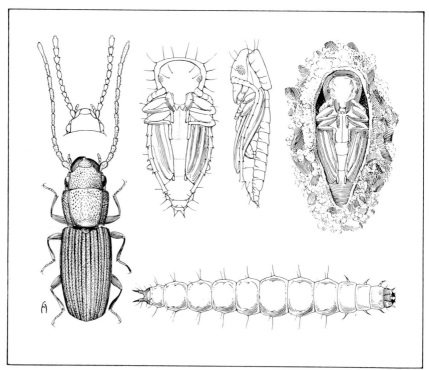

Fig. 16-24. The adult, pupal and larval stages of the flat grain beetle.

stored and milled products. It is commonly found in stored corn in the South and on dried dates and shelled peanuts in California. The flat grain beetle is often associated with the heating of grain and it is for the most part a secondary pest which is usually found with other grain-infesting insects. Davies (1949) showed whole grains are immune from attack, but any grain with even minute damage can be infested.

Back (1931) states the small white eggs are placed in the crevices of the grain or are dropped loosely in farinaceous materials and the larvae injure the germ of the grain. The larva pupates in cocoons of gelatinous material to which food particles adhere. Payne (1946) studied the life history of this insect at 78° F/ 25.5° C and found the egg stage to be eight to 10 days, the larval stage 26 to 45 days and the pupal stage six to nine days. The female may live as long as a year. Bishop (1959) studied the life history of the species *ferrugineus* (Steph.) *C. turcicus* (Grouville), and *C. pusilloides* (Steel and Howe). Ahsby (1961) and Lefkovitch (1964) studied the biology of insects in this genus.

Gould (1948) found the flat grain beetle and *C. punctatus* LeC., as well as *C. ferrugineus* (Steph.) present in corn processing plants in Indiana, where they acted as scavengers.

Freeman (1952) obtained control of insects in this genus with chloropicrin and carbon tetrachloride plus ethylene dichloride.

THE RUSTY GRAIN BEETLE

Cryptolestes ferrugineus (Steph.) (Family Cucujidae)

This beetle is similar to *C. pusillus,* but since it is more resistant to cold than the flat grain beetle, it is found more commonly in the North. Back (1931) states the two species may be readily distinguished by the fact that in the flat grain beetle the antenna of the male is about as long as the body and about twice as long as that of the female beetle, whereas in the rusty grain beetle the antenna of the male is short.

This beetle is cosmopolitan and is usually found in stored grain and other foodstuffs, as well as under the bark of trees. It is less common than *C. pusillus.* Rilett (1949) studied this species in Canada. He notes this insect became a major pest in western Canada in stored grain during World War II. The eggs were deposited "through cracks in the wheat grains in the germ region below the outer layers of the seed coat, between the grains, and among detritus." The larvae feed on the germ and endosperm, and the adults on damaged grains and wheat dust. The adults fly well at higher temperatures. At 80° F/27° C eggs hatch in four to five days and the adults emerge 69 to 103 days after hatching. The insects develop most rapidly at 90° to 100° F/32° to 38° C, and a relative humidity of 75 percent or higher. The newly-emerged larvae can enter wheat grains through microscopic breaks in the bran layers.

THE SMALLEYED FLOUR BEETLE

Palorus ratzeburgi (Wissm.) (Family Tenebrionidae)

This small reddish brown beetle is but three mm long. It is cosmopolitan and common in flour mills in the United States and breeds in grain and milled products, particularly when these are situated in flour-mill basements.

THE DEPRESSED FLOUR BEETLE

Palorus subdepressus (Woll.) (Family Tenebrionidae)

This insect is closely related to the smalleyed flour beetle, but is much less common than *P. ratzeburgi* in the United States except perhaps in the Great Plains region.

THE LONGHEADED FLOUR BEETLE

Latheticus oryzae Waterh. (Family Tenebrionidae)

The longheaded flour beetle is slightly less than ⅛ inch/three mm long. Although somewhat similar in shape to the confused flour beetle, it is pale yellow in color. It can also be separated by the peculiar shape of the antennae and by the minute canthus behind each eye. It feeds on wheat, rice, corn, barley, rye and flour. It is of little importance in the United States and occurs for the most part in the southwestern states.

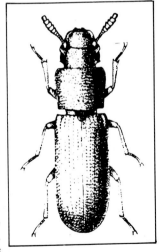

Fig. 16-25. The longheaded flour beetle.

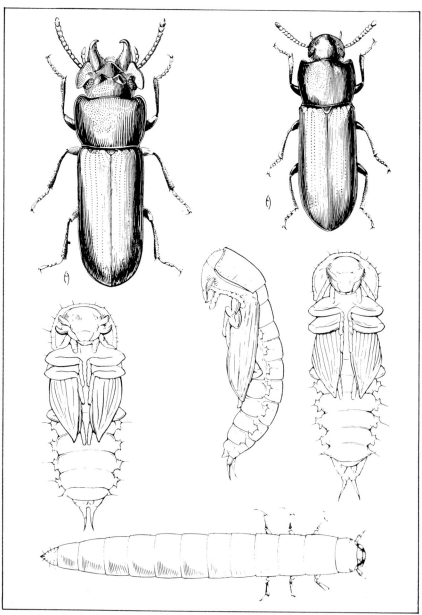

Fig. 16-26. The broadhorned flour beetle. Upper left, adult male; upper right, adult female; below, pupae and larva.

According to Patton (1931), the eggs are laid either in the flour or wheat and the larvae, which resemble that of *Tribolium*, can be separated by the two fleshy processes at the end of the ninth abdominal segment which are narrow and smaller than those of *Tribolium*. The insect molts six to nine times and has a life cycle of 25 to 40 days.

THE BROADHORNED FLOUR BEETLE

Gnathocerus cornutus (F.) (Family Tenebrionidae)

This insect is cosmopolitan in distribution and prominent as a pest of cereal products on the Pacific Coast. The adult is a red brown beetle, about ¹/₆ inch/ four mm long. The male has a pair of broad, stout horns on the mandibles.

Pimentel (1949) studied the life history of this insect. He found at optimum temperatures of 75° to 85° F/24° to 29° C, seven days were required for the incubation period, 40 days for the seven molts of larval development and 10 days for the pupal stage. A total of 77 days was required from egg to egg. The female will deposit up to 400 eggs during a five-month period.

Shepherd (1924) found the larva to feed on flour, corn meal, germea, dog biscuits, corn, pancake flour, yeast cakes, bran and farina. It prefers flaky materials. Patton (1931) states it feeds on sugar cane.

Control. According to Shepherd, the chief means of control of the broadhorned flour beetle are, "the use in Germany, of *Bacillus thuringiensis:* in England hermetical sealing; in Canada, of chloropicrin, and in the United States and elsewhere, the elimination of breeding places, fumigation with carbon bisulfide and hydrocyanic acid gas and the heating of mills to a temperature of 120°-130° F for a period of 12 hours."

THE SLENDERHORNED FLOUR BEETLE

Gnathocerus maxillosus (F.) (Family Tenebrionidae)

This beetle is confined much more closely to tropical and subtropical areas than is *Gnathocerus cornutus*. In the United States, it occurs for the most part in the South. It is smaller than the broadhorned flour beetle, being but ⅛ inch/ three mm long. The shape of the horns on the mandibles of the male is slender and incurved. In habits, the slender-horned flour beetle is similar to *Gnathocerus cornutus*.

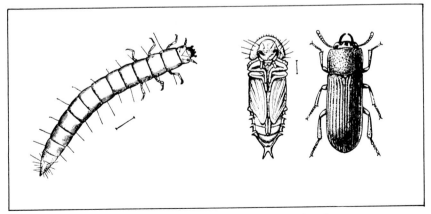

Fig. 16-27. Larva, pupa and adult of slenderhorned flour beetle.

THE COFFEE BEAN WEEVIL

Araecerus fasiculatus (DeG.) (Family Anthribidae)

The coffee-bean weevil is about ⅛ inch/three mm long, dark brown with light and dark pubescence. The antennae and the legs are reddish brown. It is cosmopolitan in distribution. This insect is common in biscuit and other factories, but is primarily a tropical and subtropical pest. It is of minor economic importance. It becomes noticeable due to its habit of leaping on window panes. Patton (1931) states it is a pest of shade plants in the Near East and the cause of black tip on bananas in Europe, a pest of seeds, corn, dried fruits and other foodstuffs, such as stored sweet potatoes and coffee beans.

Back (1931) notes it is very common in the South, where it breeds in dried fruit, coffee berries, cornstalks, corn, seeds and seed pods. It lays its eggs in the soft kernels of corn and breeds in it after it has been harvested, causing little damage since the corn is too hard for it. The coffee bean weevil is also a pest of drugs and spices. Concha (1956) controlled the larvae and adults with carbon disulfide.

DeFigueiredo (1957) states this insect is an important pest of coffee in Brazil. It has a life cycle of 30 to 45 days from egg to adult, and there may be as many as eight to 10 generations a year. Methyl bromide is used in the control of this insect. Childers and Woodruff (1980) prepared a bibliography of this weevil.

THE TWOBANDED FUNGUS BEETLE

Alphitophagus bifasciatus (Say) (Family Tenebrionidae)

This handsome-looking beetle is slightly less than ⅛ inch/three mm long, reddish brown, with two broad black bands across the wing covers. According to Back (1931), it is cosmopolitan in distribution and a general feeder on fungi and molds, as well as a scavenger in waste grain and milled products. The larvae occur in moist corn meal and in spoiled cereals. These insects are often found in the holds of grain ships, in wet or damaged grain. They are of little economic importance.

THE SQUARENECKED GRAIN BEETLE

Cathartus quadricollis (Guer.)
(Family *Cucujidae*)

This beetle is closely related to the sawtoothed grain beetle which it resembles. It is a flattened, oblong, polished, reddish-brown beetle, about ¹/₁₀ inch/2.5 mm long and according to Back (1931), it differs from the sawtoothed grain beetle by having a square-shaped thorax that lacks the toothlike projections. It is principally a pest in the South, although it is practically cosmopolitan in distribution. It is very common in stored corn and in the cornfields. The larva will devour the germ of the seed in which it feeds. It develops from egg to adult in about three weeks.

THE LESSER MEALWORM

Alphitobius diaperinus (Panz.)
(Family Tenebrionidae)

This is a minor and cosmopolitan household pest. Cotton and Good (1937) state it feeds on damp and moldy grain, milled products and spoiled foods. Hard-

Fig. 16-28. The squarenecked grain beetle, *Cathartus quadricollis* (Guer.).

ing and Bissell (1958) found these insects infesting corn cob litter in brooder houses. Anon. (1961) found them migrating from cemeteries into houses. Barke and Davis (1969) studied the life cycle of this insect.

THE BLACK FUNGUS BEETLE

Alphitobius laevigatus (F.) (Family Tenebrionidae)

The black fungus beetle is a shiny black beetle whose ventral surface is brown or rust red. The antennae and legs are brownish. This insect has been reported from flour, bread, maize, groceries, etc. Gould (1948) frequently found this beetle and the lesser mealworm in corn processing plants in Indiana.

THE BLACK FLOUR BEETLE

Tribolium audax Halstead (Family Tenebrionidae)

This beetle occurs throughout the northern and western states, southern Canada, central and northern Europe and Egypt. Usually it is found under the bark of trees, but may occasionally be taken from flour, meal, seeds, and grain.

THE FALSE BLACK FLOUR BEETLE

Tribolium destructor (Uttenb.) (Family Tenebrionidae)

Linsley and Michelbacher (1943) studied this insect as a stored product pest in California. The adult beetle is very similar to the confused flour beetle, "but is slightly larger, usually about ¼ inch long and is black instead of reddish brown." This insect has been found to infest animal food pellets, flour and other stored products. Habits and control are similar to that of the confused flour beetle.

THE LARGER GRAIN BORER

Prostephanus truncatus (Horn.) (Family Bostrichidae)

The larger grain borer is a dark brown beetle about ⅙ inch/four mm long. It is a tropical insect, and occurs for the most part in warmer areas of the United States, where it lives in warehouses and breeds in corn, other stored grains and on tubers.

THE HAIRY CELLAR BEETLE

Mycetaea hirta (Marsh.) (Family Endomychidae)

This minute, pale brown beetle, which has long erect hairs and shallow round punctures on the head, measures about 1.5 mm in length. It is found in Europe and North America, where it is recorded boring into the corks of bottles stored in wine cellars. Hinton (1945) believes the damage may actually have been done by a moth larva (*Tinea* sp).

THE SIAMESE GRAIN BEETLE

Lophocateres pusillus (Klug.) (Family *Trogositidae*)

This elongated, flat, reddish-brown beetle prefers warmer areas and is for the most part confined to the southern states. It received its name by first appearing in this country in exhibits of rice and cereals from Siam, Liberia and Ceylon at the World's Columbian Exposition. It breeds in all kinds of stored grain and is also recorded from flour, seeds, spices, dried apples, beans and macaroni.

THE HAIRY FUNGUS BEETLE

Typhaea stercorea (L.) (Family Mycetophagidae)

This brownish beetle with black eyes is 2.2 to three mm in length. Cotton and Good (1937) note this insect is found in dwellings, warehouses, stores, flour mills and outdoors. It is a pest of stored grain and seeds, tobacco, peanuts, cocoa, etc.

THE PLASTER BEETLES

Microgramme filum (Aube) (= *Cartodere*) and *Microgramme* (= *Cartodere) argus* Reitter (Family Lathridiidae)

Hinton (1941 and 1945) made a study of the members of this family that were of economic importance and the following notes are from his papers.

Microgramme filum is 1.2 to 1.6 mm in length, brownish in coloration and has a two-segmented antennal club. This insect feeds on fungus spores and in

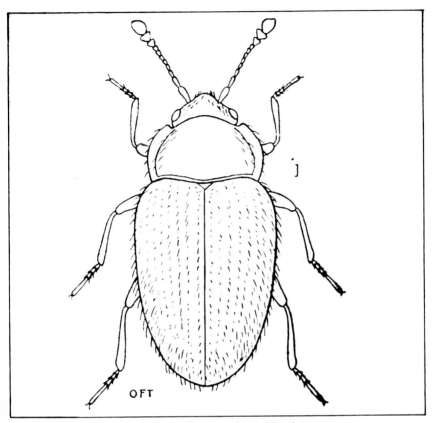

Fig. 16-29. The hairy cellar beetle, *Mycetaea hirta* (Marsh.).

Germany it has been found "in damp houses, on damp walls near water taps and badly closing windows, on mouldy wall-paper and in large numbers in a laboratory on mouldy papier-mache-dishes." In England, it has been found on wheat in warehouses and in soap powder. The life cycle requires some 36 days at 73° to 77° F/23° to 25° C, the insects undergoing three instars. This insect is found in Europe, N. Africa and in North and South America.

   M. argus Reitter is somewhat similar to the above species and has a three-segmented antennal club. It has been found in ground cereals in Oregon, in Ohio on the walls of a house, in a drug store and on heads of wheat in a field in Texas. In Pennsylvania, the author encountered an infestation in an old brick house that had been remodeled recently into apartments. The infestation was most prominent in early June after a period of heavy rainfall. The insects were found on the walls which had been plastered some three months prior to the infestation, as well as around the windows. About 30 of the beetles were found clustered around a light fixture in the ceiling. The owner attempted to control these insects with a pyrethrum oil-base spray and was not successful in this respect. An application of a six percent DDT oil-base spray was only partially successful. Some three months later, in later summer, the infestation had disappeared. It is possible the new plaster was not completely dried at the beginning of the infestation. In any event, the heavy rainfall at that time may also have been responsible for keeping the walls from drying out. This condition may have encouraged the growth of molds, and since *M. filum* can feed on fungus spores, in all likelihood, *M. argus* can do likewise and the growth of molds encouraged the infestation. With the coming of hot weather, the walls dried out and the molds and insects disappeared.

Kerr and McLean (1956) studied the biology and control of several species of Lathridiidae. They controlled these insects with DDT, chlordane or dieldrin sprays. E. Witt treated an infestation of *C. constricta* (Gyll.) with silica aerogel and an 0.5 percent lindane spray.

THE KHAPRA BEETLE

Trogoderma granarium Everts (Family Dermestidae)

The khapra beetle, unlike most dermestids, prefers grains and cereal products to animal products. In many parts of the world, particularly India, it is an extremely important pest in granaries and warehouses. For this reason, the discovery of the khapra beetle in Fresno County, California, in 1953 has been responsible for great activity among entomologists and quarantine officials. Apparently, the khapra beetle was found in the United States as early as 1946 when it was confused with another dermestid. Eradication programs eliminated the infestations in California, Arizona, Texas and New Mexico, as well as the Republic of Mexico. New infestations were found in 1980 and 1981 in some eastern states, as well as in Michigan, Texas and California. These were also eliminated.

"During the past several years, scientists at the USDA's Science and Education Administration (SEA); University of Wisconsin-Madison; College of Environmental Science and Forestry, Syracuse, N.Y.; and Max Planck Institute in West Germany have cooperated in research that has led to improved monitoring for the presence of khapra beetles by the USDA's Animal and Plant Health Inspection Service (APHIS). Officials of APHIS are now using a synthetic sex attractant to search for adult beetles. Researchers also have discovered that larval khapra beetles are attracted to wheat germ oil and this oil is being mixed with the pheromone."

According to Harper (1955), "The word Khapra is from an Indian word which means brick. This name described its habits in India of aestivating in the pores of the bricks used in the construction of storage warehouses." Many studies have been devoted to the biology and control of this species, especially by workers in India. Lindgren et al. (1955), Hadaway (1956) and Burges (1962) have completed some of the most recent studies on the biology of the khapra beetle. This insect is so similar in appearance to some of the other species of *Trogoderma*, particularly *T. inclusum* LeConte, it can only be separated from it by a highly trained specialist. Howe and Burges (1956), Beal (1956) and Strong and Okumura (1966) prepared papers showing anatomical details that separate the larvae and adults of the khapra beetle from some of the other *Trogoderma*.

The beetle is about ⅛ inch/three mm in length. The larva is yellowish-brown and about ¼ inch/six mm long. Although the beetle prefers grains such as wheat, barley and rice, it will also feed on dried blood, milk and fish-meal. Noon (1958) showed it could also feed on pollen and dead insects. Beal (1960) and Spangler (1961) tells us how to recognize the larva.

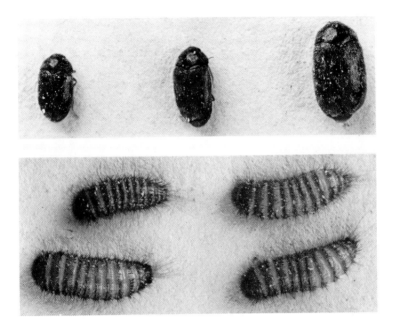

Fig. 16-30. Adults and larvae of the khapra beetle.

Lindgren et al. (1955) note a typical infestation is characterized by large numbers of larvae and their cast skins. The young larvae feed largely on damaged kernels and the other larvae feed on whole grain or seeds. This beetle usually restricts its activity to the top 12 inches/30 cm of the grain. The larvae, like most dermestid larvae, are very resistant to starvation. As the larvae become older, they are repelled by light. Pupation occurs in the top layer of the food material. "The adults usually mate immediately upon emergence and egg laying begins several days after, the interval depending on the temperature. Adults may live from a few days to several months, according to the temperature." The adult has not been observed to fly.

According to Lindgren et al. (1955), at 90° F/32° C the average time for the various stages is eight days for the egg stage, 27 days for the larval stage, six days as a pupa and 16 days as an adult, with an average of 37 days from egg to adult.

Shepherd (1957) notes its habit "of secreting itself in cracks and crevices of storage facilities, the ability of the larvae to resist starvation for long periods time and its resistance to normal sanitation measures make it a formidable pest of the grain and seed industries."

Control. Warehouses are fumigated for 24 hours with a 4 to 12 pound/1.8 to 5.4 kg dosage of methyl bromide per 1000 cu. ft/28 cubic meters, the dosage rate depending on the temperature and the commodity. The fumigated buildings are wrapped in gastight tarpaulins and tightly sealed. In addition, according to Shepherd (1957), "an area of 100 feet in diameter surrounding the site is sprayed three times at three-day intervals with a solution consisting of five pounds of actual malathion to 100 gallons of diesel oil." Anon. (1960) fumigated infested burlap-wrapped steel from Europe.

LITERATURE

ALEXANDER, P., J. A. KITCHENER & H. V. A. BRISCOE — 1944. Inert dust insecticides. I. Mechanism of action. Ann. Appl. Biol. 31(2):143-149. Biol. Abst. 19:800-801, 1945.

1944. Inert dust insecticides. II. The nature of effective dusts. Ann. Appl. Biol. 31(2):150-156. Biol. Abst. 19:801, 1945.

1944. Inert dust insecticides. III. The effect of dusts on stored products pests other than *Calandra granaria*. Ann. Appl. Biol. 31(2):156-159. Biol. Abst. 19:801, 1945.

ALFIERI, A. — 1931. Les insectes de la tombe de Tautankhamon. Bull. Soc. R. Ent. Egypte. fasc. 3-4:188-189.

ANNAND P.M. — 1944. Report of the Chief of the Bureau of Entomology and Plant Quarantine, Agricultural Research Administration, 1943. 58 pp. U.S.D.A. Washington 25, D.C.

ANONYMOUS — 1943. Protect home-cured meat from insects. U.S.D.A. Ext. Serv. Bureau of Ent. & P. Q. AWI-32. 1944. Microanalysis of food and drug products. Food & Drug Circular No. 1. Food and Drug Administration. 1945. Seed protection with DDT. Soap & S.C. 21(12):86. 1946. DDT in flour mills. Soap & S.C. 22(4):157. 1947. Food and Agriculture Org. U. N. studies use of chemicals to protect stored foods. Agr. Chem. 2(9):30-31, 66. 1949a. Concerning the use of Pyrenone concentrates for the control of common grain insects that attack grain and cereals in storage. U.S.I. Insecticide Bull. 1949b. Bug barrier. Chem. Indust. 65(3):341-342. 1954. Mealworm U.S.D.A. Leaflet No. 195. 1955. Home-cured meat; How to protect it from insects. U.S.D.A. Leaflet 485. 1956. PCO inspection check list for food grain elevator. Pest Control 24(5):22. 1958. Insect control in flour mills. U.S.D.A. Leaflet 485. 1956. PCO inspection check list for food grain elevator. Pest Control 24(5):22. 1958. Insect control in flour mills. U.S.D.A. Agr. Handbook No. 133. 1961. Lesser mealworm *(Alphitobius diaperinus)*. U.S.D.A. Coop. Econ. Ins. Rpt. 11(37):873, Sept. 15. 1964. Insect resistance to pesticides. U.S.D.A. Marketing, pp. 4-5, March. 1965. Film packages foil insects. Agr. Marketing, 10(4):12-13. 1966. Control of insects that attack dry beans and peas in storage. U.S.D.A. Agr. Inform. Bull. 303. 1972. National Pest Control Association Sanitation Committee. Sanitation and pest control floor-level inspection manual. NPCA, Inc., Vienna, Virginia. 22 pp.

APT, A. C. — 1950. A method of rearing the flat grain beetle and the grain mite. J. Econ. Entomol. 43(5):735.

ARMITAGE, F. D. & B. VERDCOURT — 1947. The preference of *Stegobium paniceum* (L.) (Col. Anobiidae) for certain drugs. Entomol. Monthly 83:133.

ARMITAGE, H. M. — 1956. The khapra beetle problem in California. J. Econ. Entomol. 49(4):490-493.

ARMITAGE, H. M. & J. B. STEINWEDEN — 1945. The fumigation of California dates with methyl bromide. Bull. Calif. Dept. Agric. 34(3):101-107.

ARMSTRONG, J. W. & E. L. SODERSTROM — 1975. Malathion resistance in some populations of the Indian meal moth infesting dried fruits and tree nuts in California. J. Econ. Entomol. 68(4):505-507.

ASHBY K. R. — 1961. The life-history and reproductive potential of *Cryptolestes pusillus* at high temperatures and humidities. Bull. Entomol. Res. 52(2):353-361.

AUSTEN, E. E., A. W. McKENNY HUGHES & H. STRINGER — 1935. Clothes moths and house moths. British Mus. Econ. Series 14.

BACK, E. A. — 1920. Angoumois grain moth. U.S.D.A. Farmers' Bull. 1156. 1930. Weevils in beans and peas. U.S.D.A. Farmers' Bull. 1275. 30 pp. 1931. Stored-grain pests. U.S.D.A. Farmers' Bull. 1260. 46 pp. 1939. The cigarette beetle as a pest of cotton seed meal. J. Econ. Entomol. 32:739-49. 1942. Cause of an infestation by *Tribolium confusum* of rolled oats packaged in three-pound cardboard cartons. J. Econ. Entomol. 35(6):957.

BACK, E. A. & R. T. COTTON — 1924. Relative resistance of the rice weevil, *Sitophilus oryza* L. and the granary weevil, *S. granarius* L. to high and low temperatures. J. Agr. Res. 28:1043-4. 1924. Effect of fumigation upon heating of grain caused by insects. J. Agr. Res. 11:1103-1115. 1926. The cadelle U.S.D.A. Bull. 1428. 1926. Biology of the saw-toothed grain beetle, *Oryzaephilus surinamensis* Linne. J. Agr. Res. 33(5):435-52. 1926. The granary weevil. U.S.D.A. Bull. 1393. 1938. Stored grain pests. U.S.D.A. Farmers' Bull. 1260.

BACK, E. A. & A. B. DUCKETT — 1918. Bean and Pea Weevils. U.S.D.A. Farmers' Bull. 983.

BACK, E. A. & W. D. REED — 1930. *Ephestia elutella* Hubner, a new pest of cured tobacco in the United States. J. Econ. Entomol. 23:1004-1006.

BALZER, A. I. — 1942a. The life-history of the corn sap beetle in rice. J. Econ. Entomol 35(4):606-607. 1942b. Insect pests of stored rice and their control. U.S.D.A. Farmers' Bull. 1906. 22 pp.

BANG, Y. H. & E. H. FLOYD — 1962. Effectiveness of malathion in protecting stored polished rice from damage by several species of stored grain insects. J. Econ. Entomol 55(2):188-190.

BANSODE, P.C. AND W. V. CAMPBELL — 1979. Evaluation of North Carolina field strains of the red flour beetle for resistance to malathion and other organophosphorus compounds. J. Econ. Entomol. 72(3):331-333.

BARE, C. O. — 1948. The effect of prolonged exposure to high vacuum on stored-tobacco insects. J. Econ. Entomol 41(1):109-110. 1948. Laboratory tests with fumigants for insects infesting stored tobacco. J. Econ. Entomol. 41(1):13-15.

BARE, C. O. & J. N. TENHET — 1950. Tests with acrylonitrile-carbon tetrachloride and hydrogen cyanide as fumigants for insects in cigarette tobaccos. U.S.D.A. B.E. & P.Q. E-794.

BARE, C. O., J. N. TENHET & W. D. REED — 1946. Effect of the thermalvacuum process on insects in stored tobacco. J. Econ. Entomol. 39(5):612-613.

BARE, C. O., J. N. TENHET & R. W. BRUBAKER — 1947. Improved techniques for mass rearing of the cigarette beetle, *Lasioderma serricorne* and the tobacco moth, *Ephestia elutella*. U.S.D.A. B.E. & P.Q. ET-247. 11 pp.

BARKE, H.F. & R. DAVIS — 1969. Notes on the biology of the lesser mealworm, *Alphitobius diaperinus* (Coleoptera: Tenebrionidae). J. Ga. Entomol. Soc. 4(3):46-50.

BARNES, D. F. & D. L. LINDGREN — 1947. Progress of work on beetle infestation in dates. Date Growers' Inst. Rpt. (1947) 24:3-4.

BARNES, D. F., E. L. MAYER, E. F. SAWALL & G. W. REILLY — 1956. Enclosures for fumigating stored raisins. U.S.D.A. Agr. Market Serv. AMS-131, July.

BEAL, R. S., JR. — 1956. Synopsis of the economic species of *Trogoderma* occurring in the United States with description of a new species (Coleoptera: Dermestidae). Ann. Entomol. Soc. Amer. 49(6):559-566.

BEAL. R. S. — 1960. Description, biology, and notes on the identification of some *Trogoderma* larvae (Coleoptera, Dermestidae). U.S.D.A. Tech. Bull. 1228.

BECKLEY, V. A. — 1948. Protection of grain against weevil. Nature 162(4123):737-738.

BENDER, E. — 1941. Investigations on the biology and morphology of the microlepidoptera that inhabit wine-cellars. Rev. Appl. Entomol. A. 35(3):71-72, 1947.

BERNS, R. E. — 1958. Straw meal moths. Pest Control. 26(11):16.

BIRCH, L. C. — 1944. Two strains of *Calendra oryzae* L. (Coleoptera). — Aust. J. Exp. Biol. Med. Sci. 22(4):271-275. Rev. Appl. Entomol. A. 34(1):271-275. 1946. 1945. A contribution to the ecology of *Calendra oryzae* L. and *Rhizopertha dominica* Fab. (Coleoptera) in stored wheat. Trans. Roy. Soc. S. Aust. 60(1): 140-149. Rev. Appl. Entomol. A. 34(7):22, 1946. 1946. The movements of *Calendra oryzae* L. (small strain) in experimental bulks of wheat. J. Aust. Inst. Agric. Sci. 12(1-2):21-26. Rev. Appl. Entomol. A. 45(6):21-26. 1947.

BIRCH, L. C. & J. G. SNOWBALL — 1945. The development of the eggs of *Rhizopertha dominica* Fab. (Coleoptera) at constant temperatures. Australian J. Exp. Biol. and Med. Sci. 23(1):37-40. Biol. Abst. 19:1949, 1945.

BISHOP, G. W. — 1959. The comparative bionomics of American *Cryptolestes* (Coleoptera-Cucujidae) that infest stored grain. Ann. Entomol. Soc. Amer. 52(6):657-665.

BISSELL, T. L. — 1943. The confused flour beetle living in bait mixtures containing cryolite. J. Econ. Entomol. 36(4):634-635.

BISSELL T. L. & M. DuPREE — 1946. Insects in shelled peanuts in relation to storage and bagging. J. Econ. Entomol. 39(4):551-552.

BOND, E. J. & C. T. BUCKLAND — 1976. Control of insects with fumigants at low temperatures: Toxicity of mixtures of methyl bromide and acrylonitrile to three species of insects. J. Econ. Entomol. 69(6):725-727.

BOND, E. J. & H. A. U. MONRO — 1954. Rearing the cadelle *Tenebrioides mauritanicus* (L.) (Coleoptera: Ostomidae) as a test insect for insecticidal research. Canad. Entomol. 86(9):402-408. 1961. The toxicity of various fumigants to the Cadelle, *Tenebrioides mauritanicus.* J. Econ. Entomol. 54(3):451-454.

BOVINGDON, H. H. S. — 1931. Pests in cured tobacco. The tobacco beetle, *Lasioderma serricorne* Fab., and the Cacao moth, *Ephestia elutella* Hb. Tobacco, 1st Aug. London.

BRANNON, L. W. & W. D. REED — 1943. Survey of tobacco carriers for stored-tobacco pests. J. Econ. Entomol. 46(2):295-299.

BRINDLEY, T. A., J. C. CHAMBERLIN, F. G. HINMAN & K. W. GARY — 1946. The pea weevil and methods for its control. U.S.D.A. Farmers' Bull. 1971.

BRISCO. H. V. A. — 1943. Some new properties of inorganic dusts. J. R. Soc. Arts 91(4650):593-607. Rev. Appl. Entomol. A. 32:38-40, 1944.

BROOKS, P. — 1934. The bookworm vanquished. Phillip. Agr. 23:171-173.

BROWN, W. B. — 1937. V. Distribution of ethylene oxide in barges containing dried fruit. J. Soc. Chem. Ind. 56:116T-122T. Rev. Appl. Entomol. A. 32:334-435, 1944.

BROWN, W. J. — 1940. A key to the species of Ptinidae occurring in dwellings and warehouses in Canada (Coleoptera). Canad. Entomol. 72:115-122.

BRUBAKER, R. W. & W. D. REED — 1943. Fumigation of tobacco at reduced pressures. J. Econ. Entomol. 36(2):300-303.

BURGES, H. D. — 1962. Studies on the dermestid beetle *Trogoderma granarium* Everts. V. — Bull. Entomol. Res. 53(1):193-213.

CANDURA, G. S. — 1943. Italy. Further damage caused by the Indian meal moth *(Plodia interpunctella)* and the Angoumois grain moth *Sitotroga cerealella)*. Int. Bull. Plant Prot. 17(2):19M-20M. Rev. Appl. Entomol. A. 32:179-180, 1944.

CHAPMAN, R. N. & H. H. SHEPARD — 1932. Insects infesting stored food products. Minn. Agr. Exp. Sta. Bull. 198.

CHILDERS, C.C. & R. E. WOODRUFF — 1980. A bibliography of the coffee bean weevil, *Araecerus fasciculatus* (Coleoptera: Anthribidae) Bull. Entomol. Soc. America 26(3):384-394.

CHILDS, D. P. — 1958. Warehouse fumigation of flue-cured tobacco with HCN to control the cigarette beetle. J. Econ. Entomol. 51(4):417-421. — 1967. Cigarette beetle control in warehouses with HCN and dichlorvos. J. Econ. Entomol. 60(1):263-265.

CHILDS, D. P. & J. E. OVERBY — 1976. Ethylene oxide fumigation of tobacco at various pressures for control of the cigarette beetle. J. Econ. Entomol. 69(1):47-50.

CHITTENDEN, F. H. — 1896. The white-marked spider beetle. U.S.D.A. Bur. Ent. Bull. 4. 1897. Some little-known insects affecting stored vegetable products: A collection of articles detailing certain original observations made upon insects of this class. U.S.D.A. Div. Entomol. Bull. (N.S.) 8. 1911. Papers on insects affecting stored products. A list of insects affecting stored cereals. The Mexican grain beetle; the Siamese grain beetle. U.S.D.A. Bur. Entomol Bull. 96. 1911. Papers on insects affecting stored products. The larger grainborer *(Dinoderus truncatus* Horn). U.S.D.A. Bur. Entomol. Bull. 96. 1911. The fig moth U.S.D.A. Bur. Entomol Bull. 194. 1917. The two-banded fungus beetle. J. Econ. Entomol. 10:282-287.

CHIU, S. F. — 1939. Toxicity studies of so-called "inert" materials with the bean weevil, *Acanthoscelides obtectus* (Say). J. Econ. Entomol. 32:240-8. 1939. Toxicity studies of so-called "inert" materials with the rice weevil and the granary weevil. J. Econ. Entomol. 32:810-821.

COLLINS, H. E. — 1963. New data on insect infestation. Modern Packaging. 37(2):132-135, 195.

CONCHA, A. C. — 1956. Biology and control of *Araecerus fasciculatus* in Barranguilla, Columbia. Chem. Abst. 50(21):16020, 1956.

CORBET, A. S. & W. H. T. TAMS — 1943. Keys for the identification of the Lepidoptera infesting stored food products. Proc. Zool. Soc. Lond. (B) 113(3):55-148, 5 pls., 287 figs. 1943. Notes on the nomenclature of the moths hitherto known as *Ephestia kühniella* Zeller, 1879 (Lepidoptera; Pyralidae) and *Endrosis lactella* (Schiffer-Müller & Denis, 1775) Lepidoptera. Oecophoridae). Entomologist 75:15; Rev. Appl. Entomol. A., 31:310. 1943.

COTTON, R. T. — 1920. Rice weevil, *(Calandra) Sitophilus oryza*. J. Agr. Research 20:409-422. 1921. Four Rhyncophora attacking corn in storage. J. Agr. Res. 20. 1922. The broad-nosed grain weevil. U.S.D.A. Prof. Paper 1085. 1941. Insect pests of Stored Grain and Grain Products. Burgess Publishing, Co. 1942. Control of insects attacking grain in farm storage. U.S.D.A. Farmers' Bull. No. 1811. 1944. Protection of stored and dried processed foods and seed supplies. J. Econ. Entomol. 39(3):380-384. 1946. The insect problems of dried food storage. Pests 14(10):50, 52, 54. 1947. Pests in stored products.

U.S.D.A. Yearbook 1943/47:874-878. 1947. Insects of stored grain and its products. Pest Control & Sanitation. 2(12):8-13, 22-24. 1950. Insect damage to stored grain and its prevention. Pest Control. 18(10):8-10. 1950. Notes on the almond moth. J. Econ. Entomol. 43(5):733.

COTTON, R. T., A. L. BLAZER & J. C. FRANKENFELD — 1944. Control of mill insects. Amer. Miller and Processor. Apr. 1944, pp. 41, 99, 100, 110.

COTTON, R. T., A. L. BALZER & H. D. YOUNG — 1944. The possible utility of DDT for insect-proofing paper bags. J. Econ. Entomol. 37:140.

COTTON, R. T. & J. C. FRANKENFELD — 1946. Insect infestation in powdered hand soap. J. Econ. Entomol. 39(3):419-420. 1949. Insect-proofing cotton bags. U.S.D.A., B. E. & P. Q., E-783. 1949. Silica aerogel for protecting stored seed or milled cereal products from insects. J. Econ. Entomol. 42(3):553.

COTTON, R. T., J. C. FRANKENFELD & G. A. DEAN — 1945. Controlling insects in flour mills. U.S.D.A. Circ. No. 720.

COTTON, R. T., J. C. FRANKENFELD & L. M. REDLINGER — 1950. The treatment of railway boxcars with insecticidal sprays — a preliminary test. Northwest Miller (Miller Prod. Sect.) 241(11):1a. Mar. 14.

COTTON, R. T., J. C. FRANKENFELD & W. B. STRICKLAND — 1949. Insect-proofing cotton bags. U.S. B. E. & P. Q. E-783. July.

COTTON, R. T., J. C. FRANKENFELD, H. H. WALKDEN & R. B. SWITZGEBEL — 1945. Tests of DDT against the insect pests of stored seed, grain, and milled cereal products. U.S.D.A., B. E. & P.Q. E-641, 7 pp. mimeo.

COTTON, R. T. & N.E. GOOD — 1937. Annotated list of the insects and mites associated with stored grain and cereal products, and their arthropod parasites and predators. U.S.D.A. Misc. Publ. 258.

COTTON, R. T. & H. E. GRAY — 1948. Preservation of grains and cereal products in storage from insect attack. Preservation of grains in storage. Food & Agriculture Organization of the United Nations, pp. 35-71.

COTTON, R. T. & R. A. ST. GEORGE — 1929. The meal worms. U.S.D.A. Tech. Bull. 95.

COTTON, R. T. & G. B. WAGNER — 1938. Practical methods for insuring the production of insect-free flour. U.S.D.A. E-419. Control of insect pests of grain in elevator storage. U.S.D.A. Farmers' Bull. 1880.

COTTON, R. T., G. B. WAGNER & T. F. WINBURN — 1942. Insecticidal value of certain dusts for the protection of stored grain. Kans. Entomol. Soc. 15(1):1-6.

COTTON, R. T., G. B. WAGNER & H. D. YOUNG — 1947. The problem of controlling insects in flour warehouses. Amer. Miller 65:22.

COTTON, R. T., H. H. WALKDEN, G. D. WHITE & D. A. WILBUR — 1953. Causes of outbreaks of stored-grain insects. Kans. Agr. Exp. Sta. Bull. 359.

CRESSMAN, A. W. — 1935. Control of an infestation of the cigarette beetle in a library by the use of heat. J. Econ. Entomol. 26:294-5.

CROMBIE, A. C. — 1944. On intraspecific and interspecific competion in larvae of graminivorous insects. J. Exp. Biol. 20(2):135-151. Rev. Appl. Entomol. A. 32:315-316. 1944.

CRUMB, S. E. & F. S. CHAMBERLIN — 1934. The effect of cool temperatures on some stages of the cigarette beetle. Florida Entomol. 18:11-14, 1936. Laboratory tests on comparative effectiveness of fumigants against cigarette beetle in cigars. J. Econ. Entomol. 28:883-992.

CURRAN, C. H. — 1946. Insects in the house: spider beetle. Natural History 55(1):46. Jan.

DAVID, M. H. & R. B. MILLS — 1975. Development, oviposition, and longevity of *Ahasverus advena*. J. Econ. Entomol. 68(3):341-345.

DAVIES, R. G. — 1949. The biology of *Laemophloeus minutus* Oliv. (Col. Cucujidae). Bull. Entomol. Res. 40(1):63-82.

DAVIS, J. J. — 1946. DDT to control household and stored grain insects. J. Econ. Entomol. 39(1):59-61. 1947. Almond moth. Pests 15(7):38.

DAVIS, W. J. — 1945. Effectiveness of DDT in flour mills. Northwestern Miller 221(7):14a. Biol. Abst. 19:1262-1263, 1945.

DEAN. G. A. — 1947. The lesser grain borer in wheat in the field. J. Econ. Entomol. 40(5):751.

DEAY, H. O. & J. M. AMOS. — 1936. Dust treatments for protecting beans from the bean weevil. J. Econ. Entomol. 29:498-501.

DeCOURSEY, J. D. — 1931. A method of trapping the confused flour beetle *Tribolium confusum* Duval. J. Econ. Entomol. 24:1079-81.

DEFIEL, F. — 1922. An experimental investigation of the supposed poisonous qualities of the granary weevil, *Calandra granaria*. Amer. J. Trop. Med. 2:199-211.

DE FIGUEIREDO, E. R., JR. — 1957. Control of *Araecerus fasciculatus*. Rev. Appl. Entomol. A. 47(3):95. 96.

DE FRANCOLINI, J. — 1935. L'emploi du bromure de methyle pour le traitment de graines de semence. Rev. Path. Veg. 22:1-8.

De ONG, E. R. — 1921. Cold storage control of insects. J. Econ. Entomol. 14:444-7. 1948. Damage to coffee by the drug store beetle. J. Econ. Entomol. 41(1):124-125.

De ONG, E. R. & C. L. ROADHOUSE — 1922. Cheese pests and their control. Calif. Agr. Exp. Sta. Bull. 343, pp. 399-424.

DICKE, R. J. — 1949. Sources of insect and rodent contamination in food processing industries. Pest Control 17(10):13-18.

DOBSON, R. M. — 1954. The species of *Carpophilus* Stephens (Col. Nitidulidae) associated with stored products. Bull. Entomol. Res. 45(2):389-402.

DONOHOE, H. C. — 1937. Indian-meal moth in California. J. Econ. Entomol. 30:680-681. 1939. Notes on Coleoptera found in raisin storages. Proc. Ent. Soc. Wash. 41:154-162. 1946a. Notes on dried fruit insects I. Pests. 14(9):32, 34. 1946b. Notes on Dried Fruit II. Pests 14(10):14-16. Oct. 1946.

DONOHOE, H. C. & D. F. BARNES — 1934. Notes on field trapping of *Lepidoptera* attacking dried fruits. J. Econ. Entomol. 27(5):1067-1072.

DONOHOE, H. C., D. F. BARNES, C. K. FISHER & P. SIMMONS — 1934. Experiments in the exclusion of *Ephestia figulilella* from drying fruit. J. Econ. Entomolol. 27:1072-1075.

DONOHOE, H. C., P. SIMMONS, D. F. BARNES, G. H. KALOOSTIAN & C. K. FISHER — 1943. Preventing damage to commercial dried fruits by the raisin moth. U.S.D.A. Leaflet 236.

DONOHUE, H. C., P. SIMMONS, D. F. BARNES, G. H. KALOOSTIAN, C. K. FISHER & C. HEINRICH — 1949. Biology of the raisin moth. U.S.D.A. Tech. Bull. 994.

DOVE, W. E. & SCHROEDER, H. O. — 1955. Protection of stored grain with sprays of pyrethrins-piperonyl butoxide emulsion. Agr. & Food Chem. 3(11):932-936.

DUFFY, E. A. J. — 1953. Lead cable severely damaged by *Ptinus tectus* Boieldieu (Coleoptera, Ptinidae) Bull. Entomol. Res. 44(1):83-84.

DURANT, J. H. & W. W. O. BEVERAGE — 1913. A preliminary report on the temperature reached in army biscuit during baking especially with reference to the destruction of the imported flour moth, *Ephestia kuhniella* Zeller. J. Reserve Army Med. Corps 20:614-634.

EASTHAM, L. E. S. & S. B. McCULLY — 1943. The oviposition response of *Calandra granaria* Linn. J. Exp. Biol. 20(1):35-42. Rev. Appl. Entomol. A. 32:81, 1944.

EBELING, W. — 1975. Urban Entomology. Univ. of Calif. 695 pp.

ECKERT, J. E. — 1947. Beekeeping in California. Calif. Agr. Ext. Serv. Circ. 100.

EDEN, W. G. — 1953. Control of rice weevil in corn with protectant dusts and sprays. J. Econ. Entomol. 46(6):1105-1106.

EICHLER, W. — 1943. *Sitodrepa panicea* as a pest of wood. Rev. Appl. Entomol. 31(A):53, 1943.

EICHMAN, R. D. — 1943. Commercial pea warehouse tests of fumigants used against the pea weevil in the Palouse Region. J. Econ. Entomol. 36(6):843-849.

ESSIG, E. O. — 1915. The dried-fruit beetle *Carpophilus hemipterus* (L.) J. Econ. Entomol. 8:396-400. 1920. Important dried fruit insects in California. Calif. Dept. Agr. Bull. No. 9 (sup. 3):119-124. 1926. Insects of Western North America. Macmillan. 1929. Origin of the bean weevil, *Mylabris obtectus* Say. J. Econ. Entomol. 22:858-61. 1933. *Ptinus tectus* Boieldieu. J. Econ. Entomol 26:734-5. 1940. Mediterranean flour moth breeding in comb of honeybee. J. Econ. Entomol. 33:949-50.

ESSIG, E. O., W. M. HOSKINS, E. G. LINSLEY, A. E. MICHELBACHER & R. F. SMITH — 1943. A report on the penetration of packing materials by insects. J. Econ. Entomol. 36(6):822-829.

EVANS, B. R. & J. E. PORTER. — 1965. The incidence, importance, and control of insects found in stored food and food-handling areas of ships. J. Econ. Entomol. 58(3):479-481.

EWER, R. F. — 1945. The effect of grain size on the oviposition of *Calandra granaria* Linn. (Coleoptera, Curculionidae). Proc. R. Entomol. Soc. Lond. A. 20 (4-6):57-63. Rev. Appl. Entomol. A. 34(10):297, 1946.

FARRAR, M. D. & W. P. FLINT — 1942. Control of insects in fourteen thousand corn bins. J. Econ. Entomol. 35:615-619.

FARRAR, M. D. & R. H. REED — 1942. Insect survival in drying grain. J. Econ. Entomol 35(6):923-928.

FARRAR, M. D. & J. M. WRIGHT — 1946. Insect damage and germination of seed treated with DDT. J. Econ. Entomol 39(4):520-521.

FISHER, C. D. — 1945. Controlling storage insects by fumigation. Food Indus. 17(10):1176-1178.

FLINT, W. P. & W. E. McCAULEY — 1937. Fabric insects — How to combat them in the home. Ill. Agr. Expt. Sta. Circ. 473.

FLOYD, E. H. & J. D. POWELL — 1958. Some factors influencing the infestation in corn in the field by the rice weevil. J. Econ. Entomol 51(1):23-26.

FRAENKEL, G. & M. BLEWETT — 1943. The natural foods and the food requirements of several species of stored products insects. Trans. R. Entomol. Soc. Lond. 93(2):457-490. 1944. Rev. Appl. Entomol. A. 32:185-186, 1944. 1944. The utilization of metabolic water in insects. Bull. Entomol. Res. 35(2):127-139. 1943. The basic food requirements of several insects. J. Exp. Biol. 20(1):28-34. Rev. Appl. Entomol. A. 32:81, 1944.

FRANKENFELD, J. G. — 1948. Staining methods for detecting weevil infestation in grain. U.S. B.E. & P.Q. ET-256, 4 pp. July.

FREEMAN, J. A. — 1952. *Laemophloeus* spp. as major pests of stored grain. Plant Pathol. 1(3):69-76.

FREEMAN, J. A. & E. E. TURTLE — 1947. The control of insects in flour mills. Ministry of Food. 84 pp. London.

FRENCH, J. C. — 1964. Protection of small quantities of beans, peas or grain from insects. Univ. Georgia Extension Leaflet No. 12.

FRINGS, H. — 1948. Inorganic salts as repellents for package-penetrating insects. J. Econ. Entomol. 41(3):413-416.

GERBERG, E. J. — 1946. DDT to control *Mezium americanum*. J. Econ. Entomol. 39(5):676.

GERHARDT, P. D. & D. L. LINDGREN — 1954. Penetration of various packaging films by common stored product insects. J. Econ. Entomol. 47(2):282-287.

GIBSON, A. — 1924. The occurrence of the Ptinid beetle. *Niptus hololeucus* Fald., in North America. Canad. Entomol. 56:74-6.

GIBSON, A. & C. R. TWINN — 1929. Household insects and their control (with a chapter on animal pests other than insects). Can. Dept. Agt. Bull. 112.

GODKIN, W. J. & W. H. CATHCART — 1949. Effectiveness of heat in controlling insects infesting the surface of bakery products. Food Technology 3:254-257.

GOOD, N. E. — 1933. Biology of the flour beetle, *Tribolium confusum* Duv. and *T. ferrugineum* Fab. J. Agr. Res. 46:1327-1334. 1936. The flour beetles of the genus *Tribolium*. U.S.D.A. Tech. Bull. 498.

GOODWIN-BAILEY, K. F. & J. M. HOLBORN — 1952. Laboratory and field experiments with pyrethrins/piperonyl butoxide powders for the protection of grain. Pyr. Post 2(4):7-17.

GORMAN, J. J. — 1952. Red-legged ham beetles. Pest Control 20(10):6, 86.

GOULD, G. E. — 1948. Insect-problems in corn processing plants. J. Econ. Entomol. 41(5):774-778.

GRAY, H. E. — 1934. Some stored product pests in Canada with special reference to the hairy spider beetle, *Ptinus villiger* Reit. Entol. Soc. Ont. Rept. 65:59-66. 1935. The hairy spider beetle, *Ptinus villiger* Reit. in Canada. Proc. World's Grain Exhibit. & Conf. 1933. 2:55-61. 1942. Spider beetles. Pests 10(7):10-13. 1952. Packaging of cereals and some chemical treatments, to increase resistance to penetration by insects. Am. Association of Cereal Chemistry 10(1):53-58.

GREEN, A. A. & J. KANE — 1959. Comparison of lindane, DDT, and malathion for the protection of bagged peanuts from infestation by *Tribolium castaneum*. Trop. Sci. 1:290-295.

GRIFFITH, C. F. — 1946. Biometric studies on the drug-store beetle, *Stegobium paniceum* L. (Col., Anobiidae). Entomologist. Monthly 82:186-191.

GROSS, J. R. — 1948. Spider beetles may be a sign of rats. Pests 16(6):42. 1949. Find the leaky roof. Pest Control 17(10):52.

GROSSMAN, E. F. — 1931. Heat treatment for controlling the insect pests of stored corn. Fla. Agr. Exp. Sta. Bull. 239.

GUNN, D. L. & R. H. KNIGHT — 1945. The biology and behaviour of *Ptinus tectus* Boie. (Coleoptera, Ptinidae), a pest of stored products VI. Culture Conditions. J. Exp. Biol. 21(3-4):132-143. Rev. Appl. Entomol. A. 34(2):61-62, 1946.

HADAWAY, A. B. — 1956. The biology of the dermestid beetles, *Trogoderma granarium* Everts and *Trogoderma versicolor* (Cruetz.). Bull. Entomol. Res. 46(4):781-796.

HALL, D. W. & R. W. HOWE — 1953. A revised key to the larvae of the Ptinidae associated with stored products. Bull. Entomol. Res. 44(1):85-86.

HAMLIN, J. C. & C. BENTON — 1925. Control of the saw-toothed grain beetle in raisins: A preliminary report. J. Econ. Entomol. 18:790-795.

HAMLIN, J. C. & W. D. REED — 1926. Metal barriers as protective devices against the saw-toothed grain beetle. J. Econ. Entomol. 19:618-624.

HAMLIN, J. C., W. D. REED & M. E. PHILLIPS — 1931. Biology of the Indian-meal moth on dried fruits in California. U.S.D.A. Tech. Bull. 242.

HARDING, W. C., JR. & T. L. BISSELL — 1958. Lesser mealworms in a brooder house. J. Econ. Entomol. 51(1):112.

HARRIS, K. L. — 1943. Some applications of insect separation methods to entomology. Proc. Entomol Soc. Wash. 45(1):19-25. 1950. Identification of insect contaminants of food by the micromorphology of the insect fragments. J. Assoc. Off. Agr. Chem. 33:898-933. Aug. 15, 1955. Additional bibliography of methods for the examination of foods for filth (1946-1954). J. Assoc. Off. Agr. Chem. 38(4):1016-1019.

HARRIS, K. L., J. F. NICHOLSON, L. K. RANDOLPH & J. L. TRAWICK — 1952. An investigation of insect and rodent contamination of wheat and wheat flour. J. Assoc. Off. Agr. Chem. 35(1):115.

HATCH, M. H. — 1933. *Ptinus tectus*, Boieldieu in America. Bull. B'klyn Entomol. Soc. 28:200-2. 1943. *Ptinus tectus* damaging furs in Alaska. J. Econ. Entomol. 36:353. 1942. The biology of stored grain insects. Bull. Assoc. Operative Millers pp. 1207-1211.

HENDERSON, L. S. — 1955. Insect prevention and control in food plants. Pest Control 23(5):18, 20-22. 23(6):18, 20-22.

HERMS, W. B. — 1915. Medical and Veterinary Entomology. Macmillan Co. 1917. The Indian meal moth, *Plodia interpunctella* Hubn. in candy and notes on its life-history. J. Econ. Entomol. 10:563.

HEUERMAN, R. F. & O. L. KURTZ — 1955. Identification of stored products insects by the micromorphology of the exoskeleton. I. Elytral patterns. J. Assoc. Off. Agr. Chem. Aug. 1955.

HICKIN, N. E. — 1942. Infestation of foodstuffs by *Ptinus tectus* Boield. (Col. Ptinidae). Entomol. Mon. Mag. 78(932):14. Rev. Appl. Entomol. A. 30:420, 1942.

HINDS, W. E. & W. F. TURNER — 1911. Life-history of the rice weevil, *Calandra oryza* Linn, in Alabama. J. Econ. Entomol. 4:230-236.

HINTON, H. E. — 1940. The Ptinidae of economic importance. Bull. Entomol. Res. 31:331-381. 1941. The Lathridiidae of economic importance. Bull. Entomol Res. 32(3):191-247. 69 figs. 1943. Natural reservoirs of beetles of the family Dermestidae known to infest stored products, with notes on those found in spiders' webs. Proc. Roy. Entomol. Soc. London Ser. A. 18(4/6):33-42. 1945. A Monograph of the Beetles Associated with Stored Products. Vol. I British Museum (Natural History). 443 pp. 505 illust. 1948. A synopsis of the genus *Tribolium* Macleay, with some remarks on the evolution of its species-groups (Coleoptera, Tenebrionidae). Bull. Entomol. Res. 39(1):13-55.

HINTON, H. E. & A. S. CORBET — 1949. Common insect pests of stored food products. Brit. Mus. (Nat. Hist.) Econ. Ser. 15. 2nd Ed.

HOFFMAN, W. A. — 1933. *Rhizopertha dominica* as a library pest. J. Econ. Entomol. 26:293-294.

HOLBORN, J. M. — 1957. Susceptibility to insecticides of laboratory cultures of an insect species. J. Sci. Food Agr. 8:182-188. Chem. Abst. 51:11643, 1957.

HORN, W. — 1934. A second contribution on insects that bore into lead, especially lead sheathing of aerial cables. Rev. Appl. Entomol. (A) 23:121, 1935.

HOWE, R. W. — 1943. Life history data for *Ptinus tectus* Boie. (Coleoptera Ptinidae) at 70% relative humidity at 21° C and 25° C. Proc. R. Entomol. Soc. Lond. (A) 18(7-9):63-65. Rev. Appl. Entomol. A. 32:35, 1944. 1949. Studies on beetles of the family Ptinidae. I. Notes on the biology of species in Britain. Entomol. Mon. Mag. 85:137-139. Rev. Appl. Entomol. A. 38(12):477-478, 1950. Studies on beetles of the family Ptinidae. III — A two-year study of the distribution and abundance of *Ptinus tectus* Boield. in a warehouse. Bull. Entomol. Res. 41(2):371-394. 1956. The biology of two common storage species of *Oryzaephilus* (Coleoptera, Cucujidae). Ann. Appl. Biol. 44(2):341-355. Biol. Abst. 31(1):273, 1957. 1957. A laboratory study of the cigarette beetle, *Lasioderma serricorne* (F.) (Col. Anobiidae) with a critical review of the literature on its biology. Bull. Entomol. Res. 48(1):9-56. 1959. Bull. Entomol. Res. 50(2):287-326.

HOWE, R. W. & H. D. BURGES — 1951. Studies on beetles of the family Ptinidae. VI. The biology of *Ptinus fur* (L.) and *P. sexpunctatus* Panzer. Bull Entomol. Res. 42(3):499-513. 1952. Studies on beetles of the family Ptinidae. VII. The biology of five ptinid species found in stored products. Bull. Entomol. Res. 43(1):153-186. 1953. Studies on beetles of the family Ptinidae. IX — A laboratory study of the biology of *Ptinus tectus* Boield. Bull. Entomol. Res. 44(3):461-516. 1956. *Trogoderma arfum* Priesner, a synonym of *T. granarium* Everts and a comparison with *T. versicolor* (Cruetz.). Bull. Entomol. Res. 46(4):773-780.
Bull. Entomol. Res. 46(4):773-780.

INCHO, H. H., E. J. INCHO & N. W. MATHEWS — 1953. Insect-proofing paper: Laboratory evaluation of pyrethrins-piperonyl butoxide formulations. Agr. & Food Chem. 1(20):1200-1203.

JAY, E. G., H. L. MUSEN & G. C. PEARMAN — 1973. Damage to stored soybeans by the cowpea weevil, *Callosobruchus maculatus*. (F.) Ga. Entomol. Soc. 8(3)164-167.

JONES, B. M. — 1947. An experiment with DDT against pests of stored products. Bull. Entomol. Res. 38(2):347-352.

JONES, C. R. — 1940. A new household pest for Colorado, *Ptinus fur* L. Pests 8(5):22.

JORDAN, K. H. C. — 1930. Cigars attacked by *Tenebrio molitor*. Rev. Appl. Entomol (A) 18:438, 1930.

JOUBERT, P. C. — 1962. The toxicity of contact insecticides to seed-infesting insects. Pyr. Post 8(2):6-14.

JUDD, W. W. — 1949. The red-legged ham beetle on imported copra. Can. Entomol. 81(2):52.

KELLOGG, V. L. — 1894. Insects injuring drugs at the University of Kansas. Insect Life 7:31.

KEMPER, H. — 1938. Biology of *Gibbium psylloides*. Rev. Appl. Entomol. (A) 26:519, 1938.

KENAGA, E. E. — 1957. Evaluation of grain fumigants. Down to Earth 13(3):10-13.

KENAGA, E. E. & F. W. FLETCHER — 1942. Effects of high temperature on several household and storage grain pests. J. Econ. Entomol. 35(6):944.

KERR, T. W., JR. & D. L. Mc LEAN — 1956. Biology and control of certain Lathridiidae. J. Econ. Entomol. 49(2):269-270.

KING, J. L. — 1918. Notes on the biology of the Angoumois grain moth, *Sitotroga cerealella* Oliv. J. Econ. Entomol. 11:87-93.

KROHNE, H. E. & D. L. LINDGREN — 1958. Susceptibility of life stages of *Sitophilus oryza* to various fumigants. J. Econ. Entomol. 51(2):157-158.

KULASH, W. M. — 1954. Save stored grain from insect pests. North Carolina Agr. Expt. Sta. Bull. 389.

KUNIKE, G. — 1941. Investigations on the protection of packaging material against penetration by pests of foodstuffs. Mitt. Biol. Reichsanst. 65:42. Rev. Appl. Entomol. A., 31:371, 1943.

KURTZ, O. L. & K. L. HARRIS — 1955. Identification of insect fragments: Relationship to the etiology of the contamination. J. Assoc. Off. Agr. Chem. 38(4):1010-1015. 1962. Micro-analytical Entomology for Food Sanitation Control. Assoc. of Official Agr. Chemists. Washington, D.C.

KURTZ, O. L., N. A. CARSON & H. C. VAN DAME — 1952. Identification of cereal insects in stored grain by their mandible characteristics. J. Assoc. Off. Agr. Chem. 35(4):817-826.

LARSON, A. O. — 1924. Fumigation of bean weevil, *Bruchus obtectus* Say and *B. quadrimaculatus* Fab. J. Agr. Res. 28:347-356.

LARSON, A. O., T. A. BRINDLEY & F. G. HINMAN — 1938. Biology of the pea weevil in the Pacific northwest with suggestions for its control on seed peas. U.S.D.A. Tech. Bull. 599.

LARSON, A. O. & C. K. FISHER — 1938. The bean weevil and the southern cowpea weevil in California. U.S.D.A. Tech. Bull. 593.

LARSON, A. O. & P. SIMMONS — 1923. Notes on the biology of the four spotted bean weevil, *Bruchus quadrimaculatus* Fab. J. Agr. Res. 26:609-16. 1924. Insecticidal effect of cold storage on bean weevils. J. Agr. Res. 27:99-105.

LATHROP. F. H. & L. K. KEIRSTEAD — 1946. Black pepper to control the bean weevil. J. Econ. Entomol. 39(4):534.

LAUDANI, H. & D. F. DAVIS — 1955. The status of Federal research on the development of insect-resistant packages. Tappi 38(6):322-326.

LAUDANI, H., D. F. DAVIS & G. R. SWANK — 1958. Improved packaging methods can cut insect infestation. Sanit. & Bldg. Maintenance. March, 1958.

LAUDANI, H., H. B. GILLENWATER, B. H. KANTACK, & M. F. PHILLIPS — 1959. Protection of citrus pulp against insect infestation with surface applications of pyrethrum-piperonyl butoxide wettable powder. J. Econ. Entomol. 52(2):224-227.

LECATO, G. L. & B. R. FLAHERTY — 1974. Description of eggs of selected species of stored-product insects. J. Kansas Entomol. Soc. 74(3):308-317.

LEECH, H. B. — 1943. Black flour beetle, *Tribolium madens* Charpi in British Columbia (Coleoptera, Tenebrionidae). Can. Entomol. 75(2):(40).

LEFKOVITCH, L. P. — 1964. The biology of *Cryptolestes pusilloides*. Bull. Entomol. Res. 54(4):649-656.

LERGENMULLER, E. — 1958. Ecological investigations on *O. surinamensis*. Z. angew. Zool. 45(1):31-97. R.A.E. A. 48(12):545, 1960.

LESNE, P. — 1941. Quelques remarques sur le *Rhizopertha dominica* F. (Col. Bostrychidae). Rev. Franc. Entomol. 7(4):145-151. Rev. App. Entomol. A. 34(11): 328, 1946.

LINDGREN, D. L., H. E. KROHNE & L. F. VINCENT — 1954. Malathion and chlorthion for control of insects infesting stored grain. J. Econ. Entomol. 47(4):705-706.

LINDGREN, D. L. & E. V. LLOYD — 1953. Nitidulid beetles infesting California dates. Hilgardia 22(2):97-118.

LINDGREN, D. L. & H. H. SHEPARD — 1932. The influence of humidity on the effectiveness of certain fumigants against the eggs and adults of *Tribolium confusum* Duv. J. Econ. Entomol. 25:248-53.

LINDGREN, D. L., L. E. VINCENT & H. E. KROHNE — 1954. Relative effectiveness of ten fumigants to adults of eight species of stored-product insects. J. Econ. Entomol. 47(5):923-926.

LINDGREN, D. L., L. E. VINCENT & H. E. KROHNE — 1955. The khapra beetle, *Trogoderma granarium* Everts. Hilgardia 24(1):36 pp.

LINSLEY, E. G. — 1942. Insect food caches as reservoirs and original sources of some stored products pests. J. Econ. Entomol 35(3):434. 1942a. Woodboring habit of the drugstore beetle. J. Econ. Entomol 35(3):452. 1942b. A further note on wood-boring by the drugstore beetle. J. Econ. Entomol. 35(5): 701. 1943. The spelling of scientific names of some stored food product pests. J. Econ. Entomol 36(1):125-126. 1943. The dried fruit moth breeding in nests of the Mountain Carpenter Bee in California. J. Econ. Entomol 36(1):122-123. 1944. Natural sources, habitat, and reservoirs of insects associated with stored food products. Hilgardia 16(4):187-224. 1944. Protection of dried packaged foodstuffs from insect damage. J. Econ. Entomol. 37(3):337-379.

LINSLEY, E. G. & A. E. MICHELBACHER — 1943. Insects affecting stored food products. Calif. Agr. Expt. Sta. Bull. 676. 1943. A report on insect infestation of stored grain in California. J. Econ. Entomol 36(6):829-831.

LLOYD, C. J. & P. S. HEWLETT — 1958. The relative susceptibility to pyrethrum in oil of Coleoptera and Lepidoptera infesting stored products. Bull. Entomol. Res. 49:177-185.

LOSCHIAVO, S. R. — 1976. Food selection by *Oryzaephilus mercator* (Coleoptera: Cucujidae). Can. Entomol. 108;827-831.

LUCAS, C. E. & T. A. OXLEY — 1946. Study of an infestation by *Laemophloeus sp.* (Coleoptera, Cucujidae) in bulk wheat. Ann. Appl. Biol. 33(3):289-293. Rev. Appl. Entomol. A 36(12):418-419, 1948.

LYLE, C. — 1936. Long survival of *Gibbium psylloides* Czemp. J. Econ. Entomol. 29:1026.

MACKIE, D. B. — 1932. Annual report of the Division of Entomology and Pest Control. Monthly Bull. Dept. Agr. Calif. 21:474-488.

MacNAY, C. G. — 1950. Yellow mealworm. 80th Ann. Rept. of Entomol. Soc. of Ontario, 1949, p. 77.

MAMPE, C. D. — 1976. Cigarette beetles. Pest Control 44(9):18.

MANTON, S. M. — 1945. The larvae of Ptinidae associated with stored products. Bull. Entomol. Res. 35:341-365.

MARCOVITCH, S. — 1929. The control of the tobacco beetle in upholstered furniture. J. Econ. Entomol. 22:602. 1934. Control of weevils in stored beans and cowpeas. Univ. Tenn. Agr. Exp. Sta. Bull. 150. 1935. Control of the bean weevil and the cowpea weevil. J. Econ. Entomol. 28:796-7.

MATHLEIN, R. — 1941. Investigations on pests of stored products. II. The grain moths. *T. secalella* and *T. granella*. Medd. Vaxtskyddsanst. 34:56 pp.

MAYER, E. L. & H. D. NELSON — 1955. Fumigation of dry beans and cowpeas on the packaging line. U.S. Agr. Mktg. Serv. AMS-4.

McCOLLOCH, J. W. — 1922. Longevity of the larval state of the cadelle. J. Econ.

Entomol. 15:240-3.

McLAINE, L. S. — 1943. The war activities of the Federal Division of Entomology and plant protection since 1939. 73rd Rpt. Entomol. Soc. Ont. 1942, pp. 7-16. Rev. Appl. Entomol. A., 31:456, 1943.

MERRILL, E. D. — 1948. On the control of destructive insects in the herbarium. J. of the Arnold Arboretum. 29:103-110.

METCALF, C. L. & W. P. FLINT — 1939. Destructive and Useful Insects. McGraw-Hill.

MICHELBACHER, A. E. — 1953. Insects attacking stored products. Advances in Food Research 4:281-358.

MICHEL, C. E. & J. STANDISH — 1946. Susceptibility of edible soya products in storage to attack by *Tribolium confusum* Duv. Tech. Bull. Minnesota Agric. Expt. Sta. 175, 1-28. 9 pl.

MILLS, H. B. & J. H. PEPPER — 1939. The effect on humans of the ingestion of the confused flour beetle. J. Econ. Entomol. 32:874-5.

MONRO, H. A. U. — 1942. Tests with a pyrethrum aerosol against cockroaches and stored product pests. 73rd Annual Report of the Entomological Society of Ontario, pp. 61-63. — 1964. Manual of Fumigation for Insect Control. Food & Agr. Org. United Nations. 1969. Second Edition.

MONRO, H. A. U. & E. UPITIS — 1956. Selection of populations of the granary weevil *Sitophilus granarius* L. more resistant to methyl bromide fumigation. Canad. Entomol. 88:37-40.

MONTGOMERY, V. H. — 1939. Spider beetles. Ext. Log 4:20.

MOORE III, S. — 1959. Malathion : A new insect protectant for stored grain. Pest Control. 27(7):40.

MORGAN, W. L. & G. PASFIELD — 1943. Dusts for protecting bean seed against *Bruchus obtectus* Say. Rev. Appl. Entomol. 31(A):262.

MUNRO, J. W. — 1942. Infestation of manufactured food by insects. Rev. Appl. Entomol. A. 30:91, 1942. 1966. Pests of Stored Products. Hutchinson, London.

NELSON, H. D. & C. K. FISHER — 1952. Control of insects that attack dried beans and peas in storage. U.S.D.A. EC-27.

NOON, Z. B., JR. — 1958. Food habits of the khapra beetle larva. J. Econ. Entomol. 5(4):465-469.

O'FARRELL, A. F., B. M. JONES & G. A. BRETT — 1949. The persistent toxicity under standardized conditions of pyrethrum, DDT and "Gammexane" against pests of stored food. Bull. Entomol. Res. 40(1):135-148.

OKUMURA, G. T. — 1966. The dried-fruit moth, *Vitula edmandsae serratilineella* Ragonot. Bull. Calif. Dept. Agr. 55(4):180-186.

OKUMURA, G. T. & I. E. SAVAGE — 1974. Nitidulid Beetles most commonly found attacking dried fruit in California. Nat. Pest Control Operator News. 34(3):2-7.

OKUMURA, G. T. & R. G. STRONG — 1965. Insects and mites associated with stored food and seeds in California. Part II. Bull. Calif. Dept. Agr. 54(1).

OSMUN, J. V. — 1954. Protection of stored shelled corn with a protectant dust in Indiana. J. Econ. Entomol. 47(3):462-464.

OSMUN, J. V. & W. H. LUCKMANN — 1964. How to identify and control the picnic beetle. Pest Control. 32(4):32, 34.

PADDOCK, F. B. — 1918. The beemoth or waxworm. Texas Agr. Exp. Sta. Bull. 231.

PALMER, E. D. — 1946. Intestinal canthariasis due to *Tenebrio molitor*. J. Parasitol. 32(1):54-55.

PAPP, C. S. & G. T. OKUMURA — 1959. A preliminary study of the Ptinidae. Bull. Calif. Dept. Agr. 48(4):228-248.

PARK, T. — 1945. Life tables for the black flour beetle, *Tribolium madens* Charp. Amer. Nat. 79:436-444. Sept./Oct. 1945.

PARKIN, E. A. — 1944. Control of the granary weevil with finely ground mineral dusts. Ann. Appl. Biol. 31(1):84-88. 1946. The toxicity of certain aliphatic chlorinated hydrocarbons to *Calandra granaria* L. and other insects infesting grain. Ann. Appl. Biol. 33(1):97-103.

PATTON, W. S. — 1931. Insects, Ticks, Mites and Venomous Animals. H. R. Grub, Ltd.

PAYNE, N. M. — 1925. Some effects of *Tribolium* on flour. J. Econ. Entomol. 18:737-744. 1946. Life history and habits of the flat grain beetle *(Laemophloeus minutus* Oliv.) J. N.Y. Entomol. Soc. 54:9-12.

PEARL, R, T. PARK & J. R. MINER — 1940. Experimental studies on the duration of life. XVI. Life tables for the flour beetle *Tribolium confusum* Duval. Anat. Rec. 78 Suppl. p. 170. Rev. Apply. Entomol. A. 32:389, 1944.

PEPPER, J. H. & A. L. STRAND — 1935. The importance of surface temperatures in heat sterilization. J. Econ. Entomol. 28:242-4.

PHILLIPS, G. L. — 1959. Control of insects with pyrethrum sprays in wheat stored in ship's holds. J. Econ. Entomol. 52(4):557-559.

PHILLIPS, G. L., W. K. WHITNEY, C. L. STOREY & H. H. WALKDEN — 1959. Bulk cocoa bean fumigation for tobacco moth. Pest Control 27(6):39-42.

PIMENTEL, D. — 1949. Biology of *Gnathocerus cornutus.* J. Econ. Entomol. 42(2):229-231.

PIRIE, H. — 1951. Insecticides in industry and public health. Chem. & Indust. Feb. 10, 1951.

POTTER, C. — 1936. The biology and distribution of *Rhizopertha dominica* (Fab.) Trans. A. Entomol. Soc. Lond. 85:449-82.

POTTER, C. & F. TATTERSFIELD — 1943. Ovicidal properties of certain insecticides of plant origin. (Nicotine, Pyrethrins, Derris Products). Bull. Entomol. Res. 34(3):225-244.

POWELL, T. E. JR. — 1931. An ecological study of the tobacco beetle, *Lasioderma serricorne* Fab. with special reference to its life history and control. Ecol. Monog. 1:333-393.

PRESS, A. F., JR. & D. P. CHILDS — 1966. Control of the tobacco moth with dichlorvos. J. Econ. Entomol. 59(2):264-265.

QUINLAN, J. K. & R. F. MILLER — 1958. Evaluation of synergized pyrethrum for the control of Indian-meal moth in stored shelled corn. U.S.D.A. Agr. Mkt. Serv. Rpt. No. 222.

RAU, P. — 1915. Duration of pupal and adult stages of the meal worm, *Tenebrio obscurus* Linn. (Coleop.) Entomol. News 26:154-157.

REDDY, D. B. — 1950. Ecological studies of the rice weevil. J. Econ. Entomol. 43(2):203-206. 1951. Determination of sex in adult rice and granary weevils. Pan-Pacific Ent. 27(1):13-16.

REED, W. D. & E. M. LIVINGSTONE — 1937. Biology of the tobacco moth and its control in closed storage. U.S.D.A. Cir. 422.

REED, W. D., E. LIVINGSTONE & A. W. MORRILL, JR. — 1933. A pest of cured tobacco, *Ephestia elutella* Hubner. U.S.D.A. Cir. 269. 1934. Experiments with suction light traps for combating the cigarette beetle. J. Econ. Entomol. 27:796-801.

REYNOLDS, J. M. — 1944. The biology of *Tribolium destructor* Uytt. I. Some

588 *Handbook of Pest Control*

effects of fertilization and food factors on fecundity and fertility. Ann. Appl. Biol. 31(2):132-142. Abst. 19:642, 1945.

RICHARDS, O. W. & G. V. B. HERFORD — 1930. Insects found associated with cacao, spices and dried fruits in London warehouses. Ann. Appl. Biol. 17: 367-395.

RICHARDS, O. W. & N. WALOFF — 1946. The study of a population of *Ephestia elutella* Hubner (Lep. Phycitidae) living on bulk grain. Trans. R. Entomol. Soc. Lond. 97 pt. 11 pp. 253-298, 11 figs., 20 refs.

RICHARDSON, C. H. — 1945. Fumigants for the cadelle in shelled corn. J. Econ. Entomol 38(4):478-481.

RILETT, R. O. — 1949. The biology of *Laemophloeus ferrugineus* (Steph.) — Canad. J. Res. D. 27(3):112-148. Rev. Appl. Entomol. A. 38(4):158-159.

RILEY, C. V. — 1874. The red-legged ham beetle — *Corynetes rufipes* (Fabr.). Rept. State Entomol. Mo. 6:96-102.

RILEY, C. V. & L. O. HOWARD — 1889. Beetles in a pin-cushion. U.S.D.A. Insect Life 2:148. 1892. Damage to boots and shoes by *Sitodrepa panicea*. Insect Life 4:403. 1893. Cigarette beetle eating silk. Insect Life 6:40.

ROBINSON, W. — 1926. Low temperature and moisture as factors in the ecology of the rice weevil, *Sitophilus oryza* L. and the granary weevil, *Sitophilus granarius* L. Minn. Agr. Exp. Sta. Tech. Bull. 41. 43 pp.

RUNNER, G. A. — 1922. The tobacco beetle and how to prevent damage by it. U.S.D.A. Farmers' Bull. No. 846.

SAMSONIYA, K. P. — 1936. Sur la nature de l'influence des carbonates sur le charancon de grenier. *(Calandra granaria* L.) Bull. Soc. Nat. Moscow Sect. Biol. 4:307-311.

SAUNDERS, J. P. & E. C. BAY — 1958. Resistance of some rodenticidal baits to infestation by *Tribolium confusum* Duv. J. Econ. Entomol. 51(3):299-302.

SCHOENHERR, W. H. — 1972. A guide to good manufacturing practices for the food industry. Lauhoff Grain Co., Danville, Illinois. VIII sections.

SCHWITZGEBEL, R. B. & H. H. WALKDEN — 1944. Summer infestation of farm-stored grain by migrating insects. J. Econ. Entomol. 37(1):21-24.

SHAPIRO, L. — 1941. *Ptinus tectus* Boield. Rev. Appl. Entomol. A. 32:101, 1944.

SHEPARD, H. H. — 1939. Insects infesting stored food. Minn. Agr. Expt. Sta. Bull. 341. 1940. Insects infesting home foods. Minn. Agr. Ext. Serv. Bull. 210.

SHEPARD, H. H. & D. L. LINDGREN — 1934. The relative efficiency of some fumigants against the rice weevil and the confused flour beetle. J. Econ. Entomol. 27:842-845.

SHEPHERD, D. — 1924. Life history and biology of *Echocerus cornutus* (Fab.) J. Econ. Entomol. 17:572-7

SHEPHERD, D. R. — 1957. Khapra beetle eradication. Agr. Chem. May 1957. pp. 32-33.

SIMMONS, P. — 1927. The cheese skipper as a pest in cured meat. U.S.D.A. Dept. Bull. 1453, 55 pp. 1943. Preventing insect damage in home-dried fruits. U.S.D.A. Leaflet 253, 4 pp. 1947. Preventing insect damage in home-dried fruits. Pest Control & Sanitation. 2(7):25, 27. 1960. Malathion sprays for reducing dried-fruit packing house infestations. J. Econ. Entomol. 53(5):969-970.

SIMMONS, P., D. F. BARNES, H. C. DONOHOE & C. K. FISHER — 1936. Progress in dried fruit insect investigations in 1935. 1/ U.S.D.A. E-382.

SIMMONS, P. & H. C. DONOHOE — 1938. Changes in the insect population of stored raisins. U.S.D.A. E-437.

SIMMONS, P. H. C. DONOHOE, D. F. BARNES & C. K. FISHER — 1937. Infestations in raisins and its control. U.S.D.A. E-414.

SIMMONS, P. & G. W. ELLINGTON — 1924. Biology of the Angoumois grain moth-progress report. J. Econ. Entomol. 17:41-45. 1925. The ham beetle, *Necrobia rufipes* DeGeer. J. Agr. Res. 30:845-63. 1933. Life history of the Angoumois grain moth in Maryland. U.S.D.A. Tech. Bull. 351.

SIMMONS, P. & C. K. FISHER — 1945. Ethyl formate and isopropyl formate as fumigants for packages of dried fruits. J. Econ. Entomol. 8(6):715-716.

SIMMONS, P., W. D. REED & E. A. McGREGOR — 1931. Fig insects in California. U.S.D.A. Circ. 157.

SIVIK, F. P. & W. M. KULASH — 1956. Treated cloth bags to control the rice weevil in corn. J. Econ. Entomol. 49(1):64-65.

SIVIK, F. P., J. N. TENHET & C. D. DELMAR — 1957. An ecological study of the cigarette beetle in tobacco storage warehouses. J. Econ. Entomol 50(3):310-316.

SMALLMAN, B. N. — 1945. Dust insecticide. Amer. Miller and Processor 73(7):52, 56, 58. Biol. Abst. 19:2214, 1945. 1945. Relation for insect damage to thiamine content of biscuits. J. Econ. Entomol 38(1):106-110. 1948. Residual insecticides for the control of spider beetles in cereal warehouses. J. Econ. Entomol. 41(6):869-874.

SMALLMAN, B. N. & T. R. AITKEN — 1944. Susceptibility of biscuits to insect damage. Cereal Chem. 21(6):499-510.

SPENCER, G. J. — 1945. On the destruction of all stages of insects in pulverized cereals and spices. Proc. Entomol. Soc. B. C. 42:16. Rev. Appl. Entomol. A 35(2):60-61, 1947.

SPITLER, G. H., J. A. COFFELT & P. L. HARTSELL — 1976. Malathion as a protectant against storage insects of inshell walnuts. J. Econ Entomol. 69(4):539-541.

STAMATINIS, N. C. — 1935. The enemies of tobacco in warehouses. *Ephestia elutella* Hb. and *Lasioderma serricorne* Fab. The biology and measures for their control. Rev. Appl. Entomol (A) 23:259, 1935.

STAMPFEL, J. — 1944. Experiments on the feeding of tenebrionids. Acta Soc. Ent. Bohem. 41(1-4):4-12. Rev. Appl. Entomol. A. 34(11):328, 1946.

STANLEY, J. — 1939. Time required for the development of *Tribolium* eggs at 27° C. Ann. Entomol. Soc. Amer. 32:654-9.

STATELER, E. S. — 1943. Mechanical treatment destroys insects in foods. Food Indust. 15(7):82-83.

STELLWAAG, F. — 1924. The fauna of deep-lying wine cellars. Rev. Appl. Entomol. (A) 13:210, 1925.

STONE, P. C. — 1949. Drug store beetle infesting calf starter pellets. J. Econ. Entomol 42(2):371.

STRACENER, C. L. — 1934. Insects of stored rice in Louisiana and their control. J. Econ. Entomol. 27:767-71.

STRONG. R. G. & G. T. OKUMURA — 1958. Insects and mites associated with stored foods and seeds in California. Bull. Calif. Dept. Agr. 47(3):233-249.

STRONG, R. G. & G. T. OKUMURA — 1966. *Trogoderma* species found in California. Bull. Calif. Dept. Agr. 55(1):23-30.

STRONG, R. G. & D. E. SBUR — 1963. Protection of wheat seed with diatomaceus earth. J. Econ. Entomol 56(3):372-374.

SURTEES, G. — 1965. Effect of grain size on development of the weevil, *Sitophilus granarius* (L.) Proc. Roy. Entomol. Soc. London (A) 40(1-3):38-40.

SWANK, G. R., D. F. DAVIS & S. I. GERTLER — 1957. N-pentylphthalimide as a repellent for possible use on insect-resistant packaging. J. Econ. Entomol. 50(4):515-516.

SWEETMAN, H. L. & A. I. BOURNE — 1944. The protective value of asphalt laminated paper against insects. J. Econ. Entomol. 37(5):605-609.

SWINGLE, M. C. — 1938. Low temperatures as a possible means of controlling cigarette beetle in stored tobacco. U.S.D.A. Circ. 462.

TARRY, D. W. — 1967. Control of ptinid beetle infestation in a canteen building. International Pest Control. 9(4):16-17.

TENHET, J. N. — 1947. Effect of sublethal dosages of pyrethrum on oviposition of the cigarette beetle. J. Econ. Entomol. 40(60):910-911. Control of insects in stored tobacco with pyrethrum oil sprays. U.S.B.E. & P. Q. E. C-9. 1955. Timing of sprays to control the cigarette beetle. Agr. Mkt. Serv. AMS-49. 1959. Pyrethrum mists and aerosols for control of insects in tobacco warehouses. U.S.D.A., A.M.S. Marketing Research Rpt. 334. 1961. Controlling the cigarette beetle in the tropics. U.S.D.A. AMS-439.

TENHET, J. N. & C. O. BARE — 1946. Redrying of tobacco and its effect on insect infestation. J. Econ. Entomol 39(5):607-609.

THOMAS, E. L. & H. H. SHEPARD — 1940. Influence of temperature, moisture, and food upon the development and survival of the saw-toothed grain beetle. J. Agr. Res. 60(9):605-615.

THOMPSON. J. A. & J. S. PERRY — 1956. Storage of pea beans in Michigan and Indiana. U.S.D.A. AMS-123.

THOMSSEN, E. G. & M. H. DONER — 1943. Insect control in food production. Soap 19:94-97. Febr. Part. I.

TREHAN, K. N. & S. A. RAJARAO — 1945. A note on the predatory habit of *Tribolium* beetles. Curr. Sci. 14(8):209-210. Rev. Appl. Entomol. A. 33:393, 1945.

TUCKER, E. S. — 1909. New breeding records of the coffee-bean weevil *(Araecerus fasciculatus* DeGeer). U.S.D.A. Bur. Entomol. Bull. 64:61-64.

ULLYETT, G. C. — 1945. Oviposition by *Ephestia kühniella* Zell. J. Ent. Soc. So. Afr. 8:53-59. Rev. Appl. Entomol. A. 37(2):54, 1949.

VANSELL, G. H. — 1943. The wax moth as a household pest. J. Econ. Entomol. 36(4):626-627.

VERCAMMEN-GRANDJEAN, P. H. — 1946. De la destruction continue, par voie, mecanique, des insects parasites des graines, de leurs produits et sous-products. Rep. 1st Int. Congr. Plant Prot. Heverlee 1946. Rev. Appl. Entomol. A. 36(9):282-283.

VINCENT, L. E. & D. L. LINDGREN — 1957. Laboratory evaluation of contact insecticides on three species of stored-product insects. J. Econ. Entomol. 50(3):372-373.

VINZANT, J. P. & W. D. REED — 1941. Type of wire screen required for excluding cigarette beetles and tobacco moths from warehouses. J. Econ. Entomol. 34(5):724.

WAGNER, G. B. — 1957. Sanitation in flour mills. Pest Control 25(19):34-35.

WAGNER, G. B., R. T. COTTON & H. D. YOUNG — 1936. The machinery-piping system of flour-mill fumigation. U.S.D.A. E-396.

WALKDEN, H. H. & H. D. NELSON — 1959. Evaluation of synergized pyrethrum for the protection of stored wheat and shelled corn from insect attack. U.S.D.A., AMS Rpt. 322.

WALKER, J. & D. H. MITCHELL — 1944. The fumigation of dates. Rept. Date

Growers' Inst. 21:4-6.

WARD, I. J. — 1941. The bean weevil, *Acanthoscelides obtectus* (Say) in stored white beans. Canad. Entomol. 73(11):216.

WATTERS, F. L. — 1961. Effectiveness of lindane, malathion, methoxychlor, and pyrethrins-piperonyl butoxide against the hairy spider beetle. *Ptinus villiger.* J. Econ. Entomol. 54(2):397.

WATTERS, F. L. & R. A. SELLEN — 1956. Further tests with DDT and pyrethrins-piperonyl butoxide against the hairy spider beetle. J. Econ. Entomol. 49(2):280-281.

WATTS, C. N. & F. G. BERLIN — 1950. Piperonyl butoxide and pyrethrins to control rice weevils. J. Econ. Entomol. 43(3):371-373.

WEISS, H. B. — 1944. Note on the death-feint of *Bruchus obtectus* (Say) J. N.Y. Entomol. Soc. 52:262.

WHITE, G. D. & H. E. McGREGOR — 1957. Epidemic infestation of wheat by a dermestid. *Trogoderma glabrum* (Herbst). J. Econ. Entomol. 50(4):382-385.

WHITNEY, W. K. & J. R. PEDERSEN — 1962. Physical and mechanical methods of stored product insect control. Pest Control, I. 30(10):42. II. 30(11):16.

WIGGLESWORTH, V. B. — 1947. The site of action of inert dusts on certain beetles infesting stored products. Roy. Entomol. Soc. London, Proc. Ser. A:Gen. Ent. 22:65-69. Sept. 20.

WILLIS, E. R. & L. M. ROTH — 1950. The attraction of *Tribolium castaneum* to flour. J. Econ. Entomol. 43(6):927.

WILSON, F. — 1946. The use of mineral dusts for the control of wheat pests. Bull. Conn. Sci. Industr. Res. Aust. No. 199. Rev. Appl. Entomol. A. 35(9):302-303.

WINBURN, T. F. — 1952. Fumigants and protectants for controlling insects in stored grain. Pest Control 20(8):9-11, 32.

WOODROFFE, G. E. — 1962. The status of the foreign grain beetle, *Ahasverus advena* as a pest of stored products. Bull. Entomol. Res. 53(3):537-540.

ZACHER, F. — 1941. Beobachtungen über "Kornmotten." Zeitschr. angew. Ent. 28:465-476. Biol. Abst. 20:821, 1946.

ZECK, E. H. — 1943. Pests of dried fruits. Agric. Gaz. N.S.W. 54(2):67-71. Rev. Appl. Entomol. 31 A., 391. 1943.

ROBERT J. SNETSINGER

Dr. Robert Snetsinger is a native of Illinois and received bachelor's, master's and Ph.D. degrees in entomology from the University of Illinois, Urbana in 1952, 1953 and 1960 respectively. Prior to receiving his Ph.D., he was employed as an orchard foreman, a farm manager and as a research assistant at the Illinois Natural History Survey, Urbana. In 1960, he joined The Pennsylvania State University as an assistant professor of entomology to conduct research on mushroom and greenhouse pests. In 1961, he was placed in charge of educational programs for the Pennsylvania structural pest control industry.

Snetsinger has conducted research on spider mites and other pests of greenhouse and ornamental crops, on the biology and control of mushroom pests, on spiders and ticks and on household pests. He has authored numerous scientific papers in these subject areas. In addition, he is co-author of a series of eight correspondence courses on structural pest control. But Snetsinger's writing goes beyond pest control subjects. He is the author of the books _Kiss Clara for Me_, a historical account of the American Civil War, based on letters written by his great-grandfather; _Diary of A Mad Planner_, an account of urban planning; and _Frederich Valentine Melsheimer — Parent of American Entomology_, an account of America's first entomologist.

His most recent book, soon to be published by Franzak and Foster Co., is _The Ratcatcher's Child_, a history of the structural pest control industry. This latest book appeals not only to those who are part of the pest control industry, but to students of history, business and all who, like Snetsinger, are interested in human progress.

Snetsinger belongs to numerous professional associations including the Entomological Societies of America, Pennsylvania and Canada and the Arachnological Society of America.

CHAPTER SEVENTEEN

Lice

Revised by R.D. Snetsinger[1]

"The cootie is the national bug of France, parley-voo?
The cootie is the national bug of France, parley-voo?
The cootie is found all over France, parley-voo?
No matter where you hang your pants.
Hinky-dinky, parley-voo!"
                    World War I song.

FLIPPANTLY, ONE CAN say a bald, non-promiscuous, black, nudist is un-likely to have problems with lice. The factors that support this assertion relate to the nature of louse problems. Head lice usually attach their eggs to hair; pubic lice are usually transferred during non-conjugal, sexual relationships; American blacks rarely have head lice because their hair structure is not conducive to the attachment of louse eggs; and body lice deposit their eggs on clothing.

Orkin (1974) projected three million cases of pediculosis (lice) of all species for the United States in 1974; 50 percent more than in 1973. Elzweig and Frishman (1977) suggested three to five million cases of head and pubic lice for the United States in 1976. In the future, perhaps changing hair styles and a greater concern for bathing may reduce the resurgence of pediculosis of the 1970s, but lice in the early 1980s are a more common problem than most people suspect.

Lice are surface parasites of birds and mammals. Most authorities divide lice into two orders, *Mallophaga* (chewing or bird lice), and *Anoplura* (sucking lice). Hopkins (1949) and others place both groups in one order — *Pthiroptera;* this name is still seen in the literature. Mallophaga are found on birds and mammals; they feed on feathers, dermal scales, and dried blood around wounds, and cause discomfort if present in large numbers. Some species are pests of domestic animals. However, it is the *Anoplura* which are of greater concern to man. Booklice, barklice, or psocids (order Psocoptera or Corrodentia), according to Zinsser (1935) and others, represent the ancestral stock from which lice evolved. Booklice, however, are free-living, though some species are pests, feeding on paste and glues and causing damage to books and other items.

Infestations of lice on humans of the order *Anoplura* are ordinarily treated

[1]*Professor of Entomology, The Pennsylvania State University, University Park, Pa.*

by physicians and other health professionals. Those on domestic animals and pets are treated by veterinarians and animal owners, using appropriately labelled insecticide formulations. The role of the pest control operator in louse control is a secondary one. Mampe (1979) observed that public concern by school boards and other local health officials when louse problems arise, sometimes involves PCOs and may require them to take action. He recommends that an attempt first be made to convince authorities that premise treatment is not usually required. Mampe also recognizes the reality that there may be situations when residual sprays may be required or are demanded by school boards or local health officials who do not understand louse control or believe they must do everything possible to reassure the public of their official concern.

General characteristics. Lice are wingless insects characterized by simple metamorphosis. The species associated with man have powerfully developed legs; each provided with a claw and an opposing thumb-like process that is adapted for clinging and gives the insect a strong grip. Most species are fairly active insects and crawl rapidly. They require warm temperatures (around 80° F/27° C) for optimum development and do not survive long removed from their host. Lice are flattened dorsoventrally; antennae are composed of three to five segments, and the tarsi of the legs have one or two segments. Specificity of host and even site of habitation on the host are a feature of louse behavior. Lice are most successful in sites of the least molestation by the host, because scratching and rubbing tend to dislodge them. Three species of *Anoplura* are major human pests — the head louse, the body louse (family Pediculidae), and the pubic louse (family Phthiridae). Busvine (1978) and others have studied double infestations on one host (head louse and body louse). There are also records of double infestations of head lice and pubic lice. Because of the possibility of double, or perhaps triple species infestations, and because of the social implications of louse infestations, care must be taken in making correct species identification.

HEAD LOUSE

Pediculus capitis DeGeer

This species infests the hair of the head, most commonly the area above the ears and the back of the head, though the entire scalp may be infested. This species rarely is found on eyelashes and other hairy portions of the body. The head of this louse is rounded in front and bluntly pointed. There is a constriction at the insertion of the antennae and a narrow "neck" where the head joins the thorax. Females are ⅛ inch/2.4 to 3.3 mm and males 1/12 inch/two mm in length. The whitish eggs or nits are about 1/30 inch/0.8 mm in length.

The communal use of combs, hair brushes and head apparel fosters the spread of the head louse and is generally regarded by United States health authorities as an important means of transfer. Young girls tend to practice communal hair care and this may tend to spread louse infestations and may explain, in part, a higher incidence of infestations among girls than boys of the same age. Donaldson (1979) tends to downgrade transfer by means of combs, hair brushes, towels, bedclothes, personal clothing, etc., as a means of spreading head lice. He contends close personal contact, such as occurs during play at school provides the major opportunities for head lice to spread. Other specialists cite the general rise in head louse incidence as being attributable to additional factors, including increased communal activities, increased population mobility, ignorance of the problem, relaxation in detection and control programs and increased popularity of day-care centers and summer camps (Lance Scholdt, personal communication, 1981).

Gratz (1976) found very high rates of head lice infestations in Canada and the United States. In a survey of three cities, infestations ranged from three percent to 20 percent of the children checked. An infestation rate of two percent to three percent of school children is general in most grade schools in the United States and England. White children are more commonly infested than black children. Roy and Ghash (1945) found a maximum infestation of 1,434 adult and 4,260 juvenile head lice. This is an extreme condition, and lice infestation levels in the hundreds rather than the thousands are more likely. Unlike body lice, head lice are not usually involved as vectors of disease. However, disease transmission cannot be totally ruled out.

Life history. Female lice deposit 270 to 300 eggs which are cemented to hairs. The eggs hatch in five to nine days and the young insects have three instars before becoming mature. The complete life cycle requires about 16 days; adults live 30 days. Nuttall, a major contributor to louse research, has estimated that a single female and her daughters could have 112,778 offspring in 48 days.

Almost immediately after hatching, a nymph takes its first blood meal. The nymph is similar to the adult except for size. Two days after reaching maturity, the female is ready to lay eggs. A fertilized female lays about six to eight eggs

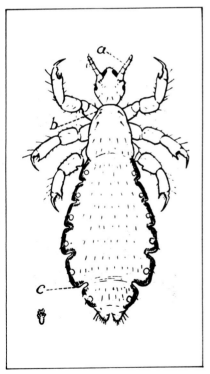

Fig. 17-1. Head louse.

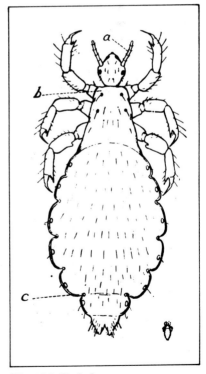

Fig. 17-2. Body louse.

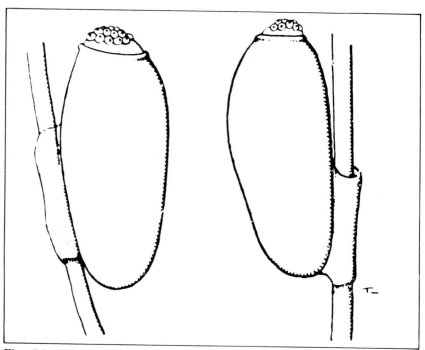

Fig. 17-3. Eggs of the body louse fastened to a hair.

every 24 hours, usually during the night. For all practical purposes, the head louse feeds only on humans and does not infest dogs or cats. The male louse has slightly larger front legs and claws than the female; these are used to hold the hind legs of the female during mating. The female produces a glue-like substance, similar to human hair which attaches the egg to hair.

BODY LOUSE OR COOTIE

Pediculus humanus L.

This species infests principally the body of man and generally oviposits on his clothing. In the United States, single, homeless men more than 50 years old have most of the problems with body lice. This is because this group of individuals tends to be less concerned with sanitation and personal hygiene. To a degree, girls are more frequently infested than boys. This may be due to the tendency of young girls to swap clothing. Counselors at girls camps often see the same blouse or skirt on three or more girls during the course of a day. In times of war or other disasters, where crowding occurs and sanitation levels are low, body lice outbreaks are frequently severe.

During the American Civil War, prisoners of war suffered severely from body lice. In fact, the Northern soldiers called lice "Graybacks" and the Southerners retaliated with "Bluebellies." Prisoners at Andersonville reported removing a quart or more of lice from clothing of a dead comrade. However, this is likely an exaggeration because even under the most squalid conditions, 400 lice is an unusually large infestation. One investigator allowed 700 to 800 body lice to feed on himself twice daily, and reported developing a tired, irritable, pessimistic state of mind which quickly disappeared when feeding was discontinued. The bites of body lice cause small spots accompanied by itching, usually followed by scratching and secondary infections. Like the head and crab louse this insect feeds on human blood.

Diseases transmitted by body louse. Epidemic or louse typhus is an acute infectious disease caused by *Rickettsia prowazkei* and is transmitted by the body louse. Severe epidemics occurred during World War I, and in World War II outbreaks threatened in Italy and North Africa, but were checked by a massive delousing program using DDT. The virus of typhus fever may be injected by the bite of an infected louse, or the infected louse, as well as its feces may be mashed into skin that is scratched and excoriated. See Pratt and Littig (1961) for further information on lice as vectors of human diseases. Trench fever, relapsing fever, and other diseases are vectored by body lice. Louse-born diseases are of mostly historical concern in the United States, because of high levels of sanitation, pest control programs and other factors.

Life history. On the average, the body louse is slightly larger ($^1/_7$ inch/three to four mm) than the head louse. About the only distinguishing characteristics are the sites where each species is found and the preference of the body louse to lay eggs on clothing versus hair for the head louse. Busvine (1978) considers each to be a valid species; others prefer a one species concept with varietal or subspecies status for each form.

The female body louse lays about 50 to 300 eggs, which hatch in six to nine days. There are three instars and maturity is reached in about 16 days. Wool clothing worn more or less continuously provides the best oviposition sites for body lice. An adult louse may live 30 to 40 days on its host, and can survive eight to 10 days after being removed from the host.

PUBIC OR CRAB LOUSE

Pthirus pubis (L.)

This species usually lives on the pubic or perianal regions of the body, but may spread to the hair of the chest, armpits and even eyebrows. The species gains one of its common names from its crab-like appearance. It is small, $^1/_{16}$ inch/1.5 to two mm in body length. The claws of the second and third pair of legs are greatly enlarged. The body shape of the crab louse resembles a flattened strawberry which is grayish white in color. For the most part, these lice remain immobile upon the host, grasping pubic hair with their hind legs so as to remain fast. Feeding in one position continues intermittently for hours or days; pubic lice are blood feeders. As it feeds, it defecates frequently, voiding blood and wastes, creating a filthy area near the site of attachment.

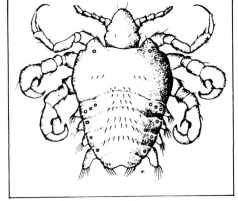

Fig. 17-4. Pubic or crab louse.

Nuttall (1918) reported female lice deposit 50 or more eggs which are fastened to the coarse hairs. Eggs are small, $1/50$ inch/0.5 mm, oval, whitish and attached near the skin junction; as the hair grows the nits become further removed from the skin (Ackerman, 1968). Incubation requires six to eight days. The young nymphs attach to feed a few hours after hatching. There are three instars, the first completed in five to six days; the second in nine to 10 days; the third in 13 to 17 days. The complete cycle requires 34 to 41 days. Removed from the host the pubic louse can survive only 12 hours to two days. Infestation is said to result from contact during coitus, but many health workers contend that crowding, clothing in locker rooms in gymnasiums and other sources cannot be ruled out. Pubic lice are restricted to humans, although there are some records of dogs being infested. (Please don't ask questions.) Children prior to puberty are not infested.

The prevalence of pubic lice world wide is on the rise (Ackerman, 1968; Grantz, 1973). Matheson (1950) believes this louse is quite widely present in the United States. Fisher and Morton (1970) report a 2.2 percent infestation rate for a limited study in Great Britain. Since the pubic louse is not a known vector of disease, it is therefore of less concern to United States public health officials. Many infestations are self-diagnosed and self-treated by those involved and do not come to the attention of physicians and other health officials. The rate of infestation is probably much higher than two percent, particularly in the age group of 15 to 30 years.

CONTROL OF LICE

Certain general sanitation and public health procedures are often established when head louse and body louse infestations are discovered. Since people abhor lice and their bites, and secondary infections cause irritability and insomnia, a single recognized infestation is often regarded as a public health threat. Notably when school children are infested by head lice, there may be a local public outcry for action. Nitzkin (1979) recommends that to avoid reinfestation all family members be treated on the same day. Body louse infestations with the potential for associated disease, become an important concern in time of war or other disaster. Pubic louse problems are generally regarded as socially improper and an indictment of a teenager's chastity or a spouse's fidelity. Thus, the presence of these lice can cause dire family problems. In general, the treatment of lice problems is largely confined to the domain of physicians and local health authorities.

Head louse control. Caution is required in diagnosing head louse infestation from the seeming presence of the eggs or nits alone. Droplets of hairsprays resemble the nits and some poorly trained officials have caused needless embarrassment and alarm by mistaking hardened droplets of hairspray for eggs or nits.

Infested individuals should, however, wash their hair and head often and carefully. Their combs, brushes, towels and other similar items should be kept clean and isolated from other family members. Their bedding and clothing should be washed and changed regularly. It should be noted that Donaldson (1979) regards walking from one host's head to another as the main mode of transfer. Further, he considers North American authorities who advised sterilization of combs, brushes and other personal items as outmoded in their recommendations. He also considers the comb and its regular use as a major preventive measure.

Donaldson (1979) regards DDT and lindane as obsolete chemical treatments for head lice and malathion and carbaryl as insecticides of choice. However, resistance of head lice to chlorinated hydrocarbons such as lindane has only been confirmed in laboratory tests in England (Maunder, 1971) and the Netherlands (Blommers and Lennep, 1978). Lotions of 0.5 percent malathion or carbaryl are available. Slightly greasy or oily hair appears to enhance the residual effect of the treatment. The hair and scalp should be thoroughly moistened and allowed to dry naturally, because heat from a hairdryer tends to degrade the insecticide. Shampoo treatments which contain about one percent insecticide are generally effective and are often more acceptable to patients. The hair should be thoroughly wetted with warm water and then the special insecticide-shampoo formulation should be applied and rubbed into all parts of the scalp. The treatment should continue for four to five minutes before the hair is rinsed with clean water; the procedure should then be repeated. Since the family may represent a reservoir of infestation, all family members frequently may be involved in treatment. Usually one treatment with the lotion provides control. Sometimes retreatment is required with the shampoo formulations; the hair can be re-washed 12 to 24 hours after treatment to remove insecticide residue. Because of secondary infections of the scalp, antibiotic therapy may be required following treatment for lice.

Body louse control. Removal and washing of all clothing and the washing of bedding is required. In situations of crowding and where louse-borne typhus and other diseases vectored by body lice may be a concern, dramatic control and preventive methods are required to protect health and prevent spread of the diseases. Such situations involve mass delousing programs and fumigation of clothing. In the United States, where body louse problems are uncommon and louse-borne diseases are not serious problems, control usually involves treatment of one or a few individuals with lotions of 0.5 percent malathion or carbaryl. In World War II and until the early 1950s, DDT was widely used in body louse control. Resistance has been reported for lindane and malathion.

Control of crab lice. There is a great deal of propinquity between an increased incidence of crab lice and an increasing incidence of venereal disease in the United States during the last decade. The spread of crab lice, like syphilis and gonorrhea is particularly associated with increased sexual activities among unmarried males and females under 25 years of age. Unlike venereal diseases, crab lice can be self-treated, using readily available products. Thus, there is no reliable information as to the number of cases. Also, *Phthirus pubis* is not a vector of diseases and does not invoke public health concern as do the other two species found on humans. The pubic louse is generally absent from children prior to puberty and is usually transferred during coitus. Unmarried teenagers usually disclaim the actual contact source of crab louse infestations and seek to suggest other more mundane sources, such as toilet seats. Since survival for more than 12 hours off a host is unlikely and most school bathroom facilities are unused for long periods of time (nights and weekends), such sources as toilet seats are not likely. However, crowding of clothing in locker rooms and gymnasiums, wearing of unwashed clothing of others and even toilet seat transfer of crab lice cannot be totally ruled out. Thus, school boards and local health officials sometimes do demand residual control measures from pest control operators. Sanitation, removal and laundering of clothing in lockers and even residual treatment with insecticides may provide window-dressing for a more medically sound program of treating infested individuals. Lindane in a one percent cream, lotion or alcohol, or a 0.5 percent malathion lotion or a 0.16 percent pyrethrum plus two percent piperonyl butoxide appear to be the pesticides of choice. Insecticide resistance has not been reported for crab lice.

The role of PCOs. The role of the pest control operator in louse control is an ambiguous one. This is because public concern about louse problems frequently requires public health officials in responsible positions to take all possible actions to rid a school or other institution of lice. Thus, these officials, either because of inadequate knowledge of lice or because of a desire to provide a decisive solution, demand that a pest control operator provide control. A thorough knowledge of louse biology is generally the best service that can be rendered by a PCO. When chemical residual treatment is required or demanded, a registered formulation of pyrethrum and lindane, which does not cause stains, may be used (Elzweig and Frishman, 1977).

Insecticide applications to inanimate surfaces by pest control operators, whether using contact or residual insecticides, may on some occasions be a useful supplement to measures aimed at the host. After all, body and head lice can survive off the host for days, and they are highly mobile. For instance, it may be appropriate to treat lockers where lousy clothing has been stored. In all cases, however, such measures should form part of a program under the guidance of public health experts.

LITERATURE

ACKERMAN, A.B. — 1968. Crabs — The resurgence of *Phthirus pubis*. New Engl. J. Med. 278(17):950-951.

BLOMMERS, L. and M. VAN LENNEP — 1978. Head lice in the Netherlands: Susceptibility for insecticides in field samples. Entomol. Exp. Appl. 23:243.

BUSVINE, J.R. — 1978. Evidence from double infestations for specific status of human head lice and body lice (Anoplura). Syst. Entomol. 3:1-8.

DONALDSON, R.J. — 1979. Parasites and western man. MTP Press Limited, Bath, England. pp. 57-77.

ELZWEIG, J. and A.M. FRISHMAN — 1977. The role of the PCO in controlling human lice. Pest Control 45(1):33-34.

FISHER, I. and R.S. MORTON, 1970. *Phthirus pubis* infestation. Brit. J. Vener. Dis. 46:326.

GRATZ, N.G. — 1973. The current status of louse infestations throughout the world. *In:* Proceedings, International Symposium on the Control of Lice and Louse-borne Diseases. Washington PAHO Sci. Pub. No. 263:23.

GRATZ, N.G. — 1976. The Epidemiology of Louse Infestations. Geneva, World Health Organization.

HOPKINS, G.H.E. — 1949. The host-associations of the lice of mammals. Proc. Zool. Soc. Lond. 119:387-604.

MAMPE, C.D. — 1979. Answers — treatment for human lice? Pest Control 47(5):64-65.

MAUNDER, J.W. — 1971. Resistance to organochlorine insecticides in head lice, and trials using alternative compounds. Med. Officer, 125:27.

MATHESON, R. — 1950. Medical Entomology. Comstock Publishing Company, Inc., Ithaca, New York. p. 203.

NITZKIN, J.L. — 1979. Head lice and hygiene. Lancet (8148):910.
NUTTALL, G.H.F. — 1918. The biology of *Phthirus pubis*. Parasitology 10:383-405.
ORKIN, M. — 1974. Pediculosis Today. Minn. Med. October.
PRATT, H.D. and K.S. LITTIG, — 1961. Lice of public health importance and their control. USPHS CDC Training Program.
ROY, D.N. and S.M. GHASH — 1945. Studies on the population of head-lice, *Pediculus humanus* var. *capitis* De G. Parasitology 36:69-71.
ZINSSER, H. — 1935. Rats, lice, and history. Little, Brown and Co. Boston. 301 pp.

NORMAN R. EHMANN
KEITH O. STORY

Norm Ehmann has a bachelor's degree from Occidental College and a master's degree in public health from the University of Michigan. His experience in structural pest control spans 28 years. He is currently general manager of Pest Control Supplies for Van Waters & Rogers, San Mateo, Calif.

Ehmann is particularly well known for his development of training courses and supportive literature for pest control technicians. These programs have made a significant contribution to improving standards of pest control in the West.

The contributions of Ehmann to PCO education and to the industry as a whole have been widely recognized. He is past president of Pi Chi Omega and of the United Pesticide Formulators and Distributors Association. Ehmann is a registered urban entomologist and registered sanitarian and is active in many professional associations. These include the National Pest Control Association, American Public Health Association and the Entomological Society of America.

Keith Story was born and raised in England and obtained his bachelor's degree in natural sciences and his master's degree in zoology and entomology from Cambridge University, England. His interest in natural history was kindled in 1955 by Dr. Carroll Williams, a Harvard professor visiting England. Later at Cambridge University his focus on entomology owed much to the inspiration of Dr. Vincent Wigglesworth, founder and acknowledged leader of modern insect physiology.

After graduation, Story embarked on a 14 year corporate career with Fisons, an international developer and marketer of specialty chemicals. After an initial period of on-the-job training in both technical and commercial areas, including working for a pest control contractor in Africa, Story began a seven year period of pesticide research and development. This culminated in his development of bendiocarb, a major new insecticide for the structural pest control market. With this achievement he quickly progressed to become manager of technical and commercial development for this sector.

In 1980, Story established his own consulting group which focuses on the development of new pest control products and supporting marketing and servicing programs. In this role he has worked with most major manufacturers. With his insider's knowledge of pesticides Story is also in strong demand for PCO training and, in addition to his overseas lectures, his United States lectures have attracted audiences of many thousands.

CHAPTER EIGHTEEN

Fleas

Revised by Norm Ehmann[1] and Keith O. Story[2]

Great fleas have little fleas upon their backs to bite 'em
And little fleas have lesser fleas, and so ad infinitum
And the great fleas themselves, in turn have greater fleas to go on
While these again have greater still, and greater still, and so on.

Augustus De Morgan.

ALTHOUGH FLEAS have a certain measure of renown for their prodigious efforts in the flea circus, they are even more notorious for their specialized talents in tormenting mankind. But by far, their chief fame rests in their ability to transmit disease, and thus it is that they must be treated in something other than a jocular vein.

The body of the flea appears to be well adapted for forward movement. Dark colored adults are flattened from side to side with many bristles or ctenidia that point backwards, and facilitate forward movement through fur, hair, or feathers. Fleas are wingless creatures, with strongly developed legs and hind legs that are especially adapted for jumping. They have sucking mouthparts and are external parasites on many animals and birds. Their life cycle undergoes what is known as a "complete metamorphosis"; that is, they have an egg, larval, pupal, and adult stage. The "pincushionlike" dorsal sense organ at the end of the abdomen is characteristic. The male is usually smaller than the unfertilized female and much smaller than the fertilized female. Moreover, the abdomen of the male is much more convex ventrally and almost straight dorsally, and the distal end is slightly tilted.

Some individuals become accustomed to fleas and are not disturbed by them. Others are extremely irritated by the mere presence of fleas, let alone their painful bites. Lunsford (1949), who studied the problem in northern California, notes fleas are common in the United States because of the even temperatures and high humidity. Newcomers are especially susceptible to flea bites, although some individuals have to be bitten repeatedly to become sensitive to the bites,

[1]*General Manager of Pest Control Supplies, Van Waters & Rogers, San Mateo, Calif.*
[2]*President, Winchester Consultants, Winchester, Mass.*

and often become immune after being bitten over a long period of time. The bite of a flea varies with the species of the flea and the person bitten. Since the bite may be mistaken for a rash, Patton's (1931) description of a flea bite should be of interest. "There is a central small, red spot where the mouthparts (mandibles) have penetrated the skin, and around it there is a red halo; there is never much swelling. The bite of the bedbug, on the other hand, produces a large wheal-like bump. This wheal or bump usually disappears in a few hours." The reaction to the flea bite is due to a secretion of the salivary glands. It has been found that various cooling preparations will often give relief to those who are sensitive to the bite. Menthol, camphor, carbolated vaseline, a three percent solution of carbolic acid in water, or calamine lotion, is recommended for this purpose. Individuals who are especially sensitive to flea bites and have allergic reactions may obtain relief through the administration of flea antigens. Feingold and Benjamini (1961) prepared antigens from the cat flea, one of the most common fleas. McIvor and Cherney showed that although the immunized person could be bitten, he was unaware of the bite. Lunsford (1949) notes that taking quinine by mouth, or thiamine hydrochloride by mouth or injection may be of some value in protecting sensitive individuals.

LIFE HISTORY AND HABITS

Eggs. The eggs are sufficiently large to be seen with the naked eye. They are $1/50$ inch/0.5 mm long, smooth, translucent, glistening, and oval. The eggs are not attached to the body of the host. Some species of fleas do not even lay their eggs on the hairs of the host, but rather in the immediate vicinity. In any event, when the eggs are laid on the body of the host, they either fall from the body or are shaken and scratched off. Therefore, we find the eggs in crevices in the floor, in dust along the edge of the wall, in the bedding of pets, under the edge of carpets, etc. The eggs are laid singly and are extruded with some force. The fleas usually oviposit from three to 18 eggs at one laying. Bacot (1914) observed *Pulex irritans* Linn. to lay 448 eggs during a period of 196 days. Temperatures around 70° F/21° C and a relative humidity of 70 percent or more are conducive to increased egg production. The eggs hatch in from one to 12 days.

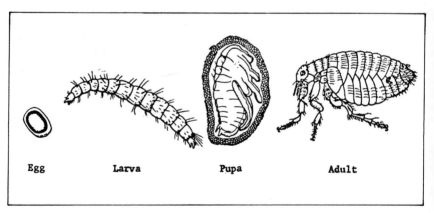

Egg   Larva   Pupa   Adult

Fig. 18-1. Life history of a flea.

Larvae. The emerging larva cracks the egg shell by means of a tooth on its head, which disappears with the first molt. The eyeless and legless larva is maggot-like, whitish, with a single row of bristles around each segment. The larva has a distinct head, three thoracic segments, and 10 abdominal segments, the head usually being a brownish color. The active larva has no legs and moves by means of the bristles which encircle each segment.

The larval period extends from seven to 15 days after hatching, but may be prolonged over six months under certain unfavorable conditions. There are usually three larval instars. The larva readily shams death upon being disturbed.

The small whitish larvae feed on any organic material found in the nest of the host or in the house where the eggs are laid. Stewart (1939) has this to say concerning the feeding habits of the larvae: "As a whole, hosts with nests are more commonly preferred to those without them. This is particularly true in the case of those species such as the European rat flea, *Nosopsyllus fasciatus* (Bosc.), whose larvae appear to require a meal of dried blood derived from the excreta of the adult flea for their development. Also in such cases, we observe that the adults spend a great deal more time in the nests than on the hosts, which is a provision of nature to supply the larvae with the necessary food. Contrasted with fleas of this type are those such as the oriental rat flea, *Xenopsylla cheopis* (Rothsch.), whose larvae apparently do not require dried blood for their subsistence. Adult fleas of this latter type, consequently, are found more frequently on the hosts and are, therefore, of greater potential danger as disease vectors and more liable to become pests of man." The larvae of the bird-feeding species live on the broken-down sheaths of the feathers of the young birds and also on their epidermal scales.

Since the larvae have a thin body-wall, their environment must of neccessity be a moist one. Like most of the adults, the larvae are light-shunners. Upon crawling in sand and debris, they bury themselves rapidly. When the debris is placed on paper, the larvae may be detected by their movements. According to Bishopp (1915), fleas occur commonly in sandy regions because the sand "maintains moisture more uniformly, and thus permits stages of the flea to develop with greater success. The sand also offers some protection to the adults and renders heavy rains less destructive to all stages of the flea present on the soil."

Pupa. Just prior to the termination of the larval stage, the larva commences to weave a silken cocoon, which is spun from its own saliva. This cocoon incorporates small pieces of debris and organic sediment, camouflaging the cocoon in its natural surroundings. The cocoon may be spun from four days to several months after hatching. The pupae in the cocoons are extremely sensitive to mechanical disturbances of any kind, and Waterston (1937) notes: "Persons entering a long-deserted house sometimes have cause to complain of hordes of fleas appearing 'suddenly' after a short time. It is probable that in such cases, fleas resting in the cocoon, beneath floors, in cracks, etc., have come out in response to the vibrations caused by people moving in their proximity."

Leeson (1936) found fleas would not feed on the day of emergence from the pupal case, but on the second day all would feed.

Adult. The flea, on emerging from the cocoon, is fat and can live for several months without food. The males usually emerge first and are less numerous than the females. Even though mated, the female flea will not lay eggs unless she has obtained a blood meal. The adults of most species of fleas visit their hosts for only a short time in order to obtain their blood meal. Others such as the adult sticktight flea, *Echidnophaga gallinacea* (Westw.), and the adult chigoe, *Tunga penetrans* (Linn.), remain attached to their host for most of their lives. Most species of fleas have definite host preferences, but do not necessarily feed exclusively on these hosts. This point is of prime importance from a health viewpoint, since the tendency of the flea to feed on several different species of animals enables it to carry disease (e.g., bubonic plague) from one animal to another, as for example, from rat to man.

The time the adult flea lives varies with the species and environmental conditions. According to Bishopp (1931), in hot weather and with no animal to feed upon, they may live but from two to five days, whereas when they feed upon blood, they may live from a month to almost a year. "During the summer, probably the average longevity of the human flea without food is about two months, of the dog flea somewhat less, and of the sticktight flea still less." Bishopp observes further that in the northern United States almost all fleas pass the winter in the immature stages, while in the South the adults may often be present on the hosts throughout the winter. Moreover, in his opinion, fleas are not as abundant in the winter and spring as in the summer and fall.

Strickland kept the northern rat flea, *Nosopsyllus fasciatus* (Bosc.), alive for 17 months without food, but the flea had to have rubbish in which to bury itself in order to keep alive. Bacot kept the human flea, *Pulex irritans* Linn., alive for 513 days, by feeding it blood daily. Mitzmain (1910) found six males of *N. fasciatus* (Bosc.) averaged 8½ days, with a maximum life of 17 days. Fifteen females he bred had an average life span of 32 to 45 days, with a maximum of 160.

Fleas are often referred to humorously as "jumping dandruff," and they are indeed famed because of their saltatorial talents. Mitzmain (1910) found a human flea was able to jump vertically 7¾ inches/20 cm and horizontally 13 inches/33 cm, but Waterston (1937) rarely observed them to jump vertically any higher than four inches/10 cm. Since fleas can traverse smooth surfaces only with difficulty, they usually progress by jumping. Also, Waterston noted fleas bury themselves when placed on cloth, fur, or feathers. "They shun daylight and are attracted by warmth. If disturbed, they frequently sham death and rest with their feet tucked up tightly to the body, in which state they are easily blown about."

As mentioned previously, fleas prefer warmth and are repelled by cold. Thus the rat, which is supposedly the first to leave a sinking ship, is likewise the first to be deserted by its flea when it dies. This last item should be noted with special care, for bubonic plague may result from infected fleas leaving a dead plague rat and biting man. Patton (1931) further elaborates on this point as follows: "The time spent by any flea on its host varies with the species. Two extremes may be mentioned. *Pulex irritans* spends only a small proportion of its life on man, whereas the plague flea, *X. cheopis* spends most of his time on its normal host, and hence it is that a rat, dying of plague, brings its fleas with it into the house, and when it dies they leave and attack man."

According to Bacot and Martin (1914), *X. cheopis* (Roth.) is a very persistent biter and it will return many times if it is disturbed during the course of feeding. If it is not disturbed, it will gorge itself in two to three minutes. Moreover, it wanders around and makes several punctures before it feeds. Mitzmain (1910) found the species do not attack with equal avidity. *Pulex irritans* Linn., the normal parasite of man, is insatiable in its blood craving. It differs in its relation to man in being more fastidious in its feeding than the rodent fleas. Although its bite is painful, it does not voluntarily feed in one spot for any great length of time." He continues with the statement that *"Pulex irritans* differs from all other species in that it squirts blood per anum during the act of biting." The flea larva is affected easily by low atmospheric humidity since it is unable to control rapid loss of water from its body during respiration. Thus, Waterston (1937) is of the opinion it is probably due to the death of the larvae that fleas are uncommon in areas where the temperature is high and the humidity is low. In any event, a hot dry summer reduces the number of fleas, and a humid, rainy summer is favorable to them.

Dispersal of fleas. Fleas are transported for the most part through the agency of man, the rat, and other animal and bird hosts. Thus, the eggs, as well as the adults may be carried in such cargoes as gunny sacks, cotton, grain, rags, hides, etc., when these are suitably moist. The eggs are also scattered when they fall from the animal host or when the bed of the host is moved. When such hosts as rats, for instance, die, the fleas seek other hosts. The flea may be carried on rats during the process of shipping or when the rats migrate.

Natural enemies of fleas. Ants have been mentioned as predators on fleas and staphylinids or rove beetles have been reported in the nests of ground squirrels in Russia where they were said to have destroyed many fleas.

Diseases caused by the flea. The flea is related so intimately to the rat in the transmission of most rat-borne diseases that these are discussed in greater detail in the chapter on rats. Here we will only summarize some of the flea-borne diseases. Fleas are concerned in the transmission of bubonic plague from rat to rat and from rat to man. In California, ground squirrels may be concerned in the transmission of plague. In India alone, plague caused more than 7,-000,000 deaths between 1896 and 1911. For further information on fleas and disease, see Jellison (1959).

Another disease caused by the rat is known as murine typhus or Brill's disease. According to Craig and Faust, this disease is a mild form of *Rickettsia prowazeki,* which causes the usual form of typhus that is spread readily by the human louse, *Pediculus humanus* Linn. The fatality of murine typhus is considerably less than the parent disease. Often it occurs where workers handle food infested with rats. Irons et al. (1944) note a number of infections acquired from handling kittens harboring infected fleas. At present, murine typhus is most common in southwestern and Gulf States and is spreading northward through the migrations of Norway rats. The disease may be transmitted from rat to rat by the tropical rat mite, *Liponissus bacoti* (Hirst), and the rat louse, *Polyplax spinulosa* (Burm.), or from man to man by the human louse, *Pediculus humanus* Linn., but is commonly carried by the flea, *Nosopsyllus fasciatus* (Bosc.) and *Xenopsylla cheopis* (Rothsch.).

Fleas and other insects are vectors of *Pasteurella tularensis,* which is the bacterium responsible for tularemia. They are also the carriers of a tapeworm which is found in the adult stage in dog and man. *Tunga penetrans* (Linn.) forms dangerous sores, which through enlargement may result in the self-amputation of the toes.

SPECIES OF FLEAS MOST COMMONLY ENCOUNTERED

Stewart (1939) states the fleas most commonly encountered in pest control work are the cat flea, *Ctenocephalides felis* (Bouché); the human flea, *Pulex irritans* Linn.; the dog flea, *Ctenocephalides canis* (Curtis); occasionally the northern rat flea, *Nosopsyllus fasciatus* (Bosc.) the oriental rat flea, *Xenopsylla cheopis* (Rothsch.); the mouse flea, *Ctenopsyllus segnis* (Schon.); and the stick-

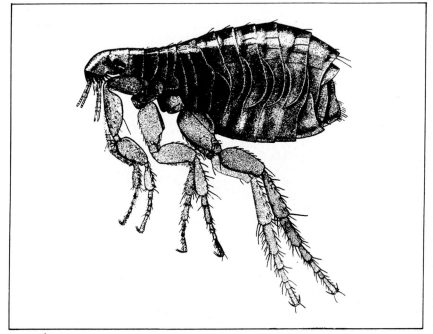

Fig. 18-2. Adult cat flea.

tight flea, *Echidnophaga gallinacea* (Westw.), which attacks poultry and may infest animals. The western chicken flea, *Ceratophyllus niger* Fox, and the European chicken flea, *Ceratophyllus gallinae* (Schrank), often leave their poultry hosts and attack men and other animals coming in contact with poultry. Mention may be made of *Tunga penetrans* (Linn.), since it is so prominent a pest in other parts of the world. Pratt and Wiseman (1962) note the squirrel flea, *Orchopeas howardii,* occasionally leaves the nest of the squirrel in attics or in hollow trees and bites man.

For the key to fleas by Pratt (1957) see Figure 18-4.

HOUSE-INFESTING FLEAS

THE CAT FLEA

Ctenocephalides felis (Bouche). The cat and dog fleas are so alike that only recently have they been separated by Rothschild. These two fleas may be distinguished from *Pulex irritans* Linn. by the ctenidia found on the head and pronotum. There are eight spines on the general comb and 16 spines on the pronotal comb. According to Herms (1939), the head of the female cat flea is twice as long as high when seen from the side and has seven to 10 bristles on the inner side of the hind femur. Whereas, the head of the female dog flea is less than twice as long as high when seen from the side and has 10 to 13 bristles on the inner side of the hind femur.

The cat flea is the most common pest in homes and lawns. It is abundant in the East and far West. It is most common during the summer months partic-

Fig. 18-3. Scanning electron micrograph of an adult cat flea (Mag. ca. 75×).

ularly when homes are reoccupied after summer vacation. The most common hosts are man, dogs, and cats but it occurs on a wide variety of animals.

THE DOG FLEA

Ctenocephalides canis (Curtis). According to Trembley and Bishopp (1940), the dog flea is a less important pest than the cat flea, but is still fairly common. This flea is most abundant during the warm summer months. It is found in all parts of the country other than the Rocky Mountain and the Intermountain regions. The most common hosts are dog, man, and cat. However, it occurs on many other animals as well.

THE HUMAN FLEA

Pulex irritans Linn. This cosmopolitan flea is a pest in buildings of all types, especially in the Middle West, South, and on the Pacific Coast. It is considered by some entomologists to be the most common flea in homes and upon man on the Pacific Coast. It has neither oral nor pronotal ctenidia. Where dust and organic debris are allowed to accumulate in the house, this flea may breed and become a pest. Moreover, when such domestic animals as pigs are in the immediate vicinity, an infestation is likely. Essig (1926) notes in the rural sections of California, the human flea is the most common flea. It is found on human beings, domestic animals, and poultry, as well as on skunks, badgers, foxes, coyotes, and other wild animals.

Patton (1931), in discussing conditions in India relative to the human flea, *Pulex irritans* Linn., states in houses that are empty during the winter, the adults hatch in large numbers during the spring and early summer. He observes

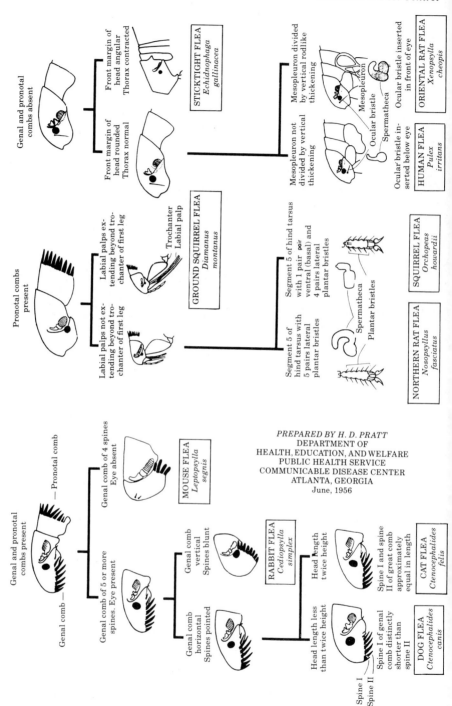

PREPARED BY H. D. PRATT
DEPARTMENT OF
HEALTH, EDUCATION, AND WELFARE
PUBLIC HEALTH SERVICE
COMMUNICABLE DISEASE CENTER
ATLANTA, GEORGIA
June, 1956

Fig. 18-4. Pictorial key to some common fleas in the United States.

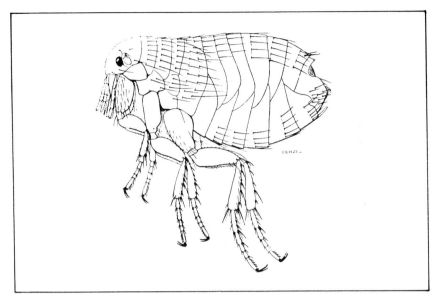

Fig. 18-5. Female human flea.

the inhabitants must walk about the rooms with sticky fly paper wrapped around their legs in order to dispose of the large number of fleas.

NORTHERN RAT FLEA

Nosopsyllus fasciatus (Bosc.). Essig (1926) observes Rucker found this flea to be the most common one in San Francisco. This is the usual flea of the house mouse, but is found also on the Norway rats, black rats, and on man. It occurs, for the most part, in Europe and America. This flea has but one ctenidium and this occurs on the pronotum and consists of 18 to 20 spines.

THE CHICKEN FLEA

Genus *Ceratophyllus*

Chicken fleas are widely distributed and at times attack man. The European chicken flea, *Ceratophyllus gallinae* Schrank, is the common flea of fowl, but has also been found in the nests of birds. This flea may be identified by a lateral row of four to six bristles on the inner surface of the hind femur. It occurs commonly in the New England states, along with the species *C. gibsoni* Fox. The western chicken flea, *C. niger* Fox, occurs in the western United States.

ORIENTAL RAT FLEA

Xenopsylla cheopis (Rothsch.) This extremely important flea is the vector of bubonic plague from rat to man. Although it is not really a household pest, it has been reported biting people in buildings. Like its rat hosts, it is found most commonly in seaport towns. According to Trembley and Bishopp (1940), the most common hosts are rats, particularly the Norway rat, *R. norvegicus;* cotton

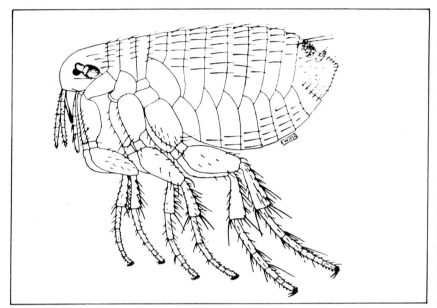

Fig. 18-6. Female oriental rat flea.

rat, *Sigmodon h. hispidus;* roof rat, *R. rattus rattus;* house mouse, *Mus musculus;* man; cottontail rabbit, and the California ground squirrel. It is reported from every part of the United States other than some of the Rocky Mountain and Intermountain states. Both the oral and pronotal ctenidia are absent and the ocular bristle is in front and just above the middle of the eye.

THE STICKTIGHT FLEA

Echidnophaga gallinacea (Westw.)

This flea is primarily a pest of poultry and only secondarily a pest on cats, dogs, rats, and horses. However, it has been found embedded in the skin of human beings and often attacks persons working with poultry. The sticktight flea is a pest particularly prominent in the South and Southwest. It is somewhat similar to the chigoe flea, *Tunga penetrans* (Linn.), and like this species, it has no combs. The sticktight flea is larger than the chigoe and, according to Patton (1931), it can be distinguished from it "by noting that it has a patch of from 18 to 20 minute, pointed, coneshaped bristles (spines) on the antero-ventral surface of each hind coxa. The bristles are usually arranged in three rows." *Tunga penetrans* does not have these bristles.

Before mating, this species is as active as any other flea and hops about in a similar manner. Once fertilized, the female attaches herself to the host for the remainder of her life. On poultry, the fleas are found most commonly in the vicinity of the eyes and comb. With dogs and cats, this flea occurs usually on the ears, especially along the edges.

Beach and Freeborn (1933) studied the life history of this flea: "The eggs laid by these attached females fall to the ground and hatch, or in case the eggs are retained in the burrows or ulcers caused by the attached females, the larvae upon hatching usually fall to the ground but may remain in the tissue, where they produce a condition resembling the infestation by blowfly maggots. These larvae are tiny white caterpillar-like organisms with chewing mouthparts. They subsist on the debris or manure on the floors of houses and yards. When they have become full-grown as larvae, they spin a white cocoon in which they change from the larval stage to that of an adult, after which they attach themselves to the skin of their host to suck blood. The constant irritation, particularly in spots where they are present in large numbers, together with a slight burrowing activity on the part of the fleas, causes the formation of ulcers, so extensive at times that blindness and subsequent death is produced. It has been shown recently in Texas that Brewer blackbirds and English sparrows are severely attacked and could easily serve as carriers from one flock to another." Badly infested young chicks may be killed and even the older and more resistant birds may die. In any event, these fleas curtail egg production seriously.

Bishopp states: "When dogs and cats are infested, the immature stages develop largely in the material used by them for beds. They require comparatively dry material in which to breed, but a large amount of air moisture is favorable to them. Adults of this species continue to emerge from infested trash for four or five months after all hosts have been removed. Hence, it is easy to understand why chicken houses may still have many fleas in time after being unused for considerable periods." The larva becomes full grown in about two weeks, pupates for another two weeks, and the entire life history is completed in approximately four weeks.

CONTROL OF FLEAS

Flea control in and around man-made structures be it homes, restaurants, warehouses, food processing plants, etc. lends itself very well to the concept of Integrated Urban Pest Management (IUPM). The IUPM concept embraces cultural, mechanical, physical, biological, regulatory and chemical control. Each of these facets of IUPM can be readily brought to bear on a flea infestation and when thoughtfully practiced will result in long range flea control.

Cultural control. Cultural control implies a sanitation inspection of the premises involved and the elimination, in so far as possible, of the environmental conditions that nurture the presence and build-up of a flea population. This may include the elimination of vegetation which provides a rodent harborage or the restacking of a lumber pile for the same reason. The careful sanitation inspection will reveal at least the "hot" points of the infestation that need special and careful attention.

Mechanical control. Mechanical control might include the screening of foundation vents to keep pets out from under the house. Many times when dogs and other pets are allowed to seek the shade under the house on a hot summer day, this can lead to a rapid build-up of a heavy concentration of fleas. Mechanical control can also include rodent proofing of the structure if the rodents are the source of the flea infestation.

Physical control. Physical control should be carried out by the building occupant prior to the time the pest control technician arrives. Physical control can mean the washing of the pets' blankets and pillows. It can also mean the vacuuming of pet resting and sleeping areas. Since flea eggs are usually laid on the host, but not attached to the hairs of the host, they fall off the host rather easily, especially when the pet scratches itself or shakes its body. Concentrations of eggs, larvae and pupae may therefore be found more often around pet resting areas. If the pet rests on the overstuffed chair or sofa, the sofa seat cushions should be removed and the peripheral cracks below, where the pencils, hairpins, and coins are usually found, should be vacuumed thoroughly. Flea eggs gravitate to this area so larvae and pupae may be found there in abundance. Particular attention should be paid to areas between the baseboard and the floor. This represents a haven to the flea larvae — chock-full of good food! Vacuuming can be helpful in closets, behind and under furniture, and in other so-called "quiet zones" where larvae can thrive unmolested. After vacuum cleaning flea-infested premises the vacuum bag should be disposed of outside the premises to avoid it becoming a flea reservoir leading to re-infestation. When wettable powder or dust formulations are used, vacuum cleaning after the treatment should be delayed for at least 10 days to avoid premature removal of the insecticide.

In the case of the sticktight flea, the floors of poultry houses should be cleaned. Moreover, all poultry manure should be turned under by plowing or spading, since the larvae of the sticktight flea, as well as the larvae of the dog flea, human flea, and the rat flea may continue to develop in this material, and for want of poultry may attack man and animals.

Biological control. Biological control can mean the use of such agents as insect growth regulators. One such agent, new on the market, is methoprene, developed and manufactured by the Zoecon Corporation. This material affects the larval stage of the flea in such a manner that it passes into the pupal stage, but the adult never completely forms and therefore, never hatches from the pupal case. This is an example of breaking the chain of the life cycle at the larval-pupal stage, never allowing an adult population to be produced (Chamberlain, 1979, and Rambo, 1980).

Regulatory control. Regulatory control makes use of the local, state or federal statutes pertaining to rodent proofing and is applicable when a rodent problem is the cause of the flea infestation.

Chemical control. Chemical control, the sixth and equally as important ingredient of IUPM, means the safe application of EPA registered pesticides for flea control. It is important to point out that before the use of any pesticide, the label should be thoroughly read and thoroughly understood by the person performing the pesticide application. Insecticides that are currently available for flea control inside or outside of structures or both inside and outside of structures include: bendiocarb (w.p. and dust); carbaryl (w.p., flowable and dust); chlorpyrifos (e.c. and granule); DDVP (e.c., aerosol and resin strip); diazinon (e.c., w.p. and microencapsulated); dioxathion (e.c.); malathion (e.c. and dust); methoprene (e.c. and aerosol); propetamphos (e.c.); propoxur (e.c., w.p. and aerosol); pyrethrum (dust, aerosol, e.c. and microencapsulated); and silica aerogel (dust, aerosol — alone and with pyrethrum). Use of DDT in the United States against fleas is now confined to control of rodent fleas in emergency programs aimed at preventing the spread of plague and murine typhus (David, 1945 and 1947). Ronnel has been used for flea control both as a spray for buildings and as a pill for dogs (Anon., 1961). Other insecticides which have been used for treatment of both buildings and pets include dusts of pyrethrins, malathion and rotenone (Anon., 1967), silica aerogel dust (Tarshis, 1959) and carbaryl dust (DeVries, 1966).

It must be emphasized the role of the pest control operator does not include treatment of pets. However, when he proposes to treat the building with insec-

ticides which inhibit cholinesterase, he may suggest the pet be treated with products which do not cause this effect.

Linduska et al. (1946) recommended the use of benzyl benzoate or 2-phenylcyclohexanol as a flea repellent on clothing. Five grains of either material in acetone were used to impregnate each square foot of clothing.

It cannot be stressed thoroughly enough that the applicator of pesticides should consult the label for proper dilution and application rates and to determine if the particular material he wishes to use has indoor or outdoor clearance or both. Successful application indoors depends upon thoroughness, including treatment of the entire surface of rugs. In this context it should be remembered that of all the long lasting residual adulticides labeled for flea control, at this time only bendiocarb, diazinon (microencapsulated), and propetamphos are cleared for treating entire rugs. Control efforts also should be directed to those areas frequented by pets, cracks of floor boards, behind baseboards, in ventilators, under edges of carpets, and similar places. In the summer, lawns and sandy soil may become infested with fleas, and the fleas will attack any warm-blooded animal passing through such areas. Outdoor treatments should focus on those areas most frequented by pets; dusts can be blown under porches and low unfinished areas where spraying might be difficult, especially "bathroom" areas of pets. Reese (1981) gives valuable guidelines on flea control.

It is very important, in addition to treating thoroughly indoors and outdoors, to have the pet involved treated. This can be accomplished by a veterinarian, a pet shampoo parlor or by the pet owner himself, but should be done at the same time the premises are being treated. Pets should not be allowed to re-enter the premises until the pesticide applied therein has completely dried. Where the insecticide treatment is administered to the pet by the veterinarian, pet owner or others may have a carryover effect on the body chemistry of the pet (e.g. a temporary reduction of the enzyme cholinesterase). Therefore, it may be wise to delay for a longer period the re-entry of the pet into premises treated with an insecticide capable of producing a similar effect. Pet treatments likely to produce a carryover effect and thus render the pet more susceptible to additive effects from similar insecticides applied in the home, include flea collars impregnated with carbamate or organophosphate insecticides. Considering the potential public health menace of fleas it would be irresponsible not to mention that perhaps the most effective and least expensive means of reducing flea infestations in the home is to forego the luxury of pet dogs and cats. But considering the value of working dogs and the companionship of such pets it is likely to be a long time before man's best friend is a fish, whether it be a catfish or a dogfish!

LITERATURE

ANONYMOUS — 1947. Fleas flee. Chem. Industries 61(6):976.

1955. Insecticides and repellants for the control of insects of medical importance to the armed forces. USDA Circ. 977.

1961. Now there's a pill to get rid of dog fleas. Chem. Week 88(7):92.

1967. Controlling fleas. USDA Home and Garden Bull. 121.

BACOT, A. — 1914. A study of the bionomics of the common rat fleas and other species associated with human habitations, with special reference to the influence of temperature and humidity at various periods of the life history of the insect. J. Hyg. 13: Plague Supplement II:447-654.

BACOT, A.W. and C.J. MARTIN — 1914. Observations on the mechanism of the transmission of plague by fleas. J. Hyg. 13: Plague supplement 3:423-439.

BEACH, J.R. and S.B. FREEBORN — 1933. Diseases and parasites of poultry in California. Calif. Agr. Ext. Serv. Circ. 8. 98p.

BISHOPP, F.C. — 1915. Fleas. USDA Bull. 248. 1931. Fleas and their control. USDA Farmers Bull. 897. 1957. How to control fleas. USDA Leaflet 152.

BROOKS, J.E. and T.D. PECK — 1969. Community Pest and Related Vector Control — Pest Control Operators of California, Inc.

BROWN, A.W.A. — 1958. Insecticide Resistance to Arthropods. World Health Org. 240 pp.

BYRNE, K.V. — 1948. Control of stickfast flea *(Echidnophaga gallinacea)* by BHC & DDT dusts. Inst. Stock N. S. Wales Ybk. 1948:103-106.

CARPENTER, S.J., R.W. CHAMBERLAIN and R. BAKER — 1945. Flea collections at army installations in the fourth service command. J. Econ. Entomol. 38(5):600-602.

CHAMBERLAIN, W.F. — 1979. Methoprene and the Flea. Pest Control 47(6):22-26.

COLE, L.C. — 1945. The effect of temperature on the sex ratio of *Xenopsylla cheopis* recovered from live rats. U.S.P.H.S. Rpts. 60(45):1337-1342.

DAGGY, H.H. — 1946. Fleas as household pests. Minn. U. Agr. Ext. Folder 81, rev. 6 p. University Farm, St. Paul, Minn.

DAVIS, D.E. — 1945. The control of rat fleas *(Xenopsylla cheopis)* by DDT. U.S.P.H.S. Reports 60(18):485-489. 1947. The use of DDT to control murine typhus fever in San Antonio, Texas. U.S.P.H.S. Rpts. 62(13) 449-463.

DeVRIES, M.L. — 1966. Control of human lice and rat fleas with Sevin. International Pest Control 8(2):10-12, 15.

EBELING, W. — 1976. Urban Entomology. University of California.

EDNEY, E.B. — 1945. Laboratory studies on the bionomics of the rat fleas, *Xenopsylla brasiliensis* Baker, and *X. cheopis,* Roths. I. Certain effects of light, temperature and humidity on the rate of development and on adult longevity. Bull. Entomol. Res. 36(4):339-416. 1947. Laboratory studies on the bionomics of the rat fleas, *Xenopsylla brasiliensis,* Baker, and *X. cheopis,* Roths. Bull. Entomol. Res. 38(2):263-280.

EHMANN, N.R. — 1981. Structural Pest Control Library — Insects that Attack Man. Van Waters & Rogers.

ESSIG, E.O. — 1926. Insects of Western North America. Macmillan.

EWING, H.E. and I. FOX — 1943. The fleas of North America. USDA Misc. Publ. 500.

FEINGOLD, B.G. and E. BENJAMINI — 1961. Allergy to flea bites; clinical and experimental observations. Ann. Allergy 19(11):1275-1289.

FOX, I. — 1940. Fleas of Eastern United States. Iowa State College Press, Ames, Iowa.

FREEMAN, R.B. — 1946. The pig as a host of *Pulex irritans* L. (Siphonaptera, Pulicidae). Entomol. Monthly Mag. 82:19-21.

GOUCK, H.K. — 1946. DDT to control rat fleas. J. Econ. Entomol. 39(3):410-411.

HARWOOD, F.H. and M.T. JAMES — 1979. Entomology in Human and Animal Health. Macmillan Co.

HERMS, W.B. — 1969 Medical Entomology. Macmillan Co.

HICKIN, N.E. — 1964. Household Pests. Benham & Co., Ltd.

HOLLENBECK, A.H. — 1946. A practical method for mass production and transfer of *Xenopsylla cheopis*. J. Parasit. 32:463-464. Oct.

HUBBARD, C.A. — 1947. Fleas of Western North America. Iowa State College Press. 524 pp.

HUNDLEY, J.M. — 1944. Anti-plague measures in Tacoma, Wash. Public Health Reports 59(38), Sept. 22, 1944, pp. 1239-1254.

IRONS, J.V., S.W. BOHLS, D.C. THURMAN, JR. and T. McGREGOR — 1944. Probable role of the cat flea, *Ctenocephalides felis,* in transmission of murine typhus. Amer. J. Trop. Med. 24(6):359-362. Biol. Abst. 19:808-809.

JELLISON, W.L. — 1959. Fleas and disease. Ann. Rev. Entom. 4:398-414.

LEESON, H.S. — 1936. Further experiments upon the longevity of *Xenopsylla cheopis* Roths. (Siphonaptera) Parasitology 28:403-409.

LINDQUIST, A.W., A.H. MADDEN and C.N. WATTS — 1944. The use of repellents against fleas. J. Econ Entomol. 37(4):485-486.

LINDUSKA, J.P., J.H. COCHRAN and F.A. MORTON — 1946. Flea repellents for use on clothing J. Econ. Entomol 39(6):767-769.

LINDUSKA, J.P. and J.H. COCHRAN — 1946. A laboratory method of flea culture. J. Econ. Entomol. 39(4):544-545.

LUNSFORD, C.J. — 1949. Flea problem in California. Arch. Derm. & Syph. 60(6):1184-1202. Biol. Abst. 24:2058. July, 1950.

MATHESON, R. — 1950. Medical Entomology. Comstock Co.

MITZMAIN, M.B. — 1910. Some new facts on the bionomics of the California rodent fleas. Ann. Entomol. Soc. Amer. 3:61-82.

MORLAN, H.B. — 1948. Ectoparasites of domestic rats and mice and the diseases they transmit. Pests 16(1):16, 18, 20.

PATTON, W.S. — 1931. Insects, ticks, mites and venomous animals of medical and veterinary importance. Part II, H.R. Grubbs, Ltd.

PINTO, L. — 1981. Plague. National Pest Control Association, TR037141-44.

PRATT, H. — 1957. It's easy to identify fleas with new CDC pictorial key. Pest Control 25(10):28, 30-32, 34, 36.

PRATT, H.D., J.S. WISEMAN — 1962. Fleas of public health importance and their control. USPHS Publication 272. Part VII.

RAMBO, G. — 1980. Precor. National Pest Control Association, TR-A-6140132.

REESE, D. — 1981. Guidelines for Flea Control. Pest Control 49(5):19-23.

RODRIGUEZ, J.L. and L.A. RIEHL — 1961. Sticktight flea control on chickens with malathion dust self treatment. J. Econ. Entomol. 54(6):1212-1214.

SMITH, C.N. and D. BURNETT, JR. — 1948. Laboratory evaluation of repellents and toxicants as clothing treatments for personal protection from fleas and ticks. Amer. J. Trop. Med. 28:599-607.

SPEAR, P. — 1978. Update on General Pest Control. National Pest Control Association, TR037130-44.

STEWART, M.A. — 1939. The control of fleas and tropical rat mites. Pests 7(5):6-8.

TARSHIS, L.B. — 1959. Use of sorptive dust on fleas. Calif. Agr. 13(3):13-14, March.

TREMBLEY, H.L. and F.C. BISHOPP — 1940. Distribution and hosts of some fleas of economic importance J. Econ. Entomol. 33:701-703.

WALDEN, B.H. — 1943. Fleas and their control. Circ. Conn. Agr. Exp. Sta. 97:37-42.

WATERSTON, B.D. — 1937. Fleas. British Mus. Econ. Series 3.

WILSON, H.G., J.C. KELLER and C.N. SMITH — 1957. Control of fleas in yards. J. Econ. Entomol. 50(3):365-366.

D. E. "Mike" HOWELL

Raised in Southern California, Dr. D. E. Howell gained his bachelor's, master's and Ph.D. in entomology and parasitology from the University of California, Berkeley. He carried out post-doctoral work at the University of Brisbane, Australia, and also attended the Naval War College.

Starting as a teaching assistant in entomology at the University of California, Berkeley, his California experience also included insect surveys of California and insect borne plant disease surveys of California. Then began a long and distinguished academic career at Oklahoma State University where he served as assistant professor, associate professor, professor and head of the Entomology Department. For one year he was acting dean of Graduate School at Oklahoma State University.

In World War II, Howell served in the South Pacific Theater, including New Guinea, Malaysia and the Philippines, and had first hand experience of insect vectors of disease. He attained the rank of captain, Medical Service Corps in the United States Naval Reserve.

An authority on flies, Howell has done consulting work for numerous corporations and for the Southeast Asian Institute of Tropical Biology. Howell was a member of the National Academy of Sciences team to evaluate AID activities in Africa and was a member of evaluation teams of Mississippi State University, University of Nebraska and Colorado State University.

As major advisor for more than 50 graduate students earning Ph.D. degrees in entomology, Howell has helped ensure a steady supply of professionals. In addition, Howell has authored approximately 150 publications in scientific journals and experiment station reports.

Howell belongs to numerous professional associations and honor societies. He is an honorary member of the Entomological Society of America, National Pest Control Association, Oklahoma Pest Control Association and is past president and honorary member of the Oklahoma Academy of Science. He is a member of the National Research Council and the American Society of Tropical Medicine and Hygiene; he is also a Fellow of the American Association for Advancement of Science. A member of Alpha Zeta, Gamma Alpha, Delta Omega and Phi Sigma. Howell is also listed in Who's Who in America and Dictionary of International Biography.

CHAPTER NINETEEN

Flies, Gnats, and Midges

Revised by D.E. Howell[1]

Our aim: to tackle houseflies cycle
Measure their pleasure
Requite their appetite
Splurge their urge
Find the dish they wish
The whiff they sniff
The wall they crawl
We're none too soon!
If we're to have a million
Nay, a trillion
Healthy, happy houseflies
Here by June!

—Hope Stoddard

ALTHOUGH THE AVERAGE individual has little interest in insects, of necessity he quickly becomes acquainted with the applied aspects of entomology in order to discourage the attention of flies, gnats and midges.

The insects in this order have one pair of wings when winged, the hind pair being represented by two short organs known as balancers or halteres. Most flies have large compound eyes and usually three simple eyes. Their mouthparts are of the lapping or piercing type. Their metamorphosis is complete with egg, larval, pupal and adult stages.

FLIES IN THE HOME

THE HOUSE FLY

Musca domestica L.

Although the house fly is one of the most familiar of all insects, the individual who can separate it from other common flies is rare. The house fly is 1/6 to 1/4 inch/four to 7.5 mm in length, with the female usually larger than the male,

[1]*Professor and Head of Entomology Department (Emeritus), Oklahoma State University, Stillwater, Okla.*

PICTORIAL KEY TO COMMON DOMESTIC FLIES
(for use with CDC fly grill record)
Harold George Scott, Ph.D.

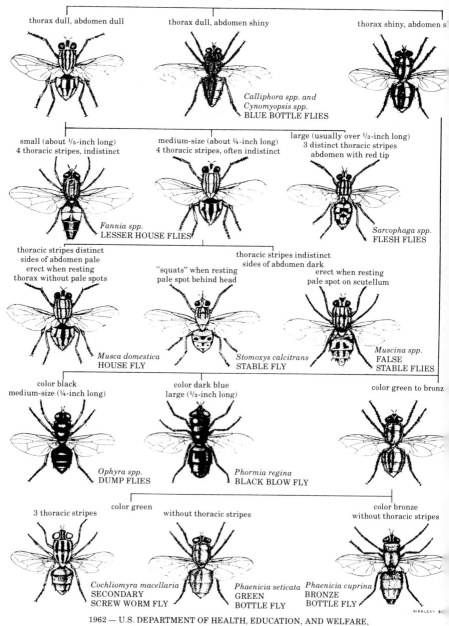

thorax dull, abdomen dull   thorax dull, abdomen shiny   thorax shiny, abdomen s

Calliphora spp. and
Cynomyopsis spp.
BLUE BOTTLE FLIES

small (about ⅕-inch long)   medium-size (about ¼-inch long)   large (usually over ⅓-inch long)
4 thoracic stripes, indistinct   4 thoracic stripes, often indistinct   3 distinct thoracic stripes
                                                                              abdomen with red tip

Fannia spp.   *Sarcophaga spp.*
LESSER HOUSE FLIES   FLESH FLIES

thoracic stripes distinct   thoracic stripes indistinct
sides of abdomen pale   sides of abdomen dark
erect when resting   "squats" when resting   erect when resting
thorax without pale spots   pale spot behind head   pale spot on scutellum

Musca domestica   *Stomoxys calcitrans*   *Muscina spp.*
HOUSE FLY   STABLE FLY   FALSE
                                                         STABLE FLIES

color black   color dark blue   color green to bronz
medium-size (¼-inch long)   large (⅓-inch long)

Ophyra spp.   *Phormia regina*
DUMP FLIES   BLACK BLOW FLY

                      color green   color bronze
3 thoracic stripes   without thoracic stripes   without thoracic stripes

Cochliomyra macellaria   *Phaenicia seticata*   *Phaenicia cuprina*
SECONDARY   GREEN   BRONZE
SCREW WORM FLY   BOTTLE FLY   BOTTLE FLY

1962 — U.S. DEPARTMENT OF HEALTH, EDUCATION, AND WELFARE,
Communicable Disease Center, Atlanta, Georgia

Fig. 19-1. Pictorial key to common domestic flies.

the size of both sexes being dependent to some extent on the availability of food in the larval stage and whether the abdomen is distended with food. The thorax bears four narrow black stripes and there is a sharp upward bend of the fourth longitudinal vein. (Fig. 19-2). The sexes can be readily separated by noting the space between the eyes, which in females is almost twice as broad as in males, or by applying pressure to the abdomen, which results in the protrusion of an ovipositor in the case of the female. (Fig. 19-3). West (1951), Hewitt (1914), and Howard (1911) considered the house fly in some detail in their books. Dethier (1963) wrote an amusing and informative book about flies that is worth reading.

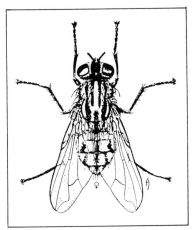

Fig. 19-2. Female house fly, *Musca domestica* L.

This insect is cosmopolitan in distribution and is called the house fly because it is very often the fly that occurs most commonly in the home. Howard (1900) collected flies in buildings in various parts of the United States and 98 percent of them were house flies, *Musca domestica* L. With the advent of the "horseless carriage," the house fly lost some of its preeminence as a household pest, although it is still very important, especially in rural areas.

There is a difference of opinion on what is the most important fly in and around homes and other buildings. This difference is evidently caused by special environmental conditions favoring certain species of flies. DeLong and Boush (1952) fogged supermarkets throughout Ohio and made a survey of the species of flies killed by the spray. They found the house fly, *Musca domestica* L., to account for 28 percent of the flies, whereas the greenbottle fly, *Phaenicia sericata* Meigen, accounted for 66.1 percent, and the black blowfly, *Phormia regina* Meigen, for five percent. Schoof and Savage (1955) noted in a survey of flies in city areas that *M. domestica* L. is most prevalent in the Southwest, while the blowflies, *Phaenicia* and *Phormia*, predominate in the Northeast. According to Haines (1953), 99 percent of the flies in houses in two cities in Georgia were *M. domestica* L.

Fig. 19-3. The sex of a house fly can be readily separated by the width of space between the eyes. The female (left) has a much wider space between the eyes than the male (right).

Food of the house fly. Fermenting, fresh horse manure is a favorite breeding place of the house fly. This manure must be less than one day old to be attractive to the egg-laying adult. However, Leikina (1942) showed the larvae of the house fly develops best in human and pig manure and not so well in horse manure. According to Coffey (1951), who made a study of fly breeding substances in Texas, the most important breeding places in descending order are horse manure, human excreta (privies), cow manure, fermenting vegetable refuse and kitchen garbage. Howell (1975) found swine barns, horse barns, sheep barns, cattle barns and poultry buildings in descending order had the most house flies. The sanitation in each barn was approximately equal. Yates (1952) noted house flies are attracted to the ammonia emanating from horse manure. Chapman (1944) observed the house fly to complete its development in a mattress wetted with human urine. Miller and Mallis (1945) found flies were causing much annoyance in the vicinity of a dump filled with incinerated garbage, particularly in citrus rind and pulp that was not completely burned. The flies bred in incinerated garbage if it was fresh and wet, but not in incinerated garbage that was scattered or two to 10 months old. Sawicki and Holbrook (1961) made a survey of rearing methods for house flies.

Entomologists in the U.S. Public Health Service have made careful studies of the breeding places of common flies both in urban and rural areas. Schoof et al. (1954) showed that in city communities *M. domestica* L. bred largely in scattered garbage, whereas the blow flies, *Phaenicia sericata* Meigen and *Phaenicia pallescens,* bred in confined garbage. *Sarcophaga* spp. use dog feces as a medium. *M. domestica* L. was the predominant species in accumulated fowl excrement. Schoof and Siverly (1954) found *M. domestica* L to be the dominant species in human feces, followed by *Sarcophaga, Muscina* and *Hermetia.* Savage and Schoof (1955) stated *Phaenicia sericata* Meigen was the predominant species in garbage dumps in Michigan and New York. *M. domestica* L. was by far the most common species in hog farms, horse stables and poultry ranches in Kansas and Arizona. Siverly and Schoof (1955a) found *M. domestica* L. to breed in suitable media all year long in Phoenix, Arizona. "Chicken and pig excrements, garbage, melons and stock feed displayed the highest production potential."

The author has been acquainted with several cases where the use of Milorganite, a human manure, on lawns has resulted in many house fly larvae migrating into cellars and basements.

In laboratories that raise house flies for the Peet-Grady method of fly spray testing, the modified Richardson larval medium is used to rear the flies. This consists of a solid material with the following components:

Percent by weight
Soft wheat bran 33.3
Dried brewers' grains 33.3
Alfalfa meal 33.3
Total 99.9

Approximately 1000 ml of an aqueous suspension containing 20 g moist cake yeast and 15 ml non-diastatic Diamalt (a maltose material) is added to 450 g of this dry larval medium.

Life cycle. The house fly passes through four stages in its life cycle: egg, larva or maggot, pupa and adult. The females, usually in clusters of 20 to 50, can be seen depositing their eggs on suitable material. The white eggs, which are about one mm long, are laid singly but pile up in small masses. The female deposits

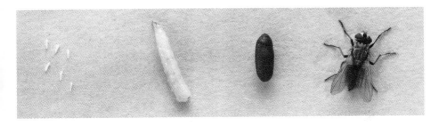

Fig. 19-4. Eggs, larva, pupa and adult of house fly.

from 75 to 150 eggs and during her lifetime she may lay five or six batches at intervals of several days between each batch. All in all, she may deposit from 350 to 900 eggs, and one female has been known to deposit 2387 in 21 batches (Dunn, 1923). Ordinarily the house fly commences to lay her eggs from four to 12 days after emerging from the pupal case and tends to favor moist materials for egg deposition. Eggs that become too dry during the incubation period will fail to hatch.

The tiny white footless maggots emerge from the eggs in warm weather in from eight to 20 hours and they immediately begin feeding. Completion of the larval stage requires three to seven days at temperatures of 70° to 90° F/21° to 32° C, although six to eight weeks may be required at lower temperatures. The larva undergoes three instars. The full-grown maggot has a greasy, cream-colored appearance and is seven to 10 mm long. It can be separated from the blowfly's maggot with which it may be confused by the sinuous slits in the posterior spiracles.

When the larva becomes fully grown, it seeks a dry, cool place to pupate and migrates from its food source to the soil beneath or into the ground beneath nearby boards, stones, etc. Barber (1919) observed house fly larvae to migrate 150 feet/45.7 m to a sewage manhole rather than pupate in the warm soil beneath a manure pile. The larvae may be in the prepupal or migrating stage for three to four days.

The pupa is the stage wherein the mature maggot transforms into the adult. It is passed in a pupal case formed from the last larval skin which varies in color as the pupa ages from yellow, red, brown, to black. Ordinarily, the size of the maggot determines the size of the pupa which in turn determines the size of the adult fly. Moreland and McLeod (1957) noted that the smaller pupae are predominantly male while the larger ones are mostly female. Pupae reared in the laboratory of Gulf Research & Development Company weighed between 18 and 24 mg with an average weight of 21 mg, and 1350 pupae weighed approximately one ounce/28.4 g. The pupa transforms into an adult in three days to four weeks or longer, depending primarily on temperature and humidity.

The emerging fly escapes from the pupal case through the use of an alternately swelling and shrinking sac or ptilinum on the front of its head which it uses like a pneumatic hammer. Bucher et al. (1948) noted at about 80° F/26.7 ° C "about one hour is required after emergence for expansion of the wings and general hardening of the body. At this temperature the adults remain relatively quiescent for about the first four hours. Normal activity is reached in about 15 hours." During the warm weather of summer, when conditions are generally optimum for the development of the house fly, it may require as little as six days (Metcalf et al., 1951) to complete the cycle from egg to adult, and there may be as many as 10 to 12 generations in one summer.

Simanton and Miller (1938), while making a study of the house fly relative to the Peet-Grady method of testing, found that in the laboratory (at 80° F/26.7° C and relative humidity 55 percent) copulation begins on the second day after emergence and almost all the females are fertilized by the end of the third day. It is heaviest from the third through the ninth day and then gradually tapers off. Patterson (1957) observed the house fly to mate twice: once when she was three days old and again when she was seven to 10 days of age.

Barber and Starnes (1949) noted the males are much more excitable than the females as indicated by their greater walking and flying activity. In the laboratory, the wings of the males are often torn or broken. This condition may be caused by the powerful blows of the middle legs of the female fending off the male. According to Patterson (1957), the metathoracic legs of the female are largely responsible for breaking the males' wings. A. Goddin observed the house fly very often is capable of flight even with one-half or more of the wing area broken off.

The usual longevity of the adult house fly is two to three days without food and from a few days to 54 days with food (Hutchinson, 1916). According to Hewitt (1914), adult flies have been kept alive in the laboratory by investigators for as long as 16 weeks. Bucher et al. (1948) found that for a short time immediately following emergence, house flies did not consume food and they began feeding when they first became active, which in turn was dependent upon environmental conditions. Howard and Bishopp (1926) observed that flies remain close to the breeding place if the food is sufficient. On the other hand, they noted a marked house fly was recaptured 13 miles from the place where it was libereated. Schoof et al. (1952) released house flies tagged with P-32 in Phoenix, Arizona and found 88 percent of the flies were captured within one mile of the point of release. Yates et al. (1952) also released house flies tagged with radioactive phosphorus and captured one 28 miles from where it was liberated.

Schoof (1959) summarized the studies of many investigators on the distances some of the common flies travel. He noted the flight of flies has been traced in recent years by means of radioisotopes. The flies are tagged with radioactive phosphorus, P-32, by feeding the adults milk-honey solutions or sugar solutions (one millicurie of P-32 per liter of milk) for a 24-hour period. These investigators showed that *M. domestica* L. can move four to 6.3 miles within 24 hours. The blowfly *Phormia regina* Meigen disperses more rapidly than any of the common flies studied. Individual specimens of *M. domestica* L. have been captured 7.3 to 20 miles from the point of release. Certain species of blowflies were captured 28 miles from the point of release.

According to Schoof (1959), ". . . from the control standpoint, the wanderings of a few individuals are of minor significance compared to the movement of the bulk of the population." Fifty-five to 96 percent of *M. domestica* L. were recovered within a one mile/1.6 km range and 77 to 100 percent within the two mile/3.2 km range.

Schoof (1959) noted that Parker described the house fly "as a migrating insect, the movement of which is partly instinct and partly due to its response to stimuli from breeding and feeding sites." Ordinarily fly control from 0.5 to one mile/ 0.8 to 1.6 km around a municipality will prevent ingress of the house fly into the municipality. In the case of blowflies, such as *Phormia regina* Meigen, "control treatments at abattoirs, rendering plants, and similar problem sites three to four miles beyond the city may be necessary."

Hewitt (1914) observed the house fly at a height of 80 feet/24 m in the air. According to Hewitt, the flies that hibernate are the young and vigorous recently-emerged flies. "On dissection it is found that the abdomens of these hibernating individuals are packed with fat cells, the fat-body having developed enormously. The alimentary canal shrinks correspondingly and occupies a very small space; this is rendered possible by the fact that the hibernating house fly seeks a dark place such as crevices as well as areas behind pictures, books, curtains and behind loose wallpaper." He also observed some of them hibernating in stables.

Hutchinson (1918) found the house fly overwinters in Washington, D.C., Columbus, Ohio and in Texas by breeding throughout the winter in warm places where food and media for deposition are available. These include protected stables, attics, heated buildings and even outdoors. Dove (1916) noted house flies overwinter as larvae and pupae in or under manure and that the adults emerge during mild weather in the winter, but the percentage that survives the winter is small. Matthysse (1945) also observed that flies overwinter by continuous breeding indoors throughout the winter. He observed thousands of house flies breeding in manure and bedding in barns. Of the flies present, 95 percent were *Musca domestica* L. and five percent were *Stomoxys calcitrans* (L.). According to Mail and Schoof (1954), *M. domestica* L. bred in cat feces and garbage sludge in basements of buildings during midwinter in Charleston, West Virginia.

The potential reproductive capacity of flies is tremendous, but fortunately can never be realized. Hodge (1911) estimated that "A pair of flies beginning operations in April may be progenitors, if all were to live, of 191,010,000,000, 000,000,000 flies by August. Allowing one-eighth of a cubic inch to a fly, this number would cover the earth 47 feet deep." According to Dr. H.A. Ambrose, it really is not quite as bad as it sounds because his calculations show the earth would be covered to a depth of only 2.5 feet.

The house fly and disease. Although the house fly is often an unbearable nuisance, we are primarily concerned with it as a carrier of disease organisms. The house fly feeds on fecal material, vomit and sputum, after which it might alight on human food. It is very well adapted by both structure and behavior for the transmission of disease. Its body is covered with fine hairs and bristles which readily pick up dirt particles. At the base of each of the two claws at the end of each leg there is a cushion-like structure, the pulvillus, which is covered with glandular hairs. The sticky secretions from the glandular hairs gather bacteria and other organisms. These sticky secretions also enable the fly to climb vertical surfaces and to walk upside down. The house fly excretes and regurgitates wherever it comes to rest. Moreover, flies commonly fall into liquid foods and contaminate them in this fashion. According to Barber and Starnes (1949), fly regurgitation "is a process of digestion during which the food is brought up from the crop bit by bit and is mixed with saliva before being passed on to the digestive tract."

House flies have been shown to carry the disease organisms causing typhoid fever, cholera, summer diarrhea, dysentery, tuberculosis, anthrax, ophthalmia, etc., as well as parasitic worms. Watt and Lindsay (1948) noted that fly control definitely reduced infection, disease and death due to diarrhea in Texas. Frison (1925) recorded the regurgitation of house fly larvae by a sick boy. Greenberg (1967) prepared an interesting article on house flies as vectors of *Salmonella* bacteria which are responsible for food poisoning and gastric infections. Experimental work has demonstrated the presence of the virus of poliomyelitis in

PICTORIAL KEY TO COMMON DOMESTIC FLIES IN SOUTHERN U.S.

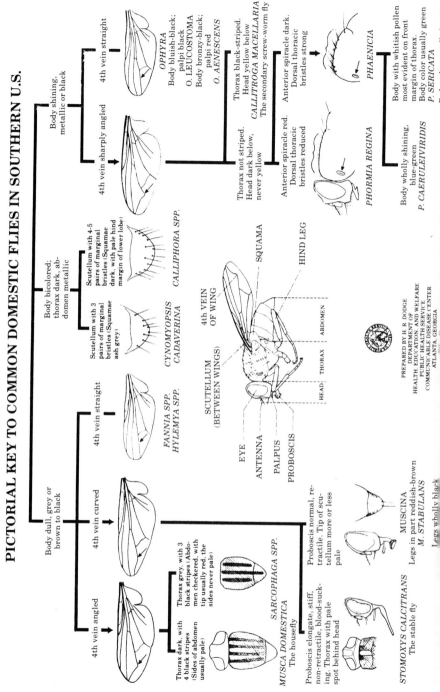

Body shining, metallic or black

4th vein straight

OPHYRA
Body bluish-black; palpi black
O. LEUCOSTOMA
Body bronzy-black; palpi red
O. AENESCENS

4th vein sharply angled

Thorax black-striped. Head yellow below
CALLITROGA MACELLARIA
The secondary screw-worm fly

Anterior spiracle dark. Dorsal thoracic bristles strong

PHAENICIA
Body with whitish pollen most evident on front margin of thorax.
Body color usually green
P. SERICATA

Thorax not striped. Head dark below, never yellow

Anterior spiracle red. Dorsal thoracic bristles reduced

PHORMIA REGINA

Body wholly shining, blue-green
P. CAERULEIVIRIDIS

Body bicolored; thorax dark, abdomen metallic

Scutellum with 4-5 pairs of marginal bristles (Squamae dark, with pale hind margin of lower lobe)
CALLIPHORA SPP.

Scutellum with 3 pairs of marginal bristles (Squamae ash grey)
CYNOMYOPSIS CADAVERINA

Body dull, grey or brown to black

4th vein curved

4th vein straight

FANNIA SPP.
HYLEMYA SPP.

SCUTELLUM
(BETWEEN WINGS)

4th VEIN OF WING

SQUAMA

HIND LEG

ABDOMEN

HEAD THORAX THORAX

EYE
ANTENNA
PALPUS
PROBOSCIS

PREPARED BY H. R. DODGE
DEPARTMENT OF
HEALTH, EDUCATION AND WELFARE
PUBLIC HEALTH SERVICE
COMMUNICABLE DISEASE CENTER
ATLANTA, GEORGIA

4th vein angled

Thorax dark, with 4 black stripes (Sides of abdomen usually pale)

Thorax grey, with 3 black stripes (Abdomen checkered, with tip usually red, the sides never pale)
SARCOPHAGA SPP.

MUSCA DOMESTICA
The housefly

Proboscis normal, retractile. Tip of scutellum more or less pale

MUSCINA
Legs in part reddish-brown
M. STABULANS

Proboscis elongate, stiff, non-retractile, blood-sucking. Thorax with pale spot behind head

STOMOXYS CALCITRANS
The stable fly

Legs wholly black

Fig. 19-5. Pictorial key to common domestic flies in the southern United States.

RECOGNITION OF SPECIES OF MUSCA

| *Musca domestica* Linnaeus (house fly) | *Musca autumnalis* DeGeer ("face fly") |

Both species with the familiar habitus of *Musca:* 4 black stripes on thorax, same wing venation, etc.

Both sexes

1. Typically slightly smaller, lighter in color; common indoors, on walls of stables, houses, etc., also found outside.

1. Typically slightly larger than *domestica,* and darker; an outdoor fly, on animals, especially about eyes and nostrils of cattle, or sitting on nearby rocks, fenceposts, etc.; may be indoors in fall and winter.

2. Propleuron-haired.

2. Propleuron bare.

3. No tympanic tuft of bristles.

3. Strong tympanic tuft of bristles at base of calypteres (often seen by lifting upper (alar) calypter with a needle or insect pin).

Males only

4. Eyes well separated (frontal stripe broad and parallel-sided).

4. Eyes almost touching.

5. Dorsum of abdomen usually yellowish at sides, or at least narrowly so toward base, *rarely* all gray-black.

5. Dorsum of abdomen entirely black in ground color, with strong gray-and-black pattern.

Females only

6. Parafrontals often yellowish-tinted anteriorly, posteriorly narrow, each about one-third as wide as median frontal stripe.

6. Parafrontals (sides of front) bright gray, wide, nearly as wide as median frontal stripe.

Curtis W. Sabrosky
Insect Identification and Parasite Introduction Research Branch, U.S. Dept. of Agriculture

Cooperative Economic Insect Report, Vol. 9(45): 11-6-59

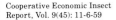

Fig. 19-6. A pictorial guide to recognizing species of musca.

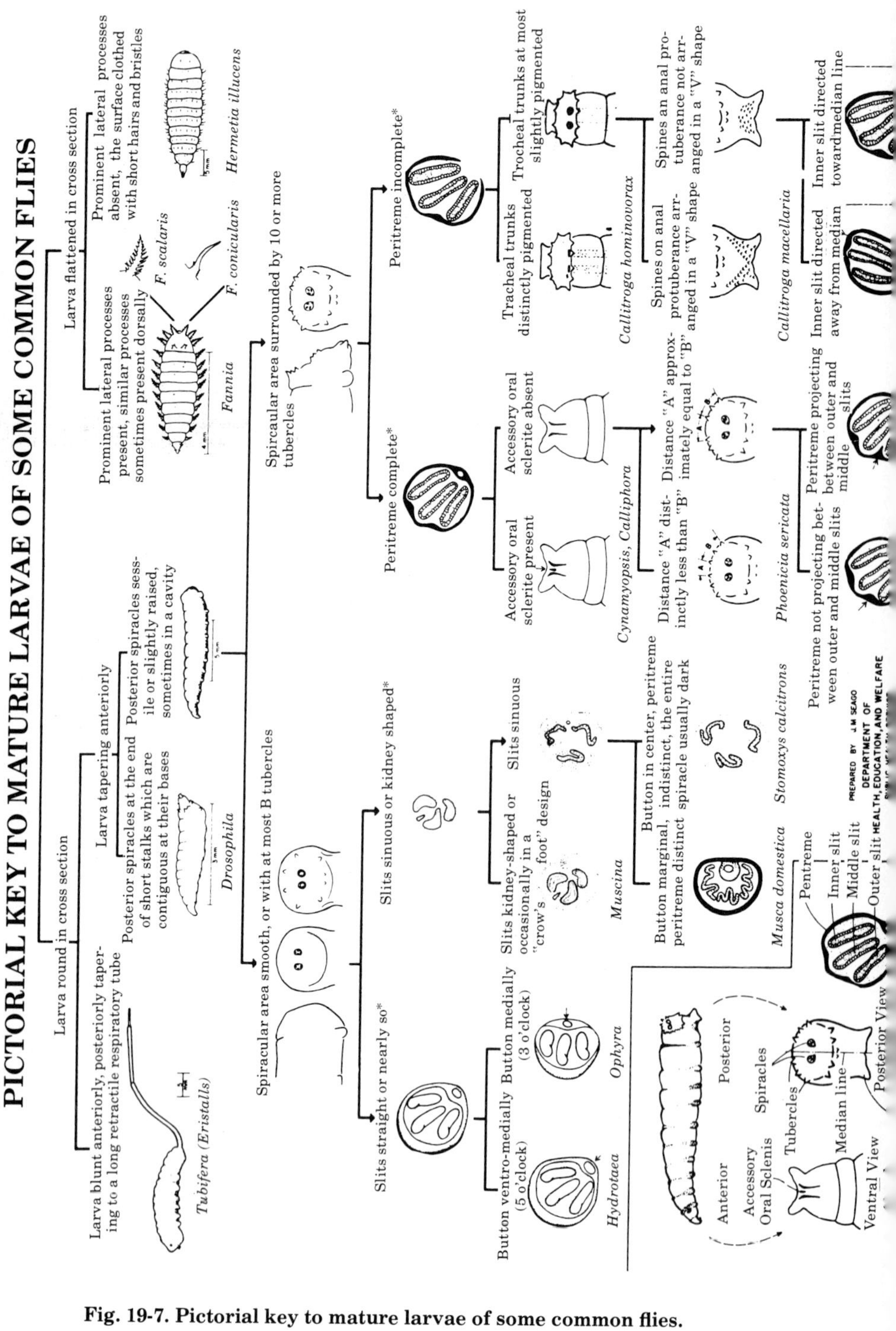

PICTORIAL KEY TO MATURE LARVAE OF SOME COMMON FLIES

Fig. 19-7. Pictorial key to mature larvae of some common flies.

or on flies trapped in epidemic areas, and the flies in homes of polio patients can contaminate food with the virus. Power and Melnick (1945), in studying the fly population in an epidemic poliomyelitis area, showed the house fly was not the dominant fly and that flies of other genera such as *Phaenecia sericata* Meigen were more common.

THE FACE FLY

Musca autumnalis DeGeer

This species is a European and Asiatic fly that hibernates in buildings in the fall and causes annoyance similar to that of the cluster fly. It was first reported in Nova Scotia in 1952 and in Long Island, New York in 1953. At present it is widely distributed throughout southern Canada and in the United States north of the 33rd parallel, where it is especially annoying on the faces of cows as it laps exudations from the eyes, nostrils and mouth. According to Teskey (1960), the adult fly feeds on animal secretions, nectar and moist manure. The adult passes the winter in the walls and attics of buildings. Decker (1961) noted in the spring it deposits eggs in fresh manure, and the egg has a dark stalk at one end. The three larval instars require three to 10 days for completion. The pupal case is dirty white in color. Teskey stated pupation takes place near the edge of the droppings and the egg to adult stage may require about two weeks. Wang (1964) found the outdoor life cycle to be 17 to 18 days at 77 to 86° F/25° to 30° C.

The two species of *Musca,* house fly and face fly, are not easily separated without magnification, but with the aid of a 10X or greater hand lens and Fig. 19-6 they may be quickly and accurately identified.

The eggs of the face fly are easily separated from those of the house fly by the presence of a respiratory stalk on the egg of the face fly which are deposited just below the surface. These may be readily found by lightly abrading the surface of the droppings with a float. This will expose the eggs, or if disturbed a little later, the yellowish, newly-hatched larvae may be seen trying to reenter the dung. In contrast, the larvae of *Musca domestica* L. are creamy white and tend to penetrate much of the droppings. The pupae of the face fly are off-white in contrast to the reddish brown color of those of the house fly.

The habits of the face fly and the house fly are quite distinct. *Musca autumnalis* DeGeer normally passes the winter as an adult in attics, little used upstairs rooms, and at times in basements. When the temperatures become warmer in the spring, the overwintering females deposit their eggs in freshly dropped manure, usually from bovines. The manure is not attractive to face flies unless it is undisturbed and less than a day old. Even trampling by animals quickly decreases the attractiveness of the manure as an oviposition site. In contrast, the house fly seldom deposits its eggs in undisturbed manure.

Cheng (1967) showed pinkeye in cattle to be correlated with face fly abundance. Adult face flies are normally closely associated with animals and are controlled by insecticides sprayed on the hosts of the face flies or by elimination of the breeding areas mentioned above. Shugart et al. (1979) evaluated the effects of feeding by *Musca autumnais* DeGeer on the eyes of cattle and indicated that 75 percent of the cattle exposed to an average of one face fly/eye for 33 days developed petechial lesions. Unexposed cattle developed almost no observable lesions. After 43 days exposure to face flies, 38 percent of the lesions had progressed from a petechial to an ecchymosis stage. This data suggests one face fly per day per eye reaches the economic level of loss.

Removal or alteration of the breeding areas is possible in the barn lots but is very difficult in the pastures. With dairy cattle, pyrethroids, pyrethrum and other toxicants and repellents can be used effectively, but they are not useful with beef cattle.

While the face fly adults normally only lap exudations from the bovine eyes, nostrils and mouth, the house flies feed on many sources.

CONTROL OF HOUSE FLIES IN THE HOME

Exclusion has been an effective method of fly control for many years. Properly maintained copper, bronze, aluminum, or plastic screens are helpful. Screen 18 mesh and smaller, if properly maintained, provide excellent control of almost all flies, mosquitoes and other insects. If control of very small gnats, flies and mosquitoes is needed, it may be worthwhile to use smaller mesh screen. Unfortunately, air circulation will be reduced and this may be undesirable.

In some situations wind screens also may be very helpful, such as when produce must be moved into buildings and the open doors would allow flies to enter. Adequate distribution of moving air must be carefully maintained and city codes followed.

CHEMICAL CONTROL

The control of flies with chemicals has changed very rapidly during the past 20 years. Major factors have been the development of resistance to pesticides by the insects and the increasing regulation of the use of pesticides. These are outlined in the *Federal Register* Vol. 40, No. 173, pp. 41175 to 41177, Sept. 5, 1975 and in more recent addenda. State and city regulations may also restrict the use of pesticides. Guidelines for the use of chemicals for fly control are published in many places and it is essential the pest control operator continue to be aware of changes. It is necessary to read the label on the pesticide container very carefully. The regular reading of professional journals, information put out by commercial companies, and recent federal and state publications will help the user to remain current. State university pesticide coordinators are available to provide help in interpreting the bewildering series of changes in the field of chemical control.

It is interesting to note that in the recently published *Federal Guidelines for the Control of Insect and Mite Pests of Foods, Fibers, Feeds, Ornamentals, Livestock, Households, Forests and Forest Products,* (Anon. Agriculture Handbook No. 571, 1980) only the following are listed for use in fly control: Allethrins, lindane, malathion (premium grade), propoxur, dichlorvos, malathion, pyrethrins plus piperonyl butoxide, pyrethroids and ronnel. Many of the insecticides which were formerly widely used are no longer labeled for fly control in the United States, though they are used against other pests. These include DDT, chlordane, dieldrin, aldrin, heptachlor and endrin. It is probable the future will bring many more changes.

SPACE, SURFACE, COMBINATION SPRAYS AND AEROSOLS

Fly sprays and household sprays are synonymous since a spray that will control house flies is usually used as a contact insecticide against a wide range of insects and other arthropods.

Three kinds of sprays are on the market: The space spray, which includes aerosols, the surface or residual spray, which may be pressurized, and the combination space and residual spray. Space sprays are applied as a mist into the air and must contact the insect at the time of spraying. They provide quick knockdown and fast results, but temporary control. A spray containing pyrethrins and small quantities of other toxicants is an example of a space spray. An aerosol is another form of a space spray whose insecticidal ingredients are usually dispersed by means of the vapor pressure of a liquified gas rather than the pressure of compressed air (e.g., the usual hand sprayer) and whose droplets are of a much smaller particle size than those formed by the usual hand sprayer.

Many pest control operators provide their commercial customers with aerosols providing from one lb. to 30 lbs. pressure for fly and other insect control in supermarkets, warehouses, etc. Odeneal (1959) described an automatic dispenser which releases a "shot" of pyrethrins/piperonyl butoxide every 15 minutes. Fisher (1955) discussed a portable barn fogging unit that uses 0.1 percent pyrethrins and one percent piperonyl butoxide. Similar units have provided excellent control of flies in barns if air movement is controlled.

Surface or residual sprays may have a petroleum or a water base. They are applied on surfaces as a wet spray rather than a mist and they leave a toxic layer of either crystals or film on the evaporation of the carrier. Ordinarily, surface sprays provide relatively slow knockdown, but usually long-lasting control against nonresistant flies, and do not have to contact the insect at the time of spraying in order to be effective. Residual sprays are ordinarily applied to surfaces frequented by house flies such as light fixtures, window screens and walls.

When a residual spray is used as a space spray, the large amount of toxicant becomes a hazard to the user and may contaminate food as well.

Space sprays. Gilbert (1946) observed the use of fly sprays is of fairly recent origin and dates from about 1919 when they were used by the British armed forces in the Middle East and other nearby areas which are notorious for fly, gnat and mosquito infestations.

MacNair (1940), however, noted the first kerosene extract of pyrethrum flowers "sold as a household insecticide was 'Walker's Devilment,' developed by a patent medicine manufacturer of that name located at Thomasville, Georgia, and sold there first in 1916." Space sprays showed a tremendous surge in sales from 1930 to 1940.

SPACE SPRAY TOXICANTS

The toxicants most widely used in current fly sprays are pyrethrum, allethrin and the pyrethrum or allethrin synergists.

Pyrethrum. This is one of the most common ingredients in a fly spray. At recommended dosages it is one of the safest toxicants and it has exceptionally fast "knockdown" (ability to quickly paralyze many insects). A few users and occupants of treated buildings may become allergic to pyrethrum sprays. Pyrethrum also has the disadvantage of not killing all the insects knocked down. The pyrethrins are used in space sprays at approximately 40 to 200 mg per 100 ml of finished spray, the usual concentration at this writing being 50 to 100 mg pyrethrins per 100 ml. It can be seen from the above figures that the pyrethrins are effective at low concentrations. Recently developed pyrethroids are very effective at low concentrations also.

Pyrethrins are usually sold as 2.5 percent and 20 percent concentrates in petroleum distillates which are readily soluble in base oils. The 2.5 percent concentrate (20 mg pyrethrins per ml) is mixed at the rate of one part concentrate (five ml) to 19 parts oil base (95 ml) to give a 100 mg pyrethrins per 100 ml spray. The 2.5 percent concentrate is known to the trade as a 20 to one pyrethrins concentrate. The 20 percent concentrate (20 percent pyrethrins by weight) is finding its widest usage in aerosols.

Simanton and Miller (1938), working with pyrethrum, demonstrated it was easier to paralyze, but more difficult to kill very young flies than older flies. They also showed female flies are approximately twice as difficult to kill and more difficult to paralyze than male flies. Miller (1946) noted: "There is no exact explanation as to why the females should be less susceptible to certain toxicants, although they are somewhat larger and heavier than the males. Also, the fact that they are somewhat less active has been shown to result in their picking up fewer droplets of spray."

David (1946) and David and Bracey (1946) showed a house fly or yellow fever mosquito flying through an insecticidal mist accumulated a large proportion of the dose on the wings. "Factors which increased the fly activity of the insects increased the dose accumulated, and *vice versa.* The quantity accumulated on the wings is about three to four times that collected on the body in the case of *Musca,* and four to six times as much in the case of *Aedes.* It is shown that the material collected on the wings contribute materially to the kill."

Miller and Simanton (1938) showed pyrethrum sprays in the Peet-Grady Chamber must be evaluated on equal numbers of the sexes to avoid highly variable ratings. A study of 18,400 flies from 11 culture jars showed that 51.4 percent of laboratory-reared *Musca domestica* L. were male. The first 10 percent of flies emerging from culture jars in the laboratory are about 65 percent males and the last 10 percent are about 40 percent males.

McGovran and Gersdorff (1945) showed feeding sugar to adult flies lowered their resistance to pyrethrins.

Pyrethrum synergists. An insecticidal synergist is a chemical which when combined with an insecticide increases the killing power of the combination beyond what might be expected by the simple use of each alone. Since pyrethrins on a pound basis are expensive, a number of synergists have been developed which produce more effective sprays using lower concentrations of pyrethrins. A common pyrethrum synergist finding commercial usage at this time is piperonyl butoxide at various concentrations in the finished spray.

Allethrin. This chemical, the allyl homolog of cinerin I, is often referred to as "synthetic pyrethrins." It has knockdown and mortality characteristics similar to natural pyrethrins when used against flying insects in space sprays. It is definitely inferior to pyrethrins against crawling insects such as cockroaches. Activation of allethrin by synergists is not necessarily the same as for pyrethrins.

Odor masking. Crocker (1949) discussed the principles of odor masking in fly sprays. The masking of odors is the covering of odors which are not wanted and the end result is to produce "an odor of minimum obviousness rather than one of distinct pleasantness. The ingredients chosen for this purpose must have a volatility similar to or less than that of the original odor and be equally stable in the air so that the odor correction attained will last long enough to be practicable. Present-day fly-sprays start with relatively low odor kerosenes but still need masking to cover the odor of the active insecticides. Insecticide odorants have to be breathed, so should be as free as possible from toxic and irritating substances. Such simple food and candy synthetics as methyl salicylate, safrol, coumarin, or vanillin are commonly used, together with a touch of oil of geranium or the geranyl esters."

Crocker recommended masking ordinarily be done with a mixture of odorants since a mixture is less evident than single odorants. "In general, masking agents have to be used with great restraint around food products, for eating the food may be a most sensitive method for detecting the presence of a masking agent in or around the container of the food. Even cooking may not succeed in driving off all of the accumulated odor, particularly when the container becomes scented with some of the modern chlorinated organics used for insecticides, fungicides, and antiseptics.

"The quantity of agent necessary to neutralize an objectionable odor, but not to create a perfume, is generally less than one part in 1000 and may be as little as one part in 20,000." This amounts to approximately 38 ml to two ml per 10 gallons/37.9 l. Besides methyl salicylate (oil of wintergreen), other odors that are commonly used are lavender, pine and cedar. Anon. (1966) also reviewed the subject of masking agents and perfumes.

RESIDUAL FLY SPRAYS

The remarkable fly control achieved with DDT sprays in such difficult places as barns, poultry houses, hog pens and similar areas in 1946 and 1947 will be long remembered by the entomologists — and the farmer — for the season-long control of house flies with one or two applications is now apparently a thing of the past! In late 1947 and 1948 the first rumbling of failure began to roll in from various sections of the country and it was soon established the house fly was not very susceptible to DDT in these areas. Within a year or two it became evident these DDT-resistant flies could not be satisfactorily controlled with other chlorinated insecticides such as chlordane, benzene hexachloride, etc. In some sections where resistant flies are present, repeated applications of residual insecticides may give some degree of control. In other areas the use of residual sprays is now of little value. At the present time, the trend is toward emphasis on sanitation, proper screening, mechanical control measures, the use of space sprays, aerosols, insecticide baits and insecticide cords. Unless some new residual insecticides or combination of insecticides make their appearance, it can be stated that for all practical purposes the problem of chemical fly control has completed a cycle from space sprays to residual sprays and back to space sprays. Malathion and diazinon sprays were worthy substitutes for DDT for a time, but now the house fly is becoming resistant to these and other organic phosphates. Since the resistant house fly is the basis for our troubles, it may not be amiss to devote some additional space to it in this section.

Resistant house flies. Brown (1958) and Brown and March (1959) prepared extensive reviews of resistance of flies to insecticides. It has been noted insects can develop physiological as well as behavioristic resistance. In physiological resistance, poisons may be absorbed at different rates, stored in nonsensitive tissues, excreted and detoxified. Some insects may even bypass organs incapacitated by poisons. In behavioristic resistance insects survive because they live in out-of-the-way habitats or because they avoid surfaces or baits treated with insecticides. Missiroli (1947) reported DDT-resistant house flies and mosquitoes, *Culex pipiens* L., which he found in the field in Italy in 1945 and 1946. Sacca (1947) also encountered flies that were not susceptible to DDT in the Tiber

Valley of Italy. Wiesmann (1947) conducted tests on DDT-resistant flies found in Sweden in 1946. Lindquist and Wilson (1948) commenced rearing DDT-resistant flies in the USDA Bureau of Entomology and Plant Quarantine laboratory in Orlando, Florida in 1946. The flies began to show resistance from the third generation on. In 1948 reports of the failure of DDT to control house flies became commonplace in the United States. King and Gahan (1949) reported DDT-resistant wild strains from the South and West. March and Metcalf (1951) noted many complaints regarding the ineffectiveness of DDT on house flies in Southern California in the spring of 1948. Hansens et al. (1948) found DDT-resistant flies in New Jersey during the summer of 1948. Shortly thereafter reports on the ineffectiveness of DDT sprays against house flies began to come from the Midwest and elsewhere, and Upholt (1950) reported an unsuccessful search in 1949 in the United States for a strain of flies non-resistant to DDT, although a number of localities were found where the flies were only slightly resistant.

Blickle et al. (1948) accidentally developed in the laboratory a strain of house flies resistant to benzene hexachloride, and Gahan and Weir (1950) found *Musca domestica vicina* Macq. to be resistant to benzene hexachloride in the field in Egypt. Wilson and Gahan (1948) developed a stock of house flies in the laboratory resistant to DDT residues, as well as to chlordane, toxaphene, pyrethrins, piperonyl cyclonene, rotenone and Thanite®. Barber and Schmitt (1949) and Hansens and Goddin (1949) reported a methoxychlor resistant strain that was non-resistant to DDT. Knipling (1950) noted house flies in Florida became highly resistant to dieldrin in one season. He also stated the field strains were resistant to DDT, methoxychlor, chlordane, toxaphene, lindane and aldrin. Busvine (1951) reported a Sardinian strain to be resistant to practically every one of the insecticides mentioned by Knipling. He also reported an Italian strain resistant to DDT and pyrethrins. Miller (1949) showed methoxychlor, chlordane and benzene hexachloride-resistant flies were susceptible to pyrethrins space sprays. Davies et al. (1958) found house flies in Sweden resistant to chlorinated hydrocarbons, parathion, diazinon, allethrin and synergised pyrethrins. Bruce (1950) developed a laboratory strain of house flies resistant to pyrethrins/piperonyl butoxide combinations. Knipling (1950) noted a pyrethrins/piperonyl butoxide spray was slow in knocking down resistant flies resulting in unsatisfactory control of the insects in a dairy in Florida. Hansens and Morris (1962) showed the house fly in New Jersey to be resistant to diazinon, ronnel, malathion, lindane and DDT.

In most instances where house flies were resistant to DDT, they showed some degree of resistance to other chlorinated insecticides, but this was not always the case. Barber and Schmitt (1949), as was mentioned previously, noted that laboratory-reared flies resistant to methoxychlor were susceptible to DDT. March and Metcalf (1949) found two highly resistant strains in Southern California, the "Bellflower," which was resistant to DDT and closely related insecticides, and the "Pollard," which was resistant to DDT, chlordane, lindane and other chlorinated insecticides. Knipling (1950) and Pimentel et al. (1950) showed DDT-resistant flies in the field were also resistant to the other chlorinated residual insecticides. Bruce and Decker (1950) demonstrated this universal resistance by DDT-resistant flies in the laboratory.

Keiding (1956) found the house fly in Denmark to be resistant to such organic phosphate insecticides as parathion and diazinon. Labrecque and Wilson (1957), Fay et al. (1958), Schoof and Kilpatrick (1958), and Hansens and Morris

(1962) showed house flies to be resistant to organic phosphate insecticides in the United States. Keiding (1965) showed in Denmark that house flies were three to seven times resistant to topical application of DDVP and 11 to 16 times to DDVP vapor.

Where house flies are resistant, each sex is resistant (Wilson and Gahan, 1948). R. A. Harrison (1951) noted that female DDT-resistant house flies were, in general, much more resistant to benzene hexachloride and chlordane than the males.

There apparently is a divergence of opinion among entomologists as to whether or not resistant house flies lose their resistance when they are not exposed to chlorinated insecticides over a number of generations in the laboratory. Two laboratories reported no loss in DDT resistance over 30 and 35 generations respectively when not exposed to DDT. However, several other laboratories have demonstrated that flies maintained under similar conditions showed a definite loss in resistance over 10 or 11 generations. Upholt (1950) stated although the house fly may lose its resistance under laboratory conditions over 10 or 11 generations, this fact is of little practical significance in the field since the flies were continuously exposed to chlorinated insecticides and were becoming increasingly difficult to control with any of these residual sprays. Georghiou and Bowen (1966) showed in California there was still a high level of DDT resistance after more than 10 years of non-use. Mallis and Miller (1967) maintained a wild strain of house flies that were still resistant to DDT after five years of non-exposure.

According to Pimentel et al. (1950), larvae from an area where DDT-resistant flies occurred were also resistant to DDT. Bruce and Decker (1950) noted the use of chlorinated insecticides as larvicides was in part responsible for the rapid development of resistant flies in an area. According to Rao (1957), larval medium with two to five ppm of DDT permitted flies to retain a high level of resistance after 10 generations. DDT resistance decreased the percentage of pupation of the flies.

Pimentel et al. (1950 and 1951) showed in the laboratory the larval periods for DDT-resistant house flies were longer than those of the nonresistant strain. Harrison (1952) noted four to five day old flies were less susceptible to DDT than older flies.

Physiologists and toxicologists are investigating the reasons for the development of resistance in house flies and other insects. Apparently the more light that is shed on the subject, the more complex it appears to be. Wiesmann (1947) believed the thickness of the cuticle may have prevented DDT from penetrating into the body of the insect. However, March and Metcalf (1949) and Sternburg et al. (1950) showed DDT-resistant flies tolerate large injected dosages and this showed fly resistance was not due to lack of penetration by DDT. March and Lewallen (1950) demonstrated thickness of cuticle and general vigor of flies are not of consequence in DDT-resistance. Fine et al. (1967) did not agree entirely with the previous authors, for in their opinion part of the resistance of flies to pyrethrins that are also highly resistant to DDT is the poor penetration of Pyrethrins I, possibly due to thicker fly cuticle in the DDT-resistant flies. Sternburg et al. (1950) and Perry and Hoskins (1950) showed DDT-resistant flies "change a large portion of the absorbed DDT into the relatively harmless ethylene derivative, DDE." Barker (1957) found the total degradation of DDT in treated flies was unchanged by age. Sternburg et al. (1953) demonstrated the presence of enzyme systems which rapidly change DDT to DDE. This same enzyme system was not found in normal CSMA flies. According to Perry and Hoskins (1950), the synergist piperonyl cyclonene inhibits DDE formation and increases the toxic effects of DDT. However, surviving DDT-resistant flies still have much unchanged DDT. In fact, an amount sufficient to kill non-resistant flies. Perry and Hoskins (1951) noted piperonyl cyclonene also acts as a synergist with TDE and methoxychlor on DDT-resistant flies. Sacktor (1950) showed DDT-resistant flies have greater cytochrome oxidase activity than normal flies. According to Fullmer and Hoskins (1951), susceptible flies on exposure to DDT have a greater respiration rate than DDT-resistant flies. Absorption of DDT in certain areas increases respiration and detoxification of DDT reduces it. Babers (1949) and Babers and Pratt (1951) reviewed the problem of resistant flies in some detail.

Harrison (1951) noted the knockdown of resistant flies is apparently inherited on a Mendelian basis. Bruce and Decker (1951) stated both sexes carry the DDT-resistant characters and crosses produce what could be called physiological blends. March and Metcalf (1951) noted there appears to be no clear-cut evidence at this time of a simple Mendelian inheritance pattern based on mortality. Pimentel et al. (1954) stated, "Rigid selection and inbreeding of a highly DDT-resistant strain of flies did not result in a homogeneous population." DDT-resistance disappeared when the resistant flies were not exposed to DDT for 20 generations.

Factors to be considered in the use of residual sprays. It is only natural that the previous discussion of resistant flies should lead us to be pessimistic regarding the use of residual insecticides. In fact, we may well ask why discuss them at all as a practical control measure? There are several reasons for our continued interest in the subject.

- DDT was once known as the "miracle" insecticide and we can hope that another "miracle" insecticide to which flies are not resistant may yet make its appearance.
- Prolonged control of resistant flies with parathion and other organic phosphates showed the responses of resistant flies differ radically with different toxicants, according to March and Metcalf (1952).
- In some sections of the country where the flies are not too resistant, some control can be obtained by repeating applications several times during the fly season.
- Synergists used with the chlorinated insecticides have at times been effective against resistant flies although eventually the flies appear to become resistant to such combinations. Antiresistant DDT, a combination of DDT and N,N,-di-n-butyl-p-chlorobenzene sulfonamide, supposedly interferes with the resistance mechanism in flies and other insects.

Factors of importance in fly control. Scudder (1949) reviewed some of the factors that are of importance in fly control, particularly as they apply to the successful application of residual sprays. Flies breed only in moist media, and "moist rich manure, stock feed, restaurant scraps, offal, or other material with a high protein or available carbohydrate content will turn out a surprising number of flies, often several hundred per pound.

"House flies are very domiciliary, for they tend to stay within a restricted locale and to go inside buildings, especially to rest there overnight in cool weather. They leave their home areas only by accident or by being forced out through the competition of their own numbers. For this reason, house fly control may be effective on as small a scale as a single farm residence or business."

According to Scudder, flies are inactive at night and will rest for several hours on one spot and thus are susceptible to residual insecticides because of the prolonged contact time.

"Studies of resting flies in various areas at night have revealed an interesting variety of preferences. At night during warm summer weather, house flies as well as blow flies rest outdoors, sharing similar resting places near their major daytime centers of activity about garbage, manure, privies, stables and kitchen entries. In cooler weather, the house flies tend to rest inside buildings on ceilings and other overhead structures, while the blow flies continue to choose outside resting places.

"Where residual sprays are applied, the emphasis should be on the treatment of overhead structures, such as light wires, cords, edges of beams and other woodwork, and all irregularities in the ceiling composition.

"House flies are so very gregarious during their oviposition that there is a definite tendency for a group of egg-laying females to infest certain chosen points as their breeding sites to the fullest, disregarding certain other adjacent areas which appear equally suitable."

Where the fly population is heavy "moist dirty linen from restaurant tables, scraps of food hidden in boxes in restaurant storerooms, collections of food trash on the floor, and soil damp with garbage liquors have been found to create phenomenal fly populations, the source of which is often very difficult to discover."

Scudder made the following observations regarding fly control in restaurants: "Any garbage station that is not kept clean and is irregularly serviced will be a prime source of flies, but an apparently well-kept garbage station may also be a source of a great number of flies, even though all litter be removed and the area washed daily. If liquid food wastes, or wash water containing food, is allowed to seep into soil, whether between bricks of a street, under platforms, or under a few inches of gravel, requisite conditions for fly breeding are set up.

"It has been observed that restaurants routinely keeping their garbage inside, even uncovered, have fewer flies than those that place it in the alley and use the alley door frequently. With proper handling of alleys by both the restaurateurs and the municipal disposal services, there would be few fly problems in restaurants.

"Ventilation systems may produce such a strong negative pressure at the back door that every time an employee uses this entrance, the flies are literally blown in from the alleyways even in spite of a double door system.

"Placing of intake fans on a level with the street is not advised as many flies may be drawn in with the air stream.

"Air-conditioning systems often keep out flies during their operation simply because the restaurant doors are then allowed to remain closed and the cool dining room is no longer as attractive to the flies; however, the kitchen of such a restaurant may continue to be as infested as ever."

Sprayers and spray application. The homeowner ordinarily uses handsprayers for space application. Such handsprayers usually vary from one cup to one gallon/3.8 l capacity, the cup and quart/0.9 l sizes being the most popular. Many of these sprayers will deliver approximately one fluid once/28 ml of spray in 50 *complete* strokes (push and pull) of the spray handle. Such sprayers are suitable for space application in small areas such as single rooms. Most of these small hand sprayers can be used only with oil base household insecticides. Water base materials, if left in the sprayer, quickly cause the sprayer to rust. Most hand sprayers are not suitable for the application of residual sprays because of their low delivery rate. Also, after a limited number of strokes, this type of sprayer tires the hand and very often leaks badly. One basic reason handsprayers have a poor reputation is the consumer ordinarily will not purchase the higher quality, more expensive models.

It can be observed from the above remarks that the usual type handsprayer is unsuitable for the application of residual sprays. Special handsprayers with a delivery rate of six to nine fluid ounces/170 to 255 ml per 50 complete strokes of the spray handle have been marketed. These are definitely superior to the usual type space sprayer for residual or wet applications. However, since they are slightly more expensive than a space sprayer, they have not been accepted by the consumer and are now practically withdrawn from the market. Therefore, the application of residual sprays by the homeowner is either limited to the usual space-type handsprayer or to a paint brush. In some instances, the adjustable sprayers on vacuum cleaners may be used.

The professional pest control operator utilizes a great variety of spray equipment for the application of space sprays. Some of the space sprayers vary from the electric sprayers of one quart/0.9 l capacity to the small mechanical aerosol generators that use centrifugal force of a whirling disc or compressed air and a special nozzle, to the large insecticide fog generators.

There are also many types of equipment for the application of residual sprays. Several companies make special handsprayers that can be switched from a wet residual spray to a space spray by flicking a nozzle pin or turning a nozzle cap. Those used by the pest control operator usually have non-removable galvanized reservoirs with approximately three quart to two gallon/1.4 to 7.6 l capacity.

There are also on the market a number of hand-pumped compressed-air galvanized and stainless steel tanks up to 3.5 gallon/13 l capacity equipped with a special oil-resistant rubber hose that are widely used for spraying surfaces, particularly small indoor areas. These sprayers have rods or wands of varying lengths which are provided with special nozzles that form a fantype spray, especially suited for the application of residual sprays. Many of the sprayers are provided with pressure gauges.

Knote (1949) described the wands that are often used with these sprayers by the pest control operator: "One unit equipped with a two-foot wand and angular nozzle body was used to treat walls. For ceilings, a separate unit with a straight nozzle body and a four-foot and two-foot wand was used for extra high ceilings. These wands are made of ⅜ inch O.D. hard tempered brass tubing threaded for ⅛ inch pipe threads."

The nozzle is an extremely important component of the residual sprayer of the type described above. Tee-jet nozzles that form a fan-type spray are manufactured by the Spraying Systems Company.

Henry (1958), discussing fly control in supermarkets, noted: "As for nozzles, we prefer the Spraying System Series 8001, 8002, 8004, and 8006. We carry all four on a fly job. The 8001 and 8002 sizes appear to be most practical for carefully applying a fine spray on painted interior surfaces where staining or runs could be a problem. For exterior applications, nozzles with larger openings, such as the 8006, are most desirable."

Several pest control supply companies also have electrically operated residual applicators which are suitable for fly application as well as for moth and carpet beetle control. These applicators have many feet of oil-resistant hose, intake hose that can be placed in any type of container, and an electrically operated pump.

Some operators use a variety of high-pressure power sprayers for large outdoor application of residuals. Knote (1949) noted that a type that is quite suitable for the pest control operator is one that has a tank capacity of 30 to 50 gallons/113 to 189 l, an adjustable pressure from zero to 150 pounds and a pumping capacity of three gallons/11.4 l per minute. Henry (1958) recommended a non-pulsating compressed air sprayer having a range of 80 to 110 pounds pressure.

Nelson (1949) and other members of a symposium on sprayer maintenance emphasized the importance of cleaning sprayers immediately after use.

Knote (1949) discussed residual applications as follows: "In residual applications, the types of surfaces affect the amount of insecticide to be applied. The absorptive power of unfinished wood, brick, accoustical tile, plastered surfaces, surfaces covered with wallpaper, metal surfaces and painted surfaces all differ. Each type of surface should require a definite quantity as well as type of insecticide (wettable powder, emulsion, or oil solution) to give maximum control without objectionable visible stains or streaks. As a possible guide in the absorptive power of these different surfaces, research workers for the U.S. Public Health Service found that in treating poplar wood with five percent DDT emulsion using a xylene solvent and five percent DDT oil solution, that only 38 percent and 30 percent respectively of the DDT applied was recovered from the top 0.001 inch of the wood which is considered to contain all the toxicant which will have insecticidal action. When surfaces are treated with emulsions, penetration seems to be related to the volatility of the solvent. The higher the volatility, the smaller the amount of penetration. With these differences in surfaces, it is difficult to set up a standard rate of application. PCOs will find when using a five percent DDT or two percent chlordane solution that wetting a surface to a point just short of run-off will give a satisfactory rate of application.

"In the application of residuals the pressure requirements of equipment are quite low. Nozzle pressures not exceeding 40 to 60 pounds per square inch will produce less misting of insecticide and consequently less possibility of direct contamination of food."

According to Knote, foods and surfaces can be protected from spray mists in the following way: "Possibly one of the most important pieces of fly control equipment, yet one that has received the least emphasis, is the drop cloth. One PCO found that heavy unbleached muslin in cloths of 72" × 72" was an ideal size to use. One side of this cloth should be stenciled "THIS SIDE UP" and the service man instructed to fold the other side in so that the possibility of contamination would be cut to a minimum. These cloths should receive frequent washings to eliminate the danger of materials rubbing or flaking directly or indirectly into food."

Baits. Before World War II, poisoned baits often were used for the control of house flies and a popular one contained formaldehyde mixed with milk and water. Because of resistance, baits made a "come-back" but LaBrecque and Wilson (1961) showed house fly resistance to baits containing malathion and trichlorfon.

Thompson et al. (1953) used 0.15 percent lindane plus TEPP in corn syrup. Gahan et al. (1954) employed 0.1 percent malathion or diazinon in 10 percent molasses plus water. Kilpatrick and Schoof (1955) used 0.1 percent DDVP. Keller et al. (1956) employed two percent malathion, Smith et al. (1960) one percent dimetilan and Howell (1961) dibrom and DDVP in various baits.

The USDA (Anon., 1955) recommended baits be made using either malathion or diazinon. A liquid bait can be prepared with ¾ pint/340 ml of either molasses syrup or sugar using one tablespoon of 25 percent emulsifiable concentrate or two tablespoonfuls of 25 percent wettable powder. These ingredients are added to water to make one gallon/3.8 l of finished liquid bait. This bait is distributed at the rate of one gallon to 1000 sq. ft./3.8 l to 93 sq. meters.

USDA (Anon., 1955) recommends dry baits and cornmeal baits be prepared as follows:

- Dry-bait. "To make a dry sugar bait, take one pound of granulated sugar and add three tablespoonfuls of a 25-percent wettable powder malathion. Add about ¼ teaspoonful of lampblack to color the sugar. Stir with a paddle until all the grains of sugar are coated with the poison and colored a dirty gray. Coloring the poisoned sugar with lampblack prevents mistaking it for ordinary sugar."
- Cornmeal bait. "A cornmeal bait is recommended for use on moist surfaces where a dry sugar bait would dissolve. While stirring one pound of coarsely ground cornmeal, slowly add the following: one tablespoonful of peanut oil; six tablespoonfuls of a 25 percent wettable powder such as malathion; two ounces of powdered sugar. Stir with a paddle until all the meal particles are coated with the sugar and the wettable powder. Five minutes' stirring insures proper mixing of quantities of one to five pounds. Mixing of larger quantities by hand is not recommended." Distribution of baits must follow current regulations and preparations of baits must be permitted by labeling.

MEASURES TO CONTROL FLY BREEDING

The average homeowner can help reduce the amount of fly breeding by having a garbage can with a tight lid. Spraying the inside and outside of the can, as well as the immediate vicinity of the garbage can with malathion insecticide, is an effective measure.

Another remedy to keep fly maggots out of a garbage can is to suspend paradichlorobenzene (PDB) crystals in a cloth bag from the cover of the garbage can (Quarterman and Mathis, 1952). Small pieces of DDVP resin strips attached to the inside of the cover also help reduce fly infestation of garbage.

Often, veritable scourges of flies appear in the home and very often this may be due to fresh manure used for lawn fertilizers. This source of annoyance will for the most part be temporary since once the manure dries, the flies cease breeding in it.

In rural areas, fresh animal manure is one of the chief sources of fly breeding. According to Bishopp and Henderson (1946), manure can be handled in several ways to reduce fly breeding:

"1) **Scattering manure on fields.** By spreading the manure thinly, the fly eggs and young maggots that are present are killed by heat, cold, or drying.

2) **Storing manure in boxes or pits.** These preferably should be made of concrete and fitted with tight lids which will reduce access of flies to manure.

3) **Compact ricking of manure.** Rectangular ricks several feet wider than the wagon are compacted by driving a wagon on top. The sides are pounded flat with shovels. The heat generated in the manure destroys many of the maggots and drives the rest to the surface, where they may be killed by applying borax or by sprinkling the edge of the pile and adjacent soil heavily with used crankcase oil, or better, with crude petroleum.

4) **Chemical control.** The chemical control of fly larvae is a separate prob-

lem that cannot be handled in the same fashion as the control of adult flies. In general, fly larvae do not appear to be susceptible to the same degree to insecticides commonly used in adult fly control. Larval toxicants are used to poison the substrate in which the larvae live, and from the wide variety of insecticides employed, it is possible these act as stomach, contact, or fumigant poisons, and in some instances, as a combination of these. Paradichlorobenzene (PDB) and orthodichlorobenzene (ODB) have found wide usage in the control of house fly and other larvae." Klassen and Williams (1951) observed PDB crystals suspended in closed garbage cans to kill fly larvae. Manis et al. (1942) showed PDB was effective against third-instar fly larvae at the rate of approximately 2.5 to 6.25 g per 450 g of material in open and closed garbage pails. They claimed the PDB acts as both a fumigant and a stomach poison. McDuffie et al. (1946) observed PDB and ODB to give good kill of larvae and fly eggs at the rate of 10 to 20 g PDB and 15 ml ODB per sq. ft./0.09 sq meters. It may also have a repellent effect on flies and it is capable of destroying adult flies that enter the treated pit.

Kilpatrick and Schoof (1956) noted since untreated privies produce few house flies, chemical treatment should be avoided. BHC, chlordane, aldrin and dieldrin, especially increase house fly production by killing the other fly species that usually breed in privies.

Sanders (1942) greatly reduced the number of flies breeding in the grounds of a racetrack. He treated the soil upon which manure was usually piled with one pound/0.45 kg of sodium fluosilicate for each square yard/0.8 sq. meters of ground. Midgely et al. (1943) prevented fly breeding by applying 2.5 to three lbs/1.1 to 1.4 kg boric acid in the bottom of a cleaned barn gutter. Bishopp and Henderson (1946) recommended the following chemical treatment of manure: "For each eight bushels of manure, one-half pound of fresh hellebore is stirred into 10 gallons of water; after standing 24 hours, it is sprinkled over the pile with a watering pot. Borax is used at the rate of 11 ounces to each eight bushels of manure."

Bishopp and Henderson also noted flies "often breed in accumulations in corners of feed boxes and mangers as well as soft dirt floors and between floor planks." Excreta in outside privies should be treated by "scattering enough borax over the excreta every three or four days to make it white."

Starr and Calsetta (1954) found a five percent suspension of ryania to be effective as a larvicide.

H. Katz controlled larvae and adult house flies in a garbage receiving bin in an apartment house using a silica aerogel.

MISCELLANEOUS METHODS OF CONTROL

Many research personnel by 1981 report strains of the house fly that show rapidly developing resistance to many of the insecticides that are still recommended for fly control. Many physiological mechanisms have been shown to be involved and all of these must be considered in the development and use of insecticides.

Liquid insecticides have largely superseded such fly control remedies as the sticky fly papers and ribbons, and the formaldehyde poison bait formula.

One formula for making sticky fly paper is as follows:

Castor oil	5 parts
Powdered resin	8 parts

These two ingredients are heated together until the resin is dissolved. They should not be boiled together. The material once prepared can be stored indefinitely in cans but is always heated before being used in order that it may be hot when applied. This mixture is spread *in as thin a coat as possible* to paper, preferably coated paper. Recently, sticky tubes which feature a light reflecting surface and pictorial fly decoys have been marketed and they are claimed to be more effective than conventional fly paper.

Quite a number of chemicals have been used in liquid baits to control flies. However, one of the safest is the formaldehyde formula. Two or three teaspoonfuls of the concentrated (40%) formaldehyde are mixed with a pint/0.47 l of water and milk mixture, the latter two in equal parts. To this are added two or three teaspoonfuls of sugar. This is then poured onto a piece of bread in a saucer and the saucer is placed in locations frequented by flies, as on the window sills. Such a bread formulation may be dangerous to use where children or pets are present and extreme caution is mandatory.

Elmore and Richardson (1936) noted the formalin should not have a marked acid reaction. They recommended another method of exposure whereby a glass tumbler is filled nearly full with the above solution. Then a circular piece of blotting paper which has a greater diameter than the tumbler is placed upon the top of the tumbler. Then, upon the blotting paper is placed an inverted dish. When the device is turned upside down, a small matchstick is inserted under the edge of the tumbler in order to admit air. The liquid bait is imbibed by the flies from the moist blotting paper.

A number of electrical devices have been successfully used for the control of flies. One of these is an electric fly screen which consists of metal bars that are spaced $9/32$ inch/7.1 mm apart. The screen is energized by a transformer and the flies are electrocuted when they attempt to fly between the bars since the moisture in the body of the insect completes the electric circuit. This electric fly screen can be inserted into a door in place of the usual screen. A portable electric fly trap which employs the same principle as the electrical fly screen has also found some usage. A bait is placed inside the fly trap and the dead flies are removed by means of a slide. The electric fly trap is preferably used immediately outside locations that cannot be protected by screening.

Zich (1944) described a method whereby air currents covering door openings are utilized to keep flies out of a dairy bottling room. A vent opening downwards all the way across was installed over each door. The blast of air from this vent prevented the flies from entering the room. Senn (1966) stated fly fans over the door must have a velocity of 1,700 feet/518 m per minute. Mathis et al. (1967) found 20 percent of flies penetrated an air curtain with a velocity of 1,700 ft/ 518 m per minute.

Bishopp (1937) and Bishopp and Henderson (1946) recommended the use of a fly trap in addition to the usual fly control measures for the reduction of flies in an area. According to Bishopp and Henderson, the conical trap is an effective fly trap. It is a 12 to 18 inches/30 to 47 cm in diameter cone with sides and top made of screen and with a screen cone inside reaching nearly to the top. The frame may be made of barrel hoops, and laths and the legs should be about one inch/2.5 cm long. (Fig. 19-8). "The bait should be placed beneath the trap in a shallow pan about four inches less in diameter than the base of the cone and one inch/2.5 cm deep. Any substance attractive to the house fly may be used as bait. A mixture of blackstrap molasses (one part) and water (three parts) makes a convenient and attractive bait. Milk and fruit waste also may be employed.

Fig. 19-8. Bishopp fly trap.

The traps should be set where flies naturally congregate. This is usually on the sunny side of a building (except in very hot weather) and out of the wind. The bait pan should be kept well filled and should be washed out occasionally. The catch is reduced when the flies become piled more than a fourth of the way up the cone. At such times the trap should be emptied."

LaBrecque (1961) showed the chemosterilants, aphoxide, aphomide and apholate to be effective in sterilizing male house flies. According to LaBrecque et al. (1962), aphoxide (tris(l-aziridinyl)phosphine oxide) used at 0.5 percent in cornmeal baits on a refuse dump in the Florida Keys drastically reduced house flies in four weeks. LaBrecque et al. (1963) showed that 0.5 percent metepa baits reduced hatching rate of fly eggs below 10 percent.

Brundrett (1953) described a homemade fly deterrent as follows: "It has become commonplace in some parts of the country to see balls of cotton stuck on screen doors." According to Gross (1949), this practice may be explained by the following ditty:

> "Paste a wad of cotton on the
> middle of the door
> And those doggone flies'll pester
> you no more."

Fig. 19-9. Spraying fairgrounds for fly control with a pyrethrum-synergist mixture by means of a mobile fog machine unit. This was done early in the morning before the grounds were opened to the public.

Gross (1948) and Wall and Bailey (1952) have shown such cotton balls have no practical value in repelling flies.

Wright (1975) indicated the insect growth regulator diflubenzuron (1-(4-chlorophenyl)-3-(2,6-difluorobenzoyl) urea) formulated in mineral blocks greatly reduced the number of house flies developing in the feces of cattle consuming the salt blocks.

OTHER FLIES IN THE HOME AND THEIR CONTROL
THE STABLE FLY

Stomoxys calcitrans (L.)

The stable fly is distributed throughout Europe and North America. At times it may invade the home and may then be confused with the house fly, which it superficially resembles. Its bite, however, immediately distinguishes it from the house fly, for the stable fly has a bayonet-like proboscis with which it pierces the skin. For this reason it is often referred to as the "biting house fly." Richardson (1936) presented the following characteristics whereby the stable fly can be distinguished from the house fly: "A slender beak projects forward from the lower part of the head; there is a prominent light-colored area between the longitudinal bands in the middle of the thorax. The upper surface of the abdomen bears a number of nearly round dark spots; when at rest, the wings are held at an angle to the body instead of projecting nearly straight backward as in the house fly."

Patton (1931) noted this fly at times becomes very annoying, especially in warm rooms when it bites through socks or stockings in the region of the ankle. "The bite is like a sharp stab with a darning needle."

It is called the stable fly because this is an environment in which it is frequently encountered. It is also found in the field, especially where cattle and horses occur. The fly is commonly seen basking in the sun on light-colored walls, fences, etc., when the temperature is moderate. During hot weather it is more frequently in the shade.

According to Hansens (1951), this fly is also known as "the beach fly or dog fly-beach flies because they are found in large numbers in seashore areas and dog flies because they annoy dogs and are most abundant in the dog days."

Both the male and female are bloodsucking, and they are important pests of domestic animals, particularly horses and cattle. They worry the animals continuously, weakening them by sucking their blood, and a number of workers have shown them to be responsible for reduced milk production in dairy cows. Stable flies probably act as vectors of disease, especially of anthrax.

Their favorite breeding place is in stacks of straw, but they readily breed in such material as decaying hay, alfalfa, grain, onions, leeks, chicken manure, lawn clippings, seashore vegetation, etc. The maggots very often may be encountered in feeding troughs.

Simmons (1944) and Simmons and Dove (1942 and 1945) concerned themselves with the stable fly or "dog fly" in Florida along the Gulf Coast where it often becomes a serious pest. The flies breed on bay grass on the shore as well as in waste celery and peanut litter.

Bishopp (1931 and 1939) and Simmons (1944) studied its life history in some detail. The ovoid, cream-colored eggs are about $1/125$ inch/0.2 mm long and hatch in one to three days after being deposited. The yellowish white maggots complete their larval growth in 11 to 30 days. The larva passes through three instars and pupates in the last larval skin which hardens to become a puparium or pupal case. The pupal stage is completed in six to 20 days in warm weather. The life history from the deposition of the egg to the emergence of the adult is completed in 21 to 25 days, although it may vary from 13 days to several months

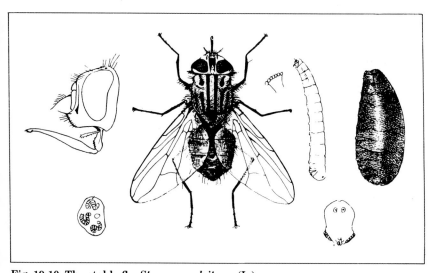

Fig. 19-10. The stable fly, *Stomoxys calcitrans* (L.).

depending on the temperature and other conditions. According to Harris et al. (1966), the female mates only once. Hansens (1951) observed the heaviest breeding on New Jersey seashores to occur at temperatures of 68 to 78° F/20° to 25° C. The average adult lives for 20 days and is most common during the summer and autumn. In northwestern Florida the third instar larvae overwintered in peanut litter and the flies emerged in warm periods. Simmons (1944) noted the feeding stable flies often punctured the skin of the host several times and became engorged in 1.5 to eight minutes, and under semi-field conditions these flies fed more readily on human beings early in the morning and late in the afternoon, and became sluggish during the heat of mid-day. Richardson (1936), discussing the behavior of the fly in Iowa, stated it frequently enters the home, particularly when the sky is overcast, where it causes much annoyance to the inhabitants. When unmolested it can imbibe more than its weight of blood at one feeding.

Hansens (1951) studied this fly as a seashore pest in New Jersey. The flies appeared within a few minutes when the wind shifted to the west and they disappeared quickly when the wind became unfavorable. The stable fly especially was attracted to dark colors and prone to attack individuals dressed in such colors. The vegetation on the bayshore was the breeding source. Seaweed out of the water for a day or two was more attractive for breeding purposes than seaweed out of the water for two or three weeks.

Campau et al. (1952) developed a technique for rearing this fly in the laboratory. The adults were fed citrated bovine blood since blood is essential for egg production. At a temperature of 80° to 84° F/26.7° to 28.9° C and a relative humidity of 50 percent ± five percent the life cycle is:

Days after Seeding Eggs

Appearance of first pupae 9.0
Emergence of first adult 14.5
Peak emergence 17.5
First egg laying 21.5

Control. In the home, this fly may be controlled by methods recommended for the house fly. Materials in which the stable fly breeds, such as straw, grass-clippings, etc., either should be burned or spread thinly since the maggots are susceptible to desiccation. Simmons and Dove (1942 and 1945) sprayed the bay grass on the shore with equal parts of a light fraction of gas condensate (a by-product of manufacture of cooking and heating gas) and a No. 2 fuel oil. Then they used a combination of one part creosote to three or four parts of diesel oil, and finally a creosote-water emulsion. These are effective against the eggs, larvae, and pupae.

THE LITTLE HOUSE FLIES

Fannia spp.

The five main species of these flies are quite similar in appearance and habitat to the house fly, but differ in being appreciably smaller and having distinctive wing venation. The males are commonly found inside buildings, particularly those that are air conditioned, when temperatures outside go above 80° F/26.7° C. They fly about in the rooms for extended periods of time and seldom rest, while the females rarely enter buildings and typically fly for only short periods of time.

Eggs are usually deposited on decaying vegetable matter or excrement of several species of animals. The eggs hatch in 24 to 36 hours and the resulting larvae appear as flattened elongate-ovate organisms with lateral protuberances which are feathered in *Fannia scalaris* (F.). The other species are not as clearly marked.

These flies may cause intestinal myiasis in man and possibly may transfer pathogenic organisms.

Smart (1943) noted the flat spiny maggots "are very characteristic and are to be found in the dried parts of manure heaps and other places where there is decaying organic matter. They are often found in the soil in hen-runs or in the urine-impregnated sawdust or other litter in rabbit hutches or the cages of laboratory animals; they may also be found under the wrappings of cheeses, bacon, etc., and feeding on dried fish and other forms of stored protein foods of animal origin."

Hewitt (1914) found the little house fly to commonly breed in human excrement.

Lewallen (1954) studied the life history of this fly at 80° F/26.7° C and 65 percent relative humidity. The egg stage requires 36 to 40 hours, the larval stage eight to 10 days, and the pupal stage eight to 12 days. The complete life cycle is 24 to 29 days. Steve (1960) had results somewhat similar to the above.

The latrine fly, *Fannia scalaris* (F.), is very similar in appearance and life history to the lesser house fly and shows a preference for human excrement. It is known to cause intestinal myiasis in man. Hewitt (1914) showed how to separate this species from *F. canicularis* by the rounded protuberance of the median point of the middle pair of legs on *F. scalaris*.

Ogden and Kilpatrick (1958) controlled the little house fly in dairy barns using parathion/diazinon-coated fly cords as discussed under house flies. Males flying in rooms may be controlled by pyrethrum and pyrethroid aerosols.

Hansens (1963) controlled this fly around a poultry farm by using a mist blower. He applied a spray consisting of eight gallons/30 l of DDT 25E and two quarts/1.9 l of malathion per 100 gallons/378 l of spray. Stone and Brydon (1965) used diazinon dust and gypsum which they stirred in poultry manure.

THE FALSE STABLE FLY

Muscina stabulans (Fallen)

Richardson (1936), James (1947) and Ware (1966) have published excellent descriptions of the biology and morphology of the false stable fly, *Muscina stabulans* (Fallen), and the closely related *Muscina assimilis* (Fallen). They pointed out that the false stable fly, which resembles the house fly in general appearance, is somewhat larger and the third wing vein which is bent slightly upward does not meet the second vein. (Fig. 19-11.)

The eggs are deposited on many kinds of feces, decaying vegetation and similar sites. The larvae may also be predacious and at times may cause intestinal myiasis in man.

Adults may enter dwellings and deposit their eggs in raw or cooked meats. Milk and related foods also may be contaminated. The flies may be found in the adult stage year round in the warmer states. Each female may deposit from 140 to 200 eggs which she scatters over the substrate.

James (1947) presented the following notes on its life history: "The females frequently enter houses and may oviposit on foods, particularly those that are

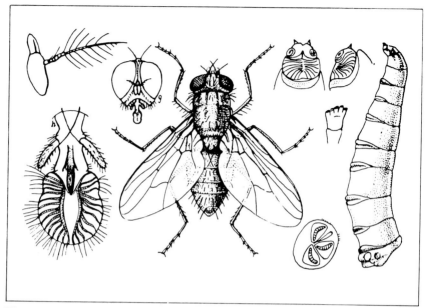

Fig. 19-11. Adult, larvae and anatomical detail of the false stable fly, *Muscina stabulans* (Fall.).

slightly tainted. It is probably in this way that man occasionally becomes parasitized. The first larval stage is of very brief duration, the second somewhat longer and the third, in which the carnivorous habit becomes the most highly developed, is by far the longest. The length of the larval life is from 15 to 25 days; higher temperatures and an abundance of animal food will make it tend toward the lower figure. Several generations may be produced in a summer. Normally hibernation takes place in the pupal stage, although larvae produced late in the season, if they survive, may remain dormant over the winter. A similar period of semi-dormancy may explain some of the protracted cases of intestinal parasitism in man."

Ware (1966) found this fly and the stable fly breeding in matted clippings lodged in power mowers.

Adequate exclusion by good screening will provide good control of the false stable fly.

THE BLOWFLIES, GREENBOTTLE FLIES, BLUEBOTTLE FLIES

Family Calliphoridae

These are the common large flies with a metallic blue or green sheen that at times make a buzzing sound, especially when flying up and down in a window. The blowflies are primarily scavengers and deposit their eggs on meat and fish, particularly if these are in decaying condition, on dead animals, in garbage and in fecal material. Entomologists in the USPHS in Savannah, Georgia, showed the stools of dogs were important breeding places of blowflies. At times the eggs may be oviposited in wounds of animals and man, and when the larvae develop, they cause the painful and often serious condition known as "myiasis." Norris (1965) reviewed the biology of blowflies in some detail.

The two common bluebottle flies are *Calliphora vomitoria* (L.) and *Calliphora vicina* R. & D., formerly *erythrocephala* (Meig.). These flies are about ²/₅ inch/10 mm long and have a metallic blue abdomen. There are also two common greenbottle flies, *Lucilia caesar* (L.) and *Phaenicia sericata* (Meig.). (Fig. 19-13). The latter was in the genus *Lucilia* until a recent change in nomenclature. They are a metallic blue/green in color. Dodge (1954) prepared a key to these and other common flies. Richardson (1936) stated in Iowa, *C. vicina* R. & D. appears the latter part of March and is not seen after December 1 and the winter is passed in the larval, pupal and adult stages. Hagmann and Barber

Fig. 19-12. Greenbottle fly, *Phaenicia sericata* (Meig.).

(1948) and Green (1951) noted the larvae of *Phaenicia sericata* (Meig.) hibernate in the soil of New Jersey and pupate a short while before emerging. *Cyanomyia cadaverina* (Desv.) has similar habits to those of the bluebottle flies and in Iowa is most common in the spring and autumn.

Where these flies occur in unusual numbers in the home, they may be breeding either in the house proper or in the immediate vicinity, although Lindquist, Yates, et al. (1951) captured *P. sericata* as far as four miles/6.4 km from the point of release.

Davis (1944) recorded the occurrence of *Phaenicia* all through a house due to the presence of dead birds in a register in the ceiling used for ventilation purposes. In 1953 the same author found dead birds in the chimney responsible for maggots throughout the home. Morse (1911) found the larvae migrating over oriental rugs, the source of the infestation being a dead squirrel in the chimney. Griswold (1942) noted the presence of a large number of maggots of *Phaenicia sericata* in a bedroom. They fell from the ceiling onto a bed. At a later date adult flies were present in this room. Although the source of infestation was not found, she notes: "It seems probable, however, that the maggots were feeding on a mouse or a squirrel that had died in the space between the ceiling of the bedroom and the floor of the room above. It has been suggested that the animal might have been injured and that eggs might have been deposited on the wound before the animal went into the wall to die." Fleischer (1957) found greenbottle flies entering a gas flue which was leaking gas. The onion smell of the escaping gas attracted the flies into the flue, whereby they entered the house. In another case the presence of blowfly maggots in a housing development was traced to larvae migrating from garbage pits. R. M. Bohart related an instance where some duck wings which had been packed away and completely forgotten served as a source for what at first appeared to be mysterious swarms of blowflies.

Careful screening of the chimneys of vacation cabins in recreation areas is often necessary to prevent their invasion during the winter months when the buildings are less frequently used.

Richardson (1936) found *Phormia regina* (Meig.) to be one of the most common blowflies in Iowa. He noted it is "present throughout the summer and venturing outdoors on warm days during the winter. It is slightly more robust in appearance than the green bottle fly and darker in color, being dark metallic green on the thorax and abdomen. The black blowfly will sometimes enter the house to lay its eggs on meat and fish or to feed upon milk and table refuse. It breeds in garbage, animal carcasses and offal. Complete development requires from about 10 days to more than three weeks." Dicke and Eastwood (1952) found this species to be the most common blowfly in Madison, Wisconsin.

Baker and Schoof (1955) outlined the useful procedures for protecting animal carcasses from fly blowing.

Control. Infested meat, either in the house or close by, must be found and destroyed by burning. Fly spray will adequately cope with the flies in the house. Several workers have shown blowflies are even more susceptible to the usual fly sprays than are house flies.

Spraying with the usual fly sprays is still effective in the control of these flies in some areas, but in others the flies are highly resistant. In these areas the pyrethrins with piperonyl butoxide, resmethrin, or malathion are most likely to succeed.

Strickland (1945) presented a novel means for ridding a screened house of blowflies: "Towards sunset, particularly when the nights are inclined to be cool, blowflies have the habit of squeezing themselves into small crevices, such as those around the edge of doors and flyscreens and in badly fitted eaves, and many thus find their way into screened homes. As they are attracted to light, they eventually reach the windows during daylight, where the screen prevents their escape. When crawling on the screens, they always move upwards, and having reached the upper edge, they crawl along it horizontally. If a pencil is pushed through the screen at one of the top corners, all blowflies will sooner or later leave through the hole, provided the window is partly open and the awnings, if any, are raised. Since the flies, when in the open, are inactive at dusk, none will return through the hole when the rooms are lighted. If it is feared mosquitoes will enter through the hole, though the author has never observed them to do so, a cork can be used to close it during the season when they are active. Several houses have been kept practically free from blowflies for a number of years by this method and it also has been used with excellent results to reduce a large population of blowflies in the buildings of a military camp."

THE CLUSTER FLY

Pollenia rudis (F.)

These flies are widely distributed in Europe, Canada, and the United States, except for the states bordering on the Gulf of Mexico. They enter houses in the fall, one by one, and may collect together like a swarm of bees. This insect is slightly larger than the house fly and has a nonmetallic, dark gray color. The thorax is without distinct stripes, but has characteristic golden hairs and the dark gray abdomen has irregular lighter patches. The wings overlap at the tips when not in use and the sharp bend in the longitudinal vein of the wing is characteristic. (Fig. 19-14). When crushed, this fly has an odor like buckwheat honey. Few insects give the housewife the "creeps" as these larger sluggish flies, appearing one-by-one in late fall or during the winter when the weather warms.

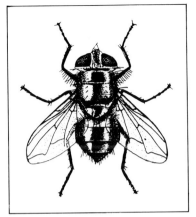 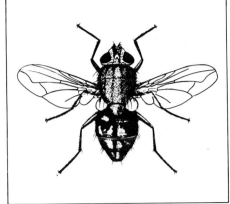

Fig. 19-13. Blow fly, *Phormia* sp.   **Fig. 19-14. Cluster fly, *Pollenia rudis* (F.).**

Herrick (1936) noted these flies normally live outdoors where they frequent flowers and fruits. With the approach of cool weather they enter homes to over-winter. Here they hide, often in a cluster, in nooks and dark corners, under clothing in closets, beneath curtains, in angles of the walls, behind pictures and furniture, in hats, under the edges of closets, etc. Austen (1920) stated in England "it frequently happens that swarms of cluster-flies, sometimes accompanied by large numbers of *Musca autumnalis,* take up their abode for the winter in the roofs of churches. On days on which the latter are warmed during winter and spring, the flies cause great annoyance by issuing from their hiding places and falling to the floor in a semi-comatose condition." These flies also swarm on windows on warm sunny days. They do no damage, but are annoying merely by being present. The author collected cluster flies on a south wall in March when the temperatures were in the fifties. The day was sunny and clear. DeCoursey (1927) noted in the spring "the flies often become a nuisance in houses. The increasing warmth of the spring days induces activity and if the hibernating quarters have been in the walls of a house or around the window frames, the flies may emerge inside the house instead of on the outside. Just as they become a nuisance in the fall while seeking hibernating quarters, they are also bothersome in the spring as they try to escape. Once inside the house, the flies crawl sluggishly over the walls and often drop into the food on the table. They are often rolled up under the windowblinds and leave unsightly greasy spots where they are crushed."

It was Keilin (1915) who first found the larvae were parasitic in the earth worms, *Allolobophora chlorotica* and *A. rosea,* and then investigated the life history of the fly in Europe. Webb and Hutchinson (1916) studied the life history of this fly in the United States and found that the eggs are laid singly in cracks in the soil. During the summer, the eggs hatch in three days and the young, emerging maggot can penetrate practically any part of an earthworm. The larval stage may last from 13 to 22 days and the pupal stage from 11 to 14 days, with a total development period of 27 to 39 days. There are four generations during the summer. Pimental and Epstein (1960) found the pupal period to be seven days at 80° F/26.7° C and 100 percent relative humidity.

Since the cluster flies breed in earthworms, no practical means has been found of directing control measures against their breeding place. In the house, the use of an ordinary vacuum cleaner, or the application of an effective fly spray, will often prove adequate. Spray it in the attic between the walls and, if possible, around the windows and all the places where these flies can hide. DDVP resin strips also have been recommended for the control of cluster flies. Rachesky (1972) sprayed an upper third of a house with one percent malathion in late summer or fall to prevent flies from entering. He noted the spray may leave white residue on dark houses.

In one case, the flies were prevented from entering the uppermost apartments of an apartment house by screening the openings to the penthouse on the roof. In another case, the problem of cluster flies issuing from the register of an air conditioner was solved by making a fresh air fiend downstairs close his window in this air-conditioned building. The author has on several occasions found cluster flies to enter offices by means of the air intake registers in air conditioners. Either close the outside of these registers with fly screen or use DDVP resin strips in the filters.

No control methods will be permanent unless the openings by which these flies enter the house are closed which often is an impossible job.

HORSEFLIES

Family Tabanidae

These flies, which are usually large and heavy-bodied, are important pests of domestic and wild animals and, at times, of man. They rarely occur indoors. Horseflies are known under a variety of names such as greenheads, deerflies, and others.

Only the females are bloodsucking, the males feeding for the most part on plant juices. The eggs are deposited on plant foliage, rocks, etc., usually in damp situations. Most of the larvae are aquatic or live in moist soil. The larvae often occur in the mud and at the bottom of ponds and ditches. Pupation usually occurs in the mud at the edge of these areas. Some of the larvae are predacious on immature insects and other small animals. The entire life cycle may require from approximately three months to as much as two years.

Gerry (1949) controlled the salt marsh greenhead fly, *Tabanus nigrovittatus* (Macquart), a pest of seashore recreational areas in Massachusetts, with DDT sprays. The adult flies emerge during June, July and August, breeding in salt marsh areas near the beach. Two treatments of DDT oil sprays, applied at the rate of one lb. DDT per acre/1.1 kg per ha provided effective control of this insect. Aluminum stearate when added to the solution at the rate of one percent by weight prolonged the residual effect of the spray. Hansens (1956) and Jamnback (1957) controlled these flies with one to 2.5 percent dieldrin granules. But under current regulations, DDT and most of the other chlorinated hydrocarbon insecticides that have been used against horseflies are no longer available in the United States.

Many other species of horse flies are very difficult to control because of their habits. Those which may spend weeks in the larval stage in dry soil may be controlled by drenching the soil with pyrethrins, but the cost is excessive. Those that live under rocks in running streams cannot be killed without danger to the environment and even the larvae which normally live in marshy areas are difficult to poison.

The pyrethrins and pyrethroids applied to the host animal may be effective repellents and toxicants, but the long flight range and resistance to most pesticides prevent good control. Fly traps are available that reduce the numbers of flies, but these seldom provide adequate control. Hansens (1981) found that resmethrin and permethrin provided good control of *Chrysops atlanticus* when applied at the rates of 0.015 to 0.112 kg/ha with a mist blower into the wooded barrier around cultivated fields.

THE FRUIT, VINEGAR, OR POMACE FLIES

Drosophila spp.

These small flies can pass through ordinary screening and are common in homes, restaurants, fruit markets, canneries and similar places. An authoritative volume on fruit flies was written by several specialists interested in genetics (Demerec, 1950).

Spencer (1950) noted that *D. funebris* (F.) is attracted to human and animal excrement and also will feed on fruits and uncooked food, and thus may act as a vector of disease. Dove (1937) noted *D. funebris* larvae are at times responsible for human intestinal myiasis. Patton (1931) observed a form of diarrhea, *"vin cochylise,"* is common among workers in vineyards and has been attributed to contamination of grapes by the fruit fly. The genus contains a large number of flies and *Drosophila melanogaster* Meigen is one of the species frequently occuring in homes. Although Ditman et al. (1936) found the above species to be the most prevalent around tomato canneries, *D. repleta* Wollaston was another important pest. A third common species is *D. funebris* (F.). Collins (1956) noted next to *D. melanogaster,* the fruit fly, *D. busckii* is secondary in importance. As *D. melanogaster* is the best known species of the fruit flies, it will be used as an example of the group.

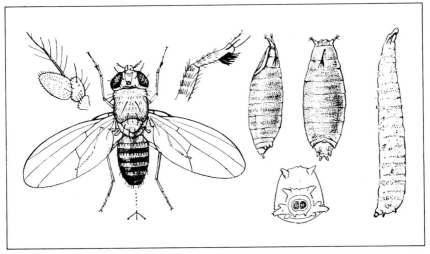

Fig. 19-15. Larvae, pupae, adult and anatomical details of the fruit fly.

The adult *D. melanogaster* Meigen is approximately ⅛ inch/three mm long, including the wings and has bright red eyes and a tan-colored head and thorax, with a blackish abdomen, the under surface of which is grayish. (Fig. 19-14). The fly is cosmopolitan in distribution and is very common wherever fruit and materials of a like nature are permitted to rot and ferment. Harrison et al. (1954) noted *D. melanogaster* and associated fruit flies are primarily attracted to fresh and fermenting fruits and vegetables caused by the multiplication of yeasts. Where decay is caused by bacteria and fungi, the fruits and vegetables begin to lose their attractiveness as a source of food to the adult and as a medium for oviposition.

The larvae are small, legless, eyeless maggots, pointed at the head end. They are common in canned fruits and vegetables that are imperfectly sealed. The maggots occur for the most part near the top of the jars and live in the briny or vinegar-like liquids; hence the name "vinegar flies." The adults deposit their eggs around the edges of the covers or jars as around spigot and bung holes of vinegar and cider barrels, and the minute larvae emerge and enter the barrels.

The larvae feed principally on yeast in the fermenting fluids. Common sources of infestation are rotting bananas, pineapples, tomatoes, mustard pickles, potatoes, etc. In cellars they have been found in such fermenting liquids as wine, cider, vinegar and beer, stale beer being particularly attractive. They have been reported from flour paste and fecal matter. Gould (1948) observed fruit flies to breed in seepage from starch vats in corn processing plants. Spieth (1951) observed *Drosophila lacicola* Patterson breeding in the rotting phloem of felled aspen.

Life history. The eggs, which are difficult to see with the naked eye, are deposited near the surface of the fermenting material. A pair of filaments that are attached to the eggs protrude above the surface of the liquid. As a rule, the larvae emerge from the eggs 30 hours after being laid. The larvae feed near the surface of the fermenting mass. The average length of the larval period is from five to six days. Prior to pupation, the larvae crawl to the drier portions of the food, or even out of it. The larva transforms into the pupa in the last larval skin, or puparium, which bears a conspicuous pair of filaments on the anterior end.

The newly emerged flies are attracted to light and become sexually active in about two days. The adults mate more than once and deposit an average of approximately 500 eggs. The life cycle from adult to adult may be completed in eight days at 85° F/29° C.

Control. Where these insects are an annoyance in homes or stores, one should carefully search for an accumulation of rotting fruits or vegetables and destroy this source by means of fire. Opened jars of fruits and vegetables should be kept in the ice box or refrigerator. Restaurants are often bothered by a plague of these flies and investigation will often reveal the source of infestation to be some rotting fruit or vegetable that was carelessly swept beneath a counter or similar place. A.C. Miller found the source of an infestation in a restaurant to be the partially empty tomato catsup bottles stored in the basement.

Gross (1948) made the following observations regarding the breeding places of *Drosophila* in buildings: "Fermenting fruits, vegetables, or juices are the most obvious sources of breeding. But repeatedly we find heavy infestations where there are no such violations of sanitation, not even one rotting potato in the bottom of the bin or lost under the fountain. What then?

"Dish-water from sinks, drain water from refrigerators or ice boxes and the slop water from floor scrubbings are saturated with food particles. Under the

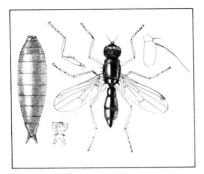

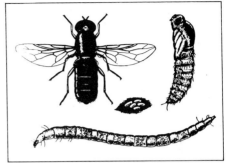

Fig. 19-16. Pupa, adult and anatomical details of *Sepsis violacea* (L.).

Fig. 19-17. Eggs, larva, pupa and adult of *Scenopinus fenestralis* (L.).

proper conditions they become fermented and make the ideal breeding place for the fruit or vinegar fly.

"One foul, soured mop can breed many thousands. The same is true of brooms, rags or paper. Garbage-laden water from sinks, coolers and ice boxes often clogs a floor drain and spreads over the floor. There it forms a thin, sour gelatinous layer, paradise for the fruit fly.

"When kitchen floors are scrubbed or mopped, the slop seeps into crevices, under the mold boards and other hidden places. There it quickly ferments into a perfect soup for the flies."

In food establishments fruit flies can be controlled by frequent application of pyrethrum-synergist sprays or aerosols. Bills (1965) found DDVP resin strips to control fruit flies in taverns.

Ditman et al. (1936) recommended canning plants be protected from these flies by the removal of debris, peelings, decaying fruit, etc., and that floors should be thoroughly cleaned and washed daily. Bickley et al. (1956) believed most of the adults in factories are brought in from the fields. Stombler et al. (1957) found a 0.11 percent pyrethrins dust applied to boxes of tomatoes in the field or to pallets of tomatoes in the receiving stations of the canneries resulted in protection from egg deposition for 24 hours. The pyrethrum dust is believed to act largely as a repellent. Anon. (1961) recommended that tomatoes in the field be dusted with 0.1 percent pyrethrins-one percent piperonyl butoxide dust. Penrose and Womeldorf (1962) found *Drosophila* infestations in wineries associated with leaky wooden storage tanks and poor sanitation. Yerington (1967) found the use of DDVP thermal aerosols to be effective in wine cellars.

DeCoursey (1925) recommended the use of a trap consisting of a pint/0.47 l or quart/0.95 l mason jar with a paper funnel. The trap was baited with bananas sprinkled with yeast and was effective for two weeks before rebaiting.

W.R. Sweadner and G.E. Wallace observed *Drosophila* sp. to emerge from the overflow drain in a sink. It is possible that the flies breed in the gelatinous layer that lines old pipes.

Procter (1941) reported the presence of the puparia of *Drosophila busckii* in milk bottles where the eggs apparently were laid in the sour milk at the bottom of neglected milk bottles. "The puparia were firmly attached to the sides of the bottle and although killed by one percent caustic detergent at 165° F/74° C were not dislodged. When the bottle was refilled they looked like hay seeds. The oc-

currence of these puparia is indicative of carelessness on the part of the consumer in not adequately cleaning and taking care of the bottle after it had been
emptied, rather than of insufficient care by the producer."

OTHER SPECIES FOUND IN HOUSES

Richardson (1923) mentioned several species that are pests in New Jersey
and undoubtedly are pests in other areas. *Sepsis violacea* (L.), a small slender
black fly (Fig. 19-16), often is found on the window panes of houses. It breeds
in excrement. *Scenopinus fenestralis* (L.), the window fly, often occurs in dwellings. This fly is "about ¼ inch in length and has a flattened abdomen. The larvae
live in cracks in the floor and in accumulations of animal and vegetable matter.
Howard (1911) stated the long, white, snake-like larva feeds on the larvae of
stored grain insects and when found in woolens or carpets it is searching for
clothes moth larvae. Howard concludes, "It is a pleasure to state that at least
one of the flies found in houses is probably beneficial rather than injurious, and
that this species is *Scenopinus fenestralis.*"

THE CHEESE OR HAM SKIPPER

Piophila casei (L.)

In the larval stage, this insect is referred to as the cheese or ham skipper
because it appears to move by skipping. The cheese skipper is an Old World pest
that is recorded in the earliest literature. This is the insect used by Redi to
disprove the theory of spontaneous generation.

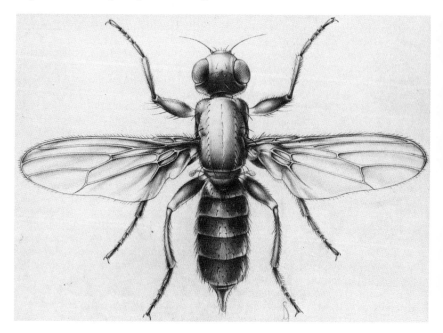

Fig. 19-18. Female cheese or ham skipper.

The adult of the skipper is a black fly with bronzy tints on the thorax, reddish brown eyes and slightly iridescent wings which lie flat over the body when the insect is at rest. The insect is ¹/₃ to ³/₅ the size of the common house fly and has lapping mouthparts like the house fly. (Fig. 19-17).

It is very likely the cheese skipper has become almost cosmopolitan in distribution because it is a scavenger in carrion. The cheese skipper has been recorded from Europe, India, the West Indies, Greenland, Alaska and throughout the United States mainland.

The cheese skipper is the principal pest of cheese, but as a rule it is a much more serious pest in meat. L.O. Howard (1900) noted at one time more than one million dollars worth of meat was condemned annually because of infestation by this insect.

Simmons (1927) noted that other than cheese, the larvae have been found in ham, bacon, human excrement, rotten fungus, human cadavers, dried bones, moist dog hair, powdered rhubarb and cooking salt. Infestation of the latter two are probably due to larval migration.

Overripe and moldy cheese, as well as lean and fat beefsteaks having a slightly putrid odor, are the preferred foods. Lean and fat ham and bacon are also common sources of infestation. Simmons reared 52,627 skippers from one dry-cured ham.

Badly infested cheese will show slightly sunken areas which, when opened, are soft and waxy. The lean portions of the meat are much more attractive for oviposition than are the fatty portions. The insects are extremely troublesome in hams and shoulders which have been cured and smoked, and they do the greatest damage to the connective and muscular tissues.

In regard to the presence of these maggots in cheese, Simmons quotes Swammerdam that the worms "were generally held in detestation, though some eat them voluptuously with the rest of the cheese, from a vulgar notion, that they are formed out of the best parts of it." And to this very day these maggots in cheese are greatly relished by some of our most fastidious gourmands.

The cheese skipper as a health menace. The larvae may feed on filth and the adults probably can transmit diseases in a fashion similar to that of the house fly. There are many records of intestinal irritation resulting from swallowing the maggot in cheese. In fact, it is believed this is the larva most commonly found in the intestines of man. The larvae also have been reported from the nose and chest of human beings.

Life history in meat. Much of the following information on the life history of this insect in meat is from Simmons (1927), who studied this insect in great detail.

The newly emerged female is beset by male flies, often before her wings have had a chance to expand. The white, shiny, slightly curved eggs are laid within 24 hours after mating. The eggs may be rapidly scattered over the surface of the meat or they may be deposited in masses in the cracks and crevices in the wrappings. Ordinarily, 140 (the maximum being 500) are deposited during a period of three to four days. The egg stage varies from one to four days, depending on the temperature.

The larva is the destructive form of the insect. It is repelled by light and aggregates with other larvae when feeding. The larva moves by skipping, as well as through the peristaltic movements of the body wall, abetted by the ambulatory teeth on the ventral surface. The skip may propel the insect 10 inches/25 cm horizontally, or six inches/15 cm vertically. In skipping, the insect curves its body into a ring by hooking its oral claws onto the end of the abdomen; it quickly releases its hold and the sudden snap propels it into the air. As a rule, however, the larva progresses by creeping. Simmons, quoting Wille, noted progression by leaping occurs only in the third instar larva.

Since the larva is slender and pointed anteriorly, it can readily penetrate meat, cheese, etc. Thus, the tiny newborn larva very often enters wrapped meat, particularly where the eggs are laid on greasy spots on the outside of the wrapper. At optimum conditions the larval stage is completed in five days, but under adverse circumstances this stage may be prolonged to eight months.

When ready to pupate, the larva leaves the food and migrates to some dark, dry crevice. It may undergo a pre-pupal period of 48 hours. The yellowish or reddish puparium, within which the larva pupates, is formed from the last larval skin. The pupal stage is completed in five days at 80° to 90° F/27° to 32° C.

The adults have an average longevity of three to 4.3 days during the late spring and summer. The odor of meat and cheese is very attractive to the adults and they feed on the juices of the larval food. They are reported to have the interesting habit of fighting with one another, apparently as a release for excess energy.

Simmons found the minimum life cycle of this insect on ham is 12 days, including the pre-oviposition period. "This brief cycle is divided about as follows: Preoviposition period, one day; larval stage, five days; pupal stage, five days. The majority of insects which are produced in hot weather take a day or two longer, and it is safe to say that two generations per month represents the normal rate of summer increase at Washington, D. C."

Life history in cheese. The fly frequents storage and curing rooms and prefers cheese that is three or more months old. DeOng and Roadhouse (1922) investigated the life history of this insect in cheese and the following facts are from their studies: The adults copulate two or three days after emergence. The eggs are deposited on the surface or in cracks in the cheese; the larvae burrow into the cheese and remain there until full grown. At 65° F/18° C the egg hatches in 30 to 48 hours, and at 65 to 95° F/18° to 35° C the larval stage extends from eight to 15 days. The larvae migrate just prior to pupation and the pupae are found in dark areas under cheese cloth covering. At 65 to 95° F/18° to 35° C the pupal stage extends from seven to 12 days.

Control in cheese. DeOng and Roadhouse (1922) used the following procedure to cope with *Piophila casei* (L.):

- Keep the place clean; remove accumulations of grease.

- Utilize cold storage to protect the cheese; slow cure the cheese at 30° to 36° F/−1° to 2° C or even up to 50° F/10° C.

- Coat cheese in paraffin heated to at least 200° F/93° C; the cheese is then protected by a thin coating of wax. This paraffin coating prevents shrinkage in weight through loss of moisture, preserves the palatibility of the cheese, and protects it from pests. The paraffin coating must be kept intact.

- Fumigation with hydrocyanic acid gas has proved successful when repeated several times. Milk, butter, water and other materials that might absorb dangerous quantities of hydrocyanic acid gas should be removed. The cheese skipper is not affected by the hydrocyanic acid gas to any great extent when it is buried in the cheese. It is most likely to be controlled when it is on the surface of the cheese, as when it prepares for pupation.

Thus, it may be necessary to fumigate more than once, at intervals of 12 to 18 days, in order to kill the cheese skippers that were deep in the cheese at the first fumigation.

Billings et al. (1942) controlled the cheese skipper adult with a pyrethrum/ sesame oil aerosol. At the present time, pyrethrum sprays and aerosols containing such synergists as piperonyl butoxide, sulfoxide and others are widely used for the control of this insect in food plants.

Sanitation is of the utmost importance in eliminating the ham or cheese skipper, and the methods recommended by the USDA Bureau of Entomology and Plant Quarantine (Anon., 1943) for protecting home-cured meat also largely applies to cheese.

"Eliminate breeding places of meat-house pests. Such insects feed and breed on any form of animal product. Even grease and crumbs of meat or cheese lodged in cracks in shelves, walls and floors harbor them.

"Brush and scrub thoroughly all places where meat has been stored. Smokehouses free of meat for any length of time should be cleaned as soon as the last meat is removed.

"Keep in tight containers all scraps of meat until they can be rendered.

"Keep insects out of meat storerooms as they can fly and carry mites.

"Use 30-inch screen or finer. Fit all doors and windows tightly.

"Slaughter, cure, and wrap meat before insects start to work in the spring.

"Wrap each piece separately and securely in waxed or other grease-proof paper. See that no insect life is on the meat when wrapped.

"Store meat in a clean, well-ventilated, dark smokehouse."

GNATS AND MIDGES

These insects, like mosquitoes, are most satisfactorily handled by the community through its medical entomologists and sanitary engineers. The individual often can cope with gnats and midges by fine mesh screening, sprays, aerosols, DDVP resin strips, repellents and smudges.

Kline and Roberts (1981) evaluated the duration of control provided by applications of several concentrations of promising insecticides. Chlorpyrifos, fenthion, malathion and propoxur were tested at one, three, five and eight percent weight/volume in evaluation of their effectiveness as residual screen treatments to retard or prevent entry of *Culicoides mississippiensis* Hoffman adults into screened areas. Propoxur at eight percent was the most effective with 72 to 92 percent quick (30 minute) knockdown for 35 days and 97 to 100 percent overall mortality. Malathion (eight percent) was effective only 14 days and fenthion ineffective at any concentration tested.

BUFFALO GNATS OR BLACK FLIES

Family Simuliidae

Black flies are small blood-sucking insects that are from 1/25 to 1/5 inch/one to five mm long. Although they are for the most part black, some are gray and a few are red. The thorax appears somewhat hump-backed, the antennae are 11-segmented, the wings lack scales and hairs, and the mouthparts are piercing. The females alone are the bloodsuckers.

These insects occur in enormous numbers in the late spring and early summer, particularly in the more northern latitudes. Their bite is extremely painful and their mouthparts are somewhat similar to those of the horsefly. Hearle (1938) stated: "On human beings they crawl into the sleeves, under the neck-band, around the tops of boots and other vulnerable places; they especially favor the head just beneath the rim of the hat."

As a result of the poison injected into the skin by the bite, a swelling and numb-soreness results, and this condition may persist for many days. There are a number of authentic records on hand where domestic animals and even human beings have been killed in a few hours through the venomous bites and the resultant loss of blood. These flies are diurnal and usually bite in shaded or partially shaded localities and may fly as far as 15 miles/24 km to feed on warm-blooded animals, including man. Davies (1951) showed dark blue cloth attracted the most and white cloth the least black flies. Black flies are known to transmit a disease of filarial worms, onchocerciasis, which cause blindness in human beings in Mexico, Central America and Africa.

The black fly, *Simulium venustum* Say, and other species of the genus *Simulium, Eusimulium* and *Prosimulium* are pests in the United States and Canada. *Simulium vittatum* Zett. is an important pest in Europe and the United States. *Simulium columbaczense* (Schiner) is a severe pest in middle and southern Europe, where on occasion it has killed thousands of domestic and wild animals.

Life history. According to Herle (1938), the life history of the black fly *Simulium venustum* Say, is as follows: These insects breed in shallow, fast-running water in rivers and in streams. They may also breed in slowly-moving water, as irrigation canals and in ditches. The small yellow eggs are attached in compact masses to stones and vegetation in the water. In some instances the fly has actually been observed to be ovipositing a foot below the surface of the water. The female oviposits approximately 500 eggs. Upon emerging from the eggs, the larvae retain their position in the water by means of the sucker-like discs and tiny hooks at the tip of the abdomen. They may also spin a fine thread which aids in anchoring them. They pupate in a cocoon which is open at one end. The adult flies take flight on emerging from the water. The winter may be passed as a larva. The life history spans about six weeks. There may be four generations a year.

Control. Effective control can only be realized by large-scale efforts of a governmental agency skilled in coping with flies, mosquitoes and similar pests. Such control necessitates the treatment of large areas where the larvae breed by airplane or helicopter, by employing crews to spray streams and other bodies of water, and by using fog machines and mist blowers to kill the adults. Jamnback and Collins (1955) prepared a monograph on the subject. Hocking (1952) reviewed the problem of protection from northern biting flies. Jamnback (1952) noted control measures must be directed against each species according to its life cycle.

Treatment of infested areas with DDT or TDE sprays from the air was recommended by numerous workers. Travis et al. (1951) sprayed a 20 percent DDT oil solution at the rate of 0.1 lb. DDT per acre/0.11 kg per ha in 100-foot wide swaths at intervals of ¼ mile/0.4 km in the Adirondacks of New York state. This resulted in effective reduction of the larvae. Goulding and Deonier (1950) controlled the emerging adults in parts of eastern Pennsylvania through airplane applications of 12 percent DDT in kerosene/xylene solution at the rate of one lb. DDT per acre/1.1 kg per ha. Large streams and open bodies of water were treated lightly or not at all to minimize injury to fish. Brown et al. (1951) reduced black fly and *Aedes* infestations for several weeks by spraying DDT from Aircraft. A 4.2 percent DDT-fuel oil solution was applied at an average dosage of 0.165 lb. DDT per acre/0.18 kg per ha by means of a Dakota C-47 plane. The USDA Bureau of Entomology and Plant Quarantine (Anon., 1952) recommended the application by plane of 0.2 lbs. DDT or TDE per acre/0.23 kg per ha when using five percent solutions. "The best way to apply these sprays is to fly the plane across the stream at intervals of about 800 feet. A plane spraying a 100-foot swath would thus make six or seven swaths for each mile of stream."

The Bureau of Entomology and Plant Quarantine (Anon., 1952) discussed treatment with ground equipment as follows: "To control black fly larvae from the ground, apply a five percent solution of DDT or TDE in oil to the water surface at about four points in every mile of stream. An ordinary compressed-air sprayer is satisfactory. Because the width and rate of flow of streams differ widely, it is not practical to recommend definite amounts to apply. However, you can estimate the area of a ¼-mile sector of the stream and then apply the solution at each point at the rate of ½ pint per acre. Allow about five to 10 minutes to apply the spray at each point. . . . . ."

The above source also noted that the use of 0.2 pound of DDT per acre/0.23 kg per ha by means of oil sprays applied from fog machines or mist blowers have at times been successful in reducing adult populations of black flies. Such treatments usually must be repeated every few days to kill the flies coming from untreated areas. Fairchild and Barreda (1945) found a four percent DDT emulsions at concentrations of one part DDT to 10 million parts water kill *Simulium* larvae in streams for a distance of several miles. Davis et al. (1957) controlled black fly larvae for a distance up to 2.8 miles/4.5 km using 0.5 and one ppm parathion. Aerial application of dieldrin at 0.04 lb. per acre/0.045 kg per ha gave 90 percent or more control for four weeks. Because of the potential adverse impact the use of DDT and other insecticides have, the widespread use of larvicides is not encouraged except where human health is seriously threatened by these flies.

As a whole, the individual can merely resort to household sprays, aerosols, repellents and smudges in order to cope with this pest. Travis et al. (1951) reported repellents in order of effectiveness against black flies in the Adirondacks to be Repellent 6-12, 6-2-2 mixture, dimethyl phthalate and Indalone. DeFoliart (1951) also conducted repellent tests in the Adirondack Mountains against black flies. He concluded as follows: "Propyl N, N-diethylsuccinamate gave the best protection, with 2,2'-thio-diethyl diacetate, Repellent 6-12 and cyclohexyl acetoacetate next in line. The other standard repellents, Indalone, dimethyl phthalate and Mixture 6-2-2 were inferior to Repellent 6-12." Repellents with diethyl-toluamide also are recommended against black flies.

The author was unable to obtain adequate protection at the margin of the hair line and other less hirsute areas on top of the head with a 6-2-2 cream in the Adirondacks despite heavy application of the cream.

MOTH FLIES, DRAIN FLIES, FILTER FLIES, OR SEWAGE FLIES

Family Psychodidae

The adults of the moth flies are at times annoying in homes, apparently appearing mysteriously from sinks and bathtub drains. Moreover, they may breed in tremendous numbers in sewage filter plants and then be carried from there by the wind to nearby homes where they penetrate through ordinary flyscreening.

Fig. 19-19. Pacific drain fly, *Psychoda pacifica* Kincaid.

Satchell (1949) noted they may be carried one mile/1.6 km from the filters by the prevailing wind. Since these flies originate from so filthy a source, any disease contracted by the inhabitants dwelling within a reasonable or unreasonable radius of the sewage works is attributed to these flies. Thus, the municipality concerned is often the victim of law suits. As a rule, the aroused citizenry who dwell close to the sewage works have some justification for their anger. For not only is there the possibility of the transmission of disease by the flies, but the flies also stick to clothes, fall into food, darken lamps, mar fresh paint, and in some instances, make breathing without swallowing them difficult. Ordman (1946) and Satchell (1949) noted cases of bronchial asthma caused by inhaling the dust resulting from the disintegration of filter flies.

Psychoda alternata Say is the common psychodid of the East. This species is 1/13 inch/two mm long with a light tan-colored body and lighter wings that are faintly mottled with black and white. The body and wings are densely covered with long hairs which give the fly a moth-like appearance, hence the name "moth-fly." The antennae are 13-segmented, each segment having a bulbous swelling with a whorl of long hairs. *Psychoda cinerea* Banks, another common Eastern species appears a little later, in April.

Habits and life history. The larvae and pupae of *Psychoda alternata* Say live in the gelatinous film that covers the filter stones. The living components of this organic film aid in purifying the sewage water. According to Headlee and Beckwith (1918), the eggs are laid upon the surface of the film in irregular masses of from 30 to 100. The eggs hatch in 32 to 48 hours at 70° F/21° C. The larvae and pupae live in the gelatinous film with the breathing tube or tubes projecting through the film.

Turner (1923) stated: "It is difficult to drown the larvae as the spiracles may even if pushed below the water, entrap a bubble of air in the hairy circle about the opening and carry this down with them. Specimens have been submerged for nearly two days without harm to them, but they will eventually drown."

The larvae feed on sediment, decaying vegetation and microscopic plants and animals. Headlee and Beckwith (1918) found the larval stage to extend from nine to 15 days at 70° F/21° C and the pupal stage from 20 to 40 hours at the same temperature. Turner (1923) noted during the brief pupal period, the front end of the pupal case bursts and the fly emerges. The life history extends from eight to 24 days at 58° to 86° F/14° to 30° C. The positively phototropic adult is sexually mature on emergence and copulates within a few hours.

In the western United States the Pacific drain fly, *Psychoda pacifica* Kincaid (Fig. 9-18), emerges from the drain of sinks and bathtubs, particularly in the spring, and is often mistaken as a moth by the householder. According to Essig (1926), this insect is distributed throughout the western United States from southern California to Alaska. The adult is two to 2.3 mm long with brownish-gray wings; otherwise it is very similar in appearance to the moth fly, *Psychoda alternata* Say.

Mallis and Pence (1941) found the immature stages of this fly to dwell in the gelatinous lining of the water-free portion of the ordinary wash basin. The flies were found emerging from the drain of a laboratory sink, down which many toxic insecticides had been poured.

E. J. Pinigis found filter flies infesting a hospital in Homestead, Pennsylvania. These flies were believed to be breeding on materials in walls wetted by a faulty air-conditioner. The flies were identified as *Telmatoscopus albipunctatus* (Will.). These same flies were found breeding in a sewage filter plant near Pittsburgh and the flies were invading buildings over 100 yards/91 m away.

Usinger and Kellen (1955) studied the role of insects in sewage disposal beds.

Control. Where the adults occur in enormous numbers because of breeding in sewage filter works, their control is then a problem for the municipality concerned. Satchell (1947) stated the warmth of the sewage beds in winter results in early emergence of the flies and infestation of natural breeding sites, and may lead to heavy infestation later in the year.

Headlee (1919) found submergence for 24 hours was destructive to the larvae and pupae. Since the eggs are not affected, this flooding of sewage filter works may have to be repeated at intervals.

Tomlinson (1945) found DDT applied at the rate of 75 lbs. per acre/85 kg per ha in percolating sewage filters controlled *P. alternata* Say and *P. severini* Tonn., as well as the fly *Anisopus fenestralis* Scop. At this concentration the DDT was not toxic to bacteria and fungi. Carollo (1946) applied DDT at the rate of 30 lbs. per acre/34 kg per ha over the surface of the filter and did not obtain adequate control. However, when he applied a Triton/xylene emulsion at the rate of "one to five ppm based on the 24 hour sewage flow and applied over the 24 hour period to the filter influent" practically all the larvae were destroyed. Warrick and Bernauer (1948) could not control adult filter flies with DDT. The former showed flooding for two hours washed out larvae and left better effluent. Brothers (1946) recommended DDT at the rate of 15 lbs. per acre/17 kg per ha or one ppm for the control of *P. alternata* Say and *Telmatoscopus albipunctatus* (Will.). This dosage was based on the total daily flow of sewage to the filter. Warwick and Bernauer (1946) found that DDT applied continuously at one ppm as emulsion eliminated larvae in three days. Simpson (1948) used a five percent DDT emulsion for the control of the filter fly larvae by applying it at the rate of one ppm DDT in the trickling filter influent. This five percent DDT spray also controlled the adult filter fly when applied at a dosage of one quart/0.9 l per 250 square feet/23 square meters. Bruce and Decker (1950) noted these filter flies are becoming resistant to DDT and chlordane. Livingston (1951) found DDT and BHC emulsions to be ineffective in filter fly control. Spiess (1952) noted trichlorobenzene added in conjunction with filter flooding provides good control of filter flies. Tetrault (1967) recommended either diazinon or malathion emulsions in combination with methoxychlor for residual control of these flies.

Mallis and Pence (1941) obtained control of the Pacific drain fly in a wash-basin drain by disconnecting the drain and removing the gelatinous lining. This was accomplished by forcing a paper swab with an outer coating of wax paper through the drain. A curved wire or a plumber's probe may be used to ram the swab down the drain. Considerable time elapses before the gelatinous lining accumulates in sufficient quantity to permit reinfestation.

The operator who attempts to clear psychodids from filter beds or other large areas is urged to determine the federal, state and municipal regulations that must be followed in his area as adverse environmental reactions may result. While several of the insecticides may be effective, it may be more desirable to use physical removal of the larvae from the surfaces where the larvae are attached. This applies to smaller areas such as the kitchen drains which can be scrubbed or treated with copious quantities of very hot water or the use of copious quantities of water under high pressure, as from a fire hose.

SAND FLIES

Family Ceratopogonidae

These small bloodsucking gnats, measuring one to three mm in length are often confused with black flies because of similarity in their mode of biting. Their bite is out of all proportion to their size and the Indians in Maine have very appropriately designated these tormentors as "no-see-ums." They are also known as "sand flies," "punkies" and "black gnats."

Adults of the genus *Culicoides* are of great economic importance in the East. Although the different species of sand flies vary in the date of emergence, they may be extremely numerous from spring through autumn when they make some areas barely habitable, discourage tourists, and in certain instances, have driven large resorts into the hands of the receivers.

These insects breed in salt marshes, mud, plant debris, fresh water areas and damp rot holes of trees. Anon. (1938) noted a hole "in a shade tree can furnish enough midges to annoy the members of an entire household." Dove et al. (1932) studied *Culicoides canithorax* Hoffman, *C. melleus*. Coq. and *C. dovei* Hall in the vicinity of Charleston, South Carolina. The breeding areas for these flies consisted of "fresh water inlets of the seacoast and on streams leading to these inlets." They found warm temperatures are not favorable for the development of sand flies in this area, and that although the pupal stage may extend for only four to seven days, the larval period may be of a duration of at least six months or longer. The females must have three to four blood meals before they deposit eggs. Breeding occurs in the decaying vegetation in tidal areas and in moist marsh areas.

These authors noted: "The evidence on breeding places of sand flies shows early development of the larvae commonly found at Charleston, South Carolina takes place in densely shaded areas at the edge of grass marshes. These locations have decaying leaves and humus which are protected from the heat of the sun. In these places concentrations of larvae are reached by brackish water of the unusually high tides. Since these tides and drainage from rainfall cause dissemination of the larvae to low areas and into fiddler-crab holes, any treatments for the destruction of the larvae should be made before dispersions of the larvae take place. In the vicinity of Charleston the most opportune time appears to be in the autumn or late spring, just after the season of sand fly activity."

In Arizona and New Mexico, *Ceratopogon stellifer* Coq. is an important pest, and the valley black gnat, *Leptoconops torrens* Townsend, and the Bodega black gnat, *Holoconops kerteszi* Kieff, are of great concern in the western United States. These gnats are carried by the winds from the river and marsh areas inland where they bite so severely that farmers and town dwellers may be forced indoors for the day. Both species are vicious biters of man, domestic animals and birds.

Smith and Lowe (1948) studied these gnats in some detail. They summarized their investigations of the valley black gnat in California as follows: "Adults of the valley black gnat occur for four to six weeks, beginning usually in the middle of May. Females feed only once; males do not feed. Unfed gnats live only six hours in captivity; with a blood meal, females live a maximum of five days. The larvae occur in clay adobe soils at a depth of 15 to 39 inches. Egress and entrance are dependent upon the drying and cracking of the soil. The larval period is at least two years in length. Larvae spend the summers in immobile aestivation. If the soil does not crack on schedule, the mature larvae enter a diapause. Some evidence is given to indicate that larvae may diapause for at least three years. Larvae are found in summer in soil with moisture content of 17 to 20 percent, salt concentration of 400 ppm, a pH of 9.6, and a temperature of 65 to 68° F." Fontaine et al. (1957) studied the ecology of *L torrens* Townsend in northern California.

The bite of these flies leaves an irritating and moist lesion which may take several weeks to heal. Tinkham (1951) noted there were a number of hospitalized cases in the Coachella Valley of California because of severe swelling of the neck and face due to the bites of *Simulium* or *Leptoconops* gnats.

Control. Smith et al. (1959) found the salt-marsh sand fly, *Culicoides furens,* resistant to chlordane, dieldrin, heptachlor and lindane, and susceptible to parathion, malathion and DDT. Jamnback (1961) killed *Culicoides obsoletus* for three weeks on screens treated with a 7.7 percent malathion/ethyl alcohol solution. Walls and ceilings around lights should also be treated because sand flies are attracted to lights. Jamnback (1963) controlled *C. sanguisuga* for four to six weeks by painting screens with either six percent deodorized malathion or six percent propoxur.

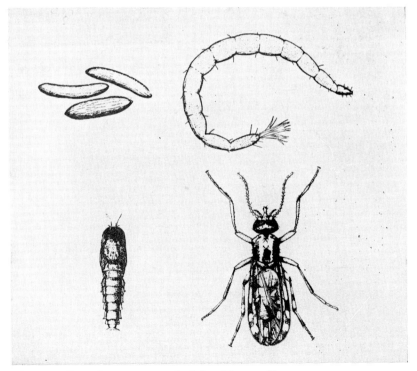

Fig. 19-20. Eggs, larva, pupa and adult of the sandfly.

Madden et al. (1946) obtained control for four days and some control for two weeks throughout an area of 200 acres/81 ha with a five percent DDT spray which was dispersed by airplane spraying at the rate of two quarts per acre/4.7 l per hectare. Jacusiel (1947) applied oil-base DDT spray to walls and ceiling at the rate of 100 to 200 mg DDT per square foot/0.09 square meter and found the control of sand flies to be very satisfactory. Bruce and Blakeslee (1948) used DDT solutions to control salt-marsh sand flies in and around a large resort near Savannah, Georgia. A five percent DDT oil solution was used to mop all the window and door screens. A 2.5 percent DDT/xylene emulsion was applied to garage, kennels and understructure of the boat dock. The outside "wall surfaces of the hotel and servants' quarters and surrounding shrubbery were sprayed to a height of about 12 feet with a water suspension containing one percent of DDT." Thermal aerosol fogs were applied at intervals over the grounds, golf course, and adjacent wooded areas. The formulation used consisted of 3¾ pounds/1.7 kg of technical DDT in one gallon/3.8 l of commercial xylene and four gallons/15 l of SAE no. 10 motor oil. When flights of sand flies were borne into the area by prevailing winds, the area was refogged. Later the management of the resort purchased a high-speed, blower-type sprayer and it was recommended a five percent DDT oil solution or emulsion be used at bi-weekly intervals during the period of annoyance.

Jamnback et al. (1958) controlled sand flies by applying DDT emulsions at the rate of one lb. per acre/1.1 kg per hectare. Goulding et al. (1953) obtained control of saltmarsh sand fly larvae with one lb. dieldrin per acre/1.1 kg per hectare. Anon. (1953) recommended two lbs. heptachlor per acre/2.2 kg per hectare.

Rees and Smith (1950 and 1952) obtained about 90 percent control of *Leptoconops* larvae and pupae in Utah by treating an infested area with a water emulsion of DDT at the rate of one pound per acre/1.1 kg per hectare. A five percent DDT emulsion applied on buildings and other resting places of the gnats was effective against the adults. A 7.5 percent DDT emulsion in fuel oil applied as a thermal aerosol fog resulted in immediate but only temporary relief from the pests. Whitsel and Schoeppner (1966) controlled *L. torrens* by applying malathion as a pre-emergence spray at the rate of two lbs. per acre/2.2 kg per hectare.

Hill and Roberts (1947) found gamma BHC applied at the rate of 100 mg per square foot/0.09 square meter in late May "not only destroyed the generations due to emerge, but also prevented the development in the treated soil of the generation that would provide the potential adult population of the following summer." Hull and Shields (1939) recommended either one of the following mixtures be applied evenly and thoroughly to doors and windows with a brush or rag. The flies already present inside the house are killed with fly spray.

Formula 1
  1 part pyrethrum extract concentrate (20 to 1)
  20 parts lubricating oil (SAE 5)
           or
Formula 2
  1 part pyrethrum extract concentrate (20 to 1)
  6 parts kerosene
  12 parts lubricating oil (SAE 10)

Screening as a means of control is unsatisfactory since these flies can pass through the smallest practical mesh. Where the sources of infestation are tree holes, these can be corrected by tree surgery or by spraying with five percent DDT household sprays. Some of the newer repellents on the market will give relief from these pests for several hours when applied to the exposed parts of the body. Travis and Norton (1950) noted the standard repellents such as 6- to 12, Indalone, dimethyl phthalate and combinations of these will keep *Culicoides* gnats from biting for 1.5 to two hours. These repellents are discussed in greater detail at the end of this chapter.

DDT and gamma BHC have provided satisfactory control of gnats of the family *Ceratopogonidae,* but they may be utilized now only with special permission of the federal government. Parathion and malathion have been quite useful and effective for large areas and will probably be available for some time. Pyrethrum and pyrethroids should be available for some time. Many of the repellents provide good protection and should be used for personal protection. Spencer et al. (1975) found Deet to be the most effective repellent.

THE MIDGES

Family Chironomidae

Homeowners very often are alarmed by large swarms of flies which they mistakenly call "mosquitoes." These are chironomids and they may be distinguished from mosquitoes by the absence of a long proboscis, the absence of scales on the wings and by the three to four-jointed palpi. These flies do not bite. Ali (1980) prepared a review of nuisance chironomids and their control.

These insects are often very important pests in and around areas where they can breed in water. Polluted water apparently favors their multiplication and emergence. Hansens and Hagman (1964) wrote about them as follows: "Generally speaking, eggs are laid in masses over open water or attached to aquatic vegetation. Under summer conditions eggs hatch in about 72 hours and the young larvae drop to the bottom where they often build tube-like structures of bottom debris held together by strands of silk. The larvae are scavengers, feeding on bottom detritus. Under favorable conditions the larval stage takes about four weeks and then pupation follows, usually lasting about 48 hours. Just before emergence the pupa rises to the water surface, emerging from the pupal skin in the same fashion as a mosquito.

"Emerged adults seek shelter until they are fully hardened. One of the noteworthy activities of many species of midges is the swarming of the males, especially at dusk. Mating occurs when females enter the swarm. So far as is known the adults do not take food and have a rather short life span, reported to be about 10 days. In New Jersey, adults of some species appear by mid-April and four or five generations may be expected during the summer. Often several species occur in the same location and produce overlapping populations. Midges overwinter in the larval stage.

"During peak emergence immense numbers of midges move into adjacent residential or industrial areas causing annoyance and damage. They completely cover houses and create a problem by their sheer numbers. The housewife who hangs out a wash may find her clothes covered with midges and, worse still, covered with small black spots of excretory waste which require that the clothes be re-washed.

"In industrial areas midges invade factories contaminating fabrics, plastics, pharmaceuticals and packaged materials. Housewives seem to object to plastic combs and brush handles with imbedded midges, although amber with similar fossil fragments is highly prized. Control measures in such situations may be difficult and expensive. Similarly, economic loss to motels and eating places may result when abundant midges attracted to lights cause patrons to go elsewhere."

In addition to the above, Grodhaus (1963) noted thick swarms of chironomid midges constitute traffic hazards and also cause allergic responses in susceptible individuals. According to Bay (1967), *Chironomus plumosus* L., the lakefly in California, defaces property by laying its egg masses on surfaces.

Bay (1967) recommended the lakefly nuisance in California can be abated somewhat by avoiding the use of unnecessary lights until 45 minutes after sundown because 90 percent or more of flight activity takes place before that time. Hagmann (1963) controlled chironomid and other midges by fogging with malathion at 1.5 to three percent in No. 2 fuel oil. He used about ½ lb. malathion per acre/0.56 kg per hectare. Patterson and Wilson (1966) controlled chironomid midges on Florida lakefronts by using one percent fenthion in a granular formulation at 0.2 lbs per acre./0.22 kg per hectare. He also fogged with malathion at two to four ounces per acre/0.14 to 0.28 kg per hectare. Excellent control was obtained for as long as four days against the adults. Bay and Anderson (1965) controlled *Chironomus* sp. in small bodies of water by stocking them with carp and goldfish at the rate of 150 to 500 pounds of fish per acre/170 to 566 kg per hectare.

Flentje (1945) controlled chironomid larvae in water reservoirs in Virginia which were appearing at customers' taps by using a one percent DDT emulsion which was sprayed on the surface of the reservoirs from a motor boat. The final concentration was 0.02 ppm of DDT. The larvae disappeared from the distribution system in 3.5 days. The availability of DDT for this purpose should be checked before using.

Gerry (1951) controlled chironomid-like larvae of the genera *Tendipes* and *Globiferus* breeding in polluted water in Massachusetts with 0.6 percent pyrethrins water emulsions. Lieux and Mulrennan (1956) studied the biology and control of *Glypototendipes paripes* in Florida. None of the control measures were entirely satisfactory.

HIPPELATES FLIES OR EYE GNATS

Family Chloropidae

The family chloropidae is composed of many species of small gnats. Some of the important agricultural insect pests belong to this family, the most important being the fruit fly and the wheat stem maggot. Only one small group, *Hippelates* spp., have medical importance. They are called "eye gnats" or "eye flies" because of their frequent feeding around the eyes. They also feed on sores, pus and at times blood. The major cause of injury is the scratching of the eyes by spines on the labellum which not only cause physical injury, but encourage the introduction of pathogenic organisms. The constant annoyance and irritation of the eyes is a serious problem much of the year and can give rise to some of the cases known as "pink eye" in human beings.

Burgess (1951) studied the life history and breeding habits of *Hippelates pusio* Loew in the Coachella Valley in California. He found the gnats to be most abundant in the fall and spring. Experimentally, 75 percent of oviposition occurred in the fall. Although the winter can be passed in the adult stage, it is usually spent as a larva in the soil. Breeding does not occur in corrals or manure heaps. The favorite breeding grounds are "freshly plowed, well irrigated sandy or gravelly soil with a heavy cover crop or application of organic fertilizer turned under."

Sabrosky (1940) found swarms of the chloropid, *Chloropisca annulata* (Walker), in window casings in a hotel in Michigan. These swarms may have been mating or ovipositing. In the fall of the year the swarms often enter the home to hibernate. Household sprays were used to disperse them.

Tinkham (1951) reduced gnat production by spraying between rows of cultivated date palms, an important breeding place for the gnats, with two lbs. aldrin per acre/2.2 kg per ha in emulsion form. Dow and Willis (1959) recommended aldrin at one to two lbs. per acre/1.1-2.2 kg per ha, chlordane at four lbs. per acre/4.5 kg per ha, and lindane at 2.5 lbs. per acre/2.8 kg per ha. Again it must be pointed out that these may no longer be available for use in the United States and some other countries.

Jones and Magy (1951) failed to control these gnats with airplane applications of thermal aerosols using DDT in oil at 0.1 and 0.3 lb. DDT per acre/0.11 and 0.34 kg per ha or as a spray at 0.4 lb. DDT per acre/0.45 kg per ha. Ground application of DDT at one lb. per acre/1.1 kg per ha as a spray or dust resulted in a small degree of control for approximately one month. Mulla et al. (1960) also obtained initial control with DDT but found soil manipulation more effective.

THE CLEAR LAKE GNAT

Chaoborus astictopus D. & S.
Family Chaoboridae

These gnats are of local importance in and around Clear Lake, Lake County, northern California, but other chaoborid gnats are becoming a problem throughout California where water is impounded (Cook, 1967). The gill-breathing, transparent larva is so difficult to see that it is often referred to as a "phantom larva." The adult is a brown-colored gnat with straw colored hairs. It does not suck blood. The flies are extremely annoying since they are attracted to lights, often in immense numbers. Herms (1937) stated R. W. Burgess found these gnats in such tremendous numbers as to practically smother one. They got into the eyes, nose, mouth and ears, and if an individual desired to be rid of them, he had to sit in the dark. The gnats appear early in May and are present until mid-October.

Deonier (1943) studied the life history of the immature stages. He found the adults to begin to emerge in large numbers in the spring. The adults emerged between 11 p.m. and 6 a.m. and spent 36 to 48 hours or more on the shore before ovipositing. Egg-laying begins at dusk and the eggs are deposited on the surface of the water where they sink to the bottom. The larvae usually emerge in 20 to 24 hours. The larval stage is not completed until the following year and the larva undergoes four instars. In the spring the pupal stage requires about two weeks. The emerging adults fly shoreward.

Brydon (1956) studied the Clear Lake gnat and its control in 1954.

A certain amount of control is realized through the use of partially concealed light traps placed at strategic points. It is necessary to use partially shielded traps or the gnats may be attracted from all over. The lights in the traps must be of greater intensity than those used in the house in order to function.

Lindquist and Bushland (1944) showed DDT to be very toxic to *Chaoborus punctipennis* Say in a lake in Florida.

Lindquist and Roth (1950 and 1951) and Lindquist et al. (1981) obtained complete control of the Clear Lake gnat for one year by spraying with one part TDE emulsion to 50 to 100 million parts of water. They applied 14,000 gallons/53,00 l of a 30 percent TDE emulsion concentrate over a 41,600 acre/16,848 ha lake. Surface craft made swaths 700 feet apart across the lake. Dolphin and Peterson (1960) noted TDE can no longer be used because of fears of accumulation of insecticides in fish and fish-eating birds. In recent years it has been realized that such apparent bioaccumulation of chlorinated hydrocarbons was often over emphasized. The early studies often failed to distinguish between non-insecticide sources (including natural sources of chlorinated hydrocarbons) and insecticide sources. Moreover, the prevalent exposure levels have not been proven detrimental to bird breeding, though nest site disturbance by people has been very harmful. According to Hazeltine (1962), methyl parathion can be substituted for TDE.

FUNGUS GNATS

Family Fungivoridae (formerly Mycetophilidae)

According to Sollers (1950), fungus gnats at times occur in the home where ferns and other house plants containing soil rich in humus may serve as a breeding source. Outdoors the larvae live in fungi, damp soil, or decayed vegetable matter. These flies, which are harmless to human beings, are often attracted to lights in the house at night. The usual type household or fly sprays are used for their control in the home. Electric fly traps are also effective in killing great numbers of them at night.

MISCELLANEOUS SMALL FLIES

Several species of minute black flies occasionally are important pests in the house. Ebeling (1964) considered them to be largely accidental invaders. Scatopsid flies breed in decaying vegetables, animal matter and excrement. Phorid flies and fungus gnats occur in rotting organic material. Decaying fruits and vegetables may be responsible for them. March flies breed in decaying vegetable material. Ebeling found them originating in highly fertilized lawns. Davis (1965) has used naled or malathion in fogging devices to control these pests. Granular insecticides may also be effective against them.

A larger fly, the stratiomyid *Hermetia illucens* (L.), has on occasion been found breeding in and under houses in rotting fruit and garbage.

RECENT DEVELOPMENTS

As resistance to pesticides developed in many species of insects, new approaches to control were initiated.

Insect growth regulators have been shown to have promise in research carried on by the USDA personnel on control of stable flies and house flies. The most promising insect growth regulators provide excellent control of both house flies and stable flies and present little evidence of long term contamination of the areas treated.

The use of biological control procedures look promising, but as yet do not provide high levels of control over long periods of time.

Wright et al. (1976) have worked with biological control agents that have provided excellent control in some areas of entomology. Recent work by Rutz and Axtel (1981), when they tested seven house fly pupal parasites as biological control agents in caged layer poultry facilities in North Carolina, suggests that several species of flies can be controlled by manipulation of known fly parasites. Combined releases of *Muscidifurax raptor* and *Spalangia cameroni* may prove to be highly effective in suppressing fly populations under the habitat and climatic conditions of that state.

Olton and Legner (1975) released three parasitoids during December to April 1969 to 1970 in an enclosed poultry house near Riverside, California. They obtained up to 46 percent parasitization of *Musca domestica* L., but only 16 percent of *Fannia* spp. Results appear promising.

Additional areas that may be productive include environmental alteration, such as soil manipulation to reduce development, more effective traps and further emphasis on biological control.

Knapp and Herald (1981) recently studied the reduction of face flies on beef cows and calves in a herd with one eight percent fenvalerate ear tag per ear during early spring. This resulted in an average of 90 percent reduction of the face flies *Musca autumnalis* DeGeer over a 21 week period. In another test of seven weeks duration, part of the herd were treated and the remainder were left untreated. The flies on the treated animals were reduced by 86 percent.

Breeden et al. (1975) added methoprene to formulations of chicken feed to control house flies breeding in chicken manure. A rate of 50 ppm of methoprene provided excellent control of house flies after the mixture was fed for three days. Feed containing five ppm provided good control after being fed for eight days.

Frazar and Schmidt (1979) compared the susceptibility of laboratory-reared stable flies to selected insecticides using adults that were no more than 24 hours old. Coumaphos and pyrethrum were most effective while the chlorinated hydrocarbons were much less toxic.

Riner et al. (1981) have found fair results from the use of Rabon® boluses in controlling face flies in cattle feces.

Williams and Westby (1980) made an evaluation of pyrethroid compounds impregnated in ear tags for control of the face fly, *Musca autumnalis* De Geer, and the horn fly, *Haematobia irritans* (L.), on pastured cattle. Ear tags tested included five and 10 percent ear tags, and 1.5 percent decamethrin ear tags. All three tags provided at least a 95 percent reduction of horn flies but less than 50 percent reduction of face flies over the 13-week study.

LITERATURE

ALI, A. — 1980. Nuisance chironomids and their control Bull. Entomol. Soc. America 26(1):3-16.

ANONYMOUS — 1938. "Sand flies" and "punkies." USDA BE&PQ. E-441. 1943. Protect home-cured meat from insects. USDA BE&PQ Ext. Serv. AWI-32. 1952. USDA Agr. Handbook no. 571. Guidelines for the control of insect and mite pests of foods, fibers, feeds, ornamentals, livestock, households, forests, and forest products. 1966. How PCOs use masking agents and perfumes. Pest Control 34(1):13-15. 1955. The housefly — how to control it. U.S.D.A.

Leaflet No. 390. 1961 Drosophila control. National Canners Assoc. Control black flies. BE&PQ. EC-20. 1980.

AUSTEN, E.E. — 1920. The house-fly as a danger to health. British Museum Econ. Ser. No. 1.

BABERS, F.H. — 1949. Development of insect resistance to insecticides. USDA BE&PQ. E-776. 1951. Not published. The current status of the physiology of insects resistant to insecticides.

BABERS, F. H. and J. J. PRATT, JR. — 1951. Development of insect resistance to insecticides. II. US BE&PQ. E-818.

BAKER, W. C. and H. F. SCHOOF. — 1955. Prevention and control of fly breeding in animal carcasses. J. Econ. Entomol. 48(2):181-183.

BARBER, G. W. — 1919. A note on migration of larvae of the house fly. J. Econ. Entomol. 12:466.

BARBER, G. W. and J. B. SCHMITT. — 1949. A line of houseflies resistant to methoxychlor. J. Econ. Entomol. 42(5):844-845.

BARBER, G. W. and E. B. STARNES. — 1949. The activities of house flies. J. N.Y. Entomol. Soc. 57(4):203-214.

BARKER, R. J. — 1957. DDT absorption and degradation in houseflies of varied age. J. Econ. Entomol. 50(4):499-500.

BAY, E. C. — 1967. What to do about lakeside property defacement by chironomid egg masses. Pest Control. 35(4):10-13.

BAY, E. C. and L. D. ANDERSON — 1965. Chironomid control by carp and goldfish. Mosquito News. 25(3):310-316.

BICKLEY, W. E., F. P. HARRISON and L. P. DITMAN — 1956. *Drosophila* as a pest of canning tomatoes. J. Econ. Entomol. 49(3):417-418.

BILLINGS, S. C., L. D. GOODHUE, and W. N. SULLIVAN — 1942. A pyrethrum-sesame oil aerosol used against cheese skipper adults. J. Econ. Entomol. 35:289.

BILLS, G. T. — 1965. Kepone and dichlorvos for controlling cockroaches and vinegar fly. International Pest Control. 7(5):8-11.

BISHOPP, F. G. — 1931. The stable fly; how to prevent its annoyance and its losses to livestock. USDA Farmers' Bull. 1097. 1937. Flytraps and their operation. USDA Farmers' Bull. 734. 1939. The stablefly: How to prevent its annoyance and its losses to livestock. USDA Farmers' Bull. 1097.

BISHOPP, F. C., and L. S. HENDERSON. — 1946. Housefly control. USDA Leaflet 182.

BLICKLE, R.L., A. CAPELLE, and W. J. MORSE — 1948. Insecticide resistant houseflies. Soap & S.C. 24(8):139, 141, 149.

BREEDEN, G. C., E. C. TURNER, and W. L. BEANE — 1975. Methoprene as a feed additive for control of the house fly breeding in chicken manure. J. Econ. Entomol. 68(4):451-452.

BROTHERS, W.C. — 1946. Experiments with DDT in filter fly control. Sewage Works. J. 18:181-207.

BROWN, A. W. A. — 1958. Insecticide resistance in arthropods. World Health Organization.

BROWN, A. W. A., and R. B. MARCH — 1959. Insecticide resistance in arthropods of medical importance. Misc. Publ. ESA Vol. #1, No. 1.

BROWN, A.W. A., R. P. THOMPSON, C. R. ETWINN, and L. K. CUTKOMP — 1951. Control of adult mosquitoes and black flies by DDT sprays applied from aircraft. Mosquito News 11(2):75-84.

BRUCE, W. G., and E. B. BLAKESLEE — 1948. Control of salt-marsh sand flies and mosquitoes with DDT insecticides. Mosquito News 8(1):26-27.

BRUCE, W. N. — 1950. Current report on housefly resistance to insecticides. Pest Control 18(4):9-10, 19.

BRUCE, W. N., and G. C. DECKER — 1950. Housefly tolerance for insecticides. Soap & S.C. 26(3):122-125, 145. 1951. Here are some solutions to insecticide-resistant houseflies. Pest Control 19(4):9-11.

BRUNDRETT, H.M. 1953. A homemade fly trap. ET-312.

BRYDON, H. W. — 1956. The Clear Lake gnat and its control in Clear Lake, California during 1954. J. Econ. Entomol. 49(2):206-209.

BUCHER, G. E., J. W. MacB. CAMERON and A. WILKES — 1948. Studies on the housefly (*Musca domestica* L.) III. The effects of age, temperature, and light on the feeding of adults. Can. J. Res. 26(1):57-61.

BURGESS, R.W. — 1951. The life history and breeding habits of the eye gnat, *Hippelates pusio* Loew, in the Coachella Valley, Riverside County, Calif. Amer. J. Hyg. 53(2):164-177. Biol Abst. 25(9):2600, 1951.

BUSVINE, J. R. — 1951. Mechanism of resistance to insecticide in house flies. Nature 1698 (4266):193-195.

CAMPAU, E. J., G. J. BAKER and F. D. MORRISON — 1952. Rearing the stable fly for laboratory tests. Proc. 38th Annual Meeting Chem. Spec. Mfr. Assoc. pp. 83-85.

CAROLLO, J.A. — 1946. Control of trickling filter flies (Psychodidae) with DDT. Sewage Works J. 18:208-211. Mar. 1946 Abst. in Chem. Abst. 40(19):5872. Oct. 10.

CHAPMAN, R. K. — 1944. An interesting occurrence of *Musca domestica* L. larvae in infant bedding. Canadian Entomol. 76:230-232. Nov.

CHENG, T.H. — 1967. Frequency of pinkeye in cattle in relation to face fly abundance. J. Econ. Entomol. 60(2):598-599.

COFFEY, J. H. — 1951. Location and community fly control. Pest Control 19(5):18, 20, 36.

COLLINS, W. E. — 1956. On the biology and control of *Drosophila* on tomatoes for processing. J. Econ. Entomol. 49(5):607-610.

COOK, S.F. — 1967. The increasing chaoborid midge problem in California. Calif. Vector Views 14(6):39-44.

CROCKER, E. C. — 1949. Principles of odor masking. Chem Ind. 64(6):948,950,952.

DAVID, W. A. L. — 1946. The quantity and distribution of sprays collected by insects flying through insecticidal mists. Ann. Appl. Biol. 33(2):133-141.

DAVID, W. A. L. and P. BRACEY — 1946. Factors influencing the interaction of insecticidal mists on flying insects. Bull. Entomol. Res. 37(2):177-190.

DAVIES, D. M. — 1951. Some observations of the number of black flies (Diptera, Simulidae) landing on colored cloths. Canadian J. Zool. 29(1):65-70. Biol. Abst. 25(9):2601, 1951.

DAVIES, M., J. KEIDING and C. G. VON HOFSTEN — 1958. Resistance to pyrethrins and to pyrethrins-piperonyl butoxide in a wild strain of *Musca domestica* L. in Sweden. Nature 182(4652):1816-1817.

DAVIS, A. N., J. B. GAHAN, J. A. FLUNO and D. W. ANTHONY — 1957. Larvicide tests against blackflies in slow-moving streams. Mosq. News 17(4):261-265.

DAVIS, J. J. — 1944. Scavenger maggots. Pests 12(11):22. 1953. Scavenger maggots. Pest Control 21(11):16.

DAVIS, L. — 1965. Small pest flies: Perennial PCO headache. Pest Control

33(4):18-20, 56.

DeCOURSEY, R. M. — 1925. A practical control for the pomace fly. J. Econ. Entomol. 18(4):626-629. 1927. A bionomical study of the cluster fly *Pollenia rudis* (Fabr.) Ann. Entomol. Soc. Amer. 20:368-381.

DeFOLIART, G. R. — 1951. A comparison of several repellents against black-. flies. J. Econ. Entomol. 44(2):265-266.

DeLONG, D. M. and G. M. BOUSH — 1952. Is the housefly being replaced by other Diptera as the major insect pest of food markets? Ohio J. Sci. 52(4):217.

DEMEREC, M. — 1950. Biology of Drosophila. 632 pp. Illus. Wiley.

DeONG, E. R. and C. L. ROADHOUSE. 1922. Cheese pests and their control. Calif. Agr. Exp. Sta. Bul. 343.

DEONIER, C. C. — 1943. Biology of the immature stages of the Clear Lake gnat *(Chaoborus astictopus)* (Diptera, Culicidae). Ann. Entomol. Soc. Amer. 36(3):383-388.

DETHIER, V. G. — 1963. To Know A Fly. Holden-Day, Inc.

DICKE, R. J. and J. P. EASTWOOD — 1952. The seasonal incidence of blowflies at Madison, WI (Diptera-Calliphoridae). Biol. Abst. 27(3):739-740, 1953.

DITMAN, L. P., E.N. CORY and A. R. BUDDINGTON — 1936. The vinegar gnats or pomace flies . . . . their relation to the canning of tomatoes. Maryland Agr. Expt. Sta. Bull. 400.

DODGE, H. R. — 1954. Fly identification is easy with CDC pictorial keys. Pest Control 22(4):10-12, 14, 16.

DOLPHIN, R. E. and R. N. PETERSON — 1960. Developments in the research and control program of the Clear Lake gnat, *Chaoborus astictopus,* D & S. Proc. 28th Ann. Conf. Calif. Mosq. Control Assoc. 90-94.

DOVE, W. E. — 1916. Some notes concerning overwintering of the housefly, *Musca domestica* Linn. J. Econ. Entomol. 9:528-538. 1937. Myiasis of man. J. Econ. Entomol. 30:29-39.

DOVE, W. E., D. G. HALL and J. B. HULL — 1932. The salt marsh sand fly problem. Ann. Entomol. Soc. Amer. 25(3):505-522.

DOW, R.P., and M. J. WILLIS — 1959. Evaluation of insecticides for the control of *Hippelates pusio* in soil. J. Econ. Entomol. 52(1):68-71.

DUNN, L. H. — 1923. Observations on the oviposition of the housefly, *Musca domestica* L. in Panama. Bull. Entomol. Res. 35:53-67.

EBELING, W. — 1964. Midges and gnats as accidental invaders of homes. PCO News. 24(11):28-30.

ELMORE, J. C., and C. H. RICHARDSON — 1936. Toxic action of formaldehyde on the adult housefly, *Musca domestica* L. J. Econ. Entomol. 29(2):426-433.

ESSIG, E.O. — 1926. Insects of Western North America. Macmillan Co.

FAIRCHILD, G. B., and E. A. BARREDA — 1945. DDT as a larvicide against *Simulium.* J. Econ. Entomol. 38(6):694-699.

FAY, R. W., J. W. KILPATRICK and G. W. MORRIS, III — 1958. Malathion resistance studies on the housefly. J. Econ. Entomol. 51(4):452-453.

FINE, B. C. et al. 1967. Resistance to pyrethrins and DDT in a strain of house-flies. J. Sci Food Agric. 18:220-224.

FISHER, E. H. — 1955. A dairy barn fogging method for fly control. J. Econ. Entomol. 48(3):330-331.

FLEISCHER, A.N. — 1957. Flue flies. Pest Control 25(11):42.

FLENTJE, M. — 1945. Elimination of midge fly larvae with DDT. J. Amer. Water Works Assoc. 37(10):1053. Biol. Abst. 20:443, 1946.

FONTAINE, R. E., D. H. GREEN and L. M. SMITH — 1957. Ecological observations of the valley black gnat, *Leptoconops torrens* Townsend. J. Econ. Entomol. 50(6):764-767.

FRAZAR, E. D., and C. D. SCHMIDT — 1979. Susceptibility of laboratory-reared horn flies and stable flies to selected insecticides. J. Econ. Entomol. 72(6):884-886.

FRINGS, H. — 1947. A simple method for rearing blowflies without meat. Science 105(2731):482. May 2.

FRISON, T. H. — 1925. Intestinal myiasis and the common housefly, *Musca domestica* Linn. J. Econ Entomol. 18:334-336.

FULLMER, O. H. and W. M. HOSKINS — 1951. Effects of DDT upon the respiration of susceptible and resistant houseflies. J. Econ. Entomol. 44(6):858-870.

GAHAN, J. B. and J. M. WEIR — 1950. Houseflies resistant to benzene hexachloride. Science III (2894):651-652.

GAHAN, J.B., H. G. WILSON and W. C. McDUFFIE — 1954 Organic phosphorous compounds as toxicants in housefly baits. J. Econ. Entomol. 47(2):335-340.

GAHAN, J. B., H. G. WILSON, J. C. KELLER and C. N. SMITH — 1957. Organic phosphorus insecticides as residual sprays for the control of houseflies. J. Econ. Entomol. 50(6):789-792.

GEORGHIOU, G. P. and W. R. BOWEN — 1966. An analysis of housefly resistance to insecticides in California. J. Econ. Entomol. 59(1):204-214.

GERRY, B. I. — 1949. Control of a salt marsh tabanid by means of residual DDT-oil spray. J. Econ. Entomol. 42(6):888-890. 1951. Some mosquito-like nuisance pests and their economic significance. Mosquito News. 11(3):141-144.

GILBERT, C. L. — 1946. Domestic fly sprays. I. Introduction. Petroleum, Feb. 1946. p. 34.

GOULD, G. E. — 1948. Insect-problems in corn processing plants. J. Econ. Entomol. 41(5):774-778.

GOULDING, R. L., R. F. CURRAN and G. C. LABRECQUE — 1953. Insecticides for control of salt-marsh sand flies in Florida. J. Econ. Entomol. 46(1):37-43.

GOULDING, R. L., JR., and C. C. DEONIER — 1950. Observations on the control and ecology of black flies in Pennsylvania. J. Econ. Entomol. 43(5):702-704.

GREEN, A. A. — 1951. Blowflies in slaughterhouses. J. R. Sanit. Inst. 71(2):138-145. Rev. Appl Entomol. B 39(12):207. 1951. The control of blowflies infesting slaughter-houses. I. Field observations of the habits of blowflies. Ann. Appl. Biol 38(2)475-494, Rev. Appl. Entomol. B 39(12):205-207, 1951.

GREENBERG, B. — 1967. Flies and disease. Sci. Amer. 213(1):92-99.

GRISWOLD, G.H. — 1942. An unusual experience with *Lucilia sericata*. J. Econ. Entomol. 35(1):73.

GRODHAUS, G. — 1963. Chironomid midges as a nuisance. Calif. Vector Views, 10(5):27-37.

GROSS, J. R. — 1948. Where the slop water seeps look for the fruit fly. Pests. 16(7):30. 1949. King cotton: super fly chaser. Pest Control 17(9):14.

HAGMANN, L. E. — 1963. The midge problem. Soap & Chem. Spec. 39(2):89, 91, 104.

HAGMANN, L. E., and G. W. BARBER — 1948. Overwintering habits of *Phaenicia sericata* (Mg.) J. Econ. Entomol. 41(3):510.

HAINES, T. W. — 1953. Breeding media of common flies. I. In urban areas. Amer. J. Trop. Med. & Hyg. 2:933-940.

HANSENS, E. J. — 1950. House fly control in dairy barns. J. Econ. Entomol. 43(6):852-858. 1951. The stable fly and its effect on seashore recreational areas in New Jersey. J. Econ. Entomol. 44(4):482-487. 1951. Stable flies at the seashore. Pest Control. 19(4):12, 14. 1956. Control of house flies in dairy barns with special reference to diazinon. J. Econ. Entomol. 49(1):27-32. 1956. Granulated insecticides against greenhead *(Tabanus)* larvae in the salt marsh. J. Econ. Entomol. 49(3):401-403. 1963. Area Control of *Fannia canicularis*. J. Econ. Entomol. 56(4):541. 1981 Resmethrin and permethrin sprays to reduce annoyance from a deer fly, *Chrysops atlanticus*. J. Econ. Entomol. 74(1):3-4.

HANSENS, E. J., and A. H. GODDIN — 1949. Reaction of certain fly strains to DDT and methoxychlor deposits. J. Econ. Entomol. 42(5):843-844.

HANSENS, E. J. and L. E. HAGMANN — 1964. The problem associated with midges and related Diptera in New Jersey. Proc. 51st Ann. Meeting NJ Mosq. Ext. Assoc. pp. 172-176.

HANSENS, E. J. and A. P. MORRIS. — 1962. Field studies of house fly resistance to diazinon, ronnel, and other insecticides. J. Econ. Entomol. 55(5):702-708.

HANSENS, E.J., J. B. SCHMITT and G. W. BARBER — 1948. Resistance of house flies to residual applications of DDT in New Jersey. J. Econ. Entomol. 41(5):802-803.

HARRIS, R. L., P. D. GROSSMAN and O. H. GRAHAM — 1966. Mating habits of the stable fly. J. Econ. Entomol. 59(3):634-636.

HARRISON, C. M. — 1951. Inheritance of resistance to DDT in the housefly, *Musca domestica* L. Nature 167:855-856. 1952. DDT resistance in an Italian strain of *Musca domestica* L. Bull. Entomol. Res. 42(4):761-768.

HARRISON, F.P., L. P. DITMAN and W. E. BICKLEY — 1954. Habits of *Drosophila* with reference to animal excrement. J. Econ. Entomol. 47(5):935.

HARRISON, R.A. — 1951. Tests with some insecticides in DDT-resistant house flies. New Zealand J. Sci. Tech. 32B(6):5-11. Chem. Abst. 46:1698, 1952.

HAZELTINE, W. — 1962. Safety of wildlife is important to planners of gnat-control programs. Agr. Chem. 17(2):12-14, 76.

HEADLEE, T. J. — 1919. Practical application of the method recently discovered for the control of the sprinkling sewage filter fly. J. Econ. Entomol. 12:35-41.

HEADLEE, T.J. and C. H. BECKWITH — 1918. Sprinkling sewage filter fly, *Psychoda alternata* Say. J. Econ. Entomol. 11:395-401.

HEARLE, E. — 1938. Insects and allied parasites injurious to livestock and poultry in Canada. Can. Dept. Agr. Farm. Bull. 53.

HENRY, S. J. — 1958. Fly control that works . . . in supermarkets. Pest Control 26(4):9,11,71.

HERALD, F. AND F. W. KNAPP — 1980. Effects of Monensin on development of the face fly and the horn fly. J. Econ. Entomol. 73(6):762-763.

HERMS, W. B. — 1937. The Clear Lake gnat. Calif. Agr. Expt. Sta. Bull. 607. 1950. Medical Entomology. The Macmillan Co. 643 pp.

HERRICK, G. W. — 1936. Insects Injurious to the Household and Annoying to Man. Macmillan.

HEWITT, G. C. — 1914. The House-fly, Its Structure, Habits, Development, Relation to Disease and Control. University Press, Cambridge, Eng.

HILL, M. A. and E. W. ROBERTS — 1947. An investigation into the effects of "Gammexane" on the larvae, pupae, and adults of *Culicoides impunctatus* Goethhebuer and on the adults of *Culicoides obsoletus* Meigen. Ann. Trop. Med. Parasit. 41(1):143-163. Rev. Appl. Entomol. B 38(37):49-50, 1950.

HOCKING, B. — 1950. Further tests of insecticides against black flies (Diptera: Simulidae) and a control procedure. Sci. Agr. 30(12):4489-4504. Rev. Appl. Entomol. B. 330:(10):162, 1951. 1952. Protection from northern biting flies. Mosq. News 12(2):91-102.

HODGE, C. — 1911. Nature and Culture, July.

HOWARD, L. O. — 1900. A contribution to the study of the insect fauna of human excrement. Proc. Wash. Acad. of Sci. 2:541-604. 1911. The House-fly Disease Carrier. Frederick A. Stokes Co.

HOWARD, L. O. and F. C. BISHOPP — 1926. The housefly and how to suppress it. USDA Farmers' Bull. 1408.

HOWELL, D. E. — 1961. Fly baits. Pest Control 29(4):9-11. 1975. Okla. Agr. Sta. Rept. pp. 54-55.

HULL, J. B., and S. E. SHIELDS — 1939. Pyrethrum and oils for protection against saltmarsh sandflies (Culicoides). J. Econ. Entomol. 32:93-94.

HUTCHINSON, R. H. — 1916. Notes on the preoviposition period of the house fly, *Musca domestica* L. USDA Bull. 345. 1918. Overwintering of the house fly. J. Agr. Res. 13(3):149-170.

JACUSIEL, F. — 1947. Sandfly control with DDT residual spray. Field Experiments in Palestine. Bull. Entomol. Res. 38(3):479-488. Rev. Appl Entomol. 36(4):68-69.

JAMES, M. T. — 1947. The flies that cause myiasis in man. USDA Misc. Publ. 631. 175 pp.

JAMNBACK, H. — 1952. The importance of correct timing of larval treatments to control specific blackflies (Simuliidae). Mosq. News 12(2):77-78. 1957. Control of salt marsh *Tabanus* larvae with granulated insecticides. J. Econ. Entomol. 50(4):379-382. 1961. The effectiveness of chemically treated screens in killing annoying punkies, *Culicoides obsoletus*. J. Econ. Entomol. 54(3):578-580. 1963. Further observations on the effectiveness of chemically treated screens in killing biting midges, *Culicoides sanguisuga* J. Econ. Entomol. 56(5):719-720.

JAMNBACK, H. and D. L. COLLINS — 1955. The control of blackflies (Diptera: Simuliidae) in New York. Albany, Univ. of State of NY. 113 pp.

JAMNBACK, H., W. J. WALL and D. L. COLLINS — 1958. Control of *Culicoides mellens* (Coq.) (Diptera:Heleidae) in small plots, with brief descriptions of the larvae and pupae of two coastal *Culicoides*. Mosq. News 18(2):64-70.

JONES, R. W. III and H. I. MAGY — 1951. Mosquito control techniques for control of *Hippelates* spp. (Diptera: Chloropidae). Mosquito News 11:102-107.

KEIDING, J. — 1956. Resistance to organic phosphorus insecticides of the housefly. Science 123(3209):1173-1174. 1965. Dichlorvos. Denmark Govt. Pest Infestation Lab. Ann. Rpt. p. 44.

KEILIN, D. — 1915. Recherches sur les larvaes de Dipteraes Cyclorraphes. Bul. Sci de la France et de Belgigue. Vol. 49, Dec.

KELLER, J.C., H.G. WILSON and C. N. SMITH — 1956. Bait stations for the control of house flies. J. Econ. Entomol. 49(6):751-752.

KILPATRICK, J. W. and H. F. SCHOOF — 1955. DDVP as a toxicant in poison baits for house fly control. J. Econ. Entomol 48(5):623-624. 1956. The use of insecticide cords for housefly control. USPHS Rpts. 71:144-150. 1956. Fly production in treated and untreated privies. USPHS Rpts. 71(8):787-796. 1959. The effectiveness of ronnel as a cord impregnant for house fly control. J. Econ. Entomol. 52(4):779-780.

KING, W. V. and J. B. GAHAN — 1949. Failure of DDT to control house flies *(Musca domestica)*. J. Econ. Entomol. 42(3):405-409.

KLASSEN, C. W. and J. D. WILLIAMS — 1951. Four years of fly control at the Illinois State Fair. Pub. Works 82(5):43-45, 76.

KLINE, DANIEL L. and R. H. ROBERTS. 1981. Effectiveness of chlorpyrifos, fenthion, malathion, and propoxur as screen treatments for control of *Culicoides mississippinesis*. J. Econ. Entomol. 74(3):331-333.

KNAPP, F. W. and F. HERALD — 1981. Face fly and horn fly reduction on cattle with fenvalerate ear tags. J. Econ. Entomol. 74(3):295-296.

KNIPLING, E. F. — 1950. Some personal observations on the treatment of Clear Lake, California for the control of the Clear Lake gnat. Mosq. News 10(1):16-19. 1950. Insecticidal resistant flies and mosquitoes. Off. Proc. 36th Mid-Year Meeting. CSMA Soap & S.C. Spec. Ed. pp. 87-88.

KNOTE, C. — 1949. Some equipment for fly control . Pests. 17(5):13, 14, 22.

LaBRECQUE, G. C. — 1961. Studies with three alkylating agents as house fly sterilants. J. Econ. Entomol. 54(4):684-689.

LaBRECQUE, G. C., M. C. EVERS and D. W. MEIFERT — 1966. Control of house flies in outdoor privies with larvicides. J. Econ. Entomol. 59(1):245.

LaBRECQUE, G. C., D. W. MEIFERT and R. L. FYE — 1963. A field study on the control of house flies with chemosterilant techniques. J. Econ. Entomol. 56(2):150-152.

LaBRECQUE, G.C., C. N. SMITH and D. W. MEIFERT — 1962. A field experiment in the control of house flies with chemosterilant baits. J. Econ. Entomol. 55(4):449-451.

LaBRECQUE, G. C., and H. G. WILSON — 1957. Housefly resistance to organophosphorus compounds. Agr. Chem. 12(9):46-47, 147, 149. 1961. Development of insecticide resistance in three field strains of house flies. J. Econ. Entomol. 54(6):1257-1258.

LEIKINA, L. I. — 1942. The role of different substrates on the breeding of *M. domestica*. Biol. Abst. 20:654, 1946.

LEWALLEN, L. L. — 1954. Biological and toxicological studies of the little house fly. J. Econ. Entomol. 47(6):1137-1140.

LIEUX, D. B. and J. A. MULRENNAN — 1956. The biology and control of midges in Florida. Mosq. News 16:201-204.

LINDQUIST, A. W. and R. C. BUSHLAND — 1944. Toxicity of DDT and pyrethrum to *Chaoborus punctipennis*. J. Econ. Entomol. 37(6):842.

LINDQUIST, A. W., A. H. MADDEN and H. G. WILSON — 1947. Pre-treating house flies with synergists before applying pyrethrum sprays. J. Econ. Entomol. 40(3):426-427.

LINDQUIST, A. W., and A. R. ROTH — 1950 Effects of dichlorodiphenyl dichloroethane on larvae of the Clear Lake gnat in California. J. Econ. Entomol. 43(3):328-329. 1951. Equipment and materials used in control of gnats in Clear Lake, Calif. Mosq. News 11(3):161-163.

LINDQUIST, A. W., A. R. ROTH and J. R. WALKER — 1951. Control of the Clear Lake gnat in California. J. Econ. Entomol. 44(4):572-577.

LINDQUIST, A. W. and H. G. WILSON. — 1948. Development of a strain of houseflies resistant to DDT. Science 107(2776):276. March 12.

LINDQUIST, A. W., W. W. YATES and R. A. HOFFMAN — 1951. Studies of the flight habits of three species of flies tagged with radioactive phosphorus. J. Econ. Entomol. 44(3):397-400.

LIVINGSTON, A. M. — 1951. Trickling filter fly control. Sewage and Industrial Wastes 23:241-244.

MacNAIR, I. P. — 1940. A brief history of household insect sprays. Soap & S.C. 16(6):104-107.

MADDEN, A. H., A. W. LINDQUIST, O. M. LONGCOY, and E. F. KNIPLING — 1946. Control of adult sand flies by airplane spraying with DDT. Florida Entomol. 29(1):5-10. Biol. Absts. 22(1):209, 1948.

MAIL, G. A., and H. F. SCHOOF — 1954. Overwintering habits of domestic flies at Charleston, West Virginia. Ann. Entomol. Soc. Amer. 47(4):668-676.

MALLIS, A. and A. C. MILLER — 1967. Prolonged resistance in the house fly and bed bug. J. Econ. Entomol. 57(4):608-609.

MALLIS, A. and R. J. PENCE — 1941. The Pacific drain fly in homes. J. Econ. Entomol. 34:586.

MANIS, H. C., A. L. DUGAS and I. FOX — 1942. Toxicity of paradichlorobenzene to third-instar larvae of the housefly. J. Econ. Entomol. 35(5):662-664.

MARCH, R. B. and L. L. LEWALLEN — 1950. A comparison of DDT-resistant and non-resistant house flies. J. Econ. Entomol. 43(5):721-722.

MARCH, R. B. and R. L. METCALF — 1949. Laboratory and field studies of DDT-resistant house flies in southern California. Bul. Calif. State Dept. Agr. 38:93-101. 1949. Development of resistance to organic insecticides other than DDT by houseflies. J. Econ. Entomol. 42(6):990. 1951. Summary of research on insects resistant to insecticides. Citrus Experiment Station Mimeog. 1952. Insecticide research for the control of resistant house flies. Pest Control 20(4):12,14,16,18.

MATHIS, W., E. A. SMITH and H. F. SCHOOF — 1967. Influence of air velocities in preventing entrance of house flies. S. E. Branch ESA.

MATTHYSSE, J. G. — 1945. Observations on housefly overwintering. J. Econ. Entomol. 38(4):493-494.

McDUFFIE, W. C., A. W. LINDQUIST and A. H. MADDEN — 1946. Control of fly larvae in simulated pit latrines and in carcasses. J. Econ. Entomol. 39(6):743-749.

McGOVRAN, E. R. and W. A. GERSDORFF — 1945. The effect of fly food on resistance to insecticides containing DDT or Pyrethrum. Soap & S.C. 21(12):165-169.

METCALF, C. L., W. P. FLINT and R. L. METCALF — 1951. Destructive and Useful Insects. 3rd Ed. McGraw-Hill.

MIDGELEY, A. R., W. O. MUELLER and D. E. DUNKLEE — 1943. Borax and boric acid for control of flies in manure. J. Amer. Soc. Agron. 35(9):779-785.

MILLER, A. C. — 1946. Housefly rearing. Gulf Research & Develop. Co. Unpublished. 1947. Tests conducted to determine the final resting place of flies killed through contact with DDT-treated walls & ceilings. Unpublished. 1949. Insecticide resistant houseflies. Proc. 4th Annual Meeting. No. Central States Branch. Amer. Assoc. Econ. Entomol. pp. 91-92. Mimeo. 1951. *Psychoda* fly nuisance. Unpublished.

MILLER, A.C. and A. MALLIS — 1945. Presence of flies in Pittsburgh dump. Gulf Research & Development Co. Unpublished.

MILLER, A. C., A. MALLIS and R. V. SHARPLESS — 1952. Aerosol insecticides. Their evaluation against house flies, cockroaches. Soap and S.C. 28(2):153,181; 28(3):143.147,149.

MILLER, A. C. and W. A. SIMANTON — 1938. Biological factors in Peet-Grady results. Soap & S.C. 14(5):103, 105,109,111, 113.

MISSIROLI, A. — 1947. Riduzione o eradicazione degli anofeli. Riv. Parasitol. 8(2/3):141-169. Biol. Abst. 22(6):1483, 1948. 1948. *Anopheles* control in the Mediterranean area. Proc. 4th Int'l. Congr. Trop. Med. & Malaria Vol. II pp. 1566-1576.

MORELAND, C. R. and W. S. McLEOD — 1957. Studies on rearing the house fly on a bran-alfalfa medium. J. Econ. Entomol. 50(2):146-150.

MORSE, A. P. — 1911. *Lucilia Sericata* as a household pest. Psyche. 18:89-92.

MOUNT, G. A., C. S. LOFGREN, AND J. B. GAHAN — 1966. Malathion, naled, fenthion, and Bayer 39007 thermal fogs for control of the stable fly. Fla. Ent. 49(3):169-173.

MULLA, M. S. — 1958. Recent developments in the biology and control of *Hippelates* eye gnats. Proc. Papers Calif. Mosq. Control Ass., 26:78-82. 1959. Some important aspects of *Hippelates* gnats, with a brief presentation of current research findings. Proc. Papers Calif. Mosq. Control Ass., 27:48-52. 1965. Biology and control of *Hippelates* eye gnats. Proc. 33rd. Annual Conf. Calif. Mosq. Control Ass., pp. 26-28.

MULLA, M. S., M. M. BARNES and M. J. GARBER — 1960. Soil treatments with insecticides for the control of the eye gnats *Hippelates collusor* and *H. hermsi*. J. Econ. Entomol. 53(3):362-365.

MULLA, M. S., R. W. DORNER, G. P. GEORGHIOU and M. J. GARBER — 1960. Olfactometer and procedure for testing baits and chemical attractants against *Hippelates* eye gnats. Ann. Entomol. Soc. Amer., 53:527-537.

MULLA, M. S., G.P. GEORGHIOU and R. W. DORNER — 1960. Effect of aging and concentration on the attractancy of proteinaceous materials to *Hippelates* gnats. Ann. Entomol. Soc. Amer., 53:835-841.

MULLA, M. S. and R. B. MARCH — 1959. Flight range, dispersal patterns and population density of the eye gnat *Hippelates collusor* (Townsend). Ann. Entomol. Soc. Amer., 52:641-646.

NELSON, J. A. — 1949. Sprayer maintenance forum. Pests and Their Control 17(5):17-18,20.

NORRIS, K. R. — 1965. The bionomics of blow flies. Ann. Rev. Entomol. 10:47:68.

ODENEAL, J. — 1959. A new method of control for flying insects. Aerosol Age 4(3):20-21,61.

OGDEN, L. J. and J. W. KILPATRICK — 1958. Control of *Fannia canicularis* (L.) in Utah dairy barns. J. Econ. Entomol. 51(5):611-612.

OLTEN, G. S. and E. F. LEGNER — 1975. Winter inoculative releases of parasitoids to reduce houseflies in poultry manure. J. Econ. Entomol. 68(1):35-37.

ORDMAN, D. — 1946. Bronchial asthma caused by the trickling sewage filter fly *(Psychoda):* inhalant insect allergy. Nature (London) 157:441. Apr. 6, 1946.

PATTERSON, R.S. — 1957. On the causes of broken wings of the house fly. J. Econ. Entomol. 50(1):104-105.

PATTERSON, R. S. and F. L. WILSON — 1966. Fogging and granule applications are teamed to control chironomid midges on Florida lakefronts. Pest Control 34(6):26,29,30,32.

PATTON, W. S. — 1931. Insects, Ticks, Mites and Venomous Animals, H. R. Grubb, Ltd.

PENROSE, R.D. and D. J. WOMELDORF — 1962. *Drosophila* Infestation in wineries. Calif. Vector Views 9(1):1-3.

PERRY, A. S. and W. M. HOSKINS — 1950. The detoxification of DDT by resistant houseflies and inhibition of this process by piperonyl cyclonene. Science III(2892):600-601. 1951. Detoxification of DDT as a factor in the resistance of house flies. J. Econ. Entomol. 44(6):850-857. 1951 Synergistic action with DDT toward resistant house flies. J. Econ. Entomol. 44(6):839-849.

PIMENTEL, D. and B. EPSTEIN — 1960. The cluster fly, *Pollenia rudis* (Diptera, Calliphoridae). Ann. Entomol. Soc. Amer. 53(4):553-554.

PINMENTEL, D., H. H. SCHWARDT and J. E. DEWEY — 1954. The inheritance of DDT-resistance in the house fly. Ann. Entomol. Soc. Amer. 47(1):208-213.

PIMENTEL, D., H. H. SCHWARDT, J. E. DEWEY and L. B. NORTON — 1950. Studies on insecticide resistant houseflies in New York. Soap & S.C. Off. Proc 37th Ann. Meeting. Dec. pp. 94-96.

PIMENTEL, D., H. H. SCHWARDT and L. B. NORTON — 1951. New methods of housefly control. Soap & S.C. 27(1):102,103,105,112A, 1950. Housefly control in dairy barns. J. Econ. Entomol. 43(4):510-515.

POWER, M. E. and J. L. MELNICK — 1945. A three-year survey of the fly population in New Haven during epidemic and non-epidemic years for poliomyelitis. Yale J. Biol. and Med. 18(1):55-69. Biol. Abst. 20:829, 1946.

PROCTER, F. — 1941. An insect pest affecting milk distribution. Dairy Indust. 6(3):70-71. Biol. Abst. 17(10):2454, 1943.

QUARTERMAN, K. D. and W. MATHIS — 1952. Field studies on the use of insecticides to control fly breeding in garbage cans. Amer. J. Trop. Med & Hyg. 1(6):1032-1037.

RACHESKY, S. — 1972. Cluster flies. Pest Control 40(10):47.

ROA, A. V. K. M. — 1957. The effect of DDT in the production of tolerant and sensitive strains of the house fly *(Musca domestica* L.) Biol. Abst. 32(4):1192.

REES, D. M. and J. V. SMITH — 1950. Effective control methods used on biting gnats in Utah during 1949 (Diptera:Ceratopogonidae). Mosq. News 10(1):9-15. 1952. Control of biting gnats in North Salt Lake City, Utah (Diptera:Heleidae). Mosq. News 12(2):49-52.

RICHARDSON, C. H. — 1923. The domestic flies of New Jersey. N.J. Agr. Expt. Sta. Bull. 307. 1936. Flies! As household pests in Iowa. Iowa Agr. Expt. Sta. Bull. 345.

RINER, JOHN L., RONNIE L. BYFORD and JAKIE A. HAIR — 1981. Sustained-release rabon bolus for face fly control in cattle feces. J. Econ. Entomol. 74(3):359-362.

RUTZ, DONALD A. and RICHARD C. AXTELL — 1981. House Fly *(Musca domestica)* control in broiler-breeder poultry houses by pupal parasites (Hymenoptera: Pteromalidae): indigenous parasite species and releases of *Muscidifurax raptor*. Environ. Entomol. 10(3):343-345.

SABROSKY, C. W. — 1940. Chloropids swarming in houses. J. Econ. Entomol. 33:946-947. 1951. Nomenclature of the eye gnats *(Hippelates* spp.) Amer. J. Trop. Med. 31(2):257-258. Biol. Abst. 25(10):2863. 1951.

SACCA, G. — 1947. Sull'esistenza di mosche domestiche resistanti al DDT. Riv. de Parassitol 8(⅔):127-128.

SACKTOR, B. — 1950. A comparison of the cytochrome oxidase activity of two strains of houseflies. J. Econ. Entomol. 43(6):832-837.

SANDERS, G. E. — 1942. Housefly control in relation to poliomyelitis. Pests 10(3):22-26.

SATCHELL, G. H. — 1947. The ecology of the British species of *Psychoda* (Diptera: Psychodidae). Annals Appl. Biol 34(4):611-621. 1949. The ecology of the British species of *Psychoda* (Diptera: Psychodidae). Ann. Appl. Biol. 34(4):611-621. Rev. Appl. Entomol. B. 37(7):141-142.

SAVAGE, E. P. and H. F. SCHOOF — 1955. The species composition of fly populations at several types of problem sites in urban areas. Ann. Entomol. Soc. Amer. 48(4):251-257.

SAWICKI, R.M. and D. V. HOLBROOK — 1961. The rearing, handling and biology of house flies (*Musca domestica* L.) for assay of insecticides by the application of measured drops. Pyr. Post 6(2):3-18.

SCHOOF, H. F. — 1959. How far do flies fly? Pest Control 27(4):16,18,20,22,66.

SCHOOF, H. F. and J. W. KILPATRICK — 1957. House fly control with parathion and diazinon impregnated cords in dairy barns and dining halls. J. Econ. Entomol. 50(1):24-27. 1958. A field strain of malathion-resistant house flies. J. Econ. Entomol. 51(1):18-19.

SCHOOF, H. F., G. A. MAIL and E. P. SAVAGE — 1954. Fly production sources in urban communities. J. Econ. Entomol. 47(2):245-253.

SCHOOF, H. F. and E. P. SAVAGE — 1955. Comparative studies of urban fly populations in Arizona, Kansas, Michigan, New York, and West Virginia. Ann. Ent. Soc. Amer. 48:1-12.

SCHOOF, H.F. and R. E. SIVERLY — 1954. Privies as a source of fly production in an urban area. Amer. J. Trop. Med. & Hyg. 3(5):930-935.

SCHOOF, H. F., R. E. SIVERLY and J. A. JENSEN — 1952. House fly dispersion studies in metropolitan areas. J. Econ. Entomol. 45(4):675-683.

SCUDDER, H. L. — 1947. A new technique for sampling the density of house fly (*Musca domestica*) populations. USPHS Rpts. 62:681-686. 1949. Some principles of fly control for the sanitarian. Amer. J. Trop. Med. 29(4):609-623. Biol. Abst. 24(4):1062, 1950.

SENN, C. L. — 1966. Fly control in food establishments. PCO News 26(4):34-35.

SHUGART, J. I., J. B. CAMPBELL, D. B. HUDSON, C. M. HIBBS, R. G. WHITE and D. C. CLANTON — 1979. Ability of the face fly to cause damage to eyes of cattle. J. Econ. Entomol. 72(4):633-635.

SIMANTON, W. A. and A. C. MILLER — 1938. Housefly age as a factor in susceptibility to pyrethrum sprays. J. Econ. Entomol. 30(6):917-921.

SIMMONS, P. — 1927. The cheese skipper as a pest in cured meats. USDA Bull. 1453.

SIMMONS, S. W. — 1944. Observations on the biology of the stablefly in Florida. J. Econ. Entomol. 37(5):680-686.

SIMMONS, S. W. and W. E. DOVE — 1942. Waste celery as a breeding medium for the stable fly, or "dog fly," with suggestions for control J. Econ. Entomol. 35(5):709-715. 1945. Experimental use of gas condensate for the prevention of fly breeding. J. Econ. Entomol. 38(1):23-25.

SIMPSON, R. W. — 1948. Filter fly larvae control. Soap & S.C. 24(5):143.

SIVERLY, R.E. and H. F. SCHOOF — 1955a. Utilization of various production media by muscoid flies in a metropolitan area. I. Adaptability of different flies for infestation of prevalent media. Ann. Entomol. Soc. Amer. 48(4):258-262. 1955b. II. Seasonal influence on degree and extent of fly production. Ann. Entomol. Soc. Amer. 48(5):320-324.

SMART, J. — 1943. A Handbook for the Identification of Insects of Medical Importance. pp. 58-59. British Museum.

SMITH, A.C. — 1958. Fly-cord studies in California. Calif. Vector Views. 5(9):58-61.

SMITH, C. N., A. N. DAVIS, D. E. WEIDHAAS and E. L. SEABROOK — 1959. Insecticide resistance in the salt-marsh sand fly *Culicoides furens*. J. Econ. Entomol. 52(2):352-353.

SMITH, C. N., G. C. LaBRECQUE, H. G. WILSON, R. A. HOFFMAN, C. M. JONES and J. W. WARREN — 1960. Dimetilan baits, fly ribbons and cords for the control of house flies. J. Econ. Entomol. 53(5):898-902.

SMITH, L.M. and H. LOWE — 1948. The black gnats of California. Hilgardia 18(3):157-183.

SOLLERS, H. — 1950. Fungus gnats, various species Family Fungivoridae (formerly Mycetophilidae). Pest Control Tech. NPCA. p. 355.

SPENCER, W. P. — 1950. Biology of *Drosophila*. M. Demerec. Editor. Wiley.

SPENCER, T.S., R.K. SHIMMIN and R.F. SCHOEPPNER — 1975. Field tests of repellants against the valley black gnat. Calif. Vector News 22(1):5-7.

SPIESS, R. — 1952. Control of filter flies by chemical treatment. Water & Sewage Works. 99:250-253.

SPIETH, H.T. — 1951. The breeding site of *Drosophila lacicola* Patterson. Science. 113(2931):232.

STARR, D.F., and D. R. CALSETTA — 1954. Ryania as a house fly larvicide. Agr. Chem. 9(11):50-53.

STERNBURG, J., C.W. KEARNS and W. N. BRUCE — 1950. Absorption and metabolism of DDT by resistant and susceptible houseflies. J. Econ. Entomol. 43(2):214-219.

STERNBURG, J., E. B. VINSON and C. W. KEARNS — 1953. Enzymatic dehydrochlorination of DDT by resistant flies. J. Econ. Entomol. 46(3):513-515.

STEVE, P. C. — 1960. Biology and control of the little house fly, *Fannia canicularis* in Massachusetts. J. Econ. Entomol. 53(6):999-1004.

STOMBLER, V., C. D. PELEKASSIS and E. S. DOYLE — 1957. *Drosophila* control on harvested tomatoes for processing in California. 1956. J. Econ. Entomol. 50(4):476-480.

STONE, R. S., and H. W. BRYDON — 1965. The effectiveness of 3 methods for the control of immature *Fannia* species in poultry manure. J. Med. Entomol. 2(2):145-149.

STRICKLAND, E. H. — 1945. A method for permanently reducing the number of blowflies in screened houses. Bull. B'klyn Entomol. Soc. 40(2):59-60.

TETRAULT, R. C. — 1967. Moth fly control at sewage treatment plants. Pa. Pest Control Quarterly, October — p. 5.

THOMPSON, R. K., A. A. WHIPP, D. L. DAVIS and E. G. BATTE — 1953. Fly control with a new bait application method. J. Econ. Entomol. 46(3):404-409.

TINKHAM, E.R. — 1951. Desert flies. Personal communication. 1952. Hyman Newsletter.

TESKEY, H. J. — 1960. A review of the life history and habits of *Musca autumnalis* DeGeer: (Diptera: Muscidae). Can. Entomol. 92(5):360-366.

TOMLINSON, T. G. — 1945. Control by DDT of flies breeding in percolating sewage filters. Nature 156(3964):478-479 Rev. Appl. Entomol. B. 35(7):102, July, 1947.

TRAVIS, B. V., D. L. COLLINS, G. DeFOLIART and H. JAMNBACK — 1951.

Strip spraying by helicopter to control blackfly larvae. Mosq. News 11(2):65-98.

TRAVIS, B. V. and F. A. MORTON — 1950. Insect repellents and nets for use against sand flies. Proc. 37th Ann. Meeting. N.J. Mosq. Exterm. Assoc. pp. 154-156.

TRAVIS, B. V., A. L. SMITH and A. H. MADDEN — 1951 Effectiveness of insect repellents against black flies. J. Econ. Entomol. 44(5):813-814.

TURNER, C. L. — 1923. The Psychodidae (Moth-like flies) as subject for studies in breeding and heredity. Biol. Bul. 57:545-558.

UPHOLT, W. M. — 1950. Significance to the insecticide industry of fly resistance. Soap & S.C. Off. Proc. 37th Ann. Meeting. CSMA Dec. pp. 69-71.

USINGER, R. L., and W. R. KELLEN — 1955. The role of insects in sewage disposal beds. Hilgardia 23(10):263-321.

WALL, W. J., JR. and S. F. BAILEY — 1952. Repellency of cotton tufts to flies. J. Econ. Entomol. 45(4):749-750.

WANG, C. H. — 1964. Laboratory observations on the life history and habits of the face fly, *Musca autumnalis*. Ann. Entomol. Soc. Amer. 57(5):563-569.

WARE, G. W. — 1966. Powermower flies. J. Econ. Entomol. 59(2):477-478.

WARRICK, L. F., and G. F. BERNAUER — 1946. How to use DDT for insect control. Water and Sewage Works 93:329-332. 1948. Spraying with DDT. Soap & S.C. 24(1):157.

WATT, J. and D. R. LINDSAY — 1948. Diarrheal disease control studies. I. Effect of fly control in a high morbidity area. USPHS Rpts. 63(41):1319-1334.

WEBB, J. L. and R. N. HUTCHINSON — 1916. A preliminary note on the bionomics of *P. rudis* (Fabr.) in America. Proc. Entomol. Soc. Wash. 18:197-199.

WEST, L. S. — 1951. The Housefly: Its Natural History, Medical Importance, and Control. Comstock Publ. Co., Inc.

WIESMANN, R. — 1947. Untersuchungen uber das physiologische. Verhalten von *Musca domestica* L. versichiedener Provenienzen. Schweiz. Ent. Gesell. Mitt 20(5):484-504.

WHITSEL, R. H. and R. F. SCHOEPPNER — 1966. Summary of a study of the biology and control of the valley black gnat, *Leptoconops torrens* Townsend. Calif. Vector Views. 13(3):17-26.

WILLIAMS, RALPH E. and ERIC J. WESTBY — 1980. Evaluations of pyrethroids impregnated in cattle ear tags for control of face flies and horn flies. J. Econ. Entomol. 73(6):791-792.

WILSON, H. G. and J. B. GAHAN — 1948. Susceptibility of DDT-resistant house flies to other insecticidal sprays. Science 107(2276):276-277. 1957. Control of housefly larvae in poultry houses. J. Econ. Entomol. 50(5):613-614.

WRIGHT, J. E. — 1975. Insect growth regulators: development of house flies in feces of bovines fed TH 6040 in mineral blocks and reduction in field populations by surface dichlorvos at larval breeding areas. J. Econ. Entomol. 68(3):322-324.

WRIGHT, J. E., G. E. SPATES and M. SCHWARZ — 1976. Insect growth regulator A 13-36206. Biological activity against *Stomoxys calcitrans* and *Musca domestica* and its environmental stability. J. Econ. Entomol. 69(1):79-82.

YATES, W. W., A. W. LINDQUIST and J. S. BUTTS — 1952. Further studies of dispersion of flies tagged with radioactive phosphorus. J. Econ. Entomol. 45(3):547-548.

YERINGTON, A. P. — 1967. Control of *Drosophila* with dichlorvos aerosols. J. Econ. Entomol. 60(3):701-704.

ZICH, J. — 1944. Giving *Musca domestica* the air. Milk Plant Monthly 33(3):60-61. Mar.

STANLEY G. GREEN

Dr. Stanley Green received his bachelor's and master's degrees from the University of Colorado. For his disserttation on the taxonomy of the family Oribatulidae he received his Ph.D. from The Colorado State University, Fort Collins, Col. in 1969.

Green is responsible for vector control and urban pest control programs in Pennsylvania, providing training to public health professionals and pest control operators. Green teaches a number of broad-ranging courses, including basic entomology and structural pest control. In addition, he teaches in-depth courses for specialists, including mosquito and termite control courses. Green's courses and workshops have played an important role in developing professionalism in the pest control industry by encouraging a more scientific approach to pest control. A <u>Vector Control</u> and <u>Pest Control</u> newsletter are published by Green to help accomplish this goal.

As an extension entomologist, Green is responsible for the preparation of the popular <u>Extension Entomological Fact Sheets</u> on household arthropod pests. These are aimed at the general public to help them better understand arthropod pests and how to deal with them.

Green is well known for his selfless attitude toward those who need advice on pest problems. Many professionals have benefitted form his wise counselling and no problems are too big or too small to capture his interest. Green's availability and ability to communicate have made him a key resource for the news media when pest stories need professional input.

Green's achievements have been recognized by numerous honor societies, including Sigma Xi, Epsilon Sigma Phi and Pi Chi Omega. In addition, he is active in many professional associations including the Entomological Society of America, the American Mosquito Control Association, the American Registry of Professional Entomologists and the American Associaiton for the Advancement of Science. Green is also active in many state and regional associations and is currently president of the Mosquito and Vector Control Association of Pennsylvania.

CHAPTER TWENTY

Mosquitoes

Revised by Stanley G. Green[1]

PLEA TO A MOSQUITO

Pray transfer your air attack
To my husband's slothful back.
Puncture him with malice keen
Then perhaps he'll fix that screen!

—Helen Gorn Sutin.

MOSQUITOES ARE IMPORTANT vectors of diseases such as malaria, yellow fever, filariasis, dengue and encephalitis. Besides their disease-carrying abilities, biting mosquitoes can be terribly annoying to people working, camping or relaxing outdoors.

Knight and Stone (1977) catalog 2,960 species of mosquitoes known from around the world. These are grouped in 34 genera. About 130 species occur in North America (Gillett, 1972). A number of publications now exist which deal regionally with the North American species. Wood et al. (1979) consider the 74 species found in Canada. An indispensible book for anyone involved in mosquito control is Carpenter and LaCasse (1955). This work covered all mosquitoes known from North America up to 1955. Since then updates of *North American Mosquitoes* have been published by Carpenter (1968, 1970, 1974). The mosquitoes found in the United States are covered by the following state publications: Alaska, 27 species — Gjullin et al. (1961); California, 47 species — Bohart and Washino (1978); Colorado, 42 species — Harmston and Lawson (1967); Connecticut, 38 species — Wallis (1960); Illinois, 55 species — Ross and Horsfall (1965); Indiana, 51 species — Siverly (1972); Iowa, 43 species — Knight and Wonio (1969); Minnesota, 47 species — Barr (1958); New York, 58 species — Means (1979); Ohio, 49 species — Anon. (1968); Pennsylvania, 41 species — Rutschky et al. (1958). (nine more species have been added recently bringing the number of species in Pennsylvania to 50).

As a rule, the mosquito problem is one that must be dealt with by public health agencies since mosquito control is essentially a problem concerned with

[1]*Associate Professor of Entomology, The Pennsylvania State University, Cooperative Extension Service, Philadelphia, Pa.*

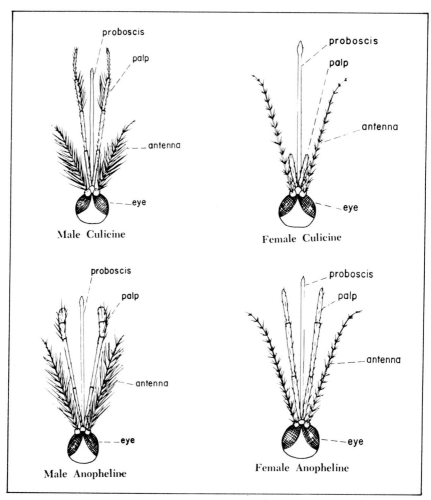

Fig. 20-1. Key to identifying Anopheline and Culicine mosquitoes.

removing, controlling or treating bodies of water in which mosquitoes breed. These agencies attack the mosquito problem through draining swamps and other bodies of water, water-level management, and by spraying standing or moving bodies of water, etc.

The author is primarily concerned with the control of mosquitoes in homes or in their immediate vicinity. The amount of effort to control mosquitoes in or around the home is largely determined by the control program undertaken by the mosquito control agencies in the area where the home is situated. The home-owner can obtain some degree of protection from mosquitoes through the use of screening, household insecticides, aerosols and repellents. He also can remove or treat with insecticides small bodies of water that breed mosquitoes in the vicinity of the home.

Mosquito characteristics and development. Mosquitoes are flies which belong to the family Culicidae. They are characterized by scales on their wings, legs and parts of their bodies. They also have a long proboscis which in the female is able to pierce the skin of animals and suck blood. Males do not feed on blood, but on nectar and plant juices. Males differ from females. They have very bushy antennae and have palpi as long as their proboscis. Female Anopheles also have long palps, but they are not swollen and hairy at their tips. (Fig. 20-1).

Mosquitoes are classified in three subfamilies:

- Toxorhynchitinae. This subfamily contains the genus *Toxorhynchites*, mosquitoes which, as adult females, do not feed on blood. Their larvae are beneficial since they feed on other mosquito larvae.
- Anophelinae. Anophelinae contains the important genus *Anopheles*, many species of which transmit malaria to man.
- Culicinae. Contains the bulk of the mosquito genera and species. Important genera in this subfamily are *Aedes, Culex, Culiseta, Psorophora* and *Mansonia*. Many of the species in these genera are known vectors of such diseases as encephalitis, yellow fever, filariasis and heartworm.

Mosquitoes undergo complete metamorphosis. They develop through four stages (egg, larvae, pupa and adult). Complete development from egg to adult usually takes from 10 to 14 days (Matheson, 1950), but varies according to the species and the temperatures.

Mosquitoes, as do many insects, have a tremendous potential for reproduction. Harrison (1978) gives the following example: A female mosquito, depending on species, can produce about 50 to 500 eggs in her first brood. She produces fewer eggs in subsequent ovipositions (of which there may be eight to 10). If we consider an average of 200 eggs per brood (of which half would be female) and that a mosquito can complete its development in less than two weeks, we can obtain in only five generations some 20 million mosquitoes. If we could get 20 million offspring from only one female what could thousands of females give rise to?

Mosquitoes of the genus *Anopheles* deposit their eggs singly on the surface of water. Their eggs are boat-shaped and have "floats" on their sides which contain trapped air. The eggs of *Culex* and some species of *Mansonia* are joined together at their sides and stand on their tips in groups called rafts (Fig. 20-2). Other species of *Mansonia* are submerged and clustered on the underside of aquatic vegetation. *Aedes* and *Psorophora* eggs are without the floats and are usually placed at the edge of a body of water or in a damp area just above the water's surface. Their eggs hatch following a period of dryness and for some species a period of freezing is required. Some of their eggs may not hatch for years. The eggs of other mosquitoes usually hatch within 48 hours.

Eggs hatch into larvae commonly called wigglers, wrigglers or wiggletails because of the way they swim. They lash their abdomen about from side to side and move wriggling tail first through the water. They also can move slowly head first propelling themselves by movement of their mouth brushes.

Larvae feed by beating their mouth brushes which draws water containing food such as algae, protozoans and minute organic debris into their mouths. There are four larval stages or instars. Larvae of all but the genus *Anopheles* have an air tube or siphon through which they breathe atmospheric air at the water's surface. Species of *Anopheles* do not have an air tube but rest horizontally at the water's surface where they breathe through an opening (spiracle)

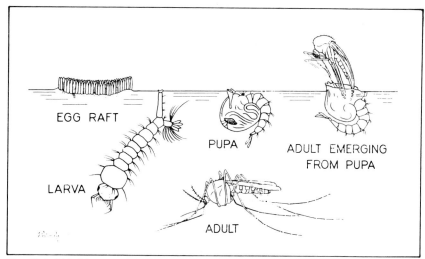

EGG RAFT

PUPA

ADULT EMERGING
FROM PUPA

LARVA

ADULT

Fig. 20-2. Life history of the house mosquito.

on their eighth abdominal segment. These mosquitoes do not feed on organic
matter below them. Gillett (1972) says they feed from the surface of the water
while at the same time breathing air at the water's surface. This would not be
possible, he says, except "for the remarkable ability of the larva to swivel its
head round 180° and so bring its mouth brushes, which normally point downward, directly in contact with the surface film. It is as if a man could float in
water back uppermost but be able to turn his head round so that he could look
upwards at the sky."

Species of the genus *Mansonia* differ from other mosquitoes by not breathing
atmospheric air. Their larvae have a short siphon which is toothed and spur-
like. They insert this into the roots of submerged plants and from them derive
their oxygen. Because *Mansonia* larvae are not found at the water's surface they
are not taken by dipping (the usual sampling method for larvae). The larval
stage lasts about seven days (depending on temperature and species).

Fourth stage larvae molt, becoming pupae which do not feed but are mobile.
They spend much of their time at the water's surface breathing air through a
pair of respiratory trumpets. When disturbed they move in a tumbling fashion,
somersaulting head over tail. Pupae are frequently called "tumblers" because
of this type of locomotion. *Mansonia* pupae use their modified pointed respiratory trumpets to clamp onto aquatic plants and in this way obtain their oxygen
supply. The pupal stage normally exists from one to three days at which time
the adult mosquito emerges.

King et al. (1944) discuss adult mosquito biology and their remarks are reproduced verbatim: "The length of life of adult mosquitoes under natural conditions is difficult to determine, but for most of the southern species it is probably only a few weeks during the summer months. Some of the northern species
of *Aedes* that emerge early in the spring apparently live much longer. Daily
observations on abundance following the emergence of a large brood of certain
species of *Anopheles* and *Aedes* have shown a marked reduction in numbers in
two weeks. The southern house mosquito probably lives longer than this and

the yellow-fever mosquito may live, on an average, a month or more, with a maximum of several months."

The above applies to female mosquitoes; males are rather short lived. James and Harwood (1969) say males only live six to seven days, but some have lived more than a month in the laboratory when kept at high humidities and given sufficient carbohydrates. They also mention some females (*Aedes* sp.) living 52 to 113 days.

How do mosquitoes survive the winter? In the northern latitudes most species of *Culex* and *Anopheles* overwinter as adult females. Russell et al. (1963) state in Siberia one species of *Anopheles* hibernates in basements where the temperature ranged from 48.2° to −0.4° F/9° to −18° C. "The insects frequently became frozen to walls and ceilings, but after gradual thawing revive and are capable of laying eggs. Most species of *Aedes* and *Psorophora* overwinter as eggs, while our species of *Mansonia, Toxorhynchites* and *Orthopodomyia*, spend the winter in the larval stage.

King et al. (1944) describe the female's proboscis as follows: "The piercing organs of the female mosquito consist of six elongated parts enclosed in a flexible sheath called the labium. When the mouth parts are inserted in the skin for bloodsucking, the sheath is bent backward in the middle like a bow. There are two pairs of slender cutting organs, the mandibles and the maxillae, and two additional organs called the hypopharynx and the labrum-epipharynx. The latter is channeled and the last two organs, when pressed together, form a tube through which blood and other liquids are drawn. A very small separate duct is found in a ventral thickening of the hypopharynx, through which is injected the secretion from salivary glands. This salivary secretion is responsible for the itching sensation caused by mosquito bites. Not all species of mosquitoes have bloodsucking females. In the genus *Toxorhynchites* the proboscis of the female is not adapted to piercing and some of the species in other genera are not known to take blood meals."

An interesting group of mosquitoes (of the genus *Malaya*) is mentioned by Gillett (1972). These mosquitoes fly in front of certain ants which have just collected honeydew from aphids. The mosquito thrusts the tip of its proboscis (which is swollen at its tip) into the jaws of the ant and sucks out the honeydew. The ant does not appear to be disturbed at the loss of food.

The great majority of female mosquitoes feed on blood which is a source of protein used in egg production. However, some species do produce a brood shortly after they emerge from the pupal case without having to take a blood meal. They obviously have enough protein accumulated from their larval feeding. They must, however, have blood to produce subsequent broods.

Gillett (1972) and Harrison (1978) reported the following on the amount of blood mosquitoes ingest: "Although a mosquito takes only some two to eight milligrammes of blood per meal, depending largely on the size of the mosquito, the total amount taken when thousands upon thousands are involved can be considerable. There are cases on record of dogs and even cattle dying following mass attack by mosquitoes; reference has already been made to human death following persistent and wholesale attack. Brian Hocking has shown the attack rate by north Canadian mosquitoes on a single exposed forearm may be as high as 289 bites per minute. Thus, a totally unprotected man could receive more than 9,000 bites per minute, which would result in a loss of about half his blood in less than two hours. But death following mass mosquito attack is not always due to loss of blood *per se*.

People and other hosts develop an allergy to mosquito bites and this response can on occasion by very severe. Fortunately, such cases are rare; usually the allergic response involves no more than the familiar itching red wheal." — (Gillett, 1972).

Fig. 20-3. Key to identifying mosquitoes.

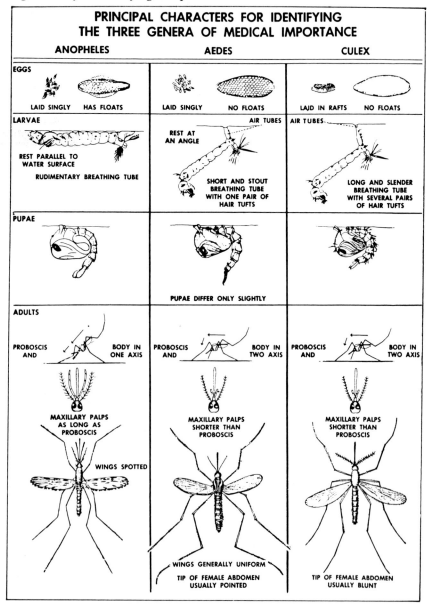

PRINCIPAL CHARACTERS FOR IDENTIFYING THE THREE GENERA OF MEDICAL IMPORTANCE

ANOPHELES	AEDES	CULEX
EGGS		
LAID SINGLY HAS FLOATS	LAID SINGLY NO FLOATS	LAID IN RAFTS NO FLOATS
LARVAE		
REST PARALLEL TO WATER SURFACE RUDIMENTARY BREATHING TUBE	AIR TUBES REST AT AN ANGLE SHORT AND STOUT BREATHING TUBE WITH ONE PAIR OF HAIR TUFTS	AIR TUBES LONG AND SLENDER BREATHING TUBE WITH SEVERAL PAIRS OF HAIR TUFTS
PUPAE		
	PUPAE DIFFER ONLY SLIGHTLY	
ADULTS		
PROBOSCIS AND BODY IN ONE AXIS MAXILLARY PALPS AS LONG AS PROBOSCIS WINGS SPOTTED	PROBOSCIS AND BODY IN TWO AXIS MAXILLARY PALPS SHORTER THAN PROBOSCIS WINGS GENERALLY UNIFORM TIP OF FEMALE ABDOMEN USUALLY POINTED	PROBOSCIS AND BODY IN TWO AXIS MAXILLARY PALPS SHORTER THAN PROBOSCIS TIP OF FEMALE ABDOMEN USUALLY BLUNT

"A female mosquito's reproductive exuberance demands quantities of protein to build and nourish her young. Hence her voracious appetite for blood. Her gorged gut may contain from one to more than four cubic millimeters of blood and she may suck twice that much in order to concentrate the haemoglobin solids from which she takes nourishment for herself and her eggs. From our blood store of some five million cubic millimeters, the loss is minute. For the mosquito the gain is immense. An average meal comes to two and a half times her unfed weight. That, in human terms translates as 400 pounds of beefsteak for a light eater. There are records of even more prodigious capacities, of virgin females taking in 15 times their own weight in blood. One glutton mosquito from Sumatra has been said to go on sucking long after her stomach is full, excreting not just plasma, but whole blood while she eats." — (Harrison, 1978).

Most female mosquitoes are able to produce fertile eggs for life after a single mating. Sperm is stored inside the female in a special organ called the spermatheca. It remains viable for months within the overwintering female.

It appears even male mosquitoes have sexual problems. Gillett (1972) notes: "When the male emerges from the pupa his sex organs are the wrong way round and mating is impossible. This curious state of affairs lasts for about a day, the actual time depending on the species and on the mean ambient temperature. Fortunately for the male (and for the female) he does not find his position particularly frustrating since he is unreceptive to the female for much of the first day anyway. The problem is solved very neatly, if in rather a surprising manner; the terminal segments of the abdomen bearing the male organs rotate through 180°. The dorsal surface comes to lie ventrally and vice versa; back becomes front and front becomes back." The above rotation is permanent for the male's life.

There are two major groups of mosquitoes, the culicine and the anopheline mosquitoes. The culicine mosquitoes include those of the genera *Culex, Aedes, Psorophora,* etc. Among the culicine mosquitoes we find some species that Bishopp (1939) has designated as "domestic mosquitoes." He states, "This grouping is made because these mosquitoes breed in and about houses, are always associated with man and have a restricted flight range. Thus, by individual effort the householder can do much to control them. It should be borne in mind that other kinds of mosquitoes may breed near houses and may enter buildings and attack man, as in the case of malaria mosquitoes." Three of the most common species are the northern house mosquito, *Culex pipiens* L., the southern house mosquito, *Culex quinquefasciatus* Say, and the yellow-fever mosquito, *Aedes aegypti* (L.) The eastern salt-marsh mosquito, *Aedes sollicitans* (Walk.), and the black salt-marsh mosquito, *Aedes taeniorhynchus* (Wied.), may migrate many miles from their breeding places. Ordinarily the flight range of the common malaria mosquito, *Anopheles quadrimaculatus,* is not more than one mile/1.6 km from the breeding source, although on occasion it may be as much as eight miles/12.9 km (Clarke, 1943).

MOSQUITO IDENTIFICATION

Mosquito identification is of primary importance. Only by knowing the correct identification of the mosquitoes you collect can you determine the following:

- The important mosquito species (those which bite man or are potential vectors of disease) versus those which are not. Obviously the control of disease vectors would take priority.

- The source of annoying mosquitoes. This is determined by the type of water they breed in and the distance they normally travel from their breeding area (i.e. flight range).
- The correct pesticide and adequate concentration for control.
- The timing for control efforts.

Figure 20-3 illustrates the key features of an adult mosquito.

To enable identification of the common mosquito genera refer to Siverly (1972).

To identify mosquitoes to species refer to publications pertaining to or near the state where they occur. These are mentioned at the beginning of this chapter. In addition the following are useful:

- For North America (Carpenter & LaCasse, 1955).
- For northeastern North America (Stojanovich, 1961)
- For southeastern U.S.A. (Stojanovich, 1960)
- For Canada (Wood et al., 1979)

Harback and Knight (1980) define and illustrate the anatomical terms used in mosquito classification.

Structures used in identifying mosquito larvae. (modified from Siverly, 1972). Figure 20-4 shows several of the structures which are used in identifying Culicine larvae. Three pairs of hairs on the top of the head are important in identifying larvae. Two pairs are close together near the middle of the head. These are numbered hair five and hair six. Number five is called the "upper head hair" and number six the "lower head hair." Number five is posterior to number six. Head hair number seven is known as the preantennal hair because

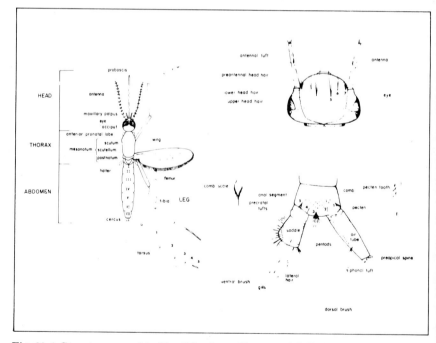

Fig. 20-4. Structures used in identification of larvae of Culicine mosquitoes.

it is near the antenna at the side of the head. The relative position and number of branches these hairs have is important in identification.

The eighth abdominal segment bears spines called comb scales and their shapes, numbers and arrangements in rows are good identifying characteristics, especially in the *Aedes*. Five hairs called the pentad hairs on the eighth segment have some diagnostic value (Fig. 20-4). The air tube or siphon is used a great deal in identification. The ratio of length to basal diameter is used (e.g. 4:1, 3:1, etc.). This ratio is known as the siphonal index. The double row of spines at the base of the anal tube is called the pecten and the arrangement of these spines is useful. Other useful structures on the siphon are the siphonal tuft and the preapical spine (see Fig. 20-4).

The anal segment in both Culicines and Anophelines bears the dorsal saddle, a dorsal and ventral brush, and anal gills (Fig. 20-4). The saddle is sclerotized and may or may not encircle the anal segment. A lateral hair or tuft arises from the posterior margin of the saddle.

The dorsal brush consists of an upper caudal tuft and a lower caudal hair. Tufts of the ventral brush which arise from the barred area or grid are called cratal tufts. Those arising anterior to these (directly under the saddle) are called precratal tufts (Fig. 20-4).

Members of the genus *Anopheles* have some of the same structures as Culicine larvae, with variations in nomenclature. Head hair five is called the inner frontal hair (it is closer to the midline); head hair six, the middle frontal; and head hair seven, the outer frontal. All the structure mentioned above are clearly pictured in Siverly's excellent diagrams.

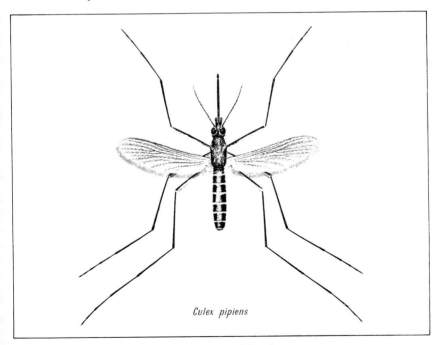

Culex pipiens

Fig. 20-5. Northern house mosquito, *Culex pipiens* L.

REPRESENTATIVE COMMON MOSQUITOES

The northern house mosquito, *Culex pipiens* L. This mosquito (Fig. 20-5) commonly is found in the northern states and Canada. It is a cosmopolitan mosquito that has been introduced into the United States and is pale brown with whitish bands across the abdomen. This mosquito also is known as "the common house mosquito" since it breeds in tin cans, rain barrels, tubs, hollow stumps, flatroofs, roadside and sewer ditches, sluggish streams and cesspools, and in fact, practically anywhere around the house where water is found. Kon (1944) found this species breeding in a Moscow subway. At times, the fertilized females may hibernate in tremendous numbers in hollow trees, street drain traps, cellars, etc. They are usually inactive in these locations, but may breed during the winter when it is sufficiently warm.

The hibernating females emerge early in the spring and lay their eggs in raft-like masses, each egg standing on end in this mass. There are 50 to 400 eggs in such a raft and the female may lay several such masses.

The larvae or "wigglers" emerge from the eggs during the summer and molt four times before becoming mature. Under optimum conditions they pupate in one week. The pupae are active but do not feed. The pupal stage may be completed in one day. When conditions are optimum the life cycle may require seven to 10 days, but often it is extended far beyond this.

Culex pipiens L. breeds continuously during the summer and is known to fly

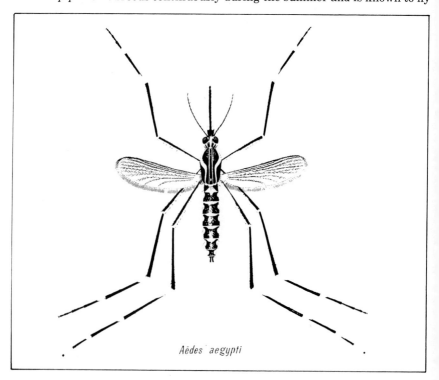

Aëdes aegypti

Fig. 20-6. Yellow fever mosquito, *Aedes aegypti* L.

up to a distance of 14 miles/22.5 km (Clarke, 1943) from the breeding place. When present in persistent numbers the common house mosquito is probably breeding in the immediate vicinity of the house. This mosquito is known to carry parasitic worms that infect man. It usually bites at dusk and after dark, hiding in some dark place during the day. Moreover, its persistent singing hum is very annoying to many individuals.

The southern house mosquito, *Culex quinquefasciatus* Say. This mosquito is very similar in appearance, life history and habits to the northern house mosquito. King et al. (1944) note it is "generally the most abundant night-biting house mosquito in the cities and towns of the southern states. In the North it is replaced by its very near relative, *Culex pipiens*, and the range of the two overlap in Virginia, northeastern Tennessee, North Carolina and other intermediate states."

The yellow-fever mosquito, *Aedes aegypti* (L.). The yellow-fever mosquito (Fig. 20-6) easily is recognized by the white or silver-colored lyre on its thorax and the silver stripes on its abdomen and legs. At one time this was a common house mosquito in the South. While *Aedes aegypti* is not common in the United States presently, it is included in this section because it is a most important disease vector and has been reintroduced into the United States many times in the past. It continues to be a potential health threat. The U.S. Department of Health (1965) lists 10 southern states in which *Aedes aegypti* can still be found including Oklahoma, Texas, Louisiana, Arkansas, Mississippi, Tennessee, Alabama, Georgia, North Carolina and Florida.

Bishopp (1939) states, ". . . the yellow-fever mosquito prefers relatively clear water and seldom, if ever, breeds in natural ponds or pools. Its wrigglers are found in tanks or cisterns, and in cans, bottles, pots, jars and other vessels that will hold even very small quantities of water. They often occur in unexpected places, such as discarded automobile casings, flower vases, holy water and baptismal fonts, urns in cemeteries, water troughs, water pans for chickens, grindstone cans, dishes of water place under legs of refrigerators and unused toilet bowls or tanks." This mosquito usually breeds close to the house. Morland and Hayes (1958) captured 78 percent of these mosquitoes within 75 feet/23 m of the release point. Berner (1947) found this species breeding not only in artificial containers, but in a number of natural situations such as in pools in interrupted drainage ditches. Christopher (1960) wrote a monograph on *A. aegypti*.

Aedes aegypti was once common in other parts of the United States, but now has been eliminated from the northern temperate region (Busvine, 1980).

The life cycle of the yellow-fever mosquito is somewhat similar to that presented for *Culex pipiens* L. except for the egg stage. According to Bishopp (1939), "the black, oval eggs of the yellow-fever mosquito are laid on the sides of water containers, usually above the water surface. If placed on the water at once, they hatch in about two days. If not moistened, however, they may remain dormant for weeks or even months. When immersed in water they promptly hatch." According to James and Harwood (1969), they may remain dormant up to a year. Hatchett (1946) showed the eggs hatch during the winter in the southern part of the United States when the mean temperature is over 70° F/21° C.

Fay (1964) has summarized the biology of *A. aegypti*. He notes the optimum water temperature is 77° to 84° F/25° to 29° C. Adults are killed in 24 hours at 43° F/6° C and by prolonged exposures at 45° to 48° F/7° to 9° C. The best temperature for the adult is 79° F/26° C. The female may take as many as 17 to 40 blood meals.

Whereas house mosquitoes are mostly night biters, the yellow-fever mosquito bites during the day, especially in early morning or late afternoon. Gjullin (1947) showed black, blue and red clothes are most attractive to several species of *Aedes*. He suggests mosquitoes choose colors on the basis of spectral reflectance.

The common malaria mosquito, *Anopheles quadrimaculatus* Say. This mosquito belongs to the genus of mosquitoes that transmits malaria to human beings, and the commom malaria mosquito is the most important carrier of malaria in the southern, central and eastern United States where it feeds on man and a great variety of animals. *A. freeborni* Aitken is the principal vector on the Pacific Coast. *A. quadrimaculatus* is most readily distinguished from the other *Anopheles* mosquitoes by the four dark spots near the center of the wing. Edwards (1931) characterizes *Anopheles* mosquitoes in the following ways: "1) the resting position: the whole body extends *in a straight line* at an angle with the surface on which the mosquito has settled; in other mosquitoes the body-line is bent, owing to the more rounded or 'humped' shape of the thorax; 2) in the great majority of the *Anopheles* the wings are spotted, whereas in almost all the others they are not; 3) in the female *Anopheles* the palpi are as long as the proboscis, while in most of the other genera they are much shorter; 4) when viewed under a microscope most mosquitoes are seen to have the abdomen covered with scales like those on a butterfly's wing, but in nearly all *Anopheles* these scales are absent; 5) the larvae of *Anopheles* when at rest lie parallel with and touching the surface of the water, being held to the surface-film by a number of remarkable rosette-shaped tufts; the breathing organ is small. Other mosquitoes have larvae which hang head downwards in the water when at rest, and possess no rosettes, but on the other hand have the breathing organ developed into a tube."

Anopheles quadrimaculatus Say breeds in bodies of fresh water containing vegetation. Keener (1945) made a detailed study of the life cycle of this mosquito at air temperatures of 76° to 80° F/24° to 27° C, water temperature of about 74° F/23° C, and a relative humidity of 70 to 80 percent. The average number of days required for completion of oviposition until emergence of the adults was 21 days with a minimum of 14 and a maximum of 27 days. The egg stage averaged two days, the four larval instars averaged five, five, five and eight days, and the pupal period averaged two days. The average adult life of the female was 21 days and that of the male was seven days. The maximum life of the female was 62 days and that of the male was 21 days. "Mating occurred as early as the first day of emergence and either before or after the first blood meal. Only one insemination was necessary for a female to continue to produce viable ova throughout her life span. Mating activity took place most frequently during the first few hours of darkness. Females were observed to take blood meals as early as 18 to 20 hours after emergence. The total number of blood meals per female is 9.4. Some blood feeding occurred during the day but rapidly increased immediately after darkness occurred, reaching a peak near the end of the night. The female deposited eggs while at rest on the water surface or on some object near the edge of the water. The eggs were released individually from the tip of the abdomen and were at first a pearly white, gradually turning to a glossy black in approximately 45 minutes. Oviposition occurred as soon as 80 hours after emergence and usually took place 60 hours after a blood meal. The average number of eggs deposited by one female was approximately 200 with a maximum of 300. No egg-laying occurred during the day, but it reached a peak immediately after the onset of darkness and then gradually diminished during the remainder of the night." There may be as many as 10 generations per year in the South. The adults are for the most part active at dusk or during the night, hiding during the day in dark, humid places such as outhouses, animal sheds and places of a similar nature.

Anopheles mosquitoes begin to hibernate in the fall in a variety of situations such as "cellars, stables, cow sheds, outbuildings of all types, caves, tree holes and similar locations." Hess and Crowell (1949) studied the winter habits of *Anopheles quadrimaculatus* in northern Alabama and noted the fertilized females hibernate in caves during October, reached a peak in late November and disappeared by the first of February. Adults may emerge during periods of warm weather at any time of the winter to take blood meals but do not develop eggs until they leave the caves in early February.

The eastern salt-marsh mosquito, *Aedes sollicitans* (Walk.). This is the most important of the salt-marsh mosquitoes and breeds along the Atlantic and Gulf Coasts (Fig. 20-7). This species can migrate for long distances, often in enormous swarms. The black salt-marsh mosquito, *Aedes taeniorhynchus* (Wied.), is another important salt-marsh mosquito pest which is found widely distributed throughout the United States, particularly in the South.

The above information about some of our common mosquito species is necessarily brief. For a more comprehensive account of these and many other species, Horsfall (1955 and 1972) is recommended. This work covers a wealth of information about 1,800 species of mosquitoes (from 29 genera). The biology of the egg, larva, pupa and adult are discussed, giving information on such aspects as distribution, development, feeding habits, dispersal, longevity, reproduction, parasites, resistance and disease implications.

MOSQUITOES AND DISEASE

Although we are primarily concerned with mosquitoes because they are transmitters of disease, we cannot overlook the annoyance, injury and economic losses caused by their presence and bites. Most people are susceptible to their bites and in some individuals they cause severe itching, swelling, pustule formation, restlessness and loss of sleep. Moreover, they prevent the enjoyment of many outdoor areas.

History indicates

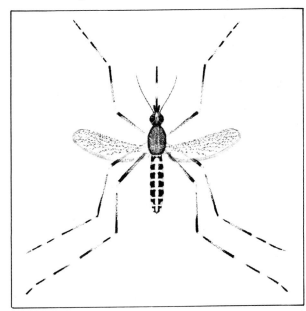

Fig. 20-7. Eastern salt-marsh mosquito.

the mosquito has long been one of the greatest scourges of man and other warm-blooded animals. The decline of Rome has been traced by some historians to the mosquito for it is believed malaria played an important role in the debilitation of the once virile Roman people. There are some 100 species of mosquitoes in the United States, but fortunately only a few are important vectors of disease organisms.

Malaria is world-wide in occurrence and is considered by some individuals as the most important disease of mankind. Malaria is caused by a microscopic animal parasite that lives within the red blood corpuscles. The malaria parasite is introduced into the blood stream by the bite of an *Anopheles* mosquito. These parasites destroy the red blood corpuscles. Pierce (1975) relates some of the history of this ancient disease. The name malaria means bad air and dates to Rome in 325 B.C. when Appius Claudius built the Appian Way through the marshes of Rome and attempted to drain them. A number of emperors attempted to drain these marshes as they were recognized as unhealthful. It was believed the air from them was poisoned.

In the United States, two of the important carriers of malaria are *Anopheles quadrimaculatus* Say and *Anopheles freeborni* Aitken. Bradley (1966) reviews the subject of malaria control in the United States. Harrison (1978) covers the history of malaria from 1880, and Russell et al. (1963) present a comprehensive treatment on malaria, covering the disease, its vectors and its control throughout the world.

The numbers of people affected with malaria is staggering. Gillett (1972) states that in 1957 it was estimated there were 250 million cases of human malaria each year, with more than two million deaths. Russell (1959) reports that in the 1930s six to seven million cases of malaria occurred each year in the United States. There are presently no malarial cases endemic to the United States. The few cases reported each year are due to blood transfusions from donors who are carriers and from travellers infected by mosquitoes abroad but developing symptoms after arriving home. However, a potential threat of malaria reoccuring in the United States still exists, since the Anopheline vectors are here. Benach et al. (1972) reported on an outbreak of malaria from a children's camp on Long Island, New York. A counsellor from the Ivory Coast was the carrier and mosquitoes, probably *Anopheles quadrimaculatus,* passed the disease to another counsellor.

Harrison (1978) relates why constant efforts are necessary if malaria is to be controlled. He reports in India the number of malarial cases had been brought down from 75 million in 1951 to about 50,000 in 1961. The use of DDT played a major role in this success, undoubtedly saving millions of lives. However, after a lessening of control efforts, the number of cases increased. From 1961 to 1963 there were less than 100,000 cases; in 1967 and 1968 there were 275,000 cases; in 1969-350,000; 1970-800,000; and in 1971-1,600,000 cases. In 1977 there are estimates of at least 30 million and perhaps 50 million cases!

Yellow fever and dengue fever are transmitted by *Aedes aegypti* (L.). The filarial worm which causes elephantiasis in man is carried by some species of mosquitoes and the southern house mosquito, *Culex quinquefasciatus,* is the most common carrier in many parts of the world.

Mosquitoes are the principal vectors of the virus diseases known as "encephalitis." Wild birds are the principal reservoirs of infection. Kandle (1964) notes that rodents, rabbits and deer may also be carriers of encephalitis. Goldfield et al. (1966) implicate cats and dogs, too. In human beings, encephalitis is also known as "sleeping sickness."

There are four major types of encephalitis affecting humans in North America. There are Eastern Encephalitis (EE); Western Encephalitis (WE); St. Louis Encephalitis (SLE) and California Encephalitis (CE). The following table compares these diseases:

TABLE 20-1

Human Encephalitis

Type	Principal Vectors*	Reservoirs**	Hosts	Disease Severity
Eastern Encephalitis	Aedes sollicitans Aedes vexans Culiseta melanura	Various species of birds. Reptiles?	Man Horse Pheasants & other birds	High mortality in man (60%); often severe neurologic damage particularly in children. Mortality in horses & pheasants almost 100%.
Western Encephalitis	Culex tarsalis Culex pipiens Culex melanimon	Various species of birds. Rodents and reptiles?	Man Horse	Low mortality in adults (5-15%). Severe in children and horses.
St. Louis Encephalitis	Culex pipiens Culex tarsalis Culex nigripalpus Aedes taeniorhynchus	Various species of birds. Horses. Bats?	Man	Mild except in old people; mortality 2-10% in young persons.
California Encephalitis	Aedes canadensis Aedes triseriatus	Small mammals such as rabbits and squirrels. Horses.	Man	Severe in children. Very mild in adults (in fact the disease may go unnoticed in adults).

*Many other species of mosquitoes are suspected of being vectors of encephalitis. However, causing them to be experimentally infected or finding them naturally infected, does not mean they can transmit disease.

**Reservoirs are animals from which the mosquito vectors can become infected. The reservoir animal normally does not become sick. The host on the other hand is affected by the disease.

CONTROL OF MOSQUITOES IN AND AROUND THE HOME

Mosquito control is a problem that is best handled by the community and public health agencies rather than by the individual because the breeding places of mosquitoes are often situated at considerable distances from the place of annoyance. However, there are some methods that the homeowner and experienced pest control operator may utilize in reducing the mosquito pest problem. See Darsie, et al. (1976) for detailed information.

Screening. There is seldom any serious mosquito problem in houses protected with 16 to 18 mesh wire screening. Bacon (1946) noted laboratory tests with several species of mosquitoes showed that 18 × 18 mesh was superior to 18 × 16 or 16 × 16 mesh. Herms and Gray (1944) discuss the screening of homes in some detail and the following is from this source.

"Four fundamental ideas must be kept in mind and applied in the case of screening: 1) The kind and size of screen must be suited to the exposure or climatic conditions; 2) The screen must be protected or reinforced where necessary against mechanical breakage due to ordinary usage; 3) All apertures, except those effectively screened, must be closed so as to prevent access by insects; 4) Ventilation with screening must be so good that the inhabitants will not prefer in warm weather to remain outdoors in the evening and at night.

"In general, extra heavy 16-mesh screen is to be preferred to 18-mesh standard screen, as the differences in effective aperture size (about 0.0019 inch) is inappreciable and the greater mechanical strength of the larger size renders such screen less liable to mechanical breakage or to deterioration under adverse atmospheric conditions."

Where the aperture size is greater than 0.0475 inch/1.2 mm, the screen may not keep out all species of mosquitoes. A smaller mesh size than that specified for the mosquitoes is necessary to keep black flies and other gnats from entering the home. Ordinarily, an electrogalvanized screen is satisfactory. In regard to the use of copper screen, Herms and Gray (1944) note "that paints containing zinc set up a reaction with copper screen, causing the screen to split at contacts with zinc-painted wood trim. Therefore, only lead base paints should be used on trim in contact with copper screen. For a similar reason galvanized nails should not be used in contact with copper screen, nor should galvanized protective screen be allowed to come in contact with copper screen.

"Chimneys and other vent flues should be screened during the mosquito season, but the screens should be removed from chimneys when winter comes and the fireplaces or stoves are used frequently or steadily, as the smoke rapidly corrodes any kind of screen material." Wesley and Morrill (1956a and 1956b) showed that an 18 × 18 mesh screen is the coarsest that can be used to exclude *Aedes aegypti* (L.). This mesh does not interfere too much with ventilation.

Breeding places around the home. Mosquitoes such as *Culex pipiens, Culex quinquefasciatus* and *Aedes aegypti* and other species may breed in the immediate vicinity of the home. Annoyance by such species can be minimized by a little diligence on the part of the homeowner.

Source reduction by water management is an essential part of mosquito control. Water-containing receptacles such as barrels and tin cans and automobile tires should be emptied and accumulations of tin cans should be removed, and if necessary buried in the ground. Leaking plumbing that results in the formation of pools on the ground or in the underareas of the home often are responsible for mosquito breeding and ice box drains are particularly important in this respect. Tunnels or underground vaults used by gas, power, light and other public utility companies will often have accumulations of water that may be responsible for local mosquito infestations. Other places that frequently harbor mosquito breeding pools are sewer inlets and catch basins, storm drains and street gutters, and clogged and defective roof gutters. In rural areas, faulty leaching cesspools may result in the breeding of large numbers of mosquitoes.

In all the above situations where the condition responsible for the accumulation of water cannot be readily corrected, it is then necessary to treat these mosquito breeding places with larvicides.

Ornamental garden ponds frequently breed mosquitoes. Where these ponds are stocked with goldfish and the mosquito eating top minnows, *Gambusia*, there is often no mosquito problem. If the fish are not fed to excess during the mosquito breeding season, they are more likely to feed on the larvae.

Herms and Gray (1944) discuss mosquito infestations in building sumps and wet basements as follows: "In large buildings with deep basements, the basements drainage is frequently too low for gravity discharge into the city sewer system. In such cases the drainage is conducted to a sump, from which it is discharged into the street sewer by an automatic ejector. Occasionally the sump may be found to be breeding mosquitoes.

"A large department store became infested with mosquitoes (*Culex pipiens*). The office force on the second floor was attacked first and in a short time the insects became so numerous that the customers were being attacked. When the matter was finally reported to the mosquito abatement district, the infestation was very severe. The source was found to be a sump and the channel leading to it, under the basement floor. The channel had several inches of water in it, caused by accumulation of muck on the bottom of the channel. The sump and channel were treated with larvicide to kill the larvae and pupae, after which the channel was cleaned and ordered flushed at monthly intervals."

Armstrong (1958) found *Culex pipiens* (?) biting in a post office in East Boston in January. The mosquitoes were breeding in a large sump beneath the furnace. Anon. (1956) notes in Georgia the mosquitoes become so bad in winter the catch basins have to be treated.

Dense shrubbery and vines around a home encourage mosquito infestations since they provide excellent resting places for a number of mosquito species. If such shrubbery and vines are removed or pruned to permit the entry of sunlight, these mosquito harborages can be eliminated or reduced to a minimum. There are several species of mosquitoes, particularly of the genus *Aedes* that breed in tree holes containing water. These tree hole mosquito breeding places can be permanently eliminated by tree surgery aimed at cleaning out and filling rotten areas. In the warmer sections of the United States, *Aedes aegypti* may occasionally breed in the home, in flower pots and their drainage saucers. The latter should be checked and emptied.

Bishopp (1951) summarizes measures for the elimination of the breeding places of domestic mosquitoes as follows: "1) Inspect your premises frequently. Remove all temporary water containers. Punch holes in the bottoms of tin cans or remove both ends and flatten them. See that cesspools, septic tanks, rain barrels and tubs in which water is stored are tightly covered and that open cisterns are covered or screened and the spouts screened; 2) Once a week empty and thoroughly wash bird baths and pans used for watering chickens; 3) Examine eaves troughs occasionally to find out whether they are stopped up or sagged, and after rains examine flat roofs to see that no water is on them; 4) Do not permit water that may be drained under the house from the refrigerator to stand in puddles or water from any source to form pools in the yard or in the street gutters; 5) Interest your neighbors and local officers in the prevention of mosquito breeding."

Larviciding. (Control of mosquito larvae). Larvae may breed wherever suitable water is available. If it is not possible to eliminate this water, insecticides should be used. The following larvicides are widely used in the United States: chlorpyrifos, fenthion, pyrethrum and temephos. Check the labels of these products for specific uses and rates. Rates vary according to specific environmental situations (catch basin; clear water; highly polluted water). In the past DDT was used with great success as a larvicide, but insect resistance is a major problem with some species in some areas, and in some countries, including the United States, it is not labeled for this use.

Oils are also used to kill larvae. Fuel oil no. 2, Flit MLO®, ARCO larvicide, diesel oil no. 2 or kerosene may be used. Only Flit MLO or temephos should be used if fish are present.

Some newer chemicals which show promise in controlling larvae are methoprene (a growth inhibiting substance) and *Bacillus thuringiensis* variety *israelensis* (a bacterium).

Methoprene does not kill mosquito larvae. It does, however, prevent them from becoming normal adults. Schaefer and Mulla (1980) state that "hand-sprayer applications with an emulsifiable concentrate of methoprene showed that *Aedes nigromaculis* larvae could be controlled with as little as $1/80$ pound active ingredient (a.i.) per acre, but that timing was extremely critical; larvae had to be treated late in the fourth stage, because the active ingredient persisted only a few hours under field conditions." Since this was written, research with slow release formulations has resulted in a sustained release briquet product. This product is claimed to release effective levels of methoprene over a 30-day period.

The bacterium Bacillus thuringiensis variety *israelensis* (BTI) is discussed by Garcia et al. (1980).

Adulticiding. (Control of mosquito adults). Adulticides can be applied in three ways:

- Thermal aerosols (fogging). Insecticides such as chlorpyrifos, fenthion, malathion and naled are used. The fog is produced by heating fuel oil in which the insecticide is diluted. Disadvantages of thermal fogging are the hazard of reduced visibility for passing motorists from the smoke, the cost of using fuel oil as a carrier and the pollution of the environment with fuel oil. This method, once widely used to kill adult mosquitoes, has been almost entirely replaced by ULV application.

- Ultra-Low Volume Application (ULV). Special machines are used with nozzles which break up an undiluted insecticide into microscopic droplets. With ULV application very small amounts of chemical are used (two to four oz per acre/140-280 g per hectare). Fuel oil is not burned and smoke is not seen coming out of the nozzle. Insecticides applied by ULV include chlorpyrifos, fenthion, malathion, naled, pyrethrins and resmethrin. Bendiocarb shows promise in this type of application. Both thermal fogging and ULV application are very temporary in nature. They kill only those adults which pass through the fog which disperses quickly.

- Mist Blower and Spraying. This method places the insecticide on the foliage of trees, shrubs, tall grass, tree trunks and the sides of structures where resting mosquitoes may contact it. Dilute amounts of pesticides are used including: carbaryl, chlorpyrifos, malathion, methoxychlor, naled, pyrethrins, propoxur and resmethrin. Mount (1970) found the optimum droplet size for adult mosquito control with space sprays or aerosols to be five to 10 u mass median diameter.

Resistance to insecticides. The resistance of both larvae and adults to insecticides is in a fluid state. Some species of mosquitoes are resistant and some are not. Further, some of the resistant species may be non-resistant in some areas. Resistance to both old and new insecticides is being constantly demonstrated by students of mosquito control. Brown (1958 and 1960) reviews the resistance of mosquitoes to insecticides. Micks (1966) is the lead paper in a symposium on resistance in mosquitoes.

All insects have the biological capability of developing resistance to all insecticides. Busvine (1980) gives a frightening account of mosquitoes and their present resistance to insecticides: "Among culicine mosquitoes, the members of the widely dispersed *Culex pipiens* group have developed resistance to various groups. Thus, *C. p. pipiens* (including probably *molestus*) is resistant to all organochlorines throughout the warmer north temperate region and also to organophosphorus compounds in France, Egypt and Israel. *C. p. fatigans* is resistant to organochlorines everywhere and in many places to organophosphates too. In the United States, the following species have become resistant to all groups: *Culex tarsalis, C. peus, Aedes nigromaculis, Ae. melanimon.* Those resistant to both organochlorines: *C. salinarius, C. restuans, Psorophora confinnis, Culiseta inornata* (and *Ae. cantator* in Canada)."

Only through a constant search for new control agents and methods can we hope to keep ahead of our dynamic insect enemies.

Carson (1962) wrote a provocative book on the effect of insecticides on the environment. Many of her statements, though not based on fact, did lead through political action to the almost complete banning of DDT in the United States, though fortunately not in countries with more rational decision processes. Mrak (1968), in a more scholarly study of pesticides and environmental health submitted to the Secretary of the Department of Health, Education and Welfare, puts the subject of pesticides and the reproductive success of eagles in better perspective. Mrak stated there was a reasonable doubt that chlorinated hydrocarbon pesticides occured in the natural feed of these birds at levels required to adversely affect reproduction. Whitten (1966) presents the insecticide issue from the viewpoint of the farmer and scientist who are trying to feed an ever-growing number of mouths in a hungry world. Borlaug (1971), winner of the 1970 Nobel Peace Prize, emphasized the continuing value of pesticides not only to save lives but, through increasing crop yields, to avoid the need to cultivate wildlife lands. Rogers (1972) examined the much publicized theory that reproductive failures of birds, especially birds of prey, are due to pesticides. He found little evidence for the theory and much evidence against it. Rogers, citing the shooting from planes of 500 Bald and Golden Eagles in 1971 in Wyoming and Colorado, concluded that shooting, collecting of birds and their eggs, encroachment of suburbia on eagle habitats, accidental electrocution and disturbance of nesting birds by ornithologists were more important factors adversely affecting birds of prey. This conclusion is supported by records showing a decline in eagles before DDT or other persistent pesticides were ever used.

Natural control by Gambusia fish. Top minnows of the genus *Gambusia,* particularly *Gambusia affinis,* often are placed in ornamental ponds and pools to control mosquito breeding. *Gambusia* fish occur naturally in the South and thrive where the winters are not too severe. These are small fish, the female being about 2½ inches/6.3 cm long and the male about an inch/2.5 cm shorter. Jordan (1927) notes that in Cuba, ". . . such little fishes are called 'Gambusinos,' hence their scientific name *Gambusia.*" When an angler returns without fish, the Cubans say he has been fishing for "Gambusinos." The eggs are hatched in the body of the female which ordinarily has six to 10 in a brood. There may be one to five broods in a season. Jordan observes the very young fish feed on algae, desmids and organisms of a similar nature, and the older fish feed on mosquito larvae and adults, as well as flies in general. The *Gambusia* will prey on mosquito larvae in shallow water less than one inch/2.5 cm in depth, a favorite breeding place of some mosquitoes. This prolific little fish will effectively control mosquitoes in some, but not all of the areas. Rees (1958) recommends it for small ponds. Herms and Gray (1944) discuss it in some detail.

Mice, fish and frogs have been shown to develop some resistance to insecticides. Boyd and Ferguson found *Gambusia* to be resistant to several chlorinated insecticides.

Kiker (1948), speaking of water-level management for mosquito control, notes a direct effect of this type of treatment is the stranding of mosquito larvae along the shore-line which exposes them to top-feeding minnows and aquatic insects. Travis (1957) discusses mosquito control by removing vegetation along the shore. The larvae are killed by wave action.

Repellents. Where the home is well screened and where a household spray or an aerosol is at hand, ordinarily there is no need for a repellent. However, when the "bug season" is on, the unscreened porch and the immediate grounds may be made untenable due to biting insects. Moreover, seaside areas, summer camps and cabins, and similar situations are notable places for conclaves of blood-imbibing insects. It is in locations such as these that effective repellents and other toxicants are invaluable. Most repellents act as contact materials which keep the insect from biting when they touch the protective chemical coating with their mouthparts. Some repellents may be sufficiently volatile so the insect refrains from coming close to the skin coated with the vaporizing chemical.

Fortunately, there are repellent liquid and cream formulations that do offer protection for several hours against biting insects that are currently on the market. The most obvious imperfection of the current formulations is the necessity of applying the repellent from the top of the head to the soles of the feet, particularly when there are large numbers of biting insects present.

Granett (1943) and Wilkes (1946) review the desirable characteristics of a repellent, a goal which some of the current repellent materials approach, but do not as yet achieve. Wilkes' observations on this subject are as follows: "A substance to be generally acceptable as a repellent should possess durability or effectiveness over a relatively long period of time against one or more of such biting or blood sucking pests as mosquitoes, flies, fleas, chiggers and the like. To this end, it should be relatively stable chemically and not readily dissipated by evaporation or vaporization although it is believed that at least some degree of volatility is essential. For dermal application, a repellent should be non-irritating and easy to apply with the hands or a swab. For use under wet-skin conditions, that is, skin which may be moist or wet as by perspiration, the repellent preferably should be relatively insoluble in water.

"Repellents which may affect clothing by staining, bleaching or weakening of the fiber, or which leave an objectionable 'oily' appearance or feel on the skin are limited in their usefulness. Preferably, the repellent should be free of odor, especially such odors as may be regarded as unpleasant or disagreeable and difficult to mask. Preferably, the repellent substance should have little or no solvent action on various finishes, paints, varnishes, lacquers and the like."

Repellents usually are applied to the skin by shaking a few drops of the liquid or squeezing a small portion of the repellent cream into the palm of the hands, rubbing them together and then smearing the repellent over the exposed skin and unexposed places such as the ankles, shoulders and shoulder blades. Aerosol sprays are now very popular for the application of repellents. Summer clothes, particularly when wet by perspiration, offer little protection against biting insects. Where chiggers (mites) and ticks are the cause of annoyance, it usually is necessary to apply the repellent material lightly over socks, clothing and such places as the neckbands, fly and cuffs of trousers. Travis et al. (1949a) recommend the use of benzyl benzoate, benzil or diphenyl carbonate in emulsion form for impregnation of clothes as protection against mites. In fact, emulsions can be made of most of the repellents and utilized for immersion of garments. Garments so treated may remain effective for days against a wide variety of pests.

Some of the disadvantages of the current repellents are listed below:

- They may cause irritation, particularly if they get into the eyes or on mucous membranes.
- They must be renewed every few hours or sooner.
- They are not effective against all biting insects, ticks or mites.
- They are solvents of paints, varnishes, plastics (synthetic watch crystals, etc.), synthetic fibers (rayon, nylon, etc.).
- Some people are sensitive or allergic to them.

Granett (1940) used the following procedure for the field evaluation of repellents against *Aedes* mosquitoes. In general, this is the technique still largely used for this purpose: "Application of the test material is made to one arm or leg of the tester. The protection afforded by a repellent is determined by the time in minutes to the first bite on these treated areas and is designated as the repellent time of the material under test. Determination also is made of the insect biting frequency during the test as indicated by observation of the bites per minute on corresponding untreated areas."

Granett & Pepper (1949) describe the technique used in the laboratory for evaluating the relative effectiveness of insect repellents. "These are tested by applying a measured quantity (one ml) uniformly to the forearm of the tester who then inserts his treated arm into a 30-inch cube cage containing approximately 2,000 *Aedes aegypti* adults. If no bites are received in a two-minute exposure period, the arm is withdrawn and inserted again after one hour. This procedure is followed until a bite is received. The time from application of the chemical until time of first bite is considered the repellent protection time. Usually a chemical of known repellency is used on the other arm of the tester so that a direct comparison can be made with the new material."

Travis (1950) studied the factors that caused variations in the results of insect repellent tests. His observations may be summarized as follows:

- Repellents are more durable on some subjects than others.
- Different species of mosquitoes are not readily repelled by the same repellent.
- Seasonal broods of the same species of mosquito vary in their reaction to the same repellent.
- Different samples of a chemical may vary in their repellency effectiveness.
- Perspiration reduces the repellent time.
- Low biting rates (relatively small biting populations) lengthened the repellency time.
- When first disturbed, mosquitoes are more avid in their biting.

Gilbert et al. (1966) showed women were protected by deet repellent for a longer time than men. Smith et al. (1963) found sweat and CO_2 were not responsible for the loss of effectiveness of deet. "Differences in protection periods between repellents appear to be due principally to differences in MED [minimum effective dosage] and secondarily to differences in the rate of loss, where resistance to loss by abrasion would be important factors."

A few repellents or repellent combinations available are fairly effective. They were for the most part widely used by the armed forces and have been accepted by the public. Travis et al. (1949b) tested hundreds of repellent materials and list the more effective ones in their paper. Lesser (1952) also prepared a general review of the subject and Dethier (1947) authored a book on attractants and repellents.

Citronella, the old standby, has been replaced by more effective repellents. Travis et al. (1949a) note the following repellents are safe and effective when used alone and in combination with one another: dimethyl phthalate, dimethyl carbate, Indalone, and Rutgers 612. One of the most effective repellents is diethyltoluamide (deet).

Description of the more common repellents. Dimethyl phthalate (dimethyl benzene orthodicarboxylate) is a colorless and practically odorless liquid with a specific gravity of 1.19. It is only 0.5 percent soluble in water or mineral oil.

Dimethyl carbate or dimethyl endolene or dimelone (dimethyl ester of cis-biscyclo [2,2,1]-5-heptene-2,3-dicarboxylic acid) is a water-white crystalline solid with a melting point about 104° F/40° C. It has a mild and pleasant odor and is soluble in Indalone and dimethyl phthalate.

Indalone (3,4-dihydro-2,2-dimethyl-4-oxo-butyl ester of 1,2H-pyran-6 carboxylic acid) is a light yellow to reddish brown liquid with a mild odor and a Sp. G. 1.056 to 1.062.

According to Granett and Haynes (1945), Rutgers 612 (2-ethyl-1,3-hexanediol) is a "slight viscous colorless liquid having a mild odor somewhat like witch-hazel. It is stable under extremes of storage conditions and unlike many other recognized repellents, its solvent action in contact with various surfaces is relatively weak." However, it does have a slight effect on spar varnish and it will soften shellac.

Diethyltoluamide or deet (N,N-diethyl-m-toluamide) is a liquid repellent that has a broad range of effectiveness. It is one of the best repellents for use against mosquitoes, fleas, biting flies, chiggers and ticks. Khan and Maibach (1972) showed deet to be the most effective repellent against the yellow fever mosquito. This repellent was synthesized by chemists in the United States Department of Agriculture. The technical grade contains 70 percent to 75 percent *m*-diethyltoluamide. The *meta* isomer was found to be the best repellent. This chemical does not have to be mixed with other repellents. However, it is usually diluted with ethyl alcohol or other alcohols.

Diethyltoluamide is resistant to rubbing and removal by perspiration and water. It may be applied as a lotion or as an aerosol or pressurized spray to both the skin and clothing. The usual precautions should be taken not to get it into the eyes or on the lips. This repellent may affect certain synthetic fibers, plastics, and painted or varnished surfaces. Gilbert et al. (1955 and 1957), Hall et al. (1957), and Smith et al. (1957) discuss this repellent in some detail.

Deet repellents were widely used in combat areas in Vietnam. Holway et al.

(1967) showed the standard deet repellent cannot be detected unless 12 inches/ 30 cm from the nose.

Three of the commonly recommended formulations which were primarily developed for military use are as follows:

TABLE 20-2

	By weight
1. Dimethyl phthalate	3 parts
Indalone	1 part
Rutgers 612	1 part
(The original 6:2:2 formula)	
2. Dimethyl phthalate	1 part
Indalone	1 part
Rutgers 612	1 part
3. Dimethyl phthalate	3 parts
Indalone	1 part
Dimethyl carbate	1 part
(Dimelone)	

Travis, Smith and Madden (1951) note mixtures of "three repellents are effective against a wider range of insects and for longer periods than mixtures containing only two repellents."

Indalone is especially effective against stable flies and ticks; dimethyl phthalate is best for the common malaria mosquito, *Anopheles quadrimaculatus* Say, for sand flies (Travis and Morton, 1950) and one of the preferred repellents for chiggers (mites); and Rutgers 612 and dimethyl carbate are the most desirable repellents for *Aedes* and other pest mosquitoes.

The Rutgers 612 formula is on the market as a liquid preparation. Granett and Haynes (1945) have shown it to offer protection against *Anopheles* and *Aedes* mosquitoes, stable flies, black flies, sand flies (*Phlebotomus* spp.), fleas and chiggers. Rutgers 612 also is available in aerosol and pressurized foams.

Repellent creams consisting of 92.5 percent liquid repellent and 7.5 percent cream base were available. This repellent mixture contained six parts dimethyl phthalate, two parts Indalone and two parts dimethyl carbate.

McAllister (1949) notes the following type of cream base has been satisfactory for use with several repellents:

Stearic acid	40.0 g
Potassium carbonate	0.6 g
Glycerine	12.0 ml
Water	68.0 ml
Liquid repellent	80.0 ml

Travis et al. (1946) state dimethyl phthalate, Rutgers 612, and Indalone gave six to 10 hours protection against the southern buffalo gnat, *Eusimulium pecuarum* (Riley), in Mississippi. They also found these repellents "prevented biting by moderate populations of *Culicoides* spp. for at least four hours. Sometimes, however, these species are very numerous and cause severe annoyance, even when repellents are used, because they land on the face, in the eyes, and in the ears and the mouth."

Smith (1950), in a review of repellents, has the following to say about benzyl benzoate, a particularly outstanding clothing repellent: "For personal protection from chiggers, benzyl benzoate is generally available and has been recognized for several years as the standard material. Clothing impregnated at a rate of about two grams per square foot, or 2½ ounces for a complete outfit of trousers, shirt or jacket, and socks, gives complete protection from chiggers until the clothing has been washed two or three times. The benzyl benzoate may be applied as a five percent solution in a volatile dry-cleaning fluid or as a five percent emulsion." He also notes mixtures containing 30 percent each of 2-butyl-2-ethyl-1,3-propanediol, N-butylacetanilide and benzyl benzoate, and 10 percent of Tween 80 emulsifier were effective as clothing repellents against mosquitoes, ticks and chiggers in Alaska. King (1951) has also prepared a list of outstanding clothing repellents for mosquitoes, fleas, chiggers and ticks. Applewhite and Cross (1951) showed 2- [2-(2-ethylhexyloxy)-ethoxy] ethanol was superior to all other repellents in impregnation of clothing against mosquitoes and sand flies in Alaska.

Travis and Morton (1946) found the 6:2:2 mixture of dimethyl phthalate, Rutgers 612 and Indalone when applied at the rate of 100 to 200 ml for shirt or trousers prevents mosquito bites through clothing from several days to several weeks.

Horsfall (1959) and Lopp and Buchanan (1959) discuss a commercial aromatic repellent which when scattered over shrubbery or grass is effective in keeping *Aedes* and *Psorophora* mosquitoes away for several days. The carrier is 30- to 40-mesh vermiculite which is soaked in an oil which is a cut between kerosene and fuel oil. It is scattered at the rate of 10 lbs. per acre/11 kg per hectare.

Gerberg (1966) found TMPD (2,2,4-trimethyl-1,3-pentanediol) to be an effective repellent.

LITERATURE

ANON. — 1956. House mosquitoes breed during Georgia winter. U.S.P.H.S. Rpts. 71(8):812. 1965. U.S. Dept. Hlth. Handbook of General Information, *Aedes aegypti*. Handbook Series No. 1. Center for Disease Control, Atlanta, GA. 16 pp. 1968. Ohio Mosquito Control Manual. Univ. Ohio Coop. Ext. Svce. and Ohio Dept. Health, Columbus, Ohio.

APPLEWHITE, K. H. and H. F. CROSS — 1951. Further studies of repellents in Alaska. J. Econ. Entomol. 44(1):19-22.

ARMSTRONG, B. — 1958. Top this one. Mosq. News 18(1):41.

BACON, R. W. — 1946. Effectiveness of insect wire screening. Mosq. News 6:85-88, June.

BARR, A. R. — 1958. The mosquitoes of Minnesota. Univ. Minn. Agr. Expt. Sta. Tech. Bull. 228. 154 pp.

BEADLE, L. D. — 1958. Status of mosquito-borne encephalitis in the United States. U.S.P.H.S. Rpts. 74(1):84-90. 1966. Epidemics of mosquito-borne encephalitis in the United States, 1960-1965. Mosq. News 26(4):482-486.

BENACH, J. L., J. J. HOWARD, T. F. BAST & A. R. HINMAN — 1972. Possible introduced malaria — New York State. Center for Disease Control: Morb. and Mort. Wkly. Rep. 21:423 (Dec. 9, 1972).

BERNER, L. — 1947. Notes on the breeding habits of *Aedes (Stegomyia) aegypti* (Linnaeus). Entomol. Soc. Amer. Ann. 40:528-529. Sept.

BISHOPP, F. C. — 1939. Domestic mosquitoes. U.S.D.A. Leaflet No. 186. 1951. Domestic mosquitoes. U.S.D.A. Leaflet No. 186.

BJORNSON, B. F. — 1961. The role of mosquitoes in the transmission of human disease. Pest Control 29(4):19-20, 22, 24, 26, 30, 32.

BOHART, R. M. & R. K. WASHINO — 1978. Mosquitoes of California. Univ. Calif. Press, Berkeley. 153 pp.

BORLAUG, N. E. — 1971. Mankind and Civilization at Another Crossroad. McDougall Mem. Lecture, FAO, United Nations.

BOYD, C. E. & D. E. FERGUSON — 1964. Susceptibility and resistance of mosquito fish to several insecticides. J. Econ. Entomol. 57(4):430-431.

BRADLEY, G. H. — 1966. A review of malaria and eradication in the United States. Mosq. News. 26(4):462-467.

BROWN, A. W. A. — 1958. Insecticide Resistance in Arthropods. World Health Org. 1960. Past, present and future in insecticide-resistance of mosquitoes. Mosq. News. 20(2):110-115.

BUSVINE, J. R. — 1980. Insects and Hygiene. Chapman and Hall, London. 568 pp.

CARPENTER, S. J. — 1968. Review of recent literature on mosquitoes of North America. Calif. Vector Views. 15:71-98. 1970. Review of recent literature on mosquitoes of North America, Supplement I. Calif. Vector Views. 17:39-65. 1974. Review of recent literature on mosquitoes of North America, Supplement II. Calif. Vector views. 21:73-99.

CARPENTER, S. J. & W. J. LA CASSE — 1955. Mosquitoes of North America. Univ. Calif. Press, Berkeley. 360 pp., 129 plates.

CARSON, R. — 1962. Silent Spring, Houghton Mifflin Co.

CHAPMAN, H. C. — 1966. The mosquitoes of Nevada. U.S.D.A. and Univ. Nevada. Bull. T2. 43 pp.

CHRISTOPHER, R. — 1960. *Aedes aegypti* The Yellow-Fever Mosquito: Its Life History, Bionomics and Structure, Cambridge University Press. 732 pp.

CLARKE, J. L. — 1943. Studies of the flight range of mosquitoes. J. Econ. Entomol. 36:(1):121-122.

DARSIE, R. F. — 1949. Pupae of the anopheline mosquitoes of the Northeastern United States (Diptera: Culicidae). Rev. Entomol. 20:509-530. 1951. Pupae of the Culicine mosquitoes of the Northeastern United States (Diptera, Culicidae, Culicini). Cornell Univ. Agr. Exp. Sta. Memoir 304, 67 pp.

DARSIE, R. F. et al. — 1976. Mosquito abatement for pest control specialists. Pest Control. 44(4):A-W.

DETHIER, V. G. — 1947. Chemical Insect Attractants and Repellents. Blakiston Co.

DODGE, H. R. — 1963. Studies on mosquito larvae I. Later instars of eastern North American species. Canad. Entomol. 95:796-813. 1966. Studies on mosquito larvae II. The first stage larvae of North American Culicidae and of world Anophelinae. Canad. Entomol. 98:337-393.

EDWARDS, F. W. — 1931. Mosquitoes and their relation to disease. British Museum Econ. Series, No. 4.

FAY, R. W. — 1964. The biology and bionomics of *Aedes aegypti* in the laboratory.

Mosq. News. 24(3):300-308.

GARCIA, R., B. A. FEDERICI, I. M. HALL, M. S. MULLA & C. H. SCHAEFER — 1980. BTI — a potent new biological weapon. Calif. Agr. 34(3):18-19.

GERBERG, E. J. — 1966. Field and laboratory repellency tests with 2,2,4-trimethyl-1,3-pentanediol (TMPD). J. Econ. Entomol. 59(4):872-875.

GILBERT, I. H., H. K. GOUCK & C. N. SMITH — 1955. New mosquito repellents. J. Econ. Entomol. 48(6):741-743. 1957. New insect repellent. Soap & Chem. Spec. 33(5):115-117, 129, 131, 133. 1957. Diethyltoluamide: new insect repellent. II. Clothing treatments. Soap & Chem. Spec. 33(6):95, 97, 99, 109. 1966. Attractiveness of men and women to *Aedes aegypti* and relative protection time obtained with deet. Fla. Entomol. 49(1):53-66.

GILLETT, J. D. — 1972. Mosquitoes. Weidenfeld & Nicholson, London. 274 pp.

GJULLIN, C. M. — 1947. Effect of clothing color on the rate of attack of *Aedes* mosquitoes. J. Econ. Entomol. 40(3):326-327.

GJULLIN, C. M., R. I. SAILER, A. STONE & B. V. TRAVIS — 1961. The mosquitoes of Alaska. USDA. ARS. Agr. Handbook No. 182:1-98.

GOLDFIELD, M., O. SUSSMAN & R. P. KANDLE — 1966. A progress report on arbovirus studies in New Jersey. Proc. N.J. Exterm. Assoc. 53rd Ann. Meet. pp. 47-51.

GRANETT, P. — 1940. Studies of mosquito repellents. 1. Test procedure and method of evaluating test data. J. Econ. Entomol. 33(3):563-565. 1943. The significance of the development of mosquito repellents for the protection of military and civilian populations. Proc. 30th meeting. N.J. Mosq. Exterm. Assoc. pp. 203-211.

GRANETT, P. & H. L. HAYNES — 1945. Insect-repellent properties of 2-ethylhexanediol-1,3. J. Econ. Entomol. 38(6):671-675.

GRANETT, P. & B. B. PEPPER — 1949. Industrially sponsored search for better insecticides and repellents. Proc. 36th Ann. Meet. N.J. Mosq. Exterm. Assoc. pp. 183-187.

HALL, S. A., N. GREEN & M. BEROZA — 1957. Insect repellents and attractants. Agr. & Food Chem. 5(9):663-667, 669.

HARBACK, R. E. & K. L. KNIGHT — 1980. Taxonomists Glossary of Mosquito Anatomy. Plexus Publ. Inc., Marlton, N.J. 415 pp.

HARMSTON, F. C. & F. A. LAWSON — 1967. Mosquitoes of Colorado. U.S. Dept. Hlth. Educ. Welf. 140 pp.

HARRISON, G. — 1978. Mosquitoes, Malaria and Man. E.P. Dutton, N.Y. 314 pp.

HATCHETT, S. P. — 1946. Winter survival of *Aedes aegypti* (L.) in Houston, Tex. U.S.P.H.S. Rpts. 61(34):1134-1144. R.A.E. (B) 7(6):99-100. 1949.

HERMS, W. B. & H. F. GRAY — 1944. Mosquito Control. 419 pp. The Commonwealth Fund.

HESS, A. D. & R. L. CROWELL — 1949. Seasonal history of *Anopheles quadrimaculatus* in the Tennessee Valley. J. Nation. Malaria Soc. 8(2):159-170. Biol. Abst. 24(4):1062, 1950.

HOLWAY, R. T., A. W. MORRILL & F. J. SANTANA — 1967. Mosquito control ctivities of the U.S. Armed Forces in the Republic of Vietnam. Mosq. News. 27(3):297-307.

HORSFALL, W. R. — 1955. Mosquitoes: Their Bionomics and Relations to Disease. Ronald Press Co., N.Y. 723 pp. Reprinted 1972, Ronald Press, N.Y. 1959. New use for an old formulation: Spot suppression of mosquitoes. Pest Control. 27(4):24.

JAMES, M. T. & R. F. HARWOOD — 1969. Herms's Medical Entomology. MacMillan Co., London. 484 pp.

JORDAN, D. S. — 1927. The mosquito fish *(Gambusia)* and its relation to malaria. Smithsonian Inst. Ann. Rpt. 1927. pp. 361-368.

KANDLE, R. P. — 1964. Continued arbovirus research activities in New Jersey — 1963. Proc. N.J. Mosq. Exterm. Assoc. 51:15-18.

KEENER, G. G., JR. — 1945. Detailed observations on the life history of *Anopheles quadrimaculatus*. J. National Malaria Soc. 4(3):263-270.

KHAN, A. A. & H. I. MAIBACH — 1972. A study of insect repellents. J. Econ. Entomol. 65(5):1318-1321.

KIKER, C. C. — 1948. Management of water to control anopheline mosquito breeding. Proc. 4th Intern'l. Cong. of Trop. Med. & Malaria. pp. 865-872, May 10-18.

KING, W. V. — 1951. Repellents and insecticides available for use against insects of medical importance. J. Econ. Entomol. 44(3):338-343.

KING, W. V., G. H. BRADLEY & T. E. McNEEL — 1944. The mosquitoes of the southeastern states. U.S.D.A. Misc. Publ. 336.

KING, W. V., G. H. BRADLEY, C. N. SMITH & W. C. McDUFFIE — 1960. A Handbook of the mosquitoes of the Southeastern United States. U.S.D.A. Agr. Handbook No. 173:1-188.

KNIGHT, K. L. & A. STONE — 1977. A catalog of the mosquitoes of the world (Diptera: Culicidae). Entomol. Soc. Am. (Thomas Say Found.), Wash. 6(2nd Ed.):1-611.

KNIGHT, K. L. & M. WONIO — 1969. Mosquitoes of Iowa (Diptera: Culicidae). Iowa State Univ. of Sci. & Technol., spec. rept. No. 61, Ames, Iowa. 79 pp.

KON', Y. S., D. I. DOBROSMAISLOR & Z. L. GINZBURG — 1944. Mosquito larvae in the tunnels of the Moscow underground railway. Rev. Appl. Entomol. Ser. B. 32(6):121, 1944.

LESSER, M. A. — 1952. Insect repellents. Soap & S.C. 28(3):136-137, 141, 142.

LOPP, O. V. & W. J. BUCHANAN — 1959. How granular mosquito repellent performed in the field. Pest Control 27(4):25-26.

MATHESON, R. — 1944. The Mosquitoes of North America. 2nd Ed. Comstock Publ. Co. 314 pp. 1950. Medical Entomology. Comstock Publ. Co., Ithaca, N.Y. 612 pp.

McALLISTER, W. G. — 1949. Insect repellents as cosmetics. Soap, Perfumery & Cosmetics. 22:848-850, 882. Aug. 1949.

MEANS, R. G. — 1979. Mosquitoes of New York. Part I. The genus *Aedes* Meigen with identification keys to genera of Culicidae. Bull. No. 430a. Univ. of the State of N.Y., State Education Dept., Albany, N.Y. 221 pp.

MICKS, D. W. — 1966. Symposium on insecticide resistance in mosquitoes. Mosq. News. 26(3):299-300.

MOORE, J. B. — 1966. New repellents. Aerosol Age 11(12):63-65, 161.

MORLAN, H. B. & R. O. HAYES — 1958. Urban dispersal and activity of *Aedes aegypti*. Mosq. News 18(2):137-144.

MORTON, F. A. — 1946. Emulsifiers for dimethyl phthalate — a preliminary investigation. N.J. Mosq. Exterm. Assoc. Proc. (1946) 33:69-71.

MOUNT, G. A. — 1970. Optimum droplet size for adult mosquito control with space sprays or aerosols of insecticides. Mosquito News 30(1):70-75.

MRAK, E. M., et al. — 1968. Report of the Secretary's Commission on Pesticides and Their Relationship to Environmental Health. U.S. Dept. Hlth. Educ. & Welf. Washington D.C.

NIELSEN, L. T. & D. M. REES — 1961. An identification guide to the mosquitoes of Utah. Univ. Utah, Biol. Ser. 12(3):1-58.

OWEN, W. B. & R. W. GERHARDT — 1957. The mosquitoes of Wyoming. Univ. Wyo. Pub. 21:71-141.

PIERCE, W. D. — 1975. The Deadly Triangle. John G. Shanfelt, Jr., Orange, Calif. 138 pp.

REES, B. E. — 1958. Attributes of the mosquito fish in relation to mosquito control. Proc. Calif. Mosq. Control Assoc. 26:71-75.

REES, D. M. & L. T. NIELSON — 1952. Control of *Aedes* mosquitoes in two recreational areas in the mountains of Utah. Mosq. News 12(2):43-49.

ROGERS, A. J. — 1972. Eagles, Affluence and Pesticides. Mosq. News 32(2).

ROSS, H. H. & W. R. HORSFALL — 1965. A synopsis of the mosquitoes of Illinois. Illinois Natural History Survey Biol. Notes No. 52. Urbana, Ill. 50 pp.

RUSSELL, P. F. — 1959. Insects and the epidemiology of malaria. Ann. Rev. Entomol. 4:415-434.

RUSSELL, P. F., L. S. WEST, R. D. MANWELL & G. MacDONALD — 1963. Practical Malariology. 2nd Ed. Oxford Univ. Press, London, 750 pp.

RUTSCHKY, C. W., T. C. MOONEY, JR. & J. P. VANDERBERG — 1958. Mosquitoes of Pennsylvania. Penn. St. Univ. Agr. Exp. Sta. Bull. 630. 26 pp.

SCHAEFER, C. H. & M. S. MULLA — 1980. Conventional and non-conventional chemicals for mosquito control. Calif. Agr. 34(3):28-29.

SIVERLY, R. E. — 1972. Mosquitoes of Indiana. Indiana State Board of Health, Indianapolis, Ind. 126 pp.

SMITH, C. N. — 1950. New materials for use as insect repellents. Soap & S.C. Off. Proc. 37th Ann. Meet. C.S.M.A. Dec. pp. 80-81.

SMITH, C. N. et al. — 1963. Factors affecting the protection period of mosquito repellents. USDA Tech. Bull. No. 1285.

SMITH, C. N. & D. BURNETT — 1949. Effectiveness of repellents applied to clothing for protection against salt-marsh mosquitoes *(Aedes taeniorhynchus; A. sollicitans)*. J. Econ. Entomol. 42(3):439-444. June.

SMITH, C. N., I. H. GILBERT & H. K. GOUCK — 1957. Use of insect repellents. USDA Agr. Res. Serv. ARS-33-26. Revised March 1957.

STOJANOVICH, C. J. — 1960. Illustrated key to common mosquitoes of southeastern United States. Privately published. 1961. Illustrated key to common mosquitoes of northeastern North America. Privately published. 49 pp.

TRAVIS, B. V. — 1949. Studies of mosquito and other biting insect problems in Alaska. J. Econ. Entomol. 42(3):451-457. 1950. Known factors causing variation in results of insect repellent tests. Mosq. News 10:126-132. 1957. Present status and future possibilities of biological control of mosquitoes. Mosq. News 17(3):143-147.

TRAVIS, B. V. & F. A. MORTON — 1946. Treatment of clothing for protection against mosquitoes. Proc. 33rd meet, N.J. Mosq. Exterm. Assoc. pp. 65-69. 1950. Insect repellents and nets for use against sand flies. Proc. Ann. Meet. N.J. Mosq. Exterm. Assoc. 37:154-156.

TRAVIS, B. V., F. A. MORTON & J. H. COCHRAN — 1946. Insect repellents used as skin treatments by the armed forces. J. Econ. Entomol. 39(5): 627-630.

TRAVIS, B. V., F. A. MORTON & C. N. SMITH — 1949a. Use of insect repellents and toxicants. U.S.D.A. B. E. & P.Q. E-698 (Revised).

TRAVIS, B. V., F. A. MORTON, H. A. JONES & J. H. ROBINSON — 1949b. The more effective mosquito repellents tested at the Orlando, Fla., Laboratory, 1942-1947. J. Econ. Entomol. 42(4):686-694.

TRAVIS, B. V., A. L. SMITH & A. H. MADDEN — 1951. Effectiveness of insect repellents against black flies. J. Econ. Entomol. 44(5):813-814.

WALLIS, R. C. — 1960. Mosquitoes in Connecticut. Conn. Agr. Exp. Sta. Bull 632:1-30.

WESLEY, C., JR. & A. W. MORRILL, JR. — 1956a. Air and insect penetration of insect screens. Mosq. News 16(3):204-206. 1956b. Effect of various insecticide solutions on different kinds of insect screens. Mosq. News 16(3): 206-208.

WHITTEN, J. L. — 1966. That We May Live. Van Nostrand.

WILKES, B. G. — 1946. Insect repellents. U.S. Patent office, 1946. No. 2,407,205.

WOOD, D. M., P. T. DANG & R. A. ELLIS — 1979. The insects and arachnids of Canada, Part 6, The Mosquitoes of Canada (Diptera: Culicidae). Biosystematics Research Institute, Ottawa. 390 pp.

WOODRUFF, R. E. — 1958. Mosquitoes may be the primary vectors of St. Louis encephalitis. Pest Control 26(4):16-18, 20, 24.

ROBERT J. SNETSINGER

Dr. Robert Snetsinger is a native of Illinois and received bachelor's, master's and Ph.D. degrees in entomology from the University of Illinois, Urbana in 1952, 1953 and 1960 respectively. Prior to receiving his Ph.D., he was employed as an orchard foreman, a farm manager and as a research assistant at the Illinois Natural History Survey, Urbana. In 1960, he joined The Pennsylvania State University as an assistant professor of entomology to conduct research on mushroom and greenhouse pests. In 1961, he was placed in charge of educational programs for the Pennsylvania structural pest control industry.

Snetsinger has conducted research on spider mites and other pests of greenhouse and ornamental crops, on the biology and control of mushroom pests, on spiders and ticks and on household pests. He has authored numerous scientific papers in these subject areas. In addition, he is co-author of a series of eight correspondence courses on structural pest control. But Snetsinger's writing goes beyond pest control subjects. He is the author of the books Kiss Clara for Me, a historical account of the American Civil War, based on letters written by his great-grandfather; Diary of A Mad Planner, an account of urban planning; and Frederich Valentine Melsheimer — Parent of American Entomology, an account of America's first entomologist.

His most recent book, soon to be published by Franzak and Foster Co., is The Ratcatcher's Child, a history of the structural pest control industry. This latest book appeals not only to those who are part of the pest control industry, but to students of history, business and all who, like Snetsinger, are interested in human progress.

Snetsinger belongs to numerous professional associations including the Entomological Societies of America, Pennsylvania and Canada and the Arachnological Society of America.

CHAPTER TWENTY-ONE

Spiders

Revised by Robert Snetsinger[1]

twas an elderly mother spider
grown gaunt and fierce and gray
with the little ones crowded beside her
who wept as she sang this lay
curses on these here swatters
what kills off all the flies
for me and my little daughters
unless we eats we dies

—Don Marquis

AN EARLY EXAMPLE of "Innard-grated" Pest Management — "There was an old woman who swallowed a fly. She swallowed a spider to catch the fly, but I don't know why she swallowed the fly. She swallowed a bird to catch the spider, etc., etc."

Spiders are viewed uniquely by most people — they are both feared and esteemed. From ancient to modern times, among the diverse folktales, beliefs and thoughts concerned with spiders, there seems to be one common idea — that it is unlucky or unwise for mankind to kill spiders, unnecessarily. This seems somewhat surprising in light of the general apprehension many people have about spiders. In the Middle Ages, the presence of spiders was a sign of a healthful household. While most modern homemakers regard cobwebs as a sign of messy housekeeping, they nonetheless commonly pick up a spider in their home with a piece of tissue paper and carefully release it out-of-doors. Those who bring live spiders in for identification will often take them home and release them outside once they have been assured the specimen is harmless. This protective mythology concerning spiders is alive today and pest controllers need to take this factor into account when involved in control programs.

The diabolical or fearsome nature of spiders also involves a mixture of myth and reality. The ancient Greeks tell of a maiden so skilled in weaving that she challenged and surpassed the Goddess Athene in producing a fine tapestry. The angered Goddess tore the weaving and caused the maiden to become a spider with poisonous jaws.

[1]*Professor of Entomology, The Pennsylvania State University, University Park, Pa.*

The poisonous nature of the spider's bite and its swiftness in paralyzing its prey in part account for the abhorrence of spiders by many people. Morris and Morris (1965) conducted an interview of 80,000 childen as to the animals they most dislike; spiders ranked second to snakes. Jones and Jones (1928) concluded the fear of snakes (presumably spiders as well) "is not innate or inherited, but learned by each individual." However, Haslerud (1938) found a variety of objects including live and stuffed snakes and toy spiders were fear-producing stimuli for young chimpanzees. Morris and Morris (1965) suggest if chimpanzees, man's nearest living relative, respond with a withdrawal reaction to certain danger signals, such as certain kinds of movement patterns, then man may share this response. Spiders, probably more than other animals, have a complex duality. They are neither "good" like a dog or "bad" like a rat.

COMMON SPECIES OF SPIDERS FOUND IN DWELLINGS

There are some 35,000 described species of spiders in the world. In most states in the United States, it is possible to collect between 400 and 700 species. Kaston (1976) reported 688 species from New England and 477 species from Connecticut. The spider faunas of most states are poorly known and species identification is very difficult. A good dissecting microscope with luminosity twice that used by most people for insects, is required to make spider identifications. Spiders are generally stored in and kept submerged while being viewed in 70 percent alcohol. Positive species identifications usually require specimens to be either mature males or females. Mature males may be recognized by their palps, located between the chelicerae (jaws) and first pair of legs; palps are swollen (resembling boxing gloves). In mating, a male deposits a droplet of sperm on a special web, then sucks it into an opening in the palps. Later the sperm is injected into the epigynum of the female. Mature females have an opening on the underside of the abdomen, behind the pedicel (restricted connection between the head-thorax and the abdomen). This opening is called the epigynum. Just prior to maturity, some semblance of the male and female reproductive organs may be observed. This last instar before maturity is called penultimate. Species are recognized essentially by the palpal and epigynumal characters. Only a very experienced spider expert will venture to identify immature spiders.

Keys to spider families, genera and species are difficult to use, unless an experienced arachnologist is present to help beginners learn the pitfalls that invariably confuse the novice. For those wishing to seriously perfect their ability to identify spiders in the eastern United States, *Spiders of Connecticut* by B. J. Kaston (1948) and its supplement by the same author (Kaston, 1976) should be obtained. For those in other parts of the United States, Kaston's *How to Know the Spiders* (Kaston, 1978) is valuable. When available, regional spider publications also are an important addition. However, for most people who occasionally want to identify spiders to family or to a common species, *Spiders and Their Kin* by H. W. Levi and L. R. Levi, Golden Press, New York is very useful. *Spiders and Their Kin* is a well illustrated picture book in which you can search for an illustration that resembles your specimen.

Most spiders that cause concern to the public and which pest control specialists are called upon to identify are either large or have striking markings. A large spider is feared to be a tarantula and a colorful one is suspected to be a black widow by much of the public. Only a few species of spiders reside in dwellings and a somewhat larger number on occasion stray or are carried into houses.

About a dozen groups or species represent more than 90 percent of the spiders of public concern. The following comments supplemented with Levi and Levi's book should permit most laypeople to make many of their own identifications.

Recognition of order Araneae. Spiders have two body regions connected by a restricted waist or pedicel. The fused head/thorax or cephalothorax has two jaws or chelicerae, two palps (small leg-like appendages), eight legs and eight or fewer eyes. The abdomen is unsegmented and has openings of the book-lungs and/or tracheae. Anteriorly, on the underside of the abdomen, is the epigynum in mature females, and posteriorly the spinnerets.

Spider mites, Tetranychidae, (order Acari) are not spiders. The common name red spider or spider mite refers to the ability of these acarines to spin silk. Harvestmen or daddy-long legs (Order Opiliones) are also related arachnids. Both Acari and Opiliones have the cephalothorax and the abdomen fused. The abdomen of harvestmen are distinctly segmented and the eyes (usually two) are located on a raised tubercle on the cephalothorax. Legs are often very long and thread-like. Without close inspection, daddy-long legs and cellar spiders, Pholcidae, may be confused.

Tarantulas. The much maligned tarantulas are mygalomorph spiders with jaws that project forward and move parallel to the long axis of the body, while with the "true" or aranemorph spiders these mouthparts project downward, perpendicular to the body. An easy way to visualize this is to lay on your stomach with your chin on the floor. Place your wrists on your cheeks and clap your hands several times. This is the position of the jaws in tarantulas. If you bring your chin down to your Adam's apple and clap your hands, this is the "jaw" position of "true spiders." In addition, typical tarantulas have two pairs of book-lungs, while "true" spiders have one pair, sometimes supplemented with tracheae.

Wolf spiders, Lycosidae. Large, "hairy", running spiders, often confused with "tarantulas," are in most cases really wolf spiders. Part of the reason for this confusion is that some of the lycosids of southern Europe are also called "tarantulas." In the Middle Ages, it was believed these wolf spiders were poisonous and the victims of the bites used wild, exotic dances as a remedy. These dances continued night and day with music from pipers, fiddlers, and Turkish drums, until the victim collapsed exhausted and temporarily "cured." The dance was the tarantella, taken from the Italian town of Taranto where the belief and dancing started.

Wolf spiders have eyes in three rows (most spiders have two eye rows). The first row in lycosids has four relatively small eyes; the second and third rows each have two quite large eyes. Wolf spiders are hunting spiders which do not construct webs. Larger species are one to 1½ inches/25 to 38 mm in body length and have a leg span of three to four inches/76 to 101 mm. *Lycosa helluo* is commonly found in cellars in the eastern United States. Many other species of lycosids enter dwellings.

Swamp spiders, Dolomedes spp, Pisauridae. Spiders of the genus *Dolomedes* are about ½- to one-inch/13 to 25 mm in body length with a leg span of two to three inches/50 to 76 cm. Most members of this genus are hunters and divers, associated with streams and ponds. However, *Dolomedes tenebrosus* is commonly found in homes in firewood. This species and others are commonly found under tree bark. Many members of the genus *Dolomedes* have a lyre-shaped marking on the head-thorax.

Fig. 21-1. Male of common jumping   Fig. 21-2. Common jumping spider. Note
spider, *Phidippus audax.*   the battery of eyes.

Funnel weavers, Agelanidae. In the late fall, the webs of funnel weavers are quite common and particularly apparent on mornings when the webs are moist with dew. *Agelenopsis pennsylvanica* and other members of this genus commonly construct their webs in windows. *Coras medicinalis* often constructs its webs in cellars of houses. In former times the silk of this spider was used to cover wounds to check bleeding.

Spiders of this family have eyes in two rows of four each. The spinnerets are long and more obvious than for most other families. The webs of *Agelenopsis* and most other members of the family form a sheet or silk platform with a tube or funnel leading off from the center or to one side. The body length of the larger species is about ½ inch/13 mm.

Jumping spiders; Salticidae. Members of this family have their eyes in three rows. The four eyes of the first row are usually very large. The third row, which has two eyes, is about halfway back on the thorax and the eyes are moderately large. The second row, which has two eyes which are often quite small, is located about midway between the first and third rows. Jumping spiders are compact in shape with relatively short legs. They are hunting spiders that are capable of jumping from a few inches up to six inches/15 cm in one leap, depending upon the species. *Phidippus audax* and other members of this genus often stray into dwellings. *P. audax* is found in most of the United States and *P. johnsoni is found* in the West. Both have red or white markings on the dorsal surface of the abdomen. *Metacyba undata* is commonly carried into houses on firewood. They measure about ½ inch/13 mm in body length. *Salticus scenicus,* the Zebra spider, is about ¼ inch/six mm in body length with white stripes and blackish/brown background. They are common on the walls of houses. Because most jumping spiders have colorful markings, many people fear they are blackwidows. Actually, they are beneficial because they pounce on flies and other insects.

Crab spiders, Thomisidae and Philodromidae. Most crab spiders found in dwellings are carried in accidently with bouquets of wild flowers or in firewood. Crab spiders are camouflaged so they match the color of the flowers in which they hide or the surface bark of trees where they wait for prey. They are easily recognized by the positioning of their legs which give them a crab-like appearance. Most species are less than ½ inch/13 mm in body length. Crab spiders do not make webs.

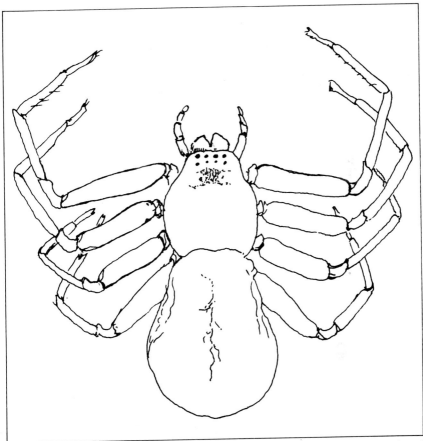

Fig. 21-3. Crab spiders, *Thomisidae* and *Philodromidae*. First and second pairs of legs laterigrade giving a crab-like appearance.

Sac spiders, Clubionidae. There are no good spot characteristics that allow the recognition of a spider as a clubionid. A few species of the family are black with white strips and red spots on the abdomen. They belong to the genus *Castianeira*. Because of these markings, some people become concerned, thinking they are black widows. Clubionids are wandering spiders and do not construct webs. Some species construct silken retreats in which they hide during daylight hours. The bites of some species of the genus *Chiracanthium* have been reported to be poisonous to man.

Gnaphosids, Gnaphosidae. Also, with this family there are no obvious spot characteristics which would easily allow layman recognition. The parson spider, *Herpyllus ecclesiastica*, quite commonly strays in houses during the summer months. The body length is about ¼ to ½ inch/six to 13 mm and the abdomen is grey/brown with a whitish band above which suggests black widow to some people. Parson spiders are hunting spiders that do not spin webs, but do construct retreats or nests.

Fig. 21-4. Sac spider, *Castianeira,* Clubionidae. This spider is characterized by a black abdomen with white stripes or red spots on some species. It measures ¹/₃ inch/eight mm long.

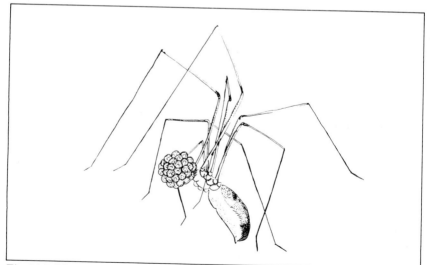

Fig. 21-5. The cellar spider is readily identified by its long thread-like legs. Note the egg cluster.

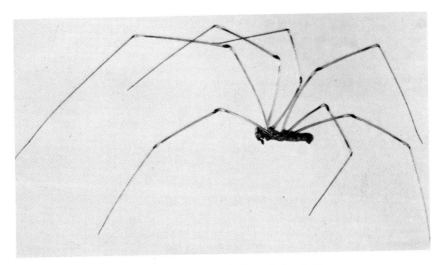

Fig. 21-6. The long-bodied cellar spider, *Pholcus phalangioides.*

Cellar spiders, Pholcidae. These spiders are common in barns, cellars and damp warehouses. Their legs are quite long. Two species are commonly encountered, the long-bodied cellar spider, *Pholcus phalangioides,* and the short-bodied cellar spider, *Spermophora meridionalis.* The former has a body length of ⅓ inch/eight mm, legs up to two inches/five cm, and eight eyes. The latter, a body length of about ¹⁄₁₀ inch/2.5 mm, legs about ⅓ inch/eight mm, and six eyes. The long-legged cellar spider hangs in its web with the abdomen upwards and, when alarmed, shakes its web violently or spins around rapidly.

Orb-weavers, Araneidae. In the fall many orb-weavers mature; some species are about one inch/25 mm in body length with bright colorful markings. Some orb-weavers construct elaborate and handsome webs. The webs and the colorful spiders attract much attention. Very commonly people bring orb-weavers to authorities for identification without any interest in having the spiders controlled. *Argiope aurantia, A. trifasciata, Araneus trifolium* and many others interest the layman.

Fiddleback spiders, Loxoscelidae. Although loxoscelines have been known since 1842, necrotic arachnidism, the disease associated with the bite of these spiders, was unverified until 1934. It wasn't until about the 1960's that bites by fiddlebacks became a matter of concern in the United States.

Fiddleback spiders have six eyes in three diads, the median diad positioned on the front of the carapace and the lateral diads positioned along the sides of the carapace, about halfway back. A fiddle-shaped

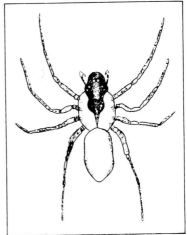

Fig. 21-7. Fiddleback spider.

marking, darker than the surrounding tan color of the carapace, covers the eye region and extends back to the pedicel. The body of the violin marking is forward and the fingerboard is posterior. Females are about ½-inch/13 mm in body length and the males are ⅓ inch/eight mm; the legs are long in both sexes, one inch/25 mm or more. The third pair of legs are the shortest.

Comb-footed spiders, Theridiidae. The black widow, the house spider and *Steatoda borealis* are comb-footed spiders of the most concern to urban pest controllers. A row of bristles located on the tarsi of the fourth pair of legs form the comb. The tarsus is the last or distal segment of the leg. The six to 10 slightly curved, serrated bristles are difficult to see, even with the aid of a microscope, unless they are pointed out to the novice. On male specimens the comb is less obvious than on female specimens.

SPECIES OR GROUPS OF MAJOR CONCERN

Black widow spiders, *Latrodectus*. No other group of spiders causes so much apprehension. The taxonomic status of black widow species in the United States has been under dispute for some time. Levi and Randolf (1975) lists *L. mactans, L. bishopi, L. geometricus, L. variolus,* and *L. hesperus.* However, these

Fig. 21-8. The black widow spider, *Latrodectus mactans* (Fabr.). View from above — (left) female and (right) male. Twice natural size.

PIDERS: KEY TO SOME IMPORTANT UNITED STATES SPECIES
Harold George Scott & Chester J. Stojanovich

ings projecting horizontally (Fig. 1A). (abdomen without tergites; tarsus with claw tufts and 2
claws) . . . .*Dugesiella hentzi* and others, TARANTULAS

ings projecting vertically (Fig. 1B) 2

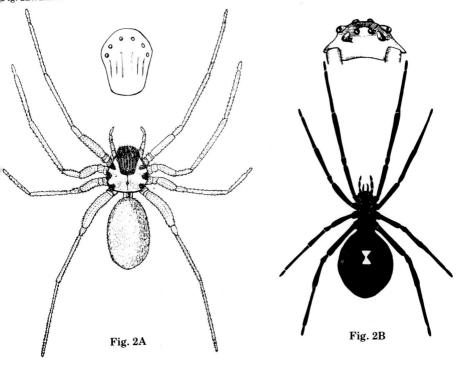

U.S. DEPARTMENT OF HEALTH, EDUCATION, AND WELFARE
PUBLIC HEALTH SERVICE, Communicable Disease Center
Atlanta, Georgia
1963

Fig. 1A

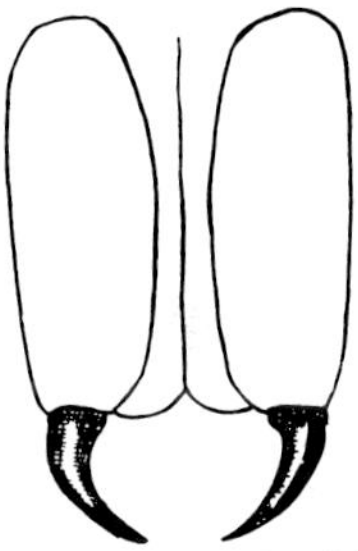

Fig. 1B

ix eyes in 3 pairs; fiddle-shaped marking on cephalothorax (Fig. 2A) . . . .
oxosceles spp. . . . . .BROWN SPIDERS

ight eyes (shiny black with red spots; usually with red hourglass on underside of abdomen)
ig. 2B). *Latrodectus mactans* . . . . BLACK WIDOW SPIDER

Fig. 2A

Fig. 2B

Fig. 21-9. Identification key to some important spider species in the United States.

authors doubt *L. hesperus* is a valid species and suggest *L. mactans* and *L. variolus* are species complexes. Although Levi doubts the occurrence of *L. mactans* in New England, Kaston (1976) holds *L. variolus* and *L. mactans* both are found in New England. It is apparent museum taxonomy (Kaston, 1970; Levi, 1975) alone will not be able to settle the nomenclatural status of black widows. While there is much variation in color and certain other characters among populations of widows, the definitive male palpal and the female epigynumal characters of matures are too much alike in the related species.

Distribution. Black widows are present in every state in the United States, in southern Canada, Caribbean Islands, Mexico, Central and South America, Mediterranean countries, southern USSR, New Zealand, Australia and other warmer parts of the world. Widows frequently are found in railroad boxcars, shipholds, interstate freight vans and other transport. Thus, northward import of south species and the introduction of exotic species are fairly common. In northern states, outbreaks or "blooms" of black widows occur erratically. Some years an area may have thousands of widows and the next year they may be gone. Certain kinds of habitats such as sand dune areas, e.g., Illinois Beach State Park (about 30 miles north of Chicago), have black widows every year. Apparently, alternating warm spells and cold snaps in the winter and spring are detrimental to survival. In much of California, the Southwest, and the South, widows do extremely well.

Preying habits. The female constructs an irregular, tangled, criss-cross web of rather coarse silk. The core of the web is a silken tunnel, in which the female spends most of the daylight hours. The web usually projects in all directions for a few inches to a few feet. The web is regularly renewed, altered, and expanded and is capable of securing large insects.

Once a prey is caught in the web, it is wrapped-up in silk which is spun out in great abundance. As the prey becomes covered with silk, it is turned over and over by the widow with her legs as more silk is applied. After the victim is secured, it is killed by a bite of the chelicerae. After the poison is injected, the prey may be fed upon or be reserved for a later feeding. During feeding a powerful saliva is released from the foregut which dissolves the protein tissue of the prey. After the prey is fed upon and the body fluids are sucked from the victim, the carcass is cut loose and allowed to drop to the ground. A variety of insects and other arthropods are eaten or imbibed.

Life History. The eggs are laid in silken cocoons or sacs which are globular in form and about ½ inch/13 mm in diameter. When first constructed, the sacs are white, but after a while turn pale brown. One to three hours are required to construct the egg sacs. About 300 to 400 eggs per sac is common, although Lawson (1933) reported a sac with 917 eggs. From four to nine egg sacs may be produced by a female during a summer. The female guards the egg sacs and moves them as necessary to repair her web. After laying eggs, the females are hungry and more likely to bite a human. The eggs hatch in about eight to 10 days. The spiderlings molt once before emerging from the sac. The period between the depositing of the eggs and emergence may be two weeks to a month. The spiderlings force an opening in the sac, usually at night. At first, the second instar spiderlings remain near the sac, but after a few hours to several days, they climb to a promontory point, where air currents are suitable, spin silk strands and float out on the breeze, like a kite on a silk thread. This ballooning scatters the spiderlings to whatever fate awaits them.

After they land, the spiderlings construct a web, obtain prey, and if successful, pass through some seven more instars when they become mature adults. Depending upon the availability of prey, maturity comes after about four months. A generation is completed in about one year. In warmer climates, development continues throughout the year. In the North, winter is spent in preadult stages with mating occurring in about May and egg laying following shortly thereafter. Deevey and Deevey (1945) found the average male matures in about 71 days after emergence from the egg sac and lives about 30 days longer if alone. The average female matures 92 days after emergence and lives about 179 days longer. The maximum longevity was 160 days for males and 550 for females. Kaston (1948) reported the survival of one female for 1,063 days after emergence from the egg sac.

Mating. Popular belief has the male eaten after one encounter with a female. Ordinarily, if the females are well fed, most males get away to mate another day. The observations in which males are eaten usually occur under laboratory conditions which are not natural and may not represent what is normal. Ross and Smith (1979) suggest males and females have chemo-receptive hairs on the tarsi and pedipalps and complementary contact pheromones are produced which are incorporated into the silk. These substances function in mate location, sex identification and courtship. Web vibration by the male and reciprocating movements by the female enter into the courtship. But the ferocity of the female towards her mate has been much over-played. Actually, males live

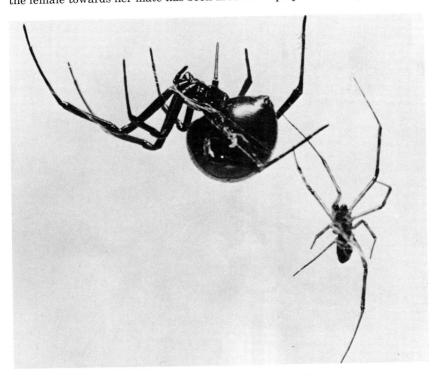

Fig. 21-10. Male (right) courting female black widow spider.

much longer when associated with a mate, because he is a kind of parasite on the food captured in the female's web. The females support their free-loading husbands. What more can any male ask?

Latrodectism. For the most part, widows are shy and retiring, except perhaps when guarding their eggs and immediately after laying eggs; at this time the females are emaciated and vigorously attack prey. Both the male and female have venom sacs. As the male reaches maturity development of the venom sac stops and becomes inactive. Incidentally, the mature male does not attack prey. *Latrodectus* venom is a neurotoxin. Frontali and Grasso (1964) separated two invertebrate neurotoxins from the venom of *Latrodectus tridecimguttatus,* one producing rapid knockdown and the other producing maximum knock down after 24 hours. The bite may at first not be felt and there is usually little evidence of a lesion, although a slight local swelling around two tiny red spots may sometimes be observed. Pain, at the site of the bite, occurs almost immediately and becomes most intense after about three hours. General aching of the body, especially the legs, are common reactions. Rigidity of large muscle groups with spasms also are a common reaction. Headache, elevated blood pressure, nausea, shock, difficulty in breathing and profuse perspiration may occur in severe cases. The condition is self-limiting and in most cases symptoms disappear in two or three days. Thorp and Woodson (1945) tabulated cases of black widow bite in the United States from 1926 to 1943. They recorded 1,291 cases and 55

Fig. 21-11. Adult female black widow spider. Hourglass mark on ventral side of abdomen is reddish in color, variable in shape and at times may be absent.

deaths. Parrish (1963) reported 63 fatal black widow bites from 1950 to 1959. Russell et al. (1979) reported the symptoms of widow bites were greatly lessened five minutes after treatment with antivenin and all symptoms were gone 20 minutes after treatment. Calcium gluconate is used intravenously to relieve and relax muscle spasms produced by widow venom. Calcium appears to block the action of the venom at nerve endings.

Recognition of black widows. Since several species are involved and only an expert has the ability to make a definitive species identification, satisfactory identification is to genus. The adult female is usually jet black above, with two reddish triangular markings, joined to form an "hour glass" shape on the venter or underside of the abdomen. Mature females are nearly ½ inch/13 mm in body length and males are about half this size. The markings of immature spiderlings and mature males are similar. Young spiders are orange and white and acquire more black in successive instars as they mature. The males have whitish streaks, bars or dots on the dorsal surface of the abdomen and two triangular red markings beneath. The immatures have one or two red markings beneath, but these may be quite variable and are not usually joined. As there are at least four native species of *Latrodectus* in the United States, and the introduction of exotic species in materials arriving from overseas is always a possibility, variations from the classification descriptions found in most publications can be expected.

Fiddleback spiders, *Loxosceles*. The identification of members of this genus to species is sufficiently difficult to require the expertise of an arachnologist to make a definitive determination. *Loxosceles* are small tan to brown spiders with six eyes in three groupings and a darker fiddle-shaped marking on the carpace or upper surface of the head-thorax. Gertsch (1958 and 1967) notes some 17 species from temperate South Africa and southern parts of Europe, 18 species from North America, Central America and the West Indies, and some 30 species from South America. Since species are readily transported by commerce, fiddlebacks are found in many parts of the world.

In the United States *L. reclusa* Gertsch, is native and in the past 20 years has greatly extended its distribution range. However, *L. unicolor, L. devia,* and *L. arizonica* are native to the Southwest; *L. caribbaea* is found in Puerto Rico and Mona Island; *L. panama* and *L. rufipes* are known from the Panama Canal Zone; *L. rufescens* has become established in the United States and other exotic species may be introduced on overseas shipments *L. laeta* was established in the basement of the Museum of Comparative Zoology at Harvard University, Cambridge, Mass., but has since been eradicated. All species must be considered dangerous to man.

Distribution. *L. reclusa* is generally found in midwestern and south central states, but has been found spreading into many other states including New York, California, Pennsylvania, New Jersey, Florida, North Carolina, Wyoming and Washington, D.C. *L. rufescens* has been reported from Georgia, Texas, and in Pennsylvania, on the campus of The Pennsylvania State University, where it dwells in heat tunnels and basement storage areas. Apparently, it has been present for at least 30 years.

Preying habits. The female constructs an irregular web in places which are undisturbed. The spiders feed on a variety of insects; *L. rufescens* in the Pennsylvania State University heat tunnels feed mostly on firebrats and in laboratory cultures readily feed on flies and honeybees. Most of the feeding activity occurs at night.

Life history. Hite et al. (1966) observed mating of *L. reclusa* in Arkansas, from February to October, but mostly in June and July. Egg laying was reported from May through August with a maximum of five egg sacs produced. Total seasonal egg production varied from 31 to 300. The maximum number of spiderlings to emerge from an egg sac was 41. The average emergence time from the egg sacs for the second instar spiderlings was 33 days. From 266 to 444 days, average 336, were required to reach maturity. The average length of life after emergence from the egg sac varied from 30 to 796 days for males and 356 to 894 for females. During the mating season, males wander about seeking females. With *L. rufescens* this wandering phase occurs at all seasons indoors or in heating tunnels. Young spiderlings also may do some wandering because they were frequently captured in firebrat traps placed in heat tunnels.

Necrotic arachnidism. The venom of *Loxosceles* spiders can produce a gangrenous slough of tissue near the site of a bite. A mild "stinging" sensation has been reported at the time of the bite, but many victims are not aware of being bitten. Mild to severe pain follows two to eight hours later. A welt of dead tissue usually forms at the site of the bite. The area becomes dark and dry, and after seven to 14 days this separates from the surrounding tissue, leaving an open ulcer which lasts for several weeks. Healing time depends upon the size of the ulcer and, if large, may require skin grafts. Spiderlings are capable of biting humans if they can get to a tender spot where they can insert their chelicerae. Most bites occur on hands and arms. Foil et al. (1980) reported two neurotoxins; one producing rapid knockdown and one producing maximum knockdown 24 hours after injection in house flies. Elgert et al. (1974) recommended antivenin be administered within 30 minutes after a bite to inhibit local necrosis in humans. Currently, there is a great deal of research being conducted on the venom of *L. reclusa*. Much of this research is published in the journal *Toxicon*. A number of drugs have been advocated for treatment of loxoscelism, including phentolamine hydrochloride, antihistamines, corticosteroids and ACTH. It is apparent a completely satisfactory treatment for the treatment of necrotic arachnidism has not been developed.

TARANTULAS

There is much diversity among species of tarantulas. Such genera as *Atypus* and *Antrodiaetus* are only a ½ inch/13 mm or so in body length and are distributed in many scattered localities in the United States. The largest tarantulas, "mygales" or "bird spiders," are tropical and have a body length of 3½ inches/8.9 cm and a span of 9½ inches/24 cm with legs extended. The largest native species in the Unites States have a body length of about two inches/five cm and a leg span of about six inches/15 cm. There is an uncertainty as to the number of native species in the United States, but the number probably is around 100. Many species are quite local or rare in distribution and the destruction of habitat may have caused the extinction of some species. However, taxonomists are still describing new species and unravelling taxonomic problems.

Most species of tarantulas are large enough so as to be capable of biting people. However, tarantulas are usually sluggish, can be handled with ease, bite only rarely and have venom generally of little harm to most people. Within recent years, tarantulas have become acceptable pets, are sold by pet stores and championed by the American Tarantula Society and their newsletter, *Tarantula Times*. Such species as Mexican redlegs, Mexican blacks, Haitians and others

Fig. 21-12. The tarantula, *Eurypelmus californica.*

are now widely sold, traded and kept in houses, apartments and school dormitories. Occasionally, these pets escape, sometimes creating panic among those who fear spiders.

Among the native species of tarantulas, the males wander about in search of females, usually during the early summer months. They sometimes stray into homes and other places where they are unwanted. In a suburb of Philadelphia, several hundred male *Atypus snetsingeri* were collected from a swimming pool where they were trapped in the pool's filter (Sarno, 1973).

Some species of tarantulas may live for 20 years or more. Except for mate-seeking males, tarantulas are secretive with many curious habits. Trap-door spiders live in burrows in the soil and have highly camouflaged doors constructed of silk and local debris. *Atypus* or purseweb spiders construct silk tubes up the base of trees or other objects.

Tarantulas are much publicized, but are rarely encountered by most people, except on the screens of movie houses and television sets. In times past when bananas were shipped as large bunches on stalks, tarantulas commonly were accidentally imported. The banana spider (not a tarantula, but a giant crab spider) was similarly found in banana bunches. Since bananas are now shipped as "hands" or small bunches, encounters with such exotics are now rare.

THE HOUSE SPIDER

Achaearanea tepidariorum (Theridiidae)

The house spider is so much a cosmopolitan and so widely distributed that no one is certain of its original homeland. It has been spread by ships, rail and other forms of transport, including its own system of ballooning. Valerio (1977) described second instar house spiderlings as "the *floating population* or dispersion stage." Second instar spiderlings can resist starvation for 25 days or longer and balloon long distances (Valerio, 1974). The mortality of these spiderlings is greater than 98 percent. Because of their small size, it is difficult for them to capture food.

The established population of house spiders includes newly established spiderlings with webs, more mature instars and adults of both sexes. With every molt, the spiderlings construct larger webs, require more food and consequently the number of individuals in the population is reduced, if prey is a limiting factor. The house spider selects web sites at random. If the sites do not yield prey, the webs are abandoned and a new site is chosen and a new web constructed (Turnbull, 1964). Air currents play an important role because they direct the movement of the prey and eventually the spiders end up constructing and maintaining webs where food is most available. The dirty webs that concern most homemakers are largely abandoned ones that have failed to yield enough prey. Because most American homes have low humidities, which causes high mortality to spiders, and relatively few flying insects enter dwellings, survival of house spiders is often low. Mature house spiders are much more commonly observed under bridges, in barns, greenhouses and warehouses. The house spider is a quite effective predator of small insects such as fruit flies, but it also can handle larger prey. Bristowe (1958) fed a house spider another species of spider nearly three times larger, and the small spider enswathed its victim in silk and paralyzed it with a fatal bite. The house spider is also credited with the capture of small snakes and of mice. The early spider authority, H.C. McCook, described the capture of a mouse by its tail. The house spider bit the mouse's tail, enswathed it with silk, bit the tail again, etc.

Montgomery (1903) described the mating of the house spider in detail. Briefly, the male acts with great caution, for he is much smaller than the female. The female, usually heavy with eggs, signals the male of her desire by

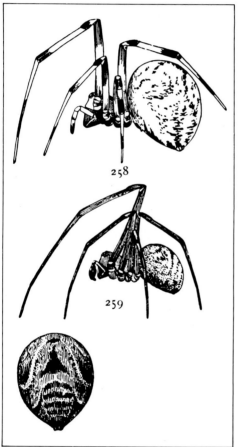

258

259

Fig. 21-13. The house spider. *Achaearanea tepidariorum* **K. Top, female; middle, male; and bottom, abdomen of female as seen from behind.**

vibrating the web and advancing toward the male. The male then eagerly approaches the female and inserts his palps, injecting his sperm. Any sign of aggressiveness by the female causes the male to drop from the web. Multiple mating is common for both sexes. Shortly thereafter, the female lays about 250 eggs in a silken sac which is flask-shaped, about ¼ to ⅓ inch/six to eight mm long, and usually placed near the center of the web. The egg sacs are moved to warmer or cooler sites, as necessary. Under favorable conditions, two or more egg sacs may be present in a web at the same time, and nine sacs may be produced in a season. The eggs hatch after about a week to 10 days. The first instar spiderlings remain in the sac for one molt. The second instar spiderlings emerge and spin silk threads and become the *floating population*.

SPIDER BITE

One hears much about spider bites. Physicians commonly tell patients with red welts or local swellings they have been bitten by spiders. In general, spiders are credited with far more bites than they are responsible for. Bites of fleas and bedbugs and stings by bees, wasps, and allergic reactions to certain substances are often mis-diagnosed as "spider bite" by physicians. Do not be cowed by the title which physicians prefix to their name. Their knowledge of arthropods may be quite limited.

The vast majority of spiders do not bite people, either because their chelicerae cannot penetrate skin or because most people rarely handle spiders. Arachnologists are the people most frequently bitten by spiders, because they collect them. The following eastern United States species have bitten the author of this section or his students. Antrodiaetidae; *Antrodiaetus unicolor;* Atypidae, *Atypus snetsingeri;* Theridiidae, Steatoda borealis; Araneidae, *Argiope aurantia, A. trifasciata, Aranea diadematus, A. nordmanni;* Tetragnathidae, *Tetragnatha elongata;* Agelenidae, *Agelenopis pennsylvanica;* Pisauridae, *Dolomedes tenebrosus;* Lycosidae, *Schizocosa crossipes, Geolycosa pikei, Lycosa rabia, L. helluo;* Gnaphosidae, *Herpyllus ecclesiastica;* Anyphaenidae, *Aysha gracilis;* Salticidae, *Phidippus audax, P. clarus, Eris marginata,* and *Metacyrba undata.* The pain from all was slight and lasted for only a few minutes, except for *Dysdera crocata,* which lasted several days. A note in *American Arachnology* (Newsletter of the American Arachnological Society) 22: 5-6 (Nov., 1980) mentions experiences of others who were bitten by about 20 species of spiders. In general, pain was only modest and the reactions mild. Many other cases of spider bites go unreported each year or the species causing the bite are not identified. Only the bites of species of *Latrodectus* and *Loxosceles* are serious matters of concern. Except for those who show unusual reactions, spider bites do not require medical treatment.

With bites of *Latrodectus* species, treatment with intravenous injections of calcium gluconate generally provide rapid relief from pain by relaxing muscle spasm. Magnesium sulphate is also used. Antivenins are effective, but results are less rapid. Such drugs as phentolamine hydrochloride and antihistamines have been used in the treatment of bites of *Loxosceles* spiders. Dillaha et al. (1964) favor a prompt treatment with ACTH and corticosteroids. Specific antivenins are not readily available in the United States. Treatment for bites by these groups of poisonous spiders should be attended to promptly by a physician. In the case of any spider bite, the spider causing the bite should be preserved, preferably in 70 percent alcohol.

Arachnophobia. Most people have never been bitten or harmed in any way by a spider, yet, fear of spiders and their kin is one of the most common animal fears expressed by those who seek psychiatric or psychological treatment. Some people have a general fear of "creepy-crawly things" and others specifically fear spiders. The popular media and nervous mothers do much to reinforce such fears. A respect for black widow spiders and fiddlebacks is healthy, but a small proportion of the public "sort of freak out" over spiders. This irrational behavior is more common in small children, young girls and women. If a person is unable to operate in society or wishes to overcome his or her fear of spiders, psychiatric or psychological help is often required. Such phobias are usually treated by nonstressful kinds of exposure to the animal, photograph of a spider, etc. Thus, by degrees the person is helped to overcome the fear. Such treatment requires trained professionals and well-meaning amateurs should stay clear of attempting to treat phobias.

CONTROL OF SPIDERS

It should be recognized that spiders are predators and require prey to survive. However, spiders frequently stray into dwellings or other indoor habitats, or may be accidentally introduced on firewood, laundry hung out to dry and on flowers. In windows and near outdoor lighting, web-building spiders frequently construct webs because insect prey may be attracted at night by the lights and by air currents. The removal of cover such as debris, lumber piles and materials may reduce the presence of some species. Care needs to be taken in using clothing, tents, sleeping bags and other items left unused for long periods in areas where spiders may be common. Improved storage, use of air tight boxes and bags, elevation of materials off the floor or ground, discarding unwanted items and periodically sweeping or vacuuming under furniture and behind mirrors and pictures are all helpful control measures. Thick leather or rubber gloves should be worn when cleaning out areas which may harbor black widows or fiddleback spiders. After the housekeeping has been improved, chemical control should be considered, based on the species or group of spiders involved and their habits.

Pesticide labels relating to spider control may simply refer to spiders as an order (i.e., all spiders or to a specific species or group of spider, i.e., the long-bodied cellar spider or black widow spider). Most tests which have been conducted to obtain efficacy information for a label for spider control have been conducted with only a few species of spiders. Since spiders have diverse habits, the label recommendations must in some cases be adjusted to meet specific problems. Web-building spiders need to be handled differently than hunting spiders. Spiders that are brought into a dwelling on firewood or on some wild flowers should be treated differently than a resident house spider problem. To obtain satisfactory control requires observation of the spider problem, before chemicals are applied, if they are required at all.

After World War II, DDT was widely used for spider control. Some of the recommendations of the time, suggested a 10 percent DDT-kerosene spray applied to the web. From 1949 into the 1970s, lindane was widely recommended and used for spider control; dilution rates varied from 0.75 to two percent. Evans (1955) recommended a two to three percent chlordane emulsion spray applied to overhangs, weeds, and the outside walls of buildings. Others have recommended a five percent chlordane dust to webs. Generally, chlordane, when applied to an area of spider activity, gave excellent results. For a time in the early 1970s dieldrin sprays, 0.5 percent and two percent dusts were used. However, most spider control in times past represented overkill or ineffective treatment, because the control was not directed against the species or group of spiders of concern. Part of this was due to a lack of knowledge of spiders and because the applicator had a basic abhorrence of spiders and wanted to kill them "deader" with more insecticide.

When applied appropriately, spiders are quite susceptible to most modern insecticides, particularly the organophosphates and carbamates. Spiders are, however, easily disturbed and drop from their webs and seek cover when sprays are applied. One consideration which is important when web-building spiders are being treated is their habit of recycling silk. They chew-up the old web and consume the silk. They also have powerful digestive juices which dissolve the silk. This means a dust formulation, lightly applied to a web can be quite effective.

A wide variety of insecticides are labelled for spider control and are effective when directed to the area of spider activity. Dust formulations are best directed against web-building spiders and to entry cracks and crevices for spiders entering from the outside of the structure. Space sprays are useful for spider probems in confined areas when there is no hazard to people using the structures (i.e., a small crawl space or storage shed; space sprays are effective for fiddleback spider control when they become established in an enclosed area; retreatment about 30 days later is necessary to kill young spiderlings which emerge from egg sacs). Residual sprays are generally applied to infested windows, corners, door frames, roof overhangs, eaves, downspouts and storage areas. Restrictions concerning food service, preparation and manufacturing must be considered when selecting a formulation and an insecticide.

Dust formulations labeled for spider control include products containing bendiocarb, diazinon, malathion, and pyrethrum. Space sprays are DDVP, pyrethrum, and resmethrin. Residual sprays include formulations based on bendiocarb, bromone, chlorpyrifos, DDVP diazinon, malathion, propetamphos, propoxur, pyrethrum, resmethrin and ronnel. Since all the materials are effective, control depends upon using a formulation that is effective against a spider species or group that gives the best results based on the habits of the spider.

LITERATURE

BRISTOWE, W.S. — 1958. The world of spiders. Collins, St. James Place, London. 304 pp.
DEEVEY, G.B. and E.S. DEEVEY, JR. — 1945. A life table of the black widows. Trans. Conn. Acad. Arts and Sci. 36:115-131.
DILLAHA, C.J., G.T. JANSEN, W.M. HONEYCUTT and C.R. HAYDEN — 1964. North American loxoscelism. J. Amer. Med. Assoc. 188(6):33-36.
ELGERT, K.D., M.A. ROSS, B.J. CAMBELL and J.T. BARNETT 1974. Immunological studies of brown recluse spider venom. Infect. Immun. 10:1412.
EVANS, D. — 1955. Black widow outbreak. Pest Control 23(10): 6 & 90.

FOIL, L.D., H.W. CHAMBERS and B.R. WORMENT — 1980. *Loxosceles reclusa* venom component toxicity and interaction in *Musca domestica*. Toxicon 18:112-117.

FRONTALI, N. and A. GRASSO — 1964. Separation of three toxicological different protein components from the venom of the spider, *Lactrodectus tridecimguttatus*. Archs. Biochem. Biophys. 106:213.

GERTSCH, W.J. — 1958. The spider genus *Loxosceles* in North America, Central America and the West Indies. Amer. Mus. Novitates. 1907:1-46.

GERTSCH, W.J. — 1967. The spider genus *Loxosceles* in South America (Araneae, Scytodidae). Bull. Amer. Mus. Nat. Hist. 136(3):1-173; 11 plates.

HASLERUD, G.M. — 1938. The effect of movement of stimulus objects upon avoidance reactions in chimpanzees. J. Comp. Psychol. 25:507-528.

HITE, J.M., W.J. GLADNEY, J.L. LANCASTER, JR. and W.H. WHITCOMB — 1966. Biology of the brown recluse spider. Univ. Arkansas Agric. Exp. Sta. Bull. 711:1-26.

JONES, H.E. and M.C. JONES — 1928. Maturation and emotion: fear of snakes. Childhood Education 5:136-143.

KASTON, B.J. — 1948. Spiders of Connecticut. Bull. Connecticut Geol. Nat. Hist. Survey 70:1-874.

KASTON, B.J. — 1970. Comparative biology of American black widow spiders. Trans. San Diego Soc. Nat. Hist. 16:33-82.

KASTON, B.J. — 1976. Supplement to spiders of Connecticut. J. Arachnol. 4:1-72.

KASTON, B.J. — 1978. How to know the spiders. W.C. Brown Co., Dubuque, Iowa.

LAWSON, P.B. — 1933. Notes on the life history of the hour-glass spider. Ann. Entomol. Soc. Amer. 26:568-574.

LEVI, H.W. and L.R. LEVI (most recent). Spiders and their kin. Golden Guide. Western Pub. Co. Inc., Racine, Wisc. 160 pp.

LEVI, H.W. and D.E. RANDOLPH — 1975. A key and checklist of American spiders of the family Theridiidae North of Mexico (Araneae). J. Arachnol. 3:31-51.

MONTGOMERY, T.H., JR. — 1903. Studies on the habits of spiders, particularly those of the mating period. Proc. Philadelphia Acad. Nat. Sci. [LV]: 59-149.

MORRIS R. and D. MORRIS — 1965. Men and snakes. McGraw-Hill, NYC. 224 pp.

PARRISH, H.M. — 1963. Analysis of 460 fatalities from venomous animals in the United States. Amer. J. Med. Sci. 245:129-141.

ROSS, K. and R.L. SMITH — 1979. Aspects of the courtship behavior of the black widow, *Lactrodectus hesperus* (Araneae: Theridiidae) with evidence for the existence of a contact sex pheromone. J. Arachnol. 7:69-77.

RUSSEL, F.E., P. MARCUS and J.A. STRONG — 1979. Black widow spider envenomation during pregnancy. Report of Case. Toxicon 17:188-189.

SARNO, P.A. 1973. A new species of *Atypus* (Araneae: Atypidae) from Pennsylvania. Entomol. News 84:37-51.

THORP, R.W. and W.D. WOODSON — 1945. Black widow. Univ. North Carolina Press, Chapel Hill. 220 pp.

TURNBULL, A.L. — 1964. The search for prey by a web-building spider *Achearanea tepidariorum* (C.L. Koch). Can. Entomol. 94:568-579.

VALERIO, C.E. — 1974. Feeding on eggs by spiderlings of *Achaearanea tepi-*

doriorum (C.L. Koch) (Araneae, Theridiidae) and the significance of the quiescent instar in spiders. J. Arachnol. 2:57-63.

VALERIO, C.W. — 1977. Population structure of the spider *Achaearanea tepidariorum* (Araneae, Theridiidae). J. Arachnol. 3:185-190.

STANLEY G. GREEN

Dr. Stanley Green received his bachelor's and master's degrees from the University of Colorado. For his disserttation on the taxonomy of the family Oribatulidae he received his Ph.D. from The Colorado State University, Fort Collins, Col. in 1969.

Green is responsible for vector control and urban pest control programs in Pennsylvania, providing training to public health professionals and pest control operators. Green teaches a number of broad-ranging courses, including basic entomology and structural pest control. In addition, he teaches in-depth courses for specialists, including mosquito and termite control courses. Green's courses and workshops have played an important role in developing professionalism in the pest control industry by encouraging a more scientific approach to pest control. A <u>Vector Control</u> and <u>Pest Control</u> newsletter are published by Green to help accomplish this goal.

As an extension entomologist, Green is responsible for the preparation of the popular <u>Extension Entomological Fact Sheets</u> on household arthropod pests. These are aimed at the general public to help them better understand arthropod pests and how to deal with them.

Green is well known for his selfless attitude toward those who need advice on pest problems. Many professionals have benefitted form his wise counselling and no problems are too big or too small to capture his interest. Green's availability and ability to communicate have made him a key resource for the news media when pest stories need professional input.

Green's achievements have been recognized by numerous honor societies, including Sigma Xi, Epsilon Sigma Phi and Pi Chi Omega. In addition, he is active in many professional associations including the Entomological Society of America, the American Mosquito Control Association, the American Registry of Professional Entomologists and the American Associaiton for the Advancement of Science. Green is also active in many state and regional associations and is currently president of the Mosquito and Vector Control Association of Pennsylvania.

CHAPTER TWENTY-TWO

Mites

Revised by Stanley G. Green[1]

There is a little chigger
That isn't any bigger
Than the point of a very small pin.
The lump that he raises
Just itches like the blazes
And that's where the rub comes in!

—Anon.

THE CLASS ARACHNIDA includes spiders, scorpions, harvestmen, pseudoscorpions, windscorpions and whipscorpions. Borror et al. (1981) says this class of arthropods has about 65,000 species.

The subclass Acari (ticks and mites) is characterized by Busvine (1980) as follows: "Most of them are small (less than five mm long), but a few mites and many ticks are larger. Except for a few primitive forms, they have no trace of segmentation and the body is not divided up by constrictions (as in insects and spiders). The larva has three pairs of legs, but the (one or more) nymphal stages and the adults have four pairs. The mouthparts consist of a pair of chelate chelicerae and two short palps."

The body is divided into two regions. They are the gnathosoma (consisting of the mouthparts) and the rest of the body, called the idiosoma.

Their development consists of four basic stages — the egg, larva, nymph and adult. The number of nymphal stages in mites varies from one to three. Males usually are distinct from females.

Sheals in Smith (1973) recognizes seven orders of Acari. Representatives from three of these (*Mesostigmata* contains the bird mites, rat and house mites; *Astigmata* includes mites of stored foods, dust and human itch mites; and *Prostigmata* includes the straw itch mite, follicle mites and chiggers) will be considered here.

Mites are much smaller than ticks and most are barely visible to the naked eye. Table I shows the differences between mites and ticks.

[1]*Associate Professor of Entomology, Pennsylvania State University; Cooperative Extension Service, Philadelphia, Pa.*

TABLE 22-1

Mites	Ticks
Soft body	Leathery body
Most are much less than 3 mm in length.	Usually 3 mm or more in length.
Lack Haller's organ.	Have a sensory organ on the top of the tarsi of legs I.
Most are free living; some are parasitic on vertebrates and invertebrates.	All are ectoparasites on mammals, birds and some on reptiles and amphibians.
Hypostome (one of the mouth parts) not barbed.	Hypostome barbed — it is this structure which makes the removal of the tick difficult.
Have 1-3 nymphal stages.	Have 1-8 nymphal stages.

From one to five simple eyes may be present, although these are often absent, and the animal is dependent on sensory setae or hairs which are present on the legs and body. Hughes (1948) notes mites breathe through tracheal openings or stigmata in the surface of the body or by the diffusion of gases through the integument.

Heal (1956) discusses the presence of mites in and around structures as follows: "The mite problems coming to the pest control industry are, with the one exception of the clover mite, usually the unusual problems. Very often the problems are quite baffling to the customer, frequently also baffling to the pest control operator and on many occasions baffling even to the entomologist who may be consulted in an advisory capacity. The pests are small and very often may not be present in large numbers, although this is by no means always the case. The identity of the pest may be difficult to establish and frequently under practical control conditions the identity may have to be determined tentatively on the basis of the presence of a host animal or of the presence of vegetable matter that may be the source of the infestation. The problem of identification becomes more acute because of the dearth of entomologists and of operators in the pest control field who can identify mites to species."

Ordinarily, the life cycle of a mite consists of an egg, larval, three nymphal stages and the adult stage. However, the number of nymphal stages varies. At times, the nymphal mites of some species may undergo a transformation into the hypopus stage which may or may not be capable of locomotion; the non-mobile hypopus is ordinarily resistant to dryness. The larva usually has only three pairs of legs. The nymphs and adult have four pairs. The length of the life cycle, under favorable conditions, is two to three weeks. The mouthparts are varied, but the mites with which we are concerned usually have piercing mouthparts. Most of the species deposit eggs, but in the case of the straw itch mite, the eggs hatch in the body and the female gives birth to living young. Control of mites in the household is important because these pests attack and annoy man, are carriers of certain diseases, damage foodstuffs and may be responsible for dermatitis and allergic reactions due to the handling or consumption of mite-infested materials. Flynn (1972) discusses parasites of laboratory animals.

Baker (1955) places household mites into five groups according to their habitat:

- Nest-inhabiting mites parasitic on birds and rodents and occasionally biting man.
- Mites parasitic on animals other than man (not household) and occasionally biting man.
- Mites parasitic on man.
- Plant-feeding mites.
- Food-infesting mites.

Baker and Wharton (1952), the National Pest Control Association (1956) and Hughes (1961) have prepared informative texts on mites. The mites covered in this handbook are discussed as follows:

1) Food inhabiting mites.
2) Parasitic mites of birds and rodents.
3) Plant feeding mites.

TYROGLYPHID MITES
CHEESE, GRAIN, OR FLOUR MITES

A number of very informative papers have been published on the tyroglyphid mites by Solomon (1943), Nesbitt (1945), Robertson (1946), Muggeridge and Dolby (1946), Hughes (1948 and 1961) and Krantz (1955). According to Robertson (1946), "The *Tyroglyphid* mites are a group of great economic importance, a number of species being common pests of stored products in many different countries. A wide variety of foodstuffs and other substances is susceptible to attack (e.g. cheese, flour, grain, seeds, meals, bulbs, straw, and also wallpaper, furniture, etc.). These mites are capable of subsisting on any fragments of organic debris in floor cracks, shelving, etc., are so readily spread by insect and human agencies, and have such wide tolerance to temperature that they are almost invariably to be found in small numbers wherever the above materials are stored. Such factors as the accumulation of debris, prolonged storage of material without cleaning or shifting, insufficient insulation against high temperatures and abnormally high humidity are important contributing factors to the development of serious infestations in stored products."

Besides the materials mentioned above, these mites are pests on dried fruits, dried meats, cereal foods, drugs, jams and jellies. Many species of mites feed on fungi and their presence on some foods may be due to the fungus growths cultured in these foods. Solomon (1946) notes tyroglyphid mites cannot penetrate wheat grains if the grain-coat is intact.

Hughes (1961) notes many tyroglyphid mites have been recorded from bird and small mammal nests. She speculates, "It seems possible, therefore, that these mites were originally the denizens of nests, where they fed on scraps of food and organic detritus. The invasion of human warehouses and larders did not necessitate any great structural change, but merely assured a more ample supply of food and a more stable environment."

Mites infest some foodstuffs to such an extent the entire mass may appear to be in motion. Chapman and Shepard (1932) state that "when present in large numbers they give off a sweetish, musty odor which is so characteristic that one who has had experience can detect their presence without having seen them.

If some of the flour which is suspected of containing mites is piled in the light, the mites will crawl away and the pile will usually flatten out."

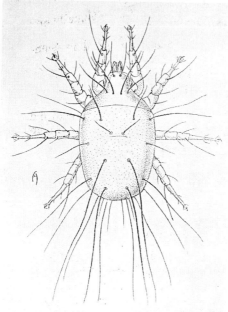

Freeman and Turtle (1947) observe the "minty" smell imparted to flour is due to a fatty surface secretion of the pests. These authors also state the mites attack the germ of wheat and feed within it. Bread made of heavily infested flour has a sour taste, poor color and does not rise properly.

A brown powder in the crevices of cheese indicates the presence of the mites. The powder consists of dead bodies, molded skins, excreta, tiny particles of uneaten cheese, as well as living mites.

The development of tyroglyphid mites usually requires from two to three weeks for completion and includes an egg and larval stage. From here the development varies. Some members of the family have three nymphal stages; others have only two. In those that have three nymphal stages the second nymph is known as a hypopus. Most workers are of the opinion the hypopus stage occurs under unfavorable conditions because it is supposed to protect the mite from dryness. According to DeOng and Roadhouse (1922), the hypopus "is like a minute tortoise. It is extremely small, pink in color and has a hard, shelly back of chitin. The legs are short, the mouth parts rudimentary and there is no evidence that it feeds. On its ventral side, it has a sucker plate by means of which it attaches itself to other mites, flies, and moths which alight on cheese, the skin or clothes of human beings. It is thus carried about until it finds a suitable place, where it drops off, molts to become a second stage nymph and commences feeding. Thus, the hypopus stage tides the mite through unfavorable conditions, and also aids in the dispersal of the mite since it can be carried around by flies, mice, men or it may be transported by air currents." Eales (1917) showed flies in cheese rooms carried mites on their legs.

Although mites lack or have poorly developed simple eyes, they can distinguish between light and darkness, and like to hide in crevices. Solomon (1943), in his review of the literature, notes high humidities are favorable for the development of tyroglyphids. Also, he notes they are fairly resistant to freezing temperatures, the hypopus is able to live at approximately 95° F/35° C and the optimum temperatures for development are between 64° and 77° F/18° and 25° C. Solomon (1946) found *T. farinae* (Linn.) most abundant in the winter and spring, and observed that grain "stored in bags has a much larger area exposed to atmospheric moisture, so that infestations tend to be more widespread than

Fig. 22-1. *Tyroglyphus* mite.

in bulk grain. Grain on concrete floors tends to get moist at floor level and encourage *Tyroglyphus* infestation." Ordinarily, the mites penetrate but only a few centimeters below the surface of the flour.

These mites may at times get on the hands, migrate under the scales of the horny layer of the skin and produce a temporary dermatitis or irritation known as "grocer's itch." This may be overcome by washing the affected member with a kerosene/soapsud emulsion which will kill these non-bloodsucking pests. According to Hase (1929), some individuals are allergic to the mites, their excrement, etc., and these individuals may get grocer's itch without being bitten by the mites. Hinman and Kampmeier (1934) note when the mites are swallowed with food, the numerous spines on the mites may irritate delicate tissues and cause diarrhea.

THE FLOUR OR GRAIN MITE
Acarus siro L. (= *Tyroglyphus farinae* De Geer)

This species is cosmopolitan and very common in cereal foods, dried vegetable materials, feed, cheese, corn, flour and dried fruits. Heavily infested flour deteriorates rapidly, may become sour and will contain enormous quantities of dead mites and their feces.

Solomon (1962) studied the ecology of this mite. He says these mites require a relative humidity of 62.5 percent. Below this the infestation will die out. The optimum environmental conditions for their rapid multiplication are 77° F/25° C and a relative humidity of 90 percent.

Munro (1966) quotes the life cycle of this mite as reported by Newstead and Duvall (1918 and 1920). They say female mites lay from 20 to 30 eggs over the food. Eggs hatch in three to four days, larvae feed for three days, then are motionless for two days. The complete life cycle at 64° to 71° F/18° to 21° C takes 17 days. In the winter, when the temperature is between 50° to 60° F/10° to 16° C, 28 days are required.

THE CHEESE MITE OR SUGAR MITE
Tyrophagus longior (Gervais)

This mite is cosmopolitan and attacks grains, cereal products, drugs, seeds, cheese and dried fruit. In the household it is found most commonly in cheese and cereals, although it does not present as serious a problem as does *Acarus siro* L. Another species, *T. putrescentiae* (Schrank, 1781) (= *T. castellani* Hirst), infests copra and causes "copra itch" in human beings handling this product. According to Essig (1926), this mite is encountered in products coming into western ports of the United States. *T. longior* has been recorded feeding on cucumber plants, beetroots, tomatoes and cyclamen seed (Hugh, 1961). The NPCA (1956) reported *T. longior* has been found in the respiratory system, urinary tract and the intestines of humans.

THE CHEESE MITE
Tyrophagus casei (= *Tyrolichus casei* Oud.) (= *Tyroglyphus siro* Michael)

Although the cheese mite is most common on cheese, it also attacks flour, cereal products, dried meats, mattresses and hair-filled upholstery. Workers handling vanilla pods may develop a rash called "vanillism" which is due to this mite.

THE MUSHROOM MITE

Tyroglyphus lintneri (Osborn)

This mite is a serious pest on mushrooms, cheese, grain and cereal products.

THE DRIED FRUIT MITE

Carpoglyphus lactis (L.) (= *Acarus passularum* Hering)

Hughes (1961) says this mite frequently is found on dried fruit, but also on honeycomb, rotting potatoes and flour. It is attracted to many fermenting substances. It can cause "dried fruit dermatitis" (Baker et al., 1956).

Control of tyroglyphid mites. Solomon (1944 and 1946) found the most effective means of preventing tyroglyphid infestations of stored products was to keep them moderately dry. He observed tyroglyphid mite infestations varied directly with the humidity and the mites did not survive in media with a moisture content less than 12 percent (corresponding to a relative humidity of 55 percent to 60 percent). A temperature range of 64° to 77° F/18° to 25° C is most favorable to mite infestation. Although the mites are resistant to low temperatures, they are destroyed by temperatures of 104° F/40° C and higher. Solomon also observed the mite *Cheyletus eruditus* (Schr.) preys on *Acarus siro*.

Munro (1966) gives the following account from Newstead and Duvall on this predator: "*Cheyletus eruditus* occurs in most grain samples which contain Tyroglyphid mites and its distribution is probably quite as wide as theirs. Like most predatory creatures it is solitary and does not herd together in masses as do its Tyroglyphid victims.

"It is very doubtful if it is ever sufficiently abundant in nature effectively to reduce Tyroglyphids when they occur in such numbers as to form a pest. Experiments carried out in the laboratory with small quantities of wheat have demonstrated that the presence of *Cheyletus* is ineffectual to prevent the rapid multiplication of the flour mite under favorable conditions of temperature and moisture for the latter. All the same, *Cheyletus* is a veritable tiger among the peaceful browsing herds of Tyroglyphids and its rapid purposeful motions are in marked contrast to the aimless hurry of *Glycyphagus* and the leisurely crawl of *Aleurobius* (now *Acarus*).

"When feeding, *Cheyletus* seizes any part of the victim between its curved pedipalps, which then close like pincer-blades with a grip from which there is no escape. It is noticeable that the prey always becomes paralyzed a few moments after capture, probably because some poison secreted by the so-called 'salivary glands' is injected. In captivity, *Cheyletus* is a shocking cannibal.

"Males are rarely found in *Cheyletus* and parthenogenetically born mites (developing from unfertilized eggs) are all female."

The ecology of *Cheyletus eruditus* has been studied by Solomon (1962) and by J. D. Norris (1958). In his most recent studies, Solomon confirmed the efficiency of *Cheyletus* as a predator of the flour mite was highest in late summer and autumn, and lowest in winter and spring. This also accords with Norris' results. Solomon concludes that as a controlling agent, *Cheyletus* has the disadvantage its success depends on physical conditions which are often lacking, namely moderately high temperature. Thus, from 10° C to 2° C *A. siro* can increase while *Cheyletus* remains static.

Olds (1942), who studied tyroglyphid mites in Canada, notes that grain "containing more than 14 percent moisture content is likely to become heavily infested unless it is stored in separate bins and turned at regular intervals. If infestation is slight, mites are often present only for a few feet down from the top and up from the bottom of the grain, and the moisture content of the grain is usually normal." He found methyl bromide used at the rate of one lb per 1,000 cu. ft./0.45 kg per 28 cu. meters penetrated 90 feet/27 meters in a bin in 36 hours and gave very satisfactory control of the mites.

Although the optimum temperature for fumigation is about 65° F/18° C, methyl bromide can be used successfully at slightly lower temperatures.

Muggeridge and Dolby (1946) and Rice (1948) recommend the use of dichloroethyl ether applied at the rate of one lb. per 100 sq. ft/0.45 kg per nine sq. meters of shelving. Rice sprayed the dichloroethyl ether into factory curing rooms at the rate of one lb. per 1,000 cu. ft./0.45 kg per 28 cu. meters for 48 hours. After the room was aired, a second treatment was applied for 24 hours. It is necessary to repeat the fumigation at regular intervals. There was no taint of odor in the body of the cheese after the fumigations.

Some control of cheese mites can be obtained by the use of pyrethrum sprays containing piperonyl butoxide and other synergists. These petroleum base sprays or water emulsions are applied into building or container crevices. The sprays can be used in infested cupboards in the home, after throwing away all infested foodstuffs. Marzke and Dicke (1959) showed 10 mg pyrethrins and 100 mg piperonyl butoxide per sq. ft./0.09 sq. meters to have residual activity against cheese mites.

Dicke et al. (1953) obtained control of cheese mites by using a four percent concentrate of sodium o-phenylphenate. Practical control also was obtained by wiping lightly infested cheese blocks with mineral or cotton seed oil.

Busvine (1980) outlines the control of mites in foodstuffs as reported in a British publication of the Ministry of Agriculture: To prevent an infestation keep grain dry (the water content should be below 12 to 13 percent). Contact acaricides such as malathion, pirimiphos-methyl, fenitrothion, iodophenphos and bromophos can be applied to the walls of empty storage rooms." Some of these pesticides may not be cleared for use in the United States. Check the label for permitted uses.

Chemicals mentioned (Busvine, 1980) which can be mixed with grain are malathion, pirimiphos-methyl or bioresmethrin (a pyrethroid). Fumigation also is recommended, but mites (especially their eggs) are somewhat resistant. Methyl bromide or phosphine can be used but Busvine states doses at 50 percent greater than for insects may be needed.

THE FURNITURE MITE

Glycyphagus domesticus (DeGeer)

This white mite (also called the house mite), with characteristic feathery hairs, may be extremely common in pantries. It is found on cereals, sugar, cheese, etc., and in furniture stuffed with green Algerian fibre, where it feeds on the fungi growing on the fibre. The presence of these mites is more annoying than harmful. When present in large numbers, the furniture may become white with them. However, they do not damage the textiles, nor do they bite. They can, however, cause grocer's itch in allergic individuals.

The furniture mite feeds on dried vegetable and animal matter that is not too hard to eat. It thrives in dampness, 80 percent to 90 percent relative humidity and thus may appear when furniture is moved from a dry locality into a damp locality. Mass infestation by the furniture mite is associated with furniture that is stuffed with a material known to the trade as Algerian fibre. The mite may spread from the furniture to all parts of the house.

Control. Furniture upholstered with Algerian fibre and infested with this mite should have this fibre removed and be restuffed with some other material. Because of the hypopus stage, which is resistant to fumigation, fumigants are not always effective. The use of hydrocyanic acid gas in a vacuum chamber is undoubtedly the most effective method of fumigation.

Glycyphagus destructor (Schrank), according to Hughes (1948), is a "rapid and jerkily moving mite commonly found in association with *T. farinae (Acarus siro)* on wheat, flour, linseed, ground nuts, etc. More rarely it occurs in large numbers by itself or mingled with other species of the genus *Glycyphagus*." It is believed this mite feeds strictly on fungi and does not feed on the grain in which it is found (Hughes, 1961).

THE STRAW ITCH MITE

Pyemotes (= Pediculoides) ventricosus (Newport)

This mite is normally beneficial since it attacks such grain and cereal insects as the *Angoumois* grain moth, rice and granary weevils, and bean and pea weevils, as well as many other injurious insects (Fig. 22-2). However, at times it devotes itself to man rather than insects.

In the active stage, this grey or yellowish mite is not visible to the eye. The females, however, when distended with eggs, can be seen. Hughes (1948) notes the average length of the non-gravid female is 223 microns, the gravid female two mm, and the male 160 microns. It is the non-visible stage of the mite that attacks humans. Since the host insects of the mite have been distributed throughout the world, the mite also has become cosmopolitan.

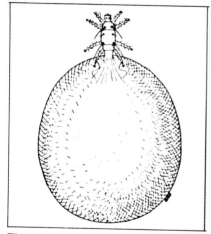

Young mites attack insect larvae, puncture the skin and suck the larval juices. Mites prey upon larvae in grain by entering the minute hole made by the larvae. A single mite and its progeny are ample for the destruction of the larva. After attacking the larva, the posterior portion of the mite's abdomen becomes greatly enlarged within a day or two and the mite becomes stationary on its victim.

Fig. 22-2. Gravid female of the straw itch mite, *Pyemotes ventricosus* (Newport).

Hughes (1948) states the male spends its life wandering over the distended abdomen of the gravid female, upon which it feeds parasitically. Also, he notes "the eggs inside the female develop there until the adult stage is reached, when their sex can be determined by characteristics which can be seen through the transparent chitin. When a young female is about to be born, it moves until it comes opposite the parent genital opening and begins to emerge. The parasitic male appears to be aware of what is happening and also moves to the opening.

With the aid of its powerfully built hind pair of legs, it seizes the young female and completes its birth by dragging it through the opening. Copulation between the two then takes place. It is said that the male never assists in the birth of young males and that even the young female can emerge without his assistance. The female can reproduce parthenogenetically, but in this case only males are produced."

Webster (1910) counted 40 to 50 young and eggs in the abdomen of the female. The eggs are forming continually and the mites pass through their metamorphosis within the abdomen of the mother, emerging as soon as they become mature. Webster quotes Wildermuth to the effect mites may produce young as early as six days after emerging from the abdomen of the mother mite. The female may give birth to 270 young, 52 being the largest number produced in a single day. The optimum temperature for existence of *Pyemotes ventricosus* (Newport) is 70° to 80° F/21° to 27° C. The mites live less than a day without food.

Injury to man and animals by mites. Man and animals are subject to attack by the straw itch mite when they come in contact with grain or straw infested with the larvae of the Angoumois grain moth, as well as other host insects. Attacks by these mites are referred to as grain itch, acarodermatitis urticarioides, barley itch, grain mite dermatitis, straw itch, mattress itch and hay itch. Sailors sleeping on straw matresses, as well as men working with straw or hay, have suffered badly from the attacks of this mite. The mites attack practically any animal when their favorite hosts disappear or are scarce.

Scott and Fine (1967) examined a house infested with the furniture beetle, *Anobium punctatum* (DeGeer), which was overrun with the straw itch mite.

The author was badly infested with the straw itch mite after investigating a barn where a horse appeared to be infested by mites. Some relief was obtained by taking a hot bath and soaping the skin. The soap suds were permitted to dry on the skin. The bites on the skin which were in evidence all over the body, disappeared after about 30 days.

David (1950) recorded an epidemic of straw itch mite in Indiana beginning in the latter part of August, 1950. The increase of this mite is attributed to the overwintering of Angoumois grain moth in stored grain, as well as to large populations of the wheat joint worm on both of which the mite is predacious. Besides the use of the warm soapy water and talcum treatment mentioned by Larson, below, Davis recommends the use of ½ percent gamma benzene hexachloride ointment or a sulfur ointment, as well as dimethyl phthalate as a repellent.

Larson (1925) found the attack of mites on human beings caused wheals and pustules with accompanying red blotches. Fever, followed by cold sweat, makes sleep impossible. Relief against this mite was obtained by bathing in warm soapy water and then applying talcum. Recommended ointments have included mercurial ointment, beta naphthol, sulfur, benzoate and lard.

Rogers (1943) notes the onset of symptoms may ensue 20 minutes to 24 hours after exposure to the mites. The mites may inject an irritating material in the process of sucking liquid substances from the skin. However, they do not burrow into the skin. The lesions "may be situated anywhere on the body, are extremely itchy, are pale pink to bright red and vary in size from pinhead to two or more inches in diameter. Many lesions are surmounted by a tiny vesicle, but no puncture wound is visible." Individuals such as farmers who are in frequent contact with the mites, may become desensitized and show no symptoms.

According to Rogers, *"Pediculoides* cannot thrive on human blood. Nevertheless, when its normal food supply is cut off, it will feed on any flesh. Hence, it remains attached to the human skin for only a short time, and so treatment with the view of destroying the mite is useless. Remedies for relieving the subjective symptoms are all that are necessary. This may be accomplished by the use of warm demulcent baths such as oatmeal or starch and a mild anti-pruritic lotion, such as phenol 15 minims (one cc); zinc oxide, one ounce (30 gm); glycerin, one drachm (four cc); lime water, four drachms (15 cc); and rose water to four ounces (120 cc)." A wider range of anti-pruritic materials now is available.

Control of the straw itch mite. Anon. (1948) states the employees of a broom factory in Baltimore, Maryland developed a rash chiefly about the waist that itched intensely for 24 hours. *Angoumois* grain moths were present on the broom corn. Control was obtained by the application of a pyrethrum and piperonyl butoxide spray. The control of the grain moths by the methods already discussed under tyroglyphid mite control will result in the elimination of straw itch mites as well.

Gay and Greaves (1942) controlled the spreading of these mites in grain weevil insect cultures by wiping a two percent solution of dinitro-ortho-cyclohexylphenol on the sides and cloth covers of the culture jars in the laboratory.

At times *Pyemotes* and *Tyroglyphus* mites, as well as others invade insect cultures. Dahm and Bauer (1949) recommended a two percent by weight dust of DMC or chlorfenethol (1,1-di-(4-chlorophenyl)ethanol) be applied to the cultures. A wide variety of insects dusted with chlorfenethol were not affected, whereas the mites were completely eliminated.

Fisk (1951) controlled mites in a roach culture with a five percent dust and a five percent spray prepared from a 50 percent wettable powder of chlorfenson (4-chlorophenyl 4-chlorobenzenesulfonate). Brown (1965) eliminated *Histiosoma laboratorium* from *Drosophila* cultures by rinsing the bottles in solutions of dicofol.

CHIGGER AND HARVEST MITES

These are the familiar red-colored mites (Fig. 22-4) that are troublesome pests, particularly in the South. Although they do not infest the household, they merit some consideration since they may be present on the grounds in the immediate vicinity of the home. Wharton and Fuller (1952) have prepared a monograph on the biology, classification, distribution and importance of chiggers. Also see Sasa (1961) and Vercammen-Grandjean and Langston (1971, 1975 and 1976). Chiggers belong to the family Trombiculidae which has worldwide distribution. The common chigger, *Trombicula alfreddugesi* (Oudemans), is widely distributed from Canada to South America. *Acariscus masoni* (Ewing) and *Trombicula batatas* (Linn.) are two other chigger pests in the South, although this does not, by any means, exhaust the number of species in the United States. Incidentally, the chigger mites *Leptotrombidium akamushi* and *L. deliensis* are proven vectors of scrub typhus or tsutsugamushi disease in Asia, Australia and Oceania.

The chigger mites are parasitic during the larval state, but free-living during the nymphal and adult stages. Rodents, birds, poultry, rabbits, livestock, snakes, toads and other animals, as well as man, are attacked by these pests. Six-inch squares of black paper placed on edge in the grass become covered with chiggers if they are present in an area.

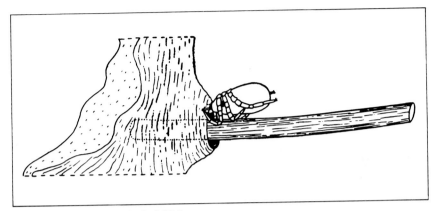

Fig. 22-3. Chigger in a hair follicle.

Stone and Haseman (1941) described the life cycle and habits of the common chigger, *Trombicula alfreddugesi* (Oudemans), in Missouri as follows: "The adult of the common chigger spends the winter in an earthen cell, about an inch to an inch and a half below the surface of the soil. These small, red, adult mites may be seen traveling slowly over the surface of the ground, when one is spading the garden and with the first warm days of spring, eggs are deposited. The adults then soon die and the eggs hatch into the chiggers, which crawl about, and in due time, make contact with man or other hosts.

"While the chigger is very small, it is swift on foot and may readily be seen scurrying about on white shoes and stockings. Chiggers reach a person's flesh by crawling onto his shoes, moving upward and even penetrating the meshes of his clothing. After coming in contact with the flesh, they may encounter such obstacles as a tight garter or belt and rather than pass over or under it they frequently settle down and begin to feed. Most of the chiggers will attack around the ankles or under the knees, although some go higher up to attack about the crotch, under the belt and occasionally in the armpits. If a person sits on infected ground, more of the chiggers are liable to attack the upper parts of the body. The chigger usually will run about a person's legs for several hours before it begins to feed.

"The active feeding stage or larva has three pairs of legs. After becoming fully fed, it drops from the host, goes into the ground and enters a quiescent stage, which some writers describe as the nymphal stage. Later in the fall, it changes from the resting nymphal stage to the adult, which has four pairs of legs and spends the winter in the ground.

"This mite attaches to man and other animals only in the larval stage. In the later nymphal and adult stages it feeds on fecal droppings of insects and other arthropods and dead organic matter. Apparently, however, the young chigger requires the blood or lymph of a living host."

Michener (1946) studied the life history of *Trombicula batatas* (Linn.) in great detail. According to his studies, the stages in the life cycle of this mite are the egg, deutovum, larva, protonymph, nymph, preadult and adult.

Seigel (1947) notes the chigger does not burrow into the skin, but inserts its mouthparts and injects a fluid into the blood vessels which prevents the blood

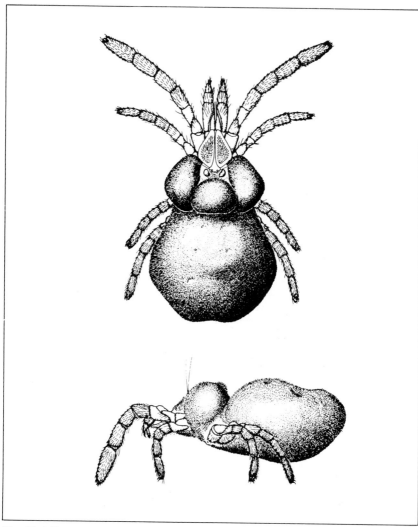

Fig. 22-4. Dorsal and side view of a chigger mite.

from clotting. This fluid then causes the typical "red blotch, at the tip of which a small water blister is formed within a day. While this itching bump is easily seen, the tiny red chigger at its top is effectively camouflaged by the inflamed surface." Matheson (1950) states the mouthparts are inserted in the skin, frequently about a hair follicle. The mite then injects a fluid into the tissues and these tissues become hardened to form a tube. The mite then feeds upon the liquified tissues. The effects of this digestive fluid results in the severe itching and dermatitis.

The USDA (Anon., 1956) recommends the preparation of the following formula by a pharmacist:

	Percent
Benzocaine	5
Methyl salicylate	2
Salicylic acid	0.5
Ethyl alcohol	73
Water	19.5
Total	100

"Apply the material to each welt with a piece of cotton. One treatment gives relief for an hour or longer. Repeat the treatment as often as necessary."

Stone and Haseman (1941) recommend the use of household ammonia, a one percent solution of "Lysol," rubbing alcohol, camphor, carbolized vaseline, chloroform, iodine or nail polish to relieve irritation, destroy the chigger and also prevent secondary infection.

Sutton (1942) found two grams of ethyl aminobenzoate in 15 grams of flexible collodium, applied with a glass applicator, soothes the irritation.

Control. Stone and Haseman (1941) recommended control of this pest by treatment of breeding areas around the home and through remedial gardening methods: "The favorite breeding places of chiggers are among briars, weeds and other thick vegetation, where there is an abundance of moisture and shade. The mowing of weeds and briars, the close clipping of lawns and the elimination of unnecessary shade in infested areas will help reduce the number of chiggers."

Repellents offer relief from chiggers. Use either (deet) diethyltoluamide, ethyl hexanediol or dimethyl phthalate applied to the clothing and exposed skin. Do not get any repellent into the eyes or mouth. Cross (1948) showed clothes dusted with five percent benzil powders resulted in effective protection.

Outdoor areas where chiggers are known to be a problem can be sprayed with carbaryl, chlorpyrifos or diazinon, paying attention to ground litter and soil, as well as grass. DeLong and Smith (1951) note the shrubbery around an infested lawn may be infested and it is necessary to treat this.

THE TROPICAL RAT MITE

Ornithonyssus (= Liponissus) bacoti (Hirst)

The bite of this grayish yellow mite is extremely annoying to man (Fig. 22-5). Dove and Shelmire (1932) have shown endemic typhus can be transmitted from one guinea pig to another through the bite of this mite. Although the tropical rat mite has been found infected naturally with murine typhus fever, there is some question in regard to its importance as a natural vector. Philip and Hughes (1948) showed this mite can transmit rickettsialpox but is not an efficient carrier.

The tropical rat mite is found commonly in the warmer parts of the United States, but its range may extend into Alaska. It also is reported in New South Wales, Western Australia, Egypt, Abyssinia and Argentina. According to Shelmire and Dove (1931), this mite was first recorded as infesting rats in Egypt. Its principal host is the Norway rat, *Rattus norvegicus*.

The mites very often become a serious problem where rat campaigns are undertaken, for under these circumstances, the mites must attack man and other animals or starve. The mites also may infest used or abandoned nests of rats. Heal (1956) states attacks by this mite occur more frequently in industrial areas than in homes.

Ebeling (1960) notes when the mites are abundant "they may be found anywhere in the home. Both nymphs and adults may attack man. Their bites may produce irritation and sometimes painful dermatitis that may continue for two or three days, leaving red spots (maculae) on the infested parts of the body. Scratching may result in secondary infections. In some families, some individuals are affected and others are not. Sometimes much time and money are spent on ineffective medication and usually it has been difficult for the infested person to obtain a correct diagnosis. This acariasis cannot be distinguished from flea bites and is sometimes misidentified as scabies." Also, Larsen (1973) discusses rat mite dermatitis.

Shelmire and Dove found rat mites in stores, restaurants, theaters and other places. The active mites usually feed at night or in semi-darkness and after feeding retreat to cracks and other dark places until their next meal. They also noted the mites accumulate in walls, particularly where there was some source of heat. Bishopp (1923) notes the mites were found on the top floor of a building 10 stories high, the mites having passed from one floor to another along the heat pipes. They also were found to drop from the ceilings. Besides favoring hot water pipes, Mills (1947) notes these mites may be present in heat installations around stoves, kitchens and bathrooms. Garman (1937) cites one case where these mites crawled over the woodwork and beds of a home in Connecticut. Large numbers of the pests were found in the empty nest of a fly catcher, *Sayornis phoebe,* which was situated directly above the window of one of the infested rooms. After the nest was removed, there was no further trouble from the mites.

Bishopp (1923) states both the nymphs and adults of the rat mite may attack man, the former being the more troublesome. Mites need not bite to be irritating, since their mere presence is annoying. They attack all parts of the body; their bite is painful, causing itching and small red spots. The irritation may continue for several hours and the spots usually last for two days.

Bertram et al. (1946), who studied the life history of this mite, note at approximately 68° to 72° F/20° to 22° C the adult mite emerges from the second nymph stage, mates and engorges within three days. The first eggs may be laid within two days and the larvae emerge within one and one-half days. The larvae molt to the first nymph within one day harden; they feed and drop from the host within two days. They then molt to the second nymph in one day and in another day molt once again into the adult stage. Approximately 11½ days are required to complete the egg to adult cycle. Unfertilized females reproduce parthenogenetically. According to Shelmire and Dove (1931), the mites require four or five blood meals to complete the life cycle and the larvae will starve unless a host is found within 12 days. Scott (1949) showed the mites may live as long as 63 days without nourishment of any kind. Also, Skaliy and Hayes (1949) studied the life history of this mite in some detail.

Control. Ebeling (1975) recommended dusting with silica aerogel. Because rats had been present in the attic, the dust was applied here as well as throughout the house. The dust also was applied "under the ticks of mattresses, on the spring supports of beds, along edges and in four corners of bed frames" and onto sofas, lounges and under pillows, as well as along floor boards and ceiling moldings. Olson and Dahms (1946) observed mites infesting laboratory animals in large numbers imbibed sufficient blood from the animals to kill them. Silica gel

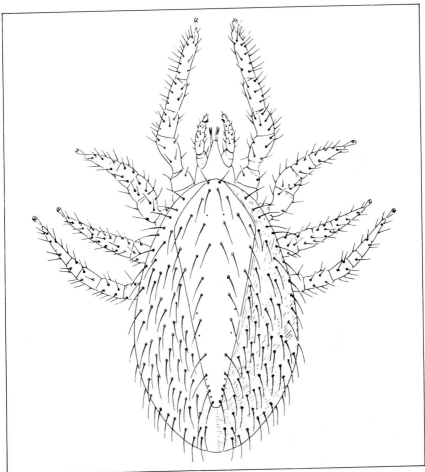

Fig. 22-5. Tropical rat mite, *Bdellonyssus bacoti* (Hirst).

can be used on mite infested animals. Fisk (1951) controlled this mite on laboratory mice with a 10 percent chlorfenson dust as well as by dipping the cages in Lysol. Scott (1958) sprayed mouse cages with a 0.3 percent pyrethrins — 1.5 percent piperonyl butoxide solution.

In searching for mite infested areas, a light should be used, checking particularly warm areas as those near hot water and steam pipes. Woodwork facings also should be examined carefully.

THE NORTHERN FOWL MITE

Ornithonyssus sylviarum (C. & F.)

NPCA (1956) notes this mite is a pest of domestic fowl and many wild birds throughout the temperate regions of the world. The northern fowl mite at times attacks man and causes itching by its bite and by crawling over the skin. According to Heal (1956), this mite "normally lives on the host and thus is not so apt to be left behind in large numbers with a migration of the host from the nest. The death of a bird in the nest area, however, can release a great number of mites which will migrate into adjacent areas of the structure and be equally annoying as the chicken mite." Meyer and Eddie (1960) note poultry parasites have been shown to have the ornithosis virus and possibly can be carriers.

This mite goes through a life cycle somewhat similar to the tropical rat mite. For information on the life history of this mite see Sikes and Chamberlin (1954), and Combs and Lancaster (1965).

Ornithonyssus bursa (Berlese), the tropical fowl mite, replaces the northern fowl mite in the warmer parts of the world. Its life cycle and habits are somewhat similar to the northern fowl mite. Anon. (1971) notes it infested an apartment in Maryland where pigeons nested on air conditioners.

Vincent et al. (1954) and Harding (1955) controlled the northern fowl mite in litter with four percent malathion dusts at the rate of ½ to one lb./0.23 to 0.45 kg per 20 sq. ft./1.9 sq. meters. Kraemer (1960) found 0.25 percent ronnel or carbaryl emulsion and five percent carbaryl dusts to be useful in the control of these mites. Knapp and Krause (1960) showed one percent ronnel dusts to be effective. Tarshis (1964) eliminated this mite with silica aerogel. He applied the dust to the nests of mite-infested birds as well as to rugs, clothing, patios, attics, subfloor areas, etc. He used one pound/0.45 kg of dust to 1000 sq. ft./28 sq. meters.

THE CHICKEN MITE
Dermanyssus gallinae (DeGeer)

This mite is a parasite of chickens and many other domesticated and wild birds. The mites may invade a house from the roosts of chickens or the nests of wild birds and then bite man. Migration of the mites from the nests of birds often, but not always, occurs when the nestlings have left the nest. Berndt (1952) and Brown (1953) incriminate sparrows, Laird (1950) starlings, and Duncan (1957) and Hantsbarger and Mast (1959) pigeons, as being the bird hosts of mites that bite man. Baker (1955) notes these mites must feed on their primary hosts, birds, or will die within two or three weeks. NPCA (1956) discusses the life history of the chicken mite and notes eggs are deposited after each blood meal. The life cycle from egg to adult can be completed in as little as seven days. The adults can resist starvation for four to five months.

Matheson (1950) claims this mite is a natural vector of St. Louis encephalitis among birds and has transovarial transmission to its young.

Furman et al. (1955) controlled these mites by spraying the chicken house with one percent malathion emulsion and by applying a two percent malathion dust to the litter at the rate of one lb. per 20 sq. ft./0.45 kg per 1.9 sq. meters.

THE HOUSE MOUSE MITE
Liponyssoides sanguineus (Hirst)

This mite, which is reported as carried by rats and mice, has been shown to be responsible (Heubner et al., 1946) for the human disease known as rickettsialpox. (Fig. 22-6). Pomerantz, a pest control operator, discovered these mites crawling on the external walls of the basement incinerators in a housing development in New York City that was infested with house mice and ravaged with rickettsialpox (Anon., 1947).

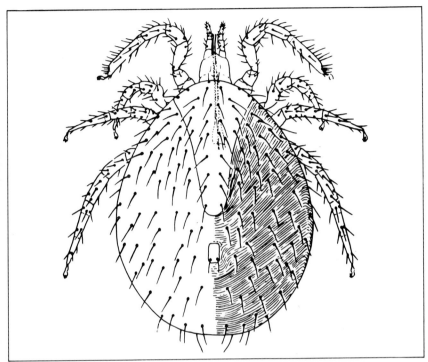

Fig. 22-6. The house mouse mite is the carrier of rickettsialpox from mouse to man.

Fuller (1954) studied the life history of this mite. The entire life cycle at 76° F/125° C required 17 to 23 days. Unfed females may live for 51 days. After obtaining a blood meal, the mite leaves the host and may be found in the nests of mice, in runways or on walls.

Control of these mites can be achieved by extermination of their common hosts (i.e., house mice) and by the application of miticides such as chlorfenethol, chlorfenson, malathion, pyrethrins and silica aerogel.

To successfully control the four mites discussed above (tropical rat mite, northern fowl mite, chicken mite and house mouse mite) proper identification is extremely important since effective control ultimately requires the control on, or of, the host animal.

THE HUMAN ITCH MITE

Sarcoptes scabiei var. *hominis* (Hering)

This mite is of greater interest to the doctor than to the pest control operator since its treatment and control is usually accomplished under a physician's care. It is considered briefly at this point, however, since the symptoms due to the human itch mite may be confused with those caused by insects or other mites.

The variety *S. scabiei hominis* attack man. Other varieties attack birds and mammals. The activities of the human itch mite result in intense itching in its victims and a rash develops after two months. This pathological condition is called scabies. The cause of scabies was discovered in 1687 by an Italian physician and pharmacist (Bonomo, 1687). The scabies skin condition occurs most commonly where there is crowding or uncleanliness.

The human itch mite is from 0.25 to 0.5 mm in length in the adult stage and infests tunnels in the upper epidermis of the human skin. The adult female has suckers on her front pair of legs. Males have suckers at the ends of the first, second and fourth pairs of legs. His third pair of legs has a bristle.

The female itch mite tunnels into the skin where she deposits her eggs. Busvine (1980) reports on the study of Mellanby (1943). He presents data showing the places on the body where the mites commonly infest. Eight hundred and eighty-six men and 119 women were examined; 63.1 percent of the men and 74.3 percent of the women had mites on the hands and wrists; 9.2 percent of the men and 9.8 percent of the women had mites on the feet and ankles; 10.9 percent of the men and 5.9 percent of the women had them on the elbows. The other parts of the body showed little involvement. It is interesting that none of the men had mites on their palms, whereas 7.5 percent of the women had mites at that site. The mites do not suck blood but feed on skin and the secretions of the hair follicles.

Eggs hatch into larvae in the subcutaneous burrow. These molt into the eight legged nymphs. Males have only one nymphal stage which molts into the adult; females go through a second nymphal stage before the adult stage is reached. The entire life cycle from egg to adult takes from nine to 11 days for the males and from 14 to 17 days for the females. However, Hand (1946) found six weeks may be required for development from egg to adult. Heileson (1946) studied the developmental stages of *Sarcoptes scabiei*.

According to Bolton (1947), the infested individual is frantic from itching and pale and haggard from loss of sleep. The most prominent symptom is itching, chiefly at night just as the infested individual gets comfortably settled in bed. The warmth of the bed makes the mites active. Beside the burrowing activity, a highly irritating fluid is released by the mite and some individuals develop an allergy to the secretion.

How long do itch mites live off the host? Mellanby (1943) found female mites removed from the skin walk about fairly rapidly at 68° F/20° C. He found they could climb polished surfaces such as glass easily. At 90 percent relative humidity and 55° F/13° C, about 50 percent of the mites survived for one week off the host and five percent lived for two weeks. Warm dry air causes rapid death. At 82° F/28° C most mites die within 24 hours and none survived longer than two days. Transmission of mites is almost entirely by intimate contact.

Control. Several prescription medicines are available for treatment of scabies including:

- Eurax cream or lotion. This contains 10 percent crotamiton (N-ethyl-o-crotonotoluide). It is manufactured by Geigy Pharmaceuticals. To use, first take a shower or bath. Then apply either the cream or lotion thoroughly to the skin. Do not use on the head. Do not shower after treatment, but repeat application 24 hours later. Then, 48 hours later, shower. Linen and clothing can be washed on the hot cycle in the washing machine before the second treatment.

- Kwell lotion or cream. This contains one percent lindane, and is made by Reed and Carnrick. Kwell also is available as a shampoo which is used to control lice. It should not be used for scabies since these mites are never found above the neck. To use, rub lotion or cream into the skin thoroughly.

About one ounce/29 ml is adequate for an adult. Leave on for eight to 12 hours and then remove by a shower or bath. One treatment is normally effective.

- Benzyl benzoate. Sheals in Smith (1973) says, "The acaricide most frequently used in the treatment of scabies is benzyl benzoate. A 25 percent emulsion is painted on the body from the neck down with a soft flat paint brush about two inches wide and the patient allowed to dry in a warm room for 10 to 15 minutes. When properly used this substance is extremely effective and one treatment is usually sufficient to kill all the mites. However, it may not be well tolerated by young children."

- Sulphur ointment. Sheals continues, "Sulphur ointment BPC and certain organic sulphur preparations, in particular dimethyl diphenyl disulphide ('Mitigal') and tetraethylthiuram monosulphide ('Tetmosol'), are also effective. Being greasy, sulphur ointment is rather unpleasant to use and it is liable to produce a sulphur dermatitis. Two treatments, either on succeeding days within the period of one week, have been recommended, and approximately three ounces of the ointment are required per treatment."

DUST MITES

Family Pyroglyphidae

Traver (1951) reports a scalp and head infestation by the rare mite, *Dermatophagoides scheremetewskyi* Bogdanov. Local physicians were unsuccessful in discovering the causal agent and diagnosed the ailment as "psychoneurotic." Also, see NPCA (1956). Traver, a zoologist, using a microscope, finally found the "imaginary" mites.

Species of *Dermatophagoides* have been reported as causing allergenic asthma by Voorhorst et al. (1964). *Dermatophagoides pteronyssinus* (Trouessant) is common to the old world and *D. farinae* Hughes, to the United States. Both these mites produce asthma symptoms in sensitive people. These mites feed on human skin debris commonly found on mattresses and on floors.

The house dust mite and house dust allergy is reported by Van Bronswijk (1973), Van Bronswijk et al. (1971), Van Bronswijk and Sinha (1971), Van Bronswijk and Koekkoek (1971 and 1972), Sinha and Paul (1972) and Keh (1973).

Control. No adequate control presently is known. However, Heller-Haupt and Busvine (1974) have studied a number of pesticides which may be of value against dust mites. They found the most promising compounds were lindane and pirimiphos-methyl. However, any chemicals used near people who are allergic to the mites may present a similar problem in sensitivity toward the pesticides.

Ebeling (1975) reviews research dealing with controlling the mites by modifying their environmental conditions.

THE CLOVER MITE

Bryobia praetiosa (Koch)

Up until the end of World War II, the clover mite (Fig. 22-7), also known as the brown mite, was considered an occasional invader and little attention was paid to it. From World War II to the present, the clover mite has become progressively a more important invader of houses and is now considered a serious pest.

Fortunately, the clover mite is a plant feeder and does not attack man. It is a pest of a great variety of herbaceous plants and deciduous trees, as well as some conifers. Just a few of its important host plants are apple, apricot, arbor vitae, beans, cherry, clover, dandelion, grasses, iris, ivy, mallow peach, pear, plum, poplar, shepherd's purse, strawberry, sweet pea, sycamore, tomato, violet and zinnia. However, Anderson and Morgan (1958) limit the clover mite largely to herbaceous plants and another species, the brown mite, to trees.

According to Snetsinger (1967), there is great confusion regarding the identity and host plant of clover mites. He lists seven clover mite forms.

These mites are pests because they invade the house, often in enormous numbers. Spear (1954) emphasizes the problem when he notes a single bedroom had an estimated 250,000 mites on the floor. If crushed, these mites leave red spots and therefore deface walls, drapes, window shades and other surfaces. Desperate homeowners often spot their walls with hundreds of stains from squashed bodies.

Spear (1954) and English and Snetsinger (1957) believe the emergence of the clover mite problem since the war is associated with the housing "boom" and the installation of well-fertilized lawns growing close to the foundation of the house. Spear notes, "The fact that greater populations of clover mite seem to be present on old lawns which have been heavily fertilized seems to argue for the importance of a soil nutrient as a determining factor."

Description. The brownish or reddish clover mite is one of the largest plant-infesting mites and is distributed throughout the country. According to Essig (1926), it is 0.75 mm in length and may be recognized by its large size, characteristically long front legs and the featherlike plates that are present on its body. Morgan and Anderson (1957) are of the opinion the clover mite is actually two species, *Bryobia praetiosa* (Koch) and *Bryobia arborea* Morgan and Anderson. They state the two species can be separated by the shape of the setae in the larvae. The larva of *arborea* has club-shaped setae, whereas the larva of *praetiosa* has setae that are much more slender in appearance. According to Morgan and Anderson, *arborea* lives in orchard trees while *praetiosa* lives on grasses, weeds and shrubs, and is the species that invades the house.

Biology and behavior. English and Snetsinger (1957), who studied the clover mite in Illinois, discuss its life history as follows: The species is partheno-genetic — no male mites having been found during their investigations. "The dark mature female mite lays bright red eggs in cracks and faults in concrete foundations, in mortar crevices, on building paper between the walls of buildings and on the underside of the basal bark of trees. In these places an unbelievable accumulation of eggs, egg shells and molt skins can often be found."

As mentioned previously, Morgan and Anderson (1957) believe the clover mite to be a complex of at least two species, consisting of the brown mite, *Bryobia arborea,* which lives only in the aerial parts of orchard trees and does not migrate, and the clover mite, *Bryobia praetiosa.*

According to Anderson and Morgan (1958), the clover mite "lives on a wide range of herbaceous plants, from which it often moves into dwellings; seldom is it found in the aerial portions of trees. This mite is considerably more tolerant of cold than the brown mite and may become active at temperatures slightly above freezing. Although it overwinters chiefly in the egg stage under bark scales of tree trunks, in cracks in fence posts, under sheathing of buildings, or in other dry, protected sites, it may overwinter in any of the other stages. The number of active mites overwintering and the percentage in each stage depend upon the fall weather conditions. The winter eggs begin to hatch very early in

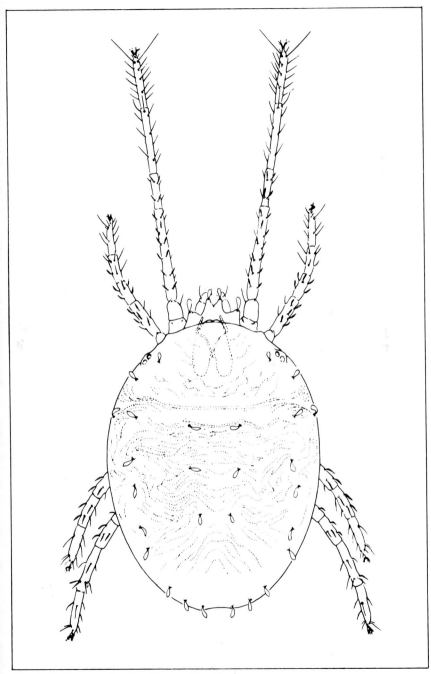

Fig. 22-7. Clover mite, *Bryobia praetiosa* Koch.

the spring, usually a month or more before those of the brown mite. One generation is completed during the spring and early summer months. Most of the eggs deposited by this generation aestivate until September, but a small number hatch in early summer and give rise to a succession of summer generations. Aestivated eggs begin to hatch about September. Only one generation is completed during the fall."

Busvine (1980) says eggs cease to hatch at temperatures between 44.6° and 35.6° F/7.0° and 2.0° C. He says eggs will not hatch when the temperature is over 86° F/30° C.

Wheeler (1955) notes radiant-heated slab foundations and other forms of modern heating create a "mild winter" for clover mites in the East.

According to Spear (1954), the mites "may or may not appear on the interior of the building during early winter, but where the clover mites have thoroughly infiltrated a building, they can be expected to be active and therefore a nuisance to the housewife during any warm period. This nuisance activity is most severe in the springtime. In the New York areas, calls will start coming to entomologists and pest control operators during any warm spell from mid-January until early summer." Also, Spear notes clover mites are sensitive to small changes in temperature and in the fall they tend to move upwards, particularly when the sun is warming the surface above them. Like some other mites they appear to be able to survive temperatures below freezing. According to Spear, the mites can be found in infested houses from November until May or June, and in late fall thousands of the mites can be found on vegetation immediately surrounding the house, as well as on the foundations and sides of the house.

Control. Control of the clover mite consists of immediate destruction of the mites inside the house to give relief to the distressed homeowner, and control outside the house to eliminate their entry into the house. Anon. (1958) recommends the use of contact sprays or pressurized sprays or aerosols containing two percent malathion. Roselle (1954) suggests the use of a vacuum cleaner to remove the mites from surfaces without crushing them on the surfaces.

Spear (1954) recommends the use of a 0.5 percent malathion emulsion to control the mites outside the house. "Spray the lawn and shrubbery to a distance of 15 to 20 ft. from the house with ½ percent malathion emulsion. It should be used at a rate of about 40 gallons per 1,000 sq. ft. This is a heavy rate of application for the average pest control operator and where large developments have had to be treated, it has led to the use of such equipment as street-watering wagons. Experience has shown, however, that success usually attends the use of ½ percent malathion at this relatively high rate and lesser amounts often fail. The foundation and exterior walls at least to the first-floor windows should be sprayed to run off with the same formulation." Anon. (1958) notes four percent of the five percent malathion dusts applied at the rate of one pound per 1,000 sq. ft/0.45 kg per 93 sq. meters may be used instead of the emulsion.

MacCreary and Connell (1962) found dicofol at the rate of ¼ pint/0.12 ml 18 percent emulsifiable per five gallons/18.9 l of water effective in eliminating the mites. The outside walls were treated up to the first floor windows, together with an adjacent 10 foot/three meter band of lawn. The spray was applied at the rate of one gallon per 100 sq. ft./0.38 l per nine sq. meters. Reinfestation occurred in some cases the following spring or fall. Dicofol also was effective when used to treat the lawn in a five foot/1.5 meter band around the house. Then the spray was applied at the rate of two gallons per 100 sq. ft./7.6 l per nine sq. meters of lawn. See page 768 for further information.

KEY TO SOME COMMON SPECIES OF FEMALE ACARI (MITES)

Harry D. Pratt and Chester J. Stojanovich

   1) Last segment of first leg with a depression known as Haller's organ; most species with a toothed hypostome on capitulum; size usually over four mm. Ticks in the suborder Ixodides (Fig. 1 A).

   Last segment of first leg without such a depression known as Haller's organ; hypostome not toothed; most species less than four mm long (Fig. 1 B). Mites 2

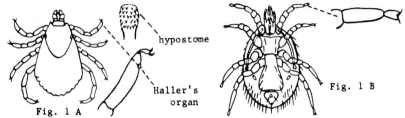

Fig. 1 A   hypostome   Haller's organ   Fig. 1 B

   2) Respiratory system with a spiracle on each side opening lateral to the bases of the 3rd or 4th pair of legs, frequently spiracles leading into slender tubes that extend forward laterally to the bases of the 1st or 2nd pairs of legs (Fig. 2 A). Mesostigmatid Mites 3

   Respiratory system without spiracles, or with spiracles opening near bases of the chelicerae (Fig. 2 B) 13

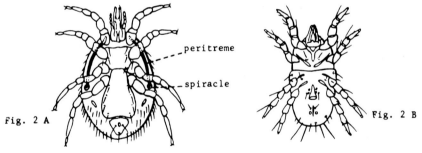

Fig. 2 A   peritreme   spiracle   Fig. 2 B

   3) Anus surrounded by a plate bearing only three setae, one on each side and one behind the anal opening; first tarsus bearing caruncle and claws at tip (Fig. 3 A) 4

   Anus surrounded by a plate bearing more than three setae; first tarsus without caruncle and claws (Fig. 3 B)   Many species of *Macrocheles*.

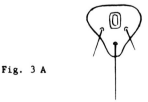

Fig. 3 A   Fig. 3 B

Fig. 22-8. **Key to some common species of female mites.**

4) Anal opening more than its length behind anterior margin of anal plate; chelicerae strongly narrowed apically, needle-like, movable absent or extremely small (Fig. 4 A). Genus *Dermanyssus* 5

Anal opening less than its length or about its length, behind anterior margin of anal plate; chelicerae not narrowed apically and needle-like, shear-like, bearing conspicuous shear-like chelae at tip which may or may not bear teeth (Fig. 4 B) 7

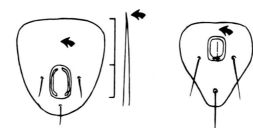

Fig. 4 A   Fig. 4 B

5) Dorsal surface of body with a single plate (Fig. 5 A) 6

Dorsal surface of body with two plates, a large anterior plate and a small posterior plate (Fig. 5 B). *Liponyssoides sanguineus*   HOUSE MOUSE MITE

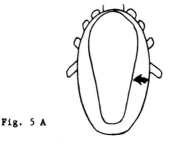

Fig. 5 A   Fig. 5 B

6) Peritreme tube somewhat sinous and extending anteriorly to a point opposite coxa 2 (Fig. 6 A). *Dermanyssus gallinae*   CHICKEN MITE

Peritreme tube short, extending forward for a distance less than half the diameter of coxa 3 (Fig. 6 B). *Dermanyssus americanus* AMERICAN BIRD MITE

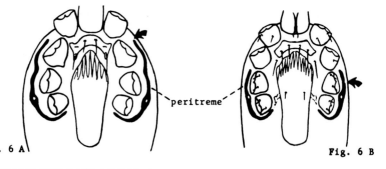

peritreme

Fig. 6 A   Fig. 6 B

7) Dorsal plate not covering entire dorsal surface of mite; genito-ventral plate typically narrowed posteriorly behind 4th coxae; chelae on chelicerae without teeth or setae (Fig. 7 A). Genus *Ornithonyssus* 8

Dorsal plate almost covering entire dorsal surface of mite; genito-ventral plate typically expanded posterior to 4th coxae; one or both chelae of chelicerae with teeth and a seta (Fig. 7 B). Family Laelaptidae 10

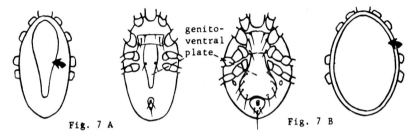

Fig. 7 A   Fig. 7 B

8) Sternal plate with anterior and middle pairs of sternal setae on the plate, posterior pair usually just off the plate (Fig. 8 A). On birds *Ornithonyssus sylviarum*   NORTHERN FOWL MITE

Sternal plate with the usual three pairs of setae on the plate (Fig. 8 B) 9

Fig. 8 A   Fig. 8 B

9) Dorsal plate narrowed posteriorly; setae in middle dorsal row of plate longer than the distance between their bases (Fig. 9 A). Normally on mammals or man *Ornithonyssus bacoti*   TROPICAL RAT MITE

Dorsal plate broader posteriorly; setae in middle dorsal row of plate much shorter than the distance between their bases (Fig. 9 B). Normally on birds *Ornithonyssus bursa*   TROPICAL BIRD MITE

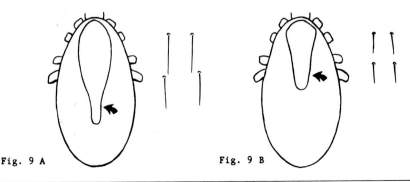

Fig. 9 A   Fig. 9 B

10) Genito-ventral plate with many fine setae; anal plate transverse, wider than long (Fig. 10 A). On domestic rats and a wide variety of wild mammals.   *Eulaelaps stabularis*
  Genito-ventral plate with one to four pairs of setae; anal plate longer than wide (Fig. 10 B) 11

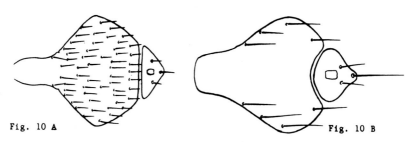

Fig. 10 A   Fig. 10 B

11) Genito-ventral plate with only a single pair of setae (Fig. 11 A). On domestic rats and mice and a wide variety of mammals and birds *Haemolaelaps glasgowi*   COMMON RODENT MITE
  Genito-ventral plate with four pairs of setae (Fig. 11 B). Normally on domestic rats 12

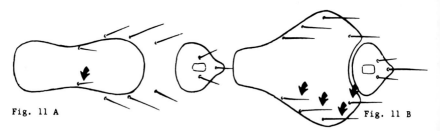

Fig. 11 A   Fig. 11 B

12) Anal plate contiguous with the genito-ventral plate, anterior margin rounded and fitting into a strong concavity in genito-ventral plate; larger species averaging 1-2 mm long. (Fig. 12 A). *Laelaps echidnina*   SPINY RAT MITE
  Anal plate somewhat separated from genito-ventral plate, anterior margin almost straight with definite anterior-lateral corners; small species averaging 0.5 to one mm long (Fig. 12 B). *Laelaps nuttalli*   DOMESTIC RAT MITE

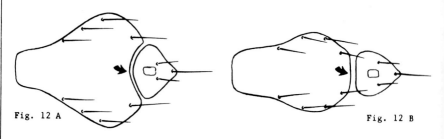

Fig. 12 A   Fig. 12 B

13) First pair of legs very long, much longer than other three pairs; anterior margin of body with four distinct flattened scales and somewhat flattened scales on other dorsal surfaces of body (Fig. 13 A). Plant feeders which invade buildings but do not bite man. *Bryobia praetiosa* CLOVER MITE
First pair of legs not markedly longer than the other three pairs of legs; no flattened scales on body (Fig. 13 B) 14

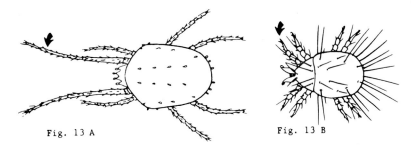

Fig. 13 A   Fig. 13 B

14) Surface of body without fine parallel lines or folds; tarsi without stalked suckers (Fig. 14 A). Adults never true parasites (Cheese or Flour mites) .. 15
Surface of body with fine parallel lines or folds; tarsi often provided with stalked suckers (Fig. 14 B). Scabies or mange mites parasitic in all stages, chiefly on vertebrates 16

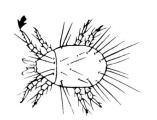
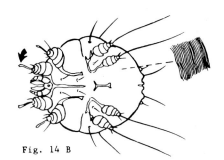

Fig. 14 A   Fig. 14 B

15) Tarsi tapering markedly to tip (Fig. 15 A)   *Glycyphagus prunorum*
Tarsi not tapering markedly to tip (Fig. 15 B). Many cheese and flour mites which are difficult to separate except with very specialized literature and a reference collection.   Genus *Tyrophagus*, Genus *Caloglyphus*, Etc.

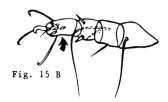

Fig. 15 A   Fig. 15 B

16) Body elongate, somewhat cigar-shaped and prolonged behind; the abdomen somewhat ringed; legs very short, apparently three-segmented; tiny species less than one mm (Fig. 16 A). In hair follicles or sebaceous glands of mammals *Demodex folliculorum*   PORE OR FOLLICLE MITE
   Body not prolonged behind and cigar-shaped (Fig. 16 B). Occasionally female grain itch somewhat balloon-shaped; larger species not found in hair follicle or sebaceous glands of mammals 17

Fig. 16 A

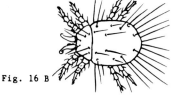

Fig. 16 B

17) A club-shaped or clavate hair between bases of first and second pairs of legs. Body divided into cephalothorax and abdomen, the latter often enormously enlarged (Fig. 17 A) *Pyemotes ventricosus* formerly *Pediculoides ventricosus*   STRAW ITCH MITE
   Setae on cephalothorax normal, no club-shaped or clavate hair between bases of first and second pairs of legs; no distinct division into cephalothorax and abdomen (Fig. 17 B) 18

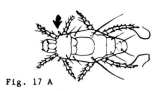

Fig. 17 A

Fig. 17 B

18) Legs short and stubby (Fig. 18 A) 20
   Legs longer and more slender (Fig. 18 B) 29

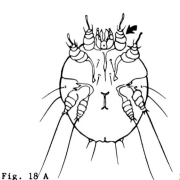

Fig. 18 A

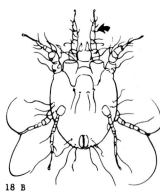

Fig. 18 B

19) Suckers of tarsi with segmented pedicels (Fig. 19 A). Non-burrowing itch mites on mammals in the genus *Psoroptes,* a common species causing scabs and crusts in the ears of rabbits in the *Psoroptes cuniculi*   RABBIT EAR MITE
   Suckers of tarsi without segmented pedicels (Fig. 19 B) Genus *Dermatophagoides*   HOUSE DUST MITE

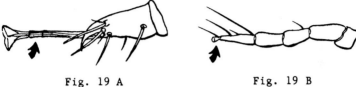

Fig. 19 A   Fig. 19 B

20) Anal opening on the dorsal surface of the body; dorsal surface of the body with only short, sharp setae (Fig. 20 A)   *Notoedres*
   Anal opening at tip of body or slightly on ventral side; dorsal surface of body with pointed scales and blunt stout spines (Fig. 20 B). *Sarcoptes scabiei*
                                                   SCABIES OR MANGE MITE

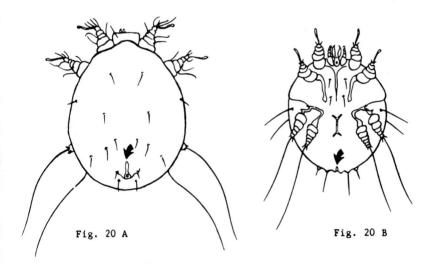

Fig. 20 A   Fig. 20 B

Katz (1965) has prepared an interesting paper on clover mites. He notes the mites may enter spaces between studs, cracks in the sill or around window frames. He recommends drenching the soil with combinations of adulticides and ovicides.

According to English and Snetsinger (1957), rain within a few days after treatment may necessitate retreatment because "the deposit of acaricide is not only depleted by the rain, but the conditions provided by the rainfall are favorable to the mites. If acaricides are necessary, they should be applied about April 1 and October 1 in the latitude of central Illinois."

English and Snetsinger (1957) also recommend leaving a six to 24-inch/15 to 61 cm grass-free band all around the house because such grass-free bands act as barriers to the house-invading clover mites.

LITERATURE

ANDERSON, N.H. & C.V.G. MORGAN — 1958. Life histories and habits of the clover mite, *Bryobia praetiosa* Koch, and the brown mite *B. arborea* M & A. in British Columbia (Acarina Tetryanychidae). Canad. Entomol. 90(1):23-42.

ANONYMOUS — 1946. Control of mite infestations. Soap & S. C. 22(5):149. 1947. Charles Pomerantz addresses the N.J.P.C.A. on mites. Pests 15(1):26. 1948. First "grain itch" outbreaks in Baltimore stopped by aid of new insecticide material. Advt. U.S.I. Chem. News. Chem. & Eng. News. 26(15):1057-1058. 1951. Chigger control. U.S.D.A. Leaflet No. 302. 1956. Chiggers; How to fight them. U.S.D.A. Leaflet No. 403. 1958. Clover mites — How to control them around the home. U.S.D.A. Leaflet 443. 1971. Tropical fowl mite. Coop. Econ. Insect Rpt. 21(3):26.

BAKER, E.W. — 1955. Mites as household pests. Pest Control 23(9):9-10, 12.

BAKER, E.W. & G.W. WHARTON — 1952. An Introduction to Acarology. Macmillan.

BERNDT. W.L. — 1952. The chicken mite attacking children. J. Econ. Entomol. 45(6):1098-1099.

BERTRAM, D.S., K. UNSWORTH & R.M. GORDON — 1946. The biology and maintenance of *Liponyssus bacoti* Hirst, 1913, and an investigation into its role as a vector of *Litomosoides carinii* to cotton rats and white rats, together with some observations on the infection in the white rats. Ann. Trop. Med. and Parasitol. 40(2):228-254. Biol. Abst. 21(3):741. 1947.

BISHOPP, F.C. — 1923. The rat mite attacking man. USDA Dept. Circ. 294.

BOLTON, W.W. — 1947. Itch, itch, itch. Hygeia 25:364-365. 404. May.

BONOMO, G.C. — 1687. Epistola che contiene osservazioni intorno a Pellicelli del corpo umano. Firenze. 16 pp., 1 plate.

BORROR, D.J., D.M. DE LONG & C.A. TRIPLEHORN — 1981. Introduction to the study of insects. 5th Ed. Saunders College Publishing. 827 pp.

BROWN, J.H. — 1953. A chicken mite infestation in a hospital. J. Econ. Entomol. 46(5):900.

BROWN, R.V. — 1965. Control of *Histiosoma laboratorium* in *Drosophila* cultures. J. Econ. Entomol. 58(1):156-157.

BUSVINE, J.R. — 1980. Insects and Hygiene. Chapman & Hall, London. 568 pp.

CARPENTER, C.C., J.A. HEINLEIN, M.B. SULZBERGER & R.L. BAER — 1946. Scabies and pediculosis treated with benzyl benzoate, DDT, benzocaine emulsion (NBIN), including a comparison with other methods used at U.S. Naval Hospital, Brooklyn, N.Y., from November 1943 to May 1945. J. Invest. Dermatology 7:93-98.

CHAPMAN, R.N. & H.H. SHEPARD — 1932. Insects infesting stored food products. Minn. Agr. Exp. Sta. Bull. 198.

COMBS, R.L., JR. & J.L. LANCASTER, JR. — 1965. The biology of the northern fowl mite. Arkansas Agr. Exp. Sta. Report Series 138.

CROSS, H.F. — 1948. Use of powders on clothing for protection against chiggers. J. Econ. Entomol. 41(5):731-734.

DAHM, P.A. & C.L. BAUER — 1949. The miticidal properties of di(p-chlorophenyl) methyl carbinol in laboratory insect rearings. Science 109 (2821):69.

DAVIS, J.J. — 1950. The 1950 epidemic of the straw itch mite. Proc. Indiana Acad. Sci. 60:183-184.

DELONG, D.M. & J.F. SMITH — 1951. The chigger *(Eutrombicula alfreddugesi)* problem in city lawns with suggestions for their control. Ohio J. Sci. 51:203-204.

DEONG, E.R. & C.L. ROADHOUSE — 1922. Cheese pests and their control. Calif. Agr. Exp. Sta. Bull. 343.

DICKE, R.J., K.D. IHDE & W.V. PRICE — 1953. Chemical control of cheese mites. J. Econ. Entomol. 46(5):844-849.

DOETSCHMAN, W.H. & D.P. FURMAN — 1949. A tropical chigger, *Eutrombicula batatas* (Linn.) attacking man in California. Rev. Appl. Entomol. B 40(7):115, 1952.

DORADO, F.G. & J.M.C. VIDAL — 1946. Cutaneous parasitism by *P. ventricosus*. Rev. Appl. Entomol (B). 36(8):138.

DOVE, W.E. & B. SHELMIRE — 1932. Some observations on tropical rat mites and endemic typhus. J. Parasitology 18:159-168.

DUNCAN, S. — 1957. *Dermanyssus gallinae* (DeGeer) attacking man. J. Parasitol. 43(6):637.

DUSTAN, G.G. — 1957. The effects of temperature and certain chemicals on cheese mites. Rept. Entomol. Soc. Ont. 68:60-7.

EALES, N.B. — 1917. The life history and economy of the cheese mites. Ann. appl. Biol. 4(1&2):28-35. Rev. Appl. Entomol. (A). 5:516, 1917.

EBELING, W. — 1975. Urban Entomology. Univ. Calif. Press, Berkeley. 695 pp.

EBELING, W. — 1960. Control of the tropical rat mite. J. Econ. Entomol. 53(3):475-476.

ENGLISH, L.L. & R. SNETSINGER — 1957. The habits and control of the clover mite in dwellings. J. Econ. Entomol 50(2):135-141.

ESSIG, E.O. — 1926. Insects of Western North America. Macmillan.

EWING, H.E. — 1921. Studies on the biology and control of chiggers. USDA Bull. No. 986.

EWING, H.E. & H.H.S. NESBITT — 1942. Some notes on the taxonomy of grain mites. (Acarina, formerly Tyroglyphidae) Proc. Biol. Soc. Wash. 55:121-124 Rev. Appl. Entomol (A.) 30:588, 1942.

FIGGE, F.H.J. & G.F. WOLFE — 1945. Use of Beta beta dithiocyano diethyl ether (RID-O) to control mite infestations in mice. Proc. Soc. Exptl. Biol. and Med. 60(1):136-138. Biol. Abst. 20:456, 1946.

FISK, F.W. — 1951. Use of a specific mite control in roach and mouse cultures. J. Econ. Entomol. 44(6):1016.

FLUNO, H.J., H.A. JONES & F.M. SNYDER — 1946. Emulsifiers for liquid acaricides. J. Econ. Entomol 39(6):810-811.

FLYNN, R.J. — 1972. Parasites of Laboratory Animals. Iowa State Univ. Press. Ames. 884 pp.

FREEMAN, J.A. & E.E. TURTLE — 1947. The control of insects in flour mills. Ministry of Food, London.

FULLER, H.S. — 1954. Studies of rickettsialpox Ill. Life cycle of the mite vector, *Allodermanyssus sanguineus*. Amer. J. Hyg. 59:236-239.

FURMAN D.P., L.E. VINCENT & W.S. COATES — 1955. External parasites of poultry. Calif. Agr. 9(1):13.

GARMAN, P. — 1937. Bird mites in a dwelling house. Bull. Conn., Agr. Exp. Sta. 396.

GAY, F.J. & T. GREAVES — 1942. The control of *Pediculoides ventricosus* (Newport) in insect cultures. Australia Counc. Sci. and Indust. Res. J. 15(4):315-317.

HAND, E.A. — 1946. Diagnosis of infestation with *"Sarcoptes scabiei"* var. *"hominis."* Discussion of life cycle of organism. U.S. Naval Med. Bull. 46(6):834-844, 9 fig.

HASE. A. — 1929. Zur pathologisch-parasitologischen und epidemiologisch-hygienischen Bedeutung der Milben, insbesondere der Tyroglyphinae (Käsemilben) sowie uber den sogenannten "Milbenkäse," *Zeitschr. Parasitenkunde* 1 (4-5):765-821.

HANTSBARGER, W.M. & G. MAST — 1959. Chicken mite, *Dermanyssus gallinae.* Coop. Econ. Insect Report 9(8):121. Febr. 20.

HARDING, W.C., JR. — 1955. Malathion to control the northern fowl mite. J. Econ. Entomol. 48(5):605-606.

HEAL, R.E. — 1956. Experiences with mites by the pest control industry. Review of Current Mite Problems. ESA, Eastern Branch, Nov. 20, 1956.

HEILESEN, B. — 1946. Studies on *Acarus scabiei* and scabies. Acta. Derm. Venereol. 26 Suppl. 14, 370 pp. Copenhagen, Rosenkilde & Bagger. Rev. Appl. Entomol. (B). 35(7):120, July, 1947.

HELLER-HAUPT, A. & J.R. BUSVINE — 1974. (Acaricides for house dust mites), J. med. Entomol. 11, 551.

HINMAN, E.H. & R.H. KAMPMEIER — 1934. Amer. J. Trop. Med. 14(4):355-362.

HILL, M.A. & R.M. GORDON — 1945. An outbreak of dermatitis amongst troops in North Wales caused by rodent mites. Ann. Trop. Med. Parasit. 39(1):46-52. Abst. Rev. Appl. Entomol (B). 35(1):7. 1947.

HODES, H. — 1946. House full of red mites *(Dermanyssus gallinae)* — they dropped from the room. Smallholder 75 (1891):12. June 14, 1946.

HUEBNER, R.J. — 1947. Rickettsial pox, a new disease. Amer. J. Clin. Path. 17:790-971. Dec.

HEUBNER, R.J., W.L. JELLISON & C. POMERANTZ — 1946. Isolation of rickettsia apparently identical with a causative agent of Rickettsial-pox from *Allodermanyssus sanguineus,* a rodent mite. U.S. Pub. Health Rpt. 61(47):1677-82.

HUGHES, A.M. — 1948. The mites associated with stored food products. Ministry of Agriculture & Fisheries. 168 pp. London. 1961. The Mites of Stored Food. Ministry Agr., Fisheries, Food. London Tech. Bull. 9.

HUGHES, T.E. — 1943. The respiration of *Tyroglyphus farinae.* J. Exp. Biol. 20(1):1-5. Rev. Appl. Entomol (A). 32:80, 1944.

HUNGERFORD, H.B. — 1943. The tropical rat mite. *(Liponyssus bacoti)* in Kansas. Kans. Entomol. Soc. J. 16:154. Oct.

KATZ, H. — 1965. How to control clover mites. Pest Control 33(6):16-17, 34.

KEH, B. — 1973. The common housedust mites of the genus *Dermatophagoides.* Calif. Vector Views. 20(5):37-45.

KELLER, J.C. & H.K. GOUCK — 1957. Small-plot tests for the control of chiggers. J. Econ. Entomol 50(2):141-143.

KNAPP, F.W. & G.F. KRAUSE — 1960. Control of the northern fowl mite, *Ornithonyssus sylviarum.* J. Econ. Entomol 53(1):4-5.

KRAEMER, P. — 1960. Relative efficacy of several materials for control of poultry ectoparasites. J. Econ. Entomol 52(6):1195-1199.

KRANTZ, G.W. — 1955. Some mites injurious to farm-stored grain. J. Econ. Entomol. 48(6):754-755.

LAIRD, M. — 1950. Notes on the infestation of man by the chicken mite *Dermanysus gallinae* (DeGeer) in New Zealand. Biol. Abst. 31(10):2990, 1957.

LARSEN, B. — 1973. Some observations on rat mite dermatitis. Calif. Vector Views 20(4):34-35.

LARSON, A.O. — 1925. Further notes on human suffering caused by mites, *Pediculoides ventricosus* Newp. Pan-Pac. Entomol. 2:93-95.

LETHWAITE, R. — 1945. Scrub-typhus: A disease of man transmitted by mites. Brit. Med. Bul. 3(9/10):227-228. Biol. Abst. 20:1252, 1946.

LINDUSKA, J.P., F.A. MORTON & W.C. McDUFFIE — 1948. Tests of materials for the control of chiggers on the ground. J. Econ. Entomol 41(1):43-47.

MACKIE, D.B. — 1937. Entomological service Bull. Dept. Agric. Calif. 25:455-481.

MAC NAY, C.G. — 1955. Control of mites in and about the home. Canad Dept. Agri. Publ. 934.

MATHESON, R. — 1950. Medical Entomology 2nd Ed. Comstock Publishing Co.

MacCREARY, D. & W.A. CONNELL — 1962. Clover mite control studies. J. Econ. Entomol. 54(5):1062-1063.

MARZKE, F.O. & R.J. DICKE — 1959. Laboratory evaluations of various residual sprays for the control of cheese mites. J. Econ. Entomol. 52(2):237-240.

McLEOD, W.S. & R.W. BROWN — 1948. Dichlor ethyl ether in the control of cheese mites. Canad. Dairy & Ice Cream J. 27(3):80. Mar.

MELLANBY, K — 1943. Scabies. Oxford Univ. Press. 81 pp. 1944. The development of symptoms, parasitic infection and immunity in human scabies. Parasitology 35:197-206. Mar.

MELVIN, R., C.L. SMITH & O.H. GRAHAM — 1943. Some observations on chiggers. J. Econ. Entomol. 36:940.

MEYER, K.F. & B. EDDIE — 1960. Feather mites and ornithosis. Science 132(3442):200.

MICHENER, C.D. — 1946. Observations on the habits and life history of a chigger mite, *Eutrombicula batatas* Ann. Entomol. Soc. Amer. 39(1):101-118.

MICHENER, C.D. — 1946. A method of rearing chigger mites (Acarina, Trombiculinae) Amer. J. of Trop. Med. 26(2):251-256.

MICHENER, M.H. & C.D. MICHENER — 1947. Chiggers. Nat. Hist. 56(5): 231-235.

MILLS, E.M. — 1947. The tropical rat mite and pest control operator. Pests 15(9):26, 28.

MORGAN, C.V.G. & N.H. ANDERSON — 1957. *Bryobia arborea* n.sp. and morphological characters distinguishing it from *B. praetiosa* Koch (Acarina: Tetranychidae). Canada. Entomol. 89(11):485-490.

MORLAN, H.B. — 1948. Ectoparasites of domestic rats and mice and the diseases they transmit. Pests 16(12):16, 18, 20.

MUGGERIDGE, J. & R.M. DOLBY — 1946. Control measures against the cheese mites, *Tyrolichus* (i.e., *Tyrophagus) casei* Ouds and *Tyrophagus longior* Gerv. New Zeal. J. Sci. & Technol. 28(1) Sec. A:1-30.

MUNRO, J.W. — 1966. Pests of Stored Products. Rentokil Library. Hutchinson, London. 234 pp.

NESBITT, H.H.J. — 1945. A revision of the family Acaridae (Tyroglyphidae), Order Acari, based on comparative morphological studies. Part I Historical, morphological and general taxonomic studies. Canad. J. Res. D. 23(6): 139-188.

NEWSTEAD, R. & H.M. DUVALL — 1918. Bionomics, Morphological and Economic Report on the Acarids of Stored Grain and Flour. R.S. Grain Comm. Rept. No. 2. Royal Soc., London.

NEWSTEAD, R. & H.M. DUVALL — 1920. Bionomic, Morphological and Economic Report on the Acarids of Stored Grain and Flour. Part II. R.S. Grain Comm. Rept. No. 8. Roy. Soc., London.

NORRIS, J.D. — 1958. Observations on the Control of Mite Infestations in Stored Wheat by *Cheyletus* spp. Ann. Appl. Biol. 46, 411.

NPCA. — 1956. A Manual of Parasitic Mites of Medical or Economic Importance. Various authors, E.W. Baker, T.M. Evans, D.J. Gould, W.B. Hull, H.L. Keegan. Technical Publication, National Pest Control Assoc. 170 pp.

OLDS, H.S. — 1942. The results of further work done on the control of grain mites in British Columbia. Proc. Entomol. Soc. B.C. 39:29-32. Rev. Appl. Entomol (A), 31:213, 1943.

OLSON, T.A. & R.G. DAHMS — 1946. Observations on the tropical rat mite, *Liponyssus bacoti,* as an ectoparasite of laboratory animals and suggestions for its control. J. Parasitol. 32(1):56-60.

PHILIP. C.B. — 1949. Scrub typhus, or tsutsugamushi disease. Sci. Mo. 69(5):281-289.

PHILIP, C.B. & L.E. HUGHES — 1948. The tropical rat mite, *Liponyssus bacoti* as an experimental vector of Rickettsialpox. Amer. J. Trop. Med. 28(5): 697-705.

PIERCE, W.D. — 1975. The Deadly Triangle. John G. Shanafelt, Jr. Orange, Calif. 138 pp.

PRATT, H.D. — 1975. Mites of Public Health Importance. U.S. Dept. Hlth, Educ. & Welf. Atlanta, Georgia. DC Publ. No. 76-8297. 38 pp.

PRATT, H.D., J.E. LANE & F.C. HARMSTON — 1949. New locality records for *Allodermanyssus sanguineus* vector of rickettsialpox. J. Econ. Entomol. 42:414-415. June.

RICE, E.B. — 1948. Queensland cheese production. Qd. Agric. J. 66(2):107-111. Rev. Appl. Entomol (A). 36(11):363, 1948.

ROBERTSON, P.L. — 1946. Tyroglyphid mites in stored products in New Zealand. Trans. Roy. Soc. New Zealand 76(2):185-207.

ROGERS, G. K. — 1943. Grain itch. J. Amer. Med. Assoc. 123(14):887-889.

ROKSTAD, I. — 1943. Endemic outbreaks of skin eruption caused by mites from foodstuffs (figs) and tobacco. Acta Dermato-Venereol 24(2):113-129. Biol. Abst. 19:809, 1945.

ROSELLE, R.E. — 1954. Clover mites *(Bryobia practiosa)*. Nebr. Agr. Col. Circ. 1570.

SASA, M. — 1961. Biology of chiggers. Ann. Rev. Entomol 6:221-224.

SAUNDERS, T.S. — 1944. Dermatitis from Tyroglyphidae in handlers of straw. Arch. Dermat. and Syph. 50:245. Oct.

SCOTT, E.W. — 1949. Longevity of tropical rat mites kept without food. J. Parasit. 35(4):434. Rev. Appl. Entomol (B). 39(9):146, 1951.

SCOTT, H.G. — 1958. Control of mites on hamsters. J. Econ. Entomol. 51(3): 412-413.

SCOTT, H.G. & R.M. FINE. — 1967. Straw itch mite dermatitis. Pest Control 35(7):19-20, 22-23.

SCOTT, J.A. — 1948. An apparatus for removing tropical rat mites *(Liponyssus bacoti)* from large quantities of bedding materials. J. Parasitol. 34:132-133, Apr. 1948.

SEIGEL, V. — 1947. What makes a good chigger? Hygeia 25:592-593. Aug.

SHANAHAN, G.J. & A.B. SHELTON — 1948. Cheese mite control. Use of dichloroethyl ether as a fumigant. Agri. Gaz. N.S.W. 59(7):381-383. Rev. Appl. Entomol (A). 37(1):24-25, 1949.

SHELMIRE, B. & W.E. DOVE — 1931. The tropical rat mite, *Liponyssus bacoti* (Hirst). J. Amer. Med. Assoc. 96:579-584.

SIKES, R.K. & R.W. CHAMBERLIN — 1954. Laboratory observations on three species of bird mites. J. Parasitol. 40:691-697.

SINHA, R.N. & T.C. PAUL — 1972. Survival and multiplication of two stored-product mites on cereals and processed foods. J. Econ. Entomol. 65(5):1301-1303.

SKALIY, P. & W.J. HAYES, JR. — 1949. The biology of *Liponyssus bacoti* (Hirst). Amer. J. Trop. Med. 29:759-772.

SLEPYAN, A.H. — 1944. A rapid treatment for scabies. J. Amer. Med. Assoc. 124(16):1127-1128.

SMITH, C.N. & H.K. GOUCK — 1944. DDT, Sulfur, and other insecticides for the control of chiggers. J. Econ. Entomol 37:131-132.

SMITH, K.G.V. — 1973. Insects and Other Arthropods of Medical Importance (Sheals, Chap. 14). British Museum of Natural History, London. 561 pp, 12 plates.

SNETSINGER, R. — 1967. Some answers to the clover mite problem. Pest Control 35(7):15-16, 18, 56.

SNYDER, F.M. & F.A. MORTON — 1946. Materials as effective as benzyl benzoate for impregnating clothing against chiggers. J. Econ. Entomol. 39(3): 385-387.

SOLOMON, M.E. — 1943. Tyroglyphid mites in stored products. Dept. of Sci. & Indust. Res., London, England. 1944. Behavior of tyroglyphid mite populations in stored grain and flour. Ann. Appl. Biol. 31(1):81. R.A.E. A. 32:319, 1944. 1946. Tryglyphid mites in stored products. Nature and amount of damage to wheat. Ann. Appl. Biol. 33(3):280-289. R.A.E. A. 36(12):417-418, 1948. 1946. Tyroglyphid mites in stored products. Ecological studies. Ann. Appl. Biol. 33(1):82-89. R.A.E. A. 35(7):209-210. 1962. Ecology of the flower mite, *Acarus siro* L. = *(Tyroglyphus farinae DeG.)*. Ann. Appl. Biol. 50(1):178-184.

SNYDER, F.M. & F.A. MORTON — 1947. Benzyl benzoate-dibutyl phthalate mixture for impregnation of clothing. J. Econ. Entomol 40(4):586-587.

SPEAR, P.J. — 1954. Some problems in the control of clover mites. Chem. Spec. Mfr. Assoc. Proc. Ann. Meeting 41:123-124.

STONE, P.C. & L. HASEMAN — 1941. The chigger and its control in Missouri, Missouri Agr. Exp. Sta. Circ. 214.

SUTTON, R.L., JR. — 1942. Trombidiosis (chigger bites). Relief of itching with ethyl aminobenzoate in flexible collodion. J. Amer. Med. Assoc. 120(1):26-27.

TARSHIS, I.B. — 1964. A sorptive dust for control of the northern fowl mite, *Ornithonyssus sylviarum,* infesting dwellings. J. Econ. Entomol. 57(1): 110-111.

TRAVER, J.R. — 1951. Unusual scalp dermatitis in humans caused by the mite *Dermatophagoides.* Proc. Entomol. Soc. Wash. 53(1):1-25.

VAN BRONSWIJK, J.E.M.H. — 1973a. *Dermatophagoides pteronyssinus* (Trouessart, 1897) in mattress and floor dust in a temperate climate (Acari: Pyroglyphidae) J. Med. Entomol. 10:63-70. 1973b. Hausstaub-Oekosystem und Hausstaub-Allergene. Acta Allergologica 27:219-28.

VAN BRONSWIJK, J.E.M.H. & H.M.M. KOEKKOEK — 1971. Nipagin (p-methyl hydroxy benzoate) as a pesticide against a house dust mite: *Dermatophagoides pteronyssinus.* J. Med. Entomol. 8:748. 1972. Effects of low temperatures on the survival of house dust mites of the family Pyroglyphidae (Acari: Sarcoptiformes). Netherlands J. Zool. 22:207-212.

VAN BRONSWIJK, J.E.M.H., J.M.C.P. SCHOONEN, M.A.F. BERLIE & F.S. LUKOSCHUS — 1971. On the abundance of *Dermatophagoides pteronyssinus* (Trouessart, 1897) (Pyroglyphidae: Acarina) in house dust. Res. Popul. Ecol. 13:67-79.

VAN BRONSWIJK, J.E.M.H. & R.N. SINHA — 1971. Pyroglyphid mites (Acari) and house dust allergy. J. of Allergy 47:31-52.

VENABLES, E.P. — 1943. Observations on the clover or brown mite, *Bryobia praetiosa.* Koch. Can. Entomol. 75(2):41-42.

VERCAMMEN-GRANDJEAN, P.H. & R. LANGSTON — 1971. The chigger mites of the world. Vol. III. *Guntherana* complex. Section A. Genus *Guntherana.* G.W. Hooper Foundation, San Francisco: 153 pp. + vii.

VERCAMMEN-GRANDJEAN, P.H. & R. LANGSTON — 1975. The chigger mites of the world. Vol. III. *Leptotrombidium* complex. Section C. Iconography. G.W. Hooper Foundation, San Francisco: 298 pp.

VERCAMMEN-GRANDJEAN, P.H. & R. LANGSTON — 1976a. The chigger mites of the world. Vol. III. *Leptotrombidium* complex. Section A. *Leptotrombidium* sensu stricto. G.W. Hooper Foundation, San Francisco: pp. 1-612.

VERCAMMEN-GRANDJEAN, P.H. & R. LANGSTON — 1976b. The chigger mites of the world. Vol. III. *Leptotrombidium* complex. Section B. *Trombiculindus, Hypotrombidium* and *Ericotrombidium,* plus heterogenera. G.W. Hooper Foundation, San Francisco: 613-1061.

VINCENT, L.E., D.L. LINDGREN & H.E. KROHNE — 1954. Toxicity of malathion to the northern fowl mite. J. Econ. Entomol. 47(5):943-944.

VOORHORST, R., F.T.M. SPIERSMA & H. VAREKAMP — 1969. House-Dust Atopy and the House-Dust Mite *Dermatophagoides pteronyssinus* (Trouessart 1897). 159 pp. Leiden: Stafleu's Scientific Publishing Company.

WEBSTER, F.M. — 1910. A predacious mite proves noxious to man. USDA Bur. Entomol. Circ. 118.

WHARTON, G.W. & H.S. FULLER — 1952. A Manual of the Chiggers. Memoirs

Entomol. Soc. Wash. No. 4.

WHEELER, E.H. — 1955. Clover mite, a household pest. Univ. Mass. Agr. Ext. Leaflet 289.

WILLIAMS, R.W. — 1944. A bibliography pertaining to the mite family Trombidiidae. Ameri. Midland Nat. 32(3):699-712. 1946. The laboratory rearing of the tropical rat mite, *Liponyssus bacoti* (Hirst) — J. Parasit. 32(3):252-256. Rev. Appl. Ent.(B). 37(7):132-133. 1946. A contribution to our knowledge of the bionomics of the common North American chigger, *Eutrombicula alfreddugesi* (Oudemans) with a description of a rapid collecting method. Amer. J. Trop. Med. 26(2):243-250. Biol. Abst. 20:1253. 1946.

WOMERLEY, H. — 1943. A revision of the Tyroglyphidae. Rev. Appl. Entomol. (A). 31:157.

STANLEY G. GREEN

Dr. Stanley Green received his bachelor's and master's degrees from the University of Colorado. For his disserttation on the taxonomy of the family Oribatulidae he received his Ph.D. from The Colorado State University, Fort Collins, Col. in 1969.

Green is responsible for vector control and urban pest control programs in Pennsylvania, providing training to public health professionals and pest control operators. Green teaches a number of broad-ranging courses, including basic entomology and structural pest control. In addition, he teaches in-depth courses for specialists, including mosquito and termite control courses. Green's courses and workshops have played an important role in developing professionalism in the pest control industry by encouraging a more scientific approach to pest control. A <u>Vector Control</u> and <u>Pest Control</u> newsletter are published by Green to help accomplish this goal.

As an extension entomologist, Green is responsible for the preparation of the popular <u>Extension Entomological Fact Sheets</u> on household arthropod pests. These are aimed at the general public to help them better understand arthropod pests and how to deal with them.

Green is well known for his selfless attitude toward those who need advice on pest problems. Many professionals have benefitted form his wise counselling and no problems are too big or too small to capture his interest. Green's availability and ability to communicate have made him a key resource for the news media when pest stories need professional input.

Green's achievements have been recognized by numerous honor societies, including Sigma Xi, Epsilon Sigma Phi and Pi Chi Omega. In addition, he is active in many professional associations including the Entomological Society of America, the American Mosquito Control Association, the American Registry of Professional Entomologists and the American Associaiton for the Advancement of Science. Green is also active in many state and regional associations and is currently president of the Mosquito and Vector Control Association of Pennsylvania.

CHAPTER TWENTY-THREE

Ticks

By Stanley G. Green[1]

Ticks ARE EXTERNAL parasites on mammals, birds, reptiles and on some amphibians. Both males and females feed on blood. There are two important families of ticks: the Ixodidae (hard ticks) and the Argasidae (soft ticks). The hard ticks are characterized as follows: only one nymphal stage is present, their mouth parts project forward and are visible from above, respiratory openings or stigmata are located on the fourth pair of legs and a scutum or shield-like plate is present on the back of the body. This plate covers the entire back of the male, but only the anterior part of the back of the female. The coxa or basal segment of each leg usually has a spur (projection). Festoons (rectangular areas separated by grooves) are generally present along the posterior margin of the body. Palpi are four segmented with the fourth segment very small. Hard ticks feed on blood once for each of the stages (unless interrupted).

The soft ticks or Argasidae have more than one nymphal stage (some have up to eight). Their mouth parts (in the nymphal and adult stages) are ventral and covered by the front margin of the body. They are not visible from above. Stigmata are between legs three and four. They lack a scutum and their skin appears wrinkled and leathery. Their coxae lack spurs and festoons are absent. Their palpal segments are equal in size. Soft ticks in all their stages feed periodically. The plate on page 778 identifies some members of the above families. families.

Most species of ticks are encountered in the woods and fields, and relatively few species occur as pests in the home. Some species of ticks are venomous and may cause paralysis. Others are vectors of such important diseases as Rocky Mountain spotted fever, relapsing fever, tularemia, Texas cattle fever and other diseases.

[1]Associate Professor of Entomology, The Pennsylvania State University, Cooperative Extension Service, Philadelphia, Pa.

PICTORIAL KEY TO GENERA OF ADULT TICKS IN UNITED STATES

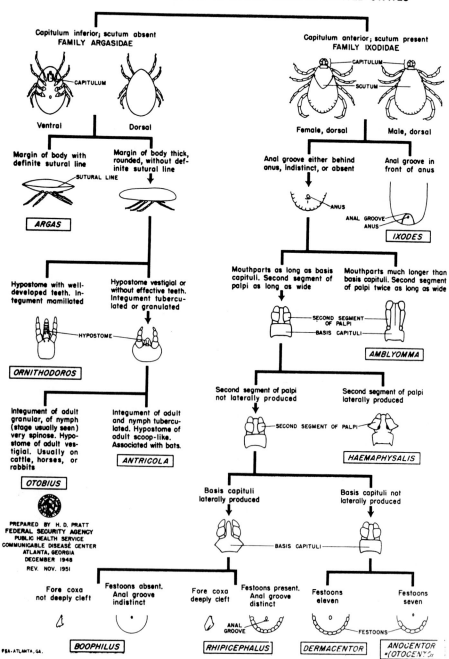

Fig. 23-1. Pictorial key to genera of adult ticks in the United States.

A vector is an animal which is capable of transmitting a pathogen from one organism to another. The pathogen is an organism which causes a disease in the animal receiving it. The following pathogens and the diseases they cause are transmitted by arthropod vectors: *Viruses* (yellow fever, encephalitis, Colorado tick fever); *bacteria* (relapsing fevers, plague, tularemia, shigella — which causes dysentery, Salmonella — cause of typhoid fever and food poisoning, anthrax and yaws); *protozoans* (malaria, Chagas disease, African sleeping sickness, piroplasmosis and anaplasmosis); *round worms* (filariasis — called elephantiasis when the worms block the lymphatics and heartworm when the worms lodge in the heart, and onchocerciasis); and *rickettsiae* (Rocky Mountain fever, louse borne typhus, rickettsialpox, Q fever, scrub typhus and trench fever).

Rice and Pratt (1972) classify the ways by which arthropods can vector disease, following the general outline of Huff (1931):

A) Mechanical transmission. Here the vector carries the pathogen in its digestive tract or on its body hairs, mouth parts or legs. It is transferred to the host organism directly without any biological changes. Examples of this type of transmission are: anthrax, typhoid fever, tularemia, yaws, cholera and food poisoning bacteria.

B) Biological transmission. The pathogen (within the vector) either multiplies or undergoes developmental changes. Three possibilities exist:

- Propagative transmission. The pathogen merely multiplies with no change in form (examples are the viruses of yellow fever, encephalitis and the rickettsiae of Rocky Mountain spotted fever).
- Cyclo-developmental transmission. No multiplication of the pathogen occurs, but it does undergo a change in form (e.g. filarial worms).
- Cyclo-propagative transmission. Not only do developmental changes occur, but there is a multiplication of the pathogens within the vector's body (e.g. malaria, Chagas disease).

Ticks and mites are important vectors of disease since they are capable of transmitting many organisms (See Table 23-1).

The number of cases of Rocky Mountain spotted fever recently has been on the increase. In 1947 there were 596 cases reported to the U.S. Public Health Service, Center for Disease Control, Atlanta, Georgia. This number dropped to 199 cases in 1959. The number of cases increased during the 60s and in 1977 a record 1,115 cases were reported (Anon., 1978).

There are two factors which contribute greatly to the successful spread of a pathogen. If the pathogen can be passed through the stages of the vector's development, (i.e. from larva to nymph to adult-transstadial transmission) and if it can enter the eggs of the vector and thus be passed onto the next generation (transovarial transmission) it will have a better chance of being passed to a vertebrate host. Ticks and mites have transstadial and transovarial transmission for most of the pathogenic organisms they carry.

The importance of transovarial disease transmission is expressed by Pierce (1975): "When the vector, like a tick, is capable of transmitting the organism hereditably through its offspring, that disease is permanent in any locality it reaches, until every vector offspring of infected offspring is exterminated." This would be quite a task since ticks produce thousands of eggs and the infection can be carried to several generations.

TABLE 23-1

Diseases vectored by ticks in North America

Disease	Pathogen	Principal Vectors*	Remarks
Colorado Tick Fever	Virus	*Dermacentor andersoni, D. occidentalis*	
Western Equine Encephalitis	Virus	*Liponyssus sylviarum Dermanyssus gallinae?*	Suspected vector
St. Louis Encephalitis	Virus	*Dermacentor variabilis*	
Powassan Encephalitis	Virus	*Ixodes cookei, I. spinipalpus. I. marxi, Dermacentor andersoni*	
Tick Paralysis	Neurotoxin	*Dermacentor andersoni, D. variabilis, Amblyomma americanum, A. ovale*	It is believed that the neurotoxin is produced by the female as she feeds. It moves from her ovaries to her salivary glands.
Rocky Mountain Spotted Fever	*Rickettsia rickettsi*	*Dermacentor andersoni, D. variabilis, D. occidentalis, Rhipicephalus sanguineus Amblyomma americanum*	
Endemic Typhus	*Rickettsia mooseri*	*Ornithonyssus bacoti*	
Benign Typhus	*Rickettsia muricola*	*Ornithonyssus bacoti*	
Human Babesiosis	*Babesia microti* (a protozoan)	*Ixodes* sp.	
Q Fever	*Coxiella burneti*	*Dermacentor andersoni Amblyomma americanum*	It is believed man becomes infested by breathing dust. Ticks have not been proven to be direct vectors to man.
Rickettsial-pox	*Rickettsia akari*	*Liponyssoides sanguineus*	

(continued)

TABLE 23-1 (cont.)

Tularemia	*Francisella tularensis* (a bacterium)	*Dermacentor andersoni, D. variabilis, Amblyomma americanum*	Deer flies also vector this disease and infection can occur from drinking contaminated water.
Relapsing Fever	*Borrelia hermsi, B. parkeri, B. turicata* (bacterial spirochetes)	*Ornithodoros hermsi, O. parkeri, O. turicata*	

*There are other vectors (not listed) which may be important in disease transmission. For example, the rabbit tick *(Haemaphysalis leporispalustris)* serves as a vector for Rocky Mountain spotted fever and tularemia among rabbits from which the principal vectors of the human disease can then become infective.

THE BROWN DOG TICK

Rhipicephalus sanguineus (Latr.) (Family Ixodidae)

The brown dog tick, *Rhipicephalus sanguineus* (Latr.), is not only a pest of dogs, but very often an unbearable nuisance in the home. The adult ticks occur in the ears and between the toes of the dog, and the larvae or seed ticks, as well as the nymphs are found in the long hair on the back. These ticks suck the blood of the dog, greatly reducing the animal's vitality, as well as making it irritable. On occasion, the brown dog tick may transmit a disease to the dog known as canine piroplasmosis. Thurman and Mulrennan (1947) found the brown dog tick on rats and mice in Florida. In the home, these ticks often emerge in great numbers from behind baseboards, molding, window and door trim, openings around window-cord pulleys, as well as from curtains and furniture. According to Pomerantz (Anon., 1946), dogs taken on vacations to tick infested areas bring the ticks back with them and eventually infest the house. The brown dog tick rarely attacks man (see Nelson, 1969). It is a vector of Rocky Mountain spotted fever in the western hemisphere and boutonneuse fever in Europe (Sheals in Smith, 1973, and Busvine, 1980).

Bishopp (1939) reports the dog tick as a pest of tropical and subtropical areas and Bishopp (1950) notes it to be most common in the South, where it may occur "under porches, in fence posts, and wood sheds," and wherever dogs are present. In the North, it is rarely found outdoors. Anon. (1961) reports it in lawns. Kohls and Parker (1948) note the brown dog tick is present in the eastern and central states, and in most of the western states. Nelson (1968) reports the tick in 50 states, but four states (Minnesota, North Dakota, Vermont and Alaska) have no preserved specimens.

Life history. Bishopp (1939) records its life history as follows: "The females, when fully engorged, are about one-third of an inch in length and bluish gray. They release their hold on the dog and seek a hiding place near by. They have a strong tendency to crawl upward and hence are often found hidden in cracks in the roofs of kennels or in the ceilings of porches. In these hiding places they deposit from 1,000 to 3,000 eggs, which hatch in from 10 to 60 days into minute, active six-legged seed ticks.

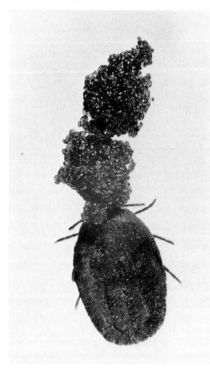

"When opportunity offers, these ticks attach to a dog and fill with blood in from three to six days. These engorged seed ticks are bluish, and about the size of a No. 8 shot. They drop from the dog and hide in cracks, and in from six to 23 days they molt their skins and become eight-legged reddish-brown nymphs. After a few days of inactivity, these nymphs are ready for attachment to dogs. After such attachment, they become engorged in from four to nine days. At this time, they are oval, about the size of a No. 5 shot, and dark gray. Again they leave the host, hide and molt their skins in from 12 to 29 days. They are now adult males and females, reddish-brown and very active when disturbed. In this stage they attach to various parts of the dog and the females become engorged in from six to 50 days, then drop off as explained previously. In each of the ungorged stages, this tick is capable of living for long periods without food. For instance, some adults have lived in confinement for over 200 days." This tick does not dwell in the woods, but instead has its habitat wherever dogs occur.

Fig. 23-2. Gravid female brown dog tick and eggs.

Control. Persistence is required if the brown dog tick is to be controlled. Both the dog and the home should be treated. Dogs should be dusted with either five percent carbaryl or three to five percent malathion. The powder should be rubbed well into the fur, particularly around the ears, the back of the neck and between the toes.

Anon. (1978) recommends the following insecticides in treating brown dog tick infested homes: 0.25 percent bendiocarb, 0.5 percent diazinon, one to two percent malathion and two percent ronnel. Propoxur, chlorpyrifos and dioxathion also are recommended, but dioxathion is labelled for outdoor use only, although use in a kennel is allowed. This is a good example of why the pest control operator should carefully read the pesticide label. He is legally responsible for the correct use of pesticides!

For the control of the ticks in the house, apply the spray to all observable crevices, such as around baseboards, trim, furniture and underneath rugs, behind pictures and on curtains and drapery.

Tarshis and Dunn (1959) recommend SG-67 silica aerogel dusts containing

CARBARYL

MALATHION

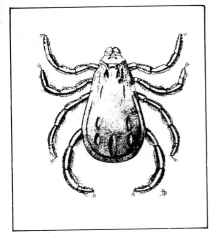

Fig. 23-3. Male brown dog tick.

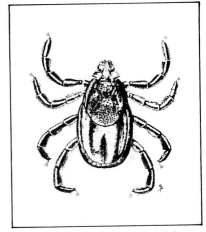

Fig. 23-4. Female brown dog tick.

two percent naled for dusting kennels. They also found ¹/₆ oz/five ml of naled e.c. per gallon/3.8 l of water to be an effective dip for dogs.

Bishopp et al. (1946) state: "Infested dogs should be kept in one place, especially during their sleeping hours. This more or less confines the ticks to that place and makes the treatment easier.

"Fumigation of infested houses is seldom advisable because the ticks are usually present in entryways, around porches and in outbuildings, where they cannot be reached with a fumigant. Furthermore, the tick is very resistant to fumigants."

Barnett and Parsons (1963) controlled a variety of ticks in fodder and bedding using two lbs/0.9 kg methyl bromide per 900 cubic feet/25 cubic meters.

THE GROUNDHOG TICK

Ixodes cookei (Packard)

Bishopp (1950) notes the nymphs and adults attack man. Occasionally they occur in summer cottages frequented by groundhogs. This species of tick is most common in the New England states.

According to the National Pest Control Assocation, this tick is controlled in the home by excluding groundhogs and other animals from the house, by spraying with appropriate solutions or emulsions of residual insecticides under the house and through the use of residual insecticides in the house.

THE RELAPSING FEVER TICK

Ornithodoros hermsi Wheeler

This tick infests cabins in the mountains of California. Herms and Wheeler (1935) found it to transmit relapsing fever. The adult female is five to six mm long, sandy in color unengorged and at first reddish, and later grayish blue when engorged.

Herms (1939) recommends cabins be made rodent proof so as to exclude chipmunks and other small animals that carry the ticks. Once the ticks are inside

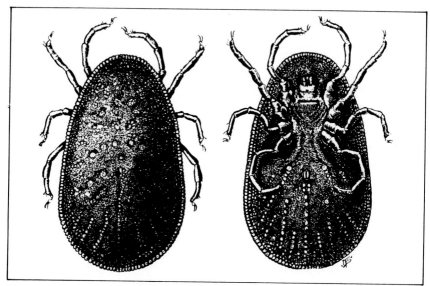

Fig. 23-5. Female fowl tick, *Argas persicus* (Oken).

the cabin, methods of control recommended for the brown dog tick may be utilized here.

Anon. (1978) states in the summer of 1973, 27 employees and 35 overnight guests at the North Rim area of Grand Canyon National Park became ill with relapsing fever. It was believed they were bitten by *O. hermsi* ticks which were collected subsequently from rodent nesting materials inside the walls and attics of the cabins in which they stayed.

THE FOWL TICK

Argas persicus (Oken) (Family Argasidae)

This cosmopolitan fowl tick at times attacks man, although this is apparently an uncommon occurrence in the United States. Cuirea and Stephanescou (1929) record the fowl tick in bedrooms in the upper floor of a new concrete building, which had no known contact with either fowls or pigeons. The inhabitants were bitten nightly and the source of infestation was not found.

The fowl tick, *Argas persicus* (Oken), also is known as the chicken tick, abode tick and "blue bug," the latter referring to the appearance of the female when engorged. This tick may injure and kill chickens, its favorite fowl host. The fowl tick has a leathery body with minute granulations and small disks. The unengorged adult females are 8.5 mm in length, the males 6.5 mm. The ticks, when turgid with blood, are red to dark blue in color. This species is most common in the South and Southwest.

Life history. According to Bishopp (1919), the female deposits brownish eggs in and around the poultry house. The eggs hatch in 10 to 90 days, depending on the temperature and other conditions. The larvae or seed ticks have six legs. As a rule the fowl tick is nocturnal, but it may be seen crawling about during the day when the sky is overcast. Upon attaching itself to a chicken it becomes engorged in three to 10 days and then crawls away from the fowl. It then hides in cracks in the hen house, under boards and behind the loose bark of trees. On occasion, it may migrate long distances and infest nearby homes. At this point the tick is two to three mm long and either dark blue or purple. It molts four to nine days after leaving the host, in the meantime adding two more legs. The tick at this stage feeds at night and secretes itself during the day. After three more molts the adult stage is reached. The female feeds once again and then commences to oviposit. She may lay from 500 to 700 eggs in a number of batches. Under optimum conditions the life cycle may be as short as 30 days. There is but one generation a year. It is interesting to note that Abdussalam and Sarwar (1953) found this tick to infest trees used by vultures and herons in Pakistan. These trees were not used by domestic birds.

Rodriguez and Riehl (1959) controlled fowl ticks in turkey roosts with water sprays containing the indicated percentages of one of the following insecticides: 0.5 percent diazinon, one percent malathion, two percent chlorobenzilate or two percent dicofol.

THE PIGEON TICK

Argas reflexus (F.)

This tick may readily be separated from *Argas persicus* (Oken) by the fact it has the thin margin around its body flexed upwards. Striations are seen on the margins (*A. persicus* has rectangular markings). Brighenti (1935) reports that in Bologna, Italy, this tick infested furniture and beds in a house, biting the occupants. The source of infestation was the attic of a house some 20 feet/6.1 m away that had once harbored poultry and pigeons. The adults hide in cracks and feed at night, and may live two to three years without food. Kemper and Reichmuth (1941) note this tick does not bite man as readily as it bites pigeons.

Control. If this tick occurs in the home it may be controlled by the same chemicals and means recommended for the brown dog tick. Insecticide treatments should be concentrated in cracks and crevices. These openings should later be sealed.

Kraemer (1960) recommends a one percent carbaryl emulsion as a spray for premises, and 0.5 percent carbaryl emulsion on the birds at the rate of one gallon/3.8 l per 100 birds.

WOOD TICKS

The so-called "wood ticks", which occur in woods and fields such as the Rocky Mountain spotted fever tick, *Dermacentor andersoni* Stiles, American dog tick, *Dermacentor variabilis* (Say), lone star tick, *Amblyomma americanum* (L.), and black legged tick, *Ixodes scapularis* Say, are not household pests, but occur often in the immediate vicinity of home (Collins et al., 1949). These ticks are annoying pests and vectors of disease. They commonly situate themselves along pathways where hikers are likely to be attacked.

Ixodes scapularis is common to the southeastern coast of the United States. *Dermacentor andersoni* is found in many of the western states in the United States. *Dermacentor variabilis* is distributed east of the Rocky Mountains and also in a few far western states (California, Idaho, Washington). Both these latter two ticks are important vectors of Rocky Mountain spotted fever in their respective geographic areas. *Amblyomma americanum* is also a vector of spot-

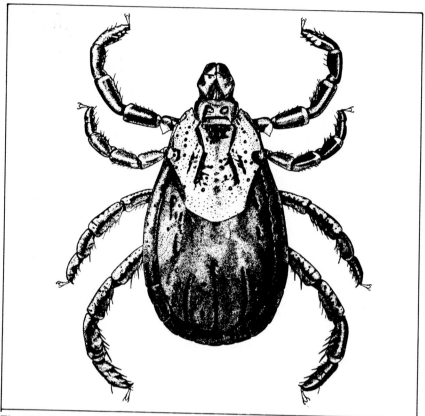

Fig. 23-5. Dorsal view of the American dog tick.

ted fever and tularemia and is easily recognized. Their palps are long and females have a white spot at the back edge of the scutum. This spot looks as if it
were painted on. This tick, once southern in distribution, has extended its range
northward, and is now found in Missouri and eastward to the coast.

The four ticks mentioned above are transmitters of a disease called tick paralysis. There is no pathogen involved. The causal agent is believed to be a
neurotoxin. This toxin is found in the ovaries of ticks as they are feeding. The
toxin then travels to the salivary glands and from there to the victim on which
the tick is attached. The toxin enters the victim as long as the tick is engorging,
which can be six to seven days. Symptoms are weakness and a progressive paralysis of the legs upwards. If the tick is not discovered and removed death can
result. A favorite feeding place of ticks on human beings is at the back of the
head, at the base of the skull. Long hair hides the tick. If you are in tick country,
it is a good idea to examine your body at least once a day for ticks. Anon. (1979)
reports on the first case of tick paralysis from the Canal Zone. The tick concerned was *Amblyomma ovale,* a new transmitter for this disease.

Control. Some protection from ticks can be achieved by the use of repellents.
Cole and Smith (1949) demonstrated the effectiveness of Indalone as a clothing

repellent for the lone star tick. Granett and French (1950) observed dibutyl adipate or n-hexyl mandelate to be an effective clothing repellent to the American dog tick when used as a solution or emulsion at the rate of two grams per square foot.

Smith et al. (1954) recommended formula M-1960 as a clothing repellent:

N-butyl-acetanilide	30%
2-butyl-2-ethyl,-1,3-propanediol	30%
benzyl benzoate	30%
emulsifier (Tween 80)	10%
Total	100%

See chapter 20 on mosquitoes for a more complete discussion on the use of repellents.

Some control will be achieved by keeping grass and weeds cut. Tall vegetation provides ticks with resting sites and protects them from desiccation. Area control of wood ticks can be accomplished using the same insecticides given for the brown dog tick. However, control efforts inside the home are not necessary.

In addition, Busvine (1980) cites the works of Hair and Howell (1970) and Dimitriev (1978) who claim that tetrachlorvinphos and fenitrothion, respectively, shows promise in area control of ticks.

LITERATURE

ABDUSSALAM, M & M.M. SARWAR — 1953. Trees as habitats of the fowl tick, *Argas persicus* (Oken). Bull. Entomol. Res. 44(3):419-420.

ANON. — 1946. Lovely surprise: about Pomerantz and ticks. Pests 14(1):26. 1961. Brown dog tick. Coop. Econ. Ins. Rpts. 11(7):70. 1978. Ticks of public health importance and their control. U.S. Dept. Hlth. Educ. & Welf. CDC Publ. No. 78-8142:47 pp. 1979. Morbidity & Mortality Weekly Report. CDC. Sept. 14 28(36):428, 433.

BAKER, G.E. — 1947. This year beware of ticks. Hygeia 25:473-475-476.

BARNETT, S.F. & B.T. PARSONS — 1963. The control of ticks on fodder and bedding using methyl bromide. Vet. Record 75(46):1213-1215.

BISHOPP, F.C. — 1919. The fowl tick and how premises may be freed from it. USDA Farmers' Bull. 1070. 1933. Ticks and the role they play in the transmission of diseases. Smithsonian Inst. Ann. Rpt. 1933 pp. 389-406, 9 plates. 1939. The brown dog tick, with suggestions for its control. USDA Bur. Entomol. & P.Q. 4 p. 1942. The brown dog tick, an important house pest. Pests 10:6-8, Sept. 1950. Ticks. Pest Control Technology. Nat'l. Pest Control Assn.

BISHOPP, F.C., C.N. SMITH & H. K. GOUCK — 1946. The brown dog tick, *Rhipicephalus sanguineus* with suggestions for its control. U.S. Bur. Entomol. & P.Q. E-292. rev.

BRIGHENTI, D. — 1935. L'*Argas reflexus* Fab. come parassita dell'uomo. Boll. Zool. 6:219-225.

BUSVINE, J.R. — 1980. Insects and Hygiene. Chapman & Hall, London. 568 pp.

COLE, M.M. & C.N. SMITH — 1949. Tick repellent investigation at Bull's Island, S.C., 1948. J. Econ. Entomol. 42(6):880-883.

COLLINS, D.L., R.V. NARDY & R.D. GLASGOW — 1949. Some host relationships of Long Island ticks. J. Econ. Entomol. 42(1):110-112.

CUIREA, I & T. STEPHANESCOU — 1929. *Argas persicus* comme parasite de l'habitation humaine. Bull. Soc. Roum. Dermat. Syph. no. 2.

DMITRIEV, G.A. — 1978. Control of Ticks. Internat'l. Pest Control 20(5):10.

EBELING, W. — 1975. Urban Entomology. Univ. Calif. Press, Berkeley. 695 pp.

GRANETT, P. & C.F. FRENCH — 1950. Field tests of clothing treated to repel American dog ticks. J. Econ. Entomol. 43(1):41-44.

GROSS, J.R. — 1949. Control of dog ticks. Pests 17(1):26.

HAIR. J.A. & D.E. HOWELL — 1970. Lone Star Ticks: their biology & control in Ozark recreational areas. Okla, Agr. Exp. Sta. Bull. B-679:1-47.

HAZELTINE, W — 1959. Chemical resistance of the brown dog tick. J. Econ. Entomol. 52(2):332-333.

HEADLEE, T.J. — 1950. The New Jersey tick problem. N.J. Agr. Exp. Sta. Circ. 395.

HERMS. W.B. — 1939. Medical Entomology. Macmillan Co., London.

HERMS, W.B. & C.M. WHEELER — 1935. Tick transmission of California relapsing fever. J. Econ. Entomol. 28:846-855.

HUFF, C.G. — 1931. A proposed classification of disease transmission by arthropods. Science 74:456-457.

JAMES, M.T. & R.F. HARWOOD — 1969. Herms's Medical Entonomogy. Macmillan Co., London. 484 pp.

KEMPER, H. & W. REICHMUTH — 1941. Biol. Abst. 20:654, 1946.

KNIGHT, K.L., D.E. BRYAN & C.W. TAYLOR — 1962. Studies on the removal of embedded lone star ticks, *Amblyomma americanum*. J. Econ. Entomol. 55(3):273-276.

KOHLS, G.M. & R.R. PARKER — 1948. Occurrence of the brown dog tick in the western states. J. Econ. Entomol. 41(1):102.

KRAEMER, P. — 1960. Relative efficacy of several materials for control of poultry ectoparasites. J. Econ. Entomol. 52(6):1195-1199.

MOORE, D. — 1950. Laboratory studies of combinations of piperonyl cyclonene, piperonyl butoxide, pyrethrins, and rotenone for the control of ticks *(Dermacentor variabilis; Rhipicephalus sanguineus)* on dogs. J. Parasitol. 36:322-325.

NELSON, V.A. — 1968. The brown dog tick in the United States. Melsheimer Entomol. Series No. 2:1-10. 1969. Human parasitism by the brown dog tick. J. Econ. Entomol. 62(3):710-712.

PIERCE, W.D. — 1975. The Deadly Triangle. John G. Shanafelt, Jr. Orange, Calif. 138 pp.

RICE, P.L. & H.D. PRATT — 1972. Epidemiology and control of vector borne diseases. U.S. Dept. Hlth. Educ. & Welf. Publ. No. (CDC) 76-8245. Publ. Hlth. Svce, Center for Disease Control, Atlanta, Georga. 52 pp.

RODRIGUEZ, J.L. & L.A. RIEHL — 1959. Fowl tick on turkeys. Calif. Agr. 13(11):11.

SMITH, K.G.V. — 1973. Insects and other arthropods of medical importance. British Museum Natural Hist., London. 561 pp., 12 plates.

SMITH, C.N., M.M. COLE, I.H. GILBERT & H.K. GOUCK — 1954. Field tests with tick repellents — 1949, 1950 and 1952. J. Econ. Entomol. 47(1):13-19.

SMITH, C.N. & H.K. GOUCK — 1946. Observations on tick repellents. J. Econ. Entomol. 39(3):375-378.

SMITH, C.N., M. MOSES & H.K. GOUCK — 1946. Biology and control of the

American dog tick. USDA Tech. Bull. 905:1-74.

SNETSINGER, R.J., E.J. WITTE & W. WILLS — 1965. Rocky Mountain spotted fever found annually in Pennsylvania. Science for the Farmer (Penn. State Univ.) 3:4-5.

TARSHIS, I.B. & M.R. DUNN — 1959. Control of the brown dog tick. Calif. Agr. 13(10):11.

THURMAN, D.C. & J.A. MULRENNAN — 1947. Occurrence of the brown dog tick on Florida rats. J. Econ. Entomol. 40(4):566-567.

WHEELER, C.M. — 1935. A new species of tick which is a vector of relapsing fever in California. Amer. J. Trop. Med. 15:435-438.

REX E. MARSH
WALTER E. HOWARD

Rex Marsh was born in South Dakota in 1929 and grew up on farms in South Dakota and California. He attended San Jose State University, majoring in biological science with a minor in physical science.

After being awarded a degree in 1953, he spent two years in the United States Army Medical Corps during the Korean conflict. He joined the staff of the Santa Clara County Department of Agriculture in 1955 and later the California Department of Agriculture, where for six years he was district supervisor for weed and vertebrate pest control. With both a knowledge of field problems and a keen interest in research, he accepted an academic position with the University of California Agricultural Experiment Station some 17 years ago and advanced to specialist in vertebrate ecology. He presently also holds a lecturer's title in wildlife and fisheries biology and is a certified wildlife manager. Vertebrate pest management has been his research area with emphasis on rodent problems of public health, agriculture, forestry and stored food commodities.

Marsh is a member of a number of organizations, including the American Society of Mammalogists, American Association for Laboratory Animal Science, The Wildlife Society, American Society for Testing and Materials, American Ornithologists' Union, Society of the Sigma Xi, and Pi Chi Omega.

Dr. Walter Howard graduated in zoology at Berkeley and received his master's and Ph.D. degrees in vertebrate ecology at the University of Michigan. He has been on the faculty of the University of California, Davis, since 1947, and currently teaches courses in wildlife ecology, principles of vertebrate control, and population problems: issues in human ecology. With assistance from Rex E. Marsh, he has many graduate students who receive their master's and/or doctorates in ecology, range management or international agricultural development.

Howard's main field of interest is what happens to vertebrate populations when man modifies ecosystems, and what is the best way of coping with those species that become pests.

Howard is a frequent overseas consultant concerning animal damage problems. He has had 12 United Nations assignments in Lebanon, Qatar, Mexico, Argentina, Bahrein, South Korea, Egypt, Malaysia and Barbados. At other times he has been a consultant in New Zealand, Malaysia, India, Taiwan (R.O.C.) and China (P.R.C.) and has spent sabbatical leaves in New Zealand, Australia, Malaysia and Taiwan.

CHAPTER TWENTY-FOUR

Vertebrate Pests

By Rex E. Marsh[1] and Walter E. Howard[2]

BIOLOGY. BATS traditionally have been associated with witchcraft, haunted houses and cemeteries, at least in Europe and North America. Those who are superstitious consider them envoys of the devil. They have been the subject of myths, fables and folklore for centuries, inevitably depicted as evil or associated with evil or the mysterious unknown. The blood-feeding habits of the vampire bat of Central and South America have contributed to many exaggerated tales on the American continent. In China, in contrast, bats were symbols of rare luck and good times; and many American Indians considered them deities (Scott, 1961a).

The fear of bats shared by a large segment of the public has possibly saved a number of lives, since rabies has been found in bats. This instilled fear of bats keeps many individuals from handling infected bats, and being bitten in consequence.

Bats, the only true flying mammals, belong to the order Chiroptera. They are distributed nearly worldwide and are among the most numerous of land vertebrates (Scott, 1961a). Bats are found throughout the United States, although certain species are more regional than others. Blood-eating (vampire) bats thrive in Central and South America but are not established in the United States.

[1]*Specialist in Vertebrate Ecology, Division of Wildlife & Fisheries Biology, University of California, Davis, Calif.*
[2]*Professor of Wildlife Biology & Vertebrate Ecologist Division of Wildlife & Fisheries Biology, University of California, Davis, Calif.*

Fig. 24-1. Scores of bats in an attic, hanging by their hind legs with their heads pointing downward.

The "wings" of bats are formed by thin membranes that extend from the greatly elongated forearm and fingers to the hind limb and body. Their bodies are furred and their eyes are small. Some are the size of a house mouse when the wings are not spread, whereas others are substantially larger. Color and head characteristics vary greatly. Examined closely, however, they are unlikely to be mistaken for any other animal.

Bats navigate by means of a sonar-like echo location system which enables them to avoid solid objects even in total darkness. That accounts for the rather erratic flight pattern. Most bats found in this country are nocturnal in habit, feeding principally on insects.

The majority of bats produce a single offspring; a few, however, have two to four young. Females of many species gather in nursery colonies to give birth to young, generally in early or midsummer. Although the birth rate is low, this is compensated for by a relatively long life span, 20 years in some species.

Some bats migrate seasonally in a true north-south direction. Others migrate more or less locally, moving their roosts seasonally, presumably for better satisfaction of their biological needs. Certain species may occupy a roost the year around and some may hibernate in winter.

Bats have adapted readily to man-made structures, with literally thousands of structures being suitable as shelters. Thus, populations of some species of bats are larger than they were under strictly natural conditions (Murray, 1968). Most colonies of bats go unnoticed or are of no particular concern to people, whereas others occasionally cause a nuisance in dwellings and other buildings.

People usually object to noise from their vocalization or activity, to odors and stains from urine and fecal material and to rejected food particles that frequently accumulate below roosting sites.

In food-processing plants and warehouses, human food may become contaminated by droppings from bats that inhabit buildings or from insects and arthropods that live in the guano deposits. Droppings are considered a food contamination by consumers, FDA and USDA. Thus, corrective measures must be taken.

Rabies is a matter of considerable concern when infected bats inhabit areas close to humans. Another recognized factor is the psychological unease felt by humans from knowing a house or place of work is infested with bats. Many people fear bats and just don't want them around, regardless of whether they are creating a disease or noise problem.

When bats roost in homes, particularly in attics, walls, chimneys or hollow floors, their droppings and urine may cause persistent stench that is insufferable to many humans. Moisture from their urine and droppings can stain the plaster in the ceiling and walls, and in some situations the only remedy is to remove and replace the stained portion. If large colonies have roosted in a building for extended periods, guano removal can be expensive, sometimes requiring the siding be removed from the building.

An excellent coverage of bat problems and solutions is provided in the National Pest Control Association (NPCA) Technical Release Number 5-75 (1975). It provides some information not included in this article and hence is a useful reference.

Aside from direct observation, other signs indicate the presence of bats, such as accumulated droppings below roosting sites. Bat droppings may be mistaken at first for mouse droppings, but the former are more irregular in shape and crumble readily. The odor and stains caused by urine and droppings may be clues, as are high-pitched squeaks and rustling sounds coming from a wall or attic. Some discoloration from body secretions is commonly found around the edges of small openings where bats enter a building.

Droppings and tell-tale insect parts often will occur in the absence of any other indication of bats. They indicate temporary roosting sites between nighttime foraging periods. These sites are likely to be less confined (porches, loading docks, etc.) than are daytime or regular roosts.

Bats are implicated in a number of human diseases, although rabies receives the greatest attention in public health. A number of excellent references are available on bat rabies. Bat rabies, first discovered in the United States in 1953, now occurs throughout most of the country, with only a couple of states other than Alaska and Hawaii having no positive reports of the disease in bats (Constantine, 1967). More than half the recognized species are known to contract the disease and the numbers of bats found infected with rabies virus in the United States increased steadily for some time after the initial discovery. Without treatment, rabies is a highly fatal virus disease of man and other mammals. Whenever sick or dead bats are found, the local public health authorities should be notified immediately.

The way in which rabies is maintained and circulated in the bat community is still poorly understood, although transmission from parent to offspring is known to occur, at least in some species. Bat rabies has been the subject of considerable research and much more is needed.

Rabid bats seldom exhibit the furious symptoms rabid carnivores do (i.e., dogs and skunks). Some typical symptoms are a more than usual erratic flight pattern, especially landing frequently in areas or on surfaces where they would not normally be expected. Appearance in the daylight hours, particularly around midday, should be viewed with suspicion. A common rabies symptom in bats is weakness or paralysis, depriving it of its ability to fly or cling to its roost, so it flops over or lies helpless on the ground near its roost. This frequently means infected bats will be found near the base of a building. There are only a few valid accounts of attacks upon humans (Murray, 1968). Most human exposure comes from being bitten when attempting to pick up infected bats. In spite of all efforts to educate the public against picking up apparently dead or sick bats, instances of such bites still occur.

Bats are also implicated in other diseases, such as Chagas' disease, relapsing fever and dermatomycoses (Scott, 1958). Several forms of encephalitis (VEE, EEE, WEE, and SLE) also have been traced to bats (NPCA, 1975), so consideration must be given to the possibility of being bitten by certain insects after they have fed on infected bats.

Histoplasmosis, an often fatal systemic fungus disease of man, may be a significant hazard to people who work at removal of bats and bat guano. The fungus, *Histoplasma capsulatum,* can be contracted by inhaling the airborne spores in the dust of bat manure, which is well suited to supporting the growth of this fungus. The bats themselves do not transmit histoplasmosis. Respirators and other protective clothing should be worn when working in bat roosts. It is important the respirators fit properly and are approved by the National Institute for Occupational Safety and Health (NIOSH) for nuisance dusts.

Anyone bitten or scratched by a bat, regardless of how healthy it appears or how small the wound, should seek medical attention. The wound should be washed with soapy or detergent water and medical attention should be obtained as soon as possible (Anon., 1969). If the bat that caused the bite can be caught or killed (preferably without crushing the skull), it should be kept for a rabies examination by public health authorities.

PCO personnel engaged in bat control or bat research can be immunized against rabies. That will provide a degree of protection, although additional rabies treatment will probably be needed for a bite by a rabid bat.

Bats carry ectoparasites, particularly the bat bug, *Cimex pilosellus,* which closely resembles the common bedbug. These may attack man when bats infest a house (Pratt, 1958). Mites, ticks and fleas, also listed among their parasites, create minor health problems at times (Scott, 1961). Bat guano attracts flies, cockroaches and other coprophagous insects. Whenever bats are controlled by exclusion or some other method, consideration also must be given to the simultaneous control of ectoparasites and other insects which may be left behind.

The habits of bats may determine the type of control method used, so it is advantageous to know the species involved before control is undertaken. For example, it is important to know if the species is highly colonial, forming large colonies, since the control approach could differ from that used on a species which roosts singly or in limited numbers (free-living bats). If the species migrates through an area and rarely stays any length of time, then control may not be warranted at all, particularly if the bats roost out-of-doors under eaves. A species not known to carry and transmit rabies would certainly not be viewed with the same alarm as a species often reported to be rabies-infected.

If you cannot identify the species, collect a specimen (avoid being bitten) for classification by the local health department, college, biology teacher or others with sufficient knowledge of bats. Identification is not enough by itself but permits you to determine its habits, life history, etc., from library references. A PCO should become familiar with the identification and biology of the bat species in his area, particularly those most apt to use buildings as roosting sites.

Control. Since bats are protected in some states and local areas, their control should be preceded by a check of local laws to determine what protection may be afforded them. Local laws or regulations may dictate the method undertaken to resolve a bat problem. For example, batproofing may be permissible and killing forbidden.

Bat control methods have not advanced much throughout the years and exclusion remains the primary means of control.

A colony of bats or a few individuals of a solitary species may be excluded from a building through batproofing. In fact, where feasible, this is the only truly permanent control. Bat colonies are easiest to dislodge from a roosting site soon after they initially take up residence. The longer a bat colony is permitted to exist at one location the more difficult it is to discourage. Additionally, the colony may grow with time. It is generally much easier to exclude bats from small buildings such as houses and churches.

They are most difficult to exclude from some large buildings, such as warehouses and factories with many beams, or from semi-open roosting situations, such as porches or roofs projecting over loading docks. A bat colony displaced from one area of a building may simply move to another section.

Some bat infestations are tolerated for a time, and because the roosts of certain bat species are strictly seasonal, the building manager is relieved when the bats leave, believing the problem has disappeared. It is unwise, however, to assume they will not return since they frequently return year after year to the same site. One of the easiest times in which to batproof a building is after the bats have left for the season.

To exclude most bat species from a building, openings larger than ⅜ inch/9.5 mm must be eliminated to prevent access. Hardware cloth (¼ inch/6.3 mm mesh) or sheet metal are the materials used most often to close entrances, although softer building materials are also useful, such as aluminum flashing, pressboard and plywood. Unlike rodents, bats will not gnaw their way through softer building materials, which are easier to work with and may match the natural texture of the building. These materials can be conveniently attached to the building with a heavy-duty staple gun.

Glass-fiber insulation blown into spaces occupied by bats has been reported to repel them (Scott, 1958). An even more effective way of filling voids within walls which may be occupied by bats is found in newer types of wall insulation which are injected as a foam through holes bored in the wall. Insulating can be done in the evening after the bats have left to feed. If foam or other loose-type insulations are used, be certain any substantial amount of droppings are removed and not just covered over, for objectionable odors may continue.

Quick-setting hard putty can be used for some small openings. Oakum, weather stripping, caulking compound or equivalent materials are effective for closing long narrow cracks. Steel wool or large stainless-steel scouring pads (which do not cause rust stains) are useful for plugging openings in Spanish-type tile roofs.

When batproofing, pay particular attention to chimneys, louvers, vents, cornices, warped siding and locations where the roof joins the sides and in the area of the eaves. Bats also have been known to take up roost sites in some rather unusual places — in one instance, the 27th floor of a 30-story building still under construction.

Species of bats which are crevice-dwellers often can be eliminated by closing up the usable crevices. However, that can be expensive and time consuming, as can eliminating access to large buildings such as warehouses. Where sanitation or public health reasons require that bats be eliminated immediately and bat-proofing is not readily achieved, the most practical solution is destruction of the colony.

While a colony still infests a building, batproofing is tricky and should be completed in two stages: seal all bat entries except one or two of the principal openings, wait for several days, and make the final closures about one-half hour after dark, presumably when all bats are out of the building. The returning bats not dispersed to other areas may cluster or flounder about the plugged ingresses. Such bats should be carefully collected (taking precautions to avoid bites) and disposed of. Bats will bite if picked up, and since they may be carrying rabies, they should always be handled with tongs or thick gloves. The building should be watched for several evenings at dusk to be sure the bats have not found an entry which may have been overlooked.

Depending on species and time of year (particularly early to midsummer), there is always the possibility young may be present in the colony and will be sealed in by batproofing. Such carcasses may create a foul odor problem.

Since bats may be attracted to a roosting site by odors left by a previous colony, destruction of a colony with toxicants may give relief only until a different group of bats reoccupies the building. Hence, batproofing measures are desirable following removal by other means. Some products designed for masking or neutralizing attracting pheromones (animal communication odors) may be useful for treating former roosting sites. They also may help mask the objectionable odor that remains from bat urine and guano even after the guano has been removed.

Bats are, to a degree, repelled by the odor of naphthalene or paradichlorobenzene, used in the same manner as discussed for chipmunks. Three to five pounds should be adequate to treat an average attic, although five to 10 pounds have been recommended by some (NPCA, 1964). These materials are often ineffective, but are worth a try as a first attempt to solve a problem. They are apparently more effective in relatively confined air spaces (attics, between walls) and are much less effective in more open situations such as louvers, porches and beneath eaves.

Sticky-type bird repellents have been used with success in a few situations where roosting surfaces can be coated. These have greater potential at sites just newly occupied and at temporary resting sites. Sticky repellents, at best, may offer some help where batproofing is impractical. Application may have to be repeated since dust can cause them to lose their tackiness.

Most people are squeamish about entering a bat-infested attic. In such cases, a person should cover himself completely with a beekeeper's net over a hard hat, along with coveralls and leather gloves.

DDT, being highly toxic to bats, was used with considerable success before its use was severely restricted. Fifty percent wettable powder was used, either as a dust or diluted with water. The material was applied to the surface of entry holes but preferably over the actual roosting area. The dust could be blown into hollow walls occupied by the bats. Bats ingested lethal amounts by grooming the DDT from their fur.

The loss of DDT has greatly hampered effective control of nuisance bats. Where bats in certain situations are designated a threat to public health, however, the use of DDT is still legally possible. According to NPCA Technical Release No. S-75, 1975, "After the presence of rabies has been established, State Public Health officials can apply to the Environmental Protection Agency for permission to use DDT. Section 18 of the Federal Insecticide, Fungicide and Rodenticide Act, as amended, provides for exemption to the DDT ban when a significant public health problem exists. Applications of DDT must be made under the direction or supervision of the state health agency." Obtaining permission to use DDT in some states involves an unbelievable amount of bureaucratic red tape even when only a few pounds of DDT are involved. The emergency procedure can hardly be considered an efficient workable system and is extraordinarily costly.

Bats also are very susceptible to other chlorinated hydrocarbons, such as chlordane, dieldrin and endrin, but such use was never common practice.

Bats have been controlled with the acute rodenticide, ANTU (Mallis, 1969). For control, the openings used by the bats are dusted with a bulb duster containing ANTU tracking powder. ANTU is registered for control of Norway rats, but not for bats. ANTU reportedly killed or drove bats away from a building for a period of three months.

Bats have been found susceptible to anticoagulant rodenticides when applied as tracking powders. Chlorophacinone in tracking powder form is the anticoagulant most often used for bat control and is presently registered for that purpose in a number of states. Other than DDT, for use under special conditions where a serious potential health problem exists, chlorophacinone is the only toxicant currently registered for bats. The chlorophacinone tracking powder contains the same percentage of active ingredient as used for the control of rats and house mice. Just as with rats and mice, the effectiveness of the anticoagulant relies on the contamination of the bats' fur and subsequent ingestion while grooming themselves or one another. The tracking powder is strategically applied according to label directions so that the bats will pick up a lethal dose. The effect of anticoagulants on the bats will be slow; results may take several weeks.

It has been suggested control with either DDT or chlorophacinone may, in fact, create a more serious potential health hazard than no control. Bats sick or dying of the toxicant, if not confined to the structure, may drop somewhere outside the roosts, where greater human contact is possible. Pets and carnivorous predators also could be exposed if they approach a dying bat which may also be rabies infected. However, there is insufficient evidence to support this contention. To do nothing about a bat infestation in populated areas may also permit rabid bats to die outside the roost and, although fewer in number, such exposure could be of a much longer duration. Even batproofing, which is so well acclaimed as the ultimate solution to bat problems, causes the dispersal of many bats which may be rabid and thus could potentially increase the health hazard.

Calcium cyanide fumigant is effective and has been used. For example, Hockenyos (1959) suggested bats in chimneys could be killed by sealing the lower opening of the chimney with wet paper and then dropping two or three tablespoons of calcium cyanide (Cyanogas A) dust down from the top. The top of the chimney is then tightly sealed off. It is important only people experienced in handling calcium cyanide attempt this. The dwelling should be vacated while such fumigation is going on. Once the bats are dead, the remaining cyanide and carcasses should be removed and disposed of. The chimney should be screened at the top to prevent reinfestation. Gases that are heavier than air, such as methyl bromide or chloropicrin, are not satisfactory (Schnurrenberger et al., 1964) because they are difficult to hold at lethal concentrations in the upper portions of buildings. When a building is tarped and being fumigated with methyl bromide for termites or other pests, however, such fumigation also will kill bats. Fumigants should always be left to professionals experienced in their proper handling.

Floodlights strung through an infested attic to illuminate all batroosting sites may cause the bats to leave and seek a new location. Laidlaw and Fenton (1971) reported the population of a bat nursery was substantially reduced by artificial lighting. This method is believed most effective if done shortly after migratory bats arrive. In some situations it is impossible to light all possible roosting locations adequately, such as in an attic, so the technique is less effective in some types of structures than in others.

High frequency sound has been advocated by some to repel bats from premises, however, there is insufficient data at present to demonstrate efficacy.

Traps have generally been considered impractical for control of bats, but it may be time we reconsidered them, along with hand removal of bats. Trapping techniques used by bat banders have become more efficient and the traps more portable. For example, the Constantine trap (1958) and modifications thereof, or the Davis hopper trap (1962), may be far more valuable in controlling bats then imagined, particularly since loss of the most effective chemical control, DDT. Even though biologically sound, trapping and hand collection are expensive approaches to resolving a bat problem.

Trapping is mentioned here as just one more method which might be used more extensively in bat control, especially when so few management techniques are available. Admittedly, trapping or hand removal of bats in any number will require a knowledge of bats and their behavior, the development of new skills and some additional investment in equipment not now needed to control bats by more traditional methods. It also may increase exposure of the trapper to rabid bats. A valuable publication for trapping bats or for routinely removing a few live bats from a building by hand-collecting is U.S. Fish and Wildlife Service Resource Publication No. 72, *"Bats and Bat Banding,"* by Greenhall and Paradiso (1968). It thoroughly covers safety, equipment and proper collection techniques.

Capturing bats at a roost with the intent of releasing them elsewhere may not only risk the spread of disease to other areas, but may be ineffective. Certain species can return to their home roosts from distances as great as 30 to 75 miles/ 48 to 120 km (Wilson and Findley, 1971; Gunier and Elder, 1971).

CHIPMUNKS

Class Mammalia
Order Rodentia
Family Sciuridae

Biology. Among the smaller members of the squirrel family (Sciuridae) are chipmunks. Two genera are recognized in the United States: The eastern chipmunk *(Tamias)* is most common in the northeastern states, from Canada westward through Michigan, and southward nearly to Florida; the western chipmunk *(Eutamias* spp.) is found from Michigan westward to the Pacific, but is not common in the Rocky and Sierra Nevada mountains. Eastern and western chipmunks overlap only in a small area around the Great Lakes. Chipmunks are found in most of the 48 contiguous states, but are rare or absent in Texas and Florida. Their distribution is relatively limited in a number of other states.

The two species closely resemble each other. The eastern type averages nine to 10½ inches/22.8 to 26.6 cm long and weighs about 2.8 to 3.1 oz/80 to 90 g; its western counterpart averages eight to 9½ inches/20.3 to 24.1 cm long and weighs 1.4 to 2.4 oz/40 to 70 g. The eastern species has three dark dorsal stripes, while the western one has five. The various species of *Eutamias* differ greatly in size. Several ground squirrels also have striped backs, but lack the definable stripes that chipmunks have running along the head.

The eastern chipmunk goes into hibernation in fall and reappears in early spring. Hibernation of the western species is not well defined. It apparently occurs in some, while in others it may not, even though periods of animal inactivity in the winter are fairly common. All species breed during spring and summer. Following a gestation period of about 29 to 31 days, litters of two to seven young are produced. The eastern chipmunk *(Tamias striatus)* often produces two litters a year, but most western species with the possible exception of the least chipmunk *(Eutamias minimus)* apparently produce only one litter annually.

Chipmunks are basically ground-dwellers, but often climb trees and other objects for food and protection. Where chipmunks are found around residences and mountain or summer cabins, it is not uncommon to find them sitting and running on porch railings, fences and rooftops. Rain-gutter downspouts may be

Fig. 24-2. The eastern chipmunk (above) is most common in the northeastern United States, from Canada westward through Michigan, and southward nearly to Florida.

used as travel routes to the rooftop, although overhanging trees are used most often. They tame quite easily and often rely on handouts provided them. They seem to accept man quite readily as a neighbor, but this friendliness and their increasing numbers are what often contribute to their status as a pest.

Chipmunks are active by day and prefer open wooded areas that have plenty of plants and trees that are food-producing. The main food is berries, fruits, nuts and small seeds, which are collected after they have fallen to the ground, or harvested by climbing trees and shrubs to pick them. Fungi, grass and leaves are also eaten. Chipmunks are sometimes carnivorous, taking slugs, snails, aphids and other insects. Small birds, eggs, mice and small snakes are also eaten.

Food that is not immediately needed is carried in the cheek pouches and cached for use in winter between periods of torpor. Chipmunks are probably the most expert hoarders of all the squirrel family, but they are selective about the things they hoard. When a chipmunk is collecting its winter store, it selects only nuts and cones, never any fruit or flesh that would go bad. For the sheer bulk of its stores chipmunks take first prize among hoarders. Reports have been made of caches containing eight quarts of acorns or a bushel of nuts. Moreover, a chipmunk may have several caches. Coniferous forests are highly preferred. Around homes and landscaped buildings they may burrow in lawns and flower beds. They often dig up and eat newly planted garden seeds and flower bulbs. Strawberries are a favorite, as are certain other cultivated fruits. Plums and apples are often taken by the eastern chipmunks, primarily for the seeds or stones.

In wooded areas, chipmunks generally nest in burrows which are quite well hidden beneath fallen logs, brush piles or vegetation. Each burrow is owned by one chipmunk who continues digging throughout its life, so the burrows may reach lengths of 30 feet/nine m or more, and have more than one entrance and perhaps several side chambers, one of which probably contains a nest of leaves and grass. Around human habitats they will nest beneath cabins or homes in burrows or other suitable locations (including basements if they can gain access). They may use outbuildings such as woodsheds, detached garages and tool-storage buildings, which are not usually rodent-proof. Rarely do they take up residence in attics or other spots which are a distance above ground. In almost all cases, chipmunks leave the building to forage and feed.

With the recent increase in the occurrence of plague in rodents and man in parts of the West, increased emphasis has been placed on chipmunk (*Eutamias* spp.) and ground squirrel control in campgrounds and other recreation areas that cater to large numbers of people. Plague is a severe disease, often fatal, caused by the bacterium *Yersina (Pasteurella) pestis.* The vectors are infected animals and their fleas. Chipmunk fleas in the West readily bite humans (Weinburgh, 1974). Where the disease is prevalent (endemic), home or property owners should be aware of the health hazards. Control of certain rodent species and their fleas is fundamental in preventing plague from spreading to humans. Also, there is always a possibility the infected wild rodents will have close enough contact with cats, dogs and suburban and urban-dwelling commensal rodents (i.e., Norway and roof rats) to infect them, thereby creating a still more serious threat to humans. The western chipmunk (*Eutamias* spp.) also is involved in transmitting relapsing fever and Rocky Mountain spotted fever, while the eastern chipmunk may be implicated in transmitting the Powassan virus (Weinburgh, 1964).

Control. Nuisance chipmunks should be built out wherever possible. Exclusion techniques should parallel those used for rat and mouse control. Essentially this involves the closure of all openings which might allow access to the building. Chipmunks and tree squirrels have been known to gain entry via chimneys.

Chipmunks and tree squirrels are said to be repelled from attics and unused cabins by paradichlorobenzene or naphthalene flakes. Generally four to five lb/ 1.8 to 2.2 kg of mothballs per 2,000 sq ft/185.8 sq m should be sufficient to drive the animal from an attic or basement, though this repellent technique is not foolproof. An animal with young in a nest in the building is unlikely to be repelled easily. In hot climates these repellents volatilize rapidly and have to be replenished often. The odor of these materials may be objectionable to some humans and thus are best placed on shallow trays for easy removal, or hung from rafters in coarse-mesh cloth sacks out of the reach of children and pets.

Where reductional chipmunk control is legal, and this should be checked out with the local fish and game official, trapping is quite effective in removing them from ground buildings. Live-catch wire-mesh traps such as Tomahawk (No. 102) or Havahart (No. 1) traps are very effective. Traps should be set in areas along pathways commonly used by the animals. Attractive baits are nutmeats, pumpkin or sunflower seeds, raisins, prunes and various grains. For maximum trapping efficacy, traps should be prebaited with the doors propped open for a few days, then the traps which have attracted the animals should be rebaited and set. This will make trapping a more decisive action.

Leg-hold traps are seldom any more effective than well-placed live-catch traps. Number 0 is the trap size generally preferred in leg-hold traps. No. 1 or 1½ is not too large to catch chipmunks and will close around the animal's head and body causing instant death. Leg-hold traps can be set in areas frequented by the animals and baited with seeds or nuts. Nutmeats, prunes and other large bait should be securely tied to the trap trigger with thread. Rolled oats may be sprinkled around the trigger as an additional attractant. Sometimes leg-hold traps are secured to ledges or on top of tree limbs to increase efficiency.

Ordinary wooden-base rat snap traps are very effective for taking chipmunks from attics, basements, rooftops and tree limbs where live traps may be difficult to place. These kill the animal almost instantly. Care should be taken to avoid setting rat traps or leg-hold traps where small dogs or kittens may be injured. If that can't be assured, then use live-catch traps.

Live-catch and leg-hold traps should be checked frequently (daily at the minimum) so non-target species accidentally trapped can be released and chipmunks humanely disposed of. Where the animals are known to be possible disease carriers, never handle them alive or dead with bare hands. Use tongs, rubber gloves or an inverted plastic bag to remove dead animals from traps. Control of ectoparasites (fleas, ticks, etc.) may be advisable prior to or simultaneously with chipmunk control.

Because chipmunks in some areas carry serious diseases, it is inadvisable to move and release live-trapped animals elsewhere. Releasing trapped animals may contribute to the spread of a disease harmful to humans as well as to chipmunks. Sacrificing a few animals may be the most humane way in the long run. In some areas it is actually illegal to move and release wild rodents into new areas.

If the burrows can be located, and often they can't with chipmunks, they can be fumigated with the same materials used for woodchucks. The burrow fumigant used most often for chipmunks, however, seems to be calcium cyanide dust. Since chipmunks are active and outside their burrows in the daytime, it is best to apply calcium cyanide after sunset, at dusk, to burrows previously located and marked. One tablespoon of calcium cyanide dust should be inserted well into the burrow and the opening then firmly closed with sod or soil to contain the gas. In such closing, make sure the calcium cyanide dust is not covered with soil since it requires contact with air to release the highly toxic hydrocyanic acid gas. Handle fumigants with caution and never use them beneath occupied buildings.

Where chipmunks are numerous, toxic bait may be the most effective and practical method of control. Strychnine baits (0.5%) have been the acute toxicant used in the past, but in more recent years oat groat or crushed oat bait treated with one percent zinc phosphide has been recommended. Baits should be placed deep in holes (tablespoon amounts) or in protected bait stations. Chipmunks are moderately susceptible to anticoagulant cereal baits designed for rat and house mouse control. Such anticoagulant baits should be exposed in suitable bait stations until feeding has ceased or until the chipmunks have been reduced to an acceptable or tolerable level.

The small size and rapid movements make chipmunks difficult to shoot. If shooting is attempted, a bird-shot cartridge fired from a .22-caliber rifle or a small-gauge shotgun may be the most appropriate. Air rifles can also be effective at close range. The morning feeding activity period is generally the best time for shooting chipmunks.

WOODCHUCKS

Class Mammalia
Order Rodentia
Family Sciuridae

Biology. There are six species of woodchucks (also called groundhogs) and marmots of the genus *Marmota* in North America. These are the largest animals of the squirrel family. "Woodchuck" and "groundhog" are common names for the species *Marmota monax,* which lives mainly in the eastern part of North America but extends into the Northwest, whereas the other five species, all in the West, have the common name "marmot." Their biology is essentially the same, although their preferred habitats and foods differ somewhat.

Woodchucks and marmots are found in open and closed forests and brushy areas. In agricultural districts they live along creeks, brush ravines, woodlot edges and pastures (Weinburg, 1964). Somé of the western species are limited to the higher elevations and, in some instances, more specialized habitats. Areas of both the southeastern coastal and Gulf States are free of these rodents, as are portions of the Southwest lowlands, including California. They are also absent in some midwestern states along the eastern slope of the Rocky Mountains.

The woodchuck is a stocky animals, weighing four to 10 lb/1.8 to 4.5 kg, with a short, stubby tail four to six inches/10.1 to 15.2 cm long. It has relatively small ears and short legs for its body size. Its short legs and chunky body give it an appearance of squatting or crouching as it moves about. Since it is a hibernating rodent, it accumulates considerable fat before winter hibernation. In the West, the yellow-bellied marmot, *M. flaviventris,* has been known to be infected with plague and in certain situations may be considered a threat to public health.

Woodchucks live in burrows, like ground squirrels and are most active in early morning and late afternoon, although at times they may lie at the burrow entrance on sunny days. Although somewhat more gregarious than wood-chucks, marmots seldom occur in large colonies and hence rarely reach very high densities. Mating takes place in the spring, with a single litter of about four to five produced each year.

Woodchucks and marmots eat a wide variety of green vegetation, from pasture forage to vegetables, and in some areas are considered agricultural pests. They occasionally invade backyard gardens, eating or destroying vegetables or succulent landscape plants. Like some ground squirrels, woodchucks are fond of tomatoes and may carry off or eat an entire backyard tomato patch. Less frequently they may damage young fruit trees by eating or clawing the bark. Woodchucks have been known to dig beneath houses and other buildings. In golf courses and the like, the burrows or the mounds of soil pushed out of burrows create a nuisance or cause an economic loss. The problem is usually confined to rural and some suburban settings, and farm animals may be injured by stepping into their burrows.

Control. Woodchucks or marmots may be protected in some areas so it is advisable to consult with the local fish and game authorities before control is undertaken. Where legal and safe, shooting may be the quickest and surest way of removing a few woodchucks (a .22-caliber rifle is suitable.) Shooting is most successful in the early morning when the animals are actively feeding. Animals spooked by the first shot will soon emerge from the burrow again. Extreme care should be taken about the direction of shots and the possibility of ricocheting bullets. Where feasible, the rifle projectiles should be short rather than long. Although liable to attract more attention, a 410 shotgun is effective and may be safer to use — that is, if the pest animal is not beyond range.

Leg-hold traps are effective. Number 1½ or 2 traps can be set just inside the burrow opening under buildings, in stone walls and similar places that woodchucks may occupy. Traps should be concealed with soil or lightly covered with a layer of dry grass or leaves and anchored securely to prevent their being dragged into the burrow. If traps cannot be placed so as to avoid accidental capture of dogs or other domestic animals (e.g., in the burrow entrance), then use livecatch traps.

With some effort, woodchucks and marmots can be live-trapped with a wire-mesh trap such as the Tomahawk (No. 103 or 105) or Havahart (No. 3 or 3a). They do not readily enter solid box traps constructed of wood or metal (Trump and Hendrickson, 1943). Traps are most effective if placed directly in their runways and baited with cut vegetables such as sweet potato, carrots, cabbage or green alfalfa. In particularly difficult situations, some individuals seed a small spot of tilled soil next to the burrow with oats or beans. The wire-mesh trap is set over the seeded area, permitting the seedlings to grow up through the trap. Since woodchucks are highly attracted to tender sprouts, better trapping success can be expected. It may take several days before an animal will become accustomed to and enter the trap. When more than one or two animals are involved, live traps are rarely a practical approach. Leg-hold traps are far more effective if properly set.

Fumigants offer a positive control for woodchucks, but cannot be used if toxic fumes will escape into occupied buildings. Calcium cyanide, carbon bisulfide, methyl bromide and gas cartridges all have been used successfully. Carbon bisulfide and methyl bromide, being phytotoxic, should not be used beneath or near plants.

Gas cartridges, sometimes referred to as "woodchuck" or "smoke" bombs, are a mixture of ingredients formed in a small cylinder-shaped cartridge with a fuse inserted. When ignited and inserted into a burrow, they give off toxic fumes and smoke. These toxic gases are retained in the burrows by closing them with sod or soil and packing them tight.

Of the available fumigants, gas cartridges are the simplest to use and do not require any previous experience. They are not as effective as carbon bisulfide or methyl bromide, although safe to use. Follow the directions on the cartridge label.

Also effective are one to three tablespoonfuls of calcium cyanide inserted well down in the burrow, or about two oz/56.7 g of carbon bisulfide placed on a wad of absorbent cotton and introduced into the burrow. All the burrow openings must be closed with sod or soil and well tamped to prevent the gas from escaping. Fumigants are most effective when the soil is damp, for moisture prevents the gas from escaping into the soil and into the atmosphere. Also effective is methyl bromide injected into the burrow. Both carbon bisulfide and methyl bromide, being hazardous to the user, should be used only by those experienced in handling them. Fumigants are not effective when the woodchuck is hibernating, for it builds an earth plug to seal itself in for its sleep. Only active burrows should be fumigated. The burrows should be checked a week after fumigation and any reopened holes should be retreated.

Baiting may be the best control where the animals are numerous or live beneath rock outcroppings, making burrow fumigation or trapping difficult. Strychnine-treated pieces of apple, carrot or sweet potato have been used as baits, as have chopped green alfalfa or dandelion greens. Because of possible hazards to children and domestic animals, the use of strychnine baits near buildings and in other areas of potential hazards is not a good practice.

Woodchucks will dig beneath ordinary woven-wire fences, but if extra wire is buried about 24 inches/60.9 cm deep and curved outward at the bottom it will normally exclude them. Although not considered very good climbers, if determined, they may even climb woven-wire fences. A one foot/30.4 cm wide sheet-metal band attached near the top of the fence will prevent this. A fence which is reasonably woodchuck-proof is expensive and is practical only for protecting small areas.

TREE SQUIRRELS

Biology. Tree squirrels belong to the rodent family Sciuridae and are represented by three genera in the United States *(Glaucomys, Sciurus,* and *Tamiasciurus).* One or more of the 10 squirrel species are found throughout the country. This distribution locally is always determined by the presence of trees, as they are never found far from forested or sparsely wooded areas. Some species are common residents in our cities, around landscaped residential and business areas where there are ample large trees. Many city parks have high squirrel populations which are fed by visitors and become favorite park attractions.

Two tree squirrels native to the eastern part of the country, the fox squirrel *(Sciurus niger)* and the eastern gray squirrel *(S. carolinensis),* have been introduced into the western United States and are now found in numerous areas. Because of their aesthetic value and some importance as game animals (especially the genus *Sciurus*), they have been deliberately introduced into areas where they formerly did not occur.

Tree squirrels are among the best known animals for there is hardly a soul who hasn't seen them scampering about in the forest or in parks. When frightened, they immediately retreat to trees to escape. Their feet are well-adapted for climbing and their bushy tails aid in balance. The loose skin along the sides of the flying squirrels *(Glaucomys)* enables them to glide great distances from one tree to another or from a tree to a rooftop. The landing thud may awaken even the soundest sleeper. Except for the flying squirrels, which are nocturnal, all others are active during the day.

Tree squirrels have two to seven young per litter and one or two litters per year, depending on the species. In natural environments the flying squirrels nearly always build their nests in hollow trees. The other tree squirrels often will use holes in trees as nests, but also build tree nests of leaves, twigs and pieces of bark.

Tree squirrels feed on a variety of materials, including fruit, bark, leaves, fungi, insects, bird eggs and occasionally other small animals. Gray and fox squirrels of the genus *Sciurus* depend on such food items as hickory nuts, acorns, pecans, beechnuts and walnuts. Tree squirrel diets vary by species and are determined by their habitat and season of the year. They do not hibernate but tend to store great quantities of food, generally in excess of that needed during winter months.

Tree squirrels, and in some regions, flying squirrels, occasionally cause damage or nuisance to the homeowner when they use buildings for nesting sites and food storage, or gnaw into attics to take up residence. They also may move into spaces between walls and floors. They often gain access through vents, broken windows, knotholes and construction gaps under eaves and gables. Occasionally the chimney and fireplace provide their entry route.

Some have a remarkable ability to destroy wooden shakes and shingles, cedar seeming to be their favorite. The amount of structural damage may at times be severe. They can be particularly destructive to summer or vacation cabins which are vacant during part of the year, since they are free to continue their activities until the owner's return. Garages, barns, stables, toolsheds and other buildings often serve as homes for tree squirrels.

Rarely do tree squirrels take up residence in an occupied building without being seen. A possible exception are the flying squirrels whose presence in the area may be unsuspected since they are only active at night, even though they may be relatively common. The typical evidence of tree squirrels includes droppings, gnawed holes, nest materials, food stores, shells, hulls, pits and other food remnants. If squirrels are in your attic or garage, you will hear them moving about, even if you can't see them. However, noise and physical evidence can sometimes be confused with that of chipmunks, woodrats or roof rats.

In backyards and in landscaped areas, squirrels will dig up vegetable and flower gardens, primarily for seeds and bulbs. Seasonally, they will clip flower buds and leaves, and often strip the bark from trees and various ornamental plants. They are particularly fond of maturing nuts and fruit. Where only a few trees exist in a backyard, they can literally strip them of a crop of walnuts, almonds, pecans, etc.

Utility companies report tree squirrels often cause loss of electrical and telephone service by gnawing into cables and stripping insulation. This damage causes shorting or grounding and can be difficult and costly to repair. Special cables with protective metal sheaths which cannot be gnawed through are available, but their cost is considerably greater than the standard cables. Over-

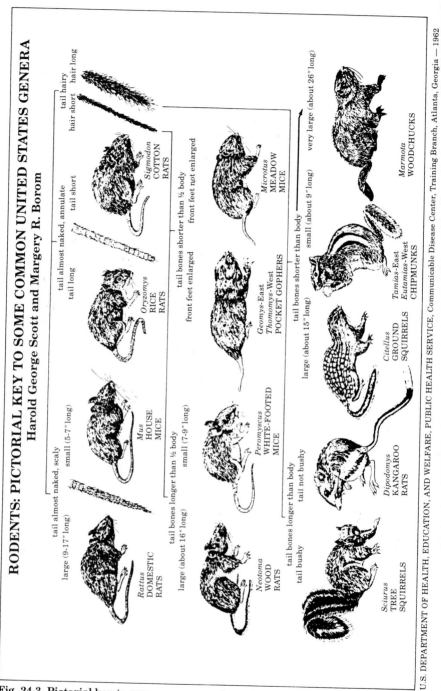

RODENTS: PICTORIAL KEY TO SOME COMMON UNITED STATES GENERA
Harold George Scott and Margery R. Borom

tail almost naked, scaly

large (9-17" long)

Rattus
DOMESTIC
RATS

small (5-7" long)

Mus
HOUSE
MICE

tail almost naked, annulate

tail long

Oryzomys
RICE
RATS

tail short

Sigmodon
COTTON
RATS

tail hairy

hair short   hair long

tail bones longer than ½ body

large (about 16" long)

Neotoma
WOOD
RATS

small (7-9" long)

Peromyscus
WHITE-FOOTED
MICE

tail bones shorter than ½ body

front feet enlarged

Geomys-East
Thomomys-West
POCKET GOPHERS

front feet not enlarged

Microtus
MEADOW
MICE

tail bones longer than body

tail not bushy

Dipodomys
KANGAROO
RATS

tail bushy

Sciurus
TREE
SQUIRRELS

tail bones shorter than body

large (about 15" long)

Citellus
GROUND
SQUIRRELS

small (about 9" long)

Tamias-East
Eutamias-West
CHIPMUNKS

very large (about 26" long)

Marmota
WOODCHUCKS

U.S. DEPARTMENT OF HEALTH, EDUCATION, AND WELFARE, PUBLIC HEALTH SERVICE, Communicable Disease Center, Training Branch, Atlanta, Georgia — 1962

Fig. 24-3. Pictorial key to some common rodents found in the United States.

head power and communication lines are frequently used by tree squirrels to travel from one area to another.

Both red squirrels *(Tamiasciurus)* and flying squirrels have been implicated in diseases of man, but direct transmission is rare. Plague has been isolated from the Sierra Nevada flying squirrel, *G. sabrinus lascivus,* in California; relapsing fever has been transmitted by contact with the blood of an infected Douglas squirrel *(T. douglasii)*; and plague also has occurred in Douglas squirrels. Colorado tick-fever virus and antibodies to Powassan virus have been found in red squirrels, *T. hudsonicus* (Weinburgh, 1964). Tree squirrels, fortunately, are not regarded as important vectors of human disease.

Tree squirrels also are subject to diseases which do not involve man. These diseases may be responsible for the rapid declines in population density observed periodically, especially following periods of great abundance. Nematode worms are common internal parasites, while fleas, ticks, mites and chiggers are common ectoparasites (NPCA, 1980a). When squirrels take up residence in attics or other parts of a dwelling, their ectoparasites may be troublesome to man.

Tree squirrels in parks often become eager or aggressive and bite the hand that feeds them. Even those that appear tame will commonly bite if attempts are made to pick them up. And they have been reported to attack pedestrians seemingly without cause (Anon., 1959). Bites should be treated by a physician.

Control. Control techniques are given for tree squirrels as a group. However, differences between species exist and the method of control must be adapted to the biology and habits of the offending animals. Proper identification of the species is necessary to comply with state statutes and regulations.

One or more tree squirrel species are considered game animals in most states and some may be entirely protected by laws and regulation. In many states, special control permits or authorizations are available when protected animals are causing property damage. It is recommended state and local laws be carefully reviewed before engaging in any tree squirrel control (NPCA, 1980b).

Exclusion (rodent proofing) is the best solution to squirrels gaining access to dwellings. Most entry points will be above eye level, but exceptions do occur. Sheet metal or wire hardware cloth are most often used to close openings. Be sure in closing all possible entry routes you do not trap the animals inside.

If the population of tree squirrels must be reduced, live-trapping is most often considered the method of choice. Live-catch traps, such as Havahart (Nos. 2 or 2A) or Tomahawk (Nos. 103 or 104), are very effective. If squirrels are entering the premises via overhead routes such as trees or power lines, traps placed on rooftops or secured to limbs may be the most appropriate settings. For best results, traps should be prebaited for several days with the doors secured in the open position; when the bait is readily taken, rebait and set the traps. A wide variety of baits may be accepted. Peanuts, walnuts, pecans, acorns, sunflower seeds and raisins have all been effective. Some tree squirrels seem to develop a liking for meat, and when other baits fail, a small piece of fresh red meat may do the trick. Fresh meat may also attract cats, but it is much less attractive to most birds. If squirrels manage to escape before the trap door completely closes, they may become very trap-shy and avoid re-entering traps for extended periods. Live-catch traps should always be checked at least once daily, preferably several times, to remove squirrels or to release other animals accidentally captured. Since tree squirrels occupy habitats favorable to many bird species, birds with a liking for the baits used also may enter the traps. Captured birds should be released immediately and, if regularly captured, traps should be moved to other locations.

The laws in some areas may not permit the killing of nuisance tree squirrels. Hence, live-trapped squirrels may have to be released elsewhere no matter how unwise such a practice is biologically.

Smaller tree squirrels, like chipmunks and wood rats, can be killed with ordinary wood snap-traps used for rats. The same baits suggested for live traps can be used. All baits should be tied to the trigger with thread or light string. Otherwise, some squirrels will become proficient at stealing baits without being caught.

Traps can be nailed or fastened to fences, tree limbs and rooftops to increase trap effectiveness, but in these outdoor situations, unless the traps are placed with considerable foresight, they will often catch birds. The novice should stick to live-catch traps and leave the snap traps to those experienced in their use. An exception to this would be the use of snap traps in attics or other indoor locations frequented by tree squirrels. If traps are set on a flat surface in a building, a strong string or light wire should be used to secure them to heavy or fixed objects to prevent them being dragged away. Like live traps, snap traps should be placed in position and baited for several days before setting. Occasionally, large gray or fox squirrels will pull free from a snap trap. Always use caution to avoid injuring pets with snap traps.

Several snap traps may be set in vacant cabins and summer homes to catch invading tree squirrels or other rodents of a similar size. A dozen or more traps might be needed if a building is vacant for extended periods.

If live-catch traps or snap traps fail, it may be necessary to resort to No. 0 or No. 1 size leg-hold traps, or a body-grip trap such as the Conibear No. 110.

Area repellents such as paradichlorobenzene crystals and napthalene flakes often are suggested for attics and similar locations. Follow the rate and method of application suggested for chipmunks. Our experiences in California with these materials have been discouraging. At least one tacky repellent is currently registered for squirrels. Non-toxic sticky repellents may be helpful in discouraging tree squirrels from walking or climbing on some surfaces.

To prevent damage to wood shingles, a mixture of one lb/0.45 kg of copper napthenate to 2½ qt/2.4 l of either mineral spirits, linseed oil, or shingle stain has been suggested (Anon., 1961) as a useful repellent.

Trapping can be used to solve most tree squirrel problems and is the method of choice for population reduction where shooting is not possible or practiced. Strychnine baits have been used in the past, but there is little justification at present for its use on tree squirrels in and about buildings. If it is in a rural area, and the rare need arises for an acute poison, zinc phosphide would be the toxicant of choice. It is reasonably effective at one or two percent concentrations on preferred foods.

Anticoagulants have been used for tree squirrel control and have been found to be fairly effective. Cereals are not always attractive to tree squirrels. However, they often will begin to feed on them if exposed for an extended period of time. Oat groats have been useful in some situations in the West, but whole corn generally seems the most preferred grain in the Midwest and East. Cereal baits prepared for rats may be accepted if used in bait boxes and placed where they intercept squirrels' travel routes. If the baits are accepted, control can be achieved, although it may take three to four weeks to accomplish the desired results. If there is evidence the squirrels are storing bait, baiting should be discontinued. Some paraffinized anticoagulant baits are accepted when they are fastened in trees, on fences or near roof eaves, which are the same types of placements used for roof rats.

In some situations where only one or two squirrels are involved, shooting may be effective, especially in the relatively rural areas. In densely populated urban or suburban areas, this approach is not usually a feasible solution, even if legal, because of the hazards associated with the use of firearms. Highpowered air rifles are sometimes useful in place of firearms.

Squirrels may be prevented from climbing trees by wide metal bands placed around the trunk at least six ft/1.8 m from the ground. Aluminum roof flashing, fitted snugly, works very well, but must allow for growth to prevent damage to the trees. These bands should be about 18 to 20 inches/45.7 to 50.8 cm wide with no gaps or rough surfaces to permit climbing. They are useless if the limbs can be reached from the ground, nearby trees, fences, or other structures accessible to squirrels because any distance of six ft/1.8 m or less can be considered an easy jump for most tree squirrels.

GROUND SQUIRRELS

Biology. The genus *Spermophilus* is represented in the United States by 17 species, having numerous subspecies. Ground squirrels, like chipmunks, woodchucks and tree squirrels, belong to the family Sciuridae, in the order Rodentia.

Probably less than half of the species of ground squirrels are considered a pest at times in some particular situation. Most of the ground squirrels are found

Fig. 24-4. California ground squirrels, *Spermophilus beecheyi.*

west of the Mississippi River, although the thirteen-lined ground squirrel *(S. tridecemlineatus)* is found as far east as Ohio. Problems with ground squirrels seem to be most severe west of the Rocky Mountains, where the number of species is greater.

Ground squirrels primarily inhabit open grassy plains and valleys and, to a lesser degree, openings in forests. They avoid highly forested or brushy habitats and are often less numerous where vegetation of any type is tall and dense. They are quite adaptable, however, and have managed to thrive very well in a variety of agricultural situations, giving them their pest status.

Depending on the species, they are brown, reddish, black, black and white, reddish and black, striped, spotted, variegated or of solid color. They vary from the size of a Norway rat to about three times as big. The adult California ground squirrel *(S. beecheyi)* often exceeds 2½ lb/1.1 kg just before going into hibernation.

All ground squirrels dig burrows for their homes. Some species are quite social, forming sizable colonies of an acre or more, with complex burrow systems; these tend to increase in size as the years pass. Squirrels seek shelter, store food, raise their young, hibernate and estivate within their burrow systems.

Ground squirrels are active in the daytime, like tree squirrels, so they do not go unnoticed in an area. If they aren't observed closely, however, damage from their digging under structures or irrigation levees can rapidly change from minor to severe in a few weeks.

In late winter and spring, ground squirrels feed on a wide variety of green herbage. In the more arid West, as green annuals begin to dry up and form seed, the squirrels will often switch to a diet nearly exclusively composed of seeds. As the season progresses they gather large amounts of seeds in their cheek pouches to carry off to bury in shallow caches or store in their burrows.

Ground squirrels hibernate in winter months and in the dry hot areas of the West they often go into a summer torpidity (estivation). The young of the species may not estivate at all and may even remain active for several months after the adults have gone into their burrows to hibernate.

The squirrels emerge from hibernation in the spring and breed shortly thereafter. The gestation period is about 28 to 30 days, with litter sizes varying with the species. Seven to nine young is about the average litter size, with only one litter produced annually. For rodents, ground squirrels have a relatively long life span, so even the single litter produced annually can result in very high densities in a few years if the habitat is favorable.

Ground squirrels and their ectoparasites are involved in transmitting a number of serious diseases to man, including plague, tularemia, relapsing fever, spotted fever and Colorado tick fever (Weinburgh, 1964). The recent increase of plague in squirrels and man in parts of the West has awakened the public to the continued danger from this dreaded disease. Plague, if not diagnosed and treated very quickly, is often fatal. Caused by the bacterium *Yersina pestis,* it can be contracted when infected squirrels are handled or infected fleas are encountered. Where plague is endemic in an area, home and property owners should be aware of the health hazards. The disease is often contracted in parks, campgrounds and other recreation areas where many people come into close association with infected chipmunks and ground squirrels or their fleas. Where the potential for the transmission of diseases is present, ectoparasite control should be undertaken before squirrel control, or at least simultaneously.

Ground squirrels are not as widely distributed as tree squirrels, thus less troublesome to most home owners. However, in many areas of the West their burrowing beneath buildings can be a problem. They are apt to dig beneath barns, granaries and other outbuildings, which provide cover as well as a food source.

Although many ground squirrels are good climbers, when frightened or disturbed they always seek their ground burrows. They do not take up occupancy in attics or in a situation away from their ground burrows, although they may temporarily invade second stories of buildings and a variety of other places to obtain food. In camping and park areas where ground squirrels thrive, they may learn to become beggars, living on handouts, just as do tree squirrels and chipmunks. Frequent results are that children are bitten and the possibility of disease transmission to people is substantially increased.

The digging done by squirrels can undermine picnic tables, barbecue pits, restrooms of recreation areas, paved road and graveled roads. They also can promote excessive water erosion on sloping terrain.

In landscaped yards and, more important, in backyard vegetable gardens and family fruit or nut orchards, they can be devastating, for their appetite and hoarding abilities seem unlimited.

Control. Before undertaking reduction control of ground squirrels, check with your local fish and game or conservation agency since some species are protected in some states.

If ground squirrels are entering buildings for food, they can be excluded by building them out, just as with Norway rats or tree squirrels. Ground squirrels are powerful gnawers, readily gnawing through one-inch/2.5 cm wood siding if they have an edge upon which to start. Well-secured heavy wire mesh or sheet metal should be used to close accesses.

Where squirrels have become dependent on handouts, they are relatively easy to live-trap. Trapping may be difficult in other situations, particularly in early spring, when they primarily eat green vegetation (i.e., grass and various herbaceous weeds and cultivated plants) rather than seeds.

Live traps, such as the Tomahawk (Nos. 103 or 104), Havahart (Nos. 2 or 2A), or traps of similar type and size, can be used effectively if only about a dozen squirrels are present. Live traps are best set near burrow entrances or along pathways used by the squirrels, the same general technique used for chipmunks or woodchucks.

Leg-hold traps are sometimes more effective than live traps on certain species of ground squirrels. Numbers 0, 1, or 1½ are used most commonly. They are best placed at, or just inside, the burrow entrance and in a concealed setting, with the trap well covered with fine soil. The traps should be secured to prevent them from being carried into the burrow. Traps set in this fashion are effective without baiting, although rolled oats sprinkled over the trap may serve as an added attraction.

Live-catch and leg-hold traps should be checked frequently (daily at the minimum) so non-target species captured accidentally can be released. In areas where the squirrels are known to be possible disease carriers, particularly in the West, the dead animal should never be handled without gloves. In disease-endemic areas, control of ectoparasites (fleas, ticks, etc.) may be advisable before or simultaneous with ground squirrel control. Your local health authorities should be consulted for the latest recommendations on ectoparasite control.

Because ground squirrels in some areas are carriers of plague and several other diseases, live-trapped squirrels should never be released, but humanely disposed of.

Some of the smaller ground squirrels can be taken with ordinary wooden rat traps. Baits of cereals, nut meats, raisins, prunes or slices of orange are quite effective, but should be tied tightly to the trigger to prevent theft without tripping the trap. Carcasses should be buried at least two ft/60.9 cm deep to prevent pets and other carnivores from digging them up.

Since most ground squirrels fear new objects, it is always wise to bait the unset traps for several days until the bait is readily accepted, and then rebait and set the traps that are being visited.

Where permitted and safe, shooting with a .22-caliber rifle can be effective for removing up to six or eight squirrels. Many species become gun-shy very rapidly, making it difficult to shoot more than a few squirrels. For safety, shooting is generally allowed only in certain rural or sparsely settled areas.

Some fumigants are very effective for controlling ground squirrels. Their use is limited to times of year when the animals are not hibernating or estivating, for during such periods the squirrels barricade themselves behind an earth plug in the burrow to exclude the outside elements, also excluding penetration by the fumigants.

Carbon bisulfide, methyl bromide and gas cartridges all have been used successfully; calcium cyanide may be effective for some species but is very poor with others. Fumigants cannot be used if there is danger that the toxic fumes will escape into occupied buildings. Since carbon bisulfide and methyl bromide are phytotoxic, they should not be used where burrows extend beneath trees or other valuable plants. Carbon bisulfide is also highly inflammable.

Gas cartridges, sometimes referred to as "woodchuck" or "smoke" bombs, are a mixture of ingredients formed in a small cylinder-shaped cartridge with a fuse inserted. When ignited and inserted into a burrow, they give off toxic fumes and smoke. These toxic gases are retained in the burrows by closing all entrances with sod or soil, packed tightly.

Of the available effective fumigants, gas cartridges are the simplest to use and can be used by anyone having no previous experience with fumigants. Both carbon bisulfide and methyl bromide are potentially hazardous to the user and should be applied only by those experienced in handling them. Label directions for use should be followed closely, regardless of the fumigant.

For further information on fumigants and their use, the reader is referred to the section on woodchucks.

When ground squirrels are numerous, poisoned baits may be the most practical method of control. The three main acute toxicants (strychnine, sodium fluoroacetate (1080), and zinc phosphide) have been used. In selecting which acute toxicant to use, one must consider the susceptibility of the particular species of ground squirrel and also the associated hazards of the toxicant to people and other nontarget species. Zinc phosphide, although not generally as effective as the others mentioned, is probably the safest of these three acute rodenticides to use in areas where access to the property by domestic animals and humans is uncontrollable. Zinc phosphide at levels of 0.8 to one percent is frequently formulated on either rolled oats or oat groats. Strychnine is not accepted well by some species of squirrels unless it is the season when they gather the seeds in their cheek pouches to cache them.

Ground squirrels are susceptible to the anticoagulant rodenticides and several firms have anticoagulant baits registered for use against some species of ground squirrels in a number of the western states. These are generally formulated at the same concentration used for rat and mouse control. Since some

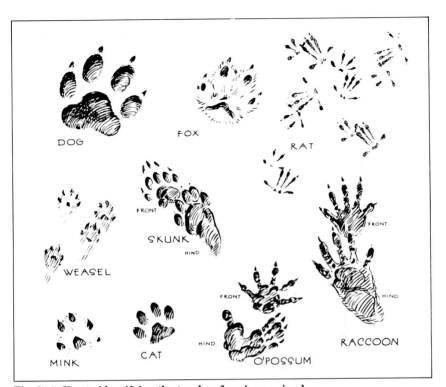

Fig. 24-5. Key to identifying the tracks of various animals.

squirrel species are far less susceptible than others to the first-generation anticoagulants, allow ample bait exposure time (three weeks or more) to achieve control. Experiments have shown the newer anticoagulants brodifacoum and bromadiolone are both highly effective. Generally the bait is exposed in bait boxes of a size suitable for squirrels. Since some species of squirrels weigh over two lb/0.90 kg, they can consume a substantial amount of bait prior to death and this can make control with anticoagulants on a large scale rather expensive.

With a little persistence, anticoagulant baits can effectively reduce a ground squirrel population to extremely low levels, even below that achieved with some of the acute toxicants. Anticoagulants also can be very effective as a follow-up treatment where acute toxicants such as zinc phosphide have been used.

No known repellents are effective against ground squirrels. Fencing to exclude them from an area is very costly and probably impractical because of their climbing ability, and they readily dig to depths of six ft/1.8 m or more.

Four species of antelope squirrels belonging to the genus *Ammospermophilus* are closely related to ground squirrels, but to our knowledge they are never pests in and about buildings; only very rarely do they become pests in agriculture or forestry.

The prairie dog, genus *Cynomys*, is another group of closely related ground-living sciurids. The habits and biology and pest status of prairie dogs closely parallel those of ground squirrels, so the control procedures suggested for ground squirrels can be used also for prairie dogs implicated as a pest beneath and about buildings. They are very sociable, but are a little more timid and less apt to be found in as close association to buildings than are some species of ground squirrels. Prairie dogs may be protected in some areas; check with your local Fish and Game or Conservation agency before taking control action against them.

WOOD RATS

Biology. Wood rats, *Neotoma* spp., also referred to as "pack rats" or "trade rats," are widely distributed over much of North America. One or more species are found in parts of most states with the exception of a few of the north-central and Great Lakes area states. They also are absent from most of the New England area.

Some eight species and many subspecies are usually restricted to a given type of habitat, but members of this genus occur from low, hot, dry deserts to the cold rocky slopes above timberline.

Wood rats apparently are attracted to small, bright, shiny objects such as spoons, small pieces of jewelry, broken bits of mirrors, coins or other items, sometimes leaving sticks, nuts or other materials in trade. The common names, "pack" or "trade" rats, were given these animals because of these antics. These characteristics are frequently the subject of exaggerated stories from which it is difficult to sift fact from fiction.

Wood rats are rat-size mammals with large ears, large dark eyes and a fairly long tail which is sparsely covered with hair or, depending on the species, well-furred with long hair. Their fur is soft; dorsal fur is colored cinnamon, brown, gray, yellowish gray, or creamy buff; feet and ventral parts are generally much lighter in color; tail is blackish or buff, paler on ventral surface. Wood rats are much larger than mice and tend to resemble the introduced Norway rat or roof rat in general size and shape. The head-and-body length is about seven to eight inches/17.7 to 20.3 cm and the tail 6½ to 7½ inches/16.5 to 19 cm long. Their clean appearance, soft fur and well-haired ears help distinguish this native species from the Norway and roof rats.

Reproductive activity is most pronounced in the spring. One to five litters a year are reported, with the number of young varying from one to four, with two about average. This species is not highly prolific and annual population increases or decreases are not great.

Wood rats have been involved in epizootics of plague and have been found infected with tularemia. They are reservoirs of the trypanosomes (parasitic, blood-infesting protozoans) of Chagas' disease (Weinburgh, 1964). Their role in transmitting disease to man is considered minor, although dead or dying wood rats should not be handled with bare hands, especially in plague areas.

Some species such as the dusky-footed wood rat *(Neotoma fuscipes)* are agile climbers and often construct bulky stick dens or nests high in the crowns of trees. More commonly, however, dens are situated on the ground. Ground dens measure three to five ft/91 to 152 cm in height and diameter; tree nests are somewhat smaller. One animal may inhabit several nests and in good feeding areas a den may be occupied for several years or a lifetime. Wood rats live alone except when mating or rearing young. The dusky-footed wood rat of the West is semiarboreal, and when traveling between tree nests, jumps squirrel-like from branch to branch.

Fig. 24-6. Wood rats are often referred to as "pack rats."

Wood rats climb readily and are chiefly nocturnal in habits, but are occasionally observed during daylight. Their food is largely determined by varying local conditions and consists mainly of a variety of green vegetation including grass, leaves, fruit, nuts, small bulbs, bark, dry seeds and fungi. They also may be attracted to human food supplies in buildings or in outdoor camps.

When this animal nests in buildings, it may utilize available foods within the building, but most often it continues to feed outside. Visible ground trails three to four inches/7.6 to 10.1 cm wide may be evident from the nest to the feeding grounds.

In forests they clip young and debark older conifer trees for food and nest building. In backyard or commercial orchards they will occasionally clip young ¼ inch/0.63 cm diameter limbs from fruit trees. Girdling of small trees and shrubs occurs, but is not common. Rarely do they become numerous enough to cause more than very limited damage to flower or backyard vegetable gardens. Generally they are considered a pest of very minor occurrence.

Wood rats are sometimes a nuisance in summer cabins or mountain or forest residences. At times these rodents gain access to homes and create problems by shredding mattresses and upholstered furniture for lining nests. They are excellent climbers and may find their way into attics or spaces between walls where they nest. Large stick nests beneath a porch, in a woodshed or attic are a giveaway to the presence of a woodrat. They may use overhanging trees to gain access to rooftops and attics.

The presence of a wood rat in the attic can be disconcerting to say the least. It is difficult to understand how such a light-footed rodent can make so much noise at night. One rat running back and forth carrying sticks, nuts, or whatever, can sound like a dozen rats wearing hiking boots. Dogs are often kept in a state of excitement every night until the rat is removed.

Control. Pack rats may be permanently excluded from buildings by the same methods used to exclude the introduced Norway and roof rats. Access by this rodent is usually at the eave or roof level, hence it is important to check for openings in attic vents, broken roof shingles or other gaps next to the eaves. In rodentproofing, be sure the rat is not unknowingly entombed in the attic.

Moth balls or crystals may be somewhat effective in repelling wood rats from an attic. (See section on chipmunks for details.)

Wood rats in buildings can be taken with the use of ordinary wooden snap rat traps. Traps can be baited with a dried prune, raisin or nut meat tied with thread to the trap trigger. Place traps across the runways of the rats. Wood rats are among the easiest rodents to trap because they show little fear of new objects placed in their environment. Live-catch traps such as the Tomahawk (No. 102) or Havahart (No. 1) are also effective in catching wood rats.

The homeowner and structural pest control operator will rarely have a wood rat infestation which cannot be handled by trapping. But where such a situation does exist, anticoagulants are the best choice of bait. Baits formulated for commensal rats and house mice have contributed to effective wood rat control. It has been found when closed box-type anticoagulant bait stations were employed, they were often filled by the rats with sticks and other debris. Open bait containers, protected by inverting a wooden lettuce crate over the bait, proved more practical. Wood rats have less tendency to carry off and store the smaller grains and bait particles. For this reason, a ground meal-type bait is preferred. Baiting sites should be located near existing rat runways, feeding sites or nests. Anticoagulant paraffin bait blocks used for commensal rodent control have proven valuable. If placed in a summer or mountain cabin, bait blocks will effectively control wood rats which may gain access while the cabin is unoccupied. The advantage of this type of bait is twofold: a more lasting bait (resistant to molds) and one which cannot readily be packed off and stored by the rats. Anticoagulant cereal baits in four oz/113 g plastic place packets are also an effective way of presenting bait.

Of the acute rodenticides, zinc phosphide is probably the most often employed. Steamed rolled oats or oat groats treated with one percent zinc phosphide adhered with vegetable oil are very effective. Tablespoonful amounts scattered on the runway near the nest are ample for control and should be applied in the late afternoon prior to their nighttime feeding.

DEER MICE

Class Mammalia
Order Rodentia
Family Cricetidae

Biology. Deer mouse is the accepted common name for *P. maniculatus,* probably one of the best known of all the *Peromyscus*. Whichever *Peromyscus* species is causing a problem, the solution is about the same. For the purpose of this chapter, all species which belong to the genus *Peromyscus* are referred to as deer mice, although each species has a different common name. Collectively, they are also called "white-footed mice."

The size and coloration of the 55 different species varies and may lead to some confusion, particularly if two different species of deer mice are caught on the same premises where they can be compared with one another. Identification of the species of this group is not always easy and is best left to the taxonomist.

In any case, if you recognize it as *Peromyscus,* that is identification enough for adopting the proper corrective measures.

P. maniculatus, used here as an example, is found throughout the United States except in areas of the South and Southeast. Every part of the country, however, has one or more species of *Peromyscus.*

Deer mice are a native species generally slightly larger than the introduced house mice, ranging from five to 15 inches/12.7 to 38.1 cm nose to tail, but easily distinguished by its relatively large eyes and ears. The tail is covered with short fine hair and is distinctly bi-colored: dark on the top and white on the bottom. Its feet and underbelly are also pure white but the rest of the body may range from sandy or gray to dark brown.

Deer mice have three to six young per litter and three or four litters per year, although those figures vary with climate and other factors. The gestation period is 21 to 24 days, with the main surge of reproduction in the spring. Since deer mice have received considerable study, information on their life history, distribution, etc., is extensive (King, 1968).

Deer mice are nocturnal (active at night) in habit and feed primarily on seeds. In the wild they are quite numerous and occupy a wide variety of habitats. Despite their numbers, they are seldom seen in the wild. They nest in hollow stumps, beneath fallen logs or in small shallow burrows. Signs of their feeding generally go undetected in the wild, leaving little evidence of their presence.

Deer mice enter homes and other buildings which are not rodentproof, but usually are considered only a minor pest because of their infrequency of occurrence. Where they gain access to mountain vacation or summer cabins, often by way of the fireplace chimney, they can do considerable damage to upholstered furniture by shredding fabric and the padding for constructing nests. Likewise, they may damage or destroy paper by building nests in drawers and file cabinets.

Nests, droppings and other signs found about structures are often attributed to house mice since they create very similar damage. Deer mice are most apt to take up residence in unoccupied cabins or outbuildings (i.e., garages, tool sheds, pump houses, barns, woodsheds, etc.) in rural settings. Mice sometimes frequent rural homes built close to the ground that are not mouseproofed. Deer mice are rare in urban residential areas unless considerable open space, parks, etc., are nearby. The exception is generally just a single animal. Recently developed subdivisions on the outskirts of cities are prone to trouble to a greater extent, at least in the West. The problem is most evident during the first year or two following a residential development.

Food-processing plants which maintain live multiple-catch traps outside will occasionally catch deer mice along with other species. Such is more likely in fall or spring when populations on adjacent uncultivated land are on the increase. Some claim deer mice will enter homes when the weather turns cold. Warmth may be the attraction, though that has not been proved.

These seed-eating mice will on occasion play havoc with seeded flower or vegetable gardens by digging up and eating the seed. They may also prevent successful reforestation by direct seeding.

Control. Deer mice in and about structures can be controlled with the same techniques used for house mice. Trapping with ordinary snap mouse traps is best where only one, and at most a few deer mice are involved. Traps can be baited with peanut butter, sunflower seed, moistened rolled oats and the like, for this species is a seed eater. For quick results, several traps should be used even if only a single mouse is believed present. Trap placement for this species is the same as used for house mice. Multiple-catch traps may be useful where mice may be entering buildings on a regular basis and where rodentproofing is inadequate. Deer mice, being very inquisitive, are easy to trap.

Anticoagulant rodenticides such as warfarin, pindone, coumafuryl, diphacinone and chlorophacinone are all quite effective on deer mice, but, as with other species, death requires multiple feedings. The second-generation anticoagulants, bromadiolone and brodifacoum, are also effective. Baiting for house mice will normally control deer mice. Deer mice may carry more bait away from bait stations than do house mice. Zinc phosphide and especially strychnine baits also are effective on deer mice.

Exclusion from buildings is the best solution to the deer mouse problem. Exclusion requires the same effort as with house mice. Openings should be closed with hardware cloth of ¼ inch/0.623 cm mesh or with sheet metal. Coarse steel wool can be packed tightly in holes and around pipes where gaps permit entry. Any gap big enough to slip a pencil through will permit a deer mouse or house mouse to enter. Unfortunately, many summer cabins and vacation homes are not constructed rodent-tight or have lost that feature through the process of aging. Some may be beyond rodentproofing without major renovation.

In these instances, some other permanent type of corrective measures may be warranted. This involves the maintenance of set traps, in particular the multi-catch traps that will not be set off by pets, or relying on anticoagulant baits. Stations or containers suitable for holding small amounts (two to six oz/ 56.7 to 170 g) of anti-coagulant baits can be placed throughout a cabin, especially if unoccupied for a period, to control the occasional invading deer mouse. Particularly useful and effective for this purpose are small paraffin bait blocks sold for the control of rats and house mice; 10 or 12 four oz/113 g baits are sufficient for a cabin of three or four rooms.

Possible damage by rodents in cabins also can be reduced by following a few simple procedures before leaving for any extended period. Loose cushions from sofas should be removed from the normal horizontal position and stored on edge, separate from one another, preferably off the floor. This makes them less attractive for nesting to any rodent species. Food items should be stored in metal or glass containers with tight-fitting lids.

MEADOW MICE

Class Mammalia
Order Rodentia
Family Cricetidae

Biology. Meadow mice (*Microtus* spp.), sometimes referred to as voles or field mice, usually require dense grass cover. They occasionally enter buildings which have floors at ground level. Being poor climbers, they cannot enter buildings via many of the routes used by house mice, deer mice or rats, but blunder into the buildings more by accident. Meadow mice are always restricted to buildings built near ground level. Although sometimes found in basements, they are never found on the second floor. They are sometimes found in stables or barns, involuntary hitchhikers on bales of hay recently removed from the field. This rodent has a blunt nose, with small furry ears and a scantily haired tail and dense fur that is blackish brown to grayish brown. Adults measure about four to five inch/10.1 to 12.7 cm in head-and-body length, with a 1¾ to 2¾ inch/4.4 to 6.9 cm tail.

Species of meadow mice are found throughout the country, generally preferring dense grassy habitat where they develop a well-defined surface runway system. Nests are constructed of grass or other vegetation which may be built either on the surface or below ground. Many litters of three to nine young may be produced each year, with the peak breeding occurring in the spring and, to a lesser extent, the fall. Females are mature when three weeks old and they take five to six weeks to produce each litter.

Meadow mice also may move from farm or uncultivated land into adjacent home gardens and landscaped areas. This is more common when crops are harvested adjacent to home sites. They have a long history of girdling orchard and homeyard fruit trees, in addition to digging shallow burrows and feeding on a variety of plants. If numerous in gardens, sometimes they do considerable damage to the exposed portions of growing beets, artichokes, low-hanging tomatoes and strawberries. However, they are mainly grass eaters and, when abundant, can cause serious damage to pastures and cereal crops.

Meadow mouse populations fluctuate dramatically, usually reaching high numbers every three or six years. As would be expected, they are reported more often in years of high population. In California and presumably elsewhere, mice may fall into swimming pools and drown because they are unable to escape. This may occur at some of the most luxurious homes built in high-value estate areas where the widely spaced homes have sufficient favorable habitat to support a meadow mouse population.

Control. Occasionally meadow mice enter buildings, but almost never become established and reproduce. They are best captured with snap mouse traps set in pairs at right angles to the wall. Triggers of traps can be made larger in area with cardboard or other suitable material and set without baiting. Within buildings, meadow mice may be attracted to baited traps, but more often run

Fig. 24-7. Meadow mice, sometimes referred to as voles or field mice, usually require dense grass cover.

into traps placed in their line of movement whether baited or not. Sherman-type live-catch mouse traps (3 × 3.5 × 9 inches/7.62 × 8.89 × 22.86 cm) work quite well when placed next to walls. The mouse's relatively large size makes ineffective some multiple-catch traps designed for the much smaller house mouse. Glue boards can be effectively used in some situations. The best solution to the meadow mouse problem is to rodentproof the building as you would to exclude house mice.

Meadow mice in gardens or landscaped areas can be poisoned with zinc phosphide baits or with an anticoagulant rodenticide (warfarin, coumafuryl, pindone, diphacinone or chlorophacinone). Meadow mice are less susceptible to anticoagulant baits than Norway rats, but concentrations normally used for commensal rats and mice can achieve control with the first-generation anticoagulants where bait acceptance is continued over a period of a week or more. Second-generation anticoagulants, bromadiolone and brodifacoum, require fewer feedings. Cardboard or plastic tube-type bait stations are effective for application of anticoagulant baits if placed in their runways or next to burrows.

Baits prepared with one or two percent zinc phosphide are quite effective for meadow mice. The lower percentage is effective if teaspoon amounts are placed outdoors in runways or near burrow openings. The higher concentration of toxicant is considered more effective if the bait is broadcast, and should be applied at five to 10 lb/2.2 to 4.5 kg of bait per acre. Meadow mice, unlike ground squirrels and deer mice, will not stray far from their normal travel routes to seek out bait, regardless of how attractive.

Meadow mouse burrow systems have so many openings that fumigants are ineffective, and no effective repellents are known.

The destruction of grassy areas adjacent to gardens where meadow mice thrive sometimes is effective in reducing the source of mice which are invading gardens and landscaped areas or buildings. Weed-free strips can serve as a peripheral buffer around an area to be protected. A weed-free buffer strip less than 10 feet/three m wide will be of little value unless used with a low mouseproof fence. Generally, the wider the strip the less likely mice will cross and become established.

Low wire or metal barriers (at least 12 inches/30.4 cm high) will exclude most meadow mice from gardens and prevent them from falling into swimming pools. To prevent the mice from digging under such fences, the bottom edge should be buried at least six inches/15.2 cm deep and preferably 12 inches/30.4 cm. Meadow mice rarely climb fences.

COTTON RATS

Biology. The cotton rat, *Sigmodon* spp., belongs to a different subfamily (Cricetinae) of the Muridae family than *Rattus* spp. and *Mus musculus*. It is a small rat-size native rodent with coarse grizzled-gray fur and a tail about half the body length, scantily covered with short hair. The adults are a light gray. The young are generally darker, sometimes almost black. The cotton rat probably is confused most often with the meadow mouse, and vice versa. They can be distinguished from one another mainly by size and tail length. The confusion arises when you have only a single animal of unknown age and nothing to compare it with. Since the control does not differ greatly between the species, an error in identification is of little practical importance. Cotton rats captured in buildings are sometimes assumed to be Norway rats.

Although sometimes found infected with plague and murine typhus, cotton rats are not generally considered much of a threat to human health. By far the most important species is *S. hispidus,* the Hispid cotton rat, for all other species have very limited distribution in this country. The Hispid cotton rat is found only in the southern portion of the United States, its distribution including the southeastern and southwestern states from Florida to California and north into Virginia, Kansas and Missouri. However, its range is gradually extending northward, especially in Kansas and California. Several other species of very limited distribution extend mainly into Mexico (Hall, 1981).

The gestation period is 27 days, with multiple litters of two to 10 produced annually. The average litter size is five to seven (Hall, 1955). Their reproductive propensity enables them to reach high population levels rather rapidly, with peak densities occurring every two to five years in a cyclic pattern. The pest status of cotton rats increases in years of peak population.

Cotton rats build shallow tunnels and create surface runways in grassy habitat. The runways, two to three inches/five to 7.6 cm wide, extend in all directions to available foods. They not only contain droppings but small piles of grass and other vegetation clipped about two to four inches/five to 10 cm long. Droppings, runways and clippings are larger than those made by meadow mice, although the signs of both animals are very similar at a distance. Nests are generally underground cavities lined with soft grass.

Cotton rats mainly eat plant material such as grass stems, roots and seeds. They are at times a pest to agriculture because of their liking for grain, alfalfa, sugar cane, cotton and row crops such as melons, sweet potatoes and tomatoes (Clark, 1972).

Control. Cotton rats occasionally enter buildings, but rarely become established and reproduce there. Their infrequent presence in fruit and vegetable packing houses partly may result from access gained as stowaways on incoming produce. Rodent problems not previously experienced are created by mechanical harvesters, which sometimes pick up rodents, and by the bulk hauling and handling of produce. The incidence of certain species such as meadow mice and cotton rats in processing plants has increased. Cotton rats can best be captured with ordinary wooden base snap rat traps. Since cotton rats are susceptible to the anticoagulant rodenticides, baiting with cereal-base anticoagulant rat baits as in Norway rat control is effective, although control may take longer than with Norway rats or house mice. Second-generation anticoagulants (brodifacoum and bromadiolone) would be expected to produce deaths with fewer feedings than with earlier anticoagulants (Gill and Redfern, 1980).

As meadow mice often do, cotton rats may move from farmland or uncultivated grassy or weedy land into adjacent home vegetable gardens and landscaped areas. If too numerous to trap, they can be poisoned with zinc phosphide or strychnine baits. Zinc phosphide is preferred to strychnine because it is less hazardous to children and domestic animals and usually equally effective or more so.

Baits prepared with one to two percent zinc phosphide are quite effective. Grains commonly used are oat groats, steam-rolled oats, whole corn, wheat or milo maize. Teaspoon amounts should be placed in outdoor runways or near burrow openings. Other effective baits are cut carrots or sweet or white potato (½ inch/1.27 cm) cubes treated with one percent zinc phosphide. For good acceptance, vegetable baits must be prepared fresh daily. The cubes are placed in the runways.

Burrow fumigants are generally ineffective because cotton rat burrow systems are relatively shallow and burrow openings may be very numerous.

If feasible, alteration of habitat in infested areas may prove effective in reducing their numbers. Herbicides, short mowing, disking, or plowing of weed and grass cover to make an adjacent area less suitable for cotton rats often limits the source of reinfestation. Weed-free buffer strips, as discussed for meadow mice, may have value if wide enough. Removal of weeds around buildings will tend to reduce the instances of rats entering those buildings.

TREE PORCUPINE

Class Mammalia
Order Rodentia
Family Erethizontidae

Biology. One of the largest native rodents is the tree porcupine *(Erethizon dorsatum)*, which is distinguished by its coat of bristling quills. According to present classification, only a single species of this family is found in the United States. It is distributed throughout Canada and the northeastern states as far south as Tennessee, the Great Lakes states, Great Plains states westward to

Fig. 24-8. Porcupines are heavy-bodied vertebrate pests which are readily identified by the more than 20,000 barbed quills which cover most of their upper body and tail.

the Pacific, and from Alaska southward through the Rocky Mountains to southern Arizona. The southeastern states and portions of the southern states are free of porcupines.

The porcupine is chunky, heavy-bodied, brown to black in color, with a small head, short legs and a short, thick tail. Its sharp barbed quills, 1½ to four in/ 3.8 to 10.1 cm long and numbering over 20,000, cover most of the upper body surface and tail. Only the face, the underparts of the body and the underpart of the tail are without quills. The quills can be raised and lowered at will, but cannot be thrown (as some believe), although they do come out easily. When disturbed, the animal will raise its quills and take a defensive stand. It may whirl or lunge at its opponent and any attacker will receive a sharp slap from the porcupine's tail. The barbed quills are driven into the flesh of the attacker, pulling free of the porcupine. The victim's own movements then embeds the quills more and more deeply, resulting in intense irritation and even death if quills that pierce the nose, mouth, tongue or other areas about the head are not removed. The affected parts may become infected and swollen. When embedded, quills prevent eating and of course the victim starves.

Porcupines mostly sleep during the daytime, in dens or trees, and do most of their feeding and moving about at night. The small, black and beady eyes reflect no light at night. The ears are small and almost hidden in the fur. Adults weigh 10 to 20 lbs/4.5 to 9.0 kg or more and measure two to three ft/60.9 to 91.4 cm long.

Porcupines reproduce once a year, giving birth to a single offspring in the spring. On rare occasions, twins are born. For a rodent, the gestation period is unusually long — seven months. The young rely on their mother for a very brief period following birth (about 10 days). They can climb trees when only two days old and become sexually mature in their second year.

Porcupines are more common in coniferous or mixed forests, but also are found where no evergreens exist. Excellent climbers, they spend considerable time in trees, especially in winter. In summer, they have been known to travel several miles from the nearest forest to feed in open meadows and fields and along the banks of streams or lakes.

In winter, they feed primarily on the inner bark (cambium and phloem tissue) of a wide variety of forest trees including white, ponderosa and pinon pine; hemlock, spruce, elm and poplar. In spring, they may eat flowers, buds and young leaves of trees. A wide variety of plant foods are eaten including some cultivated crops or home gardens, especially if near forested or wooded areas.

The porcupine makes its den in hollow trees or logs, crevices of rocky ledges or beneath buildings and the like if he can gain access. It is not uncommon for six to 10 animals to share the same den in winter. At other times they are less social in their habits.

Signs of porcupines are fairly easy to detect. In snow, they leave a troughlike trail about 10 inches/25.4 cm wide. In soft soil or mud their tracks tend to toe-in and impressions from the long claws often are evident. Quills are commonly found in feeding areas but appear also along trails. Generally there are abundant droppings (about ¾ inch/1.9 cm long and ½ inch/1.2 cm in diameter) in areas where they feed and near their dens. In winter, the droppings look like compressed sawdust or excelsior (the result of feeding on bark and twigs), but in summer the droppings may be harder to identify because more succulent vegetation is being consumed.

Damage to trees from bark-gnawing, and gnawing around buildings or objects which contain salt, can quite easily be identified as the work of porcupines. Damage to succulent garden plants may puzzle the casual observer, however, unless the porcupine is actually seen.

Porcupines will invade landscaped yards and family fruit and vegetable gardens and eat a wide variety of succulent plants (e.g., rose bushes and lily pads), which can make up much of their summer diet. They feed on berry bushes or corn, in the process often trampling other garden plants in their path. A small family orchard within the range of porcupines may go without damage for years and then suddenly be invaded by one or two animals which may cause serious damage in just a few nights.

Porcupines can become troublesome near homes, summer cabins, logging camps, campgrounds and other outdoor recreation areas. They gnaw the wooden underpinnings of buildings and even the siding or door frames of buildings. They have a particular liking for some types of plywood, apparently because of the glue used in the laminating process. The animal's strong craving for salt leads it to visit human habitations, where it chews or gnaws on tool handles, wooden wheelbarrow handles, boat oars, porch furniture, toilet seats, saddles or anything else which may contain traces of salt from perspiration or other sources.

Porcupines may be regarded as pests simply because they are an attractive nuisance to inquisitive dogs. Dogs often end up with a nose full of quills, leading to a sizable veterinary bill for their removal. Some dogs learn to avoid porcupines from a single experience, but others cannot resist confrontation whenever the opportunity arises, particularly if some time has elapsed since the last experience.

Porcupines are known to harbor some diseases, such as Colorado tick fever (Hull, 1963), which can infect man, but close contact between porcupines and man is relatively uncommon. As a result, porcupines are not considered a public health threat. People who handle them, either dead or alive, should use the same precautions advised for handling any wild rodent.

Control. Porcupines may be protected in some areas, so check with your local game or conservation agency officials before initiating control.

When they are living beneath buildings, the best solution is to exclude them by building them out. Sheet metal or heavy wire mesh may discourage them from gnawing on building timbers or siding — though not always. We know of no repellent that will effectively prevent a porcupine from gnawing at favorite spots. Electric fences have been used (Spencer, 1948), but many situations do not lend themselves to that approch.

Live-catch traps are effective around buildings and gardens if placed along routes the animals use in entering or leaving the area. Place the traps directly in their trail, if it can be identified. Homemade box traps, about 15 inches square/38.1 cm and 36 inches long/91.4 cm, are effective, as are Tomahawk (Nos. 108 or 109) and Havahart (Nos. 3 or 3A) live traps. They can be baited with apples or carrots. In some situations, bait acceptance is improved if freshly sectioned apples are sprinkled with salt. Small amounts of the bait should be placed outside the trap to entice the animal inside. Wire wings, about two to three ft/60.9 to 91.4 cm long, made of chicken wire and attached on each side of the trap, will make the trap more effective by guiding the animal inside.

Where safe to use, effective traps are No. 1½ to 2 single-spring leghold traps placed along travel routes or in den entrances. If other animals will not be endangered, bury the traps in the center of porcupine trails flush with the ground surface. Cover the trap pans with an appropriate pad and then lightly cover them with fine soil or leaves.

Since porcupines are attracted by the offensive fetid scents used by commercial fur trappers, scent-baited traps can be placed to the side of trails.

All traps should be visited first thing in the morning to recover the porcupine or to release any non-target species accidentally caught.

Because of the potential hazards to domestic pets and livestock and to other non-target species, porcupine control with acute rodenticides such as strychnine is rarely, if ever, justified in and about buildings. In some situations in the western states where forests are severely damaged, those experienced in porcupine control have used a strychnine-salt formulation in a manner which avoids danger to non-target species. Another acute rodenticide, sodium arsenite, reportedly has been used on porcupines (Faulkner and Dodge, 1962), but has not come into general use. Some of the anticoagulant rodenticides should be effective, but have received only limited consideration (Marsh et al., 1974).

Shooting with a .22-caliber rifle, effective for eliminating one or two troublesome animals, is best done within the first four hours after sunset. A spotlight or strong flashlight is a must in locating them. Shooting should be done only where it does not endanger people and is lawful.

Because porcupines are relatively slow and sluggish, it is possible to spot them at night with a light and to herd them into a hand-held wire-mesh trap or cage. Caution must be used to avoid being slapped by the tail, for at times they can move surprisingly fast. A tennis racket-size plywood paddle with a 30 inch/76.2 cm handle is helpful in directing the animal. A large fish-landing net also works well, but don't expect the net to be undamaged.

SKUNKS

Class Mammalia
Order Carnivora
Family Mustelidae

Biology. Skunks belong to the weasel family and are a well-known furbearer. Being a frequent subject of stories and cartoons, the skunk has characteristics and habits that most people are familiar with even if they have never seen the animal in the wild. Since skunks are not particularly disturbed by man's presence and activities, they will frequently move in from their native habitat and take up residency beneath buildings.

Of the five species of skunks found in this country, only three are of any economic or public-health importance: the striped skunk *(Mephitis mephitis)* and the western *(Spilogale gracilis)* and eastern *(S. putorius)* spotted skunks. The hooded *(Mephitis macroura)* and hog-nosed *(Conepatus mesoleucus)* skunks may be implicated as pests in localized areas.

As for distribution, the striped skunk is found throughout the country except the desert area of the Southwest. The western and eastern spotted skunks are distributed throughout about three-quarters of the country, being absent along most of the eastern seaboard, the northeastern states and the region around the Great Lakes.

Hooded and hog-nosed skunks are confined primarily to areas of the Southwest and southward into Mexico. Their biology and habits do not vary greatly

Fig. 24-9. Dogs are usually the big losers in an encounter with a skunk, along with their owners.

from those of the more widely distributed skunks. Any problems caused by those two species are solved in the same way as problems from the striped or spotted skunks.

The striped skunk is about the size of an adult house cat, about 30 inches/76 cm including tail, and its fur is mostly black with white on top of the head and neck. In most animals the white extends posteriorly, usually separating into two white stripes. All-black or nearly all-black individuals are sometimes seen.

The striped skunk often is found in wooded areas, close to stream banks and irrigation levees, etc. Litters of four to seven (occasionally up to 10) young are born in late spring or early summer. Some females may produce a second litter. The young are weaned in six to seven weeks and the family breaks up by fall. They become mature in about a year and in captivity have lived for 10 to 12 years.

Spotted skunks, as their name implies, are black with white spots or short streaks of white. They are smaller than the striped skunk, about half the size of a house cat. They are found in a variety of habitats, including brushy areas, stream beds, rocky outcrops, road culverts, industrial yards and around rural homes and farmyards, and in suburban areas which are within about ½ mile/0.8 km of their natural habitat.

Spotted skunks are much better climbers than striped skunks and may occasionally have dens in tree holes. Because of their climbing ability, they are more capable than striped skunks of entering open unscreened windows of a cabin or house or other suitable openings above ground level.

Spotted skunks breed in early spring and have two to six young, born in early summer. In more temperate climates they may have a second litter in the fall.

Skunks are most commonly known for their scent glands, which can eject a potent liquid for six to 10 ft/1.8 to three m. The secretion is acrid enough to cause nausea and can produce severe burning or temporary blindness if it strikes the eyes. The persistence of skunk odor on anything touched by the fluid is remarkable.

The omnivorous diet of skunks comprises grubs and adult insects of various species, small rodents, snakes, frogs, carrion, fruit, berries, mushrooms, bird eggs, nestlings and garbage, if available.

Although they occasionally dig dens in a fresh-cut bank, their dens are most often enlargements of burrows of other animals. Rock piles or outcroppings also are used in some areas, as are hollow logs. When no natural sites are available, they may establish their dens beneath a house or other building if they can gain access. More than one family may occupy a den, although that is not common.

The odor of skunk may be strong or only faintly evident at inhabited dens. Occupied dens will often show signs of fresh digging, at least in spring. Droppings will be evident and usually contain numerous insect fragments. Hair and rub marks can also be found.

Skunks are nocturnal in behavior; their occasional presence may go unnoticed around buildings for awhile until they take up occupancy beneath a building or have a confrontation with some other animal, leaving the tattletale scent of their presence. A faint lingering skunk odor occasionally is detected where skunks have fed or traveled, even though the animals have not scent-sprayed the area.

Skunks have been found infected with an array of diseases that may or may not affect man. These include histeriosis, mastitis, distemper, Q fever, histoplasmosis, microfilaria and, by far, the most important, rabies (Maynard, 1965).

Rabies, an infectious disease caused by a virus organism, is found in the saliva of infected animals about the time they become ill. It is transmitted from animal to animal or animal to man by a bite. The disease, once contracted, is usually fatal to the animal (except bats). If medical attention is given in time to humans bitten by rabid animals, death can be prevented.

More often than any other species of wild carnivore, skunks are implicated in human exposure to rabies through bites. All skunk bites should be attended to by a physician and the local health department should be notified of the incident.

Skunks that seem tame or listless and wander around in the daytime should be treated with suspicion, since such behavior is symptomatic of rabies. And also to the contrary, if they exhibit no fear of humans or a tendency to be aggressive, the chances are quite high they are rabid. In an endemic rabies area, mothers rightfully are fearful their small children will attempt to pet a diseased skunk and be bitten.

As with porcupines, skunks are also a sort of attractive nuisance to dogs, for the inexperienced, and even experienced, dog will investigate the trail of a skunk. The dog nearly always comes out the loser in the encounter. When the dog tangles with a skunk in the woods away from home, your only concern is decontaminating the dog; but if the encounter occurs under your porch or patio deck, the decontamination problem is substantially greater.

Control. Since skunks are furbearers, they are harvested seasonally in some states. Hence, before any reductional control is undertaken, be sure to check with officials of the local fish and game or conservation agency. Legal provisions normally permit action to remove skunks where a health threat exists or where damage occurs.

As with many other vertebrate pests, the best solution to skunk problems beneath buildings is to screen or block them out. Such sealing is essential for all entrances or openings in the foundations of homes and other outbuildings. Spaces beneath porches, stairs and mobile homes should be closed off to keep skunks out.

Once skunks have made their home beneath a building, the problem is a little more difficult, for you have to be sure the animals have left before you close the entrance. That can sometimes be accomplished by sprinkling a smooth layer of flour, about ⅛ inch/0.31 cm thick, on the ground at the suspected entrance to form a tracking patch and then examining the area for skunk tracks soon after dark. When tracks lead out of the entrance, the opening usually can then be safely closed off.

If you are unsure of the number of skunks present, the tracking patch can be supplemented by hanging a section of ½ inch/1.2 cm hardware cloth over the opening, hinged at the top and left loose on the other three sides. It must be larger than the opening, so it cannot swing inward. The skunks will push it open to leave but cannot reenter.

Also, one or two lb/0.45 or 0.9 kg of mothballs (or equivalent in paradichlorobenzene or naphthalene crystals) can be divided into four to six equal parts and placed beneath the infested building away from the entrance. That is sometimes effective in driving skunks out and also keeping them out if replenished occasionally. Ordinary household ammonia has been effective for the same purpose if placed in shallow open-top watertight containers to permit the vapors to penetrate the space beneath the building.

The placement of several floodlights under the floor of the opposite side of the building from their normal entrance has been useful in driving skunks out from beneath a building or from the interior of an outbuilding.

When skunks dig in lawns in search of grubs and other insects, control of the insects will generally remedy the problem though perhaps not immediately.

Live-catch box-type traps constructed of wood, sheet metal or wire mesh probably offer the best method of removing skunks from beneath or around buildings. The traps should be about 10 × 10× 30 inches/25.4 × 25.4 × 76.2 cm or slightly larger. Place them where the animals are entering the building or in trails they are known to use. Bait can be fish (canned or fresh), fish-flavored cat food, raw or cooked bacon or chicken parts.

Skunks are relatively easy to trap and, if the trap is handled with minimum shaking and disturbance, can be carried in the trap to a suitable location for disposal. With live-catch traps constructed of wire mesh, a piece of plastic, canvas or burlap placed carefully over the trap prior to moving it will keep the skunk in the dark and it will be less apt to release its scent. If the trap is covered when set, it may help encourage skunks to enter the trap. There is no assurance a trapped skunk will not scent-spray both you and the area when it is live-trapped or moved. The odds of being sprayed while handling trapped skunks decreases as you gain experience in handling them.

Chloroform, carbon monoxide or carbon dioxide can be used to kill a skunk in the trap, or the trap can be immersed in water to destroy the animal. Because of the potential of spreading rabies, releasing trapped skunks elsewhere is not advised.

Skunks can be caught with No. 1 or 1½ leg-hold traps at the entrance to dens or in regularly used trails. The trap should be set and covered for concealment. It can be left unbaited in some situations or baited just beyond the trap or between two trail-set traps. Fetid scents, some of which are commercially available, often are used to attract them to the traps. The same baits suggested for live-catch box traps can be used also for leg-hold traps.

If the trap is set at the opening to a den or building, the trap stake should be driven the full length of the trap chain away from the opening. That will keep the trapped animal out of the den so it can be removed without tugging on the animal and exciting it into a defensive attitude. It is often desirable to extend the trap chain by eight to 10 ft/2.4 to three m, using wire or chain. This permits pulling the stake and moving the animal with less chance that it will spray the area.

The animals can be disposed of in the same manner as used for box-type traps, though handling a skunk without being sprayed requires more experience with a leg-hold trap than with a box-type trap. There are many stories of various methods of handling skunks without exciting them, though none is 100 percent effective. To be safe rather than sorry, wear old clothes which can be discarded and goggles to protect your eyes.

Since skunks are nocturnal, they can in some situations be spotlighted and shot at night as a control method, assuming the shooting can be done legally and safely.

Since an infestation beneath or around buildings normally involves only a few skunks, trapping should be the control method of choice. In years past, however, skunks were occasionally controlled with acute poisons such as strychnine, particularly where a rabies epidemic was involved in rural areas. Rarely does the situation warrant any toxicants for skunks beneath or around urban or occupied buildings.

Dens found away from buildings can be fumigated with the same fumigants used for woodchucks. Beneath an occupied building, however, safety does not permit fumigation.

Fencing can also exclude skunks from landscaped areas, backyard gardens, school yards, etc. A three ft/0.9 m high/one inch/2.5 cm hexagon wire fence, extending about six inches/15.2 cm beneath the ground and then six inches/ 1.52 cm outward horizontally beneath the earth's surface will discourage most skunks from digging beneath it. Spotted skunks will occasionally scale such a fence, though that is rare.

A chemical called neutroleum-alpha is probably one of the most useful odor neutralizers available for getting rid of the unpleasant odor of skunk scent. About a tablespoonful of the water soluble form in a water bath can be used to decontaminate dogs and humans (Cummings, 1965). It also can be used to scrub basements, garages, floors, walls, outdoor furniture, etc. At a higher concentration (two oz/56 cc to one gal /3.8 l water), it can be sprayed on the soil in a contaminated area.

It is unfortunate neutroleum-alpha is not always readily available when needed. PCO's involved with skunk control should keep some on hand; others may find the only local source to be a hospital supply house. Neutroleum-alpha is used in a variety of deodorizing products. As a last resort, liberal applications of canned tomato juice can be used on contaminated dogs.

RACCOONS

Class Mammalia
Order Carnivora
Family Procyonidae

Biology. The raccoon *(Procyon lotor),* often called "coon," belongs to the family Procyonidae of the order Carnivora. It is one of the best known of all the furbearers. With its distinctive shaggy, grayish-brown fur, unique black "bandit" face mask, black-and-white ringed bushy tail and great popularization in children's books and cartoons, description is almost unnecessary. It is a huskily built animal; the average raccoon weighs 10 to 16 lb/4.5 to 7.2 kg, although an occasional adult male may reach 50 lb/22.6 kg or more. The raccoon is found through most of the United States, although its distribution is limited in Montana, Wyoming, Utah and Nevada.

The raccoon is omnivorous, feeding upon fish, frogs, small mammals, birds and their eggs, mollusks, crustaceans, fruits, nuts, small grains in storage and in fields and certain vegetables such as corn. The long slender toes ("fingers") on its front feet are used with great dexterity in grasping small objects and searching for food, while the larger surfaces of its hind feet allow it to balance easily on its feet. Tracks in the mud or dust, which look like a small human hand print, often reveal that raccoons may be quite common in an area even though only occasionally seen because they are nocturnal and seldom move about in daylight.

Raccoons are usually found near a marsh, stream, lake or pond, but many venture away from natural water. They prefer trees or brush as natural cover. They do not dig dens of their own, preferring hollow logs, hollows in trees and other natural shelters. Sometimes they use rock dens or abandoned burrows of other species.

The single litter per year, in the spring, varies from two to eight (average four). The gestation period is 60 to 70 days. The newborn are almost hairless and their eyes open in 18 days. After about 10 weeks they emerge from the nest and take progressively longer foraging trips, staying with their mother for about one year.

Even though they are normally easily frightened from one's garden, they can be fierce fighters when cornered. In such instances, they have been known to inflict fatal wounds on even relatively large dogs. Raccoons are very inquisitive and, on rare occasions, will gain access into attics, basements or crawl space beneath houses, sometimes taking up residence in barns, stables and various outbuildings. Since they need substantial foraging space and cover, they are less common in areas densely settled by humans unless a stream with wooded banks is nearby. They may become troublesome to those living in rural or sparsely settled suburban areas that have a suitable raccoon habitat.

For the purpose of this chapter, comments will be restricted for the most part to raccoons as a nuisance in and about homes, outbuildings and their immediate vicinity, including landscaping and home gardens.

One of the most common problems with raccoons is their turning over garbage cans and scattering the contents in search of food. The commotion they make in the process can be most disturbing at night. Dogs confined near raiding raccoons often begin barking and howling. By raiding food supplies, backpacks and garbage cans, raccoons are a common problem in both public and private camping and recreational facilities.

Raccoons and their relation to human health are not of great concern, although they often are infested with various fleas and ticks. They have been implicated in several infectious diseases transmissible to humans, including leptospirosis, rabies, Chagas' disease and tularemia (Johnson, 1970). Being more resistant to rabies than are many other carnivores (e.g., skunk and fox), they do not play an important role in its spread, although some exceptions have occurred (Johnson, 1970). Raccoons which are apparently sick and show no fear of man should never be approached or handled and if humans are bitten, medical care should be sought immediately.

Fig. 24-10. Raccoons are readily identified by their distinctive shaggy, grayish-brown fur and unique black "bandit" face mask.

Control. In most states raccoons are considered furbearing animals and are harvested under seasons set by law and regulation. Since the protection offered them varies from state to state, consult your local game warden or conservation agency before any control measures are undertaken.

Where feasible, troublesome raccoons should be discouraged by exclusion. They are usually easy to exclude from dwellings by blocking entrances, although exclusion for some outbuildings is not always practical.

Lids of garbage cans should be secured to prevent removal and the cans should be placed in racks or otherwise anchored to prevent being toppled. On some backyard garden pools, removable frames over which wire mesh has been stretched can be submerged horizontally just below the surface of the water to discourage raccoons from entering and catching goldfish and other species.

In a few situations, special fences can be useful for protecting some backyard vegetable gardens and family-size fruit orchards, but ordinary fences will not exclude raccoons, which are agile climbers. When electric fences are feasible, wire fences of one inch/2.5 cm mesh, 36 inches/91.4 cm high, work well if a single strand of electric wire charged with a cattle-type fence charger is attached on the outside of the fence eight to nine inches/20.3 to 22.8 cm above the ground and about eight inches/20.3 cm outward from the fence. Raccoons, being very sensitive to electric shock, rapidly learn to avoid charged fences.

Another type of electric fence has only three strands of smooth wire spaced 4½, nine and 13 inches/11.4, 22.8 and 33.0 cm above the ground, making a low raccoon barrier. Generally installed as temporary fences, they are effective in keeping raccoons from corn patches and away from poultry houses.

Where safe and legal, spotlighting raccoons and shooting them will eliminate a few animals.

Live-trapping is probably the best method of control for the novice. Tomahawk live traps (Nos. 108, 109, 207, 608, or 609.5), Havahart (Nos. 3 or 3A) and other traps of similar size and style are effective. Raccoons are clever, however, so some are difficult to lure into a live trap. If the trap is the type with a door on each end, set the trap with only one door open. Traps should be secured to a tree, fencepost or stake driven alongside the trap to prevent the animal from tipping it over and stealing the bait without entering the trap. Live-catch traps large enough for a raccoon are quite expensive which greatly increases the cost of control.

Traps should be set to intercept a raccoon as it approaches the garden or garbage cans. They usually follow established trails, fence lines, building or other cover as they move from the woods. In soft or sandy soil, push the trap back and forth until the soil tends to cover the wire mesh on the bottom of the trap.

Chunks of corn-on-the-cob in the milk stage make good bait, as do pieces of melon, prunes, peanut butter and syrup or honey on bread. Also effective in many situations are baits of smoked fish, fish-flavored canned cat food, sardines, cooked fatty meat or fried bacon. The bait should be placed at the rear or closed end of the trap, but protected so the animal cannot reach in through the side of the trap and steal the bait. Live traps should first be tied open for several days to permit the animals to become accustomed to the traps and to feeding on the bait. Once the bait is being accepted, then set the traps.

Experienced trappers find raccoons relatively easy to trap with leg-hold traps. Probably best suited for this purpose is the No. 2 double-coil spring. Several kinds of "sets" are commonly used successfully for raccoons, with the "dirt-hole" set being one of the best. Set the trap about ½ inch/1.2 cm below the ground, one to two ft/30.4 to 60.9 cm from the side of a trail and cover lightly with sifted soil. Cover the trap pan with a piece of tissue, plastic, or canvas to prevent dirt from getting under it and locking it open. Dig a small hole — about six inches/15.2 cm deep and three inches/7.6 cm across — at a slant just behind the trap. The baits are the same as suggested for live-catch traps and are placed in the hole. The trap chain should be securely wired to a stake or a drag.

The "cubby" set especially is useful in winter because it protects the traps from snow and rain, but it can be used in any season. First, a triangular "shelter" should be made with large sticks or small logs about one ft/30.4 cm high and two ft/60.9 cm deep. Scrap lumber can be used if trapping is to be conducted in and around buildings. Hide the trap just inside the open end and place the bait behind it. A cubby set also can be made with two open ends, with a trap in each end and bait in the middle. Using a drag instead of a stake for securing the trap is desirable so that the cubby set will not be destroyed by a trapped animal. A cubby leg-hold trap-set is the best substitute when live or box traps are unavailable.

"Water" sets are popular in many places, particularly along streams. The main advantage of a water set is it is easy to make one which does not arouse the suspicions of a raccoon. Place the trap two or three inches/five or 7.6 cm beneath the surface, in a riffle or a similar shallow place, five or six inches/12.7 to 15.2 cm from shore, or at the entrance to a water cubby. Attach a shiny object (such as a piece of aluminum foil, tin, or a bright button) to the pan of the trap. The raccoon's curiosity causes him to investigate the shiny object on the trap pan.

Metal guards 18 inches/45.7 cm or wider wrapped around trees five to six ft/
1.5 to 1.8 m above the ground can sometimes be used to deny raccoons access
to the roof of a house or other building.

ARMADILLOS

Class Mammalia
Order Edentata
Family Dasypodidae

Biology. Armadillos belong to a primitive group of mammals which includes
anteaters and sloths. The nine-banded armadillo *(Dasypus noveminctus)* has a
protective "armor" of horny material that covers its body, tail, head and legs,
hence its common name. One plate covers the head down to the tip of the nose,
one covers the rear of the animal and another covers the neck and front shoulders. These are connected by nine flattened narrow plates which cover the rib
section. The tail and outside portion of the legs also are covered with protective
plates. Only the ears and underside of the armadillo are without horny covering.
This well-armored animal is unique and should not be easily mistaken for any
other species. The nine-banded armadillo is the only species found in the United
States.

The head and body is about 15 to 17 inches/38.1 to 43.1 cm long, and the
average weight is eight to 17 lb/3.6 to 7.7 kg. The front feet are well adapted
for digging, and the tracks it leaves generally appear to be left by a three-toed
animal.

In spite of its bulky appearance, the armadillo can run rapidly when alarmed.
Also, it takes readily to water and swims very well. Its eyesight is not good, but
is compensated for by a keen sense of smell and hearing. Even though armadillos often may be seen during the day, they are generally nocturnal.

Only one litter a year is produced, generally in February or March. The litter
always consists of identical quadruplets. Gestation is about five months. The
young look like miniature adults, although their protective plates are soft and
leathery.

The armadillo was found only in Texas in the mid-1800's, but since then it
has extended its range northward and eastward and is now found in New Mexico, Arkansas, Oklahoma, Missouri, Kansas and all of the Gulf states (Hall,
1981 and Humphrey, 1974). On the eastern coast it ranges into Georgia and
South Carolina. Its presence in Florida is the result of deliberate introductions.
The reason for this species extending its range into the other states is unknown.

The armadillo commonly is found in dense shady cover, such as brush, woodland or pine forest. It digs rather simple burrows, seven to eight inches/17.7 to
20.3 cm in diameter and up to 15 ft/4.5 m long, usually in rock piles, brush piles,
levees, dikes or around tree stumps. It also may dig beneath orchard trees and
may cause root damage or excessive aeration, weakening or killing the tree. It
often constructs several burrows and may occupy different burrows from day
to day.

Individual armadillos sometimes become a nuisance by rooting in backyard
vegetable gardens or landscaped areas in search of their favorite diet of worms,
insect larvae, pupae and other soil insects. They are fond of rooting and digging
in leaf mold and other rich humus material containing high concentrations of

insects and other soil invertebrates. Like skunks, they can be particularly damaging to lawns of golf courses, cemeteries, parks and playing fields by digging through the turf in search of grubs. Fungi, young plant roots and shoots also make up a small portion of their diet. Consumption of cantaloupes, watermelons, peanuts and tomatoes has been reported (Fitch et al., 1952), though plant matter is not the mainstay of their diet.

The construction of den burrows in these same areas also may be as destructive as their feeding activities. Burrows beneath farmhouses or other rural buildings are generally viewed with some concern. The armadillo's pronounced odor is objectionable to many people. Although a trace of the odor may be evident at all times, it is most evident when the animal is excited or has been engaged in a struggle or conflict. With some exceptions, armadillos are only a very minor pest around dwellings and structures. Structural pest control operators are sometimes called on to handle armadillo problems; hence the need for control information.

Armadillos can be infected with several diseases important to man including leptospirosis and Chagas' disease. More important is the relatively recent discovery that some animals are naturally infected with a leprosy-like disease, but at present the relationship of armadillos in the transmittal of human leprosy is unknown (Walsh et al., 1975).

Control. Indirect control by eliminating much of their food may prove effective in some situations. That can be accomplished by treating the soil with an insecticide to control the insects and invertebrates, banishing the armadillos to better feeding grounds.

Where armadillos are a constant problem in a garden, they can be fenced out. The type of fence suggested for skunks is effective for armadillos too.

Armadillos can be live-trapped in Tomahawk (Nos. 108, 109, 207, 208), Havahart (No. 3), or wooden box traps of similar size (30.4 × 30.4 × 91.4 cm/1 × 1 × 3 ft). Traps should be set near dens, burrows or on trails next to a fence or beside a building. They can be made more effective on some trails by placing two long boards on edge to form a funnel that herds them into the trap. The trap can be baited with overripe fruit such as pears or apples. A small amount of ground meat (marble size) or an egg also is effective in some situations (Chamberlain, 1980).

Number 1 or 1½ leg-hold traps are also effective and should be set at the mouth of the burrow. Leg-hold traps require more skill for success than do box traps. The Conibear trap (size 110) can also be used, if set so as to avoid harming nontarget animals.

Burrow fumigation with carbon bisulfide, methyl bromide, calcium cyanide or smoke cartridges is effective if all of the burrows can be located. The general fumigation procedures used are the same as for woodchucks. Since the armadillo is most likely to be in its burrow during midday, that is the best time to use fumigants.

Poison baits prepared using small eggs or marble-size balls of ground meat have been used. However, only rarely will a situation occur around buildings that will justify the use of toxic baits.

Where legal, armadillos can be hunted at night, using a light to spot them. Shining a light into their eyes at night will often cause them to "freeze" for a moment, facilitating clubbing or shooting. Armadillos are edible and, in fact, are hunted for food in some localities. The meat is quite tasty when properly prepared and fried or barbecued.

OPOSSUMS

Class Mammalia
Order Marsupialia
Family Didelphidae

Biology. The Virginia opossum, *Didelphis virginiana,* is the only marsupial native to the United States. Existence of the genus in ancient times is evidenced by fossils in Pliocene deposits of South America.

The opossum (familiarly known as a "possum") is found throughout most of the eastern half of the country, although absent in the extreme northern latitudes. It is gradually extending its range westward into the Rocky Mountain region. Except for isolated locations, it is absent in large areas just west of the Rockies. However, because of introductions, it is found all along the Pacific Coast (Gardner, 1973).

Opossums are about 12 inches/30.4 cm high and 33 inches/83.8 cm long, including a 12 inch/30.4 cm tail. Their underfur is dense with long guard hair; they have

Fig. 24-11. The opossum is found throughout most of the eastern half of the United States.

five toes on each foot, with an opposable and clawless hallux on the hind feet which functions as a sort of thumb. They are generally light gray in color, although phases of darker color exist. Well-known characteristics of the animal are a naked prehensile tail and a well-developed marsupium (pouch) in the female. Whereas rodents only have two upper incisors and the carnivores six, opossums have 10 upper incisors, making their skulls easy to identify.

Litters may number 18 or more; the young are about the size of a kidney bean and quite undeveloped at birth. Following emergence, the young work their way through the mother's fur to the opening of the pouch, which they enter. Attaching themselves to one of the 13 mammary glands, they continue to develop for four to six weeks in the pouch. All young greater than this number of mammary glands are doomed from the beginning and only about six young usually survive the first few weeks outside the pouch. The young become independent when about 14 weeks old and breed before they are one year old. Females normally produce one litter a year; in the southern regions, however, a second litter may be born.

Opossums prefer living adjacent to streams, lakes, swamps or other bodies of water and in wooded areas where they can find suitable cover. Where food is available, their nightly travel may be limited to a few hundred yards.

Nocturnal in habit and sluggish in its movement, the opossum seems to wander aimlessly throughout an area in search of food. Each opossum usually has a home range of six to seven acres, but it may be much larger. It may move to a new area but stay only a short time, or go a long distance and establish a new homesite. It dens in almost any sheltered location; hollow logs or trees, rock crevices, drainage pipes, culverts, crawl spaces beneath buildings, attics of houses, various accessible outbuildings or holes dug by other animals.

The opossum is quite varied in food habits, eating almost anything available. A true omnivore, it eats garbage, fruit, vegetables, green plants, insects, snakes, frogs, fish, crustaceans, mushrooms, eggs, birds and small mammals. Since opossums eat both fresh meat and carrion, they are often hit by vehicles on highways while attempting to feed on other road-kill animals.

Since opossums have been involved in the transmission of tularemia to humans, they should not be handled or skinned without protective gloves. Additionally, they have been reported to be infected with and may be carriers of a number of other diseases including leptospirosis, relapsing fever, murine typhus and Rocky Mountain spotted fever (Barr, 1963). Fortunately, very few cases of rabies have been reported in opossums and laboratory evidence suggests adult opossums are quite resistant to rabies infections, although the young are more susceptible. Opossums are often heavily infested with ectoparasites.

Occasionally opossums become a nuisance or pest in and around buildings. Where numerous, they may den beneath dwellings or porches or take up residence in attics or outbuildings. They frequently raid uncovered garbage cans and tear open plastic garbage bags set out for disposal. Their nighttime prowling about a premise often arouses kenneled or tied dogs, causing them to bark. In rural areas, poultry or eggs are sometimes preyed upon.

Control. Since opossums may be classified as game or furbearers in your state, check with local fish and game or conservation officials before undertaking reductional control.

The best way of correcting opossum problems resulting from their living under buildings or in attics or outbuildings is to "build them out." That will permanently solve the problem. Paradichlorobenzene or naphthalene crystals have been used with some success as repellents in enclosed areas.

Opossums can be fenced out of an area with a fence similar to that recommended for raccoons, although expensive and rarely justified. Remember, opossums are excellent climbers.

Opposums are readily taken with wooden box traps or with other live traps such as the Havahart (No. 3) or Tomahawk (Nos. 106, 107, 205, 206). Traps should be placed in trails or at entrances to dens. Excellent bait is raw red meat, poultry, fish or moist canned dog food. Leg-hold traps (No. 1½ or 2) can take them, but traps should be set to avoid catching nontarget wild species and pets. Since they are quite sluggish in their movements, they can, if it is legal, be hunted at night with a spotlight and grabbed by the tail, netted, or shot. Dogs are a great aid in tracking them at night.

Strychnine-treated eggs have been used to control opossums in rural areas where their numbers have reached high levels and predation is severe (Merrill, 1962), but strychnine is no longer legal for such use. Rarely will a PCO meet a situation that warrants the use of toxic baits for opossum control.

SNAKES

Biology. Snakes, along with lizards, belong to the order Squamata of the class Reptilia. Actually, snakes evolved from lizards. They are one of the most successful vertebrate forms, living in almost every conceivable habitat: forests, swamps, grasslands, deserts and both fresh and salt water. Some are nocturnal, others diurnal. The majority of snakes are territorial; some are arboreal and others essentially fossorial or aquatic.

Fish stories must take second place to snake yarns. Unfortunately for the story tellers, in America there are no hoop snakes, dreaded puff adders with poisonous breath, snakes with a poisonous sting-tail or species that suck milk from cows. Also, snakes are not "slimy." Snakes, being deaf, live in a world of silence and are not charmed by snake charmers' music. Few creatures are the subject of so much ignorance. The only really dangerous snakes in the United States are the rattlesnakes (massasaugas), copperheads and water moccasins (cottonmouths), all of the pit viper family (Crotalidae). In addition, coral snakes of the family Elapidae can be dangerous.

Snakes utilize various methods in procuring prey, but usually swallow it head first, dead or alive. A water snake seizes a fish, overpowers it and swallows it, often alive. The constrictors, such as gopher snakes and king snakes, utilize their many teeth to capture prey, then coil about it to hold or kill it. The "hypodermic needle" fangs of vipers inject venom into their prey. The venom of pit vipers is mainly hemotoxic, i.e., destroys blood cells and affects heart action, whereas the venom of the coral snakes is primarily neurotoxic, affecting the victim's nervous system but causing little or no swelling or discoloration.

Since snakes have recurved teeth and can dislocate their jaws, they are able to swallow prey perhaps twice the diameter of their own heads. Even so, many snakes also feed on small frogs, salamanders, crawfish, earthworms, slugs, insects and spiders, although the major diet of the larger snakes consists of rodents, lizards, other snakes, eggs and occasionally birds. King snakes can even swallow snakes longer than themselves.

Most snake species produce a large number of eggs enclosed by a tough but flexible membrane. The eggs may grow slightly larger as the embryos mature; baby snakes extricate themselves from the eggs with a temporary egg tooth. Other species, such as garter snakes, water snakes and rattlesnakes, give birth to live young. Garter snakes have been known to produce many dozens of young in a single brood. Once born or hatched, the young are able to take care of themselves. A newborn rattlesnake six to eight inches/15.2 to 20.3 cm long, can strike within minutes, although it is still too small and young to have much venom. Most snakes begin breeding when two or three years old.

Rattlesnakes, water moccasins (cottonmouths) and copperheads are called pit vipers because they have heat sensors located in visible loreal (facial) pits between each eye and the nostril. These sense organs mainly enable them to seek out and capture warmbodied prey, even in darkness. It is even dangerous to pick up a recently killed rattlesnake or the head of a decapitated rattler, for example, because the warmth of the hand will cause the snake's mouth to open reflexively, with the fangs ready to inject venom if the hand can be bitten.

Practically all other snakes commonly seen in the United States belong to the largest family, Colubridae. This includes bull snakes, various racers, gopher snakes, king snakes, garter and water snakes, the horn stinging or thunder snake, coachwhip snake, blowsnake and pilot black snake, to name some of the more common ones.

Being cold-blooded, snakes can regulate their body temperature only by moving to a place where the exposure provides a suitable temperature. In temperate climates all snakes hibernate.

Snakes have a superolfactory organ in their forked tongue which allows them to detect minute odor concentrations in air, water, and soil. For example, after a rattlesnake strikes a cottontail or rodent and injects venom, it then uses its tongue to trail the prey to the place where the poison finally immobilizes or kills it.

Snakes have a "strong" stomach with potent gastric juices. They can digest feathers, fur and bones. Snakes, having no eyelids, never close their eyes and the yellow eye lens sharpens their vision by filtering out ultraviolet rays. Being legless, snakes use their scales to propel themselves forward in a quiet gliding motion. For greater speed, and in sand, they may undulate laterally in a wriggling fashion.

Most snakes have teeth in the upper and lower jaws, and in venomous (poisonous) snakes, some of the teeth are modified with a groove or are hollow to serve as fangs.

Almost any snake in a home garden may, unfortunately, invoke mild hysteria in those who have no affection for snakes. One reason for this concern is the lay public has difficulty in identifying snakes correctly. Only about 16 percent of the species in the United States are dangerous. Several good references on identification of snakes in the United States are Barker (1964), Conant (1975), Leviton (1971) and Stebbins (1966).

Klussman and Ramsey (1976) describe the main distinguishing characteristics of the poisonous snakes found in the United States.

Rattlesnake
- Rattle present, if not lost by mishap.
- Head distinctly wider than neck and somewhat triangular.
- Stout body; adults range in length from 18 inches/45.7 cm in the pygmy rattlesnake to about seven ft/2.1 m in the western diamondback.
- Presence of facial (loreal) pits.
- Vertically elliptical pupils or "cat-eyes."
- Colors vary with species but include shades of brown, black, gray, chalky white, dull red, tan and olive green. Rattlesnakes frequently have definite diamond, chevron or blotched markings of particular colors.
- Large, hollow, movable fangs.

Copperhead
- Reddish-brown crossbands on a lighter background color.
- Head distinctly wider than neck and somewhat triangular.
- Vertically elliptical pupils, or "cat-eyes."
- Usually small and rather slender; adults range from 20 to 40 inches/50.8 to 101.6 cm in length.
- Hollow and movable fangs.
- Presence of facial (loreal) pits.

Cottonmouth or Water Moccasin
- Head distinctly wider than neck and somewhat triangular.
- Usually dark olive, black, or dark brown in color, with 10 or 15 wide dark crossbands. The jaw area below the eye is light compared with the dark color on top of the head. Young snakes are vividly marked, strongly resembling copperheads in pattern, coloration and yellow-tipped tail.
- Vertically elliptical pupils or "cat-eyes."
- Adults rather large and heavy, sometimes reaching five ft/1.5 m long.
- Large, hollow, movable fangs.
- Presence of facial (loreal) pits.

Coral Snake
- Small and slender; adults usually less than 30 inches/76.2 cm long.
- Color pattern consists of yellow, red and black rings encircling the body.

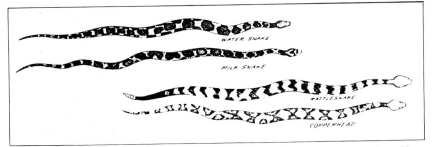

Fig. 24-12. Water snakes and milk snakes are examples of harmless snakes. Venomous snakes found in the United States include rattlesnakes and copperheads.

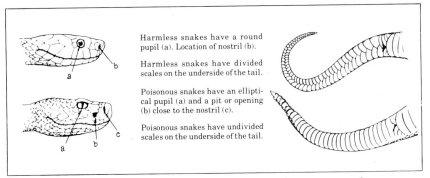

Fig. 24-13. How to differentiate harmless snakes from venomous snakes.

Some nonpoisonous snakes, such as the milk snake, have similar markings, but in the coral snake the yellow and red rings always touch each other. Thus the warning, "Red on yellow kills a fellow."
- Round pupils.
- Head not distinctly wider than neck.
- Fangs short and permanently erect.

Control. Since many people dread even non-venomous snakes, these species are often unwanted around homes. Very few people are willing to tolerate poisonous snakes in their gardens or their yards, especially if children frequently play in the area. For these reasons, snake control is often sought about homes and suburban housing areas. Other situations where some controls may be justified are recreation areas, around farmhouses, during the nesting season at bird sanctuaries, in duck nesting marshes and at fish hatcheries to protect loss of fry.

The first step in control is to find out what kind of snake is creating the problem and to learn something of its habits. Snake control is usually expensive, aside from a single snake trapped in an empty swimming pool or some other place where it readily can be located and removed or killed. Unfortunately there is usually no simple or single method of eliminating snakes.

It is often quite difficult to identify snakes, especially on the basis of a description provided by a distraught mother frightened by a snake in her yard.

Fortunately, some control methods do not require positive identification, although it is usually desirable to know whether the snake in question is venomous or harmless. Except for the somewhat dangerous coral snake of the South, all of the really venomous snakes in the United States have a sensory loreal pit between their eyes and nose and have vertical eye pupils. Most people, however, will not even examine a dead snake closely enough to identify these characteristics.

Actually, rather than attempting to control snakes in mountainous or other wild areas, it is easier to take sensible precautions to prevent accidents with venomous snakes when camping or in the field. Select clean and open campsites and wear trousers and adequate footwear.

If snakes are gaining access into a building, a thorough search should be made for cracks in the foundation and holes in basement screens. Check clearance under door sills and look for improper sealing where plumbing or other conduits enter the base of the building. Vents under buildings should not have screens larger than ¼ inch/0.63 cm mesh since snakes can pass through very small openings.

A fairly sure way of excluding venomous snakes from around a house or a child's play area is to construct a snakeproof fence. Rattlesnakes, copperheads, cottonmouths and most harmless snakes are not likely to climb a fence 18 to 36 inches/45.7 to 91.4 cm high. It should be made of galvanized ¼ inch/0.63 cm mesh (screen) hardware cloth with the bottom edge buried several inches to make sure the snakes cannot go under it.

To make the fence more snakeproof, put the posts on the inside of the fence and make certain the gate fits snugly and has a strong reclosing spring. A fence that slopes outward about 30° is probably better, though much harder to construct. No fence will be effective, however, if lumber or other items are piled against the outside of the fence. In fact, a narrow strip along the outside of the fence also should be kept free of weeds and other vegetation.

Some biological control methods can be used to prevent snake problems. Do not let the adjacent areas become favorable habitats for snakes. In the country, where venomous snakes are common, children's play areas can be made much safer by removing all piles of rock, tall weeds or brush, lumber or other items that might attract snakes to search for prey or to escape from the sun. Effective rodent control around households and the closing of all entrances to rodent burrows makes an area less attractive to snakes. It also helps if one is persistent in controlling rats, mice and field rodents around the household and farm buildings.

Snakes that have entered or escaped in a house may be difficult to find, as many people have learned when a pet snake escaped. One recommendation is to place a wad of damp cloth covered with a dry one at different places along walls (Stickel, 1953). They seem to like the moisture. Each pile must be large enough so the snake can crawl under it. Check the cloth piles daily.

Funnel traps along drift or lead fences capture many snakes (Imler, 1945). The method is time-consuming and expensive, however, because the traps must be examined daily to remove a great number of other kinds of animals which also enter such traps. Den trapping in early spring, where feasible, can be effective regionally and permits nonpoisonous snakes to be released unharmed.

Chorinated hydrocarbon insecticides such as endrin, dieldrin and chlordane are quite toxic to snakes and other reptiles. At one time DDT was dusted in rock piles, cracks, crevices and burrows to control venomous snakes around buildings. With the ban of that insecticide, however, such is not now legal. Where chlorinated hydrocarbon insecticides are used in accordance with directions as foundation treatments for termites, for grubproofing lawns, or for the control of other soil insects through ground spraying or dusting, some side benefits are undoubtedly derived in controlling unwanted snakes and lizards. No chemicals are presently registered by EPA for snake control and no known effective snake repellents are now available.

TOADS AND FROGS

Biology. Toads, frogs and other amphibians are represented by a number of orders in the subclass Salientia and belong to the class Amphibia. Their distribution is correlated closely with the distribution of water. Most are terrestrial as adults, but many frogs, toads and salamanders require free water in which to pass through egg and larval stages. The moist or humid warmer climates have greater and more varied numbers of amphibians than do colder or drier climates.

Amphibians show a wide diversity of tolerance to urbanization, with some surviving very well where others do not (Schlauch, 1976; Campbell, 1973).

Toads and large frogs are sometimes unwanted; some people fear them just as they fear snakes. Under certain conditions they can become quite numerous, crushed by cars in driveways or under foot on garden or park walkways. Such sporadic population increases are usually brief, lasting a few weeks at most.

Frogs occasionally may emerge from ponds in great numbers and disperse into the surrounding area. The adjacent street may become slippery with crushed frogs, creating a local problem.

Frogs and toads also become a problem when they are numerous enough to clog water outlets or filter systems connected to reservoirs supplying domestic water. Small frogs work their way into lawn sprinkler systems and into disconnected garden hoses, which clogs them when they are connected.

Adult amphibians eat live, moving animals such as insects, crustaceans, worms and small mollusks. Some of the large aquatic species, such as bullfrogs, eat small fish, birds and small mammals. The aquatic larvae of toads and frogs feed mainly on algae and bits of dead animal matter in the water.

Toxic acrid secretions of the skin glands on the back of toads protect them from some predators and partly discourage others, although some snakes eat them. The secretions of some species are toxic enough to kill dogs and other predators that gulp their prey (Fitzwater, 1974). Raccoons and skunks are reported to roll toads in the dirt to get rid of the secretion before eating them (Storer, 1943). Raccoons also learn to eat around the head and shoulders, avoiding the poisonous glands (Fitzwater, 1974). Young dogs and cats may be slow in learning of this defense mechanism, although many adult dogs apparently learn to avoid toads after a single encounter.

Control. Indirect control by eliminating the food supply (i.e., insects and other soil invertebrates) with insecticides has been partially effective in reducing the number of resident frogs and toads occupying a yard or garden. When chlorinated hydrocarbon insecticides were extensively used, some direct toxicity to amphibians resulted (Rudd, 1964). Amphibians are especially susceptible to endrin, dieldrin, aldrin, chlordane and toxaphene (Kaplan and Overpeek, 1964; Vinson et al., 1963; Sanders, 1970; Ferguson and Gilbert, 1968). When copper sulfate is used in ponds for the control of algae or other plant growth, it often has a dramatic adverse effect on frogs (Kaplan and Yoh, 1961).

No pesticides, unless specifically registered for that purpose, should be used where there is a possibility of contaminating water that supports fish or water used by other nontarget wildlife. Do not use pesticides near open water supplies destined for human consumption or livestock use.

No chemicals are currently federally registered for the control of amphibians, nor are there any known effective chemical repellents useful for frogs or toads (Fitzwater, 1974).

The presence of toads and frogs generally reflects a healthy habitat and environment. Hence, most people favor their presence and are reluctant to get rid of them even if they are causing minor problems. In those rare situations where a problem becomes unbearable, however, some temporary corrective measure may be needed.Some species of amphibians are protected by law so toad and frog problems generally should be resolved without killing them.

Control through habitat modification may be possible in some situations if a long-term solution is desired. To be effective, the biological requirements of the species involved must be understood, but most often this involves pond or water management (National Academy of Sciences, 1970).

Low barriers, temporary or permanent, as the situation warrants, will remedy most migration problems by redirecting their movements. Low fences about 14 to 18 inches/35 to 45 cm high, constructed of sheet metal, ¼ inch/0.63 cm mesh, or even wood, tile, or plastic can be used to surround an ornamental fish or lily pond or a water trap on a golf course, thereby restricting mass dispersal of young frogs from those ponds. Such a barrier also can protect a swimming pool from becoming contaminated with frog or toad carcasses.

The problem of frogs or toads clogging water-intake screens or filter systems generally can be resolved permanently by correcting the placement of intakes or by installing specially designed screened intakes which do not readily clog with extraneous matter.

PIGEONS

Class Aves
Order Columbiformes
Family Columbidae

Biology. Feral pigeons, *Columba livia,* are descendants of the rock dove or blue rock of Europe, Asia and Africa, and belong to the Columbidae family of birds. They have a long history of being raised and kept by man. As far back as 3,000 BC the Egyptians raised pigeons for food. The pigeon was believed first brought to this country as a domestic bird in about 1606 and it is now found in a feral (wild) state in virtually every city. Many consider feral pigeons interesting and even attractive, however, when they become numerous they invariably take the status of a pest when they conflict with man's interest or when they present health problems.

Since feral pigeons are one of the most common city birds and are recognized by everyone, they require no description. At times, however, native doves and band tail pigeons are mistakenly thought to be feral pigeons.

Feral pigeons adapt well to man-made environments and are the most troublesome bird pest in cities and suburbs. The abundance of shelter provided by the design of many buildings assures pigeons will have ample places to roost, loaf and nest. Food and water is often in adequate supply but, when it isn't, the

Fig. 24-14. Feral pigeons adapt well to man-made environments and are the most troublesome bird pest in cities and suburbs.

birds seek these necessities from nearby rural or undeveloped areas which are generally within flight range.

Pigeon droppings deface and accelerate deterioration of statues, buildings and equipment, besides fouling areas where people may walk or work. Pigeon droppings and nests clog drain pipes and air intakes, mar window sills and render fire escapes hazardous. Their droppings and feathers can contaminate large quantities of stock feed and food destined for human consumption.

The serious and constant public health problems they create are unmatched by any other bird species. They are known to carry or transmit pigeon ornithosis, encephalitis, Newcastle disease, histoplasmosis, cryptococcosis, toxoplasmosis, pseudotuberculosis, pigeon coccidiosis and salmonella food poisoning. Pigeon ectoparasites include a number of bugs, fleas, ticks and mites, many of which bite man (Scott, 1961b).

Pigeons lay one to two eggs about eight to 12 days after mating, which hatch about 18 days later. The young squabs are fed predigested food (pigeon milk) until the young fledge at four to six weeks of age. Additional eggs may be laid before the first brood is off the nest. Pigeons are monogamous, mating for life. Breeding occurs year-round, although the peak reproduction occurs in the spring and summer. Wild pigeons can have a long life span, living for 15 years. Occasionally, in captivity, they live for 30 years or more. In one urban population study, however, 34 percent of the adult birds died each year (Murton et al., 1972), indicating few pigeons under urban conditions live more than three to four years. The same study revealed only about one-third of the adult feral

Fig. 24-15. Pigeons quickly become accustomed to visual repellents such as plastic owls, rubber snakes, etc. Thus, they are generally ineffective.

pigeons reproduced during the year. Juvenile mortality from the fledgeling stage to adult is relatively high (43%), according to Murton et al. (1972).

The diet of pigeons normally consists of seeds, grain, some fruit and green feed. They also feed on garbage, livestock manure, insects and a relatively wide range of other foods when the more preferred foods become scarce.

Control. There are a number of control options available for resolving pigeon problems including sanitation, stoppage (barriers), trapping, repellents, toxicants, etc., yet often the problems are such that it taxes available control measures to the maximum. To solve many bird problems requires a substantial amount of ingenuity and imagination, simply because bird problems vary so greatly in their environmental and physical settings and in their biological complexity. Feral pigeons are not protected by federal laws, but state and local laws should be checked before reductional control through shooting, trapping or poisoning is attempted.

Good sanitation, in the way of removal of spilled food or refuse, can do much to reduce the attractiveness of an area to pigeons. This depends on the situation and whether or not the food source can be effectively and economically limited. The removal of nests and nest sites also may be included as part of a sanitation program.

The best permanent solution to pigeons and other birds that roost or nest in or on buildings, is to build them out by making the site pigeon proof (stoppage). This is easily said but often difficult and/or expensive to accomplish. Openings in buildings, exposed rafters on overhanging dock roofs, bridge bracings, etc. can be screened with rust-proof wire of ¾-inch/1.9 cm mesh, which also will keep out sparrows and starling (Anon., 1960). One half inch/1.3 cm mesh would be needed if rats are also to be excluded. In the last two decades, various types of commercial bird netting has become available and this netting has become more popular for pigeon stoppage. Such netting is less expensive and, because of its light weight, can be installed over expansive areas at much less expense than wire mesh. It is also less visible than wire screening.

Nest destruction can be helpful in preventing pigeon populations from increasing, but to be effective, the nest and eggs must be destroyed at two-week intervals. Nest removal is best when used in conjunction with other types of reductional control. By itself, unless carried out during a long period, it has little effect on localized pigeon populations. To be efficacious, nest destruction must be continued until natural mortality accounts for the surviving adults.

Sharp pointed wires or projections (wire prongs, sheet metal spikes, etc.) can be highly effective as physical barriers in preventing pigeons. It also prevents other birds from perching on ledges and beams of buildings. The temporary discomfort inflicted by the spikes causes the birds to avoid lighting on these surfaces. Several kinds of these devices are commercially available in strip form that can be installed with clip fasteners, wire ties or an appropriate adhesive. Strips of sharp projections can be installed permanently on ledges, rafters, window sills or other locations where birds might roost, loaf or nest. Wide surfaces may require two or more parallel rows of projections. The manufacturers provide instructions for effective installation. The expense of the devices and their installation can be substantial, but their permanent efficacy often justifies the cost.

Installation of grounded electrical wires on roosting surfaces can be highly effective. Such an installation is expensive but offers a relatively permanent solution in keeping pigeons off building ledges. The wires carry high voltage but low amperage current, similar to cattle-type electric fences and intermittently shock the birds without killing them. Such installations are not without problems, for they can be shorted out by an accumulation of dirt on insulators and by sticks and debris which may fall on the wires.

Chemical bird repellents, which are sticky to the touch, are available for application to ledges and beams where pigeons may roost or nest. These non-toxic tacky materials cause the birds to avoid the treated surfaces, but do not entrap them. While effectively repelling birds in many situations, the tackiness is lost in time, principally through an accumulation of dust. Some also are adversely affected by temperature extremes.

The several sticky repellents are sold in a gel form in cartridges for application with caulking guns. The material is laid down in a wavy bead on the edges of roosting surfaces. While the gel form is the most popular, some tacky repellents are available in viscous liquid form to be sprayed or brushed on surfaces. Small squeeze tubes and aerosol cans also are marketed for convenient application over relatively small areas.

Frightening devices in the way of auditory or visual repellents are, in general, ineffective for pigeons. Pigeons quickly become accustomed to plastic owls, rubber snakes, aluminum twirlers, flashing lights and the like. Auditory-type repellents, such as automatic exploders (gas cannons), electronically produced noises, bioacoustics (alarm calls), pyrotechnics and shellcrackers, although effective for some other bird species, rarely give desired results with pigeons.

Live trapping of pigeons can be a very effective method of control. A variety of traps have been used, including loft traps, funnel traps and bob-type traps; with few exceptions, the bob-type trap is the most effective. The sizes of these

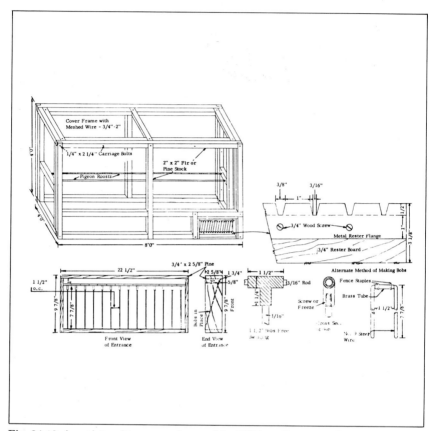

Fig. 24-16. A variety of traps can be used by PCOs to live trap pigeons.

traps vary considerably, from being large enough to walk into to only 10 inches/ 25.4 cm high and 18 to 24 inches/45.7 to 61 cm in width and length. Regardless of size, the bob-type traps all work on the same principle. The door or entrance through which the pigeons are lured is the key feature of the trap and consists of a row of evenly spaced individual, one-way, free swinging bobs. These bobs permit the pigeons to push them upward and inward to enter the trap, but prevents their exit. Grain (e.g. wheat and cracked corn) is scattered at the door entrance to entice the birds into the trap. One to three live decoy pigeons in the trap greatly improve trapping efficiency. Water and food has to be provided for the decoys. Live-catch traps should be serviced frequently to remove captured pigeons. Various methods are used for disposing of trapped pigeons, but in no case should they be taken away from the area and released, for the pigeon's homing ability can defeat any trapping and release program.

The conibear kill trap (Nos. 110 or 120) is an excellent means of removing a few birds from nests. The double wire trigger is spread about one inch/2.5 cm and the trap carefully set on an actively used nest. The trapped birds are killed instantly. The conibear trap and ordinary snap-type rat trap will also take pigeons when baited with pieces of bread or other suitable food and set at a roof-top roosting or loafing area.

Shooting, where legal, is another option in pigeon control, but it is very time consuming if more than just a few birds are involved. Where permissible, shooting with a .22 rifle with bird shot or short range "BB" caps can be effective. Shotguns and high-powered air rifles also have been effectively used. Careless shooting can be hazardous to people and can damage structures.

Toxic perches can be highly effective for pigeon control, but require considerable knowledge of bird behavior for proper and effective placement. The toxic perch is essentially a hollow metal tube two feet/61 cm long with a lengthwise wick that contacts the perching birds' feet and permits transfer of the toxic solution from inside the tube to the surfaces of the feet. The chemical, either endrin or fenthion, is absorbed through the skin or groomed from the skin in amounts which cause death. The birds may die at the exposure site or some distance away at their roosting site (Jackson, 1978). The perches are not placed in a casual way wherever the pigeons might happen to light, but only in select locations where pigeon roosting is highly predictable. In this way, well placed perches are capable of controlling hundreds of birds. Toxic perches also are used in conjunction with roof-top feeders which help attract the pigeons. Perch installations must be serviced at regular intervals.

An aza-steroid (20, 25 diazacholesterol dihydrochloride) is the only reproductive inhibitor (chemosterilant) registered for bird control. It has been registered for pigeon control for a number of years. Success has been reported where the pigeon populations have been thoroughly baited twice a year for three or more years over large areas, such as an entire city. Chemosterilants have been viewed favorably by those who have been faced with an ever increasing opposition to any killing of pigeons. The chemosterilant bait is applied where it will be fed upon repeatedly for about 10 days, which will then inhibit female fertility for about six months. To effectively inhibit reproduction necessitates two baiting periods per year, with the first application suggested for February or March when reproduction is naturally at its lowest. Even with prebaiting, the likelihood of reaching all the pigeons with repeated doses to totally inhibit reproduction in the population is not great. If all reproduction were stopped, the rate the pigeon population would be reduced would equal the death rate of the mature birds. Using a natural mortality rate of 30 percent, a pigeon population of 1,000 birds would be reduced to 343 birds following three consecutive years of treatment. If this chemosterilant is consumed in excessive amounts, it is toxic and can result in some direct mortality of a pigeon population. Even though it is a useful control method, situations where it can be considered an effective solution to a pigeon problem are limited. Treatments involving a single premise, seaport dock or railroad yard are predictably unsuccessful because of the intermingling and immigration of non-sterile birds into the treated area from adjacent areas.

The chemical frightening agent 4-aminopyridine, sometimes referred to as a psychochemical, is available as a bait or concentrates and is quite effective for pigeon control. It is lethal to the birds that ingest sufficient quantities, but prior to death, the affected bird, depending on the species, may display erratic behavior and emit distressing cries which, in turn, frighten the other birds of the flock. The treated bait is diluted with clean bait to limit the number of birds which actually will consume a biologically active dose. In this way, by dosing a relatively small number of birds, the material is capable of producing flock-alarm reactions which repel the rest of the birds from the area. Pigeons are not easily frightened and a high degree of mortality may be needed to appreciably effect the population. Pre-baiting is recommended prior to baiting. Repeated application of bait may be required until the population ceases to return to the area or until an acceptable population limit is attained. After an initial success, bait need only be applied periodically following pre-baiting to keep pigeons from returning to the area.

Several narcotizing chemicals have been used in bait form to immobilize pigeons. The stupefied pigeons then are recovered and disposed of. A number of experimental trials were conducted using this approach but, in spite of its appeal to those who generally oppose pigeon control by more direct means, the method has many shortcomings. The use of narcotizing chemicals for pigeon control has not become a routine approach for pigeon problems in the United States and no chemicals are presently registered for this purpose.

In many situations, strychnine-treated baits remain one of the most economical ways of reducing large pigeon populations in industrial areas, but only where hazard to nontarget species is minimal. Strychnine, in varying degrees, is poisonous to all birds and mammals and its selectivity in control operations relies on the type of bait used and the place and method of exposure. Pigeons killed by strychnine can cause secondary poisoning if eaten in sufficient amounts by other animals.

Pre-baiting with untreated bait (preferrably whole corn) is essential to achieve effective control with strychnine. This conditions the pigeons to feeding on the bait and also permits selection of the sites where feeding is most intense. Once the pigeons take the pre-bait for several days, the poison bait is substituted, but only for one day. The poison bait should be removed at the end of the day and the dead pigeons recovered and disposed of. If additional treatment is necessary, wait at least two weeks to repeat the process after prebaiting again.

HOUSE OR ENGLISH SPARROWS

Class Aves
Order Passeriformes
Family Ploceidae

Biology. The house sparrow, *Passer domesticus,* was introduced from England into the United States in New York in 1850, hence, it received the common name, English sparrow. The bird is not a true sparrow and is a member of the weaver finch family, Ploceidae. There are many species of native sparrows (family Fringillidae) in the United States. A few, such as crowned sparrows (*Zonotrichia* spp.), have become pests, as have English sparrows to certain agricultural crops, but none of these are considered pests in and about buildings. This chapter will primarily address the English sparrow as a city or suburban pest and not the English sparrow problems in agricultural crops, as the remedies differ rather substantially.

As with feral pigeons, the house sparrow is at home in man-modified environments and has adapted well to living in close association with man and his domestic animals. They are found throughout the country, seeming to prefer to reside in cities and around rural farm buildings.

The house sparrow is a small, but stocky bird of 5½ to 6½ inches/14 to 16.5 cm in length. The male house sparrow may be distinguished from all common native sparrows by its black throat and upper breast and ash-gray crown. In addition, a conspicuous chestnut color cape extends from the eyes along the sides and back of the neck. The female lacks the black throat and has a grayish-brown head and rump. Its back is streaked with black and reddish-brown and the breast and under parts are a dirty white with a brownish tinge. Immature birds of both sexes resemble the female.

They are a relatively social bird, nesting in close proximity to one another and flying and feeding in small flocks. Their nests often are built in or on buildings. These nests of twigs, grass, paper and string are built in gutters, on rafters, ledges of buildings and on almost any other conceivable elevated place where they can anchor a nest in a semi-protected spot. They frequently build nests in protected areas such as inside warehouses, airport hangers and stadium roofs.

Sparrows are prolific breeders, raising at least two and up to five broods per year, with the peak breeding season occurring in the spring. Three to eight eggs (averaging 4.5 to five) are laid per clutch, taking 11 to 17 days (average 14) to hatch. The young fledge at about two weeks of age. The annual natural mortality rate of mature English sparrows has been calculated at 54 percent (Anderson, 1978).

The sparrow is primarily a granivorous species, the exception being the breeding season when they feed mostly insects to their nestlings. Sparrows feed on a wide variety of cereals and seeds. However, their diet also includes young seedlings, buds and flowers of a variety of plants. Small soft fruit (cherries, grapes, berries) at times make up part of their diet. Many sparrows thrive at cattle feed lots, dairies, and hog and poultry farms since food is plentiful. The house sparrow has become largely dependent on man for both food sources and nesting sites.

In addition to their messy nests and the contamination and defacement caused by droppings, sparrows damage Styrofoam® and other soft insulation in warehouses and in poultry and hog raising facilities. In electrical substations their nests have been know to cause short circuits and fires. The nesting, roosting and feeding activities all may contribute to the sparrows' pest status.

English sparrows are implicated in the transmission of more than 25 diseases of humans and domestic animals (Weber, 1979) including psittacosis, salmonellosis and several forms of encephalitis. Sparrows in and around poultry and hog farms, because of their disease carrying potential, are of particular concern to farmers. Beside the diseases they may transmit, a number of ectoparasites are associated with the bird and its nests.

Control. The English sparrow is not protected by federal or state laws. However, check with the local game officials concerning any local laws which might prohibit their control, at least by certain methods.

Restricting access, whenever possible, by the eliminating of nesting and roosting sites offers the best permanent solution to the sparrow problem (Anon. 1966). Sparrow stoppage or exclusion may involve screening windows and openings, replacing broken panes of glass, plugging gaps beneath a corrugated metal roof or other similar measures. Open rafters can be netted on the underside to restrict access. Exterior building ledges can sometimes be structurally modified so the birds can no longer roost on the surfaces. Bird proofing a building ranges all the way from a relatively simple and inexpensive process to what is nearly impossible at any expense.

While the breeding period is an extended one, the systematic destruction of nests and eggs at 10 to 12 day intervals will reduce reproduction and often eventually move the birds from a building. This can be done with long poles

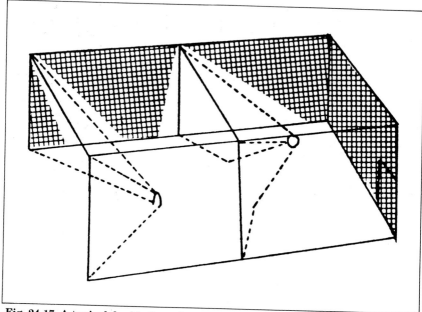

Fig. 24-17. A typical double funnel trap for controlling sparrows.

equipped with a large hook at the end. A strong stream of water also can be used to destroy nests under eaves and on ledges. Nests in vine-covered buildings are more difficult to remove and consideration should be given to severely trimming vines or removing them.

All nests that are knocked down should be cleaned up and destroyed to prevent the birds from reusing the material and to prevent the spread of nest parasites.

Since the natural mortality of sparrows is relatively high, nest destruction, if carried out over two years, will have a noticeable effect on the nesting population. The adults generally return to the same nesting sites year after year, if not disturbed, while juveniles may wander in search of unoccupied nesting sites. Recolonization by the young adults often will occur on buildings previously cleared unless some other corrective action is taken.

Sharp pointed wires or projections used as physical barriers on ledges and rafters are as useful against sparrows as they are against pigeons, except more may be required because sparrows can light on ledges too narrow for use by pigeons.

Electrical wires installed on roosting ledges, as mentioned for pigeon control, are effective for sparrows and starlings also, as are the sticky type bird repellents. However, the ability of sparrows to light on small projections makes the use of sticky repellents more expensive in time and materials. Certain sticky-type bird repellents are suitable for spraying on trees where sparrows, starlings or blackbirds may be roosting.

Most visual and audible scaring devices commonly used for scaring some other species of birds are of only temporary value at best. The use of frightening devices, such as shotguns, shell crackers or pyrotechnics, are a possible exception with sparrows roosting in trees if used before the birds have habituated to the site. The longer a roosting site is used, the more difficult it is to drive the birds elsewhere. It often requires a great deal of persistence to drive sparrows from a site.

A variety of traps have been used to control sparrows, but the funnel trap has been the most popular and, all around, the most effective. Funnel traps are variety of sizes, ranging up to 24 inches/61 cm wide and six to eight feet/1.8 to 2.4 m long, but their basic trapping principle is the same.

The wire mesh trap of ½ or ¼ inch/1.3 or 0.63 cm hardware cloth consists of two compartments; the birds enter the first chamber through a funnel entrance that is at floor level. In their effort to escape the first compartment, they inadvertantly find their way through another small opening which is at the apex of the second funnel that takes them into the holding compartment of the trap. Escape from this compartment rarely occurs. The birds can be removed through a small door. The traps should be serviced regularly and the removed birds destroyed in an acceptable manner. By providing food and water, several birds can be left in a trap as decoys.

Traps of the funnel type can be effectively used in areas where sparrows are in the habit of feeding. They can be baited with canary grass seed, cracked wheat, milo or corn or with chick-scratch feed. Although it is time consuming, a great many sparrows can be trapped from some locations. Trapping success often varies dramatically with the season of the year and the availability of food in the vicinity.

Sparrow traps, other than the funnel trap, are available on the market. They vary in effectiveness, but few can match a good funnel trap.

Shooting to remove a relatively few sparrows from inside large buildings, such as aircraft hangers, is sometimes effective, depending on the situation. Where legal, a .22 rifle loaded with dust (bird) shot or short range "BB" caps can be effective. Shotguns used with #9 shot also are effective, as are highpowered air rifles. The situation must be appraised carefully to avoid hazard to people, equipment and buildings.

Toxic (endrin or fenthion) perches can be highly effective for house sparrow control. Their use parallels that given in the section on pigeons. The key to success is determined through critical observations of flight and perching habits of the bird so perches can be placed in the locations where they will be used to their fullest. This maximizes control with a minimum number of perches.

The frightening agent 4-aminopyridine is registered for sparrow control, but sparrows are not easily frightened. Successful results with sparrows often rely on the associated mortality in the chemically affected birds to reduce the population. Both prebaiting and subsequent baiting should be conducted in those areas where feeding has been observed. Sparrows are ground feeders, but they will feed from "V" shaped troughs and flat feed trays strategically placed in or near sparrow nesting or roosting sites.

Poisoning sparrows with strychnine bait, where legal and feasible, is one of the most effective and economical methods of control, and is comparable to toxic perches. However, each and every situation must be assessed to determine if strychnine baits can be used without hazard to humans or to nontarget species. Consideration must also be given to possible adverse public reaction to poisoning birds.

Flat bait trays or "V" shaped troughs should be located in spots where they will attract the most sparrows. If the trays and troughs are well situated, acceptance of the prebait by house sparrows should be well established within one week. Remove the remaining prebait and then place strychnine bait in the trays or troughs, but do not leave the poison bait exposed more than two days. If further population reduction is required, repeat the process by prebaiting again for one week prior to exposing the poison bait.

STARLINGS
Class Aves
Order Passeriformes
Family Sturnidae

Biology. The common or European starling, *Sturnus vulgaris,* is a member of the starling family Sturnidae and was deliberately introduced from Europe into New York in the 1890s (Kalmbach, 1931). After living in that part of the country for about 30 years, they rapidly extended their range and now are found from coast to coast in most areas of the United States, extending to Alaska and Hawaii.

The starling is a stocky 8½ inch/21.6 cm, short-tailed bird with dark adult plumage which is iridescent blue-black with the feathers tipped in light tan. The flecked feathers are most pronounced in their winter plumage; the beak is a blackish color. As they approach the spring breeding season, the beak changes to yellow. The young, immature birds are a plain brownish-gray.

The starling produces a number of characteristic rasping, squawking or squeaking calls, most of which are harsh and unpleasant to hear. They will, however, mimic or imitate other species such as robins and bobwhite.

Starlings mate in the spring and build their nests, for the most part, in cavities such as tree hollows, woodpecker holes, bird houses, palm trees and crevices or confined structural spaces on buildings. The nest is constructed of fibrous material lined with fine grass and other soft material which may be available. From three to eight (averaging four to five) pale greenish-blue eggs are laid per clutch; these hatch in about 12 days. The young remain in the nest for two to three weeks. One to three broods per year may be produced.

After young birds leave the nest, they form small flocks. As the summer progresses, the flocks increase in size to hundreds or thousands of birds. These large flocks may come together from miles around to establish large communal roosts. In winter, the communal roosts usually consist of a larger number of birds than the summer roosts. Since starlings are somewhat migratory in their habits, in the fall thousands of starlings may move from the northern breeding areas to more southerly regions.

Starlings, although predominantly insectivorous, are omnivorous in their feeding habits. They feed their young entirely on insects. Starlings are very fond of various soft fruits and berries such as grapes, figs, cherries and apples (Howard, 1959). They also will feed on seed and grain, and in the winter, livestock feeds may constitute a large portion of their diet. Starlings are highly adaptive and in time of food scarcity they will feed on almost anything, including garbage. Much of the success of starlings is due to their wide diet, together with their adaptability in using resting sites and their aggressiveness.

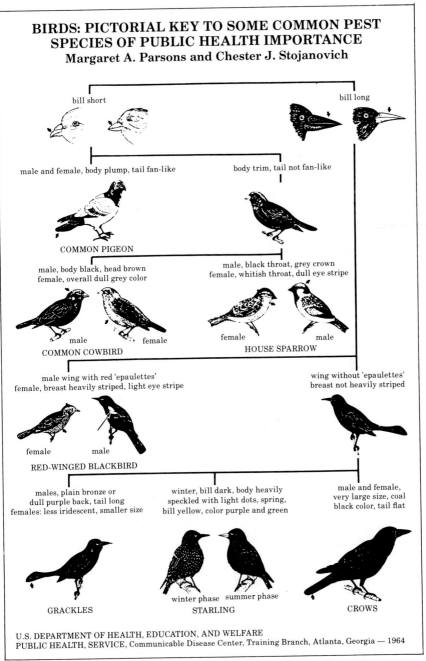

BIRDS: PICTORIAL KEY TO SOME COMMON PEST SPECIES OF PUBLIC HEALTH IMPORTANCE
Margaret A. Parsons and Chester J. Stojanovich

bill short

bill long

male and female, body plump, tail fan-like

body trim, tail not fan-like

COMMON PIGEON

male, body black, head brown
female, overall dull grey color

male, black throat, grey crown
female, whitish throat, dull eye stripe

male   female
COMMON COWBIRD

female   male
HOUSE SPARROW

male wing with red 'epaulettes'
female, breast heavily striped, light eye stripe

wing without 'epaulettes'
breast not heavily striped

female   male
RED-WINGED BLACKBIRD

males, plain bronze or
dull purple back, tail long
females: less iridescent, smaller size

winter, bill dark, body heavily
speckled with light dots, spring,
bill yellow, color purple and green

male and female,
very large size, coal
black color, tail flat

GRACKLES

winter phase summer phase
STARLING

CROWS

U.S. DEPARTMENT OF HEALTH, EDUCATION, AND WELFARE
PUBLIC HEALTH, SERVICE, Communicable Disease Center, Training Branch, Atlanta, Georgia — 1964

Fig. 24-18. Pictorial key to common bird pests of public health importance.

In cities, starlings are disliked because of the vocalizations made at roosting time and because of the filth they leave behind. Problems of starlings nesting on or in buildings is of relatively minor concern compared to the roosting problem. The starling has adapted well to cities where winters are severe, utilizing a multitude of roosting sites that provide shelter and warmth. Building ledges, lighted signs, marquees and billboards with ample structural bracing or ledges all make excellent roosting sites, and thousands of starlings may invade a city at dusk and roost side by side forming solid rows of birds. Starlings also roost in trees, taking on a pest status when the trees are in a city park or close to human habitation. Not only is the roost obnoxious to the senses, but the birds are known to transmit diseases such as encephalitis, ornithosis and histoplasmosis. Starlings, in addition to consuming large amounts of poultry, hog and cattle feed, are implicated in the spread of diseases of livestock, for example, hog cholera.

Since this book is concerned with pests in and around buildings and structures, none of the many problems of starlings in growing cereal, fruit and vegetable crops will be included. Remedies for starling crop depredation problems differ from those employed in a city roosting problem.

Starlings are not protected by federal or state laws. However, local ordinances may preclude their control by certain methods (i.e., use of sound). Other bird species, such as blackbirds or robins, may roost with starlings. Therefore, permits may be needed before certain control measures are instigated.

Control. Starling control in urban roosts is best undertaken as city or community programs. It is the winter starling roost that presents the biggest problem, for the birds may return to the same site night after night from about mid-November to about mid-March. As with other bird problems, exclusion is the only permanent solution and, when feasible, ledges and irregular ornamental architecture that provide starling roosts should be screened off with wire, nylon or plastic netting. Support beams and braces of large lighted signs and billboards can be blocked off, although this is often very expensive.

Habitat modification is sometimes of value where starlings are roosting in street or park trees. Select pruning, to open up the canopy of the trees, may make the trees unsuitable for roosting cover.

Starlings can be repelled from night roosts with sound such as pyrotechnics, exploding cracker shells and recorded starling distress calls. For best results, such distress calls or noise repellents should be initiated as soon as the birds begin using the location and should be continued until they leave. Three to four consecutive evenings is generally adequate to move the birds to another roost, hopefully less objectionable. However, if after six to seven days, the birds have not moved, the technique should be reevaluated or discontinued. Scaring must begin early in the evening when the birds first begin to arrive and when there is sufficient light for the starlings to find alternate roosts. When repelling large numbers of starlings, there is always a risk that the new or alternative site selected by the birds will also be objectionable to humans.

Sharp pointed projections and grounded electrical wires also can be used to repel starlings from ledges, as they are described for pigeons and sparrows. Roosts with innumerable perching sites to accommodate the large flocks of starlings greatly increase the cost of using wire projections or sticky repellents, even though they can be effective. Reapplication of sticky repellents is usually necessary in order to maintain maximum effectiveness (refer to section on pigeons).

The surfactant called PA-14, sometimes referred to as a stressing agent, has been effectively used to control starlings and blackbirds in communal roosts. Sprays containing recommended amounts of PA-14 in water are applied either by ground equipment or by aircarft onto the roosting birds. This nonionic surface-active agent (detergent) lowers the surface tension of water and, thereby, enhances the feather wetting ability of water. This enables the spray solution to penetrate the oily feathers, reducing their insulation properties.

To be effective, applications must be limited to nights when at least ½ inch/ 1.3 cm of rain and temperatures below 45° F/7° C can reasonably be expected to occur following treatment and before the birds leave the roost area at dawn. The cold temperature causes the wet birds to lose body heat and mortality results through stress induced by chill.

PA-14 is registered by EPA and can only be used under the direction of the U.S. Fish and Wildlife Service. For several reasons, stressing agents are most suitable for roosts located in wooded situations in rural areas.

The frightening agent, 4-aminopyridine, does not lend itself to use in most urban roosting situations because it must be consumed in baits and starlings do not feed at the roost site. It is, however, highly effective for repelling birds from feeding sites such as cattle feed lots, dairies and hog and poultry farms.

The modified Australian crow trap with live decoys makes an effective starling trap at feed lots, orchards and vineyards where starlings are feeding (Zajanc and Cummings, 1965). Traps are not the solution to problems stemming from winter roosts.

The lethal avicide 3-chloro-p-toluidine hydrochloride is effective in some situations, but, as with 4-aminopyridine or trapping, its effectiveness is limited primarily to feeding sites, although some baiting success has been achieved in the birds' preroosting staging area.

Toxic perches using endrin or fenthion are registered for starlings and can be effective if the roosting situation lends itself to perch placement. To be effective, a large number of perches may be needed and the presence of dead birds in the area must be tolerable from a public relations point of view. Various limitations make toxic perches a rather impractical technique for ridding moderate or large numbers of starlings using roosts.

WOODPECKERS

Class Aves
Order Piciformes
Family Picidae

Biology. Woodpeckers belong to the family Picidae which also includes the flickers and sapsuckers. Various species are found throughout the country. Some are migratory while others are not.

Occasionally, and generally quite locally, certain species of woodpeckers and flickers cause damage to wooden structures by their pecking or drilling of holes. The species involved include the Pileated woodpecker, *Dryocopus pileatus;* Acorn woodpecker, *Melanerpes formicivorus;* and Red-shafted flicker, *Colaptes cafer.* Several others also may be implicated regionally.

Because of their sap feeding habits, the Yellow-bellied sapsucker, *Sphyrapicus varius varius;* Red-breasted sapsucker, *S. v. ruber;* and the Williamson's sapsucker, *S. thyroideus,* have all been considered pests to trees (Oliver, 1968). Sapsuckers are not, however, a problem in structural damage.

Woodpeckers are essentially arboreal in their habits and obtain the greater part of their food from trees. Their physical conformation, including short legs with two sharp clawed backward pointing toes and stiff tail feathers enables them to cling easily to the trunks and branches of trees or to the wood siding of a house to strike effective blows with a stout chisel-shaped beak. Their specially developed long tongue can be extended a considerable distance and is used most effectively to dislodge larvae or ants from their burrows in wood or bark.

Sapsuckers differ strikingly from other genera of woodpeckers in both feeding habits and tongue structure. They feed on cambium, phloem and sap of trees. Their tongue is well adapted for this purpose and, instead of the long barb-tipped tongue of other woodpeckers, the sapsucker has a tongue tipped with stiff hairs to which sap droplets readily cling.

The characteristic feeding of sapsuckers leaves a distinct pattern of a uniform series of holes drilled through the bark and cambium and often into the wood. These are arranged in neat parallel rings or partial rings around a tree trunk or limb.

Problems with woodpeckers damaging buildings are relatively infrequent and tend to be localized. Wooden houses or buildings in the suburbs or in the more rural wooded settings are most apt to be damaged, for this is the habitat where the woodpeckers thrive. The numbers of woodpecker always are relatively low and they do not flock in large numbers, although some species live in small social groups. Generally, damage to a building involves only one or two birds, but possibly up to six or eight, at any one period.

They can be particularly destructive to summer or vacation homes which are vacant during part of the year, since they are free to continue their activities until the owners return.

Damage to wooden buildings may take one of several forms. Holes may be drilled into wooden siding or eave facing boards and, if the accessible cavity is suitable, it may be used as a nesting site. The acorn woodpecker is responsible for drilling rows of holes just large enough for each to accommodate an acorn. Acorns also may be wedged between or beneath roof shakes, with a few holes actually being drilled. Woodpecker damage to utility poles can be severe and widespread in some regions, necessitating frequent pole replacement (Jorgensen et al., 1957). Contrary to one rather common belief, some species of woodpeckers readily peck holes in sound wooden fence posts, utility poles and in wood sidings of homes and outbuildings. The wood need not be insect infested, as many believe.

Drumming is the term given to the noise made by woodpeckers pecking in rapid rythmic succession on wood. This is a springtime activity of males proclaiming their territories. Drumming may occur a number of times during a single day and the activity may go on for some time. When occupied houses are used for drumming, the noise can become quite annoying and considerable disfiguring is done to exterior surfaces.

As mentioned previously, sapsuckers do not damage buildings or wooden structures, but they are responsible for rows of holes pecked in the bark of healthy trees. Where this is occurring to trees in landscaped areas, it becomes of concern to the owner. Repeated attacks may cause sufficient damage to girdle the tree, causing a tree or limb to die above the injured region. Fungi and insects can enter the sapsucker wounds, resulting in further injury or death of the tree. In some regions, foresters experience substantial losses from sapsucker damage due to preharvest mortality and/or reduced timber quality (Caslick, 1976).

Control. All members of the woodpecker family are protected by federal, as well as by the laws of most states. A permit is required from the U.S. Fish and Wildlife Service to kill damage-causing birds. Non-lethal methods of resolving the problem are always preferred over having to kill the birds.

Sheet metal or hardware cloth of ¼ or ½ inch/0.63 to 1.3 cm mesh can be fastened over areas of wood buildings which are being damaged.

The eaves or wood siding can also be netted with nylon or plastic netting to exclude the birds. The netting should be attached so there is at least four inches of space between the netting and the damaged building.

Sticky or tacky bird repellents are sometimes effective to prevent building damage, but may have to be reapplied periodically. When these materials are used for pigeon, sparrow or starling control, they are generally applied to flat surfaces, but for repelling woodpeckers the material must be applied to vertical surfaces, often in areas of the building where they are very visible. This sometimes presents problems as some of these materials tend to run in hot weather and others may stain some wood surfaces or finishes. As a precaution against unforseen problems, before using any of these materials to cover a large area of a building, apply some to a small test area in an inconspicuous place.

Where permits are granted for taking the one or two woodpeckers or flickers involved in doing the damage, shooting may be the quickest method of dispatching the birds. At close range, air rifles or .22 caliber rifles loaded with dust shot or "BB" caps can be effective. The offending birds may also be taken by nailing a wooden-base rat trap just below the hole or pecked area. Results are better if the trap trigger is enlarged with a square of cardboard or wire screen.

Sapsuckers can be discouraged from shade or backyard trees by wrapping hardware cloth or burlap around the area being tapped. The sticky type bird repellents are also sometimes effective.

LITERATURE

ANDERSON, T.R. — 1978. Population studies of European sparrows in North America. Museum of Natural History, University of Kansas, Occasional Papers No. 70:1-58.

ANONYMOUS. — 1959. Control of tree squirrels. U.S. Department of Interior, Fish and Wildlife Service, Washington, D.C. Wildlife Leaflet 403. 4 pp.

ANONYMOUS — 1960. Controlling birds — vagrant domestic pigeons. U.S. Department of Interior, Fish and Wildlife Service, Washington, D.C. Wildlife Leaflet 413. 4 pp.

ANONYMOUS. — 1961. Controlling tree squirrels. U.S. Department of Interior, Fish and Wildlife Service, Lafayette, Indiana. Leaflet 324. 2 pp.

ANONYMOUS. — 1966. Controlling sparrows. U.S. Department of Interior, Fish and Wildlife Service, Lafayette, Indiana. Leaflet 207. 2 pp.

ANONYMOUS — 1969. Bats. U.S. Department of Health, Education and Welfare, Public Health Service. Health Information Series No. 141. 4 pp.

BARKER, W. — 1964. Familiar Reptiles and Amphibians of America. Harper and Row, New York. 220 pp.

BARR, T.R.B. — 1963. Infectious diseases in the opossum: A review. J. Wildlife Management 27(1):53-71.

CAMPBELL, C.A. — 1973. Survival of reptiles and amphibians in urban environments. In: Wildlife in an Urbanizing Environment. Monograph — Planning and Resource Development, Series No. 28. Massachusetts Cooperative Extension Service, University of Massachusetts, Amherst, Massachusetts. pp. 61-66.

CASLICK, J.W. — 1976. Biodeterioration by birds. Proceedings Third Inter. Biodegradation Symposium (J.M. Sharpley and A.M. Kaplan, eds.) pp. 309-315. Applied Science Publication, London. 1138 pp.

CHAMBERLAIN, P.A. — 1980. Armadillos: problems and control. Proceedings Ninth Vertebrate Pest Conference (J.P. Clark, ed.), Fresno, California, March 4-6, 1980. University of California, Davis. pp. 163-169.

CLARK, D.O. — 1972. The extending of cotton rat range in California — their life history and control. Proceedings Fifth Vertebrate Pest Conference (R.E. Marsh, ed.), Fresno, California. March 7-9, 1972. University of California, Davis, pp. 7-14.

CONANT, R. — 1975. A Field Guide to Reptiles and Amphibians of Eastern and Central North America. Houghton, Mifflin, Boston. 429 pp.

CONSTANTINE, D.G. — 1958. An automatic bat-collecting device. J. Wildlife Management 22(1):17-22.

CONSTANTINE, D.G. — 1967. Bat rabies in the western United States. Public Health Reports 82(10):867-888.

CUMMINGS, M.W. — 1965. Skunks and their control. University of California Extension Service. OSA No. 126. 2 pp.

DAVIS, R., C. HERREID & H. SHORT. — 1962. Mexican free-tailed bats in Texas. Ecological Monographs 32:311-346.

FAULKNER, C.E. & W.E. DODGE. — 1962. Control of the porcupine in New England. J. Forestry 60(1):36-37.

FERGUSON, D.E. & C.C. GILBERT. — 1968. Tolerances of three species of anuran amphibians to five chlorinated hydrocarbon insecticides. J. Mississippi Academy of Science 13:135-138.

FITCH, H.S., P. GOODRUM & C. NEWMAN. — 1952. The armadillo in the southeastern United States. J. Mammalogy 33(1):21-37.

FITZWATER, W.D. — 1974. Reptiles and amphibians — a management dilemma. Proceedings Sixth Vertebrate Pest Conference (W.V. Johnson, ed.), Anaheim, California. March 5-7, 1974. University of California, Davis. pp. 178-183.

GARDNER, A.L. — 1973. The systematics of the genus *Didelphis* (Marsupialia: Didelphidae) in north and middle America. Texas Tech University, The Museum, Special Publications No. 4. 81 pp.

GILL, J.E. & R. REDFERN. — 1980. Laboratory trials of seven rodenticides for use against the cotton rat *(Sigmodon hispidus)*. J. Hygiene 85:443-450.

GREENHALL, A.M. & J.L. PARADISO. — 1968. Bats and bat banding. U.S. Department of Interior, Fish and Wildlife Service, Bureau of Sport Fisheries and Wildlife Resource Publication 72. 47 pp.

GUNIER, W.J. & W.H. ELDER. — 1971. Experimental homing of gray bats to a maternity colony in a Missouri barn. The American Midland Naturalist 86(2):502-506.

HALL, R.E. — 1955. Mammals of Kansas. University of Kansas. Museum of Natural History Publication No. 7. pp. 136-138.

HALL, R.E. — 1981. The Mammals of North America. Volume II. John Wiley & Sons, New York. 576 pp.

HOCKENYOS, G.L. — 1959. Control of chimney swifts and bats in chimneys. National Pest Control Association, Technical Release No. 20-59, lp.
HOWARD, W.E. — 1959. The European starling in California. The Bulletin, California Department of Agriculture. 48(3):171-179.
HULL, T.G. (ed.). — 1963. Diseases transmitted from animals to man. Fifth Edition. Charles C. Thomas, Publisher, Springfield, Illinois. 967 pp.
HUMPHREY, S.R. — 1974. Zoogeography of the nine-banded armadillo *(Dasypus novemcinctus)* in the United States. BioScience 24(8):457-462.
IMLER, R.H. — 1945. Bullsnakes and their control on a Nebraska wildlife refuge. J. Wildlife Management 9:265-273.
JACKSON, W.B. — 1978. Rid-A-Bird perches to control bird damage. Proceedings Eighth Vertebrate Pest Conference (W.E. Howard, ed.), Fresno, California. March 7-9, 1978. University of California, Davis. pp. 47-50.
JOHNSON, A.S. — 1970. Biology of the raccoon *(Procyon lotor varius* Nelson and Goldman) in Alabama. Auburn University Agricultural Experiment Station Bulletin 402. 148 pp.
JORGENSON, R.N., H.T. PFITZENMEYER & W.C. BRAMBLE. — 1957. Prevention of woodpecker damage to wooden utility poles. Pennsylvania State University Agricultural Experiment Station, Progress Report 173. 4 pp.
KALMBACH, E.R. — 1931. The European starling in the United States. U.S. Department of Agriculture, Washington, D.C. Farmers' Bulletin No. 1571. 26 pp.
KAPLAN, H.M. & J.G. OVERPECK. — 1964. Toxicity of halogenated hydrocarbon insecticides for the frog, *Rana pipiens*. Herpetologica 20(3):163-169.
KAPLAN, H.M. & L. YOH. — 1961. Toxicity of copper for frogs. Herpetologica 17(2):131-135.
KING, J.A. (ed.). — 1968. Biology of *Peromyscus* (Rodentia). American Society of Mammalogists, Special Publication No. 2. 593 pp.
KLUSSMAN, W. & C.W. RAMSEY. — 1976. Poisonous snakes of Texas. Texas Agricultural Extension Service, Texas A & M University, College Station. 4 pp.
LAIDLAW, G.W.J. & M.B. FENTON. — 1971. Control on nursery colony populations of bats by artificial light. J. Wildlife Management 35(4):843-846.
LEVITON, A.E. — 1971. Reptiles and Amphibians of North America. Doubleday, New York. 250 pp.
MALLIS, A. — 1969. Handbook of Pest Control. Fifth Edition. MacNair-Dorland Company, New York. 1158 pp.
MARSH, R.E., P.C. PASSOF & W.E. HOWARD. — 1974. Anticoagulants and alphanaphthylthiourea to protect conifer seeds. In: Wildlife and Forest Management in the Pacific Northwest (H.C. Black, ed.). Oregon State University. pp. 75-83.
MAYNARD, R.P. — 1965. The biology and control of skunks in California. California Vector Views 12(4):17-20.
MERRILL, H.A. — 1962. Control of opossums, bats, raccoons, and skunks. Proceedings Vertebrate Pest Control Conference, National Pest Control Association, Elizabeth, New Jersey, pp. 79-97.
MURRAY, K.F. — 1968. Bats in California. California Vector Views 15(10):109-113.
MURTON, R.K., R.J.P. THEARLE & J. THOMPSON. — 1972. Ecological studies of the feral pigeon. *Columba livia* var. J. Appl. Ecol. 9:835-874.
NATIONAL ACADEMY OF SCIENCES. — 1970. Amphibians and reptiles as

pests. In Vertebrate Pests: Problems and Control, Volume 5, National Academey of Sciences, Washington, D.C. pp. 42-57.

NATIONAL PEST CONTROL ASSOCIATION. — 1964. Bats. NPCA Technical Release No. 17-64, 2 pp.

NATIONAL PEST CONTROL ASSOCIATION. — 1975. Bat control. NPCA Technical Release No. 5-75, 8 pp.

NATIONAL PEST CONTROL ASSOCIATION. — 1980a. Tree squirrels, Part I Biology and habits. Technical Release ESPC 043201A. 4 pp.

NATIONAL PEST CONTROL ASSOCIATION. — 1980b. Tree squirrels, Part II Control. Technical Release ESPC 043205A. 4 pp.

OLIVER, W.W. —. 1968. Sapsucker damage to ponderosa pine, J. Forestry 66(11):842-844.

PRATT, H.D. — 1958. Ectoparasites of birds, bats and rodents and their control. Pest Control 26(10):55-56, 58, 60, 94, 96.

RUDD, R.L. — 1964. Pesticides and the living landscape. The University of Wisconsin Press, Madison. 320 pp.

SANDERS, H.O. — 1970. Pesticide toxicities to tadpoles of the western chorus frog *Pseudacris triseriata* and Fowler's toad *Bufo woodhousii fowleri,* Copeia 2:246-251.

SCHLAUCH, F.C. — 1976. City snakes, suburban salamanders. Natural History 85(5):47-53.

SCHNURRENBERGER, P.R., J.R. BECK & F. BURSON. — 1964. Bat rabies — a discussion of problems existing in Ohio. Ohio State Medical Journal 60(4):361-364.

SCOTT, H.G. — 1958. Bats — public health importance and control. U.S. Department of Health, Education and Welfare. Public Health Service CDC Training Leaflet. 4 pp.

SCOTT, H.G. — 1961a. Bats — public health importance, identification and control. Pest Control 29(8):4 pp.

SCOTT, H.G. — 1961b. Pigeons — public health importance and control. U.S. Department of Health, Education and Welfare, Public Health Service, Communicable Disease Center. CDC Training Program. 18 pp.

SPENCER, D.A. — 1948. An electric fence for use in checking porcupine and other mammalian crop depredations. J. Wildlife Management 12(1):110-111.

STEBBINS, R.C. — 1966. Field Guide to Western Reptiles and Amphibians (Peterson Field Guide Series, 16). Houghton Mifflin, Boston. 279 pp.

STICKEL, W.H. — 1953. Control of snakes. U.S. Fish and Wildlife Service. Wildlife Leaflet 345. 8 pp.

STORER, T.I. — 1943. General Zoology. First Edition, McGraw-Hill Book Company, New York. 798 pp.

TRUMP, R.F. & G.O. HENDRICKSON. — 1943. Methods for trapping and tagging woodchucks. J. Wildlife Management (4):420-421.

VINSON, S.B., C.E. BOYD & D.E. FERGUSON. — 1963. Aldrin toxicity and possible cross-resistance in cricket frogs. Herpetologica 19(2):77-80.

WALSH, G.P., E.E. STORRS, H.P. BURCHFIELD, E.H. COTTRELL, M.F. VIDRINE & C.H. BINFORD. — 1975. Leprosy-like disease occurring naturally in armadillos. J. Reticuloendothelial Society. 18(6):347-351.

WEBER, W.J. — 1979. Health Hazards From Pigeons, Starlings and English Sparrows. Thomson Publicatins, Fresno, Calif. 138 pp.

WEINBURGH, H.B. — 1964. Field rodents, rabbits and hares — public health importance, biology, survey, and control. U.S. Department of Health, Education and Welfare. Public Health Service, Communicable Disease Center, Atlanta, Georgia. 87 pp.

WILSON, D.E. & J.S. FINDLEY. — 1972. Randomness in bat homing. The American Naturalist 106(949):418-424.

ZAJANC, A. & M.W. CUMMINGS. — 1965. A cage trap for starlings. University of California Agricultural Extension Service. OSA No. 129.

ARNOLD MALLIS

Arnold Mallis was born in New York City on October 15, 1910, the son of Russian immigrants to the United States. He attended grade school and high school in Brooklyn and Long Island. In 1927, at the age of 16, he moved with his parents to Los Angeles, California, where he completed his last year of high school.

In 1929, he entered the University of Southern California and began pre-dental courses. Since the Great Depression began at about this time, he dropped out of college and worked in a Los Angeles garment factory. After two years on the job he decided to return to college and, because of his interest in trees and natural history, began a two year course in forestry at Pasadena Junior College (now Pasadena City College).

He entered the University of California at Berkeley in 1932, where he studied under such outstanding entomologists as Professor E. O. Essig in economic entomology, Dr. E. C. Van Dyke in forest entomology, Professor W. B. Herms in medical entomology and insect ecology, and Dr. W. M. Hoskins in insect toxicology. These teachers and others instilled in him a life-long interest in insects and their control. He received a bachelor's degree in entomology in 1934.

After two years graduate work he left the University in 1936 and sought employment as an entomologist in southern California. Since there were no jobs open in entomology he took the state examinations for a license in structural pest control and received a Class A license (all categories, including fumigation).

Mallis worked two years for pest control firms in Los Angeles, Hollywood and Bakersfield and then in 1938 obtained a job as a field aide in the USDA Bureau of Entomology, doing research on vegetable insects in southern California. This job lasted for six months. He then returned to Berkeley for two purposes, to complete his master's degree on the ants of California and to collect information for the first edition of the Handbook of Pest Control. In 1939, he returned to southern California and became entomologist-in-charge of pest control for the Buildings and Grounds Department, UCLA. In 1942, shortly after the United States entered WWII, he became associated with the USPHS, implementing malaria control programs around military camps in Louisiana. In 1944, he began to screen compounds as pyrethrins substitutes for Hercules Powder Company at the University of Delaware. The screening resulted in the discovery of Compound 3956, later known as toxaphene.

In 1945, Mallis commenced his employment in the entomology laboratory of Gulf Oil Corporation in Harmarville, Pennsylvania. Here he worked as an entomologist on household insects and household insecticides for 23 years until the company closed its entomology laboratory in 1968. He then became an extension entomologist for The Pennsylvania State University and retired as an associate professor in 1975, at the age of 65.

CHAPTER TWENTY-FIVE

Miscellaneous Household Pests

By Arnold Mallis[1]

ANYONE WHO HAS worked in pest control for some time comes up with unusual household pests that make life interesting and often present new control problems. These pests are referred to in a number of ways such as "miscellaneous pests", "occasional invaders" or "minor pests." It should be noted that although it may be a "minor" pest to the pest control operator, it is a "major" pest to the homeowner or tenant. Moreover, what is a common pest in one part of the country may be a rare or introduced pest in another area.

EARTHWORMS

Phylum Annelida
Class Oligochaeta
Order Terricolae

   Earthworm biology. Earthworms in the soil are almost always beneficial. Occasionally however, they may become unduly plentiful, particularly after heavy rains, when they will crawl about walks, patios and into the cellar and other accessible parts of the house. Ordinarily, no control is necessary since the invasion ceases with the disappearance of the excess soil moisture. Walton (1928) discusses the economic aspects of earthworms and much of the following is from his paper.

   The earthworms in which we are interested belong to the family *Lumbricidae*. These worms have a great number of common names such as earthworms, angleworms, dewworms, night crawlers, fishworms, rainworms, etc.

   Earthworms are most common in moist soils containing considerable organic matter. They are considered beneficial animals since they aerate the soil and triturate or divide decaying vegetable matter into organic material, thus making it available to the plants.

   Although earthworms feed to a large extent on the soil, they will feed on meat, fat and sugar. At times they will also consume young seedlings and occasionally

[1]*Associate Professor (Retired) Extension Entomologist, The Pennsylvania State University, University Park, Pa.*

may occur in such numbers in lawns that they loosen the roots of grass, and their hillocks or heaps of castings may render the soil lumpy.

The most common earthworms belong to the genus *Helodrilus.* The rain-worm, *Lumbricus terrestris,* is one of our largest worms, reaching a length of 10 inches.

Earthworms have no eyes, nevertheless they are sensitive to light as demonstrated when they have a light flashed on them at night. They crawl by the "alternate contraction and expansion of the muscular rings forming their bodies" and are assisted in these movements by the short, stiff bristles on the side of their bodies.

Walton discusses the reproduction of earthworms as follows: "Every earthworm is both male and female. Each worm produces eggs from the forward half of the body near the flesh-colored, swollen band that may be observed there. The eggs are deposited within a ring of gelatinous matter that quickly solidifies and is then slipped over the head and assumes the shape of a capsule. Such capsules contain several eggs, together with a quantity of albumin to serve as food. The young worms emerge from the capsule fully formed and ready for their struggle for existence. The egg capsules of the common, large earthworms are about the size of a grain of wheat." The young worms emerge from the cocoon in one to five months and are ready to reproduce after another six to 18 months. In captivity *L. terrestris* has lived six years.

Control. In the past DDT and other chlorinated insecticides were used where earthworms became a pest. At this writing we are unaware any pesticide is registered for earthworm control. However, diazinon as used for the control of sowbugs in the lawn should be effective against earthworms. In addition, certain benzimidazolecarbamate fungicides, especially benomyl, have produced substantial incidental kills of earthworms when used against fungal infections of crops or turf.

SOWBUGS AND PILLBUGS

Phylum Arthopoda
Class Crustacea
Order Isopoda

Biology. Sowbugs and pillbugs are the only crustaceans that have become completely adapted to spending their whole life on land. They have oval bodies, convex above and flat or hollow beneath. They never reach more than ¾ inch/ 19 mm long. The head and abdomen are small, but the thorax is comparatively large, composed of seven hard individual but overlapping plates. There are seven pairs of legs, the large pair appearing only after the first molt.

The common pillbug, *Armadillidium vulgare* (Latreille), and the dooryard sowbug, *Porcellio laevis* Koch, and *P. scaber* (Latreille) are world wide in distribution. In certain instances they become pests of young plants. They like moist locations and are found under objects on the damp ground, as well as under vegetable debris of all kinds. At times they may even bury themselves several inches in the soil.

The sowbugs can be separated from the pillbugs by the fact they cannot roll up into a tight ball like the pillbug. Moreover, the sowbug has two prominent tail-like appendages the pillbug does not have. Sowbugs and pillbugs will at times invade damp basement areas as well as the first floors of houses. When

Fig. 25-1. Upper left, *Armadillidium vulgare* (Latreille); upper right, the pillbug rolled up; lower left, *Porcellio laevis;* lower right, side view of sowbug.

this occurs, they are almost always present in considerable numbers in the soil and under plants immediately outside the house.

Sowbugs and pillbugs normally feed on decaying vegetable matter. The female carries her young in a pale-colored vivarium or marsupium on the underside of the body. She gives birth to many very white living young. There are usually one or two generations a year, varying with environmental conditions. They become inactive during the winter months, although this may not be the case in artificially heated buildings such as greenhouses.

Hatchett (1947) studied the biology of several species of isopods in Michigan. He notes that "isopods are considered to be gravid when they are carrying eggs, embryos or young in their marsupia. The period of gravidity, therefore extends from the time the eggs are first deposited in the marsupium until the young leave it." However, this period is quite variable since the young may leave of their own volition. Thus, *Porcellio scaber* and *Armadillidium vulgare* may carry the brood for an average of 44 and 43 days, respectively. The females may bear two or more broods per year. *P. scaber* averaged 28 young and *A. vulgare* 24 young per brood.

According to Hatchett (1947), an isopod molts for the first time within 24 hours after leaving the marsupium. Second and third molts occur during the second and third weeks respectively. Beginning with the fourth instar, molts occur every two weeks until the animal is 20 weeks old. After this age the period between molts is irregular. Many sowbugs live to be two years old.

Control. Leaves, grass clippings, mulch, boards, stones and similar materials close to the building should be removed since these may harbor sowbugs and pillbugs.

For chemical control outdoors one of the following may be used:

■ Bendiocarb — Mix one packet or one scoop of bendiocarb per gallon water

(0.25%) and apply outside around doors and other places where they can enter the building.

- Diazinon 4E — Apply four fl. oz/114 ml in three gallons/11.4 l water per 1,000 sq. ft./93 square meters.
- Propoxur — 70 percent wettable powder — Mix two ounces/57 g to one gallon/3.8 l water and apply as a residual spray. Do not treat plants (lawns, flowering plants, shrubs, trees, etc.) with this concentration.

For indoor use — Apply contact or residual sprays usually used to control crawling pests indoors.

AMPHIPODS

Phylum Arthropoda   "HOUSE HOPPER"
Class Crustacea
Order Amphipoda   TALITRIDAE
Talitrus sylvaticus Haswell ✕

Fig. 25-2. Young emerging from vivarium of mother pillbug, *Armadillidium vulgare* (Latreille).

Homeowners in Los Angeles are very often startled by the appearance of small red shrimp-like animals in their homes. These are dead when found and thus are a nuisance merely because of their presence. This amphipod is related to the beach flea. C.R. Shoemaker, in correspondence with the author, states, "This species was first taken in California in 1918 and has been taken there several times since. It is probable that it has been observed on many occasions which have not been recorded. The species was first described from New South Wales and afterwards reported from several of the Pacific islands, including Hawaii, and in 1918 it was first observed in California, which constituted the first record for the Western Hemisphere. It has been taken also in Louisiana."

Mallis (1942) found the living amphipods, which are brownish-black in color, to live under ivy used as a ground covering. Here they jump about like fleas and are captured only with great difficulty. The amphipods were present in the soft ground up to a depth of ½inch/13 mm.

It is very likely rain or excess moisture forces these animals to seek a somewhat drier environment. They then invade the house by jumping through windows, or crawling over thresholds, and here they die because of insufficient moisture, their dark-brown bodies turning shrimp red. Ivy, as a ground cover, and leaf mold beneath shrubbery apparently offer a suitable habitat for these moisture-loving animals. The entry of this amphipod into the home is a rare occurrence and on no occasion reported has it been necessary to resort to control methods after the initial complaint. However, Ebeling (1975) notes the use of an 0.5 percent diazinon emulsion was effective in a troublesome situation.

R86   - SEP
✕ R87394 - MAY

Les Escargots
My heart doesn't harden
At snails in a garden
I'm fond of their leisurely gait
But here at the table
I wish they were able
To run, like gazelles, off my plate
Milton Bracker

SLUGS AND SNAILS

Phylum Mollusca
Class Gastropoda

Slugs and snails ordinarily are of little interest to individuals concerned with household pests. Nevertheless, occasionally they may assemble in large numbers in basements, on walls and doorways and along walks making these areas unsightly. Moreover, they may be crushed underfoot, resulting in an unsightly

Fig. 25-3. Amphipods that invade the home.

and odorous mess. However, not everyone regards snails with distaste for they have been prized as food by the Stone Age people and when prepared by modern chefs they are consumed with gusto by many gourmets.

The common slugs in and around the home are the spotted garden slug, *Limax maximus* L., the tawny garden slug, *L. flavus* L., and the true garden slug, *Deroceras agrestis* L. The brown garden snail, *Helix aspersa* (Muller), and the cellar or greenhouse snails of the genus *Oxychilus* are among the more common house-invading snails. Lange (1944) lists 19 species of slugs from California alone.

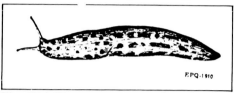

Fig. 25-4. The common spotted garden slug.

White and Davis (1942) have prepared the following notes on the habits of these common creatures: "Snails and slugs are mainly nocturnal, but they come out of their hiding places and feed in the evening or on dark days. Their favorite hiding places are under old decaying boards and logs, under board walks, in cellars, creameries and springhouses, in rock piles, along hedgerows and beneath damp refuse. Snails are less particular in this respect than slugs, as they have the power, when confronted with unfavorable living conditions, of sealing the opening of the shell with a mucous sheet, the operculum, which soon hardens to a leathery texture. The snails then become dormant, and some may exist for as long as four years. When conditions again become favorable, the 'door' of the shell is rasped away, and the snail resumes its normal activity.

"It is probable that a few slugs survive the winter out of doors in the colder regions, but they are able to survive in such places as drain pipes, cellars, greenhouses, storage pits and well walls. They are said to perish quickly when exposed to temperatures below freezing. Snails seem to be more hardy. Greenhouse snails have successfully passed the winter out of doors as far north as Washington, D.C., by sheltering themselves beneath trash piles.

"Soon after emerging from the eggs, the young slugs and snails begin to move about in search of food. This consists of such material as is near at hand, since they do not wander far, remaining for four or five weeks in a colony near the place where the eggs were deposited. In some of the introduced species this colonial habit persists throughout life, but the native species tend to wander farther and farther afield until the colony is broken up. The homing instinct is well developed in some slugs, each individual returning to its particular hiding place night after night, unless disturbed or unless the place becomes too dry. The route taken in returning is usually the same as that taken in going out. Snails and slugs will, if possible, avoid all dusty, dry or sharp objects."

According to White and Davis (1953), the life history of a slug is as follows: Slugs lay oval translucent eggs, about 25 or more under boards, trash, flowerpots, compost, earth clods, stones and similar materials.

The eggs are laid from spring to fall. However, in greenhouses slugs may deposit eggs in the winter. Incubation of the eggs require about one month at 60° to 70° F/16° to 21° C. Slugs reach adult size in three to 12 months. In temperate climates they may live less than a year. Temperature and moisture are of prime importance in regard to slug activity.

Natural enemies of slugs, and for that matter snails too, are toads, some predacious beetles and their larvae, wild birds, and chickens and ducks.

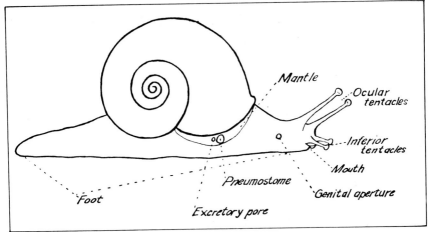

Fig. 25-5. Diagram showing the external anatomy of a snail.

White and Davis (1953) note the life history of snails and slugs are similar in some respects. They present the life history of the white garden snail, *Theba pisana,* as somewhat typical of the common snails:

"The eggs are round and white and have a calcareous, or limy, shell. They are laid in a cavity hollowed out by the parent snail about one inch beneath the surface of the ground. From 10 to more than 200 eggs may be laid in one mass, depending on the time of the year and the size of the parent. As in the case of the slugs, weather conditions control the time of incubation, but the average is about 18 to 20 days. The young snails are very small when newly emerged from the egg and for several months they remain close to the place of hatching. They grow slowly, adding coils to the shell as they grow. Probably as with the slugs more than a year is required for them to attain full size. The European species of *Helix* require two or three years."

Control of snails and slugs. Sanitation is important in the control of snails and slugs. Check the places where these pests are in the greatest numbers as in accumulations of leaves and cut grass, under loose boards, flat stones and similar places and remove such habitats where possible.

Methiocarb, a carbamate pesticide, 4-(methylthio)3,5-xylyl methylcarbamate, has largely replaced metaldehyde as a pesticide for the control of snails and slugs. It is available as a two percent granular bait and is broadcast around buildings at the rate of one lb per 1,000 sq. ft./0.45 kg per 93 sq. meters. Boards tilted on stones can be used to cover the bait where children and dogs are present because both have been known to eat the bait.

Metaldehyde baits consisting of bran, additional materials and added toxicants may still be on the market although it appears mexacarbate baits and sprays are no longer available. Metaldehyde baits and their application are described as follows:

The bait is distributed as a wet mash in small piles (one tablespoonful) on the ground immediately around the house. The bait should be put out towards evening. Snails and slugs are more active on foggy, moist nights, or immediately after a rain, and control should be conducted against them at such times. It is often necessary to repeat the treatment several times at semi-weekly intervals. Once the bran is moistened, it will not keep, and for this reason small amounts, sufficient for one treatment, should be prepared for immediate distribution.

Woglum (1941), speaking of some tests conducted with baits containing metaldehyde, calcium arsenate and combinations of the two in southern California, notes: "Metaldehyde baits appear to be less effective when broadcast as compared to being placed in small piles; arsenical baits were about equally effective both ways during dry weather, but piles gave a slightly higher kill during rainy weather. During cool weather, as in January, metaldehyde alone gave no kill at all in the shade. In commercial practice, metaldehyde baits are often placed near the tree trunks and growers are sometimes greatly impressed by the spectacular accumulation of snails. However, our observations show many of these snails merely *stupefied* and not killed. It is possible that the percent kill of snails by metaldehyde baits gradually improves over a period of time but for immediate and certain results, metaldehyde baits are more effective when placed where sunshine has access to the piles of material."

Thomas (1948) recommends one part metaldehyde to 30 parts bran for the control of slugs.

Floyd F. Smith, an entomologist with the USDA, has found stale beer to be very attractive to slugs. "Placed in shallow pans, the beer attracts 30 times as many slugs as metaldehyde, a standard bait for decades. The slugs crawl into the pans and drown, apparently finding it a wonderful way to go."

He said fresh beer is equally effective, a recommendation not likely to gain widespread approval!

MILLIPEDES

Phylum Arthropoda
Class Diplopoda

Back (1939) studied millipedes as household pests. These creatures normally live outdoors where they feed on damp and decaying wood and vegetable matter, as well as on tender roots and green leaves on the ground. Their slow-crawling, rounded bodies have two pairs of legs on each body segment except for the first three which have but one pair of legs on each segment. Millipedes protect themselves by means of glands which secrete an unpleasant odor. There are about 1,000 species of millipedes in the United States and only a few experts can tell them apart.

As there is a dearth of information on the life histories of some of the common millipedes, the author wrote to N.B. Causey, an authority on millipedes, for information on these animals. The answer was as follows: "In brief, the life history of most of them is as follows: fertilization is internal; eggs are deposited in clusters in the soil, sometimes in a capsule prepared for them by the female; the first larvae have three pairs of legs in most species; the larvae pass through a series of molts; during each molt the number of legs and segments is increased, the number of molts, legs, and segments varying with the species. Usually there are from seven to 10 molts. When sexual maturity is reached, the molting is usually discontinued. Many species reach sexual maturity the second year; others spend four or five years in the larval stage and then live several years after that." Cloudsley-Thompson (1958) discusses the life-history of millipedes in some detail.

According to Back, the two species most annoying to homeowners are *Parajulus venustus* Wood and *P. impressus* (Say). They are brownish, one to 1½ inches/25 to 38 mm long, nocturnal and hide during the day beneath various objects located on damp soil.

Back has the following to say regarding their injurious habits: "At certain times millipedes become restless for some reason and then leave the soil and crawl into houses, sometimes swarming over basements and first-floor rooms. Millipedes are most troublesome during the fall of the year, when hordes may crawl into area-ways, into cellars, up foundation walls, and through small openings and reach the interior of living rooms. They may crawl up side walls and drop from ceilings. More often they are confined to the basement and to floors of rooms over unexcavated soil. The fall migrations are possibly concerned with a natural urge to seek hibernation quarters. Sometimes continuous or heavy rains raise the water level in the soil, force the millipedes out of their natural abode, and send them in search of shelter elsewhere. In cottages in wooded areas, or in newly developed areas with virgin soil filled with decaying vegetation, as many as 700 millipedes may enter a room during an evening."

An infestation of the millipede *P. impressus* (Say) in a suburb of Pittsburgh was brought to the attention of the author in November, 1949. The millipedes were climbing in large numbers over the threshold into the house and underneath the garage door into the garage much to the consternation of the housewife. The house had been built a year or two previously on some newly subdivided land. Strangely enough, other homes along the street were rarely invaded. In approximately one week, the infestation abated without any effort on the part of the owner. In the late fall of 1950, the house once again underwent an invasion of the millipedes; the infestation was not nearly as severe as the previous year. The owner applied a suspension of 50 percent DDT wettable powder around the entire house and suffered no further annoyance from these pests.

Another millipede epidemic occurred in western Pennsylvania from November 13 through November 18, 1958. The author's laboratory received 15 telephone calls and pest control operators in Pittsburgh and vicinity probably received hundreds more. This small millipede, ½ to one inch/13 to 25 mm in length, which was later determined as *Iulus* sp., was by the hundreds, climbing the walls of houses and invading the house itself. The epidemic occurred during a period of unusually warm weather for that time of the year, the temperature rising above 75° F/24° C during the day. A cold snap November 18 immediately put a stop to the complaints from householders.

Snetsinger (1967) noted 90 percent of the millipedes that invaded buildings in central Pennsylvania were *Uroblaniulus jerseyi* Causey. He writes about this

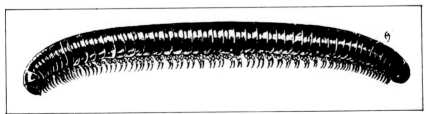

Fig. 25-6. Millipedes normally live outdoors where they feed on damp and decaying wood and vegetable matter, as well as on tender roots and green leaves on the ground.

invasion as follows: "In October, *U. jerseyi* did not come out until shortly after sunset. At first the millipedes were observed crawling in the forest cover debris and grass. Quickly they migrated to the trunks of trees, walls of dwellings, etc. Often 20 or 30 millipedes could be collected on a tree trunk during times of heavy migration. They climbed up the tree trunks to the height of nine feet maximum. The millipedes remained on the trunk for four to six hours on the nights during the study. Toward morning the millipedes migrated down the tree trunks and walls into the soil where they spent the daylight hours. The following night the same procedure occurred. Mating apparently occurs when the millipedes are on the tree trunks or on the walls.

"The millipedes overwinter in the soil near the foundations of dwellings and under debris near the trunks of trees. Invasions of dwellings appear to be accidental, associated with the search for a humid site for overwintering. In most basements millipedes shortly die because the environment is too dry. New millipedes replace those that die if the weather is warm enough so that millipedes remain active in the evenings."

Bennett and Kerr (1973) noted the two most important millipede pests in Florida were the greenhouse millipede, *Oxidus gracilis* Koch, and the tropical millipede, *Orthomorpha coarctata* (Saussure). The greenhouse millipede breeds in decaying leaf litter and the tropical millipede in lawngrass thatches. They were most active through the night. In laboratory tests, carbaryl, propoxur and methomyl were effective against them.

None of the aforementioned invasions compare with that of the millipede, *Fontaria,* recorded by Brooks (1919) on farms in West Virginia: "Mr. McDougle stated that on the morning following the coming of the army to his home he opened a screen door between his kitchen and back porch and that the door on swinging back swept up a heap of millipedes a foot in height. He immediately got a shovel and cleaned up two washtubsful from the porch and from a small ditch that extended along the side. Every morning thereafter for two weeks he collected ½ bushel or more about his.home and carried them away."

Similar incidents have been recorded periodically in Europe. For instance, in 1878 a train was brought to a halt in Hungary by a mass of millipedes that carpeted the ground and made the wheels slip on the rails. Trains were again stopped in this way in northern France in 1900.

Control. Since millipedes thrive in damp organic matter, accumulations of leaves, plant debris, mulch and similar materials should be inspected and removed if necessary. Check for wood imbedded or buried in the soil. Sugerman (1960) and others have found this an important source for millipedes.

For chemical control see sowbugs and pillbugs. Thurston et al. (1959) applied sprays or dusts for millipedes in a 15-foot/4.6 m band around the building. He also applied them to the lower part of the outside walls.

Story (1981) recommends the following insecticide control methods: "Perimeter spraying is the best method of control but the spray should be very thorough to ensure soaking the soil in a five- to 20-foot band around the building. Since millipedes can travel from a few inches to several feet in a minute it is important to use insecticides that produce a kill after a few minutes exposure. Carbamate insecticides are faster acting on millipedes than organophosphates and I recommend you use Baygon®, Ficam® or Sevin®. Wettable powder formulations of these insecticides will provide better residual control on soil or concentrate than other formulations. Particular attention should be paid to treating across doorways and other openings."

SCORPIONS

Phylum Arthropoda
Class Arachnida
Order Scorpionida

Stahnke (1950) has been a student of scorpions and scorpion venoms for many years and much of the following is from his work.

Scorpions are most common in the southern states from the Atlantic to the Pacific, but they are found as far north as British Columbia. Most of the scorpions that invade the house and sting when accidentally crushed or contacted are relatively non-venomous and their sting is no more poisonous than a bee or wasp. However, there are extremely venomous scorpions in North Africa, India, North and South America and other parts of the world (Roch, 1941). In the United States one deadly species of scorpion is to be found for the most part in southern Arizona and adjacent areas of California, New Mexico and nearby Texas. Stahnke reports a total of 64 deaths due to scorpion stings, many of them small children, during a 20-year period. The importance of the problem in the Southwest can be observed from the fact that physicians reported 1,573 cases of "scorpionism" during a 10-month period.

Scorpions have eight legs, a pair of large pincers or pedipalps and a pair of mouth pincers or chelicerae. Underneath at the base of the last pair of legs are the comb-like pectines which are organs of touch. The body of the scorpion consists of a cephalothorax, seven other segments of the trunk and six segments of the tail. The last bulbous segment of the tail can be moved like a universal-

Fig. 25-7. Side view of the scorpion, *Hadrurus sp.*

SCORPIONS—PICTORIAL KEY TO SOME COMMON UNITED STATES SPECIES

Chester J. Stojanovich and Harold George Scott

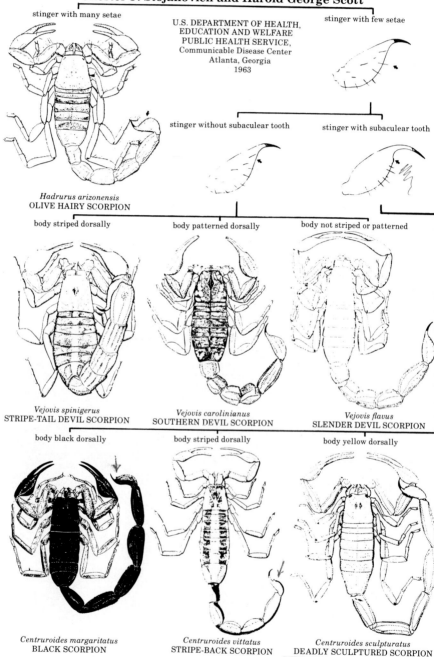

stinger with many setae

U.S. DEPARTMENT OF HEALTH, EDUCATION AND WELFARE PUBLIC HEALTH SERVICE, Communicable Disease Center Atlanta, Georgia 1963

stinger with few setae

stinger without subaculear tooth

stinger with subaculear tooth

Hadrurus arizonensis
OLIVE HAIRY SCORPION

body striped dorsally

body patterned dorsally

body not striped or patterned

Vejovis spinigerus
STRIPE-TAIL DEVIL SCORPION

Vejovis carolinianus
SOUTHERN DEVIL SCORPION

Vejovis flavus
SLENDER DEVIL SCORPION

body black dorsally

body striped dorsally

body yellow dorsally

Centruroides margaritatus
BLACK SCORPION

Centruroides vittatus
STRIPE-BACK SCORPION

Centruroides sculpturatus
DEADLY SCULPTURED SCORPION

Fig. 25-8. Pictorial key to some common species of scorpions found in the United States.

joint and "contains two poison glands with ducts leading into the terminal stinger." Although scorpions possess from two to 12 eyes, their sight is very poor and their pincers are used as feelers.

The common striped scorpion, *Centruroides vittatus* (Say.), is typical of the group, has two broad, dark longitudinal bands on the dorsal side of the abdomen, with yellowish brown appendages and postabdomen. The male is 6.7 centimeters long and the female reaches a length of 5.9 centimeters. It is distributed widely in the South.

Ewing (1928) notes: "Out of doors it is found very abundantly under the loose bark of large trees and logs, and under logs and stones on the ground. About human habitations it prefers probably above all else the woodpile and crumbling stone or brick foundations. In some parts of Texas the writer has found it infesting back yards, and reports of its infestation of houses have been frequent." This scorpion dwells under objects on the ground. When these are turned over, the scorpion will sting if touched.

Smith (1927) studied the life history of this species. The mother produces living young which climb on her back and remain there for five to 15 days. They molt in three to six days. Maturity is probably attained in three to four years. The female produces an average of 32 young, ranging from 25 to 39 young. Smith has the following to say in regard to the food habits of this animal: "Scorpions are easily kept in captivity if provided with water and food. They will eat small insects such as grasshoppers and roaches, but refuse caterpillars. They will also eat raw lean beef. Young scorpions feed readily upon termites, apparently in preference to anything else." Carr (1946) notes the scorpion "seizes its prey with its lobster-like pincers and either subdues it through the "squeeze" system or poisons it by means of the needle-sharp stinger."

There is one extremely venomous scorpion in the United States, *Centruroides sculpturatus* Ewing. At one time it was thought there was another very poisonous species, *C. gertschi* Stahnke, but later it was shown to be a color variant of *C. sculpturatus*. This scorpion reaches a maximum length of three inches/7.6 cm although two inches/five cm is the usual size. The "tail" is relatively thin, being 1/16 inch/1.6 mm in diameter. At the base of the stinger is a characteristic blunt thorn. The pincers are narrow, about six times as long as the widest part.

According to Ennik (1972), *C. sculpturatus* occurs "in western New Mexico, Arizona, in adjacent Mexico and sporadically along the west bank of the Colorado River in California." Ennik also notes scorpions can be separated into two groups, those that occur *above* ground level and those that burrow into the ground and are at ground level. *C. sculpturatus* usually occurs above ground level under tree bark, in palm trees and crevices in rocky cliffs. This scorpion often clings to loose objects and people are stung when they startle or accidently touch it.

The scorpions responsible for mild poisoning cause local swelling and pain at the site of the sting. In regard to the highly poisonous *C. sculpturatus*, Ennik (1972) notes it contains in its venom "protein fractions which have a marked effect on neuromuscular impulse transmission and perhaps with functioning of other systems." He describes the medical symptoms as follows: "The sting of *C. sculpturatus* usually produces a localized burning sensation *without* local tissue response, numbness at the site of the sting and sometimes around the mouth and over the face, hyperactivity, profuse salivation, syncopy, dysphonia, dysphagia, and in some cases, convulsions. Respiratory distress may follow."

Fig. 25-9. (Left) *Hadrurus sp.* and (right) the poisonous *Centruroides* sp.

Children under 12 and elderly persons are especially vulnerable and should be kept under observation for at least 24 hours to make certain there are no serious complications. Ennik further states, "There is no first aid measure of real value in the treatment of scorpion stings. Placing a piece of ice over the wound site may reduce pain and cause some desirable local vasoconstriction." Antivenins for treatment of stings by *C. sculpturatus* are available only at the larger medical institutions.

These poisonous scorpions are attracted by their desire for moisture to the vicinity of the home and the house proper. Stahnke notes around the home scorpions will thrive in accumulations of lumber, bricks, brush, etc., and in the house they may occur from the cellar to the attic. The venomous scorpions can readily enter the house through cracks ¹/₁₆-inch in height. Occasionally scorpions are carried into the house, particularly in firewood.

Stahnke recommends individuals living in scorpion-infested areas observe some simple precautions. Objects should be picked up carefully and inspected on the underside. Also, shoes and clothing should be shaken vigorously before putting them on. The editor knows of a fellow entomologist who, putting on his undershorts in darkness, found a scorpion had climbed in before him — a case of judgment striking in the end. The legs of a crib should be placed in wide-mouthed glass jars since scorpions cannot climb clean glass. Beds should be removed from the walls. Moreover, one should not walk barefooted at night since it is at this time that scorpions become most active.

Control. According to Stahnke, scorpions can live for six months without food and water, and they may hide for two to three months after feeding. Moreover, very venomous species require three to five years to reach maturity. Thus, continued and thorough control is essential to rid a dwelling of them.

In controlling scorpions, the greatest success has been obtained through the use of residual insecticides. Stahnke notes contact insecticides such as pyrethrum in oil base will kill scorpions if they are wetted with the spray. However, under practical conditions it is not possible to reach all their hiding places in a building with such a spray.

Since scorpions can hide in innumerable places in the house, a thorough application of residual sprays that wet the surface is essential. All crevices in woodwork, closets, around plumbing and similar openings should be sprayed. On the outside of the house, areas immediately above the foundation, around windows, doorways, pipe openings and other possible avenues of entry are treated. Where it is not convenient to remove accumulations of lumber, firewood and other materials, such accumulations should be sprayed.

The following insecticides may be used for the control of scorpions:

- Propoxur 70 percent wettable powder — Mix two ounces/57 g to one gallon/ 3.8 l of water and apply as a residual spray or with a paint brush to surfaces of buildings, porches, screens, window frames, patios and garages. Do not treat plants (lawns, flowering plants, shrubs, trees, etc.) with this concentration.
- Bendiocarb — Mix one packet or one scoop of bendiocarb w.p. per gallon of water (0.25%) and apply outside around doors and other places where they can enter the building.
- Diazinon 4E — Apply a ½ percent spray around windows, doors and along baseboards. Make spot applications to other areas over which pests may crawl.

For treatment of crawl spaces dust formulations of residual insecticides can be useful.

The Centipede
I objurgate the centipede
A bug we do not really need
At sleepy time he beats a path
Straight to the bedroom or the bath
You always wallop where he's not
Of, if he is he makes a spot!
—Ogden Nash

CENTIPEDES

Phylum Arthropoda
Class Chilopoda

Biology. Centipedes are flattened, elongated animals with one pair of legs on most of their body segments. They have many-jointed antennae and poison jaws connected to poison glands which kill insects and other small creatures for food.

Ordinarily, centipedes dwell outdoors in damp localities, beneath accumulations of leaves, stones, boards, etc., and when disturbed they run swiftly. They may crawl into the house and conceal themselves in various articles, and if crushed they may bite, causing some pain and swelling.

The house centipede. *Scutigera coleoptrata* (Linnaeus). — Back (1935) notes the house centipede "has a wormlike body, an inch or more in length, with a pair of very long, slender antennae springing from the head, and with 15 pairs of long legs arranged along the sides of its body, the last pair being much more than twice the length of the body in the female." The body of this grayish-yellow centipede is marked with three longitudinal dark stripes. The legs are encircled with alternating dark and white bands. The last pair of hind legs are modified to lasso and hold its victims. There are six larval instars and four post-larval instars before maturity.

The house centipede is now found throughout the country; it originally came from Mexico. It lives outdoors as well as indoors and in the home it occurs in moist cellars, damp closets and in bathrooms where it feeds on insects and spiders. One house with a sump pump in the basement near Pittsburgh, had more than 200 house centipedes in one year.

Marlatt (1930) notes this species "may often be seen darting across floors with very great speed, occasionally stopping suddenly and remaining absolutely motionless, presently to resume its rapid movements, often darting directly at inmates of the house, particularly women, evidently with a desire to conceal itself beneath their dresses, and thus creating much consternation." The weak jaws of the house centipede can penetrate the skin only with difficulty. Curran (1946) states there are several records where the bite resulted in swelling and pain, but usually it is no worse than that of a bee sting.

The larger centipedes, *Scolopendra heros* Gir. and *S. morsitans* (L.) of the southern states, as well as *S. polymorpha* Wood of the southwestern states and Mexico, reach a length of six inches/15 cm. These can cause severe pain and swelling by their bite.

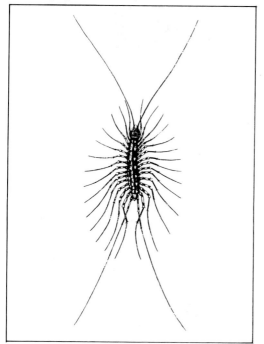

Fig. 25-10. The common house centipede.

The author was involved in a case where a centipede of the above genus crept across the face of a movie actress while she slept in a Hollywood hillside home. The face of this actress allegedly was scratched by the dragging claws of the centipede. Despite the efforts of a pest control operator, every now and then this famous Hollywood household discovered one or more of these formidable creatures. After months of effort the pest control operator accidentally discovered a nest of centipedes in the ground, beneath boulders, forming a rock wall. It seemed likely this was the source of the invasion.

Control. Centipedes that occur in the house require moist habitats. Wherever possible, such conditions should be corrected. Outdoors, remove accumulations of materials near the house that provide places for them to hide and breed. The following insecticides have been found effective against centipedes:

- Bendiocarb w.p. — Mix one packet or scoop of bendiocarb w.p. per gallon of water (0.25%) and apply around doors and windows, and other places where these pests may enter premises. Spray baseboards, storage areas and other locations where these pests are found.
- Propoxur — The National Pest Control Association recommends propoxur residual sprays against centipedes.

INSECTS AS OCCASIONAL INVADERS

At times, plant-feeding insects invade a building in sufficient numbers to become pests merely by their presence for they rarely do any harm in the building. Indoors, they are usually controlled by contact and residual insecticides. The careful use of insecticides recommended for outdoor use will curtail their numbers and help keep them from entering houses and similar structures.

BEETLES

As can be seen from the following review, there are many species of beetles that are occasional invaders. We will discuss some of the most important, although it should be noted their prominence as pests varies from year to year and from place to place.

The strawberry root weevil, *Brachyrhinus ovatus* (L.), according to Wilcox et al. (1934), becomes a pest in the spring and fall in many localities in the United States. These authors quote E. M. Patch of Maine as follows: "We have been overrun with a hateful pest. I killed more than 400 one evening in the front room. They travel all over the house and crawl from the baseboard to the ceiling only to drop to the carpet and try it over and over again. They hide under any protection, carpet, clothing, bedding and are a general nuisance."

McDaniel (1941) notes in Michigan these insects invade the house in summer and late fall where they frequently may be found in sinks, bathtubs, wash bowls and other places where moisture is available.

Spencer (1942) observes in British Columbia, besides the strawberry root weevil, *Brachyrhinus ovatus* (L.), the black vine weevil, *B. sulcatus* (F.) and *B. singularis* (L.), enter houses for shelter in the fall. Some other pest species are *B. rugostriatus* (Goeze) and *B. cribricollis* (Gyll.). According to Ebeling (1975), the latter insect was first noticed in Los Angeles County in 1928 and has since spread throughout the state. Unfortunately, only a skilled taxonomist can separate the different species of *Brachyrhinus*. Breakey (1952) separated the strawberry root weevil from the black vine weevil by noting the former is 1/5 inch/five mm long and the black vine weevil is 2/5 inch/10 mm long and is often marked with small flecks of yellow or white.

Root weevils are pests of a great variety of plants including strawberries, blueberries, rhododendrons, azaleas, yews, camellias and other plants. The larvae feed on the roots of the plants and the adults feed on the leaves and between the two of them injure or kill the plants. When these plants are badly infested, they produce many adults which invade the house and other structures. In the house sprays containing one percent diazinon, one percent propoxur and 0.5 percent chlorpyrifos have been successful against them. Doorways and windows should be carefully treated. Some pest control operators spray the foundation walls outdoors.

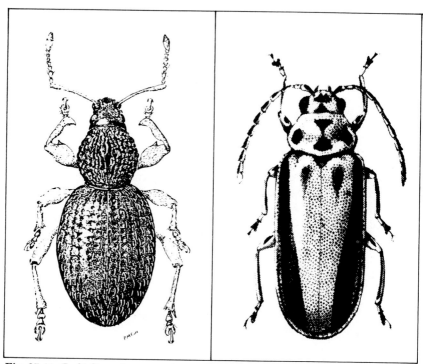

Fig. 25-11. (Left) Strawberry root weevil. (Right) Elm leaf beetle.

In the past, the soil about the evergreens was treated with two percent chlordane emulsions to kill both the larvae and adults. At this writing the USDA recommends for the control of the strawberry root weevil and the black vine weevil one tablespoonful of 20 percent lindane emulsifiable concentrate per gallon/3.8 l of water be applied in May or early June to trunks and soil, thoroughly wetting the surface of infested plants to control both adults and larvae. For adults of the black vine weevil alone, the USDA recommends seven oz/198 g acephate 75 percent w.p. per 20 gallons/76 l of water to be applied to infested evergreens when the first feeding damage occurs. The application should be repeated at four-week intervals until the first heavy frost.

The elm leaf beetle, *Pyrrhalta luteola* (Muller), formerly known as *Galerucella xanthomelaena* (Schrank) is a common and important pest of elm trees. These yellow to olive green beetles with a dark stripe along the edge of each wing cover are approximately ¼ inch/six mm in length. In the fall of the year they enter houses and other buildings to hibernate and are found throughout the building. When the insects first become abundant during the summer they once again may become invaders. The USDA recommends spraying with methoxychlor, three tablespoonfuls of 25 percent emulsifiable concentrate per gallon/3.8 l of water, on the undersurface of the leaves when the larvae are first noticed. Carbaryl is labeled for use as a 0.1 percent high volume spray against elm leaf beetle.

Indoors diazinon or propoxur can be used in the same manner as for cockroach control. For control of these insects in the immediate vicinity of the building apply these sprays at concentrations used for the control of lawn insects.

The imported long-horned weevil, *Calomycterus setarius* Roelofs, according to Johnson (1944), at times becomes a problem as an invader of homes. These insects, which are native to Japan, have become established in the Northeast and Midwest. The adults emerge from the soil at the end of June and become most abundant during July and August. The wingless and parthenogenetic adults feed on a wide variety of foliage. Anon. (1968) reports them as pests on red clover in Iowa and alfalfa in Wisconsin. When they become abundant they are found crawling on people and vehicles. They are also found in dwellings and are present on walls, ceilings, furniture and even on food on dining tables. Contact household sprays will control these pests in houses; outdoors they can be controlled with diazinon and other organophosphate insecticides at concentrations used for lawn insect control.

The Asiatic oak weevil, *Cyrtepistomus castaneus* (Roelofs), is a pest somewhat similar to the imported long-horned weevil. It infests alfalfa and red, scarlet, white and pin oaks. The adults are attracted to lights in houses. It has been reported invading houses in Delaware and Pennsylvania, but is probably present in other states. Superficially the imported long-horned weevil resembles the Asiatic oak weevil. It differs from the latter by having white scales and short blunt hairs, whereas the Asiatic oak weevil has iridescent blue to green scales and pointed hairs. In the house, the control for the Asiatic oak weevil is similar to that for the imported long-horned weevil.

Anon. (1938) mentions the weevil, *Trachyphloeus bifoveolatus* Beck as a pest in houses in the eastern United States and Anon. (1954) notes this weevil entered houses in large numbers in Oregon.

Lady beetles, which are almost entirely beneficial insects due to their feeding on plant lice and other harmful insects, at times make a nuisance of themselves by invading homes. The author is acquainted with a small invasion of the two-

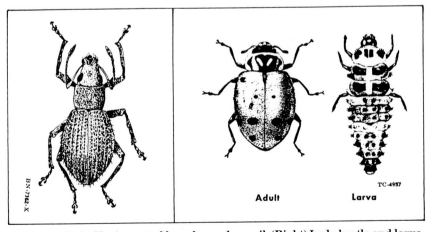

Adult   Larva

Fig. 25-12. (Left) The imported long-horned weevil. (Right) Lady beetle and larva.

spotted beetles, *Adalia bipunctata* (L.), which occurred in January and February, 1946, in a home near Pittsburgh. A woman claimed the beetles were throughout the house. Debris such as leaves had accumulated on a small roof near the second floor windows and the lady beetles found this to be a suitable place for hibernation. Whenever the weather warmed-up, the beetles came out of hibernation and entered the house through cracks.

Other entomologists such as Curran (1946) and Hatch and Houk (1948) report similar invasions. The latter authors note a woman on removing a coal shed from her yard in the spring found a mass of two-spotted lady beetles six to eight inches/15 to 20 cm deep and more than one yard/0.9 m square. "This same person reported the beetles would gather in great numbers on the outer wall of the building, perhaps attracted by the warmth, and enter the building through any available crevices, and this seemed in general to be the source of the beetles rather than the hibernating masses within the building." Svihla (1952) notes he was bitten by the two-spotted lady beetle prior to their going into hibernation. Unless these lady beetles become extremely objectionable, it is preferable not to kill them since they are normally considered to be beneficial insects.

The larvae and adults of carabid or ground beetles are beneficial since they are predacious on many injurious insects, snails, etc. During the summer in many parts of the country, ground beetles and many other insects are attracted in tremendous numbers to bright lights at night. Blue neon lights, which are so commonly used around cafes, taverns, drive-in restaurants and similar places are especially attractive, and the ground beetles and other insects become very annoying by crawling in these establishments. Mampe (1977) recommends applying bendiocarb to a band of soil six to 10 ft/1.8 to three m wide around the building as well as on foundation walls.

The stink beetle, *Nomius pygmaeus* (Dej.), is another carabid that occasionally occurs in homes in the western United States and Canada. Hinton (1945) states this insect has an offensive smell and articles that have been in contact with the beetle may retain the disagreeable odor for several weeks.

According to Spencer (1942), the stink beetle is an accidental intruder which comes into buildings in summer. Speaking of British Columbia, he notes this insect is "probably the most concentratedly malodorous animal for its size in the world. For some years now I have checked up reported flights of this beetle and find that their presence in towns can be associated with forest fires; when the horizon is blotted out by smoke, these stinkers may arrive. Apparently they are forest dwellers which are driven out by the fires and may then travel long distances, scattered by the smoke."

The tule beetle, *Agonum maculicolle* Dejean, is a ground beetle that occurs in the Sacramento and San Joaquin valleys of California. According to Essig (1926), after fall rains, these beetles may enter houses in large numbers. Ebeling (1975) recommends for their control the installation of a light bulb above a container filled with water and detergent, about 50 feet/15 m from the house. The beetles are attracted to the light and drown upon falling into the water.

French (1944) states *Anthocomus bipunctatus* Harrer, one of the malachiid beetles, were present in dwellings in Richmond, Virginia and Summit, New Jersey. H. Katz found the click beetle, *Aeolus mellilus* (Say), to be an occasional pest in houses in Pittsburgh, Pennsylvania. *A. dorsalis* (Say), known in Pennsylvania as the "corn wireworm," also is a common pest in buildings in the fall of the year. Dorsey (1957) notes a rhinoceros beetle, *Dynastes tityus* (L.), became a nuisance in houses in West Virginia by invading them through the chimney. According to Michelbacher (1953), the acorn weevil, *Balaninus uniformis* LeConte, occasionally enters houses in California. The above beetles can be controlled in buildings with insecticides ordinarily used against cockroaches and other crawling insects.

The tulip tree weevil, *Odontopus calceatus* (Say), is a small black weevil about 1/10 inch/2.5 mm long. It is a feeder on the foliage of magnolia, sassafras and tulip poplar trees. The adults are attracted to lights in buildings. They are most common in July and August. In one case, the beetles were found on the 37th floor of a skyscraper in Pittsburgh, Pennsylvania. This weevil occurs throughout the eastern and southern states. Both the adults and the larvae feed on the leaves. For their control in the home, use contact sprays with pyrethrins, resmethrin or DDVP. Wallace et al. (1968) discuss this insect in some detail.

The lathridiid beetles of the genera *Enicmus, Coninomus, Cryptophagus, Cartodere* and others are at times pests in the home. In England, Anon. (1940) notes such insects are often referred to as "plaster beetles" since they may occur in newly-built and reconditioned homes and apartments, especially those that are freshly plastered. The beetles and their larvae or grubs can exist only under damp conditions, since they feed on the molds that grow on moist surfaces. Thus, they are found "in damp cellars where packing materials are stored and occur also on such articles as cheese, jam, fibres and carpeting." Damp plaster and wet straw or vegetable fibers support molds and mildews, particularly in damp weather. Thus, these insects often become particularly prominent after a period of rainy weather. Wallpaper placed on fresh plaster is another likely source of mold which supports these plaster beetles. Once the infested area becomes dry, the mold and the associated beetles disappear. Contact sprays such as pyrethrum when applied in crevices under baseboards and similar places will kill many of these pests, but they will not eradicate them. If the source of infestation cannot be found, periodical spraying is necessary. Hinton (1945) has treated these insects in great detail in his book.

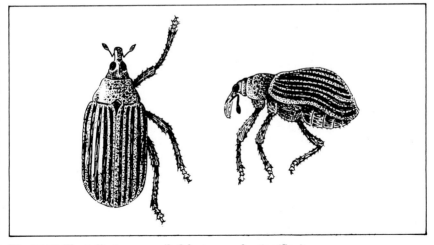

Fig. 25-13. The tulip tree weevil, *Odontopus calceatus* (Say).

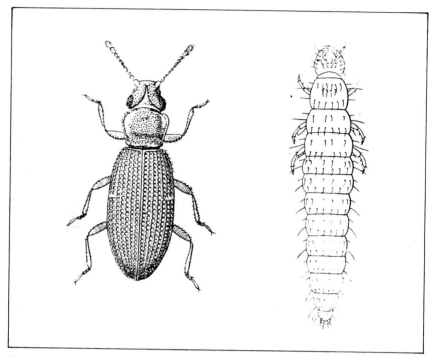

Fig. 25-14. Plaster beetle, *Cartodere filum* (Aubé), adult and larva.

An infestation of an apartment near Pittsburgh, Pennsylvania, by the plaster beetle, *Cartodere argus* Reitter, in June, 1947, was investigated by the author. The infestation was in an old brick house which had been remodeled into several apartments. A recently renovated apartment with newly plastered walls showed a moderate number (20 to 30) of the beetles. The beetles were found on the ceiling, around windows and other electric light fixtures. They were also present in the cellar and to a lesser extent in the apartment next door. The insects crawled about and made no attempt to fly. The infestations became noticeable after three consecutive rainy days. A pyrethrum contact spray did not control the infestation effectively, but a six percent DDT oil-base spray resulted in fair control, although a few were still present several days after the DDT spraying. Several months later, the insects disappeared entirely and have not reappeared after several years. It is quite possible mold on newly plastered walls supported these insects and the heat of summer curtailed the mold and the fungus-feeding insects. For the control of plaster beetles use insecticides used for cockroach control.

MAYFLIES, CADDIS FLIES AND STONEFLIES

Mayflies, also known as shadflies and lakeflies, are as their common names imply, aquatic insects. The eggs are deposited in water and the nymphs live at the bottoms "of streams, pools and lakes, feeding upon small aquatic plants, animals and organic material. They lead a very precarious life, being the chief food of many aquatic insects and fishes." The adults are attracted to lights at night and often occur in tremendous numbers during the summer when they make a pest of themselves by their mere presence in and around houses. Some idea of the great swarms of these insects may be derived from Anon. (1947) reporting that three electric locomotives were put out of commission by millions of these insects that "covered the box cars, clogged overhead power apparatus and short-circuited electric motors." Leiux and Mulrennan (1955) controlled the mayfly, *Hexagenia munda orlando,* in Florida by applying 0.24 lb/0.1 kg gamma BHC per acre/0.4 ha of lake. This treatment did not injure aquatic life.

According to Munroe (1951), adults of the caddis fly *Hydropsyche bifida* Wlk. and other caddis flies become important pests in homes along the St. Lawrence drainage system by swarming around trees and lights and entering homes where they settle in large numbers on walls and other surfaces. Also see Peterson (1952) and Fremling (1960).

Stoneflies emerge during the winter and spring, some as early as February. They may then alight in great numbers on buildings and fences where they disturb the inhabitants. Lloyd Adams found a stonefly, *Allocapnia* sp., swarming on a milkhouse in Pennsylvania on February 1. Here they contaminated milk cans and other equipment.

THRIPS

Thrips are very small, slender insects that often are barely visible to the eye. Although small in size, they loom very large as pests of many crops. Bailey

Fig. 25-15. Flower thrips.

(1936) states the onion thrips, *T. tabaci* Lind., particularly the larvae, are apt to bite people, and the bite produces a pricking sensation which results in a slight itching but no swelling. At times these insects invade the home from adjacent fields. Peairs and Davidson (1956) note that the grass thrips or "oat bugs," *Anaphothrips obscurus* (Mull.), feed on grains and grasses. From there they migrate into houses. People have also been bitten by thrips brought in on blankets hung in the yard. On occasion, pear thrips, *Taeniothrips inconsequens* (Uzel), will bite. Aerosols for flying insects may give some relief from these insects.

CATERPILLARS

Caterpillars of many species, particularly those that infest trees and shrubbery outside the home, often enter dwellings. Curran (1945), speaking of the fall webworm, *Hyphantria cunea* (Drury), states: "There are two generations of webworms each year, one in June and July, which is usually small, and one in August and September which may be very large. When the caterpillars mature they wander about in search of a suitable place in which to build a cocoon, and unless a building is tightly screened, they may crawl in and distribute themselves over the walls and ceilings." These insects are controlled by spraying trees and shrubs early in the summer with labeled residual insecticides such as carbaryl and chlorpyrifos.

The eastern tent caterpillar, *Malacosoma americanum* (F.), may become an important pest in and around the house. Anon. (1956) found the larvae migrating from wild cherry and fruit trees annoying the residents of a Baltimore suburb. J. O. Rowell, in the previous reference, noted that in Virginia the larvae left their tents and crawled over the lawns to become a great nuisance to the homeowner. In 1959, in Pittsburgh, Pennsylvania, tent caterpillars were leaving wild black cherry and other trees to find places to pupate. People complained about them getting on walks, porches, sides of buildings, etc. Sprays containing lindane and malathion killed the migrating larvae on vegetation, and oil-base household sprays and aerosols were used against the larvae when on other surfaces. Carbamate and organophosphate insecticides, such as carbaryl and chlorpyrifos respectively, are effective against tent caterpillars and are labeled for spraying trees and ornamentals.

The gypsy moth, *Portheria dispar,* heavily infested 11 million acres of forest in the Northeast in 1981 and this led to numerous homes being invaded by caterpillars, with some residents developing rashes from contact with their hairs. Non-pesticide control measures, such as moth trapping and tree banding, were of some value in isolated cases, but on a large scale tree spraying with insecticides was essential to prevent defoliation of trees and invasion of homes. The carbamate insecticide carbaryl was the most widely used material, but three organophosphates, acephate, malathion and trichlorfon were also used to good effect, provided they were applied early enough. A chlorinated hydrocarbon, methoxychlor, is also labeled for tree spraying against gypsy moth. Sprays based on *Bacillus thuringiensis,* a bacterial insecticide, proved effective against gypsy moth, particularly on a forest treatment scale.

Although stinging caterpillars with poisonous or urticating hairs often injure people handling them, these caterpillars rarely occur in the house. Their are about 25 species of caterpillars with stinging hairs. The puss caterpillar, *Megalopyge opercularis*, is one of the worst and sometimes the victims are confined

STINGING CATERPILLARS
:TORIAL KEY TO SOME IMPORTANT UNITED STATES SPECIES
Harold George Scott & Chester J. Stojanovich

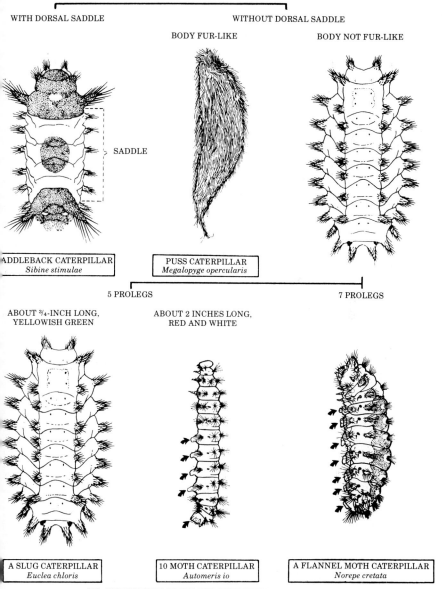

WITH DORSAL SADDLE   WITHOUT DORSAL SADDLE

BODY FUR-LIKE   BODY NOT FUR-LIKE

> SADDLE

ADDLEBACK CATERPILLAR
Sibine stimulae

PUSS CATERPILLAR
Megalopyge opercularis

5 PROLEGS   7 PROLEGS

ABOUT ¾-INCH LONG,   ABOUT 2 INCHES LONG,
YELLOWISH GREEN   RED AND WHITE

A SLUG CATERPILLAR
Euclea chloris

10 MOTH CATERPILLAR
Automeris io

A FLANNEL MOTH CATERPILLAR
Norepe cretata

U.S. DEPARTMENT OF HEALTH, EDUCATION, AND WELFARE
PUBLIC HEALTH SERVICE, Communicable Disease Center
Atlanta, Georgia

Fig. 25-16. Pictorial key to some important United States species of caterpillars.

for hospital treatment. These caterpillars, as well as the other stinging species, feed on a variety of shrubs and trees. Anon. (1964) recommends spraying the trees with malathion, DDT or toxaphene to kill the foliage-feeding larvae. Anon. (1961) and Scott (1964) discuss these caterpillars in some detail.

Pence and Hogue (1957) record the fungus moth, *Aglossa caprealis* Hubner, as a pest in houses infested with the dry rot fungus, *Poria incrassata*. The larvae feed on the mycelia of the fungus. To control the fungus moth permanently, it is necessary to correct the conditions causing decay. They also feed on carpets, clothes and shoes.

HORNTAILS

These large insects, about one inch/25 mm in length, are related to wasps but do not sting or bite. Occasionally they are of some concern in the house because of the emergence holes (¼ inch/six mm in diameter) they make in lumber and adjoining materials such as hardwood floors, carpeting, linoleum, wallboard, etc. Horntails most commonly occur in the poorer grades of lumber such as is used in studs, joists and subflooring; they attack both coniferous and deciduous woods.

The female is called a "horntail" because of the long hornlike projection that protects the ovipositor. According to Ebeling and Wagner (1963), the female "uses this ovipositor to deposit eggs deeply into the wood of coniferous trees that have been weakened or that are dying as the result of fire, disease or other injury. The larvae make cylindrical holes in the wood, packing the tunnels left behind them with frass from their borings.

"The duration of the life cycle of the wood wasp ranges from one to three years. Therefore, the adults may not emerge for as long as two or three years after a house is built. Pupation takes place at the end of the larval burrow, but the emerging adult chews through nearly an inch of wood and bark in forest trees."

The larva of the horntail can be readily recognized by the spine on the last abdominal segment. Its jaws are so powerful it can chew through lead sheeting. Chandler (1959) notes *Sirex cyaneus* F. is the species he most commonly receives for identification at Purdue University, Indiana. Ebeling and Wagner (1963) reports *Sirex longicauda* Middlekauf to be a common pest in California.

Although the horntail does not reinfest wood in the house, it is a nuisance because of its presence and because of its emergence holes. There is no practical method of control. If a piece of wood is suspected to be infested, it may be removed if the owner wishes to undertake the expense.

ANTS

Mallis (1973) reports an unusual problem with stinging ants, *Ponera pennsylvanica* Buckley, that occurred in a complex of large commercial greenhouses near Pittsburgh, Pennsylvania. The supervisor complained he and the other greenhouse workers were being bitten by flies. The author collected some flying midges the supervisor thought were responsible for the annoyance. When these were checked in the laboratory, none of them had biting mouthparts.

A few weeks later the supervisor sent the author some insects he was sure were responsible for the bites. Upon examination, these proved to be not flies, but ponerine ant queens, armed with long stings and biting mouthparts. Colonies of this predacious ant were breeding in corn cob mulch used to grow roses. The ants were feeding on fly larvae and other insects in the mulch.

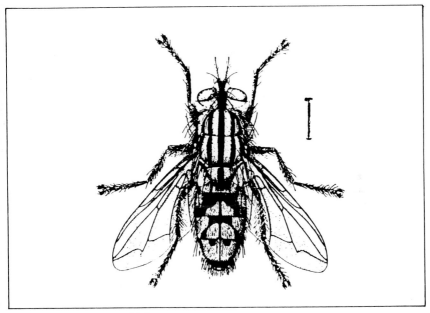

Fig. 25-17. A sarcophagid fly.

What probably happened is the flying ponerine ant queens became enmeshed in the hairs and perspiration on the skin of the workers and then stung and bit them. The problem ended when the mulch was treated with chlordane. Probably any of the insecticides now used for ant control would be effective against the ant colonies.

FLIES

Sarcophagid flies are at times household pests. *Sarcophaga aldrichi* Park. occurs throughout southern Canada and much of the eastern and midwestern United States. *S. houghi* Aldrich is a pest in the southern United States. Mallis (1973) found *S. aldrichi* Park. to be a very annoying pest in resort areas in Somerset County, Pennsylvania, in late June. Some individuals referred to these flies as "red-eyed" flies because of their prominent red eyes. The flies pestered golfers on the course, invaded buildings and parked cars with open windows. Although they did not bite, they sat on the skin and hovered around in annoying numbers.

Somerset County is famous for its sugar maple and its sugar maple trees. The larvae of these flies are important predators of the pupae of the forest tent caterpillar, a common pest of sugar maples. When there is a heavy infestation of forest tent caterpillars, the adult sarcophagid flies may appear in great numbers. This is a case of a beneficial insect becoming a pest because of sheer numbers. *S. houghi* Aldrich parasitizes the forest tent caterpillar and the elm spanworm in the southern United States and has habits similar to *S. aldrichi* Park.

Usually, after several weeks, the flies disappear without any treatment. If necessary, spraying or fogging nearby trees with malathion may be helpful.

The author has on several occasions identified an onion fly, *Tritoxa flexa* (Wied.), as a pest in houses. It infested potted chives.

Pence (1955) names some other occasional pests in houses in California amongst which are the beetle, *Podabrus tomentosus* (Say), the oak moth, *Phryganidia californica* Packard, and the fly, *Fannia benjamini*.

MAUSOLEUM PESTS

A great variety of insects occur in and around mausoleums, especially beetles and flies. NPCA (1977) notes hide beetles, *Dermestes sp.*, are prominent in such situations since they feed on cadavers. For further information on dermestids see the chapter on hide and carpet beetles. Ebeling (1975) found the following beetles to be present in mausoleums in California: *Necrobia rufipes* (DeGeer), *Necrobia ruficollis* (F.) *Alphitobius piceus* (Olivier) and *A. diaperinus* (Panzer).

Many species of flies and gnats, especially fungus flies, feed on decaying vegetation and cadavers in mausoleums. Phorids, scatopsids and similar fungus gnats are common in such situations. H. Katz found the phorid, *M. Scalaris* abundant in mausoleums in Pittsburgh. Under optimum conditions many of these flies complete their life cycles in a month or less.

Bendiocarb is labeled for use against other pests in mausoleums and it may be effective in controlling flies that breed in cadavers. However, bendiocarb is not labeled for control of their flies. Periodical fogging with space sprays and insect electrocutors should be helpful in reducing fly populations.

The control of beetles in mausoleums leaves much to be desired. Repeated applications of residual insecticides used for the control of cockroaches and other crawling insects may result in some control. H. Katz, who studied the problem in some detail, states the crypts are quite damp and the moisture weakens and detoxifies most of the dusts and sprays used against crawling insects. Possibly, bendiocarb, may be of some help.

OTHER OCCASIONAL INVADERS

According to Becker and Sweetman (1964), larvae of the sawfly, *Macremphytus tarsatus* (Say), crawled about buildings and dooryards of houses in Massachusetts. The larvae also burrowed into wooden structures to construct their pupal cells.

Farrier (1958) found a lygaeid bug, *Ischnodemus falicus*, in North Carolina to swarm about houses in early November; Knowlton (1947 and 1958) reports leafhoppers and tingids biting indoors.

Hackberry gall psyllids, *Pachypsylla sp.*, consisting of a number of species, invade houses in the East and Midwest in the fall. These gallmakers are restricted to the hackberry trees. Smith and Taylor (1953) studied the biology and control of these insects. Frequent application of organic phosphate sprays to the trees controlled these gall insects.

Grass bugs of the genus *Crassus* are at times reported to invade houses in California in great numbers. Anon. (1966), speaking of *Arhyssus crassus*, notes overwintering adults spot materials in the house.

LITERATURE

ANONYMOUS — 1938. USDA newsletter for October. 1940. Plaster beetles. British Museum (Natural History). 1947. May flies turn brakeman. Pests 15(9):36. 1954. A weevil *(Trachyphloeus bifoveolatus)*. Coop. Econ. Insect Rpt. 4(40):92. 1956. Eastern tent caterpillar. Coop. Econ. Insect Rpt. 6(22):496. 1961. Puss caterpillars. Pest Control 29(5):15, 16, 18. 1964. Florida researchers list latest controls for stinging caterpillars. Pest Control. 32(9):60. 1966. A coreid bug *(Arhyssus crassus)*. Coop. Econ. Insect Rpt. 16(49):1117. 1968. A Japanese weevil, *Calomycterus setarius*. Coop. Econ. Insect Rpt. 18(31):725. 1980. Guidelines for the control of insect and mite pests of food, fibers, feeds, ornamentals, livestock, households, forests, and forest products. USDA Agr. Handbook No. 571.

BACK, E.A. — 1939. Centipedes and millipedes in the house. USDA Leaflet 192.

BAILEY, S.F. — 1936. Thrips attacking man. Canadian Entomol. 68:95-98.

BECKER, W.B. & H.L. SWEETMAN — 1946. Leaf-feeding sawfly larvae burrowing in structural wood. J. Econ. Entomol. 39(3):408.

BENNETT, D.R. & S.H. KERR — 1973. Millipedes in and around structures in Florida. Florida Entomol. 56(1):43-48.

BREAKEY, E.P. — 1952. Control of strawberry root weevils. Wash. Agr. Expt. Sta. Mimeograph Circular No. 151.

BROOKS, F.E. — 1919. A migrating army of millipedes. J. Econ. Entomol. 12:462-464.

CARR, W.H. — 1946. The truth about scorpions. Natural History 55(2):80-86.

CHANDLER, L. — 1959. Home building speed-up helps make horntails a pest in structural timbers. Pest Control 27(6):46, 48-53.

CLOUDSLEY-THOMPSON, J.L. — Spiders, Scorpions, Centipedes, and Mites. Pergamon Press, N.Y. 228 pp.

CURRAN, C.H. — 1945. The fall webworm. Insects in the house. Natural History 54(7):332. 1946. Uninvited guests in the house, the house centipede *(Scutigera forceps)*. Natural History 55:240, May. 1946. Insects in the house: lady beetles. Natural History 55(9):437, Nov.

DORSEY, C.K. — 1957. A rhinoceros beetle *(Dynastes tityus)*. Coop. Econ. Insect Rpt. 7(9):16, March 1.

EBELING, W. — 1975. Urban Entomology. Univ. Calif. Div. Agr. Sciences pp. 695, illustrated.

EBELING, W. & R.E. WAGNER — 1963. Biology and control of the siricid wood wasp. P.C.O. News 23(9):16-17.

ENNIK, F. — 1972. A short review of scorpion biology, management of stings, and control. Calif. Vector Views. 19(10):69-80.

ESSIG, E.O. — Insects of Western North America. Macmillan.

EWING, H.E. — 1928. The scorpions of the western part of the United States with notes on those occurring in Northern Mexico. Proc. U.S. National Museum 73:1-24.

FARRIER, M.H. — 1958. Coop. Econ. Insect Rpt. 8(13):245, March 28.

FREMLING, C.R. — 1960. Biology and possible control of nuisance caddis flies of the upper Mississippi River. Iowa State University Research Bull. 483.

FRENCH, G. — 1944. *Anthocomus bipunctatus* (Harrer), a new household insect. J. Econ. Entomol. 37:103.

HATCH, M.H. & H. HOUK — 1948. *Adalia bipunctata* in buildings in Washington. J. Econ. Entomol. 41:412.

HATCHETT, S.P. — 1947. Biology of the Isopoda of Michigan. Ecological Monographs 17:47-79, Jan.

HINTON, H.E. — 1945. A Monograph of the Beetles Associated with Stored Products. Vol. I. British Museum (Natural History).

JOHNSON, J.P. — 1944. The imported long-horned weevil, *Calomycterus setarius* Roelofs. Conn. Agr. Expt. Sta. Bull. 479.

KNOWLTON, G.F. — 1947. Leafhopper bites man. Bull. B'klyn Entomol. Soc. 42(5):169. 1958. Tingidae are biters. Bull. B'klyn Entomol. Soc. 53(3):73.

LANGE, W.H., JR. — 1944. Land slugs in California. Bull. So. Calif. Acad. Sci. 43(1):33-40.

MALLIS, A. — 1942. An amphipod household pest in California. J. Econ. Entomol. 35:595. 1973. Unusual insect problems in Pennsylvania. PA Pest Control Quarterly. 16(2):3, 6. 1979. Some unusual household pests in Pennsylvania. Pest Control 47(10):36, 37, 44, 45.

MAMPE, C.D. — 1977. Ground beetles. Pest Control 45(3):32.

MARLATT, C.L. — 1930. The house centipede. USDA Farmers' Bull. 627.

McDANIEL, E.I. — 1941. Strawberry root weevils and crickets as household pests. Michigan State College Ext. Bull. 230.

MICHELBACHER, A.E. — 1953. Dooryard visitors. Pest Control 21(8):32-33.

MUNROE, E.G. — 1951. Pest Trichoptera of Fort Erie, Ontario. Canadian Entomol. 83(3):69-72.

NATIONAL PEST CONTROL ASSOCIATION — 1977. Mausoleum pests. NPCA Tech. Release, May 27, 1977.

PEAIRS, L.M. & R.H. DAVIDSON — 1956. Grass thrips or "oat bugs." Insect Pests of Farm, Garden, and Orchard. Wiley. p. 190.

PENCE, R.J. — 1955. Invasion of strangers. Pest Control 23(10):30, 32, 38, 40.

PENCE, R.J. & C.L. HOGUE — 1957. A new record of a fungus moth as a household pest in California. P.C.O. News 17(12):6.

PETERSON, D.G. — 1952. Observations on the biology and control of pest Trichoptera at Fort Erie, Ontario. Canadian Entomol. 84(4):103-107.

ROCH, M. — 1941. Les pigures des scorpions. Biol. Abst. 20:663, 1946.

SCOTT, H.G. — 1964. Stinging caterpillars. Pest Control 29(8): 23-26.

SMITH, F.R. — 1927. Observations on scorpions. Science 65:64.

SMITH, R.C. & R.S. TAYLOR — 1953. The biology and control of the hackberry psyllids *(Pachypsylla)* in Kansas. Kansas Entomol. Soc. J. 26:103-115.

SNETSINGER, R. — 1967. Millipedes in dwellings. PA Pest Control Quarterly 10(6):5.

SPENCER, G.J. — 1942. Insects and other arthropods in buildings in British Columbia. 39:23-29.

STAHNKE, H.L. — 1950. Scorpions. Arizona State College. 1971. Some observations of the genus *Centruroides* Marx (Buthidae, Scorpionida) and *C. sculpturatus* Ewing. Entomol. News 82(11):281-307.

STORY, K.O. — 1981. Questions and Answers. Pest Control Technology 9(3):50.

SUGERMAN, B.B. — 1960. Millipedes. Coop. Econ. Insect Rpt. 10(40):9.

SVIHLA, A. — 1952. Two-spotted lady beetles biting man. J. Econ. Entomol. 45:134.

THOMAS, D.C. — 1948. Use of metaldehyde against slugs. Ann. Appl. Biol. 35:202-227. Chem. Abst. 43:9339-9340, 1949.

THURSTON, R., G.M. BOUSH, & K.J. STARKS — 1959. Control of millipedes infesting new homes in Kentucky. Pest Control 27(10):40, 98-99.

WALLACE, G.E., S.G. GESELL, & A. MALLIS — 1968. The tulip weevil as a household pest. PA Pest Control Quarterly 11(4):3, 5.

WALTON, W.R. — 1928. Earthworms as pests and otherwise. USDA Farmers' Bull. No. 1569.

WHEELER, A.G., JR. — 1975. Birch catkin bug a nuisance insect. PA Pest Control Quarterly, Spring, p. 4.

WHITE, W.H. & A.C. DAVIS — 1942 and 1952. Land slugs and snails and their control. USDA Farmers' Bull. No. 1895.

WILCOX, J., D.C. MOTE, & L. CHILDS — 1934. The root weevils injurious to strawberries in Oregon. Oregon Agr. Expt. Sta. Bull. 330.

WOGLUM, R.S. — 1941. Tests on the relative effectiveness of snail baits. Calif. Fruit Growers Exchange Pest Control Circular No. 74, pp. 270-272.

C. DOUGLASS MAMPE
KEITH O. STORY

Dr. C. Douglass Mampe received degrees in entomology from Iowa State University, North Dakota State University, and a Ph.D. from North Carolina State University. Upon completing his university education, he joined the staff of the National Pest Control Association (NPCA). During his 10 years there he was responsible for keeping abreast of all technical and regulatory actions related to the structural pest control industry. This included reviewing research proposals and overseeing research projects related to the biology and control of termites, cockroaches, commensal rodents, and other related pests.

Mampe left NPCA as its technical director and joined Western Termite and Pest Control, Inc. and its sister company Residex, a pest control industry supply house. He served as Western's technical director for six years and developed practical procedures for field operations. He eventually became general manager of Residex, which included developing new pesticide registrations and industry training programs, including certification training programs for a number of the northeastern and middle Atlantic states.

In 1980, Mampe started his own consulting firm for the urban and structural pest control industry. His firm provides technical consultation, training and training programs, evaluates new pesticides and equipment, custom develops personnel management programs and provides expert testimony for legal cases.

Keith Story was born and raised in England and obtained his bachelor's degree in natural sciences and his masters degree in zoology and entomology from Cambridge University, England. His interest in natural history was kindled in 1955 by Dr. Carroll Williams, a Harvard professor visiting England. Later at Cambridge University his focus on entomology owed much to the inspiration at Cambridge of Dr. Vincent Wigglesworth, founder and acknowledged leader of modern insect physiology.

After graduation, Story embarked on a 14 year corporate career with Fisons, an international developer and marketer of specialty chemicals. After an initial period of on-the-job training in both technical and commercial areas, including working for a pest control contractor in Africa, Keith began a seven year period of pesticide research and development. This culminated in his development of bendiocarb, a major new insecticide for the structural pest control market.

CHAPTER TWENTY-SIX

Chemicals Used in Controlling Household Pests

Revised by C. Douglass Mampe[1] and Keith O. Story[2]

Elmer was a chemist
But Elmer is no more
For what Elmer thought
Was H_2O was H_2SO_4

—Anonymous

IN THIS CHAPTER the chemicals used in the control of household pests are briefly reviewed. In many cases the insecticides considered here were treated in greater detail in previous chapters where they were pertinent to the control of the pest concerned. Frear (1948), Shepard (1951), Brown (1951), Bailey and Smith (1951), Martin (1961) and O'Brien (1967) have authored books on insecticides that should be consulted by the reader who desires further information.

Lists of manufacturers and distributors of chemicals and equipment used in the control of household pests can be obtained from the *Pesticide Handbook and Entoma,* published by the Entomological Society of America, *"Blue Book,"* published by *Soap and Chemical Specialties,* as well as from the monthly issues of the periodicals, *Pest Control, Pest Control Technology, Soap and Chemical Specialties, Journal of Economic Entomology* and *Mosquito News.* The above publications are available in many public libraries.

Chemicals used in controlling insects are grouped in a number of ways. They may be classified according to the *method of application* such as sprays (aerosols, space, or residual sprays), dusts, baits, paints or fumigants; by *mode of action,* such as stomach, contact, residual and respiratory (fumigant) poisons, or as to

[1]*President, DM Associates, Westfield, N.J.*
[2]*President, Winchester Consultants, Winchester, Mass.*

895

chemical nature such as inorganic, organic and botanical. Since many of the poisons have one or more modes of action, the author has grouped the insecticides according to their chemical nature.

If one should compare the insecticides in the chapter on chemicals in the first edition of this book with those in the present chapter, it will be observed that many of the standbys of the pest control operator have been largely replaced by the newer organic chemicals. Now that house flies, mosquitoes, lice, cockroaches and other insects show resistance to some organic insecticides, it may be necessary, as in the case of house flies, to return to some of the older insecticides for effective control. It is for this reason both the older inorganic and newer organic insecticides have been considered throughout the book. Those currently registered, available and commonly in use are considered in the"current insecticides" section. Others are considered in the "historical pesticides" section.

THE INSECT IN RELATION TO CHEMICAL CONTROL

The integument. The integument of insects acts as a case to hold the body fluids and prevent the escape of moisture, as a shield to prevent the entry of foreign materials, as an armor against attack by insects of similar size and as an external skeleton which also provides places for muscle attachment. This integument has certain properties which must be considered in the application of insecticides.

The integument of arthropods, including insects, consists of a cuticle, a single layer of epidermal cells that secrete the cuticle, and a basement membrane that separates the epidermal cells from the body cavity. The surface layer of the cuticle is the epicuticle which is a thin layer covered with a lipoid or wax layer. It is this fatty or waxy layer that enables the insect integument to prevent ready entrance of water. Hairs, as well as the waxy layer of insects, often hinder the entry of sprays and dusts. Petroleum distillates of the fly-spray base type, as well as other materials are capable of acting as solvents of the waxy coating, thereby permitting the entry through the integument of the oil base and toxicant. The inner portion of the epicuticle and the outer portion of the next layer of the cuticle, the exocuticle, contain chitin, a carbohydrate which is responsible for much of the durability of the insect integument. The endocuticle is the third and thickest layer of the cuticle, and the one closest to the epidermal cells. Pore canals, dermal glands and sense organs originate in the epidermal layer and perforate the cuticle to the surface of the integument. The dermal glands secrete waxes or greases in some insects, e.g., cockroaches. The insect cuticle consists of hard and darkened processes known as sclerites and soft areas situated between the sclerites. These soft areas are flexible and permit limited growth of the insect body. Insects secrete molting fluids which permit shedding of the integument and subsequent growth. Richards (1951) has prepared a monograph on the integument of arthropods.

Respiratory system. The respiratory system consists primarily of *tracheae*, which are air filled tubes of small diameter which grow finer and finer in size. The finest tubes are called *tracheoles* and these are sufficiently minute to aerate the individual cells. The openings of the tracheae at the surface of the integument are known as *spiracles*. Some insects are capable of regulating the opening or closure of the spiracles, a factor which is of economic importance since the effectiveness of fumigants may thereby be determined. Spray and dust particles may enter the insect body through the spiracles. In most adult insects the spiracles are arranged so there is a pair in every segment. However, the larvae of some insects may have but one or two pairs. Cockroaches and some other insects are capable of "pumping" air in and out of the spiracles so they appear to "breathe."

Circulatory system. The circulatory system of an insect is relatively simple. It consists of a heart that is a simple tube open at both ends and which is situated along the dorsum or back of the insect. The body cavity of the insect is bathed by blood cells which are generally colorless and have no oxygen-carrying cells. The blood is pumped forward through the heart toward the head, then circulated backward through the body cavity and once again pumped through the heart. Small pulsating organs or supplementary hearts aid in pumping the blood into appendages and other relatively inaccessible portions of the bodies of insects.

Digestive system. The digestive system consists of three main sections — the foregut, midgut and hindgut. The mouthparts of the insect which are essentially of the chewing, lapping, or sucking type convey the food to the mouth, the first opening into the foregut. If the mouthparts are of the chewing type, they can carry solid particles into the mouth. If of the sucking or lapping type, liquid food can be taken up. Usually, salivary glands discharge their secretions in the mouth. Some of these salivary secretions are anticoagulant in nature which prevent clotting of the blood and blocking of the narrow channels, in the sucking or lapping insect mouthparts. The food passes through the mouth usually into the crop, a temporary storage chamber, and then into the midgut. The midgut contains the digestive enzymes and actual digestion of the food occurs here. The blood absorbs the food from the cells of the midgut. The hindgut prepares the food for excretion. The Malphighian tubules which empty into the intestine at the junction of midgut and hindgut are largely responsible for removing nonusuable food. If the insect is a sap-imbiber (e.g., an aphid), the excretory product may be of a liquid nature; if it is an insect such as a dry-wood termite or clothes moth which consume relatively dry food and must retain body moisture, the feces are excreted as dry pellets.

Nervous system. The nervous system consists of a central nervous system of nerve groupings which lie along the floor of the body cavity. Nerves are distributed to various parts of the body from the central nervous system. Such sensory organs as the eyes and rod-like organs of taste and smell which project through the cuticle, serve as external receptors for stimulation of the insect nervous system.

MODE OF ACTION OF INSECTICIDES

Fundamental contributions to a knowledge of the mode of action of chemicals on insects and related animal life are made by the physiologist and toxicologist. Metcalf (1948a), Wigglesworth (1950), Richards (1951), Brown (1951), Roeder, et al. (1953), Metcalf (1955) and Dethier (1963) have summarized much of the available information on the subject. The most recent and comprehensive book on the subject is by Hayes (1975). The following remarks present a few of the highlights relating to the chemical control of household pests.

Stomach insecticides. Stomach insecticides for the most part enter through the midgut, the part of the alimentary canal where food is absorbed. Ant baits, certain mothproofing materials and insecticidal dusts which the insect may carry into its mouth while cleaning itself, are typical of the stomach insecticides used in the control of household pests. Some stomach insecticides such as sodium fluoride also may be absorbed through the integument and thus act as contact insecticides.

Fumigants or respiratory insecticides. Fumigants or respiratory insecticides usually enter the insect body through the spiracles and then through the tracheae and tracheoles which supply the insect body with air. To a limited extent fumigants can also penetrate the softer portions of the integument such as those lying between the sclerites. Fumigants are more readily transported through the insect respiratory system at higher temperatures. Since some insects can close their spiracles and thereby hinder the entry of fumigants into the body, carbon dioxide, which at certain concentrations forces the insect to open its spiracles, is often mixed with commercial fumigants. Atmospheric oxygen is made available to the insect by the cytochrome present in the tissues and muscles. The cytochrome carrier under the influence of oxidase takes up and releases oxygen. HCN and some other fumigants combine with the cytochrome enzymes, paralyze the oxygen-carrying system and prevent cellular respiration.

Contact insecticides. Contact insecticides may penetrate directly through the insect integument, through the tracheae, or through sensory glands and other organs that perforate the insect cuticle. Oil sprays wet the lipoid or waxy covering of the cuticle and permit the entry of the insecticide through the integument, tracheae and other openings. Fly spray base oils by themselves are toxic to some species of insects. The toxic principles of dry dusts such as sodium fluoride, pyrethrum and others apparently are soluble in the lipoid layer of the insect cuticle since they pass through the insect integument. Organic compounds and botanicals such as pyrethrum, rotenone, etc., are more commonly applied than inorganic chemicals as contact sprays since the former apparently have a greater ease of penetration through the waxy covering of the insect cuticle. Hairs, spines, and other structures on the integument may interfere with the wetting. A contact insecticide such as pyrethrum acts on the nervous system ultimately destroying many of the nerve cells. Some workers believe pyrethrum synergists interfere with the pyrethrum detoxifying mechanism in the insect by preventing the repair of injured nerve cells.

Residual insecticides. Residual insecticides are long-lasting insecticides which leave either a crystalline or non-crystalline layer of insecticide on a surface. Adult insects ordinarily take up the insecticide through their tarsi. Larvae may be affected through the legs (if present) or through other parts of the body such as the mouthparts. Many residuals act as nerve poisons and show the ability to penetrate the chitin in the cuticle. They also may be dissolved in surface secretions which are then absorbed through the integument. Lindane, chlordane and chloropyrifos not only act as residuals but also show some vapor toxicity.

CURRENT INSECTICIDES

The chemicals discussed in this section are currently registered, available and commonly used by the pest control industry. Descriptions of other materials which are little used or no longer available are preserved in the "historical" section of this chapter. A good reference on the properties of each insecticide is Spear and Pinto (1980).

INORGANIC INSECTICIDES

Sodium fluoride. Sodium fluoride (NaF) is a white powder which is tinted blue or green when used for insecticidal purposes. It has a solubility in water of about four percent at ordinary temperatures. This insecticide is used in dusts for roach, silverfish and ant control. In the past, it was widely used as a dust in combination with pyrethrum. Several investigators have investigated the action of sodium fluoride against roaches, and it apparently acts both as a stomach and contact insecticide.

Borax. Borax ($Na_2B_4O_7 \cdot 10H_2O$) is a relatively slow acting insecticide that was widely used as one of the ingredients of cockroach and ant powders, and is occasionally used to control fly maggots in manure. It is a white powder that is readily soluble in water. Boric acid (N_3BO_3) is used for roach and ant control as a finely ground dust which acts by contact and stomach action. Boric acid is relatively ineffective in solution.

Silica gel and diatomaceous earth. Commonly mixed with pyrethrins, these dusts interact with the integument, causing water loss and dehydration of insects. They remain effective for long periods of time, in dry situations.

ORGANIC INSECTICIDES — ORGANIC PHOSPHATES (OP'S)

There are apparently an infinite number of these compounds available and each year more are shown to be insecticidally active. All of them inhibit the cholinesterases and vary greatly in their toxicity to warm-blooded animals. Some of the less toxic compounds are finding wide usage in the control of household pests.

Malathion. Malathion, O,O-dimethyl thiophosphate of diethyl mercaptosuccinate is manufactured by the American Cyanamid Co. This insecticide is slightly soluble in petroleum oils, but readily soluble in many organic solvents. It is a yellowish liquid which is sold in a variety of formulations. Because of its relatively low warm-blooded animal toxicity it is used in the household usually as a two percent concentration.

Diazinon. Diazinon O,O-diethyl O-(2-isopropyl-4-methyl-6-pyrimidyl) phosphorothioate is primarily manufactured by the Ciba-Geigy Corp. The technical product is a pale to dark brown liquid which is soluble in petroleum oils, alcohol, xylene and acetone. It is of moderate toxicity to warm-blooded animals. This insecticide is effective against a wide range of household insects and is used in the house usually at one percent concentrations. Diazinon is also available in a new microencapsulated formulation used at 0.5 to one percent concentrations. Microencapsulation reduces dermal toxicity while enhancing residual life. There is evidence that this formulation is effective against diazinon-resistant German cockroaches.

Ronnel. Ronnel, O,O-dimethyl 0-2,4,5-trichlorophenyl phosphorothioate is a product of the Dow Chemical Co. This insecticide is readily soluble in most organic solvents including fly spray base oils. Until recently it was available in a variety of formulations. Ronnel is of relatively low toxicity to warm-blooded animals and is used in the house at 0.5 to two percent concentrations primarily for fleas and cockroach control.

DDVP. DDVP, O,O,-dimethyl 2,2-dichlorovinyl phosphate, first synthesized and tested by the USPHS, is used in space and surface sprays at from 0.2 to 0.5 percent by weight. DDVP, which is a liquid, has a relatively high vapor pressure and thus in some respects acts like a fumigant. It remains active for a short time, having little residual effect. Like pyrethrins this insecticide quickly paralyzes flies, cockroaches and many other crawling pests. It is a more potent killer than pyrethrins.

At the recommended concentrations it is of low mammalian toxicity. Nevertheless, special respiratory masks should be worn where there is prolonged exposure to the vapors or to DDVP concentrates. DDVP is soluble in fly spray base oils at 0.5 percent concentrations. DDVP is being used at a 20 percent concentration in resin strips against flying insects. If the DDVP vapors are not quickly removed, the resin strips are slowly effective over a long period.

Fenthion. Fenthion, O,O, Dimethyl O-[4-(methylthio)-m-tolyl] phosphorothioate is of moderate mammalian toxicity. It is used in various formulations and is effective against cockroaches and other insects as a residual insecticide. Parr (1962) recommends a two percent concentration.

Dimethoate. Dimethoate, O,O-Dimethyl S (N-methylcarbamoylmethyl) phosphorodithioate is used in fly sprays. It is most useful for outdoor fly control since indoor use is limited by its strong odor.

Chlorpyrifos. Chlorpyrifos, O,O-diethyl O-(3,5,6-trichloro-2-pyridyl) phosphorothioate is manufactured by Dow Chemical Co. This residual insecticide is effective against some resistant strains of the German cockroach. Its broad spectrum of activity and moderate toxicity permit its wide use in the industry. It is most commonly applied as a 0.5 percent spray, but is also formulated as a paint-on lacquer and as granules.

Acephate. Acephate, O,S-dimethyl acetylphosphoramidothioate is a new compound developed and manufactured by Chevron Chemical Co. This residual insecticide is just appearing on the market and shows great promise for cockroach control.

Fenitrothion. Fenitrothion, O,O,-Dimethyl O-(4-nitro-m-tolyl) phosphorothiate is a residual insecticide widely used in structural pest control in Japan and Europe and under development in the United States by Stauffer Chemical Co.

Propetamphos. Propetamphos, (E)-1-methylethyl 3-[[(ethylamino) methoxyphosphinothioyl] oxy]-2-butenoate is a new low odor broad spectrum contact and residual insecticide developed and manufactured by Sandoz, Inc. It is particularly effective against fleas and German cockroaches.

ORGANIC INSECTICIDES — CARBAMATES

Carbaryl. Carbaryl, 1-Naphthl *N*-methylcarbamate is a white crystalline solid soluble in many organic solvents. It is a contact insecticide with moderate residual properties and is effective against fleas, millipedes, brown dog ticks and other pests, particularly wasps and bees.

Dimetilan. Dimetilan, 2-dimethylcarbamyl-3-methylpyrazolyl (5)-dimethylcarbamate is used in fly bands.

Propoxur. Propoxur, O-isopropoxyphenyl methylcarbamate is a white odorless crystalline powder. It is not soluble in oil. It is used by pest control operators as a one percent emulsion against cockroaches. Propoxur has good knockdown and killing properties.

The emulsifiable concentrate has an exotic solvent system which is unstable in water below 40° F, which limits its use in the North during the winter. It is also available as two percent bait and a 70 percent wettable powder, but the prime use of propoxur is in consumer aerosols.

Bendiocarb. Bendiocarb, 2,2-dimethyl-1,3-benzodioxyl-4-ol-methylcarbamate is a broad spectrum contact and residual insecticide. It could not be easily formulated into a liquid concentrate so a special wettable powder formulation designed for compressed air hand sprayers was developed. It is also available as a one percent dust. It has virtually no odor or vapor, permitting its use in sensitive areas such as hospitals.

ORGANIC INSECTICIDES — CHLORINATED HYDROCARBONS

Lindane. Lindane is the gamma isomer of benzene hexachloride of not less than 99 percent purity with only traces of the other isomers. It is practically free of the musty odor of technical benzene hexachloride. "Chemically, lindane is the gamma isomer of 1,2,3,4,5,6-hexachlorocyclohexane. It is a white crystalline substance having a melting point of 233° F/112° C. In the presence of strong alkaline solutions, lindane has the tendency to break down." At 68° F/ 20° C it is two percent soluble in fly spray base oil and 10 percent soluble in pine oil, and almost insoluble in water. It can be prepared in solution, emulsion, wettable powder or dust formulations. Hensill (1951) reviewed the characteristics and uses of lindane for insecticidal purposes.

Lindane acts as a contact, stomach, fumigant and residual insecticide. It quickly paralyzes most insects and is considered to have fast "knockdown" for a chlorinated insecticide. In the author's laboratory, Peet-Grady tests on house flies with 80 mg lindane per 100 ml base oil (approximately 0.1 percent lindane by weight), showed knockdown for three and 10 minutes of 46 and 94 percent, respectively. This is to be compared with a knockdown of approximately 90 and 98 percent for pyrethrum (100 mg pyrethrins per 100 ml), one of our fastest paralytic agents, at three and 10 minutes, respectively.

Lindane remains as a residue on surfaces in the form of minute crystals which at room temperature continually evolve lindane vapors. Lindane, being much more volatile than DDT, has a shorter residual life than DDT. Chemically stable, it may be volatilized by heating without decomposition, such as in electrically heated vaporizers.

Lindane is very effective against a wide variety of non-resistant household pests such as the adults and immature stages of house flies, cockroaches, carpet beetles, clothes moths, grain insects, lice and fleas. It is used in space sprays at 0.1 percent lindane by weight; however, up to 0.5 percent lindane by weight may be required for residual deposits.

Hornstein and Sullivan (1953) developed a lindane-resin (chlorinated polyphenyl) mixture in the laboratory which keeps the lindane from crystallizing and slows down the evaporation of lindane. Aroclor 5460, from the Monsanto Chemical Co., has been used as the resin type material in the following emulsion formulation:

Parts by weight
Lindane 20
Aroclor 5460 40
Xylene 100
Triton X-100 3

Lindane is nearly twice as toxic as DDT to warm-blooded animals as an acute poison and only one-fourth as toxic as DDT as a chronic poison. It is eliminated rapidly from the body, disappearing one to two weeks after cessation of intake.

It does not accumulate in fatty tissues like DDT. Horton et al., (1948) on applying 2.0 gm of gamma benzene hexachloride per square foot of cloth found it to be highly effective against mites and to withstand repeated launderings. However, they observed bodily motion apparently increases the absorption of gamma benzene hexachloride crystals through the skin and is quite toxic to animals at the dosage of 2.0 gm per square foot on human clothing when applied to rabbits. Lindane dusts, however, are apparently not toxic to human beings since they are being used for the control of human lice and fleas, as well as other insect pests by some military agencies, particularly in the case where lice are resistant to DDT.

Chlordane. Chlordane is an insecticide that was developed during World War II in laboratories of the Velsicol Corp. It is a very effective household insecticide and has found wide use in the control of roaches, ants, carpet beetles, termites and other pests. It is a slow-acting toxicant and may take as long as six days to kill roaches paralyzed by it. Chlordane is the common name assigned to it by the United States Department of Agriculture. This insecticide is offered commercially as a mixture containing 60 to 75 percent of the pure compound and 25 to 40 percent of related compounds occurring in the manufacturing process. The active insecticidal compound has the empirical formula $C_{10}H_6Cl_8$, (1068) and the chemical name for this compound is 1,2,3,4,5,6,7,8,8-octachloro-2,3,3a,4,7,7a-hexahydro-4,7-methanoindene. Chlordane is a viscous, amber-colored liquid which has a density of 1.5-1.63 at 25° C (13.0 to 13.5 lbs. per gal.). It is insoluble in water, but soluble in a great many organic solvents such as kerosene and refined petroleum fractions used for household spray purposes. Chlordane acts as a stomach, contact, residual and fumigant poison. Its volatility ranges between DDT and lindane. Unlike DDT sprays, which leave a crystalline deposit upon evaporation of the carrier, chlordane sprays remain as a non-crystalline film. Chlordane can be used in oil solution, although pest control operators may apply chlordane sprays in homes either as oil solutions or as water emulsions. Unfortunately, house flies, German roaches and some other insects developed resistance and in the United States label restrictions have limited its use in and around buildings to control subterranean termites. For this use it is the dominant insecticide.

Bussart and Schor (1948) recommended the "soluble oil" method for the preparation of emulsion concentrates whereby the hydrocarbon solvent, which is usually kerosene, is added to the chlordane and an emulsifier is added to the oil solution. The water emulsion is then prepared by adding the emulsion concentrate to the required amount of water which may contain such stabilizers as starch or dextrine to retard "creaming." The mixture is then vigorously agitated. One formulation recommended by these authors in order to make one gallon of emulsion concentrate containing 47 to 49 percent by weight chlordane is:

Chlordane	4.0 lbs.
Emulsifier	0.6 lb.
Kerosene	0.6 gal.

The wettable powders are used for spraying rough interior areas where the white residue is not objectionable, as well as the ground around structures for the control of certain insects such as some species of ants.

Chlordane dusts, usually at a five percent chlordane concentration, were widely used by pest control operators for controlling ants and other pests on the ground, and thereby preventing their establishment in the home. Chlordane also was sold as granules.

Heptachlor. Heptachlor, which is related to chlordane and has the empirical formula $C_{10}H_{15}Cl_7$, has many of the properties of chlordane, but because of its warm-blooded animal toxicity was used primarily as an agricultural insecticide. It was widely used against agricultural and fire ants and is now used as a termiticide.

BOTANICAL INSECTICIDES

Pyrethrum. Pyrethrum is a "botanical" or plant product. Pyrethrum extracts are widely used in liquid household sprays, aerosols, livestock and garden sprays. Prior to the advent of DDT and some of the other residual insecticides, pyrethrum powder "insect powder" and mixtures containing pyrethrum powder were commonly used in household dusts, and are now finding wide usage as dusts for stored grains. Gnadinger (1933 and 1945) has devoted two volumes to the subject of pyrethrum. Moore (1966) discusses the chemistry of pyrethrins in some detail.

Since pyrethrum supposedly was first used by the Persians, it has, in the past, often been referred to as "Persian Powder." There are many stories on the initial use of pyrethrum powder against insects, but this tale from Kruhm (1939) is as good as any:

"Around 1840 a woman of Dubrovnik, Dalmatia, discovered the dried flowers of *Pyrethrum cinerariaefolium* to be death to insects. Attracted by the singular beauty of these wild flowers on the surrounding hills, so the story goes, she picked a bouquet of them to take home. When they withered she threw the faded bunch into a corner where, several weeks later, she found it surrounded by dead insects."

Pyrethrum flowers are grown commercially in Kenya, Tanzania, and Rwanda in East Africa, New Guinea and Ecuador, and formerly were exported from Japan, the Belgian Congo and Dalmatia. Two thirds of the worldwide production of about 16,000 metric tons of dried flowers comes from Kenya, which, with the other two East African nations account for 90 percent of total world output. Before World War II, Japan was a major producer of pyrethrum flowers, but with the advent of war, the previously mentioned countries took over the world market.

The flower grown for the commercial production of pyrethrum is *Chrysanthemum cinerariaefolium,* formerly in the genus *Pyrethrum,* and hence the designation "pyrethrum." The insecticidal principles occur for the most part in the achenes of the flower, but some of the ingredients are also present in other parts of the flower such as the petals, receptacles, scales and disc florets. The flowers are picked when the disc florets are three-fourths open since at this stage the pyrethrum content is greatest.

Simanton (1939), speaking of experimental plantings of pyrethrum flowers in the United States, notes: "The plant is similar to the common field daisy in growth habits and appearance. Being a perennial in suitably cool and well-drained regions, a planting will last for six to 10 years, producing a half crop the second year and three full crops the following years. After the fifth year the planting gradually deteriorates until the yield no longer justifies continuance."

Pyrethrum was introduced into Kenya in 1928 and commercial production was initiated in 1932. The flowers are raised at altitudes of 6,500 to 9,500 feet above sea level. The growing season is nine to 10 months long and the properly-ripened flowers are picked by hand during this period. They are then artificially dried in trays at 130° F/54° C. The dried flowers are pressed in bales of 448 pounds each for export.

When the flowers arrive in the United States they are ground and either used in dusts or the pyrethrins are extracted from the ground flowers with petroleum distillates and organic solvents. During the last few years 25 to 35 percent pyrethrum extracts have been exported, and less of the flowers.

Kenya and other leading pyrethrum growers have decided advantages in growing pyrethrum flowers, such as high pyrethrum content flowers varying from 1.3 to more than two percent pyrethrins per pound of flowers, a very long growing season, and relatively low labor costs.

Pyrethrum has been grown commercially in the United States as early as 1876 and sold widely as "Buhach" powder (Essig, 1931). A number of companies have since attempted to raise pyrethrum in this country but have not found it commercially feasible. It is interesting to note the latest strains of pyrethrum raised in the United States are of an erect flowering variety that matures within a week or two (usually during June and again in September) and lends itself to harvesting with a mechanical harvester. Nevertheless, because of the great investment and many problems entailed in cultivating pyrethrum, there are now no commercial pyrethrum plantings in the United States.

The active principles of pyrethrum are pyrethrin I, pyrethrin II, cinerin I, cinerin II, jasmolin I, and jasmolin II (Godin et al., 1965). Gersdorff (1947) studied the toxicity to house flies of the pyrethrins, cinerins and their derivatives and found that pyrethrin I was 4.3 times as toxic as pyrethrin II; cinerin I was four times as toxic as cinerin II; pyrethrin I was 1.4 times as toxic as cinerin I, and pyrethrin II was 1.3 times as toxic as cinerin II. Pyrethrins are usually sold as concentrates containing 2 g, 10 g, or 20 g, pyrethrins per 100 ml. base oil. The 2 g pyrethrins formulation is commonly known as a 20-1 concentrate to the trade. The concentrate is then diluted with the petroleum distillate to the desired pyrethrin content. Aerosol brand pyrethrum concentrates are specially refined to remove Freon insoluble waxes and resins that are irritating and clog the small orifices in the aerosol valve. Pyrethrum sprays readily deteriorate when exposed to light and air, and are packaged in either amber bottles or cans. According to Peet-Grady, tests on house flies conducted in the author's laboratory, pyrethrum in oil base when packaged in amber bottles with a minimum of air space showed very little if any deterioration over a 10-year period.

Weed noted a satisfactory fly spray for that time contained "approximately 45 mg of Pyrethrin I and 55 mg of Pyrethrin II, or a total of 100 mgm of pyrethrins per 100 cc., yielding knockdowns in excess of 95 percent in 10 minutes and kills lying between 65 and 70 percent in 24 hours" by the Peet-Grady test method. At the present time small amounts of synergist, DDVP, diazinon, etc., may be added to raise the 24-hour mortality on flying insects to 99 percent, plus since pyrethrins alone do not give 100 percent kill.

Wilcoxson and Hartzell (1933) showed pyrethrum acts by contact through the integument, paralyzing the nerve ganglia. Weed (1943) noted the active principles of pyrethrum "gain entrance into the insect through openings in the body wall, and by their apparent solubility in body excretions, particularly in the vicinity of the joints of appendages and segments of the body, where the exoskeleton is thin." Hurst (1945) believes pyrethrins interfere with the oxidative enzyme systems essential in carrying oxygen to and from the cells.

Wilson (1949) showed house flies are especially susceptible to pyrethrins when these contact the thin membranes of the "neck" and between the coxae, whereas the top of the thorax and the membranous parts on the bottom of the abdomen are relatively resistant to pyrethrum applications.

David (1946) found by increasing the activity of flies and mosquitoes a greater dosage of spray was accumulated, resulting in higher kill. David and Bracey (1946) note pyrethrum sprays accelerate the activity of mosquitoes and flies much more than DDT sprays. David (1946) showed the quantity of spray accumulated on the wings of *Musca domestica* was three to four times, and of *Aedes aegypti,* four to six times that accumulated on the body.

Pyrethrum sprays have such wide usage because pyrethrins are practically non-toxic to warm-blooded animals in the amounts used in household sprays when taken orally or applied externally. Nevertheless, pyrethrum can cause a dermatitis or an asthmatic attack on susceptible individuals. Martin and Hester (1941) showed it was not the pyrethrins, but other undetermined ingredients present in the flowers that caused the dermatitis and attendant irritation. Zucker (1965) showed there were no allergic reactions to the latest refined pyrethrins.

Pyrethrum synergists. It was previously noted pyrethrum is widely used in sprays, dusts and aerosols because it is effective at low concentrations, provides fast knockdown for both flying and crawling insects and is relatively non-toxic to human beings. However, at times, pyrethrum is in short supply and may become relatively expensive as the demand for it increases.

When certain chemicals are combined with pyrethrum the total effectiveness of the pyrethrum-combination against some insects will be greater than is to be expected from the mere additive effect of each chemical alone. Such chemicals are called *pyrethrum synergists* or *activators.* They currently are widely used in household insecticides because they provide less costly and more effective pyrethrum formulations. The synergist, itself, may or may not have insecticidal properties, but is usually insecticidally inert at the commonly employed concentrations. Often, such combinations of pyrethrum and synergist are quite specific in their action. For example, the combination may be very active against flying insects such as flies and mosquitoes and relatively inactive against crawling insects such as cockroaches, or in any event, show a toxicity only slightly greater than that which is to be expected from the pyrethrum alone.

Some of the chemicals that have been used as pyrethrum synergists are sesame oil extractives, IN 930 or isobutyl undecylenamide, Synergist 264, piperonyl cyclonene, piperonyl butoxide, n-propyl isome and sulfoxide. Ethylene glycol ether of pinene and terpene diacetate are synergists of lesser importance. At this writing, piperonyl butoxide, sulfoxide and Synergist 264 are finding the widest commercial application.

Linquist, et al. (1947) observed that flies showed no signs of injury from synergists alone and these synergists were highly effective even though applied 30 minutes to four hours before exposure to pyrethrins. It is their belief synergists may disarrange or injure nerve or other tissue so that the pyrethrins later are highly effective in producting knockdown. Moreover, they may affect the absorption of the toxicant through the cuticle. David and Bracey (1947) showed a similar delayed effect of pyrethrum synergists on the yellow fever mosquito 15 minutes before application of the pyrethrum spray.

Wilson (1949) elaborated further on pyrethrum-synergist interaction in a study of the physiological action of pyrethrins and piperonyl butoxide and piperonyl cyclonene. He advances the hypothesis piperonyl butoxide prevents recovery of flies from knockdown by pyrethrins by inhibiting the detoxification of the pyrethrins in the fly. According to Wilson, the piperonyl compounds damage the nerve cell bodies and pyrethrins damage the nerve fibers, and the piperonyl compounds interfere with repair of damage caused by the pyrethrins.

David and Bracey (1947) demonstrated that IN 930 (isobutyl undecylenamide) and sesame oil are relatively non-volatile, and their inclusion in spray formulae decreases the volatility of the spray droplets and leads to increase in persistence and particle size of mists. They conclude the increased kill with the use of the adjuvants is due to the increased particle size of the mist. This increase in particle size can also be obtained through the use of a heavy lubricating oil. The authors note this may be a partial explanation since other unknown factors may be involved.

Piperonyl butoxide. Piperonyl butoxide (the technical product containing 80 percent of pure compound (3,4-dioxymethylene-6-proply benzyl) (butyl) diethylene glycol ether) and piperonyl cyclonene, formerly called piperonyl cyclohexenone (condensation product of the alkyl-3, 4-dioxymethylenestyryl ketones with ethyl acetoacetate) are used as pyrethrum synergists in dusts, sprays, emulsions, wettable powders and aerosols. These formulations are finding application in the fields of household, stored product and agricultural pest control. It is claimed these piperonyl compounds also are capable of acting as rotenone synergists.

McAlister, Jones and Moore (1947) discuss the development of these pyrethrum synergists as follows: "The linking by Haller and coworkers (1942) of the increased effectiveness of a methylene-dioxyphenyl group contained in sesamin to the discovery by Eagleson (1940) that the addition of sesame oil to pyrethrum extracts increased their effectiveness against houseflies was a promising approach. Hedenburg (1946) had independently shown that certain compounds containing the methylenedioxy-phenyl group were effective as insecticides, and from this research a technical product has been developed that has found prompt commercial acceptance under the name piperonyl cyclohexenone (piperonyl cyclonene). An intensified search for a product of increased activity and of complete miscibility with *Freon* and with petroleum hydrocarbons led to the synthesis by Wachs (1947) of (3,4-methylenedioxy-6-propyl benzyl (butyl) diethylene glycol ether, the technical grade of which is widely used commercially under the name of piperonyl butoxide."

Dove (1947) describes the physical properties of piperonyl butoxide as follows: "Piperonyl butoxide, as an industrial product, is practically odorless, is a pale yellow oily liquid and has a specific gravity at 25° C of about 1.06. In its pure state it is clear. It possesses a faint bitter taste that is evident several seconds after it is exposed to the tip of the tongue. This insecticide is soluble in all dilutions in mineral oils commonly used as solvents, forms clear solutions with liquified gases that are used as propellents for aerosols and is neutral to litmus."

Piperonyl butoxide, besides increasing the mortality, is known to accelerate the rate of knockdown of flies. These pyrethrin-piperonyl combinations are sold by Fairfield Chemical Division, FMC Corp., and the ratio of pyrethrins to piperonyl butoxide usually varies from 1:2.5 to 1:10. Piperonyl butoxide has been shown to be no more toxic to rats than the deodorized base oil alone. This synergist has slight insecticidal activity of its own. Piperonyl cyclonene, due to its insolubility in petroleum distillates, and Freon (for aerosol use) must be dissolved in auxiliary solvents and is most widely used in agricultural dust formulations.

Sulfoxide (n-octyl sulfoxide of isosafrole). Sulfoxide is a pyrethrins synergist of the same order of effectiveness as piperonyl butoxide. Starr (1950) discusses it in some detail. This chemical as well as n-propyl isome (condensation product of n-propyl maleate with isosafrole), another pyrethrins synergist which was developed at the Boyce Thompson Institute, is marketed by S. B. Penick & Co. Both sulfoxide and n-propyl isome are also used as allethrin synergists.

Sesame oil extractives. Sesame oil extractives, including sesamin (Eagleson, 1942), were used as pyrethrins synergists in aerosol cans during World War II. The concentration of these sesame oil extractives required to effectively synergize pyrethrins is greater than that of most of the synergists currently on the market. (Mallis et al., 1952). However, since sesame oil extractives are less expensive than some of the other commercial synergists, they are marketed in combination with pyrethrins.

Synergist 264 ((n-octyl bicyclo-heptene dicarboximide), (Moore, 1950a). Synergist 264 is a pyrethrins and rotenone synergist. It has been shown to be one of the better allethrin synergists and is superior to most synergists in raising allethrin effectiveness on cockroaches since allethrin is definitely inferior to pyrethrins in cockroach knockdown and mortality. Synergist 264 or MGK-264 is sold in allethrin combinations by McLaughlin Gormley King Company.

Allethrin. Allethrin is often referred to as "synthetic pyrethrins." It is not a botanical, but a synthetic compound. However, since its insecticidal activity and characteristics are similar to pyrethrum, it has been placed in this section. Allethrin is the common name assigned to the completely synthetic allyl homolog of cinerin I by the U.S. Department of Agriculture. The patent for the synthesis of allethrin was issued to Schechter and LaForge of the Bureau of Entomology and Plant Quarantine who demonstrated the natural active principles of pyrethrum flowers are four compounds which have been isolated and identified as pyrethrin I, cinerin I, pyrethrin II and cinerin II. Schechter, Green and LaForge were the first to synthesize compounds closely analogous to cinerin I which is the least complex of the four components of pyrethrum flowers. Allethrin has the formidable chemical name *dl*-2-allyl-4-hydroxy-3-methyl-3-cyclopenten-1 – one esterified with a mixture of cis and trans *dl*-chrysanthemum monocarboxylic acids.

McNamee (1950) describes some of the physical and chemical properties of allethrins as follows: "It is a clear liquid with a specific gravity of 1.005 at 20° C and a refractive index of 1.5050 at 20° C. It is completely miscible with petroleum distillates ordinarily employed in the formulation of insecticidal sprays. The color of the solution is very light. Less than 0.1 percent of this product is insoluble in the "Freon" used in aerosols. Moreover, it stays that way in storage and does not develop polymers and gunks which clog orifices." The present commercial product contains about 92 percent of the actual synthetic allyl homolog of cinerin I. Starr et al. (1950) and other workers have shown allethrin is safe to use in aerosols and sprays. Also, it appears to be no more toxic to warm-blooded animals than pyrethrins.

According to Nash (1950), Peet-Grady tests show allethrins to be approximately as toxic as pyrethrins to house flies. This same paper demonstrates allethrin to be definitely inferior to pyrethrins on cockroaches. Approximately twice the concentration of allethrin is necessary to equal pyrethrins by the U.S.D.A. liquid roach testing method against German cockroaches and two to four times in the case of American cockroaches. Granett et al. (1951) found allethrin to be one-third as toxic as pyrethrins to larvae of the yellow fever mosquito. Jones et al. (1950) conclude from their studies on the effect of synergists on allethrins that the effects are "not equivalent to those obtained when the same synergists are used with pyrethrins. In general, combinations of allethrin with synergists show a lower order of insecticidal effectiveness than similar combinations of pyrethrins with synergists." Several laboratories have shown that synergist 264 and n-propyl isome are the better allethrin synergists, particularly when the allethrin-synergist combination is used against house flies and cockroaches.

Bioallethrin. Bioallethrin or d-trans-allethrin is a newer synthetic pyrethroid with superior insecticidal activity to allethrin. A corresponding isomer known as (s)-bioallethrin developed by the French company Roussel-Uclaf has faster knockdown than bioallethrin. Both pyrethroids are yellow viscous liquids, practically insoluble in water but miscible with most organic solvents including refined kerosene. As with natural pyrethrins and allethrin, their metabolic detoxication by insects is delayed by the addition of synergists such as piperonyl butoxide and MGK 264.

Tetramethrin. Tetramethrin is another synthetic pyrethroid with a strong knockdown action on flies, mosquitoes and some other public health or nuisance pests, but with little effect on cockroaches. It is a white crystalline solid with a pyrethrum-like odor. It is soluble in many organic solvents and it is formulated alone and with other insecticides as space or residual sprays and dusts. It is synergized by piperonyl butoxide.

Resmethrin. Resmethrin, also known as 'SBP-1382' and 'NRDC 104,' is one of the most widely used synthetic pyrethroids in the structural pest control industry. It is a white, waxy solid made up of two isomers, 20 percent to 30 percent bioresmethrin and 70 percent to 80 percent cismethrin. Like most marketed pyrethroids, it is somewhat more stable than natural pyrethrins, but still degrades fairly rapidly on exposure to air and light. It is insoluble in water, but soluble in all the common organic solvents. Knockdown and flushing are not as great as with natural pyrethrins or bioallethrin, but it has excellent killing properties with little recovery after knockdown.

Extensive use sometimes results in odor problems but its "chrysanthemate odor" (described by many as akin to cat urine) quickly dissipates and is more a nuisance to the professional user than returning occupants.

Resmethrin is formulated with and without other pyrethroids and synergists in aerosols and also as emulsifiable concentrates, ULV concentrates and wettable powders. With increasing prices and periodic shortages of natural pyrethrins it has become an important alternative for space and contact sprays.

Permethrin. Permethrin, introduced in the last decade under the code name 'NRDC 143,' is representative of the newest generation of synthetic pyrethroids. Already extensively used in United States agriculture and overseas in public health and structural pest control, these new pyrethroids are currently being developed for control of household pests in the United States. Permethrin is a solid, freely soluble in most organic solvents, but barely soluble in water.

Permethrin is a contact insecticide with much greater photostability than natural pyrethrins or earlier synthetic pyrethroids. This permits its use as both a contact and moderately persistent residual insecticide. It shows a broad spectrum of activity against pests of buildings, ranging from flies to cockroaches.

It is more toxic to mammals than most other pyrethroids and like them is characterized by higher toxicity to fish than most insecticides.

ORGANIC INSECTICIDES — GROWTH REGULATORS

Methoprene (Isopropyl(E.E)-11-methoxy-3,7,11-trimethyl-3,4 dodecadienoate). Methoprene is the first representative of a new class of insecticides now available for household pest control. Methoprene is a growth regulator developed by Zoecon Corp. It has been available for some years for control of mosquitoes and various agricultural pests, but it was only in September 1980 that a clearance was obtained for its use in the United States for indoor control of fleas (Rambo, 1980).

Methoprene works by inhibiting the life cycle of the flea at the pupal stage. After exposure to this material the larvae pupate, but adults never emerge from the pupal cocoon. Methoprene may also play a part in inhibiting eggs from hatching, but this is of secondary importance.

If a high population of adult fleas is present, the use of adulticides is necessary for immediate relief. Since many adulticides such as propetamphos and diazinon also kill larvae, the use of methoprene provides insurance against survivors from standard insecticide treatments.

The greatest advantages of methoprene are its absence of odor and its very low toxicity to non-target species including pets, which allows reoccupation of treated areas within 30 minutes of use. Methoprene is available as an emulsifiable concentrate which is applied as a 0.1 percent spray and as a 0.15 percent total release aerosol.

Methoprene is under development in the United States for control of Pharaoh ants, a use which is already established in Europe. Other so called biorational compounds, including growth regulators and pheromones are expected to become available for household pest control in the future. An excellent review of growth regulators is presented by Kiess (1981).

HISTORICAL INSECTICIDES

The chemicals discussed in this section were all developed more than 20 years ago and many of them are no longer available, or if available are severely restricted in their use and little used in household pest control.

INORGANIC INSECTICIDES

White arsenic. During the Middle Ages the Borgias kept the conversation at the dinner table at a high level of interest by eliminating the more tiresome guests with white arsenic. Its success in this respect may have suggested its use for the eradication of four-legged and six-legged pests. Roark (1935) notes Worlidge recommended arsenical baits for ant control as early as 1669. White arsenic, arsenic trioxide, or arsenious oxide (As_2O_3) is a heavy, white, odorless and tasteless powder. According to Sennewald (1947a), white arsenic is "sparingly and extremely slowly soluble in cold water. About ½ of one percent technical arsenic trioxide will dissolve in water in five days' time. It is more soluble in boiling water; however, when the solution cools, the As_2O_3 does not remain in solution." This same author also separates white arsenic into several categories including:

- Commercial white arsenic — This is the common chemical.

- Micronized and Microfine arsenic — This is white arsenic that is ground
  in a micronizer mill. It is a very fine powder containing microscopic particles of arsenic trioxide.
- Soluble arsenic trioxide — This is a commercial brand of arsenic that contains an organic arsenical compound and is sold as a soluble white arsenic.

Finely ground white arsenic is preferable for use in rodent control because
it has greater solubility in water than the coarser material. Water soluble arsenic also is more readily soluble in the body fluids of the rat. White arsenic is
used in ant pastes and rat poisons.

Sodium arsenite. Sodium arsenite ($NaAsO_2$) is a heavy white powder in the
chemically pure form (99%) and a grayish-white powder in the commercial
grade (95%). It is readily soluble in water, and solutions of sodium arsenite are
alkaline and have a bitter taste. Sodium arsenite in the chemically pure form
is used in ant syrups. In the commercial grade it is used as soil poisons for
termites.

Paris green. Paris green ($Cu(C_2H_3O_2)_2 \cdot 3 Cu (AsO_2)_2$ is a copper aceto arsenite. It is sold commercially as a heavy green powder and has found some use
as dust for the control of drywood termites and agricultural ants. London purple
is a mixture of calcium arsenate and calcium arsenite and also has been used
for the control of agricultural ants.

Acid lead arsenate. Acid lead arsenate ($PbHAsO_4$) is a white powder that
must be colored pink so as not to be mistaken for flour or other edibles. It occasionally is used as a soil poison in termite control.

Calcium arsenate. Calcium arsenate consists for the most part of tricalcium
meta arsenate ($Ca_3(AsO_4)_2$) and a lesser amount of acid calcium arsenate
($CaHAsO_4$). It is a white powder which is colored pink and used in solid baits
for snails, sowbugs, and pillbugs.

Sodium fluosilicate. Sodium fluosilicate (Na_2SiF_6) also is known as sodium
"silicofluoride." Marcovitch and Stanley (1942) studied this and other fluoride
compounds in some detail. They describe sodium fluosilicate as "a white, granular, nonvolatile, odorless powder with an acid reaction. At room temperature
it dissolves in water at the rate of one part to 154. It is insoluble in the common
organic solvents. When the fluosilicate is swallowed, it usually is more toxic
than any of the fluorides, since it contains 60 percent fluorine." Sodium fluosilicate is used in solution for the control of such household insects as clothes
moths and as a dust against earwigs and crickets. It must be tinted blue or green
like sodium fluoride.

Barium fluosilicate. Barium fluosilicate is used in silverfish baits. Magnesium fluosilicate, sodium aluminum fluosilicate and other fluosilicates are
water soluble fluosilicates used in mothproofing solutions.

Thallium sulfate. Thallous sulfate (Tl_2SO_4) is commonly known as thallium
sulfate. Sennewald (1947c) notes thallium sulfate is available as a heavy white
powder or as a white crystalline material. Sennewald also notes the Federal
Rodenticide Act requires thallium sulfate be tinted a nile green color. It is approximately five percent soluble in cold water and much more soluble in warm
water. The commercial grades have a purity of 99 percent or more.

Since solutions of thallium sulfate have a slightly acid reaction, they should
be prepared in glass or enamel ware. The vapors of boiling solutions containing
thallium are poisonous and must not be inhaled. Thallium sulfate is odorless
and tasteless and quite effective against rats, mice and some species of ants.
Since the compound also is odorless and tasteless to human beings, as well as

a cumulative poison, it must be used with care. *Children and pets are especially likely to consume sweet ant baits containing it.* The USDA barred thallium sulfate for household use in 1965.

White phosphorus. Sennewald (1947b) discusses phosphorus and phosphorus pastes as follows: "When pure and recently manufactured, phosphorus is a colorless semi-transparent solid. It has a waxy luster and consistency about like that of beeswax. It fumes when exposed to the air, has a disagreeable odor and is luminous in the dark. With age it becomes yellowish, or acquires a white coating.

"The correct name for the kind of phosphorus in phosphorus paste is white phosphorus. In former days because it was not pure, or had developed a yellow color, it was called yellow phosphorus. White and yellow phosphorus are the same.

"White phosphorus ignites spontaneously in air and therefore must be packed, stored, shipped and handled under water. It is supplied in 2½ pound cakes, or wedges and in sticks about ½ inch in diameter and 10 inches long.

"White phosphorus has a specific gravity of 1.82. It is almost twice as heavy as water. It melts at 111 degrees F. and ignites spontaneously in air at about 122 degrees F. It is almost insoluble in water, but imparts to it an odor. It is slightly soluble in alcohol and ether, more soluble in benzene, chloroform and certain oils, and very soluble in carbon disulfide.

"Every pest control operator has seen the white clouds or 'smoke' formed when phosphorus paste is stirred. These clouds or 'smoke' are phosphorus pentoxide. Phosphorus pentoxide is a white, practically odorless powder. It is not toxic. The odor of phosphorus paste is due to phosphorus trioxide and phosphorus itself, and not to phosphorus pentoxide.

"In a properly prepared phosphorus paste, the particles of phosphorus are so small that they cannot burn and produce a flame. When the paste is stirred, they are changed to 'smoke' or clouds of phosphorus pentoxide.

"The toxic action of phosphorus is due to the fact that it is absorbed by the fats and oils present in the body. It is interesting to note that the toxicity of phosphorus, like the toxicity of arsenic trioxide, depends upon the particle size of the poison."

One phosphorus paste on the market contains about two percent white phosphorus. It is effective against rats and the American and oriental cockroaches. Phosphorus pastes are used in damp places where dusts or sprays cannot be used. Baits and pastes containing white phosphorus are extremely poisonous and great care should be taken by pest control operators and others to place them so they are not accessible to children and pets.

Mercuric chloride. Mercuric chloride, corrosive sublimate, bichloride of mercury ($HgCl_2$), is a white crystalline material that is also available commercially in the granular or powder state. It is soluble in water and alcohol. In tropical countries it has been applied to wood, book bindings, paper, etc., for protection against termites, cockroaches, silverfish and other pests. It also has been used to impregnate tapes used as barriers against ants. Mercuric chloride and its solutions should be handled with great care.

ORGANIC INSECTICIDES — CHLORINATED HYDROCARBONS

DDT. In 1874, Zeidler of the University of Strasbourg, prepared and described DDT for his doctor's thesis. The insecticidal properties of DDT were first discovered by Paul Müller, a research chemist of the Swiss dye firm, J.R. Geigy,

S.A., who synthesized this chemical and demonstrated its effectiveness against house flies and mosquitoes. Müller also discovered the residual effect of DDT and prepared formulations of DDT solutions, emulsions, dusts and wettable powders. For these achievements, he was awarded a Nobel prize in 1948.

Knipling (1945) notes samples of DDT were received in the Orlando laboratory in Florida in November, 1942, and were found to be very effective against the body louse, as well as against mosquitoes, house flies, bed bugs, fleas and cockroaches. The compound was identified in the laboratory by H.L. Haller. English workers studied the insecticidal properties of DDT at about the same time as American investigators. Their research paralleled that of the Americans.

Haller and Busby (1943-1947) discuss DDT as follows: "The symbol DDT is a contraction for dichloro-diphenyl-trichloroethane, the generic name of the active insecticidal principle. Theoretically, there are 45 possible dichloro-diphenyl-trichloroethanes, excluding stereo isomeric forms. However, the term DDT has been confined to the product obtained on condensation of chloral (on its alcoholate or hydrate) with chlorobenzene in the presence of sulfuric acid. The product thus obtained is termed technical DDT. It is a white to cream-colored powder, possessing a fruit-like odor. Its major constituent is 1,1,1 trichloro-2,2-bis (p-chlorophenyl) ethane, which has the formula $(ClC_6H_4)_2 CHCCl_3$ and is called p,p'-DDT.

"As the name applies, technical DDT is a commercial grade. Since it melts over a range of several degrees, the solidification point rather than the melting point is used to give an indication of its purity."

The p,p'-DDT in technical DDT varied from 65 percent to 75 percent. The major impurity was o,p'-DDT from 19 percent to 21 percent and others, including TDE (DDD) from 0.007 percent to four percent.

Solubility of DDT. For all practical purposes, DDT is insoluble in water, slightly soluble in petroleum oils and very soluble in some organic solvents. Jones, Fluno and McCollough (1945) have prepared extensive tables on the solubility of DDT in various solvents at 80° to 86° F/27 ° to 30° C, and the accompanying solubility table for DDT has been compiled from their paper.

TABLE 26-1
Solubility of DDT in Some Common Solvents
At 27° to 30° C per 100 grams of solvent

Acetone	74
Benzene	89
Carbon Tetrachloride	28
Cyclohexanone	122
o-Dichlorobenzene	45
Ethyl alcohol (94 percent)	2
Ethyl ether	39
Aliphatic Petroleum Fractions	
Gasoline	13
Stoddard Solvent	12
Kerosene	10-12
Fuel Oil No. 1	10-14
Fuel Oil No. 2	8-12
Lubricating Oil, S.A.E. 30	6
Kerosene (refined, fly-spray base)	5

Aromatic Petroleum Fractions

PD-544-C, broad fraction
  (Socony-Vacuum Culicide Oil B
  chiefly methyl- and polymethyl-naphthalenes)   42
APS 202 (refined, high boiling
  fraction of S/V Culicide Oil B)   44
Velsicol AR-50
  (chiefly mono- and dimethyl naphthalenes)   56
Velsicol AR-60
  (chiefly di- and trimethylnaphthalenes)   58

Miscellaneous

Xylene, 10-degree   61
Linseed Oil, raw   12

Although crude kerosenes can be used to prepare a five percent DDT solution, fly-spray base oils may require 13 percent to 20 percent auxiliary solvents, particularly when it is desirable to keep DDT in solution at 32° F/0° C or lower temperatures.

Certain of the above solvents are inflammable and may form highly explosive combinations in enclosed areas. For this reason, they are usually used only in water emulsion concentrates or with wettable powders.

INSECTICIDAL PREPARATIONS OF DDT

Solutions. Solutions of DDT on the market for use in residual or surface sprays are usually five percent or six percent DDT by weight in a fly-spray base oil. As previously noted, auxiliary solvents such as Socony-Vacuum PD-544-C and the Velsicols, to name but a few, must be added as auxiliary solvents to the fly-spray base oil to prevent the DDT from coming out of solution at low temperatures. Because the auxiliary solvents are quite odorous, pest control operators may use only three percent DDT in a fly spray base oil to avoid the use of these solvents.

Fleck and Haller (1946) showed kerosene obtained from naphthenic-base crude oils dissolve more DDT than do those obtained from paraffin-base crude oils. The aniline point of kerosene may be used as a general guide to its solvent power for DDT. In general, the solubility of DDT increases as the aniline point decreases.

Iron has been shown to decompose DDT. Also, DDT will lose its insecticidal activity more rapidly on iron than on copper screens.

Emulsions. Haller and Busby (1943-1947) have the following to say in regard to water emulsions of DDT: "Two types of emulsions have been used — those in which the DDT is dissolved in a volatile solvent, such as xylene, which evaporates after spraying to leave a deposit of DDT crystals, and those in which the DDT is dissolved in a relatively nonvolatile solvent, such as a petroleum oil, which leaves the sprayed surface coated with a solution of DDT in oil after evaporation of the water. A great variety of emulsifying agents are available for the preparation of DDT emulsions. The use of excessive amounts of emulsifier should be avoided in order to prevent excessive runoff of the spray and to avoid coating the DDT deposit by the emulsifier after evaporation of the water and solvent." Jones and Fluno (1947) have prepared a detailed list of emulsifiers and the results obtained when they are used in DDT-xylene emulsions.

An emulsion concentrate that was developed by workers in the Bureau of Entomology and Plant Quarantine of U.S.D.A. as a residual spray and larvicide is presented below:

	By weight
DDT	25%
Triton X-100	10%
Xylene	65%
Total	100%

According to U.S.D.A. Miscellaneous Publication 606, "Emulsions are made by adding the required amount of concentrate slowly to water with continuous stirring. To prepare a five percent emulsion of DDT, one volume of the concentrate is mixed with four volumes of water (the density of the concentrate is close enough to one gm per milliliter to make this dilution by volume satisfactory). A one percent DDT emulsion made from this concentrate is cheaper than one percent DDT solutions in fuel oil or kerosene." The disadvantages of the above formula is its low flash point due to the xylene. Also, the DDT will come out of solution at low temperatures.

Commercial emulsion concentrates on the market usually contain 25 percent or 30 percent DDT by weight, an auxiliary DDT solvent (either Socony-Vacuum PD-544-C, the Velsicol DDT solvent, or any of the other available DDT solvents) emulsifiers such as the Triton, Span, or Tween series, an emulsion stabilizer, anti-corrosion agent and anti-foam agent. These concentrates are diluted with water to make the desired solutions.

Suspensions (Wettable or dispersible powders). Simple suspensions can be prepared by mixing powdered DDT with some dust material like talc in addition to a small amount of a wetting agent.

Chisholm (1948), in a review of DDT formulations, discusses the preparation of wettable powders containing 90 percent or more of DDT. The compound is micronized and the particles are coated to prevent caking. Anti-caking agents are usually used with these. They are "low-bulk-density diluents, the probable function of which is to separate the DDT particles and thus prevent packing in the mill during grinding or coalescence during storage. These are used with wetting-dispersing agents, auxiliary-dispersing agents, polymeric film-forming agents or wetting agents."

The same author notes the diluents for powders containing 50 percent or less DDT are of low or high bulk density (i.e., where the weight is equal materials of low-bulk density take up more space than those of high-bulk density). The anti-caking agents are examples of low bulk density. Pyrophyllite, talc, calcite and clays are examples of high-bulk-density. "Mixtures of both types often are used to prepare products which have practical bulk-density values and also will resist caking on storage at high temperatures. It is possible to use either type alone for products containing 50 percent DDT, but diluents of low-bulk density may be impractical because of the excessive space occupied by the product. The use of high-bulk-density diluents alone may give products that become packed or lumpy on storage. The possible exceptions are Kaolin-type clays, which are intermediate in bulk density. For products containing 10 percent or less of DDT the high-bulk density diluents are satisfactory." Wetting agents are usually added to the DDT and diluent before grinding.

The DDT wettable powders are used at from 0.1 percent to five percent concentrations, but ordinarily at 2½ percent concentration or less to avoid clogging of the sprayer nozzles. During the application, it is essential the suspension be constantly agitated to prevent settling out.

Woodruff and Turner (1947) showed a reduction of DDT particle size in water suspension causes an increase in toxicity both in the laboratory and field.

Dusts. DDT dusts are prepared by grinding technical DDT with talc, pyrophyllite, kieselguhr, gypsum, chalk, pyrethrum marc, clay, sulphur, etc. Technical DDT softens in the mill at 192° F/89° C. For this reason, overheating during the grinding process must be avoided. A 10 percent DDT dust is widely used for louse and flea control.

Precautions in use of DDT. DDT has an excellent safety record. Nevertheless, it is a poison and certain safety factors should be observed while working with it. Neal and von Oettingen (1946) present a summary on the use of DDT relative to its safety to man and animals. The precautions to be observed for DDT are somewhat similar to those for other chlorinated insecticides.

"DDT in *dust form* is not absorbed through the skin unless greases, oils or greasy skin lotions are already present on the skin. Nevertheless, DDT powders should not be allowed to remain on the skin and excessive inhalation of the powder should be avoided.

"DDT in *oil solution* is readily absorbed through the intestine and also is absorbed through the skin. Therefore, DDT-oil solutions should not be allowed to remain on the skin or saturate clothing. Wash the hands and exposed skin with warm soapy water; and if oil solutions or concentrates are spilled on the clothes, change them promptly. Avoid inhaling the mist and contaminating food with the spray. *Never use it on the skin or coat of animals.*

"It should be pointed out that many of the solvents (kerosene, etc.) used in preparing DDT insecticides in themselves may cause irritation of the skin and other harmful effects when handled carelessly. By proper precautions and cleanliness, these can be avoided.

"If a good deal of spraying is to be done, it is advisable to wear gloves, goggles and a respirator to avoid excessive contact with and inhalation of DDT and its solvents."

Miscellaneous notes on DDT. Schmitz and Goette (1948) studied the penetration of DDT into wood surfaces (poplar wood). According to their tests, "The DDT which remained within 0.001 of an inch of the top surface was considered to include all which was available biologically; all DDT which penetrated further than 0.001 of an inch of the top surface was probably of no value biologically. About 38 percent of the DDT applied in an emulsion and about 30 percent of the DDT applied in a kerosene solution were deposited within 0.001 of an inch of the top surface. Better recovery of DDT from the emulsion was probably due to a more rapid rate of DDT crystallization from the emulsion than from the kerosene solutions."

Parkin and Green (1947) showed that on wallboard a mechanical stimulus such as movement of flies on treated surface may cause the DDT present in supersaturated solutions to crystallize with a resultant increase in toxicity of the film.

Parkin and Hewlett (1946) demonstrated the low toxicity of films of DDT on limewash. Barlow and Hadaway (1947) noted that DDT and "Gammexane," when applied as oil solution or emulsion to limewash and mudwalls, resulted in a loss of the insecticide due to absorption. Also, they showed wettable powders show less loss due to this factor.

Lindquist, Jones and Madden (1946) demonstrated DDT residues exposed "to both ultraviolet and sunlight reduced the effectiveness of the treatments in most cases, as shown by the times required to give a knockdown of house flies, *Musca domestica* L.," and DDT deposited in emulsion or suspension did not decompose as rapidly as those in solution. For political reasons use of DDT is no longer permitted in the United States, except in public health emergencies. However, DDT continues to play a vital role in disease vector control programs overseas.

TDE or DDE. TDE or DDD or dichlorodiphenyl-dichloroethane, 1,1-dichloro-2,2-bis (*p*-chlorophenyl) ethane was synthesized by German research workers during World War II (Frear, 1948). This light-colored granular solid is less toxic to warm-blooded animals than DDT, but is also less effective insecticidally to most species of insects. It is widely used in mosquito larviciding work. Ginsburg (1948) notes it is "nearly as toxic as DDT to mosquito larvae, but is less injurious to fish and warm blooded animals."

Cristol, Haller and Lindquist (1946), in a study of toxicity of several DDT isomers, came to the following conclusion: "The p,p'-DDT isomer is several times as toxic as the o,p'-DDT isomer to both goldfish and mosquito larvae. The methoxy analogue equals DDT in toxicity to fish, but is less toxic to mosquito larvae. On the other hand, the DDD analogue ranks about as high as DDT in toxicity to mosquito larvae, but appears to be less toxic to goldfish."

Methoxychlor. Methoxychlor is the approved generic name for 1,1,1-trichloro-2,2-di-(4-methoxyphenyl)ethane. Methoxy DDT, methoxy analog of DDT and dianisyl trichloroethane are some of the other names by which it is known. Technical methoxychlor is a cream-colored amorphous solid consisting of 90 percent methoxychlor mixed with petroleum hydrocarbons and one percent wetting agent. Smith et al. (1946) show that methoxychlor is one of the chlorinated compounds that is least toxic to warm-blooded animals. A. Goddin, in an unpublished work, showed methoxychlor to have a much faster knockdown against flies than DDT. Methoxychlor has some residual activity but it is not as long as that of DDT. Krister (1951) notes that it is effective against fleas on dogs as a 10 percent dust or 0.5 percent dip.

Anon. (1948a) found that a concentration of 0.6 percent methoxychlor based on the weight of garment when applied in perchlorethylene solution gave protection for about one year against carpet beetles and clothes moths. Methoxychlor oil-base sprays at 0.3 percent to three percent by weight are used against flies as space or surface sprays, and its use is recommended where DDT sprays and residues are objectionable. Unfortunately, DDT-resistant flies are usually resistant to methoxychlor. A five percent methoxychlor dust was effective against silverfish. Methoxychlor also was used in mothproofing aerosols. Interest in methoxychlor has been recently re-kindled because of its usefulness in gypsy moth control in urban areas.

Perthane or Q-137. The chemical name of this compound is 1,1-dichloro-2,2-di-(4-ethylphenyl) ethane and it was developed in the laboratories of Rohm & Haas Co. This chemical is as toxic as methoxychlor to house flies, is of a low order of mammalian toxicity and is more readily soluble in fly spray petroleum distillates. It was often combined with DDT as a mothproofing agent or used alone for control of clothes moths and carpet beetles.

Benzene hexachloride, HCH or BHC. The insecticidal properties of benzene hexachloride were discovered by French and English workers from 1940 through 1942. The chemical name of this toxicant is 1,2,3,4,5,6-hexachlorocyclohexane, consisting of several isomers and containing 12 percent to 14 percent of the gamma isomer, and the formula $C_6H_6Cl_6$ led to its being called "666." It has been shown only the gamma isomer is highly toxic to insects, and the alpha, beta, delta and epsilon isomers are of relatively low insecticidal activity. In the United States, the gamma isomer of benzene hexachloride of not less than 99 percent purity is known as *lindane*. It is lindane that is used in homes for the control of household pests. "Gammexane" is approximately the British equivalent of lindane. All the early studies were made with benzene hexachloride. It and lindane will be treated separately. Crude BHC used for the control of agricultural pests has a musty, disagreeable and persistent odor. It is sold in dusts with a variable amount of gamma BHC; 10 percent to 12 percent gamma BHC is a common dust mixture.

Benzene hexachloride acts as a stomach, contact and fumigant poison. Like DDT and chlordane, benzene hexachloride can be formulated as a dust, wettable powder, solution or an emulsion.

Dustan et al. (1947) showed residual deposits of benzene hexachloride and chlordane lost their effectiveness when exposed to air due to fumigant action, whereas chlorinated camphene (toxaphene) was hardly affected and DDT was without fumigant action.

DeMeillon (1946) showed when benzene hexachloride was fed to a rabbit in small dosages, toxic effects become evident on *Cimex lectularius* and *Aedes aegypti* when the rabbit had a total dosage of 100 mg "Gammexane." The African relapsing fever tick, *Ornithodorus moubata,* also was affected. Barlow (1947), reporting the work of Wilson, notes "the blood of cattle which has received benzene hexachloride by oral ingestion was toxic to tsetse flies for a considerable time after treatment."

Benzene hexachloride has been known to taint with its odor fruits and vegetables upon which it was applied. In this respect, it is interesting to observe that Hixson and Muma (1947) note a wettable benzene hexachloride containing five percent of the gamma isomer gave the meat of poultry a distasteful flavor and objectionable odor when applied to chickens, chicken house or the litter.

Dieldrin ($C_{12}H_8OCl_6$). Dieldrin has the chemical name 1,2,3,4,10,10-hexachloro-exo-6,7-epoxy-1,4,4a,5,6,7,7,8a-octahydro-1,4-endo,exo-4,8-dimethanonaphthalene. This insecticide is a product of the Shell Chemical Corporation. It is slightly soluble in fly spray base oils and readily soluble in aromatic solvents. Dieldrin is sold as an oil base concentrate, emulsifiable concentrate, wettable powder and in granular formulations. This insecticide finds wide use overseas against household insects at a concentration of 0.5 percent. It also is used as a soil insecticide against termites and agricultural ants, including fire ants.

Aldrin. Aldrin is closely related to dieldrin and also is sold by the Shell Chemical Corporation in formulations somewhat similar to that for dieldrin. Because of its greater warm blooded animal toxicity, it is not used like dieldrin for the control of insects in the house. It was widely used against termites and agricultural ants, including fire ants, and is still used as a soil insecticide for termite control.

Chlordecone (Kepone). Chlordecone has the chemical name decachloro-octahydro-1,3,4-metheno-2H, 5H-cyclobuta(cd)pentalen-2-one. It is used in baits for control of cockroaches and ants, including fire ants and Pharaoh ants. It is no longer manufactured but existing stocks can still be used.

Mirex. Mirex has the chemical name dodecachloro-octahydro-1,3,4-metheno-2H-cyclobuta(cd) pentalene. Like chlordecone, mirex is a stomach insecticide with little contact activity and was mostly used in baits for control of ants. It was particularly valuable in control of fire ants and harvester ants but is no longer used in the United States.

Orthodichlorobenzene ($C_6H_4Cl_2$). Orthodichlorobenzene is a clear colorless liquid. The technical grade which contains 90 percent of the ortho- and 10 percent of the para-isomer is used for insecticidal purposes. It is soluble in oil and miscible with most organic solvents, as well as being a good solvent itself. Fletcher (1945) describes its physical and insecticidal properties. Orthodichlorobenzene is still used for the control of termites and powder post beetles.

Paradichlorobenzene or PDB ($C_6H_4Cl_2$). PDB is a white crystalline substance which vaporizes more rapidly than naphthalene to form gas with an ether-like odor. It is soluble in kerosene, carbon tetrachloride and other organic solvents. Roark (1947) notes that Arnold Erlenbach of Germany in 1911 patented the use of paradichlorobenzene against clothes moths and other insects that feed on woolens and animal materials. The use of PDB as a deodorizer was developed by Dr. R. C. Roark in 1928. Paradichlorobenzene is used in the same way as naphthalene as a fumigant to control and repel clothes moths and carpet beetles. It is available in deodorant blocks and in moth cakes, flakes and nuggets.

Toxaphene ($C_{10}H_{10}Cl_8$). Toxaphene is a chlorinated camphene having a chlorine content of 67 percent to 69 percent which was developed as compound "3956" by chemists of the Hercules Powder Co., and screened on such household insects as cockroaches, house flies and mosquitoes by A. Mallis at the University of Delaware. It is a yellowish, waxy solid which is soluble in most of the organic solvents. It was used in combination with knockdown agents at one percent or two percent in oil base sprays and in aerosols at five percent concentration for the control of household insects. Strobane, a terpene polychlorinate with a chlorine content of approximately 66 percent, has properties somewhat similar to toxaphene and was used in household insecticides like toxaphene. Strobane is a brownish, viscous liquid which is readily soluble in fly spray base oil. Parker and Beacher (1947) prepared an introductory paper, and Roark (1950) summarized the literature on toxaphene. Toxaphene is still available for agricultural pest control but its use is rapidly declining.

ORGANIC INSECTICIDES — ORGANIC PHOSPHATES (OP's)

Dibrom. Dibrom, which has the chemical name 1,2-dibromo-2,2-dichloroethyl dimethyl phosphate, is an oily liquid. It is chemically related to DDVP, but has a lower vapor pressure. Dibrom is of moderate mammalian toxicity. It was used for fly control in barns, in fly baits and for mosquito control.

Dicapthon. Dicapthon has the code name "4124" and the chemical name 0-(2-chloro-4-nitrophenyl)0,0-dimethyl phosphorothioate. It was a product of the American Cyanamid Company and was used against resistant cockroaches as a one percent spray. Dicapthon is of moderate toxicity to warm-blooded animals.

Dioxathion. Dioxathion has the chemical name 2,3-p-dioxanedithiol s,s-bis (O,O-diethyl phosphorodithioate). It is a highly toxic insecticide and acaricide introduced more than a quarter of a century ago by Hercules, Inc. It primarily was used for ectoparasite control on livestock and against phytophagous mites. Recently it has been resurrected for use by certified pest control applicators for control of ticks and fleas outdoors.

DPN, Parathion and TEPP. DPN, parathion and TEPP are organic phosphate insecticides of high toxicity to warm blooded animals. They are sometimes used outdoors by public health workers in the control of mosquito larvae. Parathion and TEPP also were used in the preparation of fly cords.

Trichlorfon. Trichlorfon, introduced by Bayer Leverkusen under the code number "Bayer L 13/59," has the chemical name dimethyl 2,2,2-trichloro-1-hydroxyethylphosphonate. It is a broad spectrum contact and stomach insecticide of moderate to low toxicity to warm blooded animals. In addition to agricultural and veterinary uses it is used for control of household pests, particularly flies and cockroaches. Fly control formulations include sugar baits and sprays. For cockroach control a water soluble powder is mostly used because its lack of odor is particularly suited to use in special situations such as hospitals. The insecticidal activity of trichlorfon is attributed to its metabolic conversion to dichlorvos.

BOTANICAL INSECTICIDES

Rotenone. Rotenone is another botanical. Derris, a source of rotenone, is produced from the roots of the plant *Derris elliptica*, grown for the most part in the Far East in such countries as Malaya, East Indies and the Philippines. Cubé is from the roots of a plant of the genus *Lonchocarpus* from Peru and Brazil. The leaves of these plants are much richer in rotenone than the roots. Both of these sources of rotenone have been used for centuries as fish poisons by the natives.

Apparently, the most important toxic principle in the roots of the plants is rotenone which has the formula $C_{23}H_{22}O_6$. Rotenone is colorless, odorless, crystalline, only very slightly soluble in petroleum distillates of the fly-spray base type and not soluble in water. Other crystalline compounds related to rotenone and found in the same roots are dequelin, tephrosin, toxicarol and sumatrol. These non-rotenone compounds also have some insecticidal activity.

Derris and cubé are finely ground and mixed with various dusts such as talc or pyrophyllite to produce a mixture containing 0.75 percent to one percent rotenone by weight.

Rotenone and its related compounds are extremely toxic to cold blooded animals, but relatively harmless to man and domestic animals. Rotenone acts primarily as a contact insecticide on household insects. In man, at times, rotenone may produce a sensation of numbness in the mucous membranes of the mouth which may last for several hours. Since fish are extremely susceptible to rotenone, they should be removed from a room where a rotenone spray or dust is applied.

Rotenone and extracts of derris, cubé and timbo, another rotenone source, are occasionally used in household sprays. Due to the low solubility of rotenone and related compounds in petroleum oils, it is necessary to add an auxiliary solvent such as safrol, butyl phthalate or similar chemicals to keep the rotenone from precipitating.

Rotenone is effective against fleas, lice, ticks, mites, house flies, cockroaches, bed bugs, ants, etc. It also is widely used as an agricultural and veterinary insecticide. At the present time, rotenone dusts find their widest use in the household for the control of fleas.

Ryania. Ryania is derived from the South American plant, *Ryania speciosa*. The stem is the principal source for the insecticidal material. A 100 percent

Ryania dust has been found to be effective for several months against German and American cockroaches. W.D. Bedingfield, a Texas pest control operator, found it effective against chlordane-resistant roaches.

MISCELLANEOUS INSECTICIDES

Many of the following items were considered elsewhere and the reader should refer to the index for more detailed information.

Temporary and permanent-type mothproofers. These were discussed in greater detail in the section on clothes moths. They consist of the water and oil-soluble agents, the temporary mothproofers that must be reapplied one or more times a year and the more permanent-type dye-bath agents that are applied in the mill.

Sodium aluminum silicofluoride is soluble in water. Other water soluble temporary mothproofing compounds are magnesium silicofluoride and ethanolamine silicofluoride. DDT and methoxychlor are oil soluble temporary mothproofers.

The dye-bath mothproofers are permanent-type mothproofers which are, as the name implies, applied in the dye bath. *Eulan* CN and CN *extra*, which are of German origin, are sulphonic acid derivatives of triphenylmethane. *Lanoc* (sodium dihydroxy-pentachloro-triphenyl-methane-sulphonate) is related to the Eulans and is manufactured by Imperial Chemical Industries, Ltd. *Mitin FF* is a dichloro-diphenyl ether and sulphonic group compound which is the product of Ciba-Geigy, A.G. *Mystox*, manufactured by Catamonace, Ltd. is primarily a pentachlorophenol compound.

Fumigants and rodenticides. Fumigants and rodenticides are discussed in some detail in their respective chapters.

Wood preservatives. Wood preservatives also are discussed in the chapters on termites and decay fungi.

Creosote. Anon. (1958) notes creosote is a coal tar distillate made by high temperature treatment of bituminous coal. Its principal constituents are liquid and solid aromatic hydrocarbons and appreciable quantities of tar acids and tar bases. It is heavier than water and does not evaporate readily. It is toxic to molds, fungi, marine borers, termites and wood-borers. Timbers treated with some coal tar creosotes are somewhat fire resistant since on ignition the dense, heavy smoke may extinguish the flame.

The American Wood Preservers' Association and the American Railway Engineering Association have set up specifications for the chemically complex crosote.

Highly refined creosotes are obtained from coal tar by fractional distillation followed by refrigeration and filtering. This material is described as follows: "It is completely fluid at ordinary temperatures, is only slightly soluble in water, does not corrode metals and is extremely toxic to wood destroying agencies such as molds, fungi, marine borers and termites. It is brown in color and imparts a rich brown color to wood." This form of creosote is suitable for the non-pressure treatment of timber products such as by the "Hot and Cold Bath Process" (Open Tank Treatment), dipping, brushing and spraying. It also was used in poultry houses to eliminate mites, ticks, larvae, etc.

Such timber products as cross ties, poles, bridge timbers and miscellaneous lumber may undergo pressure treatment in vacuum cylinders or retorts for maximum retention of creosote in the wood cells. Open tank creosote treatment of timber is conducted at atmospheric pressure and the timber is dipped in the preservative.

Chlorinated phenols. According to Anon. (1946), "The oil soluble chlorinated phenols have found extensive commercial use in the millwork industry, and five percent solutions of these materials in suitable oil carriers have been widely adopted as standard treatments for the preservation of sash, frames, doors and other millwork products. For this purpose, the following are employed: pentachlorophenol, tetrachlorophenol and chlor-orthophenyl-phenol."

Pentachlorophenol. Pentachlorophenol, technical, is a solid material usually consisting of dark-colored flakes and sublimed needle crystals and has a characteristic odor. The volatility of pentachlorophenol is of a low order and thus is not readily lost.

Anon. (1946) has prepared a table of pentachlorophenol solvents showing the solubility in grams per 100 grams solvent at 77° F/25° C. Some of the important solvents in this table are as follows:

TABLE 26-2

Coal Tar Creosote (AWPA No. 1)	54
Deobase	4
o-Dichlorobenzene	12
Fuel Oil	12
Pine Oil	89
Stoddard Solvent at 40° C	4
Turpentine	10

Pentachlorophenol is widely used as a wood preservative for timbers, posts and poles.

Tetrachlorophenol. The technical grade wood preservative, Dowicide 6, 2, 3, 4, 6-tetrachlorophenol (Anon., 1946), varies from brown flakes to a sublimed solid with a strong characteristic odor. It is extremely soluble in most organic solvents, and Anon. (1946) lists the solvents for this chemical in grams per 100 grams solvent at 77° F/25° C:

TABLE 26-3

Coal Tar Creosote (AWPA No. 1)	127
Fuel Oil	37
Pine Oil	203
Stoddard Solvent at 41° C	38
Turpentine	42

Tetrachlorophenol is one of the ingredients used along with pentachlorophenol and chlor-orthophenyl phenol.

Chlor-orthophenyl phenol. According to Anon. (1946), this is a "clear, colorless to straw-colored, viscous liquid with a faint characteristic odor." It is soluble in most organic solvents and is used along with pentachlorophenol and tetrachlorophenol as a millwood preservative.

Proprietary preservatives. The following are some of the chemicals and combinations of chemicals which form the basis of numerous proprietary products which protect wood against insects either directly or indirectly by resisting decay:

- Sodium fluoride + sodium chromate + anhydrous di-sodium arsenate + dinitrophenol.
- Sodium fluoride + potassium bichromate + dinitrophenol.

- Zinc meta-arsenite.
- Zinc chloride.
- Chromated zinc chloride.
- Copper sulfate + potassium bichromate + acetic acid.

Naphthalene ($C_{10}H_8$). Napthalene is a white crystalline solid, which as naphthalene flakes or as moth balls, is one of our most familiar insecticides. It is an important constituent of coal tar and is slowly volatile. It is soluble in a number of organic solvents. Flake naphthalene is prepared by sublimation from refined naphthalene and ball naphthalene is prepared by pressing by mechanical means crushed or chipped naphthalene. Naphthalene is widely used for clothes moth control ordinarily being confined in a trunk, garment bag, etc.

Lethane. Lethane 384 Regular and Lethane 384 Special belong to the group of insecticides known as organic thiocyanates and are used in space sprays of the household type. These two insecticides are produced by the Rohm and Haas Company and are readily soluble in petroleum distillates of the fly spray base type.

Lethane 384 Regular is a liquid insecticide concentrate containing 50 percent by volume beta butoxy beta thiocyano diethyl ether, standardized with petroleum distillate. Lethane 384 Special is a "liquid insecticide concentrate containing 12.5 percent by volume beta butoxy beta thiocyano diethyl ether and 37.5 percent by volume beta thiocyano ethyl esters of aliphatic acids containing 10 to 18 carbon atoms, standardized with petroleum distillate."

Lethane sprays that are effective against house flies and other flying insects consist of four percent or more by volume of Lethane 384 Special, two percent or more of Lethane 384 Regular or 1½ percent Lethane 384 Special plus 1½ percent Lethane 384 Regular. The Lethane sprays show rapid knockdown of flies by the Peet-Grady method of testing and the addition of 0.5 to one percent by weight DDT ensures 99 percent or more kill of non-resistant flies in 24 hours. Lethane sprays are not as effective in knockdown speed or kill as pyrethrum when used against cockroaches.

Thanite. Thanite is an insecticide concentrate produced by the Hercules Powder Company. It is a 100 percent active toxicant which consists of at least 82 percent isobornyl thiocyanoacetate and 18 percent other active terpenes. It has no other diluents and is readily soluble in petroleum distillates of the fly spray base type. It is used in space sprays at two percent to four percent thanite and varying amounts of chlorinated insecticides. Although this is an effective spray against flying insects with fast knockdown, it is not as effective as pyrethrum against cockroaches.

INSECTICIDE FORMULATIONS

Insecticides are rarely used or available for use as the technical grade, which is the form in which they are first manufactured. Instead, the technical insecticide is mixed (formulated) with other components which may increase efficacy, reduce toxicity and facilitate handling and dilution to the usable concentration.

The main types of formulation used are oil solution concentrates, ready-for-use oil solutions, emulsifiable concentrates, wettable powders, granules, dusts and baits. In addition, more complex technology has made available microencapsulated insecticides as well as insecticide aerosols, lacquers, tapes and strips (Bennett et al., 1980).

The choice of formulation is often as important as the choice of insecticide ingredient. Where insect control is dependent on residual action (in other words, applying the insecticide to surfaces which will subsequently be contacted by the insect), rather than direct contact of the insect, the reaction between the insecticide and the surface treated must be known. For instance, many insecticides are quickly degraded on highly alkaline surfaces such as concrete or whitewashed walls. Some organic phosphate insecticides are quickly broken down on stainless steel because of the catalytic action of trace elements in the steel. Wet surfaces will often break down an insecticide faster than dry surfaces. High light intensities, especially ultra-violet light, and high temperatures will accelerate the degradation of insecticides on surfaces.

The effects of surfaces are not confined to degradation of the insecticide. Many surfaces absorb insecticides making them wholly or partially unavailable for killing insects which subsequently walk on the surface. This is particularly true of porous surfaces which may quickly absorb 80 percent of an insecticide in oil solution and most of an insecticide emulsion. On porous surfaces, use of wettable powders, dusts or granules is usually preferable where they will not be unsightly. Absorption, however, is not confined to porous surfaces and many insecticides, even if formulated as wettable powders, are absorbed by fresh oil paint, latex paint and vinyl tiles. In some cases, residual insecticides applied to such surfaces are totally absorbed and unavailable to kill insects within a few days of application. Since few product labels warn of such degradation or absorption, the user must learn by experience or from specialists which products and surfaces are most compatible for the longest residual action.

In selecting formulations, efficacy is not the only consideration. Safety to the user, occupants and property treated are even more important. In all cases the precautionary statements on product labels must be followed. These precautions may range from avoiding contamination of non-target areas, such as food and food preparation surfaces, to avoiding staining sensitive surfaces or causing electrical short circuits.

OIL SOLUTION CONCENTRATES

Oil solution concentrates are made by dissolving technical insecticide in a solvent which, under the likely conditions of transportation and storage, will hold the insecticide in solution. Such solutions are typically 20 percent or higher in strength. Prior to use, the oil solution concentrate is diluted with base oil by the user to achieve the correct application strength. For simplicity in use and to minimize dilution mistakes by semi-skilled workers, some insecticides are marketed as ready-for-use oil solutions. In these cases, the insecticides are already diluted in base oil.

Base oils for household sprays are nonstaining, colorless, odorless petroleum distillates (Anon. 1952). According to Scoggin et al. (1946), the term "deodorized" petroleum oil "means an oil carried to or beyond the neutral odor stage, thus removing all evidences of mineral spirits, kerosene and distillate origin." Thus, the olefins, benzenes, sulfur compounds and acidic and basic components are absent. The author's laboratory (Gulf R. & D. Co) made a survey of 14 of these highly refined commercial petroleum distillates suitable for use as a carrier for household insecticides.

Table 26-4 represents the average, as well as the minimum-maximum range of the physical and chemical specifications of insecticidal base oils for household sprays as determined by API tests.

TABLE 26-4

Physical and Chemical Specification of Base Oils for Household Sprays

	Average	Minimum	Maximum
Gravity API	48.9	45.4	53.6
Sp. Gr. 60/60° F	0.7846	0.7645	0.7999
Visc. centi. 100° F	1.70	1.19	2.14
Color	+29 to 30	—	—
Flash PM° F	*160.5	140	178
Aniline Point	*172.5	160.9	184.3
% Olefins	0.29	0.0	1.5
ES-45% Aromatics	0.52	0.0	1.9
Br$_2$ No.	0.28	0.0	1.4
Refractive Index N^D 20° C	1.4353	1.4250	1.4431
% Sulfur	0.014	0.00	0.024
Doctor Test	Sweet	—	—
Distillation Range			
IBP	373	344	398
5%	396	349	417
10%	401	350	420
50%	419	361	436
95%	458	388	480
EP	488	424	511

*Only 13 samples tested.

In recent years, isoparaffinic solvents have to some extent replaced the highly refined petroleum distillates commonly used in household insecticides. Fiero (1964) has prepared a very useful paper on these isoparaffinic solvents. He notes those prepared by Esso Research are manufactured synthetically. He states such solvents "insure an odorless product with quick evaporating qualities and low surface tension to bring about rapid spreading over the body of the insect. The lower solvent power of these carriers makes the insecticide less liable to affect surfaces and result in stains. Their high flash point insures their safety from a flammability standpoint."

Oil solutions of an insecticide usually provide a faster knockdown and kill of insects on contact than other formulations. This is because the solvents themselves have insecticidal action, as well as speed uptake of the insecticide by dissolving the waxy covering of the insect integument. Other advantages of oil solutions include better penetration of cracks, better adhesion to greasy surfaces and, because they are non-conductors of electricity, increased safety around electrical equipment.

Disadvantages of insecticides in oil solution include the high cost of the diluent; solvent damage to asphalt, rubber, plastic, (including sprayer components and insulation on electrical wires), some paints, etc.; phytotoxicity; increased odor; skid risk on tile floors; and flammability if used near open flames or on hot surfaces. In addition, as mentioned earlier, oil solutions are quickly absorbed and rendered ineffective by porous surfaces.

EMULSIFIABLE CONCENTRATES

Emulsifiable concentrates are formulations made up of the technical insecticide, solvent(s) and emulsifying agents. The use of emulsifying agents allows insecticides which are not soluble in water to mix with water to form an emulsion. The emulsion is made up of small droplets of insecticide dissolved in solvent and dispersed in suspension in the water diluent.

By using water instead of base oil as a diluent, the disadvantages of using insecticides in a total oil solution are partially avoided. However, insecticide emulsions conduct electricity and require agitation to maintain dispersion. Moreover, because some organic solvent as well as water are present in the emulsion, some surfaces may be damaged.

WETTABLE POWDERS

Wettable powders are formulations made by impregnating or coating an appropriate powder with the technical insecticide and adding wetting and dispersing agents to facilitate dilution of the product with water. Wettable powders do not dissolve in water — they simply form suspensions in water. These suspensions usually are much less stable than emulsions and require more frequent agitation to avoid settling out of the insecticide particles.

The advantages of wettable powder formulations arise from the lack of solvents. Thus, they will not damage plastics, asphalt, etc., and will not cause solvent "burn" on plants. Wettable powder formulations are preferable to emulsions and oil solutions for treating porous surfaces because after the water diluent has evaporated or soaked in, the insecticide remains on the surface. This efficacy benefit also can be a practical disadvantage from the viewpoint of visible deposits and many professionals confine their use of wettable powders to surfaces which are not easily seen.

GRANULES

Granular formulations are manufactured by coating or impregnating coarse particles of an appropriate substrate with the insecticide. The granules may be based on inorganic material such as clay pellets and limestone chips or organic material such as corn-cob.

Granules can be used to control occasional insect invaders of buildings by scattering them around the perimeter.

DUSTS

Dusts are ready-for-use formulations, usually made up of combinations of an insecticide with a dust diluent, which is usually inert. In the case of boric acid powder, the insecticide often comprises all of the dust, with no other component. In other cases an organic insecticide such as pyrethrum is combined with an inorganic dust diluent with desiccant properties, such as diatomaceous earth or silica aerogel. In those cases where the diluent is inert its purpose is the same as with liquid diluents, namely to facilitate application.

Dusts have the big advantage that they can penetrate hidden spaces such as wall voids and pipe ducts better than sprays. The ease of void penetration and the extent to which dusts adhere to vertical surfaces varies with the dust. Shepard (1951) and Hansberry (1943) consider the characteristics of dust diluents including such factors as particle shape, particle density, particle size, hardness, electrostatic charge, absorption and adsorption, bulk and particle density, flowability and sticking properties.

Hansberry notes the shape of the particle is important in the following ways: "Round or crystalline particles flow more freely, but do not adhere as well as flat, fibrous or irregular particles."

According to Shepard, particle density is the actual density of the dust particles, whereas bulk or apparent density comprises the weight of a specific volume of dust. Bulk density is dependent upon the physical properties of the dust, as well as handling, container size and packing.

Particle size is an important factor in dust diluents. Most dusts are composed of particles in the 250 to 350 mesh range, but silica aerogels have the extremely small particle size of less than 400 mesh. It has been shown that the toxicity of a number of dusts to insects increases as the particle size decreases. Hansberry's table (Table 26-5) indicates one way in which the fineness of a dust can be determined. Gooden (1944) also made a careful study of size specifications of fine powders. He notes when "the particle size is reduced, the dust floats better, has better sticking qualities and becomes lighter, but tends to flow from the duster less freely." The hardness of dusts is of importance when abrasion of equipment must be considered.

TABLE 26-5*

How to Judge the Fineness of a Dust

Particle size in microns	Screen mesh	Other description
1200	16	Particles will just pass ordinary window screen
420	40	The size of fine, gritty sand
150	100	Smallest size where individual particles can be seen without magnification
75	200	Ordinary flour is this size
44	325	Finest commonly used screen
3-5	—	Particles dance in water when viewed with high power microscope
0.2-4	—	Particles float in air like smoke
Less than 0.2	—	Particles smaller than wave-length of light, therefore invisible.

*From Hansberry (1943)

According to Shepard, electrostatic charges are caused "by friction between the particles themselves and between the particles and dusting equipment." Electrostatic charges influence adhesion to surfaces, as well as dust distribution. Permanent adhesion is not realized since the charge is soon lost.

The physical characteristics of particles such as surface, size, shape, etc., affect the absorption and adsorption properties as well as the flowability of dusts. Dusts that have a tendency to absorb moisture will cake.

Hansberry notes particles "with internal surfaces adsorb and absorb liquid insecticides more thoroughly, but smooth, hard particles may permit a greater percentage of the liquid to contact the insect as liquid."

Watkins and Norton (1947) classified insecticide dust diluents and carriers in two major categories, botanical flours and minerals, the latter being subdivided into elements, oxides, carbonates, sulfates, silicates, phosphates and indeterminate. Weidhass and Brann (1955) discuss dust diluents and carriers at some length.

Hansberry presents the characteristics of some of the common dusts and dust diluents in Table 26-6.

BAITS

Insecticide baits are based on a combination of insecticide with an attractant carrier, usually a preferred food of the target insect. Baits may take the form of granules, pellets, paste or liquids. In the case of sugar baits for flies, the sugar is not an attractant, but an arrestant which helps ensure that an arriving fly stays long enough to pick up a lethal dose of insecticide.

Because of the food content of baits great care must be taken to avoid harming non-target animals. In general they should not be used in food processing areas.

MICROENCAPSULATED INSECTICIDES

Technology is now available to microencapsulate pesticides by suspending the pesticide particles or droplets in polymers of various types. By altering the chemistry of the polymer or by changing factors in the processing, microcapsules can be formed of various sizes, solubilities, wall thickness and penetrability. These factors govern the speed with which the active ingredient contained within is released, which affects the residual life, odor, safety and efficacy.

Microencapsulated formulations appear as flowable suspensions and are generally handled in the same manner as wettable powders. Microencapsulation reduces the toxicity of the formulation. Some formulations are water-activated. They do not begin releasing the active ingredient until dry *after* mixing with water. This permits active ingredients that hydrolyze to remain stable in the spray tank. Rewetting halts chemical release until the residue drys again. This could be an advantage or disadvantage, depending upon the situation. Currently available microencapsulated products for use in buildings are based on pyrethrum and diazinon.

AEROSOL INSECTICIDES

Insecticidal aerosols are extremely fine mists produced by dispersing insecticides dissolved in liquid gases. Anon. (1950a) explains the operating principle as follows: "It is a simple principle. Insecticide solutions are combined with a liquified gas held under pressure in a metal container. The only exit from the container is through a pinpoint aperture controlled by a valve. As a valve is opened, the gas, in liquid form, rushes from the relatively high pressure of the bomb's interior to the lower atmospheric pressure outside. The insecticidal solution is, of course, carried with the gas in its dash to freedom. Rushing through the restrictive opening, the gas-insecticide solution forms a fine mist. On contact with the lower pressure of the atmosphere, the gas expands rapidly and evaporates. The insecticidal ingredients, being relatively nonvolatile, are left in the form of miniscule droplets. Due to their small size, they remain suspended in the air for comparatively long periods, comprising, in effect, millions of tiny "barrage balloons" which destroy the unsuspecting fly or mosquito that comes in contact with them in its flight."

TABLE 26-6*

Characteristics of Some Common Insecticidal Dusts*

Material	Chemical Nature	Apparent Density	Particles
Talc	Hydrate of magnesium silicate	10-40 lbs. per cu. ft.	Flat, long or irregular
Clay	Hydrate of aluminum silicate	15-25 lbs. per cu. ft.	Irregular
Diatomaceous earth	Silicate	6-10 lbs. per cu. ft.	Crushed diatom shells
Bentonite	Hydrate of aluminum silicate	20-40 lbs. per cu. ft.	Pumice-like
Pyrophyllite	Hydrate of aluminum silicate	20-40 lbs. per cu. ft.	Flat, irregular
Gypsum	Calcium sulfate	25-40 lbs. per cu. ft.	Irregular
Walnut shell flour	Cellulose, lignin,	28 lbs. per cu. ft.	Round, smooth
Lime	Calcium hydroxide	15-30 lbs. per cu. ft.	Irregular
Calcium carbonate (chalk)		20-40 lbs. per cu. ft.	Irregular
Calcium carbonate (mineral)		50-75 lbs. per cu. ft.	Irregular
Lead arsenate		20-30 lbs. per cu. ft.	Irregular
Sodium fluoride		20-50 lbs. per cu. ft.	Crystalline
Pyrethrum (ground flowers)		10-15 lbs. per cu. ft.	Irregular
Derris or Cubé (ground root)		10-15 lbs. per cu. ft.	Irregular

*From Hansberry (1943)

Anon. (1951) explains the principles of space spray aerosol operation in a little different manner which further clarifies the ways in which insecticidal aerosols operate.

"These aerosols consist of a solution of active ingredients and propellent in a sealed container with a specially designed valve and a standpipe. The "Freon" propellent is put into the container, usually under refrigeration to keep it liquified. Part of the "Freon" remains in solution while the rest changes to a gas and fills the head space in the container. Pressure, supplied by the vapor pressure of the "Freon" propellent, is tailored to suit different products.

"It is this pressure that serves as the propelling force, and as the valve is opened, the solution is pushed up the standpipe, through the valve and out the discharge nozzle. As the material passes through the nozzle and valve, the "Freon" still in solution instantly changes to a gas and literally blasts the active ingredient into very minute particles."

Some of the earliest investigations on insecticidal aerosols were conducted by workers in the U.S. Bureau of Entomology and Plant Quarantine, and L.D. Goodhue and W.N. Sullivan were granted in 1943, U.S. Patent 2,321,023 assigned to the Secretary of Agriculture, covering their method of producing an aerosol. The insecticidal aerosol proved so effective as a space spray against flying insects such as mosquitoes and flies that nearly 30,000,000 aerosol dispensers were used by the U.S. Armed Forces during World War II. Since 1946, insecticidal aerosols have been merchandised in hundreds of millions of dispensers. Insecticidal aerosols, as a means of combating flying insects, has been accepted by the public because they offer a convenient and effective means of dispersing insecticides against flying insects. However, the advantages are not all on the side of the aerosols, since they are relatively more expensive than liquid sprays and less effective against crawling insects such as roaches than liquid sprays applied in hand sprayers. Miller et al. (1952) showed the dosage of typical aerosols when used as space sprays for the control of American, oriental and German cockroaches is approximately 15 to 25 times as great as that required for house flies.

For information on aerosols see Fulton (1957a) Anon. (1957), Herzka and Pickthall (1961) and Shepherd (1961). Anon. (1958) presents U.S. Government specifications for aerosols and liquid sprays.

The formulation. Insecticidal aerosols consist of three primary components — the formulation, valve, and dispenser.

An aerosol formulation contains two major components: the "volatiles" and the "nonvolatiles." The volatile is the propellent gas such as Freon-12 (dichlorodifluoromethane) which has a gage pressure of 70 psi at 70° F/21° C and a boiling point of $-22°$ F/$-30°$ C; Freon-11 (trichloromonofluoromethane), which has a boiling point of 75° F/24° C; methylene chloride has a boiling point of 103° F/39° C. Freon-12 is mixed at a ratio of 50:50 with F-11 to obtain 35-38 psi at 70° F/21° C. Or the formulation may contain by weight 42.5 percent F-12, 21.25 percent F-11 and 21 to 25 percent methylene chloride. Methyl chloroform is also mixed with F-12 to reduce the pressure. All the propellents used in household aerosols must be "nonflammable, nonexplosive, nonirritating, of low order of toxicity, colorless and essentially odorless." In addition, they should be noncorrosive to metals, good insecticidal solvents and relatively inexpensive.

It should be noted the first widely used propellents were the Freon gases, and these gases were referred to as F-11, F-12, etc. Freon is being phased out as a propellant because of concerns about theoretical effects on the upper atmosphere. Other propellants are replacing Freon, although they are either much more expensive, more hazardous or less effective as propellants. Carbon dioxide is one substitute, but better substitutes will undoubtedly be found in the future.

The nonvolatiles are usually associated with the insecticidal agents, solvents and petroleum distillates in the formulation, and are often called the "actives," whereas the propellents are termed "inerts." Nelson et al. (1949) note technically the terms volatiles and nonvolatiles may be misleading since "only a small proportion of the ingredients when dispersed as an aerosol may be classified as nonvolatile." Fulton (1957b) discusses formulas in some detail.

The insecticides used in aerosols contain the "knockdown" agents such as pyrethrins, resmethrin, DDVP. Pyrethrins are used in aerosols at concentrations ranging from 0.1 percent to 0.6 percent in commercial formulations. Pyrethrins in aerosols are usually combined with synergists such as piperonyl butoxide, sulfoxide and others. These pyrethrins-synergist combinations are commonly used in aerosol formulations since they not only provide knockdown and good kill of flying insects, but are superior to other permissible insecticides for knockdown and kill of crawling insects. Moore (1950b) and Fales et al. (1951) and other workers have shown allethrin to be as effective against flying insects in aerosols as equal concentrations of pyrethrins. Synergist 264 may be used with allethrin to increase the effectiveness of the formula. Maughan et al. (1951) demonstrated that two to four percent Lethane 384 may be substituted for part of the more costly pyrethrins in some formulations.

Where pyrethrins are used in aerosols, they are usually from specially refined 10 percent or 20 percent concentrates with a minimum of Freon insolubles. The amount of Freon insolubles in pyrethrum concentrates is of importance since these insoluble waxes and resins may be a source of irritation and may also clog the valve.

Thus, we have noted the pyrethrins and other contact insecticides in the formula provide "knockdown" and some kill; the synergists aid in increasing the knockdown and kill. The petroleum distillates are responsible for the remainder of the nonvolatiles and the ratio in which they are combined with the volatiles is partially responsible for the particle size of the aerosol. The size and structure of the valve, particularly the nozzle orifice and the pressure in the container, also are responsible for particle size.

The first formulations used by the army contained one percent pyrethrins, two percent sesame oil, seven percent petroleum distillates, and 90 percent Freon-12. Fulton and Rohwer (1949) note this formula was finally modified to contain 0.4 percent pyrethrins (two percent of a 20 percent pyrethrum extract), three percent DDT, five percent to seven percent cyclohexanone or 10 percent to 12 percent methylated naphthalene and the remainder Freon-12. This formula was widely used in the high pressure aerosols.

Fales et al. (1946) showed a ratio by weight of 15 percent nonvolatiles to 85 percent volatiles provided aerosols that had a high insecticidal toxicity to flying insects. Goodhue and Riley (1946) observed the particle size of an aerosol increased as the proportion of nonvolatiles in the formulation increased. Fulton and Rohwer (1949) define insecticidal aerosols as "a system of particles suspended in air where 80 percent of the particles are less than 30 microns in diameter and no particles larger than 50 microns." Goodhue (1946a) notes particles of 10 to 20 microns are the most satisfactory particle size for confined spaces. Others (Bennett et al. 1975) have confirmed this and have shown that smaller particles do not impinge well on insects and larger particles fall out too quickly (this is the principle upon which "ULV" applications operate). The nonvolatiles in the aerosols may range from 10 percent to 20 percent. As soon as the formulation is sprayed, a fine mist forms and the liquefied gas evaporates immediately so the insect is actually contacted by high concentrations of the insecticide in the nonvolatile droplets.

The low pressure formulations have 15 percent to 20 percent nonvolatiles — 85 percent to 80 percent volatiles and contain varying amounts of some of the following ingredients; pyrethrins, resmethrin, piperonyl butoxide, sesame oil extractives, n-propyl isome, sulfoxide, 264 and other synergists, and may or may not include a residual insecticide.

A typical formulation was:

	% by Weight
Pyrethrins	0.4
Piperonyl butoxide	1.0
Petroleum distillates	13.6
Propellents	85.0

Lee (1967) notes the "introduction of isobutane, n-butane and propane as propellants on a commercial basis revolutionized a large segment of the aerosol industry within a few years." The reason for this is these propellants introduced the era of water-based insecticides. "A water emulsion toxicant system with hydrocarbon propellants in place of the fluorocarbons brought a marked reduction in the cost of pressurized household sprays. A water-base system with hydrocarbon propellants was a natural, since the flammability of the finished product was sharply reduced."

Glynne Jones and Weaning (1966) list the advantages and disadvantages of the water based formulas:

ADVANTAGES

- "Saving of cost of propellant, i.e. replacing 80 percent F 11/12 by 30 percent Butane (plus 3% to 10% F 12 in some formulations).
- "Reduction in oil solvent costs.
- "Less phytotoxic — hence dual purpose: House and Garden resulting in increased overall sales.
- "Easier perfuming.
- "Claimed less irritancy."

DISADVANTAGES

- "A more stringently controlled filling technique due to the care required in preparing emulsions and handling highly inflammable propellants.
- "Greater can corrosion risks.
- "Long-term storage trials are required.
- "Cold-filling methods are inappropriate due to freezing of the emulsion that would occur.
- "Slower delivery rate than for conventional aerosols."

Another evident disadvantage is the customer must shake the can before using in order to obtain the proper insecticidal spray. In general, water-base insecticides are not as effective as oil-base sprays against German cockroaches because the oil, itself, has insecticidal activity against German cockroaches.

The following formulation is typical of water-base insecticidal aerosols:
- 0.25% pyrethrins
- 1.25 % piperonyl butoxide
- 0.10% perfume
- 1.00 % petroleum distillate
- 67.40% deionized water, emulsifiers and additives
- 30.00% propellent — combination of isobutane and propane

A number of manufacturers use propellent gases to disperse surface or wet sprays. These usually contain diazinon, propoxur or malathion. The nonvolatiles range from 25 to 70 percent. Thus, a typical formula of this type which is used in industrial establishments contains:

Diazinon	0.5%
Methylated solvents	13.0%
Petroleum distillates	26.5%
Propellents	60.0%

The valve. The aerosol valve is one of the three essential components of the aerosol. Tuttle (1951) discusses aerosol valves as follows: "The dictionary describes a valve as: 'A contrivance or arrangement that is used to open or close a passage to permit or stop the flow of a liquid, gas, vapor or loose material.' An aerosol-type valve is actually much more than this. It also conducts the liquid from the bottom of the container to the actual valve, preventing the escape of the gas alone; it acts as a valve as defined; it transmits the operating motion to the valve without leakage; it reduces the flow to a very small rate; it regulates this small flow accurately; it causes the fluid mixture and the liquefied compressed gas to change into a mixture of finely divided liquid and gas, and it causes this mixture to discharge into the atmosphere so that the suspended particles in the air are in a certain size range effective for the purpose intended."

Fulton et al. (1951) offer some additional information concerning aerosol valves: "The proper design of nozzles and valves is important to the production of aerosols of the desired particle-size. Particle-size may be changed by varying the size of the orifice and the structure of the valve through which the material must pass. The most practical type of nozzle is one that incorporates an expansion chamber between two orifices. The most satisfactory ratio of size of the inside orifice to the outside one is approximately 2:3. The restricted inside orifice may be designed as a metering rod, loose-fitting pin, or any type of device that will give a satisfactory drop in pressure.

"In the valves now being marketed, a rubber-like material or nylon is used for the seat of the valve. The sealing material has been one of the most costly items in the development of valves. The solvents that are necessary in aerosol formulations usually cause swelling or some other deleterious action on the valve seat. The performance of a valve can be determined only after a long test program. Our standard test program includes containers stored at temperatures from 32° to 120° F, with the valves in two positions, up and down." Some of the earlier valves were machined from brass, the latest valves are molded from nylon. Molded nylon valves are less expensive and are less likely to react with the ingredients than brass valves.

The earliest aerosol valves were of a design adopted from tire valves. Later, a screw-cap type that worked on a needle-valve principle was widely used in the high-pressure aerosols. Several of the earlier insecticidal aerosol manufacturers suffered large monetary losses due to the valves clogging or leaking or leaks occurring where the valve assembly was joined to the dispenser.

The ease with which the valves can be clogged can be readily understood when it is noted some passages have clearances that are only several thousandths of an inch in diameter. Sharpless (1950), working in the author's laboratory, made an investigation of 590 defective aerosols and summarizes his studies on clogged orifices as follows: "Failure was the direct result of foreign material such as metal burs, particles of lithographing, flux, sealing compound, lint and insolubles. The presence of the first four items listed appeared necessary to act as a nucleus around which the lint and insolubles collected."

Precipitates or sediments such as the Freon-insoluble waxes and resins in pyrethrum concentrates must be guarded against in formulations. Every effort must be made in insecticidal aerosol formulations to prevent the entry of moisture, since this may result in corrosion in the container or deleterious changes in the formulation with resultant precipitates or sediments that may clog the valve orifices.

Tuttle (1951) lists eight points for the evaluation of aerosol valves:

- Loss of volatiles — only a few grams per year.
- Loss of oils — almost zero.
- Valve operation — small ratio of flow with minimum clogging tendency.
- A metering accuracy of 25 percent.
- Aerosol size particles with a low-pressure propellent.
- Self-reclosing or push-button operation.
- Upward or downward spraying.
- High cost-efficiency.

Aerosol valves are constantly being developed or redesigned to improve ease of operation and promote greater efficiency. In this respect, A.J. Samuel of the Gulf Oil Corporation invented a valve that permits spraying in an inverted position without loss of gas. The latest valves are adjustable to produce a pinstream or space spray.

The aerosol dispenser. Like the formulations and the valve, the dispenser is another vital consideration in the manufacture and merchandising of aerosols. The high pressure aerosol formulations (70 pounds per square inch/4.92 kg per square cm at 70° F/29° C) necessarily must be confined to heavy drawn steel containers which are either welded or brazed. In this respect, Peterson (1947) notes: "While a pressure of 70 pounds is, in itself, not too excessive, this pressure is increased by a rise in temperature (e.g., 185 pounds pressure is developed at 130° F by Freon-12). These containers are covered by the Interstate Commerce Commission's Specifications Nos. 9 and 40." To counteract the high initial cost of the high pressure aerosol container, some manufacturers offered a refillable type which "enabled the consumer to save about half the original purchase price by returning the empty container." Cylinders containing five pounds and 30 pounds of either low or high pressure aerosols are now being used in large commercial establishments.

In 1947, the Interstate Commerce Commission amended its regulations permitting "the shipment of non-inflammable aerosols with less than 40 pounds gage pressure at 70° F in light containers." This resulted in the marketing of the beer can-type aerosol dispenser with the low-pressure formulation of 35 to 38 psi/2.46 to 2.67 kg per sq. cm at 70° F/21° C. As previously described, this lowering of pressure was accomplished by mixing Freon-11 with Freon-12 and other suitable gases.

The propellents are compressed gases. Thus, the aerosol filling "must be done either with pressurized equipment at room temperature or with the ingredients at sub-zero temperature so that the propellent may be handled as a liquid."

Anon. (1951) describes the cold or refrigerated procedures, the method commonly used for packaging low pressure aerosol insecticides, as follows: "By far the bulk of the low pressure aerosol products on the market today are filled by the refrigerated method. This method of loading has several advantages. It is usually faster because refrigerated filling permits the ingredients to be placed in the container and the 'Freon' added directly rather than through the valve. Also, crimping the top or bottom of the can in place is done after filling; thus, sufficient "Freon" is permitted to evaporate between the filling operation and the crimper to displace all water vapor and air before the container is sealed. In certain formulations, it may not be necessary to remove the air from the containers, but air has a tendency to increase the pressure in the container, and if this pressure is already calculated to be close to the allowable Interstate Commerce Commission limits, air may be definitely unwanted and dangerous in the container.

"An additional advantage in cold filling is the fact that automatic machinery is available which is capable of sorting, loading, capping, temperature testing, and labeling containers at the rate of 15 to 90 cans per minute."

Pressure filling recently has replaced a great deal of the cold-fill procedure. This has taken place because of the use of water-base formulas which naturally freeze when refrigerated. Whitmire (1958) describes the pressure fill method as follows: "In the pressure fill method the concentrate material is filled into the open can at room temperature. This is followed by crimping on the valve. The propellant gas is forced into the can through the valve stem. Where an appreciable amount of gas is forced into the can with this method, the air is forced to the top and builds up a false pressure as the gas itself will liquefy after its pressure is reached at set temperatures. This is not the case with the air. Of course there are ways of inverting the can and purging this air, but no method has been devised that will do a thorough job of removing all the air from the can.

"With the new combination pressure and cold fill, the concentrate is run in at room temperature and is then followed by the addition of a small percentage of the propellant gas. Next the valve is crimped and the remainder of the gas is pressure filled. The advantage here is that the first addition of the propellant gas with an open head drives out the air before the valve is crimped and the remaining gas can be filled and still maintain a true pressure. The advantage of this new method is that it would eliminate the expensive refrigeration equipment necessary to chill the concentrated material since only the propellant gas would require refrigerations."

Aerosol generators. Aerosol generators that utilize hot air, steam and compressed air currently are available. This type of apparatus is used for fogging extensive areas outdoors and even indoors for insect control. The repeated application of these aerosol fogs in an area very often results in the effective control of such flying insects as mosquitoes, flies, gnats and similar pests. The American Mosquito Control Association (1952) reviews some of these aerosol or fog generators.

In the hot air type, the machine consists of a gasoline motor, compressor and a spark combustion chamber. A six-inch/15 cm duct or pipe is present in the rear. Hand-operated gears permit the duct to be turned 360 degrees in a horizontal fashion and 180 degrees in a vertical position. The liquid insecticide issues from the inner area of the duct and the superheated air from a narrow slit around the periphery. The superheated air blasts the insecticide into aerosol-size droplets and disperses them. This superheated air cools on exposure to the atmosphere and forms a dense fog, the liquid insecticide particles serving as a nucleus for the condensation of the vapor. The particle size can be regulated from two to 60 microns. When used outdoors, control of the deposition of the fog is largely dependent on the direction and velocity of the wind, and thus can be used only on relatively quiet days. Another machine utilizes a jet-type engine and the insecticide is introduced into the exhaust gases. A third type of generator utilizes steam to atomize and disperse the insecticidal solution.

Whiting and Bartlett (1963) note the high operating temperatures of thermal foggers reduce the effectiveness of pyrethrins.

On the market at the present time there are so-called mechanical aerosol generators that do not use gas, hot air or steam as propellents. One mechanism utilizes the centrifugal force of a whirling disc to atomize and disperse the aerosol particles; another uses compressed air and a special nozzle with several perforated discs. These mechanical aerosol generators are used for large indoor areas such as cafeterias and food plants and employ highly concentrated pyrethrins-synergist combinations.

The more recent development of ultra low volume equipment (ULV) has permitted the application of highly concentrated insecticides at reduced cost and increased effectiveness. This equipment uses air and venturi to break up the insecticide particles into the desired range (10 to 20 microns). Special formulations should be used to obtain the proper droplet size. The carrier is usually Klearal rather than a deodorized kerosene.

Yeomans makes the following recommendations in applying oil base formulations: "Owing to the explosion hazard when oil solutions are used indoors, not more than one gallon of these solutions should be used per 100,000 cubic feet and they should not be released in the vicinity of an open flame. All work indoors should be done while wearing a proper respirator. The pyrethrum formula is recommended for use around exposed foodstuffs." Yeomans (1960) has upped the rate to not more than two gallons per 100,000 cu. ft./7.6 l per 2,800 cubic meters.

"The dosage will vary according to the insect and the insecticide used. Because of better dispersion, rooms with high ceilings require a smaller dosage than those with low ceilings. In closed warehouses there are indications that one pound of DDT in one gallon of solution per 6,500 square feet of floor space applied about every two weeks will provide excellent protection of commodities against insect infestation."

Yeomans and Van Leeuwen (1954) showed the presence of tetrachloroethylene in the insecticidal solution helped to prevent explosions.

Anon. (1952) carefully analyzed the very few cases of fires and explosions in indoor fogging operations and concluded that inexperienced operators, carelessness and faulty maintenance are largely responsible for such disasters.

LACQUERS, TAPES AND STRIPS

Various formulation techniques have been developed with the prime objective of prolonging the period of effectiveness of the insecticide.

In Europe, insecticide lacquers have been in use for more than 20 years and a few years ago such formulations were introduced in the United States. The insecticide is dissolved in a clear lacquer which is applied by paintbrush or sprayer to surfaces not damaged by the rather aggressive solvents. The lacquer dries forming a waterproof film through which the insecticide migrates to form a surface "bloom" of crystals. As insects or cleaning measures remove the surface layer of insecticide, more migrates to the surface to replace it. Under normal circumstances such lacquered surfaces may continue to kill insects which walk on them for six months or more. The original concentration of insecticide in the lacquer is five or more times greater than in conventional sprays and this helps sustain the reservoir of insecticide. Such lacquers are useful in areas which are frequently washed down, but solvent odor and damage to some surfaces has greatly limited their use. Moreover, while such lacquer deposits are resistant to water they have limited use outdoors because ultra-violet light degrades the insecticide.

In Europe, the use of insecticide lacquers is minimal, partly because of the progressive darkening of the lacquer to form unsightly brown bands. This has led to their use being largely confined to ships' galleys which are either too dirty to matter or which are re-painted every six months to prevent rust.

Insecticide tapes have proved useful in extremely sensitive situations such as inside telephones, computers or under animal cages where use of conventional formulations might have caused problems. The tapes consist of a sandwich of an adhesive backing and a surface protective layer with an insecticide reservoir layer in-between. The tapes are cut into appropriate lengths and stuck on concealed surfaces over which insects will walk. As with the lacquers, insecticide migrates from within to form a mono-molecular layer on the protective surface. Insects walking over the tapes pick up the insecticide and die. The surface layer is then replenished from the insecticide reservoir layer.

Insecticide strips based on DDVP impregnated into vinyl have been available for some years. Unlike the lacquers and tapes, which require the insects to walk on them, the DDVP strips emit the insecticide in vapor form. In confined areas, these vapors attain levels that kill insects, especially small insects. These insecticide strips have proved very effective against flying insects and have an excellent record of safety.

TOXICOLOGY OF SOME COMMONLY USED INSECTICIDES

Hayes (1975) has prepared tables on the toxicology of some common insecticides and the following table (Table 4) is from his book. Gleason et al. (1963) and Morgan (1977) prepared useful volumes on the toxicology of commercial products with information on the procedures to use in case of poisoning. Anyone using pesticides should have a copy of Morgan's book. Negherbon (1959), and more recently Hayden (1975), compiled very valuable and detailed works on the toxicology of insecticides.

The mean lethal dose in these tests is the average dosage necessary to kill all of the test population. LD_{50} is the average dosage necessary to kill 50 percent of the test population. Test rats usually weigh from 150 to about 200 grams, the latter weight being that of the adult (Table 26-7).

ANTIDOTES

The following notes are based on the recommendations of the American Medical Association (1957):

First-aid measures for poisoning. The aim of first-aid measures is to help prevent absorption of the poison. Speed is essential. First-aid measures must

TABLE 26-7*

Acute Oral & Dermal LD$_{50}$ Values of Insecticides (technical grade)*** for Male and Female White Rats

Compound	Oral LD$_{50}$(mg/kg)		Dermal LD$_{50}$(mg/kg)	
	Males	Females	Males	Females
Aldrin	39*	60*	98*	98*
Bendiocarb	179**	—	566	—
Carbaryl	850	500	>4000	—
Chlordane	335*	430*	840*	690*
Chlorpyrifos	97-276	—	—	—
DDT	113*	118*	—	2510*
DDVP	80	56	107	75
Diazinon	108	76	900	455
Dieldrin	46*	46*	90*	60*
Dimethoate	215	—	400	—
Fenthion	215	245	330	330
Heptachlor	100	162	195	250
Lindane	88	91	1000	900
Malathion	1375	1000	>4444	>4444
Propoxur	100	—	>1000	—
Parathion	13	3.6	21	6.8
Piperonyl Butoxide	+7,500	—	—	—
Pyrethrins	200-1500	—	>1800	—
Resmethrin	4240	—	>3040	—
Ronnel	1250	2630	—	—
Trichlorfon	630	560	>2000	>2000

*With the exception of the dermal LD$_{50}$ for dimethoate, these values were determined by the Toxicology Section under standardized conditions.

**LD$_{50}$ for the 76% wettable powder formulation.

N.B. Chronic toxicity data are not presented since most modern insecticides such as the organo-phosphates and carbamates are quickly metabolized and not stored in the body. This is not true of some of the inorganics and the chlorinated hydrocarbons.

***From Hayes (1975)

be started at once. If possible, one person should begin treatment while another calls a physician. When this is not possible, the nature of the poison will determine whether to call a physician first or begin first-aid measures and then notify a physician. Save the poison container and material itself if any remains. If the poison is not known, save a sample of the vomitus.

MEASURES TO BE TAKEN BEFORE ARRIVAL OF PHYSICIAN

I. **Swallowed poisons.** Many products used in and around the home, although not labeled "Poison," may be dangerous if taken internally. For example, some medications which are beneficial when used correctly may endanger life if used improperly or in excessive amounts.

In all cases, *EXCEPT THOSE INDICATED BELOW,* remove poison from the patient's stomach immediately by inducing vomiting. This cannot be over-emphasized, for it is the essence of the treatment and is often a lifesaving procedure. Prevent chilling by wrapping patient in blankets if necessary. Do not give alcohol in any form.

A. Do Not Induce Vomiting If:
   - Patient is in coma or unconscious.
   - Patient is in convulsions.
   - Patient has swallowed petroleum products (i.e., kerosene, gasoline, lighter fluid).
   - Patient has swallowed a corrosive poison (symptoms: severe pain, burning sensation in mouth and throat, vomiting).

CALL PHYSICIAN IMMEDIATELY

a) Acid and acid-like corrosives: sodium acid sulfate (toilet bowl cleaners), acetic acid (glacial), sulfuric acid, nitric acid, oxalic acid, hydrofluoric acid (rust removers), iodine, silver nitrate (styptic pencil).

b) Alkali corrosives: sodium hydroxide — lye (drain cleaners), sodium carbonate (washing soda), ammonia water, sodium hypochlorite (household bleach).

If the patient can swallow after ingesting a *corrosive poison,* the following substances (and amounts) may be given:
   - For acids: milk, water, or milk of magnesia (one tablespoon to one cup of water).
   - For alkalis: milk, water, fruit juice, or vinegar.
   - For patient one to five years old — one to two cups.
   - For patient five years and older — up to one quart.

B) Induce Vomiting When Non-corrosive Substances Have Been Swallowed:
   - Give milk or water (for patient one to five years old — one to two cups; for patient over five years — up to one quart).
   - Induce vomiting by placing the blunt end of a spoon or your finger at the back of the patient's throat.

When retching and vomiting begin, place patient face down with head lower than hips. This prevents vomitus from entering the lungs and causing further damage.

II) Inhaled poisons.
   - Carry patient (do not let him walk) to fresh air immediately.
   - Open all doors and windows.
   - Loosen all tight clothing.
   - Apply artificial respiration if breathing has stopped or is irregular.
   - Prevent chilling (wrap patient in blankets).
   - Keep patient as quiet as possible.
   - If patient is convulsing, keep him in bed in a semidark room; avoid jarring or noise.
   - Do not give alcohol in any form.

III) Skin contamination.
   - Drench skin with water (shower, hose, faucet).
   - Apply stream of water on skin while removing clothing.
   - Cleanse skin thoroughly with water; rapidity in washing is most important in reducing extent of injury.

IV) Eye contamination.
- Hold eyelids open, wash eye with gentle stream of running water *immediately*. Delay of few seconds greatly increases extent of injury.
- Continue washing until physican arrives.
- *Do not use chemicals;* they may increase extent of injury.

V) Injected poisons (scorpion and snake bites).
- Make patient lie down as soon as possible.
- Do not give alcohol in any form.
- Apply tourniquet above injection site (e.g., between arm or leg and heart). The pulse in vessels below the tourniquet should not disappear, nor should the tourniquet produce a throbing sensation. Tourniquet should be loosened for one minute every 15 minutes.
- Apply ice-pack to the site of the bite.
- Carry patient to physician or hospital; *do not let him walk.*

VI) Chemical burns.
- Wash with large quantities of running water (except those caused by phosphorus).
- Immediately cover with loosely applied clean cloth.
- Avoid use of ointments, greases, powders and other drugs in first-aid treatment of burns.
- Treat shock by keeping patient flat, keeping him warm and reassuring him until arrival of physician.

MEASURES TO PREVENT POISONING ACCIDENTS:
- Keep all drugs, poisonous substances, boric acid and other household chemicals out of reach of children. This includes all insecticides and rodenticides.
- Do not store nonedible products on shelves used for storing food.
- Keep all poisonous substances in their original containers; do not transfer to unlabeled containers. Material in unlabeled containers should be destroyed.
- Do not use beverage bottles for pesticides.
- When medicines are discarded, destroy them. Do not throw them where they might be reached by children or pets.
- When giving flavored and/or brightly colored medicine to children, *always* refer to it as medicine — *never* as candy.
- Do not take or give medicine in the dark.
- Read labels before using chemical products and dispose of empty containers according to label directions.

LITERATURE

AMERICAN MEDICAL ASSOCIATION — 1957. First-aid measures for poisoning. J. Amer. Med. Assoc. Vol. 165, Oct. 12.

AMERICAN MOSQUITO CONTROL ASSOCIATION — 1952. Ground equipment and insecticides for mosquito control. AMCA Bull. No. 2.

ANNAND, P.N. — 1943. Aerosol method patented. Soap & S. C. 19(10):117-119.

ANONYMOUS — 1942. What specifications for insecticide base oil? Soap & S.C. 18(11):87-89. 1946. Dowicides. The Dow Chem. Co. 1946. Testing of roach sprays. Soap & S.C. 22(7):145, 147, 148E. 1948a. Methoxychlor bibliography: results of tests. E. I. Du Pont de Nemours & Company. Mimeographed. 1948b. Official antidotes. Calif. State Board of Pharmacy. 1949. The tentative N.A.I.D.M. aerosol test method for flying insects. Soap & S.C. 24(5):114-117. 1950a. Easy does it. The Orange Disc (Gulf Oil Corp.) July-Aug. pp. 8-10. 1950b. Change in TOTA procedure. Soap & S.C. 26(9):135. 1951. Package for profit. E.I. du Pont de Nemours Company. 1952. Fire and explosion hazards of thermal insecticidal fogging. Appendix on spray flammability and flammable liquids terminology. Nat'l. Board of Fire Underwriters. 1957. CSMA Aerosol Guide. CSMA, 50 E. 41st St., N.Y. 17. 1958. U.S. Gov't Specifications. Soap & Chem. Spec. 1958. Blue Book.

BAILEY, S.F. & L.M. SMITH — 1951. Handbook of Agricultural Pest Control. 191 pp. Industry Publ.

BARLOW, F. & A.B. HADAWAY — 1947. Preliminary notes on the loss of DDT and Gammexane by absorption. Bull. Entomol. Res. 38(2):335-346.

BENNETT, G.W. & E.S. RUNSTROM — 1980. Efficacy of New Insecticide Formulations in Urban Pest Control. Pest Control 48(12):19-24.

BOOCOCK, D. — 1952. Use of pyrethrum with the Todd insecticidal fog applicator. Pyr. Post 2(4):18-22.

BROWN, A.W.A. — 1951. Insect Control by Chemicals. 998 pp. John Wiley & Sons, Inc.

BUSSART, J.E. — 1950. Recent developments on chlordane and heptachlor. Soap & S.C. Off. Proc. 37th Ann. Meeting C.S.M.A. Dec. p. 76.

BUSSART, J.E. & A. SCHOR — 1948. Chlordane. Soap & S.C. 24(8):126-128.

CHISHOLM, R.D. — 1948. A review of DDT formulations. U.S. B.E.&P.Q. E-742.

CRISTOL, S.J., H.L. HALLER & A.W. LINDQUIST — 1946. Toxicity of DDT isomers to some insects affecting man. Science 104(2702):233-234. Oct. 11.

DAVID, W.A.L. — 1946. The quantity and distribution of sprays collected by insects flying through insecticidal mists. Ann. Appl. Biol. 33(2):133-141. Biol. Abst. 20(9):2002, 1946.

DAVID, W.A.L. & P. BRACEY — 1946. Factors influencing the interaction of insecticidal mists on flying insects. Bull. Entomol. Res. 37(2):177-190. 1947. Factors influencing the interaction of insecticidal mists and flying insects. Part IV. Some experiments with adjuvants. Bull. Entomol. Res. 37(3):393-398.

DeMEILLON, B. — 1946. Effect on some blood-sucking arthropods of "Gammexane" when fed to a rabbit. Nature 158(4023):839.

DETHIER, V.G. — 1963. The Physiology of Insect Senses. John Wiley, 266 p.

DOVE, W.E. — 1947. Piperonyl butoxide, a new safe insecticide for the household and field. Pests 15(9):30.

DUSTAN, G.G., H. ARMSTRONG & W.L. PUTMAN — 1947. The influence of air currents on the insecticidal action of DDT, Benzene Hexachloride, Hercules Toxicant 3956, and Velsicol 1068. Can. Entomol. 79(3):45-50.

EAGLESON, C. — 1940. U.S. Patent 2,202,145. May 28, 1940. Sesame oil. 1942. Sesame in insecticides. Soap & S.C. 18(12):125-127.

ESSIG, E.O. — 1931. A History of Entomology. Macmillan.

FALES, J.H., E.R. McGOVRAN & L.D. GOODHUE — 1946. "Aerosol toxicity." Effect of nonvolatile content of a DDT aerosol on mortality of houseflies. Soap & S.C. 22(6):157-158.

FALES, J.H., R.H. NELSON, R.A. FULTON, & O.F. BODENSTEIN — 1951. Insecticidal effectiveness of sprays and aerosols. J. Econ. Entomol. 44(1):23-28.

FIERO, G.W. — 1964. Isoparaffinic solvents as bases for pyrethrum insecticides. Pyr. Post 7(4):3-6, 8.

FLECK, E.E. & H.L. HALLER — 1946. Solubility of DDT in kerosene: effect of auxiliary solvents at subzero temperatures. Ind. & Eng. Chem. 28(2):177-178.

FLETCHER, F.W. — 1945. Ortho-dichlorobenzene as an insecticide. Pests 13(3):15-28.

FREAR, D.E.H. — 1948. Chemistry of Insecticides, Fungicides and Herbicides. Second Edition. D. Van Nostrand Company, Inc. 417 p.

FREAR, D.E.H. & M.T. HILBORN — 1950. Pest control materials. 1950. Pa. Agr. Expt. Sta. Prog. Rpt. 20, 148 p.

FULTON, R.A. — 1948. A chronological list of publications on liquefied-gas aerosols. U.S. Bur. Ent. & P.Q. E-754. 1951. Supplemental list of publications on liquefied-gas aerosols. U.S.D.A. E-814. 1957a. A list of publications on liquefied-gas aerosols 1933-1956. USDA Agr. Res. Service ARS-33-37. 1957b. Liquefied-gas insecticide formulas — 1947, present and future. CSMA Proc. Mid-Year Meeting 43:63-66.

FULTON, R.A., R.H. NELSON & A.H. YEOMANS — 1951. Evaluation of liquefied-gas aerosols. Soap & S.C. 27(8):129-131.

FULTON, R.A. & S.A.ROHWER — 1949. Liquefied gas insecticidal aerosols. Soap & S.C. 25(2):122-124, 148C.

GERSDORFF, W.A. — 1947. Toxicity to house flies of the pyrethrins and cinerins, and derivatives, in relation to chemical structure. J. Econ. Entomol. 40(6):878-882.

GINSBURG, J.M. — 1948. An evaluation of toxicants for mosquito control. Agr. Chem. 3(5):23-24.

GLEASON, M.N., R.E. GOSSELIN, H.C. HODGE — 1963. Clinical Toxicology of Commercial Products: Acute poisoning (Home and Farm) Williams & Wilkins. 1160 pp.

GLYNNE JONES, G.D. & A.J.S. WEANING — 1966. The formulation and activity of water-based aerosols containing pyrethrum. Aerosol Age 11(3):21-23.

GNADINGER, C.B. — 1933. Pyrethrum Flowers. 380 pp. McLaughlin Gormley King Co., Minneapolis, Minnesota. 1945. Pyrethrum Flowers Supplement 1936-1945. McLaughlin Gormley King Co.

GODIN, P.J., J.H. STEVENSON & R.M. SAWICKI — 1965. The insecticidal activity of Jasmolin II. J. Econ. Entomol. 58(3):548-551.

GOLDBERG, M. — 1949. Low-pressure aerosols become big-volume sellers. Chem. Indust. 65(3):374, 376, 378.

GOODEN, E.L. — 1944. Size specifications for fine powders. J. Econ. Entomol. 37(2):204-208.

GOODHUE, L.D. — 1946a. Aerosols and their application. J. Econ. Entomol. 39(4):506-509, 1946b. The evaluation of liquefied-gas aerosol formulations. Soap & S.C. 22(2):133, 135, 165.

GOODHUE, L.D. & R.L. RILEY — 1946. Particle-size distribution in liquefied gas aerosols. J. Econ. Entomol. 39(2):223-226.

GRANETT, P., D.P. CONNOLA & J.V. LEMBACH — 1951. Laboratory tests of allethrin for stability, residual action and toxicity. J. Econ. Entomol. 44(4):552-557.

HALLER, H.L. — 1946. Insecticides and synergists containing the 3,4-methylene dioxyphenyl group. Proc. 25th Annual Meeting No. Central States

Branch Amer. Assoc. Econ. Entomol. p. 71, March 27-29.
HALLER, H.L. & R.L. BUSBEY — 1947. The chemistry of DDT, U.S.D.A. Yearbook 1943-1947, pp. 616-622.
HALLER, H.L., E.R. McGOVRAN, L.D. GOODHUE & W.N. SULLIVAN — 1942. The synergistic action of sesamin with pyrethrum insecticides. J. Org. Chem. 7:183-184.
HANSBERRY, R. — 1941. How insecticides kill. Pests. 9(2):20-22. 1942. About stomach insecticides. Pests. 10(3):14-15. 1943. Some concepts of particle size in insecticidal dusts. Pests. 11(4):26-28.
HARDER, F.K. — 1953. You can make money in neighborhood fogging. Pest Control 21(6):9, 10, 12.
HAYES, W.J.H. — 1963. Clinical Handbook on Economic Poisons. U.S. Public Health Service Publication No. 476. 144 pp.
HAYES, W.J.H. — 1975. Toxicology of Pesticides. 580 pp. The Williams & Wilkins Co., Baltimore, Maryland.
HEDENBURG, O.F. — 1946. Patent Pending. See also U.S. Patents 2,520,930, April 22, 1949; 2,521,366, Sept. 5, 1950.
HENSILL, G.S. — 1951. Lindane. Soap & S.C. 27(9):135, 137, 159.
HERZKA, A. & J. PICKTHALL — 1961. Pressurized Packaging (Aerosols) 2nd Ed. Academic Press Inc. pp. 509.
HIXSON, E. & M.H. MUMA — 1947. Effect of benzene hexachloride on the flavor of poultry meat. Science 106 (2757):422-423. Oct. 31.
HORTON, R.G., L. KAREL & L.E. CHADWICK — 1948. Toxicity of gamma-benzene hexachloride in clothing. Science. 107(2275):246-247.
HORNSTEIN, I. & W.N. SULLIVAN — 1953. The role of chlorinated polyphenyls in improving lindane residues. J. Econ. Entomol. 46(6):937-940.
HURST, H. — 1945. Nature 145:462.
JONES, H.A. & H.J. FLUNO — 1947. DDT-xylene emulsion for use against insects affecting man. J. Econ. Entomol. 39(6):735-740.
JONES, H.A., H.J. FLUNO & G.T. McCOLLOUGH — 1945. Solvents for DDT. Soap & S.C. 21(11):110-115, 155.
JONES, H.A., H.O. SCHROEDER & H.H. INCHO — 1950. Allethrin with synergists. Soap & S.C. 26(8):109, 133, 135, 137, 139.
KERR, R.W. — 1948. The effect of starvation on the susceptibility of houseflies to pyrethrum sprays. Australian J. Sci. Res. Ser. B. 1(1):76-92. Biol. Abst. 24(1):243.
KIESS, A. — 1981. Uses of Insect Growth Regulators. Pest Control 49(4):27-28.
KINNARD, V. — 1946. Pyrethrum: Lethal flowers generate significant trade. Foreign Commerce Weekly 24(6):5-7, 33, 34, Aug. 10.
KNIPLING, E.F. — 1945. DDT insecticides developed for use by the armed forces. J. Econ. Entomol. 38(2):205-207.
KRISTER, C.J. — 1951. Methoxychlor. Agr. Chem. 6(7):39-41. 93.
KRUHM, A. — 1939. Use of dried flowers to kill insects dates back to Persians. Pests. 7(12):20.
LEE, J.E. — 1967. Aerosol insecticide formulations up to date. Soap & Chem. Spec. 43(4):62.
LINDQUIST, A.W., H.A. JONES & A.H. MADDEN — 1946. DDT residual-type sprays as affected by light. J. Econ. Entomol. 39(1):55-59.
LINDQUIST, A.W., A.H. MADDEN, & H.C. WILSON — 1947. Effect of pretreating house flies with synergists before applying pyrethrum sprays. J. Econ. Entomol. 40(3):426-427.

MALLIS, A., A.C. MILLER & R.V. SHARPLESS — 1952. Effectiveness against house flies of six pyrethrum synergists alone and in combination with piperonyl butoxide. J. Econ. Entomol. 45(2):341-343.

MARCOVITCH, H.S. & W.W. STANLEY — 1942. Fluorine compounds useful in the control of insects. Tenn. Agr. Exp. Sta. Bull. 182.

MARTIN, H. — 1961. Guide to the Chemicals Used in Crop Protection. Canada Dept. of Agr. 4th Ed. Science Service Lab., London, Ontario.

MARTIN, J.T. & K.H.C. HESTER — 1941. Brit. J. Dermatol. Syphilis. 53:127.

MAUGHAN, F.B., F.M. MIZELL & J.P. NICHOLS — 1951. Further studies on aerosol formulations with Lethane and other toxicants. Soap & S.C. 27(2):125, 127, 131.

McALISTER, L.C., JR., H.A. JONES & D.H. MOORE — 1947. Piperonyl butoxide with pyrethrins in wettable powders to control certain agricultural and household insects. J. Econ. Entomol. 40(6):906-909.

McNAMEE, R.W. — 1950. General nature of allethrin. Soap & S.C. 26(8):106.

METCALF, R.L. — 1948a. The mode of action of organic insecticides. National Research Council. Review No. 1. 1955. Organic Insecticides: Their Chemistry and Mode of Action. Interscience Publishers, Inc., N.Y. 392 pp.

MILLER, A.C., A. MALLIS & R.V. SHARPLESS — 1952. Aerosol insecticides . . . their evaluation against house flies, cockroaches. I. Soap & S.C. 28(2):151. 153, 181, II, 28(3):143, 145, 147, 149.

MONRO, H.A.U. — 1949. Insect resistance to insecticides. Pest Control 17(9):16, 18, 20.

MOORE, J.B. — 1950a. Synergist "264." Soap & S.C. Off. Proc. 36th Mid-Year Meet. p. 72. 1950b. The place of allethrin in the aerosol program. Soap & S.C. Off. Proc. 37th Ann. Meeting C.S.M.A. Dec. pp. 89-91. 1950c. Relative toxicity to insects of natural pyrethrins and synthetic allyl analog of Cinerin I. J. Econ. Entomol. 43(2):207-213. 1966. Chemistry and biochemistry of pyrethrins. Pyr. Post 8(4):27-31.

MORGAN, D.P. — 1977. Recognition and Management of Pesticide Poisonings. U.S. Government Printing Office. 75 pp.

NASH, K.B. — 1950. Biological tests of allethrin without a synergist. Soap & S.C. 26(9):127, 129.

NEAL, P.A. & F.W. VON OETTINGEN — 1946. Toxicity of DDT: a report on experimental studies. Soap & S.C. 22(7):135, 137, 138, 141, 143.

NEGHERBON, W.O. — 1959. Handbook of Toxicology. Volume III: Insecticides. W.B. Saunders Co. 845 pp.

NELSON, R.H., R.A. FULTON, J.H. FALES & A.H. YEOMANS — 1949. Low pressure aerosols: their efficiency in relation to formulation and particle size. Soap & S.C. 25(1):120-121, 123, 125, 166.

O'BRIEN, R.D. — 1967. Insecticides. Action and Metabolism. Academic Press. N.Y. 322 p.

PARKER, W.L. & J.H. BEACHER — 1947. Toxaphene: A chlorinated hydrocarbon with insecticidal properties. Del. Agr. Exp. Sta. Bull. 264.

PARKIN, E.A. & A.A. GREEN — 1947. DDT residual films. I. The persistence and toxicity of deposits from kerosene solution on wall board. Bull. Entomol. Res. 38(2):311-325.

PARKIN, E.A. & P.S. HEWLETT — 1946. The formation of insecticidal films on building materials. I. Preliminary experiments with films of pyrethrum and DDT in a heavy oil. Ann. Appl. Biol. 33(4):381-386. Abst. in Biol. Abst. 21(6):1538-1539. 1947.

PARR, T. — 1962. Entex. Pest Control 30(4):16.

PETERSON, H.E. — 1947. Low pressure aerosols. Continental Can Co. Bulletin No. 14.

RAMBO, G. — 1980. Precor. National Pest Control Association, TR-A-6140132.

RICHARDS, A.G. — 1951. The Integument of Arthropods. 411 pp. Univ. of Minn. Press.

ROARK, R.C. — 1935. Household insecticides. Soap & S.C. 11:101, 1947. Roark gives para facts. Soap & S.C. 23(8):165. 1950. A digest of information on toxaphene. U.S.D.A. & B.E.&P.Q. E-802.

ROEDER, K.R., Editor — 1953. Insect Physiology. John Wiley & Sons. 1100 pp.

SCHMITZ, W.R. & M.B. GOETTE — 1948. Penetration of DDT into wood surfaces. Soap & S.C. 24(1):118-121.

SCHROEDER, H.O. — 1950. Piperonyl butoxide in low pressure aerosol insecticides. Soap & S.C. 26(6):145, 147, 148A, 148C.

SCHROEDER, H.O., H.A. JONES & A.W. LINDQUIST — 1948. Certain compounds containing the methylene dioxyphenyl group as synergists for pyrethrum to control flies and mosquitoes. J. Econ. Entomol. 41(6):890-894.

SCOGGIN, B.J., W.L. STEINER & C.C. ALLEN — 1946. Deodorized petroleum base oils. Soap & S.C. 22(4):149-151, 1953.

SEIFERLE, E.J. & D.E.H. FREAR — 1948. Insecticides derived from plants. Indust. & Eng. Chem. 40(4):683-691.

SENNEWALD, E.F. — 1947a. Arsenic trioxide. Senco News. 13(3):2. 1947b. White phosphorus. Senco News 14(6):2-3. 1947c. Thallium sulfate. Senco News. 14(4):2-3. 1947d. Sodium arsenite and sodium arsenate. Senco News. 13(6):2-3. 1947e. Strychnine alkaloid. Senco News. 14(3):2-3.

SHARPLESS, R.V. — 1950. Examination of 590 defective aerosol bombs. Gulf Research & Development Company (mimeographed).

SHEPARD, H.H. — 1951. The Chemistry and Action of Insecticides. 504 pp. McGraw-Hill Book Co.

SHEPHERD, H.R., editor. — 1961. Aerosols: Science and Technology. Interscience Publishers, Inc. 548 p.

SIMANTON, W.A. — 1939. The pyrethrum problem for Gulf. Gulf Research & Development Company Research Project MO 71. Mimeo. 1942. Growing pyrethrum in the United States. Gulf Research & Development Company. Chemistry Division Report. MO 71.

SMITH, M.L., H. BAUER, E.F. STOHLMAN & R.D. LILLIE — 1946. J. Pharm. & Exptl. Therapeutics. 88:359-365.

SPEAR, P.J. & PINTO, L.L. — 1980. Technical Data for Pesticides of the Structural Pest Control Industry. 124 pp. National Pest Control Assoc.

STARR, D.F. — 1950. Recent advances on the allethrin-synergist combination. Soap & S.C. Off. Proc. 37th Ann. Meeting C.S.M.A. Dec. pp. 82-83. 1950 Sulfoxide, Soap & S.C. Off. Proc. 36th Mid-Year Meet. pp. 70-71.

STARR, D.F., P. FERGUSON & T.N. SALMON — 1950. Toxicity of a synthetic pyrethrin. Soap & S.C. 26(3):139, 141, 143.

TUTTLE, W. — 1951. Aerosol valves. Soap & S.C. 27(1):107, 109.

USPHS — 1952. 1956. Clinical memoranda on economic poisons. Communicable Disease Center, U.S. Public Health Service, Savannah, Ga.

WACHS H. — 1947. Synergistic insecticides. Science 105(2733):530-531. May 16.

WATKINS, T.C. & L.B. NORTON — 1947. A classification of insecticide dust diluents and carriers. J. Econ. Entomol. 40(2):211-214.

WEED, A. — 1938. New insecticide compound. Soap & S.C. 14(6):133, 135. 1943. Insect kill: The action of insecticides on insect tissue. Soap & S.C. 19(6):117-121.

WEIDHAAS, D.E. & J.L. BRANN, JR. — 1955. Handbook of Insecticide Dust Diluents and Carriers. 2nd Ed. Dorland Books, Caldwell, N.J. 233 pp.

WHITING, R.W. & P.G. BARTLETT — 1963. High temperatures cut effectiveness of thermal foggers. Food Processing 24(8):105-106, 108.

WHITMORE, B.J. — 1958. Aerosols. Soap & Chem. Spec. 34(9):89.

WIGGLESWORTH, V.B. — 1950. The Principles of Insect Physiology. 544 pp. E. P. Dutton & Co.

WILCOXON, F. & A. HARTZELL — 1933. Some factors affecting the efficiency of contact insecticides. III; further chemical and toxicological studies of pyrethrum Contrib. Boyce Thompson Inst. 5:115.

WILSON, C.S. — 1949. Piperonyl butoxide, piperonyl cyclonene, and pyrethrum applied to selected parts of individual flies. *(Musca domestica; Calliphora vomitoria)*. J. Econ. Entomol. 42(3):423-428.

WOODRUFF, N. & N. TURNER — 1947. The effect of particle size on the toxicity of DDT diluents in water suspension. J. Econ. Entomol. 40(2):206-211.

YATES, W.W. & A.W. LINDQUIST — 1950. Exposure of house flies to residues of certain chemicals before exposure to residues of pyrethrum. J. Econ. Entomol. 43(5):653-655.

YEOMANS, A.H. — 1952. Directions for industrial use of aerosols. B.E. & P.Q. E-835. 1960. Directions for industrial use of aerosols. USDA ARS-33-58.

YEOMANS, A.H. & E.R. VAN LEEUWEN — 1954. Explosion characteristics of insecticidal aerosols. Agr. Chem. 9(4):75,77.

ZUCKER, A. — 1965. Investigation of purified extracts. Annals of Allergy 23:335-339.

C. DOUGLASS MAMPE

Dr. C. Douglass Mampe received degrees in entomology from Iowa State University, North Dakota State University, and a Ph.D. from North Carolina State University. Upon completing his university education, he joined the staff of the National Pest Control Association (NPCA). During his 10 years there he was responsible for keeping abreast of all technical and regulatory actions related to the structural pest control industry. This included reviewing research proposals and overseeing research projects related to the biology and control of termites, cockroaches, commensal rodents, and other related pests.

Mampe left NPCA as its technical director and joined Western Termite and Pest Control, Inc., and its sister company Residex, a pest control industry supply house. He served as Western's technical director for six years and developed practical procedures for field operations. He eventually became general manager of Residex, which included developing new pesticide registrations and industry training programs, including certification training programs for a number of the northeastern and middle Atlantic states.

In 1980, Mampe started his own consulting firm for the urban and structural pest control industry. His firm provides technical consultation, training and training programs, evaluates new pesticides and equipment, custom develops personnel management programs and provides expert testimony for legal cases.

Mampe served as editor of NPCA's _Approved Reference Procedures for Control of Subterranean Termites_, and editor and co-author of the _Manual for Structural Wood Decay_. He writes a monthly column called "Answers" for _Pest Control_ magazine. This column, which deals with a variety of questions, has consistently been one of the best read sections of the magazine since the column began at the beginning of 1975.

Mampe is a member of the NPCA, Pi Chi Omega (a fraternity of industry professionals involved in training), the Entomological Society of America, and a number of state pest control associations. He is currently licensed as a certified pesticide applicator in a number of northeastern states. In 1979, he was chosen "Pest Control Operator of the Year" by the New Jersey Pest Control Association for his contributions to the industry in that state. He is currently serving as a resource person for the United States House of Representatives Committee on Agriculture.

CHAPTER TWENTY-SEVEN

Household Fumigation

Revised by C. Douglass Mampe[1]

FUMIGATION as a method of insect and animal control has been known for many centuries, the fumes of sulfur having been used as early as the 12th century B.C. Cotton (1941) notes: "The burning of incense or aromatic substances was closely associated with the early religious ceremonies of primitive man. It was used to counteract the disagreeable odors arising from the slaughter and burning of the animals offered in sacrifices, to impart a pleasing odor to the sacrifice and to mystify and exert a benign physiological influence over the religious devotees. Later, it was used in churches to purify the air in time of public sickness and to dispel the foulness caused by large congregations of poisonous gases arising from poorly constructed vaults under the church floors."

To a large extent, pest control operators have supplanted fumigation with residual sprays, since their use results in long-lasting control without the danger, expense, and inconvenience often associated with fumigation. Nevertheless, fumigation is still employed since it is usually an effective means of ridding a house of vermin, and when used in combination with residual sprays, supplies the quick kill often lacking in residual sprays. Monro (1969) prepared a useful manual on fumigation.

One may ask, why fumigate? This is a pertinent question since ordinarily there are so many alternative methods of controlling household pests. Fumigation may be used in lieu of contact, stomach poisons, or other means of control for any one of the following reasons:

- The most commonly used fumigants are toxic to all forms of life (e.g. insects and rodents) and thus it is possible by means of one fumigation to control the animal pests in a building.

[1]President, D.M. Associates, Westfield, N.J.

947

- Fumigation is usually the quickest method for controlling animal pests.
- Fumigants frequently, but not always kill insects where ordinarily contact or residual sprays and stomach poisons may not reach them (e.g. behind baseboards, in hollow spaces in the walls, and in building timbers.)
- In specific cases, where it may be too dangerous to have poisonous dusts or powders in the vicinity of food, or where sprays may taint such food with their odors, certain fumigants can be used to kill the pests within or in the vicinity of food.
- In specific instances, it is less expensive to apply a fumigant than to control the pests with repeated treatments by sprays or dusts.

There are a number of good reasons why the pest control operator may forego fumigation, preferring some other method of control. Some of these reasons are:

- Practically every fumigant that is effective against insect and animal pests is also very toxic to man. Rowe (1957) and Torkelson et al. (1966) reviewed the toxicity of the common grain fumigants.
- Since fumigants are poisonous, special equipment, (e.g., gas mask, halide lamp, etc.) is required to safeguard the operator against the effects of these fumigants.
- The building must be completely sealed. This necessitates much labor and sealing material.
- As a rule, the application of fumigants requires more technical skill than other methods of control.
- The fumigants and the labor factor involved make the initial cost of fumigation high compared with other control methods.
- It is not permissible for one man alone to fumigate, whereas only one individual need be involved in the application of sprays and dusts.
- Residents are inconvenienced since they must absent themselves from the structure from one to several days.
- It is less expensive and often more practical to treat a localized infestation by some method of control other than fumigation.
- Fumigants may give an almost immediate and complete kill of animal pests, but a new infestation, from outside sources, may commence immediately after completion of the fumigation.
- It is necessary to obtain licenses and to observe many legal restrictions before fumigation may be undertaken.
- Special insurance coverage is necessary.

How do fumigants affect insects? Insects breathe air through the openings of spiracles which are present on the sides of their thoracic and abdominal segments. The larger respiratory tubes in the body are known as the tracheae, and these attenuate into smaller tubes known as tracheoles. It is through the thin walls of the tracheoles that the oxygen diffuses to enter the cell proper. Active insects "breathe," that is, the air sacs and larger tracheae in the thorax and abdomen are alternately compressed and relaxed. The more rapidly the fumigant enters this tracheal system, the more readily is the insect affected.

It is believed fumigants greatly affect the enzymes concerned with oxidation and respiration in the insect body, and thereby prevent the assimilation of oxygen by its tissues. Carbon dioxide accelerates the rate and amplitude of respiration in insects as well as in warm-blooded animals, and is believed to affect the tracheal valve control in insects. For this reason carbon dioxide is mixed with a number of fumigants.

Many insects are inactive at low temperatures and their rate of respiration is correspondingly reduced, making them more resistant to fumigants. With an increase in temperature, up to a certain point, they become more susceptible to fumigants. Gases are more active at high temperatures than at low temperatures and are able to penetrate warm commodities at a more rapid rate.

Cotton (1932) found the susceptibility of an insect to a fumigant varies with the rate of respiratory metabolism. The three most important factors that increase the susceptibility of the insects to a fumigant are: (1) an increase in temperature, 2) an increase in the carbon dioxide content of the fumigation chamber, and 3) a decrease in the oxygen content of the fumigation chamber.

All stages of the insect breathe (egg, larva, pupa, and adult), but the egg and pupa, being physically inactive, do not breathe as readily as the larva and the adult. Therefore, as a rule, they are not as readily affected by equal concentrations of gas as are the larva and adult. Monro (1960 and 1969) reviews the use of various fumigants. Hoyle (1961), Sheehan (1961), Whitney (1961), and Torkelson et al. (1966) discuss fumigation hazards that should be studied by those interested in fumigation.

HOUSE FUMIGATION

The principles of fumigation are presented here, and they apply to most types of fumigation. Specific characteristics of certain fumigants are noted in the section on fumigants.

Sealing the building. The more tightly the building is sealed, the longer the pests will be confined with toxic amounts of the gas. The material most useful in sealing windows, doors, etc., in homes is masking tape, which is sold in varying widths, two inches/five cm and four inches/10 cm being the most useful. Masking tapes have the great advantage of adhering to surfaces of all types and can usually be removed without damaging them. Upon removing masking tape from wall paper and similar materials, care should be used since the masking tape may strip the finish from the surface.

It should be noted that within two hours after the start of the fumigation, much of the gas may be lost, for no matter how tightly the building is sealed, the gas leaks into the atmosphere.

Vents and openings of a like nature can be sealed off by covering them with four mil polyethylene and then using masking tape along the edges.

The fumigator should make certain there are no conduits, openings, pipes, etc., which lead from the building being fumigated to an adjoining building. The gas might travel to the adjoining building. Since it is often difficult to determine if two or more houses are connected to one sewer, the sewer connection in the basement or house proper should be sealed with tape.

In warehouses and places of a like nature where appearance is not a primary consideration, flour paste and newspaper strips may be used for sealing. Flour mills use a sealing paste made up of a putty-like mixture of flour and lubricating oil, which is smeared into all crevices.

Rather than sealing, an alternate method is to cover the building with tarps. Four or six mil polyethylene may be used, but vinyl laminated nylon is reusable and more manageable in the wind. Tarps are usually laid out and the adjoining edges rolled together and clamped. Sand snakes are used to effect a seal to the soil.

Fires. At the concentrations at which modern fumigants are used there is practically no possibility of it exploding. However, all fires, such as pilot lights, oil burners, etc., should be extinguished since they cannot be supervised during the fumigation.

Provision for ventilation. The rear and front doors on the main floor should be locked so that they may be opened from the outside when ready to ventilate. Opening these doors will create a draft which will help to dissipate the gas. If the building is tarped, windows should be opened.

Temperature in the structure. The temperature within the structure should be at least 65° F/18° C. Insects are less active and usually more resistant to the fumigants at temperatures lower than 65° F/18° C. It is advisable to commence the fumigation in the morning because rising temperature will be encountered as the day progresses and until the structure is ready for ventilation.

Removal of foods and articles. Some fumigants such as sulfuryl fluoride should not contact foods because of label restrictions. Methyl bromide reacts with some products to create a permanent off-odor. Anon. (1971) has a list of such commodities. Living plants must be removed as most fumigants will kill them.

Diffusion and penetration. Where large rooms or areas are being fumigated, it is often necessary to release gas at a number of locations, otherwise the gas may diffuse unevenly. Fans can also be used to circulate air and the fumigant. Fumigants lighter than air are usually released on lower floors and heavier-than-air fumigants are released in upper areas. When possible, the gas should be released from outside the area being fumigated so the fumigator is not exposed to toxic concentrations.

Respiratory protection. Respiratory equipment is required if a fumigator is likely to be exposed to toxic concentrations of a fumigant.

The ultimate protection is a self-contained unit such as the Scott Air Pak. Such units are heavy and bulky, which prevents fumigators from easily entering ship holds and other confined spaces. The usual alternative is a full-face gas mask with a canister. Canisters are designed for specific types of gases and the proper one must be used for the gas in question. NIOSH now approves canisters for fumigants. The manufacturer or suppliers of the fumigant can supply information on the proper canister for specific applications.

Canisters have expiration dates beyond which they should not be used. They also have uselife limits, depending upon the concentration of the fumigant and size of the canister. Spent canisters should be crushed or otherwise rendered non-usable to prevent someone from using a spent canister in error.

Fumigators test their masks by placing their hands over the inlet of the mask and inhaling. If the mask is absolutely tight, it collapses against the face. Then the canister is placed on the mask, after removing the tape covering its inlet, and the process repeated.

When entering an area to be gassed or under gas, fumigators should work in pairs so they can assist one another in case of an emergency. The two should never be separated. When releasing gas from within the building a plan should be made so the gas is released in an orderly fashion, starting from the point most distant from the exit and working towards the final exit. In multi-storied structures, the gas is usually released on the top floor first and then the fumigators work downwards towards the exit. Under no circumstances should fumigators use interior elevators once gas is released, as a malfunction of the elevator could trap them in a toxic atmosphere for a period longer than the life of their canisters or air supply.

Ventilation. In accordance with arrangements made previously, the front and rear doors should first be opened from outside, opening first the door opposite the direction from which the wind is blowing. Gas masks should be worn during the entire period of opening the building.

After this preliminary ventilation has been in progress for 30 minutes more or less, depending upon weather conditions, it should be safe for the operators to enter, wearing gas masks. Additional windows should be opened. Operators should not remain in the building until it has been thoroughly aired. It is important to have guards at entrances during ventilating to keep anyone from entering the building.

Specific information as to length of time required for ventilation cannot be given to meet all cases. Much depends upon the movement of air currents, temperature, humidity of the air and rate of gas leakage from the building during the hours of fumigation. It may require anywhere from three to 24 hours for complete aeration. To be absolutely safe, the building should be ventilated until there is no odor of gas before persons are allowed to enter for normal resumption of activities.

It is obviously the fumigator's responsibility to make sure the premises are completely and thoroughly aerated before turning them over to the tenants. He should take all necessary precautions, such as testing with available detection equipment to determine safe levels or absence of the fumigant.

Detection equipment. Most modern fumigants have no detectable odor at toxic levels. Warning agents such as chloropicrin have been used, but they often settle out and usually aerate at a rate different than that of the primary fumigant. Thus, they are not reliable indicators of the amount of fumigant remaining in a fumigated atmosphere.

Hassler (1961) discusses the use of a Fumiscope for measuring concentrations of methyl bromide. This equipment can also measure ethylene oxide and sulfuryl fluoride. Equipment of this type is not sensitive enough at low concentrations to determine if an area is free of gas.

Heseltine (1959) shows methyl bromide can be detected at low concentrations using detector tubes and a halide leak detector. Tubes of this type, but with different reagents, can be used to detect low concentrations of nearly all modern fumigants. Dumas & Munro (1966) provide a good review of detector tubes and their use in the field.

HYDROCYANIC ACID GAS (HCN)

Hydrocyanic acid gas is a fumigant that was used by many pests control operators or public officials in household fumigation.* Since hydrocyanic acid gas will cause death in a few minutes, and unconsciousness with just a few breaths, it is used only by experienced and duly licensed or authorized individuals. Before we delve into a discussion of this effective, but deadly fumigant, for safety's sake, it must be repeated that *HYDROCYANIC ACID GAS SHOULD BE USED ONLY BY THOROUGHLY EXPERIENCED INDIVIDUALS.*

*At present, cyanide is not readily available, but new formulations and registrations are being developed.

The ancients were well aware of the poisonous effects of hydrocyanic acid gas, and Cotton (1941) notes, "the Egyptian priests used decoctions made from peach kernels to get rid of people who were too curious about their religious secrets." This gas was first used on a practical scale by Coquillet in 1886 for the fumigation of citrus trees under tents.

Langhorst (1947) notes liquid HCN must be kept cool below its boiling point of 79° F/26° C. It is approximately $7/10$ as dense as water. HCN volatilizes rapidly and has the characteristic odor of bitter almonds. Ordinary commercial liquid HCN is 96 to 98 percent pure and slightly acid. According to Langhorst, liquid HCN "is inflammable and burns with a lavender colored flame. The products of the combustion are harmless, being nitrogen, carbon dioxide, and water vapor. Hence, one of the simplest means for removing danger from HCN which has been spilled, or otherwise must be disposed of, is to set fire to it and let it burn." A concentration of 10 percent HCN in air must be reached before the combination is inflammable and explosive. Such concentrations are rarely reached in the usual type of HCN household fumigation.

HCN METHODS OF FUMIGATION

1) Discoids. A fibrous absorbent material is saturated with liquid HCN and kept in sealed and airtight containers. The employment of discoids is one means of household fumigation in the United States.

2) Dusts. Calcium cyanide on exposure to atmospheric moisture forms hydrocyanic acid gas.

3) Liquid Cyanide. Liquid cyanide kept under pressure in cylinders is pumped through pipes into the building or machinery to be fumigated. Flour mills, warehouses, and other industrial buildings may use this type of HCN fumigation.

4) Pot Method. HCN gas is generated in a container from a mixture of sodium cyanide, sulfuric acid, and water. This method for evolving HCN has been largely replaced by the aforementioned methods.

The following table indicates the amounts of other HCN fumigants required to equal one pound of liquid HCN:

HCN discoid	1 pound
calcium cyanide (Cyanogas 25% CN)	4 pounds
sodium cyanide 52% CN)	2.5 pounds

DISCOID METHOD OF CONTROL

HCN discoids are an efficient, clean, and convenient method of fumigation. No equipment is necessary other than a special can opener and a gas mask. Once the discs have been used they may simply be burned or thrown into a garbage can since they are harmless.

HCN discoids consist of porous, absorbent fibre discs that are saturated with liquid hydrocyanic acid gas and packed in gas-tight containers. When these discs are exposed to the atmosphere, the liquid hydrocyanic acid volatilizes into the toxic gas. The discoids come in approximately two- and five-pound cans and contain 16 ounces and 40 ounces HCN, respectively. These cans are opened with a special can opener. Since the gas evolves immediately, one must wear a gas mask when working with discoids.

The building is prepared in the usual manner with the usual precautions. Rubber gloves should be worn by the fumigators, who, of course, work in pairs. The discoids are scattered in thick layers of newspapers or wrapping paper to prevent staining. The individual discs are not handled one by one, but scattered from the container over the papers. For ordinary household fumigation the discoids are used at the rate of eight ounces/226 g of HCN (eight ounces of discoids) for every 1,000 cubic feet/28 cubic meters at a temperature of at least 65° F/18° C with an exposure of 12 hours or more.

Since the gas evolves so quickly upon opening the can, it is often desirable, especially in warm weather, to slow the evolution of the gas. This is accomplished by pre-cooling the cans several hours before the fumigation. Cans can be pre-cooled by placing dry ice or ordinary ice on the tops and around the base of the cans three to four hours before they are to be opened.

The author is indebted to the HCN *Discoids Manual* distributed by the American Cyanamid Company for much of the following information regarding the use of HCN discoids for fumigation.

Determination of the amount of fumigant that is necessary. Estimate the number of cubic feet in the area to be fumigated. The HCN *Discoids Manual* recommends the following useful form:

Date of Fumigation ——————————————————————————

Building ——————————————————

Address ————————————————

——————————————————————————————————————

——————————————————————————————————————

——————————————————————————————————————

Type of Building ——————————————————————————

Parts to be fumigated:

Location	Dimensions	Cubic Feet	Gas lbs.	Windows	Doors (to be sealed)
Basement	150×40×10	60,000	40		1
First Floor	150×40×14	84,000	42	12	3
Second floor	150×40×12	72,000	40	10	
Third floor	100×40×10	40,000	20	8	4
	etc.	etc.	etc.	etc.	1

Remarks:
Description of other openings to be sealed:

Totals: Floors ——————— Cubic Feet ——————— Gas (lbs.) ———————

   Windows ——————— Doors ——————— Warning Signs ———————

Fig. 27-1. PCO opens discoid can for hydrocyanic gas fumigation.

Since the resistance of different insects to the same fumigant varies greatly, one must consult the manufacturer's recommendations for the correct dosage. The shorter the exposure, the more gas will be necessary. The tightness with which the premises can be sealed, as well as temperature and outside wind velocity are other factors that must be considered.

Removal of foods and household articles. Hydrocyanic acid gas is absorbed by commodities during fumigation and rapidly given off (desorbed) after 24 to 72 hours aeration. As the HCN is desorbed during the aeration period, it continues to kill the insect stages that survived the initial exposure. Liquid materials release the HCN much more slowly than dry commodities.

Liquid and fatty foods not in sealed containers, such as jams, preserves, milk, butter, and others should be removed before fumigation since these moist ingredients readily absorb HCN. Fresh fruits and vegetables are also in this class. Flour, cereals, and other finely ground materials, such as ground chocolate, coffee concentrates, etc., absorb HCN and release it relatively slowly. Thus, it is advisable not to expose these materials to fumigants if they are not infested.

The emulsions on unexposed photographic film, the lubricating oils in clocks and the coating on inexpensive mirrors may be affected by HCN and should be removed prior to fumigation.

Sherrard (1942), speaking of tests with such fumigants as HCN, chloropicrin, methyl bromide, ethylene oxide — carbon dioxide 1:9, and ethylene dichloride — carbon tetrachloride 3:1, notes that none of these in a gaseous state affected the color or texture of animal or vegetable fabrics or corroded metals. Sherrard (1945) makes an exception to his 1942 notes since he found fragile fabrics, such as white linen window curtains, may take an orange tinge at high HCN concentrations. He also states if HCN discoids contact rayon, a gummy mass may form due to the solvent action of the liquid HCN on fibers or dye. These discoids have also been shown to damage linoleum "through two layers of heavy brown paper, but that four layers gave protection." Other floor coverings and highly finished floors are subject to "burns" if contacted by HCN discoids.

Staining with HCN. When fumigating with hydrocyanic acid gas, greasy areas around light switches, enameled gas ranges, etc., may become stained. Such discoloration may be removed with ordinary soap and water. Fumigation should not be undertaken where there are freshly painted walls, since these may be discolored. Where such discoloration occurs it is often removed with hydrogen peroxide.

Arrangement of the contents in the room. The contents in the room should be arranged so they facilitate penetration by the gas. Closets should be opened and clothing hung loosely. Trunks and other tight containers should be unlocked and the contents draped over bedsteads, etc. Blankets and other bedding may be draped over a temporary line erected across the room. Carpets and rugs should have their edges rolled back. Books in cases should be loosely arranged.

The actual fumigation. Immediately preceding the fumigation, fumigators should inspect the house thoroughly to see that no human beings, as well as plants and animals, are left on the premises. After the fumigator is satisfied that all openings are properly sealed, fires are out, etc., warning signs should be placed on the doors. All the doors should be locked except the one to be used as an exit by the fumigators. A large warning sign, some two feet square, should also be attached to each door. The sign should have the words FUMIGATION, DEADLY GAS, DANGERS, the name of the gas, and preferably a skull and crossbones. As an additional precaution, a guard should be placed at the main entrance to prevent anyone from entering. Another individual, in the meantime, should check the building room by room to make certain that neither human beings nor pets are present.

Distribute the required number of cans of *Discoids* (unopened) throughout the building, at the places where it is desired to apply the fumigant. As a precaution against staining or marring floors, place some suitable protective material alongside each can so at the time of fumigation the discs may be scattered on to these surfaces. Some pest control operators use wire baskets. Suitable baskets may be obtained in almost any department store. They are of the "dish drain" type and are sold for about $1 each. They are approximately 12 inches/ 30 cm by 18 inches/45 cm. These baskets have short legs which keep the basket clear of the floor and it is only necessary to shake the *Discoids* from the can, spreading them in the basket. The bottom of the basket being raised off the floor permits circulation of air around the *Discoids,* resulting in quicker diffusion of the gas.

Another equally practical procedure is to obtain a suitable quantity of 36 inches/91 cm square sheets of corrugated paper upon which to scatter the *Discoids.* One sheet of this size is ample for one 16-ounce/454 g can of *Discoids* or two sheets for a 40-ounce/1.13 kg can. The sheets are light in weight, making it easy to distribute them, as well as to gather them up after fumigation. The corrugations promote evolution of gas from the discs. Newspapers may also be used.

When all the unopened cans of *Discoids* have been distributed, and the operator in charge is satisfied that all is in readiness, a guard should be placed at the exit door with orders to allow no one to enter, or the door should be locked from the inside.

With an assistant (both operator and assistant wearing gas masks) the operator starts the process of fumigating from the top floor, beginning with the can or cans farthest from the exit and working toward the exit. One man opens the cans, while the other follows along scattering the discs. They should work together and time their action so the man opening the cans does not get too far ahead of his co-worker.

Another method of procedure is to first open each can and immediately cap it with the fibre cap (each can of *Discoids* is equipped with a fibre cap on top and bottom as a protection during shipment). Then, after all the cans on the floor have been opened, the contents of each may be scattered.

Still another method is described as follows: After proceeding as above by spotting the cans in the desired places, each one of them should be opened and up-ended on the floor with the open end down. This plan eliminates the use of the fibre cap, saves time, and will hold the gas in the can equally as long as the caps. In opening the cans and scattering the discs, the operators should always work in a direction *away* from the gas and *toward* the exit.

Operators should *never* re-trace their steps while scattering the *Discoids,* even though equipped with gas masks. It is always dangerous to re-enter a fumigated space. Operators should work quickly, but should not rush.

Having finished the top floor, the operators proceed immediately to the next lower floor to repeat the operation, and so on down the line skipping the main or street floor, in order to fumigate the cellar. The main floor should be fumigated after the cellar. After the work is completed, operators should lock and seal the doors *from the outside.* One or more guards should patrol outside of premises during the entire operation.

Testing after ventilation. When the fumigator has aerated the building, he should than test for HCN. Detector tubes are available. He should also test with methyl-orange papers for the presence of HCN in bedding, mattresses, and pillows. Finally, if there is any doubt as to whether or not the building *and the furnishings* are completely free of HCN, then the fumigator should not permit occupancy, even if it is necessary to keep the tenants out of the premises overnight.

Removing the paper seals from fireplaces, ventilators, and heat registers will considerably speed up the process of ventilation. Also, in the case of industrial buildings with sub-basements and other places that are difficult to aerate, the use of a few properly placed office fans is suggested.

Aeration of bedding, pillows, mattresses, etc. Ventilation of a fumigated home by opening windows, doors, etc., is only one phase of ridding the premises of the fumigant. It is also *absolutely essential to beat and air all pillows, mattresses, bedding, clothing and bedroom rugs before residents are permitted re-entry.*

Williams (1938) states the danger from mattresses arises from the fact that "hydrocyanic acid gas retained in the mattress condenses into a liquid and thereafter dries out very much more slowly than it would if it had remained as gas. Thus, a person sleeping on a mattress is exposed to the emanation of the gas for a number of hours." Moreover, during cold weather, the in-rushing cold air during ventilation chills the mattresses, pillows, blankets, etc., impairing the volatilization of the gas from them. The warmth of the human body is sufficient to hasten the release of hydrocyanic acid gas from the interior of the mattress. Young children are the ones most often affected, since they often sleep under bed clothes with their heads covered.

The majority of fumigation accidents from mattresses occur during cold weather, since the gas is more likely to be retained in the mattress. Moreover, in cold weather the windows are more likely to be closed. Thus, there is little ventilation. During warm weather mattresses should be ventilated for at least eight hours and preferably overnight before being slept on. Where the weather is cold and the humidity is high, the proper ventilation of the house is made more complicated. Low temperatures and high humidity make the gas less volatile, and thus it clings to furnishings and walls. Many deaths have resulted from overlooking this factor.

A good method of ventilating is to air the premises thoroughly, then close the windows, and heat the rooms to 75° to 85° F/24° C to 29° C for an hour or longer. Then open all windows and ventilate again. It should be remembered not only do the furnishings absorb the gas, but the walls do too.

Hubbel (1941), speaking of ventilation, says, "In many cases, where a heavy charge of hydrocyanic acid gas was applied in cool, damp weather, I have found it necessary to open, air, close up, and heat up, twice a day for as many as three days and nights before I considered it safe for people to inhabit. And in all these cases two or more very large (27 inches) ventilating fans were used for hours each day." Generally ventilation can be hastened by placing a number of office fans at a point farthest from the window and directed towards the windows.

Disposal of spent discoid. The used discoids should be gathered up and burned or buried immediately after the ventilation is completed.

How to test for cyanide. Horsfall (1941), in discussing the use of methyl orange-mercuric chloride test papers (originally yellow in color), notes these turn a deep red in 10 seconds if the concentration is 890 parts per million, or one ounce per 1000 cubic feet, deep pink in one minute if the concentration is 89 parts per million. Either of the above concentrations of hydrocyanic acid gas are too high for occupancy, as 300 parts HCN per million will quickly kill human beings and animals. These test strips may be placed between pillows and mattresses or between two pillows. If after eight minutes exposure the paper turns only slightly pink, the concentration is approximately nine parts per million and the room is safe for occupancy.

The American Cyanamid Company, which distributes these papers in bottles, states they can be made as follows: 10 grams mercuric chloride dissolved in 500 cc distilled water; 2½ grams methyl-orange dissolved in 250 cc distilled water. "It is necessary to warm these mixtures somewhat until the solution is complete. Cool to 30° C or lower. Mix with 50 cc glycerine U.S.P. Allow solution to stand overnight and filter before use.

"Sheets of filter paper are immersed in this solution and hung up to dry in air entirely free of HCN. At end of drying period, paper should still be soft. At this point the paper should have a moisture content of 55 to 60 percent on the wet basis. Over-drying results in loss of sensitivity. The dried paper is cut into strips and preserved in glass containers, tightly stoppered.

"Any high grade filter paper may be used, but the paper selected should have a neutral reaction and should not be more than 0.020 inches in thickness.

"These test papers are not likely to be effective if dried out. They should therefore be kept in a moist condition in a tightly sealed bottle, away from an atmosphere of hydrocyanic acid gas. A fruit jar, having a small sponge fastened inside the cover, makes a very satisfactory container. The sponge should be kept moistened.

"A ready method of determining if the supply of test paper is in good condition is to hold one of them over an open can of Discoids for a second. The paper should turn red immediately."

Effect of HCN on human beings. Cyanide, either in solution or as a gas, may be absorbed by the skin and pass from there into the blood. Williams (1935), discussing the effect of HCN on human beings, makes these pertinent remarks: "In concentrations of eight ounces hydrocyanic acid per 1000 cubic feet, a man may be dangerously poisoned by skin absorption alone in five or 10 minutes." Cyanogen (CN), the essential component of all cyanides, "prevents the body cells from removing the oxygen from the blood, it stops the normal process of life, and if its effect is maintained for more than a short time, causes the cells to die.

"Nerve cells are the first affected, which fact accounts for the very early appearance of confusion, loss of consciousness and cessation of breathing. The heart will continue to beat for several minutes after breathing has stopped."

Gas masks. Gas masks are used when fumigating with any toxic gas. The materials in the canister remove the gases in the air so the individual wearing the mask inhales the air without the gas for which the canister acts as a filter. These canisters are filled with materials which absorb or react with the gases for which they are specifically made. Glidden (1964) notes there are "four ways in which a gas mask canister or respirator cartridge works: filtration, adsorption, absorption, and catalization." For this reason, when working with hydrocyanic acid gas, a canister that filters hydrocyanic gas must be used. Thus, where chloropicrin is applied, a canister that will filter this organic vapor is utilized. Where both HCN and chloropicrin are used in fumigation, the latter as a warning gas, a canister that filters both of these vapors is necessary. Pearce (1961) and Fulton et al. (1962) discuss masks.

The U.S. Bureau of Mines has a standard color scheme for canisters relative to the gas or gases for which they are used. According to Wagner (1942), the canisters differ as follows:

- The white canister affords protection against low concentrations (less than two percent in air) of acid gases such as hydrocyanic acid and sulfur dioxide. This canister contains soda lime, caustic pumice or caustite (a sodium hydroxide preparation), and activated charcoal.
- The yellow canister affords protection against low concentration (less than two percent in air) of a combination of organic vapors and acid gases, such as a combination of hydrocyanic acid and chloropicrin. This canister contains activated charcoal and soda lime or other alkaline granules.

Back and Cotton (1932) discuss the serviceability of canisters as follows: "A canister will last for only a certain period, which depends upon the concentration of the gases to which the canister is subjected, the length of the exposure, and the manner in which it is stored when not in use. As supplied by the manufacturers, the canister is sealed by a cap over the inlet valve and by a cork in the nipple. Stored in this condition and without use, the canister should last a year. If air is allowed to enter, the contents of the canister deteriorate rapidly. A fresh canister is good for several hours of continuous use. When it begins to give out, the gas fumes will penetrate the mask. When this occurs, the wearer should immediately go into fresh air and replace the canister with a new one." Unused canisters supposedly last indefinitely when stored in a cool dry place away from gases.

The HCN *Discoid Manual* makes the following interesting note: "Medical science indicates that there is a definite possibility that an individual with punctured ear drums can, while wearing a gas mask in the presence of HCN, "breathe in" HCN vapors by reason of the fact that the mask creates a slight vacuum on inhalation and this, in turn, will allow gas to come in through the ears, which is then exhaled by the operator into the gas mask. Individuals having this trouble have solved the difficulty by using ear plugs of cotton smeared with oil."

First aid. (HCN *Discoid Manual*). In case of accident, keep cool. Poisoning by the gas should not be fatal if prompt action is taken. The following action should be taken if a worker is overcome by fumes.

1) Get the person out of the gas. When a person is overcome by gas, the first thing to do is to get them into fresh air quickly. Fresh air does not mean out of doors in cold weather. Many people have walked from a warm room containing gas only to collapse in the cold outside air. Take the patient to a room free from gas and comfortably warm. Be quick, but not unnecessarily rough.

2) If the patient is not breathing, start artificial respiration at once by the Schafer Method. This technique must be followed verbatim in all cases of unconsciousness.

3) Do not rush an unconscious person to a hospital. Prompt action on the spot is essential.

4) Send for a physician, but in the meantime, if the patient is breathing, keep him in fresh air, but do not permit him to exert himself. Have patient inhale from bottle of ammonium carbonate. Do not leave patient alone until he is normal.

5) Do not neglect immediate and continued first aid treatment to call a doctor. Have someone else call a doctor, but continue first aid until doctor arrives.

6) If patient is fully conscious and recovery is still delayed, give 16 drops of aromatic spirits of ammonia in a half glass of water.

7) Do not breathe gas yourself even for a short time. If it does not overcome you, it will cut down your strength. If you have to go into the gas to get a person out, remember that nobody is immune. Protect yourself, wear a gas mask.

Domestic fumigation practices recommended by the NPCA.

1) No crew should be allowed to fumigate unless they are properly equipped with gas masks approved by the U.S. Bureau of Mines and which have been properly tested before entering into the gas.

2) All fumigating crews shall be required to carry to all jobs sufficient warning signs and such safety devices as necessary.

3) No fumigation of any building and re-occupancy should take place within the same 24-hour period, and in no event should a fumigated space be reoccupied before an eight-hour ventilation period. This does not include a treatment for mice or rats in burrows and harborages.

4) No partial fumigation of any building should be allowed unless the entire structure is vacated.

5) On every job, one or more watchmen should be on guard outside the fumigated structure during the entire period from the time of the fumigation until it is safe for experienced fumigators to enter without masks.

6) All greasy or damp foodstuffs should be removed. This includes milk, butter, green vegetables, eggs, opened and not corked bottles of liquids, unexposed film, fine clocks, and all plants and pets.

7) All toilets should be flushed at the completion of the fumigation. If toilets are used for the disposal of any fumigant residue, several flushings should be made.

8) Any premise to be fumigated should be sealed in such a manner as to confine the fumigant to the space intended to be fumigated. Also, careful examination should be made of all parts of the enclosed space to determine that no persons or domestic animals remain before fumigation material is distributed and the final exit made.

9) All doors and accessible windows should be locked and warning signs should be placed on all entrances before the fumigant is released.

10) Two or more experienced men should be on every fumigation job, both at the time of releasing the fumigant and at the time of initial ventilation. Under no circumstances should anyone but an experienced fumigator be permitted to conduct or supervise a fumigation.

11) Fumigation regulations in effect in the city, municipality or state in which fumigation operations are to be performed should be strictly complied with. In the event of no local ordinance or law pertaining to fumigation operations, notifications should be made at least to the fire department and to the police department or any sub-division of the local municipal government with jurisdiction.

12) All bedding and overstuffed material, as well as absorbing articles, such as woolens and furs should be placed for easy penetration and ventilation of the fumigant. It is recommended that the structure be heated when necessary to assist driving the fumigant out of the materials. The use of fans or any other mechanical means of ventilation is recommended.

13) Extreme caution should be taken to make sure all tenants are notified of the fumigation as to the time to vacate and when to re-enter the premises. Such notification should be made in writing to each individual or a notice be posted in a conspicuous place.

14) Each crew should be outfitted with a safety kit containing the following:
- A diagram showing the Schafer Prone method of resuscitation.
- Smelling salts (ammonium carbonate).
- Amyl nitrite. 5 minims.
- Gauze bandage, Band-Aid, adhesive tape, and antiseptic.
- Turkish towel.
- New canister.
- Permit if such is required.
- Aromatic spirits of ammonia (well stoppered in a dark bottle).

15) No fumigation project should be carried out unless the operators are familiar with the Schafer Prone method, especially on cyanide fumigations.

16) No person should be permitted to enter the fumigated premises before the fumigator has satisfied himself, by personal inspection without gas mask, that it is safe for occupancy.

Miscellaneous observations. The following questions were asked of applicants for a fumigators license in one state, and interestingly reveal what safeguards must be taken in two unusual circumstances:

"Describe in detail what you would do if you were hauling a cylinder of liquid HCN through the Liberty Tunnel, and if traffic became congested and you were stopped in the tunnel, and at the same time an irreparable large leak occurred at the cylinder valve."

Answer — "Don a gas mask, quickly place cylinder on floor of tunnel, and ignite fumes of liquid issuing from cylinder. The liquid or gas will burn as fast as it issues from leak and the products of combustion; namely, carbon dioxide, nitrogen and water vapor are harmless."

"Suppose a fire started in the house? What would you do?"

Answer — "Ring fire alarm. Locate policeman on beat. Have him stand watch while I opened doors and windows on main floor. Tell fire chief the danger of entering the building unless his men are equipped with 'all service' masks."

Some deaths because of fumigation may have been due to:

1) Lack of a proper understanding of construction, particularly an understanding of the porous walls that divide old buildings. In such cases, neighbors are often stricken because of gas leaking through these walls.

2) Not enough consideration given to checking that all people are out of a structure. Written and verbal notices are insufficient.

3) Where there are children, someone should stand guard at the entrances. A child does not read signs, and acts impulsively. One child is alleged to have thought that chairs placed before an entrance amounted to nothing more than a joke. The chairs were brushed to one side and the child was stricken after climbing a flight of stairs.

4) Insufficient heating of over-stuffed articles.

5) Too much haste in allowing tenants to return.

LIQUID HYDROCYANIC ACID

This colorless, volatile liquid (96 to 98 percent HCN) was marketed in 30- to 75-pound/13 to 34 kg cylinders. The cyanide cylinder was placed on scales, and the number of required ounces were weighed out and discharged from the cylinders by a small portable compressor. This method was utilized for the most part in large buildings such as warehouses, flour mills, etc. Only fumigation experts used the liquid HCN method.

The building was prepared in the usual way, the cylinders containing the liquid were placed outside the building, and the gas was forced through ⅜ inch/nine mm flexible copper piping into the building by compressed air. Synthetic and pressure rubber tubing may also be utilized. Spray nozzles are attached to the end of the tubing. In large institutions, this piping may be of a permanent nature. Cotton et al. (1945) consider the application of liquid HCN in flour mills in some detail. At present, liquid hydrocyanic acid is not available.

CALCIUM CYANIDE

Calcium cyanide dust on exposure to atmospheric humidity generates hydrocyanic acid gas according to the following formulas:

$$Ca(CN)_2 + 2H_2O \rightarrow Ca(OH)_2 + 2HCN \quad or$$
$$CaH_2(CN)_4 + 2H_2O \rightarrow Ca(OH)_2 + 4HCN.$$

The dust is applied on strips of paper on the floor. The gas evolves more slowly than when HCN discoids are used.

According to Williams (1939), "The amount of HCN produced is approximately one-half the weight of the calcium cyanide entering into the reaction. Therefore, when *Cyanogas* is used, the HCN produced will be between one-fifth and one-fourth of the weight of the raw material." *Cyanogas* contains from 40 to 50 percent calcium cyanide.

In using calcium cyanide, one proceeds in a manner similar to that indicated for the methods discussed previously. Strips of paper, preferably several layers of wrapping paper, are spread across the room. Then the cans containing the material are distributed according to plan. The operator, wearing a mask equipped with a cyanide canister, removes the top of the cans and scatters the dust on the strips of paper, always working from the farthest end of the room toward the door. The dust should not be more than ⅛ inch/three mm thick at any point.

Since from *Cyanogas* there evolves only one-half the amount of the HCN that an equal amount by weight of sodium cyanide generates, it must be used in dosages twice as great as sodium cyanide.

In addition to the usual hydrocyanic acid gas precautions, certain special precautions should be observed with calcium cyanide. These include:

- Wear a mask when cleaning the structure of residue, since some hydrocyanic acid gas may still be emanating.
- Use care when opening up the structure for ventilation, since a breeze may blow the residue about the room.
- Use plenty of wrapping paper or newspaper to prevent staining of floor, etc.

Cyanogas G fumigant, which is somewhat sand-like, allows the slow evolution of the gas. For insects, use two pounds per 1000 cubic feet/0.9 kg per 28 cubic meters and for rats and mice one-half pound per 1000 cubic feet/0.23 kg per 28 cubic meters. *Cyanogas* A dust is a finely powdered material, which upon exposure to the moisture in the air releases HCN more rapidly than *Cyanogas* G fumigant. Cyanogas A is applied usually in rat burrows with a special foot pump.

POT METHOD OF FUMIGATION

In this method of generating hydrocyanic acid gas the chemicals are placed in earthenware crocks, barrels, or similar type containers. Sodium cyanide eggs, which weigh approximately one ounce/28 g each are mixed with sulfuric acid and water. In the reaction hydrocyanic acid gas is formed as follows:

$$NaCN + H_2SO_4 \rightarrow NaHSO_4 + HCN.$$

The greatest advantage of this method of fumigation is it is an inexpensive method for generating cyanide. On the other hand, it requires crocks, pots, and materials of a like nature. The sulfuric acid may spatter and injure the woodwork, rugs, furnishings, etc. The sulfuric acid is also dangerous to handle. The sodium cyanide is deadly poisonous if it penetrates through any wound in the skin. This method has been replaced largely in household fumigation by the discoid procedure.

The formula for the pot method of cyanide generation is as follows:

Sodium cyanide (96% to 98% pure)	1 pound/0.45 kg
Sulfuric acid (66% Baume)	1.5 pints/0.7 l
Water	3 pints/1.4 l

First pour the water into the crock and add the acid; *never pour the water into the acid because the reaction is so violent that the resultant spattering will injure the operator and damage the surrounding materials.* Upon the addition of the sodium cyanide, hydrocyanic acid is generated immediately. One pound/0.45 kg of sodium cyanide, when used in the above formula, will be sufficient to fumigate 1000 cubic feet/28 cubic meters of space.

For rooms of 2,000 to 3,000 cubic feet/56 to 84 cubic meters, earthenware or stone crocks of four gallon/15.1 l capacity are used. For smaller rooms, smaller crocks are practical. According to Back and Cotton (1932), where large houses are to be treated, "a few large containers are sometimes preferable to many small ones. In such cases 50-gallon/189 l wooden barrels can be used. They are large enough to handle safely a charge of 30 pounds/13.6 kg of sodium cyanide or sufficient material to fumigate 30,000 to 40,000 cubic feet/approximately 800 to 1,100 cubic meters of space. Such barrels must be watertight. They should be thoroughly scrubbed both within and without, filled with water, and allowed to stand for at least 12 hours previous to their use. Ordinary oil or molasses barrels are usually satisfactory if all hoops are sound. Flour barrels are not heavy enough. Before the chemicals are added, the barrels should be set in galvanized-iron washtubs, in each of which has been placed a pailful of water containing several handfuls of ordinary washing soda. This is a precaution against leaking barrels and provides for catching and neutralizing any small quantities of the acid-water mixture that may work out of the barrel. Galvanized-iron tubs containing a similar soda solution should also be used, if possible, when crocks are used as generators."

Back and Cotton give further detailed directions in the use of the pot method of fumigation as follows: "After the room or house to be fumigated has been prepared, the generators should be distributed and the water and the acid added. The acid should be added to the water slowly, since this operation produces heat and a too-sudden change in temperature is likely to cause the crocks to crack. If the floors are highly polished and will be damaged by the heat from the jars, the jars may be set on bricks set in each tub. If crocks are used, the water and acid should be added outdoors or in a room with a concrete or tile floor and with drain(s) close by so that the breaking of a crock will not result in an acid-burned floor. It is even more advisable to fill barrels outside if labor is plentiful and the charge for carrying them in is not too heavy.

"The proper quantity of sodium cyanide, which should be weighed out beforehand, should be placed in a paper sack or small piece of newspaper, in a compact bundle, and set near the generator to which it belongs. If several generators are employed, the generation of the gas can be delayed by wrapping the cyanide in several thicknesses of paper, or by using two paper bags of different size and placing the smaller inside the larger. The gas will not generate until the acid has eaten through the paper, so the operator will have a longer time in which to complete his work and escape, than if only one thickness of paper were used.

"When all is ready, the fumigator should start upstairs in the room farthest from the stairway, gently lower the bag of cyanide into the generator containing the acid and water, leave the room, and close the door. He should repeat the operation in the other rooms on that floor, proceed to the downstairs rooms and start the generators there, then leave by a pre-determined exit. If all the rooms on the first and second floors are fumigated as a single unit by using one or two large containers on the first floor, all the inside doors are left open and the containers are located close by the exit door.

"In buildings equipped with fire-alarm systems which are set off by heat, the generation of hydrocyanic acid from large charges of cyanide in barrels directly beneath the wires may cause the alarm to ring. This can be avoided by properly placing the barrels.

"If the cyanide is well wrapped, one person will have sufficient time to start the generators in all the rooms in an ordinary house before there is danger of being overcome by the gas. In larger houses, two or more persons can work together. A definite plan should be made and rehearsed in advance so that each man knows just what he has to do and when he is to do it. The work should be done rapidly but not recklessly or nervously. If a charge is overlooked, let it go; to go back is dangerous."

Also note the following:

- Wear rubber gloves and goggles when handling sulfuric acid and cyanide.
- Do not place a paper sack containing a charge of cyanide on the floor since it may contact some stray acid which may have been spilled.
- Wear mask at all times.

Protect the floor, furnishings, furniture, etc., by placing plenty of newspapers beneath and around the crock and over the nearby furnishings, etc. In order to prevent spattering, use a four-gallon crock for a three-pound charge, a three-gallon crock for a two-pound charge, etc.

A 12-hour exposure is preferred when using the pot method. If it is necessary to reduce the time for exposure, the amount of hydrocyanic gas generated per 1,000 cubic feet should be increased.

The residue in the containers may be dumped down street drains, flushed down toilets, or buried in a hole in the ground. Where the residue has crystallized and must be broken up, the operator should exercise care so as not to inhale any gas arising from the crock.

Cyanide generators. Some years ago, mechanical cyanide generators which had fumigating capacities up to 25,000 cubic feet/700 cubic meters were available. These generators could be hooked up in multiple units. Sodium cyanide eggs were used in this process. The generator method had the advantage of permitting the operator to remain outside the premises and did away with the large number of crocks needed.

To overcome some of the problems of hydrogen cyanide, Dr. H.W. Houghton (Houghton, 1974) developed a briquette which, when immersed in muriatic acid, released cyanogen chloride. This gas has penetrating and performance characteristics similar to HCN, but has other characteristics that are more desirable.

It is a lachrymator (tear gas) from the beginning of gas evolution, is almost the same weight as air, and fumigation can be completed within five hours. Its characteristics caused it to be widely used to fumigate mattresses and ships. Aeration is quick and simple, and the odor is detectable at very low concentrations. It is still being manufactured and used for the above and for hide and powderpost beetle work where tarping is impractical.

METHYL BROMIDE (CH_3Br)

Methyl bromide is a colorless, odorless, volatile liquid which at room temperature is a gas approximately 3.3 times as heavy as air. It may be obtained in one pound/0.45 kg cans, 24 to the case, or in 10-, 50-, 100-, and 200-pound cylinders.

According to Dow Chemical Company, the cylinders "are pressured with 60 pounds of air before leaving the factory. Thus, the gauge pressure on the full cylinder varies from 60 pounds per square inch at 40° F to approximately 120 pounds per square inch at 110° F. No air pressure is applied to the one-pound cans so their gauge pressure varies from zero pounds per square inch at 40° F to 48 pounds per square inch at 110° F. Cylinders are provided with fusible safety plugs, which will melt out at 165° F. Cans will distort at 150° F and burst at 185° to 190° F. Containers must never be subjected to temperatures as high as the above."

Methyl bromide is not used as widely as sulfuryl flouride in household fumigation since it produces persistent odors in certain materials. Searls et al. (1944) state that sponge rubber, white kid leather, certain iodized salts, and some woolens are affected in this manner. Silver polishing papers, rubber mattresses, and rubber pads are also reacted on by methyl bromide. Anon. (1977) has a list of materials that may react adversely.

Liquid methyl bromide should not be permitted to come in contact with painted or varnished materials since it acts as a solvent on these. Where methyl bromide fumigation is undertaken, flames should be extinguished, since the combustion of the methyl bromide results in products that are corrosive to metals.

Since methyl bromide is a very penetrating fumigant it has found wide use in mill, warehouse, boxcar, and similar type gas treatments.

Methyl bromide, like all the available fumigants, is toxic to man and unlike HCN is considered a cumulative poison. It is necessary to wear a gas mask during the fumigation, during aeration, and whenever it is essential to enter areas containing the gas. A black *canister* or *yellow all purpose canister* is used in the mask for methyl bromide. Smith (1952) and Cardiff (1953) review the recommended safety practices for fumigating buildings with methyl bromide. Thompson (1966) has prepared a review of the use of methyl bromide as a fumigant. Dow provides a variety of manuals on the use of methyl bromide.

According to Anon. (1943), at times liquid methyl bromide will cause burning upon contacting the skin. Some individuals are more sensitive than others to skin contact with this fumigant.

Methyl bromide gas can be detected with a halide gas detector which operates with an acetylene gas flame. According to Anon. (1950a), the detector "consists of an acetylene torch which heats a copper cone and an air tube through which the air to be tested is passed over the hot copper." Air containing methyl bromide will show a green to blue flame, the latter when high concentrations of the gas are present.

Latta (1953), Monro et al. (1953), Phillips (1957), Hassler (1961) and Dumas and Monro (1966) discuss the use of thermal conductivity gas analyzers. These instruments accurately measure the concentrations of methyl bromide in air.

Monro and Delisle (1943) note insects are often alive and active immediately after exposure to methyl bromide, but die in time. "An interesting phenomenon is the large percentage of survivors living some time after exposure, but which subsequently died after failing to progress to the next stage of their life history. This effect of methyl bromide has been noticed especially with lepidopterous larvae. Many survivors manifest nervous and uncoordinated movements which may persist for many weeks before they finally succumb."

Warehouse fumigation with methyl bromide. Because of its remarkable penetrative powers, methyl bromide finds wide use in warehouse fumigation. It is believed this penetrating power is due to its slight solubility in water and the fact it is barely absorbed by most other materials. Monro (1947), in discussing the fumigation of plant products in steel barges and the holds of ships, notes methyl bromide penetrated for a distance of 10 to 20 feet/three to six m through solid piles of peanut bags.

The warehouse is sealed in the manner discussed previously. However, since methyl bromide is so penetrative, it is quickly lost through leaks. Uncoated paper, stuffed burlap bags, and similar materials used in HCN fumigation will not retain methyl bromide sufficiently long to result in effective fumigation. Masking tape should be used. If paper is used as a sealing material, it should be coated with oil.

Methyl bromide in one pound/0.45 kg cans is under pressure at any temperature above 40° F. For this reason it must be opened with a special puncturing instrument. The Dow Chemical Company discusses this as follows: "When the can is inserted into the steel band of the applicator and the lever pulled, the can is punctured and the hole automatically gasketed, so that the liquid flows through tubing into the fumigation chamber, box car, or other space. The application from outside the structure provides maximum safety and frequently eliminates the necessity of a gas mask during the application. A gas mask should, however, be available for emergency use.

"So that the pressure within the can will cause the liquid to flow through the tubing, the point of puncture should be the lowest point on the can, but never on the side seams. If the can is punctured high, methyl bromide vapors rather than liquid flow out causing the contents to become cooled and reducing the pressure to a point where it may be quite difficult to get the liquid out of the can."

Saran or copper tubing may be fitted to the applicator for use in box car and tarpaulin fumigations. Moreover, the applicator can be modified for use in fumigation chambers.

Should the applicator fail or break, the operator should stay away from the can until the gas has dissipated and been carried away. When empty cans are removed "the can should be tilted at an angle so that no liquid can run out through the puncture and contact the hands. An occasional drop or two may still remain in the can. After removal, the can should be thrown a few feet to one side." The can may also be cooled with ice and salt or dry ice to temperatures below its boiling point (40.1° F/4.5° C). The can is then held firmly on a level surface and opened with a beer can opener and the contents are poured into evaporating pans. All precautions must be observed. Gas masks should be worn, skin contact should be avoided, and clothing wetted with the liquid methyl bromide should be removed immediately. The Dow Chemical Company recommends the use of fans where such use is applicable. "Because of its rapid rate of evaporation, methyl bromide gas leaves the cylinder at a very low temperature, and like cold air stratifies near the floor. This heavy concentration in the lower portion of the fumigation area results in an overdose at the base of the load and none at the top. To overcome such tendencies, methyl bromide should always be released near the top of the load or near the ceiling of the fumigation chamber and (except for fumigations under tarpaulin or in a box car) kept from stratifying with electric fans."

Hassler (1955a) discusses the use of heat exchangers in methyl bromide fumigation. The steam-operated heat exchangers hasten the formation of the gas, increase penetration, and completely vaporize the gas.

Aerating the building. The Dow Chemical Company recommends the following procedure: "At the end of the exposure period *most buildings will contain a considerable amount of gas.* Proper planning will allow the operators to vent the building with a minimum exposure to the gas. The amount of gas remaining in the building at this time will depend on several factors such as original dosage, type of building, type of sealing, length of exposure, and weather conditions during exposure. A quick check with the gas detector will indicate the approximate concentration of gas remaining in the building, which would play an important part in the plans for aeration.

"The important thing is to get a few windows or doors open on opposite sides of the building on each floor to allow for cross ventilation. The same high diffusion rate that gives methyl bromide extra power to penetrate to the center of large masses of stock, makes possible its quick and easy removal from the building. Start the aeration by opening all the doors on the first floor from the outside. In order to reduce the hazard to operators, allow the building to stand in this condition until the gas detector fails to show a blue color. The circulating fans should be started as soon as possible and allowed to run during the aeration period. *Do not try to open all the windows on each floor the first time through.* Two men equipped with gas masks and *fresh* canisters should work together during the aeration as well as in releasing the gas. The operators should take rest periods in the open air free from any gas after opening each floor. After a few windows have been opened on each floor, the building should be allowed to aerate for one hour or longer. At the end of this aeration period, the gas detector will indicate which floors require additional ventilation."

For stored product insects such as *Sitophilus granarius* (L.) and the Angoumois grain moth, *Sitotraga cerealella* (Oliv.), a dosage of one pound/0.45 kg per 1000 cubic feet/28 cubic meters for 1.5 hours or more is satisfactory. Hassler (1954) recommends methyl bromide be used at the rate of two lbs/1000 cu. ft. for two hours; one lb./1000 cu ft. for five hours, and five ozs. for 24 hours.

Mackie (1938) states, "In experimental fumigation of cheese in an air-tight room, in which an electric fan was kept running, a complete kill of mites and their eggs was obtained in 24 hours with the mixture of methyl bromide and carbon dioxide at rates varying from eight to 20 pounds per 1000 cubic feet and at temperatures of from 58° to 63° F." A slight taint to the cheese in flavor and odor disappeared after two days. Searls et al. (1944) studied methyl bromide as a fumigant in dairy factories. They found this fumigant in no way affected the dairy products or dairy equipment.

Monro and Upitis (1956), through selection in the laboratory, developed a strain of *Sitophilus granarius* (L.) resistant to methyl bromide.

"Picride" consists of 80 percent methyl bromide and 20 percent chloropicrin by weight. It is used in the fumigation of buildings that are "too loosely constructed to hold methyl bromide long enough for a successful fumigation." This mixture is applied at the rate of one to two pounds per 1000 cubic feet/0.45 to 0.9 kg per 28 cubic meters.

CHLOROPICRIN (CCl$_3$NO$_2$)

This fumigant was used as a tear gas in World War I. It is a slightly yellowish liquid with a liquid density of 1.65, which weighs 13.75 pounds/6.24 kg per gallon/3.8 l. Its vapor is 5.7 times heavier than air. At room temperatures, chloropicrin is non-explosive and non-inflammable, it vaporizes slowly and has a decided lachrymatory effect.

Chloropicrin is preferred to hydrocyanic acid by some household fumigators principally because human beings cannot enter a room when this gas is present even in small quantities. It is used in flour mills and against stored product insects, since it penetrates commodities well. In the home, it has the disadvantage of persisting on fabrics and furnishings for some times.

Sherrard (1939) used chloropicrin discoids in ship fumigation to bring "stowaways" into the open some 30 minutes prior to the introduction of hydrocyanic acid gas.

In household fumigation, the house is prepared in a manner similar to that for hydrocyanic acid gas. The gas should not be used at temperatures below 60° F/15.6° C. The operators who work in pairs are equipped with a mask provided with a canister protecting them from chloropicrin. The easiest way to apply the liquid fumigant is with a sprinkler bottle made from a 28-ounce ginger ale bottle equipped with a sprinkler top stopper.

The chloropicrin is poured upon a burlap sack crumpled in shallow 12 inch/ 30 cm square pans about one inch/25 mm high. About ½ pint/0.24 l of chloropicrin is used on each sack. The sacks and accompanying pans should be well distributed throughout the house and placed in closets and similar places. About ½ to one pound/0.23 to 0.45 kg of the material is used in each pan. The time of exposure is 24 hours. If it is desired to hasten the volatilization of the gas, it may be necessary to spray the chloropicrin. Sprinkling cans and compressed air sprayers may be used for this purpose. Care should be exercised so the liquid does not fall on polished metal, shellac, varnish, stain, or painted woodwork for all these may be damaged.

Cotton (1941) states that where chloropicrin has splashed upon the skin it should be washed off with alcoholic di-sodium sulfite. Eyes affected by the gas are bathed with boric acid or a two percent solution of sodium bicarbonate.

The manufacturer recommends the following dosages: "For bedbugs, clothes moths, roaches, use at rate of one pound per 1000 cubic feet; for carpet beetles use at rate of 1.25 pounds per 1000 cubic feet, with an exposure of 24 hours at 60° F or higher. The chloropicrin dosage for rodents is .25 pounds per 1000 cubic feet. Against insects in a tight warehouse, apply one pound per 1000 cubic feet, and in vaults with sacked material 1.5 to three pounds per 1000 cubic feet." It should be noted that one pint/0.47 l of the material weighs 1.75 pounds/0.8 kg, and thus will fumigate under standard conditions 1750 cubic feet/49 cubic meters. Where the material is partially sprayed, the exposure can be shortened to six or eight hours. Chloropicrin and carbon tetrachloride mixtures have been marketed as fumigants. Larvabrome-20 is a mixture of 80 percent methyl bromide and 20 percent chloropicrin.

Under ordinary conditions, the life of a chloropicrin canister is at least 10 to 20 operating hours, and replacements should be available after 10 hours usage. Chloropicrin is shipped in steel cylinders of 25, 50, 100, and 180 pounds. It is also sold in one-pound glass bottles packed in individual metal cases, 12 bottles to a case.

There should be at least 15 hours ventilation after the fumigation, and the area treated should be warmed to facilitate the removal of the gas. Electric fans used to blow the gas out of the windows are quite useful in ventilation.

Grayson (1948) showed chloropicrin applied for the fumigation of peanuts at a dosage of one to three pounds/0.45 to 1.36 kg per 1000 cubic feet/28 cubic meters did not harm germination as long as the moisture content of peanuts was not more than 10 percent.

ETHYLENE OXIDE (C₂H₄O)

This fumigant is a gas at ordinary temperatures and is a colorless liquid below 50° F/10° C. It is not nearly as toxic to human beings as are some of our more common fumigants. Ethylene oxide is stored and transported in steel cylinders. The ordinary dosage is three pounds/1.36 kg per 1000 cubic feet/28 cubic meters. Ethylene oxide is effective at low temperatures and is toxic to insect eggs. It is used in combination with carbon dioxide because of its inflammable vapors. The ethylene oxide (one part) is passed over crushed dry ice (carbon dioxide-nine parts).

Carboxide is a commercial product that comes in cylinders under pressure and consists of nine parts carbon dioxide and one part ethylene oxide. The carbon dioxide renders the mixture non-inflammable, causes the insect to respire more rapidly, thereby enabling the fumigant to enter it more readily, and reduces the absorption of the toxic gas by the commodities being fumigated. This gas is popular in vacuum fumigation and in atmospheric vault fumigation. In the latter case, the dosage varies from 10 to 30 pounds per 1000 cubic feet/4.5 to 13.6 kg per 28 cubic meters for 12 to 48 hours.

Cotton and Roark (1928) used ethylene oxide against household insects at the rate of two pounds/0.9 kg per 1000 cubic feet/28 cubic meters with an exposure of 24 hours. They found ethylene oxide to be effective at low temperatures of 60° to 70° F/15° to 21° C. It was found capable of killing insects buried in overstuffed furniture, in sealed packets of cereals, etc. Shepard and Lindgren (1932) used *Carboxide* for the control of carpet beetles, *Anthrenus sp.,* and obtained satisfactory control. Brown (1933) used *Carboxide* against bedbugs and cockroaches at the rate of three pounds/1.36 kg and four pounds/1.81 kg respectively, per 1000 cubic feet/28 cubic meters. This mixture has been used to kill insects infesting valuable books and papers. Fulton et al. (1963) conducted tests on insects with this fumigant.

SULFUR DIOXIDE (SO₂)

The burning of sulfur produces sulfur dioxide and a small amount of sulfur trioxide. At times, it may result in the deposition of a film of elemental sulfur on smooth surfaces. The gas is toxic to insects and rapid in action. It is not as effective as most of the fumigants; it is non-inflammable, slightly heavier than air, and very irritating to animals. Liquefied sulfur dioxide in steel cylinders is a clean, but somewhat expensive method of fumigation.

The primary advantages of fumigating with sulfur dioxide by burning sulfur are it is economical and comparatively safe. Its disadvantages are its tendency to tarnish and corrode metals, to bleach and rot fabrics, and poor penetrability. In regard to the rotting of fabrics, the damage is most severe when the humidity is high. The injury often does not show up until the fabric is washed, when it simply falls to pieces. It is used at the rate of four to eight pounds per 1000 cubic feet/1.8 to 3.6 kg per 28 cubic meters.

Hockenyos (1940) gives the following procedure for the burning of sulfur in sulfur dioxide fumigation: "The mechanics of burning sulfur are simple and the chief precautions are to avoid fire hazards and see that the sulfur burns out completely. Prepared sulfur candles may be used, but it is cheaper to burn powdered sulfur. The sulfur is usually placed in a metal pan such as a stew pan, which in turn is placed in a larger pan such as a dish pan, and this dish pan is in turn set on some support such as sticks or bricks to keep the heat off the floor. Sulfur melts before it burns, and at times specks of burning sulfur may spatter out of the small pan. It is sometimes recommended that water be placed in the larger pan to quench any spattered burning sulfur. This practice is, however, not only unnecessary, but undesirable in that water cools the inner pan and may result in the sulfur not being burned out completely. The sulfur should never be heaped up in the pan so that it might overflow when melted, and it is not advisable in any case to place more than three pounds of sulfur in any one container.

"The burning of the sulfur may be started by pouring on a few ounces of alcohol and touching a match. Be sure the alcohol is straight 95 percent and has not been diluted with water. Some operators line the small pan with a double thickness of cloth before placing the sulfur therein, and others place some inert filler such as sand or kieselguhr in the space between the smaller and larger pans. The purpose of both of these procedures being to help hold the heat into the inner pan of sulfur and thus assure complete burning." Swisher (1944) notes that sulfur dioxide-acetone mixtures were effective against bedbugs at a dosage of one pound per 1000 cubic feet/0.45 kg per 28 cubic meters.

SULFURYL FLUORIDE (SO$_2$F$_2$)

Kenaga (1957) describes sulfuryl fluoride as a non-inflammable, colorless, odorless compound which boils at $-55.2°$ C at 760 mm pressure. The high vapor pressure of this fumigant favors gaseous penetration of commodities and structural materials. Sulfuryl fluoride is heavier than air in the gaseous state. Many of the materials injured by methyl bromide are not affected by sulfuryl fluoride.

Stewart (1957) showed this fumigant to be useful for the control of the drywood termite, *Incisitermes minor* (Hagen). Stewart and Meikle (1964) discuss aeration after fumigation with this gas. Bartlett (1967) used this fumigant in large scale tent fumigation.

Kenaga (1957) shows that sulfuryl fluoride is very effective against many adult and larval forms of insects, but not as effective against eggs. This is not important when fumigating for the control of social insects such as drywood termites, as the eggs will not develop without care by nymphs and workers. However, this is not true when fumigating for powderpost beetles and old house borers. Dow (1978) specifies that four times the concentration used for termites be used for old house borer and 10 times the standard dosage be used for powderpost beetles. These high concentrations are necessary to assure kill of eggs that exist.

Because sulfuryl fluoride is expensive, especially when used at high concentrations to control beetle eggs, Stewart (1962 and 1966) devised calculators to determine dosage needed and half-loss times to minimize the amount of fumigant needed for a particular situation. Using these calculators with a gas analyzer can reduce costs and assure complete kill.

Unlike HCN, sulfuryl fluoride readily penetrates frame construction. Thus, successful fumigation using this gas usually requires tarping the structure.

Detection of sulfuryl fluoride is difficult at low concentrations. Detector tubes are available for sulfur dioxide. Sulfuryl fluoride must be passed through a furnace, which breaks it down and sulfur dioxide is formed. The sulfur dioxide is then measured with the tubes.

HYDROGEN PHOSPHIDE (H₃P)

Lindgren et al. (1958) studied this fumigant which has been used in Europe for a number of years. According to these authors, the "proprietary compound is manufactured as a high compressed tablet composed of ammonium carbamate and aluminum phosphide, which on exposure to moisture decomposes to hydrogen phosphide, aluminum hydroxide, ammonia, and carbon dioxide. This decomposition is slow and dependent on the moisture content of the atmosphere and on the temperature. Each tablet weighs three grams and upon decomposition produces one gram of hydrogen phosphide." This fumigant has been shown to be effective against a wide range of stored product pests. Rutledge (1967) discusses the use of this fumigant in some detail. A variety of pellets and tablets are available, each being developed for specific uses.

Hydrogen phosphide penetrates very well and leaves little or no residue on foodstuffs. It is widely used in the grain industry. Phosphine is released slowly, which permits the placement of tablets or pellets without respiratory safety equipment. This "advantage" is also a disadvantage in that fumigation times range from 48 to 96 hours, depending upon temperature. This long period makes the compound impractical for some types of fumigation work.

It is not used extensively for space fumigations because it sometimes reacts with and corrodes copper and zinc and can render motors, telephones, electrical wiring, and plumbing inoperable.

Hydrogen phosphide has a garlic odor which is detectable at concentrations toxic to man. Very low concentrations can be confirmed using the proper detector tubes.

The manufacturers have comprehensive manuals and instructions for the use of their products. These should be consulted for details on the specific use of their products.

Newer formulations may eliminate the long time periods currently required, but the safety benefits of the current products may disappear if the gas is generated more quickly.

ETHYLENE DICHLORIDE (C₂H₄Cl₂)

Ethylene dichloride is used usually in combination with carbon tetrachloride for the control of such household pests as moths in closets and grain insects in cupboards. It is a colorless liquid with an odor similar to chloroform. The gas is approximately three times as heavy as air. Three volumes of ethylene dichloride are usually combined with one volume of carbon tetrachloride to make a non-inflammable mixture. Ethylene dichloride is one of the gases that is less toxic to man than HCN. It is slow to vaporize and often requires 24 to 72 hours exposure to kill insects. This gas damages paints and varnishes. Fatty foods retain the gas for long periods, and it leaves an unpleasant flavor in tobacco.

Ethylene dichloride comes in cans and is used at the rate of 10 to 18 pounds per 1,000 cubic feet/4.5 to 8.2 kg per 28 cubic meters for a period of 24 hours. Since it is heavier than air, the liquid is placed in pans on the uppermost shelf of the space to be fumigated. In bin fumigation, it is used at the rate of 2.5 gallons/9.5 l per 1,000 bushels/35,000 l.

Cotton and Roark (1927) found the mixture of ethylene dichloride and carbon tetrachloride killed *Tineola bisselliella* (Hum.), *Anthrenus vorax* (Waterh.), *Attagenus piceus* (Oliv.) at a concentration of six pounds per 1000 cubic feet/2.7kg per 28 cubic meters at 85° F/29° C for 24 hours exposure. At 65° F/18° C one must use 12 pounds/5.4 kg of the mixture. They also state, "A dosage of 14 pounds to 1,000 cubic feet at 80° F, which is the strength recommended for general fumigation, gave a perfect kill in special tests against *Tribolium confusum* (Duval), *Sitophilus oryza* (Linn.), sealed in cartons of cereals, *Plodia interpunctella* (Hubner) and *Silvanus surinamensis* (Linn.) buried in boxes of sweets. The fumigant is cheap, not injurious to furniture or fabrics, is simple to use, and not dangerous unless inhaled in high concentrations for a long period."

Richardson and Casanges (1942) note that ethylene dichloride has a delayed killing action and the confused flour beetle could lay viable eggs before death. They also state that carbon tetrachloride in the mixture added little or nothing to the toxicity.

CARBON DISULFIDE (CS₂)

Carbon disulfide or carbon bisulfide is a colorless volatile liquid with a specific gravity of 1.26, which forms a heavy vapor that is very toxic to insects. Unfortunately, it forms explosive mixtures with air that have a tendency to ignite spontaneously, and for this reason other fumigants are preferred. Moreover, commercial carbon disulfide usually contains some hydrogen sulfide, which gives the gas a very unpleasant odor. It is rarely used in the home, but does have a place in killing certain soil inhabiting insects, such as ants, where the material may be poured into the nest, as well as for the fumigation of grain in bins. According to Winburn (1952), CS₂ is one of the most effective grain fumigants from a cost and efficiency viewpoint.

There are also mixtures of carbon disulfide with either carbon tetrachloride or sulfur dioxide on the market that reduce the inflammability of the vapors of carbon disulfide, making this gas safer to use. The vapors of carbon disulfide are poisonous if breathed for extended periods, and use of the gas often results in slight giddiness, which quickly disappear if the affected individual leaves the vicinity.

ETHYLENE DIBROMIDE (C₂H₄Br₂)

Ethylene dibromide is a colorless liquid with a specific gravity of 2.17. It is claimed this material gives no odor or taste to the grain, is noninflammable, nonexplosive, results in good penetration, and is toxic to grain insects. A five percent by volume mixture of ethylene dibromide in a chlorinated solvent is used in treating the surface layer of stored grain. The above mixture is used at the rate of two gallons per 1,000 bushels of grain in tightly constructed bins. For loosely constructed bins, three to four gallons of the mixture are applied. Another mixture containing 70 percent by volume ethylene dibromide is used as a local mill machinery or spot fumigant. Hassler (1955) discusses ethylene dibromide (EDB) as a soil fumigant for the control of subterranean termites.

CARBON TETRACHLORIDE (CCl₄)

Carbon tetrachloride is a colorless, volatile, non-flammable, heavy liquid which is an excellent solvent. It has a specific gravity of 1.59 at 70° F/21° C, weighs 13.28 pounds/six kg per gallon/3.78 l and boils at 170° F/77° C.

Ordinarily, it is not used alone, but combined with other volatile liquids as ethylene dichloride to decrease the inflammability of the latter. Jefferson (1943) states when carbon tetrachloride is combined with methyl bromide, methyl formate, or ethylene dichloride, a slight decrease in toxicity of the mixture is indicated when used against such test insects as *Tribolium castaneum* (Hbst.). At times, carbon tetrachloride has been used in grain fumigation and as a substitute for sodium cyanide in insect-killing cyanide bottles. American roaches, as well as other species of roaches, are killed quickly when wetted with carbon tetrachloride. Due to its toxicity to man, both in the liquid state and as a fumigant, it is rarely used for the control of household pests.

Richardson (1946) recommends the use of two gallons/7.6 l of CCl_4 per 1,000 bushels/35,000 l of corn for 24 hours at temperatures of 70° F/21° C or higher, for the control of several species of grain insects.

ETHYL FORMATE ($C_3H_6O_2$)

Ethyl formate is used for the fumigation of individual packages of dried fruit. Simmons and Fisher (1945) discuss its use in the dried fruit industry. It is a colorless liquid which weighs approximately 7.6 pounds per gallon/3.4 kg per 3.78 l at 68° F/20° C and has a boiling point of 129.4° F/54° C. At concentrations of 2.5 percent and above it is explosive. Like most fumigants, care must be taken not to inhale it. Simmons and Fisher describe its use as follows: "The process, which is in sequence with the other operations of the packing line, consists in pumping a small quanity of the fumigant into the package just before the dried fruit is put in. The package is sealed immediately thereafter, and the resulting exposure to the gas kills any insects or their eggs that may have escaped removal during cleaning and washing operations."

The dosage is four milliliters in hot weather and seven milliliters in cold weather for a 25 pound/11 kg box of raisins. Mayer and Nelson (1955) discuss ethyl formate and 12 other fumigants in the fumigation of dry beans and cowpeas on the packaging line.

Isopropyl formate is very similar in its physical characteristics and fumigant action to ethyl formate. It is used also for dried fruit package fumigation. Frisselle (1948) notes that although this fumigant is much less toxic than some of the common fumigants, "the rate at which the fumigant leaves the package is very slow, resulting in a long-term fumigation which is adequate to destroy insect life in properly designed packages."

ACRYLONITRILE (C_3H_3N)

Acrylonitrile is a colorless liquid with a specific gravity of 0.801 at 77° F/25° C. Since the gas at high concentrations is inflammable, it is usually mixed in equal volumes with carbon tetrachloride. Cotton and Young (1943) found it to be effective in the laboratory against the rice weevil and the confused flour beetles. It was used in the spot treatment of elevator boots, spouts, and other mill units. Bare and Tenhet (1950), in fumigation against insects in cigarette tobaccos, recommend a 1:1 mixture of acrylonitrile-carbon tetrachloride in atmospheric fumigation chambers at a dosage of 20 ounces per 1,000 cubic feet/0.56 kg per 28 cubic meters for 72 hours. In warehouses, the dosage is 40 ounces per 1,000 cubic feet/1.13 kg per 28 cubic meters for 72 hours. Tenhet (1955) used a 34-66 mixture of acrylonitrile and CCl_4 at the rate of four lbs. per 1,000 cu ft./1.8 kg per 28 cubic meters for the fumigation of cigars.

Trichloracetonitrile (CCl_3CN) is a yellowish liquid with a specific gravity of 1.44 at 77° F/25° C. Cotton and Young (1943) note: "It is nonflammable and is highly corrosive to iron and steel. It is relatively toxic to human beings, but the vapors in low concentrations produce great irritation to the mucous membranes of the nose, throat, and eyes so that anyone entering an appreciable concentration is forced to leave." Bovingdon and Coyne (1944) note trichloracetonitrile compares favorable with ethylene oxide for fumigating wheat. The toxicity was increased when 10 percent or more CO_2 was added.

PROPYLENE DICHLORIDE ($C_3H_6Cl_2$)

This gas is used in mixtures with carbon tetrachloride. According to Hutson (1933), mixtures of this material are about as efficient as carbon disulfide for fumigation against grain-infesting insects, without the fire hazard attending the use of the former material. Cotton (1941), on the other hand, states it has never proved successful as a grain fumigant.

CARBON DIOXIDE (CO_2)

This fumigant is a gas at room temperature and comes in cylinders under pressure or is sold as "dry ice." It is mixed with other fumigants since one of the properties of carbon dioxide is to increase the respiratory rate of many insects. Moreover, it reduces the fire hazard when used with the more flammable fumigants. Dry ice has been used as a fumigant against rats in large refrigerators.

NAPHTHALENE ($C_{10}H_8$)

Naphthalene is a white crystalline material, the common "moth balls" or "flakes" of commerce. It vaporizes slowly with a pungent odor. It must be used at the rate of one pound/0.45 kg in 10 to 100 cubic feet/0.28 to 2.8 cubic meters for pests of household fabrics and insect collections. For detailed information of its use against fabric pests see the chapter on clothes moths.

PARADICHLOROBENZENE ($C_6H_4Cl_2$)

This material is a white crystalline substance which vaporizes slowly, but more rapidly than naphthalene to form gas with an ether-like odor. It is used at the rate of one pound/0.45 kg in 10 to 100 cubic feet/0.28 to 2.8 cubic meters. It acts as a repellent against fabric and museum pests and is effective against moths only in tightly closed chests or closets as in naphthalene. This chemical is also discussed in greater detail in the chapter on clothes moths.

VAULT FUMIGATION AT ATMOSPHERIC PRESSURE

Furniture and stored product warehouses, furriers, etc., may have vaults made especially for the fumigation of various commodities. The best vaults are constructed of metal, although they may also be made of concrete, brick, etc. Plywood vaults with metal lining are quite common. The vaults are equipped with air-tight doors and a ventilating system, as well as with a circulation fan. In some instances the vault may contain an electrical heating unit so that the temperature is high during the course of the fumigation. The fumigant is piped into the vault by compressed air, or in the case of the heavier than air fumigants, they reach an evaporating pan by means of gravity. Usually 24 hours are required for fumigation in atmospheric vaults. Vault design and construction are discussed by Hill (1964) and Monro (1969).

The Dow Chemical Company, speaking of vault fumigation with methyl bromide, notes that the "dosage for chamber fumigation will depend upon many factors, such as temperature, exposure period, type of commodity to be treated, and others. For example, as the temperature decreases, the dosage must be increased; as exposure period is increased, the dosage may be decreased. More finely divided commodities require slightly higher dosages than is the case with those of larger particle size.

"For bagged grain, rice, or other commodities of similar particle size, any one of the following dosage and exposure period combinations may be used, provided the temperature is above 60° F:

- 3 lb. per 1,000 cu. ft. for 4 hr.
- 2 lb. per 1,000 cu. ft. for 6 hr.
- 1 lb. per 1,000 cu. ft. for 12 hr.

Milled grain products may have to be treated with 1.25 pounds methyl bromide per 1,000 cubic feet/0.57 kg per 28 cubic meters with an 18-hour exposure.

The requirements for vault fumigation with HCN, methyl bromide, and other gases vary with the fumigant. It is essential to know the dosages and other recommendations made by the manufacturers of these gases.

VACUUM FUMIGATION

In vacuum fumigation the commodities to be fumigated are placed in a gas-tight steel chamber. The air is exhausted and replaced by the fumigant. Vacuum fumigation is one of the most effective means of fumigation, since the exhaustion of the air aids in rapid penetration of the commodity, and once the oxygen is reduced, insects as a general rule are more susceptible to the effect of the fumigant. By this method the dosages and time for exposure can be lessened.

Monro (1941) notes that "an adequate industrial fumigation outfit should be capable of producing two inches or better of absolute pressure in not more than 10 minutes, and of maintaining the initial vacuum with but little loss during the fumigation period of two or three hours." Monro discusses vacuum fumigation as follows: "There are two methods of vacuum fumigation commonly practiced, the so-called 'sustained' and 'dissipated' vacuum treatments. In the sustained vacuum, after the removal of the air and the introduction of the fumigant into the vault, no further alteration is made to the pressure during the period of exposure, and atmospheric pressure is not restored within the vault until the time for the 'air washing' process. This is now conceded to be the most effective method for fumigating many commodities, but it cannot be applied at low pressures to those plants, fruits, vegetables, and other materials which are too delicate to withstand the reduced pressures.

"In the 'dissipated' vacuum, after the vault is evacuated, the fumigant is introduced either slightly before, or at the same time, that atmospheric air is allowed to flow into the chamber until normal pressure is again reached, or until a very low vacuum, sufficient to ensure closure of the door of the vault, is attained. In this way the pressure during the exposure period is approximately that of the atmosphere. In this technique, circulation of the fumigant is usually effected by some system of ducts and fans."

Due to the decreased exposure time and increased sorption of the fumigant by commodities undergoing vacuum fumigation, it is necessary to increase the fumigant dosage. Thus, whereas one pound methyl bromide per 1,000 cubic feet/ 0.45 kg per 28 cubic meters is usually a satisfactory dosage under atmospheric conditions, two to three pounds/0.9 to 1.5 kg of methyl bromide per 1,000 cubic feet/28 cubic meters is used with a sustained vacuum of 25 to 27 inches with an exposure of from 1.5 to several hours. Monro (1952 and 1956) considers vacuum fumigation in some detail.

Fig. 27-2. Vacuum fumigation chamber.

Monro makes this interesting statement concerning vacuum fumigation, which is undoubtedly appropriate to other methods of insect control: "The one serious drawback to the successful fumigation of foodstuffs, under vacuum or atmospheric conditions, is that the dead insects are left inside the treated commodities. Indeed, some consumers have expressed the opinion that they "would rather find healthy living insects than dead ones" inside their food. Peas infested with pea weevils, dead or alive, have altogether discouraged some people from eating pea soup."

TARPAULIN TREATMENT

Warehouse men often resort to the fumigation of bagged grain, boxed foodstuffs and similar materials, through the use of air-tight rubberized (neoprene) or plastic-treated tarpaulins. Dawson (1944) notes the plastic, ethyl cellulose, is not affected by methyl bromide vapors. Phillips and Nelson (1957) made a study of the permeability of plastic films and plastic and rubber-coated fabrics to methyl bromide. According to these workers, polyethylene and vinyl films and synthetic rubbers, and materials coated with them effectively retain methyl bromide. The tarpaulin edges are sealed with "sand snakes," which are elongated bags filled with sand, or with some of the bagged material itself.

The Dow Chemical Corporation discusses the procedure for this type of fumigation when using methyl bromide: "If the material to be fumigated must be unloaded from a railroad car or truck, it should be stacked in a square area to a height of five or six feet, allowing for complete tarpaulin coverage with the necessary margin of two feet on all sides. It is preferable that the fumigation be performed on a concrete floor, or other air-tight surface free of cracks. Where floors are not air-tight, cracks should be properly caulked, or the area first covered with a relatively gas-tight paper.

"After the material is stacked, four sacks should be centered upright on top of the pile to form the gas expansion dome. Copper or saran tubing is used to connect the cylinder or the jiffy can puncture to a point near the center of this dome. The tube should be placed so that the liquid fumigant does NOT come in contact with the tarpaulin, because the plastic coating on the cloth may be damaged or washed off if exposed to liquid methyl bromide.

"The floor should be swept clean surrounding the sacked material to be fumigated and the tarpaulin should be unfolded or pulled over the stack, taking particular care to leave the necessary two or three foot margin on the floor for sealing. The saran or copper tube through which the gas is injected runs out under the edge of the tarpaulin which should be folded at the corner, thus eliminating folds at the floor. The stack is then sealed by laying a row of bagged materials completely around it. "Canvas snakes," canvas tubes about four inches in diameter and filled with sand or other heavy material, may also be used."

The manufacturers' of methyl bromide recommend a dosage of 1.5 pounds methyl bromide per 1,000 cubic feet/0.68 kg per 28 cubic meters with an exposure of 12 to 18 hours at 60° F/15.6° C or higher, depending on how finely ground the fumigated material is.

RAILROAD CAR FUMIGATION

Many commodities are fumigated in the box car. For the most part, steel cars of tight construction are suitable for fumigation. These cars are sealed tightly with masking tape and are then treated with such gases as methyl bromide or hydrogen phosphide. The former is especially favored for box car fumigation. Anon. (1956a) recommends papers laminated with asphalt such as T475M50 of the Technical Association of the Pulp and Paper Industry be used to cover the car floor. Four mil polyethylene is also used.

The box cars are sealed with masking tape. A small hole is drilled through the bottom of the car opposite the door and then a saran tube with suitable apertures is pushed through this opening so it is near the roof of the car. The tube is connected to the fumigant cylinder and the valve is opened. Upon completion of the fumigation, the tube is withdrawn, and the hole in the car is plugged.

Where methyl bromide is used, the methyl bromide applicators with one pound methyl bromide cans can be used in place of the cylinders. Saran tubing leads from the cans into the cars. Monro and Delisle (1945) used several applicators to apply several pounds of methyl bromide at one time. The above authors note that saran plastic tubing becomes brittle when cooled to 32° F/0° C, as is often the case in methyl bromide fumigation. For this reason, it must be handled carefully.

Monro and Delisle (1945) recommend the use of 1.5 pounds methyl bromide per 1,000 cubic feet/0.68 kg per 28 cubic meters with an exposure period of 16 to 24 hours at temperatures of 60° F/15.6° C and above. Dean and Cotton (1943) used up to four pounds per 1,000 cubic feet (12 pounds per car) for fumigating cars loaded with flour. Anon. (1956b) gives the following dosages for 12 to 18 hours exposure at 60° F or higher:

- 10 lbs/4.5 kg in steel cars in good condition.
- 12 lbs/5.4 kg in steel cars in fair condition.

- 12 lbs/5.4 kg in wooden cars in good condition.
- 15 lbs/6.8 kg in wooden cars in fair condition.

In regard to ventilation, Monro and Delisle (1945) note when the fumigation is completed, "both doors were opened wide and the cars were not entered for at least two hours. Under most summer conditions, this was sufficient for the dissipation of the gas from the body of the car, and only mild reactions for methyl bromide of approximately 50 parts per million were recorded among the bags towards the end of the cars. As might be expected, weather conditions influenced the ventilation of the gas considerably and a number of different effects were observed, the more important of which are listed herewith:

1) On warm windless days the gas dissipated rapidly from all parts of the cars and most quickly in the absence of extensive cloud cover.

2) On very windy days, either cold or warm when a strong draught of air crossed the middle of the cars, the gas was sometimes "pocketed" at one or both ends of the cars.

3) In cool damp weather and during rain, the gas often took considerably more than six hours to dissipate completely from the free air space in the cars."

They conclude that there are no "hard and fast rules for natural ventilation."

SHIP FUMIGATION

Hydrocyanic acid gas and methyl bromide were the gases used most commonly in ship fumigation, but HCN currently is not readily available. HCN is applied either by means of discoids or from cylinders.

According to the HCN *Discoid Fumigation Manual,* the recommended dosage for rats and mice is two ounces HCN per 1,000 cubic feet/56 g per 28 cubic meters of space with an exposure of two to three hours. Where roaches, bedbugs, and other insects are concerned, the dosage is usually eight ounces/227 g HCN per 1,000 cubic feet, with an exposure period of at least eight hours. Monro et al. (1952) compared HCN and methyl bromide as fumigants for insect control in empty cargo ships and found them to be about equally effective, although HCN was more effective than methyl bromide against the cadelle. On the other hand, methyl bromide was more effective than HCN against the adult and pupa of the granary weevil.

Monro (1947 and 1951) studied the use of methyl bromide in the fumigation of ships in great detail. He discussed the dosage as follows: "At least one pound per thousand cubic feet of air space, for a period of ten hours, is required for satisfactory control of insects at temperatures above 60° F. Methyl bromide has shown promising results inside holds at temperatures as low as 30° F, if the amount is increased, and the exposure period is extended to 12 hours. As this gas is about 3.25 times as heavy as air, it is inclined to stratify toward the bottoms of the holds when first introduced. The use of circulating fans to improve distribution is strongly recommended. These fans are usually operated for 30-60 minutes from the beginning of the exposure. Once good distribution of the gas has been obtained, the fumigant will not settle again toward the bottom of the space." Since methyl bromide will readily escape from an unsealed area, special care must be taken to close all openings.

Monro (1951) recommends forcing air into the holds by means of blowers to aerate the hold.

LITERATURE

ADKISSON, P.L. — 1957. The relative susceptibility of the life history stages of the rice weevil to certain fumigants. J. Econ. Entomol. 50(6):761-764.

ANON. — 1931. Carbon tetrachloride, Roessler & Hasslacher Chem. Co., N.Y. 1935. Proxate fumigation handbook. Liquid Carbonic Corp., Chicago, Ill. 1938. Methyl Bromide, Dow Chemical Co., Midland, Mich. 1938. The Larvacide Log. Various issues, Innis, Speiden, & Co., New York, N.Y. 1938. Zyklon discoids fumigation manual. American Cyanamid & Chemical Corp. 1939. How to be safe and sure in home fumigation work and other fumigation jobs. Larvacide (Chloropicrin). 1940. Sealing materials. Pests 8(8):26. 1940. Domestic fumigation practices. Pests 8(2):21-22. 1940. Methyl Bromide. Neal A. Maclean Co., San Francisco, Calif. 1943. Methyl bromide fumigation. U.S. Bur. Ent. & P.Q. Circ. E-601. 1943. Tarpaulin fumigation for bagged and packaged foods with methyl bromide. The Dow Chemical Co. 1944. Fumigation manual: Aero Brand HCN Discoids. Amer. Cyanamid Co. 1945. Effect of hydrocyanic acid on foodstuffs. Pests 13(9):26. 1945. Insect and rodent control. War Dept. Tech. Manual TM 5-632. 1947. Fast, thorough, low cost fumigation of railroad cars with Dow methyl bromide. Dow Chem. Co. 1947. Methyl bromide applicators and supplementary equipment. Dow Chem. Co. 1947. Manual for construction and operation: Atmospheric fumigation chamber. Dow Chem. Co. 1947. Reference manual for mill and warehouse fumigations. Dow Chem. Co. 1948. Rodent Control Manual. U.S. Public Health Service. 1948. Dowfume EB-5 controls insects pests in stored grain. Dow Chem. Co. 1948. Reducing losses in farm-stored grain. Down to Earth (Dow Chemical Co.) 4(2):14, 1948. Kill farm and garden pests with Cyanogas. American Cyanamid Co. 1948. Kill rats with Cyanogas. American Cyanamid Co. 1949. Methyl bromide box car fumigation. Pest Control 17(11):11. Tarpaulin fumigation 17(12):3. 1950a. Precautions for methyl bromide fumigations. Dow Chem. Co. 1978. Dowfume EB-15. Dow Chem. Co. 1950. New fumigation method. Pest Control 18(10):54. 1956a. USDA and A.A.R. urge improvement procedures on fumigation of commodities in freight cars. USDA & Assoc. Amer. R.R. Mimeog. 22036.2, Feb. 15, 1956b. Use Dow methyl bromide for box car fumigation. Dow Chemical Co. Form No. 132-89-77. 1959. 1977. Phostoxin — new grain fumigant. Pest Control 27(3):38. Commodities unsuited for methyl bromide fumigation. Dow Chemical Co. Form No. 132-141-77. 1978. Vikane technical manual for structural fumigation. Dow Chemical Co. Form No. 132-136-78.

BACK, E.A. & R.T. COTTON — 1935. Industrial fumigation against insects. U.S.D.A. Circ. 369. 1932. Hydrocyanic acid gas as a fumigant for destroying household insects U.S.D.A. Farmers' Bull. 1670.

BACK, E.A., R.T. COTTON & G.W. ELLINGTON — 1930. Ethylene oxide as a fumigant for food and commodities. J. Econ. Entomol. 23:226-231.

BARE, C.O. & J.N. TENHET — 1950. Tests with acrylonitrile — carbon tetrachloride and hydrogen cyanide as fumigants for insects in cigarette tobaccos. U.S. Bur. Ent. & P.Q. E-794.

BARTLETT, A. — 1967. Tent fumigation of 4-acre building challenges Orkin. Pest Control 35(2):21-23.

BOVINGDON, H.H.S. & F.P. COYNE — 1944. Trichloracetonitrile as a fumigant. Ann. Appl. Biol. 31(3):255-259. Biol. Abst. 19:1264, 1945.

BROWN, E.W. — 1933. Carboxide gas: a new insecticidal fumigant for bedbugs and cockroaches. U.S. Nav. Med. Bull. 31:253-268. 1934. The efficiency of carboxide gas as an insecticidal fumigant for naval and merchant vessels. U.S. Nav. Med. Bull. 32:294-317.

CARDIFF, D.G. — 1953. Suggested standards for fumigation with methyl bromide. P.C.O. News 13(2):1, 7.

COTTON, R.T. — 1930. Carbon dioxide as an aid in the fumigation of certain highly absorptive commodities. J. Econ. Entomol. 23:231-233. 1932. The relation of respiratory metabolism of insects to their susceptibility to fumigants. J. Econ. Entomol. 25:1088-1102. 1941. Insect Pests of Stored Grain and Grain Products. Burgess Publishing Co.

COTTON, R.T., J.C. FRANKENFELD & G.A. DEAN — 1945. Controlling insects in flour mills, U.S.D.A. Cir. 720.

COTTON R.T., & H.E. GRAY — 1948. Preservation of grains and cereal products. Food & Agr. Org. of U.N. Agr. Studies No. 2. pp. 35-71.

COTTON, R.T. & R.C. ROARK. — 1927. Ethylene dichloride — carbon tetrachloride mixture; a new non-burnable, non-explosive fumigant. J. Econ. Entomol. 20:636-639. 1928. Ethylene oxide as a fumigant. Indus. and Engin. Chem. 20:805.

COTTON, R.T. & G.B. WAGNER — 1941. Control of insect pests of grain in elevator storage. U.S.D.A. Farmers' Bull. No. 1880.

COTTON, R.T. & H.H. WALKDEN — 1947. Fumigation of grains and other stored foods. Agr. Chem. 2(1):33-35.

COTTON, R.T., H.H. WALKDEN & R.B. SCHWITZGEBEL — 1944. The role of sorption in the fumigation of stored grain and milled cereal products. J. Kans. Entomol. Soc. 17(3):98-103.

COTTON, R.T. & H.D. YOUNG — 1943. Acrylonitrile and trichloroacetonitrile in admixture with carbon tetrachloride as possible fumigants for stored grain. J. Econ. Entomol. 36(1):116-117.

COTTON, R.T., H.D. YOUNG & G.B. WAGNER — 1936. Fumigation of flour mills with hydrocyanic acid gas. J. Econ. Entomol 29(3):514-23.

DAWSON, J.C. — 1944. Methyl bromide fumigation. Soap & S.C. 20:96-98. March.

DEAN, G.A. & R.T. COTTON — 1943. Fumigation with methyl bromide. Pests 11(9):14-16.

DUDLEY, H.C. — 1941. Studies on foodstuffs fumigated with methyl bromide. Pests 9(7):20.

DUMAS, T. & H.A.U. MONRO — 1966. Detector tubes for field determination of fumigant concentrations. Pest Control 34(7):20-23, 52.

FISK, F.W. & H.H. SHEPARD — 1938. Laboratory studies of methyl bromide as an insect fumigant. J. Econ. Entomol. 31(1):79-84.

FRISSELLE, P. — 1948. Line preservation. Down to Earth (Dow Chemical Co.) 4(2):15.

FULTON, R.A., F.F. SMITH & R.L. BUSBEY — 1962. Respiratory devices for protection against certain pesticides. USDA ARS-33-76.

FULTON, R.A., A.H. YEOMANS & W.N. SULLIVAN — 1963. Ethylene oxide as a fumigant against insects. J. Econ. Entomol. 56(6):906.

GERSDORFF, W.A. — 1932. Bibliography of ethylene dichloride. U.S.D.A. Misc. Publ. No. 17.

GLIDDEN, G.M. — 1964. How your gas mask canister works. Pest Control

31(1):9, 11-12, 14, 16.

GRAYSON, J.M. — 1948. Germination of fumigated peanuts. J. Econ Entomol. 41(5):816-817.

HAMBLIN, D.O. — 1946. Toxicology of insecticides and fungicides. Agr. Chemicals 1(6):28-31.

HASSLER, K. — 1947. Methyl bromide fumigation. Pest Control and Sanitation 2(7):16-17. 1951. On "Death of a PCO." Pest Control 19(2):6, 1954. Outline of fumigation. 3rd Cal. Poly Pest Control Conference. Dec. 1954, Calif. State Polytechnic College, San Dimas, Calif. 1955. Methyl bromide fumigation. Heat exchangers. Pest Control 23(8):6. 1955. Termite fumigation in California. Pest Control 23(2):14-16.

HASSLER, R.K. — 1961. How a measuring device makes for better methyl-bromide fumigation. Pest Control 29(7):12, 14.

HESELTINE, H.K. — 1959. The detection and estimation of low concentrations of methyl bromide in air. Pest Technology; July/August.

HINDS, W.E. — 1925. Carbon disulfide as an insecticide. U.S.D.A. Farmers' Bull. No. 799.

HOCKENYOS, G.L. — 1940. Sulfur dioxide fumigation. Pests 8(11):23-25.

HORSFALL, J.L. — 1938. Common problems arising in relation to household fumigation. Pests 6(12):10-13. 1940. Case histories as related to fumigation procedures. Pests 8(4):19-23. 1941. Fumigants, Pests 9(3):12-15.

HORSFALL, W.R. — 1934. Some effects of ethylene oxide on the various stages of the bean weevil and the confused flour beetle. J. Econ. Entomol. 27(2):405-409.

HOUGHTON, C.W. — 1974. Fumigants: a history of one. Pest Control Technology (4):30-33.

HOYLE, H.R. — 1961. Control of health hazards during application of grain fumigants. Pest Control 29(7):25-26, 28, 30.

HUBBEL, F.D. — 1941. Public relations — Fumigation Service. Domestic fumigation practices recommended by N.P.C.A. Pests 9(6):6-7.

HUTSON, R. — 1933. Propylene dichloride as a fumigating material. J. Econ. Entomol. 26:291.

JEFFERSON, R.N. — 1943. Influence of carbon tetrachloride on the toxic efficiency of certain volatile organic compounds. J. Econ. Entomol. 36(2):253-259.

JOHNSON, C.C. — Chloropicrin: Its widening commercial use, characteristics and advantages. Innis Speiden & Co.

JONES, R.M. — 1935. The toxicity of carbon dioxide-methyl formate mixtures to the confused flour beetle (*Tribolium confusum* Duv.) J. Econ. Entomol. 28:475-485. 1938. Toxicity of fumigant-CO_2 mixtures to the red flour beetle. J. Econ. Entomol. 31(2):298-309.

KENAGA, E.E. — 1957. Some biological, chemical and physical properties of sulfuryl fluoride as an insecticidal fumigant. J. Econ. Entomol. 50(1):1-6.

KITCHEL, R.L. & W.M. HOSKINS — 1935. Respiratory ventilation in the cockroach in air, in carbon dioxide and in nicotine atmospheres. J. Econ. Entomol. 28(6):924-925.

KNIGHT, K.L. — 1940. Fumigation of sacked grain with chloropicrin. J. Econ. Entomol. 33(3):536-539.

LANGHORST, H.J. — 1947. Hydrocyanic acid gas fumigation. Agr. Chem. 2(4):30-33, 69.

LATTA, R. — 1953. BE & PQ-perfected device measures methyl bromide concentrations. Pest Control 21(2):30.

LINDGREN, D.L., L.E. VINCENT & R.G. STRONG — 1958. Studies on hydrogen phosphide as a fumigant. J. Econ. Entomol. 51(6):900-903.

MACKIE, D.B. — 1938. Methyl bromide — its expectancy as a fumigant. J. Econ. Entomol 31:70-79.

MACKIE, D.B. & W.B. CARTER — 1937. Methyl bromide as a fumigant: a preliminary report. Calif. Dept. Agr. Bull 26:153-162.

MACLEAN, N.A. — 1950. Warehouse fumigation with methyl bromide. Dow Chem. Co. Down to Earth 5(4):17.

MARLATT, C.L. — 1906. Sulphur dioxide as an insecticide. U.S.D.A. Bur. Entomol. Bull. 60.

MAYER, E.L. & H.D. NELSON — 1955. Fumigation of dry beans and cowpeas on the packaging line. U.S. Agr. Mktg. Serv. AMS-4.

MONRO, H.A.U. — 1941. Vacuum fumigation for insect control. Sci. Agr. 22(3):170-177 (Canada). 1945. Low temperature fumigation. Can. Entomol. 77:192-196. Oct. 1947. Methyl bromide fumigation of plant products in steel barges and the holds of ships. Sci. Agr. 27(6)267-283 (Canada). 1951. Insect pests in cargo ships. Can. Dept. Agr. Publ. 855. 1952. Some aspects of research and development in fumigation in Europe and North Africa. 83rd Ann. Rpt. Entomol. Soc. Ontario. pp. 3-19. 1956. Theory and practice of vacuum fumigation. In 1951-1956 Report of Science Service Labortory, Univ. of Western Ontario, London, Ontario, pp. 29-31. 1960. Modern fumigants for the control of pests. Ann. Conf. Public Health Inspectors. Sept. 1960, Scarborough, England. 1969. Manual of Fumigation for Insect Control. Food & Agr. Org. of United Nations. 381 pp.

MONRO, H.A.U., C.T. BUCKLAND & J.E. KING — 1953. Preliminary observations on the use of the thermal conductivity method for the measurement of methyl bromide concentrations in ship fumigation. 84th Ann. Rpt. Entomol. Soc. Ontario. pp. 71-76.

MONRO, H.A.U., C.R. CUNNINGHAM & J.E. KING — 1952. Hydrogen cyanide and methyl bromide as fumigants for insect control in empty cargo ships. Sci. Agr. 32:241-265. May.

MONRO, H.A.U. & R. DELISLE — 1943. Further applications of methyl bromide as a fumigant. Sci. Agr. 23(9):546-556 (Canada). 1945. Methyl bromide fumigation of plant products in railroad freight cars with special reference to work supervised by the Dominion Department of Agriculture during 1944. Sci. Agr. 25(12):794-816. Aug. 1945. 1946. Methyl bromide fumigation of plant products in railroad freight cars. Pests 14(9):14, 16-17, 20-21.

MONRO, H.A.U. & E. UPITIS — 1956. Selection of populations of the granary weevil *Sitophilus granarius* L. more resistant to methyl bromide fumigation. Canad. Entomol. 88(1):37-40.

MOORE, W. — 1918. Fumigation with chloropicrin. J. Econ. Entomol. 11:357-362. 1932. Reactions of sulphuric acid on sodium cyanide. J. Econ. Entomol. 25:729-30.

PEARCE, S.J. — 1961. Types and colors of gas mask canisters to be used for respiratory protection against specific fumigants. Pest Control 29(7):28.

PHILLIPS, G.L. — 1957. Current use of thermal conductivity gas analyzers. Pest Control 25(7):18, 20, 22, 24, 26.

PHILLIPS, G.L. & H.D. NELSON — 1957. Permeability to methyl bromide of plastic films and plastic- and rubber-coated fabrics. J. Econ. Entomol. 50(4):452-454.

PIPER, W.R. & R.H. DAVIDSON — 1938. Methyl bromide vapor against five species of stored product insects. J. Econ. Entomol. 31(3):460-61.

QUAYLE, H.G. — 1928. Fumigation with calcium cyanide dust. Calif. Agr. Exp. Sta. Hilgardia 3:207-32.

REED, W.D., E.M. LIVINGSTON & A.W. MORRILL, JR. — 1934. The fumigation of tobacco warehouses. E-325. USDA, B.E.&P.Q.

RESSLER, I.L. — 1938. Hazards in relation to household fumigation. Pests 6(12):14-16.

RICHARDSON, C.H. — 1946. Efficiency of carbon tetrachloride, ethylene dichloride and certain other fumigants in shelled corn. J. Econ. Entomol. 39(5):598-607.

RICHARDSON, H.H. & A.H. CASANGES — 1942. Toxicity of acrylonitrile, chloroacetonitrile, ethylene dichloride and other fumigants to the confused flour beetle. J. Econ. Entomol. 35(5):665-668.

ROWE, V.K. — 1957. Toxicology hazards and properties of commonly-used grain fumigants. Pest Control 25(9):18, 20, 22, 24, 26-27.

RUTLEDGE, J.H. — 1967. Potential uses of phosphine by pest control operators. Pest Control 35(7):11-14.

SCHUENEMAN, J.J. — 1951. Case History: death of a PCO. Pest control 19(1):24.

SEARLS, E.M., F.W. FLETCHER & E.E. KENAGA — 1944. Methyl bromide as a fumigant for dairy factories. J. Econ. Entomol. 37(6):822-829.

SHEEHAN, J.M. — 1961. Prevention of fire or explosion in storage and use of fumigants. Pest Control 29(7):34, 36, 38.

SHEPARD, H.H. & A.W. BUZICKY — 1939. Further studies of methyl bromide as an insect fumigant. J. Econ. Entomol 32(6):584-589.

SHEPARD, H.H. & D.L. LINDGREN — 1932. Ethylene oxide — liquid carbon dioxide mixture in house fumigation. J. Econ. Entomol 25(1):138-139.

SHEPARD, H.H., D.L. LINDGREN & E.L. THOMAS — 1937. The relative toxicity of insect fumigants. Minn. Agr. Exp. Sta. Tech. Bull. 120.

SHERRARD, G.C. — 1939. Chloropicrin as a prewarning gas in ship fumigation. U.S.P.H.S. Rpts. 54(42):2297-2302. 1942. Five fumigants for disinfestation of bedding and clothing: a comparative study of insecticidal properties. Pests 10(9):14-18. 1945. A report of damage to fabric by liquid hydrocyanic acid gas in fumigation. U.S.P.H.S. Rpts. 60(44):1308-1309. 1946. A report of damage to fabric by liquid hydrocyanic acid gas in fumigation. Pests 14(2):30.

SIMMONS, P. & C.K. FISHER — 1945. Ethyl formate and isopropyl formate as fumigants for packages of dried fruits. J. Econ. Entomol. 38(6):715-716.

SMITH, H.C. — 1952. ABC's of safety in fumigating a house with methyl bromide. Pest Control 20(8):20, 28.

SMITH, ROGER C. — 1926. House fumigation with calcium cyanide. J. Econ. Entomol. 19:65-77.

STADHOLZ, B. & G.W. JARMAN, JR. — 1946. Fumigation and reconditioning process for damaged foodstuffs. Pests 14(5):42.

STEWART, D. — 1957. Sulfuryl fluoride — a new fumigant for control of the drywood termite *Kalotermes minor* Hagen. J. Econ. Entomol. 50(1):7-11. 1962. Precision fumigation for drywood termites with Vikane. Pest Control 50(2). 1966. Balanced fumigation for better termite control. Down to Earth 22(2):8-10.

STEWART, D. & R.W. MEIKLE — 1964. Post-fumigation aeration of Vikane. Down to Earth 20(2):2, 14-16.

STRAND, A.L. — 1926. Preliminary experiments on the use of chloropicrin as an insect fumigant in flour and cereal mills. J. Econ. Entomol. 19:504-510.

SULLIVAN, W.N., E.R. McGOVRAN & L.D. GOODHUE — 1942. Fumigation action of a mixture of orthodichlorobenzene and naphthalene applied by a new method. Pests 10(4):16-17.

SWISHER, E.M. — 1944. Sulfur dioxide-acetone as a household fumigant. J. Econ. Entomol. 35(5):694-697.

TENHET, J.N. — 1955. Fumigation of cigars. Biol. Abst. 30(3):875, 1956.

THOMPSON, R.H. — 1966. A review of the properties and usage of methyl bromide as a fumigant. J. Stored Prod. Res. 1(4):353-376.

TORKELSON, T.R., H.R. HOYLE & V.K. ROWE — 1966. Toxicological hazards and properties of fumigants. Pest Control 34(7):13-16, 18, 42.

TRAUTMAN, J.A. — 1934. Methylene blue in the treatment of HCN gas poisoning. Ext. Log. 2:15.

WAGNER, G.B. — 1942. Insect control manual for flour millers. Millers' National Federation, Chicago, Ill. Gas masks pp. 40-47.

WAKELAND, C. — 1925. Fumigation for the control of household insects. Univ. Idaho Ext. Circ. 50.

WHITNEY, W.K. — 1961. Fumigation hazards as related to physical, chemical, and biological properties of fumigants. Pest Control 29(7):16, 18-21.

WILLIAMS, C.L. — 1934. Fumigation of foodstuffs. Ext. Log. 2(5):9. 1935. The treatment of cyanide poisoning. Ext. Log 3(11):16-17. 1938. Hydrocyanic acid gas absorbed in bedding. Pests 6(11):15-17. 1939. Successful fumigations. Pests 7(8):10-11.

WINBURN, T.F. — 1952. Fumigants and protectants for controlling insects in stored grain. Pest Control 20(8):9-11, 32, 42.

YOUNG, H.D., R.H. CARTER & S.B. SOLOWAY — 1943. Bromine residues from methyl bromide fumigation of cereal products. Cereal Chem. 20(5):572-578. Rev. Appl. Entomol., A 32:351, 1944.

VON OETTINGEN, W.F. — 1946. The toxicity and potential dangers of methyl bromide with special reference to its use in the chemical industry, in fire extinguishers, and in fumigation. U.S.P.H.S. Bull No. 185.

HARRY L. KATZ

Harry Katz was raised in Canonsburg, Pennsylvania, a small town 20 miles southwest of Pittsburgh. It was a family friend, Fred Pollock, who started a pest control supply and service business in 1928 in East Liberty, Pittsburgh, who got Katz interested in pest control in the 1930s.

After service in World War II, Katz took over the Elco Manufacturing Company operation. The impact of DDT in 1946 altered Elco's direction in favor of pest control service and at the same time Katz started to build an extensive library of pest control. In 1961, Elco acquired a one acre site in Sharpsburg, near Pittsburgh, and built a warehouse, office and a training room where classes for PCOs have been conducted since 1971.

In 1953, he helped form the Western Division of the Pennsylvania Pest Control Association and served until 1980 as recorder (secretary) when he sold the pest control service portion of his business.

The National Pest Control Association has been a consuming interest for Katz and he served on the Technical Committee (Council) for two years; as regional chairman, Wood Destroying Organisms Committee for one term; and as regional director 1975-78.

Locally, Katz was busy in the Sweadner Entomological Society (successor to Western Pennsylvania Entomological Society, established in 1910), serving as secretary for more than 10 years and as program chairman for 20 years. He is also a research associate in the Carnegie Museum of Natural History, Pittsburgh.

Statewide, the Entomological Society of Pennsylvania, oldest in the United States, captured Katz's attention, and he has held office and served on committees for this Society. In 1975, he established an annual Elco award for the best entomology student in graduate school in Pennsylvania. In 1980, he was recipient of the Frederich V. Melsheimer Distinguished Service Award of the Entomological Society of Pennsylvania.

Pest control and entomology are not Katz's only interests. He also is noted for his sincerity and the time he devotes to helping others. In the 1930s, he organized and chaired a Log Cabin Preservation Committee in Canonsburg, Pa. (1st school of the Presbytery West of the Alleghenies and forerunner of W & J University.)

In 1955, Katz founded the Parkway Jewish Center in a suburb near Pittsburgh and he is honorary president of this congregation. He is also a life trustee and assistant treasurer of the Hebrew Free Loan Association of Pittsburgh. Here he devotes two hours a week interviewing people unable to borrow at banks and helping them get on their feet with no interest loans.

Lecturing before diverse audiences is another one of Katz's loves. This includes PCOs and sanitarians at conferences, as well as university and high school classes. Katz has also contributed many articles to trade journals such as _Pest Control_ and is contributing editor to _Pest Control Technology_ magazine.

CHAPTER TWENTY-EIGHT

Equipment

By Harry L. Katz[1]

AS THE PEST control industry matured, various trade customs and practices developed in different areas of the United States. Out of this maturation process a kind of speciation of devices also developed!

A brief description of the systems used by PCOs in the recent past should be chronicled. "What is past is prologue" also may apply to some of the ideas our predecessors developed. Many thought the references to dusts in the Fifth Edition of this book in 1969 were superfluous.

Robert Snetsinger's imminent book, *The Ratcatcher's Child*, describes the state of the art of pest control before the 1930s. This chapter takes up from there to describe the panorama of equipment into the 1980s.

There may be devices in use regionally which may not be described. If a product is identified by name, superiority is not implied over other similar brands. Equipment intended primarily for horticulture will be only incidentally mentioned in this chapter.

Before the age of liquid spray residuals, PCOs used a variety of devices to dispense liquid pesticides. These included the hand pump atomizers, hand operated hydraulic pistol sprayers, electric paint sprayer and various home-made adaptations.

The developing pest control industry modified existing devices which were available in the home and garden market.

One of the earliest devices was the intermittent hand pump atomizer which emits a puff of spray with each stroke of the plunger. This was followed by the

[1]*President, Elco Manufacturing Co., Pittsburgh, Pa.*

continuous hand sprayer which provides a buildup of air pressure in the space above the liquid. The pressure forces the spray through a tube continuously into the air stream where it is broken up into airborne particles.

The writer recalls using a "Jacques" continuous sprayer with hollow needle attachment to inject pesticides into upholstered furniture in the late 30s and early 40s.

More popular in those early days was the hand-operated hydraulic pistol sprayer. This device was patterned after the familiar trigger oil can in which pressure on the trigger drives out the liquid displaced by the piston. The size of the spray particle is dependent on finger pressure. Pressure is zero at the start of the stroke, rises to a max-

Fig. 28-1. Original B&G sprayer, 1950.

imum on the squeeze and immediately drops to zero. This causes variability in particle size of the cone spray (Spear, 1955). The sturdy metal sprayers are equipped with pinstream and spray nozzles and are still used in some parts of the country, but they have been largely supplanted by the low cost plastic window spray bottle with adjustable nozzles. A PCO may want to use a specific material for a specific problem and by using such pistol sprayers avoid the chore of emptying and cleaning larger sprayers.

In the early 40s, Jack Benmosche and Phil Meyer developed a pressure sprayer. It was a modified blow torch with a hand pump built into it to pressurize the liquid. Snetsinger's *Ratcatcher's Child* records this development. These units were equipped with extensions and nozzles for pinstream and spray.

Later, William Brehm and George Gilmore, two students at Purdue University, modified a garden sprayer to create the prototype of the compressed air sprayer so popular today. This model is now in the Archives of Purdue University.

Since then, Brehm's firm and several other manufacturers have improved the unit in several ways. Their sprayers feature a hand operated brass pump with check valve all inserted from above. This forces the liquid through a tube, hose, lever valve and precision nozzle.

When pumped 25 strokes, a gallon/3.8 l of liquid with about three pints/1.4 l of air space will register 40 psi/2.8 kg per sq cm. This was the pressure designed by the nozzle engineers to deliver, "while walking," a spray which will wet a floor or wall to the point of "runoff" (just short of running). A multi-tip nozzle allows a variation from a heavier spray for more absorbent surfaces (cinder blocks and stucco) to a less absorbent (ceramic and stainless steel). Numbers on the tip are coded to indicate the flow in gallons per minute and the angle of the spray. The figure is based on 40 psi/2.8 kg per sq cm pressure. Some manufacturers attach a pressure gauge or can supply it. This is becoming increasingly popular. Regulatory agencies prefer no more than 20 psi/1.4 kg per sq cm pressure in order to avoid unnecessary drift into the air.

Also, Brehm introduced a modification of the unit in which the air is delivered with a pump, powered by a chargeable battery — a system which maintains a constant 20 psi/1.4 kg per sq cm pressure.

Some sprayers have wide mouth openings for easy cleaning. The lid in this case is kept in place by the air pressure.

The bottom of the tank could bulge downward or upward. The latter resists bulging from pressure, but allows liquid in a nearly empty tank to deteriorate the juncture if long neglected. There should be little difference if used and cleaned regularly.

Most of the tanks are made of stainless steel type 302 or 304 gauge. They will last indefinitely if used and cleaned regularly. Otherwise, long standing organophosphorus emulsions or oil solutions will degrade into an acidic condition which will eat pinholes in the steel. Solutions will react with the moisture in the air above it. This condenses on the walls of the sprayer and runs into the solution. Brehm (1968) comments on sprayer care.

Tanks have rubber washers often referred to as "soft goods" which need to be replaced every six months. The solvent element of emulsions tends to swell the rubber or synthetic washers. Viton® check valves are now available which resist this deterioration. Most manufacturers sell kits which contain the necessary replacement parts. PCOs are urged to become familiar enough with their equipment to replace the soft goods periodically. A list follows which describes problems and solutions associated with the compressed air sprayer.

SOLVING PROBLEMS WITH COMPRESSED AIR SPRAYERS

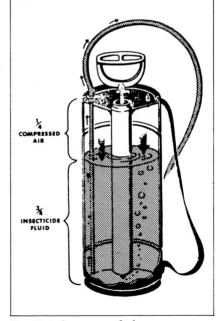

- If it does not shut off, the shut-off valve is not working properly.

Answer: Open the sprayer, clean and replace washers if necessary.

- Liquid comes out of the top in a stream after the cylinder is pumped. Cylinder fills up with liquid.

Answer: Check to see if the valve is faulty. It may be worn, dirt may be under it or there may be a chemical buildup under the check valve. Replace after cleaning.

- A few minutes after pumping, liquid leaks over the top.

Answer: Dirt or chemical may have built up under check valve. Clean and replace.

- No pressure develops after repeated pumping; tank holds no pressure.

Answer: Tank gasket is worn and needs replacing. Sealer ring needs attention, perhaps replacement or tightening. Cup leather needs oiling or replacing.

- Spray pattern is inadequate, despite good pressure.

Fig. 28-1. Compressed air sprayer.

Answer: Tip of nozzle is worn or dirty; clean or replace.

- Pesticide dripping as sprayer is being used, even with good pressure.

Answer: Replace "O" ring in nozzle if worn or missing. Straighten tip if not secured properly. Replace seat gasket if worn, replace spring if broken or clean dirt under seat gasket.

Preventive Maintenance prolongs sprayer life and avoids problems. Here are some tips.

- Sprayer should be rinsed at the end of each day's use.
- Sprayer should be cleaned once a week.
- When cleaning, use ammoniated detergent with warm water.
- When not in use for an hour or more, release pressure. A full tank at full pressure left in the sunlight on a hot day can explode. Hoses also can burst.
- When emptying tank, also empty hose. Do this by releasing the air pressure, holding the nozzle high and squeezing the trigger.
- Avoid plugging nozzle by using clean water and using the proper size filter. Most importantly, never place pump cylinder on ground while filling the sprayer.
- In the absence of mixing directions on pesticide labels, when mixing concentrates first add water, then concentrate and then water. If adding emulsifiable concentrate to wettable powder, mix the powder first. Follow all label directions.

Some tanks are ½ gallon/1.9 l in size, to fit into a carrying case. By far the most popular is the one-gallon/3.8 l sprayer, the total weight being comfortable for most technicians. However, larger sizes are available with shoulder straps.

A variety of attachments are available for the compressed air container. These include an inspection light — attached to the sprayer extension, a pressure gauge and pump handle cradle.

Plastic models of the compressed air sprayer also have been appearing in the marketplace. These still are basically for the garden trade, sometimes with adaptations of nozzle and valve for the structural pest control industry.

These plastic units have also been marketed through janitor supply distributors. The plastic material is designed to tolerate the aromatic elements and solvents in emulsifiable concentrates and oil solutions.

It should be noted that do-it-yourselfers may not appreciate the distinction between garden compressed air sprayers and the professional models. The nozzle of the former is designed to deliver quantities of liquid for vegetation far in excess of the amount needed for pest control indoors. Until they learn the differences the excess spray produced by gardening models can damage property or injure occupants in addition to wasting money.

Most compressed air sprayers use a nine inch/ 23 cm extension between the valve and the nozzle. This is in contrast with a few which attach the nozzle directly to the valve; or those which use an 18 inch/46 cm extension. A telescoping spray extension set also is available.

Fig. 28-3. Typical compressed air sprayer.

The disadvantages of the stubby valve-nozzle assembly is the back splash of droplets on the hands. An extension largely eliminates this problem. A disadvantage of the 18 inch/46 cm length is the inability to get into close quarters. Its principal value is in the ease of treating elevated sites.

Most PCOs have abandoned the original "Trigger Tee Jet" valve for the extended valve containing a no-drip feature. This type shuts off the flow at the nozzle rather than at the valve using a flexible wire inside the extension tube. Several manufacturers supply excellent no-drip nozzles with their compressed air sprayers. These include B&G Equipment Co., H.D. Hudson Manufacturing Co. and Root Lowell.

Multi-tip nozzles feature pinstream openings. These permit treatment of cracks and crevices. But for those whose aim may wander, a modification in the multi-tip assembly permits the use of a crack and crevice tip extension. This is a six inch/15 cm plastic tube which easily fits in the nozzle assembly. It makes possible treatment of cracks and crevices without getting toxicants on surrounding surfaces. Some of the labels, as well as specific USDA regulations forbid any surface treatment.

In treating with the crack and crevice tube, the air pressure should be reduced to as little as three psi/0.21 kg per sq cm (Spear, 1955). This is because some of the voids treated may be cul de sacs — closed spaces — and a splash back will occur if excess air pressure is used.

Another amendment to the basic compressed air sprayer is the pure bristle brush attachment. This allows a steady supply of residual insecticide (e.g. chlorpyrifos formulated in lacquer base) through the brush.

Brushing residual toxicants on surfaces has long been used to control pests. Several over-the-counter products have a brush furnished with the pesticide.

Synthetic hoses have replaced the rubber ones in order to withstand the aromatic solvents in some of the pesticides. However, cold weather can make most of the synthetics less pliable and subject to cracking.

A recent development by Tom Parker is a three gallon/11.4 l sprayer on wheels. It has an electric powered diaphragm pump which delivers at an even pressure of up to 50 psi/3.5 kg per sq cm. With a 26 ft./7.9 m range it requires convenient electrical outlets.

American PCOs are an ingenious lot. Many have created devices and systems for themselves which have merit. One such system is a back-pack sprayer devised by John Becker of Pittsburgh in the 50s.

This was a two gallon/7.6 l oval steel tank, strapped to the back with hose, standard valve and nozzle. Charged with liquid and air from an independent air source, the technician's arms were free to hold a flashlight or to move objects.

A variety of sprayers have been developed which require air from an independent source of air pressure (service station, plant air compressor).

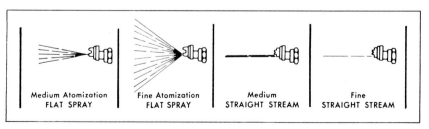

| Medium Atomization
FLAT SPRAY | Fine Atomization
FLAT SPRAY | Medium
STRAIGHT STREAM | Fine
STRAIGHT STREAM |

Fig. 28-4. Examples of multi-tip nozzle patterns.

In 1960 Alvin Burger fitted a regular one-gallon/3.8 l compressed air sprayer with an air chuck.

In the early 30s, Milwaukee Sprayer made a quart/0.9 l size steel sprayer with handle which was pressurized with an air chuck. Along with other manufacturers' models, it still is available with jet stream and mist spray nozzles.

CO_2 pressure bomb filler units are supplied where compressed air is unavailable. Several manufacturers make similar units.

For PCOs willing to work with CO_2 as a source of pressure, B & G Equipment Co. makes a dual purpose spraying unit. Mounted on a two-wheel cart, a one-gallon/3.8 l compressed air tank powered with a 5 lb/2.3 kg CO_2 cylinder, produces either a 10 to 15 micron aerosol for flying insects or a cylinder crack and crevice treatment, each with different controls. The CO_2 cylinder lasts an hour.

Industrial plants or other institutions equipped with convenient compressed air outlets sometimes find it possible to use a portable atomizer as described by Ebeling (1975).

PCOs who have easy access to compressed air supply sometimes use a stainless steel one pint/0.4 l can which is fitted with an air intake valve and an adjustable nozzle with crack and crevice tip. They can refill the aerosol can at low cost. Of course, the advantage of non-repellency of toxicant in propellant is lost.

SIZING UP AEROSOL PARTICLES

A very important tool in the management of flying and crawling insect pests is the generation and application of pesticide aerosol particles.

This is accomplished with devices which fall within three general categories: mechanical aerosol generators (MAGs), thermal generators and pressurized aerosol containers (PACs), (Katz, 1981).

The function of all these systems, of course, is to convert liquid pesticide into airborne particles. The systems vary in the range of sizes of the particles they produce and in the composition of the particle.

The size of the particle has much to do with the effect on the target. Particles that are too small may be too light to impinge on the target. Ebeling (1975) elaborates on particle impaction, distribution and size. For instance, particles may be too small to wet a leaf of shrubbery in order to kill adult mosquitoes resting thereon. If too small, the slightest wind will carry it beyond the target area outdoors. Indoors, however, the tiny particles can travel throughout a habitat which can be traversed with a blast of air. These include pipe chases, skids, aisles, false ceilings with panels removed from opposite ends, sewers and other such habitats.

Particles that are too large may not float long enough to kill enough of the flying insects and will wet interior surfaces excessively. The large particle is responsible for excessive use of pesticide.

According to Bals (1978), if a droplet of 50 um contains enough chemical to kill an insect, the 500 um droplet (dosage variation of 1:1,000) is wasting 999 units of chemical, thus creating an unneccessary level of environmental contamination.

Particle size is measured in microns. A micron is 1/1,000 of a millimeter (1/25,000 of an inch). A sheet of paper is 100 microns thick.

According to the American Mosquito Control Association (Anon., 1952), the following table shows how far a particle falls or drifts.

TABLE 28-1

Rate of fall in still air and drift on a four-mile wind, of droplets of different diameters

Diameter of Particle	Rate of fall per second (about)	Time of fall one foot/ 0.3 m (about)	Distance carried while settling four feet/1.2 m (about)
Microns			
100.0	30.0 cm (12 in.)	1 second	$23^7/_{15}$ feet/7.1 m
10.0	3.0 mm (1/8 in.)	$1^2/_3$ minute	$2346^2/_3$ feet/715 m
1.0	30.0 microns (1/800 in.)	2¾ hours	44 miles/70 km
0.1	0.3 microns (1/80,000 in.)	11½ days	4404 miles/700 km

These figures assume the particle remains the same size throughout its flight. Of course, evaporation reduces the particle according to the vehicle, temperature, pressure and relative humidity. Some formulations in pressurized containers contain no solvent, so the particles often are more stable than those in water or oil carriers, thus holding their integrity longer.

With water or oil solutions, the size of the particles can change rapidly. According to Dean Collins of Root Lowell, a 20-micron, oil based particle shrinks to 10 microns in five minutes if the relative humidity is 50 percent and the temperature is 75° F/24° C. If water based, 90 percent of the particle will evaporate after being thrust forward 35 ft./10.7 m under the same time and conditions.

Another aspect which bears on the integrity of particle size is the surface area exposed to the elements, as shown in the table below.

TABLE 28-2

Relation of droplet diameter to number of droplets and to total surface area (1 gram of water)

Type of Dispersion	Droplet Number	Mm	Microns	Total Surface Area of Droplets Sq. Cm
				15
Rain	30	4.0	4000.0	(3 postage stamps)
				1,500
Spray	30,000	0.4	400.0	(12 by 17 inches)
				150,000
Coarse aerosol	30,000,000	0.04	40.0	(10 by 14 feet)
				15,000,000
Medium aerosol	30,000,000,000	0.004	4.0	(2 city lots)
				1,500,000,000
Fine aerosol	30,000,000,000,000	0.0004	0.4	(37 acres)

With such a range of surface area exposed to the elements, one can appreciate the list of factors compiled by Dan Stout, Whitmire Research Laboratories, Inc., which alter the effectiveness of the insecticide particle: size, velocity, surface tension, electrostatic charge, ability to penetrate the pest, compatibility with other materials, temperature and environment.

Insecticides dispersed in the air are classified according to particle size, as follows:

TABLE 28-3

Classification of different types of dispersed insecticides according to size range in microns

Coarse spray	400 and larger
Fine spray	100 to 400
Mist	50 to 100
Aerosols and fogs	0.1 to 50
Fumes and smokes	0.001 to 0.1
Vapors	Less than 0.001

The classification of aerosols by particle size can be confusing, according to Truman et al. (1976). In some devices, the particle size can be adjusted to either a fog or a mist classification. Also, he says the names of several devices can be misleading, having either "fog" or "mist" as part of their brand name. Other authorities consider any particle under 100 microns to be an aerosol.

If this were not a book for pest control operators, aerosols could refer to shaving cream, spray paint or even a sneeze, which is the most universal aerosol of all.

COLD AEROSOL GENERATORS

Mechanical aerosol generators (MAGs) started with the electric paint sprayer in the 30s. They were then adapted to pest control work. This was the most widely used piece of equipment of PCOs in those early years. It was used to mist a room and the residue of kerosene and pyrethrins literally coated every horizontal surface, producing a high body count of insect pests. Populations of cockroaches, bedbugs and other vermin may have been worse in the decade before the residual chlorinated hydrocarbons came into use, and the harvest was usually spectacular. The electric sprayer was used to wet surfaces just as PCOs use compressed air hand sprayers today.

Suppliers to the pest control industry have since produced a variety of MAGs, sometimes called cold aerosol generators. The aerosol particles could be produced by spinning discs and rotors through swirl chambers or through a combination of both. The harder the hit, the smaller the droplet.

In some of the units, blowers blast the particles into the air where they remain at the mercy of air currents. In the absence of a breeze, particles will fall at a rate shown previously in this book.

At the upper range of particle sizes are the mists. Portable mist blowers have been widely used since the 30s. As described by Ebeling (1975), their range is between 15 and 60 microns and their thrust is 75 ft./23 m.

For large scale mosquito control, a standard tool has been the motorized mist blower. Few PCOs are involved with this generator. Mist blowers use dilute sprays. PCOs generally work with MAGs below the mist classification.

Mechanical aerosol generators can be regulated to change the rate of flow and particle size within a specified range. Rotating platforms are available to spread the particles evenly throughout the room.

The ability to control the range of particle sizes is more refined in the ULV-ULD mechanical aerosol generators.

The expression ULV originated with agriculture and by definition limits the term to a device which uses no more than ½ gallon/1.9 l of spray per acre/0.4 ha with no reference to particle size or indoor use. A letter from the EPA by F.S. Bishop, dated October 20, 1980, confirmed this position.

The term ultra low dosage (ULD) for indoor situations is accepted by the industry as one synonymous with ultra low volume (ULV) to accommodate the concept of generating a more concentrated liquid pesticide with a narrow range of small particle sizes.

A study of manufacturers' claims for ULV-ULD shows a variance with numbers ranging from five to 20, 10 to 20 and five to 30 microns. The optimum average would range from nine to 15 microns.

The ULV-ULD devices are available in hand held models, on wheels and on skid-mounted set-ups for large scale applications.

Claims abound concerning the efficacy of aerosols with various particle sizes (e.g. different impingement on insects or penetration of habitats).

Researchers at Purdue University found no significant penetration by the particles into a six inch/15 cm opening in doors of a cabinet housing a cage of insects (Bernhard and Bennett, 1980).

In a talk to a group of PCOs at a Whitmire seminar in August, 1981, Alvin Burger claimed no significant benefit from mechanical aerosols generators in his operation. But other industry leaders use them regularly on a routine basis with satisfaction.

A Pittsburgh sanitarian, Gene Becker, reported extraordinary success with the following technique. He hand carries a small ULV-ULD device directing it into equipment and habitat sites with the dial turned to the lowest possible flow rate. This takes about 15 minutes for a large room. The velocity of the air blast exceeds 70 mph, and can reach otherwise inaccessible habitats. He then turns the device to a medium setting and walks around the room for about 10 minutes. He never aims at a wall or equipment, but *alongside* them. Becker feels the tiny micron range brings out well-entrenched cockroaches and the large particles that follow "do them in."

Research is in progress which points to the value of the low micron range particle. In one instance, a room was filled with this aerosol and then evacuated. Hours later caged insects were brought in and died quickly. Clearly, there is much more to be learned about the merits of both small and large particles in MAG.

Another phenomenon occurs with MAG which should be noted. Oil based particles, especially on the larger end of the size range, tend to collide and stick, a phenomenon known as agglomeration. These get larger and fall faster, sometimes wetting the surfaces, especially close to the device.

Operators are warned not to look for visible clouds with ULV. Only in certain refractive light can anything be seen. Turning to the highest setting may make the particles too large for the purpose. Shining a flashlight in a darkened room will more easily reveal the particles floating. Aerosols in excess amounts are more visible.

Exceeding the label directions accomplishes several things including:

- It kills insects a few moments sooner.
- It increases the chances of leaving a contaminating residue.
- It increases the chances of an oil slick with oil-based solution.
- It increases the chance of fire.
- It costs more.
- It violates the law and leaves the operator without legal support in any customer law suit.

Dan Stout, of Whitmire Research Laboratories, Inc., sums up these facts about ULV which PCOs should know about.

- Water sprays fall faster than oils and most solvents.
- Lateral dispersion is accomplished by air currents after terminal velocity is reached.
- Particles are not conveyed into dead end cracks or into materials in which air does not circulate.
- Small particles move laterally to greater distances than do larger particles because they stay suspended in the air longer.
- In unheated buildings, air currents are at a minimum.
- Heating sets up convection currents that are a great aid to particle dispersion.
- Fans are extreme aids to dispersion of particles.
- Ninety-five percent of particles settle on horizontal surfaces and five percent on walls and ceilings.
- Surfaces colder than the air temperature attract small particles.
- Many structures are too open to successfully and economically use ULV with the equipment presently available. Select a mist blower that will give a particle range above 50 microns.

The succession of concepts — from spraying dilute pesticides to more concentrated ones, to ultra low volume may be progressing to a newer concept — ultra low dosage with a more controlled particle size. Thus, a new term "CDA," meaning controlled droplet application has been coined. Matthews (1978) discusses the concept.

At present this type of applicator uses grooved, spinning discs to deliver precisely sized particles, designed to avoid drift and minimize evaporation. For structural PCOs Micron West's portable CDA surface spraying device is well suited to flea and tick control and for weeds. This device, developed in the United Kingdom, uses "D" batteries for 100 hours and it is claimed that a five pt/2.4 l bottle will cover ¾ of an acre/0.3 ha with 250 micron droplets. It can also dispense wettable powders.

Using more than ½ gal./1.9 l per acre, this is not ULV but it is ULD and can result in considerable savings in pesticide. However, with 250 micron size very little drift is possible compared with more conventional applicators.

Similarly designed is a hand carried space spraying MAG which is also a British invention. It produces 30 to 40 micron particles for mosquitoes and flies. A CDA device, it has a 40 ft/12 m thrust and it operates with a portable 12-volt battery.

Other systems are used to generate aerosol particles. A superheated dry steam aerosol generator was popular in California and, according to the Van

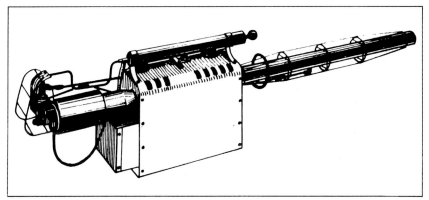

Fig. 28-5. A hand carried thermal aerosol fogger (gasoline powered).

Waters and Rogers Division of Univar, it is still in use. Heated with electric coils, the emerging steam breaks up the insecticide into particles from 10 to 100 microns.

Also, in California as well as other areas, some veteran PCOs favor a high pressure aerosol generator. As Truman (1976) describes it, the solution in such machines is atomized when forced through a turbulence chamber by compressed nitrogen gas.

HOT AEROSOL GENERATORS

Only thermal aerosol generators are true foggers. With these devices, oil-based insecticide is heated to vaporization with heat from electricity or petroleum. Upon contact with ambient air, the vapor condenses into a true fog consisting of particle sizes from 0.2 to nine microns in diameter. Some units involve steam ejection; still others utilize the internal combustion engine exhaust pipe.

Large foggers for large outdoor mosquito work and large warehouses can use insecticide at 30 to 60 gallons/113 to 227 l per hour. Smaller foggers, utilizing gasoline, use from five to 15 gallons/19 to 57 l per hour. Small electric foggers are best suited for out-of-doors because of the possibility of fire, although manufacturers of recent models claim this has been corrected.

Truman et al. (1976) states that indoors fog will not usually move against cold exterior walls. Indoors, using more than one gallon/3.8 l of oil-based pesticide per 50,000 cu ft/1,400 cubic meters may be explosive if ignited, even with an electric switch. This situation also can occur with non-thermal aerosol generators if the carrier is petroleum distillate.

Carbon forms in the heat chambers of the thermal units and must be removed after each use before it has had time to harden and accumulate in a thick layer.

An advantage of fogs outdoors is the ability to penetrate between the leaves of vegetation and to float long enough to kill mosquitoes. However, winds in excess of five miles/eight km per hour will reduce the effectiveness.

One advantage of this very visible fog is the public relation benefit of taxpayers seeing their tax money being spent to kill mosquitoes. On the other hand, persons obsessed with any pesticide dispersed on public and private property will sometimes make problems for the public health agency conducting the application. A cloud of ULV spray applied to an area may escape the eye of many citizens. It is hard to have it both ways.

Claims are made that pyrethrins are degraded by the heat of thermal generators. However, the Pyrethrin Board reports the degradation to be minimal.

CANNED INSECTICIDES (AEROSOLS)

Americans must love canned insecticides. About 200 million cans of pressurized aerosol containers (PACs) of insecticide were packed by CSMA members in 1979. Apparently, PCOs favor them too, particularly because of ease of use.

The problems attendant on oil slick or contamination are practically eliminated. Some formulations have a minimum of solvents, others have none, the toxicant being dissolved in the propellant.

According to Richard Vega, of the Whitmire Research Laboratories, propellants fall into two classes: liquified gases and compressed gases. These he explained as follows, citing references from Johsen and Dorland (1972) and Sanders (1979).

Aerosol propellants. "A lot of confusion currently exists regarding the different types of chemicals that are used as aerosol propellants. Basically, propellants can be divided into two classes:

1) **"Liquified gases.** The broad classification of "liquified" gases used as aerosol propellants defines those chemicals that exist as liquids under pressure in a normal aerosol container at room temperature (70° F). The great advantage offered by the liquified gases is 1) they maintain a constant vapor pressure throughout the use of an aerosol, and, 2) provide tremendous breakup of the spray as the liquid gas vaporizes at the nozzle expanding tremendously in the air. As a result, almost all true aerosols producing particles in the zero to 50 microns range are propelled by liquified gases. The liquified gas propellants include:

- **Fluorocarbons.** Commonly called Freons (E.I. DuPont's tradename for non-flammable fluorocarbons), the most common fluorocarbons used as propellants are dichlorodifluoromethane (F-12), trichlorofluoromethane (F-11) and dichlorotetrafluorethane (F-114). These propellants are no longer in general use because of the ban resulting from the ozone depletion theory.
- **Hydrocarbons.** The hydrocarbons most commonly used as propellants include propane, butane, and isobutane. All of these compounds are flammable; however, since the fluorocarbon ban, they are the principle propellants for 90 percent of all aerosol or pressurized products.

Chlorocarbons and ethers. While these chemicals are included in the liquified gas propellants class, they are not in widespread use because of problems involving toxicity, odor, and extreme flammability. Vinyl chloride is the only chlorocarbon used to any extent as a propellant. Among the ethers, only dimethyl ether is able to function as a true propellant. Since it is very soluble in water, it is currently receiving some attention as a potential propellant for aqueous-based products.

2) **Compressed gases.** The term "compressed gas" is used in the aerosol industry to denote a gas which can be liquified only by the application of very low temperatures or very high pressures. The compressed gases most often used as propellants are carbon dioxide, nitrous oxide, and nitrogen. These compounds all exist as gases inside normal aerosol products. As a result, as the product is

used, the vapor pressure in the aerosol gradually decreases. A typical CO_2-propelled unit may start out with 90 PSIG vapor pressure initially and contain only 40 PSIG when the can is empty. This loss in vapor pressure means that it is hard to maintain a constant delivery rate with aerosols propelled by compressed gases. The other disadvantage using compressed gas propellant is that they do not aid in spray breakup like the liquified gases. Since the propellant is already a gas in the container, it cannot vaporize at the nozzle and no expansion occurs to produce small particles. As a result, the compressed gases are usually used in products where a coarse wet spray is desired.

Fig. 28-6. Disposable aerosol dispenser.

"An exception to this is . . . . large 15 lb. cylinders of products designed for space treatment. In these products, carbon dioxide is used as the propellant and small ULV-size particles (in the 20 to 30 micron range) are produced. This is accomplished by using higher percentages of carbon dioxide at higher pressures than are found in typical small aerosol cans and by using a proprietary spray nozzle. This nozzle contains a special chamber that swirls and shears the liquid stream mechanically breaking up the spray into small particles."

Particle sizes in small conventional retail PACs, as well as in the total release cans are not in the ULV-ULD range. The dispenser valve is relatively inefficient, as one can observe fallout on newspapers under the emptied can.

Generally, the larger packages are provided with finely engineered applicator devices. These insure particle size in the ULV-ULD range, essentially doing what mechanical generators can do — breaking a drop of insecticide into a cloud of optimum particle sizes.

One advantage of the propellant system is the tendency for particles to expand and separate, rather than agglomerate as with oil-based particles.

On the safety of aerosols, Vega comments as follows: "I would like to comment on an accident that recently occurred when, as I understand it, an aerosol product fell into a deep-fat fryer and exploded. I believe, accidents like this are caused by careless storage practices. Every aerosol or pressurized spray product in use in this country today clearly warns the user not to store or use the product near heat or open flame. Whenever a product is packaged under pressure, the potential hazard of the container bursting always exists. However, if proper cautions are followed in storage and use, this hazard is minimal. As evidence of how minimal this hazard is, statistics for the period from 1966 to 1974 indicate that only 93 incidents were reported which involved exploding aerosol cans. During this time period, literally billions of aerosols were produced or, in other words, only one can in every 300,000,000 produced was involved in an accident. This kind of safety record speaks for itself . . ."

Relative to the advantages of aerosol insecticides, Vega has this to say: "Probably the greatest advantage that aerosol insecticides offer is that they minimize environmental contamination. This is accomplished in two ways:

- Less product is used (when compared to liquid sprays).
- The amount of aerosol product that is used contains less oil than the same amount of liquid spray.

"Comparing the formulas and label use directions for typical aerosols versus liquid spray products quickly points out how environmental contamination is minimized with aerosols. When using aerosols for fly control, the label will typically tell you to spray for a maximum three seconds per 1,000 cubic feet of space. Based on the "delivery rate" of a typical aerosol, that three seconds of spray will deliver about three grams of insecticide into the 1,000 cubic feet space. Considering the fact that the typical aerosol formulation contains at least 80 percent volatile solvents and propellant gas, only about 0.6 grams of insecticide spray (20 percent × three grams) will fall out of the air and land on horizontal surfaces (including food processing surfaces) causing environmental contamination.

"In comparison, the label of a typical liquid spray will call for one ounce of spray per 1,000 cubic feet for fly control. Since the typical liquid contains the active insecticide mixed in a non-volatile oil such as kerosene, the entire ounce of spray will fall causing extensive contamination of horizontal surfaces. This means that the liquid spray produces about 50 times more environmental contamination than the aerosol product.

$$\frac{(1 \text{ oz.} \times 28.35 \text{ gms/oz.}}{0.6 \text{ gms}} = \text{(47 times more oil with liquid)}$$

This oil fallout will cause slippery floors and oily food processing equipment.

"The oil-based fog from a mechanical sprayer also can cause a fire and explosion hazard particularly in areas where there are a lot of ignition sources such as pilot lights and ovens. This hazard is not nearly so great when using aerosols simply because of the small amount of oil dispensed by aerosols when used according to directions.

"The final disadvantage with liquid spray products is that they require the purchase of a mechanical sprayer. Often mechanical fogging equipment can be expensive ($200 to $300) to purchase and maintain.

"Aerosol products provide a convenient package form that does not require mixing of product or clean-up and maintenance of spray equipment. And of course, the effective control of flying insects using aerosol insecticides should not be minimized. Laboratory and field tests prove again and again how effective aerosols are in controlling flies."

On the other hand, when considering the economics of canned aerosol insecticides one must remember the tendency of users to over-use an easy to use product. Moreover, the can, when empty, is normally thrown away and thus wasted.

The expansive character of propellants serves another useful purpose. Propellants expand to 240 times their space, which exerts a pressure in a small airtight room of 0.89 psi/0.06 kg per sq cm according to Whitmire engineers (personal communication). Thus, particles of toxicant can be carried to hidden habitats by these high or low pressure air movements. Results are sometimes spectacular when flushing insecticides are injected into crevices leading to habitats.

Particles from these pressurized aerosol containers are different from those from conventional generators. The toxic particle is quickly stripped of its carrier and floats as virtually 100 percent toxicant.

PCOs should be alerted to vagaries in the EPA registration of pesticide labels for PACs as shown by Katz (1981b).

One label can recommend as much as 131 times as much pyrethrins as another label, both designed to kill flies. This data was supplied by E. Paisley. Cost per 1000 cu. ft. should be calculated and systems' cost compared.

Circulating fans (not exhaust) can keep particles floating longer, being somewhat more useful with CO_2 and hydrocarbon propellant than with Freon propellants, according to Vega.

Widely distributed throughout the United States are timed, dispensing devices which hold 12 oz/340 g cans of PACs. These are timed to release enough pyrethrin or similar botanical at 15 minute intervals to keep an area of 6,000 cu. ft./168 cubic meters free of flying insects. While generally not permitted in small kitchens, they have been a mainstay in the fly control program of thousands of establishments. The dispenser can be operated by battery or house current. Most of the dispensers can accept most brands of the product.

PAC systems are available in which a series of nozzles are placed throughout a structure and connect with a central PAC. Timers periodically release the material.

Economies can be gained in some cases if illumination is left for a time in a corner, especially in the food preparation area of a darkened room. The PCO releases the hand held aerosol container as he or she approaches the lighted area. Dosage should be increased somewhat to compensate for some of the particles which spread to untreated areas.

Treatment for crawling insects requires as much as 10-fold the amount needed for flying insects. Because cockroaches do not fly through a cloud of particles, the air at floor level must be loaded with sufficient particles to kill the insects.

One of the biggest innnovations in pest control in the 70s was the crack and crevice concept with pressurized aerosol containers. While the concept of treating restricted habitats rather than indiscriminate baseboard spraying was used by many PCOs (See H. Katz, 1965), the expression "crack and crevice" together with a formulation without solvent or emulsifier must be credited to the Whitmire staff.

Embodying the principles of Integrated Pest Management (IPM) long before the term was promoted, Whitmire used the propellant essentially as a solvent. This made possible the application of a precise amount of technically pure toxicant at a precise spot, to which insects entering the habitat would be exposed.

Longevity in this protected environment is greatly enhanced, free of the degrading influence of alkali cleaners, organic contaminating debris, strong lights and air currents.

The availability of such aerosols, together with other technical advances (including improved liquid and dust products and application equipment) helped the industry grow into the IPM program naturally.

Scientific calculations explaining the basis for the effectiveness of crack and crevice treatment is documented by Shore (1974).

Systems using a combination of space and crack and crevice outputs have been developed for technicians to professionalize the use of this concept. In one system a belt with two pouch holders permits the technician to hold one PAC of residual pesticide and one PAC with a space treatment, each with a specific coiled hose, safety valve, clamp adapter and a nozzle for void injection or fogging. The technicians is free to move objects and carry a flashlight. Even more impressive are low cost PACs with adjustable valves which permit one aerosol can to be used for controlled or total release space spraying, as well as pinstream spraying.

For industrial accounts, the PCO can carry the same toxicants in a 15 lb/6.8 kg cylinder. A ULV jet and hose assembly and a crack and crevice injection gun and hose assembly were designed for more efficient distribution. Or walking around a room with a five lb/2.3 kg PAC would deliver the aerosol efficiently and economically. It produces 75 minutes of aerosol.

DUST TO DUST IN FORTY YEARS!

Dusting was one of the early PCOs best methods for controlling pests. When the residual emulsifiable concentrates were popularized dusting was all but forgotten. Limitations of the liquid residual insecticides eventually surfaced and many PCOs are returning to dusts (Loftus, 1980).

Dusts are not absorbed into the treated surface, but are instead available to the insect. Dust particles travel well and can be relocated with air turbulence. Moreover, dusts are even more economic than canned aerosols since they require no throwaway can and no solvents or propellants.

Excellent tools are becoming available to use dusts more effectively and with ease. Several hand dusters are marketed for treating small areas or crevices and these are based on rubber cylinders or bulbs. The original Getz duster has been improved and is quite popular because of its small size.

The bulb type is still popular. A modification of the original bulb type was made by H. Katz which stopped the blow-back on the hand and the plugging in the tube.

An advance in design similar to the Getz duster has been developed by Woodstream. Truman et al. (1976) recommend placing pebbles or ball bearings inside these small hand dusters to keep the dust from clumping.

The garden store plunger duster is still in use for dusting attics and above false ceilings. It is all plastic, light weight, and holds two quarts/1.9 1. A plug prevents spilling.

Many dust products are in small applicator packs. These often take the form of a plastic cylinder with an applicator tube for crack and crevice injection.

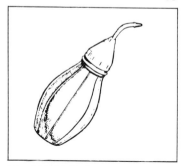

Hand cranked turbine dusters, often supported by a neck strap, are useful for dusting large areas. Electric dusters are also of great help to PCOs in attics, unexcavated areas, pipe chases and utility tunnels. Backpack high capacity models are available for quickly treating very large areas.

According to Engineer Dean Collins, of Root-Lowell, dust moving rapidly over a plastic surface created a positive charge in the metal nozzle which in turn is picked up by the dust as it leaves the nozzle. Since insects and surfaces are negative, such posi-

Fig. 28-7. Bulb duster.

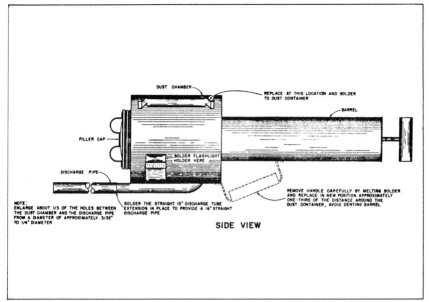

Fig. 28-8. Side view illustrating the mechanics of a hand pump duster.

tively charged particles hold on to the target more readily. Holding the metal nozzle negates the charge. Thus, an operator should hold the plastic rather than the metal when directing the dust stream.

Fire extinguishers with air chuck fittings were converted to dusters and have been popular for years in California and other areas. They are still being used according to some western PCOs for cockroach control in sewers and for bees, wasps and other uses.

More professional compressed air dusters are available. These units must be filled with outside compressed air. Made of stainless steel they can hold up to 100 psi/seven kg per sq. cm pressure, contain up to 10 lbs/4.5 kg dust and have safety features and a pressure gauge. However, care in releasing the pressure from the can is needed because the four to 10 lbs/1.8 to 4.5 kg of dust it holds can be dispersed very quickly.

Foot dusters for calcium cyanide were not available for several years but they are now available again since 'A' dust has been reintroduced to the marketplace. These dusters are specially designed so the calcium cyanide can be discharged with one or two strokes of the handle, then an air baffle can be moved and air alone pumped in for four or five strokes to disperse the gas arising from the dust throughout the rodent burrow.

TOOLS FOR TERMITE TREATMENT

Ingenuity and resourcefulness of the American PCO created a wide variety of equipment for conducting termite work in the early decades of pest control.

Some of these early systems involved compressed air in heavy tanks, (which can be dangerous to use), proportioner pumps, gear pumps, roller pumps, twin piston and centrifugal pumps. Tools to inject the toxicant were often home made.

Fortunately, the modern PCO easily can obtain well built and finely engineered units completely assembled for use with electricity or gasoline. For

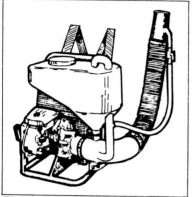

PCOs who would compare the power unit of gas versus electricity, M. Koistinen, a Hypro Corporation engineer says, as a rule of thumb, it is two to one; that is, a three horsepower gasoline engine would be equivalent in power to a 1.5 horsepower electric motor.

Pumps operate these units which serve different functions. Roller pumps are vane-type pumps in which resilient rollers replace flat sliding vanes. The four roller pumps develop pressures of 75 to 150 psi/5.3 to 10.5 kg per sq. cm depending on size of motor and rpm and yielding up to nine gallons/34 l per minute and as little as two gallons/7.6 l per minute with the smaller motors and with lower rpm.

Fig. 28-9. Portable power duster/ mister.

They can handle wettable powders, as well as emulsions.

Larger roller pumps can produce more than 20 gallons/76 l per minute. These are used by companies that do termite, lawn and horticultural work. Roller pumps are made of iron and nickel alloy. Twin piston pumps also are popular in some areas. These are positive displacement pumps and also can handle wettable powders readily. The small twin piston pumps three gallons/11.4 l per minute at up to 500 psi/35 kg per sq. cm at 1800 rpm. Its big brother, pumps 10 gallons/38 l per minute at 400 psi/28 kg per sq. cm at 600 rpm. Standard rubber seals should not be used for pesticides in these pumps. Pest control supply houses all have literature with pump data.

Hypro Corporation products are described because they dominate the small pump market, according to a survey of PCO supply firms. No superiority over other brands of small pumps is implied.

Myron Koistinen anticipates that with the EPA inspired trend to lower pressure in termite control applications, centrifugal pumps may become more popular. He claims that with the Hypro 9300 pedestal, a cast iron unit, using five horsepower gas engine will produce 10 gallons/38 l per minute at 175 psi/12.3 kg per sq. cm pressure. Less equipment is needed to operate the centrifugal, pressure can be reduced, maintenance is easier and it will last longer, even with wettable powders.

Indeed, lower pressures seem to be mandated by the new termiticide labels (Katz, 1981a). A California label specifies 20 psi/1.4 kg per sq. cm pressure when the surface at the perimeter of unexcavated areas is treated. Otherwise, no specific figure is designated as "low pressure."

The apparent reason for these regulations is that higher pressure means more small particles can become airborne and get off target, possibly contaminating non-target sites.

One advantage of the use of low pressure four-roller pumps in corrective termite work is there is minimal risk of creating voids in the backfill which could divert ground water toward the foundation wall in basement houses. However,

PCOs in many areas need a pump with enough pressure and volume for pretreatment and corrective work. Pretreatment requires large volumes quickly. Slab work requires good pressure for lateral distribution in the opinion of veteran termite operators.

Pressure and volume also are needed for lawn work. In the South, considerable pressure is needed to penetrate the thick thatch which protects target pests, according to Bruce Sibson, a Florida PCO. Failure to control such pests as chinch bugs, earwigs, millipedes, sowbugs, pillbugs and other occasional invaders has often resulted from superficial treatment.

Regional pump preferences vary. In California, Van Waters and Rogers report their customers greatly prefer the small twin piston to the roller pumps with the new proportioner pump unit gaining favor.

In the Northeast, Essco reports that 60 percent of PCOs use rollers, of which 75 percent are four rollers and 25 percent use six and eight rollers. Of the twin piston adherents, 25 percent use the small and 75 percent use the bigger version. Essco, too, reports an increase in proportioner pumps.

Cal Stevenson, Jr. reports that in the Southeast, roller pumps outsell twin pistons 20 to one, and the large rollers outsell the small ones 10 to one. However, Francis Rossbach's experience in south Florida is the reverse. PCOs there are active in lawn and ornamental work. He says wettable powders are hard on nylon rollers; twin pistons are easier to repair. In the central states, Mercury sells 10 rollers to one twin, while Crown sells five rollers to one twin piston, with both categories evenly divided between small and large.

While the label refers to pressure at the pumps, one must recognize the loss of pressure at the point of treatment. According to M. Koistinen, a 75 psi/5.3 kg per sq. cm flow of three gallons/11.4 l per minute through the 100 feet/30.5 m of ⅜ inch/nine mm hose loses 50 psi/3.5 kg per sq. cm, netting 25 psi/1.75 kg per sq. cm at the nozzle.

M. Koistinen further comments that an additional drop in pressure derives from the resistance in some types of quick-connect couplings.

Because of pressure loss, Doug Palmer, of Broward Junior College, teaches that calibration with a pressure gauge be made at the nozzle. He recommends that nozzle pressure for such pests as chinch bug be under 90 psi/6.3 kg per sq. cm; but for weeds it should be under 10 psi/0.7 kg per sq. cm at the nozzle.

And of course the high pressure of 250 to 450 psi/17.6 to 31.6 kg per sq. cm which was specified by architects in their early literature is out of the question for two reasons.

- It is not possible to "pressure treat" wood in place according to the Approved Reference Procedures for Termite Control issued by the National Pest Control Association, 1980 (except perhaps for a species such as red oak.)
- It is forbidden by most product labels to treat wood for termite control by drilling and injecting termiticide. However, a new label for bendiocarb permits drilling and injection into existing openings for temporary control.

Buying a complete unit with tank, hose, fittings and reel is of course the preferred way for most PCOs. Many firms in their early years built their rig out of separate elements. Pump units without tank are readily available. Some PCOs still use the portable pump unit with a 55 gallon/208 l drum.

Complete units are available which include the pump mounted on a steel base, a pressure relief valve, pressure gauge (usually glycerine filled), surge tank to minimize pulsation in pumps, bypass pressure regulator with unloader valve on some units, suction hose with strainer assembly, bypass overflow, chemical resistant discharge hose, quick disconnect snap fittings and applicator tools. The hose is ⅜ inch/nine mm I.D. in small units and ½ inch/13 mm I.D. and even ⅝ inch/16 mm I.D. in the larger ones.

Tanks now are built which are rectangular and compact for use in the smaller trucks. They are usually made of fiberglass, although polyethylene is available at lower cost where sunlight is not a factor as in northern climates.

The plastic composition is designed to withstand the aromatic oil component in emulsifiable concentrates. Most tanks are translucent so the liquid level is apparent. A strip indicating level of gallonage is often affixed. Sizes usually vary from 100 to 250 gallons/378 to 946 l. Openings to fill are six to 12 inches/ 15 to 30 cm. Baffle plates are often installed on the larger tanks. An agitator booster fitting at the end of the fill tube causes increased turbulence.

Round plastic tanks are supported by saddles, steel plates curved to the shape of the tank. Many tanks have a small depression and outlet at the bottom to collect sediment.

An agitator is a necessity in 250 gallon/946 l tanks, especially where 72 percent chlordane emulsifiable concentrate is used. This product weighs more than 12 pounds a gallon/1.4 kg per l and when poured into 250 gallons/946 l of water, it may not mix thoroughly, especially if cool, even if circulated with the bypass. A buildup of the concentrate can accumulate on the end opposite the hose connection. The lower concentration formulations would have no problem mixing, being closer to the weight of the water. The writer uncovered this condition after a Pittsburgh PCO experienced failures during a cool spring. M. Eldridge suggests rolling the drum of concentrate before use.

Proportioners have been used for decades on a small scale but not until Tom Parker of Philadelphia engineered his model did this device come into more common use.

The power of the micro injection proportioner comes from a twin piston pump which forces water from a house faucet or tank into a treating hose. As it moves at the rate of no less than 2.3 gallon/8.7 l per minute it asperates emulsifiable concentrate out of the original pesticide container. Two dials determine the flow, one is set at the percent desired (⅛, ¼, ½ or one) and the other set at the labeled percentage of concentrate. There is no way the pesticide could flow back into the water lines of the structure. A preheater warms the pesticide in colder weather so that the flow is accurate for the thicker concentrate. Parker points out that with this proportioner, there is no "left over" pesticide in the tank to dispose of. The meter accurately notes how much material is being used; the operator can thus follow the label more accurately.

If the treatment is to be done in an unexcavated, odor-prone area, and if the operator has data that indicates that less than label dosage would suffice, the adjustment on the dial makes the change easy to do. Operators are cautioned, however, to use labelled dosages everywhere else on the structure, lest the extraordinary good record of the industry be marred.

The unit comes equipped with soil treating tools. If homemade tools are used in which the apertures are too small, the device will not work properly. A flow of less than 2.3 gallons/8.7 l per minute will not give a true mix.

At the time of publication, other units were appearing on the marketplace and no superiority is implied of one brand over the other.

Proportioners have other uses. With a five degree angle fan spray and a 26 ft/7.9 m vertical reach, treatment could be made without a ladder of the crevices under the eaves of some two-story buildings for the control of cluster flies, wasps, bees and other insects which may enter. Foundation walls and turf around the perimeter of a home could be treated for clover mites.

Handling 100 feet/30.5 m or more of hose several times a day is a time consuming chore. Reels make it much easier and faster. The most common unit with a hand crank rewind is constructed of a heavy steel frame cradle holding a steel spool turned on self-aligning bearings. A swivel joint inlet permits liquid to flow while the hose is being used.

While some models are turned manually with a crank, others are turned by battery to an electric rewind motor. Hose reels can hold up to 325 ft/99 m of ½ inch/13 mm PVC hose and even more of ⅜ inch/nine mm hose. PVC high pressure hose is the most common type. It largely has replaced the much heavier rubber hose of previous years. Working pressure is recommended at 600 psi/42 kg per sq. cm, although the tested burst strength is about 1850 psi/130 kg per sq. cm. Most of this plastic hose has been imported. It is commonly colored tan, orange or yellow. It does not stiffen in cool weather.

Many tools are available for injecting termiticide into the target site.

For treating the soil, soil probes, similar to root feeders, are made by several manufacturers. They are made of a heavy metal handle with quick action shut off valve attached for convenience. Centered in the handle is a four foot/1.2 m tube with a treating tip at the end. The outside diameter is ½ or ¾ inch/13 or 19 mm. For the ½ inch/13 mm size, extra heavy ¼ inch/six mm iron pipe schedule 80 pipe had been often used in the 50s and 60s with good success. There is also heavy aluminum tubing available.

Four foot/1.2 m extensions of these tubes with couplings are available if treatment is to be made laterally through the foundation wall, and immediately below the slab. Multiples of these extensions can reach situations otherwise difficult to treat.

For treating under a slab, and into wall voids, a popular tool is a sub-slab injector. This locks a nozzle with expansion ring tightly into opening and permits pressure to be applied under the slab. A short model for walls was increased 10 inches/25 cm in length to make it more comfortable to treat floors. Such a tool is especially necessary when treating concrete slab covered with a wood floor on screeds. A combination for slab, soil, and void is available.

Another model without the expanding gasket is available, but depends rather on downward pressure into a cone-shaped rubber washer. All such devices have a heavy handle with valve attached for quick shutoff. Some valves close with rotating handle, others with a lever.

Termite control specialists need specialized tools for their work. Holes must often be drilled through floor tile. One of the most popular tools for this is a floor tile plug cutter kit which includes pilot plate, replaceable plug cutter head, threaded shank and a plug cutter sharpening stone. Any standard electric drill could handle this cutter.

A less expensive drill is an Arch Punch, available in ½, ¾ or one inch/13, 19 or 25 mm or larger. It is sharpened with a sharpening stone tip that fits in the drill. A heater, propane torch or hair dryer and a hammer complete the tile cutting kit.

Competition has provided PCOs with a wide choice of excellent hammer drills. A typical hammer operates at 3,000 blows and 500 revolutions per minute. The clutch is adjustable for a safe torque level. Smaller hammer drills are intended, for wall void drilling and occasional floor drilling. However, PCOs who use the small drill largely for floor drilling will soon wear it out.

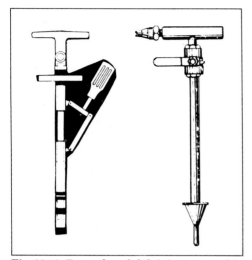

Fig. 28-10. Examples of slab injectors used in termite control work.

Reading the instructions and following maintenance schedules will improve performance and extend hammer life.

Problems arise when the extension cord is too light. Tips break when they hit an iron rod or flint pebble in the concrete. According to M. Eldridge, PCOs use a smaller star drill to break the obstructing pebble. Operators should guide, not force, the electric hammer, letting the tool do the work. Hammers should (if directions indicate it) be oiled and cleaned before, during and after use. Further, he suggests that drill bits be changed every ½ hour of continuous use to cool off.

Some PCOs prefer longer bits to save workmen from bending. Others, such as Tom Cox, say the chance of breaking a long bit is increased, with the risk of injury to the feet of the operator. This accident has occurred. In this case, short bits are used first, followed by longer ones where the concrete is thicker. Economies are gained with the use of short bits, but tapered bits cost even less.

Smaller holes require smaller size injector tools. Moreover, the holes are drilled more quickly, make less dust and are less conspicuous.

Diagnostic tools are important to a termite control specialist. Good eyes and a good flashlight are still the most important tools to spot termite tubes or specks of termite tube material which they use to plug a tiny hole in the wood. A good sounding tool is as essential as a cheerleader's plastic baton. A small sharp knife blade may leave less tell-tale damage than a large ice pick, though some PCOs prefer to leave visible evidence of the thoroughness of their inspection.

Excess moisture conditions often lead to infestations of carpenter ants, termites, other insects and also rot organisms. Lawsuits have been increasing dramatically over errors of omission in inspection reports of such conditions.

According to Cassens (1979), a vital tool for PCOs is the electric moisture meter or register. These small hand-held devices measure moisture below the surface of wood. Two probing needles are inserted into wood parallel to the grain. Moisture between the needles can be measured accurately.

More accurate and more expensive models which measure below 15 percent moisture content are used by the lumber industry, but for PCOs the need for that degree of accuracy is not so critical. Danger points are above 20 percent moisture content. In making inspections however, one wood member showing 18 percent or 19 percent moisture content, when all the others indicate below 15 percent, could uncover an incipient infestation.

Pat Butler, a Pittsburgh PCO, spotted active termites in several inspections solely with a moisture meter. No other evidence indicated termites, save the elevated moisture content of the infested wood members. Of course the inspector

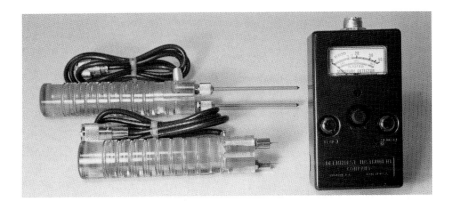

Fig. 28-11. Electric moisture meters measure moisture levels in wood.

would check to see if there is another reason for moisture.

Elevated readings mean trouble for the homeowner one way or another. Such warnings are a service to the client.

As the pest control industry becomes more professional, the use of moisture registers will become more widespread.

RODENT CONTROL EQUIPMENT

The Pied Piper of Hamelin would drool over the assortment of weapons available to his modern counterpart and he would probably ask why, with all these tools, we still have so many rats and mice.

The most universal tool at present is poison bait, which is available in bulk or in "place" packs. The term "throw" packs" is frowned upon by many professionals.

Most experts recognize that bait acceptance is vastly increased in a protected bait station. Legislation has mandated use of these bait stations in most cases. But there are still areas where bait cups can be placed. Within these areas are spots inside pallets in non-food sections, restricted sections or in unexcavated inaccessible areas.

For these places, bulk bait is placed in uncovered bait cups which are now largely plastic, having largely displaced the fibre ones previously used.

Most of the plastic cups are four inch/10 cm square, some with dividers. While intended for liquid and bait, the divider serves to stiffen the cup.

Mini bait cups, two inches/five cm square and plastic, also are available for the more potent rodenticides such as brodifacoum, bromadialone and zinc phosphide.

Prospects of lawsuits have forced more PCOs into using bait stations. Here too, plastic models have replaced many of the fibre boxes.

The fibre bait boxes are still widely used because of cost. There is a wax coated corrugated box, but the solid fibreboard, also waxed, is more popular. These boxes are packed flat and are quickly assembled. They hold about two lbs/0.9 kg of bait. Three inch/7.6 cm holes on each side are cut about one inch/2.5 cm off the bottom. Plastic ties are available to comply with "tamper proof" requirements.

Neither unit is intended for outdoor conditions for very long, although the

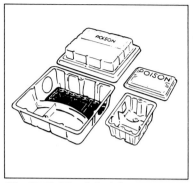

Fig. 28-12. Plastic bait boxes are useful in controlling rodents.

solid fibreboard will outlast the corrugated. In protected environments, they quickly acquire the ambient odor and are readily entered.

Careless workers in industrial warehouses destroy many bait stations. The reason the fibre boxes are still used is probably because such losses are less costly than with plastic boxes.

As with the fibre, these plastic boxes are available flat, and knocked down for economy in shipping. However, forklift cowboys took their toll of these plastic boxes, so manufacturers came up with a "memory plastic." This unit could be reset after being crushed.

More popular than these, however, are the two piece plastic boxes made by several manufacturers including J.T. Eaton & Co. and Bell Laboratories, each contributing an advantage in their product. Plastic boxes have either dome or flat lids. The dome lid resists storage above it, and the flat lid fits into tight spots. The flat lid model is also available in metal.

Eaton added a clear plastic box to their line. This has the advantage of saving time for a technician who can observe activity without opening the station, but is disadvantaged in a well-lit warehouse; rodents prefer to eat in dark protected places. Sherman Technology came out with a heavy well-constructed plastic box with a sliding funnel to load and dividers within to contain the bait and avoid spillage on the floor outside the box.

Bell's most recent product is a 'memory' plastic body of 150 mil polyethylene. They claim the plastic softens the shock of cold and heat for rodent paws. The bait station features a locking, sliding metal lid for rigidity; other improvements include a baffle system for child or tamper resistance.

Even before the fibre boxes were popular, nested, galvanized, metal boxes were available for mice and for rats. These sturdy "cafeterias" were widely used. The larger units can contain a water fount, but some have no lip to contain loose bait, as does the mice station. The covers are water proof for outdoor use.

Miniatures of these also are available for mice. They are usually small enough to place inside a pallet. Where pallets are stacked, multi-tier placement is a useful technique.

Helland's liquid rodent bait dispensers are small enough to fit into most bait stations. This is a nine oz/255g plastic reservoir snapped into a trough. Larger dispensers are available by several firms.

While the large metal bait stations are sufficient for outdoor rodent infestations, there are situations where a much larger bait supply is desired. This could be a meat packer near a river, grain elevator, zoo, dump or farm. For this reason, Rocon Industries came out with a large plastic bait station consisting of a domed lid over a tray with six openings and a holder for 20 lbs/9.1 kg of bait.

Of course, some PCOs will continue the practice, learned decades ago, to use inverted five gallon/19 l drums on bricks or cutting 55 gal./208 l drums in half to cover large containers of bait.

An increase in mouse populations, some of resistant strains, has stirred up more interest in live trapping. Wind-up traps are available which take advantage of the mouse's curiosity. If placed along a wall, an entering mouse is swept into a hopper by a "revolving door." Winding the spring six to eight times will catch as many as 15 mice. Overwinding can cut a mouse in half. They can also be drowned in an attachment to the trap. If not removed, the mice will die and decompose. When cleaned with strong solvents and degreasers, the odor will repel mice for a time.

Another trap for catching mice alive is low enough to fit under skids; it can trap up to 30 mice and does not require wind up. Mice enter through a trap door, the big advantage of the two live traps and glue boards is the ability to catch several mice with one setting.

Live rat traps are available, but few PCOs use them. Ketchall, Havahart and Tomahawk are amoung the manufacturers of these traps. They often are used to trap rats for testing resistance.

One of the earliest rat and mouse traps is the snap trap. It is still widely used by the public and also by PCOs.

Improvements have been made in the basic trap by Woodstream Corp. A professional model features an oversize plastic trigger pedal on a wood base with an option for firm or sensitive settings. Placing the trap next to the wall at right angles in a spot traversed by the rodent will usually work. Bait need not be used. Once the trap has acquired the odor of the area, it will most likely do its job. Using a sufficient number of traps is the mark of a professional.

Also available are small easy set metal mouse traps. Jaws close on the mouse and are easily opened to drop the body.

A four-hole choker type trap made of plastic with metal wire trap and a spring also is used. These latter two are hardware items.

Electrocuter rat traps were shown at the National Pest Control Association Convention in 1980. It is a rather large and elaborate arrangement in which the rat enters, is electrocuted and then bagged for easy disposal.

Another form of trap is an entrapment glue. Its use has increased phenomenally in the recent past, although it has been available for decades in bulk.

Formulated from adhesive resins, the product will not run in warm weather and will still work if cool. It will not work well if placed near wet spots, wet paws not being susceptible.

Ready-to-use glue boards were pioneered by J.T. Eaton & Co. and are available for mice and for rats. These are professional and easy to use. They can be mounted vertically near a pipe or near an opening in the attic to trap a rat. The ready-to-use boards are usually packed two to a set. Some glue boards are ready-made to fit into a bait box.

People who would make their own glue boards can buy a gallon/3.8 l of glue and apply it to sections of scrap roof paper or heavy plastic sheets or even lumber. One need only wet the hands, scissors and knife in detergent in water. Scoop out a handful, wet the hands again and smear it over the base.

For a large serious problem, one PCO cut hundreds of pieces of 1¼ inch/32 mm plastic tubing in eight inch/20 cm section and slipped into it a narrow strip of cardboard coated with glue. Mice can scarcely resist entering the opening. The strips are prepared by laying them on a table, then dipping a brush into detergent, and dipping the wet brush into the can of warm glue which had been held in a bucket of hot water. A dab of the glue will keep the tube from rolling.

Placing these close together in badly infested premises will trap many mice. Dust and water will not disturb these pipe stations. Three to four inch/7.5 to 10 cm pipe would work with rats; the glue must be spread thicker.

One warehouse with huge doors that were open all night and exposed to a rodent infested area, controlled rodents by placing large expendable cardboard boxes between the parked trailers. Several three inch/7.6 cm holes were cut near the bottom on each side, and the floor of the carton covered with a glue-coated cardboard. Incoming transient rats eagerly sought the protection of the box and became attached to it. This turned out to be an excellent rat trap.

Where bait is not working because of food preference or shyness, and glue traps are not desired, tracking powder is very much indicated. Whether the tracking powder is an anticoagulant or an acute rodenticide, it should be placed inside a protected environment so as to avoid contamination or poisoning by pets or humans.

A triangle shaped 12 inches/30 cm long plastic tunnel called a mouse tracking station has a hole in the center into which the tracking powder is dusted. Felt pads at the ends help remove excess dust. A less expensive fibre station for zinc phosphide powder is also available. It has a lid in the top center to introduce a ¼-teaspoon of ZP tracking powder in the center.

Confusion and controversy exists over the extent of rodent control with ultrasonic and electromagnetic systems. The ultrasonic device transmits high frequency sound waves (above human hearing range). Manufacturers claim that rats within a 1,500 to 3,000 sq. ft./139 to 279 sq meter area are repelled by these sounds. Apparently, rats can hide behind objects in a cluttered room and may cope with the annoyance. The Environmental Protection Agency instituted a one-year study in 1981 and takes no position. However, others have worked with the system, and found benefits, including top authorities such as W.B. Jackson. Charles Knote best sums up the results in *Pest Control* magazine, August 81:" the astute and correct use of ultrasonic devices in very select situations such as exterior doorways has proven helpful." Use and satisfaction with ultrasonic devices is growing, largely as a result of training by manufacturers such as Impex Industries.

With electromagnetic units, however, there is no confusion. The EPA's position is that they are ineffective. The electromagnetic device is buried in the ground or attached to a pipe driven into the ground. Claims are made that electromagnetic waves disorient animals throughout an area as large as 30 acres/12 hectares. An NPCA Government Affairs Bulletin 12/79 describes EPA actions against the promoters.

To detect rodent urine, black light hand lamps are available. Spectronics makes kits to detect rodent urine contamination including a chemical test of the rodent urine (other contaminants can confuse the reading).

OUTDOOR/YARD EQUIPMENT

The modern PCO is blessed with an incredible array of tools. Veteran PCO Maurice Oser, of Denver, often said that any piece of equipment is worth owning even if its use is limited. The possession of a variety of tools is a mark of a good technician.

Many of these devices originated from the agricultural and home gardening industries and were adapted to PCO needs. According to Snetsinger (1981), much of the PCOs equipment originated with agriculture, but pesticides for structural pest control were often initiated by the structural pest control industry.

PCOs need a variety of outdoor equipment. They are often called upon to treat outdoor sites adjacent to structures. Some of the target pest species include:

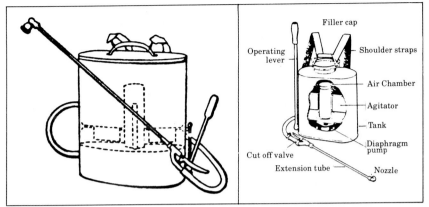

Fig. 28-13. Exterior and interior views of a knapsack sprayer.

outdoor biting or stinging insects such as mosquitoes, biting flies, bees, wasps; ornamental pests which support the wasp population; turf pests; fleas and ticks; cluster flies, clover mites, earwigs, scorpions, spiders, sowbugs, millipedes; wood roaches and other occasional pests.

These occasional tools will be listed even though they may be used only in limited areas of the United States.

Backpack equipment such as the Hudson Porta Pak ULV, Solo Mist Blower, Echo DM 9 and others fill a niche between the large scale mosquito project and the very small one. Warehouse, public parks, stadia and difficult terrain can be readily treated with back packs.

These are usually gasoline powered mist blowers which can generate fine aerosols for mosquito control and also larger particles to wet vegetation for residual control of many outdoor pests. Because these are ULV devices and can use more concentrated formulations, they can cover more acres without refilling. They can use emulsifiable concentrates or wettable powders. Some gas engine models can use dust and granules.

These devices have powerful radial fans which break up the liquid into airborne particles and propel them about 50 feet/15 m. The weight of one unit is 45 to 60 lbs/20 to 27 kg loaded.

Closely related to the backpacks is the knapsack sprayer. These are strapped to the back and manually operated with a short lever. A similar model for dust and granules has limited use by PCOs in trailer and mobile home parks, weed control, mosquito larviciding, treatment of attics and sewers, mole cricket baiting for lawns and applying insecticide granules in the lawn.

Another occasional tool is the rotary duster. Held in front of

Fig. 28-14. Trombone sprayer.

the operator for treating small areas, it also dispenses insecticide granules or bait as well as herbicides.

Another tool with only occasional use is the trombone sprayer. Using trombone-like strokes, a spray 25 ft./7.6 m high can be made. Long a garden and home tool, it has aided some PCOs when a more basic device was not available. Generally PCOs are reluctant to use equipment available at a garden store for fear it will hurt their professional image. As a result much time is often wasted through use of unsuitable equipment.

Along these lines, a more common hardware store item, which some PCOs use is the garden hose end sprayer. For years, a Pittsburgh PCO offered customers a choice of a lower price if he used his hose-end sprayer. This was only offered when the area to be treated was small. He saved time and money and the clients were satisfied enough that he was called back annually to repeat.

Hosing 20 to 30 gallons/76 to 113 l into a small patch of millipedes-infested mulch with such a device is certainly preferable to skimming the surface with a gallon/3.8 l of spray with a compressed air sprayer.

Of course, a very professional approach would also be made with a proportioner device described under the termite heading. This would give an adequate drench, put up a good appearance and save considerable time.

There also are small power sprayers on wheel available to PCOs. Some models hold up to 30 gallons/113 l. These are useful for a wide variety of outdoor pests PCOs are called upon to control.

TOOLS FOR HOLISTIC PEST CONTROL

Many systems are available in addition to chemical pesticides to support a holistic approach to pest control (Katz, 1979).

Holistic pest control is a more comprehensive approach to pest control which is dedicated to removing causes rather than treating symptoms. It implies total commitment of both the PCO and the client.

Integrated Pest Management (IPM) assumes there is a population level to be tolerated. As with IPM, chemicals are to be used judiciously where needed, selection always to be of the least offensive to the environment. Non-chemical tools are preferred where there are choices.

Insect electrocutors have become an important IPM tool. For food factories, they can serve as monitors helping sanitarians determine incidence of the intruders. Knowing this, an effort to correct the source is possible. Moreover, these devices are useful in the reduction of flying pest species. In 1981, a new organization of manufacturers of electrocutors was formed called the Electronic Pest Control Association.

Insect electrocutors take advantage of photopositive attraction of insects to light. The black light lamps emit wave lengths in the near ultra violet range of 300 to

Fig. 18-15. Garden hose sprayer.

400 nanometers which are particularly attractive to flies and many other flying insects. A transformer converts 100 volts to about 4,000 plus or minus 500 volts, but the United States testing organization does not permit the milliamp to exceed 9.5. This is the "let go" limit which will not injure humans.

Claims are made about how far the unit will reach. Insect species vary greatly in their perception of black light. Indian meal moths see nothing and are not at all attracted. *Drosophila* see only five to 10 feet/1.5 to three m. But some harmless outdoor moths can see long distances and so the claims are valid.

Installation is critical. Units should not be set where the sun shines on them or in the path of a strong air current, or behind a partition when the view is obstructed. They should be placed as close as possible to the "action" where insects are emerging or entering.

Situations where the author has found them useful include above false ceilings to attract cluster flies and in mausoleums to attract phorid flies. However, they do not attract dermestids in mausoleums.

They may not give good control of flies in a food facility unless a sufficient number of these electronic devices are installed. They cannot replace sanitation and exclusion. But they are an important aid in suppressing populations of flying insects if placed properly.

Other flying insect traps are available which attract with light, but without electrocuting the insects. A sticky surface traps them; with others, detergent and water drown them. With still others, a small fan draws them into the trap.

As indicated earlier, moisture registers are a vital tool to an IPM termite specialist. But these should not be restricted to the termite department. With such a tool, a technician can determine if the plaster on an outside wall contains more moisture than an inside wall. This could mean 1) a leak, 2) a condensation fault and 3) a problem with insulation.

An elevation of the moisture content in the plaster on the outside wall over the reading on the inside wall indicates a potential problem. Many insects seek such an environment. These include psocids, collembola, silverfish, carpenter ants, other ants and cockroaches. Alerting the client to a moisture

Fig. 28-16. A typical electric insect control unit.

condition is a valuable service which can avoid problems for a homeowner.

Control of birds by non-chemical means sometimes is possible by exclusion with netting. Many other remedies are advertised. All work at times, but PCOs

Fig. 28-17. Sticky traps for cockroach monitoring and control are growing in popularity among PCOs and homeowners alike.

know the pest birds have learned to live with their "nuisance" and return after a time. These include flashing lights, noise makers and anguish calls. Biologists well trained in bird behavior, however, can sometimes get good results with some of these systems.

Other times, stainless steel strips of spikes, such as made by Nixalite, can keep the animals from alighting if they are placed properly.

Rid-a-perch, utilizes the bird's tendency to perch on limbs to kill them. A band of absorbent material lines the upper side of the perch. When saturated with a toxic material, the alighting bird picks up a lethal does. The toxicant is in a closed system (i.e. it is restricted to perching species). The label specifies placement only where pest species are roosting. Unfortunately, there is a possibility that registration may be denied the toxicants currently used, primarily because of emotional protests about killing birds.

Another IPM tool is a crack filler in a pressurized container, Polycel® 100, a urethane product. This effectively hardens and seals crevices against insect pest intrusion. When hardened, occasional wetting is tolerated. Constant wetness is not recommended. Rodents are not kept out by this.

Air doors are another useful IPM tool. These are giant blowers that direct a wall of air from an overhead unit through louvers to the floor. The air moves at about 15 mph/24 km per hour. Depending on the size of the door, motors can be one to two horsepower or more.

Air doors to keep out the cold have smaller motors. Maintenance of the unit is important. If the louvens directing the blast of air are deflected even slightly, the current can change direction sufficiently to suck insects into the building near the floor.

Tubes and paper strips coated with tacky resins are available to reduce fly populations without pesticides. Such traps can be hung in sensitive areas such as hospital operating rooms.

Sticky traps for cockroach control are becoming popular with the general public as well as with PCOs. These serve principally as monitors, but some "harvest" is assured, with the body count obvious. Control is dependent on the number of traps placed and where they are placed. Generally, PCOs use many more traps than a lay person is likely to use. Some traps have strong odors which can be absorbed by fatty foods, if they are exposed.

Odors can cause much grief for PCOs. Ailments not easily diagnosed could wind up as a law suit because of a persistent odor. If cross ventilation is not available, miniature exhaust fans are available. These can be installed in unexcavated and odor-prone areas. PCO storage rooms, if not ventilated, should have such a low power exhaust fan installed. A typical blower fan uses 32 watts, has $1/100$ hp, has a two inch/five cm opening and handles 60 cu. ft./1.68 cubic meters a minute. This is ideal to install in a damp unexcavated area. It will keep the area relatively dry, at least during the summer months, and it is a good ancillary service to clients for PCOs to tend to this inexpensive device for seasonal servicing.

SAFETY EQUIPMENT

Various check lists for safety equipment have been published by the National Pest Control Association and other groups. The following is a compilation derived from these lists.

- **First aid kit.** This should be placed in every vehicle and in company offices and store rooms. The company doctor should be consulted to review the items in the kit. He may want to add a specific medication or device. The name and number of the physician and poison control center should be prominently displayed in the kit.
- **Fire extinguisher.** This should be an all-purpose type for oil, paper and electric fires. For some operations, the technician should be alert to the possible need of the extinguisher which is normally in the vehicle. If the vehicle is parked a block away, a fire could get out of hand before the extinguisher can be used. On certain fogging operations, an extinguisher should be nearby. Most industrial premises have extinguishers and these should be noted.
- **Flashlights and extra batteries and bulbs.** For bird work, spotlights are needed. For unexcavated areas, large hand lanterns are important. For general pest control, a bargain flashlight can be expensive by not allowing adequate detection of evidence of harborages. A dome light should be over chemicals and equipment in vehicle. This should be checked to be in good operating order. Store rooms should have adequate lighting also.
- **Racks to store material and equipment.** There should be a place for every item. Periodically the vehicle should be washed of chemical residues with degreaser or strong detergent. The vehicle should not have cans rolling around. The store should have shelves on racks to properly segregate combustible from non-combustible containers. The storage area must be locked. Consult the licensing inspector for state regulations which affect storage of pesticides.
- **Vehicle safety.** The vehicle is very much a part of the equipment of a PCO. The unit should be inspected periodically for safety features, such as brakes, lights of various kinds, wipers, horns, flares, mirrors, locks and exhaust system; also separation between driver and chemicals.
- **Containers.** They should be unbreakable and impervious to pesticides and should be labeled properly.
- **Measuring devices.** While the trend is to use premeasured packages, there is still a need to measure from a larger container. A funnel and measuring cup should be kept in a sealed container to avoid contamination of air and vehicle. Excellent pre-measuring devices are furnished by some pesticide manufacturers and container molders.
- **For spillage.** Clean water in clean containers should be available with soap and towels. Wiping rags and a rubber or plastic squeegee for some cases are needed. Carry paper towels for small spills.

Wiping rags should be stuffed into basement drains before a termite job is started. In case of an accident, there will be no pollution into the sewer. Carry a bag of absorbant granules, 25 or 50 lb. bag, the kind garages use to clean oil slicks. Janitor supply outlets sell it.

- **Clothing.** Extra sets of clothing, including underclothes and shoes and socks, should be carried for service people. A splash on the garment is equivalent to a patch test for the technician for the rest of the day. Don't wear cuffed pants since such pants may retain pesticide dusts and granules.

Coveralls are needed if extensive misting is to be done, or for crawling in unexcavated areas or in attics. Exposing skin to excessive abrasion, chemicals or to organisms such as "creeping eruption" in a crawl space is dangerous. Knee guards also are needed.

Aprons, when pouring concentrates, are mandatory on several labels. These should be made of neoprene or similar material.

For termite work, soles of shoes should be of a chemical resistant rubber-like material which is not affected by oil and pesticides. For bird work, on roofs, non-skid shoes are needed. Steel toe shoes are also useful. When working with, or treating new, electrical equipment, wear shoes with non-conductive soles and heels.

Gloves should be the kind which do not absorb pesticide, such as neoprene or a canvas impregnated with a plastic material. These also help to avoid a shock with ordinary electric tools. Don't use gloves lined with cloth. Chemicals can splash down the wrist and be absorbed by the cloth. Gloves periodically should be washed inside and out. Remember, the skin is a barrier for most substances, but it can be a funnel with some chemicals. Ideally, gloves should extend to near the elbow and be loose enough to shake off. Gloves or tongs should be used when handling dead rodents.

Bump hats are necessary in crawl spaces, in attics, in meat packing establishments and other industrial settings where they are mandatory. PCOs are expected to bring their own hats. In bird work, technicians should wear hard hats.

Some labels specify protection for eyes and face. Goggles or face shields should be donned if there's any chance for overhead spraying to splash onto the face. Outdoors, driven mist must be kept from eyes with chemical splash goggles. Goggles should have vents to resist fogging. For termite work and in certain industrial facilities, safety goggles should be worn.

For bee work, a bee veil, sleeve protectors and gloves are necessary.

Respirators are one of the most important and under-utilized safety tools among many PCOs. Many labels make them mandatory, but hot weather, indifference and fear of what the customer might think are partly responsible for the under-utilization of respirators.

They are more important than ever, what with aerosol particle sizes being produced in the low micron range. Particles under five microns can be more easily absorbed into the lungs, and more readily enter the blood stream.

During any volumetric treatment with aerosol pesticides, the pesticide respirator used should be jointly certified by the Mine Safety and Health Administration of the Department of Labor (MSHA), and by NIOSH under provisions of 30 CFR Part 11. Pesticide supply houses can furnish the approved mask for aerosol pesticides in common use. Information in this section was confirmed by R. Grunberg and R. Markle of the Mine Safety Appliances Company.

The most common pesticide respirator is a soft rubber halfmask facepiece with a rolled cushion edge to fit most faces. It is available in three sizes.

The two cartridges, each containing a combination of activated charcoal to absorb organic vapors which do not exceed 0.1 percent of the air volume, and a particulate pre-filter to remove dust and mist particles, is generally appropriate for most aerosol pesticides. They are not acceptable for some greenhouse uses and of course should not be used with fumigants.

Particles less than one micron in size are effectively removed by this type of respirator which is certified by NIOSH/MSHA.

Not many PCOs appreciate the difficulty in controlling leaks where the masks fit the face. As much as a 10 percent leak can occur even with a smooth shaven face. Face shapes and sizes in both sexes vary greatly — it is essential to pick one of the three sizes available which fits.

Beards interfere seriously. Anyone required to wear a respirator should not wear a beard or long sideburns. They will interfere with the facial seal. Whatever advantages there are in the efficiency of filtering particles and vapors through the cartridges can be partially lost with a bad fit with the technician's physiognomy.

PCOs can determine the extent of leaks with comparatively simple qualitative fit tests which can be made with a procedure described in Mine Safety Appliances Bulletin 1,000-16.

For persons whose eyes are sensitive to any of the aerosol pesticides, a full face chin style pesticide mask is necessary. This unit contains one larger canister and can also handle concentrations of vapors of 0.5 percent by volume. With this unit, the fit is somewhat easier. But beards and glasses create problems. Spectacle kits are available for full face pieces for those who must wear corrective lenses. Contact lenses should never be worn under a full face piece since contaminants could get under the contact lenses and also they could be dislodged.

For PCOs who do extensive industrial aerosol applications, an even larger cartridge is available than the chin style. A universal harness holds the canister against the body. This unit can handle vapor concentrations up to two percent of the air volume.

All masks should be used only by the technician for whom it was fitted. They should be kept in a plastic bag and washed and air dried after each use.

Technicians are warned to replace the cartridge or canister when breathing resistance is encountered, when odor or taste of contaminant is detected, or if nausea or illness is experienced. If the cartridge or canister feels hot, this is an important warning that high concentrations have been encountered. The canister should be discarded after each use.

PCOs working for extended periods in higher concentrations of gas or confined spaces deficient in oxygen should use an air supplied respirator with an egress cylinder or a self contained breathing apparatus. If in doubt, one should have this equipment available for use.

PCOs involved in fumigation require special training and licensing and no attempt is made here to cover fumigation equipment. Indeed, this is one subject where incomplete knowledge is dangerous for the technician and possibly the client. Suffice it to say that fumigation equipment ranges from sand snakes to gas meters and includes such special items as oiled ear plugs for technicians with punctured ear drums.

Technicians using dusts should use a dust respirator certified by NIOSH/MSHA for toxic dusts and mists. Dust respirators have a soft rubber face cushion to seal from contamination. It also is appropriate to use the combination chemical filter cartridge containing a particulate prefilter. A double headband keeps the mask in place.

The operator always should be aware of the possibility that the filter capacity may be reached, as indicated by excessive breathing resistance. When this happens, inhaled air becomes contaminated by the excess deposits of toxic particles.

The technician should have an ample supply of spare refills and a clean plastic bag to store the mask.

For termite crews, anti-siphoning equipment or procedures are mandatory according to the termiticide labels. A back-flow preventer is readily available which fits onto a hose connection at a source of water. Electric tools must be properly grounded. Most homes have grounding plugs. They should be checked with testing devices before using. If not working, a ground wire to a water line is necessary.

Devices are available at pest control supply houses which can cut off the power if the house wiring is improper, or if the tool develops a short, or if the bit strikes metal inside the concrete. These are known as ground fault interrupters and drill stoppers.

Of course, only three wire extensions should ever be used. And the size of the wire should be heavier than the capacity of the tools. Again suppliers will help with the selection. Cords should be checked for breaks periodically. Since so much of the termite work is outdoors, only the round extension cord for outdoor use should be used.

A stepladder or small stool should be part of the equipment. Customer's property should not be used.

Drop cloths and tarps are often a necessity if law suits are to be avoided. Dust masks and bump hats should be available when needed and safety goggles should be worn to prevent eye damage from stone chips when drilling. Signs and bridges for hoses laid across sidewalks are needed to accommodate pedestrian traffic.

Small tools such as a pen knife, pliers, and electrical tape should be readily available.

A broom and a dustpan to leave the clients property clean and uncontaminated is important for reasons of safety and good public relations. This is often more impressive to some homeowner than the work done to stop the termites.

Despite having a full array of safety equipment accidents will happen unless the pest control technician is constantly safety conscious. The management of pest control companies have a key role to play in this by adopting procedures that help eliminate human error. Good employee selection, training and supervision are essential to maintain a good safety record. After all, the main causes of accidents include lack of training, poor supervision, lack of alertness, poor judgment, indifference and laziness.

LITERATURE

ANON. — 1952. Characteristics in the Dispersion of Insecticide Particles. AMCA Bull. No. 2, Mar. 1980. Approved reference procedures for subterranean termite control. Nat'l. Pest Control Assn. Vienna, Virginia. 1981. Pressurized Products Survey, U.S., 1980. CSMA.

BALS, E.J. — 1978. Controlled droplet application today. World Crops, July/ Aug.

BERNHARD, K.M. and G.W. BENNETT — 1980. Ultra Low Volume Applications of Synergized Pyrethrin for Stored Products Pest Control. Thesis completed June 1980. JEE and Pest Control (In press).

BREHM, W.L — 1968. Care & Maintenance of the One Gallon Stainless Steel Sprayer. Pest Control. 36(5).

CASSENS, D.L. — 1979. Using Electric Moisture Meters. Pest Control Technology. 7(2):15, 16, 46.

EBELING, W. — 1975. Urban Entomology. Univ. of Calif. 695 pp.

JOHSEN, M. and E. & W. DORLAND — 1972. The Aerosol Handbook.

KATZ, H. — 1965. How to Treat Veterinary Hospitals. Pest Control. 33(10):16. 1979. Integrated Pest Management, Then and Now. Pest Control. 47(11):16-18, 36, 38, 40, 81. 1981a. New Labels Mean Changes in Termite Control Procedures. Pest Control Technology. 9(4):34-36. 1981b. Aerosols, Taking the Fog out of Fogging. Pest Control. 49(9):30.

LOFTUS, R.D. — 1980. The Trend Blows Toward Dust. Pest Control Technology. 8(11):10-12.

MATTHEWS, G.A. — 1976. New Spraying Techniques for Field Crops. World Crops, May-June.

SANDERS, P — 1979. Handbook of Aerosol Technology. 2nd Ed.

SHORE, J. — 1974. Calculations Crack and Crevice Treatment. Pest Control. 42(7).

SNETSINGER, R. — 1982. The Ratcatcher's Child. Franzak and Foster (In press).

SPEAR, P.J. — 1955. Serviceman's Manual. Nat'l. Pest Control Assn.

TRUMAN, L.C., G.W. BENNETT and W.L. BUTTS — 1976. Scientific Guide to Pest Control Operations. 3rd Ed. Purdue University/Harvest Publishing Co. 276 pp.

WILLIAM E. BLASINGAME

Bill Blasingame graduated from the University of Georgia in 1950 with a bachelor's degree in entomology. Since graduation he has been continuously associated with the pest control industry in three different capacities — a regulatory official, a member of the industry, and since 1974, as a consultant and teacher in pest control.

Immediately after graduation, Blasingame began working with the Georgia (State) Department of Entomology as assistant state entomologist. He was appointed state entomologist in 1955, and served in that capacity until 1964 when he resigned to join Getz Services. Blasingame worked with Getz as vice-president and technical director until 1974. Since 1974, he has been director of Stephenson Services, a division of Stephenson Chemical Company.

The Georgia Structural Pest Control Act was passed in 1955 and Blasingame served as the first chairman of the Georgia Structural Pest Control Commission. He served on the Commission until he left state service in 1964. As director of the Entomology Division of the Georgia Department of Agriculture he was also responsible for the enforcement of this law. He was one of the organizers of the Association of Structural Pest Control Regulatory Officials and served one term as chairman.

Blasingame has served as president or chairman of the following professional organizations: National Pest Control Association, Georgia Pest Control Association, Georgia Entomological Society, Georgia Council of Entomology, Georgia Plant Pathology Association, Southern Plant Board and Pi Chi Omega.

As Director of Stephenson Services, Blasingame has participated in training courses that have involved over 5,000 pest control technicians. In addition to the regularly scheduled courses at the Stephenson Training Center in College Park, Georgia, he has conducted many courses in various states for state and local pest control associations and for individual pest control companies.

CHAPTER TWENTY-NINE

Legislation

By William Blasingame[1]

WITHIN THE STRUCTURAL PEST control industry, the broad area of pesticide legislation has been a matter of discussion and, to a degree, controversy for many years. Early activity at both state and federal levels was restricted to "labeling" laws, but beginning in 1930, states, primarily in the South, began enacting laws that regulated the use of pesticides. By the end of 1960, at least 17 states had passed legislation regulating the practice of structural pest control.

It was not until the passage of the Federal Environmental Pesticide Control Act (FEPCA) in 1972 that Federal legislation moved from purely labeling laws to a regulatory scheme that encompassed the use of pesticides. The passage of this amendment effectively put an end to the long-running debate concerning self-regulation vs. government regulation. However, its major changes relating to pesticide registrations and the completely new pesticide use regulations have certainly stimulated the controversy that historically has been associated with this type of regulatory legislation.

STATE STRUCTURAL PEST CONTROL LAWS

In any consideration of legislation affecting structural pest control, it is important to understand the basic differences in concept and design between these earlier state applicator laws and the use provisions of the current amended Federal Insecticide, Fungicide and Rodenticide Act (FIFRA).

The flurry of state legislation during the 1940s and 50s was not due to concern regarding the use of pesticides, but rather was because of public concern about

---

[1]*Director of Stephenson Services, Stephenson Chemical Company, Inc., College Park, Ga.*

the pest control industry. A major part of this concern dealt with one particular structural pest, the termite. Several problems inherent with termite control were important contributors, but two virtually unique factors spawned most of the concern: a) the difficulty of determining incipient infestations, and b) the inability to evaluate control efforts. These two factors created a situation that permitted unskilled, untrained individuals to engage in termite control. The activity of these so-called "fly by night" operators, coupled with the absence of generally accepted control standards, created consumer suspicion and mistrust that eventually surfaced as laws regulating the practice of pest control. The laws and resulting regulations that emerged clearly reflect the primary concern with termites. Most of the regulations and by far most of the enforcement activity related to termite control. An important part of the regulations in many of these states was the establishment of required minimum standards for termite control.

FEDERAL LAWS

Insecticide Act of 1910. Federal regulation of pesticides began with the enactment of the Federal Insecticide Act of 1910. The basic purpose of this Act was to protect the consumer from substandard products by preventing the manufacture, sale or transportation of adulterated or misbranded insecticides and fungicides. Two insecticides, Paris green and lead arsenate, were mentioned by name.

The complexity and number of pesticides changed very little from this period until the era of World War II. Spurred in large measure by the threat imposed by such pest-borne diseases as typhus and malaria and the increased demands in agriculture, there was a major acceleration in pesticide research and development. By 1945 a number of new synthetic pesticides had been developed, including one of the better known compounds, DDT. The number and variety of chemical pesticides manufactured and used increased rapidly and in 1946 the Congress began holding hearings on more comprehensive federal legislation.

Federal Insecticide, Fungicide and Rodenticide Act. In June of 1947, the Insecticide Act of 1910 was completely replaced by the Federal Insecticide, Fungicide and Rodenticide Act (FIFRA). This regulatory legislation, though still a labeling law, greatly expanded the simple provisions of the 1910 Act. Important features of the Act required:

- Registration of all economic poisons moving in interstate commerce. The Act defined "economic poisons" a) as any substance or mixture of substances intended for preventing, destroying, repelling, or mitigating any insects, rodents, nematodes, fungi, weeds and other forms of plant or animal life or viruses, except viruses on or in living man or other animals, which the Secretary of Agriculture shall declare to be a pest, and b) any substance or mixture of substances intended for use as a plant regulator, defoliant or desiccant.
- The display of poisonous warning statements on labels of highly toxic compounds.
- The inclusion of warning statements on the label to prevent injury to people, animals and plants.
- Instructions for use to provide adequate protection for the public.
- The coloring of certain white insecticide powders to prevent their being mistaken for foodstuffs.

FIFRA was amended in 1954 to include several new types of agricultural chemicals within the definition of economic poisons. Amendments in 1964 required for the first time that each pesticide carry a license number identification and further expedited procedures for suspending the marketing of previously registered pesticides which were found to be unsafe.

Miller Amendment. In 1954, pesticide regulations were expanded further by an amendment to the Food and Drug Act authorizing its administrator to set tolerance limits for the residues of pesticides on foods. This provision, known as the "Miller Amendment," required the pretesting of a chemical pesticide before its use on food crops to show its usefulness to agriculture and how much residue, if any, would remain on a food crop after application. This has some significance to pest control operators who apply pesticides in and around raw agricultural crops.

The Environmental Protection Agency. Prior to 1970, the two basic federal laws, FIFRA and the Federal Food, Drug and Cosmetic Act, were administered by the Departments of Agriculture and Health, Education and Welfare. In December 1970, the Environmental Protection Agency was established and all legal functions relating to pesticide control were transferred to this new agency. The creation of this new agency resulted from increased public and governmental awareness and concern regarding environmental safety. With regard to pesticide regulations it signaled a change in primary emphasis from efficacy to safety. Critics of FIFRA before 1972 claimed too much attention was given to the benefits of using pesticides. Critics since the FEPCA amendments have claimed too much attention to potential risks. The debate continues.

Federal Environmental Pesticide Control Act. In October of 1972, significant amendments were made to FIFRA by the passage of the Federal Environmental Pesticide Control Act (FEPCA). The consumer protection features of FIFRA remained, but FEPCA additionally required broader consideration of the need to protect the public from potentially harmful effects of pesticide use. Of primary importance was the requirement that EPA refuse to register a pesticide unless it is determined that "when used in accordance with widespread and commonly accepted practice it will not cause unreasonable adverse effects on the environment." Many changes were made in the registration process, but of primary importance to users of pesticides were those provisions that established a completely new area of federal control over the use of pesticides. The label became the law and a compulsory system of certification was established governing the use of certain restricted pesticides.

Provisions of the Act relating to registrations primarily impact manufacturers, formulators and distributors, but certain of the new registration requirements directly affect the user of pesticides. Examples of such changes include the requirement that:

- All pesticides be registered with the Environmental Protection Agency.
- All existing products be reregistered.
- Pesticides be classified for general use, restricted use or both.

Reregistration. According to EPA, there were approximately 34,000 products registered when FIFRA was amended in 1972. Reregistration of these products is required to determine if the data supporting these registrations meets the new standards imposed by the 1972 amendments. If during this review it is determined certain criteria of safety to man and the environment have not been met or exceeded, EPA issues a *rebuttable presumption against registration* (RPAR). The RPAR is used by EPA to initiate a study of the benefits and risks of the continued use of the pesticide. If the review is unfavorable, suspension or cancellation hearings are initiated.

The delay in completing reregistration continues to have a major impact on pest control operators and other groups. Immediately, on the effective date of FEPCA in 1972, it became a violation of the law to use a pesticide in a manner inconsistent with the label. This created a problem in that existing labels were not written to provide this type of complete use direction. For the most part, product labels provided only general information as to the pests controlled, application sites, application techniques and, in some cases, even use concentrations. Historically, specific instruction pertaining to use was obtained from recommendations of authoritative sources such as experiment stations, universities, agricultural extension services and professional organizations like the National Pest Control Association (NPCA). One of the major problems to pest control operators was the original requirement the target insect had to be specified on the label (changed in 1978 amendments). In hearings before the Congress the NPCA identified 17 common household pests that were not on any available pesticide label. The problem of inadequate directions for use is perhaps best illustrated by the termiticides. In October of 1972, few, if any, termiticide labels contained more than very basic directions for use. Put very simply, the pest control operator was placed in the awkward position of abiding by the law or controlling termites.

The inadequacy of existing labels was recognized but thought to be a short-term problem in that FEPCA required reregistration be completed by October 1976. When it became apparent this deadline could not be met, the EPA instituted a series of Pesticide Enforcement Policy Statements (PEPS) to correct major problem areas.

PEPS were not designed to interpret the law or otherwise define what is and what is not legal, but only to inform the public how the EPA will exercise its prosecutorial discretion. Four PEPS were issued that affected structural pest control.

PEPS No. 1-Covered the use of pesticides at less than label dosage rates.

PEPS No. 2-Covered the use of pesticides for the control of pests not named on the label.

PEPS No. 4-Covered preventive pest control treatments in the absence of target pests.

PEPS No. 6-Covered the use and labeling of service containers for the temporary storage of pesticides.

FIFRA amendments in 1978 eliminated the necessity for all the above PEPS with the exception of PEPS No. 4.

PEPS No. 4 continues in force and covers the important area of preventive pest control. Since so much of the work done in structural pest control is preventive, details of the PEPS should be clearly understood. The need for this statement arises from the fact many labels do not affirmatively provide for the use of the pesticide as a preventive treatment. The question arises that in such cases would preventive application constiue a use inconsistent with the label? This PEPS gives EPA's position as it relates to enforcement actions. The following is a summary of this statement as published in the Federal Register, Vol. 41, No. 132:

"In the exercise of its prosecutorial discretion the Agency has determined that it will not generally initiate an enforcement action against a person who uses a registered pesticide in a preventive pest control treatment in the absence of the target pest where all of the following conditions are met:

1) the label of the pesticide which is used does not affirmatively prohibit preventive treatments; and

2) the target pest is reasonably expected to infest the treated area; and

3) the pesticide is normally safe and efficacious against the target pest when used in a preventive capacity."

Another "user" problem associated with registration has been the voluntary cancellation of registrations by manufacturers. BHC, chlordecone, acrylonitrile, Perthane®, sodium arsenite and Strobane® are pesticides where the basic manufacturer has requested cancellation of some or all of the product's registrations. It is expected additional pesticides will be similarly withdrawn, or at least have some previous uses deleted. This would be especially likely with pesticides that have limited, minor use and there is a need for additional research data to support their continued registration.

As previously mentioned, the reregistration process was scheduled for completion by October 1976. Relatively few pesticides have been reregistered to date and it is difficult to predict how many additional years may be required for completion. This simply means the problem of dealing with incomplete labeling, particularly as it relates to use, will continue with many products for some period of time.

Classification. As with the reregistration process, classification as "general" or "restricted" use was originally required to be completed by October of 1976. This deadline was delayed by Congress for one year. Then, in the 1978 FIFRA amendments it was provided that classification of existing products could be by regulation. As a practical matter, this means that some products may be restricted, but current labeling would not so indicate.

1) A general use pesticide is defined as one which the Administrator (EPA) has determined that, when applied in accordance with its directions for use, warnings and cautions, will not generally cause unreasonable adverse effects on the environment.

2) A restricted use pesticide is one which the Administrator has determined that, when applied in accordance with its directions for use, warnings and cautions, may generally cause, without additional regulatory restrictions, unreasonable adverse effects on the environment.

Some pesticides may be registered for both general and restricted use. In such cases, the classification is not based on the pesticide per se, but rather the uses permitted by the label. In general, restricted pesticides must be applied by or under the direct supervision of a certified applicator. Because of its toxicity or hazard, a pesticide may be restricted to only certified applicators. Pesticides classified at the federal level for general use may be changed to restricted use by individual states. States do not have the authority to change a restricted use classification to general use.

Opinions were expressed during congressional hearings in 1971 and 1972 that relatively few pesticides would be restricted. This has to date been proven true; certainly in the area of structural pest control very few of the more commonly used pesticides have been so classified.

Certification. A major provision of Amended FIFRA is the requirement that restricted pesticides must be used by or under the direct supervision of a certified applicator. After the passage of the FEPCA amendments in 1972, the EPA took the position that structural pest control operators were not only users of pesticides, but also distributors and sellers of pesticides. This created quite a controversy within the industry, primarily because of the additional enforcement options this provided EPA, the most important being that it gave the Agency the right to inspect pest control establishments. The definition contained in the 1978 amendments redefined "certified operator" to say in effect that any applicator who applies a pesticide only to provide a service of controlling pests is *not* considered to be a dealer or seller of pesticides. With this important change, PEPS No. 6 that covered the use and labeling of service containers was no longer necessary.

Amended FIFRA provides for two types of certified applicators:

1) Private Applicator — A certified applicator who uses or supervises the use of a restricted pesticide for the purpose of producing any agricultural commodity on property owned or rented by him or his employer.

2) Commercial Applicator — A certified applicator who uses or supervises the use of a restricted pesticide for any purpose or on any property other than as provided by the definition of "private applicator."

The active involvement of the certified applicator in the use of restricted pesticides is of paramount importance to the entire concept of classification and certification. In permitting the use of restricted pesticides by or "under the direct supervision of a certified applicator," the question is raised as to what constitutes direct supervision. Amended FIFRA states that unless otherwise prescribed by the labeling, a pesticide shall be considered applied under direct supervision if it is applied by a competent person acting under the instructions and control of a certified applicator who is available if and when needed. Some states have gone further in attempting to assure the involvement of the certified person.

Commercial applicator categories. EPA has established 10 occupational categories for commercial applicators. The states are required to adopt these categories, but may add sub-categories to fit their particular needs. For example, Category 7 — industrial, institutional, structural and health related pest control — frequently contains the subcategories household pest control, wood-destroying organisms and fumigation. The 10 categories are:

1) Agricultural pest control
2) Forest pest control
3) Ornamental and turf pest control
4) Seed treatments
5) Aquatic pest control
6) Right-of-way pest control
7) Industrial, institutional, structural and health related pest control
8) Public health pest control
9) Regulatory pest control
10) Demonstration pest control

To be certified a commercial applicator must pass a written examination to determine competence to use and handle pesticides. Regardless of the particular occupational category, all commercial applicators are examined in the following general areas:

- Label and labeling comprehension
- Safety
- Environment
- Pests
- Pesticides

- Equipment
- Application techniques
- Laws and regulations

The second part of the examination covers additional questions in the above subject areas of particular importance to the occupational category.

Requirements for certification and examinations may vary from state to state. In states that have specific structural pest control laws a much higher level of skill and competence is required. The range of questions, technical and otherwise, is much greater in scope and in specific detail, making the examinations more difficult. There are often prerequisites for admission to the examination. This may take the form of required formal education and/or work experience, age of the applicant, etc.

Penalties. The Act contains provisions for both civil and criminal penalties. Civil penalties for commercial applicators may be up to $5,000 for each offense, criminal penalties up to $25,000 or imprisonment for not more than one year or both. In determining the amount of the civil penalty, the EPA must consider the size of the business of the person charged, the effect on the person's ability to continue in business and the gravity of the violation. Penalties are normally assessed against the owner of the business, but may include the person committing the act.

In states having specific structural pest control legislation, penalties may also be in the form of suspension and cancellation of applicator certification.

Enforcement. Enforcement of FIFRA from 1972 to 1978 was primarily by EPA. A review of the enforcement activity involving pest control operators during this period shows the vast majority of warnings and fines imposed by EPA involved one violation, the use of a pesticide in a manner inconsistent with the label. EPA began with a very literal interpretation of what constituted "inconsistent use." The PEPS identified certain "use inconsistencies" that would not be enforced. A more positive solution came when the 1978 amendments declared the following do *not* constitute use inconsistent with the label:

- Applying a pesticide at any dosage, concentration or frequency less than that specified on the labeling;
- Applying a pesticide against any target pest not specified on the labeling if the application is to the crop, animal or site specified on the labeling (unless the labeling specifically states that the pesticide may be used only for pests specified on the labeling);
- Employing any method of application not prohibited by the labeling.

States now have the primary responsibility for enforcing the use requirements of FIFRA (1978 amendments). The role of EPA will be to monitor state enforcement activities and otherwise provide secondary support.

To date, 48 states have completed the necessary requirements to assume this responsibility. States may additionally enter a grant program with EPA to enforce other provisions of FIFRA.

The future. In many areas of national policy our actions and reactions are almost like the swinging of a pendulum. A far swing to the left triggers reaction that may swing the pendulum completely to the right. This, in a very broad sense, is the critical issue with pesticides, with benefits of using pesticides on one side and risks on the other. How nearly the pendulum can be balanced will in a real sense be the primary influence on the future.

The issues involved in assessing these benefits and risks are varied and complex, running the gamut from technical to economic, from short-term to long-range effects. Three branches of the federal government have a direct effect on the position of the pendulum: the Congress, courts and the EPA.

It seems clear from legislative history that Congress fully supports a balanced assessment, recognizing that a certain element of risk is involved in any use of any chemical, whether it be a pesticide, pharmaceutical or detergent. Amendments to FIFRA in 1975 and 1978 reflected the feeling of Congress that not enough attention was being given to the benefits side of the equation. These amendments strengthened the role of the U.S. Department of Agriculture in the decision-making process and a new Scientific Advisory Panel was created. The latter panel was created to provide an unbiased assessment of potential health and environmental hazards.

The courts have taken a very conservative approach to their evaluation of risks. Even prior to the FEPCA amendment, the courts, in a major decision (1971), ruled that when a substantial question of safety is raised the burden of proof shifted to the manufacturer. Subsequent court decisions in cases involving aldrin, dieldrin, heptachlor and chlordane further reflect primary attention to potential risks with subordinate or secondary attention to benefits.

The position of the EPA is obviously and properly influenced by these two branches of government. Adding to the burden is the fact that several major requirements of FEPCA are not yet complete even after nine years and there have been additional amendments to FIFRA since 1972. As to the future, there is reason to believe there will be significant changes in the posture of this Agency. At Senate confirmation hearings and in subsequent speeches, the current Administrator, Anne Gorsuch, has spelled out some future changes in management principles and program objectives. From these statements one can predict a:

- Simplification and streamlining of the regulatory process.
- Closer working relationship with the states. The states will be dominant in the important areas of certification, enforcement, training and education. Further, one might expect a greater state influence in the overall regulatory process.
- More balanced approach to the question of benefits and risks. Statements of the EPA Administrator such as the following give credence to such a conclusion: "We must recognize that EPA is affected today by economic, energy and environmental considerations largely unknown when many of the laws were passed. The public is no less committed to environmental protection, but increasingly aware of the need to balance all of these interests."

The most certain prediction that can be made is regulations will increase, particularly in states that have enacted applicator laws since 1972. In these states an "official" connection with structural pest control has just begun. There is historical precedent to predict that as enforcement begins areas will be discovered where they feel additional regulations are required. For example, state structural pest control laws originated at a time when the "state of the art" was much less than today. Many of the problems and conditions that gave rise to the legislation are no longer real factors, yet the history of these laws reflects an increase in regulations. Additional predictions include:

- In states having structural pest control laws, the pattern of enforcement will change from an almost complete preoccupation with termite inspections to a more balanced consideration of inconsistent use.
- EPA enforcement activities involving pest control operators were almost

entirely due to consumer inquiries. This will undoubtedly continue with the states as they take over enforcement.

- If the certified applicator is not actually involved in the use or supervision of restricted pesticides, then all of the words pertaining to certification and classification mean nothing. Therefore, it is likely that the states will be increasingly active in attempting to tie down the requirement of direct supervision.

As indicated in the opening paragraph of this chapter, legislation has always been an issue that generates strong emotion. There are valid arguments to support the need for legislation and likewise there are convincing reasons against the regulatory approach to solving problems. I believe there is one benefit that few would dispute. More than all the components that have gone into a pesticide regulatory program, enforcement, penalties, etc., the chief value that has accrued to the public *and* the pest control industry is in the general area of education and training. With legislation have come close relationships with universities, research agencies, and extension specialists. Such associations have been a dominant force in protecting the public and advancing the professional stature of the pest control industry. "Teaching, not policing, is the key to the future."

HARRY L. KATZ

Harry Katz was raised in Canonsburg, Pennsylvania, a small town 20 miles southwest of Pittsburgh. It was a family friend, Fred Pollock, who started a pest control supply and service business in 1928 in East Liberty, Pittsburgh, who got Katz interested in pest control in the 1930s.

After service in World War II, Katz took over the Elco Manufacturing Company operation. The impact of DDT in 1946 altered Elco's direction in favor of pest control service and at the same time Katz started to build an extensive library of pest control. In 1961, Elco acquired a one acre site in Sharpsburg, near Pittsburgh, and built a warehouse, office and a training room where classes for PCOs have been conducted since 1971.

In 1953, he helped form the Western Division of the Pennsylvania Pest Control Association and served until 1980 as recorder (secretary) when he sold the pest control service portion of his business.

The National Pest Control Association has been a consuming interest for Katz and he served on the Technical Committee (Council) for two years; as regional chairman, Wood Destroying Organisms Committee for one term; and as regional director 1975-78.

Locally, Katz was busy in the Sweadner Entomological Society (successor to Western Pennsylvania Entomological Society, established in 1910), serving as secretary for more than 10 years and as program chairman for 20 years. He is also a research associate in the Carnegie Museum of Natural History, Pittsburgh.

Statewide, the Entomological Society of Pennsylvania, oldest in the United States, captured Katz's attention, and he has held office and served on committees for this Society. In 1975, he established an annual Elco award for the best entomology student in graduate school in Pennsylvania. In 1980, he was recipient of the Frederich V. Melsheimer Distinguished Service Award of the Entomological Society of Pennsylvania.

Pest control and entomology are not Katz's only interests. He also is noted for his sincerity and the time he devotes to helping others. In the 1930s, he organized and chaired a Log Cabin Preservation Committee in Canonsburg, Pa. (1st school of the Presbytery West of the Alleghenies and forerunner of W & J University.)

In 1955, Katz founded the Parkway Jewish Center in a suburb near Pittsburgh and he is honorary president of this congregation. He is also a life trustee and assistant treasurer of the Hebrew Free Loan Association of Pittsburgh. Here he devotes two hours a week interviewing people unable to borrow at banks and helping them get on their feet with no interest loans.

Lecturing before diverse audiences is another one of Katz's loves. This includes PCOs and sanitarians at conferences, as well as university and high school classes. Katz has also contributed many articles to trade journals such as Pest Control and is contributing editor to Pest Control Technology magazine.

CHAPTER THIRTY

Entomophobia

By Harry L. Katz[1]

"Well I must admit, the joke was on me.
If no bugs in the package, for you to see.
Will send another sample. Hoping you to convince,
Was not just seeing things, nor, have I been since.
Neither a skin specialist, do I yet need,
Unless to patch the punctures, where the little pests feed.
Will send hallucinations, all sealed up in a jug,
If any escape from this, they will be some bug.
Those standing on hind feet were placed in jug alive;
Don't tell me this was wicked, for goodness sakes alive.
If, in one single night these all should land on you,
By the time you caught them all, you might want to punish, too.
If hallucinations, you cannot yet see,
Consulting an eye specialist might be well with thee.

— J.J. Davis

T O SOME DEGREE PHOBIAS are responsible for some of the pest control operator's revenue. A considerable portion of the industry's income is derived from relatively harmless creatures.

Phobias are simply fears which are unusually intense. According to Marks (1978), a phobia is a special kind of fear which is out of proportion to the demands of the situation, cannot be explained or reasoned away, is beyond voluntary control and leads to avoidance of the feared situation.

It is an anxiety, he says, which is triggered by a particular situation. The phobia may be minor or it may be a handicapping disorder.

Mild fears of spiders, mice, dogs and other animals are extremely common in our culture, but such fears are rarely strong enough to be called phobias.

The term "entomophobia" was not listed in three dictionaries on Psychology (by English, Harrison and Warren). These were found at the Western Psychiatric Hospital Library in Pittsburgh, Pennsylvania, a teaching hospital of the University of Pittsburgh Medical School.

[1]*President, Elco Manufacturing Co., Pittsburgh, Pa.*

The Encyclopedia Brittanica (1974) states, "The word phobia is derived from the Greek phobos meaning dread, panic or fear; among neuroses, phobias are specific fears out of proportion to the apparent stimuli." Among the animal fears listed are those dealing with cats, dogs, horses, lizards, and snakes. There is no term listed for fear of insects.

Experienced PCOs can usually spot a potential entomophobe. Some of the typical complaints include expression of "invisible insects"; "they can be white, black, colored or clear", they "dart into the skin, blanket or clothes"; "they move so fast one can scarcely see them", and they "penetrate emollients and tooth-paste." The caller talks of the necessity of daily washing of clothes, walls and floors — an obsessive compulsion. The customer already has tried some home remedies, has seen a succession of exterminators and may have visited physicians.

As stated by Hackley (1944), even if the PCO believes the customer is imagining insects or mites, the PCO should remember that a sick person should be respected with patience and sympathy.

Further, the PCO is not qualified to make such a diagnosis. There are many legitimate reasons for itches other than from the arthropods PCOs deal with. Later in this section, case histories will illustrate how a wrong diagnosis can be made.

Dr. R. Landay, an eminent allergist (personal communication) lists 10 of the more common causes of itching which are not related to insects or arachnids.
- Allergic dermatoses (eczema, contact dermatitis, and urticaria).
- Itching may appear as a component of a generalized allergic reaction.
- Parasitic infestations (such as pinworms, trichinosis and related infections).
- Pregnancy (Itching is a fairly common manifestation in the last tri-mester of pregnancy).
- Liver disease.
- Renal disease.
- Diabetes.
- Thyroid disorders.
- Malignancy (itching is associated with some malignant disorders).
- Neurologic disorders (such as *Herpes zoster* in its pre-eruptive stage and *Tabes dorsalis*).

Howell (1961) adds to the list by mentioning the possibility of drugs which cause hallucinations.

Ophthalmologists speak of optical BLEBS or floating spots in the eyes (a definition in Dorland's illustrated Dictionary, 25th edition).

Dr. R. A. Kern decries the fact that physicians overlook minor allergies because the symptoms mimic other diseases. "When no anatomic lesion can be found", he writes, "these patients are dubbed neurotic."

Particles and gases generated in the work place also cause itches to susceptible employees (Public Health Service, 1954). Allergic reactions are sometimes started with elements in house dust, and at times it is from particles from cockroaches and other arthropods in very heavily infested premises. Cornwell (1968) reports many people react positively to cockroach allergies. Okumura (1967) describes dermestid allergy and Ebeling (1975) covers a broad range of such disorders. Perlman contributed much original work relating to the effects of insect parts on people. (1961, 1962, 1965a and 1965b)

When visiting the customer, an examination of the premises must be thorough. It should include a liberal placement of sticky insect traps. They should be placed near the beds, in the bathroom and where the victim claims to be attacked. Later, the traps should be examined and the trapped creatures identified. The insects which attack humans are well documented; it must be noted that other flying non-biters could alight on the skin and cause a slight sensation.

Sticky traps are particularly desirable in establishing presence or absence of causal organisms in what is sometimes called "Bell's syndrome." This phenomenon occurs in yet another category of imagined infestations.

The expression, "Bell's Syndrome" may have originated in large rooms where a battery of telephone operators handled phone calls. One person will feel a strong itch and scratch vigorously, The next operator observes and is fearful of being the next victim. Others search for insects and usually find some innocuous flyer. The power of suggestion is such that itching occurs in sympathetic fellow workers and can be just as intense as from a genuine bite.

Case histories. Mampe (1976) discusses a case of this type. Most often this starts on a Monday or the day after a summer holiday. People react to mosquito bites differently. Some itch and swell at once; others experience this as much as 24 or more hours later. Thus, a bite on the beach or in the camp on Sunday will suddenly itch sometime Monday. It could also activate an itch at an old bite site.

Epidemics of acute anxiety are well documented. Two cases are reported by Marks (1978) which have nothing to do with insects. In Singapore, during July 1967, there was an outbreak of Swine Fever. Much publicity about inoculating pigs to control the outbreak resulted in a rumor that eating pork from inoculated pigs would cause Koro. Koro is an imagined shrinking of the penis. For the next few days up to 100 cases a day came to the hospitals and many more individuals consulted doctors. On the seventh day, a panel of experts appeared on television and radio explaining that Koro was psychological and it was impossible for the penis to retract into the abdomen.

In another situation at a girls' school in Britain, two-thirds of 500 girls showed anxiety symptoms with one-third actually hospitalized for fainting. It seems the school girls had waited in parade formation for three hours because of the late arrival of a royal visitor. Twenty of them felt faint and had to break ranks to lie down. The next morning there was much chatter about fainting. At assembly, one girl fainted, others felt dizzy and were told to sit on the chairs in the hall. A mistress thought they should lie on the floor rather than fall down. And there they were during the morning break for all to see. The phenomenon now became epidemic. Fainting, dizziness, pins and needles in the extremities and cramps in the muscles of arms and legs affected 125 girls in the first day. By the twelfth day, the nature of the epidemic was realized and firm management prevented the problem from spreading further. Symptoms of anxiety slowly subsided over a few days.

In Pittsburgh, Pennsylvania, Regis Murphy reports the case where a woman fainted an hour after the drilling for a termite job had started. She blamed it on the chemical which had not yet been applied. Thus, we see what the power of suggestion can do.

There being many degrees of severity of pest phobias, Hackley (1944) talks of benevolent fraud for the less severe cases. He proposes a somewhat exhibitory treatment. In an article, "Handling Entomophobia Cases," (1979) it is called a placebo.

Truman (1976) and Gage (1957) agree that no treatment should be made if no insects are seen. Gage believes if such treatment will not cure the problem, it would be more difficult to handle at a later date.

In Pittsburgh, Pennsylvania, several such cases were treated with total release pressurized aerosol containers. The client and especially the spouse were told that nothing could be found, but if there was anything in the room undiscovered, it could not survive the treatment. The client also was told that should there be no relief, a physician should be consulted with one of the other possible reasons for itch in mind. Further, the home would have to be vacated for four hours, along with the pets and fish tanks covered and the aerating pump stopped.

If this procedure is followed, it would be prudent for the PCO to return in four hours to remove the empty cans and to aerate the premises; bringing along an exhaust fan to be placed in a strategic opening to speed the process. Besides being of immense psychological value, this exhibition of concern could head off a possible lawsuit from a related problem, "Chemophobia," which could be a chapter heading in a future edition of this book, if lawsuit trends continue.

Few will disagree that honesty is the best policy. But consider this case. A dentist and his wife suspected bedbugs when they were both bit on the shoulders and head every third or fourth day. A careful examination by a Pittsburgh PCO found no bedbugs, no telltale fecal marks, no bedbug odor and no blood stains. The couple spent the next several weeks with a succession of allergists, dermatologists and others without relief. The PCO was called back to try again. This time he sprayed for bedbugs in the crevices of the bedstead, mattress and baseboard areas. Two bedbugs emerged. (Sex was not determined. Were they both males?) The PCO's honesty in this case wasted money and caused aggravation for the client.

Howell closed his 1961 paper with this statement. "Many pest control operators make a decision to treat a home for insects which have not been found or to flatly tell the individual no insects can be found and treatment is not warranted, even though they are convinced the homeowner may derive relief from imagined attacks by the application of insecticides. Several lawsuits resulted from refusal to treat a home under such circumstances followed by mentioning the reasons for the refusal to other pest control operators."

Several additional case histories are described in this section, each illustrating a different aspect of the phobia phenomenon — all from a PCOs viewpoint.

A young nurse suffered from incessant itching of the scalp. A variety of medications and visits to physicians gave no relief. As stated by H. Katz (1961), while attending a patient lice crawled from his head onto the operating table. She was convinced she had become infested and pointed to the nits in her hair to prove it. It turned out these were globules of hair spray. Pointing out the difference under a microscope corrected the louse phobia and her itch subsided. Wilson (1952) wrote in some detail on the delusion of parasitosis.

In another instance a woman became hysterical when sowbugs appeared outside her home. Trying to quiet her distress with assurances was difficult. However, a good drenching with pesticide helped. But each year for 12 years the sow-bugs returned, the treatment and the lecture were repeated and the hysteria gradually lessened to a mere annoyance. Another case of a phobia controlled with patience. The lectures were just as important as the treatment.

Howell (1961) reports on a housewife complaining of bugs biting her in bed at night. Howell took a microscope to her home, examined each particle and showed her they were bits of debris. She was then convinced that she had an illusion.

Howell (1961) also described the case of a man complaining of bites on his cheeks and forehead every night. Examination showed no bites, but much scratching. He was told to smear vaseline over the affected areas with the idea of trapping the bug. In the morning, he could wipe off the vaseline and the bug. Two weeks later, he reported complete success.

It should be pointed out that PCOs are not qualified or permitted to examine a person. The aforementioned case illustrates how susceptible to cure some of the entomophobia cases actually are.

Not all calls that are apparently entomophobic in nature turn out that way. Take the celebrated case reported by Traver (1951), in which several members of a family complained of dermatitis. Examined by several physicians, they were dismissed as neurotics until one of the victims, a zoology professor, actually caught the mites and had them classified by a specialist as a species of *Dermatophagoides*.

Howell (1961) tells the story of a woman complaining of bites by very small bugs. No creatures could be found. She was asked to swab affected areas each day and to save the cotton swabs. Pollen grains from Bermudagrass were found in 30 percent of the swabs. When Bermudagrass pollen was placed on her skin, it reacted.

Similarly, Howell (1961) tells of a woman who complained of bites on her arms and chest. Samples of insects were identified as non-biting species. A second series of swabs found elongate crystalline particles. They turned out to be from a dish-washing cleaner. Miller (1954) presents an account of insect hallucinations in an elderly couple.

In another case, Howell describes a large warehouse and office building which was fumigated for rat control. After the job, mites left the harborage and attacked personnel. The building was again fumigated, but two women still complained of bites. A careful search revealed no live mites.

The women were asked to swab irritated areas with alcohol and to bottle the swabs for examination. Forty-two percent of all swabs showed fragments of parts of tropical mites. They had been sensitized by the mite and the mite particles had caused the itch. Unfortunately, not all cases end successfully.

Howell tells of a 64-year-old man who claimed persistent bites from tiny bugs which no one else could see. "He doused his home with heavy concentrations of the then available hydrocarbons. No evidence of any arthropod could be found in this environment. He treated his body with undiluted Lysol® and visited Dr. Howell weekly for six months and then died, never finding relief from hallucination."

In response to early editions of this book, letters were written by severely troubled people revealing pathetic cases of entomophobia at its worst. All were lengthy letters listing all the symptoms mentioned earlier, each with additional variations. In one case, the PCO left a residue from a mechanical aerosol generator which, together with the odor, left an additional problem. PCOs are warned to avoid chemophobia possibilities.

One woman insisted that little black flies went through the glass. Some of the writers were extremely literate, others barely understandable. Some had exhausted their savings with a succession of medical people and pest control operators. One woman shaved the hair on her head because of the bites and the large red "lumps." Another, at her wit's end, wrote that whatever was in her scalp was now in her pubic area. She said further that various medical people could find no cause, but that a psychiatrist agreed she was not psychosomatic and blamed the negligence of the medical contingent. The red blotches she said got larger and sometimes got a pustule at the top.

Some of the correspondents are sick in more ways than one. An elderly manager of a small apartment house couldn't get rid of her bugs even after bathing twice a day. She blamed negro youngsters for, "putting the bugs on white people."

Robert Snetsinger of Pennsylvania State University received a letter from a woman who claimed itching from minute dark "grits" on bed and floor; and snow-white mite-sized "grits" on furniture and in the hair. Her children, she said, sneezed constantly, her legs were bitten when she ironed and her nose and ears itched. She found the dark "grits" on the underpants and a white mite on the fold of the skin on a son's swollen penis. Her home was exterminated, she said, but it didn't help. The letter was published in part by Stan Green in Public Health Entomology, September 1969, Alert #8 of the Penn State Extension Service.

So much for the most severe cases of entomophobia, delusory parisitosis, hallucinations or whatever one chooses to call it.

Something should be said for the thousands of people who suffer from a much lesser degree of a phobia against insects, a phobia in which insects are not imagined and in which the customers clearly feel uncomfortable and are disturbed about the possibility of pest intrusion into their premises. For them, periodic treatments are important and necessary for their peace of mind, mental health and, in cases of sensitivity to insects, their physical health. These people feel "clean" when the PCO is treating the property and, depending on the intensity of their phobia, look forward to the monthly visit. The peace of mind they get is important to their well being and justifies periodic spraying.

This is particularly important as stresses mount in our society. Several private interest groups have alerted millions of people to be aware of the dangers of insects and of chemicals.

An excellent paper on the spectrum of attitudes toward arthropods was written by H. Olkowski and W. Olkowski (1976). Forrest E. St. Aubin (1982) covered many aspects of entomophobia in an exhaustive paper based on an extensive bibliography.

Whether it is the pesticide purveyors pointing to the hazards of insects or the environmental groups pointing out the dangers of chemicals in general, and sometimes pesticides in particular, each has succeeded in causing a loss of perspective for a significant part of the population. Each viewpoint can best be marketed by highlighting the worst statistics that can be developed.

LITERATURE

ANON. — 1979. Handling Entomophobia Cases. Pest Control Technology 7(6):24, 25.

CORNWELL, P.B. — 1968. The Cockroach. Vol. I Hutchinson, London. 391 pp.

EBELING, W. — 1975. Urban Entomology. Univ. of Calif. 695 pp.

GAGE, R.W. — 1957. What to do about insect phobias. Pest Control 25 (10): 42, 47.

GREEN, S. — 1969. Public Health Entomology Alert #8, PSU Extension Service, Sept. 1969.

HACKLEY, R.E. — 1944. Entomophobia. Soap and Sanitary Chemicals. 20 (3):107.

HOWELL, D.E. — 1961. Hallucinations or Illusions of Insect Attack. Proc. Okla. Acad. Sci. Vol. 41: 83-87.

KATZ, H. — 1958. Notebook — Spray on Lice. p. 92.

KERN, R. A., M.D. — 1962. Environment in relation to Allergic Disease. Arch. Environmental Health. 4:28-49, 55.

MAMPE, D.C. — 1976. Irritation in office. Pest Control. 44(11):41.

MARKS, I.M. — 1978. Living With Fear.

MILLER, L.A. — 1954. An account of insect hallucinations affecting an elderly couple.

OKUMURA, G. — 1967. A report on canthariasis and allergy caused by *Trogoderma*. Calif. Vector News 14:19-22.

OLKOWSKI, H.and W. — 1976. Entomophobia in the urban ecosystem: some observations and suggestions. Bull. Entomol. Soc. America 22(3)313-317.

PERLMAN, F. — 1961 Insect Allergens. J. of Allergy 1 Vol. 32 No. 2 p. 93-101. 1962 Insects as Allergen Injectants. Calif., Med. Jan. 96:1-10. 1965a. Drug Reactions and Arthropod Sensitivity Chap. 28. 1965b Arthropods in Respiratory Tract Allergy, Acta Allergologica XXI 241-253.

PUBLIC HEALTH SERVICE — 1954. Occupational and Related Dermatoses PHS Publ. No. 364.

TRAVER, J.R. — 1951. Unusual scalp dermatitis in humans caused by the mite *Dermatophagoides*. Proc. Entomol. Soc. Wash. 53 (1) 1-25.

TRUMAN, L.C., G.W. BENNETT and W.L. BUTTS — 1976. Scientific Guide to Pest Control Operations. 3rd. Ed. Purdue University/Harvest Publishing Co. 276 pp.

WILSON, J. W. — 1952. Delusion of parasitosis (acarophobia). Arch. Dermatol. & Syphiol. 66:577-585.

HARRY L. KATZ

Harry Katz was raised in Canonsburg, Pennsylvania, a small town 20 miles southwest of Pittsburgh. It was a family friend, Fred Pollock, who started a pest control supply and service business in 1928 in East Liberty, Pittsburgh, who got Katz interested in pest control in the 1930s.

After service in World War II, Katz took over the Elco Manufacturing Company operation. The impact of DDT in 1946 altered Elco's direction in favor of pest control service and at the same time Katz started to build an extensive library of pest control. In 1961, Elco acquired a one acre site in Sharpsburg, near Pittsburgh, and built a warehouse, office and a training room where classes for PCOs have been conducted since 1971.

In 1953, he helped form the Western Division of the Pennsylvania Pest Control Association and served until 1980 as recorder (secretary) when he sold the pest control service portion of his business.

The National Pest Control Association has been a consuming interest for Katz and he served on the Technical Committee (Council) for two years; as regional chairman, Wood Destroying Organisms Committee for one term; and as regional director 1975-78.

Locally, Katz was busy in the Sweadner Entomological Society (successor to Western Pennsylvania Entomological Society, established in 1910), serving as secretary for more than 10 years and as program chairman for 20 years. He is also a research associate in the Carnegie Museum of Natural History, Pittsburgh.

Statewide, the Entomological Society of Pennsylvania, oldest in the United States, captured Katz's attention, and he has held office and served on committees for this Society. In 1975, he established an annual Elco award for the best entomology student in graduate school in Pennsylvania. In 1980, he was recipient of the Frederich V. Melsheimer Distinguished Service Award of the Entomological Society of Pennsylvania.

Pest control and entomology are not Katz's only interests. He also is noted for his sincerity and the time he devotes to helping others. In the 1930s, he organized and chaired a Log Cabin Preservation Committee in Canonsburg, Pa. (1st school of the Presbytery West of the Alleghenies and forerunner of W & J University.)

In 1955, Katz founded the Parkway Jewish Center in a suburb near Pittsburgh and he is honorary president of this congregation. He is also a life trustee and assistant treasurer of the Hebrew Free Loan Association of Pittsburgh. Here he devotes two hours a week interviewing people unable to borrow at banks and helping them get on their feet with no interest loans.

Lecturing before diverse audiences is another one of Katz's loves. This includes PCOs and sanitarians at conferences, as well as university and high school classes. Katz has also contributed many articles to trade journals such as Pest Control and is contributing editor to Pest Control Technology magazine.

CHAPTER THIRTY-ONE

Holistic Pest Control

By Harry L. Katz[1]

RECENT PUBLICITY has introduced the concept of Integrated Pest Management (IPM) to the American public. A term promoted by environmentalists concerned with alleged abuses in agriculture, IPM principles are being adapted by proponents of urban pest management (Katz, 1979b).

This has raised the level of consciousness of some PCOs to the excess use of toxicants. Katz (1979a) lists non-chemical options in pest control which can be exercised to some extent. For instance, sales of rodent and insect entrapment systems have increased sharply according to communication with manufacturers.

These trends have not, however, made a significant change in pest populations. Thousands of apartment houses, particularly in urban settings, are plagued with increasing cockroach and ant populations. At best, restaurants, food processing plants, hospitals, nursing homes and other institutions appear to be barely holding the pests to a tolerable level, with their pest control programs.

Ted Sleek of Sandoz reports on a survey made in 1979 which shows the pest control industry is growing by 15 percent to 20 percent annually over and above inflation costs because of these problems.

Darley (1976) believes "pest control" is a complete misnomer. He thinks the term "pest treatment" is more appropriate. He points out that most treatments are made in response to complaints and that attempts to eradicate often end up with moving the population to neighboring premises.

In a talk before a Whitmire Seminar group in August 1981, Alvin Burger, president of Bugs Burger Bug Killers, Inc., Miami, Fla., said pest control is not

[1]President, Elco Manufacturing Co., Pittsburgh, Pa.

the same as pest eradication. Pest control implies that a certain level of infestation is expected and tolerated.

Herein lies the basis for a newer concept of dealing with pests. *Holistic pest control* implies *total* eradication of pest populations with a total commitment of the PCO and the customer. It is dedicated to removing causes rather than treating symptoms.

IPM concepts for agriculture hold that a minimum level of pests are acceptable. Some proponents would have us introduce parasitic wasps to seek out the capsules of cockroaches and thus control them naturally. This completely ignores the federal regulations governing food handling establishments and the attitude of most citizens about insects.

Holism defined. Holism by definition, in the new college edition *American Heritage Dictionary,* emphasizes the importance of the whole and the interdependence of its parts.

One must look at the structure from an overall viewpoint. In some advanced medical facilities, opinions of the specialists are coordinated in treatment of a patient; so must all "organs" of a structure be considered in their ability to sustain a variety of pest populations.

Indeed, a study of household ecology should be an important course in training to eradicate pests. One would learn to think of the infinite variety of habitats, micro and macro, of temperatures and humidity ranges, and availability of nourishment and shelter. These include, besides the traditional appliances and furniture entities in home and factory, the following:

- Sewer systems. A repository for rich, nutritious effluent from garbage grinders, kept warm year round by waste water and connected with miles of arteries below the frost line. They serve as a perfect haven for psychodid drain flies, American cockroaches and rats.
- Suspended ceilings and false walls.
- Electrical conduit systems spanning structures in all directions.
- Heating and air conditioning ducts, utility pipe chases and large wall openings for utility pipes.
- Hollow modular concrete and steel units, permitting pest movement from wall to wall.
- Retrofitted insulation with the vapor barrier left on the wrong side, creating superb conditions favoring insects and rot organisms.

It appears many of the systems designed for our creature comfort have unwittingly given comfort to creatures which share our habitat.

We live in an artificial semi-tropical environment of our pest-susceptible structures. There is rapid movement of goods and people between diverse regions of the world. Therefore, new species, new strains and increases in pest populations are occuring.

The expression and concept "Pest Free Zone" may be ascribed to Cliff Darley in 1971. According to a paper by Story (1977), Darley, as chief public health inspector of Birkenhead, was able to mobilize limited resources, train personnel and motivate the public to attempt eradication of pest species in the entire community.

The city was divided into zones and each structure treated, residential and commercial, together with the areas outside, and also the sewers.

This is the system often described on a smaller scale by Austin Frishman in his training programs: to fragment the premises and to consider each piece of furniture, appliance or equipment as an entity in itself.

Holistic pest control is not possible without commitment by the customer to alter the environment which supports pest populations.

Millions of dollars were poured into public rat control projects in the 70s. One needs only to ride through housing developments today to observe the results when little is done to alter the environment.

Even fogging for mosquitoes is of little benefit, according to Irving Fox of the University of Puerto Rico (Fox, 1980).

Knipling (1979) graphically illustrates how hopeless it is to control populations without controlling the source. It is not how many we can kill, he says, but how much larval breeding material is available that determines the population of such insects as the house fly, stable fly, *Drosophila* and other Diptera. He further emphasizes the importance of community effort.

In the same publication, (page 245) Knipling says: "Pests in home or industrial establishments and many of those affecting man and animals are affected little by natural biotic agents. Therefore, the use of insecticides against such pests creates little or no imbalance. For many pests of minor or sporadic importance, alternative control methods cannot be used that will be as practical as chemical insecticides".

Why not holistic pest control? There are many reasons why holistic pest control is difficult to attain. The general public does not demand it. Too many are willing to tolerate insects at various levels. The cultural background of some is such that a certain level of insects is tolerated traditionally, according to Cornwell (1968). Sanitation practices are usually ancillary to this attitude.

In addition, few are aware of the potential hazard of disease-carrying insect pests. For others, there is a social stigma for having an exterminator. (Pests mean "I'm dirty" to the neighbors). Fear of pesticides and chemical odors keeps other homeowners from permitting exterminators into their homes.

Industry and institutions are often locked into a low-bid syndrome. The award is usually based on a figure that barely covers the cost of service under the most optimum conditions.

Most PCO firms have some technician training program for the record, but few have a comprehensive program in actual practice. They do not take enough time from production to properly train their technicians. A survey by the National Pest Control Association indicated that training books were bought, but few were used properly. To correct this, the EPA financed a Train-the-Trainer Program.

Progress is being made, however, in that attendance at training courses and seminars is improving, according to several pest control supply houses sponsoring them. Association and university conferences are also well attended.

Leaders in industry are now aware there is an alternative to the "low bid" type of pest control. The spectacular success of one PCO firm specializing in holistic pest free service attests to this fact. Other progressive firms are providing comprehensive services which include consultation on all aspects which can help attain pest free status.

The following is a partial list of possible techniques a PCO in holistic pest control would consider in commercial or institutional establishments:

- Drilling into inaccessible habitats and applying dusts if dry or with pressurized aerosol containers with crevice tube.
- Removing access panels in buildings or in equipment and then treating.
- Treating motor housings.
- Treating cracks and crevices in public areas with repellent-type toxicants

and non-repellent materials in hidden habitats.
- Treating sewers with drain openers monthly and occasional treatment with a water resistant material such a lacquer-based insecticide.
- Removing vegetation cover outdoors which can harbor pests.
- Moving bright lights away from doors to attract insects to a more distant location.
- Careful inspection for building faults which permit pest entry.
- Monitoring trash distribution and collection.
- Monitoring employee lockers.
- Monitoring pest status in pipe chases.
- Monitoring moisture leaks from equipment, plumbing and roof.
- Monitoring storing practices of merchandise.
- Monitoring employee sanitation practices (in food establishments).
- Monitoring wildlife in neighborhood.
- Monitoring incoming shipments.
- Training employees in basic sanitation practices, in their native tongue if necessary.

For a municipality, holistic pest control would include flushing the sewer system with massive volumes of water from fire hydrants when the water supplies are at their peak. This would remove much of the substrate which supplies pest populations to hundreds of homes. Following up with application of dust or aerosols would destroy the few surviving pests.

If one is to adapt the principles of holistic pest control to the home of an environmentally concerned person who would be willing to pay for it, here are some additional ideas for controlling pests.
- Periodically monitor the moisture content of the plaster or panelling of inside walls. An elevated moisture content on the outer wall would indicate trouble. There could be a leak in the roof or retrofitted insulation with the vapor barrier on the cold side. Such a condition will likely result in rot and insects such as carpenter ants and others.
- A moisture meter can alert to leaks around bathtub, shower and sinks which lead to insects and rot.
- In the basement, walls should be checked with a moisture meter to determine possible leaks in roof drainage pipes below grade or drainage in the grade outside.
- Wood joists can be checked for possible elevation of moisture content. The moisture meter needle could jump at the spot where an incipient termite colony exists.
- Check gutters for stoppage to help prevent carpenter ant damage; damage by squirrels and birds also can be noted at this time.
- Screens over the chimneys and gutters also should be checked; plugged soffit vents and leaks around chimney and vents also should be noted.
- When foundation plants need to be replaced, the owner should be advised to move them further away from the foundation to allow a strip of soil to be barren.
- Wooden stakes can monitor the presence of termites. There are substances attractive to termites with which these can be treated, but plain wood is usually sufficient.
- Trash piles, additional construction, changes in grade, bad sanitation practices all can be checked each time an inspection is made.
- Door checks, screens, weather seals and blocked dryer vents reduce the

likelihood of pest intrusion.
- Limited use of pesticides in cracks and crevices in some areas may be necessary, periodically.

Such a program does not come cheap because it calls upon all the skills of highly trained technical people. But for homeowners overly concerned about environmental pollution the option of holistic pest control is welcome.

In commercial premises, once the concept of a pest free environment is understood by management and it is willing to commit resources for equipment, training and motivation of employees, there are progressive PCOs who can truly provide this service.

The tools and training are available to all PCOs. All they need is commitment to use them and courage to ask to be paid for their services. Of course, only a small percentage of the industry would ever be involved with this type of pest control because only a small percentage of customers would appreciate it. But perhaps there's something in between for PCOs who are properly trained and motivated. Remember, there is a market for PCOs who can deal with the causes of symptoms, and not with the mere suppression of symptoms.

LITERATURE

CORNWELL, P.B. — 1968. The Cockroach. Vol. I. Hutchinson, London. 391 pp.

DARLEY, C.D. — 1976. Wirral's Pest Free Zones, Pest Control. 44(7):54-56.

FOX, IRVING — 1980. Evaluation of ULV Aerial and Ground Applications of Malathion Against Natural Populations of *Aedes aegypti* in Puerto Rico. Mosq. News, Vol. 40, No. 2, June pp. 280-283.

KATZ, H. — 1979a. Non Chemical Pest Control. Pest Control Technology. 7(6):8-14.

KATZ, H. — 1979b. Integrated Pest Management Then and Now. Pest Control. 47(11).

KNIPLING, E.F. — 1979. The Basic Principles of Insect Population, Suppression and Management. USDA Agric. Handbook No. 512.

STORY, K.O. — 1977. Speculations on Application of New Methods and Materials for Pesticide Uses in Urban Entomology. Pesticide Management and Insecticide Resistance. Academic Press. pp. 455-474.

INDEX OF COMMON NAMES AND EXAMPLES OF REGISTERED TRADE NAMES

COMMON NAME	REGISTERED TRADE NAME(S)
Acephate	ORTHENE
Acrylonitrile	VENTOX
Aldrin	ALDRITE, OCTALENE, SEEDRIN
Allethrin	PYNAMIN
Aluminum phosphide	CELPHOS, DELICIA, DETIA, GASTOXIN, PHOSTOXIN
*4-aminopyridine	AVITROL
Antu	ANTU
Bacillus thuringiensis	DIPEL, THURICIDE
Bendiocarb	DEXA-KLOR, FICAM, UBICIDE
Benomyl	BENLATE
Benzene, hexachloride, BHC or HCH	666
Beta naphthol	BRUCE PRESERVATIVE
Bioallethrin	BIOALLETHRINE, D-TRANS, ESBIOL
Boric acid	ROACH PRUFE
Brodifacoum	TALON
Bromadiolone	BROMONE, MAKI
Calcium cyanide	CYANOGAS, A-DUST
Carbaryl	SEVIN
Chlordane	BELT, OCTACHLOR, ORTHO-KLOR, VELSICOL 1068
Chlordecone	KEPONE
Chlorfenethol	DIMITE, MITRAN, QUIKRON
Chlorfenson	CROTRAR, OVEX, OVOTRAN, SAPPIRON
Chlorophacinone	CALD, DRAT, LIPHADIONE, MICROZUL, QUICK, RAMICIDE, ROZOL, TOPITOX
Chloropicrin	ACQUINITE, LARVACIDE, PIC-CHLOR, PICFUME
*3-chloro-p-toluidine hydrochloride	STARLICIDE
Chlorpyrifos	DURSBAN, KILLMASTER, LORSBAN
Coumachlor	RATILAN, TOMORIN
Coumafuryl	FUMARIN, FUMASOL, KRUMKIL, LURAT, RATAFIN, RAT-A-WAY
DDT	GESAROL, NEOCID
Deet	METADELPHENE

COMMON NAME

REGISTERED TRADE NAME(S)

*20,25 diazacholesterol dihydrochloride	ORNITROL
Diazinon	BASUDIN, DIAZITOL, NEOCIDOL, NUCIDOL, SAROLEX, SPECTRACIDE
Dichlorvos, DDVP	DEDEVAP, HERKOL, NOGOS, NO-PEST, NUVAN, VAPONA, VAPONITE
Dicofol	KELTHANE
Dieldrin	OCTALOX
Diflubenzuron	DIMILIN, TH 6040
Dimethoate	CYGON, DE-FEND, DIMETATE, PERFEKTHION, REBELATE, ROGOR, ROXION
Dimetilan	SNIP
Dioxacarb	ELOCRON, FAMID
Dioxathion	DELNAV, DELTIC
Diphacinone	DIPHACIN, PID, RAMIK
Endrin	ENDREX, HEXADRIN, MENDRIN, RID-A-BIRD
Ethylene dibromide, EDB	BROMOFUME, CELMIDE, DOWFUME, E-D-BEE, FUMOGAS, NEPHIS
Fenitrothion	ACCOTHION, CYTEL, CYFEN, FOLITHION, SUMITHION, VERTHION
Fenthion	BAYCID, BAYTEX, ENTEX, LEBAYCID, RID-A-BIRD, TIGUVON
Fluoracetamide	FLUORAKIL, FUSSOL, YANOCK, COMPOUND 1081 (code name)
Heptachlor	DRINOX, HEPTAMUL
Hydrogen cyanide, HCN, or hydrocyanic acid	CYCLON
Iodofenphos	ALFACRON, ELOCRIL, NUVANOL N
Isobornyl thiocyanoacetate	THANITE
Isovaleryl indandione	INCCO, ISOVAL, PMP, VALONE
Lindane, gamma BHC or gamma HCH	GAMMEXANE, KWELL
Malathion	CYTHION, KARBOPHOS, MALAPHOS, MALATHIOZOL, MALATHON, SUMITOL
Metaldehyde	META, HELARION

COMMON NAME

REGISTERED TRADE NAME(S)

Methiocarb	DRAZA, MESUROL
Methoprene	ALTOSID, PRECOR, PHARORID
Methoxychlor	MARLATE
Methyl bromide	BROM-O-GAS, BROZONE, DOWFUME, METH-O-GAS, TERR-O-GAS
Mirex	GC 1283 (code number)
Naled	DIBROM
Norbormide	SHOXIN, RATICATE
Paradichlorobenzene, PDB	PARACIDE, PARA-DI, PARADOW
Parathion	BLADAN, FOLIDOL, NIRAN, THIOPHOS
Pentachlorophenol, PCP	DOWICIDE, PENTA, PENTACON, SANTOBRITE, SANTOPHEN
Permethrin	ECTIBAN, NRDC 143 (code number), PERMANONE
Pindone	CHEMRAT, DUOCIDE, PIVACIN, PIVAL, PIVALDIONE, PIVALYN, TRI-BAN
Piperonyl butoxide	BUTACIDE
Pirimiphos-methyl	ACTELLIC, ACTELLIFOG, BLAX
Propetamphos	SAFROTIN
Propoxur	BAYGON, BLATTANEX, SUNCIDE, UNDENE
Red squill	DETHDIET, RODENE, RODINE, SILMURIN, SQUILL, TOPZOL
Resmethrin	CHRYSON, CROSS FIRE, SYNTHRIN
Ronnel, fenchlorvos	KORLAN, NANKOR, TROLENE
Rotenone	DERRIS
Silica aerogel	DRI-DIE, SILIKIL
Silica aerogel + pyrethrum	DRIONE
Sodium aluminum silicofluoride	LARVEX
Sodium fluoride	FLOROCID
Sodium fluoracetate	COMPOUND 1080 (code name)
Strychnine	KWIK-KIL, RODEX
Sulfuryl fluoride	VIKANE
Temephos	ABATE, ABATHION, BIOTHION, NIMITEX, SWEBATE
TEPP	NIFOS, VAPOTONE
Tetramethrin	NEO-PYNAMIN
Toxaphene	ATTAC

COMMON NAME   REGISTERED TRADE NAME(S)

Trichlorfon	ANTHON, DIPTEREX, DYLOX, NEGUVON, TUGON
Warfarin	COUMAFENE, DETHMOR, DUOCIDE, WARFARICIDE
Zinc phosphide	RUMETAN, ZP

*No common name for these chemicals.

GLOSSARY OF TECHNICAL TERMS

This glossary provides definitions of technical terms used in this book. Some of these terms have additional or different meanings when used elsewhere.

Abdomen — The posterior of the three main body divisions.

Absorption — The process of one substance (usually gas or liquid) being taken into another substance (usually liquid or solid).

Acute oral LD 50 — In toxicological studies, the dose required to kill 50 percent of the test animals when given as a single dose by mouth. The dose is normally expressed as the weight of chemical per unit weight of animal (e.g. mg/kg).

Adsorption — The adhesion of substances in an extremely thin layer to the surfaces of solids or liquids with which they are in contact.

Aestivation — Dormancy during a warm or dry season.

Aliphatic — Belonging to a group of organic compounds having an open-chain structure and consisting of the paraffin, olefin and acetylene hydrocarbons and their derivatives.

Alate. — Winged.

Ametabolous — Without metamorphosis.

Anal — Pertaining to the last abdominal segment (which bears the anus); also the posterior basal part of the wing.

Antennal club — The enlarged distal segments of a clubbed antenna.

Anterior — Front; in front of.

Anus — The posterior opening of the alimentary tract.

Apical — At the end, tip, or outermost part.

Aquatic — Living in water.

Asymmetrical — Not alike on the two sides.

Basal — At the base; near the point of attachment (of an appendage).

Basal cell — A cell near the base of the wing, bordered at least in part by the unbranched portions of the longitudinal veins.

Beak — The protruding mouth-part structures of a sucking insect; proboscis.

Book lung — A respiratory cavity containing a series of leaflike folds (spiders).

Brood — The individuals arising from one mother; individuals that hatch at about the same time and normally mature at about the same time.

Cannibalistic — Feeding on other individuals of the same species.

Carnivorous — Feeding on the flesh of other animals.

Carrier — The liquid or solid material added as a diluent to a pesticide to facilitate its application.

Caterpillar — A larva with a cylindrical body, a well developed head and with both thoracic legs and abdominal prolegs (e.g. butterfly, moth, sawfly).

Cell — A unit mass of protoplasm, surrounded by a cell membrane and containing one or more nuclei or nuclear material; a space in the wing membrane partly or completely surrounded by veins.

Cephalothorax — A body region consisting of head and thoracic segments (Crustacea and Arachnida).

Cercus (pl. cerci) — One of a pair of appendages at the end of the abdomen.

Chelate — Pincerlike, having two opposable claws.

Chelicera (pl. chelicerae) — The anterior pair of appendages in arachnids.

Chitin — A nitrogenous polysaccharide occurring in the cuticle of arthropods.

Chrysalis (pl. chrysalids or chrysalides) — The pupa of a butterfly.

Class — A subdivision of a phylum or subphylum containing a group of related orders.

Cleft — Split or forked.

Clubbed — With the distal part (or segments) enlarged; clubbed antennae.

Cocoon — A silken case inside which the pupa is formed.

Comb — A row of hairs or bristles (e.g. on head of fleas).

Commensalism — A living together of two or more species, none of which is injured thereby, and at least one of which is benefited.

Compound eye — An eye composed of many individual elements or ommatidia, each of which is represented externally by a facet; the external surface of such an eye consists of circular facets that are very close together, or of facets that are in contact and more or less hexagonal in shape.

Compressed — Flattened from side to side.

Constricted — Narrowed.

Contact insecticide — An insecticide which works when the insect is directly exposed to airborne particles, droplets or vapor.

Contiguous — Touching each other.

Coxa (pl. coxae) — The basal segment of the leg.

Cross vein — A vein connecting adjacent longitudinal veins.

Ctenidium (pl. ctenidia) — A row of stout bristles like the teeth of a comb (e.g. fleas).

Cuticle — The noncellular outer layer of the body wall of an arthropod.

Desiccant — A compound which promotes loss of moisture.

Diad — A raised group of eyes (some spiders).

Diapause — A period of arrested development.

Distal — Near or toward the free end of an appendage; that part of a segment or appendage farthest from the body.

Distillate — A liquid product condensed from vapor during distillation.

Diurnal — Active during the daytime.

Dormancy — A state of quiescence or inactivity.

Dorsal — Top or uppermost; relating to the back or upper side.

Ecdysis — Molting; the process of shedding the exoskeleton.

Ectoparasite — A parasite that lives on the outside of its host.

Elbowed antenna — An antenna with the first segment elongated and the remaining segments coming off the first segment at an angle (e.g. ants).

Elytron (pl. elytra) — A thickened, leathery, or horny front wing (e.g. Coleoptera, Dermaptera and some Homoptera).

Emergence — The act of the adult insect leaving the pupal case or the last nymphal skin.

Emulsifiable concentrate — See formulation.

Emulsifier — A surface-active chemical which reduces interfacial tension and which can be used to facilitate formation of an emulsion of one liquid in another.

Emulsion — A mixture in which very small droplets of one liquid are suspended in another liquid, e.g. oil in water. When the emulsion consists of droplets of water in oil it is known as an "invert" or "mayonnaise" emulsion.

Endocuticle — The innermost layer of the cuticle.

Epicuticle — The very thin, nonchitinous, external layer of the cuticle.

Epigynum — The external female genitalia of spiders.

Epipharynx — A mouth-part structure on the inner surface of the labrum or clypeus; in chewing insects a median lobe on the posterior (ventral) surface of the labrum or clypeus.

Exocuticle — The layer of the cuticle just outside the endocuticle, between the endocuticle and the epicuticle.

Exoskeleton — A skeleton or supporting structure on the outside of the body.

Extragenital — Unusual form of insemination not involving joining of the sexual organs (e.g. bedbugs).

Eye, compound — See compound eye.

Eye, simple — See ocellus.

Family — A subdivision of an order, suborder, or superfamily, and containing a group of related genera, tribes, or subfamilies. Family names end in *-idae*.

Feces — Excrement, the material passed from the alimentary tract through the anus.

Femur (pl. femora) — The third leg segment, located between the trochanter and the tibia.

Feral — Wild, including having escaped from domestication and become wild (e.g. feral cats, feral pigeons).

Forewing — The anterior or front pair of wings.

Formulation — The process by which pesticides are prepared for practical use or a preparation containing a pesticide in a form suitable for practical use, e.g.

Emulsifiable concentrate — A type of formulation for spray application consisting of a concentrated solution of a pesticide and an emulsifier in an organic solvent, which will form an emulsion when added to water and agitated.

Granule — A type of formulation for dry application consisting of granules which serve as a carrier for the pesticide.

Wettable powder — A type of formulation for spray application in which a pesticide is mixed with an inert carrier, the product finely ground and a surface-active agent added so that it will form a suspension when agitated with water.

Frass — Plant fragments made by a wood-boring insect, usually mixed with excrement.

Funiculus (or funicle) — The antennal segments between the scape and the club.

Gaster — The rounded part of the abdomen posterior to the nodelike segment (ants).

Genal comb — A row of strong spines borne on the anteroventral border of the head (fleas).

Generation — From any given stage in the life cycle to the same stage in the offspring.

Genus (pl. genera) — A group of closely related species; the first name in a binomial or trinomial scientific name. Names of genera are Latinized, capitalized, and when printed are italicized.

Gill — Evaginations of the body wall or hindgut, functioning in gaseous exchanges in an aquatic animal.

Gnathosoma — The anterior region of ticks and mites containing the mouthparts.

Granule — See formulation.

Gregarious — Living in groups.

Grub — A thick-bodied larva with a well-developed head and thoracic legs, without abdominal prolegs, and usually sluggish but often wriggling when disturbed.

Hallux — The first or preaxial digit of the hind limb of vertebrates.

Halter (pl. halteres) — A small knobbed structure on each side of the metathorax representing the hind wings (Diptera).

Head — The anterior body region, which bears the eyes, antennae, and mouth parts.

Herbivorous — Feeding on plants.

Hibernation — Dormancy during the winter.

Honeydew — Liquid discharged from the anus of certain Homoptera.

Host — The organism in or on which a parasite (or parasitoid) lives; the plant on which an insect feeds.

Hypopharynx — A median mouthpart structure anterior to the labium; the ducts from the salivary glands are usually associated with the hypopharynx, and in some sucking insects the hypopharynx is the mouthpart structure containing the salivary channel.

Idiosoma — The body section of ticks and mites excluding the mouthparts.

Instar — The stage of an insect between successive molts, the first instar being the stage between hatching and the first molt.

Instinctive behavior — Unlearned stereotyped behavior in which the nerve pathways involved are hereditary.

Integument — The outer covering of the body.

Joint — An articulation of two successive segments or parts.

Labial — Of or pertaining to the labium.

Labium — One of the mouthpart structures; the lower lip.

Labrum-epipharynx — A mouthpart representing the labrum and epipharynx.

Larva (pl. larvae) — The immature stages, between the egg and pupa, of an insect having complete metamorphosis or the six-legged first instar of Acarina. Larvae are distinctly different from the adult.

Larviform — Shaped like a larva.

Maggot — A legless larva without a well-developed head capsule (e.g. Diptera).

Malpighian tubules — Excretory tubes that arise near the anterior end of the hindgut and extend into the body cavity.

Mandible — Jaw; one of the anterior pair of the paired mouthpart structures.

Marginal cell — A cell in the distal part of the insect wing.

Maxilla (pl. maxillae) — One of the paired mouthpart structures immediately posterior to the mandibles.

Mesonotum — The dorsal sclerite of the mesothorax.

Mesothorax — The middle or second segment of the thorax.

Metamorphosis — Change in form during development.

Metathorax — The third or posterior segment of the thorax.

Midgut — The mesenteron, or middle portion of the alimentary tract.

Molt — A process of shedding the exoskeleton. Also known as ecdysis.

Morphology — The science of form or structure.

Myiasis — A disease caused by the invasion of dipterous larvae.

Nocturnal — Active at night.

Notum (pl. nota) — The dorsal surface of a body segment (usually used when speaking of the thoracic segments).

Nymph — An immature stage (following hatching) of an insect that does not have a pupal stage; the immature stages of Acarina that have eight legs.

Ocellus (pl. ocelli) — A simple eye of an insect or other arthropod.

Olefin — An unsaturated open-chain hydrocarbon containing at least one double bond.

Omnivorous — Feeding on a wide variety of substances of both animal and vegetable origin.

Ootheca (pl. oothecae) — The covering or case of an egg mass (e.g. cockroaches).

Order — A subdivision of a class or subclass, containing a group of related families.

Oviposit — To lay or deposit eggs.

Ovipositor — The egg-laying apparatus; the external genitalia of the female.

Palp — A segmented process borne by the maxillae or labium.

Pathogen or **pathogenic organism** — An organism which causes a disease in the animal receiving it.

Parasite — An animal that lives in or on the body of another living animal (its host), at least during a part of its life cycle.

Parasitic — Living as a parasite.

Parthenogenesis — Reproducing by eggs that develop without being fertilized.

Pecten — A comblike or rakelike structure.

Pectines — Comblike organs of touch (e.g. scorpions).

Pedicel — The second segment of the antenna; the stem of the abdomen, between the thorax and the gaster (ants), or between the head-thorax and the abdomen (spiders).

Pedipalps — The second pair of appendages of an arachnid.

Penultimate — Next to the last.

pH — A measure of the acidity or alkalinity of a medium. A pH value of 7.0 indicates neutral; lower values indicate acid and higher values indicate alkaline.

Pheromone — A chemical substance produced by an animal which acts as a stimulus to other individuals of the same species for one or more behavioral responses (e.g. aggregation pheromone of German cockroaches).

Phylum (pl. phyla) — One of the dozen or so major divisions of the animal kingdom.

Phytotoxic — Toxic to at least some plants.

Pilose — Covered with hair.

Pilosity — hairiness.

Polymorphic — Having many different forms or sizes (e.g. worker ants of some species).

Posterior — Hind or rear.

Postnotum (pl. postnota) — A notal plate behind the scutellum, often present in wingbearing segments.

Preapical — Situated just before the apex (e.g. preapical tibial bristle of Diptera).

Predator — An animal that attacks and feeds on other animals (its prey), usually animals smaller or less powerful than itself.

Prepupa — A quiescent stage between the larval period and the pupal period.

Proboscis — The extended beaklike mouthparts.

Pronotum — The dorsal sclerite of the prothorax.

Prothorax — The anterior of the three thoracic segments.

Psammophore — A group of long hairs on the underside of an ant's head, used for cleaning, etc.

Handbook of Pest Control

Ptilinum — A temporary bladderlike structure that can be inflated and thrust out through the frontal suture, just above the bases of the antennae, at the time of emergence from the puparium (Diptera).

Pulvillus (pl. pulvilli) — A pad or lobe beneath each tarsal claw (Diptera).

Pupa (pl. pupae) — The stage between the larva and the adult in insects with complete metamorphosis, a nonfeeding and usually an immobile stage.

Puparium (pl. puparia) — A case formed by the hardening of the next to the last larval skin, in which the pupa is formed (Diptera).

Pupate — To transform to a pupa.

Rectum — The posterior region of the hindgut.

Residual insecticide — An insecticide which remains capable of killing insects which contact treated surfaces long after the application.

Rostrum — Beak or snout.

Scape — The basal segment of the antenna.

Scavenger — An animal that feeds on dead plants or animals, on decaying materials, or an animal wastes.

Scent gland — A gland producing an odorous substance.

Scientific name — A Latinized name, internationally recognized, of a species or subspecies. The scientific name of a species consists of the generic and specific name and the name of the describer of the species, and that of a subspecies consists of generic, specific, and subspecific names and the name of the describer of the subspecies. Scientific names (excluding authors' names) are always printed in italics.

Sclerite — A hardened body wall plate bounded by sutures or membranous areas.

Sclerotized — Hardened.

Scutellum — A sclerite of a thoracic notum.

Scutum — The middle division of a thoracic notum, just anterior to the scutellum. Also the plate on the back of ticks.

Segment — A subdivision of the body or of an appendage, between joints or articulations.

Serrate — Toothed along the edge like a saw; serrate antenna; serrate bristles.

Seta (pl. setae) — A bristle.

Setaceous — Bristlelike, (e.g. setaceous antenna).

Setate — Provided with bristles.

Simple — Unmodified, not complicated; not forked, toothed, branched or divided.

Siphon — Breathing tube of mosquito larvae.

Species — A group of individuals or populations that are similar in structure and physiology and are capable of interbreeding and producing fertile offspring, and which are different in structure and/or physiology from other such groups and normally do not interbreed with them.

Spermatheca (pl. spermathecae) — The saclike structure in the female in which sperms from the male are received and often stored.

Spinneret — A structure with which silk is spun, usually fingerlike in shape.

Spiracle — An external opening of the tracheal system; a breathing pore.

Spiracular bristle — A bristle very close to a spiracle (Diptera).

Spur — A movable spine; when on a leg segment usually located at the apex of the segment.

Sternite — A subdivision of a sternum.

Sternum (pl. sterna) — A sclerite on the ventral side of the body; the ventral sclerite of an abdominal segment.

Stigma (pl. stigmata) — A thickening of the wing membrane along the costal border of the wing near the apex. Also respiratory opening in mites and ticks.

Stylus (pl. styli) — A short, slender, fingerlike process.

Subclass — A major subdivision of a class, containing a group of related orders.

Sublimation — The process of causing a substance to pass from a solid state directly to the vapor state (by heating) and then condensing again to solid form (e.g. in production of flake naphthalene).

Subspecies — A subdivision of a species, usually a geographic race. The different subspecies of a species are ordinarily not sharply differentiated and intergrade with one another and are capable of interbreeding.

Surface-active agents or **surfactants** — Substances which, when added to a liquid, affect the physical properties of the liquid surface (e.g. for the formulation of emulsifiable concentrates and wettable powders and for increasing the wetting properties of sprays).

Suspension — Particles or microcapsules of pesticide suspended in liquid.

Synergism — The combined effect of two or more pesticides mixed together leading to a greater pesticidal effect than would be predicted from the behavior of each component when applied singly.

Synergist — A chemical that enhances the effectiveness of an active ingredient.

Tarsus (pl. tarsi) — That leg segment beyond the tibia, consisting of one or more segments or subdivisions.

Taxis (pl. taxes) — A directed response involving the movement of an animal toward or away from a stimulus.

Taxonomy — The science of classification into categories of varying rank, based on similarities and differences, and the describing and naming of these categories.

Terpenes — Any of various hydrocarbons $(C_5H_8)n$ found in essential oils and resins (esp. from conifers) and used especially as solvents.

Thigmotaxis — A movement in which surface contact (esp. with a solid or a rigid object) is the directive factor.

Tibia (pl. tibiae) — The fourth segment of the leg, between the femur and the tarsus.

Trophallaxis — Food exchange of mutual benefit (esp. in social insects).

Tubercle — A small knoblike or rounded protuberance.

Vector — An animal capable of transmitting a pathogen from one organism to another.

Venter — The ventral side.

Vitelline membrane — The cell wall of the insect egg; a thin membrane lying beneath the chorion.

Wettable powder — See formulation.

LIST OF ILLUSTRATIONS

INDEX

Bug *(continued)*
 Assassin, 332
 Bat, 326
 Backswimmers, 344
 Bed bug, 319
 Birch catkin, 343
 Boxelder, 347
 Buffalo, 401
 Burrower, 342
 Chimney swift, 328
 Chinch, 343
 Conenose, 332
 Eastern bat, 326
 Electric light, 345
 European bed bug, 326
 Giant waterbugs, 344
 Grass, 340, 890
 Green stink, 340
 Kissing, 336
 Leafhopper assassin, 336
 Lygaeid, 342, 890
 Mexican, 334
 Mexican chicken, 327
 Negro, 342
 Occasional invaders, 341
 Poultry, 327
 Reduviidae, 332
 Seed, 342
 Squash, 340
 Swallow, 326
 Swimming pool insects, 344
 Texas, 334
 Tissue paper, 413
 Toe biters, 345
 Tropical human bed bug, 325
 Two-spotted stink, 342
 Water scavenger beetle, 345
 Waterboatmen, 344
 Waterbug, 114
 Western bat, 326
 Western boxelder, 340
 Wheel, 326
Buprestids, 298
Buprestis aurulenta, 298
Burrower bug, 342

C

Caddis flies, 884
Cadelle, 519, 527
 Control, 528
Cadra figulilella, 540
Calcium arsenate, 910
Calcium cyanide, 60, 961
California death watch beetle, 287
California ground squirrel, 810

California hardwood bark beetle, 298
California harvester ant, 455
Callidium antennatum, 302
Calliphora
 vicina, 649
 vomitoria, 649
Callosobruchus maculatus, 534
Calomyceterus setarius, 881
Camel crickets, 161
Camponotus
 abdominalis floridanus, 463, 474
 clarithorax, 463
 ferrugineus, 463
 hyatti, 463
 laevigatus, 463
 modoc, 463
 nearcticus, 463
 pennsylvanicus, 463
 vicinus, 463
Campsomeris tolteca, 497
Canned insecticides (aerosols), 998
 Advantages, 999
 Aerosol propellants, 998
 Chlorocarbons, 998
 Compressed gases, 998
 Disadvantages, 1000
 Ethers, 998
 Fluorocarbons, 998
 Hydrocarbons, 998
 Safety, 999
Carbamates, 900
 Bendiocarb, 901
 Carbaryl, 900
 Dimetilan, 900
 Propoxur, 900
Carbaryl, 134, 163, 900
 Carbon bisulfide, 61
Carbon dioxide, 974
Carbon disulfide, 972
Carbon monoxide, 61
Carbon tetrachloride, 972
Carpenter ant, 463
 Black, 463
 Control, 465-467
 Florida, 463
 In the home, 464
 Red, 463
Carpenter bee, 494-496
 Control, 496
 Mountain, 494
 Valley, 495
Carpet beetle, 387-423
 Birdnest, 408
 Black, 397
 Common, 389, 400
 Furniture, 402

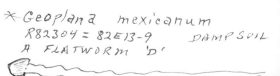
✗ Geoplana mexicanum
R82304 = 82E13-9   DAMP SOIL
A FLATWORM 'D'

I

T

Tabanidae, 652
Tabanus nigrovittatus, 652
Tabatrex, 144
Tachycines asynamorus, 161
Taeniothrips inconsequens, 886
Talitrus sylvaticus, 866 *RED SHRIMP-LIKE*
Tamias striatus, 799
Tamiasciurus
   douglasii, 807
   hudsonicus, 807
Tanolith, 273
Tapestry moth, 354
Tapinoma sessile, 458
Tarantulas, 730
   Bird spiders, 730
   Mygales, 730
   Purseweb, 731
   Trap-door, 731
Tarpaulin fumigation, 976
Tawny garden slug, 868
TDE, 916
Tear fungus, 260
Telmatoscopus albipunctatus, 663
Tenebrio
   molitor, 519, 544
   obscurus, 544
Tenebroides mauritanicus, 519, 527
TEPA, 144
TEPP, 919
Termite, 177-257
   Arid-land subterranean, 209
   As native pests, 215
   As social insects, 179
   Biological information, 188
   Caste system, 185
   Control, 212-249
   Control, Introduction, 212
   Courtship and pairing, 189
   Dampwood, 197, 199
   Dark dampwood, 199
   Desert, 212
   Desert dampwood, 205
   Desert subterranean, 211
   Distinguishing ants from termites, 195
   Distinguishing termite injury, 216
   Distribution, 192
   Drywood, 200
   Drywood, other species, 203
   Drywood termite control, 242-248
   Eastern subterranean, 210
   Food, 182
   Formosan subterranean, 211
   Fungi, 183
   Furniture, 204

Grooming, 185
Ground treatment, 226
Guests, 195
How supplementary kings and queens
   are produced, 190
King, 190
Moisture, 181
Nesting sites, 194
Pacific dampwood termite, 197
Powder post, 204
Protozoa, 183
Queen, 190
Reaction to their environment, 181, 182
Reproductive caste, 185
Resistant woods, 242
Small dampwood, 199
Soil conditions, 182
Soil insecticides, 233
Soldier caste, 191
Sound, 192
Subterranean, 205
Subterranean control, 271-242
Supplementary queens, 187
Swarming, 188
Tabulation of species, 196
Temperature, 181
Termite castes, 185
Termites and fungi, 214
Termites and roaches, 179
Termites of economic importance, 197
Western drywood, 200
Western subterranean, 206
What is a termite, 179
Worker caste, 191
Termite, Chemicals rarely used or not
   currently labelled, 235
   Creosote, 237
   Lead arsenate, 235
   Lindane, 235
   Orthodichlorobenzene, 237
   Pentachlorophenol, 236
   Sodium arsenite, 235
   Trichlorobenzene, 237
Termite, Chemicals used in soil
   treatment, 233
   Aldrin, 234
   Chlordane, 234
   Dieldrin, 234
   Foundation treatment, 238
   Handling, 233
   Heptachlor, 234
   Nature of soil, 238
   Other chemicals, 234
Termite treatment equipment, 1003
   Drills, 1007
   Electric moisture meter, 1008